T0230639

differential equations, dynamical systems, and control science

**PURE AND APPLIED MATHEMATICS**

**A Program of Monographs, Textbooks, and Lecture Notes**

## LECTURE NOTES IN PURE AND APPLIED MATHEMATICS

1. *N. Jacobson,* Exceptional Lie Algebras
2. *L.-Å. Lindahl and F. Poulsen,* Thin Sets in Harmonic Analysis
3. *I. Satake,* Classification Theory of Semi-Simple Algebraic Groups
4. *F. Hirzebruch, W. D. Newmann, and S. S. Koh, Differentiable Manifolds and Quadratic Forms*
5. *I. Chavel,* Riemannian Symmetric Spaces of Rank One
6. *R. B. Burckel,* Characterization of C(X) Among Its Subalgebras
7. *B. R. McDonald, A. R. Magid, and K. C. Smith,* Ring Theory: Proceedings of the Oklahoma Conference
8. *Y.-T. Siu,* Techniques of Extension on Analytic Objects
9. *S. R. Caradus, W. E. Pfaffenberger, and B. Yood,* Calkin Algebras and Algebras of Operators on Banach Spaces
10. *E. O. Roxin, P.-T. Liu, and R. L. Sternberg,* Differential Games and Control Theory
11. *M. Orzech and C. Small,* The Brauer Group of Commutative Rings
12. *S. Thomier,* Topology and Its Applications
13. *J. M. Lopez and K. A. Ross,* Sidon Sets
14. *W. W. Comfort and S. Negrepontis,* Continuous Pseudometrics
15. *K. McKennon and J. M. Robertson,* Locally Convex Spaces
16. *M. Carmeli and S. Malin,* Representations of the Rotation and Lorentz Groups: An Introduction
17. *G. B. Seligman,* Rational Methods in Lie Algebras
18. *D. G. de Figueiredo,* Functional Analysis: Proceedings of the Brazilian Mathematical Society Symposium
19. *L. Cesari, R. Kannan, and J. D. Schuur,* Nonlinear Functional Analysis and Differential Equations: Proceedings of the Michigan State University Conference
20. *J. J. Schäffer,* Geometry of Spheres in Normed Spaces
21. *K. Yano and M. Kon,* Anti-Invariant Submanifolds
22. *W. V. Vasconcelos,* The Rings of Dimension Two
23. *R. E. Chandler,* Hausdorff Compactifications
24. *S. P. Franklin and B. V. S. Thomas,* Topology: Proceedings of the Memphis State University Conference
25. *S. K. Jain,* Ring Theory: Proceedings of the Ohio University Conference
26. *B. R. McDonald and R. A. Morris,* Ring Theory II: Proceedings of the Second Oklahoma Conference
27. *R. B. Mura and A. Rhemtulla,* Orderable Groups
28. *J. R. Graef,* Stability of Dynamical Systems: Theory and Applications
29. *H.-C. Wang,* Homogeneous Branch Algebras
30. *E. O. Roxin, P.-T. Liu, and R. L. Sternberg,* Differential Games and Control Theory II
31. *R. D. Porter,* Introduction to Fibre Bundles
32. *M. Altman,* Contractors and Contractor Directions Theory and Applications
33. *J. S. Golan,* Decomposition and Dimension in Module Categories
34. *G. Fairweather,* Finite Element Galerkin Methods for Differential Equations
35. *J. D. Sally,* Numbers of Generators of Ideals in Local Rings
36. *S. S. Miller,* Complex Analysis: Proceedings of the S.U.N.Y. Brockport Conference
37. *R. Gordon,* Representation Theory of Algebras: Proceedings of the Philadelphia Conference
38. *M. Goto and F. D. Grosshans,* Semisimple Lie Algebras
39. *A. I. Arruda, N. C. A. da Costa, and R. Chuaqui,* Mathematical Logic: Proceedings of the First Brazilian Conference
40. *F. Van Oystaeyen,* Ring Theory: Proceedings of the 1977 Antwerp Conference
41. *F. Van Oystaeyen and A. Verschoren,* Reflectors and Localization: Application to Sheaf Theory
42. *M. Satyanarayana,* Positively Ordered Semigroups
43. *D. L Russell,* Mathematics of Finite-Dimensional Control Systems
44. *P.-T. Liu and E. Roxin,* Differential Games and Control Theory III: Proceedings of the Third Kingston Conference, Part A
45. *A. Geramita and J. Seberry,* Orthogonal Designs: Quadratic Forms and Hadamard Matrices
46. *J. Cigler, V. Losert, and P. Michor,* Banach Modules and Functors on Categories of Banach Spaces

47. *P.-T. Liu and J. G. Sutinen,* Control Theory in Mathematical Economics: Proceedings of the Third Kingston Conference, Part B
48. *C. Byrnes,* Partial Differential Equations and Geometry
49. *G. Klambauer,* Problems and Propositions in Analysis
50. *J. Knopfmacher,* Analytic Arithmetic of Algebraic Function Fields
51. *F. Van Oystaeyen,* Ring Theory: Proceedings of the 1978 Antwerp Conference
52. *B. Kadem,* Binary Time Series
53. *J. Barros-Neto and R. A. Artino,* Hypoelliptic Boundary-Value Problems
54. *R. L. Sternberg, A. J. Kalinowski, and J. S. Papadakis,* Nonlinear Partial Differential Equations in Engineering and Applied Science
55. *B. R. McDonald,* Ring Theory and Algebra III: Proceedngs of the Third Oklahoma Conference
56. *J. S. Golan,* Structure Sheaves Over a Noncommutative Ring
57. *T. V. Narayana, J. G. Williams, and R. M. Mathsen,* Combinatorics, Representation Theory and Statistical Methods in Groups: YOUNG DAY Proceedings
58. *T. A. Burton,* Modeling and Differential Equations in Biology
59. *K. H. Kim and F. W. Roush,* Introduction to Mathematical Consensus Theory
60. *J. Banas and K. Goebel,* Measures of Noncompactness in Banach Spaces
61. *O. A. Nielson,* Direct Integral Theory
62. *J. E. Smith, G. O. Kenny, and R. N. Ball,* Ordered Groups: Proceedings of the Boise State Conference
63. *J. Cronin,* Mathematics of Cell Electrophysiology
64. *J. W. Brewer,* Power Series Over Commutative Rings
65. *P. K. Kamthan and M. Gupta,* Sequence Spaces and Series
66. *T. G. McLaughlin,* Regressive Sets and the Theory of Isols
67. *T. L. Herdman, S. M. Rankin III, and H. W. Stech,* Integral and Functional Differential Equations
68. *R. Draper,* Commutative Algebra: Analytic Methods
69. *W. G. McKay and J. Patera,* Tables of Dimensions, Indices, and Branching Rules for Representations of Simple Lie Algebras
70. *R. L. Devaney and Z. H. Nitecki,* Classical Mechanics and Dynamical Systems
71. *J. Van Geel,* Places and Valuations in Noncommutative Ring Theory
72. *C. Faith,* Injective Modules and Injective Quotient Rings
73. *A. Fiacco,* Mathematical Programming with Data Perturbations I
74. *P. Schultz, C. Praeger, and R. Sullivan,* Algebraic Structures and Applications: Proceedings of the First Western Australian Conference on Algebra
75. *L Bican, T. Kepka, and P. Nemec,* Rings, Modules, and Preradicals
76. *D. C. Kay and M. Breen,* Convexity and Related Combinatorial Geometry: Proceedings of the Second University of Oklahoma Conference
77. *P. Fletcher and W. F. Lindgren,* Quasi-Uniform Spaces
78. *C.-C. Yang,* Factorization Theory of Meromorphic Functions
79. *O. Taussky,* Ternary Quadratic Forms and Norms
80. *S. P. Singh and J. H. Burry,* Nonlinear Analysis and Applications
81. *K. B. Hannsgen, T. L. Herdman, H. W. Stech, and R. L. Wheeler,* Volterra and Functional Differential Equations
82. *N. L. Johnson, M. J. Kallaher, and C. T. Long,* Finite Geometries: Proceedings of a Conference in Honor of T. G. Ostrom
83. *G. I. Zapata,* Functional Analysis, Holomorphy, and Approximation Theory
84. *S. Greco and G. Valla,* Commutative Algebra: Proceedings of the Trento Conference
85. *A. V. Fiacco,* Mathematical Programming with Data Perturbations II
86. *J.-B. Hiriart-Urruty, W. Oettli, and J. Stoer,* Optimization: Theory and Algorithms
87. *A. Figa Talamanca and M. A. Picardello,* Harmonic Analysis on Free Groups
88. *M. Harada,* Factor Categories with Applications to Direct Decomposition of Modules
89. *V. I. Istrăţescu,* Strict Convexity and Complex Strict Convexity
90. *V. Lakshmikantham,* Trends in Theory and Practice of Nonlinear Differential Equations
91. *H. L. Manocha and J. B. Srivastava,* Algebra and Its Applications
92. *D. V. Chudnovsky and G. V. Chudnovsky,* Classical and Quantum Models and Arithmetic Problems
93. *J. W. Longley,* Least Squares Computations Using Orthogonalization Methods
94. *L. P. de Alcantara,* Mathematical Logic and Formal Systems
95. *C. E. Aull,* Rings of Continuous Functions
96. *R. Chuaqui,* Analysis, Geometry, and Probability
97. *L. Fuchs and L. Salce,* Modules Over Valuation Domains

98. *P. Fischer and W. R. Smith,* Chaos, Fractals, and Dynamics
99. *W. B. Powell and C. Tsinakis,* Ordered Algebraic Structures
100. *G. M. Rassias and T. M. Rassias,* Differential Geometry, Calculus of Variations, and Their Applications
101. *R.-E. Hoffmann and K. H. Hofmann,* Continuous Lattices and Their Applications
102. *J. H. Lightbourne III and S. M. Rankin III,* Physical Mathematics and Nonlinear Partial Differential Equations
103. *C. A. Baker and L, M. Batten,* Finite Geometrics
104. *J. W. Brewer, J. W. Bunce, and F. S. Van Vleck,* Linear Systems Over Commutative Rings
105. *C. McCrory and T. Shifrin,* Geometry and Topology: Manifolds, Varieties, and Knots
106. *D. W. Kueker, E. G. K. Lopez-Escobar, and C. H. Smith,* Mathematical Logic and Theoretical Computer Science
107. *B.-L. Lin and S. Simons,* Nonlinear and Convex Analysis: Proceedings in Honor of Ky Fan
108. *S. J. Lee,* Operator Methods for Optimal Control Problems
109. *V. Lakshmikantham,* Nonlinear Analysis and Applications
110. *S. F. McCormick,* Multigrid Methods: Theory, Applications, and Supercomputing
111. *M. C. Tangora,* Computers in Algebra
112. *D. V. Chudnovsky and G. V. Chudnovsky,* Search Theory: Some Recent Developments
113. *D. V. Chudnovsky and R. D. Jenks,* Computer Algebra
114. *M. C. Tangora,* Computers in Geometry and Topology
115. *P. Nelson, V. Faber, T. A. Manteuffel, D. L. Seth, and A. B. White, Jr.,* Transport Theory, Invariant Imbedding, and Integral Equations: Proceedings in Honor of G. M. Wing's 65th Birthday
116. *P. Clément, S. Invernizzi, E. Mitidieri, and I. I. Vrabie,* Semigroup Theory and Applications
117. *J. Vinuesa,* Orthogonal Polynomials and Their Applications: Proceedings of the International Congress
118. *C. M. Dafermos, G. Ladas, and G. Papanicolaou,* Differential Equations: Proceedings of the EQUADIFF Conference
119. *E. O. Roxin,* Modern Optimal Control: A Conference in Honor of Solomon Lefschetz and Joseph P. Lasalle
120. *J. C. Díaz,* Mathematics for Large Scale Computing
121. *P. S. Milojević,* Nonlinear Functional Analysis
122. *C. Sadosky,* Analysis and Partial Differential Equations: A Collection of Papers Dedicated to Mischa Cotlar
123. *R. M. Shortt,* General Topology and Applications: Proceedings of the 1988 Northeast Conference
124. *R. Wong,* Asymptotic and Computational Analysis: Conference in Honor of Frank W. J. Olver's 65th Birthday
125. *D. V. Chudnovsky and R. D. Jenks,* Computers in Mathematics
126. *W. D. Wallis, H. Shen, W. Wei, and L. Zhu,* Combinatorial Designs and Applications
127. *S. Elaydi,* Differential Equations: Stability and Control
128. *G. Chen, E. B. Lee, W. Littman, and L. Markus,* Distributed Parameter Control Systems: New Trends and Applications
129. *W. N. Everitt,* Inequalities: Fifty Years On from Hardy, Littlewood and Pólya
130. *H. G. Kaper and M. Garbey,* Asymptotic Analysis and the Numerical Solution of Partial Differential Equations
131. *O. Arino, D. E. Axelrod, and M. Kimmel,* Mathematical Population Dynamics: Proceedings of the Second International Conference
132. *S. Coen,* Geometry and Complex Variables
133. *J. A. Goldstein, F. Kappel, and W. Schappacher,* Differential Equations with Applications in Biology, Physics, and Engineering
134. *S. J. Andima, R. Kopperman, P. R. Misra, J. Z. Reichman, and A. R. Todd,* General Topology and Applications
135. *P Clément, E. Mitidieri, B. de Pagter,* Semigroup Theory and Evolution Equations: The Second International Conference
136. *K. Jarosz,* Function Spaces
137. *J. M. Bayod, N. De Grande-De Kimpe, and J. Martínez-Maurica,* *p*-adic Functional Analysis
138. *G. A. Anastassiou,* Approximation Theory: Proceedings of the Sixth Southeastern Approximation Theorists Annual Conference
139. *R. S. Rees,* Graphs, Matrices, and Designs: Festschrift in Honor of Norman J. Pullman
140. *G. Abrams, J. Haefner, and K. M. Rangaswamy,* Methods in Module Theory

141. *G. L. Mullen and P. J.-S. Shiue*, Finite Fields, Coding Theory, and Advances in Communications and Computing
142. *M. C. Joshi and A. V. Balakrishnan*, Mathematical Theory of Control: Proceedings of the International Conference
143. *G. Komatsu and Y. Sakane*, Complex Geometry: Proceedings of the Osaka International Conference
144. *I. J. Bakelman*, Geometric Analysis and Nonlinear Partial Differential Equations
145. *T. Mabuchi and S. Mukai*, Einstein Metrics and Yang–Mills Connections: Proceedings of the 27th Taniguchi International Symposium
146. *L. Fuchs and R. Göbel*, Abelian Groups: Proceedings of the 1991 Curaçao Conference
147. *A. D. Pollington and W. Moran*, Number Theory with an Emphasis on the Markoff Spectrum
148. *G. Dore, A. Favini, E. Obrecht, and A. Venni*, Differential Equations in Banach Spaces
149. *T. West*, Continuum Theory and Dynamical Systems
150. *K. D. Bierstedt, A. Pietsch, W. Ruess, and D. Vogt*, Functional Analysis
151. *K. G. Fischer, P. Loustaunau, J. Shapiro, E. L. Green, and D. Farkas*, Computational Algebra
152. *K. D. Elworthy, W. N. Everitt, and E. B. Lee*, Differential Equations, Dynamical Systems, and Control Science
153. *P.-J. Cahen, D. L. Costa, M. Fontana, and S.-E. Kabbaj*, Commutative Ring Theory
154. *S. C. Cooper and W. J. Thron*, Continued Fractions and Orthogonal Functions: Theory and Applications
155. *P. Clément and G. Lumer*, Evolution Equations, Control Theory, and Biomathematics

*Additional Volumes in Preparation*

# differential equations, dynamical systems, and control science

*A Festschrift in Honor of Lawrence Markus*

edited by

K. D. Elworthy
*University of Warwick*
*Coventry, England*

W. Norrie Everitt
*The University of Birmingham*
*Birmingham, England*

E. Bruce Lee
*University of Minnesota*
*Minneapolis, Minnesota*

MARCEL DEKKER

MARCEL DEKKER, INC. NEW YORK • BASEL

**Library of Congress Cataloging-in-Publication Data**

Differential equations, dynamical systems, and control science : a
festschrift in honor of Lawrence Markus / edited by K.D. Elworthy,
W. Norrie Everitt, E. Bruce Lee.
p. cm. -- (Lecture notes in pure and applied mathematics ; v.
152)
Includes bibliographical references.
ISBN 0-8247-8904-0 (acid-free)
1. Control theory. 2. Differential equations. 3. Differentiable
dynamical systems. I. Markus, L. (Lawrence).
II. Elworthy, K. D. III. Everitt, W. N. (William Norrie).
IV. Lee, E. B. (Ernest Bruce). V. Series.
QA402.3.D483 1994
515'.64--dc20 93-30077
CIP

The publisher offers discounts on this book when ordered in bulk quantities. For more information, write to Special Sales/Professional Marketing at the address below.

This book is printed on acid-free paper.



MARCEL DEKKER, INC.
270 Madison Avenue, New York, New York 10016

Current printing (last digit):
10 9 8 7 6 5 4 3 2

To Lawrence Markus
on the occasion of his seventieth birthday

Professor Markus lectures on linear systems. (Photo by Sarah Knoepfler)

# Preface

Lawrence (Larry) Markus, Regents' Professor of Mathematics in the University of Minnesota in the United States of America, and Professor (sometime Nuffield Professor) of Mathematics in the University of Warwick in the United Kingdom, was born on 13 October 1922 at Hibbing, Minnesota.

The academic year 1992 to 1993 marks then Larry's 70th birthday and his friends and colleagues have taken this opportunity to prepare a "Festschrift" to acknowledge his many and manifest contributions to mathematics, and to offer him a token to represent their gratitude for his friendship and professional support over many years.

The year 1992 to 1993 sees also a programme at the Institute of Mathematics and its Applications, University of Minnesota, devoted to Control Theory and Its Applications, one of the areas of mathematics to which Larry has devoted his time and research abilities. The Director of the Institute, Professor Avner Friedman, and his colleagues willingly agreed to lend the support of the Institute to the preparation of the Festschrift. Many of the contributions are linked to this area of control theory and many are written by those who have visited and worked in the Institute.

In writing this Preface my first thought was that it should have been penned by one of Larry's colleagues in either the University of Minnesota or the University of Warwick; he has served in both Universities for many years and with such success that this responsibility could but be deemed appropriate for a member of one or the other of these Universities. However, reflection suggests that a colleague and collaborator from elsewhere but who has visited and worked with Larry at both Universities may be in a position to see more clearly the contribution he has made to both institutions. So the privilege of writing falls to me, and if I am honest I both sought the responsibility and guarded the position lest it pass to another.

It is well known that Larry began his university education at the University of Chicago. It is perhaps not so well known that his studies began not only in mathematics (B.S. 1942) but also in meteorology (M.S. 1946). In fact, Larry became an Instructor of Meteorology at the age of nineteen, and later used this professional knowledge as a scientist on the "Manhattan Project" for the development of the atomic bomb. The interest in meteorology may well have been influenced by the upheaval of the second world war; in any event these studies were interrupted with Larry spending the years 1944 to 1946 in the United States Naval Reserve, during which period he achieved the rank of Lieutenant (junior grade). Many months during these years of naval service were spent attempting to achieve good order and naval discipline whilst surviving the worst which the Atlantic Ocean could hurl against a weather ship struggling to keep station amid the billows.

In 1950 Larry Markus married Lois Shoemaker. They have remained happily married for over forty years. Two children were born to the marriage, a daughter Sylvia, and then a son Andrew.

Larry began the serious study of mathematics after the war; his *curriculum vitae* reads like a catalogue for the Ivy league; Harvard, Yale, Princeton. In 1952, he gained the Ph.D. degree from Harvard University. He was an Instructor during the years 1951 to 1955 at Harvard and then Yale Universities; and then served as a Lecturer at Princeton University for the years 1955 to 1957. This period brought the young Larry Markus into contact with many of the great and significant names in mathematics during those formative years of modern developments in the subject. I can but hope that Larry will find the time and energy to write of his experiences as a young and hopeful mathematician. Many of those with whom he had contact have now been swept away by time's ever-flowing stream, but those who were conscious of the need to help the young, and those who were not, deserve to have their names on record.

In 1957 Larry began his long and distinguished service to the University of Minnesota. He was appointed Assistant Professor in 1957, Associate Professor in 1958, Professor in 1960 and Regents' Professor in 1980. He has served as Associate Head of the Department of Mathematics, Director of the Center for Control Sciences, and, from 1980 until recently, Director of the Control Science and Dynamical Systems Center. His 35 years of service have included the award of many ONR and NSF research grants, collaboration with colleagues in mathematics and engineering, the whole range of undergraduate teaching and the supervision of many research students, some of whom are now collaborators.

This is a record of service of which Larry himself will rarely speak, but to us, who have the knowledge and experience to make a judgement, this is a record of dedication and distinction which commands admiration.

In the United Kingdom the 1960s brought a recognition that the study of control theory and differential equations was not strong enough and required special research support. The University of Warwick, founded in 1960, had appointed Christopher Zeeman (now Sir Christopher Zeeman, FRS) to the Chair of Mathematics. He persuaded the then Science Research Council to allocate considerable funds to support a research programme at the newly founded Institute of Mathematics. This led to the start of Larry's long and distinguished association with the University of Warwick in England. From 1968 onwards he has served, for varying periods, as Nuffield and Leverhulme Professors of Mathematics, and Director of the Control Theory Centre of the University. In 1985 he was appointed to the position of Honorary Professor in the University.

Since it was started in the 1960s the Mathematics Institute at the University of Warwick has been a remarkable success in the development of the mathematical sciences in the United Kingdom. The Institute now ranks in excellence with those at the Universities of Oxford and Cambridge. My colleagues at the Institute are clear that this early decision to associate Larry Markus with the University of Warwick led to a contribution which has played a very significant role in this achievement.

It is rare to find a colleague who has the time, energy and ability to make such remarkable and outstanding contributions at this level to two Universities in different countries, with such different traditions in education.

One of the many keen pleasures in helping with the preparation of this Festschrift has been in working with my colleagues Bruce Lee in the University of Minnesota and David Elworthy in the University of Warwick. In this way these two Universities are paying tribute and recognition to the long years of outstanding service which Larry Markus has given so generously and with such enthusiasm.

The association with the University of Warwick has had a remarkable effect upon the lives of Larry and Lois Markus. For many years now they have divided their time between living in Minneapolis and nearby the University of Warwick. They remain citizens of the United States of America but in their affection for their adopted country, I have often thought that they should be offered Honorary Citizenship of the United Kingdom. They delight in their two homes, an apartment in Minneapolis and for some years now a flat in Royal Leamington Spa.

This is not the place to enter into a detailed assessment of the contribution which Larry Markus has made to mathematics. However, to read his *curriculum vitae* is to see a collection of monographs and research papers with a range of contributions to analysis, algebra and geometry. It must suffice here to note that the harvest of his labours in the vineyard of mathematics produced a choice cull from differential equations, control theory, dynamical systems, Hamiltonian systems, symplectic and differential geometry, linear algebra, group theory, non-standard analysis and cosmological mathematical models.

It is appropriate to single out here the text Foundations of Optimal Control Theory, written jointly with his colleague Professor E.B. Lee of the Department of Electrical Engineering at the University of Minnesota. The first edition of this book appeared in 1967 about the time that Larry was first appointed to the University of Warwick. There was a second edition in 1986, and the book has been translated into Russian and into Japanese. This monograph has held, for twenty-five years, a leading position in the subject of optimal control theory of both linear and non-linear systems; this is a remarkable achievement attained by very few research-orientated books in a field developing as rapidly as control theory has done, and continues to do.

In preparing for this volume the Editorial Board issued an invitation to Larry Markus to write a short history of control theory. This request received a positive response and the leading article is entitled "A Brief History of Control."

There are few, if any, who could have approached this responsibility with such erudition as Larry Markus, and no other who could have written with such panache and with such impressive command of the subject.

Larry has many other interests in life as well as mathematics and care for his family.

Many who consider themselves proficient at table tennis will recall their shattered ambitions with Larry at the opposite end of the table. Once on a visit to Minneapolis in the middle 1970s I was invited one evening to the home of James and Barbara Serrin. Larry arrived late in the evening, dressed in a track suit, in the company of another visitor to the University. It emerged in conversation that Larry and the visitor had come to the Serrins' following a session of table tennis. I ventured to ask the visitor if the game had gone well; he replied briskly that he did not wish to discuss the matter, but that he would never be the same man again. Perhaps our visitor should have been informed that in 1974 Larry had played for the Minnesota State Table Tennis Team.

Conversation on topics chosen by others is one of Larry's great pleasures, but beware that your statements are correct if talk steers you into US or UK politics, or the contents of the *London Times* or *Sunday Times* newspaper (I once ventured to speak of the 1776 American Rebellion, but never again, for the sake of his late Majesty King George III). On the other hand my wife and I were fortunate when on a visit to Minneapolis in October 1991 the baseball World Series was underway at the Metrodome. By the time the Twins had vanquished the Braves and we ourselves had survived the tension and extraordinary equal fortunes of the 6th and 7th games, a combination of Larry's tuition and the sheer excitement of these events made of us baseball enthusiasts for the years to come.

Larry Markus is an international mathematician. The long list of invitations to international meetings, including the major invitation to lecture at the International Congress of mathematicians in Vancouver in 1974, reads like a world tour of all sorts and conditions of countries. I have had the good fortune to be with him on a number of these occasions; in particular in Warwick, Birmingham, Colorado Springs, Vancouver, Oberwolfach and Beijing. The visit to the People's Republic of China in 1983 is memorable; our hosts were kindness personified but the pace of the tour was gruelling and Larry, Lois and I formed ourselves into a survival unit. There is a conference photograph taken in Beijing in that year with $n$ ($n$ large) Chinese friends and colleagues. It was unspeakably hot and humid but I turned up for the photographic session wearing my Oxford (Balliol College) tie and pointed out this symbol of British imperialism to Larry. Without a moment of hesitation Larry, who was wearing his shirt open-necked, put his hand in his pocket and pulled out a bow tie; with one flowing movement of arm and hand this symbol of American diplomacy was in place. There, if you look at the photograph, you will find the two of us, the only well-dressed but suffering figures amid a host of smiling relaxed faces.

There is a videotape of Larry's lecture to the Royal Institution in London in the year 1982. The lecture was designed to give (and succeeded in doing so) young scientists at school level a glimpse of modern mathematical ideas. If you wish to see Larry playing a guitar and singing like a professional, then here is your opportunity. To this there is the added attraction of a wonderful assistant to aid Larry in his endeavours with the young: Lois Markus.

But life has its downturns for all of us. Some five years ago Larry had to undergo serious and lengthy surgery on one side of his face and neck. The operation was successful but for upwards of many months the aftermath was hard to bear. All of Larry's friends and colleagues are delighted that in due course he made an excellent recovery and that indeed there is now little or no sign of his ordeal.

The contents of this volume speak for themselves. When David Elworthy, Bruce Lee and I issued our invitation for contributions we expected an enthusiastic response. We have in fact been overwhelmed by the number and quality of the contributions. Inevitably these have flowed in part outside the immediate areas suggested by the title of the volume. However, we have been moved by the number of former students, collaborators and colleagues of Larry Markus who seized this opportunity to recognize his influence and to thank him for his inspiring work as a mathematician.

Here then is our Festschrift. From near and from afar, from today and from down the long years we pay this tribute to Larry Markus. We who have had the privilege of his friendship and the manifold advantages of his mathematics offer this tribute to him in return.

W. Norrie Everitt

# Acknowledgments


When the idea of this Festschrift was conceived the Editorial Board:

| | |
|---|---|
| Professor K.D. Elworthy | (Mathematics, University of Warwick) |
| Professor W.N. Everitt | (Mathematics, University of Birmingham) |
| Professor Avner Friedman | (Director of the IMA, University of Minnesota) |
| Professor E.F. Infante | (Vice-President and Provost, and Professor of of Mathematics, University of Minnesota) |
| Professor E.B. Lee | (Electrical Engineering, University of Minnesota) |
| Professor Sir Christopher Zeeman | (Mathematics, University of Oxford) |

agreed to appoint K.D. Elworthy, W.N. Everitt and E.B. Lee as Coordinators for the volume. We take this opportunity to thank all members of the Board for their confidence, help and support.

For our part we thank our assistants

X.-M. Li (University of Warwick)

J.P. De Jager (University of Minnesota)

for their help in preparing the manuscripts, and in the process of obtaining referee reports on the separate contributions.

The Editorial Board approached the publisher Marcel Dekker, Inc., New York, to seek inclusion of the Festschrift volume in the prestigious series Lecture Notes in Pure and Applied Mathematics. The Board is grateful to Marcel Dekker for accepting its proposal and thanks the two Editors for the Series, Professor E.J. Taft and Z. Nashed, for their support in this venture.

The Coordinators take this opportunity to thank the Staff of Marcel Dekker, Inc. for their help and cooperation. In particular we are grateful to Maria Allegra (Acquisitions Editor) and Walter Brownfield (Production Editor) for continuing advice during the preparation of the volume.

Likewise we thank the secretarial staff of the Universities of Minnesota and Warwick for their contribution.

K. D. Elworthy
W. Norrie Everitt
E. Bruce Lee

# Contents

Preface *W. Norrie Everitt* v

Acknowledgments xii

Contributors xix

A Brief History of Control *Lawrence Markus* xxv

PART I: DIFFERENTIAL EQUATIONS

1. **De Motu Arietum (On the Motion of Battering Rams)** 1
*Rutherford Aris and Bernard S. Bachrach*

2. **A Note on Continuation Algorithms for Periodic Orbits** 15
*Pavol Brunovský and Milan Kubala*

3. **Uniformly Isochronous Centers of Polynomial Systems in $\mathbf{R}^2$** 21
*Roberto Conti*

4. **Some Remarks on the Titchmarsh-Weyl *m*-Coefficient and Associated Differential Operators** 33
*W. Norrie Everitt*

5. **Periodic Solutions of Single Species Models with Delay** 55
*H. I. Freedman and Huaxing Xia*

6. **Asymptotic Estimates for a Nonstandard Second Order Differential Equation** 75
*William A. Harris, Jr., Patrizia Pucci, and James Serrin*

7. **Asymptotic Phase, Shadowing and Reaction-Diffusion Systems** 87
*Morris W. Hirsch*

8. **Numerical Methods for Studying Parameter Dependence of Solutions to Schrödinger's Equation** 101
*M. Holthaus, C. S. Kenney, and A. J. Laub*

9. **Differential Systems and Algebras** 115
*Michael K. Kinyon and Arthur A. Sagle*

10. **Limit Cycles and Centres: An Example** 143
*N. G. Lloyd and J. M. Pearson*

11. **Stability Criteria with a Symmetric Operator Occurring in Linear and Nonlinear Delay-Differential Equations** 159
*James Louisell*

12. **Approximating Piston-Driven Flow of a Non-Newtonian Fluid** 173
*David S. Malkus, John A. Nohel, and Bradley J. Plohr*

13. **A Necessary and Sufficient Condition for Exponential Stability of Large-Scale Stochastic Delay Systems in Hierarchical Form** 193
*Xuerong Mao*

14. **The Averaging Method and the Problem on Separation of Variables** 213
*Yu. A. Mitropolsky*

15. **Construction of Lyapunov Functions Using Integration by Parts** 225
*Patrick C. Parks*

16. **Remarks on Williams' Problem** 235
*W. Parry*

17. **Approximations of the Long-Time Dynamics of the Navier-Stokes Equations** 247
*Victor A. Pliss and George R. Sell*

18. **About the Solution of Some Inverse Problems in Differential Galois Theory by Hamburger Equations** 277
*J. P. Ramis*

19. **A Remark on Bessel Functions** 301
*Yasutaka Sibuya*

20. **The Broadwell System, Self-Similar Ordinary Differential Equations, and Fluid Dynamical Limits** 307
*Marshall Slemrod and Athanasios E. Tzavaras*

21. **The Statistical Mechanics of Asset Prices** 321
*Michael Stutzer*

22. **The Winding Problem for Stochastic Oscillators** 343
*Ananda P. N. Weerasinghe*

23. **On the Convexity of Carrying Simplices in Competitive Lotka-Volterra Systems** 353
*E. C. Zeeman and M. L. Zeeman*

PART II: DYNAMICAL SYSTEMS AND CONTROL SCIENCE

24. **The Shuffle Product and Symmetric Groups** 365
*A. A. Agrachev and R. V. Gamkrelidze*

25. **Optimal Control of Infinite Dimensional Systems Governed by Integro Differential Equations** 383
*N. U. Ahmed*

26. **Data Analysis of a Lumped System** 403
*Louis Auslander*

27. **Ergodic Bellman Systems for Stochastic Games** 411
*Alain Bensoussan and Jens Frehse*

28. **Some Results on Feedback Stabilizability of Nonlinear Systems in Dimension Three or Higher** 423
*William M. Boothby and Riccardo Marino*

29. **Robust Stabilization of Infinite-Dimensional Systems with Respect to Coprime Factor Perturbations** 437
*Ruth F. Curtain and A. J. Pritchard*

30. **Time-Delayed Perturbations and Robust Stability** 457
*Richard F. Datko*

31. **Positive Controllability of Linear Systems with Delay** 469
*Mohamed A. El-Hodiri and F. S. Van Vleck*

32. **Discrete Time Partially Observed Control** 481
*Robert J. Elliott and John B. Moore*

33. **Symmetries of Differential Systems** 491
*Fabio Fagnani and Jan C. Willems*

34. **Relaxation in Semilinear Infinite Dimensional Control Systems** 505
*H. O. Fattorini*

35. **An Algebraic Approach to Hankel Norm Approximation Problems** 523
*P. A. Fuhrmann*

36. **Stabilizing Solutions to Riccati Inequalities and Stabilizing Compensators with Disturbance Attenuation** 551
*Aristide Halanay*

37. **Vector Field Approximations Preserving Structural Properties** 569
*Henry Hermes*

38. **Nonlinear Boundary Stabilization of a von Kármán Plate Equation** 581
*Mary Ann Horn and Irena Lasiecka*

39. **Global Null Controllability of Linear Control Processes with Positive Lyapunov Exponents** 605
*Russell Johnson and Mahesh Nerurkar*

40. **Linear Two-Dimensional Systems with Deviating Arguments** 623
*Tadeusz Kaczorek*

41. **Boundary Controllability in Transmission Problems for Thin Plates** 641
*John E. Lagnese*

42. **Approximation of Linear Input/Output Delay-Differential Systems** 659
*E. Bruce Lee*

43. **Local Smoothing and Energy Decay for a Semi-Infinite Beam Pinned at Several Points, and Applications to Boundary Control** 683
*Walter Littman and Stephen W. Taylor*

44. **Abnormal Sub-Riemannian Minimizers** 705
*Wensheng Liu and Héctor J. Sussmann*

45. **Some Algebraic Approaches for Stability Analysis of Two-Dimensional Systems and Digital Filters** 717
*Wu-Sheng Lu*

46. **A Control-Theoretic Banach Lie Group $G_{A,B}$—The Stability Group** 737
*Dahlard L. Lukes*

47. **Min-Max Game Theory and Algebraic Riccati Equations for Boundary Control Problems with Analytic Semigroups: The Stable Case** 757
*Christine A. McMillan and Roberto Triggiani*

48. **Approximate Controllability of Linear Functional-Differential Systems: A State Space Independent Approach** 781
*Andrzej W. Olbrot*

49. **The Attainability Order in Control Systems** 793
*Emilio O. Roxin*

50. **Extending Linear-Quadratic Optimal Control Laws to Nonlinear Systems and/or Nonquadratic Cost Criteria** 799
*D. L. Russell and Xiaohong Zhang*

51. **Existence of Optimal Controls for a Free Boundary Problem** 859
*Thomas I. Seidman*

52. **Maximum Principle for Optimal Control of Distributed Parameter Stochastic Systems with Random Jumps** 867
*Shanjian Tang and Xun-Jing Li*

53. **Spillover Problem and Global Dynamics of Nonlinear Beam Equations** 891
*Yuncheng You*

54. **A Dynamical Systems Approach to Solving Linear Programming Problems** 913
*Stanislaw H. Żak, Viriya Upatising, Walter E. Lillo, and Stefen Hui*

# Contributors

**A. A. Agrachev** Steklov Mathematics Institute, Moscow, Russia

**N. U. Ahmed** Departments of Electrical Engineering and Mathematics, University of Ottawa, Ottawa, Ontario, Canada

**Rutherford Aris** Department of Chemical Engineering and Materials Sciences, University of Minnesota, Minneapolis, Minnesota

**Louis Auslander** Graduate School and University Center, City University of New York, New York, New York

**Bernard S. Bachrach** Department of History, University of Minnesota, Minneapolis, Minnesota

**Alain Bensoussan** Institut National de Recherche en Informatique et en Automatique, Le Chesnay, France

**William M. Boothby** Department of Mathematics, Washington University, St. Louis, Missouri

**Pavol Brunovský** Institute of Applied Mathematics, Comenius University, Bratislava, Slovakia

**Roberto Conti** Department of Mathematics, University of Florence, Florence, Italy

**Ruth F. Curtain** Mathematics Institute, University of Groningen, Groningen, The Netherlands

**Richard F. Datko**, Department of Mathematics, Georgetown University, Washington, D.C.

**Mohamed A. El-Hodiri** Department of Economics, University of Kansas, Lawrence, Kansas

**Robert J. Elliott** Statistics and Applied Probability Department, University of Alberta, Edmonton, Alberta, Canada

**W. Norrie Everitt** School of Mathematics, The University of Birmingham, Birmingham, England

**Fabio Fagnani** Classe di Scienze, Scuola Normale Superiore, Pisa, Italy

**H. O. Fattorini** Department of Mathematics, University of California at Los Angeles, Los Angeles, California

**H. I. Freedman** Department of Mathematics, University of Alberta, Edmonton, Alberta, Canada

**Jens Frehse** Institüt für Angewandte Mathematik der Universität Bonn, Bonn, Germany

**P. A. Fuhrmann** Ben-Gurion University of the Negev, Beer-Sheva, Israel

**R. V. Gamkrelidze** Steklov Mathematics Institute, Moscow, Russia

**Aristide Halanay** Faculty of Mathematics, University of Bucharest, Bucharest, Romania

**William A. Harris, Jr.** Department of Mathematics, University of Southern California, Los Angeles, California

**Henry Hermes** Department of Mathematics, University of Colorado, Boulder, Colorado

**Morris W. Hirsch** Department of Mathematics, University of California at Berkeley, Berkeley, California

**M. Holthaus** Center for Nonlinear Sciences and Center for Free-Electron Laser Studies, Department of Physics, University of California at Santa Barbara, Santa Barbara, California

**Mary Ann Horn** Institute for Mathematics and Its Applications, University of Minnesota, Minneapolis, Minnesota

**Stefen Hui** Department of Mathematical Sciences, San Diego State University, San Diego, California

**Russell Johnson** Dipartimento di Sistemi e Informatica, Universitá di Firenze, Florence, Italy

**Tadeusz Kaczorek** Department of Electrical Engineering, Warsaw University of Technology, Warsaw, Poland

**C. S. Kenney** Department of Electrical and Computer Engineering, Center for Control Engineering and Computation, University of California at Santa Barbara, Santa Barbara, California

**Michael K. Kinyon** Department of Mathematics and Computer Science, Indiana University at South Bend, South Bend, Indiana

**Milan Kubala** Institute of Applied Mathematics, Comenius University, Bratislava, Slovakia

**John E. Lagnese** Department of Mathematics, Georgetown University, Washington, D.C.

**Irena Lasiecka** Department of Applied Mathematics, University of Virginia, Charlottesville, Virginia

**A. J. Laub** Department of Electrical and Computer Engineering, Center for Control Engineering and Computation, University of California at Santa Barbara, Santa Barbara, California

**E. Bruce Lee** Department of Electrical Engineering, University of Minnesota, Minneapolis, Minnesota

**Xun-Jing Li** Department of Mathematics and Institute of Mathematics, Fudan University, Shanghai, China

**Walter E. Lillo** Department of Systems Evaluation, The Aerospace Corporation, Los Angeles, California

**Walter Littman** School of Mathematics, University of Minnesota, Minneapolis, Minnesota

**Wensheng Liu** Institute for Mathematics and Its Application, University of Minnesota, Minneapolis, Minnesota

**N. G. Lloyd** Department of Mathematics, University College of Wales, Aberystwyth, Wales

**James Louisell** Department of Mathematics, University of Southern Colorado, Pueblo, Colorado

**Wu-Sheng Lu** Department of Electrical and Computer Engineering, University of Victoria, Victoria, British Columbia, Canada

**Dahlard L. Lukes** Department of Applied Mathematics, University of Virginia, Charlottesville, Virginia

**David S. Malkus** Department of Engineering Mechanics and Astronautics, Center for the Mathematical Sciences, and Rheology Research Center, University of Wisconsin, Madison, Wisconsin

**Xuerong Mao** Department of Statistics and Modelling Science, University of Strathclyde, Glasgow, Scotland

**Riccardo Marino** Department of Electronic Engineering, University of Rome "Tor Vergata," Rome, Italy

**Christine A. McMillan** Department of Applied Mathematics, University of Virginia, Charlottesville, Virginia

**Yu. A. Mitropolsky** Institute of Mathematics, Ukranian Academy of Sciences, Kiev, Ukraine

**John B. Moore** Research School of Physical Sciences, The Australian National University, Canberra, Australia

**Mahesh Nerurkar** Department of Mathematics, Rutgers University, Camden, New Jersey

**John A. Nohel** Forschungsinstitüt für Mathematik, ETH-Zentrum, Zürich, Switzerland, and Department of Mathematics, University of Wisconsin, Madison, Wisconsin

**Andrzej W. Olbrot** Department of Electrical and Computer Engineering, Wayne State University, Detroit, Michigan

**Patrick C. Parks** Department of Mathematics, ACM Group, Royal Military College of Science, Shrivenham, England

**W. Parry** Mathematics Institute, University of Warwick, Coventry, England

**J. M. Pearson** Department of Mathematics, University College of Wales, Aberystwyth, Wales

**Victor A. Pliss** Faculty of Mathematics and Mechanics, University of St. Petersburg, St. Petersburg, Russia

**Bradley J. Plohr** Department of Mathematics, State University of New York at Stony Brook, Stony Brook, New York

**A. J. Pritchard** Mathematics Institute, University of Warwick, Coventry, England

**Patrizia Pucci** Department of Mathematics, Universitá di Perugia, Perugia, Italy

**J. P. Ramis** Institute de Mathématiques et d'Informatique, Université de Louis Pasteur, Strasbourg, France

**Emilio O. Roxin** Department of Mathematics, University of Rhode Island, Kingston, Rhode Island

**D. L. Russell** Department of Mathematics, Virginia Polytechnic Institute and State University, Blacksburg, Virginia

**Arthur A. Sagle** Department of Mathematics, University of Hawaii at Hilo, Hilo, Hawaii

**Thomas I. Seidman** Department of Mathematics and Statistics, University of Maryland, Baltimore County, Baltimore, Maryland

**George R. Sell** School of Mathematics, University of Minnesota, Minneapolis, Minnesota

**James Serrin** School of Mathematics, University of Minnesota, Minneapolis, Minnesota

**Yasutaka Sibuya** School of Mathematics, University of Minnesota, Minneapolis, Minnesota

**Marshall Slemrod** Department of Mathematics, University of Wisconsin, Madison, Wisconsin

**Michael Stutzer** Carlson School of Management, University of Minnesota, Minneapolis, Minnesota

**Héctor J. Sussmann** Department of Mathematics, Rutgers University, New Brunswick, New Jersey

**Shanjian Tang** Department of Mathematics and Institute of Mathematics, Fudan University, Shanghai, China

**Stephen W. Taylor** Department of Mathematics, Montana State University, Bozeman, Montana

**Roberto Triggiani** Department of Applied Mathematics, University of Virginia, Charlottesville, Virginia

**Athanasios E. Tzavaras** Department of Mathematics, University of Wisconsin, Madison, Wisconsin

**Viriya Upatising** School of Electrical Engineering, Purdue University, West Lafayette, Indiana

**F. S. Van Vleck** Department of Mathematics, University of Kansas, Lawrence, Kansas

**Ananda P. N. Weerasinghe** Department of Mathematics, Iowa State University, Ames, Iowa

**Jan C. Willems** Department of Mathematics, University of Groningen, Groningen, The Netherlands

**Huaxing Xia** Department of Mathematics, University of Alberta, Edmonton, Alberta, Canada

**Yuncheng You** Department of Mathematics, University of South Florida, Tampa, Florida

**Stanislaw H. Żak** School of Electrical Engineering, Purdue University, West Lafayette, Indiana

**E. C. Zeeman** Hertford College, University of Oxford, Oxford, England

**M. L. Zeeman** Division of Mathematics, Computer Science and Statistics, University of Texas at San Antonio, San Antonio, Texas

**Xiaohong Zhang** Department of Mathematics, Virginia Polytechnic Institute and State University, Blacksburg, Virginia

# A Brief History of Control

**Lawrence Markus** University of Minnesota, Minneapolis, Minnesota

LET THERE BE LIGHT!

So went the seminal control command that created the physical universe [Mo] [Ma 6], and gave it impetus for the explosive expansion of the original Big-Bang [Haw] [Ma 3]. From the viewpoint of control theory the primary question is:

> Was the creation an open-loop command initiating the response at a specified time-say, at $t = 0$?

or

> Was this a closed-loop command set to trigger action at a specified state of the universe–say, at a quantum-mechanical negative-vacuum?

Probably we shall never know the answer since the experiment is too difficult to repeat, and, despite scientific chutzpah, the state space is too intricate to analyze.

### 1. Pre-human Control Era.

Within the first few seconds following the Big Bang, the expanding sphere [Ma 2] [Ma 8] of intense radiant energy became less dense and hot until the radiation began to condense into protons, electrons and the elementary particles of matter–even into simple hydrogen atoms (see: modern cosmology mythology). These atoms gravitated into massive gaseous balls with central pressures and temperatures so great that the hydrogen atoms fused to produce helium in a nuclear fire. Thus a star is born; and the stage is set for the second great control experiment.

A star is intrinsically an unstable dynamical system, since the inward gravitational pull on the surface layers of the star intensifies as the radius decreases, and thus gravity acts as positive feedback. But the radius of the star can be feedback-stabilized through the outward radiation pressure generated by the nuclear fusion at the center. If the gravitational force pulls the surface layers of the star inward from the equilibrium radius, the central pressure and nuclear fusion intensifies, and then the outward radiation pressure restores the radius towards equilibrium.

But what happens to the star when all the available hydrogen at the center has been converted into helium? Why then the equilibrium fails and the radius collapses, intensifying the central pressure until the helium begins to fuse. This establishes a new stable equilibrium between inward gravitation and outward radiation. But what happens when the helium is exhausted? Why then the succeeding equilibria fail until the star collapses to the point where super-fusions of higher elements occur causing a supernova explosion, fragmenting the star and ejecting the debris of carbon, oxygen, nitrogen, phosphorous, and iron–the elements of life. Thus ultimately a star is a failed control system–unless a supernova instability is deemed a design success.

After this spectacular beginning for the astronomical universe, nothing of importance happens for about 10 billion years (see: current astrophysical mythology). Then commenced the great Gaia experiment. The goddess Gaia (or Gaea) is the Mother Earth who nurtures the biosphere of life, and the human species in particular, (see: classic Greek mythology). The Gaia Hypothesis [G] refers to the action of the biosphere as a feedback stabilizer for the geological dynamics of the Earth's atmosphere and hydrosphere.

For instance consider the control problem of stabilizing the ratio of the relative densities of $CO_2$ and $O_2$ in the atmosphere. If the $CO_2$ density rises, then the foliage of the tropical rain forests and the plankton of the antarctic oceans luxuriate in the resulting Green House effect. These plants then grow with augmented energy arising from photosynthesis, and thereby consume the $CO_2$ and release $O_2$. On the other hand, if $O_2$ density rises, then forest fires flare and more $CO_2$ is produced. By such regulator feedback mechanisms this atmospheric equilibrium has been kept quasi-stable for almost a billion years (see: concerned ecological mythology).

Popular misconceptions of these biochemical interactions have led to the belief that "Nature" has a consciousness or intelligent volition, rather than to the recognition of the system-theoretic phenomenon of negative feedback. Will the Gaia Hypothesis and the Biospheric-compensator also turn out to be a failed control system? We do not know as yet, and it seems rather preposterous and presumptuous to predict consequences, or to try to control this complicated system.

## 2. Artisans and Technicians Control Era.

Since the most antique times, artisans and technicians have skillfully employed control methods in refining ore, working glass, and most spectacularly in regulating large-scale irrigation processes by the use of water wheels, screw pumps, and dam flues [D] [Nee]. The diorite slab inscribed with the Code of Hammurabi (1700 BC) lists regulatory procedures for communal cooperation in agricultural irrigation [Ham]; and this trend can be followed through the legal Code of Justinian (and the approximately contemporaneous Talmud, about 500 AD), and thence to the Congressional Colorado River Bill of 1956 for reservoir and dam control valid to the year 2000 AD.

It is not entirely clear that such endeavors fit the framework of automatic control, but there are many individual examples of bona fide authenticity. Detailed records over centuries describe the perfection of the float value, for regulating the water level in a tank, by the Greek Ktesibios (250 BC); the development of the water clocks in Egypt and China; and the introduction of an oven temperature regulators by C. Drebbel about 1600 AD [Th].

At the beginning of the industrial revolution, with the great advances in instrumentation, we find pendulum clocks improved by Christian Huygens and reaching perfection in the chronometer of John Harrison (1763), with an intricate design of motion and temperature compensators. But certainly Watt's application of the centrifugal governor for regulating steam engines (1775) can be taken as the start of modern control engineering.

**3. Early scientific Control Era** (1850–1900).

In 1868 J. C. Maxwell published a mathematical analysis of Watt's governor and related mechanical regulators. In his famous paper "On Governors", Maxwell [Max] applied the principles of Newtonian mechanics to describe the action of such a governor by linear differential equations with constant coefficients. He recognized that a necessary and sufficient condition for (asymptotic) stability was that the corresponding characteristic polynomial should have roots only in the left-half complex plane. After resolving the stability question for third degree equations, he then declared: "I have not been able completely to determine these conditions for equations of a higher degree than the third; but I hope that the subject will obtain the attention of mathematicians".

For real polynomials the stability problem was attacked by C. Hermite and then totally resolved in a famous paper of A. Hurwitz (1894) [Hu]. The applications to engineering were emphasized by J. Vishnegradskii (1876) [V].

Yet here we are, over a century later, still at it–for partial differential equations and functional differential equations–with Hardy space $H^\infty$, Paley-Wiener theory, and matrices over Noetherian rings! (see: Abstract mathematical mythology).

The research in dynamics, especially stability theory including the work of Routh (1877), is well reported in the foundational two volumes "Theory of Sound" by Baron Rayleigh (1894) [Ra]. It is interesting to note that in the section on power line transmission in volume I, Rayleigh quotes the earlier work of Heaviside " ... circuit ... is wholly cleared of electrification and current in the time $\ell/v$". This was in connection with feedback boundary damping for the one-dimensional wave equation by what is now called "matched impedance".

The other great treatise of this era on the stability of dynamical systems was the remarkable volume "Problème général de la stabilité du mouvement" by A. Lyapunov (1892) [Ly1] [Not to be confused with the important paper on the mathematical basis of bang-bang control "Sur les fonctions-vecteurs complètement additives" by a different A. Lyapunov (1940) [Ly2]]. Here Lyapunov (the first one) introduced two methods of analyzing the stability of a vector nonlinear differential system: Method I consists of linearization and relates to the local behavior near an equilibrium; Method II depends on dissipation of an energy (Lyapunov) function and yields a global result. These two methods of stability analysis lie at the heart of many subsequent investigations over the past century.

**4. Classical Engineering Control Era** (1900–1940).

The central problem of control engineering for the first half of this century was the accurate reproduction and amplification of a signal on a telephone cable. This amplification, only practical following De Forest's invention of the triode tube, was essential for long distance communication, and the problems posed and resolved led to the classical methods of control engineering.

In the opening decades of the century O. Heaviside and C. Steinmetz pioneered the path through the algebraization of linear differential systems by means of operator calculus and complex analysis. Although these methods had been known in a mathematical context since the times of Lagrange and Laplace, they were now available in a format that matched the approaches of electrical engineers. But the most important discoveries were by H. Nyquist ("Regeneration Theory", 1932) [Ny], and shortly thereafter, H. Bode and W. Evans. These papers first introduced the transfer function for a feedback circuit, and then developed practical graphical methods for analyzing the stability of the circuit for various possible compensators. Essentially the Nyquist plot, the Bode diagram, and the Evans root-locus technique all depend on classical results in complex function theory–particularly the Principle of the Argument.

In retrospect the fundamental paper of Nyquist appears unnecessarily complicated. Instead of defining the transfer function around a feedback loop in terms of an equilibrium condition, it is constructed by considering an infinite series of regenerations of the current around the loop. The complex analysis repeats much of the classical analysis of complex integration theory with hypotheses of integrands of "bounded variation", and integration paths "over the appropriate Riemann surface". This exposition caused the importance of the basic discovery to be misjudged by many electrical engineers, who laughed at the paper as a "snow job" [Th]. Nevertheless Nyquist does come to his following:

> "Rule: Plot plus and minus the imaginary part of $A\,J(iw)$ against the real part for all frequencies from 0 to $\infty$. If the point $1 + i0$ lies completely outside this curve the system is stable; if not it is unstable".

While much of this research was accomplished at Bell Telephone Laboratories, and later at the MIT Radiation Laboratory [JNP], in connection with problems of telephone and radar operation, there is another work of that era which is little known and which has had great consequences, as explained to me by my colleague, Alfred Nier, Regents' Professor of Physics at the University of Minnesota. In 1935 Nier published a short note on the operation of the mass spectrometer [Ni], in which he describes the problem:

> "In practically all methods of positive ray analysis a combination of an electric and a magnetic field is used to separate the ions of different masses ... The fact that the deflection suffered by an ion depends upon both the magnetic field and the electric field suggests that instead of attempting to keep the magnetic field constant one could cause the electric field to fluctuate automatically along with the magnetic field in such a manner that the deflection of an ion would be independent of the magnetic field fluctuations.
>
> "This can be accomplished very simply by a vacuum tube amplifier."

This enormously important discovery by Nier made the mass spectrometer a practical instrument for guiding the separation of Uranium 235 and 238. It was the key to all future development in the production of atomic energy, and research in nuclear physics.

In summary the classical control era was concerned with the stabilizing of feedback compensating circuits, which were described mathematically in terms of linear ordinary differential equations with constant coefficients. A few attempts had been made to extend these methods to equations with time-varying coefficients, as with the describing-function of Kochenburger [Ko], and to allow for some stochastic noise and signal delays [Mi2], but such developments were largely left to future generations.

## 5. Innovations of Control in WWII Era.

During the Second World War the main stimulus to control engineering arose from the development of new technologies of weaponry, especially in aeronautics and marine navigation. These constructions, of jet fighter and long-range bomber aircraft and ballistic missiles in military offense, as well as radar and electronic detection and navigation in defense, all demanded high level innovations of science and engineering.

In the USA much of this research was conducted at MIT, particularly in the Radiation Laboratory that was mainly concerned with the perfection of radar. The control problem was to use the radar data as feedback to guide the gun tracking of enemy aircraft. Thus there were major efforts in the control theory of target-tracking and also target-prediction. New sophisticated mathematics theories involving stochastic processes and statistical methods of functional analysis came to prominence, led by the researches of Norbert Wiener [W]. Some of these topics, specially the more applicable results, were reported in the influential volume "Theory of Servomechanisms" by James, Nichols and Phillips [JNP].

In the USSR the theory of nonlinear vibrations and oscillations was advanced by Krylov and Bogoliubov [Bo K], and their followers, while strong cohorts of theoretical engineers were led by A. M. Letov [Let], A. I. Luré and many others. The prediction theory of Kolmogorov was developed independently from that of Wiener.

Any detailed study of the control advances in the era would require volumes to record, and more scholarship than I possess. Essentially this would constitute a great chapter in the history of technology, and so I leave this topic with only the few remarks offered above.

## 6. Mathematical Control Era (1950–1970).

It is always easier to write a survey about something on which your knowledge is restricted to vague generalities, rather than to report accurately on a complex collection of events in which you are an individual participant. In this latter case the report is colored by your own attitudes and aptitudes. Accordingly, this brief sketch of control science of the decades following 1950 must be regarded as a personal history, rather than an objective chronicle.

Shortly after the war Solomon Lefschetz, a world famous topologist and past President of the AMS, argued that America should have a national policy of potent strength in fields of applied mathematics. With the enthusiastic encouragement of the Office of Naval Research, Lefschetz established a center for research in nonlinear ordinary differential equations (ODE) and dynamics at Princeton University–somewhat in parallel to the new Courant Institute at NYU which specialized in partial differential equations (PDE). Lefschetz was mainly interested in nonlinear oscillations and stability theory, as evidenced by his text [Lef]. After Lefschetz retired from Princeton, he moved his research group first to RIAS (research institute of Martin Corporation) and later to Brown University where it became the Lefschetz Center for Dynamical Systems under the direction of Joseph La Salle, with the cooperation of Wendell Fleming, E. (Jim) Infante, Harold Kushner, and others.

I joined the Princeton group as Lecturer and Lefschetz' assistant (1955–1957), and helped to organize his weekly seminar. There were many senior mathematicians who participated in these programs including R. Bellman, L. Cesari, J. La Salle, J. Leray, L. S. Pontryagin, as well as more junior experts like A. Antosiewicz, R. Bass, S. Diliberto, J. Hale, R. Kalman, J. Moser, M. Peixoto, G. Reeb, and S. Smale.

Lefschetz who was then a quite elderly mathematician (but no older than I am now [Ma 5]), was born in Moscow in 1884 but began his professional career as an engineering graduate of the Ècole Centrale of Paris [Ma 7]. Later he emigrated to the United States as an industrial engineer. But in a tragic engineering accident he lost both his arms, and thereafter used artificial limbs and hands with great skill in daily life and even in his manifold writings and blackboard-chalk lectures. His discoveries in algebraic topology and algebraic geometry were among the high points of twentieth century mathematics. His enthusiasms and scientific insights were an inspiration to his many students and colleagues and he was certainly a major influence on my mathematical development.

Many of the mathematicians and theoretical engineers from the Princeton-RIAS-Brown groups later established their own centers of research. At the University of Minnesota I cooperated with my colleague Bruce Lee in organizing the Center for Control Science and Dynamical Systems. I was the first Director of this Center, and later Bruce took over that responsibility with the help of other colleagues like K. S. P. (Pat) Kumar. Our scientific cooperation culminated in the writing of our joint treatise, ***Foundations of Optimal Control Theory*** (1967) [LM], which was widely used and subsequently translated into Russian and into Japanese.

In 1968 I became associated with the University of Warwick, England, where I alternated my duties with Minnesota. At Warwick, with the encouragement of E. C. Zeeman, I established and directed the Control Theory Centre (later directed by P. C. Parks and A. J. Pritchard).

The lists of lecturers and visitors at these two Centers (Centres) incorporate many of the most distinguished control scientists and world leaders in related mathematical and engineering fields. My own research activities, and those of my doctoral students, emphasized nonlinear ordinary differential equations–especially the geometric and algebraic aspects of control theory, (lately more in the direction of PDE and hybrid control systems, see [Li M1], [Li M2], and [MY]). Bruce Lee became a world authority in delay-differential (or more general hereditary functional differential equations (FDE)) control systems, with an entire school of doctoral and post-doctoral students following his leadership.

Meanwhile, back in the USSR at the Steklov Institute in Moscow, L. S. Pontryagin had established a somewhat similar concentration of young mathematicians with strong interests in nonlinear differential equations. These included his fellow Lenin Prize winners, V. Boltyanski, R. Gamkrelidze, and E. Mischenko, who coauthored the famous seminal book, "*The mathematical theory of optimal processes*"[PBGM]. Other younger associates, now famous mathematicians, included D. Anosov and V. I. Arnol'd.

Like Lefschetz, Pontryagin was world famous as an algebraic topologist (homotopy groups, cohomology classes) and as the creator of the duality theory for topological groups. He, too, had suffered an early accident which left him blinded. Also, although it was not widely known by the mathematical establishment, Pontryagin had long pursued a creative career in applied mathematics. For instance, Pontryagin generalized the Hurwitz stability criterion to cover linear delay-differential equations [P1], which have exponential-polynomial characteristic functions. Also it was he who first introduced the concept of structural stability (robustness) for dynamical systems [P2].

Other important research centers in the USSR developed around Mitropolsky in Kiev, Krasovskii and Myshkis in Sverdlovsk, as well as other powerful scientific institutions in Leningrad and Moscow. The famous Institute of Automatic Control and Telemechanics in Moscow was directed by A. I. Luŕe and later by A. Letov (and A. Ya. Lerner) for several years, after which Letov became the director of the center for environmental control at the International Institute for Applied Systems Analysis (IIASA) near Vienna.

The build-up of research in control theory proceeded at the Universities of Oxford, Cambridge, Warwick, Manchester, and Imperial College in London. In France the Paris based IRIA was led by J. L. Lions, and other institutes soon followed in the rest of Europe as well as in Japan and in South America. It would be futile, and also presumptuous, for me to try to give any full impression of the manifold of distinguished research centers and institutes that had established scientific centers of cooperation between mathematicians and control engineers by the 1970's. My personal interactions were with A. V. Balakrishnan [Ba] and L. Neustadt in California.

As this era came to full flowering, it could be asserted that, from an overall viewpoint, the mathematical researches in the areas of control science were expanding into new areas:

(i) Distributed parameter systems (PDE) based on the revolutionary treatises of A. G. Butkovskiy [Bu] and J. L. Lions [Li 1]. The former staying closer to the engineering tradition, and the latter bringing into play the full Bourbaki apparatus of modern functional analysis and partial differential equations.

(ii) Nonlinear differential systems (ODE) involving new methods of Lie algebras and Lie groups–both in the deterministic and the stochastic cases–with generalizations stretching towards (FDE).

(iii) Linear systems theory founded on the concepts of controllability, observability, and identification. Here the interplay between the time-domain and the frequency-domain was to be exploited by a virtuosity of abstract algebra, with the lead taken by Rudy Kalman, and H. Rosenbrock.

## 7. An anecdotal history of two astonishing surprises.

At the International Congress of Mathematicians in Edinburgh in 1958 L. S. Pontryagin presented a major invited address "Optimal Processes of Regulation" [P3]. At that time Russians in the West were exotic phenomena, and the excitement of the occasion was inflated by the fame and mathematical eminence of Pontryagin–the great topologist who was renowned for his researches into cohomology by means of characteristic classes and for the duality theory of topological groups. The lecture hall was overflowing and Lipman Bers had assumed his place for an English simultaneous translation of the lecture.

The international mathematical audience, led by a concentration of abstract topologists, were flabbergasted and astonished as the lecture developed. Pontryagin seemed to be talking about some kind of engineering, leaving them feeling ignorant and confused. The Maximum Principle of Pontryagin seemed mysterious and incomprehensible, except to those relatively few who were experts on control theory and who were already familiar with the Bang-Bang principle in the linear case treated previously by Bellman [BGG]. I recognized this direction of mathematical analysis, and also I had previously met Pontryagin (as Lefschetz' assistant) and knew very well of his long-standing interest in the problems of applied dynamical systems.

After a few days of reflection and gossip at the Congress, the mathematical establishment decided that the Maximum Principle wasn't about engineering, after all, but was instead a topic in the classical calculus of variations (similar to earlier results of M. Hestenes [Hes]). Thus the consensus of the Congress came to the conclusion that control theory might be mathematically respectable, but that it was dull and boring–the position still held by many of the avant-garde of abstract mathematicians.

In 1961 R. Kalman and R. Bucy published their remarkable paper [BuK] on the optimal control of linear systems, with Gaussian disturbances, and relative to a quadratic cost functional. The mathematical result was that the optimal controller could be synthesized through the solution of a Riccati matrix differential equation. Accordingly, a reasonably elementary computer program could generate the desired optimal controller, which also served as a feedback stabilizer.

Not only was this analysis elegant mathematics, but more importantly, it led directly to practical hardware for engineering control of guidance devices. In particular, the Kalman-Bucy filter was an absolutely essential component of the guidance system for the Apollo Moon-Rocket.

The industrial engineering community were flabbergasted and astonished that any such "fancy mathematics" could lead to useful and important engineering products. Thus mathematical control theory was gradually accepted and deemed to be potentially useful and practical. For many years, whenever control theoreticians were challenged to defend their craft to corporate engineering directors or to governmental funding agencies, the automatic and parrot-like response would ring out, "Kalman filter".

## 8. New trends for control theory (1970–1990).

During the past quarter century new trends have appeared and flourished in the mathematical theory of control systems. Some of the recent branches on the tree of mathematical control are:

i) Algebraization of frequency-domain analysis, leading into the realm of ring theory and algebraic geometry. The works of R. Kalman [Kal] and his colleagues, particularly E. Sontag, are fundamental developments.

ii) Partial differential equations and functional analysis for distributed parameter systems, and systems with infinite dimensional state spaces. Pioneers in this field include H. Fattorini [Fa], J. L. Lions [Li2], D. Russell [Rus]. New researches are carried forward by their students and colleagues.

iii) Lie groups and algebras in a geometric treatment of nonlinear control systems. See papers of R. Brockett [Br], H. Hermes [HL], A. Isidori [Is], H. Sussmann [JS], et al. Also I should recall the historic role played by C. Carathéodory (1909) in his axiomation of classical thermodynamics, where the reachable states under cyclically reversible processes are described by Lie algebra techniques [Ca].

iv) Functional differential equations, especially hereditary or delay-differential equations, are an active field for investigations by J. Hale and E. B. Lee, each with schools of doctoral students.

v) Frequency-domain analysis for infinite dimensional systems, as described by PDE or FDE, leading to new types of problems in complex analysis. In particular, Hardy space $H^\infty$ is at the center of these studies. This is paradoxical and is much to the chagrin of Hardy's spirit, in view of the total scorn and hostility to applications frequently expressed in the writings of Hardy [Har]

vi) Fields so new and under such continual turbulent re-evaluation (e.g. robotics, computer logic, and bioengineering based on physiological homeostasis) that their influence on control theory or vice versa is impossible to assess at this time.

But for insights into the future research for the next quarter century, one could do no better than to review the contributions to this Festschrift volume.

While all these developments of control theory may have been of only uncertain value to control engineering, there is no question but that they have been a bonanza for mathematics.

**Acknowledgement.**

I thank the editors for inviting me to prepare this brief essay on the development of mathematical control theory, and to record some personal reminiscences, as suggested in the Preface.

## Selected References


[Ba] Balakrishnan, A. V., *Applied functional analysis*, Springer-Verlag, 1976.

[Be 1] Bellman, R., *On the application of the theory of dynamic programming to the study of control processes*, Proc. Sym. on Nonlinear Circuit Analysis, Polytech Inst. Brooklyn (1956), 199–213.

[Be 2] Bellman, R., *Control Theory*, Scientific Amer. (Sept. 1964), 186–200.

[BGG] Bellman, R., I. Glicksberg, and O. Gross, *On the "Bang-Bang" control problem*, Quart. Appl. Math., 14 (1956), 11–18.

[Be K] Bellman, R. and R. Kalaba, *Selected papers on mathematical trends in control theory*, Dover, 1964.

[Bo K] Bogoliubov, N. and N. Krylov, *Introduction to Nonlinear Mechanics*, Princeton Annals of Math. Studies (1949).

[Br] Brockett,R., *Lie algebras and Lie groups in control theory, Geometric Methods in Systems Theory*, Reidel Publ. Boston (1973), 43–82.

[Bu K] Bucy, R. and R. Kalman, *New results in linear filtering and prediction theory*, Trans. ASME, J. Basic Eng., Series 83D (1961), 95–108.

[Bu] Butkowskiy, A. G., *Theory of Optimal Control of distributed parameter systems*, Elsevier NY, 1969.

[Ca] Carathéodory, C., *Unter suchungen über die Grundlagen der Thermodynamik*, Math. Ann. **67** (1909), 355-386.

[Con] Conti, R., *Problemi di controllo e di controllo ottimale*, Turin, Union Typographico editrice, 1974.

[CP] Curtain, R. and A. J. Pritchard, *Functional analysis in modern applied mathematics*, Acad. Press NY, 1977.

[D] DaVinci, Leonardo (C. Singer, et al, ed.), see "History of Technology III", Oxford Univ. Press, London (1957).

[E] Elworthy, K. D., *From local times to global geometry*, Pitman Research Notes in Math 150, 1987.

[EM] Everitt, W. N. and L. Markus, *Nonlinear quasi-differential control systems*, Resultate der Mathematik (1992), 65–82.

[Fa] Fattorini, H., *On complete controllability of linear systems*, J. Diff. Eqs. **3** (1967), 391–402.

[Fr] Friedman, A., *Optimal control for hereditary processes*, Arch. Rat. Mech. Anal. **15** (1963), 396–416.

[G] *Gaia Hypothesis*, see Gaia: a new look at life on earth, J. E. Lovelock, Oxford U. Press, 1987.

[Ham] Hammurabi, see "The oldest code of laws in the world", sections 53–56, by C. John, Edinburgh 1903.

[Har] Hardy, G., *A mathematician's apology*, Cambridge U Press, 1967.

[Haw] Hawking, S., *A brief history of time*, Bantam Publ. NY, 1988.

[HL] Hermes, H. and J. LaSalle, *Functional analysis and time optimal control*, Acad. Press NY, 1969.

[Hes] Hestenes, M., *Calculus of variations and optimal control theory*, Wiley NY, 1966.

[Hu] Hurwitz, A., *On the conditions under which an equation has only roots with negative real parts*, Math. Annalen **46** (1895), 273–284.

[I] Infante, E., *Some results on the Lyapunov stability of functional equations*, Volterra and Functional Differential Equations (Lecture Notes in Pure and Applied Math., 81), Marcel Dekker NY (1982).

[Is] Isidori, A., *Nonlinear control systems*, (2nd edit), Springer-Velag (1989).

[JNP] James, H., N. Nichols and R. Phillips, *Theory of servomechanisms*, McGraw Hill NY, 1947.

0 [JS] Jurdjevick V. and H. Sussmann, *Control systems on Lie groups*, J. Diff. Eqs (1972), 470–476.

[Kal] Kalman, R., *On the general theory of control systems*, Proc. IFAC Congress, Butterworths, Ltd London (1961), 2020–2030.

[Ko] Kochenburger, R., *Frequency response method for analysis of a relay servomechanism*, Trans. AIEE

69 (1950), 270–283.

[La] LaSalle, J., *Time optimal control systems*, Proc. Nat. Acad. Sci **45** (1959), 573–577.

[LL] LaSalle, J. and S. Lefschetz, (Edit), *Proc. Internat. Symp. on Nonlinear Differential Equations and Nonlinear Mechanics*, Academic Press NY, 1963.

[Lee] Lee, E. B., *Optimal control of linear hereditary systems, in Distributed Parameter Control Systems: New Trends and Applications*, M. Dekker NY (1991), 233–269.

[LM] Lee, E. B. and L. Markus, *Foundations of Optimal Control Theory*, Wiley NY 1967 and Krieger Publ. 1986. (Russian translation, Moscow 1972 and Japanese translation, Tokyo 1974)

[Lef] Lefschetz, S., *Differential equations: geometric theory*, Interscience NY, 1963.

[Let] Letov, A. M., *Stability in nonlinear control processes*, (English translation) Princeton Univ. Press, 1961.

[Li 1] Lions, J. L., *Optimal control of systems governed by partial differential equations*, Springer-Verlag NY, 1971. (S. K. Mitter, Translator)

[Li 2] Lions, J. L., *Exact controllability, stabilization and perturbations for distributed systems*, SIAM Review 30 (1988), 1–68.

[Li M1] Littman, W. and L. Markus, *Exact boundary controllability of a hybrid system of elasticity*, Arch. for Rat. Mech. and Anal. (1988), 193-236.

[Li M2] ———, *Stabilization of a hybrid system of elasticity by feedback boundary damping*, Annali di Mat. Pura ed Appl. (1988), 281–330.

[Ly 1] Lyapunov, A., *Problème général de la stabilité du mouvement*, (1892), Princeton Annals of Math Study (1949).

[Ly 2] Lyapunov, A., *Sur les fonctions-vecteurs complètement additives*, Bull. Acad. Sci. USSR, Ser. Math. Izveztia Acad. Nauk SSSR **4** (1940), 465–478.

[Ma 1] Markus, L. (with N. Wagner), *Stability theorems for a class of nonlinear servomechanics*, J. Mat. and Mech. 6 (1957), 393–400.

[Ma 2] Markus, L. (with E. Calabi), *Relativistic space forms*, Ann. of Math **75** (1962), 63–76.

[Ma 3] Markus, L., *Cosmological models in differential geometry*, Lecture Notes at Univ. Minn. (1963), 1–198.

[Ma 4] ———, *What is control theory-and why?* Inaugural Lecture - Univ. Warwick, Bull. of Inst. Math and its Applications **8** (1972), 94–97.

[Ma 5] ———, *Optimal control of limit cycles, or what control theory can do to cure a heart attack – or to cause one*, Proc. Symp. Diff. Eqs., Univ. Minn. (1972), 108–134.

[Ma 6] ———, *Mathematical elements in the existence proofs of God*, Open Lecture at Univ. Warwick, Math. Inst. Report (1973).

[Ma 7] ———, *Solomon Lefschetz – An appreciation in memoriam*, BAMS **79** (1973), 663–680.

[Ma 8] ———, *Large scale structure of space-time*, (Review of book by S. Hawking and G. Ellis), BAMS **82** (1976), 805–817.

[MY] Markus, L. (with Y. C. You), *Dynamical boundary control for elastic plates of general shape*, SIAM J. Control and Optimization (1993).

[Max] Maxwell, J. C., *On governors*, Proc. of the Royal Soc. of London **16** (1868), 270–283.

[Mi 1] Minorsky, N., *Control problems*, J. Franklin Inst. **232** (1941), 451–487, 519–551.

[Mi 2] ———, *Self-excited oscillations in dynamical systems possessing retarded action*, Jour. of Appl. Mech. **9** (1942), 65–71.

[Mo] Moses, *Genesis 1*.

[Nee] Needham, J., *Science and Civilization in China*, Cambridge Univ. Press, 1954.

[Ni] Nier, A., *Device to compensate for magnetic field fluctuations in a mass spectrometer*, R.S.I. **6** (1935), 254–255.

[Ny] Nyquist, H., *Regeneration Theory*, Bell. Syst. Techn. J. **11** (1932), 126–147.

[P1] Pontryagin, L., *On some zeros of transcendental functions*, Translations Am. Math. Soc. 2, **1** (1955), 95–110.

[P2] Pontryagin, L. with A. Andronow, *Systèmes grossiers*, Doklady Akad Nauk **14** (1937), 247–251.

[P3] ———, *Optimal processes of regulation*, Proc. Internat. Math. Congress, Edinburgh, 1958.

[PBGM] Pontryagin, L. S., V. Boltyanski, R. Gamkrelidze, E. Mischenko, *The mathematical theory of optimal processes*, (English translation) Interscience Publ. Inc., NY, 1962.

[Ra] Rayleigh, Baron (J. Strutt), *Theory of Sound I, II*, Macmillan London, 1877–1894, (2nd ed. rev. 1926), Reprint Dover 1945.

[Ro] Roxin, E., (Edit), *Modern Optimal Control–Conference on Bang-Bang Principle*, Kingston, R.I., M. Dekker NY, 1989.

[Rus] Russell, D., *Controllability and stabilizability theory for linear partial differential equations*, Survey for SIAM Review **20** (1978).

[Th] Thaler, G., *Automatic Feedback Control-Past, present, future*, Lecture at COMCOM3, (Personal Comm.) (1992).

[V] Vyshnegradski, J., *Sur la théorie général de régulateure*, Compte. Rend, Acad Sci. Paris **83** (1876), 318–331.

[W] Wiener, N., *Cybernetics*, J. Wiley 1948, see also article in "Mathematics in the modern world", Readings from Scientific American, 1968, pp. 378–384.

[Z] Zeeman, E. C., *Catastrophe theory: Selected papers*, Addison-Wesley NY, 1977.

# 1 De Motu Arietum (On the Motion of Battering Rams)

**Rutherford Aris** University of Minnesota, Minneapolis, Minnesota

**Bernard S. Bachrach** University of Minnesota, Minneapolis, Minnesota

Introduction.

Of the many virtues of Larry Markus none is more widely appreciated than his willingness to transgress departmental barriers and cooperate freely and fruitfully with colleagues in other disciplines. This is natural enough in the physical sciences, where there is a mode of thought sympathetic to "the euristic vision of mathematical trance", but rare indeed in the social sciences, where the writ of natural law runs haltingly and the concepts are much less amenable to the niceties of mathematics. Nevertheless, for some years Markus ran a seminar jointly with Holt, a distinguished colleague in political philosophy, and, as his bibliography will show, has published in this area. The following paper, which essays to use mathematical modelling as an ancillary science in the study of an historical question, is dedicated to Larry Markus by two of his colleagues who esteem his disciplinary transgressions as highly as they regard his mathematical rectitude.

This is no place to launch into a full-scale defense of mathematical modelling as a valuable element in the graith of the historian, but a few words of apology are in order. For the historian of science a knowledge of mathematics is, of course, absolutely necessary and, though the general historian should not have this exacted of him in addition to the traditional ancillary disciplines (languages, law, diplomatic, palæography, etc.), some appreciation of what mathematical modelling is about may prove more enlightening than might at first be suspected. A mathematical model is a system of equations that purports to represent some phenomenon in a way that gives insight into the origins and consequences of its behavior. Being mathematical, it is patient of a precision in the statement of its assumptions and the exposition of its results that can be stimulating and suggestive to the historian. A model must not be so venerated as to be allowed to impose a spurious exactness, or to be found wanting if it does not predict values to N decimal places, but it can show the relative importance of various factors and lead the historian to examine them, using, indeed, the traditional tools of historical study, but with a fresh perception of their relevance. The model does not detract from the primacy of historical thought, but seeks to serve it as a true *ancilla,* perhaps even as an *ancilla ancillarum rerum scriptoris* - a possibility that should drive the politically correct of our day straight up the wall !

Rightly understood, a mathematical model can guard against impossibilities being considered (to use the ram as an informal example, there are natural limits to the energy that can be 'pumped' into the ram's motion), can reveal sensitivities and instabilities (the tendency of the ram to wobble if suspended in certain ways) and suggest lines of enquiry (is there any evidence as to how the ram was suspended?). Because the model has a life of its own, it can present alternatives (such as different regimes of 'pumping up' the ram) and seek out optimal strategies which might well have been discovered by the trial and error of experience. Models can give ranges to quantities, saying, for example, that if such a wall were battered down the ram must have been at least this big and, if it were manageable on such a terrain it could not have been bigger than that. Models introduce the dimensionless parameters of the situation, one of the greatest intellectual beauties of natural philosophy, in which quantities attain significance by being compared, not with an arbitrary standard or unit of mensuration, but with standards intrinsic to the problem itself.

In what follows we shall study the motion of the ram as a dynamical system - hence the title, since a more general study, *De Arietibus*, has yet to be completed - and the strategies of its effective use. We shall prescind from the question of how much energy is needed for a successful 'batter', for this is a much broader problem. It raises many questions concerning wall materials and construction which are not really mathematical and which will be more appropriately explored elsewhere.

The ram and its use.

The battering ram (*aries*) was a heavy object, usually a long cylinder with a blunt,conical tip, suspended from a beam within a 'cat', or sheltering penthouse (*pluteus*), also known as a 'tortoise' or 'testudo'. The cat would be moved forward to the wall or door to be battered and the ram set aswinging by the team of soldiers within the shelter of the cat. In the absence of a cat, a team of soldiers might run at the wall carrying the ram, but they would clearly be much more vulnerable and would either have to retire and repeat the run or be limited to a much smaller amplitude of swing. Clearly a much smaller and better protected team could be fielded if the weight of the ram were taken by the cat, and this is the case we will consider.

Let the ram be represented by a heavy cylinder suspended symmetrically by two light cords so that it swings in the plane of the cords and the center line of the cylinder. Deviations from this planar motion will be considered later, but it may well have been discovered that, by suspending the ram from a V of rope or chain, the tendency of the ram to 'wobble' out of the plane could be minimized - an important point for the safety of the team. If the cat were run up until the tip of the ram almost touched the wall, the only procedure would be to haul the ram as far as possible away from the wall, keeping the cat stationary, and let it fall back against the wall, possibly with a further push. If the single blow were successful, this tactic would have all the elements of surprise and swiftness beloved of military commanders, but it would also require a relatively large team to haul the ram back sufficiently far. A commander might prefer to put a smaller team at risk and instruct them to 'pump up' the ram by repeatedly forcing the oscillation according to some concerted program. It will be found that friction imposes a limit on the amplitude of the swing and that it can be attained in remarkably short order. Of course, room must be left between the wall and the cat for this swing to be developed and after the last forcing the crew would have to run the cat up to the wall, ideally in such a way that the velocity of the ram on impact would be as large as possible. We shall see how all these quantities can be calculated and how different tactics can be compared. It is not to be supposed that there was any such *Libellus de operandi arietum modo* in ancient or mediæval times. But, if our basic picture is a sound one, our results will give the ranges of the values for the various parameters of consequence. What is known from the ancient records will be discussed in detail elsewhere; the brief bibliography at the end of the paper will serve as an introduction for any interested reader.

Characteristic Quantities and Basic Equations.

Since every part of the ram moves in an arc of circle, the dimensions of the ram are not significant until we come to consider the full three-dimensional motion. The length scale in the basic problem is determined by L, the length of the ropes by which the ram depends from the ridge of the cat. The ram, however, determines the scale of the forces by its weight, Mg. Thus, if the force with which the team can heave or push the ram is F, the dimensionless force will be expressed by $f = F/Mg$. If the ram performs oscillations sufficiently small as to be in the region of linearity, these will be of frequency $\omega = \sqrt{(g/L)}$, so $1/\omega$ is a natural unit of time and $\tau = \omega t$ the dimensionless time.

If, as we are supposing, its motion in one plane, each particle of the ram describes an arc of a circle of radius L and the ram may be replaced by a point mass M at the end of a inextensible string of length L. Let the tip of the ram be toward the left and x,y coordinates be taken with origin at the tip and the x-axis pointing to the right, then, when the inclination of the cords to the vertical is q, the tip is at

$$x = L \sin\theta, \qquad y = L(1 - \cos\theta). \tag{1}$$

If we neglect any resistance to the motion, Newton's laws applied in the x and y-directions, give

$$M\ddot{x} = F - 2\,T\cos\theta, \qquad M\ddot{y} = 2\,T\cos\theta - M\,g \tag{2}$$

where T is the tension in the cords and F the force imposed by the handlers, assumed to be constant and purely horizontal. These may be combined into a single equation for $\theta$ as is appropriate to a system with one degree of freedom,

$$M\,L\,d^2\theta/dt^2 = F\cos\theta - M\,g\sin\theta.$$

Dividing through by Mg and defining $\tau$ as $\omega t$, and $\phi$ as $\tan^{-1}(F/Mg)$ gives

$$d^2\theta/d\tau^2 = -\sin\theta + f\cos\theta = -\sec\phi\,\sin(\theta-\phi). \tag{3}$$

We observe that by changing $\theta$ to $\theta-\phi$, $\tau$ to $\sqrt{(\sec\phi)}\tau$ that the equations reduce to the case of no external force of which the solution in terms of elliptic functions is well known (see, for example Whittaker IV,44). For, multiplying eqn. (3) by 2 $d\theta/d\tau$ and integrating,

$$(d\theta/d\tau)^2 = A - 2\sec\phi.\cos(\theta-\phi) = A' - 4\sec\phi.\sin^2[(\theta-\phi)/2].$$

Letting $z = \sin(\theta-\phi)/2$ and $k^2 = (A'/4)\cos\phi$, we have

$$z = k\,\mathrm{sn}[(\sec\phi)^{1/2}(\tau-\tau_0),\,k] \tag{4}$$

A', determined by the initial condition as $[(d\theta/d\tau)_0]^2 + 4\sec\phi.\sin^2[(\theta_0-\phi)/2]$, is small enough to keep k less than 1 or else the ram would go over the top.

When the frictional resistance to motion is taken into account a term $2\mu(d\theta/d\tau)$ must be added to the second derivative in eqn. (3), giving

$$d^2\theta/d\tau^2 + 2\mu(d\theta/d\tau) + \sec\phi.\sin(\theta-\phi) = 0 \quad (5)$$

$\mu$ is the dimensionless resistance coefficient and is probably quite small. Whilst it would be beyond the ability of mediæval mechanics to formulate in these terms, it would be a commonplace observation that some rams, when set freely swinging, would go through a greater number of oscillations than others in settling back to rest and that these rams would be easier to 'pump-up'. They would, in fact, be those with smaller values of $\mu$.

There is no longer a solution in terms of tabulated functions, but the solution for non-zero $\phi$ is still related to that for $\phi = 0$. If $\theta = \Theta(\tau;\ \mu,\ \phi;\ \theta_0,\ \psi_0)$ is the solution with $\theta = \theta_0$ and $d\theta/d\tau = \psi_0$ at $\tau = 0$, then

$$\Theta(\tau;\ \mu,\ \phi;\ \theta_0,\ \psi_0) = \phi + \Theta(\tau\surd\sec\phi;\ \mu\surd\cos\phi,\ 0;\ \theta_0-\phi,\ \psi_0\surd\sec\phi). \quad (6)$$

The equations can be linearized and are then easily soluble. It will be convenient to write

$$\psi = d\theta/d\tau,\ \mu = \sin\chi,\ \ C(\tau) = \exp(-\sin\chi.\tau)\cos(\cos\chi.\tau),\ S(t) = \exp(-\sin\chi.\tau)\sin(\cos\chi.\tau) \quad (7)$$

for then the solution of the linearized equation

$$d^2\theta/d\tau^2 + 2\mu(d\theta/d\tau) + \theta = \phi \quad (8)$$

for which $\theta = \theta_0$ and $d\theta/d\tau = \psi_0$ at $\tau = 0$ is

$$\theta(\tau) = \theta_0\,C(\tau) + \{\sec\chi.\psi_0 + \tan\chi.\theta_0\}S(\tau) + \{1 - C(\tau) - \tan\chi.S(\tau)\}\phi \quad (9)$$

$$\psi(\tau) = \psi_0\,C(\tau) - \{\sec\chi.\theta_0 + \tan\chi.\psi_0\}S(\tau) + \sec\chi.S(\tau).\phi \quad (10)$$

These solutions will be quite serviceable so long as $\theta$ and $\phi$ are sufficiently small that $\theta$ and $\sin\theta$ do not differ greatly. The linear case thus provides a useful limiting case as check on calculations done with the full equations.

<u>The single swing.</u>

To be effective the force F must be applied according to some program, the simplest scenario being for the crew to push back as far as possible and, stepping aside from the path of any handles or protuberances they might have used, to let the ram swing forward under its own weight. Thus we have to solve eqn. (5) with $\phi>0$ and initial conditions $\theta = d\theta/d\tau = 0$ at $\tau = 0$ to find $\theta$ at its maximum, say $\theta_1$, and then solve eqn. (5) with $\phi=0$ and initial conditions $\theta = \theta_1$, $d\theta/d\tau = 0$ for the free fall forwards. In this way we can calculate the velocity, and so also the kinetic energy, of the ram at the point of impact with the wall, namely $\theta = 0$. In the linear limit this can be done explicitly by using the solutions (9) and (10). With $\theta_0 = \psi_0 = 0$, $\psi = \phi.\sec\chi.S(\tau)$, which vanishes at time $\pi.\sec\chi$ and gives $\theta_1 = \phi(1 + e^{-\pi.\tan\chi})$. With $\theta_0 = \theta_1$ and $\psi_0 = 0$, eqn.(9) shows that $\theta = 0$ when the time is $(\pi/2 + \chi).\sec\chi$ and that the velocity is then

$$\psi_1 = -\phi(1 + e^{-\pi.\tan\chi}).e^{-(\pi/2 + \chi).\tan\chi}.$$

The kinetic energy is best expressed as a fraction of the greatest difference in the potential energy, namely MgL. Thus the kinetic energy is $(ML^2/2)(g/L)(\psi_1)^2 = \kappa.MgL$ and

$$\kappa = (\psi_1)^2 /2 = E.\phi^2. \tag{11}$$

The coefficient E is a function only of $\chi$, i.e. of $\mu$, in the linear limit; it will be a function of $\phi$ also in the as the value of $\phi$ increases, but a considerably weaker one.

The calculation is most easily accomplished by writing the equation of motion as a pair of first order equations

$$d\theta/d\tau = \psi \tag{12}$$

$$d\psi/d\tau = - \sin\theta - 2\mu\psi + \phi.\cos\theta.\Gamma(\theta,\psi) \tag{13}$$

where $\Gamma(\theta,\psi)$ contains the program. For example, in the mode we are considering the force is applied only in the first quadrant of the $\theta,\psi$ -plane, so

$$\Gamma(\theta,\psi) = (1/4)\{1+\mathrm{sgn}(\theta)\}\{1+\mathrm{sgn}(\psi)\}. \tag{14}$$

This can be easily accomplished with a desk top computer using Hubbard andWest's excellent MacMath program (Springer, 1991) or Stella II® (High Performance Systems) - the latter makes it easier to get numerical values, but the former is particularly valuable for "seeing" solutions of two or three first order differential equations. Since we are not concerned with the time it takes for the operation, the phase plane representation is ideal. Fig. 1 shows a typical example. The curve OPQ represents the motion, for it starts at the origin, the position of rest, the ram moves with increasing $\theta$, accelerating at first but then slowing down as the extreme position P, $\theta = \theta_1$ is reached. Its motion in the free fall, PQ, is with decreasing $\theta$, i.e. $\psi$ is negative and it passes through $\theta = 0$ with velocity $-\psi_1$. It should be noted that, because of the damping, this is not the greatest velocity, which seems to be attained at the point R in Fig.1.

This Chaucerian (for the Ram has only 'his halve cours yronne') program is minimal in terms of the effort a team can put into a single blow. Had they sufficient agility, they might change direction and urge the ram in the direction towards the wall also. To do this with the same force and for the whole of the return stroke probably represents the most effort that could be put into a single batter. Thus we have a maximal program given by

$$\Gamma(\theta,\psi) = \mathrm{sgn}(\psi), \tag{15}$$

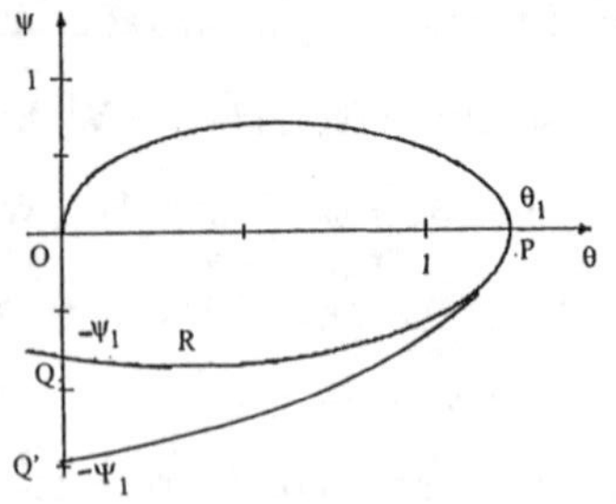


Figure 1. Paths of the ram's motion in the phase plane for the single batter.

which represents a constant force always impressed in the direction of motion of the ram. The path in the phase plane is the same as for the minimal policy over the first half OP, but for the return it maturally results in a greater velocity as shown by the trajectory PQ'. This value might be denoted by $-\Psi_1$ and all reasonable efforts will clearly produce speeds that lie in the interval $(-\psi_1, -\Psi_1)$. Table I, at the end of the paper, gives values of $\theta_1$, $\psi_1$, and $\Psi_1$ and the corresponding energy factors $\kappa_1$ and $K_1$. In terms of these dimensionless numbers, the physical quantities of interest are:

i, extent of back swing, $(FL/Mg).\sin\theta_1$;

ii, velocity at impact, $\psi_1\surd(Lg)$ or $\Psi_1\surd(Lg)$; (16)

iij, kinetic energy at impact, $\kappa_1 MgL$ or $K_1 MgL$.

The first columns (labelled $\phi$=0) are from the linear case whose equations are given above for the minimal policy and, for the maximal, $\Psi_1 = -\phi(2 + e^{-\pi.\tan\chi}).\sec\chi.S(t)$

where t is given by $C(t) + S(t).\tan\chi = 1/(2 + e^{-\pi.\tan\chi})$.

Pumping the swings.

The purpose of suspending the ram is to prevent its weight being a burden to the crew leaving them free to put their effort into getting the ram swinging. A much smaller crew than would be needed to push the ram back and let it fall forwards against the wall in a single blow could be trained to coordinate their pushing and 'pump up' the amplitude of the swings, much in the way that a child will pump itself up on the playground swings by coordinated swinging of legs and pulling back with the arms. Indeed we may imagine a minimal and a maximal mode given by the two functions $\Gamma(\theta,\psi)$ of eqns. (14) and (15). The least they could do would be to catch the ram every time it moved back through the position where the ropes are vertical ($\theta$=0) and push with force F as far as possible ($\psi$=0), letting the ram fall forwards and back to the central position, i.e. the remaining three quarters of the oscillation, under its own weight. The most they could do would be to apply the force always in the direction of motion, reversing their effort at the end of each half swing (i.e. where $\psi=0$). In practice this would demand a nimbleness and dexterity that would be difficult to attain, but it still represents a sensible upper bound on what could be expected. The minimal policy could also be generalized to the imposition of a positive force whenever $\theta > \alpha$ and $\psi > 0$.

The work done by the crew goes partly to increase the total energy (kinetic plus potential) of the ram and is partly dissipated by the resistance to motion and the damping forces. Since the latter increase with increasing speed and amplitude, there comes a point when the work done by the impressed forces is entirely dissipated by the resistive forces and the oscillation settles down into a steady limit cycle. If there is no resistance, the forcing builds up the amplitude without limit, and, in fact, the model breaks down, but, as soon as there is any damping force, there is a natural limit to the amplitude of the swing that any team can build up which is governed by the two parameters, $\phi$, the ratio of the force to the weight (F/Mg), and $\mu$, the damping constant. In practice the structure could not tolerate too great an amplitude of swing and the limit of 1.5 radians = 86 degrees, which we have imposed, is excessively generous.

As the forcing increases (for given resistance), the size of the limit cycle increases. As the damping increases (for a given forcing), the size decreases, but the rapidity with which the limit cycle is reached increases greatly. This can be seen by comparing the four swings that it takes to get into the dark ring of the limit cycle in Fig.2b,($\mu=0.1$), with the two that suffice when $\mu=0.2$ and the fact that the limit cycle is described the first time around in Fig. 2 d, $\mu=0.5$. In Fig. 2a where is only 0.05, one can count at least eight oscillations before the limit cycle is approached.

Figure 2 is drawn for the limiting case of small amplitude and the coordinates in the plane are $\theta/\phi$ and $\psi/\phi$ respectively. The quantities that characterize the limit cycle are marked on Fig. 2b and tabulated in table II. They are:

$\theta_\infty$, the ultimate angle of back-swing that can be attained;

$\psi_\infty$, the angular velocity when $\theta=0$, giving the velocity at impact; and

$\theta'_\infty$, the absolute value of the greatest forward angle, from which the space needed to pump-up the ram to its maximum energy can be calculated.

Thus to pump-up the ram to its fullest ferocity would require a space $A = L\,\theta_\infty$ behind and $A' = L\,\theta'_\infty$ before, whilst the maximum kinetic energy on impact would be $MgL(\psi_\infty)^2$. Actually the maximum velocity of the ram occurs a little before the ropes are vertical, but the difference is not great.

The maximal effort policy given by eqn. (15) likewise generates limit cycles. They will be symmetrical in the sense of being invariant under change of sign of both $\theta$ and $\psi$ so that $\Theta_\infty$ and $\Theta'_\infty$ are the same and in the linear limit, since both occur at multiples of $\pi.\sec\chi$, they are given by

$$\Theta_\infty = \Theta'_\infty = \phi.\coth[(\pi/2)\tan\chi] \qquad \text{where} \qquad \chi = \sin^{-1}\mu \tag{17}$$

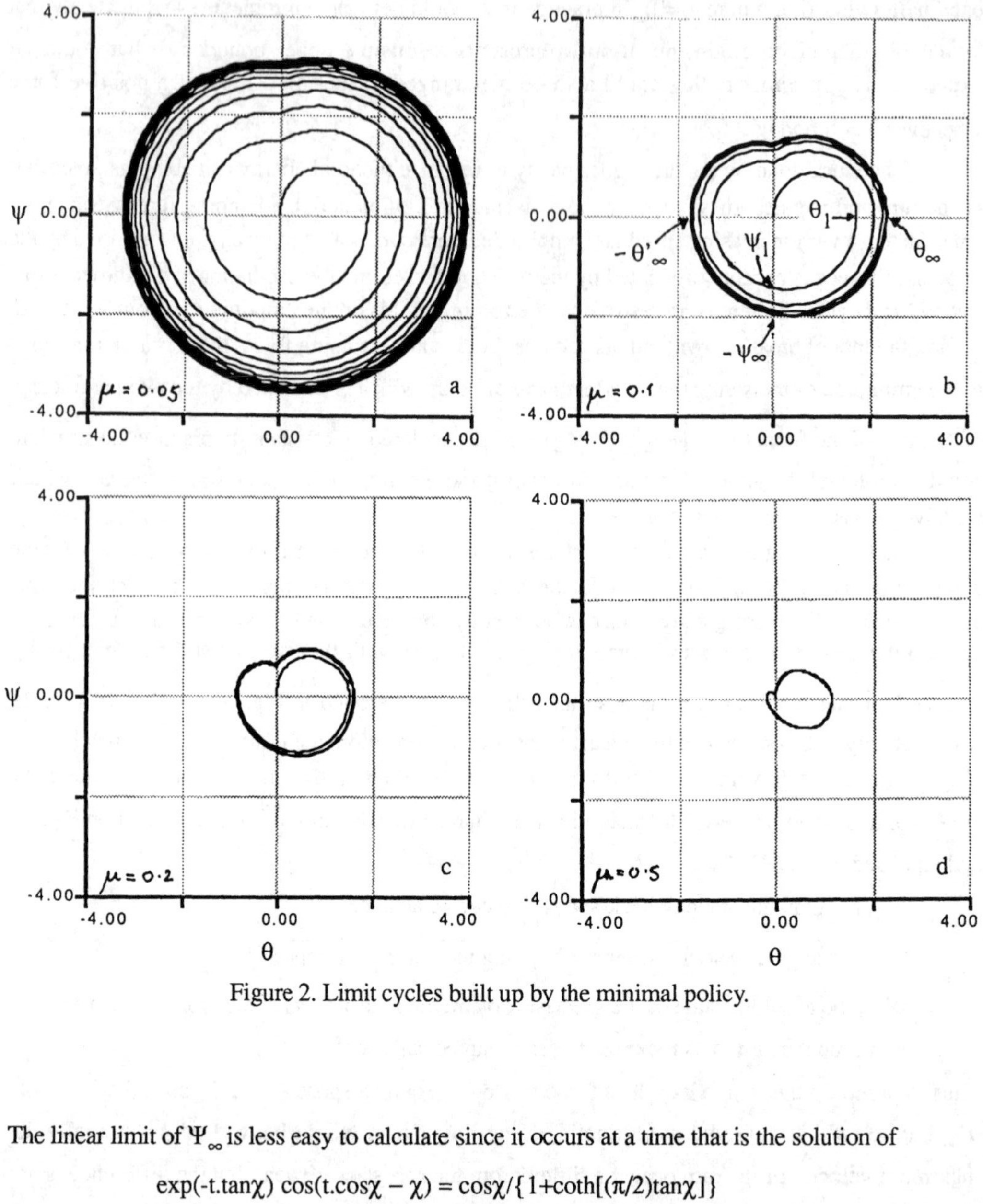


Figure 2. Limit cycles built up by the minimal policy.

The linear limit of $\Psi_\infty$ is less easy to calculate since it occurs at a time that is the solution of

$$\exp(-t.\tan\chi)\cos(t.\cos\chi - \chi) = \cos\chi/\{1+\coth[(\pi/2)\tan\chi]\}$$

and then
$$\Psi_\infty = (\Theta_\infty + \phi).\sec\chi.|S(t)|. \quad (18)$$

The linear limits under both maximal and minimal policies are tabulated in Table II, whilst the finite amplitude characteristics obtained from the solution of the full nonlinear equations are given in Table III. The omitted entries are for cases where the swing back would be greater than $\pi/2$. These can be achieved when the damping is not large, but the forcing is considerable, for the asymptotic limit cycle (as $\phi$ goes to infinity) is the pair of homoclinic curves joining the points $\theta=\pm\pi/2$, $\psi=0$, the upper one with $\Gamma=+1$, the lower with $\Gamma=-1$.

<u>The final run-up.</u>

With the progam of a single heave back with either letting the ram fall forward under its own weight or pushing it forward, the cat can be brought as close as possible to the wall so that the tip of the ram in its rest position is in contact with the wall. When the ram is to be pumped-up, however, we have seen that sufficient room must be left in front of the cat during this operation, namely $L\sin\theta'_\infty$ or $L\sin\Theta'_\infty$. To be effective the cat must be moved up to the wall after the last "pump" so that it may impinge when $\theta = 0$, since the speed is then close to its greatest value. To refine the estimate of the kinetic energy at impact we must consider this final run-up and ask what force would be necessary to move the cat into position by the time the ram has reached $\theta = 0$.

The ram may be considered as a mass suspended, not, this time, from a fixed support, but from another mass, M', which can be moved horizontally by an impressed force, H. The system now has two degrees of freedom and, in addition to $\theta$, we may take $\zeta = z/L$ as a dimensionless generalized coordinate, where z is the distance of the wall from the cat. The origin for z is taken to be the position of the cat when the tip of the ram, in its "rest" position $\theta = 0$, is just in contact with the wall. $\zeta$ is reckoned positive into the wall and hence has the opposite sign to $\theta$. Since the horizontal component of the ram's dimensionless velocity is now $d(\zeta-\sin\theta)/d\tau$, the dimensionless kinetic energy of the whole system is

$$\kappa = (1/2)\{(1+\gamma)(d\zeta/d\tau)^2 - 2\cos\theta(d\zeta/d\tau)(d\theta/d\tau) + (d\theta/d\tau)^2\}. \quad (19)$$

where $\gamma=M'/M$. Also the work done by the external forces, again made dimensionless by MgL and with $\varpi = H/Mg$, is

$$dQ/MgL = \varpi.d\zeta - \sin\theta.d\theta \quad (20)$$

and so Lagrange's equations of motion give

$$(1+\gamma)(d^2\zeta/d\tau^2) - \cos\theta.(d^2\theta/d\tau^2) + \sin\theta.(d\theta/d\tau)^2 = \varpi$$

and

$$-\cos\theta.(d^2\zeta/d\tau^2) + (d^2\theta/d\tau^2) = -\sin\theta. \quad (21)$$

It will have been noticed that we have ignored any resistance to the motion of the cat or damping of the ram's swing. These, if, as previously supposed, proportional to velocity could have been included at the expense of two extra terms, while if the resistance to the movement of the cat is constant it can be accommodated in $\varpi$. In the interest of simplicity we will treat them here after the manner of the middle dot in the Markus ellipsis convention.

Equations (21) can be written

$$
\begin{aligned}
d\zeta/d\tau &= \xi \\
d\theta/d\tau &= \eta \qquad (22) \\
d\xi/d\tau &= \{\varpi - \eta^2\sin\theta - \sin\theta.\cos\theta\}/\{\gamma + \sin^2\theta\} \\
d\eta/d\tau &= \{\varpi \cos\theta - \eta^2\sin\theta.\cos\theta - (1+\gamma)\sin\theta\}/\{\gamma + \sin^2\theta\}
\end{aligned}
$$

with initial conditions $\xi = \eta = 0$, $\zeta = -\sin\Theta$, $\theta = \Theta$. For given $\gamma$ and $\Theta$ we want to find $\varpi$ so that for the solution of (22) $\zeta$ and $\theta$ vanish at the same moment. Then the kinetic energy on impact will be

$$\kappa = (1/2)\{\xi^2 - 2\cos\theta.\xi\eta + \eta^2\} \qquad (23)$$

To the two place accuracy which we are using - and it does not seem appropriate to cultivate any greater precision in this context - the required force, $\varpi$, is proportional to the back-swing angle, $\Theta$, the constant of proportionality being a function of $\gamma$, the ratio of the mass of the cat to that of the ram.

The kinetic energy also correlates quite well with $\Theta^2$ and we have

| $\gamma$ | 1 | 2 | 5 |
|---|---|---|---|
| $\varpi/\Theta$ | .785 | 1.15 | 2.32 |
| $\kappa/\Theta^2$ | 1.612 | 1.268 | 1.012 |

Attempts to express these constants of proportionality as powers of $\gamma$ are less accurate, but rough approximations are

$$\varpi = 0.77\ \Theta\ \gamma^{2/3}, \qquad \kappa = 1.58\ \Theta^2\ \gamma^{-1/4}.$$

Table IV gives figures that have been calculated from the solution of the equations and it includes the time that the run-up takes. This, surprisingly, does not vary much with the weight of the cat, but the reason seems to be that the extra effort required to get a well-timed batter compensates for the greater weight of the cat. The kinetic energy achieved decreases as the cat becomes more massive and it is clear that experience would result in a suitable compromise between the unwieldiness of a larger structure and the vulnerability of a smaller.

Further problems.

A major problem that remains to be tackled is the stability of the planar motion of the ram. This has three degrees of freedom, since each point of attachment of the ropes must lie in a sphere, but the two points must be a fixed distance apart. This is a non-holonomic problem and, if $\theta$ and $\varphi$, with appropriate subscripts, denote the angle of the cord from the vertical and the angle between the vertical plane in which the cord lies and the vertical plane through the rest position of the ram, we have always to satisfy the relation

$$\cos\theta_1.\cos\theta_2 + \sin\theta_1.\sin\theta_2.\cos(\varphi_1 - \varphi_2) + 2\alpha(\sin\theta_1.\cos\varphi_1 - \cos\varphi_2.\ \sin\theta_2) = 1 \qquad (24)$$

where the distance between the points of attachment is $2\alpha L$. A little more symmetry can be introduced into the problem by taking $\varphi_1 - \varphi_2 = 2\theta_3$ and $\varphi_1 + \varphi_2 = 2\theta_4$ , whence $\theta_4$ can be expressed in terms of $\theta_1$, $\theta_2$ and $\theta_3$ , but the formulae are not elegant.

We can simplify the problem in various ways, for instance by considering twisting motions in which the ram stays in a horizontal plane and its centroid moves in a vertical line. For these motions have but one degree of freedom and the generalized coordinate may be taken to be $\chi$, the angle between the ram and its rest position. The height of the center of gravity is

$$y = 1 - \sqrt{(1-2\alpha^2[1-\cos\chi])},$$

whence

$$y' = dy/d\chi \ = \alpha^2.\sin\chi/\sqrt{(1-2\alpha^2[1-\cos\chi])}.$$

The dimensionless kinetic energy is

$$T/MgL = (1/2)\ \{y'^2 + \delta^2\} = \{\alpha^4.\sin^2\chi/(1-2\alpha^2[1-\cos\chi]) + \beta^2/3 + \gamma^2/4\}$$

where the ram is of length $\beta L$ and its mean radius is $\gamma L$. The last two terms on the right are the square of the radius of gyration of the ram and may be easily modified if the ram is of a different shape. The work done when $\chi$ increases by $d\chi$, is $-MgL[\ 2\mu\beta^3 + y']\ d\chi$, where $\mu$ is a damping constant, the proportionality of the resistance to the speed. Thus Lagrange's equations give

$$\{y'^2 + \delta^2\}(d^2\chi/d\tau^2) + 2\mu\beta^3(d\chi/d\tau) + y'(\chi) = 0 \qquad (25)$$

This equation must be solved for finite oscillations but for small undamped oscillations the equations clearly show that the ratio of the frequencies of the swings to that of the twists is the ratio of the distance between the points of attachment to the radius of gyration of the ram. We shall reserve the discussion of stability to a later investigation, but it seems that it would be advantageous to suspend the ram from points as distant as possible if these twisting motions are to be avoided. this presents certain problems with the design of the cat and a sensible engineering compromise is to take the points of suspension at the radius of gyration so that neither frequency dominates.

There is certainly much more to be said about the dynamics of a ram - its forced and free motions, its response to the impact, the possibility of one rope going slack and its sensitivity to imperfections in the crew's discipline. Whether these more esoteric aspects will reflect the experiences of ancient and mediæval man in this school of hard knocks remains to be seen, but the problems themselves have a certain panache that makes them worthy of mention in a paper dedicated to one who has brought to his mathematics a unique touch of élan and éclat.

Appendix. Tables of results.

Table I. Values of the backswing and impact parameters for the single batter.

| $\Theta_1, \theta_1$. | | $\phi = 0$ | 0.1 | 0.2 | 0.5 | 1.0 |
|---|---|---|---|---|---|---|
| | $\mu$=0 | 2.00$\phi$ | .20 | .39 | .92 | 1.56 |
| | .05 | 1.85$\phi$ | .18 | .37 | .86 | 1.46 |
| | .1 | 1.73$\phi$ | .17 | .34 | .81 | 1.38 |
| | .2 | 1.53$\phi$ | .15 | .30 | .71 | 1.23 |
| | .5 | 1.16$\phi$ | .12 | .23 | .59 | 0.96 |
| | | $\phi = 0$ | 0.1 | 0.2 | 0.5 | 1.0 |
| $\psi_1$ | 0 | 2.00$\phi$ | .20 | .39 | .89 | 1.41 |
| | .05 | 1.71$\phi$ | .17 | .34 | .77 | 1.22 |
| | .1 | 1.46$\phi$ | .14 | .29 | .66 | 1.06 |
| | .2 | 1.06$\phi$ | .11 | .21 | .48 | 0.82 |
| | .5 | 0.35$\phi$ | .03 | .07 | .16 | 0.26 |
| $\Psi_1$ | 0 | 2.83$\phi$ | .28 | .56 | 1.26 | 2.00 |
| | .05 | 2.54$\phi$ | .25 | .50 | 1.15 | 1.84 |
| | .1 | 2.29$\phi$ | .23 | .45 | 1.04 | 1.94 |
| | .2 | 1.90$\phi$ | .19 | .37 | 0.87 | 1.48 |
| | .5 | 1.17$\phi$ | .12 | .23 | 0.55 | 0.99 |

Table II. Characteristics of the linear limit cycle under the maximal and minimal polices.

| $\mu$ | $\theta_\infty$ | $\theta_\infty$' | $\psi_\infty$ | $\Theta_\infty$ | $\Psi_\infty$ |
|---|---|---|---|---|---|
| .05 | 3.90$\phi$ | 3.34$\phi$ | 3.39$\phi$ | 12.74$\phi$ | 12.67$\phi$ |
| .1 | 2.35$\phi$ | 1.72$\phi$ | 1.98$\phi$ | 6.39$\phi$ | 6.29$\phi$ |
| .2 | 1.63$\phi$ | 0.86$\phi$ | 0.64$\phi$ | 3.22$\phi$ | 3.03$\phi$ |
| .5 | 1.16$\phi$ | 0.09$\phi$ | 0.09$\phi$ | 1.39$\phi$ | 0.77$\phi$ |

Table III. Characteristics of the solutions of the nonlinear equations

| $\mu$ | $\phi$ | $\theta_\infty$ | $\theta_\infty'$ | $\psi_\infty$ | $\kappa$ | $\Theta_\infty$ | $\Psi_\infty$ | K |
|---|---|---|---|---|---|---|---|---|
| .05 | .2 | 0.71 | 0.60 | 0.64 | 0.20 | | | |
| .1 | .2 | 0.46 | 0.33 | 0.39 | 0.08 | 1.08 | 1.03 | 0.53 |
| .2 | .2 | 0.32 | 0.17 | 0.22 | 0.024 | 0.62 | 0.60 | 0.18 |
| .5 | .2 | 0.23 | 0.037 | 0.068 | 0.002 | 0.27 | 0.25 | 0.031 |
| .05 | .5 | 1.40 | 1.16 | 1.18 | 0.73 | | | |
| .1 | .5 | 1.03 | 0.73 | 0.82 | 0.34 | | | |
| .2 | .5 | 0.75 | 0.39 | 0.51 | 0.13 | 1.29 | 1.20 | 0.73 |
| .5 | .5 | 0.54 | 0.087 | 0.16 | 0.013 | 0.65 | 0.59 | 0.17 |
| .05 | 1 | | | | | | | |
| .1 | 1 | 1.64 | 1.10 | 1.22 | 0.75 | | | |
| .2 | 1 | 1.28 | 0.63 | 0.82 | 0.34 | | | |
| .5 | 1 | 0.96 | 0.74 | 0.26 | 0.03 | 1.12 | 1.04 | 0.54 |

Table IV. Effort required and kinetic energy achieved in the final run-up.

| $\Theta$ | $\gamma$ | $\varpi$ | $\tau$ | $\kappa$ |
|---|---|---|---|---|
| 0.1 | 1 | 0.078 | 1.60 | 0.0182 |
| 0.1 | 2 | 0.117 | 1.85 | 0.0137 |
| 0.1 | 5 | 0.234 | 2.10 | 0.0110 |
| 0.2 | 1 | 0.156 | 1.60 | 0.0724 |
| 0.2 | 2 | 0.234 | 1.90 | 0.0545 |
| 0.2 | 5 | 0.467 | 2.10 | 0.0437 |
| 0.5 | 1 | 0.394 | 1.70 | 0.4338 |
| 0.5 | 2 | 0.585 | 1.90 | 0.3326 |
| 0.5 | 5 | 1.163 | 2.06 | 0.2699 |
| 1.0 | 1 | 0.785 | 1.79 | 1.570 |
| 1.0 | 2 | 1.150 | 1.92 | 1.246 |
| 1.0 | 5 | 2.250 | 2.03 | 1.036 |

## References.


Marsden, E.W. 1969. *Greek and Roman Artillery: Historical Developments* and
1971. *Greek and Roman Artillery: Technical Treatises*. Clarendon Press, Oxford.

Sakur, W. 1925. *Vitruv und die Poliorketiker.*
*Vitruv und die christliche Antike*. Bautechnishes aus der Literatur des Alterums. Berlin.

Smith, C and J.Hawthorne. *Mappae Clavicula: A little Key to the World of Mediæval Techniques*. Trans. American Philosophical Society. **270** (1974).

Whittaker, E.T. 1904. *A Treatise on Analytical Dynamics*. Cambridge University Press. Cambridge.

# 2 A Note on Continuation Algorithms for Periodic Orbits

**Pavol Brunovský** Comenius University, Bratislava, Slovakia

**Milan Kubala** Comenius University, Bratislava, Slovakia

## 1. Introduction

In parallel with theoretical studies, problems of location, dependence on parameters and branching of periodic trajectories of dynamical systems have been extensively investigated numerically for some time. Several computer codes have been developed for this purpose [DK,HK,KKLN,K] some of which are distributed commercially at present. In most of the underlying algorithms the orbit is represented by one or several of its points. A substantial part of the algorithm consists in continuation: some of the parameters of the system are changed in small steps and, knowing a point on a periodic orbit for some value of the parameter (or, a sequence of such points for several values of the parameter) a nearby point on a periodic orbit for the next value of the parameter is found.

To determine this point a system of nonlinear equations is solved numerically. In order to have a full set of equations another condition (to be called anchor condition) is added to the periodicity condition. Typically, two types of anchor conditions are employed.

A) one of the coordinates of the anchor point is fixed

B) one of the coordinates has to attain an extreme on the trajectory at the anchor point.

In order that the algorithm works it is necessary that the resulting system of equations is not degenerate. Equally, it is obvious that this goal cannot be achieved for all values of the parameter. For instance, between a parameter value for which a periodic orbit has a transversal intersection point with the anchor hyperplane and another parameter value for which the orbit does not intersect this hyperplane at all there must be a parameter value for which the orbit is tangent to the hyperplane.

A numerical code for periodic orbit following assumes tacitly that, at least for almost all values of the parameter, the above non degeneracy requirements are met. In this note we give a certain justification of the above assumption: we prove that it holds for "almost all vector fields $f$" in the topological sense, i.e. generically.

## 2. Algorithm A

Consider the differential equation

$$\dot{x} = f(x, \mu) \tag{1}$$

on an open ball $B \subset \mathbb{R}^n$ with $\mu \in \mathbb{R}$ and $f \in \mathcal{F}: = C^r(\overline{B}, \mathbb{R}^n),\ r > 0$.

**Theorem 1.** *Fix* $c_1, \dots, c_n \in \mathbb{R}$. *Then, there exists a residual subset* $\mathcal{G}_1$ *of* $\mathcal{F}$ *such that for each* $f \in \mathcal{G}_1$

*(a) the set of points* $(x^*, \mu, T)$ *such that some periodic orbit of (1) intersects some hyperplane* $x_i = c_i$, $1 \le i \le n$, *nontransversally at* $x^*$ *has no accumulation point in* $B \times \mathbb{R} \times [0, \infty)$,

*(b) if a periodic orbit intersects some hyperplane* $x_i = c_i$ *nontransversally then it intersects all other hyperplanes* $x_j = c_j$, $j \ne i$ *transversally.*

The *proof* is an almost straighforward exercise in transversality theory.

Without loss of generality we take $i = 1$. If we denote the (in general local) flow of (1) by $\phi^\mu$ then $x^*$ is a point of nontransversal intersection of the periodic orbit $\Gamma$ of (1) with the hyperplane $x_1 = c_1$ if and only if the set of equations

$$
\begin{aligned}
\phi^\mu_{T,1}(c_1, \tilde{x}^*) - c_1 &= 0 \\
\tilde{\phi}^\mu_T(c_1, \tilde{x}^*) - \tilde{x}^* &= 0 \\
f_1(c_1, \tilde{x}^*, \mu) &= 0
\end{aligned}
\tag{2}
$$

is satisfied, where $\tilde{x} = (x_2, \dots, x_n)$, $\tilde{\phi}_t = (\phi_{t,2}, \dots, \phi_{t,n})$.

Since $\phi$ depends on $x, \mu, T, f$ in a $C^r$ way, we can write the system of equations (2) as

$$\Phi(\tilde{x}, \mu, T, f) = 0$$

where

$$\Phi(\tilde{x}, \mu, T, f) = \begin{pmatrix} \phi^\mu_{T,1}(c_1, \tilde{x}) - c_1 \\ \tilde{\phi}^\mu_T(c_1, \tilde{x}) - \tilde{x} \\ f_1(c_1, \tilde{x}, \mu) \end{pmatrix}$$

considered as a map from its (open) definition domain $D$ in $\tilde{B} \times \mathbb{R} \times (0, \infty) \times \mathcal{F}$ into $\mathbb{R}^{n+1}$, $\tilde{B} = \{\tilde{x} : (c_1, \tilde{x}) \in B\}$, is $C^r$ (note that $D$ may be a proper subset of $\tilde{B} \times \mathbb{R} \times (0, \infty) \times \mathcal{F}$ since solutions of (1) are in general not defined for all positive times).

If 0 is a regular value of $\Phi(.,.,.,f)$, then $N := [\Phi(.,.,.,f)]^{-1}(0)$ has no accumulation point in $D$. Therefore, in order to establish (a) we prove that

($i$) for a residual subset of $\mathcal{F}$, $f(c_1, \tilde{x}, \mu) \ne 0$ for $\tilde{x} \in \tilde{B}$ and 0 is a regular value of $\Phi(.,.,.,f)$

($ii$) if $N$ has no accumulation point in $D$ then it has no accumulation points in $\tilde{B} \times \mathbb{R} \times [0, \infty)$.

In order to prove ($i$) we first note that, obviously, for a residual subset of $\mathcal{F}$ we have $f(c_1, \tilde{x}, \mu) \ne 0$ for $\tilde{x} \in \tilde{B}$. To complete the proof we employ a slight extension of the "parametric" transversality theorem [$AR$, 18.2, 19.1] by which ($i$) holds provided 0 is a regular value of $\Phi$, i.e. $D\Phi(\tilde{x}, \mu, T, f)$ is surjective if $\Phi(\tilde{x}, \mu, T, f) = 0$. The extension consists in $D$ not being a product of an open subset of $\tilde{B} \times \mathbb{R} \times (0, \infty)$ and $\mathcal{F}$; the proof of [$AR$] applies to this extension without change.

We prove surjectivity of $D\Phi$ by proving that $D_f\Phi$ is surjective, i.e. that for each $\xi \in \mathbb{R}^n, \eta \in \mathbb{R}$ there exists an $h \in \mathcal{F}$ such that

$$D_f\Phi(\tilde{x}, \mu, T)h = (\xi, \eta).$$

Let $\tilde{\Phi}(\tilde{x}, \mu, T) = 0$, $T > 0$ being the minimal period of a periodic orbit $\Gamma$ through $(c_1, \tilde{x})$. We have

$$D_f\Phi(\tilde{x}, \mu, T)h = (y(T), h_1(c, \tilde{x}, \mu)),$$

where $y(t)$ is the solution of the variational equation

$$\dot{y} = D_x f(\phi_t^\mu(x), \mu)y + h(\phi_t^\mu(x), \mu)$$

satisfying $y(0) = 0, x = (c, \tilde{x})$. The variation of constants formula yields

$$y(T) = \int_0^T Y(T, t)h(\phi_t^\mu(x), \mu)dt \tag{3}$$

where $Y(t, s)$ is the solution of the homogeneous equation

$$dY/dt = D_x f(\phi_t^\mu(x), \mu)Y$$

satisfying $Y(t, s)$ =identity.

To construct $h$ we choose an arbitrary $C^r$ periodic function $p(t)$ such that $p_1(0) = \eta$ and define $h$ in the points of the trajectory $\Gamma$ of $x$ by

$$h(\phi_t^\mu(x), \mu) = p(t)$$

It is obvious that $h$ is well defined and $C^r$ on $\Gamma$ and can be extended to a function from $\mathcal{F}$.

To prove (ii) let $f \in \mathcal{G}_1$ and let $(\tilde{x}, \mu, T, f)$ be an accumulation point of $N$ in $\tilde{B} \times \mathbb{R} \times [0, \infty)$. From the basic theorem on the continuous dependence of solutions of differential equations on data it follows that $\phi_T^\mu(c, \tilde{x}, \mu)$ is defined and $\phi_T^\mu(c_1, \tilde{x}, \mu) = (c, \tilde{x})$. Since $\phi_.^\mu(c_1, \tilde{x}, \mu)$ is the limit of a sequence of nonconstant periodic orbits the periods of which are bounded below by a positive constant by [LY], we have $T > 0$. Hence, $(c_1, \tilde{x}, \mu, CalF) \in D$. This completes the proof of (a).

To prove (b) we note that if $x^\star$ is a point of non-transversal intersection with the the hyperplane $x_1 = c_1$ of a $T$-periodic orbit $\Gamma$ of (1) and if $\Gamma$ intersects nontransversally another hyperplane $x_j = c_j$ then the set of equations

$$\begin{aligned} \phi_{T,1}^\mu(c_1, \tilde{x}^\star) - c_1 &= 0 \\ \tilde{\phi}_T^\mu(c_1, \tilde{x}^\star) - \tilde{x}^\star &= 0 \\ f_1(c_1, \tilde{x}^\star, \mu) &= 0 \\ \phi_{t,j}(c_1, \tilde{x}^\star, \mu) - c_j &= 0 \\ f_j(c_1, \tilde{x}^\star, \mu) &= 0. \end{aligned} \tag{4}$$

is satisfied. If we denote the right-hand side of the set of equations (4) by $\hat{\Phi}$ and consider $\hat{\Phi}$ as a map from it's open domain $\hat{D}$ in $\tilde{B} \times \mathbb{R} \times (0, \infty)^2 \times \mathcal{F}$ to $R^{n+3}$ we can repeat the argument from the proof of (a) to show that 0 is a regular value of $\hat{\Phi}$. Since the dimension of $\tilde{B} \times \mathbb{R} \times (0, \infty)^2$ is higher than that of $R^{n+3}$, 0 being a regular value of $\hat{\Phi}$ means that it cannot be attained by $\hat{\Phi}$ at all.

*Remark.* There is a straightforward extension of Theorem 1 to the case of several anchor hyperplanes for one coordinate.

## 3. Algorithm B

**Theorem 2.** *Let $r > 1$. Then, there exists a residual set $\mathcal{G}_2$ of $\mathcal{F}$ such that for each $f \in \mathcal{G}_2$*

*(a) the set of points $(x, \mu, T)$ some of the coordinates $x_i$ of which has a degenerate critical value over some $T$-periodic orbit of (1) at $x$ does not have accumulation points in $B \times \mathbb{R} \times [0, \infty)$ except, possibly, for equilibria,*

*(b) for any $\mu$ and any periodic $\Gamma$ at most one coordinate $x_1$ has a degenerate critical point over $\Gamma$.*

*Proof.* The coordinate $x_i$ has a degenerate critical point over a $T$-periodic orbit of (1) if and only if

$$\phi_T^\mu(x) - x = 0$$
$$f_i(x, \mu) = 0$$
$$Df_i(x, \mu) f(x, \mu) = 0$$

The proof of Theorem 2 is similar to that of Theorem 1 with $\Phi, \hat{\Phi}$ replaced by

$$\Psi(x, \mu, T, f) = \begin{pmatrix} \phi_T^\mu(x) - x \\ f_i(x, \mu) \\ Df_i(x, \mu) f(x, \mu) \end{pmatrix}$$

$$\hat{\Psi}(x, \mu, T, f) = \begin{pmatrix} \phi_T^\mu(x) - x \\ f_1(x, \mu) \\ Df_1(x, \mu) f(x, \mu) \\ f_j(\phi_t^\mu(x), \mu) \\ D_x f_j(\phi_t^\mu(x), \mu) f(\phi_j^\mu(x), \mu) \end{pmatrix},$$

respectively.

**Theorem 3.** *Let $r > 2$. Then, there exists a residual subset $\mathcal{G}_3$ of $\mathcal{F}$ such that for each $f \in \mathcal{G}_3$ the global maximum and the global minimum of each coordinate over any non-constant periodic orbit is non-degenerate.*

*Proof.* If the coordinate $x_i$ has a degenerate maximum or degenerate minimum over a $T$-periodic orbit $\Gamma$ $(T > 0)$ of (1) at $x^*$ then we have

$$\frac{d}{dt}\phi_{T,i}^\mu(x) = \frac{d^2}{dt^2}\phi_{T,i}^\mu(x) = \frac{d^3}{dt^3}\phi_{T,i}^\mu(x) = 0$$

i.e.

$$f_i(x^*, \mu) = 0$$
$$D_x f_i(x^*, \mu) f(x^*, \mu) = 0$$
$$f(x^*, \mu)^T D_{xx} f_i(x^*, \mu) f(x^*, \mu) + D_x f_1(x^*, \mu) D_x f(x^*, \mu) f(x^*, \mu) = 0.$$

Proceeding similarly as in the proof of Theorem 1 again, the proof can be reduced to to the verification of the fact that for any $\eta_1, \eta_2, \eta_3 \in \mathbb{R}$ there exists a function $f \in \mathcal{F}$ such that

$$f_1(x^*, \mu) = \eta_1$$
$$D_x f_1(x^*, \mu) f(x^*, \mu) = \eta_2$$
$$f(x^*, \mu) D_{xx} f_1(x^*, \mu) f(x^*, \mu) + D_x f_1(x, \mu) D_x f(x, \mu) f(x, \mu) = \eta_3.$$

For i=1, such a function can e.g. be given by

$$f_1(x,\mu) = \eta_1 + \eta_2(x_2 - x_2^*) + \frac{\eta_3}{2}(x_2 - x_2^*)^2$$
$$f_2(x,\mu) = 1,$$
$$f_i(x^*,\mu) = 0 \text{ for } i > 2 \text{ (provided } n > 2).$$

## 4. Applications to the numerical continuation of periodic orbits

Theorems 1 and 2 suggest that, if a curve of period orbits is followed for a randomly picked function $f$, the necessity to change the anchor condition occurs at isolated points only which may accumulate if the orbit approaches the boundary of $B$, if the period tends to infinity or (in case of Algorithm B), if a Hopf bifurcation point is approached. At the points of degeneracy of some anchor condition the alternative ones are nondegenerate. Hence, any one of them can be picked as a new anchor condition.

Yet, Theorem 1 is less satisfactory than Theorem 2. Practically, in the numerical codes, at the points of degeneracy of an anchor condition a new anchor condition is not picked from a pre-chosen set of conditions. Instead, if the anchor conditions $x_i = c_i$ degenerates at the point $(x^*,\mu)$ of a periodic orbit $\Gamma$ then a new anchor condition is chosen in two ways:

(i) a $j \neq i$ is chosen and $c_j$ is taken as

$$c_j = 1/2[\max_{x\in\Gamma} x_j + \min_{x\in\Gamma} x_j],$$

(cf. [HK]),

(ii) a $j$ is found such that $f_j(x^*\mu) \neq 0$ (note that such a $j$ has to exist) and $c_j := x_j^*$ is chosen [K].

We have not been able to prove a result in the spirit of Theorem 1 for this choice of anchor conditions

Theorem 3 suggests that, generically, by following the global maximum of some coordinate along a periodic orbit one can avoid the necessity to switch to another coordinate at all. However, following the global maximum is computationally more involved.

## References

[AR] R.Abraham, J.Robbin, *Transversal mappings and flows*, Benjamin, 1967.

[DK] E.Doedel, J.P.Kernevez, *Software for continuation and bifurcation problems in ordinary differential equations* (1986), California Insistute of Technology.

[HK] M. Holodniok, M. Kubicek, *DERPER - An algorithm for the continuation of periodic solutions of ordinary differential equations*, J. of Comp. Physics **55** (1984), 254-267.

[KKLN] A. Khibnik, Yu. Kuznetsov, V. Levitin, E. Nikolaev, *Interactive local bifurcation analyser, version 1.1* (1990), Research Computing Centre, USSR Academy of Sciences, Pushchino.

[K] M. Kubala, *Numerical methods of continuation of periodic orbits*, MSc.Thesis (1991), Comenius University, Bratislava.

[LY] A.Lasota, J.Yorke, *Bounds for periodic solutions of differential equations in Banach spaces*, Journal of Diff.Equations **10** (1978,), 887-899.

# 3 Uniformly Isochronous Centers of Polynomial Systems in $R^2$

**Roberto Conti** University of Florence, Florence, Italy

0.

Let us consider a polynomial system in $\mathbb{R}^2$, i.e., a system of two ordinary differential equations

$$(0.1) \qquad \dot{x} = X(x,y) , \qquad \dot{y} = Y(x,y)$$

where $\dot{x} = dx/dt$ , $\dot{y} = dy/dt$ , $t \in \mathbb{R}$ , and $X(x,y)$, $Y(x,y)$ are two polynomials of $(x,y) \in \mathbb{R}^2$ with real coefficients. The degree $k$ of (0.1) is the maximum of degrees of $X(x,y)$ and $Y(x,y)$.

Let the origin 0 of $\mathbb{R}^2$ be a center of (0.1) and let $N_0$ denote the largest region of $\mathbb{R}^2$ such that for every point $p \in N_0 \setminus \{0\}$ the trajectory passing through $p$ is a cycle which surrounds 0 and no other singular point of (0.1).

For every $p \in N_0 \setminus \{0\}$ let $T(p)$ denote the period of the cycle through $p$ . When $T(p)$ is constant 0 is said to be an isochronous center.

$T(p)$ is the time it takes to the ray $Op$ to make a complete turn around 0 as $p$ moves along the cycle.

The fact that 0 is isochronous means that $T(p)$ is independent of the cycle, but this does not imply that the angular velocity of the ray $Op$ be the same for all the cycles in $N_0$ . When this happens we shall say that the center 0 is <u>uniformly isochronous</u>. □

Let us consider two examples.

Example 0.1

In the linear case $(k = 1)$

$$\dot{x} = ax + by , \qquad \dot{y} = cx + dy$$

0 is a center if and only if

$$a + d = 0 , \qquad \Delta = 4(a^2 + bc) < 0 .$$

The trajectories $\neq \{0\}$ are the ellipses $cx^2 - 2axy - by^2 = \text{const.}$ and the period is $4\pi / \sqrt{-\Delta}$ . Introducing polar coordinates $\rho , \theta$ by $x = \rho \cos\theta$ , $y = \rho \sin\theta$ as usual, the angular velocity is the same $\dot{\theta} = c\cos^2\theta - 2a\cos\theta\sin\theta - b\sin^2\theta$ for every ellipse, so 0 is a uniformly isochronous center.

Example 0.2 (W.S. Loud [3]).

Let us consider the quadratic system $(k = 2)$

$$\dot{x} = y - xy , \qquad \dot{y} = -x - \frac{1}{4} y^2 .$$

0 is the unique singular point and it is a center. $N_0$ is the half plane $x < 1$ and 0 is isochronous. In fact the cycles are represented by

$$\begin{cases} x = 1 - 4(1 - r)(2 - r + r\cos t)^{-2} \\ y = -2r\sin t\,(2 - r + r\cos t)^{-1} \end{cases}$$

for $0 < r < 1$ so they all have the same period $2\pi$ .

However 0 is not uniformly isochronous. In fact the angular velocity of the ray $Op$ is represented by $\dot\theta = -1 + \frac{3}{4}\rho\cos\theta\sin^2\theta$, i.e., it depends on $\rho$. □

We shall show first that for every integer $k \geq 1$, $0$ is a uniformly isochronous center for some system of degree $k$.
Next it will be shown that if $0$ is a uniformly isochronous center then $N_0 = \mathbb{R}^2$ if $k = 1$, whereas if $k > 1$ the boundary $\partial N_0$ of $N_0$ is the finite union of open unbounded trajectories whose number is $\leq k-1$.
□

1.

Let $0$ be an isochronous center of (0.1). Then, as a consequence of a more general result about the period function $p \to T(p)$ it can be shown (M. Villarini [4]) that (0.1) can be written

$$\dot x = y + P(x,y), \qquad \dot y = -x + Q(x,y) \tag{1.1}$$

where $P(x,y)$, $Q(x,y)$ are polynomials which do not contain terms of degree $\leq 1$. In particular $\rho^{-2}[x\,Q(x,y)-y\,P(x,y]$ can be constant if and only if it is zero.
From (1.1) it follows

$$\dot\theta = -1 + \rho^{-2}\,[x\,Q(x,y) - yP(x,y)]$$

If $0$ is a uniformly isochronous center then $\dot\theta = -1$, i.e.,

$$xQ(x,y) - yP(x,y) = 0, \qquad (x,y) \in \mathbb{R}^2,$$

i.e.,

$$P(x,y) = xR(x,y), \qquad Q(x,y) = yR(x,y)$$

where $R(x,y)$ is $\equiv 0$ if $k = 1$ and a polynomial of degree $k-1$ if $k > 1$.
Therefore, if $0$ is a uniformly isochronous center (1.1) has the form

$$\dot x = y + xR(x,y), \qquad \dot y = -x + yR(x,y). \tag{1.2}$$

□

Since

$$[y + xR(x,y)]^2 + [-x + yR(x,y)]^2 =$$

$$= (x^2 + y^2)[1 + R^2(x,y)] > 0 , \qquad x^2 + y^2 > 0$$

it follows that if $0$ is a uniformly isochronous center then $0$ is the unique singular point of (1.2).

Therefore no polynomial system can have more than one uniformly isochronous center, whereas there are systems with more than one isochronous center, like, for instance (cf. W. S. Loud [3])

$$\dot{x} = y - xy , \qquad \dot{y} = -x + \frac{1}{2} x^2 - 2y^2$$

with isochronous centers at $0$ and $(2,0)$ . □

If $k > 1$ , $R(x,y)$ can be written

$$R(x,y) = \sum_{1}^{k-1}{}_h \; R_h(x,y)$$

where $R_h(x,y)$ is a homogeneous polynomial of degree $h$ .

Then the trajectories of (1.2) correspond to the solutions $\theta \to \rho(\theta)$ of

$$\frac{d\rho}{d\theta} = - \sum_{1}^{k-1}{}_h \; R_h(\cos\theta, \sin\theta)\rho^{h+1} . \tag{1.3}$$

To every cycle of (1.2) there corresponds a positive $2\pi$-periodic solution of (1.3) and viceversa.

Therefore $0$ is a uniformly isochronous center if and only if (1.3) has a family of such solutions. □

2.

Let $k \geq 2$ and let $R(x,y)$ be a homogeneous polynomial of degree $k-1$, i.e., let

(2.1) $$R(x,y) = R_{k-1}(x,y) = \sum_{j+\ell=k-1} r_{j,\ell} x^j y^\ell .$$

Theorem 2.1

Let $k \geq 2$. Then $0$ is a uniformly isochronous center of

(2.2) $$\dot{x} = y + x R_{k-1}(x,y) , \qquad \dot{y} = -x + y R_{k-1}(x,y)$$

if and only if either $k$ is even, or $k$ is odd, $k = 2m+1$ , and

(2.3) $$\sum_0^{2m} {}_s\, r_{2m-s,s} \int_0^{2\pi} \cos^{2m-s}\varphi \sin^s\varphi \, d\varphi = 0$$

holds.

Proof. When (2.1) holds the corresponding equation (1.3) reduces to

(2.4) $$\frac{d\rho}{d\theta} = - R_{k-1}(\cos\theta, \sin\theta)\rho^k$$

so that the solution satisfying

$$\rho(0) = r > 0$$

is represented by

(2.5) $$\rho^{1-k}(\theta) = r^{1-k} + (k-1) \int_0^\theta R_{k-1}(\cos\varphi, \sin\varphi) d\varphi$$

as long as the right hand side is $> 0$ .

If

(2.6) $$\int_0^{2\pi} R_{k-1}(\cos\varphi, \sin\varphi) d\varphi = 0$$

then $\theta \to \int_0^\theta R_{k-1}(\cos\varphi, \sin\varphi) d\varphi$ is $2\pi$-periodic, hence bounded. Therefore the right hand side of (2.5) is $> 0$ if $r > 0$ is close enough to $0$

and for such $r$, $\theta \to \rho(\theta)$ is defined for all $\theta$ and is $2\pi$-periodic, so that $0$ is a center.

If, on the contrary

$$\int_0^{2\pi} R_{k-1}(\cos\varphi,\sin\varphi)\,d\varphi = \delta \neq 0$$

then, for $\nu = 1,2,\ldots$

$$r^{1-k} + (k-1)\int_0^{\pm 2\nu\pi} R_{k-1}(\cos\varphi,\sin\varphi)\,d\varphi = r^{1-k} \pm \nu(k-1)\delta$$

and from (2.5) it follows $\rho(2\nu\pi) \to 0$ if $\delta > 0$, $\rho(-2\nu\pi) \to 0$ if $\delta < 0$, so that $0$ is a focus.

When $k$ is even, then $R_{k-1}(\cos(\varphi+\pi), \sin(\varphi+\pi)) = -R_{k-1}(\cos\varphi,\sin\varphi)$, so that (2.6) holds.

When $k$ is odd, $k = 2m+1$, then the coefficients $r_{j,\ell}$ of $R_{k-1}(\cos\varphi,\sin\varphi)$ must satisfy (2.3). □

Remark 2.1

Verifying whether (2.3) holds or not requires the calculation of the integrals

$$I_{m,s} = \int_0^{2\pi} \cos^{2m-s}\varphi \sin^{s}\varphi\,d\varphi, \qquad s = 0,1,\ldots,2m .$$

When $s$ is odd then $I_{m,s} = 0$. When $s$ is even then $I_{m,s}$ is the finite sum of integrals

$$\int_0^{2\pi} \cos^{2n}\varphi\,d\varphi = 4\int_0^{+\infty} (1+\tau^2)^{-n-1}\,d\tau, \qquad n = 0,1,\ldots .$$

□

Remark 2.2

When (2.6) holds $\int_0^{\theta} R_{k-1}(\cos\varphi,\sin\varphi)\,d\varphi$ is a polynomial of $\cos\varphi,\sin\varphi$ of degree $\le k-1$ and from (2.5) it follows that every trajectory $\neq \{0\}$

of (2.2) is a branch of an algebraic curve whose order is $= 2(k-1)$ if $k$ is even, and $= k-1$ if $k$ is odd. □

3.

When $N_0 = \mathbb{R}^2$, $0$ is said to be a global center. This means that all the trajectories $\neq 0$ are cycles.

In the linear case $(k=1)$ if $0$ is a center, then it is global and uniformly isochronous (Cfr. Example 0.1).

In the non linear case $(k>1)$ we have

Theorem 3.1

If $k > 1$ and $0$ is a uniformly isochronous center then $0$ cannot be a global one.

Proof. $0$ is a global center of (1.2) if and only if all the positive solutions $\rho$ of (1.3) are $2\pi$-periodic.

To show that this is impossible (unless $R_h \equiv 0$, $1 \le h \le k-1$) we replace $\rho$ by $\sigma = \rho^{-1}$, $\rho > 0$, sending the solutions $\rho$ of (1.3) into the solutions $\sigma$ of

$$(3.1) \qquad \frac{d\sigma}{d\theta} = \sum_{1}^{k-1}{}_h \; R_h(\cos\theta, \sin\theta)\sigma^{-h+1}$$

so that the cycles of (1.2) correspond to the positive $2\pi$-periodic solutions of (3.1).

If $0$ is a (uniformly isochronous) center of (1.2) the family $F$ of such solutions is non-empty. We shall show that there exist solutions $\notin F$

Let

$$\sigma_o = \inf\{\sigma(0) : \sigma \in F\} \quad .$$

If $\sigma_o > 0$ then every solution $\sigma$ of (3.1) satisfying $0 < \sigma(0) < \sigma_o$ does not belong to $F$.

Let $\sigma_o = 0$ and let us consider the sequence of solutions $\sigma_n$ such that

$\sigma_n(0) = \frac{1}{n}$, $n = 1,2,\ldots$ Clearly $\sigma_n \in F$. Moreover the graphs of $\sigma_n$, $\sigma_{n+1}$ cannot intersect by the uniqueness of such solutions. Thus, since $\sigma_{n+1}(0) < \sigma_n(0)$ by definition, $\sigma_{n+1}(\Theta) < \sigma_n(\Theta)$. Therefore there exist

$$\lambda(\Theta) = \lim_n \sigma_n(\Theta), \qquad 0 \leq \Theta \leq 2\pi$$

If $\lambda(\bar{\theta}) > 0$ for some $\bar{\theta}$ then no solution $\sigma$ such that $0 < \sigma(\bar{\theta}) < \lambda(\bar{\theta})$ belongs to $F$.

Therefore let $\lambda(\theta) = 0$, $0 \leq \theta \leq 2\pi$ .

From (3.1) we have

$$\sigma_n^{k-1}(\theta) = \left(\frac{1}{n}\right)^{k-1} + (k-1)\int_0^\theta R_1(\cos\varphi, \sin\varphi)\sigma_n^{k-2}(\varphi)\,d\varphi + \ldots \tag{3.2}$$

$$\ldots + (k-1)\int_0^\theta R_{k-2}(\cos\varphi, \sin\varphi)\sigma_n(\varphi)\,d\varphi + (k-1)\int_0^\theta R_{k-1}(\cos\varphi, \sin\varphi)\,d\varphi$$

On the other hand since $\{\sigma_n\}$ is a decreasing sequence converging to a continuous function (namely $\lambda = 0$) it follows (Dini's theorem) that $\sigma_n \to 0$ uniformly.

Then from (3.2) it follows

$$\int_0^\theta R_{k-1}(\cos\varphi, \sin\varphi)\,d\varphi = 0, \qquad 0 \leq \theta \leq 2\pi$$

i.e.,

$$R_{k-1}(\cos\varphi, \sin\varphi) = 0, \quad 0 \leq \varphi \leq 2\pi, \quad \text{which is impossible.}$$

□

Remark 3.1

It can be proved (M. Galeotti - M. Villarini [2]) that no polynomial system of even degree can have a global center.

Theorem 3.1 shows that the same is true also when the degree is odd provided that the center be uniformly isochronous.

□

4.

From theorem 3.1 it follows that if $k > 1$ and 0 is a uniformly isochronous center the boundary $\partial N_0$ of $N_0$ is the finite union of open unbounded trajectories. If the number of such trajectories is $\nu$ we say that 0 is a center of type $B^\nu$. □

Theorem 4.1

If 0 is a uniformly isochronous center of a polynomial system of degree $k > 1$ then 0 is a center of type $B^\nu$ with $\nu \leq k-1$.

Proof. Given a vector field $(X(x,y),Y(x,y))$ in $\mathbb{R}^2$ and an algebraic curve C of order n in $\mathbb{R}^2$ represented by $f(x,y) = 0$, recall that a point $(x,y)$ is said to be a contact point of C with $(X(x,y),Y(x,y))$ if it is a solution of the system

$$(4.1) \qquad f(x,y) = 0\,, \qquad f_x(x,y)X(x,y) + f_y(x,y)Y(x,y) = 0$$

i.e., if either $(x,y)$ is a singular point of $(X(x,y),Y(x,y))$ on C or the vector $(X(x,y),Y(x,y))$ is tangent to C at $(x,y)$.
If C is a circle, $f(x,y) = x^2 + y^2 - r^2$, and

$$X(x,y) = y + xR(x,y)\,, \qquad Y(x,y) = -x + yR(x,y)$$

then (4.1) become

$$x^2 + y^2 - r^2 = 0\,, \quad R(x,y) = 0$$

so that by Bézout theorem the number of contacts on C is $\leq 2(k-1)$.
If r is large enough the circle C must intersect all the trajectories $\subset \partial N_0$ and the number of intersections is at least = 2 for each trajectory. Such intersections divide C into two arcs each containing one contact point at least. Therefore the number of trajectories $\subset \partial N_0$ must be $\leq k-1$.
The next example shows that the upper bound $k-1$ can be attained by $\nu$.

Example 4.1

Let

$$R_{k-1}(\cos\varphi, \sin\varphi) = \sin(k-1)\varphi \quad , \quad k > 1 \ .$$

Then (2.5) becomes

$$\rho^{1-k}(\theta) = r^{1-k} + 1 - \cos(k-1)\theta \quad , \quad r = \rho(0)$$

with cycles corresponding to $0 < r < +\infty$, whereas $\partial N_0$ corresponds to $r = +\infty$ so that it is represented by

$$\rho^{k-1} = (1 - \cos(k-1)\theta)^{-1}$$

i.e., by an algebraic curve with $k-1$ unbounded real branches. □

Remark 4.1

Examples are known (R. Conti [1]) showing that for every $k > 1$ there exist polynomial systems of degree $k$ with a center of type $B^{k-1}$, but in those examples the center is not even isochronous. □

References

[1] R. Conti, On centers of type B of polynomial systems, Archivum Math. (Brno), 26(1990), 93-100.

[2] M. Galeotti - M. Villarini, Some properties of planar polynomial systems of even degree, Annali di Matematica pura ed appl., (4) 161 (1992), 299-314.

[3] W. S. Loud, Behavior of the period of solutions of certain plane autonomous systems near centers, Contributions to Diff. Eqs., 3(1964), 21-36.

[4] M. Villarini, Regularity properties of the period function near a center of a planar vector field, J. Nonl. Analysis, 19 (1992), 787-803.

# 4 Some Remarks on the Titchmarsh-Weyl $m$-Coefficient and Associated Differential Operators

**W. Norrie Everitt** The University of Birmingham, Birmingham, England

## 1. Introduction

The Titchmarsh-Weyl $m$-coefficient for second-order ordinary differential equations was first introduced by Titchmarsh in 1941, see [21], but based on the fundamental work of Weyl in 1910, see [23]. For some discussion on the history of the $m$-coefficient see Bennewitz, Everitt [5, Section 1], and Fulton [16].

This paper, as is [16], is concerned with the definition and properties of the $m$-coefficient in the so-called (Weyl) limit-circle case. However, we remark on the (Weyl) limit-point case at the end of the paper.

As presented here, the thesis is that this theory is best seen in conjunction with the properties of the unbounded differential operators, defined in the underlying Hilbert function space, associated with the given Sturm-Liouville differential equation. This equation, written in standard form, is

$$-(py')' + qy = \lambda wy \text{ on } (a, b) \tag{1.1}$$

with real-valued coefficients $p$, $q$ and $w$, where the non-negative weight $w$ determines the required Hilbert space $L^2((a, b); w)$.

The $m$-coefficient in the limit-circle case can be defined by a limit process in terms of a sequential approach to the end-point of the interval of definition of the differential equation; see Titchmarsh [22, Chapter II], Coddington and Levinson [9, Chapter 7], Everitt [12], Hille [17, Chapter 10], Levitan and Sargsjan [18, Chapter 2]. To obtain the required meromorphic properties of the $m$-coefficient, use is made of the Vitali boundedness result, see Titchmarsh [22, Section 5.21], or some form of uniform convergence of integrals, see [22, Sections 2.83 to 2.85] or Copson [10, Sections 5.4 and 5.5], in the Cauchy theory of functions of a complex variable.

Similar methods are used here to define the $m$-coefficient, but utilize either a continuous limit process from regular points to the singular end-point, or a direct definition using the singular boundary conditions of the associated differential operator. Some of the ideas involved in this approach are inspired by the work of Fulton [16], but more recently by results on the regular approximation to singular Sturm-Liouville boundary value problems given in Bailey, Everitt, Zettl [2], and the subsequent paper with Weidmann [3].

The methods used here extend to differential equations of more general form than (1.1). For the most general scalar equations but with function coefficients, see Everitt and Zettl [15], and Everitt [13, 14]. For Sturm-Liouville type differential equations with measure coefficients, see Bennewitz [4], and Clark and Hinton [8], to which the methods of this paper would apply.

Throughout the paper we make use wherever possible of the properties of the differential operators in $L^2((a,b);w)$ generated by the differential equation (1.1). The properties of these unbounded operators required here can be found in Akhiezer and Glazman [1], and Naimark [19]. In particular, the form of the separated boundary conditions to determine these operators is discussed in some detail in [3].

Properties of the differential equation (1.1) are discussed in Sections 2 and 3; differential operators in 4 and 5; analytic properties and the $m$-coefficient in 6 and 7; the regular approximation in 8; the link with the Weyl circles in 9 and 10; concluding remarks are made in Section 11.

**Acknowledgements.** The author expresses his gratitude to Christer Bennewitz for continuing help and discussion on the theory of the Titchmarsh-Weyl $m$-coefficient over many years. Some of the ideas considered in this paper result from collaboration in the preparation of the manuscript [6], which we hope to see published in 1993.

The author is greatly indebted to Mari-Anne Hartig, Department of Mathematical Sciences, Northern Illinois University, DeKalb, USA for preparing the typed version of this paper.

## 2. Differential equations

In the equation (1.1) let $(a,b)$ be an open interval of the real line $\mathbb{R}$, let the spectral parameter $\lambda = \mu + i\nu \in \mathbb{C}$ the complex plane, and let the coefficients $p$, $q$ and $w$ satisfy:

(i) $p, q, w : (a,b) \to \mathbb{R}$ and be Lebesgue measurable

(ii) $p^{-1}, q, w \in L^1_{\text{loc}}(a,b)$ (2.1)

(iii) $w(x) > 0$ (almost all $x \in (a,b)$).

For any open set $U \subseteq \mathbf{C}$ of the complex plane, let $H(U)$ denote the family of Cauchy analytic functions which are holomorphic (differentiable) on $U$.

The standard existence theorems for the differential equation (1.1), see [9, Chapter 3], [15], [17, Chapter 10], [19, Chapter IV], and [22, Chapter I], yield the following result:

$$\text{let } c \in (a,b) \text{ and let } \xi_r(\cdot) \in H(\mathbf{C}) \quad (r = 1,2) \text{ be given} \tag{2.2}$$

then there exists a unique mapping $y : (a,b) \times \mathbf{C} \to \mathbf{C}$ with the properties

$$\begin{aligned} &\text{(i) } y(\cdot,\lambda) \text{ and } (py')(\cdot,\lambda) \in AC_{\text{loc}}(a,b) \qquad (\lambda \in \mathbf{C}) \\ &\text{(ii) } y(x,\cdot) \text{ and } (py')(x,\cdot) \in H(\mathbf{C}) \qquad (x \in (a,b)) \\ &\text{(iii) } y(c,\lambda) = \xi_1(\lambda) \quad (py')(c,\lambda) = \xi_2(\lambda) \qquad (\lambda \in \mathbf{C}) \\ &\text{(iv) } y(\cdot,\lambda) \text{ satisfies (1.1) almost everywhere on } (a,b). \end{aligned} \tag{2.3}$$

The quasi-derivatives of the solution $y$ are denoted by $y^{[s-1]} (s = 1,2)$, see [15], and defined as

$$y^{[0]}(x,\lambda) := y(x,\lambda) \qquad y^{[1]}(x,\lambda) := (py')(x,\lambda) \qquad (x \in (a,b), \lambda \in \mathbf{C}). \tag{2.4}$$

Let $y$ be a solution determined by initial conditions (2.2); let $z$ be another such solution. The Wronskian $W(y,z)$ is defined by, for $x \in (a,b)$ and $\lambda \in \mathbf{C}$,

$$W(y,z)(x,\lambda) := (y^{[0]}z^{[1]} - y^{[1]}z^{[0]})(x,\lambda). \tag{2.5}$$

It is known $W(y,z)(x,\lambda) \equiv W(y,z)(\lambda)$ is, for each $\lambda \in \mathbf{C}$, a constant function on $(a,b)$, and is not zero on $(a,b)$ if and only if the pair $\{y(\cdot,\lambda), z(\cdot,\lambda)\}$ is linearly independent and hence forms a basis of solutions.

Let $c \in (a,b)$; the particular basis $\{y_1, y_2\}$ is determined, for all $\lambda \in \mathbf{C}$, from the initial conditions

$$y_r^{[s-1]}(c,\lambda) = \delta_{rs} \qquad (r,s = 1,2). \tag{2.6}$$

For each $x \in (a,b)$ we have $y_r(x,\cdot) \in H(\mathbf{C})$, and since $p$, $q$, $w$ are real-valued on $(a,b)$ and the initial conditions (2.6) are real, it follows that the symmetry condition

$$\bar{y}_r(x,\lambda) = y_r(x,\bar{\lambda}) \qquad \overline{(py_r')}(x,\lambda) = (py_r')(x,\bar{\lambda}) \qquad (\lambda \in \mathbf{C}) \tag{2.7}$$

is satisfied, and that $\{y_r(x,\cdot)\}$ are real-valued on $\mathbf{R} \subset \mathbf{C}$. (Here the $^-$ denotes complex conjugation.)

For

$$f : (a,b) \to \mathbf{C} \text{ with } f \text{ and } pf' \in AC_{\text{loc}}(a,b) \tag{2.8}$$

let the differential expression $M$ be defined by

$$M[f](x) := -(p(x)f'(x))' + q(x)f(x) \qquad (x \in (a,b)). \tag{2.9}$$

The Green's formula for $M$ is, for any compact $[\alpha,\beta] \subset (a,b)$

$$\int_\alpha^\beta \{\bar{g}(x)M[f](x) - f(x)M[\bar{g}](x)\}\, dx = [f,g](x)\Big|_\alpha^\beta \tag{2.10}$$

where the skew-symmetric bilinear form $[\cdot,\cdot](\cdot)$ is defined by

$$[f,g](x) = f^{[0]}(x)\bar{g}^{[1]}(x) - f^{[1]}(x)\bar{g}^{[0]}(x) \qquad (x \in (a,b)) \tag{2.11}$$

for any pair $f$, $g$ satisfying the condition (2.8).

Below we make use of the Plücker identity; for further details see Everitt [11]. Let $\{f_r, g_r : r = 1,2,3\}$ be any six functions, each of which satisfies the condition (2.8); then for the $3 \times 3$ matrix $[[f_r, g_s](x)]$ $(r,s = 1,2,3)$ we have the identity on $(a,b)$

$$\det[[f_r, g_s](x)] = 0 \qquad (x \in (a,b)). \tag{2.12}$$

## 3. Classification of end-points of $(a,b)$

The end-point $b$ for the differential equation (1.1) is said to be *regular* (R) if $b \in \mathbf{R}$ and for some $\delta > 0$ such that $a < b - \delta < b$

$$p^{-1}, q, w \in L^1[b-\delta, b); \tag{3.1}$$

similarly for the end-point $a$. Clearly this classification is independent of the spectral parameter $\lambda$ in (1.1).

If $b$ is not R, then this end-point is said to be *singular*; clearly if $b = \infty$, then $b$ is singular. If $b \in \mathbf{R}$, then $b$ is singular if and only if (3.1) fails for at least one of the three coefficients $p$, $q$, $w$. There is a similar classification of singular at end-point $a$.

For the classification of singular end-points we have to introduce the Hilbert function space $L^2((a,b);w)$ which is defined by (recall property (iii) of (2.1))

$$\left\{ f : (a,b) \to \mathbf{C} \mid f \text{ is Lebesgue measurable and } \int_a^b w(x)|f(x)|^2\, dx < \infty \right\} \tag{3.2}$$

With due regard for the introduction of equivalence classes (Lebesgue measure), this family of functions may be regarded as a Hilbert space with inner-product

$$(f, g) := \int_a^b w(x) f(x) \bar{g}(x)\, dx; \tag{3.3}$$

the zero element is the equivalence class of all functions zero almost everywhere on $(a, b)$.

For any $c \in (a, b)$ there are similar definitions of the spaces $L^2((a, c]; w)$ and $L^2([c, b); w)$.

Let the end-point $b$ of the differential equation (1.1), with spectral parameter $\lambda \in \mathbb{C}$, be singular; then $b$ is said to be *limit-circle* (LC) in $L^2((a, b); w)$ if for some (and then for all) $c \in (a, b)$ the basis set $\{y_1, y_2\}$ determined by the initial conditions (2.6) satisfies

$$y_r(\cdot, \lambda) \in L^2([c, b); w) \qquad (r = 1, 2). \tag{3.4}$$

Otherwise the singular end-point $b$ is said to be *limit-point* (LP) in $L^2((a, b); w)$.

There is a similar classification at end-point $a$.

The classification of a singular end-point as LC or LP is independent of the spectral parameter $\lambda$, and depends only on the coefficients $p$, $q$ and $w$. We have

**Lemma 3.1** *Let the coefficient conditions (2.1) hold; let $c \in (a, b)$ and let $b$ be a singular point of the equation (1.1); let the limit-circle condition (3.4) hold for some point $\lambda \in \mathbb{C}$; then*

*(i)* $y_r(\cdot, \lambda) \in L^2([c, b); w) \qquad$ *($r = 1, 2$ and for all $\lambda \in \mathbb{C}$)*

*(ii) the integrals $\int_c^b w(x)\, |y_r(x, \lambda)|^2\, dx$ ($r = 1, 2$) are uniformly convergent at $b$ for all $\lambda \in K$, where $K$ is any compact set of the complex plane $\mathbb{C}$*

*(iii) for any compact $K \subset \mathbb{C}$ there exists a real, positive number $L \equiv L(K) > 0$ such that*

$$\int_c^b w(x)\, |y_r(x, \lambda)|^2\, dx \leq L \qquad (r = 1, 2 \text{ and for all } \lambda \in K)$$

*(iv) let $f \in L^2([c, b); w)$ and define $F_r : \mathbb{C} \to \mathbb{C}\, (r = 1, 2)$ by*

$$F_r(\lambda) := \int_c^b w(x) f(x) y_r(x, \lambda)\, dx \qquad (r = 1, 2\ \lambda \in \mathbb{C});$$

*then*

$$F_r(\cdot) \in H(\mathbb{C}) \qquad (r = 1, 2).$$

*Proof.*

(i) See [9, Chapter 9, Theorem 2.1].

(ii) The proof of (i) in [9] can be extended to give this result.

(iii) Again the proof in (i) can be extended for this result; see also the remark in [9, Chapter 9, Section 4, remark following display (4.7)].

(iv) This follows from the uniform convergence result in (ii).

## 4. Maximal domains

Let the coefficients $p$, $q$ and $w$ satisfy the conditions (2.1) and let the differential expression $M$ be defined by (2.9).

The maximal domain $\Delta$ determined by the coefficients $p$, $q$ and $w$ is a linear manifold of the Hilbert function space $L^2((a,b);w)$ defined by

$$\Delta := \{f : (a,b) \to \mathbb{C} \mid f, pf' \in AC_{\text{loc}}(a,b) \\ f, w^{-1}M[f] \in L^2((a,b);w)\}. \tag{4.1}$$

The sub-domains $\Delta_a$ and $\Delta_b$ are the restricted domains (here $c \in (a,b)$ and can be chosen as proves to be convenient)

$$\Delta_a(\text{resp. } \Delta_b) := \{f \in \Delta \mid \text{domain of } f \text{ restricted to } (a,c] \text{ (resp. } [c,b))\} \tag{4.2}$$

In general the bilinear form $[f,g](\cdot)$ is defined on the open interval $(a,b)$, for all $f$, $g$ satisfying (2.8). If now we restrict $f$, $g$ to belong to $\Delta$, then from Green's formula (2.10) and the definition (4.1) we can define $[f,g](\cdot)$ on the closed interval $[a,b]$ (identifying $[-\infty,c] = (-\infty,c]$ and $[c,\infty] = [c,\infty)$ when required) by

$$[f,g](a) := \lim_{x\to a^+}[f,g](x), \qquad [f,g](b) := \lim_{x\to b^-}[f,g](x), \qquad (f,g \in \Delta). \tag{4.3}$$

When the differential equation (1.1) is regular at both end-points $a$ and $b$ it follows that $[f,g](\cdot) \in AC[a,b]\,(f,g \in \Delta)$. When the equation is singular at one or both end-points, this result has to be relaxed to $[f,g](\cdot) \in C[a,b]\,(f,g \in \Delta)$.

It is known that, see [19, Section 18.3] and [22, Chapter III], the differential equation (1.1) is

$$\text{LP at } b \text{ if and only if } [f,g](b) = 0 \qquad (f,g \in \Delta) \tag{4.4}$$

$$\text{R or LC at } b \text{ if and only if there exists } f,g \in \Delta \text{ with } [f,g](b) \neq 0. \tag{4.5}$$

There is a similar result at end-point $a$.

We have

**Lemma 4.1** *Let the above definitions hold; let the differential equation (1.1) be LC at $b$; let the basis pair of solutions $\{y_1, y_2\}$ be defined by (2.6); then*

*(i)* $y_r(\cdot, \lambda) \in \Delta_b \qquad (r = 1, 2 \quad \lambda \in \mathbb{C})$

*(ii) if $g \in \Delta_b$ and $F_r(\lambda) := [y_r(\cdot, \lambda), g](b)$ for $\lambda \in \mathbb{C}$ and $r = 1, 2$ then $F_r(\cdot) \in H(\mathbb{C})$* $(r = 1, 2)$.

*Proof.*

(i) Clear.

(ii) From Green's formula (2.10)

$$[y_r(\cdot, \lambda), g](b) = y_r^{[0]}(c, \lambda)\bar{g}^{[1]}(c) - y_r^{[1]}(c, \lambda)\bar{g}^{[0]}(c) + \\ + \int_c^b w(x) y_r(x, \lambda)\bar{g}(x)\, dx \qquad (\lambda \in \mathbb{C})$$

and the required result follows from (2.3) and Lemma 3.1 (iv).

**Remarks.**

1. There is a similar result at end-point $a$.

2. For ease of notation we write sometimes

$$[y_r(\cdot, \lambda), g](b) \equiv [y_r, g](b, \lambda) \qquad (\lambda \in \mathbb{C}) \tag{4.6}$$

## 5. Differential operators

In this section we define the self-adjoint differential operators in $L^2([c, b); w)$, with so-called separated boundary conditions, which are associated with the Titchmarsh-Weyl $m$-coefficient of the differential equation (1.1) in either the regular or limit-circle case at end-point $b$.

There are similar definitions for end-point $a$.

To form the boundary condition at the regular end-point $c$, we define a new basis pair from $\{y_1, y_2\}$, where the latter satisfy the initial conditions given by (2.6).

For $\sigma \in [0, \pi)$ define, for $x \in (a, b)$ and $\lambda \in \mathbb{C}$,

$$\begin{aligned} \theta_\sigma(x,\lambda) &:= \cos(\sigma)y_1(x,\lambda) - \sin(\sigma)y_2(x,\lambda) \\ \varphi_\sigma(x,\lambda) &:= \sin(\sigma)y_1(x,\lambda) + \cos(\sigma)y_2(x,\lambda) \end{aligned} \tag{5.1}$$

(the $\theta$, $\varphi$ notation is essentially that of Titchmarsh [22, Chapter I]). Then from (2.6) and (2.7)

$$W(\theta_\sigma, \varphi_\sigma)(x,\lambda) = (\cos^2(\sigma) + \sin^2(\sigma))W(y_1, y_2)(x,\lambda) = 1 \qquad (x \in (a,b) \quad \lambda \in \mathbb{C}) \tag{5.2}$$

and

$$\left[\theta_\sigma(\cdot,\lambda), \varphi_\sigma(\cdot,\bar{\lambda})\right](x) = W(\theta_\sigma, \varphi_\sigma)(x,\lambda) = 1 \qquad (x \in (a,b) \quad \lambda \in \mathbb{C}). \tag{5.3}$$

We note that, compare with (2.7),

$$\theta_\sigma(x,\bar{\lambda}) = \bar{\theta}_\sigma(x,\lambda) \qquad \varphi_\sigma(x,\bar{\lambda}) = \bar{\varphi}_\sigma(x,\lambda) \qquad (x \in (a,b) \quad \lambda \in \mathbb{C}) \tag{5.4}$$

and that $\theta_\sigma(\cdot,\lambda)$ and $\varphi_\sigma(\cdot,\lambda)$ are real-valued when $\lambda \in \mathbf{R} \subset \mathbb{C}$.

If $y(\cdot)$ is any solution of (1.1), or if $f$ is any element of the domain $\Delta$, we find for $\sigma \in [0, \pi)$

$$\begin{aligned} [y(\cdot), \varphi_\sigma(\cdot,\lambda)](c) &= \cos(\sigma)y(c) + \sin(\sigma)(py')(c) \\ &= \cos(\sigma)y^{[0]}(c) + \sin(\sigma)y^{[1]}(c) \end{aligned}$$

and

$$[f(\cdot), \varphi_\sigma(\cdot,\lambda)](c) = \cos(\sigma)f^{[0]}(c) + \sin(\sigma)f^{[1]}(c).$$

A separated boundary condition at end-point $c$ may then be written in the form, for some $\sigma \in [0, \pi)$,

$$[y(\cdot), \varphi_\sigma(\cdot,\lambda)](c) = 0 \qquad \text{or} \qquad [f(\cdot), \varphi_\sigma(\cdot,\lambda)](c) = 0 \tag{5.5}$$

where $\lambda \in \mathbb{C}$ is a free parameter, since $\varphi_\sigma^{[0]}$ and $\varphi_\sigma^{[1]}$ are independent of this variable at $c$, from (5.1) and (2.6).

When $\sigma = 0$ we obtain the Dirichlet condition $y(c,\lambda) = 0$ or $f^{[0]}(c) = 0$, and when $\sigma = \pi/2$, the Neumann condition $(py')(c,\lambda) = 0$ or $f^{[1]}(c) = 0$.

The form of the boundary condition (5.5) is equivalent to the established boundary condition; see [9, Chapter 9, Section 2] and [22, Chapters I and II].

To obtain the boundary condition form at a regular, or singular limit-circle end-point $b$, we select two *real-valued* functions $\{\chi_1, \chi_2\}$ from $\Delta$ which satisfy the independence condition

$$[\chi_1, \chi_2](b) = 1. \tag{5.6}$$

If $b$ is R, then $\chi_1$, $\chi_2$ can be basis solutions of the differential equations satisfying initial conditions at $b$ of the form (2.6) for $\{y_1, y_2\}$. If $b$ is LC, then we can choose $\chi_1(x) = \theta_\sigma(x, \lambda)$, $\chi_2(x) = \varphi_\sigma(x, \lambda)$ $(x \in (a, b))$ for some *real* $\lambda$; in this case (5.6) follows from (5.3). However, in this LC case, it is sometimes desirable to choose $\chi_1$, $\chi_2$ from $\Delta$ which are not solutions; examples of the solution choice are given in [2, Section 6, Examples 1 to 6], and for domain examples, see the same reference Examples 7 and 8. Given then $\{\chi_1, \chi_2\}$ satisfying (5.6) define, for $x \in (a, b)$ and for a parameter $\tau \in [0, \pi)$, (compare with (5.1))

$$\begin{aligned} \kappa_\tau(x) &:= \cos(\tau)\chi_1(x) - \sin(\tau)\chi_2(x) \\ \delta_\tau(x) &:= \sin(\tau)\chi_1(x) + \cos(\tau)\chi_2(x) \end{aligned} \tag{5.7}$$

With this definition we have then

$$\begin{aligned} &\text{(i) } \kappa_\tau,\ \delta_\tau \in \Delta \text{ and are real-valued on } (a, b) \\ &\text{(ii) } [\kappa_\tau, \delta_\tau](b) = 1 \qquad (\tau \in [0, \pi)) \\ &\text{(iii) for each } \tau \in [0, \pi) \text{ there exists } c' \equiv c'(\tau) \in (c, b) \text{ with the property} \\ &\qquad [\kappa_\tau, \delta_\tau](\beta) \neq 0 \qquad (\beta \in (c', b)). \end{aligned} \tag{5.8}$$

We can now define all self-adjoint differential operators in the Hilbert function space $L^2([c, b); w)$ which are generated by:

(i) the Lagrange symmetric (formally self-adjoint) differential expression $M$ given by (2.9)

(ii) separated, real, symmetric boundary conditions at the regular end-point $c$, and at the regular or limit-circle end-point $b$.

These differential operators depend on the choice of the boundary condition parameters $\sigma$ and $\tau$. Thus, we adopt the notation $\{T_{\sigma,\tau} : \sigma \in [0, \pi), \tau \in [0, \pi)\}$ in order to describe the set of such operators. The separate domains of definition and operators are defined by

$$\begin{aligned} D(T_{\sigma,\tau}) &:= \{f \in \Delta_b \mid [f, \varphi_\sigma](c) = 0 \quad [f, \kappa_\tau](b) = 0\} \\ T_{\sigma,\tau} f &:= w^{-1} M[f] \qquad (f \in D(T_{\sigma,\tau})). \end{aligned} \tag{5.9}$$

The properties of $T_{\sigma,\tau}$ are essentially to be found in Naimark [19, Sections 18 to 21], but see also the results in [22, Chapters I to III] and [9, Chapter 9]; see also the discussion in Fulton [16, Section 2]. In this respect we have

**Theorem 5.1** *Let the differential equation (1.1) be regular at the point $c \in (a, b)$, and regular or limit-circle at end-point $b$; let the differential operators $\{T_{\sigma,\tau} : \sigma \in [0, \pi), \tau \in [0, \pi)\}$ be defined as above; then*

*(i)* $T_{\sigma,\tau}$ *is self-adjoint and unbounded in* $L^2([c,b);w)$

*(ii)* $T_{\sigma,\tau}$ *has a (real) simple, discrete spectrum*

*(iii)* $T_{\sigma,\tau}$ *has infinitely many, countable eigenvalues*

*(iv) if* $\lambda \in \mathbb{R}$ *is an eigenvalue of* $T_{\sigma,\tau}$*, then the associated eigenvectors all have the representation* $f = \alpha\varphi_\sigma(\cdot,\lambda)$ $(\alpha \in \mathbb{C})$ *and*

$$[\varphi_\sigma(\cdot,\lambda),\kappa_\tau(\cdot)](b) = 0 \qquad [\varphi_\sigma(\cdot,\lambda),\delta_\tau(\cdot)](b) \neq 0. \tag{5.10}$$

**Remarks.** The proof of results (i), (ii) and (iii) is essentially contained in [19, Sections 18 to 21].

We comment on (iv). Any eigenfunction of $T_{\sigma,\tau}$ with eigenvalue $\lambda$ has to be a solution of the differential equation $M[y] = \lambda wy$ on $[c,b)$; if the boundary condition at $c$ is to be satisfied, this solution has to be $\varphi_\sigma(\cdot,\lambda)$; if the boundary condition at $b$ is to be satisfied, then $[\varphi_\sigma(\cdot,\lambda),\kappa_\tau(\cdot)](b) = 0$. The condition $[\varphi_\sigma(\cdot,\lambda),\delta_\tau(\cdot)](b) \neq 0$ follows from the Plücker identity (2.12); let the set $\{f_r, g_r : r = 1,2,3\}$ be determined by

$$f_1, f_2, f_3 \longrightarrow \varphi_\sigma(\cdot), \kappa_\tau, \delta_\tau \qquad g_1, g_2, g_3 \longrightarrow \theta_\sigma(\cdot), \kappa_\tau, \delta_\tau.$$

Since all these functions are in $\Delta$, we can take $x = b$ in (2.12); recalling that $[\varphi_\sigma,\theta_\sigma](b) = -1$, $[\kappa_\tau,\kappa_\tau](b) = [\delta_\tau,\delta_\tau] = 0$, and $[\kappa_\tau,\delta_\tau](b) = 1$, a calculation from (2.12) shows that

$$1 + [\varphi_\sigma(\cdot,\lambda),\delta_\tau(\cdot)](b)\cdot[\kappa_\tau(\cdot),\theta_\sigma(\cdot,\lambda)](b) = 0 \tag{5.11}$$

If now $[\varphi_\sigma,\kappa_\tau](b) = 0$, then we have a contradiction.

This result (iv) shows that the differential operators in the set $\{T_{\sigma,\tau} : \sigma \in [0,\pi), \tau \in [0,\pi)\}$ are all distinct; an eigenvector of one such operator cannot be an eigenvector of any one of the other operators in the set.

Note that it follows also from (5.11) that when $\lambda$ is an eigenvalue of $T_{\sigma,\tau}$, i.e. $[\varphi_\sigma(\cdot,\lambda),\kappa_\tau](b) = 0$, then

$$[\theta_\sigma(\cdot,\lambda),\kappa_\tau(\cdot)](b) \neq 0. \tag{5.12}$$

## 6. Analytic properties

The results of the previous section have important consequences for certain analytic functions associated with solutions of the differential equation (1.1) in the regular or limit-circle end-point classifications. Let $\varphi_\sigma$ and $\kappa_\tau$ be determined as in the previous section; define $F_{\sigma,\tau}$ :

$\mathbf{C} \to \mathbf{C}$ for $\sigma, \tau \in [0, \pi)$ by

$$F_{\sigma,\tau}(\lambda) := [\varphi_\sigma(\cdot, \lambda), \kappa_\tau(\cdot)](b) \equiv [\varphi_\sigma, \kappa_\tau](b, \lambda) \qquad (\lambda \in \mathbf{C}). \tag{6.1}$$

**Theorem 6.1** *Let $F_{\sigma,\tau}$ be defined by (6.1); then for all $\sigma, \tau \in [0, \pi)$, the following given properties hold:*

*(i)* $F_{\sigma,\tau} \in H(\mathbf{C})$

*(ii)* $F_{\sigma,\tau}(\lambda) \neq 0 \qquad (\lambda \in \mathbf{C} \setminus \mathbf{R})$

*(iii)* *$F_{\sigma,\tau}$ has a countable infinity of real zeros*

*(iv)* *all the zeros of $F_{\sigma,\tau}$ are simple.*

**Remarks.** Properties (i), (ii) and (iii) follow essentially from the results of Theorem 5.1.

We comment on (iv). If $b$ is a regular end-point of the differential equation, then this result follows from arguments similar to the results given in [22, Chapter I].

If $b$ is limit-circle, then the argument is more protracted. Suppose at the point $\lambda \in \mathbf{R}$ the function $F_{\sigma,\tau}$ has a zero of order two or higher; then, for $\delta > 0$,

$$[\varphi_\sigma, \kappa_\tau](b, \lambda + i\delta) = 0(\delta^2) \qquad (\delta \to 0^+) \tag{6.2}$$

From (5.10) and Lemma 3.1 (iv)

$$\lim_{\delta \to 0+} [\varphi_\sigma, \delta_\tau](b, \lambda + i\delta) = [\varphi_\sigma, \kappa_\tau](b, \lambda) \neq 0. \tag{6.3}$$

From Lemma 3.1 (ii)

$$\begin{aligned}\lim_{\delta \to 0+} \int_c^b w(x) \left|\varphi_\sigma(x, \lambda + i\delta)\right|^2 dx &= \int_c^b \lim_{\delta \to 0+} \left\{w(x) \left|\varphi_\sigma(x, \lambda + i\delta)\right|^2\right\} dx \\ &= \int_c^b w(x) \left|\varphi_\sigma(x, \lambda)\right|^2 dx > 0 \end{aligned} \tag{6.4}$$

Now substitute in the Plücker identity (2.12) as follows

$$f_1 = g_1,\ f_2 = g_2,\ f_3 = g_3 \longrightarrow \varphi_\sigma(\cdot, \lambda + i\delta), \kappa_\tau, \delta_\tau.$$

Using Green's formula (2.10) and the properties of $\kappa_\tau$, $\delta_\tau$, the identity (2.12) at the point $b$ gives, for all $\delta > 0$,

$$\begin{aligned}2i\delta \int_c^b w(x) \left|\varphi_\sigma(x, \lambda + i\delta)\right|^2 dx - [\varphi_\sigma, \kappa_\tau](b, \lambda + i\delta)[\delta_\tau, \varphi_\sigma](b, \lambda + i\delta) \\ + [\varphi_\sigma, \delta_\tau](b, \lambda + i\delta)[\kappa_\tau, \varphi_\sigma](b, \lambda + i\delta) = 0.\end{aligned}$$

Divide this last result by $\delta > 0$ and then let $\delta \to 0^+$. From (6.2) and (6.3) we obtain a contradiction on (6.4).

**Corollary 6.1** *Let the analytic function $G_{\sigma,\tau}$ be defined for $\sigma, \tau \in [0, \pi)$ by*

$$G_{\sigma,\tau}(\lambda) := [\theta_\sigma(\cdot, \lambda), \kappa_\tau(\cdot)](b) \equiv [\theta_\sigma, \kappa_\tau](b, \lambda) \qquad (\lambda \in \mathbb{C}). \tag{6.5}$$

*Then all the results of Theorem 6.1 for $F_{\sigma,\tau}$ hold for $G_{\sigma,\tau}$.*

**Remarks.**

(i) The proof of this corollary follows the same lines as the proof for Theorem 6.1.

(ii) The identity (5.11) shows that the zeros of $F_{\sigma,\tau}$ and $G_{\sigma,\tau}$ are distinct. Additional arguments show that the zeros of these two analytic functions are interlaced on the real line $\mathbb{R}$, but we do not consider this property here in this paper.

## 7. The $m$-coefficient

Let all the conditions leading to Theorem 5.1 above hold; then corresponding to each differential operator $T_{\sigma,\tau}$ we define the Titchmarsh-Weyl $m_{\sigma,\tau}$ coefficient by the formula

$$m_{\sigma,\tau}(\lambda) := -\frac{[\theta_\sigma(\cdot, \lambda), \kappa_\tau(\cdot)](b)}{[\varphi_\sigma(\cdot, \lambda), \kappa_\tau(\cdot)](b)} \qquad (\lambda \in \mathbb{C} \setminus \mathbb{R}). \tag{7.1}$$

Then $m_{\sigma,\tau} : \mathbb{C} \to \mathbb{C}$ has the following given properties:

(i) $m_{\sigma,\tau}$ is meromorphic on $\mathbb{C}$

(ii) $m_{\sigma,\tau} \in H(\mathbb{C} \setminus \mathbb{R}) \qquad m_{\sigma,\tau}(\lambda) \neq 0 \ (\lambda \in \mathbb{C} \setminus \mathbb{R})$

(iii) $m_{\sigma,\tau}$ has a countable infinity of simple poles on the real line $\mathbb{R}$, i.e., at the zeros of $[\varphi_\sigma, \kappa_\tau](b, \cdot)$.

(iv) $m_{\sigma,\tau}$ has a countable infinity of simple zeros on the real line $\mathbb{R}$, i.e., at the zeros of $[\theta_\sigma, \kappa_\tau](b, \cdot)$.

**Remarks.** The results (i), (ii) and (iii) follow from the results in Theorem 6.1. Property (iv) requires Corollary 6.1. The poles and zeros of $m_{\sigma,\tau}$ do not coincide; see Remark (ii) after Corollary 6.1.

The formula (7.1) is similar to results given in [9, Chapter 9, Section 4, (4.8)] and [16, Section 2, (2.16)]. However, (7.1) is essentially concerned with using the maximal domain function $\kappa_\tau$ for the boundary condition at $b$, rather than a solution of the differential equation. However, it is known that the two methods are equivalent.

The use of $\varphi_\sigma$ and $\kappa_\tau$ in (7.1) relates the $m_{\sigma,\tau}$ coefficient to the differential operator $T_{\sigma,\tau}$; it is this approach to which reference was made in the introduction to this paper. In this sense, the role of $\theta_\sigma$ in (7.1) is auxiliary; however, these roles can be reversed. If in the initial conditions (5.1), given $\tau \in [0, \pi)$, we put $\rho = \tau + \pi/2 \,(\mathrm{mod}\,\pi)$, then

$$\theta_\rho = -\varphi_\sigma \qquad \varphi_\rho = \theta_\sigma \qquad \text{on } (a, b) \times \mathbf{C}$$

and (7.1) gives

$$m_{\rho,\tau} = -\frac{[\theta_\rho, \kappa_\tau]}{[\varphi_\rho, \kappa_\tau]} = \frac{[\varphi_\rho, \kappa_\tau]}{[\theta_\rho, \kappa_\tau]} = -\frac{1}{m_{\sigma,\tau}}.$$

Thus for $\sigma \in [0, \pi)$ and $\rho = \sigma + \pi/2 \;(\mathrm{mod}\,\pi)$ we have

$$m_{\sigma,\tau}(\lambda) m_{\rho,\tau}(\lambda) = -1 \qquad (\lambda \in \mathbf{C} \setminus \mathbf{R}) \tag{7.2}$$

At this point we make mention of the underlying spectral theory. The self-adjoint operator has a discrete spectrum determined by the (real) zeros of the entire function $[\varphi_\sigma, \kappa_\tau](b, \cdot)$; these discrete points of $\mathbf{R}$ are exactly the poles of the meromorphic function $m_{\sigma,\tau}$.

The boundary value problem, always considered as embedded in the space $L^2([c, b); w)$,

$$\begin{aligned} &-(py')' + qy = \lambda w y \quad \text{on } [c, b) \\ &[y, \varphi_\sigma](c) = 0 \quad [y, \kappa_\tau](b) = 0 \end{aligned} \tag{7.3}$$

can be studied *either*

(i) from the definition and spectral properties of the differential operators $\{T_{\sigma,\tau} : \sigma, \tau \in [0, \pi)\}$, *or*

(ii) from the definition and meromorphic properties of the Titchmarsh-Weyl coefficients $\{m_{\sigma,\tau} : \sigma, \tau \in [0, \pi)\}$.

The study through (i) can be seen from the results in Naimark [19, Chapters V and VI]; see also Akhiezer and Glazman [1, Appendix 1]. The study through (ii) is given in Titchmarsh [22, Chapter II]; see also Hille [17, Chapter 10].

These two theories are analytically equivalent; a specific account in this direction is to be found in Chaudhuri and Everitt [7]; see also Bennewitz and Everitt [5].

## 8. The regular approximation

Let all the previously given conditions hold; in particular, the differential equation (1.1) is R at $c$, and R or LC at $b$. The operators $\{T_{\sigma,\tau}\}$ and $m$-coefficients $\{m_{\sigma,\tau}\}$ are then obtained from the boundary value problem determined by (7.3).

Returning to (5.8) (iii), let $c' \in (c, b)$ such that

$$[\kappa_\tau, \delta_\tau](\beta) \neq 0 \qquad (\beta \in (c', b)). \tag{8.1}$$

With this condition satisfied and given any $\beta \in (c', b)$, we can consider the regular boundary value problem in the Hilbert function space $L^2([c, \beta]; w)$, for any $\sigma, \tau \in [0, \infty)$,

$$\begin{aligned} -(py')' + qy &= \lambda w y \quad \text{on } [c, \beta] \\ [y, \varphi_\sigma](c) = 0 \quad &[y, \kappa_\tau](\beta) = 0. \end{aligned} \tag{8.2}$$

In determining this problem we retain the original boundary condition at the R end-point $a$, but we *inherit* a boundary condition at the R end-point $\beta$ from the previously chosen R or LC boundary condition at end-point $b$; this is effected by retaining the function $\kappa_\tau \in \Delta_b$, noting that $\kappa_\tau, \delta_\tau \in \Delta_\beta$, and then using the pair $\{\kappa_\tau, \delta_\tau\}$ at $\beta$ with (8.1) to provide the necessary linear independence.

For the problem (8.2) we can define, using the methods in earlier sections, the self-adjoint differential operator $T_{\sigma,\tau}(\beta; \cdot)$ in $L^2([c, \beta]; w)$ and the $m$-coefficient $m_{\sigma,\tau}(\beta; \cdot)$ as follows:

$$\text{(i)} \qquad \begin{aligned} D(T_{\sigma,\tau}(\beta;\cdot)) &:= \{f \in \Delta_\beta \,|\, [f, \varphi_\sigma](c) = [f, \kappa_\tau](\beta) = 0\} \\ T_{\sigma,\tau}(\beta; f) &:= w^{-1} M[f] \qquad (f \in D(T_{\sigma,\tau}(\beta;\cdot))) \end{aligned} \tag{8.3}$$

$$\text{(ii)} \qquad m_{\sigma,\tau}(\beta; \lambda) := -\frac{[\theta_\sigma(\cdot, \lambda), \kappa_\tau(\cdot)](\beta)}{[\varphi_\sigma(\cdot, \lambda), \kappa_\tau(\cdot)](\beta)} \qquad (\lambda \in \mathbf{C} \setminus \mathbf{R}). \tag{8.4}$$

Note that in (8.3) and (8.4) the boundary condition function $\kappa_\tau$ remains unchanged; however, its function values at $\beta$ are now those which determine the boundary condition at the R end-point $\beta$.

The relationships between the operator $T_{\sigma,\tau}$ and the coefficient $m_{\sigma,\tau}$, and the families of operators $\{T_{\sigma,\tau}(\beta;\cdot) : \beta \in (c', b)\}$ and coefficients $\{m_{\sigma,\tau}(\beta;\cdot) : \beta \in (c', b)\}$, have been studied only recently. These studies were initiated by Bailey, Everitt, Zettl [2] and continued, with Weidmann, in [3].

From [3] we have the result

(i) the family $\{T_{\sigma,\tau}(\beta;\cdot : \beta \in (c',b)\}$ can be extended to a family $\{T_{\sigma,\tau}{}'(\beta;\cdot : \beta \in (c',b)\}$ of self-adjoint operators all acting in $L^2([c,b);w)$, and this extended family is norm resolvent convergent to $T_{\sigma,\tau}$; the spectral consequences of this result are reported on in [3].

In contrast, it may be seen, by direct consideration of the properties of the bilinear form $[\cdot,\cdot]$ and Lemma 3.1, that in the simple topology of $\mathbb{C}$,

$$\text{(ii)} \qquad \lim_{\beta \to b-} m_{\sigma,\tau}(\beta;\lambda) = m_{\sigma,\tau}(\lambda) \qquad (\lambda \in \mathbb{C}\setminus\mathbb{R}) \tag{8.5}$$

and that this convergence is uniform on compact subsets of $\mathbb{C}\setminus\mathbb{R}$.

In this respect it is possible to obtain this continuous limit process from $\{m_{\sigma,\tau}(\beta:\cdot)\}$, given by the regular problems (8.2), to the limit $m_{\sigma,\tau}(\cdot)$ as given by the possibly singular problem (7.3). As mentioned in the introduction to this paper, this limit process has previously been considered in sequential, rather than continuous, form; again see [22, Chapter II], [9, Chapter 9] and [17, Chapter 10]. The inherited boundary condition, in (8.2), at $\beta$ on $[c,\beta]$, prevents the need for such a sequential process, but requires initially an understanding of the possibly singular problem on the interval $[c,b)$ and the use of separated boundary conditions.

## 9. Some identities

To link the definition of the $m$-coefficient $m_{\sigma,\tau}$ in Section 7 above with the so-called circle method of the Titchmarsh-Weyl theory, we prove certain identities. In these results the complex number $\lambda = \mu + i\nu \in \mathbb{C}\setminus\mathbb{R}$ so that $\nu \neq 0$. To simplify the presentation we write, see (4.6),

$$[\theta_\sigma,\varphi_\sigma](\beta,\lambda) \equiv [\theta_\sigma(\cdot,\lambda),\varphi_\sigma(\cdot,\lambda)](\beta) \tag{9.1}$$

$$[\varphi_\sigma,\varphi_\sigma](\beta,\lambda) \equiv [\varphi_\sigma(\cdot,\lambda),\varphi_\sigma(\cdot,\lambda)](\beta) \tag{9.2}$$

along with similar formulae.

Note that from Green's formula (2.10), from (5.4) and $[\varphi_\sigma,\varphi_\sigma](c,\lambda) = 0$ $(\lambda \in \mathbb{C}\setminus\mathbb{R})$

$$\begin{aligned}
[\varphi_\sigma,\varphi_\sigma](\beta,\lambda) &= 2i\nu \int_c^\beta w(x)|\varphi_\sigma(x,\lambda)|^2\,dx \\
&= -2i(-\nu)\int_c^\beta w(x)|\varphi_\sigma(x,\bar\lambda)|^2\,dx \\
&= -[\varphi_\sigma,\varphi_\sigma](\beta,\bar\lambda) \neq 0
\end{aligned} \tag{9.3}$$

We now state, using the notation of earlier sections

**Lemma 9.1** *Let the conditions of Theorem 5.1 hold; let $c' \in (c,b)$ be determined so that (8.1) is valid; let $\lambda \in \mathbb{C} \setminus \mathbb{R}$; let $\beta \in (c',b]$ on the understanding if $b = \infty$, then a limit is to be taken; let $\sigma, \tau \in [0,\pi)$; then for all $z \in \mathbb{C}$*

$$\begin{aligned}
&[\theta_\sigma + z\varphi_\sigma, \theta_\sigma + z\varphi_\sigma](\beta,\lambda) \\
&\quad = -2i\, im\,(z) + 2i\nu \int_c^\beta w(x)\, |\theta_\sigma(x,\lambda) + z\varphi_\sigma(x,\lambda)|^2 \, dx && (9.4)\\
&\quad = [\varphi_\sigma,\varphi_\sigma](\beta,\lambda) \left[ \left| z + \frac{[\theta_\sigma,\varphi_\sigma](\beta,\lambda)}{[\varphi_\sigma,\varphi_\sigma](\beta,\lambda)} \right|^2 + \frac{1}{[\varphi_\sigma,\varphi_\sigma](\beta,\lambda)^2} \right]. && (9.5)
\end{aligned}$$

*Proof.* The result (9.4) follows from Green's formula (2.10), from (5.1), and from (2.6).

The result (9.5) is obtained by a calculation (we omit some of the details)

$$\begin{aligned}
&[\theta_\sigma + z\varphi_\sigma, \theta_\sigma + z\varphi_\sigma](\beta,\lambda) \\
&\quad = [\varphi_\sigma,\varphi_\sigma](\beta,\lambda)|z|^2 + [\varphi_\sigma,\theta_\sigma](\beta,\lambda)z \\
&\qquad\qquad + [\theta_\sigma,\varphi_\sigma](\beta,\lambda)\bar{z} + [\theta_\sigma,\theta_\sigma](\beta,\lambda) \\
&\quad = [\varphi_\sigma,\varphi_\sigma](\beta,\lambda) \left[ \left| z + \frac{[\theta_\sigma,\varphi_\sigma](\beta,\lambda)}{[\varphi_\sigma,\varphi_\sigma](\beta,\lambda)} \right|^2 \right. \\
&\qquad\qquad \left. + \frac{([\theta_\sigma,\theta_\sigma][\varphi_\sigma,\varphi_\sigma] - [\theta_\sigma,\varphi_\sigma][\varphi_\sigma,\theta_\sigma])(\beta,\lambda)}{[\varphi_\sigma,\varphi_\sigma](\beta,\lambda)^2} \right].
\end{aligned}$$

To complete the proof of (9.5) we note that

$$([\theta_\sigma,\theta_\sigma][\varphi_\sigma,\varphi_\sigma] - [\theta_\sigma,\varphi_\sigma][\varphi_\sigma,\theta_\sigma])(\beta,\lambda) = 1 \tag{9.6}$$

for all $\lambda \in \mathbb{C}\setminus\mathbb{R}$ and all $\beta \in (c',b]$. This last result (9.6) is obtained from the Plücker identity (2.12) with

$$f_1 = f_1,\ f_2 = g_2,\ f_3 = g_3 \longrightarrow \theta_\sigma(\cdot,\lambda),\ \varphi_\sigma(\cdot,\lambda),\ \varphi_\sigma(\cdot,\bar{\lambda})$$

and use of (9.3).

## 10. The Weyl circles

Consider the singular boundary value problem (7.3). To obtain further information about the $m_{\sigma,\tau}$ coefficient and to provide a means to construct the Green's function for this problem we introduce, in the Titchmarsh notation, the $\psi$ solution of the differential equation; we extend this notation and define, for $x \in [c,b)$ and $\lambda \in \mathbb{C} \setminus \mathbb{R}$

$$\psi_{\sigma,\tau}(x,\lambda) := \theta_\sigma(x,\lambda) + m_{\sigma,\tau}(\lambda)\varphi_\sigma(x,\lambda). \tag{10.1}$$

We note from the definition (7.1) of $m_{\sigma,\tau}$ that $\psi_{\sigma,\tau}$ satisfies the boundary condition in (7.3) at the end-point $b$, i.e.,

$$[\psi_{\sigma,\tau}(\cdot,\lambda),\kappa_\tau(\cdot)]\,(b)=0 \qquad \lambda\in\mathbf{C}\setminus\mathbf{R} \tag{10.2}$$

From this result and an application of the identity (2.12) with

$$f_1=f_1,\ f_2=g_2,\ f_3=g_3\longrightarrow\psi_{\sigma,\tau}(\cdot,\lambda),\ \kappa_\tau,\ \delta_\tau$$

we then obtain

$$[\psi_{\sigma,\tau}(\cdot,\lambda),\psi_{\sigma,\tau}(\cdot,\lambda)]\,(b)=0 \qquad \lambda\in\mathbf{C}\setminus\mathbf{R} \tag{10.3}$$

If now in the identity (9.5) we put $z=m_{\sigma,\tau}(\lambda)$ and $\beta=b$, we see that the left-hand side is zero and so

$$\left|m_{\sigma,\tau}(\lambda)+\frac{[\theta_\sigma,\varphi_\sigma](b,\lambda)}{[\varphi_\sigma,\varphi_\sigma](b,\lambda)}\right|^2=\left(2|\nu|\int_c^b w(x)\,|\varphi_\sigma(x,\lambda)|^2\,dx\right)^{-2} \tag{10.4}$$

on using (9.3).

Hence the complex number $m_{\sigma,\tau}(\lambda)$ lies on a circle $C(\sigma,\lambda)$ of radius

$$\left(2|\nu|\int_c^b w(x)\,|\varphi_\sigma(x,\lambda)|^2\,dx\right)^{-1} \tag{10.5}$$

and center

$$[\theta_\sigma,\varphi_\sigma](b,\lambda)/[\varphi_\sigma,\varphi_\sigma](b,\lambda) \tag{10.6}$$

This is the Weyl circle; note that the circle itself depends on $\sigma\in[0,\pi)$ and $\lambda\in\mathbf{C}\setminus\mathbf{R}$, but not on $\tau\in[0,\pi)$.

To see how the point $m_{\sigma,\tau}(\lambda)$ varies on this circle as the parameter $\tau$ varies on $[0,\pi)$, we have from (7.1) and (5.1)

$$m_{\sigma,\tau}(\lambda)=-\frac{[\theta_\sigma,\chi_1](b,\lambda)\cos(\tau)-[\theta_\sigma,\chi_2](b,\lambda)\sin(\tau)}{[\varphi_\sigma,\chi_1](b,\lambda)\cos(\tau)-[\varphi_\sigma,\chi_2](b,\lambda)\sin(\tau)}. \tag{10.7}$$

A further application of the identity (2.12) with (5.6) and

$$f_1=\theta_\sigma(\cdot,\lambda),\quad g_1=\varphi_\sigma(\cdot,\bar\lambda),\quad f_2=g_2=\chi_1,\quad f_3=g_3=\chi_2$$

leads to the determinant identity, for all $\lambda\in\mathbf{C}\setminus\mathbf{R}$

$$[\theta_\sigma,\chi_1](b,\lambda)[\varphi_\sigma,\chi_2](b,\lambda)-[\theta_\sigma,\chi_2](b,\lambda)[\varphi_\sigma,\chi_1](b,\lambda)=1$$

for the bilinear map (10.7). From known results in complex variable theory, see [20], it then follows that as $\tau$ varies over $[0,\pi)$, the point $m_{\sigma,\tau}(\lambda)$ describes the whole circle $C(\sigma,\lambda)$.

From (9.3) we obtain also the result, again using (10.4), valid for all $\sigma,\tau \in [0,\pi)$ and $\lambda \in \mathbb{C} \setminus \mathbb{R}$

$$\int_c^b w(x)\,|\theta_\sigma(x,\lambda) + m_{\sigma,\tau}(\lambda)\varphi_\sigma(x,\lambda)|^2\,dx = \frac{\mathrm{im}(m_{\sigma,\tau}(\lambda))}{\mathrm{im}(\lambda)}. \tag{10.8}$$

This is the well-known result of Titchmarsh [22, Chapter II] which shows that $m_{\sigma,\tau}(\cdot)$ is a Nevanlinna function on $\mathbb{C}$; for further details see [5, Section 4].

The Green's function for the singular boundary value problem (7.3) is then defined by

$$\begin{aligned} G_{\sigma,\tau}(x,\xi;\lambda) &:= \varphi_\sigma(x,\lambda)\psi_{\sigma,\tau}(\xi,\lambda) \qquad c \le x < \xi < b \\ &:= \varphi_\sigma(\xi,\lambda)\psi_{\sigma,\tau}(x,\lambda) \qquad c \le \xi < x < b. \end{aligned} \tag{10.9}$$

All these results (10.1)–(10.9) hold also, by inspection, for the regular approximation boundary value problem (8.2) with now, for $x \in [c,\beta]$ and $\lambda \in \mathbb{C} \setminus \mathbb{R}$

$$\psi_{\sigma,\tau}(x,\lambda;\beta) := \theta_\sigma(x,\lambda) + m_{\sigma,\tau}(\beta;\lambda)\varphi_\sigma(x,\lambda).$$

## 11. Concluding remarks

**(1)** We have considered the end-point $b$ of the differential equation (1.1) in discussing properties of the differential operators $T_{\sigma,\tau}$ and the corresponding $m$-coefficients $m_{\sigma,\tau}$. Entirely similar considerations lead to a similar theory for the end-point $a$, when this end-point is regular or limit-circle.

**(2)** If the end-point $b$ is singular and in the limit-point classification, then Weyl [23] showed that if the solutions $\theta_\sigma$ and $\varphi_\sigma$ are defined by (5.1), then for all $\lambda \in \mathbb{C} \setminus \mathbb{R}$

$$\theta_\sigma(\cdot,\lambda) \notin L^2([c,b);w) \quad \text{and} \quad \varphi_\sigma(\cdot,\lambda) \notin L^2([c,b);w).$$

However, there exists a unique $m$-coefficient $m_\sigma(\cdot)$ with the property that for all $\lambda \in \mathbb{C} \setminus \mathbb{R}$

$$\theta_\sigma(\cdot,\lambda) + m_\sigma(\lambda)\varphi_\sigma(\cdot,\lambda) \in L^2([c,b);w).$$

In this case we can define a pair $\{\kappa,\delta\}$ of real-valued elements of $\Delta_b$ with the properties

(i) for some $c' \in (c,b)$

$$[\kappa,\delta](\beta) \neq 0 \qquad (\beta \in (c',b))$$

(ii) however, from (4.4)

$$[\kappa, \delta](b) = 0.$$

This leads to a self-adjoint differential operator $T_\sigma$ in $L^2([c,b);w)$.

If we define, for all $\lambda \in \mathbb{C} \setminus \mathbb{R}$

$$m_\sigma(\beta;\lambda) := -\frac{[\theta_\sigma(\cdot,\lambda),\kappa(\cdot)](\beta)}{[\varphi_\sigma(\cdot,\lambda),\kappa(\cdot)](\beta)}$$

then it may be shown that, and this is a continuous limit,

$$\lim_{\beta \to b^-} m_\sigma(\beta;\lambda) = m_\sigma(\lambda) \qquad \lambda \in \mathbb{C} \setminus \mathbb{R}.$$

However, in this case the terms $\{[\theta_\sigma,\kappa](\beta,\lambda)\}$ and $\{[\varphi_\sigma,\kappa](\beta,\lambda)\}$ do not, in general, have separate limits as $\beta \to b^-$.

In this the limit-point case, no boundary condition is required at the singular end-point $b$ to define either the differential operator or the $m$-coefficient. Hence the notation $T_\sigma$ and $m_\sigma$ indicating dependence only on the boundary condition parameter $\sigma \in [0,\pi)$, for the regular end-point $c$ of the interval $[c,b)$.

# References


[1] N. I. Akhiezer and I. M. Glazman. THEORY OF LINEAR OPERATORS IN HILBERT SPACE: Volumes I and II (Pitman and Scottish Academic Press, London and Edinburgh, 1980; translated from the Russian edition of 1965).

[2] P. B. Bailey, W. N. Everitt, and A. Zettl. *Computing eigenvalues of singular Sturm-Liouville problems.* RESULTS IN MATHEMATICS **20** (1991), 391-423.

[3] P. B. Bailey, W. N. Everitt, J. Weidmann, and A. Zettl. *Regular approximation of singular Sturm-Liouville problems.* (To appear in RESULTS IN MATHEMATICS.)

[4] C. Bennewitz. *Spectral asymptotics of Sturm-Liouville equations.* PROC. LOND. MATH. SOC. (3) **59** (1989), 294-338.

[5] C. Bennewitz and W. N. Everitt. *Some remarks on the Titchmarsh-Weyl m-coefficient.* Proceedings of the Pleijel Conference, University of Uppsala: TRIBUTE TO ÅKE PLEIJEL (1979), 49-108. (Published by the Department of Mathematics, University of Uppsala, Sweden.)

[6] C. Bennewitz and W. N. Everitt. *The Titchmarsh eigenfunction expansion theorem: new lamps for old.* (In preparation.)

[7] Jyoti Chaudhuri and W. N. Everitt. *On the spectrum of ordinary, second-order differential operators.* PROC. ROYAL SOC. EDINBURGH (A) **68** (1969), 95-119.

[8] S. J. Clark and D. B. Hinton. *Strong nonsubordinacy and absolutely continuous spectra for Sturm-Liouville equations.* (Preprint 1992.)

[9] E. A. Coddington and N. Levinson. THEORY OF ORDINARY DIFFERENTIAL EQUATIONS (McGraw-Hill, New York: 1955).

[10] E. T. Copson. THEORY OF FUNCTIONS OF A COMPLEX VARIABLE. (Oxford University Press, 1946.)

[11] W. N. Everitt. *A note on the self-adjoint domains of second-order differential equations.* QUART. J. OF MATH. (Oxford) (2) **14** (1963), 41-45.

[12] W. N. Everitt. *Integrable-square solutions of ordinary differential equations.* QUART. J. OF MATH. (Oxford) (2) **14** (1963), 170-180.

[13] W. N. Everitt. *On the transformation theory of ordinary second-order linear symmetric differential equations.* CZECH. MATH. J. **32** (107) (1982), 275-306.

[14] W. N. Everitt. *A note on linear ordinary quasi-differential equations.* PROC. ROYAL SOC. EDINBURGH (A) **101** (1985), 1-14.

[15] W. N. Everitt and A. Zettl. *Generalized symmetric ordinary differential expression I: the general theory.* Nieuw Archief voor Wiskunde (3) **XXVII** (1979), 363-397.

[16] C. T. Fulton. *Parametrizations of Titchmarsh's $m(\lambda)$-functions in the limit circle case.* TRANS. AMER. MATH. SOC. **229** (1977), 51-63.

[17] E. Hille. LECTURES ON ORDINARY DIFFERENTIAL EQUATIONS. (Addison-Wesley Publishing Company, London: 1969.)

[18] B. M. Levitan and I. Sargsjan. INTRODUCTION TO SPECTRAL THEORY; SELF-ADJOINT ORDINARY OPERATORS. (American Mathematical Society, Providence: 1975; translated from the Russian edition of 1970.)

[19] M. A. Naimark. LINEAR DIFFERENTIAL OPERATORS: II (Ungar Publishing Company, New York: 1968; translated from the Russian edition of 1960.)

[20] E. C. Titchmarsh. THEORY OF FUNCTIONS (Oxford University Press, 1939.)

[21] E. C. Titchmarsh. *On expansions in eigenfunctions IV.* QUART. J. MATH. (Oxford) **12** (1941), 33-50.

[22] E. C. Titchmarsh. EIGENFUNCTION EXPANSIONS: I (Oxford University Press, 1962.)

[23] H. Weyl. *Über gewöhnliche Differentialgleichungen mit Singularitäten und die zugehörigen Entwicklungen willkürlicher Funktionen.* MATH ANN. **68** (1910), 220-269.

# 5 Periodic Solutions of Single Species Models with Delay

**H. I. Freedman*** University of Alberta, Edmonton, Alberta, Canada

**Huaxing Xia**† University of Alberta, Edmonton, Alberta, Canada

**Abstract**

This paper is concerned with the existence, uniqueness and global attractivity of periodic solutions of the delay differential equation $\dot{x}(t) = x(t)F(t, x_t)$, where the functional $F$ is periodic in $t$.

* Research partially supported by the Natural Sciences and Engineering Research Council of Canada, Grant No. NSERC A4832.

† Research partially supported by a Province of Alberta Graduate Fellowship

## 1. Introduction

In mathematical models of single-species growth in a closed environment, a fundamental assumption is that the growth rate is proportional to the population concentration. In continuous models, this takes the form of a differential equation of the form

$$\dot{x} = xF,$$

where $F$ is a function describing the growth law (see [7]).

In the earliest (and most common) models, $F$ is time independent leading to an autonomous differential equation. Such a model, however, is relevant only in the case of a constant environment, and has been thoroughly discussed elsewhere [7,16,21].

However, in the real world, environments tend to fluctuate, particularly with seasons. Thus it is reasonable, as a second approximation to reality, to consider the case where $F$ is time dependent and periodic in $t$ (e.g. see [1,2,3,7,9,10,22,23]).

In addition to the above, it is a fact that in real populations, even at the bacterial or protozoan levels, there are time delays in response to environmental changes and due to gestation (see [25] for the later). Incorporating such delays then leads to a functional differential equation with periodic time dependence and delays (e.g. see [1,3,5,8,10,11,19,20,24,25,27,28]).

The class of models we consider is sufficiently general to include both discrete and distributed delays. For example, it includes equations of the type

$$\dot{x}(t) = x(t)\big(a(t) - b(t)x(t-\tau)\big)$$

and

$$\dot{x}(t) = x(t)[a(t) - \int_{-\infty}^{0} k(t)x(t+\theta)d\theta]$$

as special cases.

It is the main purpose of this paper to consider such a class of models and to obtain criteria for the existence, uniqueness, and global attractivity of our class of models.

The paper is organized as follows. In the next section we describe our equation (model) and obtain some preliminary results. The existence of periodic solutions is discussed in Section 3, and the uniqueness and global attractivity in the final section.

## 2. Preliminaries

Let $C = BC\big((-\infty, 0]; \mathbb{R}^n\big)$ denote the Banach space of all bounded continuous functions, endowed with the usual supremum norm. We consider the following class of $\omega$-periodic functional differential equations

$$\dot{x}(t) = x(t)F(t, x_t) \tag{2.1}$$

where $x_t \in C$ is defined by $x_t(\theta) = x(t+\theta)$, $\theta \in (-\infty, 0]$, for a bounded continuous function $x \in (-\infty, +\infty)$, $F : \mathbb{R} \times C \to \mathbb{R}^n$ is a continuous $\omega$-periodic functional, i.e. $F(t+\omega, \varphi) \equiv F(t, \varphi)$ for all $\varphi \in C$ and $t \in \mathbb{R}$. Throughout this section, we assume the existence and uniqueness of solutions to each initial value problem of (2.1) (see [6,13-15] for details in various cases).

To obtain an $\omega$-periodic solution of (2.1), we associate with it a family of equations

$$\dot{x}(t) = x(t)F_\lambda(t, x_t), \quad \lambda \in [0, 1] \tag{2.2}$$

where $F_\lambda$ satisfies the same properties as $F$, with $F_1 = F$ and $F_0$ a function from $\mathbb{R} \times \mathbb{R}^n \to \mathbb{R}^n$. Therefore, when $\lambda = 0$, (2.2) reduces to

$$\dot{x}(t) = x(t)F_0(t, x) \tag{2.3}$$

which is an ordinary differential equation. We denote by $A$ the Poincare map of equation (2.3), i.e. $Ax_0 = x(\omega, x_0)$, $x_0 \in \mathbb{R}^n$, where $x(t, x_0)$ is the unique solution of (2.3) with the initial data $x(0, x_0) = x_0 \in \mathbb{R}^n$. By assumption, $A$ is well-defined and continuous.

Let $K := \{(x_1, \ldots, x_n) \in \mathbb{R}^n | x_i \geq 0,\ i = 1, 2, \ldots, n\}$ be the usual positive cone which is endowed with a partial ordering. We call a continuous map $T : K \to K$ a *cone expansion* if there are two numbers, $0 < r < R$, such that $Tx \ngeq x$ for all $x \in K$ with $\|x\| = r$, and $Tx \nleq x$ for $x \in K$ with $\|x\| \geq R$. It is called a *cone compression* if there exist $R > r > 0$ such that $Tx \nleq x$ for $x \in K$ with $\|x\| = r$, and $Tx \ngeq x$ for $x \in K$ with $\|x\| = R$. It is well-known that $T$ has a fixed point in $K$ if either $T$ is a cone expansion or a cone compression (see [12,17]).

We note that the Poincare map $A$ defined above for equation (2.3) maps $K$ into itself.

To go further along with our analysis, we need the following result from [4].

THEOREM 2.1 [4]. *Suppose that there are two numbers, $0 < m < M$, such that $m < \sup_{t \in [0,\omega]} |x(t)| < M$ for any positive $\omega$-periodic solution of equation (2.2) for some $\lambda \in [0,1]$. Suppose further that there is $m_1 > 0$ such that $|x(0)| > m_1$ for any positive $\omega$-periodic solution of equation (2.3). If the map $A$ is either a cone expansion or a cone compression, then equation (2.1) has at least one nontrivial positive $\omega$-periodic solution.*

To apply the above result, we need some information about the solutions to equation (2.3). For the purpose of applications, we assume that

$$F_0(t, x) = \gamma(t) g\big(x, K(t)\big). \tag{2.4}$$

Then equation (2.3) takes the form

$$\dot{x}(t) = \gamma(t) x(t) g\big(x(t), K(t)\big). \tag{2.5}$$

Throughout the rest of this section, we assume that

(i) $\gamma$, $K : (-\infty, +\infty) \to (0, \infty)$ are continuous and $\omega$-periodic;

(ii) $g : [0, \infty) \times [0, \infty) \to \mathbb{R}$ is differentiable such that $g(0, K) > 0$, $\frac{\partial g}{\partial x} < 0$, $\frac{\partial g}{\partial K} > 0$ and $g\big(K(t), K(t)\big) \equiv 0$ for all $t \in \mathbb{R}$.

Again for the convenience of applications, we give the following two simple lemmas.

LEMMA 2.2. *Assume (i) and (ii) hold. Then there is a number* $\delta > 0$ *such that* $x(t) \geq \delta$ *for all positive* $\omega$*-periodic solutions* $x(t)$ *of (2.5).*

PROOF: Choose $\delta > 0$ such that $g\big(\delta, K(t)\big) > 0$ for all $t \in [0, \omega]$. Let $x(t)$ be any positive $\omega$-periodic solution of (2.5). If $x(t) \geq \delta$ for all $t \in [0, \omega]$ is not true, we can find a $\xi \in [0, \omega]$ such that $0 < x(\xi) < \delta$ and $x'(\xi) < 0$. However, from (2.5)

$$\begin{aligned} \dot{x}(\xi) &= \gamma(\xi)x(\xi)g\big(x(\xi), K(\xi)\big) \\ &\geq \gamma(\xi)x(\xi)g\big(\delta, K(\xi)\big) > 0 \end{aligned}$$

a contradiction. This proves the lemma.

□

The next lemma indicates that the Poincare map $A$ is a cone compression.

LEMMA 2.3. *Under the assumptions (i) and (ii),* $A : K \to K$ *is a cone compression, i.e., there are two positive numbers* $r < R$ *such that* $Ax > x$ *for all* $x \leq r$ *and* $Ax < x$ *for all* $x \geq R$.

PROOF: Choose $r > 0$ such that $g\big(r, K(t)\big) > 0$ for all $t \in [0, \omega]$. Consider any solution $x(t)$ to the following equation

$$\begin{cases} \dot{x}(t) &= \gamma(t)x(t)g\big(x(t), K(t)\big) \\ x(0) &= x, \quad x \leq r\,. \end{cases} \tag{2.6}$$

We claim that $x(t) > r$ which implies $Ax > x$. Indeed, from (2.6), $\dot{x}(0) = \gamma(0)x(0)g\big(x, K(0)\big) > 0$. If $x(\bar{t}) \leq r$ for some $\bar{t} \in [0, \omega]$, we can find a $\xi \in [0, \omega]$

such that $x(\xi) = r$ and $\dot{x}(\xi) \leq 0$. But equation (2.6) gives

$$\begin{aligned}\dot{x}(\xi) &= \gamma(\xi)x(\xi)g\big(x(\xi), K(\xi)\big) \\ &= r\gamma(\xi)g\big(r, K(\xi)\big) > 0,\end{aligned}$$

a contradiction. This shows that $Ax > x$ for all $x \leq r$.

The proof for $Ax < x$ for all $x \geq R$ is similar. In fact, one can choose $R > 0$ such that $g\big(R, K(t)\big) < 0$.

□

We remark that the above two lemmas ensure the existence of an $\omega$-periodic solution for equation (2.5). However, we are more interested in the existence of an $\omega$-periodic solution for delay equation (2.1). We will treat this in the next section.

## 3. Existence of Periodic Solutions

In this section, we prove an existence result for periodic solutions to the following delay differential equation

$$\dot{x}(t) = x(t)[\gamma(t) - a(t)x(t) - b(t)\big(G(x(t)) + H(t, x_t)\big)] \tag{3.1}$$

where $b(t)$ and $\gamma(t)$ are positive, continuous $\omega$-periodic functions, $a \geq 0$, $G : [0, \infty) \to [0, \infty)$ is continuous and nonincreasing, $H : \mathbb{R} \times C \to [0, \infty)$ is a continuous functional and is $\omega$-periodic in $t$. Motivated by [3], we make the following assumption on the functional $H$.

(H) There exist continuous $\omega$-periodic functions $p(t) > 0$, $q(t) \geq 0$ and a constant $\tau > 0$ such that

$$H(t, \varphi) \geq p(t)V(\widetilde{\varphi}) + q(t)$$

for all $\varphi \in C$ with $\varphi(\theta) > 0$, $\theta \in (-\infty, 0]$ and a functional $V$ with

$$\lim_{\substack{\min |\widetilde{\varphi}(\theta)| \to \infty \\ \theta \in [-\tau, 0]}} V(\widetilde{\varphi}) = \infty,$$

where $\widetilde{\varphi} = \varphi|_{[-\tau,0]}$ for $\varphi \in C$.

We first obtain a priori bounds on the periodic solutions.

LEMMA 3.1. *Assume that (H) holds. If* $[G(0) + H(t,0)]b(t) < \gamma(t)$, *then there exist two constants* $m$ *and* $M > 0$ *such that* $m < \|x\| := \sup_{t \in [0,\omega]} |x(t)| < M$ *for any positive* $\omega$*-periodic solution* $x(t)$ *of*

$$\dot{x}(t) = x(t)[\gamma(t) - (a(t) + \lambda b(t))x(t) - (1 - \lambda)b(t)(G(x(t)) + H(t, x_t))] \tag{3.2}$$

*for some* $\lambda \in [0, 1]$.

PROOF: Choose $m > 0$ such that

$$\max\{m, G(0) + \max_{\substack{0 \le x \le m \\ 0 \le t \le \omega}} H(t,x)\}b(t) + ma(t) < \gamma(t).$$

Let $\delta = \max\{m, G(0) + \max_{\substack{0 \le x \le m \\ 0 \le t \le \omega}} H(t,x)\}$. If $\|x\| \le m$, then (3.2) gives

$$\begin{aligned}
\dot{x}(t) &\ge x(t)[\gamma(t) - (a(t) + \lambda b(t))m - (1 - \lambda)b(t)(G(0) + \max_{\substack{0 \le x \le m \\ t \in [0,\omega]}} H(t,x))] \\
&\ge x(t)[\gamma(t) - ma(t) - b(t)(\lambda\delta + (1 - \lambda)\delta)] \\
&= x(t)[\gamma(t) - ma(t) - b(t)\delta] > 0.
\end{aligned}$$

This is impossible since $x(t)$ is an $\omega$-periodic solution.

Next, we show that there exists a constant $M > 0$ such that $\|x\| < M$ for any $\omega$-periodic solution $x(t)$ of (3.2).

Suppose that it is not true. Then we can find a sequence $\{t_n\}_{n=1}^{\infty}$, $\{\lambda_n\}_{n=1}^{\infty}$ and $\{x_n(t)\}_{n=1}^{\infty}$, $x_n(t) > 0$ such that

(i) $x_n(t_n) \to +\infty, \quad n \to +\infty$

(ii) $x_n'(t_n) > 0, \quad n = 1, 2, \ldots$

where $x_n(t)$ is an $\omega$-periodic solution of (3.2) for $\lambda = \lambda_n$. We claim that $x_n(t_n + \theta) \to +\infty$ for all $\theta \in [-\tau, 0]$, as $n \to \infty$.

Indeed, by integrating (3.2), it follows that

$$\begin{aligned} x_n(t_n) &= x_n(t_n + \theta) e^{\int_{t_n+\theta}^{t_n} \{\gamma(t) - a(t)x_n(t) - b(t)\left(\lambda_n x_n(t) + (1-\lambda_n)G(x_n(t) + H(t, x_{nt})\right)\} dt} \\ &\le x_n(t_n + \theta) e^{\int_{t_n+\theta}^{t_n} \gamma(t) dt}, \quad \theta \in [-\tau, 0] \end{aligned}$$

which implies that $x_n(t_n + \theta) \ge e^{-\int_{t_n-\tau}^{t_n} \gamma(t) dt} x_n(t_n) \to +\infty$ as $n \to +\infty$. Therefore, there is an $N > 0$ such that for $n \ge N$,

$$G(x_n(t_n)) + H(t, x_{nt_n}) \ge p(t_n) V(\widetilde{x}_{nt_n}) + q(t_n) \ge \frac{\gamma(t_n)}{b(t_n)},$$

and

$$x_n(t_n) > \frac{\gamma(t_n)}{b(t_n)}.$$

Consequently,

$$\begin{aligned} \dot{x}_n(t_n) &= x_n(t_n)\big(\gamma(t_n) - a(t_n)x_n(t_n) - b(t_n)[\lambda_n x_n(t_n) \\ &\quad + (1 - \lambda_n)(G(x_n(t_n)) + H(t_n, x_{nt_n}))]\big) \\ &< x_n(t_n)\big(\gamma(t_n) - [\lambda_n \gamma(t_n) + (1 - \lambda_n)\gamma(t_n)]\big) = 0 \end{aligned}$$

which is a contradiction to (ii). This completes the proof.

□

By combining Theorem 2.1 and Lemmas 2.2, 2.3 and 3.1, we obtain the following existence result about the $\omega$-periodic solution of equation (3.1).

THEOREM 3.2. *Assume (H) holds. If* $G(0) + H(t,0) < \frac{\gamma(t)}{b(t)}$ *and* $ca(t) + b(t)\big(G(c) + H(t,c)\big) \not\equiv \gamma(t)$ *for all positive constants* $c$, *then (3.1) has at least one nonconstant* $\omega$*-periodic solution* $x(t)$ *such that* $m \leq \|x\| \leq M$, *where* $M$ *and* $m > 0$ *are given in Lemma 3.1.*

If we choose

$$H(t,\varphi) = \sum_{i=1}^{n} c_i(t)\varphi^i(-\tau_i) + q(t), \quad \tau_i > 0 \tag{3.3}$$

where $c_i(t) > 0$, $i = 1, 2, \ldots, n$, and $q(t) \geq 0$ are continuous $\omega$-periodic functions, then, clearly (H) is satisfied. Therefore, we have

COROLLARY 3.3. *Suppose that* $\sum_{i=1}^{n} c_i(t)c^i + q(t) \not\equiv \gamma(t)/b(t)$, $c > 0$ *is arbitrary and* $q(t) < \gamma(t)/b(t)$. *Then the delay equation*

$$\dot{N}(t) = N(t)[\gamma(t) - b(t)\sum_{i=1}^{n} c_i(t)N^i(t-\tau_i) - b(t)q(t)] \tag{3.4}$$

*has a positive, nonconstant* $\omega$*-periodic solution.*

PROOF: Let $G(x) \equiv 0$ and $a(t) \equiv 0$. The corollary follows immediately from Theorem 3.2.

□

In modelling the growth of a single-species population, the functional $H(t,\varphi)$ often takes the form

$$H(t,\varphi) = c(t)\int_0^T H(s)\varphi(-s)ds \tag{3.5}$$

where $c(t) > 0$ is continuous and $\omega$-periodic, $0 < T \leq \infty$ and $H : [0,T] \to [0,\infty)$ is an integrable delay kernel with $\int_0^T H(s)ds > 0$ (see [11]). It is easily checked that

$H(t,\varphi)$ in (3.5) verifies the assumption (H). Therefore, Theorem 3.2 implies the following.

COROLLARY 3.4. *If* $G(0) < \frac{\gamma(t)}{b(t)}$ *and* $ca(t) + b(t)[G(c) + cd(t)\int_0^T H(s)ds] \not\equiv \gamma(t)$ *for each constant* $c > 0$, *where* $d(t) > 0$ *is a continuous* $\omega$*-periodic function, then the equation*

$$N'(t) = N(t)[\gamma(t) - a(t)N(t) - b(t)(G(N(t)) + d(t)\int_0^T H(s)N(t-s)ds)] \tag{3.6}$$

*has at least one nonconstant* $\omega$*-periodic solution.*

## 4. Uniqueness and Global Attractivity of Periodic Solutions

In this section we study the global asymptotic stability of the periodic solutions obtained in Section 3. Since it is generally difficult to give sufficient conditions under which the periodic solution predicted in Theorem 3.2 is globally stable, we concentrate our attention on the following two typical delay models

$$\dot{N}(t) = N(t)(a(t) - b(t)N(t - \tau(t))) \tag{4.1}$$

$$\dot{N}(t) = N(t)(a(t) - b(t)\int_0^T H(s)N(t-s)ds) \tag{4.2}$$

where $a, b, \tau$ are $\omega$-periodic continuous functions with $a, b > 0$ and $\tau \geq 0$, $T > 0$ is a constant, $H : [0,T] \to [0,\infty)$ is integrable with $\int_0^T H(s)ds = 1$. Equation (4.1) is similar to one discussed in [10], whereas (4.2) is considered in [11].

If $a(t) \not\equiv b(t)$, by Theorem 3.2, equation (4.1) (resp. (4.2)) has an $\omega$-periodic positive solution. We denote it by $N^*(t)$. Let $\ell n\,(1 + x(t)) = \ell n\, N(t) - \ell n\, N^*(t)$, where $N(t)$ is a solution of equation (4.1) (resp. 4.2). The equations 4.1 and 4.2 are therefore transformed into the following form

$$\dot{x}(t) = -\gamma(t)\big(1 + x(t)\big)x\big(t - \tau(t)\big) \tag{4.3}$$

$$\dot{x}(t) = -b(t)\big(1 + x(t)\big)\int_0^T K(t,s)x(t-s)ds \tag{4.4}$$

where $\gamma(t) = b(t)N^*\big(t - \tau(t)\big)$ and $K(t,s) = H(s)N^*(t-s)$. To study the global attractivity of $N^*(t)$, we need only to show the global stability of the zero solution of (4.3) (resp. 4.4).

Let us first consider (4.3) and begin with the following lemma.

LEMMA 4.1. *Let* $\varphi$ *denote the initial data of a solution* $x(t)$ *to equation (4.3), defined on* $[-\max_{t\in[0,\omega]} \tau(t), 0]$. *If* $1 + \varphi(s) \geq 0$; $s \in [-\max_{t\in[0,\omega]} \tau(t), 0]$ *and* $1 + \varphi(0) > 0$, *then* $x(t) > -1$, *for all* $t > 0$, *and is bounded above.*

PROOF: Since $x(0) > -1$, by integrating (4.3) we have

$$1 + x(t) = \big(1 + x(0)\big)e^{-\int_0^t \gamma(s)x(s-\tau(s))ds}$$

for small $t > 0$, from which we deduce that $x(t) > -1$ for all $t > 0$.

To show the upper boundedness of $x(t)$, we separate our considerations into two cases.

CASE I: $x(t)$ is non-oscillatory, i.e. for some $N > 0$, $x(t) \neq 0$ for all $t \geq N$. Since $1 + x(t) > 0$, we know that $\dot{x}(t)$ does not change sign for $t \geq N$ and, therefore, $\lim_{t\to+\infty} x(t) = \ell$ for some constant $\ell$. Note that $x(t)$ is a solution to (4.3) and that $\gamma(t)$ is positive. We must have $\ell = 0$ and we are done.

CASE II: $x(t)$ is oscillatory, i.e. there is an unbounded sequence $\{t_n\}_{n=1}^{\infty}$ such that $x(t_n) = 0$, $n = 1, 2, \ldots$ .

Let $\xi_n \in [t_n, t_{n+1}]$ be such that $x(\xi_n) = \max_{t\in[t_n,t_{n+1}]} x(t)$. Then $\dot{x}(\xi_n) = 0$ and hence $x(\xi_n - \tau(\xi_n)) = 0$. Integrating the equation, we have

$$\begin{aligned} 1 + x(\xi_n) &= e^{-\int_{\xi_n-\tau(\xi_n)}^{\xi_n} \gamma(s)x\left(s-\tau(s)\right)ds} \\ &< e^{\int_{\xi_n-\tau(\xi_n)}^{\xi_n} \gamma(s)ds} \leq e^{\overline{A}}, \end{aligned}$$

where

$$\overline{A} = \max_{t\in[0,\omega]} \int_{t-\tau(t)}^{t} \gamma(s)ds. \tag{4.5}$$

Therefore, $x(\xi_n) < e^{\overline{A}} - 1$ which implies $x(t) < e^{\overline{A}} - 1$ for all $t \geq 0$. This completes the proof.

□

LEMMA 4.2. *Let* $0 \leq a \leq b$ *be two constants. If* $e^{1-a} \geq b$, *then for all* $x > 0$,

$$\ell n\,(1+x) > b(1-e^{-ax}).$$

PROOF: Let $f(x) = \ell n\,(1+x) - b + be^{-ax}$. Then $f'(x) = \frac{1}{1+x} - abe^{-ax}$. Note that

$$\min_{x\geq 0} \frac{e^{ax}}{1+x} = ae^{1-a} \geq ab.$$

We see that $f'(x) > 0$ for $x > 0$ with $a(1+x) \neq 1$, and, consequently, $f(x) > f(0) = 0$, as desired.

□

Let $\overline{A}$ be defined as in (4.5) and

$$\underline{A} = \min_{t\in[0,\omega]} \int_{t-\tau(t)}^{t} \gamma(s)ds.$$

Clearly, $0 \leq \underline{A} \leq \overline{A}$.

The following lemma gives a sufficient condition for the global attractivity of the zero solution of (4.3). Our proof follows the idea of Wright [26].

LEMMA 4.3. *If* $\overline{A} \le e^{1-\underline{A}}$, *then every solution* $x(t)$ *of (4.3) with the initial data* $\varphi$ *with* $1+\varphi(s) \ge 0$ *on* $[-\max_{t\in[0,\omega]} \tau(t), 0]$ *and* $1+\varphi(0) > 0$, *satisfies* $\lim_{t\to\infty} x(t) = 0$.

PROOF: By Lemma 4.1, we can assume $x(t)$ is oscillatory, so there is an unbounded increasing sequence $\{t_n\}_{n=1}^{\infty}$ such that $x(t_n) = 0$, $n = 1, 2, \dots$ . Since $x(t)$ is bounded, there are well-defined two nonnegative numbers

$$\alpha = \limsup_{t\to+\infty} x(t), \quad \beta = -\liminf_{t\to+\infty} x(t).$$

Let $\varepsilon > 0$ be arbitrarily given. For a sufficiently large $N > 0$, we have

$$-\beta - \varepsilon < x(t) < \alpha + \varepsilon, \quad t \ge N. \tag{4.6}$$

Let $\xi_n \in [t_n, t_{n+1}]$ be such that $x(\xi_n) = \max_{t\in[t_n,t_{n+1}]} x(t)$. Then $x(\xi_n - \tau(\xi_n)) = 0$ and

$$\ell\text{n}\,\big(1 + x(\xi_n)\big) = -\int_{\xi_n-\tau(\xi_n)}^{\xi_n} \gamma(s)x\big(s-\tau(s)\big)ds. \tag{4.7}$$

Choose $N_1 > 0$ so large that $t_n \ge N + 2\max_{t\in[0,\omega]} \tau(t)$ for $n \ge N_1$. From (4.6) and (4.7), we have for $n \ge N_1$

$$-\underline{A}(\alpha+\varepsilon) < \ell\text{n}\,\big(1 + x(\xi_n)\big) < \overline{A}(\beta+\varepsilon).$$

Therefore

$$e^{-\underline{A}(\alpha+\varepsilon)} - 1 < x(\xi_n) < e^{\overline{A}(\beta+\varepsilon)} - 1$$

for $n \ge N_1$. Similary we have

$$e^{-\underline{A}(\alpha+\varepsilon)} - 1 \le x(\eta_n) < e^{\overline{A}(\beta+\varepsilon)} - 1$$

for $n \geq N_1$, where $\eta_n \in [t_n, t_{n+1}]$ are such that $x(\eta_n) = \min_{t \in [t_n, t_{n+1}]} x(t)$. Therefore for sufficiently large $n$

$$\alpha - \varepsilon < x(\xi_n) < e^{\overline{A}(\beta+\varepsilon)} - 1,$$
$$e^{-\underline{A}(\alpha+\varepsilon)} - 1 < x(\eta_n) < -\beta + \varepsilon.$$

Letting $\varepsilon \to 0$, we obtain

$$0 \leq \alpha \leq e^{\overline{A}\beta} - 1$$
$$0 \leq \beta \leq 1 - e^{-\underline{A}\alpha}. \tag{4.8}$$

We consider two cases.

CASE I: If $\alpha\beta = 0$, then from (4.8), we know $\alpha = \beta = 0$ and we are done.

CASE II: If $\alpha > 0$ and $0 < \beta < 1$, it follows from (4.8) that

$$1 + \alpha \leq e^{\overline{A}\beta} \leq e^{\overline{A}(1-e^{-\underline{A}\alpha})}$$

which gives

$$\ell n\,(1+\alpha) \leq \overline{A}(1 - e^{-\underline{A}\alpha}).$$

This is impossible by Lemma 4.2, since we assume $\overline{A} \leq e^{1-\underline{A}}$. Therefore, we have shown $\alpha = \beta = 0$ and the proof is complete.

□

REMARK 4.1: We believe that the above result on the global attractivity of the zero solution of (4.3) could be improved if one uses analysis more carefully as in [26]. For the sake of simplicity, we decide not to pursue that here.

We need one more lemma on the upper and lower bounds of the $\omega$-periodic solution of (4.1). In what follows, we use the notations

$$M = \max_{t\in[0,\omega]} \tfrac{a(t)}{b(t)}, \qquad m = \min_{t\in[0,\omega]} \tfrac{a(t)}{b(t)}$$

$$a = \min_{t\in[0,\omega]} \exp\left(\int_{t-\tau(t)}^{t} a(s)ds\right), \qquad A = \max_{t\in[0,\omega]} \exp\left(\int_{t-\tau(t)}^{t} a(s)ds\right)$$

$$b = \min_{t\in[0,\omega]} \exp\left(\int_{t-\tau(t)}^{t} b(s)ds\right), \qquad B = \max_{t\in[0,\omega]} \exp\left(\int_{t-\tau(t)}^{t} b(s)ds\right).$$

LEMMA 4.4. *Let* $N(t)$ *be an* $\omega$*-periodic solution of (4.1). Then*

$$\frac{ma}{B^{MA}} \leq N(t) \leq MA.$$

PROOF: Let $\xi \in [0,\omega]$ be such that $N(\xi) = \max_{t\in[0,\omega]} N(t)$. Then $N'(\xi) = 0$ and thus $N(\xi - \tau(\xi)) = \frac{a(\xi)}{b(\xi)}$. Integrating the equation, one has

$$\begin{aligned} N(\xi) &= N(\xi - \tau(\xi)) \exp\left(\int_{\xi-\tau(\xi)}^{\xi} (a(s) - b(s)N(s - \tau(s)))ds\right) \\ &\leq \frac{a(\xi)}{b(\xi)} \exp\left(\int_{\xi-\tau(\xi)}^{\xi} a(s)ds\right) \leq MA. \end{aligned}$$

On the other hand, let $\eta \in [0,\omega]$ be such that $N(\eta) = \min_{t\in[0,\omega]} N(t)$. We have $N(\eta - \tau(\eta)) = \frac{a(\eta)}{b(\eta)}$ and

$$\begin{aligned} N(\eta) &= N(\eta - \tau(\eta)) \exp\left(\int_{\eta-\tau(\eta)}^{\eta} (a(s) - b(s)N(s - \tau(s)))ds\right) \\ &\geq \frac{a(\eta)}{b(\eta)} \exp\left(\int_{\eta-\tau(\eta)}^{\eta} (a(s) - b(s)N(\xi))ds\right) \geq \frac{ma}{B^{MA}}. \end{aligned}$$

□

Now we are in a position to prove the following theorem on the global stability of the periodic solution $N^*(t)$.

THEOREM 4.5. *If*

$$MAb^{MA}\,\ell n\,B \le e \tag{4.9}$$

*then the periodic solution* $N^*(t)$ *obtained in Section 3 is globally stable, i.e. for any positive solution* $N(t)$ *of (4.1),*

$$\lim_{t\to\infty} |N(t) - N^*(t)| = 0. \tag{4.10}$$

PROOF: Recall that in (4.3), $\gamma(t) = b(t)N^*(t-\tau(t))$. By Lemma 4.4,

$$\begin{aligned}\overline{A} &= \max_{t\in[0,\omega]} \int_{t-\tau(t)}^{t} b(s)N^*(s-\tau(s))ds \\ &\le MA \max_{t\in[0,\omega]} \int_{t-\tau(t)}^{t} b(s)ds = MA\,\ell n\,B\end{aligned}$$

and

$$\begin{aligned}\underline{A} &= \min_{t\in[0,\omega]} \int_{t-\tau(t)}^{t} b(s)N^*(s-\tau(s))ds \\ &\le MA \min_{t\in[0,\omega]} \int_{t-\tau(t)}^{t} b(s)ds = MA\,\ell n\,b.\end{aligned}$$

Now (4.9) implies that $\overline{A} \le e^{1-\underline{A}}$. By Lemma 4.3, $\lim_{t\to\infty} \frac{N(t)}{N^*(t)} = 1$ and hence (4.10) holds. This completes the proof.

□

We remark that the existence of $N^*(t)$ was first obtained by [4]. However, we proved its global stability under (4.9) which [4] did not give.

Let us now consider the global attractivity of $N^*(t)$ for (4.4). We use the following notations.

$$\widetilde{A} = \max_{t\in[0,\omega]} \int_{t-T}^{t}\int_{0}^{T} b(u)K(u,v)dvdu$$

$$\underset{\sim}{A} = \min_{t\in[0,\omega]} \int_{t-T}^{t}\int_{0}^{T} b(u)K(u,v)dvdu$$

$$A_T = \max_{t\in[0,\omega]} \exp\Big(\int_{t-T}^{t} a(s)ds\Big), \quad a_T = \min_{t\in[0,\omega]} \exp\Big(\int_{t-T}^{t} a(s)ds\Big)$$

$$B_T = \max_{t\in[0,\omega]} \exp\Big(\int_{t-T}^{t} b(s)ds\Big), \quad b_T = \min_{t\in[0,\omega]} \exp\Big(\int_{t-T}^{t} b(s)ds\Big).$$

By the same argument as before, we obtain the following lemma.

LEMMA 4.6. *If* $\widetilde{A} \le e^{1-\underset{\sim}{A}}$, *then every solution* $x(t)$ *of (4.4) with the initial data* $\varphi$ *with* $1+\varphi(s)\ge 0$ *on* $[-T,0]$ *and* $1+\varphi(0)>0$ *satisfies* $\lim_{t\to\infty} x(t) = 0$.

PROOF: The proof is essentially the same as that of Lemma 4.3, so we omit it.

□

Similarly, one obtains the upper and lower boundedness of $\omega$-periodic solutions of (4.2).

LEMMA 4.7. *Let* $N(t)$ *be an* $\omega$*-periodic solution of (4.2). Then*

$$\frac{ma_T}{B_T^{MA_T}} \le N(t) \le MA_T.$$

□

By Lemmas 4.6 and 4.7, we are now able to prove the following theorem.

THEOREM 4.8. *Under the condition*

$$MA_T b_T^{MA_T}\,\ell n\, B_T \le e \tag{4.11}$$

*the periodic solution* $N^*(t)$ *of (4.2) is globally stable.*

PROOF: Note that $K(u,s) = H(s)N^*(u-s)$ and $\int_0^T H(s)ds = 1$. Following the same proof as in Theorem 4.5, one can show that for any positive solution $N(t)$ of (4.2), $\lim_{t\to\infty} |N(t) - N^*(t)| = 0$.

□

REMARK 4.2: If $\widetilde{A} \le 1$, then $\widetilde{A} \le e^{1-\underset{\sim}{A}}$ is automatically satisfied. Note that by Lemma 4.7

$$\begin{aligned}\widetilde{A} &= \max_{t\in[0,\omega]} \int_{t-T}^{t} \int_0^T b(u)H(v)N^*(u-v)dudv \\ &\le MA_T \max_{t\in[0,\omega]} \int_{t-T}^{t} b(u)du \le \frac{a^0}{b_0} e^{a^0 T} \cdot b^0 T = \frac{b^0}{b_0} a^0 T e^{a^0 T}\end{aligned}$$

where

$$a^0 = \max_{t\in[0,\omega]} a(t), \quad b^0 = \max_{t\in[0,\omega]} b(t), \quad b_0 = \min_{t\in[0,\omega]} b(t).$$

Therefore, if we let

$$\frac{b^0}{b_0}\, a^0 T e^{a^0 T} \le 1 \tag{4.12}$$

then the $\omega$-periodic solution $N^*(t)$ of (4.2) is globally stable. Comparing our condition (4.12) with the similar one (2.12) of [11], we see that (4.12) is weaker. Consequently, the sufficient condition (4.11) for the global attractivity of $N^*(t)$ is more general.

We finally remark that the question of obtaining sufficient conditions for the existence of a globally attractive $\omega$-periodic solution of the equation

$$\dot{N}(t) = N(t)[a(t) - b(t) \int_0^\infty H(s)N(t-s)ds]; \quad t > 0 \tag{4.13}$$

is still open (see [11]). We did prove the existence of an $\omega$-periodic solution of (4.13) as already shown in Section 3, but we are unable to get sufficient conditions for its global attractivity.

## References

[1] M. Bardi and A. Schiaffino, *Asymptotic behaviour of positive solutions of periodic delay logistic equations*, J. Math. Biol. **14** (1982): 95-100.

[2] M. Bardi, *An equation of growth of a single species with realistic dependence on crowding and seasonal factors,* J. Math. Biol. **17** (1983): 33-43.

[3] M. Bardi, *A nonautonomous nonlinear functional differential equation arising in the theory of population dynamices,* J. Math. Anal. Appl. **109** (1985): 492-508.

[4] V.I. Borzdyko, *A topological method of proving the existence of positive periodic solutions of functional-differential equations*, Differential Equations **26** (1990): 1230-1236.

[5] J.M. Cushing, "Integrodiffferential Equations and Delay Models in Population Dynamics," Lecture Notes in Biomathematics, **20** Springer-Verlag, Heidelberg, 1977.

[6] R.D. Driver, *Existence and stability of solutions of a delay-differential system,* Arch. Rat. Mech. Anal. **10** (1962): 401-426.

[7] H.I. Freedman, "Deterministic Mathematical Models in Population Ecology," HIFR Consulting Ltd., Edmonton, 1987.

[8] H.I. Freedman and K. Gopalsamy, *Global stability in time-delayed single-species dynamics,* Bull. Math. Biol. **48** (1986): 485-492.

[9] H.I. Freedman, V.S.H. Rao and J.W.-H. So, *Asymptotic behaviour of a time-dependent single-species model*, Analysis, **9** (1989): 217-223.

[10] H.I. Freedman and J. Wu, *Periodic solutions of single-species models with periodic delay*, SIAM J. Math. Anal. **23** (1992): 689-701.

[11] K. Gopalsamy, X.Z. He and L.Z. Wen, *Global attractivity and oscillations in a periodic logistic integrodifferential equation,* Houston J. of Math. **17** (1991): 157-177.

[12] D. Guo and V. Lakshmikanthan, "Nonlinear Problems in Abstract Cones," Academic Press, New York, 1988.

[13] J.K. Hale, "Theory of Functional Differential Equations," Springer-Verlag, Heidelberg, 1977.

[14] J.K. Hale and J. Kato, *Phase space for retarded equations with infinite delay,* Funkcialaj Ekvac. **21** (1978): 11-41.

[15] Y. Hino, S. Murakami and T. Naito, "Functional Differential Equations with Infinite Delay," Lecture Notes in Mathematics **1473**, Springer-Verlag, Heidelberg, 1991.

[16] G.E. Hutchinson, *Circular casual systems in ecology,* Ann. New York Acad. Sci. **50** (1948): 221-246.

[17] M.A. Krasnosel'skii, "Positive Solutions of Operational Equations," P. Nourdhoff, Groningen, The Netherlands, 1964.

[18] M.A. Krasnosel'skii and P.P. Zabreiko, "Geometrical Methods of Nonlinear Analysis," Springer-Verlag, New York, 1984.

[19] Y. Kuang, B.G. Zhang and T. Zhao, *Qualitative analysis of a nonautonomous nonlinear delay differential equation*, Tôhoku Math. J. **43** (1991): 509-528.

[20] B.S. Lalli and B.G. Zhang, *On a periodic delay population model,* Quart. Appli. Math. (to appear).

[21] R.M. May, *Models for single populations,* in "Theoretical Ecology," (R.M. May, ed.), Saunders Publ. Co., Philadelphia (1976): 4-25.

[22] R.M. Nisbet and W.S.C. Gurney, *Population dynamics in a periodically varying environment,* J. Theoret. Biol. **56** (1976): 459-475.

[23] S. Rosenblat, *Population models in a periodically fluctuating environment*, J. Math. Biol. **9** (1980): 23-36.

[24] J. Sugie, *On the stability for a population growth equation with time delay,* Proc. Royal Soc. Edinburgh **102A** (1992): 179-184.

[25] B.G. Veilleux, *An analysis between the predatory interaction between Paramecium and Didinium,* J. Anim. Ecol. **48** (1979): 787-803.

[26] E.M. Wright, *A nonlinear difference differential equation,* J. Reine Angew. Math. **194** (1955): 66-87.

[27] B.G. Zhang and K. Gopalsamy, *Global attractivity and oscillations in a periodic delay logistic equation*, J. Math. Anal. Appl. **150** (1990): 274-283.

[28] B.G. Zhang and K. Gopalsamy, *Global attractivity in the delay logistic equation with variable parameters,* Math. Proc. Comb. Phil. Soc. **107** (1990): 579-590.

# 6 Asymptotic Estimates for a Nonstandard Second Order Differential Equation

**William A. Harris, Jr.** University of Southern California, Los Angeles, California

**Patrizia Pucci** Universitá di Perugia, Perugia, Italy

**James Serrin** University of Minnesota, Minneapolis, Minnesota

## §1. Introduction.

We are concerned with the asymptotic behavior at infinity of solutions of the linear differential equation

$$y'' + 2b\,t^{\alpha-1}y' + c\,t^{2\beta-2}y = 0, \tag{1.1}$$

where $\alpha$, $\beta \in \mathbb{R}$ and $b$, $c$ are non-zero real constants. Such problems arise for example in the study of radial solutions of the linear elliptic equation $\Delta u + c|x|^{2\beta-2}u = 0$ in $n$ dimensions, when $u(r) = u(|x|)$ is a solution of

$$u''(r) + \frac{n-1}{r}u'(r) + c\,r^{2\beta-2}u(r) = 0, \qquad r > 0.$$

Another case of importance is that of a damped harmonic oscillator, when $\beta = c = 1$ and $b > 0$. More generally we may think of (1.1) as a linear differential equation with a highly irregular singular point at $\infty$.

In an earlier paper [2], based on the techniques of [1,3], we showed that the rather unexpected results indicated in [4] for the special case $\beta = 1$ have the following more general manifestation: *the* $(\alpha, \beta)$ *plane is divided into sectors by the set of rays from the origin* $(0,0)$ *given by*

$$\begin{cases} \beta < 0, & \alpha = 0 \\ \beta = s_N\alpha, & \alpha > 0, \quad N = 0,1,\ldots, \quad s_N = 1 - 1/2(N+1), \\ \beta = \alpha, & \alpha > 0 \\ \beta = t_N\alpha, & \alpha > 0, \quad N = 0,1,\ldots, \quad t_N = 1 + 1/(2N+1) = 1/s_N, \\ \beta > 0, & \alpha = 0 \\ \beta = 0, & \alpha < 0 \end{cases} \tag{1.2}$$

*which represent lines of change in the asymptotic representation of* $y$ *and* $y'$.

The concept of a change in asymptotic behavior of soultions as $(\alpha, \beta)$ crosses one of the rays (1.2) is sufficiently delicate that we indicate here precisely the meaning which is intended. This is somewhat easier to state for the non-oscillatory case, so we consider this first. Here, corresponding to any open sector $S$ in $\mathbb{R}^2$ bounded by a pair of consecutive rays (1.2), there is a positive real analytic function $\phi_S(t) = \phi_S(t; \alpha, \beta)$, $(\alpha, \beta) \in S$, $t > 0$, having the property that

$$\frac{y(t)}{\phi_S(t)} \to \text{finite limit} = \ell \qquad \text{as } t \to \infty \tag{1.3}$$

for any (non–oscillatory) solution $y$ of (1.1) for which $(\alpha, \beta) \in S$. Explicit expressions for the asymptotic functions $\phi_S$ are given in [2].

Turning to the oscillatory case, we note to begin with that this can occur only when $c > 0$, and only for $(\alpha, \beta)$ in the parameter region defined by

$$\beta \geq \max\{0, \alpha\},$$

see [2]. Then, corresponding to any sector $S$ in this region, there are *two* real analytic functions

$$\phi_S(t) = \phi_S(t; \alpha, \beta), \quad \psi_S(t) = \psi_S(t; \alpha, \beta), \quad (\alpha, \beta) \in S, \quad t > 0, \tag{1.4}$$

such that

$$\frac{y(t)}{\phi_S(t)} = [A\cos(\psi_S(t) + \theta)] \cdot [1 + o(1)] \tag{1.5}$$

for any real solution $y$ of (1.1) for which $(\alpha, \beta) \in S$, where $A$, $\theta$ are appropriate real constants depending on the solution.

The function $\phi_S$ and $\psi_S$ depend continuously on $(\alpha, \beta) \in S$, and also on $b$, $c$ for $b$, $c \neq 0$. Moreover, they have *continuous extensions* in the variables $\alpha, \beta$ to the *closure* of $S$ in $\mathbb{R}^2\backslash\{(0,0)\}$. The extended functions obey the following *transition formulas* on rays $R$ of the form (1.2), separating a pair of adjacent sectors $S$ and $S'$:

*Non–oscillatory case.* For $(\alpha, \beta) \in R$

$$\phi_{S'}(t; \alpha, \beta) = t^{\lambda}\phi_S(t; \alpha, \beta), \quad t > 0, \tag{1.6}$$

where $\lambda$ is a non–zero constant, depending only on $b$, $c$ and the ray $R$.

*Oscillatory case.* For $(\alpha, \beta) \in R$

$$\phi_{S'}(t; \alpha, \beta) = \phi_S(t; \alpha, \beta), \quad \psi_{S'}(t; \alpha, \beta) = \psi_S(t; \alpha, \beta) + \mu \log t, \quad t > 0, \tag{1.7}$$

where $\mu$ is a non–zero real constant, again depending only on $b$, $c$ and the ray $R$.

The formulas (1.6), (1.7) graphically illustrate the jump discontinuity in asymptotic behavior which occurs as one crosses a ray (1.2). Note that in the oscillatory case the *amplitude* function $\phi$ *does not change* across rays, there being only a *phase shift* in the amount $\mu \log t$.

In this paper we shall refine the above results to obtain the asymptotic behavior as $t \to \infty$ for the first and second derivatives of solutions of (1.1). In particular we show in the non–oscillatory case that

$$\frac{y'(t)}{\phi'(t)} \to \ell \quad \text{as } t \to \infty, \tag{1.8}$$

with the exception of parameter values $(\alpha, \beta)$ on the following rays

$$\alpha = \beta > 0; \quad \alpha = 0, \ \beta < 0; \quad \beta = 0, \ \alpha < 0.$$

In (1.8) the function $\phi$ is given by

$$\phi(t) = \begin{cases} \phi_S(t;\alpha,\beta) & \text{if } (\alpha,\beta) \in S, \\ t^\lambda \phi_S(t;\alpha,\beta) & \text{if } (\alpha,\beta) \in R, \end{cases}$$

where in the second case $S$ is one of the two sectors adjacent to $R$.

On the other hand, in the *oscillatory case* we show that

$$\frac{[t^{1-\beta} y'(t)]^2 + c\,y^2(t)}{\phi^2(t)} \to c\,A^2 \qquad \text{as } t \to \infty,$$

again with the exception of the rays

$$\alpha = \beta > 0; \quad \alpha = 0,\ \beta < 0; \quad \beta = 0,\ \alpha < 0.$$

The behavior of the second derivatives is treated in Section 6.

For the non-oscillatory case our proofs are based on the Sturmian form of (1.1), namely

(1.9) $$(p\,y')' + c\,p\,t^{2\beta-2}\,y = 0,$$

where

(1.10) $$p = p(t) = \begin{cases} \exp\left(\dfrac{2b}{\alpha} t^\alpha\right), & \alpha \neq 0 \\ t^b, & \alpha = 0. \end{cases}$$

For the oscillatory case we work directly with the fundamental matrix $Z(t)$ related to equation (1.1).

The origin $\alpha = 0$, $\beta = 0$, corresponds to the Euler equation $t^2 y'' + 2b\,t\,y' + c\,y = 0$, whose treatment is standard and which is best set aside from the main considerations of the paper. *Thus in what follows we assume always that* $(\alpha,\beta) \neq (0,0)$.

## §2. Asymptotic behavior when $\alpha > \beta$ and $\alpha > 0$.

This case is best divided into the two subcases $\alpha < 2\beta$ and $\alpha > 2\beta$. The treatment of asymptotic behavior on the rays (1.2) themselves is slightly different than for the open sectors between the rays. For simplicity in this and the next two sections, we shall therefore only treat values $(\alpha,\beta)$ in the interior of the sectors. The discussion when $(\alpha,\beta)$ belongs to a ray will be given later at the end of Section 5.

*2.1 The case $\alpha < 2\beta$ and $b > 0$.* Here by (3.8) of [2], we have

(2.1) $$\phi(t) = \phi_N(t) = \exp\left(\sum_1^N \lambda_k \int_1^t s^{\alpha-1+2k(\beta-\alpha)} ds\right) = \exp\Sigma(t), \qquad t > 0,$$

where all the coefficients $\lambda_k$ are real and non-zero. It follows from (1.9) that

(2.2) $$\frac{y'(t)}{\phi'(t)} = -\frac{c}{p(t)\phi'(t)} \int^t p(s) s^{2\beta-2} y(s) ds.$$

By (2.1) we get

$$\Sigma(t) = \frac{\lambda_1}{2\beta - \alpha} t^{2\beta-\alpha} [1 + o(1)],$$

(2.3)
$$\frac{\phi'(t)}{\phi(t)} = \Sigma'(t) = \lambda_1 t^{2\beta-\alpha-1} [1 + o(1)],$$

and so from (1.8)

(2.4)
$$p(t)\phi'(t) \to \infty \qquad \text{as } t \to \infty.$$

We now apply L'Hospital's rule to the right hand side of (2.2) in order to determine its limit as $t \to \infty$. Suppose first that in (1.3)

(2.5)
$$\frac{y(t)}{\phi(t)} \to \ell \neq 0 \qquad \text{as } t \to \infty.$$

Then the integral in (2.2) is divergent to $\pm\infty$ as $t \to \infty$, since $\alpha < 2\beta$. L'Hospital's rule therefore gives

(2.6)
$$\lim_{t\to\infty} \frac{y'(t)}{\phi'(t)} = - \lim_{t\to\infty} \frac{c\,p(t)\,t^{2\beta-2}\,y(t)}{p(t)\phi''(t) + p'(t)\phi'(t)},$$

provided the right hand limit exists.

To investigate this, we observe first that $p' = 2b\,t^{\alpha-1}\,p$. It is therefore enough to consider the behavior as $t \to \infty$ of the function

(2.7)
$$I = \frac{\phi'' + 2b\,t^{\alpha-1}\phi'}{t^{2\beta-2}y} = \left(\frac{\phi''}{\phi} + 2b\,t^{\alpha-1}\frac{\phi'}{\phi}\right)\frac{1}{t^{2\beta-2}} \cdot \frac{\phi}{y}.$$

From (2.3) we have

(2.8)
$$\frac{\phi''(t)}{\phi(t)} = \Sigma''(t) + (\Sigma'(t))^2 = \lambda_1^2\, t^{2(2\beta-\alpha-1)} [1 + o(1)],$$

since $\alpha < 2\beta$. Thus, using (2.3),

$$\frac{\phi''}{\phi} + 2b\,t^{\alpha-1}\,\frac{\phi'}{\phi} = 2b\,\lambda_1\,t^{2\beta-2} [1 + o(1)]$$

since $\alpha > \beta$. Consequently

$$\lim_{t\to\infty} I(t) = \frac{2b\lambda_1}{\ell}$$

by (2.5).

Hence the limit on the right hand side of (2.6) exists, and so in turn

$$\frac{y'(t)}{\phi'(t)} \to \text{finite limit;}$$

by (2.5) and by L'Hospital's rule this limit must in fact be $\ell$, as required.

Next suppose that $\ell = 0$ in (1.3). Fix $\varepsilon > 0$. Then $|y(t)/\phi(t)| < \varepsilon$ for all $t$ sufficiently large, say $t \geq T$. Hence from (2.2) and (2.4)

$$\begin{aligned} \left|\frac{y'(t)}{\phi'(t)}\right| &\leq \frac{|c|}{p|\phi'|}\left|\left(\int^T + \int_T^t\right) p(s)s^{2\beta-2}y(s)ds\right| \\ &\leq \varepsilon + \varepsilon\frac{|c|}{p(t)|\phi'(t)|}\int_T^t p(s)s^{2\beta-2}\phi(s)ds \end{aligned}$$

for suitably large $t$. The use of L'Hospital's rule as in the previous calculation then gives

$$\limsup_{t\to\infty}\left|\frac{y'(t)}{\phi'(t)}\right| \leq \varepsilon + \varepsilon\frac{|c|}{2b|\lambda_1|}.$$

Since $\varepsilon > 0$ is arbitrary it follows that

$$\lim_{t\to\infty}\frac{y'(t)}{\phi'(t)} = 0,$$

as required.

*2.2 The case $\alpha < 2\beta$ and $b < 0$.* Here, instead of (2.1), we have by (3.9) of [2],

$$\phi(t) = \phi_N(t) = t^{1-\alpha}\exp\left(-\frac{2b}{\alpha}\left[t^\alpha - 1\right] + \Sigma(t)\right),$$

where the function $\Sigma(t)$ was defined in (2.1). Corresponding to (2.3) we have, as $t \to \infty$,

$$\frac{\phi'(t)}{\phi(t)} = (1-\alpha)t^{-1} - 2b\,t^{\alpha-1} + \lambda_1 t^{2\beta-\alpha-1}\left[1 + o(1)\right], \tag{2.9}$$

and then, after a short calculation,

$$\frac{\phi''(t)}{\phi(t)} = \left[\frac{\phi'(t)}{\phi(t)}\right]' + \left[\frac{\phi'(t)}{\phi(t)}\right]^2 = 4b^2t^{2\alpha-2} - 4b\lambda_1 t^{2\beta-2}\left[1+o(1)\right], \tag{2.10}$$

since $\beta < \alpha < 2\beta$.

In order to apply L'Hospital's rule, we investigate the term $p\phi'$ in (2.2). A simple calculation gives

$$p(t)\phi'(t) = 2|b|\exp\left(\frac{\lambda_1}{2\beta-\alpha}t^{2\beta-\alpha}\left[1+o(1)\right]\right) \longrightarrow \begin{cases} 0 & \text{if } \lambda_1 < 0 \\ \infty & \text{if } \lambda_1 > 0 \end{cases} \tag{2.11}$$

as $t \to \infty$.

We first consider the case when $\lambda_1 > 0$. Let (2.5) hold, that is $\ell \neq 0$ in (1.3). Then L'Hospital's rule applies to the right hand side of (2.2), since the integral also diverges. As before, it is clear that we must investigate the function $I$ given in (2.7). By (2.9) and (2.10)

$$\frac{\phi''}{\phi} + 2b\,t^{\alpha-1}\frac{\phi'}{\phi} = -2b\lambda_1 t^{2\beta-2}\left[1+o(1)\right],$$

since $\alpha < 2\beta$. Consequently

$$\lim_{t\to\infty} I(t) = -\frac{2b\lambda_1}{\ell}.$$

The limit on the right hand side of (2.6) exists, and so

$$\lim_{t\to\infty} \frac{y'(t)}{\phi'(t)} = \ell.$$

The case $\ell = 0$ in (1.3) is treated as in subsection *2.1.*

We now consider the case when $\lambda_1 < 0$. Here the integral on the right hand side of (2.2) converges and so (2.2) can be written in the form

$$\frac{y'(t)}{\phi'(t)} = \frac{c}{p(t)\phi'(t)} \left( \int_t^\infty p(s)s^{2\beta-2}y(s)ds + d \right), \tag{2.12}$$

where $d$ is a constant.

If $d \neq 0$ then $y'/\phi'$ tends to $\pm\infty$ as $t \to \infty$. In turn, since $\phi(t) \to \infty$, we obtain by integration

$$\lim_{t\to\infty} \frac{y(t)}{\phi(t)} = \infty,$$

which is a contraction. Hence $d = 0$ in (2.12).

We can now apply L'Hospital's rule directly to (2.12), and the remaining discussion is unchanged from the previous case $\lambda_1 > 0$.

It remains to treat

*2.3. The case* $\alpha > 2\beta$. Suppose first that $b > 0$. Then by (3.8) of [2], with $N = 0$,

$$\phi = \phi_0(t) = 1,$$

where we recall the agreement that $\sum_1^0 = 0$, see footnote 1 in [2]. Here the ratio $y/\phi$ tends to a finite limit as $t \to \infty$ and so (1.3) takes the simple form

$$y(t) \to \ell \qquad \text{as } t \to \infty.$$

Actually in the present case it is convenient to choose a slightly different function $\phi$, namely

$$\phi(t) = 1 - \frac{c}{2b(2\beta - \alpha)} t^{2\beta-\alpha}.$$

Of course, even with this choice we still have

$$\frac{y(t)}{\phi(t)} \to \ell \qquad \text{as } t \to \infty.$$

Now by (1.9)

$$y'(t) = -\frac{c}{p(t)} \int^t p(s)s^{2\beta-2}y(s)ds, \tag{2.13}$$

where $p(t) \to \infty$ as $t \to \infty$ since $b > 0$. If $\ell \neq 0$ then L'Hospital's rule gives

$$\frac{y'(t)}{\phi'(t)} \to \ell \qquad \text{as } t \to \infty. \tag{2.14}$$

When $\ell = 0$ the argument at the end of case *2.1* shows that (2.14) continues to hold, with $\ell = 0$.

It remains to consider $b < 0$. Here by (3.9) of [2]

$$\phi(t) = \phi_0(t) = t^{1-\alpha} \exp\left(\frac{2|b|}{\alpha}[t^\alpha - 1]\right),$$

and an easy calculation yields $p(t)\phi'(t) \to 2|b| \exp(-2|b|/\alpha) > 0$ as $t \to \infty$. Moreover the integral in (2.2) converges because of (1.3), (1.10) and the fact that $\alpha > 2\beta$. Thus finally

$$\frac{y'(t)}{\phi'(t)} \to \text{limit} = \ell \qquad \text{as } t \to \infty.$$

## §3. Asymptotic behavior when $\beta > \alpha > 0$.

This case is best divided into the two subcases $c < 0$ and $c > 0$.

*3.1. The non-oscillatory case* $c < 0$. By (3.13) in [2] we have

$$\phi(t) = \hat{\phi}_N(t) = t^{(1-\beta)/2} \exp\left(-\frac{b}{\alpha}[t^\alpha - 1] + \psi(t)\right), \tag{3.1}$$

where

$$\psi(t) = \psi_N(t) = \sqrt{|c|} \sum_0^N \mu_k \int_1^t s^{\beta-1+2k(\alpha-\beta)} ds, \qquad \mu_0 = 1.$$

First one checks, for the formula (2.2), that $p\,\phi'$ tends to $\infty$ as $t \to \infty$ and that, when $\ell \neq 0$ in (1.3), the integral diverges as $t \to \infty$.

Consequently L'Hospital's rule applies. Proceeding as in subsection *2.1*, we find after a straightforward calculation that the function $I$ in (2.7) satisfies

$$\lim_{t\to\infty} I(t) = \frac{|c|}{\ell}, \qquad \ell \neq 0,$$

since $\beta > \alpha > 0$. Hence from (2.2)

$$\lim_{t\to\infty} \frac{y'(t)}{\phi'(t)} = \ell.$$

The case $\ell = 0$ can be treated in the usual way.

*3.2. The oscillatory case* $c > 0$. The previous calculations were based on the formula (2.2). It is more convenient in the oscillatory case to work directly with the fundamental matrix $Z(t)$ given by (2.9) in [2]. The function $\lambda(t)$ here has the form, as $t \to \infty$,

$$\lambda(t) = i\sqrt{c}\,M(t) + O(t^{-1-\beta+\alpha}),$$

where $M(t) = M_N(t) = \sum_0^N \mu_k\, t^{\beta-1+2k(\alpha-\beta)} = t^{\beta-1}[1 + o(1)]$ is given in (3.10) of [2].

Now for appropriate constants $A$ and $\theta$ we have, as in (1.5),

$$\frac{y(t)}{\phi(t)} = A\cos(\psi(t) + \theta) \cdot [1 + o(1)] \qquad \text{as } t \to \infty, \tag{3.2}$$

where, by (3.14) of [2],

$$\phi(t) = t^{(1-\beta)/2} \exp\left(-\frac{b}{\alpha}\,[t^{\alpha} - 1]\right).$$

In fact (3.2) is just the linear combination, with coefficients $\frac{1}{2}Ae^{i\theta}$ and $\frac{1}{2}Ae^{-i\theta}$, of the two elements on the first row of the fundamental matrix $Z(t)$. Applying the same operation to the two elements in the second row of $Z(t)$ yields, as $t \to \infty$,

$$\begin{aligned}\frac{y'(t)}{\phi(t)} &= -b\,t^{\alpha-1}A\cos(\psi(t)+\theta)\cdot[1+o(1)] \\ &\quad +\sqrt{c}\,t^{\beta-1}A\sin(\psi(t)+\theta)\cdot[1+o(1)] + O(t^{-1-\beta+\alpha}).\end{aligned}$$

This formula can be expressed more usefully in the form

$$\frac{[t^{1-\beta}y'(t)]^2 + c\,y^2(t)}{\phi^2(t)} = c\,A^2[1+o(1)].$$

It should be observed that the oscillatory case does not have asymptotic formulas for $y'$ corresponding to those of the non–oscillatory case.

## §4. Asymptotic behavior when $\alpha < 0$.

*4.1 The case $\beta > 0$ and $c < 0$.* Here by (3.18) of [2] we have

$$\phi(t) = t^{(1-\beta)/2} \exp\left(\frac{\sqrt{|c|}}{\beta}\,[t^{\beta} - 1]\right). \tag{4.1}$$

This case is completely analogous to subsection *3.1*, and we get

$$\frac{y'(t)}{\phi'(t)} \to \ell \qquad \text{as } t \to \infty.$$

*4.2. The oscillatory case $\beta > 0$ and $c > 0$.* As before it is more convenient in the oscillatory case to work directly with the fundamental matrix $Z(t)$. Here the function $\lambda(t)$ is given by

$$\lambda(t) = i\sqrt{c}\,t^{\beta-1} + O(t^{-1-\beta+\alpha}) \qquad \text{as } t \to \infty.$$

Hence this case is exactly the same as subsection *3.2*, except that because $\alpha < 0$ the function $\phi$ now has the form

$$\phi(t) = t^{(1-\beta)/2}.$$

*4.3. The case $\beta < 0$.* From (4.2) of [2] one has immediately

$$\frac{y(t)}{t} \to \ell \quad \text{and} \quad y'(t) \to \ell \qquad \text{as } t \to \infty.$$

## §5. Behavior on the rays (1.2).

As shown in [2], in the non-oscillatory case the asymptotic behavior of a solution $y$ of (1.1) for $(\alpha, \beta)$ belonging to a ray $R$ of type (1.2) is governed by the function

$$\phi_R(t) = t^\lambda \phi(t), \tag{5.1}$$

where $\phi$ is one of the two functions corresponding to the adjacent sectors $S$ and $S'$ whose common boundary is $R$. Moreover $\lambda$ is a non-zero constant depending only on $b$, $c$ and $R$.

*Note: Formula* (5.1) *applies on all rays* (1.2) *with the exception of the three special ones*

$$\beta = \alpha,\ \alpha > 0; \qquad \alpha = 0,\ \beta < 0; \qquad \beta = 0,\ \alpha < 0;$$

*the following considerations consequently do not apply in these cases.*

For the functions $\phi_R$ in (5.1) the calculations of the previous sections carry over almost unchanged. We indicate the details only when the ray $R$ lies in the sector $\beta < \alpha \le 2\beta$.

First for $\beta < \alpha < 2\beta$ and $b > 0$, the formula (2.3) becomes

$$\frac{\phi_R'}{\phi_R} = \frac{\lambda}{t} + \frac{\phi'}{\phi} = \frac{\lambda}{t} + \Sigma' = \lambda_1 t^{2\beta-\alpha-1}\,[1 + o(1)], \tag{5.2}$$

which is exactly as before, while (2.8) takes the form

$$\frac{\phi_R''}{\phi_R} = -\frac{\lambda}{t^2} + \Sigma'' + \left(\frac{\lambda}{t} + \Sigma'\right)^2 = \lambda_1^2 t^{2(2\beta-\alpha-1)}\,[1 + o(1)], \tag{5.3}$$

again exactly as before. The rest of the argument is now unchanged, yielding

$$\lim_{t\to\infty} \frac{y'(t)}{\phi_R'(t)} = \lim_{t\to\infty} \frac{y(t)}{\phi_R(t)} = \ell.$$

The case $b < 0$ can be treated as above.

For the ray $\alpha = 2\beta$ and $b > 0$, we use (3.4) of [2] with $N = 0$ and the case of equality. Consequently by (2.9) of [2] we obtain

$$\phi_R(t) = t^{\lambda_1}, \qquad \lambda_1 = -c/2b.$$

We claim that

$$\lim_{t\to\infty} \frac{y'(t)}{\phi_R'(t)} = \lim_{t\to\infty} \frac{y'(t)}{\lambda_1 t^{\lambda_1 - 1}} = \ell.$$

Indeed by (1.9)

$$\frac{y'(t)}{t^{\lambda_1-1}} = -\frac{c}{t^{\lambda_1-1}p(t)} \int^t p(s)s^{\alpha-2}y(s)ds,$$

where $t^{\lambda_1-1}p(t) \to \infty$ as $t \to \infty$ since $b > 0$. When $\ell \neq 0$ the integral diverges as $t \to \infty$ and L'Hospital's rule gives

$$\frac{y'(t)}{t^{\lambda_1-1}} \to -\frac{c}{2b}\ell,$$

as claimed. The case $b < 0$ is treated exactly as in subsection *2.3*.

## §6. Second derivative behavior.

We determine here the asymptotic behavior of $y''$ for all non–oscillatory cases in the parameter range $2\beta > \max\{0, \alpha\}$, $\alpha \neq \beta$, and in fact show in these cases that

$$\frac{y''(t)}{\phi''(t)} \to \ell \qquad \text{as } t \to \infty. \tag{6.1}$$

For this purpose we write, using the equation (1.1),

$$\begin{aligned} \frac{y''}{\phi''} &= \frac{\phi}{\phi''}\left\{-2b\,t^{\alpha-1}\frac{\phi'}{\phi}\cdot\frac{y'}{\phi'} - c\,t^{2\beta-2}\frac{y}{\phi}\right\} \\ &= -2b\ell\,t^{\alpha-1}\frac{\phi}{\phi''}\left\{\frac{\phi'}{\phi} + \frac{c}{2b}t^{2\beta-\alpha-1}\right\}[1+o(1)]. \end{aligned} \tag{6.2}$$

When $\beta < \alpha < 2\beta$ and $b > 0$, we get from (2.1)

$$\frac{\phi'}{\phi} = \lambda_1 t^{2\beta-\alpha-1} + \lambda_2\,t^{4\beta-3\alpha-1}[1+o(1)],$$

while from (2.8)

$$\frac{\phi''}{\phi} = \lambda_1^2\,t^{4\beta-2\alpha-2}\,[1+o(1)].$$

Recalling that $\lambda_1 = -c/2b$, and inserting the above estimates into (6.2), we have

$$\frac{y''}{\phi''} \to -2b\frac{\lambda_2}{\lambda_1^2}\ell = \ell \qquad \text{as } t \to \infty.$$

When $\beta < \alpha < 2\beta$ and $b < 0$, we get from (2.9) and (2.10)

$$\begin{aligned} \frac{\phi'}{\phi} &= -2b\,t^{\alpha-1}\,[1+o(1)], \\ \frac{\phi''}{\phi} &= 4b^2t^{2\alpha-2}\,[1+o(1)], \end{aligned}$$

and the required conclusion follows at once using (6.2).

Finally, when $\beta > \max\{0, \alpha\}$ and $c < 0$, we get from (3.1) and (4.1)

$$\begin{aligned} \frac{\phi'}{\phi} &= \sqrt{|c|}\,t^{\beta-1}\,[1+o(1)], \\ \frac{\phi''}{\phi} &= \left(\frac{\phi'}{\phi}\right)' + \left(\frac{\phi'}{\phi}\right)^2 = |c|\,t^{2\beta-2}\,[1+o(1)], \end{aligned}$$

and (6.1) follows again.

*Acknowledgements.* This paper was completed while William A. Harris, Jr. was on sabbatical leave from the University of Southern California as a Senior Fellow, AHPCRC, University of Minnesota, supported in part by the United States Army under Grant DAAL 03-89-C-0038.

Patrizia Pucci is a member of the *Gruppo Nazionale di Analisi Funzionale e sue Applicazioni* of the *Consiglio Nazionale delle Ricerche*, Rome, Italy. This reseach has been partly supported by the Italian *Ministero della Università e della Ricerca Scientifica e Tecnologica.*

James Serrin was supported in part by the National Science Foundation.

## References

[1] W.A. Harris, Jr. and D.A. Lutz, *A unified theory of asymptotic integration*, J. Math. Anal. Appl. **57** (1977), 571-586.

[2] W.A. Harris, Jr., P. Pucci and J. Serrin, *Asymptotic behavior of solutions of a nonstandard second order differential equation*, to appear.

[3] N. Levinson, *The asymptotic nature of solutions of linear systems of differential equations*, Duke Math. J. **15** (1948), 111-126.

[4] P. Pucci and J. Serrin, *Asymptotic properties for solutions of strongly nonlinear second order differential equations*, Rend. Sem. Mat. Univ. Pol. di Torino, Fasc. Speciale (1989), 121-129.

# 7 Asymptotic Phase, Shadowing and Reaction-Diffusion Systems

**Morris W. Hirsch** University of California at Berkeley, Berkeley, California

## 0 Introduction

It is with pleasure and gratitude that we honor Professor Larry Markus for his contributions to mathematics. In 1956 he published *Asymptotically autonomous differential systems* [6] in a series called *Contributions to the Study of Nonlinear Oscillations*— a subject which today would be called "Dynamical Systems". The present article is a direct descendant of Markus' influential paper, through Conway, Hoff and Smoller [2].

Consider a smooth flow $\{\Phi_t\}$ having an attracting limit cycle $\gamma$. It is well known that if $\gamma$ is a hyperbolic attractor— all Floquet exponents having negative real parts— then every trajectory $\Phi_t x$ attracted to $\gamma$ is asymptotic with the trajectory of a unique point of $\gamma$. If $\gamma$ is parameterized by the interval $[0, 2\pi]$ then $y$ can be interpreted as an angle, called the *asymptotic phase* of $x$.

I abstract this notion as follows. Consider a trajectory $\Phi_t x$ attracted to some positively invariant set $A$. If $y \in A$ is such that $\lim_{t\to\infty} ||\Phi_t x - \Phi_t y|| = 0$ then I call $y$ an *asymptotic phase* for $x$. (For clarity I use $||a - b||$ to denote the distance between points $a, b$ in any metric space.) Notice that uniqueness of $y$ is not required here.

If $A$ is not negatively invariant it may happen that $x$ does not have an asymptotic phase in $A$, but that $\Phi_s x$ does, for some $s > 0$. In this case I say $x$ has an *eventual asymptotic phase* in $A$.

It is frequently incorrectly assumed that every orbit approaching an attractor has an eventual asymptotic phase in the attractor. A common situation is that of a cascade of two systems, that is, a system of the form:

$$\begin{aligned} \frac{dx}{dt} &= F(x,y) \\ \frac{dy}{dt} &= G(y). \end{aligned}$$

If $(x(t), y(t))$ is a particular solution such that $y(t) \to c$, it is often asserted without justification that $x(t)$ is asymptotic to a solution of $dz/dt = F(z,c)$. A simple counter-example in the plane is:

$$\begin{aligned} \frac{dx}{dt} &= xy \\ \frac{dy}{dt} &= -y^3. \end{aligned}$$

The goal of this paper is to find conditions ensuring existence of an eventual asymptotic phase.

# 1 Main Results

Let $a = a(q)$ denote a nonnegative real-valued function of a variable $q$ whose domain is understood to be a terminal segment of either the positive reals or the positive integers. Define

$$\mathcal{R}a = \mathcal{R}_{q\to\infty}a(q) = \limsup_{q\to\infty} a(q)^{\frac{1}{q}}.$$

Then

$$\mathcal{R}(a+b) = \max(\mathcal{R}a, \mathcal{R}b),$$

and for any constant $\kappa > 0$:

$$\mathcal{R}(\kappa a) = \mathcal{R}(a).$$

Let $F = \{F_t\}_{t\geq 0}$ be a flow (more precisely, a partial semiflow) on a metric space $X$ (ususally, a Banach space). For clarity the distance between points $x, y \in X$ is denoted $||x-y||$. In applications $X$ is usually a subset of a Banach space. I shall always assume the maps $F_t$ have the following *local Lipschitz property:* For any $t_0 \geq 0, x_0 \in X$ there exist $L \geq 0$ and neighborhoods $N \in \mathbf{R}_+$of $t_0$ and $U \in X$ of $x_0$ such that

$$||F_t x - F_t y|| \leq L||x-y||$$

for all $t \in N, y \in U$.

Denote by $A \subset X$ a closed subspace having the following properties:

(a) $A$ is positively invariant under $F$.

(b) $A$ has the structure of a Riemannian manifold without boundary homeomorphically embedded in $X$. The norm of a tangent vector $Y$ in the Riemannian metric is denoted by $||Y||$.

(c) There is a smooth ($=C^1$) tangent vector field $G$ on $A$ whose flow $\Phi = \{\Phi_t\}_{t\in\mathbf{R}}$ coincides with $F_t|A$ for $t \geq 0$.

Let $K \subset A$ denote a nonempty compact set positively invariant under $F$, so that $K$ is also $\Phi$-invariant. The *inset* of $K$ under $F$ is the set

$$In(K) = In(K, F) = \{x \in X : \lim_{t\to\infty} \text{dist}(F_t x, K) = 0\}.$$

For $x \in In(K)$ define the *rate of approach to* $K$ of $x$ to $K$ under $F$ to be the number

$$\mathcal{P}(x, K, F) = \mathcal{R}_{t\to\infty}\text{dist}(F_t x, K).$$

Evidently $0 \leq \mathcal{P}(x, K, F) \leq 1$. If $\mathcal{P}(x, K, F) < 1$ I say $x$ is *exponentially attracted to* $K$.

Fix a Riemannian metric on $A$. The closed ball in $A$ with radius $\rho \geq 0$ centered at $x \in A$ is denoted by $B(\rho, x)$.

For a diffeomorphism $h$ between open subsets of $A$, the *expansion constant of* $h$ *at* $x \in A$ is the positive number

$$EC(h, x) = ||T_x h^{-1}||^{-1} = \min_{||Y||=1} ||T_x h(Y)||.$$

Here $Y$ denotes tangent vectors to $A$ at $x$, and $||T_x h||$ denotes the operator norm of the differential of $h$ at $x$ (defined by the Riemannian metric). Thus $EC(h, x) \geq \mu$ iff $||T_x z|| \geq \mu||z||$ for all $z \in T_x$.

Now for any compact subset $K \subset A$ define

$$EC(h, K) = \min_{x\in K} EC(h, x)$$

If $EC(h, K) > \nu > 0$ then it is not hard to see that there exists $\rho_* > 0$ such that if $x \in K$ and $0 < \rho \leq \rho_*$ then

$$hB(\rho, x) \supset B(\mu\rho, h(x));$$

see Hirsch and Pugh [4].

The *expansion rate of* $\Phi$ *at* $K$ is the nonnegative number

$$\mathcal{E}(\Phi, K) = \sup_{t>0} EC(\Phi_t, K)^{\frac{1}{t}}.$$

Since $[T_x\Phi_t]^{-1} = T_{\Phi_t x}\Phi_{-t}$, we have

$$\mathcal{E}(\Phi, K) = \sup_{t>0} \min_{x\in K} ||T_{\Phi_t x}\Phi_{-t}||^{-\frac{1}{t}}$$

The expansion rate is is the largest $\mu > 0$ having the following property: If $0 < \nu < \mu$ then there exist $s > 0, \rho_* > 0$ such that

$$\Phi_s B(\rho, x) \supset B(\nu^s \rho, \Phi_s x)$$

provided $x \in K$ and $0 < \rho \leq \rho_*$.

The expansion rate depends on the dynamics and the Riemannian metric. In some cases it is possible to estimate it from a formula for the vector field, from the dynamics of its flow, or from estimates using other metrics. Here are several such estimates.

**(i)** Assume that $A = \mathbf{R}^n$ with the standard inner product $\langle\cdot,\cdot\rangle$ , and denote $T_x\Phi_t$ by $D\Phi_t(x)$. The variational equation along orbits of the reversed time flow $\Phi_{-t}$, generated by the vector field $-G$ on $A$, gives the following matrix differential equation:

$$\frac{d}{dt}D\Phi_{-t}(x) = -DG(\Phi_{-t}x)D\Phi_{-t}(x)$$

Therefore for every nonzero vector $Y \in \mathbf{R}^n$ and every $t \geq 0, y \in K$ we have, setting $y = \Phi_t x \in K$:

$$\frac{d}{dt}||D\Phi_{-t}(y)Y|| = ||D\Phi_{-t}(y)Y||^{-1}\langle -DG(\Phi_{-t}y)D\Phi_{-t}(y)Y,\ D\Phi_{-t}(y)Y\rangle$$

The inner product on the right hand side is bounded above by $-\beta||D\Phi_{-t}(y)Y||^2$ where $\beta = \beta(G, K)$ denotes the minimum over $x \in K$ and unit vectors $\xi \in \mathbf{R}^n$ of $\langle DG(x)\xi, \xi\rangle$. Equivalently, $\beta$ equals the smallest eigenvalue of the symmetric matrix $\frac{1}{2}[DG(x)+DG(x)^T]$ where $T$ denotes the transpose of a matrix. Therefore

$$\frac{d}{dt}||D\Phi_{-t}(x)|| \leq \beta||D\Phi_{-t}(x)||,$$

whence

$$||D\Phi_{-t}(x)|| \leq e^{-t\beta}.$$

This proves $EC(\Phi_t, x) \geq e^{t\beta}$ for all $t \geq 0, x \in K$. We get the convenient estimate:

$$\mathcal{E}(\Phi, K) \geq e^{\beta(G,K)}. \tag{1}$$

(ii) Another estimate is obtained by noticing that

$$|\beta| \leq M = M(G, K) = \max_{x \in K} ||DG(x)||$$

(using the Schwarz inequality) so that $\beta \geq -M$. This yields the estimate:

$$\mathcal{E}(\Phi, K) \geq e^{-M(G,K)}. \tag{2}$$

which will be used in Section 2.

(iii) A different estimate can be obtained in case all forward and backward trajectoies in K are attracted to hyperbolic periodic orbit (possibly stationary). Suppose that the real parts of the Floquet exponents of these periodic orbits are all $\geq \gamma \in \mathbf{R}$. Then it can be proved that:

$$\mathcal{E}(\Phi, K) \geq e^{\gamma} \tag{3}$$

Suppose for example that the flow in $A$ is the gradient flow of a function $g : A \to \mathbf{R}$ having a finite set of critical points, and $K$ is a compact attractor containing all the critical points. Then $\gamma$ is the minimum of the eigenvalues of the Hessian of $g$ at critical points in $K$.

(iv) More generally, it can be shown that if $L \subset K$ is a compact set containing all alpha and omega limit points in $K$, then $\mathcal{E}(\Phi, K) = \mathcal{E}(\Phi, L)$. The reason is that any semi-trajectory in $K$ spends all but a finite amount of time in any given neighborhood of $L$.

(v) If $K$ is a smooth submanifold and the flow in $K$ is isometric for some Riemannian metric, then $\mathcal{E}(\Psi, K) = 1$. This is the case, for example, when $K$ is a periodic orbit; when $K$ is a smooth submanifold consisting of stationary points; or when the $K$ is an $n$-dimensional torus and the flow is translation by a one parameter subgroup.

(vi) It seems reasonable to conjecture that if $\Psi$ is generated by a vector field $H$ on $A$ of the form $H(x) = c(x)G(x)$ where $c$ is a positive function on $A$, then $\mathcal{E}(\Phi, K) = \mathcal{E}(\Psi, K)$.

It would be very useful to know that $\mathcal{E}(\Phi, K)$ is preserved, or at least well controlled, by a smooth or continuous reparameterization of the trajectories, or by a topological conjugacy between flows. A key test case is a $C^2$ flow on a 2-torus without periodic orbits: Is the expansion rate equal to 1?

**(vii)** Clearly $\mathcal{E}(\Phi, K) \geq \mathcal{R}_{t\to\infty} EC(\Phi_t, K)$. The latter number is easier to estimate and in some ways is more natural. For example it is easy to prove that it is independent of the Riemannian metric on $A$.

The main result says roughly that $x$ *is exponentially attracted to* $K$ *at rate* $\lambda$*, while the expansion rate at* $K$ *of the flow in* $A$ *is* $\mu > \lambda$*, then* $x$ *is eventually asymptotic at rate* $\lambda$ *to a unique trajectory in* $A$*:*

**Theorem 1.1** *Let* $\mathcal{E}(\Phi, K) = \mu$*. Suppose* $x \in In(K)$ *approaches* $K$ *at rate*

$$\mathcal{P}(x, K, F) = \lambda < \min(1, \mu).$$

*Then:*

**(a)** *There exists* $r \geq 0, y \in A$ *such that*

$$\mathcal{R}_{t\to\infty} ||\Phi_{t+r}x - \Phi_t y|| = \lambda.$$

**(b)** *Let* $y$ *be as in (a). Suppose* $l > 0, z \in A$ *are such that*

$$\mathcal{R}_{t\to\infty} ||\Phi_{t+l}x - \Phi_t z|| \leq \lambda.$$

*Then* $z$ *and* $y$ *are on the same orbit of* $\Phi$*.*

This is proved in Section 3 below. The same argument yields the analogous result for mappings.

The proof of the following corollary is left to the reader:

**Corollary 1.2** *If* $\mathcal{P}(x, K, F) = \lambda < \min(1, \mathcal{E}(\Phi, K))$*, then* $x$ *has an eventual asymptotic phase* $y \in K$*. If* $\mathcal{E}(\Phi, K) \geq 1$ *then the* $\Phi$*-trajectory of such a* $y$ *is unique.*

As a simple example illustrating Theorem 1.1, consider a smooth flow in some manifold $A$ having an invariant $n$-torus $K = T^n = (\mathbf{R}/2\pi\mathbf{Z})^n$ in which the flow is quasiperiodic, the generating vector field $G$ in $T^n$ being covered by a constant vector field in $\mathbf{R}^n$. It is clear that $\mathcal{E}(\Phi, T^n) = 1$, using the Riemannian metric covered by the Euclidean metric on $\mathbf{R}^n$. Therefore by Theorem 1.1, any orbit attracted to $T^n$ at a rate of approach less that 1 has an asymptotic phase in $T^n$. It is not hard to show that the same conclusion holds if the flow in $T^n$ is generated by $gG$ where $g$ is any smooth real-valued function on $T^n$. The proof is based on the fact that orbits of the lifted flow in $\mathbf{R}^n$ stay in parallel lines.

**Remark 1.3** Suppose $K$ is a normally hyperbolic submanifold, or a hyperbolic subset, for the flow in $A$ (see [3, 4, 5, 7]). Then any point $x \in A$ attracted to $K$ belongs to the strong stable manifold of some $y \in K$. Therefore $x$ is exponentially asymptotic with $y$.

**Remark 1.4** The main results apply equally to discrete-time systems, i. e. to a mapping $f$ from an open subset $X_0 \subset X$ to $X$. Everything makes sense if $t$ is restricted to the natural numbers, $F_t$ is the $t$'th iterate of $f$, and $\Phi$ is replaced by the iterates of the map $h = f|A \cap X_0$, assumed to be a diffeomorhism from $A_0 = A \cap X_0$ onto a neighborhood of $K$ in $A$. In fact the main part of the proof of the main theorem in Section 3 consists of a proof of the discrete-time case; this is applied to the mapping $f = F_s$ for suitable $s > 0$.

# 2 Reaction Diffusion Systems

Theorem 1.1 is applied to reaction diffusion systems of the following kind. Let $\overline{\Omega} \subset \mathbf{R}^m$ be a smooth (i. e. $C^1$) compact submanifold with interior $\Omega$. We look for a continuous function $u(x,t)$, $x \in \overline{\Omega}$, $t \geq 0$ with values in $\mathbf{R}^n$satisfying for $t > 0$

$$\frac{\partial u}{\partial t} = B\Delta u + \sum_{j=1}^{m} C_j(x,u)\frac{\partial u}{\partial x_j} + f(u), \tag{4}$$

$$\frac{\partial u}{\partial \nu} = 0. \tag{5}$$

Here $\Delta$ is the Laplacean in the spatial variable $x \in \overline{\Omega}$, operating on each component $u_j$ of $u$; $B$ is a positive definite $n \times n$ matrix; each $n \times n$ matrix-valued function $C_j$ is continuous in $(x,u)$; $f$ is a smooth vector field on $\mathbf{R}^n$; $\nu$ is the inward pointing unit vector field normal to the boundary of $\Omega$.

It is known that solutions to this system form a *solution semiflow* $S = \{S\}_{t \geq 0}$ in the Sobelev space $H^1(\overline{\Omega}, \mathbf{R}^n)$: The solution taking initial values $u(x,0) = v(x)$ is $u(x,t) = (S_t v)(x)$.

Let $A \subset H^1(\overline{\Omega}, \mathbf{R}^n)$ denote the linear subspace of constant maps $\overline{\Omega} \to \mathbf{R}^n$, and identify $A$ with $\mathbf{R}^n$ in the natural way. The form of Equation (4) shows $A$ is positively invariant under $S$.

A trajectory of $S$ in $A$ defines a spatially homogeneous solution to Equations (4), (5). Such a solution has the form $u(x,t) = y(t)$ where $y$ is a solution to the autonomous system $dy/dt = f(y)$.

The restriction to $A$ of the solution flow $S$ of (4), (5) coincides for $t \geq 0$ with the flow $\Phi$ obtained by integrating the vector field $f$.

Suppose from now on that $\Gamma \subset \mathbf{R}^n$ is a compact invariant rectangle[1] (the product of $n$ nondegenerate compact intervals.) We identify $\Gamma$ with a compact subset of $A$, namely the constant functions with values in $\Gamma$. *Invariance* means that if the initial

[1] More generally, $\Gamma$ can be an invariant region as defined in Conway, Hoff and Smoller [2].

map $v : \overline{\Omega} \to \mathbf{R}^n$ takes values in $\Gamma$ then the same holds for every map $S_t v$. When $B$ is a diagonal matrix, invariance holds provided that for every $y$ on the boundary of $\Gamma$, the vector $f(y)$ does not point out of $\Gamma$.

In [2] a condition is given ensuring that $\Gamma$ attracts every initial $v \in H^1(\overline{\Omega}, \mathbf{R}^n)$ taking values in $\Gamma$, or in other words, that the set $X = H^1(\overline{\Omega}, \Gamma)$ lies in the inset of $\Gamma$. This condition is given in terms of the real parameter

$$\sigma = b\Lambda - M - c\sqrt{m\lambda} \tag{6}$$

defined in terms of the following constants: The positive number $b$ is the smallest eigenvalue of the positive definite matrix $B$; $\Lambda$ (also positive) is the smallest eigenvalue of $-\Delta$ on $\Omega$ with homogeneous Neumann boundary conditions (5); $c$ is the maximum matrix operator norm $||C_j(x, y)||$, $(1 \le j \le m,\ x \in \overline{\Omega}, y \in \Gamma)$; and as before, $M = \max_{y \in \Gamma} ||Df(y)||$.

It will also be convenient to consider the slightly different parameter:

$$\sigma_2 = \sigma - M = b\Lambda - 2M - c\sqrt{m\lambda} \tag{7}$$

For each $v \in X$ set $v_t = S_t v$, and denote by $\overline{v}_t \in \mathbf{R}$ the average of $v_t$ over $\overline{\Omega}$. Notice that $\overline{v}_t$ is a curve in $X$, but it need not be a trajectory of the flow $S$, that is, $\overline{v}_t(x)$ need not be a solution to Equations (4, 5).

Let $||\cdot||_\infty$ denote the $L_\infty(\overline{\Omega}, \mathbf{R}^n)$ norm.

The following result is a corollary of Theorem 3.1 of [2][2]

**Theorem 2.1** (Conway, Hoff, Smoller [2])
*Assume $\sigma > 0$ and let $v \in X = H^1(\overline{\Omega}, \Gamma)$. Then:*

**(a)** *There is a constant $c_1 > 0$ such that $||v_t - \overline{v_t}||_1 \le c_1 e^{-\sigma t}$ for all $t \ge 0$.*

**(b)** *If the matrices $C_1, \ldots, C_n$ are zero, or if $C_1, \ldots, C_n$ and $B$ are diagonal, then there is a constant $c_2 > 0$ such that $||v_t - \overline{v_t}||_\infty \le c_2 e^{-\frac{2\sigma}{m}t}$ for all $t \ge 0$.*

This says that when $\sigma$ is positive, in the appropriate norm trajectories of the reaction-diffusion system approach spatially homogeneous functions. In fact in [2] it is proved that the spatial averages $\overline{v_t}$ satisfy a nonautonomous system $d\overline{v_t}/dt = f(\overline{v_t}) + g(t)$ with $||g(t)||_1 \le c_3 e^{-\sigma t}$ for some constant $c_3 \ge 0$. Conway, Hoff and Smoller say that "because of a result of Markus [6] it follows that the *asymptotic* behavior of $\overline{v_t}$ is determined only by $f$".

In the terminology of Section 1 we have:

---

[2]The statements of Theorem 2.1 are proved but not stated in this form. The exponent in (b) is given as $-\frac{\sigma}{m}t$, I think incorrectly.

**Corollary 2.2** *Under the same hypothesis as Theorem 2.1:*

**(a)** $\mathcal{R}_{t\to\infty}(||v_t - \overline{v_t}||_1) \le e^{-\sigma}$.

**(b)** *If the matrices $C_1, \ldots, C_n$ are zero, or if $C_1, \ldots, C_n$ and $B$ are diagonal, then also* $\mathcal{R}_{t\to\infty}(||v_t - \overline{v_t}||_\infty) \le e^{-\frac{2\sigma}{m}}$.

While the Conway-Hoff-Smoller theorem provides much information about such systems, it leaves open the question of whether trajectories have an asymptotic phase in $A$. The following result gives a sufficient condition for this.

Let $\mu = \mathcal{E}(\Phi, \Gamma)$, the expansion rate in $\Gamma$ of the flow in $A = \mathbf{R}^n$ defined by $dy/dt = f(y)$.

**Theorem 2.3** *Assume $\sigma > 0$ and $e^{-\sigma} < \mu$. Let $v \in H^1(\overline{\Omega}, \mathbf{R}^n)$ take values in the invariant rectangle $\Gamma \subset \mathbf{R}^n$. Then the trajectory $S_t v$ in $In(\Gamma)$ of the solution flow in $H^1(\overline{\Omega}, \mathbf{R}^n)$ of the reaction-diffusion system (4),(5) has an an eventual asymptotic phase in the space $A$ of constant maps. More precisely, if $S_t v(x) = u(x,t)$ then for every sufficiently large $s \ge 0$ there is a unique solution to $dy/dt = f(y)$ such that:*

**(a)** $\mathcal{R}_{t\to\infty}(||u(\cdot, t+s) - y(t)||_1) \le e^{-\sigma}$.

*Moreover, if the matrices $C_1, \ldots, C_n$ are zero, or if $C_1, \ldots, C_n$ and $B$ are diagonal, then:*

**(b)** $\mathcal{R}_{t\to\infty}(||u(\cdot, t) - y(t)||_\infty) \le e^{-\frac{2\sigma}{m}}$.

**Corollary 2.4** *If $\sigma_2 > 0$ then the conclusions of Theorem 2.3 hold.*

**Proof** Corollary 2.2(a) implies $v$ has rate of approach $\le e^{-\sigma}$ to $\Gamma$. Therefore Theorem 2.3 follows from Theorem 1.1 (with $K = \Gamma$) and the assumption $e^{-\sigma} < \mu$.

To prove Corollary 2.4, assume $\sigma_2 > 0$. Then $\sigma > 0$ and $e^{-\sigma} < e^{-M}$ (see (7)). Since estimate (2) therefore implies $e^{-M} \le \mathcal{E}(\Phi, \Gamma)$, the corollary is a consequence of Theorem 2.3. **QED**

# 3 Shadowing

The main theorem will be derived from the results of this section. The same notations and assumptions as in Section 1 are in force, although at first the setting is quite general.

Let $X_0 \subset X$ be any subset and let $g : X_0 \to X$ be a map ($g$ = some $F_t$ in the application). Let $0 \le \lambda < 1$. I call a sequence $\{y_k\}$ in $K$ a *$\lambda$-pseudoorbit* for $g$ if

$$\mathcal{R}_{k\to\infty}||g(y_{k-1}) - y_k|| \le \lambda.$$

**Lemma 3.1** *Suppose $g$ is $\alpha$-Hölder, $0 < \alpha \leq 1$. Let $\{y_k\}$ be a sequence in $X$ which is $\lambda$-shadowed by a point $u \in X_0$. Then $\{y_k\}$ is a $\lambda^\alpha$-pseudoorbit for $h$. In particular if $g$ is Lipschitz then $\{y_k\}$ is a $\lambda$-pseudoorbit.*

**Proof** Fix $C > 0$ such that $||g(a) - g(b)|| \leq C||a - b||^\alpha$. Observe that

$$\begin{aligned} ||g(y_{m+k-1}) - y_{m+k}|| &\leq ||g(y_{m+k-1}) - g^k u|| + ||g^k u - y_{m+k}|| \\ &\leq C||y_{m+k-1} - g^{k-1}u||^\alpha + ||g^k u - y_{m+k}||. \end{aligned}$$

Therefore (see Section 1)

$$\begin{aligned} \mathcal{R}_{k\to\infty} C||g(y_{k-1}) - y_k|| &\leq \max(\mathcal{R}_{k\to\infty}||y_{m+k-1} - g^{k-1}u||^\alpha, \mathcal{R}_{k\to\infty}||g^k u - y_{m+k}||) \\ &\leq \max(\lambda^\alpha, \lambda) = \lambda^\alpha. \end{aligned}$$

**QED**

Now set $A_0 = A \cap X_0$, assume $g(A_0) \subset A$ and $g(K) \subset K$. Set $g|A_0 = h$ and assume from now on that $h$ is a $C^1$ diffeomorphism of $A_0$ onto some neighborhood of $K$ in $A$.

A point $u \in A_0$ (or its orbit) is said to $\lambda$-*shadow* the sequence $\{y_k\}$ in case $h^k(u)$ is defined for all $k \in \mathbf{N}$, and:

$$\mathcal{R}_{k\to\infty}||h^k(u) - y_{k+m}|| \leq \lambda$$

for some $m \geq 0$.

**Theorem 3.2** *Assume the expansion rate of $h$ in $K$ is $EC(h, K) = \mu > 0$. Let $\{y_k\}$ be a $\lambda$-pseudoorbit in $K$ such that*

$$0 < \lambda < \min(1, \mu).$$

*Then:*

**(a)** *There exists $z \in A_0$ which $\lambda$-shadows $\{y_k\}$.*

**(b)** *If $z, w \in A_0$ both $\lambda$-shadow $\{y_k\}$ then there exist natural numbers $l, r$ such that $h^l z = h^r w$.*

**Remark 3.3** The proof shows that $z$ in the theorem can be chosen in $K$ if $K$ is a smooth compact submanifold without boundary, or if $K$ is an attractor for $h$, or if the pseudoorbit $\{y_k\}$ is eventually bounded away from the boundary of $K$ in $A$. In any case the forward orbit of $z$ is attracted to $K$ and its omega limit set is in $K$.

**Remark 3.4** The theorem is valid under the more general hypothesis where $\mu$ denotes $\sup_{k>0} EC(h^k, K)^{\frac{1}{k}}$.

**Proof** Fix $\rho_* > 0$ so small that if $0 \leq \rho \leq \rho_*$ then

$$hB(\rho, x) \supset B(\mu\rho, h(x)) \tag{8}$$

for all $x \in K$, where $B$ refers to closed balls in $A$. Then this also holds for all $x$ in some neighborhood $N \subset A_0$ of $K$, since $K$ is compact.

Choose $\nu$ such that

$$0 < \lambda < \nu < \min(1, \mu).$$

Pick $\delta$ such that

$$\nu < \delta < \min(1, \mu).$$

I claim that for all sufficiently large positive integers $k$ we have:

$$hB(\delta^{k-1}, y_{k-1}) \supset B(\delta^k, y_k). \tag{9}$$

To see this observe that $\delta^j < \rho$ and $B(\delta^{k-1}, y_{k-1}) \subset N$ for large $j$. Therefore by (8) it suffices to prove for sufficiently large $k$ that

$$B(\mu\delta^{k-1}, h(y_{k-1})) \supset B(\delta^k, y_k). \tag{10}$$

And this last will hold by the triangle inequality provided we show

$$\mu\delta^{k-1} \geq \delta^k + ||h(y_{k-1}) - y_k||. \tag{11}$$

Because $\{y_k\}$ is a $\lambda$-pseudoorbit, for large $k$ we have

$$||h(y_{k-1}) - y_k|| < \nu^k. \tag{12}$$

Therefore it suffices to show

$$\mu\delta^{k-1} \geq \delta^k + \nu^k \tag{13}$$

or equivalently

$$\mu \geq \delta + (\frac{\nu}{\delta})^{k-1}\nu \tag{14}$$

for sufficiently large $k$. This is true, say for $k \geq m$, because $\mu > \delta > \nu$.

Therefore estimate (9) holds for $k \geq m$. This implies that for $n \geq m$ the set

$$Q_n = \cap_{i \geq 0}(h|B(\delta^n, y_n))^{-i} B(\delta^{i+n}, y_{i+n})$$

is not empty, and the orbit of any point in $Q_m$ $\lambda$-shadows $\{y_k\}$. This proves statement (a) of the theorem.

From the assumption $EC(h, K) > \lambda$ it follows easily that $Q_n$ is a singleton for every $n \geq m$. This implies (b). **QED**

**Proof of Theorem 1.1** With the notation and assumptions of Theorem 1.1, fix $r > 0$ so that

$$EC(\Phi_r, K) = \mu_0 > \lambda.$$

Set $h = \Phi_r : A_0 \to A$ where $A_0$ denotes the domain of $\Phi_r$— a neighborhood of $K$ in $A$. For $k \in \mathbf{N}$ let $y_k \in K$ be a point nearest to $h^k(x)$. It then follows from Lemma 3.1(a) with $u = x$ and $g = F_r$, and the standing assumption that each $F_t$ is Lipschitz, that $\{y_k\}$ is a $\lambda$-pseudoorbit for $h$. By Theorem 3.2 $\{y_k\}$ is $\lambda$-shadowed by the orbit of some $z \in A_0$. It follows that for some $m \geq 0$ we have:

$$\mathcal{R}_{k\to\infty}||\Phi_{k+m}x - \Phi_k z|| = \lambda \ (k \in \mathbf{N}).$$

Continuity of the flow now implies:

$$\mathcal{R}_{t\to\infty}||\Phi_{t+m}x - \Phi_t z|| = \lambda \ (t \in \mathbf{R}).$$

This proves part (a) of Theorem 1.1.

Part (b) follows similarly from part (b) of Theorem 3.2. **QED**

**Remark 3.5** The connection between asymptotic phase and shadowing is more extensive. For simplicity consider a diffeomorphism $h$. Suppose the orbit of some point $x$ is attracted to a compact invariant set $K$, not necessarily at an exponential rate. By choosing $y_k \in K$ to be a point nearest to $h^k(x)$ we obtain a sequence $\{y_k\}$ in K with the property that $||h(y_{k-1}) - y_k|| \to 0$. If $h|K$ has the property of *unique shadowing*, described below, then it is easy to see that $\{y_k\}$ is asymptotic to the orbit of a unique point $z \in K$. Such a $z$ would therefore be an asymptotic phase for $x$.

To say the map $h|K$ has unique shadowing means the following. For $\delta > 0$, $\{y_k\}$ is an *$\delta$-pseudoorbit* in case $||h(y_{k-1}) - y_k|| < \delta$. "Unique shadowing" means that for every $\epsilon > 0$ there exists $\delta > 0$ such that for every $\delta$-pseudoorbit $\{y_k\}$ there is a unique $z \in K$ such that $||y_k - h^k(z)|| < \epsilon$, or in other words $\{y_k\}$ is *$\epsilon$-shadowed* by $z$.

R. Bowen [1] showed that if $K$ is a hyperbolic invariant set, then $h|K$ has unique shadowing. Suppose for example that $V$ is a compact smooth invariant submanifold of $A$ and that $h|V$ is an Axiom A diffeomorphism in the sense of Smale [7]. If $x \in A$ is attracted to $V$ then it is easy to see that in fact $x$ is attracted to what Smale calls a basic set $K$ for $h|V$, which is by definition a hyperbolic invariant set. Therefore Bowen's theorem implies that $x$ has an asymptotic phase in $K$, hence also in $V$.

# References


[1] R. Bowen, *On axiom A diffeomorphisms*, CBMS Regional Conference Series in Mathematics, vol. 35, American Mathematical Society Publications, Providence, Rhode Island, 1978.

[2] E. Conway, D. Hoff, and J. Smoller, *Large time behavior of solutions of systems of nonlinear reaction-diffusion equations*, SIAM Journal of Applied Mathematics **35** (1978), 1–16.

[3] M.W. Hirsch, J. Palis, C.C. Pugh, and M. Shub, *On invariant subsets of hyperbolic sets*, Inventiones Mathematicae **9** (1970), 121–134.

[4] M.W. Hirsch and C.C. Pugh, *Stable manifolds for hyperbolic sets*, Global Analysis, Proceedings of Symposia in Pure Mathematics, No. 14, American Mathematical Society, Providence, Rhode Island, 1970, pp. 133–163.

[5] M.W. Hirsch, C.C. Pugh, and M. Shub, *Invariant manifolds*, Springer Lecture Notes in Mathematics, vol. 583, Springer-Verlag, New York, Heidelberg, Berlin, 1977.

[6] L. Markus, *Asymptotically autonomous differential systems*, Contributions to the Theory of Nonlinear Oscillations, Vol. III (S. Lefschetz, ed.), Princeton University Press, 1956, Annals of Mathematics Study No. 36, pp. 17–30.

[7] Smale. S., *Differentiable dynamical systems*, Bulletin American Mathematical Society **73** (1967), 747–817.

# 8 Numerical Methods for Studying Parameter Dependence of Solutions to Schrödinger's Equation

**M. Holthaus*** University of California at Santa Barbara, Santa Barbara, California

**C. S. Kenney** University of California at Santa Barbara, Santa Barbara, California

**A. J. Laub**† University of California at Santa Barbara, Santa Barbara, California

**Abstract**

Numerical methods are presented for handling two of the most important computational problems associated with the study of parameter dependence of solutions to the periodically time dependent Schrödinger equation as approximated by a finite system of ordinary differential equations (ODEs): 1) lengthy integration times due to the growth in parameters as the number of states increases and 2) repeated integrations over a grid in parameter space. The new methods are independent of the numerical algorithm used to integrate the ODE system. A first method uses a simple diagonal transformation of the solution to control the magnitude of terms appearing in the differential system, and a second method relies on matrix interpolation together with a Taylor series expansion to reduce the number of integrations in parameter space. The combination of these two procedures results in a dramatic decrease in computation time for an example problem of an electron confined to a quantum well.

*Keywords*: Schrödinger's Equation, Parameter Study, Matrix Interpolation
*AMS(MOS) subject classification*: 65L05, 65F30.
*Abbreviated title*: Integrating Schrödinger's Equation

*The research of this author was supported by a Feodor-Lynen Research grant from the Alexander von Humboldt - Stiftung.

†The research of these authors was supported by the National Science Foundation under Grant No. ECS-9120643 and the Air Force Office of Scientific Research under Contract No. AFOSR-91-0240.

# 1 Introduction

In the theoretical study of periodically forced quantum systems, such as atoms or molecules under the influence of strong monochromatic laser radiation, physicists often have to deal with an $n \times n$ matrix Schrödinger equation of the form

$$i\frac{d}{dt}\psi(t) = (D + \lambda(t)\sin(\omega t)H)\psi(t) \tag{1}$$

where $D$ is the Hamiltonian matrix of the unperturbed system and $H$ describes its interaction with an external periodic force (the laser field) of frequency $\omega$; the smooth function $\lambda(t)$ models the envelope of the forcing pulse. In order to investigate the variation of the solution $\psi$ at some terminal time $t_p$ with respect to changes in the system parameters ($\omega$, $\lambda$, etc.), it is a common practice to integrate (1) repeatedly over a grid in parameter space. Since each integration may itself be computationally intensive, this approach can lead to weeks or months of computer time or to a retreat to an unrealistic problem model in which, for example, the number of states is kept very low. In this paper, we describe two simple methods that we have found to be very useful in: 1) reducing the effort needed for each integration, and 2) reducing the number of integrations needed to get a good picture of the effects of parameter variation. For example, in the problem described below, we initially estimated that 25 years of computer time would be needed for a particular parameter study; this was reduced to 3.5 days by applying the methods described in this paper — a reduction by a factor of 2600.

## 1.1 Physical motivation

To motivate the methods we have developed it is useful to take a detailed look at a particular example. Consider an electron in an infinite quantum well [12] of width $2a$. If we work within a basis of "particle-in-the-box" states, $D$ is a diagonal matrix with diagonal elements

$$D_{kk} = \frac{1}{8}\left(\frac{\pi k}{a}\right)^2 , \quad k = 1,2,3,\ldots \tag{2}$$

and $H$ is given by

$$H_{jk} = \begin{cases} \frac{4a}{\pi^2}\left(\frac{1}{(j+k)^2} - \frac{1}{(j-k)^2}\right) & \text{for } j \neq k \bmod (2) \\ 0 & \text{for } j = k \bmod (2) \end{cases} . \tag{3}$$

(We use a system of dimensionless units such that Planck's quantum $\hbar$ is equal to 1.)

The elements $D_{kk}$ are the eigenenergies of the unperturbed states in the quantum well. Since they grow quadratically with the quantum number $k$, the energy differences $D_{kk} - D_{k-1,k-1}$ will eventually become large compared to the photon energy $\omega$. Then the external force can no longer induce transitions between neighboring states, and, therefore, it is sufficient to consider only a finite number of states: $k = 1, 2, ..., n$, where $n$ has to satisfy

$$D_{nn} - D_{n-1,n-1} \gg \omega \quad . \tag{4}$$

Despite its apparent simplicity, this model exhibits surprisingly rich dynamics. As an example, we fix the parameters $a = 1.0$, $\omega = 3.5$, and $n = 10$, and choose an envelope function

$$\lambda(t) = \lambda_{max} \sin^4 \left( \frac{\pi t}{t_p} \right) \quad , \quad 0 \leq t \leq t_p \tag{5}$$

with a pulse length of 100 cycles: $t_p = 200\pi/\omega$. We assume that at $t = 0$ the ground state is populated, i.e., we have the initial condition $\psi(0) = e_1$ ($e_k$ denotes the $k^{th}$ standard unit vector consisting of all zeros except for a 1 in the $k^{\text{th}}$ position), and solve the Schrödinger equation (1) numerically to obtain the wave function $\psi(t_p)$ after the pulse. To get an idea of how our system responds to pulses with different strengths, we plot the resulting occupation probabilities of the unpertubed states as a function of the peak field strength $\lambda_{max}$.

Figure 1 (see also [6]) shows the occupation probabilities of the ground state $k = 1$ and of the second excited state $k = 3$. Interestingly, the latter state does not seem to play any role up to $\lambda_{max} \approx 13$ when, all of a sudden, a rapid oscillation starts. The excitation energy $D_{3,3} - D_{1,1}$ is close to $2.8\,\omega$, which means that we have an example of a "multiphoton resonance."

There is a very simple explanation for this striking behavior. If one forgets the time dependence of the amplitude and instead keeps $\lambda$ fixed, the Schrödinger equation (1) has a strictly periodic time dependence, and, for each value of $\lambda$, we can calculate the corresponding Floquet multipliers by solving the matrix initial value problem

$$\frac{d}{dt} U(t) = -i(D + f(t)H)U(t) \quad , \quad U(0) = I \quad , \tag{6}$$

with $f(t) = \lambda \sin(\omega t)$ and computing the one-cycle-evolution operator $U(2\pi/\omega)$. (Because of the particular symmetries of quantum well problem, it is actually sufficient to integrate only over a quarter of the period in order to determine $U(2\pi/\omega)$. However, we will omit a discussion of this point since our goal is to develop methods which apply to general systems without this feature.) The diagonalization of $U(2\pi/\omega)$ then yields the Floquet multipliers $\exp(-i\varepsilon_k 2\pi/\omega)$; the characteristic exponents $\varepsilon_k$ are called "quasienergies" in the physical literature [17]. A plot of these quasienergies versus the driving strength $\lambda$ contains valuable information. To come back to our example, Figure 2 shows such a plot of the quasienergies for the five lowest states of the quantum well problem. The interesting feature is an "avoided level crossing" of $\varepsilon_1$ and $\varepsilon_3$ at $\lambda \approx 13$, precisely where the oscillations in Fig. 1 appear. In fact, this avoided crossing is the key to understand the oscillations. Associated with each quasienergy $\varepsilon_k$ there is a quantum Floquet state $u_k(t)$. The adiabatic principle of quantum mechanics (see, e.g., [13]) guarantees that under the influence of a pulse with a smooth envelope $\lambda(t)$, such as (5), the initial state $e_1$ is first "adiabatically shifted" into the "connected" Floquet state $u_1(t)$. But when, during the rise of the pulse, $\lambda(t)$ reaches the avoided crossing, adiabaticity breaks down and the wave function is split into a superposition of the two Floquet states $u_1(t)$ and $u_3(t)$. In the second half of the pulse, $\lambda(t)$ decreases and reaches the avoided crossing a second time. Then the two parts of the wave function interfere, and it is this quantum mechanical interference phenomenon that produces the oscillations in the occupation probabilities of the final states (see [6] for details). Oscillations of this type are not a specific property of the quantum well model, but a generic feature of pulsed quantum systems. They have recently been observed in experiments with highly excited atoms that were exposed to pulses of microwave radiation [2], [16]. It has also been argued that related effects could possibly be observed in real quantum wells made of compound semiconductors such as $Al_xGa_{1-x}As$ [5].

To sum up, a strategy to understand the dynamics of a quantum system interacting with pulses of a periodic force, where the smooth envelope $\lambda(t)$ varies between zero and some value $\lambda_{max}$, is to compute the characteristic exponents $\varepsilon_k$ in the range $0 \leq \lambda \leq \lambda_{max}$. A plot of these quasienergies versus $\lambda$ immediately allows the identification of multiphoton resonances.

## 1.2 Computational considerations

From the computational point of view, the main task then is to integrate the system (6) to determine the matrices $U(2\pi/\omega)$. In our model example with $\omega$ fixed, we have used as many as 300 equally spaced values of $\lambda$. If we were also interested in the effect of frequency variations, and if we allowed $\omega$ to vary on the same scale, we would need to integrate (6) 90,000 times! Clearly some other approach is called for if we want to explore parameter variation in any detail. This type of problem can be handled very effectively by matrix interpolation and, when combined with a Taylor expansion about $\lambda = 0$, yields an efficient method of studying the parameter dependence of the solutions to the Schrödinger equation (see Section 3).

Interpolation is helpful in reducing the number of integrations, but even one integration can be computationally demanding if we must use the form of the differential equation given by (6). For example, the quantum well problem becomes very interesting when the frequency $\omega$ is considerably larger than $D_{2,2} - D_{1,1}$. In this case, a large number of states is relevant, and the quantum dynamics approaches the semiclassical limit. On the other hand, the classical counterpart of the model exhibits chaotic dynamics [11]. Thus, for large $\omega$ the quantum well model allows us to study the influence of classical chaos on quantum mechanics [12], [7]. But then the criterion (4) implies that a basis with a large number $n$ of states must be used. However, this leads to an increase in the norm of the $D$ matrix since the $n^{th}$ diagonal element of $D$ is $\frac{1}{8}\left(\frac{\pi n}{a}\right)^2$, which grows quadratically with $n$.

This type of growth translates into a reduction in the allowable time step size for most numerical integration routines such as Runge-Kutta or Adams methods (see [1] or [15]). Since the step size reduction is usually inversely proportional to the norm of $D$, the number of time steps can grow as $n^2$ for a system like (6). Combining this with the order $n^3$ arithmetic operations needed per time step (because of the matrix multiplications used in evaluating the derivative) leads to a total computational effort on the order of $n^5$. This is unacceptable even for modest values of $n$, as illustrated in Table 1 of Section 2.

In Section 2, we address this problem by working with the transformation

$$W(t) = e^{itD}U(t) . \tag{7}$$

This gives the system

$$\frac{d}{dt}W(t) = -iA(t)W(t) \quad , \quad W(0) = I \tag{8}$$

where the entries of $A(t)$ have the same modulus as the corresponding entries of $f(t)H$. The effect of this transformation is to eliminate the large norm problems of (6) and the associated lengthy integration times. A related transformation, which is very simple to program but not as efficient, is also discussed in Section 2.

**Remark:** The transformation (7) is often used in quantum mechanical time dependent perturbation theory, where it is referred to as the "transformation to the interaction picture" (see [3]).

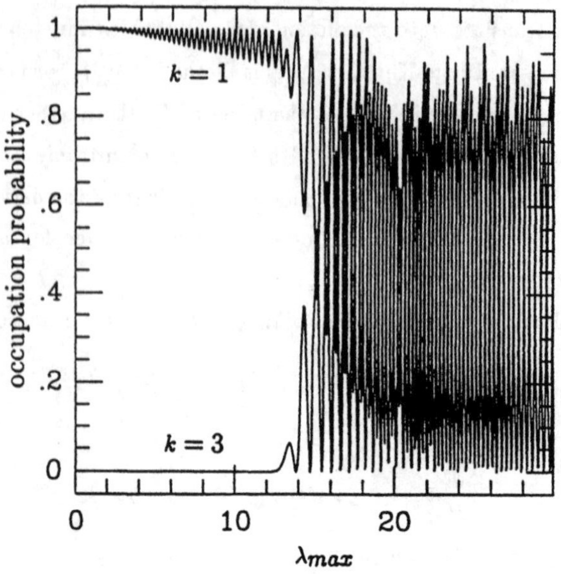


**Fig. 1** Occupation probabilities of quantum well states $k = 1$ and $k = 3$, after the system has interacted with pulses with the envelope (5).

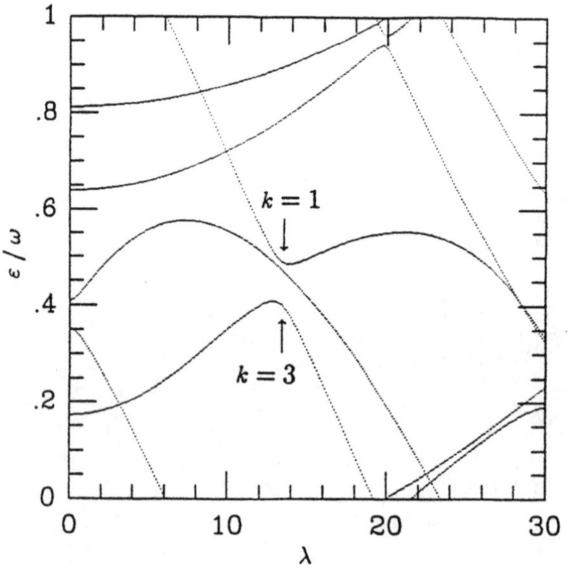


**Fig. 2** Characteristic exponents (quasienergies $\varepsilon_k$ in units of $\omega$) for this situation. Note the avoided crossing of $\varepsilon_1$ and $\varepsilon_3$ (indicated by the arrows).

## 2 Transformation to an Equivalent System

In the Introduction, we noted that the large entries of the diagonal matrix $D$ (see (2)) place restrictions on the size of time steps used by numerical integration routines. On the other hand, the elements of the matrix $H$ in (3) are bounded above by $\frac{4a}{\pi^2}$ independently of $n$. This suggests a possible remedy: get rid of $D$.

To do this write (6) as

$$\frac{dU}{dt} + iDU = -ifHU. \tag{9}$$

Now pre-multiply by $e^{itD}$ on both sides to get

$$\frac{d}{dt}\left(e^{itD}U\right) = -ife^{itD}HU. \tag{10}$$

Using the substitution $W(t) \equiv e^{itD}U(t)$ gives

$$\frac{d}{dt}W = -ife^{itD}He^{-itD}W\ , \quad W(0) = I. \tag{11}$$

Since $D$ is diagonal so is $e^{itD}$. In fact, $e^{itD}$ is unitary with $k^{\text{th}}$ diagonal entry equal to $\cos d_k t + i \sin d_k t$ where $d_k = \frac{1}{8}\left(\frac{\pi k}{a}\right)^2$ as per (2). Thus the transformation $U \mapsto W$ is well-conditioned and numerically inexpensive (i.e., requiring only $O(n^2)$ arithmetic operations). Moreover, the matrix $e^{itD}He^{-itD}$ has entries $H_{jk}e^{it(d_j-d_k)}$ where $H_{jk}$ is given by (3). Thus the entries of $e^{itD}He^{-itD}$ are bounded in norm by $\frac{4a}{\pi^2}$ independently of $n$.

In terms of numerical effort, evaluation of the derivative of $W$ in (11) requires one full matrix multiplication and two diagonal matrix scalings; in total this is about the same effort needed to evaluate the derivative of $U$ in (6). The difference, however, between (6) and (11) is that numerical integration routines can take much longer time steps for (11) than for (6).

More general problems, in which we only assume that $D$ and $H$ are symmetric, can also be treated by this method. First write $D$ as

$$D = V^T D_1 V \tag{12}$$

where $V$ is orthogonal and $D_1$ is diagonal. Then define $U_1(t) = VU(t)V^T$ and $H_1 = VHV^T$ to get

$$\frac{d}{dt}U_1 = -i(D_1 + f(t)H_1)U_1\ , \quad U(0) = I\ . \tag{13}$$

However, the main diagonal entries of $H_1$ may not be zero (e.g., this occurs in the triangular quantum well problem described in [4]) so we can improve on $e^{itD_1}$ as a transformation matrix.

Suppose that $H_1$ is split into its diagonal part $H_d$ and its off-diagonal part $H_o$:

$$H_1 = H_d + H_o. \tag{14}$$

If $H_o = 0$ then the solution to (13) is given by

$$B(t) = e^{-i\int_0^t (D_1 + f(s)H_d)\,ds}. \tag{15}$$

Even if $H_o$ is not 0, $B(t)$ defined by (15) is unitary, so that $B^{-1}(t) = B^H(t)$ where the superscript $H$ denotes the conjugate transpose. Moreover, $B$ is diagonal with

$$B_{kk}(t) = e^{-i\int_0^t (d_k + f(s)h_k)\,ds} \tag{16}$$

where $d_k$ and $h_k$ are the $k^{\text{th}}$ diagonal entries of $D_1$ and $H_d$, respectively. We are assuming that the function $f$ is integrable; in most practical problems $f$ is simple enough to integrate explicitly.

Since $B(t)$ approximates $U_1(t)$, modulo $H_o$, it is natural to try a transformation of the form

$$\begin{aligned} W(t) &= B^{-1}(t)U_1(t) \\ &= B^H(t)U_1(t). \end{aligned} \tag{17}$$

Using (13) we see that $W$ satisfies

$$\frac{d}{dt}W(t) = -if(t)B^H(t)H_oB(t)W(t). \tag{18}$$

Since $B(t)$ is diagonal and unitary, the matrix $B^H(t)H_oB(t)$ has entries whose absolute values are equal to the corresponding entries of $H_o$.

Both (11) and (18) were inspired by a similar transformation. This method uses the vector form of the Schrödinger equation. Let $e_k$ denote the $k^{\text{th}}$ standard unit vector and let $u_k$ be the vector solution to

$$\frac{d}{dt}u_k(t) = -i(D + f(t)H)u_k(t)\ , \quad u_k(0) = e_k. \tag{19}$$

Then $u_k(t)$ is simply the $k^{\text{th}}$ column vector of the matrix solution $U(t)$ to (6). If $H$ were equal to zero then $u_k(t)$ would be equal to the unit vector $e_k$ times the scalar function $e^{-itd_k}$ where we assume that $D$ is diagonal with $k^{\text{th}}$ entry equal to $d_k$. This suggests the transformation

$$w_k(t) = e^{itd_k}u_k(t)\ , \tag{20}$$

which has the effect of eliminating the $k^{\text{th}}$ diagonal entry of $D$. The effect of the unchanged diagonal entries of $D$ is partially offset by the initial condition on $u_k$; every entry is zero except the $k^{\text{th}}$ entry. However, as the solution evolves in time the other entries of $u_k$ grow in magnitude and the large uncanceled entries of $D$ can start to limit the allowable time step size. Thus the programming simplicity of this approach is attained at some cost in computational efficiency. This is illustrated in Table 1 for (2) with $a = 1/\sqrt{2}$, $\omega = 0.1$, and two values of field strength $\lambda = 0$ and $\lambda = 0.1$. Table 1 reports the total time required for both integrations from $t = 0$ to $t_p = 2\pi/\omega$ as well as the time needed to diagonalize the final unitary matrix $U(2\pi/\omega)$. (Such diagonalizations are common in many applications and do not affect the execution time significantly for this problem; however, they are included for comparison with the next section in which they play a large role.) In Table 1, the column labeled "no transformation" refers to integration of (6), "partial transformation" refers to integration of (19) after applying (20), and "complete transformation" refers to the integration of (11). All numerical integrations were done using the Adams routine DE of Shampine and Gordon [15].

| Number of States | No Trans. | Partial Trans. | Complete Trans. |
|---|---|---|---|
| 5 | $8.1 \times 10^1$ | $1.3 \times 10^1$ | $1.8 \times 10^1$ |
| 10 | $1.3 \times 10^3$ | $2.5 \times 10^2$ | $2.3 \times 10^2$ |
| 15 | $6.7 \times 10^3$ | $1.7 \times 10^3$ | $9.4 \times 10^2$ |
| 20 | $2.4 \times 10^4$ | $6.2 \times 10^3$ | $2.6 \times 10^3$ |
| 25 | $6.3 \times 10^4$ | $1.8 \times 10^4$ | $5.6 \times 10^3$ |
| 30 | $1.4 \times 10^5$ | $4.0 \times 10^4$ | $1.0 \times 10^4$ |

**Table 1 Execution Time in Seconds for Three Integration Methods.**

For this problem, the execution time for the integration of the completely transformed system grows like $O(n^{3.4})$ while both the partial and untransformed systems have about $O(n^{4.5})$ growth. This shows that while (11) has a big computational edge, there is room for improvement.

One way in which such improvement might be effected can be found in the recent work of Chu [8] on simultaneous orthogonal reduction of matrices to desired forms. For example, in our problem it would be nice to have an orthogonal matrix $Q$ such that $D_1 = QDQ^T$ is block diagonal with, say, 2 by 2 blocks or even 3 by 3 blocks, while $H_1 = QHQ^T$ has a sparse structure such as bandedness with a diagonal bandwidth $p << n$. The block structure of $D_1$ would still allow the efficient computation of $e^{itD_1}$ which inherits the same block structure. Thus a transformed system like (11) could still be used, and the sparse structure of $H_1$ would reduce the cost of the derivative evaluation to about $pn^2 \ll n^3$ arithmetic operations. The desired orthogonal matrix $Q$ is found by integrating a matrix ODE of the form

$$\frac{d}{dt}Q = -g(Q)$$

until an equilibrium point ($dQ/dt = 0$) is reached. The cost of this integration would presumably be offset by the savings in integration time for the transformed Schrödinger equation, especially if many such integrations have to be performed for different parameter values. We are currently investigating this approach.

## 3 Matrix Interpolation

The function $f$ in (6) usually depends on physical parameters such as the frequency $\omega$ and field strength $\lambda$ of the sinusoidal field in the quantum well problem.

Because of the periodic time dependence of this problem, it is the eigenvalues of the terminal matrix $U(2\pi/\omega)$ that are of interest. The dependence of these eigenvalues on the parameters is usually studied by plotting the eigenphases mod $2\pi$ (or, in physical terms, the quasienergies mod $\omega$) versus $\lambda$ for fixed $\omega$ or more ambitiously by plotting the eigenphase surfaces as functions of both $\lambda$ and $\omega$.

The most straightforward approach to generating such plots is simply to integrate (6) or its transformed version (11) for many values of $\lambda$ (and $\omega$ if surface plots are desired). The phase plots get smoother as the number of integrations goes up but unfortunately we soon encounter limitations in terms of computation time. This can be seen easily from Table 1 of Section 2 in which only two values of $\lambda$ were used and the number of states is rather small (in some cases, especially in studies of quantum dynamics in the semiclassical limit, we would like to use several hundred states).

This leads us to consider using interpolation based on the data generated from integrating the system (11) for only a few values of $\lambda$ (assuming for now that $\omega$ is fixed). Direct interpolation of the eigenvalue phases works fine as long as they are distinct (mod $2\pi$), and has the further advantage of eliminating the costly eigendecompositions of the terminal unitary matrices for many different values of $\lambda$. Unfortunately, the degenerate cases, in which some eigenvalues have the same phase, are often of great interest from a physical point of view.

Moreover, as pointed out by Overton [14], the eigenvalues of differentiable matrix functions are not necessarily differentiable at points of degeneracy, as shown by the example from [14]

$$A = \begin{bmatrix} 1+x_1 & x_2 \\ x_2 & 1-x_1 \end{bmatrix}.$$

The eigenvalues of $A$ are $\mu = 1 \pm \sqrt{x_1^2 + x_2^2}$ and are not differentiable at $(x_1, x_2) = (0,0)$. This is unfortunate because the accuracy of function interpolation depends crucially on the smoothness of the function and its derivatives.

There are ways to get around this problem but in the interest of general applicability and simplicity, we turn instead to matrix interpolation: if $f$ is a smooth function of the parameters then so is the terminal unitary matrix solution to the Schrödinger equation, i.e., $U\left(\frac{2\pi}{\omega}\right) = U\left(\frac{2\pi}{\omega}, \lambda\right)$ depends smoothly on $\lambda$ and $\omega$ (see [9]).

Matrix interpolation is the direct analog of scalar function interpolation. For example, given two matrices $M_0 = U(2\pi/\omega, \lambda_0)$ and $M_1 = U(2\pi/\omega, \lambda_1)$ we can form the linear interpolant

$$M(\lambda) = M_0 + \frac{M_1 - M_0}{\lambda_1 - \lambda_0}(\lambda - \lambda_0). \tag{21}$$

More generally, given $M_0, \ldots, M_k$ corresponding to $\lambda_0, \ldots, \lambda_k$ we first form the Newton divided differences

$$\begin{aligned} M[\lambda_1,\lambda_0] &= \tfrac{M_1-M_0}{\lambda_1-\lambda_0} \\ M[\lambda_2,\lambda_1,\lambda_0] &= \tfrac{M[\lambda_2,\lambda_1]-M[\lambda_1,\lambda_0]}{\lambda_2-\lambda_0} \\ &\vdots \\ M[\lambda_k,\ldots,\lambda_0] &= \tfrac{M[\lambda_k,\ldots,\lambda_1]-M[\lambda_{k-1},\ldots,\lambda_0]}{\lambda_k-\lambda_0}\,. \end{aligned} \tag{22}$$

Then the interpolating matrix polynomial is given by

$$M(\lambda) = M_0 + M[\lambda_1,\lambda_0](\lambda-\lambda_0) + \cdots + M[\lambda_k,\ldots,\lambda_0](\lambda-\lambda_0)\cdots(\lambda-\lambda_{k-1}) \tag{23}$$

and satisfies the interpolating conditions

$$M(\lambda_j) = M_j \quad \text{for } 0 \le j \le k. \tag{24}$$

See [1] for a discussion of scalar function interpolation. The matrix case, especially the interpolation property (24), follows by considering each matrix entry as a separate scalar function interpolation problem. The error in the approximation

$$E(\lambda) = M(\lambda) - U\left(\frac{2\pi}{\omega},\lambda\right)$$

can be gauged by measuring the difference between polynomial estimates obtained by using $k$ and $k+1$ interpolation points. More conservatively, we may find the $\lambda$ value, say $\lambda_{\max}$, where these successive polynomial estimates have a maximal difference (for $\lambda_0 \le \lambda \le \lambda_k$) and then evaluate $U(2\pi/\omega,\lambda_{\max})$ to compute the true error at $\lambda_{\max}$. However, there is a simpler error estimation procedure that is commonly used by practitioners in the field. The exact solution $U(2\pi/\omega,\lambda)$ is unitary and hence has all of its eigenvalues on the unit circle. Thus, the distance from the unit circle of the eigenvalues of an approximation to $U(2\pi/\omega,\lambda)$ gives a lower bound on the error in the approximate eigenvalues. This bound is usually very close (within a factor of 10 say) of the true error independently of whether the approximation was obtained via numerical integration or interpolation. Moreover, since we need to compute the eigenvalues of $U(2\pi/\omega,\lambda)$ in order to construct the phase plot, this method of error estimation is essentially free.

Aside from the error, another point of interest is that Hermite interpolation [1] may be used in which several values of $\lambda$ may be repeated interpolation points with the Newton divided differences replaced by the appropriate derivatives. For example, as described below it is rather easy in compute $U(2\pi/\omega,\lambda)$, $U_\lambda(2\pi/\omega,\lambda)$, and $U_{\lambda\lambda}(2\pi/\omega,\lambda)$ at $\lambda=0$ for $f(t)=\lambda\sin wt$. (Here the subscript $\lambda$ denotes the partial derivative with repect to $\lambda$.) If we also evaluate $U(2\pi/\omega,\lambda)$ at another value, say $\lambda=\lambda_3$, then we can construct the cubic interpolating polynomial

$$M(\lambda) = M_0 + M[0,0]\lambda + M[0,0,0]\lambda^2 + M[\lambda_3,0,0,0]\lambda^3 \tag{25}$$

where $M[\lambda_3,0,0,0]$ is formed as in (22) with

$$M_0 = U\left(\frac{2\pi}{\omega},0\right) \tag{26}$$

$$M[0,0] = U_\lambda\left(\frac{2\pi}{\omega},0\right) \tag{27}$$

$$M[0,0,0] = \frac{1}{2}U_{\lambda\lambda}\left(\frac{2\pi}{\omega},0\right) . \tag{28}$$

To find expressions for $U_\lambda(2\pi/\omega,0)$ and $U_{\lambda\lambda}(2\pi/\omega,0)$, we apply variation of parameters [9] to (6) to get

$$U(t) = e^{-itD} - i\int_0^t e^{-i(t-s)D} f(s)HU(s)\,ds. \tag{29}$$

Repeated substitution of the left side of (29) into the right side yields

$$\begin{aligned} U(t) = e^{-itD} \; &- \; i\textstyle\int_0^t e^{-i(t-s)D} f(s)He^{-isD}\,ds \\ &+(-i)^2 \textstyle\int_0^t\int_0^s e^{-i(t-s)D} f(s)He^{-i(s-s_1)D} f(s_1)He^{-is_1D}\,ds_1 ds \\ &+\cdots \end{aligned} \tag{30}$$

For $f(t) = \lambda\sin\omega t$, we find

$$U(t,0) = e^{-itD} \tag{31}$$

$$U_\lambda(t,0) = -i\int_0^t e^{-i(t-s)D}\sin(\omega s)He^{-isD}\,ds \tag{32}$$

$$U_{\lambda\lambda}(t,0) = 2\int_0^t\int_0^s e^{-i(t-s)D}\sin(\omega s)He^{-i(s-s_1)D}\sin(\omega s_1)He^{-is_1D}\,ds_1 ds . \tag{33}$$

Since $D$ is diagonal as per (2), we can use standard integral formulas [10] to evaluate $U(t,0)$, $U_\lambda(t,0)$, and $U_{\lambda\lambda}(t,0)$ at $t = 2\pi/\omega$ to get $M_0$, $M[0,0]$, and $M[0,0,0]$ as above.

Table 2 compares this cubic interpolation procedure with the transformation method (11) for the quantum well problem with $a = 1/\sqrt{2}$, $\omega = 10$, and 202 output points with $\lambda$ varying between 0 and 0.2. The cubic interpolation was implemented by setting $\lambda_3 = 0.2$ and then integrating (11) to obtain $U(2\pi/\omega,\lambda_3)$. At each of the 202 output points the approximation to $U(2\pi/\omega,\lambda)$ was diagonalized to obtain its eigenvalues using the EISPACK routines COMHES and COMLR2. This additional work accounts for the fact that the time ratios for these two methods did not approach the ratio 202:1 of required integrations but instead is nearer 40:1. The accuracy of both methods was comparable (max error $\leq 10^{-6}$) as established by using the numerical integration of (11) with the tolerance parameters set equal to $10^{-12}$.

| Number of States $n$ | Complete Transformation (as per (11)) | Cubic Interpolation (as per (25)) |
|---|---|---|
| 10 | $7.8 \times 10^2$ | $2.1 \times 10^1$ |
| 20 | $6.2 \times 10^3$ | $1.7 \times 10^2$ |
| 30 | $2.1 \times 10^4$ | $5.8 \times 10^2$ |
| 40 | $5.4 \times 10^4$ | $1.3 \times 10^3$ |

**Table 2 Execution Times in Seconds for Two Parametric Methods.**

# 4 Conclusion

Parametric studies of solutions to problems in quantum mechanical systems are often impeded because of 1) lengthy integration times for the associated time dependent Schrödinger equation, and 2) multiple integrations for different parameter values. In this paper we have presented two general methods for handling these problems. The problem of lengthy integration is addressed by using a technique of quantum mechanical time dependent perturbation theory, called the "transformation to the interaction picture." This is a unitary diagonal transformation associated with the unperturbed quantum system and effectively eliminates the large derivative norm problems of the untransformed system. This method can potentially yield a reduction in computational time on the order of $n^2$ where $n$ is the number of states used to describe the system. However, in practice the observed reduction was a somewhat smaller power of $n$ rather than $n^2$. The second problem of multiple integrations over parameter space was addressed by using matrix interpolation methods together with a Taylor expansion technique about the unperturbed state. For many problems this simple approach reduces the number of integations needed by several orders of magnitude. These methods work well in combination and on a sample quantum well problem reduced the time required for a parametric study from an estimated 25 years to 3.5 days.

## References


[1] K.E. Atkinson, *An Introduction to Numerical Analysis*, Wiley, New York, 1978.

[2] M.C. Baruch and T.F. Gallagher, *Ramsey Interference Fringes in Single Pulse Microwave Multiphoton Transitions*, Phys. Rev. Lett., **68** (1992), 3515-3518.

[3] G. Baym, *Lectures on Quantum Mechanics*, Benjamin/Cummings, Menlo Park, 1969

[4] F. Benvenuto and G. Casati and I. Guarneri and D. Shepelyansky, *A Quantum Transition from Localized to Extended States in a Classically Chaotic System*, Z. Phys. B – Condensed Matter, **84** (1991). pages = 159-163.

[5] B. Birnir and B. Galdrikian and R. Grauer and M. Sherwin, *Nonperturbative Resonances in Periodically Driven Quantum Wells*, Phys. Rev. B (to appear).

[6] H.P. Breuer and K. Dietz and M. Holthaus, *The Role of Avoided Crossings in the Dynamics of Strong Laser Field — Matter Interactions*, Z. Phys. D., **8** (1988), 349-357.

[7] H.P. Breuer and M. Holthaus, *A Semiclassical Theory of Quasienergies and Floquet Wave Functions*, Ann. Phys., **211** (1991), 249-291.

[8] M. Chu, *A Continuous Jacobi-Like Approach to the Simultaneous Reduction of Real Matrices*, Lin. Alg. Appl., **147** (1991), 75–96.

[9] E. Coddington and N. Levinson, *Theory of Ordinary Differential Equations*, New York, McGraw-Hill, 1955.

[10] I. Gradshteyn and I. Ryzhik, *Table of Integrals, Series, and Products*, 4th ed., Academic Press, New York, 1965.

[11] W.A. Lin and L.E. Reichl, *External Field Induced Chaos in an Infinite Square Well Potential*, Physica D, **19** (1986), 145-152.

[12] W.A. Lin and L.E. Reichl, *Spectral Analysis of Quantum–Resonance Zones, Quantum Kolmogorov–Arnold–Moser Theorem, and Quantum–Resonance Overlap*, Phys. Rev. A **37**, (1988), 3972-3985.

[13] A. Messiah, *Quantum Mechanics, Vol. II*, North Holland, Amsterdam, 1962.

[14] M. Overton, *Large-Scale Optimization of Eigenvalues*, SIAM J. Optim. (to appear, 1992).

[15] L. Shampine and M. Gordon, *Computer Solution of Ordinary Differential Equations*, Freeman and Co., San Francisco, 1975.

[16] S. Yoakum and L. Sirko and P.M. Koch, *Stueckelberg Oscillations in the Multiphoton Excitation of Helium Rydberg Atoms: Observation with a Pulse of Coherent Field and Suppression by Additive Noise"*, Phys. Rev. Lett., **69** (1992), 1919-1922.

[17] Ya.B. Zel'dovich, *The Quasienergy of a Quantum-Mechanical System Subjected to a Periodic Action*, Zh. Eksp. Teor. Fiz., **51** (1966), 1492-1495, 1966, [Note: English translation appeared in *Sov. Phys. JETP*, 24 (1967), pp. 1006–1008].

# 9 Differential Systems and Algebras

**Michael K. Kinyon** Indiana University at South Bend, South Bend, Indiana

**Arthur A. Sagle** University of Hawaii at Hilo, Hilo, Hawaii

## 1 Introduction

In 1960, Lawrence Markus wrote a paper [17] in which he introduced a new way of thinking about polynomial differential systems. To describe this point of view, we illustrate the basic idea with an example.

The Euler equations in $\mathbb{R}^3$ are of the form

$$\begin{aligned} \dot{x_1} &= a_1 x_2 x_3 \\ \dot{x_2} &= a_2 x_1 x_2 \\ \dot{x_3} &= a_3 x_1 x_2. \end{aligned}$$

The right hand side of this system defines a homogeneous quadratic vector field $p : \mathbb{R}^3 \to \mathbb{R}^3 : X \mapsto p(X)$ where $X = (x_1, x_2, x_3)^t$ . Thus, $p$ satisfies $p(aX) = a^2 p(X)$ for $X \in \mathbb{R}^3$ and for all $a \in \mathbb{R}$. Associated with $p$ is a unique, symmetric, bilinear mapping $\hat{p} : \mathbb{R}^3 \times \mathbb{R}^3 \to \mathbb{R}^3$ defined by $\hat{p}(X,Y) := \frac{1}{2}(p(X+Y) - p(X) - p(Y))$. Making the further abbreviation $XY := \hat{p}(X,Y)$, we can write the Euler equations as

$$\dot{X} = X^2. \tag{1}$$

In this setting, Markus' insight was to think of the bilinear mapping $\hat{p}$ as defining a multiplication on $\mathbb{R}^3$, thus giving $(\mathbb{R}^3, \hat{p})$ the structure of a *nonassociative algebra* [30]. This enables us to think of equation (1) as a quadratic differential equation *occurring* in a nonassociative algebra. Note that because $\hat{p}$ is symmetric, the algebra $A$ is commutative.

We can also think of more general polynomial vector fields as defining algebras. Given a vector space $V$, a polynomial mapping $p : V \to V$ has the form

$$p(X) = p_0(X) + p_1(X) + \cdots + p_m(X)$$

where each $p_k : V \to V$ is homogeneous of degree $k$. For each $p_k$, there exists a unique, totally symmetric, $k$-linear mapping $\tilde{p}_k : V \times V \to V$ satisfying

$\tilde{p}_k(X,\ldots,X) = p_k(X)$; this map is defined by $\tilde{p}_k(X_1,\ldots,X_k) := p_k^{(k)}(0)\cdot(X_1,\ldots,X_k)$ where $p_k^{(k)}(0)$ denotes the $k$th derivative of $p_k$ at 0. We think of the structure $(V,p_k)$ as being a *$k$-ary algebra*. It also causes no harm to drop the tildes and write either $p_k(X)$ or $p_k(X_1,\ldots,X_k)$, the context making clear which mapping is meant. Since each $k$-linear mapping $p_k$ is symmetric, each algebra $(V,p_k)$ is *commutative*. Unless otherwise noted, all $k$-ary algebras in this paper will be assumed to be commutative. In case $p$ is itself homogeneous of degree $m$, we think of the equation $\dot{X} = p(X)$ as *occurring* in the algebra $(V,p)$.

As another example, the $n$-species predator-prey model is given by the quadratic system

$$\dot{X} = \begin{bmatrix} \dot{x}_1 \\ \vdots \\ \dot{x}_n \end{bmatrix} = \begin{bmatrix} c_1 & & \\ & \cdots & \\ & & c_n \end{bmatrix} \begin{bmatrix} x_1 \\ \vdots \\ x_n \end{bmatrix} + \begin{bmatrix} x_1 \sum b_{1j}x_j \\ \vdots \\ x_n \sum b_{nj}x_j \end{bmatrix}$$

$$\equiv \; p_1(X) + p_2(X).$$

This example is discussed in [9], [17].

As a cubic example, the usual transformation carrying higher order ordinary differential equations into first order systems transforms the van der Pol equation $\ddot{z} = (3cz^2 + d)\dot{z} + abz$ into the cubic system

$$\dot{X} = \begin{bmatrix} \dot{x}_1 \\ \dot{x}_2 \end{bmatrix} = \begin{bmatrix} 0 & 1 \\ ab & d \end{bmatrix} \begin{bmatrix} x_1 \\ x_2 \end{bmatrix} + \begin{bmatrix} 0 \\ 3cx_1^2x_2 \end{bmatrix}$$

$$\equiv \; p_1(X) + p_3(X).$$

Many other well-known differential systems are, at worst, polynomial in their nonlinearities. Just to name a few, we mention the Lorenz and Rossler systems, and the matrix Riccati equation.

The theme of the point of view begun by Markus in [17] is that there should be an interrelationship between the structure of the algebra in which a given quadratic differential system occurs and the behavior of the solutions to the associated initial value problem; thus, an understanding of the former will help with an understanding of the latter. Markus himself proved many of the basic results, and also gave the first classification (up to affine equivalence) of homogeneous quadratic differential systems in $\mathbb{R}^2$. On page $v$ of the Preface to Volume V of *Contributions to the Theory of Nonlinear Oscillations*, one finds the following written by L. Cesari, J.P. LaSalle, and S. Lefschetz:

"The next paper by Markus deals with a curious application of abstract algebra to the classification of [non]linear differential systems."

In the years since [17] appeared, Markus' "curious application" has been coming into its own as a developing speciality in differential equations. Rather than give a complete history, we note a few contributions: Koecher [16], Kaplan and Yorke [8], Röhrl [24], [25], Gerber [2], [3], [4], Walcher [34], [35], [38], Roth [26], and Kinyon and Sagle [9], [10], [11], [12], [13].

In this paper, we will survey some of the results in this area, and we will also present new results that have not appeared elsewhere. Many of our new results will be generalizations of results appearing in [9] and [10]; we will often not pause to point this out.

## 2 Solutions

Given a polynomial vector field $p(X) = p_0(X) + \cdots + p_m(X)$ occurring in a vector space $V$ with each $p_k$ homogeneous of degree $k$, let $\widetilde{V} = V \times \mathbb{R}$ and define a homogeneous degree $m$ mapping $\tilde{p} : \widetilde{V} \to \widetilde{V}$ by

$$\tilde{p}(\widetilde{X}) := \begin{pmatrix} \sum_{k=0}^{m} u^{m-k} p_k(X) \\ 0 \end{pmatrix},$$

where $\widetilde{X} = \begin{pmatrix} X \\ u \end{pmatrix}$ with $X \in V$, $u \in \mathbb{R}$. As described in Section 1, we also use $\tilde{p}$ to denote the associated $m$-linear mapping, i.e., the associated multiplication. Then the equation $\dot{\widetilde{X}} = \tilde{p}(\widetilde{X})$ occurs in the commutative algebra $(\widetilde{V}, \tilde{p})$, and solutions to the original system are obtained from solutions $\widetilde{X}(t)$ by setting $u = 1$ in the initial condition. As noted in [9] and [35], this homogenization process allows us (at least hypothetically) allows us to restrict our attention to the homogeneous equation $\dot{X} = p(X)$ occurring in $(V, p)$. Perhaps a better way to think of it is that results can be proven in the general nonhomogeneous setting by first proving them in the homogeneous case and then "pulling back" the result to the nonhomogeneous case. We give examples of this below.

We return to the general case where the vector field $p$ is an arbitrary polynomial mapping. We follow a notation advocated by Olver [22] and denote the flow for the initial value problem $\dot{X} = p(X)$, $X(0) = Z$ by $(\exp tp)(Z)$. As Olver notes, this notation has the advantage of agreeing with the matrix exponential in the case that $p$ is a linear mapping, and is consistent with the use of the exponential mapping in Lie theory. In addition, it is justified by the semigroup property for the flow, and by the power series form of the solution, as we shall see below.

The following will be useful later.

**Lemma 2.1** *Let $(V, p)$ be a $m$-ary algebra. Then the flow for $\dot{X} = p(X)$ satisfies $(\exp tp)(aZ) = a(\exp a^{m-1} tp)(Z)$ for each $Z \in V$ and $a \in \mathbb{R}$ for which both sides are defined.*

**Proof:** Observe that both sides satisfy the initial value problem $\dot{X} = p(X)$, $X(0) = Z$.

The solution of $\dot{X} = p(X)$, $X(0) = Z$ is given locally in $t$ by a power series of the form

$$(\exp tp)(Z) = \sum_{k=0}^{\infty} \frac{t^k}{k!} p^{[k]}(Z), \tag{2}$$

where $p^{[0]}(Z) := Z$ and $p^{[k+1]}(Z) := (p^{[k]})'(Z) \cdot p(Z)$. In the case where $p$ is homogeneous of degree $m$, the terms $p^{[k]}(Z)$ can each be written in terms of the product in $(V, p)$. For example, when $p$ is quadratic and $XY \equiv p(X, Y)$, one can write

$$(\exp tp)(Z) = Z + tZ^2 + \frac{t^2}{2!}(2ZZ^2) + \frac{t^3}{3!}(2Z^2Z^2 + 4ZZZ^2) + \cdots .$$

These latter observations are essentially due to Koecher [16]; see also [9] in the quadratic case, and [35] in the $m$-ary case.

We now use the power series solution (2) to find some closed form solutions in certain special cases. We first establish some notation and terminology.

For an $m$-ary algebra $(V, p)$, we define the *multiplication map* by

$$L_p(X_1, \ldots, X_{m-1})Y := p(X_1, \ldots, X_{m-1}, Y).$$

We also use the abbreviation $L_p(X)Y := L_p(X, \ldots, X)Y$. In addition, we will also drop the subscript $p$ when it is clear which multiplication is being invoked. An algebra $(V, p)$ is said to be *power-associative* if for each $X \in V$ and each $k > 0$, all $k$-times iterated compositions

$$V \times \cdots \times V \to V$$

that can be constructed from $p$ agree when applied to $(X, \ldots, X)$. For binary power-associative algebras, see Schafer [30]. Our definition in the $m$-ary case is taken from Röhrl [24, Def. 2.16, pp. 76-77] and from Walcher [35, p. 29].

From a differential systems viewpoint, the interest in power-associative algebras stems from the fact that the associated homogeneous differential equation has solutions behaving like those in the one dimensional case. The following result formalizes this remark. Various versions of the result have appeared in the literature for binary algebras. It seems to have first appeared in Gerber [2], as Gerber himself notes in [4]. For the more restrictive class of power-associative algebras with unit element, the result can be found in Koecher [16]. The general binary case can also be found in Walcher [35, Prop. 2.7, p. 29]. For associative Banach algebras with unit element, the result was also proven (without the use of power series) by Vogt (see [32,

Prop. 4.1, p. 76] for the proof and [33, p. 226] for the Banach algebra interpretation).

To the authors' knowledge, the following is the first appearance of the result in the $m$-ary power-associative case for all $m \geq 2$ (although Walcher alludes to the possibility [35, p. 29]).

**Theorem 2.2** *Suppose the $m$-ary algebra $(V, p)$ is power-associative. Then the solution to $\dot{X} = p(X)$ through $Z \in V$ is given by*

$$(\exp tp)(Z) = (I - (m-1)tL(Z))^{-1/(m-1)}Z. \tag{3}$$

**Proof:** We begin with a

**Lemma 2.3** $(\frac{d}{ds}L(X + sY))_{s=0} = (m-1)L(X, \ldots, X, Y)$.

**Proof of Lemma:** Write $L(X+sY)Z = p(X+sY, \ldots, X+sY, Z)$, differentiate using the product rule, set $s = 0$, and use commutativity.

For $t$ in a sufficiently small interval about 0, the power series formula (2) is valid. Recall the recursion formulae $p^{[0]}(Z) = Z$, $p^{[k+1]}(Z) = (p^{[k]})'(Z) \cdot p(Z)$. We wish to establish that under the power-associativity hypothesis, we have, for each $k \geq 0$,

$$p^{[k]}(Z) = r(k)L(Z)^k Z$$

where $r(0) := 1$ and $r(k+1) := r(k)(1 - k(m-1))$. The case $k = 0$ is obvious. Assuming the induction hypothesis, we compute

$$\begin{aligned}
p^{[k+1]}(Z) &= r(k)[L(Z)^k Z]' \cdot p(Z) \\
&= r(k)\tfrac{d}{ds}\{L(Z + sp(Z))L(Z)^{k-1}Z + \cdots + \\
&\qquad L(Z)^{k-1}L(Z+sp(Z))Z + L(Z)^{k-1}(Z + sp(Z))\}_{s=0} \\
&= r(k)\{(m-1)L(Z, \ldots, Z, p(Z))L(Z)^{k-1}Z + \cdots + \\
&\qquad (m-1)L(Z)^{k-1}L(Z, \ldots, Z, p(Z))Z + L(Z)^{k+1}Z\} \\
&= r(k)(k(m-1)+1)L(Z)^{k+1}Z \\
&= r(k+1)L(Z)^{k+1}Z,
\end{aligned}$$

where the third equality uses the lemma and the fourth equality uses power-associativity. Thus in this case, the power series solution takes the form

$$(\exp tp)(Z) = \sum_{k=0}^{\infty} \frac{r(k)}{k!} t^k L(Z)^k Z.$$

If for real $\alpha$ and nonnegative integer $j$, we define $\binom{\alpha}{j} := \alpha(\alpha-1)\cdots(\alpha-j+1)/j!$, the generalized binomial coefficient, then it can easily be shown that $(-1)^k(m-1)^k\binom{-1/(m-1)}{k} = r(k)/k!$. Thus we have the formula

$$(\exp tp)(Z) = \sum_{k=0}^{\infty}\binom{-1/(m-1)}{k}(-(m-1)tL(Z))^k Z.$$

But for $t$ sufficiently small, this is just formula (3) in the operator-valued case, because the generalized binomial theorem states that $\sum_{k=0}^{\infty}\binom{\alpha}{k} z^k = (1+z)^\alpha$ for $z$ complex and $|z|$ sufficiently small. This completes the proof.

**Remark:** In the binary case ($m = 2$), the converse of Theorem 2.1 also holds; that is, if formula (3) is the solution to $\dot{X} = X^2$ through each $Z \in V$, then $(V, p)$ is power-associative; see Walcher [35, Prop. 2.7, p. 29]. The proof uses a result due to Albert that a commutative binary algebra (over a field of characteristic zero) is power-associative if and only if it is *fourth power associative*, i.e., $XXX^2 = X^2X^2$ for all $X \in V$. We conjecture that if similar characterizations turn out to hold for commutative $m$-ary algebras for $m > 2$ (i.e., if one only has to check power-associativity up to some finite order), then the converse of Theorem 2.1 will hold for such algebras.

We now illustrate how results obtained in the homogeneous case can be "pulled back" to the nonhomogeneous case; we shall only sketch the proof of the result. We shall call a not necessarily homogeneous polynomial mapping $p : V \to V$ *power-associative* if the homogenization $\tilde{p} : \tilde{V} \to \tilde{V}$ gives a power-associative $m$-ary algebra. To find a formula for the solution to $\dot{X} = p(X)$, $X(0) = Z$ in $V$, we use formula (3) in $\tilde{V}$. Thus we must first find a block matrix representation of $L_{\tilde{p}}(\tilde{Z})$. For $p(X) = p_0(X) + \cdots + p_m(X)$, we shall use the abbreviation $L_j(X)Y := L_{p_j}(X)y$. For $\tilde{Z} = \begin{bmatrix} Z \\ u \end{bmatrix}$, one can show after some calculations that

$$L_{\tilde{p}}(\tilde{Z}) = \begin{bmatrix} \sum_{j=1}^{m} \frac{j}{m} u^{m-j} L_j(Z) & \sum_{j=1}^{m-1} \frac{m-j}{m} u^{m-j-1} p_j(Z) \\ 0 & 0 \end{bmatrix}. \tag{4}$$

If we now examine (3), we find that one must invert and raise to the $1/(m-1)$-st power a block matrix of the form $\begin{bmatrix} I-A & B \\ 0 & I \end{bmatrix}$, where the matrix $(I-A)^{-1/(m-1)}$ is assumed to exist. Some tedious but straightforward linear algebra yields the formula

$$\begin{bmatrix} I-A & B \\ 0 & I \end{bmatrix}^{-\frac{1}{m-1}} = \begin{bmatrix} (I-A)^{-\frac{1}{m-1}} & -(\sum_{i=1}^{m-1}(I-A)^{1+\frac{i}{m-1}})^{-1}B \\ 0 & I \end{bmatrix}.$$

Using (4), we apply this to $I-(m-1)tL_{\tilde{p}}(\widetilde{Z})$, and then multiply the resulting block matrix by $\widetilde{Z}$. We then set $u = 1$ and examine the $V$-component of the resulting formula. This sketches the proof of the following.

**Theorem 2.4** *Let $p : V \to V$ be a power-associative polynomial mapping on a vector space $V$. Then the solution to $\dot{X} = p(X)$, $X(0) = Z$ is given by* $(\exp tp)(Z) =$

$$H(t,Z)^{-\frac{1}{m-1}}Z + (m-1)t\sum_{j=1}^{m-1}\frac{m-j}{m}(\sum_{i=1}^{m-1}H(t,Z)^{\frac{m+i-1}{m-1}})^{-1}p_j(Z),$$

*where* $H(t,Z) = I-(m-1)t\sum_{k=1}^{m}\frac{k}{m}L_k(Z)$.

**Remark:** Observe that when $p_j(X) \equiv 0$ for $0 \le j \le m-1$, Theorem 2.4 reduces to Theorem 2.2, as expected.

For our next example, we generalize a result of Walcher [35, Prop. 2.8, p. 30] about binary algebras to $m$-ary algebras. Suppose $(V,p)$ is an $m$-ary algebra ($m \ge 2$) satisfying the identities

$$p(p(X),\ldots,p(X)) = 0 \tag{5}$$

$$p(X,p(X),\ldots,p(X)) = 0 \tag{6}$$

$$\vdots$$

$$p(X,\ldots,X,p(X),p(X)) = 0 \tag{7}$$

In the case $m = 2$, if we write $XY := p(X,Y)$, these collapse to the single identity $X^2X^2 = 0$, which defines the class of binary algebras considered by Walcher.

We wish to establish the following claim:

$$L(X,\ldots,X,p(X))L(X)^jX = 0 \quad \text{for } j \ge 1. \tag{8}$$

This is equivalent to $p(X,\ldots,X,L(X)^jX,p(X)) = 0$ for $j \ge 1$. To prove this we shall establish the identities

$$p(L(X)^jX,p(X),\ldots,p(X)) = 0 \tag{9}$$

$$p(X, L(X)^jX, p(X), \ldots, p(X)) = 0 \qquad (10)$$
$$\vdots$$
$$p(X, \ldots, X, L(X)^jX, p(X), p(X)) = 0 \qquad (11)$$

for $j \geq 0$.

To begin, linearize (5) (i.e., replace $X$ by $X + Y$ and use multilinearity and (5) itself to cancel equal expressions) to obtain

$$m^2 p(L(X)Y, p(X), \ldots, p(X)) = 0.$$

Divide by $m^2$ and set $Y = L(X)^iX$ for $i \geq 0$ to obtain (9) for $j \geq 1$, while (6) is (9) for $j = 0$. Next linearize (6):

$$p(Y, p(X), \ldots, p(X)) + m(m-1)p(X, L(X)Y, p(X), \ldots, p(X)) = 0.$$

Set $Y = L(X)^iX$ for $i \geq 0$. By (9), the leftmost expression is 0. Dividing by $m(m-1)$ gives (10) for $j \geq 1$, while the identity which follows (6) is (10) for $j = 0$. Continuing in this way, we obtain (11) by induction. Finally we linearize (7):

$$(m-2)p(X, \ldots, X, Y, p(X), p(X)) + 2mp(X, \ldots, X, L(X)Y, p(X)) = 0.$$

Set $Y = L(X)^iX$ for $i \geq 0$. By (11), the leftmost expression is 0. Dividing by $2m$, we establish our claim (8).

With these preliminaries out of the way, we now turn to

**Theorem 2.5** *Let $(V, p)$ be an $m$-ary algebra $(m \geq 2)$ satisfying (5) through (7). Then the solution to $\dot{X} = p(X)$, $X(0) = Z$ is*

$$(\exp tp)(Z) = \frac{m-1}{m}Z + \frac{1}{m}\exp(mtL(Z))Z.$$

**Proof:** We must first prove that for $k \geq 1$, we have $p^{[k]}(Z) = m^{k-1}L(Z)^kZ$. The case $k = 1$ is obvious. For $k > 1$, we assume the induction hypothesis and use the proof of Theorem 2.2 and Lemma 2.3 to obtain

$$\begin{aligned}
p^{[k+1]}(Z) &= m^{k-1}(L(Z)^kZ)' \cdot p(Z) \\
&= m^{k-1}\{(m-1)L(Z, \ldots, Z, p(Z))L(Z)^{k-1}Z + \cdots + \\
&\quad (m-1)L(Z)^{k-1}L(Z, \ldots, Z, p(Z))Z + L(Z)^k p(Z)\} \\
&= m^{k-1}\{(m-1)L(Z)^{k-1}p(Z, \ldots, Z, p(Z)) + L(Z)^{k+1}Z\} \\
&= m^k L(Z)^{k+1}Z.
\end{aligned}$$

(The third equality follows from (8); all terms except the last two are zero.) Thus the power series solution (2) becomes

$$\begin{aligned}(\exp tp)(Z) &= Z + \textstyle\sum_{k=1}^{\infty} \frac{t^k}{k!} m^{k-1} L(Z)^k Z \\ &= \tfrac{m-1}{m} Z + \tfrac{1}{m} \textstyle\sum_{k=0}^{\infty} \frac{1}{k!} (mtL(Z))^k Z \\ &= \tfrac{m-1}{m} Z + \tfrac{1}{m} \exp(mtL(Z)) Z.\end{aligned}$$

This completes the proof.

Next we pull this result back to the nonhomogeneous case.

**Theorem 2.6** *Let $p : V \to V$ be a polynomial mapping on a vector space $V$ whose homogenized algebra $(\widetilde{V}, \tilde{p})$ satisfies (5) through (7). Then the solution $t\ \dot{X} = p(X)$, $X(0) = Z$ is given by* $(\exp tp)(Z) =$

$$\frac{m-1}{m} Z + \frac{1}{m} \exp t(\sum_{j=1}^{m} jL_j(Z))Z + \sum_{j=1}^{m-1} \frac{m-j}{m^2} \int_0^t \exp s(\sum_{j=1}^{m} jL_j(Z))p_j(Z)ds.$$

**Proof:** Use our earlier calculation of $L_{\tilde{p}}(\widetilde{Z})$ and the fact that the exponential of a block matrix of the form $t \begin{bmatrix} A & B \\ 0 & 0 \end{bmatrix}$ is $\begin{bmatrix} \exp tA & \int_0^t (\exp sA) B ds \\ 0 & I \end{bmatrix}$.

# 3 Structure

In this section we discuss a few results connecting algebraic structure and differential systems. For more complete treatments of some of the issues involved, see [9] and [35].

Most structural results in this area stem from an equivalence of categories which we now describe.

**Definition:** Suppose $\dot{X} = p(X)$ occurs in the vector space $V$ and $\dot{Y} = q(Y)$ occurs in the vector space $W$. A mapping $\Phi : V \to W$ is said to be a *solution-preserving map* (or *morphism*) from $(V, \dot{X} = p(X))$ into $(W, \dot{Y} = q(Y))$ if $Y(t) := \Phi(X(t))$ is a solution to $\dot{Y} = q(Y)$ for each solution $X(t)$ to $\dot{X} = p(X)$. In terms of flows, this means $(\exp tq)(\Phi(X)) = \Phi((\exp tp)(X))$ for all $X \in V$ and for all $t \in \mathbb{R}$ for which the right side is defined. The equivalent "infinitesimal" condition is $\Phi'(X) \cdot p(X) = q(\Phi(X))$ [35, Lemma 2.1, p. 18].

The connection between morphisms and algebras is given in the following lemma [35, Lemma 2.2, p. 20]. In the discussion that follows, we shall assume that polynomial maps have no constant term for clarity of exposition.

**Lemma 3.1** *Let $p(X) = p_1(X) + \ldots + p_m(X)$ be a polynomial mapping defined in a vector space $V$, and let $q(X) = q_1(X) + \ldots + q_m(X)$ be a* *polynomial mapping defined in a vector space $W$. Let $\Phi : V \to W$ be linear. Then $\Phi$ is a morphism from $(V, \dot{X} = p(X))$ into $(W, \dot{Y} = q(Y))$ if and only if (1) $\Phi \circ p_1 = q_1 \circ \Phi$, and (2) $\Phi$ is a homomorphism from $(V, p_k)$ into $(W, q_k)$ for $2 \leq k \leq m$.*

Thus for $m \geq 1$, we have on the one hand a category $Diff_m$ whose objects are pairs $(V, \dot{X} = p(X))$ consisting of a vector space $V$ and a differential equation $\dot{X} = p(X)$ occurring in $V$ with $p$ homogeneous of degree $m$, and whose morphisms are linear, solution-preserving maps. On the other hand, we have a category $Alg_m$ whose objects are $m$-ary algebras $(V, p)$ and whose morphisms are algebra homomorphisms. Röhrl [24, Theorem 1.4] observed the following (see also [35, Theorem 2.3, p. 20]):

**Lemma 3.2** *$Diff_m$ and $Alg_m$ are equivalent categories.*

More general differential equations $\dot{X} = p(X)$ with $p : V \to V$ a polynomial mapping can be viewed as living in the category $Diff = \bigcup_{m \geq 1} Diff_m$. In order to obtain structural results for such differential equations, one generalizes concepts from $Alg_m \cong Diff_m$ to $Diff$.

As examples, consider the notions of subalgebra and ideal. A *subalgebra* of an $m$-ary algebra $(V, p)$ is a subspace $W$ such that $p(W, \ldots, W) \subseteq W$. An *ideal* of an $m$-ary algebra $(V, p)$ is a subspace $W$ such that $p(V, \ldots, V, W) \subseteq W$. For $p = p_1 + \cdots + p_m$, we generalize this as follows: a *subobject* of $(V, \dot{X} = p(X))$ is a subspace $W$ such that $p_1(W) \subseteq W$ and $(W, p_k)$ is a subalgebra of $(V, p_k)$ for each $k > 1$. An *ideal* of $(V, \dot{X} = p(X))$ is a subspace $W$ such that $p_1(W) \subseteq W$ and $(W, p_k)$ is an ideal of $(V, p_k)$ for each $k > 1$. Given an ideal $W$, the quotient $V/W$ is well-defined, and the canonical mapping $V \to V/W$ is solution-preserving from $\dot{X} = p(X)$ into $\dot{\bar{X}} = \bar{p}(\bar{X})$ where $\bar{p}(\bar{X}) := p(X) + W$; see [35, Prop. 2.4].

Both of these concepts have characterizations that are independent of the decomposition $p = p_1 + \cdots + p_m$. The importance of this is that such characterizations might later be taken as *definitions* for more general classes of mappings, e.g., $C^1$-mappings. In particular, $(W, p)$ is a subobject if and only if $p(W) \subseteq W$, and $(W, p)$ is an ideal of $(V, p)$ if and only if $p'(X) \cdot W \subseteq W$ for all $X \in V$.

The importance of ideals for differential systems is given in the following result [35, Theorem 2.5, p. 23].

**Lemma 3.3** *Let $U$ be an ideal of $(V, \dot{X} = p(X))$ where $p = p_1 + \cdots p_m$. Choose a subspace $W \subseteq V$ so that $V = U \oplus W$ and let $\phi : V/U \to W$ be the natural isomorphism. Then the solution of $\dot{X} = p(X)$, $X(0) = Z$ in $V$ be obtained by finding the solution $\bar{Z}(t)$ of $\dot{\bar{X}} = \bar{p}(\bar{X})$, $\bar{X}(0) = \bar{Z}$ in $V/U$ and then solving*

$$\dot{Y} = \sum_{i=0}^{m} \sum_{k_i=1}^{i} \binom{i}{k_i} p_i(Y, \ldots, Y, \phi(\bar{Z}), \ldots, \phi(\bar{Z})) - [\phi(\bar{Z})' - p(\phi(\bar{Z}))] ,$$

$Y(0) = Z - \phi(\bar{Z})$ *in* $U$.

As formidable as it might look, this result is very useful for decomposing differential equations and obtaining solutions in special cases, as we shall see.

An algebra $(V, p)$ is called *solvable* if the sequence of subalgebras defined by $V^{(0)} := V$, $V^{(k+1)} := p(V^{(k)}, \dots V^{(k)})$ is eventually zero, i.e., $V^{(n)} \neq \{0\}$, $V^{(n+1)} = \{0\}$ for some $n > 0$. Generalizing this to $p = p_1 + \cdots + p_m$, we say that $(V, \dot{X} = p(X))$ is *solvable* if the sequence of subobjects defined by $V^{(0)} := V$, $V^{(k+1)} := p(V^{(k)}, \dots V^{(k)})$ is eventually zero, i.e., $V^{(n)} \neq \{0\}$, $V^{(n+1)} = \{0\}$ for some $n > 0$. The following generalizes Theorem 4.3 of [9]; the proof is essentially the same as in that paper.

**Theorem 3.4** *Suppose $(V, \dot{X} = p(X))$ is solvable with each $V^{(k)}$ an ideal. Then $\dot{X} = p(X)$ can be solved by solving a sequence of linear equations.*

**Proof:** Choose a sequence of subspaces $W_1, \dots, W_n$ so that $V = V^{(k)} \oplus W_k$ and use Lemma 3.3 inductively.

For an algebra $(V, p)$, let $\mathbb{R}[Z]$ denote the subalgebra generated by the element $Z$, i.e., the subalgebra consisting of all sums of products of $Z$ with itself. $(V, p)$ is said to be *nil* if for each $Z \in V$, $\mathbb{R}[Z]$ is a nilpotent subalgebra, i.e., there exists $n = n(Z)$ such that all $n$-times iterated products of $Z$ vanish. Since the power series solutions of $\dot{X} = p(X)$ eventually terminate in nil algebras, we have the following.

**Theorem 3.5** *Let $(V, p)$ be a nil algebra. For each $Z \in V$, $(\exp tp)(Z)$ is a polynomial.*

Finally, we turn to 1-dimensional subalgebras. Such subalgebras are characterized by special elements which can be found from any element of the subalgebra after a suitable normalization. An element $N \in V$ is said to be a *nilpotent* of index $m$ (briefly, a nilpotent) if $p(N) = 0$. An element $E \in V$ is said to be an *idempotent* if $p(E) = E$ and an *anti-idempotent* if $p(E) = -E$. Note that if $m$ is even, then one can normalize an anti-idempotent to be an idempotent by multiplying by $-1$.

A nilpotent is an equilibrium; we discuss these further in the next section. For $E \in V$ an idempotent, we have $(\exp tp)(Z) = (1 - (m-1)t)^{-1/(m-1)}E$, and for $E \in V$ an anti-idempotent, we have $(\exp tp)(Z) = (1 + (m - 1)t)^{-1/(m-1)}E$. These can be seen from either the power series solution or direct calculation. Notice that the trajectory through an idempotent blows up in finite positive time and approaches the origin as $t \to -\infty$, while the trajectory through an anti-idempotent approaches the origin as $t \to \infty$ and blows up in finite negative time.

For the existence of nilpotents, idempotents, and anti-idempotents, we have the following result which is essentially due to Röhrl [25] (though the case of a real algebra is in [35, Theorem 3.3, p. 35]). Earlier versions were proven by different means for $m = 2$ by Markus [17] in dimensions 2 and $2k - 1$, and Kaplan and Yorke [8] for all dimensions.

**Theorem 3.6** *Let $(V, p)$ be a real (commutative) m-ary algebra.*
*(1) If $m$ is even, then $(V, p)$ has an idempotent or a nilpotent.*
*(2) If $m$ is odd and dim $V$ is odd, then $(V, p)$ has an idempotent, an anti-idempotent, or a nilpotent.*

We shall use this result in the next section to obtain some results on stability.

# 4 Equilibria

**Definition:** An *equilibrium* of $\dot{X} = p(X)$ is a constant solution. Let $\mathcal{E}$ denote the set of all equilibria for $\dot{X} = p(X)$.

We shall focus on the case where $p$ is homogeneous of degree $m$; i.e., $\dot{X} = p(X)$ occurs in the algebra $(V, p)$. We may do this because equilibria are preserved by the homogenization process. To see this, suppose that $N \in V$ is an equilibrium of $\dot{X} = p(X)$ where $p = p_1 + \cdots + p_m$ is a polynomial mapping. Then it is easy to check that $\begin{pmatrix} uN \\ u \end{pmatrix}$ is an equilibrium for $\dot{\tilde{X}} = \tilde{p}(\tilde{X})$ in $\tilde{V}$, for every $u \in \mathbb{R}$. Conversely, if $\begin{pmatrix} X \\ u \end{pmatrix}$ $(u \neq 0)$ is an equilibrium for $\dot{\tilde{X}} = \tilde{p}(\tilde{X})$ in $\tilde{V}$, then $N := u^{-1}X$ is an equilibrium for $\dot{X} = p(X)$ in $V$: $p(N) = p_1(u^{-1}X) + \cdots + p_m(u^{-1}X) = u^{-m}(\sum_{k=1}^{m} u^{m-k} p_k(X)) = 0$. This justifies us in concentrating on the case where $p$ is homogeneous of degree $m$.

**Remarks:** (1) As observed earlier, $N \in V$ is an equilibrium if and only if $N$ is a nilpotent.
(2) If $0 \neq N \in \mathcal{E}$ for $p$ homogeneous of degree $m$, then the line $< N >:= \{sN : s \in \mathbb{R}\} \subseteq \mathcal{E}$, since $p(sN) = s^m p(N)$.
(3) No equilibrium $N \in \mathcal{E}$ is hyperbolic. This follows from computing the derivative of $p$: $p'(X)Y = (\frac{d}{ds}[p(X + sY) - p(X)])_{s=0} = mp(X, \ldots, X, Y) = mL(X)Y$; thus $p'(X) = mL(X)$. In particular, if we linearize at $N$, we find $p'(N)N = mL(N)N = mp(N) = 0$. Consequently $p'(N)$ has a zero eigenvalue so that $N$ is not hyperbolic. It follows that $N$ has a (nonunique) center manifold. This center manifold can always be chosen so as to contain $< N >$.
(4) In particular, for $N \neq 0$, linear stability analysis is applicable by considering the eigenvalues of $L(N)$. The presence of idempotents also plays a role; see [9, Corollary 3.8] for example.

With these considerations out of the way, we now turn to stability of the origin. A system $\dot{X} = p(X)$ is called *dissipative* if there exists a ball $B$ centered at the origin such that for any $Z \in V$, there exists a $T = T(Z) \in \mathbb{R}$ so that $(\exp tp)(Z) \in B$ for $t \geq T$. Thus dissipativeness is a generalization of asymptotic stability.

Part (2) of the following was also obtained by Koditschik and Narenda by a different method [15].

**Theorem 4.1** *Let $\dot{X} = p(X)$ occur in the $m$-ary algebra $(V, p)$.*
*(1) If there exists a nonzero idempotent, then the origin is unstable.*
*(2) If $m$ is even, then the origin is not asymptotically stable.*
*(3) If $m$ is even, then the system $\dot{X} = p(X)$ is not dissipative.*

**Proof:** (1) If $E \neq 0$ is an idempotent, then for each $\delta > 0$, we have $(\exp tp)(\delta E) = \delta(1-(m-1)\delta^{m-1}t)^{-1/(m-1)}E$ (by Lemma 2.1), which blows up in finite positive time. Thus for each neighborhood of the origin, there is an unbounded solution starting in that neighborhood, so that the origin is unstable.
(2) Suppose that the origin is asymptotically stable. By (1), there is no nonzero idempotent, so by Theorem 3.6, there is a nilpotent $N \in V$, and thus $< N > \subseteq \mathcal{E}$. Thus for each neighborhood of the origin, there is a trajectory starting in that neighborhood that does not approach the origin, which is a contradiction.
(3) Suppose that the system is dissipative. Then there can be no nonzero idempotent (because the trajectory passing through it would blow up), so by Theorem 3.6, there is a nilpotent $N \in V$. But then for any ball $B$ centered at the origin, there is a multiple of $N$ not in $B$, and the trajectory through that multiple never enters $B$. Contradiction.

Of course for $m$ odd, no such conclusions can be drawn because the existence of an anti-idempotent is consistent with both asymptotic stability and dissipativeness.

The stability of the origin can also be shown by Lyapunov functions. An important special case of Lyapunov functions are first integrals with compact level sets. From an algebraic perspective, an important class of such first integrals are positive definite quadratic forms, or equivalently, positive definite bilinear forms. The following result is not the most general possible, but will suffice for the discussion that follows.

**Theorem 4.2** *Let $\dot{X} = p(X)$ occur in a vector space $V$. If $V$ has a positive definite bilinear form $C : V \times V \to \mathbb{R}$ satisfying $C(X, p(X)) = 0$ for all $X \in V$, then the quadratic form $\gamma(X) := C(X, X)$ is a first integral. In particular, the origin is stable.*

**Proof:** $\frac{d}{dt}\gamma(X(t)) = 2C(X(t), \dot{X}(t)) = 2C(X(t), p(X(t))) = 0.$

**Example:** Let $\dot{X} = X^2$ occur in the Euler algebra of Section 1. Let $C(X,Y) := \sum b_i x_i y_i$ for $b_i > 0$, $i = 1,2,3$ be chosen so that $\sum a_i b_i = 0$. Then $C(X, X^2) = (\sum a_i b_i) x_1 x_2 x_3 = 0$, so that the origin is stable.

As an aside, we now discuss how variations of the Euler equations give the differential geometry of invariant Lagrangian mechanics on a reductive homogeneous space $G/H$ [1, App. 2] [6] [7] [14] [28]. Thus let $G$ be a connected Lie group and let $H$ be a closed (Lie) subgroup with Lie algebras $g$ and $h$ respectively. The homogeneous space $G/H$ is called reductive if there exists a subspace $m$ of $g$ such that $g = m + h$ (direct sum) and $(Ad\ H)m \subseteq m$, i.e., $[h, m] \subseteq m$. For example, let $g$ and $h$ be semisimple and $m = h^\perp$ relative to the Killing form of $g$. For a reductive space, there exists a bijective correspondence between the set of $G$-invariant connections $\nabla$ on $G/H$ and the set of algebras $(m, \alpha)$ with $Ad\ H \subseteq Aut(m, \alpha)$; here $\alpha : m \times m \to m$ is the bilinear multiplication on $m$ and $Aut(m, \alpha)$ is the group of automorphisms of the algebra $(m, \alpha)$ [6] [20]. This correspondence is given by $\alpha(X,Y) = (\nabla_{\tilde{X}} \tilde{Y})(\bar{e})$ with $Z \in m$ giving the invariant vector field $\tilde{Z}(\bar{a}) = \tau'(a)Z$ where $\tau(a) : \bar{x} = xH \mapsto \overline{ax} = axH$. Next a modification of notation in [6] [7] shows that a curve $\sigma(t)$ in $G/H$ is a geodesic if its tangent field $X(t) := \dot{\sigma}(t)$ satisfies the quadratic equation $\dot{X} + \alpha(X,X) = 0$. [In particular the line $tX$ in $m$ become geodesics $\sigma(t) = \pi \exp tX$ in $G/H$ if and only if $\alpha(X,X) = 0$, i.e., $(m, \alpha)$ is an anticommutative algebra.] A pseudo-Riemannian connection $\nabla$ is given by a connection algebra $(m, \alpha, C)$ where $C$ is a symmetric nondegenerate bilinear form on $m$ satisfying $C$ is $Ad\ H$-invariant and $C(\alpha(Z,X),Y) + C(X, \alpha(Z,Y)) = 0$ for all $X, Y, Z \in m$ [20] [27]. In particular, for a Riemannian connection $C(X, \alpha(X,X)) = 0$ so that for $\dot{X} + \alpha(X,X) = 0$, the origin in $m$ is stable. For example, let $g$ and $h$ be semisimple, $g = m + h$ with $m = h^\perp$ as above, and let $B(X,Y)$ be the Killing form of $g$ restricted to $m$. Then we can write $C(X,Y) = B(SX, Y)$ where the "moment of inertia operator" $S \in GL(m)$ is $B$-symmetric [27]. In this case, $\alpha(X,X) = S^{-1}[X, SX]_m$ where $[X,Y]_m$ denotes the projection of $[X,Y] \in g$ into $m$. The quadratic system for geodesics is $\dot{X} = S^{-1}[X, SX]_m$; that is, $\dot{Z} = [Z, S^{-1}Z]_m$ for $Z = SX$ which is a Lax equation occurring in $m$.

Now let $G/H$ be a configuration space for a "least action" variational problem given by a $G$-invariant Lagrangian $L : T(G/H) \to \mathbb{R}$ where $T(G/H)$ is the tangent bundle of $G/H$ [1, App. 2] [7] [28]. Using the $G$-invariance, $L$ is given by an $Ad\ H$-invariant function $K : m \to \mathbb{R} : X \mapsto K(X)$; thus $L(\bar{q}, \dot{\bar{q}}) = K(\tau'(q^{-1})(\bar{q}) \cdot \dot{\bar{q}})$. An "extended Euler field" $Y(X)$ is obtained from $K$ and the second derivative $Y^{(0)}(0)Z^{(2)}/2! = -\alpha(Z,Z)$ gives the connection algebra obtained from $L$. For example [1] [28], let $H = \{e\}$ and $K : g \to \mathbb{R} : X \mapsto (1/2)C(X,X)$ where $C$ is a nondegenerate bilinear form on $g$ which represents kinetic energy. Then a solution curve $f(t) = (\sigma(t), \dot{\sigma}(t))$ in $T(G)$

of the Euler-Lagrange equation which goes through the origin satisfies

$$C(\ddot{\sigma}, W) + C(\dot{\sigma}, [W, \dot{\sigma}]) = 0$$

for all $W \in g$. Thus using the nondegeneracy of $C$, define the Euler vector field $Y$ on $g$ relative to $L$ by

$$C(Y(Z), W) = C(Z, [Z, W])$$

for all $Z, W \in g$. Then $X(t) = \dot{\sigma}(t)$ satisfies $\dot{X} = Y(X)$. In terms of algebra, let $\alpha : g \times g \to g$ be a bilinear multiplication on $g$ so that

$$C(\alpha(Z, Z), W) = C(Z, [W, Z])$$

for all $Z, W \in g$. Then the nondegeneracy of $C$ implies $Y(Z) = -\alpha(Z, Z)$. Thus $X(t) = \dot{\sigma}(t)$ satisfies $\dot{X} + \alpha(X, X) = 0$. Also $C$ satisfies $C(\alpha(Z, X), Y) + C(X, \alpha(Z, Y)) = 0$ so that the trajectory $\sigma(t)$ is a geodesic relative to the connection given by $(g, \alpha, C)$. For $G = SO(n)$, choose the moment of inertia operator $S$ to be diagonal relative to a suitable basis to obtain the usual Euler equation [7] [23].

# 5 Periodic Solutions

**Definition:** Let $(\exp tp)(Z)$ be a solution to $\dot{X} = p(X)$. $Z$ is a *periodic point* if there exists $\tau > 0$ so that $(\exp(t + \tau)p)(Z) = (\exp tp)(Z)$ for all $t \in \mathbb{R}$, and the *period* is the smallest $\tau > 0$ satisfying this condition. In this case, $\{(\exp tp)(Z)\}$ is called a *periodic trajectory* and we let $\mathcal{P}_\tau$ denote the set of all periodic trajectories of period $\tau$.

Periodicity is another property preserved by the homogenization process; this is just a consequence of the fact that solutions to $\dot{\widetilde{X}} = \tilde{p}(\widetilde{X})$ in $\widetilde{V}$ remain in the hyperplanes $u \equiv const.$

First we obtain some nonexistence results in special cases. The following result generalizes [9, Theorem 3.11].

**Theorem 5.1** *Let $(V, p)$ be a power-associative $m$-ary algebra. Then no solution to $\dot{X} = p(X)$ is periodic.*

**Proof:** Suppose that the solution $(\exp tp)(Z)$ through $X(0) = Z$ is periodic with period $\tau$. Then using (2.2), we have $(I - (m-1)\tau L(Z))^{-1/(m-1)} Z = Z$. Multiplying both sides of this equation by $(I - (m-1)\tau L(Z))^{1/(m-1)}$ and using the equation itself, we obtain

$$(I - (m-1)\tau L(Z))^{1/(m-1)} Z = (I - (m-1)\tau L(Z))^{-1/(m-1)} Z.$$

Continuing in the same fashion, we eventually obtain $(I - (m-1)\tau L(Z))Z = Z$, and thus $Z - (m-1)\tau p(Z) = Z$. But this implies that $Z$ is an equilibrium point, a contradiction.

**Corollary 5.2** *Let $p : V \to V$ be a power-associative polynomial mapping. Then no solution to $\dot{X} = p(X)$ is periodic.*

**Theorem 5.3** *Let $(V, p)$ be an $m$-ary algebra satisfying the identities (5) through (7). Then no solution to $\dot{X} = p(X)$ is periodic.*

**Proof:** Suppose that the solution $(\exp tp)(Z)$ through $X(0) = Z$ is periodic with period $\tau$. Then using the solution formula developed in Theorem 2.5, we have $\frac{m-1}{m}Z + \frac{1}{m}\exp(mtL(Z))Z = Z$, and thus $\exp(mtL(Z))Z = Z$. But then $Z$ is an eigenvector for $\exp(mtL(Z))$ corresponding to the eigenvalue 1, and since $V$ is a real algebra, this implies that $Z$ is an eigenvector for $mtL(Z)$ corresponding to the eigenvalue 0. This implies that $Z$ is an equilibrium, a contradiction.

**Corollary 5.4** *Let $p : V \to V$ be a polynomial mapping whose homogenization $\tilde{p}$ satisfies the identities (5) through (7). Then no solution to $\dot{X} = p(X)$ is periodic.*

**Remark:** Another nonexistence result is the following: if $\mathbb{R}[Z]$ is a nilpotent subalgebra (of index $> 2$), then the solution through $Z$ is not periodic; this is because it is a polynomial (Theorem 3.5).

In the following result, $(\alpha, \omega)$ denotes the maximal interval of existence for the initial value problem $\dot{X} = p(X)$, $X(0) = Z$.

**Lemma 5.5** *Let $\dot{X} = p(X)$ occur in $(V, p)$. Suppose that there exist $Z \in V$, $a \neq 0$, and $r > 0$ so that $(\exp rp)(Z) = aZ$. Then*
*(1) If $m$ is even, then $a > 0$, in which case,*
*(a) if $0 < a < 1$, then $\frac{a^{m-1}r}{a^{m-1}-1} < \alpha$,*
*(b) if $a > 1$, then $\omega < \frac{a^{m-1}r}{a^{m-1}-1}$, and*
*(c) if $a = 1$, then $(\exp tp)(Z)$ is periodic.*
*(2) If $m$ is odd, then*
*(a) if $|a| < 1$, then $\frac{a^{m-1}r}{a^{m-1}-1} < \alpha$,*
*(b) if $|a| > 1$, then $\omega < \frac{a^{m-1}r}{a^{m-1}-1}$, and*
*(c) if $|a| = 1$, then $(\exp tp)(Z)$ is periodic.*

**Proof:** The proof hinges on the following calculation for $a \neq 1$:

$$\begin{aligned} (\exp \tfrac{a^{m-1}r}{a^{m-1}-1}p)(Z) &= (\exp(\tfrac{r}{a^{m-1}-1} + r)p)(Z) = \\ = (\exp(\tfrac{r}{a^{m-1}-1})p)(\exp rp)(Z) &= (\exp(\tfrac{r}{a^{m-1}-1})p)(aZ) = \\ &= a^{m-1}(\exp \tfrac{a^{m-1}r}{a^{m-1}-1}p)(Z), \end{aligned}$$

which is a contradiction. The rest is a case by case analysis of what exactly is being contradicted which we leave to the reader.

Since a periodic orbit cannot intersect any line through the origin more than once, we have the following well-known result in dimension 2.

**Corollary 5.6** *Let $\dot{X} = p(X)$ occur in an m-ary algebra $(V,p)$ with dim $V = 2$. Then no periodic solution exists.*

In the following result, for $a$ real and $\gamma$ a trajectory, $a\gamma := \{aP : P \in \gamma\}$.

**Theorem 5.7** *Let $\dot{X} = p(X)$ occur in an m-ary algebra $(V,p)$, and fix $a \in \mathbb{R}$. If $\gamma \in \mathcal{P}_\tau$, then $a\gamma \in \mathcal{P}_{\tau/|a|^{m-1}}$.*

**Proof:** Note that for $a > 0$, $(\exp(\tau/|a|^{m-1})p)(aZ) = a(\exp \tau p)(Z) = aZ$. For $a < 0$, use $(\exp(-\tau)p)(Z) = (\exp(-\tau)p)(\exp \tau p)(Z) = Z$.

**Corollary 5.8** *Let $\dot{X} = p(X)$ occur $(V,p)$. Then no periodic orbit is an attractor.*

**Proof:** If $\gamma$ is a periodic orbit, then for each $a \in \mathbb{R}$, the orbit $a\gamma$ is also periodic and thus does not approach $\gamma$.

**Example:** We give a 3-dimensional quadratic system with periodic solutions. Let $V$ be the 3-dimensional algebra with basis $\{X_0, X_1, X_2\}$ and multiplication table

| | $X_0$ | $X_1$ | $X_2$ |
|---|---|---|---|
| $X_0$ | $\lambda X_0$ | $cX_2$ | $-cX_1$ |
| $X_1$ | $cX_2$ | $\mu X_0$ | 0 |
| $X_2$ | $-cX_1$ | 0 | $\mu X_0$ |

where $c \neq 0$ and $\lambda$, $\mu \in \mathbb{R}$. The system $\dot{X} = p(X) = X^2$ in $V$ is given in coordinates by

$$\begin{aligned} \dot{x_1} &= \lambda x_0^2 + \mu(x_1^2 + x_2^2) \\ \dot{x_2} &= -2cx_0x_2 \\ \dot{x_3} &= 2cx_0x_1 \end{aligned}$$

and solutions are discussed in [9]. In particular, for $\lambda < 0$ and $\mu > 0$, let $Z = x_0X_0 + x_1X_1 + x_2X_2$ be on the cone $\mathcal{C} = \{\lambda x_0^2 + \mu(x_1^2 + x_2^2) = 0\}$. Then except for the origin, we obtain a periodic solution

$$(\exp tp)(Z) = -\frac{\lambda}{|\lambda|}\mu r^2 X_0 + r[\cos(2c\theta(t) + \psi)X_1 + \sin(2c\theta(t) + \psi)X_2]$$

where $r = (x_1^2 + x_2^2)^{1/2}$, $\theta(t) = -\frac{\lambda}{|\lambda|}\mu r^2 t$, $\psi = \tan^{-1}(x_2/x_1)$ if $x_1 \neq 0$ or $\psi = \pi/2$ if $x_1 = 0$, and the period is $2\pi/|\mu|r^2$. The nappe $\mathcal{C}^+ = \mathcal{C} \cap \{x_0 \geq 0\}$ is closed, invariant, and attracts nearby solutions as they spiral along a horizontal cylinder determined by the periodic solution on $\mathcal{C}^+$; i.e., the periodic solutions are limit cycles. However $0 \in \mathcal{C}^+$ is an unstable equilibrium since $V$ has a nonzero idempotent $(1/\lambda)X_0$; see Theorem 4.1. We return to this example in the next section using automorphisms of $V$.

## 6 Symmetries and Invariants

In this section, we discuss some of the connections between another of Markus' interests [18], symmetries of differential equations, and the results we have obtained so far. In this area the work of S. Lie is fundamental; see Olver [22] for historical notes. Most treatments of symmetries of differential equations (e.g. [22], [31]) are presented in full generality so that the case of partial differential equations can also be covered. A very readable treatment of symmetries for ordinary differential equations can be found in [37]; it is partially based on the dissertation of Roth [26] and avoids all the complicated machinery found in other works. For the case of ordinary differential equations, the most important aspect of the theory of symmetries is its connection with dynamical systems and bifurcation theory. Here is where *linear* symmetries take center stage. The Birkhoff normal form is a classical topic; see [5] and [36]. For equivariant bifurcation theory, the standard references is [5]. This latter work discusses the physical importance of solutions which are given by orbits of one-parameter subgroups of the linear symmetry group; see also the remarks in [10]. For the connections between symmetry theory and algebras, see [9], [10], [16], and [35] (especially Chapters 9 and 10).

**Definition:** Let $(V,p)$ be an $m$-ary algebra. Define $G(p)$ to be the set of all germs of maps which are analytic at 0, map 0 to 0, have a local inverse that is analytic at 0, and are morphisms from $(V,\dot{X}=p(X))$ into itself. Clearly $G(p)$ is a group under composition; $G(p)$ is called the *symmetry group* of $(V,\dot{X}=p(X))$. The subgroup of $G(p)$ consisting of *linear* maps is call the *automorphism* group of $(V,\dot{X}=p(X))$, and is denoted by $Aut(V,p)$. As noted in the previous lemma, $Aut(V,p)$ is identical with the automorphism group of $(V,p)$ in the sense of algebras.

For polynomial mappings $p,q : V \to V$, we define the *(Poisson) bracket* of $p$ and $q$ by $[p,q](X) := p'(X)\cdot q(X) - q'(X)\cdot p(X)$. The vector space $C(p)$ consisting of all polynomial mappings $q : V \to V$ whose bracket with $p$ vanishes is clearly a Lie algebra under the bracket $[\cdot,\cdot]$, and is called the *symmetry algebra* of $(V,\dot{X}=p(X))$. The subspace of *linear* mappings in $C(p)$ is a Lie subalgebra of $C(p)$ called the *derivation* algebra of $(V,\dot{X}=p(X))$, and is denoted by $Der(V,p)$. $Der(V,p)$ is identical with the derivation algebra of $(V,p)$ in the sense of algebras; see [35, p. 142].

The automorphism group $Aut(V,p)$ is a Lie group whose Lie algebra is $Der(V,p)$ [29], [35, Theorem 9.1, p. 142]. If $dim\ C(p) < \infty$, then $G(p)$ is a finite dimensional, local Lie transformation group whose Lie algebra is the subalgebra of $C(p)$ consisting of nonconstant maps [35, Theorem 10.2, p. 155]. In particular, if $q \in C(p)$, then $(\exp tq) \in G(p)$.

**Definition:** Let $H$ be a subgroup of $Aut\ E$. A polynomial $\gamma : \mathbb{R}^n \to \mathbb{R}$ is called *invariant* for $H$ if $\gamma(\phi X) = \gamma(X)$ for all $X \in \mathbb{R}^n$ and all $\phi \in H$. If $h$ is the Lie subalgebra of $Der\ E$ corresponding to $H$, then we also say that $\gamma$ is *invariant* for $h$; the corresponding "infinitesimal" condition is $\gamma'(X) \cdot D = 0$ for all $X \in \mathbb{R}^n$ and $D \in h$ [35].

**Remark:** The set of all $H$-invariant (or $h$-invariant) polynomials is clearly a real associative algebra. We shall denote this algebra by $\mathcal{I}(H)$ or $\mathcal{I}(h)$.

The importance of knowing the algebra of invariants for a given subalgebra of $Der(V,p)$ is suggested in the next result [35, Lemma 9.11].

**Proposition 6.1** *Suppose $h \neq \{0\}$ is a subalgebra of $Der(V,p)$ such that $\mathcal{I}(h)$ is nontrivial (i.e., $\neq \mathbb{R}$) and is generated by $\gamma_1, \ldots, \gamma_m$. Then the mapping $\Gamma : V \to \mathbb{R}^m$ defined by $\Gamma(X) := (\gamma_1(X), \ldots, \gamma_m(X))$ is a morphism from $(V, \dot{X} = p(X))$ into some polynomial differential equation in $\mathbb{R}^m$.*

It is known that the algebra $\mathcal{I}(h)$ is finitely generated when $h$ is 1-dimensional [35]. This is the situation we will consider later in this section.

Next we consider solutions to $\dot{X} = p(X)$, $X(0) = Z$ which are given by $(\exp tq)(Z)$ for $q \in C(p)$. The case of $q = D \in Der(V,p)$ is of particular importance. As it turns out, it is enough that $p$ and $q$ agree on the initial point. See [9, Prop. 5.3] for the following in the derivation case.

**Theorem 6.2** *Suppose $q \in C(p)$. Then $(\exp tq)(Z)$ is a solution to $\dot{X} = p(X)$ if and only if $q(Z) = p(Z)$.*

**Proof:** As noted earlier, $q \in C(p)$ implies $(\exp tq) \in G(p)$. Now $q(Z) = p(Z)$ if and only if $(\exp tq)(q(Z)) = (\exp tq)(p(Z))$ if and only if $q((\exp tq)(Z)) = p((\exp tq)(Z))$ if and only if $\frac{d}{dt}(\exp tq)(Z) = p((\exp tq)(Z))$.

For $q = D \in Der(V,p)$, this result was used in [9] and [10] to construct solutions of the form $(\exp tD)(Z)$ to quadratic differential systems occurring in algebras. Later we shall discuss one of these examples, but we will give a new construction based on the following observation.

**Proposition 6.3** *Suppose that $(\exp tD)P$ is a solution to $\dot{X} = p(X)$, and suppose $\gamma_1, \ldots, \gamma_m$ is a set of generators of $\mathcal{I}(D)$. Then each $\gamma_i$ is conserved on the trajectory $\{(\exp tD)P\}$.*

**Proof:** This is just a consequence of $\gamma_i((\exp tD)P) = \gamma_i(P)$.

In particular, if $\dot{Y} = q(Y)$ is the image equation under the morphism $\Gamma : V \to \mathbb{R}^m$ described earlier, then the trajectories given by automorphisms can be described by the equations $\gamma_1(X) = k_1, \ldots, \gamma_m(X) = k_m$ where the constants $k_1, \ldots, k_m$ must satisfy the equations $F(k_1, \ldots, k_m) = 0$; see the example later for details.

Recall from the previous section the definitions of $\mathcal{E}$ and $\mathcal{P}_\tau$. The following result is actually an extension of its own corollary, which appears independently in [9, Prop. 5.7]. There is a similar result, whose proof is also similar, for quasiperiodic orbits; it essentially states that the image of a quasiperiodic orbit a symmetry is also quasiperiodic with the same set of quasiperiods. The corollary to this result appears independently in [10, Prop. 1.5].

**Theorem 6.4** *Let $\dot{X} = p(X)$ occur in a vector space $V$ and suppose that $\Phi \in G(p)$ is defined in a neighborhood $\mathcal{U}$ of the origin. Then*
*(1) $\Phi(\mathcal{U} \cap \mathcal{E}) \subseteq \mathcal{E}$,*
*(2) $\Phi(\mathcal{U} \cap \mathcal{P}_\tau \subseteq \mathcal{P}_\tau)$ for all $\tau > 0$.*

**Proof:** (1) Given $N \in \mathcal{U} \cap \mathcal{E}$, we compute $p(\Phi(N)) = \Phi'(N) \cdot p(N) = 0$ so that $\Phi(N) \in \mathcal{E}$.
(2) Given $\gamma \in \mathcal{U} \cap \mathcal{P}_\tau$ and $Z \in \gamma$, we compute $\Phi(Z) = \Phi((\exp \tau p)(Z)) = (\exp \tau p)(\Phi(Z))$, so that $\Phi(Z)$ is a periodic point of period $\sigma$ for some $\sigma \leq \tau$. But since $\Phi$ is (locally) bijective, essentially the same calculation shows that $Z$ is a periodic point of period $\sigma$, so that $\sigma = \tau$. Thus $\Phi(\gamma) \subseteq \mathcal{P}_\tau$.

Now, as promised, we obtain periodic solutions of $\dot{X} = p(X)$ in terms of one-parameter subgroups of $G(p)$. The following extends [9, Prop. 5.8] from the derivation algebra to the full symmetry algebra. There is a similar result with similar proof for quasiperiodic orbits; the derivation version of this can be found in [10, Theorem 1.6].

**Theorem 6.5** *Suppose $\gamma \in \mathcal{P}_\tau$ is isolated in $\mathcal{P}_\tau$ (i.e., has a neighborhood intersecting no other trajectory of $\mathcal{P}_\tau$). Suppose $q \in C(p) \neq \{0\}$ and $Z \in \gamma$ are such that $q(Z) \neq 0$. Then there exists $a \in \mathbb{R}$, $a \neq 0$, such that $(\exp atq)(Z) = (\exp tp)(Z)$.*

**Proof:** Let $\gamma \in \mathcal{P}_\tau$, $Z \in \gamma$, and $q \in C(p)$ be as in the hypotheses. Consider the mapping $k(s) := (\exp sq)(Z)$. Since $(\exp sq) \in G(p)$, it follows from the previous proposition that $k(s)$ is periodic of period $\tau$ for all $s$ in the maximal interval of existence for the initial value problem $\dot{X} = q(X), X(0) = Z$. Since the image of $k$ is connected, $(\exp 0q)(Z) = Z \in \gamma$, and $\gamma$ is isolated in $\mathcal{P}_\tau$, we have $k(s) \in \gamma$ for all $s$. Thus there exists a function $u(s) \in \mathbb{R}$ so that

$$(\exp sq)(Z) = (\exp u(s)p)(Z) \tag{12}$$

The function $u(s)$ is differentiable, and we may assume that $u(0) = 0$ (otherwise, set $\tilde{u}(s) = u(s) - u(0)$; then $(\exp sq)(Z) = (\exp(\tilde{u}(s) + u(0))q)(Z) = (\exp \tilde{u}(s)q)(\exp u(0)q)(Z)$; set $s = 0$ to obtain $Z = (\exp u(0)q)(Z)$, and thus $(\exp sq)(Z) = (\exp \tilde{u}(s)q)(Z)$.) Differentiating (12), we obtain

$$\begin{aligned} q((\exp sq)(Z)) &= \tfrac{d}{ds}(\exp sq)(Z) = \tfrac{d}{ds}(\exp u(s)p)(Z) \\ &= u'(s)p((\exp u(s)p)(Z)), \end{aligned}$$

by the chain rule. Setting $s = 0$, we obtain $q(Z) = u'(0)p(Z)$. Since $q(Z) \neq 0$, we have $u'(0) \neq 0$. Set $a = 1/u'(0)$ to obtain $aq(Z) = p(Z)$. By Theorem 6.2, $(\exp tp)(Z) = (\exp atq)(Z)$. All that is left is to prove that the maximal interval of existence of $(\exp atq)(Z)$ is $(-\infty, \infty)$ so that the trajectories entirely agree. But this is clear, since otherwise the trajectory $(\exp atq)(Z)$ would blow up.

**Remark:** Of course the case of most interest is when $q = D \in Der(V,p)$. For a discussion of the orbital stability of periodic orbits given by exponentiated derivations, see [10]. In particular, the automorphisms themselves can be used to find the Floquet decomposition of the solution of the associated variational equation.

We now discuss an example where trajectories are given by automorphisms. This example was constructed in [9] and extended in [10].

**Example:** Suppose that $V$ is a 3-dimensional commutative binary algebra with a derivation $D \neq 0$, a real number $b \neq 0$, and a point $0 \neq Z \in V$ such that $bDZ = p(Z) = p_1(Z) + p_2(Z) =: TZ + Z^2$, and such that the corresponding trajectory $(\exp tp)(Z) = e^{tbD}Z$ is periodic. Since we desire a linear group action to have a periodic orbit, we may absorb the frequency into the parameter $b$ and assume that relative to some basis $\{X_0, X_1, X_2\}$, $D$ has the matrix representation $\begin{bmatrix} 0 & 0 & 0 \\ 0 & 0 & 1 \\ 0 & -1 & 0 \end{bmatrix}$. Working in the complexification $V_{\mathbb{C}}$ of $V$, we decompose $V_{\mathbb{C}}$ relative to $D$:

$$V_{\mathbb{C}} = V_{\mathbb{C}}(0) + V_{\mathbb{C}}(i) + V_{\mathbb{C}}(-i) \text{ (direct sum).}$$

We then use the multiplicative relations $V_{\mathbb{C}}(u)V_{\mathbb{C}}(v) \subseteq V_{\mathbb{C}}(u+v)$ between the eigenspaces of $D$. The general 3-dimensional commutative algebra $V$ having $D$ as a derivation is given by the following multiplication table [9].

| | $X_0$ | $X_1$ | $X_2$ |
|---|---|---|---|
| $X_0$ | $\lambda X_0$ | $c_1X_1 + c_2X_2$ | $-c_2X_1 + c_1X_2$ |
| $X_1$ | $c_1X_1 + c_2X_2$ | $\mu X_0$ | $0$ |
| $X_2$ | $-c_2X_1 + c_1X_2$ | $0$ | $\mu X_0$ |

It is also straightforward to determine the general $3\times3$ matrix $T : V \to V$ that commutes with $D$ (i.e., $T$ such that $D \in Der(V,T)$). It is given in the same basis by

$$T = \begin{bmatrix} a & 0 & 0 \\ 0 & \alpha & \beta \\ 0 & -\beta & \alpha \end{bmatrix}.$$

Thus the general 3-dimensional quadratic differential equation $\dot{X} =$

$p(X) = TX + X^2$ with $D \in Der(V, p)$ is, for $X = x_0X_0 + x_1X_1 + x_2X_2$,

$$\dot{x_0} = ax_0 + \lambda x_0^2 + \mu(x_1^2 + x_2^2)$$

$$\dot{x_1} = \alpha x_1 + \beta x_2 + 2x_0(c_1x_1 - c_2x_2)$$

$$\dot{x_2} = -\beta x_1 + \alpha x_2 + 2x_0(c_2x_1 + c_1x_2).$$

Now one way to find trajectories given by $(\exp tp)(Z) = (\exp tbD)Z$, is to solve the equation $bDZ = TZ + Z^2$ in $V$; this is the method advocated in [9] and [10]. Here we shall use a different method which makes use of the algebra $\mathcal{I}(Der(V, p))$ of polynomial invariants of $Der(V, p)$. The following is straightforward [10, Lemma 2.2].

**Lemma 6.6** *The polynomials $\gamma_0(X) = x_0$ and $\gamma_1(X) = x_1^2 + x_2^2$ generate $\mathcal{I}(Der(V, p))$ as an algebra.*

The invariant $\gamma_1$ can also be found by considering a trace form on the algebra $V$; see [9].

Applying Proposition 6.3, we have

**Proposition 6.7** *The map $\Gamma : V \to \mathbb{R}^2$ defined by $\Gamma(X) = \begin{pmatrix} \gamma_0(X) \\ \gamma_1(X) \end{pmatrix}$ is a morphism from the differential equation $\dot{X} = TX + X^2$ in $V$ described above to the differential equation $\dot{Y} = F(Y)$ in $\mathbb{R}^2$ described in coordinates by*

$$\dot{y_0} = ay_0 + \mu y_1 + \lambda y_0^2$$

$$\dot{y_1} = 2(\alpha + 2c_1y_0)y_1.$$

**Proof:** We compute

$$\dot{y_0} = \dot{x_0} = ax_0 + \lambda x_0^2 + \mu(x_1^2 + x_2^2) = ay_0 + \mu y_1 + \lambda y_0^2,$$

and

$$\begin{aligned} \dot{y_1} &= \tfrac{d}{dt}(x_1^2 + x_2^2) \\ &= 2x_1\dot{x_1} + 2x_2\dot{x_2} \\ &= 2x_1(\alpha x_1 + \beta x_2 + 2x_0(c_1x_1 - c_2x_2)) \\ &\quad +2x_2(-\beta x_1 + \alpha x_2 + 2x_0(c_2x_1 + c_1x_2)) \\ &= 2\alpha(x_1^2 + x_2^2) + 4c_1x_0(x_1^2 + x_2^2) \\ &= 2(\alpha + 2c_1y_0)y_1. \end{aligned}$$

Now to find those points $Z$ for which $(\exp tp(Z) = (\exp tbD)Z$ is a trajectory, we use Proposition 6.3. Since $\Gamma((\exp tbD)(Z) = \Gamma(P)$ for all $t \in \mathbb{R}$, $\gamma_0$ and $\gamma_1$ must be conserved on the trajectories $(\exp tp)(Z) = (\exp tbD)Z$.

This means that the trajectories must lie in the two parameter family of curves

$$x_0 = k_0 \text{ and } x_1^2 + x_2^2 = k_2.$$

Now we use the differential system $\dot{Y} = F(Y)$ given in the new variables $y_i = \gamma_i(X)$. In particular, $\dot{y}_i = \frac{d}{dt}\gamma_i((\exp tbD)Z) = \frac{d}{dt}\gamma_i(Z) = 0$. Thus to find those trajectories $(\exp tp)(Z) = (\exp tbD)Z$, we start by finding the *equilibria* for the system $\dot{Y} = F(Y)$. This procedure will determine an algebraic variety (described by the equation $F(\Gamma(X)) = 0$) in which such trajectories must lie. We will then interpret this as determining relations among the parameters $k_0$ and $k_1$: $F\begin{pmatrix} k_0 \\ k_1 \end{pmatrix} = 0$.

Thus we now solve the system of equations

$$\begin{aligned} ay_0 + \mu y_1 + \lambda y_0^2 &= 0 \\ 2(\alpha + 2c_1 y_0)y_1 &= 0. \end{aligned}$$

To maintain clarity of the exposition, we shall not do this in full generality. Instead, we shall make the simplifying assumption that the linear transformation $T : V \to V$ is the zero mapping, i.e. $T = 0$ (as in [9]). In particular, this means that $a = \alpha = 0$. We shall also ignore the special subcases that arise when the various parameters in this system of equations are zero. Thus we must solve the system

$$\begin{aligned} \mu y_1 + \lambda y_0^2 &= 0 \\ 4c_1 y_0 y_1 &= 0. \end{aligned}$$

We assume that $\mu \neq 0$ (otherwise it turns out that *every* solution to $\dot{X} = X^2$ is periodic and given by $(\exp tp)(Z) = (\exp btD)Z$ [9]), solve the first equation for $y_1$, and plug into the second equation to obtain $(-4c_1\lambda/mu)y_0^3 = 0$. If $y_0 = 0$, then by the first equation, $y_1 = 0$, so we have no nontrivial solutions in this case. Thus $c_1\lambda = 0$. Now we assume that $\lambda \neq 0$ (otherwise it turns out that we are in the same special case referred to previously [9]), so $c_1 = 0$. Applying this to the multiplication table for $V$, we find that our algebra is the algebra of Section 5, where $c = c_2$.

Returning to the $x_i$ coordinates, the desired trajectories must lie on the variety determined by the equation $\lambda x_0^2 + \mu(x_1^2 + x_2^2) = 0$. In the presence of our earlier assumptions, this will be nontrivial if and only if $\lambda\mu < 0$; thus the trajectories lie on the cone $C$ described in Section 5. Since the trajectories lie in the two parameter family of curves $x_0 = k_0$ and $x_1^2 + x_2^2 = k_1$, each trajectory is determined by specifying a solution $(k_0, k_1) \neq (0, 0)$ to the equation $\lambda k_0^2 + \mu k_1 = 0$.

Having selected a trajectory (i.e., specified a pair $(k_0, k_1)$), all that remains is to determine the parameter $b$ so that the trajectory is given by

$(\exp tp)(Z) = (\exp tbD)Z$ for some $Z$ in the trajectory. Now we use the equation $bDZ = Z^2$. Instead of using this equation to find $Z$ as in [9] and [10], we use it to determine $b$. For $Z = z_0X_0 + z_1X_1 + z_2X_2$, the equation is given in coordinates by the system

$$\begin{aligned} 0 &= \lambda z_0^2 + \mu(z_1^2 + z_2^2) = \lambda k_0^2 + \mu k_1 \\ bz_2 &= -2cz_0z_2 \\ -bz_1 &= 2cz_0z_1. \end{aligned}$$

The first equation is redundant. Multiplying the second equation by $z_2$, the third by $z_1$, and subtracting gives $bk_1 = -2ck_0k_1$, so (since $k_1 \neq 0$) $b = -2ck_0$.

We summarize this discussion as follows. To find a trajectory given by the action of $\exp tbD$ for some $b \in \mathbb{R}$ on an initial point, we fix a nonzero solution to the equation $\lambda k_0^2 + \mu k_1 = 0$. The trajectory is then determined by the pair of equations $x_0 = k_0$ and $x_1^2 + x_2^2 = k_1$. For $Z$ in this trajectory, the solution starting at $Z$ is described by $(\exp tp)(Z) = (\exp tbD)Z$ where $b = -2ck_0$.

**Remarks:** (1) See [10] for examples of quasiperiodic trajectories and "hyperbolic" trajectories given by the action of $Aut(V, p)$ on an initial point. (2) The authors have also been investigating examples of quadratic systems where $Der(V, p)$ is a representation of $sl(2, \mathbb{R})$, and acts irreducibly on the algebra $V$. These results will appear elsewhere.

## References

[1] V.I. Arnold, "Mathematical Methods of Classical Mechanics", Second Edition, Springer-Verlag, Berlin-Heidelberg-New York, 1989

[2] P. D. Gerber, Left alternative algebras and quadratic differential equations I and II, IBM Thomas J. Watson Research Center, New York (1973)

[3] P. D. Gerber, Linearization of Cauchy's problem for quadratic semilinear partial differential equations, IBM J. Res. Develop., July 1973, 314-323

[4] P. D. Gerber, LIP loops and quadratic differential equations, Chapter XIII of "Quasigroups and Loops: Theory and Applications", Sigma Ser. Pure Math., **8**, Heldermann, Berlin, 1990, pp. 431-443

[5] M. Golubitsky, I. Stewart, and D.G. Schaffer, *Singularities and Groups in Bifurcation Theory*, Vol. II, Applied Mathematical Sciences, vol. 69, Springer-Verlag, Berlin-Heidelberg-New York, 1988.

[6] S. Helgason, "Differential Geometry, Lie Groups, and Symmetric Spaces", Academic Press, New York, 1978

[7] R. Hermann, "Differential Geometry and the Calculus of Variations", Math. Sci. Press, 1977.

[8] J. Kaplan and J. Yorke, Nonassociative real algebras and quadratic differential equations, Nonlinear Analysis TMA, **3** (1979), pp. 49-51

[9] M. K. Kinyon and A. A. Sagle, Quadratic dynamical systems and algebras, to appear in J. Diff. Eq.

[10] M. K. Kinyon and A. A. Sagle, Automorphisms and derivations of ordinary differential equations and algebras, to appear in Rocky Mountain Math. J.

[11] M. K. Kinyon and A. A. Sagle, Quadratic dynamical systems, in "Proceedings of the International Symposium on Nonassociative Algebras and Related Topics", K. Yamaguti and N. Kawamoto (eds.), World Scientific, 1992, pp. 101-114.

[12] M. K. Kinyon and A. A. Sagle, Critical elements of quadratic systems, in "Topology, Analysis, and Applications", G.M. Rassias (ed.), World Scientific, 1992.

[13] M. K. Kinyon, H. C. Myung, and A. A. Sagle, Quadratic differential equations, in "Hadronic Mechanics and Nonpotential Interactions 5 - Part I. Mathematics", H.C. Myung (ed.), Nova Science, 1991.

[14] S. Kobayashi and K. Nomizu, "Foundations of Differential Geometry, II", Wiley, New York, 1968

[15] D. E. Koditschek and K. S. Narendra, The stability of second-order quadratic differential equations, IEEE Trans. on Automatic Control, Vol. AC-27, no. 4, pp. 783-798

[16] M. Koecher, Die Riccatische Differentialgleichung und nicht-assoziative algebren, Abh. Math. Sem. Univ. Hamburg, **46** (1977), pp. 129-141

[17] L. Markus, Quadratic differential equations and nonassociative algebras, in "Contributions to the Theory of Nonlinear Oscillations", Vol. V, L. Cesari, J.P. LaSalle, and S. Lefschetz (eds.), Princeton Univ. Press, Princeton, 1960, pp. 185-213

[18] L. Markus, Group theory and differential equations, Lecture Notes, University of Minnesota, Minneapolis, 1960.

[19] H. C. Myung and A. A. Sagle, Quadratic differential equations and algebras, Contemporary Math., **131** (1992), pp. 659-672

[20] K. Nomizu, Invariant affine connections on homogeneous spaces, Amer. J. Math., **9** (1954), pp. 33-65

[21] M. Oka, Classifications of two-dimensional homogeneous cubic differential equation systems I and II, TRU Math. 16-2 (1980), pp.19-62, and TRU Math. 17-1 (1981), pp. 65-88

[22] P. J. Olver, "Applications of Lie Groups to Differential Equations", Springer-Verlag, Berlin-Heidelberg-New York, 1986

[23] T. Ratiu, The motion of the free $n$-dimensional rigid body, Indiana Math. J., **29** (1980), pp. 609-630

[24] H. Röhrl, Algebras and differential equations, Nagoya Math. J., **68** (1977), pp. 59-122

[25] H. Röhrl, A theorem on non-associative algebras and its application to differential equations, Manuscripta Math. **21** (1977), pp. 181-187

[26] M. Roth, Zentralisator und Normalisator quadratischer Polynome, Dissertation, TU München (1992)

[27] A. A. Sagle, Jordan algebras and connections on homogeneous spaces, Trans. A.M.S., **187** (1974), pp. 405-427

[28] A. A. Sagle, Invariant Lagrangian mechanics, connections, and nonassociative algebras, Algebras, Groups, and Geometries, **3** (1986), pp. 199-263

[29] A. A. Sagle and R. Walde, "Introduction to Lie Groups and Lie Algebras", Academic Press, New York, 1973

[30] R. D. Schafer, "Introduction to Nonassociative Algebras", Academic Press, New York, 1966

[31] H. Stephani, "Differential Equations: Their Solution Using Symmetries", Cambridge University Press, 1991

[32] A. Vogt, The Riccati equation: when nonlinearity reduces to linearity, in "Nonlinear Semigroups, Partial Differential Equations, and Attractors", Proceedings, Washington, D.C., 1985, ed. T.L. Gill and W.W. Zachary, Lecture Notes in Math **1248**, Springer-Verlag, 1987

[33] A. Vogt, The Riccati equation revisited, in "Nonlinear Semigroups, Partial Differential Equations, and Attractors", Proceedings, Washington, D.C., 1987, ed. T.L. Gill and W.W. Zachary, Lecture Notes in Math **1394**, Springer-Verlag, 1989

[34] S. Walcher, A characterization of regular Jordan pairs andi its application to Ricatti differential equations, Comm. Alg. **14** (10) (1986), 1967–1978.

[35] S. Walcher, "Algebras and Differential Equations", Hadronic Press, Palm Harbor, 1991.

[36] S. Walcher, On differential equations in normal form, Math. Ann., **291** (1991), pp. 293-314.

[37] S. Walcher, Symmetries of ordinary differential equations, to appear in Nova J. Alg. Geom.

[38] S. Walcher, Algebraic structures and differential equations, preprint.

# 10 Limit Cycles and Centres: An Example

**N. G. Lloyd** University College of Wales, Aberystwyth, Wales

**J. M. Pearson** University College of Wales, Aberystwyth, Wales

## 1: Introduction

Systems of the form

$$\dot{x} = P(x, y), \quad \dot{y} = Q(x, y) \tag{1.1}$$

in which $P$ and $Q$ are polynomial functions continue to attract a great deal of interest. One of the famous problems in the theory of nonlinear oscillations is the number of limit cycles which such systems can have when $P$ and $Q$ are of specified degrees. This is part of the sixteenth of the 'problems for the twentieth century' posed by David Hilbert in 1900 [6]: though easily stated it has proved remarkably intractable, and progress has on the whole been slow and intermittent. Since little progress can be expected in general most research has proceeded by considering particular classes of equations and concentrating on limit cycles which bifurcate from structures such as critical points and homoclinic loops.

Until the last ten years or so work on polynomial systems was somewhat fitful, but since then there has been a marked quickening of interest and significant deepening of understanding. This has occurred at least in part as a result of the availability of effective Computer Algebra systems to perform the very large calculations which arise. We refer to the survey article [8] for the history of work on Hilbert's sixteenth problem and a description of progress up to 1988.

Our work has concentrated on limit cycles which bifurcate from critical points and the closely related issue of the conditions under which a critical point is a centre. The derivation of necessary and sufficient conditions for a centre is of independent interest and is, as we shall see, strongly connected with the question of the integrability of the systems concerned.

There is an extensive literature on quadratic systems and a considerable amount of work has by now been done on cubic systems. In this paper we illustrate our recent research by detailed reference to the quartic system

$$\dot{x} = y(x+1), \quad \dot{y} = -x - a_1x^2 - a_2x^3 - a_3x^4 - a(1+ux)xy - wy^2 \tag{1.2}$$

which was investigated by Cherkas in the seventies. He showed that certain polynomial systems could be transformed to Liénard form:

$$\dot{X} = Y - F(X), \quad \dot{Y} = -G(X).$$

He then used the theory which he and others had developed for Liénard systems to obtain conditions under which the origin is a centre. However calculations performed by Wang [15] strongly suggested that the list of conditions given by Cherkas for (1.2) was incomplete [1,2]. We confirm that this is indeed so and present a set of necessary and sufficient conditions. We make extensive use of the Computer Algebra system REDUCE, both in proving the necessity of the conditions and their sufficiency.

## 2: Liapunov Quantities and Limit Cycles

Our approach to the bifurcation of limit cycles from critical points (multiple Hopf bifurcation) is described in detail in [9] and has been used in several recent papers - see [10] and the references given therein, for example. Necessary conditions for a centre are obtained simultaneously but as seen in [11] severe computational difficulties sometimes arise.

We give a brief description of the technique. Suppose that the origin is a critical point of focus type for (1.1). We can write the system in the form

$$\dot{x} = \lambda x + y + p(x,y), \quad \dot{y} = -x + \lambda y + q(x,y) \tag{2.1}$$

where $p$ and $q$ are polynomials without linear terms. If the origin is not a centre when $\lambda = 0$ it is said to be a *fine focus*. The number of limit cycles that can bifurcate from the origin under perturbation of $P$ and $Q$ is determined by the multiplicity of $x = 0$ as a zero of the displacement map $H : x \mapsto h(x) - x$, where $(h(x), 0)$ is where the positive semi-orbit through $(x, 0)$ next meets the $x$-axis. The classical way to find this multiplicity is to use the fact that there is a function $V$, analytic in a neighbourhood of the origin, such that $\dot{V}$, its rate of change along orbits, is of the form $\eta_2 r^2 + \eta_4 r^4 + \eta_6 r^6 + \ldots$, where $r^2 = x^2 + y^2$.

The $\eta_{2k}$ are the *focal values* and are polynomials in the coefficients which occur in $P$ and $Q$. The stability of the origin is determined by the sign of the first non-zero focal value, and the origin is a centre if and only if all the focal values are zero. The fine focus at the origin is of *order* $k$ if $\eta_{2\ell} = 0$ for $\ell \leq k$ and $\eta_{2k+2} \neq 0$. It is easily shown that at most $k$ limit cycles can bifurcate from a fine focus of order $k$.

For a given class $\mathcal{C}$ of systems it follows from the Hilbert basis theorem that there is $K(\mathcal{C})$ such that the origin is a centre if $\eta_{2k} = 0$ for all $k \leq K+1$. The smallest value of $K$ with this property is thus the maximum order of a fine focus for systems in $\mathcal{C}$. To maximise the number of limit cycles which bifurcate a system with a fine focus of maximal order is taken and a sequence of perturbations introduced each of which reduces the order by one and reverses the stability of the origin.

To implement this approach the first task is to calculate the focal values. It is possible to do this by hand only in the simplest situations and it is therefore necessary to use an appropriate Computer Algebra system. We use a procedure which we have written called FINDETA. This is described in [9] and uses REDUCE. Having computed as many focal values as seems necessary each is reduced modulo those already known: that is, for each $\ell$ the relations $\eta_2 = \eta_4 = \ldots = \eta_{2\ell} = 0$ are used to simplify $\eta_{2\ell+2}$. The expressions so obtained are the *Liapunov quantities* and are denoted by $L(k)$; it is immediate that $L(0) = \lambda$. It usually becomes obvious what the maximum possible order of a fine focus is likely to be. To prove that this guess ($\kappa$, say) is correct we prove independently that the origin is a centre if $\eta_{2k} = 0$ for $k \leq \kappa + 1$. Thus we obtain necessary conditions for a centre but at the same time require a full knowledge of sufficient conditions. The final stage is to find a sequence of perturbations under each of which a limit cycle bifurcates. Such limit cycles are said to be of *small amplitude*.

To prove that the origin is a centre under certain conditions two well known results are available. We have a centre if the system is Hamiltonian: $P_x + Q_y = 0$, or if the system is symmetric in one of the axes. However, these two simple criteria by no means cover all cases and much effort has gone into developing other approaches, as we shall describe in Section 3.

We now turn to system (1.2) and show how the approach we have described is implemented in practice. First we derive a set of conditions one of which must be satisfied if the origin is a centre and then return to the question of limit cycles. Using the sufficiency of the centre conditions we shall show that four limit cycles - and no more - can bifurcate from a fine focus at the origin.

The first task is to compute the focal values. Using a REDUCE 3.3 version of FINDETA running on the Amdahl 5890 at the Manchester Computing Centre we computed the focal values up to $\eta_{14}$ in 48 seconds cpu. From $\eta_4$ we find that

$$L(1) = -a(u - w - a_1).$$

Since it is the sign of the Liapunov quantities that is significant, positive constants multiplying them may be omitted. If $L(1) \neq 0$ the origin is of order one, and at most one limit cycle can bifurcate. We therefore suppose that $L(1) = 0$. Now if $a = 0$ the origin is a centre by the symmetry criterion: the phase portrait is symmetric in the $x$-axis. We therefore suppose that $a \neq 0$. For a centre or a fine focus of order greater than one we require $L(1) = 0$ and we take

$$a_1 = u - w. \tag{2.2}$$

Then $L(2) = a\Phi$ where

$$\Phi = (3a_3 + a_2(w + 2 - 5u) + uw(2w + 3 - 5u))$$

and from $L(2) = 0$ we take

$$a_3 = -\tfrac{1}{3}(a_2(w + 2 - 5u) + uw(2w + 3 - 5u)). \tag{2.3}$$

We then find that

$$L(3) = aAB, \quad L(4) = -aAC, \quad L(5) = aAD \tag{2.4}$$

where

$$A = 2u^2w - u(2w^2 + 3w - 2a_2) - a_2(w + 2),$$

$$B = 35u^2 - 35u - 2w^2 - 4w + 6$$

and $C, D$ are polynomials in $a, a_2, u$ and $w$ ($C$ is of degree 4 with 25 terms and $D$ is of degree 6 with 73 terms). The relations (2.4) suggest that if $a_1$ and $a_3$ are as given above then the origin is a centre if $A = 0$ or if $B = C = D = 0$.

When $A = 0$, (2.3) simplifies to give $a_3 = a_2 u + wu^2$. Consider the possibility that $B = C = D = 0$. We use REDUCE to evaluate $R_1$, the resultant of $B$ and $C$ with respect to $u$, and $R_2$, the resultant of $B$ and $D$ with respect to $u$. We find that

$$R_1 = \gamma_1(w-1)^2(w+3)^2(3w+2)(3w+4)E$$

and

$$R_2 = \gamma_2(w-1)^2(w+3)^2(3w+2)(3w+4)F,$$

where $\gamma_1$ and $\gamma_2$ are non-zero integers, $E = 8w^2+16w+11$ and $F$ is a polynomial in $a, w$ and $a_2$ of degree 6. The coefficients of the highest powers of $u$ in $B, C$ and $D$ are non-zero, so for $B, C$ and $D$ to be simultaneously zero we require both resultants to be zero. Now $E$ is positive definite, so $R_1 = R_2 = 0$ only if $w = 1, -3, -\frac{2}{3}$ or $-\frac{4}{3}$. The corresponding values of $u$ are found from $B = 0$; the pairs $(u, w)$ are

$$(0,1), (1,1), (0,-3), (1,-3), \left(\tfrac{1}{3}, -\tfrac{2}{3}\right), \left(\tfrac{2}{3}, -\tfrac{4}{3}\right).$$

We confirm that $C$ and $D$ are both zero for these values of $u$ and $w$. Thus we have the following result in which necessary conditions for a centre are listed.

THEOREM 2.1 *Suppose that the origin is a centre for the system (1.2). Then one of the following conditions holds:*

(1) $a = 0$,

(2) $a_1 = u - w,\ a_3 = a_2 u + w u^2,\ A = 0$,

(3) $u = 0,\ w = 1,\ a_1 = -1,\ a_3 = -a_2$,

(4) $u = 1,\ w = 1,\ a_1 = 0,\ a_3 = \frac{2}{3}a_2$,

(5) $u = \frac{1}{3},\ w = -\frac{2}{3},\ a_1 = 1,\ a_3 = \frac{1}{9}a_2$,

(6) $u = \frac{2}{3},\ w = -\frac{4}{3},\ a_1 = 2,\ a_3 = \frac{8}{9}(a_2 - 1)$,

(7) $u = 0,\ w = -3,\ a_1 = 3,\ a_3 = \frac{1}{3}a_2$,

(8) $u = 1,\ w = -3,\ a_1 = 4,\ a_3 = 2a_2 - 8$.

REMARK. The conditions (4),(7) and (8) are not given by Cherkas.

We shall see in Section 3 that the conditions given in Theorem 2.1 are in fact sufficient. With this knowledge we can return to the question of the number of limit cycles which can bifurcate from a fine focus at the origin. Referring to the Liapunov quantities given above and using the sufficiency of the conditions given in Theorem 2.1 we show that the maximum order of a fine focus is four.

We see that for the origin to be a fine focus of order greater than three we require $a_1$ and $a_3$ to be given by (2.2) and (2.3) respectively, and $B = 0$. Moreover $aA$ and at least one of $C, D$ must be non-zero. Suppose that $B = 0$; we take $u = (35+v)/70$, where $v^2 = 35E$. Then

$$C = -12(w-1)(w+3)(E + (w+1)v)/35.$$

For a fine focus of order greater than four we must also have $C = 0$. If $w = 1$ then $u = 1$ or $u = 0$. For $u = 1$ we have that $a_1 = 0$ and $a_3 = \frac{2}{3}a_2$, and the origin is a centre by condition (4) of Theorem 2.1. When $u = 0$ then $a_1 = -1$, $a_3 = -a_2$, in which case the origin is a centre by condition (3) of Theorem 2.1. Similarly if $w = -3$ we find that the origin is a centre by conditions (7) and (8) of Theorem 2.1. If $w \neq -1$ we have $v = -E(w+1)^{-1}$ and for consistency we must have

$$E^2 = 35(w+1)^2(8w^2 + 16w + 11)$$

which is satisfied only if $w = -\frac{2}{3}$ or $w = -\frac{4}{3}$. In both cases the origin is a centre (conditions (5),(6) of Theorem 2.1).

It follows that the origin is a fine focus of order four or more only if $w = -1$, $u = (35 \pm \sqrt{105})/70$, $a_1$ and $a_3$ are given by (2.2) and (2.3) respectively, and $aA \neq 0$. Under these conditions $C$ is a non-zero constant, so the order is exactly four. Note that $A = (2u-1)(a_2 - u)$ so that $A \neq 0$ provided $a_2 \neq u$.

LEMMA 2.2 *The maximum order of a fine focus at the origin for (1.2) is four. It is of order four if* $w = -1$, $u = (35 \pm \sqrt{105})/70$, $a_1 = u + 1$, $a_3 = -\frac{1}{3}(1-5u)(a_2 - u)$, $a \neq 0$ *and* $a_2 \neq u$.

Having established the maximum order of the fine focus we turn to the final stage of the process, which is to find perturbations under which four limit cycles bifurcate from the origin. We start with a system satisfying the conditions given in Lemma 2.2. We take $u = u_* = (35 - \sqrt{105})/70$ for definiteness; a similar argument works if $u = (35 + \sqrt{105})/70$. First we perturb $u$ so that $L(3)L(4) < 0$; that is $BC > 0$. Now $C > 0$, so we perturb $u$ so that $B > 0$. Since we start with $u = u_*$ this is achieved by decreasing $u$. At the same time $a_1$ and $a_3$ are adjusted so that (2.2) and (2.3) still hold. The stability of the origin is reversed and its order is reduced by one, so a limit cycle bifurcates.

The next stage is to perturb $a_3$ so that $L(2)$ becomes non-zero but $L(2)L(3) < 0$. This means that $\Phi$ must have the sign of $-A$ after perturbation. Since $u_* < \frac{1}{2}$, $A > 0$ if $a_2 < u_*$ and $A < 0$ if $a_2 > u_*$. Thus we increase $a_3$ if $a_2 > u_*$ while we decrease $a_3$ if $a_2 < u_*$. While making this perturbation we adjust $a_1$ so that (2.2) is satisfied, and the second limit cycle bifurcates.

The third limit cycle is produced by perturbing $a_1$ so that $u + 1 - a_1$ has the sign of $-A$ (to ensure that $L(1)L(3) > 0$ and so $L(1)L(2) < 0$). This is achieved by increasing or decreasing $a_1$ according to whether $a_2 < u_*$ or $a_2 > u_*$. To complete the construction of four limit cycles a non-zero value of $\lambda$ is introduced (see (2.1)). Its sign is such that $\lambda L(1) < 0$, that is $\lambda a > 0$ if $a_2 < u_*$ and $\lambda a < 0$ if $a_2 > u_*$.

THEOREM 2.3 *There are systems of the form (1.2) with four small amplitude limit cycles and this is the maximum possible number.*

## 3: Sufficiency of the Centre Conditions

We now turn to the question of proving the sufficiency of the conditions given in Theorem 2.1. System (1.2) is interesting in this regard because of the variety of techniques which are required. The problem of distinguishing between a centre and a focus is a difficult one and has attracted a great deal of attention over the years. There are well-known conditions for a centre for quadratic systems and symmetric cubic systems but until recently necessary and sufficient conditions were known for few other classes of systems. The basic ideas are well established, indeed the conditions for quadratic systems date back to Dulac (1908) and Kapteijn (1911,1912) (see [16] for details). However it is only quite recently that it has been possible to approach the question at all systematically. This again is in no small measure due to the advent of Computer Algebra systems.

We exploit the fact that if the origin is a critical point of focus type for a system of the form (1.1) then it is a centre if there is a function $B$ such that $B(0,0) \neq 0$ and

$$\frac{\partial}{\partial x}(BP) + \frac{\partial}{\partial y}(BQ) = 0. \tag{3.1}$$

When such a function $B$ exists the system

$$\dot{x} = B(x,y)P(x,y), \quad \dot{y} = B(x,y)Q(x,y),$$

which has the same orbits as (1.1), is Hamiltonian. Thus the original system can be transformed to a Hamiltonian system by a position-dependent transformation of time.

A function such as $B$ is said to be an *integrating factor* or a *Dulac function.* As explained in [13] and [14] the existence of a function $B$ satisfying (3.1) is closely related to whether the system has an elementary first integral. These ideas have their roots in work of Darboux in the last century and we refer to [7] for a modern treatment.

Finding integrating factors often requires considerable ingenuity. We follow the approach described by Christopher [3] who was able to derive centre conditions for systems which had previously proved intractable. We seek functions $B$ of the form $C_1^{\alpha_1} C_2^{\alpha_2} \dots C_r^{\alpha_r}$ where each $C_k$ is a polynomial in $x$ and $y$ which is non-zero at the origin. Each curve $C_k = 0$ is invariant under the flow determined by (1.1) and the $\alpha_k$ are independent of $x$ and $y$. The significance of invariant algebraic curves has long been understood, both in the context of centre conditions (see [5], for instance) and in questions relating to the existence and uniqueness of limit cycles. Several results of this kind are given by Ye in [16] and other questions are raised by him in [17].

Suppose that $C(x, y)$ is a polynomial. We say that $C = 0$ (or simply $C$) is *invariant* with respect to (1.1) if there is a polynomial $L$ such that $\dot{C} = CL$, where $\dot{C} = C_x P + C_y Q$, and $C(0,0) \neq 0$. We call $L$ the *cofactor* and its degree is one less than that of the maximum degree of $P$ and $Q$. Let $C_1, C_2, \dots, C_r$ be invariant algebraic curves such that

$$\dot{C}_k = C_k L_k \qquad (k = 1, \dots, r). \tag{3.2}$$

Let $B = \prod_{k=1}^{r} C_k^{\alpha_k}$ where all the $\alpha_k$ are non-zero. Then

$$\frac{\partial}{\partial x}(BP) + \frac{\partial}{\partial y}(BQ) = B(P_x + Q_y + \alpha_1 L_1 + \dots + \alpha_r L_r).$$

The aim is to determine relationships between the coefficients in $P$ and $Q$ and to find $C_k, L_k$ satisfying (3.2) and $\alpha_k$ such that

$$P_x + Q_y + \alpha_1 L_1 + \alpha_2 L_2 + \dots + \alpha_r L_r = 0. \tag{3.3}$$

If the degree of $C_k$ is $m_k$ and the degree of the system is $n$ then by comparing coefficients (3.2) and (3.3) give rise to

$$\tfrac{1}{2}n(n+1)-(r+1)+\sum_{k=1}^{r}\tfrac{1}{2}(n+m_k)(n+m_k+1)$$

equations for the

$$\tfrac{1}{2}rn(n+1)+\sum_{k=1}^{r}\tfrac{1}{2}(m_k^2+3m_k)$$

unknowns ( $\alpha_k$ and the coefficients of $C_k, L_k$ ).

The invariant algebraic curves which we seek may be real or complex. If $C$ is complex, it can still be used to construct a real Dulac function by noting that $C\overline{C}$ is a real invariant curve.

We have been able to partly automate the search for invariant curves. We have written a REDUCE procedure to obtain the equations corresponding to (3.2) and (3.3) above. A start has been made to make the whole process automatic but this is a much more sophisticated computational problem.

Considering the specific system (1.2) it is immediately clear that the line $\Lambda : x+1=0$ is invariant. We observe that $(x+1)^\alpha$ is a Dulac function if and only if $\alpha = 2w-1$ and $a=0$. Thus the origin is a centre if $a=0$. This is condition (1) of Theorem 2.1, a result which we have already noted follows from the symmetry criterion.

For system (1.2) there is no advantage in considering more than one invariant curve in addition to $\Lambda$ at any one time. If $C$ is invariant and of degree $N$ let $C = 1+\sum_{k=1}^{N}\Gamma_k$ where $\Gamma_k$ is homogeneous of degree $k$;similarly $L = L_1+L_2+L_3$.

By considering the terms of degree one in (3.2) we see that if $\Gamma_1 = rx+sy$ then $L_1 = -sx+ry$.

LEMMA 3.1 *Suppose that $C$ is an invariant polynomial for system (1.2) with cofactor $L$. Then $\Gamma_N$ is a constant multiple of $x^N$ and $L_3 = 0$.*

PROOF: For (1.2), $L$ is of course of degree at most three. That $\Gamma_N$ is independent of $y$ follows from a result in [3]. For a direct proof suppose that $\ell = cx + dy$ is a linear factor of $\Gamma_N$, real or complex. Write $\Gamma_N = \ell^m K$ where $\ell$ does not divide $K$. Then

$$\ell^m KL = -a_3x^4(m\ell^{m-1}dK + \ell^m K_y),$$

from which it follows that $\ell$ divides $x^4K$. Since $\ell$ does not divide $K$, $\ell = cx$. Now comparing terms of highest degree in the relation (3.2) we have $\Gamma_N L_3 = 0$, whence $L_3 = 0$.

We first seek an invariant conic of (1.2) and then use it together with the straight line $\Lambda$ to construct an integrating factor. By Lemma 3.1 an invariant conic must be of the form

$$C = tx^2 + rx + sy + 1 \tag{3.4}$$

and its cofactor is $L_2 - sx + ry$. Comparing terms of degree 4 in (3.2) we have

$$-a_3sx^4 = tx^2L_2.$$

Hence $L_2 = \mu x^2$ where $\mu = -sa_3/t$. We then obtain six equations involving $r, s$ and $t$. One solution gives the polynomial $(x+1)^2$, which is obviously invariant. There is another invariant conic only if the relation (3.5) below is satisfied.

LEMMA 3.2 *Suppose that (1.2) has an invariant conic other than $(x+1)^2$. Then*

$$a_3 = a_2w_1 + ww_1^2, \tag{3.5}$$

*where $w_1 = w + a_1$. The conic is of the form (3.4) with $r = -w$, $t = a_3w_1^{-1}$ and $s$ satisfying the two relations*

$$s^2 - as - w^2 - w + 2a_3w_1^{-1} = 0 \tag{3.6}$$

$$s^2w_1 - aus + a_3(w+2)w_1^{-1} = 0. \tag{3.7}$$

Having established the conditions for the existence of an invariant conic $C$ we seek a Dulac function of the form $B = (x+1)^{\alpha_1} C^{\alpha_2}$. We need $\mu\alpha_2 = au$, $s\alpha_2 + a = 0$ and $\alpha_1 = 2w - 1 + w\alpha_2$. These relations require $u = w_1$, in which case (3.6) and (3.7) are satisfied provided that

$$w^2 + w - 2a_3 w_1^{-1} + a_3(w+2)w_1^{-2} = 0. \tag{3.8}$$

Thus $B$ is a Dulac function if $a_1 = u - w$, $a_3 = a_2 u + wu^2$ and (3.8) is satisfied. These are exactly the conditions (2) of Theorem 2.1.

THEOREM 3.3 *Suppose that condition (2) of Theorem 2.1 holds. Then the origin is a centre and* $(x+1)^{\alpha_1} C^{\alpha_2}$ *is an integrating factor with* $\alpha_1, \alpha_2$ *and* $C$ *as given above.*

We move on to the possibility of invariant cubics. We find $(x+1)^3$ of course, and $(x+1)C$ where $C$ is the invariant conic of Theorem 3.3. More significantly there are two (and only two) sets of circumstances in which there is an invariant irreducible cubic: these lead directly to conditions (4) and (7) of Theorem 2.1. We summarise our conclusions as follows.

THEOREM 3.4

*(1) Suppose that* $w = 1$, $a_1 = 0$, $a_3 = \frac{2}{3}a_2$ *and* $u = 1$. *Then*

$$C = a_3 x^3 + a_2 x^2 + sxy + sy + 1,$$

*where* $s^2 - as + 2a_2 = 0$, *is invariant and* $(x+1)^{\alpha_1} C^{\alpha_2}$ *is a Dulac function with* $\alpha_1 = 1$ *and* $\alpha_2 = -a/s$. *Hence condition (4) of Theorem 2.1 is sufficient for the origin to be a centre.*

*(2) Suppose that* $w = -3$, $a_1 = 3$, $a_3 = \frac{1}{3}a_2$ *and* $u = 0$. *Then*

$$C = a_3 x^3 + a_2 x^2 + sy + 3x + 1,$$

*where* $s^2 - as + 2a_2 - 6 = 0$, *is invariant and* $(x+1)^{\alpha_1} C^{\alpha_2}$ *is a Dulac function with* $\alpha_1 = -7 - 3\alpha_2$ *and* $\alpha_2 = -a/s$. *Hence condition (7) of Theorem 2.1 is sufficient for the origin to be a centre.*

We have so far shown that the sufficiency of four of the conditions for a centre given in Theorem 2.1 can be proved by constructing a Dulac function from invariant polynomials. We note that two of these - those derived from the existence of invariant cubics - were not included by Cherkas in his original list. Despite intensive efforts we have not been able to show that there are invariant polynomials of degree greater than three. It is possible to build up considerable information about the structure of possible invariant polynomials of degree $N$ by looking at the terms of degree $N+3, N+2, N+1$ in (3.2). We summarise some of our conclusions but do not give details. In the following we express the fact that a polynomial is a linear combination of certain monomials by writing them as a list within curly brackets.

LEMMA 3.5 *Suppose that* $C = 1 + \sum_{k=1}^{N} \Gamma_k$ *is invariant and that its cofactor is* $L$.

*(1) If* $4 \leq N \leq 7$, $L_2 = \{x^2\}$, $\Gamma_N = \{x^N\}$, $\Gamma_{N-1} = \{x^{N-1}, x^{N-2}y\}$ *and* $\Gamma_{N-2} = \{x^{N-2}, x^{N-3}y, x^{N-4}y^2\}$.

*(2) If* $8 \leq N \leq 11$, $L_2 = \{x^2, xy\}$, $\Gamma_N = \{x^N\}$, $\Gamma_{N-1} = \{x^{N-1}, x^{N-2}y, x^{N-3}y^2\}$ *and* $\Gamma_{N-2} = \{x^{N-2}, \ldots, x^{N-6}y^4\}$.

We have sought invariant curves of degree less than or equal to nine but have found none of degree greater than three; we believe that there are none. This strongly suggests that the approach via invariant polynomials is inadequate to prove the sufficiency of all the cases in Theorem 2.1. We prove that if condition (3) holds then there are no invariant polynomials, a result which is of independent interest in the light of the work recently reported by Odani [12].

Suppose that $a_1 = -1$, $w = 1$, $u = 0$ and $a_3 = -a_2$ with $aa_2 \neq 0$. Then the system can be written in terms of coordinates $x$ and $Y$, where $Y = y(x+1)$:

$$\dot{x} = Y, \quad \dot{Y} = -axY + q(x), \tag{3.9}$$

where $q(x) = x(x^2-1)(a_2x^2+1)$. The transformed system is of Liénard form and the origin is immediately seen to be a centre by symmetry in the $Y$-axis. Note that

(1.2) has an invariant polynomial if and only if (3.9) does. Reverting to coordinates $(x, y)$ let

$$C = f_0(x) + f_1(x)y + \ldots + f_n(x)y^n$$

where $f_n$ is not identically zero and $f_0(0) \neq 0$. Suppose that $\dot{C} = CL$, where $L = \sum_{k=0}^{4} L_k(x)y^k$. Since (3.9) is of degree five, the cofactor $L$ is of degree at most four. We then have

$$y\left(\sum_{k=0}^{n} f_k'(x)y^k\right) + (q(x) - axy)\left(\sum_{k=1}^{n} kf_k(x)y^{k-1}\right)$$

$$= \left(\sum_{k=0}^{n} f_k(x)y^k\right)\left(\sum_{k=0}^{4} L_k(x)y^k\right).$$

By comparing coefficients of $y^{n+4}, y^{n+3}$ and $y^{n+2}$ we immediately see that $L_4 = L_3 = L_2 = 0$. Considering the coefficients of $y^{n+1}$ we have $f_n' = f_n L_1$. Since $f_n$ is a polynomial $L_1 = 0$ and $f_n$ is constant: let $f_n(x) = c_n \neq 0$. Continuing with the comparison of coefficients of successively lower powers of $y$,

$$f_{n-1}(x) = c_{n-1} + c_n \int (L_0 + anx) \tag{3.10}$$

where $c_{n-1}$ is a constant, and for $1 \leq \ell < n$

$$f_{n-\ell-1}(x) = c_{n-\ell-1} + \int ((L_0 + a(n-\ell)x)f_{n-\ell}(x) - (n-\ell+1)q(x)f_{n-\ell+1}(x)) \tag{3.11}$$

where $c_{n-\ell-1}$ is constant. Finally, by considering the terms independent of $y$ we have the consistency condition

$$q(x)f_1(x) = L_0(x)f_0(x). \tag{3.12}$$

From (3.10) and (3.11) we see that if $\partial L_0 = m$ then $\partial f_{\ell-1} = \partial f_\ell + m + 1$. But from (3.12) $\partial f_0 = \partial f_1 + 5 - m$, so $m = 2$. Write $L_0 = \mu_1 x + \mu_2 x^2$. Now we find by induction from (3.10) and (3.11) that none of the functions $f_\ell(x)$ contains a term of degree one. Then comparing the coefficients of $x^2$ in (3.12) gives $\mu_2 = 0$ or $c_0 = 0$. But $f_0(0) \neq 0$, so we must have $\mu_2 = 0$.

We now observe that if $\ell$ is even, then the highest degree term in $f_{n-\ell}$ is a constant multiple of $c_n a_2^{\frac{1}{2}\ell} x^{3\ell}$ while if $\ell$ is odd it is a constant multiple of $\delta_n c_n a_2^{[\frac{1}{2}\ell]} x^{3\ell-1}$ where $\delta_n$ is a linear combination of $\mu_1$ and $a$, and [ ] denotes integer part. This can be proved from (3.11) by induction. We then consider the highest degree terms in (3.12). If $n$ is odd, comparing coefficients of $x^{3n+2}$ gives $a_2 = 0$, which is excluded by hypothesis. If $n$ is even we consider the terms of degree $x^{3n}$ and $x^{3n+1}$ and we have that $a_2 = 0$ or $a = \mu_1 = 0$, both of which are contrary to hypothesis.

We deduce the following result:

THEOREM 3.6 *System (1.2) has no invariant polynomials when conditions (3) of Theorem 2.1 hold.*

We remark that in his recent paper Odani [12] has shown that a Liénard system $\dot{x} = y$, $\dot{y} = -yf(x) - g(x)$ cannot have algebraic limit cycles if $g$ is of lower degree than $f$.

The search for invariant polynomials for (1.2) can be conducted by writing the polynomial in the form $\sum y^k f_k(x)$. A recursive set of first order linear differential equations are obtained for the $f_k$. Using this approach we rapidly recovered the (known) invariant curves leading to conditions (1),(2),(4) and (7) but no others were obtained.

We strongly conjecture that there are no invariant polynomials when conditions (5),(6),(8) hold, but it remains to prove that the origin is a centre in these cases. This has to be done in a completely different way. Christopher [4] has clarified the original work of Cherkas and shown how the sufficiency of all the conditions in Theorem 2.1 can be deduced by considering appropriate Liénard systems. Note that this method is successful for condition (8) which was omitted by Cherkas and which has not been obtained by the technique described here. Some background on Liénard systems is required and the details are given in [4]. As we have previously remarked it is interesting how so many facets of the subject are illustrated by the system (1.2).

THEOREM 3.7 *The conditions given in Theorem 2.1 are sufficient for the origin to be a centre.*

**References**

[1] L A Cherkas, Conditions for a Liénard equation to have a center, *Differentsial'nye Uraveniya* **12** (1976) 292-298.

[2] L A Cherkas, Conditions for the equation $yy' = \Sigma_{i=0}^{3} p_i(x)y^i$ to have a centre, *Differential Equations* **14** (1978) 1133-1138.

[3] C J Christopher, Invariant algebraic curves and conditions for a centre, Preprint, The University of Wales, Aberystwyth, 1989.

[4] C J Christopher, Necessary and sufficient conditions for a centre for a class of systems considered by Cherkas, Preprint, The University of Wales, Aberystwyth, 1991.

[5] H Dulac, Détermination et intégration d'une certaine classe d'équations différentielles ayant point singulier un centre, *Bull. Sci. Math.* **32(1)**, (1908) 230-252.

[6] D Hilbert, Mathematical problems, *Bull. Amer. Math. Soc.* **8** (1902) 437-479.

[7] J P Jouanolou, *Equations de Pfaff Algebriques*, Lecture Notes in Mathematics, No.708 (Springer-Verlag, Berlin, 1979).

[8] N G Lloyd, Limit cycles of polynomial systems - some recent developments, *New Directions in Dynamical Systems* (eds. T Bedford & J Swift) L.M.S. Lecture Notes **127** (Cambridge University Press, 1988) 192-234.

[9] N G Lloyd and J M Pearson, REDUCE and the bifurcation of limit cycles, *J. Symbolic Comput.* **9** (1990) 215-224.

[10] N G Lloyd and J M Pearson, Conditions for a centre and the bifurcation of limit cycles in a class of cubic systems, in: J.-P. Françoise and R. Roussarie, Eds., *Bifurcations of Planar Vector Fields* (Springer, Berlin, 1990).

[11] N G Lloyd and J M Pearson, Computing centre conditions for certain cubic systems, *J. Comp. Appl. Math.* **40** (1992), 323-336.

[12] K Odani, The limit cycle of the van der Pol equation is not algebraic, *J. Differential Equations*, to appear.

[13] M J Prelle and M F Singer, Elementary first integrals of differential equations, *Trans. Amer. Math. Soc.* **279** (1983) 215-229.

[14] M F Singer, Formal solutions of differential equations, *J. Symbolic Comput.* **10** (1990) 59-94.

[15] Dongming Wang, Centre conditions for a system considered by Cherkas, private communication (1990).

[16] Yanqian Ye, *Theory of limit cycles*, Trans. Math. Monographs **66** (Amer. Math. Soc., 1986).

# 11 Stability Criteria with a Symmetric Operator Occurring in Linear and Nonlinear Delay-Differential Equations

**James Louisell** University of Southern Colorado, Pueblo, Colorado

## Section 1

In a recent paper ( [ 6 ] ) this author has given theorems for delay - differential equations which are parallel to theorems of Markus and Yamabe on ordinary differential equations in their paper *Global Stability Criteria for Differential Systems* ( [ 8 ] ) . In this paper we review the ideas for these theorems and we give new theorems putting these ideas in a broader, more flexible, and handier framework.

In order to give an introduction to the ideas presented in this paper, we first examine a basic question found in the area of ordinary differential equations. Here we consider the linear ordinary differential equation $\dot{x}(t) = Ax(t)$ , $x(0) = x_0$ , where A is any member of $R^{nxn}$ . For vectors $x,y \in R^n$, we write « x,y » for the standard inner product of x and y . Introducing the quadratic form $V(x) = $ « Ax,Ax » , we consider the real function $V(x(t))$ , where $x(\cdot)$ is any solution of $\dot{x}(t) = Ax(t)$ . Noting that $\dot{x} = Ax$ and examining the time derivative of this function, we obtain

$$\dot{V}(x) = (A\dot{x})^T \dot{x} + (\dot{x})^T A\dot{x} = (\dot{x})^T A^T \dot{x} + (\dot{x})^T A\dot{x} = (\dot{x})^T (A^T + A)\dot{x} = 2\text{« } A\dot{x},\dot{x} \text{ »} .$$

This suggests that there may be cases where the matrix $A^T + A$ could be useful in the stability analysis of the system $\dot{x}(t) = Ax(t)$ . In fact, if $\lambda = \lambda_1 + i\lambda_2 \in C$ is any eigenvalue of A , with eigenvector $w \in C^n$, we can write $w^*(A^T + A)w = \bar{\lambda}w^*w + \lambda w^*w = (\lambda + \bar{\lambda})| w |^2 = 2\lambda_1 | w |^2$ . On the other hand, for $\lambda_{max}(A^T + A) = 2\gamma$ , we see from the Cauchy - Schwarz inequality that $z^*(A^T + A)z \leq 2\gamma | z |^2$ for each $z \in C^n$. This gives $2\lambda_1 | w |^2 \leq 2\gamma | w |^2$ , and noting that $w \neq 0$ , we know that $\lambda_1 \leq \gamma$ . Thus the real part of any eigenvalue of A is no greater than one - half the maximum eigenvalue of $A^T + A$ , and this relates the spectrum of $A^T + A$ to that of A . Particularly, if $\lambda_{max}(A^T + A) < 0$ , then the system $\dot{x}(t) = Ax(t)$ is asymptotically stable.

This fact from linear ordinary differential equations has been used by Markus and Yamabe as a starting point for an analysis of the global stability of certain nonlinear ordinary differential equations. One topic explored by these authors is the analysis of systems of the form $\dot{x}(t) = f(x(t))$ , where $f : R^n \rightarrow R^n$ is continuously differentiable. Here they use the Lyapunov function $V(x) = $ « f(x),f(x) » , just as the Lyapunov function $V(x) = $ « Ax,Ax » is used in the linear system above, and they consider the matrix $M(x) = f'(x)^T + f'(x)$ , defined for each $x \in R^n$, just as the matrix $A^T + A$ is considered above. Making fairly mild assumptions on $\det(M(x))$ and $\operatorname{trace}(M(x))$, they prove that if $M(x)$ is negative definite for all $x \in R^n$, then the system has a unique equilibrium in $R^n$, and all trajectories of the system approach this equilibrium at an exponential rate as $t \to \infty$. For further applications of this method in the area of nonlinear ordinary differential equations, one can refer to the paper by Markus and Yamabe ( [ 8 ] ) on dynamical systems defined over connected Riemannian manifolds, to the paper by Hill and Mareels ( [ 4 ] ) which includes applications to the problem of estimating stability basins, and to the book by Krasovskii ( [ 5 ] ).

This author has investigated a similar problem in the area of time - varying delayed dynamics. In this problem one considers the linear time - varying delay - differential equation (†) $\dot{x}(t) = A_0 x(t) + A_1 x(t - h(t))$, where $A_0, A_1 \in R^{n \times n}$, and $h(\cdot)$ is a function having domain $[0,\infty)$ and range contained in a bounded subset of $[0,\infty)$. Here we define the set $H_\gamma = \{ \eta \in [0,\infty) : f_\eta(s) = | sI - A_0 - e^{-s\eta} A_1 |$ is nonzero for each complex s having $\mathrm{Re}(s) \geq \gamma \}$. In the analysis of the autonomous system (*) $\dot{x}(t) = A_0 x(t) + A_1 x(t - \eta)$, it is known that if $\eta \in H_\gamma$, then for each $\phi \in C[-\eta,0]$, there is a constant $C > 0$ such that the solution $x(t) = x(\phi,t)$, having $x(u) = \phi(u)$ for $-h \leq u \leq 0$, satisfies $| x(t) | \leq C e^{\gamma t}$ for all $t \geq 0$. Examining the time - varying delay - differential system (†), a natural question is whether, or under what conditions, any similar exponential bound holds in the case that $H_\gamma \supset \operatorname{range}(h(\cdot))$. For the case $\gamma = 0$, this becomes the basic question of the stability of the time - varying system (†) or the autonomous system (*), and for this reason we denote the set $H_0$ as $H_{sta}$.

The author has given an analysis of this question of the growth and stability of the time - varying system (†) $\dot{x}(t) = A_0 x(t) + A_1 x(t - h(t))$ for functions $h(\cdot)$ having $H_\gamma \supset \operatorname{range}(h(\cdot))$. In this analysis the values of $| x(t) |^2$ are related to the *average* magnitude of $h'(\cdot)$ over the interval $[0,t]$, i.e. to $\frac{1}{t} \int_0^t | h'(\tau) | d\tau$, the average rate at which $h(\cdot)$ varies in $H_\gamma$ over $[0,t]$. The following are examples of theorems which the author has proven on this topic ( [ 7 ] ). The expression $| \cdot |$ below is used to denote Lebesgue measure, and it is useful here to recall that an absolutely continuous real function has a derivative almost everywhere in Lebesgue measure.

Let $h(\cdot)$ be any absolutely continuous function having domain $[0,\infty)$ and range contained in a compact subset of $H_\gamma$, and let $a(t) = \frac{1}{t} \int_0^t | h'(\tau) | d\tau$ for $t > 0$. Now consider the delay - differential equation

(†) $\dot{x}(t) = A_0 x(t) + A_1 x(t - h(t))$.

Theorem 1 : There exists a constant $B > 0$, and constants $\mu_1, \mu_2$, with $\mu_1 < 0 < \mu_2$, such that if $\mu_1 < h'(t) < \mu_2$ a.e. over $[0,\infty)$, then for each solution $x(\cdot)$ of (†), the following inequality holds for some constant $C > 0$: $| x(t) |^2 \leq C e^{t(2\gamma + B a(t))}$ for all $t > 0$.

Theorem 2 : If $\int_0^\infty | h'(\tau) | d\tau < \infty$ , then for $\kappa = \int_0^\infty | h'(\tau) | d\tau$ , and $\tilde{B} = e^{B\kappa}$, the inequality above can be sharpened to the following inequality : $| x(t) |^2 \leq \tilde{B} \cdot C e^{2\gamma t}$ for all $t > 0$ .

For the case $\gamma = 0$ this author has proven :

Theorem 3 : If $\liminf_{t \to \infty} a(t) = 0$ , then for each *bounded* solution $x(\cdot)$ of (†) , one has *both* of a) and b) below :

a) $\liminf_{t \to \infty} | x(t) | = 0$
b) for any $\varepsilon > 0$ and $\beta$ having $0 < \beta < 1$ , there is a sequence $\{ t_i \}$ with $t_i \uparrow \infty$ and $\frac{1}{t_i} \cdot | \{ \tau : 0 \leq \tau \leq t_i , | x(\tau) | \geq \varepsilon \} | < \beta$ .

Note that part a) of Theorem 3 states that if $a(t)$ at least *attains* arbitrarily small values as $t \to \infty$ , then $| x(t) |$ also does, while part b) states under the same hypotheses that there is a sense in which $| x(t) |$ actually *adheres* to arbitrarily small values as $t \to \infty$ , i.e. the ratio of elapsed time having $| x(t) | \geq \varepsilon$ to total time elapsed can be made as small as demanded by proper choice of the total elapsed time $t_i$ . Naturally, part a) is an easy corollary of part b) .

## Section 2

In this section we give a framework for working with velocity functionals which is analogous in the theory of delay - differential operators to the previously mentioned velocity Lyapunov function $V(x) = \langle\langle Ax, Ax \rangle\rangle$ used in the analysis of the ordinary differential equation $\dot{x}(t) = Ax(t)$ . To begin, we denote the vector space of all continuous functions taking the interval $[-h,0]$ into $R^n$ by $C[-h,0]$ , and naturally the vector space of all continuously differentiable functions taking the interval $[-h,0]$ into $R^n$ is denoted by $C^1[-h,0]$ . We will be interested in the linear transformation $D : C^1[-h,0] \to C[-h,0]$ defined by $D\phi = \dot{\phi}$ for $\phi \in C^1[-h,0]$ , where $\dot{\phi}$ is the derivative of $\phi$ . For any $t \geq 0$ and any vector function $x(t)$ with domain including $[-h,t]$ , we let $x_t$ denote the function given by $x_t(u) = x(t + u)$ for $-h \leq u \leq 0$ . Then $x_t$ is a member of $C[-h,0]$ if $x(\cdot)$ is continuous over $[t - h,t]$ . If $x(\cdot)$ is continuously differentiable over $[t - h,t]$ , then $x_t \in C^1[-h,0]$ , $Dx_t$ is well - defined, and $Dx_t = \dot{x}_t$ , where $\dot{x}_t(u) = \dot{x}(t + u)$ for $-h \leq u \leq 0$ . There will be cases in which we consider the extensions of the above spaces $C[-h,0]$ and $C^1[-h,0]$ to include functions having range in $C^n$ . In these cases we will denote these extensions as $C[-h,0,C^n]$ and $C^1[-h,0,C^n]$ , and of course $D$ can be extended to $C^1[-h,0,C^n]$ . Finally, we shall denote inner products on these spaces by $< , >$ , and the induced norm by $\| \cdot \|$ .

We now consider the familiar linear delay - differential equation $(*)\ \dot{x}(t) = A_0x(t) + A_1x(t-h)$ , where $A_0, A_1 \in R^{nxn}$ , and h is any member of $(0,\infty)$ . Given $A_0, A_1 \in R^{nxn}$ and $h > 0$ , we can define the linear delay transformation $L: C[-h,0] \to R^n$ by $L\phi = A_0\phi(0) + A_1\phi(-h)$ . For any trajectory $x(\cdot)$ of the system (*) and $t \geq 0$ , we have $Lx_t = A_0x(t) + A_1x(t-h)$ , and the delay - differential equation (*) is written as $(*)\ \dot{x}(t) = Lx_t$ . The complex function $g_h(s) = |\, sI - A_0 - e^{-sh}A_1 \,|$ is referred to as the characteristic function for the system (*) , and for any complex $\lambda$ having $g_h(\lambda) = 0$ and any nonzero null vector $w \in C^n$ of the matrix $\lambda I - A_0 - e^{-\lambda h}A_1$ , the function $x(t) = e^{\lambda t}w$ is called a characteristic solution of the delay - differential equation (*) . With these notations, we can now give the theorem immediately below. Although this will be quite straightforward, the facts contained provide a convenient framework for analyzing a variety of velocity functionals.

**Theorem 2.1 :** Let $<\,,\,>$ be an inner product on the space $C[-h,0,C^n]$ . Suppose that for each $\lambda \in C$ having $g_h(\lambda) = 0$ , there is at least one complex null vector w of the matrix $\lambda I - A_0 - e^{-\lambda h}A_1$ making $< D\phi,\phi > + < \phi, D\phi >$ strictly negative for the function $\phi(u) = e^{\lambda u}w$ defined over $[-h,0]$ . Then the system $(*)\ \dot{x}(t) = A_0x(t) + A_1x(t-h)$ is exponentially asymptotically stable.

**Proof :** Take any complex $\lambda$ having $g_h(\lambda) = 0$ . Noting the hypothesis, take any $w \in C^n$ having $(\lambda I - A_0 - e^{-\lambda h}A_1)w = 0$ and $2 \cdot \mathrm{Re}(< D\phi,\phi >) < 0$ for the function $\phi(u) = e^{\lambda u}w$ . Then $\dot{\phi}(u) = \lambda\phi(u)$ , so that $< D\phi,\phi > + < \phi,D\phi > = \lambda < \phi,\phi > + \bar{\lambda} < \phi,\phi > = (\lambda + \bar{\lambda}) \| \phi \|^2$ . Since $< D\phi,\phi > + < \phi,D\phi >$ is negative, we have $\lambda + \bar{\lambda} < 0$ , i.e. $\mathrm{Re}(\lambda) < 0$ . We have shown that for any complex $\lambda$ having $g_h(\lambda) = 0$ , one has $\mathrm{Re}(\lambda) < 0$ . Thus the system (*) is exponentially asymptotically stable. Q. E. D.

We are now able to show how the simple principle given in the above theorem can be valuable in giving interesting examples of quadratic functionals for use in analyzing the stability of delay - differential systems.

**Theorem 2.2 :** Consider the linear delay - differential equation $(*)\ \dot{x}(t) = A_0x(t) + A_1x(t-h)$ . For each Hermitian matrix $K > 0$ , define the 2n x 2n matrix $S_K$ by $S_K = \begin{pmatrix} A_0^T + A_0 + K & A_1 \\ A_1^T & -K \end{pmatrix}$ . If there exists Hermitian $K > 0$ such that $S_K$ is negative definite, then the system (*) is exponentially asymptotically stable .

**Proof :** Take any Hermitian matrix $K > 0$ , and consider the bilinear functional $<\,,\,>$ defined on $C[-h,0,C^n]$ by

$$< \phi,\psi > \;=\; \phi^*(0)\psi(0) + \int_{-h}^{0} \phi^*(u)K\psi(u)du \quad \text{for } \phi,\psi \in C[-h,0,C^n]$$

. One can easily check that this bilinear functional is actually an inner product on $C[-h,0,C^n]$ . Letting $g_h(s) = |\, sI - A_0 - e^{-sh}A_1 \,|$ , we consider any complex $\lambda$ having $g_h(\lambda) = 0$ , and take any nonzero $w \in C^n$ with $(\lambda I - A_0 - e^{-\lambda h}A_1)w = 0$ . Setting $\phi(u) = e^{\lambda u}w$ for $-h \leq u \leq 0$ , we know that $\dot{\phi}(u) = e^{\lambda u}(\lambda w) = e^{\lambda u}(A_0 + e^{-\lambda h}A_1)w = e^{\lambda u}(A_0\phi(0) + A_1\phi(-h))$ , and particularly $\dot{\phi}(0) = A_0\phi(0) + A_1\phi(-h)$ . We now write

$$< D\phi,\phi > = \dot{\phi}(0)^*\phi(0) + \int_{-h}^{0} \dot{\phi}(u)^*K\phi(u)du \quad \text{and} \quad < \phi,D\phi > = \phi(0)^*\dot{\phi}(0) + \int_{-h}^{0} \phi(u)^*K\dot{\phi}(u)du .$$

This yields

$$< D\phi,\phi > + < \phi,D\phi > = \dot{\phi}(0)^*\phi(0) + \phi(0)^*\dot{\phi}(0) + \int_{-h}^{0} \frac{d\ \phi(u)^*K\phi(u)}{du} du$$

$$= ( \phi^*(0)A_0^T + \phi^*(-h)A_1^T )\phi(0) + \phi^*(0)( A_0\phi(0) + A_1\phi(-h) ) + \phi^*(0)K\phi(0) - \phi^*(-h)K\phi(-h) .$$

In matrix form this is written $< D\phi,\phi > + < \phi,D\phi > = ( \phi^*(0) \quad \phi^*(-h) )\, S_K \begin{pmatrix} \phi(0) \\ \phi(-h) \end{pmatrix}$, and noting Theorem 2.1 and the fact that $| \phi(0) \quad \phi(-h) | = | w \quad e^{-\lambda h}w | \neq 0$, it is now clear that if $S_K$ is negative definite, then the system (*) is exponentially asymptotically stable. Q. E. D.

**Theorem 2.3 :** Consider the delay - differential equation (*) $\dot{x}(t) = A_0x(t) + A_1x(t - h)$ . Suppose there exists Hermitian $K > 0$ such that the matrix $S_K$ of Theorem 2.2 is negative definite. Then for $\lambda_{max}( S_K ) = 2\gamma$ and $\alpha = \inf \{\beta : g_h(s)$ has no zeros with $Re(s) \geq \beta \}$ , we have $\alpha ( 1 + \lambda_{max}(K) \cdot \frac{1 - e^{-2\alpha h}}{2\alpha} ) \leq ( 1 + e^{-2\alpha h} )\gamma$ .

**Proof :** To demonstrate this formula for the bound on the x -coordinate of characteristic values of the system (*) , we let $\lambda$ be any zero of the characteristic function $g_h(s)$ , and for any $w \in C^n - \{0\}$ with $( \lambda I - A_0 - e^{-\lambda h}A_1 )w = 0$ , we examine the value of the functional $< D\phi,\phi > + < \phi,D\phi >$ for the function $\phi(u) = e^{\lambda u}w$ . We then see that

$$< D\phi,\phi > = \bar{\lambda}w^*w + \bar{\lambda} \int_{-h}^{0} e^{\bar{\lambda}u}w^*Kwe^{\lambda u}du = \bar{\lambda} ( w^*w + w^*Kw \int_{-h}^{0} e^{( \lambda + \bar{\lambda} )u} du ) .$$ Integrating, we have

$$< D\phi,\phi > = \bar{\lambda} ( w^*w + \frac{1 - e^{-( \lambda + \bar{\lambda} )h}}{\lambda + \bar{\lambda}} w^*Kw ) ,$$ so that

$$< D\phi,\phi > + < \phi,D\phi > = ( \lambda + \bar{\lambda} ) w^* ( I + \frac{1 - e^{-( \lambda + \bar{\lambda} )h}}{\lambda + \bar{\lambda}} K ) w ,$$ and

noting the matrix expression for $< D\phi,\phi > + < \phi,D\phi >$ given in Theorem 2.2 , we have

$$( \phi^*(0) \quad \phi^*(-h) )\, S_K \begin{pmatrix} \phi(0) \\ \phi(-h) \end{pmatrix} = ( \lambda + \bar{\lambda} ) w^* ( I + \frac{1 - e^{-( \lambda + \bar{\lambda} )h}}{\lambda + \bar{\lambda}} K ) w .$$

Write $[ \phi(0) \quad \phi(-h) ] = [ w \quad e^{-\lambda h}w ]$ , $\lambda = \lambda_1 + i\lambda_2$ and $2\lambda_1 = \lambda + \bar{\lambda}$ , and then apply the Cauchy - Schwarz inequality to see that

$$(\phi^*(0)\quad \phi^*(-h))\, S_K \begin{pmatrix} \phi(0) \\ \phi(-h) \end{pmatrix} \le 2\gamma |\, w\; e^{-\lambda h} w\, |^2 = 2\gamma(1 + e^{-2\lambda_1 h}) |\, w\, |^2 .$$

Now $w^*Kw \le \lambda_{max}(K)\, w^*w$, by Theorem 2.2 we know $\alpha < 0$, and since $\lambda_1 \le \alpha$, we have $1 - e^{-2\lambda_1 h} < 0$. We now see that

$$2\lambda_1 w^*w + (1 - e^{-2\lambda_1 h})\lambda_{max}(K)\, w^*w \le 2\lambda_1 w^*w + (1 - e^{-2\lambda_1 h}) w^*Kw ,$$

and this expression on the right is just the right side of the expression for $(\phi^*(0)\quad \phi^*(-h))\, S_K \begin{pmatrix} \phi(0) \\ \phi(-h) \end{pmatrix}$ above. Thus

$$(2\lambda_1 + (1 - e^{-2\lambda_1 h})\lambda_{max}(K))\, |\, w\, |^2 \le (\phi^*(0)\quad \phi^*(-h))\, S_K \begin{pmatrix} \phi(0) \\ \phi(-h) \end{pmatrix} \le 2\gamma(1 + e^{-2\lambda_1 h}) |\, w\, |^2 ,$$ so that

$0 \le (2\gamma(1 + e^{-2\lambda_1 h}) - 2\lambda_1 - (1 - e^{-2\lambda_1 h})\lambda_{max}(K))\, |\, w\, |^2$. Noting that $w \ne 0$, we see that

$$0 \le 2\gamma(1 + e^{-2\lambda_1 h}) - 2\lambda_1 - (1 - e^{-2\lambda_1 h})\lambda_{max}(K) ,\ \text{i.e.}\ 2\lambda_1 + (1 - e^{-2\lambda_1 h})\lambda_{max}(K) \le 2\gamma(1 + e^{-2\lambda_1 h}) .$$

Since $\lambda_1 \le \alpha < 0$, we know that $-\lambda_1 \ge -\alpha > 0$, and thus $2\lambda_1 + (1 - e^{-2\lambda_1 h})\lambda_{max}(K) \le 2\gamma(1 + e^{-2\alpha h})$.

Noting that $\alpha = \sup \{ \lambda_1 : \text{there exists } \lambda = \lambda_1 + i\lambda_2 \text{ with } g_h(\lambda) = 0 \}$, we conclude that

$$2\alpha + (1 - e^{-2\alpha h})\lambda_{max}(K) \le 2\gamma(1 + e^{-2\alpha h}) ,\ \text{i.e.}\ \alpha\left(1 + \lambda_{max}(K) \cdot \frac{1 - e^{-2\alpha h}}{2\alpha}\right) \le (1 + e^{-2\alpha h})\gamma .$$

Q. E. D.

It will be possible to give a criterion sufficient for the existence of a positive Hermitian matrix $K$ making the above $2n \times 2n$ matrix $S_K$ negative definite. In the case that the matrix $A_1$ is symmetric this condition will also be necessary. If such $K$ exists we can give a specific choice of $K$ which does make $S_K$ negative definite, and this matrix $K$ will be real symmetric. For convenience in the proof we first recall the notion of a congruence from linear algebra. Any two members $F, G$ of $C^{nxn}$ are said to be congruent if there is a nonsingular $T \in C^{nxn}$ with $T^*FT = G$. This notion is particularly valuable in the area of quadratic forms, since, for instance, for any Hermitian $F > 0$, we will have $G > 0$ for any matrix $G$ congruent to $F$. Before proceeding, we first give the following facts from linear algebra, with several short proofs given here to provide understanding for the next theorems.

**Lemma 2.1 :** Let $A,T \in C^{nxn}$, with $A$ Hermitian and $T$ nonsingular. Then for $\hat{A} = T^*AT$, we have $\hat{A} > 0$ if, and only if, $A > 0$.

**Proof :** For $x = Tu$, we have $x^*Ax = (Tu)^*A(Tu) = u^*(T^*AT)u$. Q. E. D.

**Lemma 2.2 :** Let $A \in C^{nxn}$ be any Hermitian positive definite matrix, and let $S = A^{1/2}$ denote the unique Hermitian square root of $A$, i.e. $S = UD^{1/2}U^*$, where $U^*AU = D$, with $U$ unitary and $D$ diagonal. Now set $T = A^{-1/2} = S^{-1}$. Then we have $T^*AT = I$. Furthermore, for any $B \in C^{nxn}$ and $\hat{B} = T^*BT$, the matrix $B^*A^{-1}B$ is congruent to $\hat{B}^*\hat{B}$.

**Proof :** The fact that $T^*AT = I$ follows from simple computation. Now noting that $TT^* = T^2 = A^{-1}$, we find that $\hat{B}^*\hat{B} = T^*B^*A^{-1}BT$, so that $\hat{B}^*\hat{B}$ is congruent to $B^*A^{-1}B$. Q. E. D.

**Lemma 2.3 :** Let $A \in C^{nxn}$ be a Hermitian positive definite matrix, and let $B = B^*$ be any Hermitian matrix. Then there exists a nonsingular $T \in C^{nxn}$ for which $T^*AT = I$ and $T^*BT$ is a diagonal matrix.

**Proof :** For $J = A^{-1/2}$, we have $J^*AJ = I$. The matrix $J^*BJ$ is Hermitian, so that we have $U \in C^{nxn}$ with $D = U^*(J^*BJ)U = (JU)^*B(JU)$, where $D$ is diagonal. Set $T = JU$, noting that $T^*AT = U^*IU = I$. Q. E. D.

**Lemma 2.4 :** Let $K$ be Hermitian and negative, and let $A$ be Hermitian. If $A + K > 0$, then $A > 0$.

**Lemma 2.5 :** Let $A,B \in C^{nxn}$, with $A$ Hermitian and positive. Then for the Hermitian matrix $F = \begin{pmatrix} A & B \\ B^* & A \end{pmatrix}$, one has $F > 0$ if, and only if, $A - B^*A^{-1}B > 0$.

**Proof :** Since $A > 0$, we have $T^*AT = I$ for $T = A^{-1/2}$. Writing $\hat{T}$ = block diag $(T,T)$ and $\hat{B} = T^*BT$, we set $\hat{F} = \hat{T}^*F\hat{T} = \begin{pmatrix} I & \hat{B} \\ \hat{B}^* & I \end{pmatrix}$, and this matrix is congruent to F. Writing $\hat{B} = U\Sigma V^*$, where $U^*U = I = V^*V$ and $\Sigma = \text{diag}(\sigma_1,...,\sigma_n)$ is the matrix of singular values for $\hat{B}$, we set $W$ = block diag $(U,V)$, and find that $W^*\hat{F}W = \begin{pmatrix} I & \Sigma \\ \Sigma^* & I \end{pmatrix} = \begin{pmatrix} I & \Sigma \\ \Sigma & I \end{pmatrix}$, so that $\begin{pmatrix} I & \Sigma \\ \Sigma & I \end{pmatrix}$ is also congruent to F. Examining the Hermitian form for this matrix, we have $[x^* \; y^*]\begin{pmatrix} I & \Sigma \\ \Sigma & I \end{pmatrix}\begin{pmatrix} x \\ y \end{pmatrix} = \sum_1^n (x_i\bar{x}_i + y_i\bar{y}_i + \sigma_i x_i\bar{y}_i + \sigma_i\bar{x}_i y_i)$ for $x,y \in C^n$.

Setting $t_{2i-1} = x_i$ and $t_{2i} = y_i$ for $i = 1,..., n$, the matrix for this form is written as block diag $(C_1,...,C_n)$, where $C_i = \begin{pmatrix} 1 & \sigma_i \\ \sigma_i & 1 \end{pmatrix}$. Thus $W^*\hat{F}W > 0$ if, and only if, $1 - \sigma_i^2 > 0$ for each $i = 1,...,n$. This occurs if, and only if, one has $\lambda(I - \hat{B}^*\hat{B}) > 0$ for each eigenvalue $\lambda$ of $I - \hat{B}^*\hat{B}$, i.e. if and only if $I - \hat{B}^*\hat{B} > 0$. Since $W^*\hat{F}W$ is congruent to $F$, we see that $F > 0$ if and only if $I - \hat{B}^*\hat{B} > 0$. Finally, $I - \hat{B}^*\hat{B} = I - A^{-1/2}B^*A^{-1}BA^{-1/2}$, so that for $S = A^{1/2}$, we have $S^*(I - \hat{B}^*\hat{B})S = A - B^*A^{-1}B$, and hence this matrix is congruent to $I - \hat{B}^*\hat{B}$. Thus $F > 0$ if and only if $A - B^*A^{-1}B > 0$. Q. E. D.

**Theorem 2.4 :** Let $A_0, A_1 \in R^{n\times n}$, and set $Z_0 = \frac{1}{2}(A_0^T + A_0)$. If one has both $-Z_0 > 0$ and $-Z_0 + A_1^T(Z_0)^{-1}A_1 > 0$, then there does exist a positive Hermitian matrix K making the matrix $S_K = \begin{pmatrix} A_0^T + A_0 + K & A_1 \\ A_1^T & -K \end{pmatrix}$ negative definite. In fact, for $K = -Z_0$, one has $K > 0$ and $S_K < 0$.

**Proof :** Setting $K = -Z_0$, we have $-S_K = \begin{pmatrix} -Z_0 & -A_1 \\ -A_1^T & -Z_0 \end{pmatrix}$, and now applying Lemma 2.5, we see that $-S_K > 0$. Q. E. D.

**Theorem 2.5 :** Let $A_0, A_1 \in R^{n\times n}$, with $A_1 = A_1^T$, and set $Z_0 = \frac{1}{2}(A_0^T + A_0)$. If there exists a positive Hermitian matrix K making the matrix $S_K = \begin{pmatrix} A_0^T + A_0 + K & A_1 \\ A_1^T & -K \end{pmatrix}$ negative definite, then one has both $-Z_0 > 0$ and $-Z_0 + A_1(Z_0)^{-1}A_1 > 0$.

**Proof :** Suppose $S_K < 0$. Then $-S_K > 0$, and writing $A_0^T + A_0 = 2Z_0$, we have $-S_K = \begin{pmatrix} -2Z_0 - K & -A_1 \\ -A_1^T & K \end{pmatrix}$. Since $(x^* \ \ 0)(-S_K)\binom{x}{0} = x^*(-2Z_0 - K)x$, we see that $-2Z_0 - K > 0$, and noting Lemma 2.4, we have $-2Z_0 > 0$, i.e. $-Z_0 > 0$. Since $-Z_0 > 0$ and $K > 0$, we note Lemma 2.3, set $J = (-Z_0)^{-1/2} = J^T$, and let R be an orthogonal matrix having $R^T(JKJ)R = D$ with D diagonal. Setting T = block diag (JR,JR) and $\hat{A}_1 = R^TJ(-A_1)JR$, we note the symmetry of $\hat{A}_1$, and write $T^*(-S_K)T = \begin{pmatrix} 2I - D & \hat{A}_1 \\ \hat{A}_1 & D \end{pmatrix}$. This matrix is positive since it is congruent to $-S_K$. Thus $2I - D > 0$, so that one can set $D_1 = I - D$ and write $2I - D = I + D_1$, $D = I - D_1$. Thus $T^*(-S_K)T = \begin{pmatrix} I + D_1 & \hat{A}_1 \\ \hat{A}_1 & I - D_1 \end{pmatrix} > 0$.

Now for any $x, y \in C^n$ with $(x,y) \neq 0$, we have

$0 < (x^* \ y^*)T^*(-S_K)T\binom{x}{y} = (x^*x + y^*y + x^*\hat{A}_1y + y^*\hat{A}_1x + x^*D_1x - y^*D_1y$, and

$0 < (y^* \ x^*)T^*(-S_K)T\binom{y}{x} = (x^*x + y^*y + x^*\hat{A}_1y + y^*\hat{A}_1x) - x^*D_1x + y^*D_1y$.

From this we see that $2(x^*x + y^*y + x^*\hat{A}_1y + y^*\hat{A}_1x) > 0$ for $x, y \in C^n$ having $(x,y) \neq 0$, and thus we have shown that $\begin{pmatrix} I & \hat{A}_1 \\ \hat{A}_1 & I \end{pmatrix} > 0$. From Lemma 2.5 we see that $I - (\hat{A}_1)^2 > 0$, and this gives $J^{-1}R(I - (\hat{A}_1)^2)R^TJ^{-1} > 0$, i.e. $-Z_0 - (-A_1)(-Z_0)^{-1}(-A_1) > 0$. Q. E. D.

Examining the previous work, we obtain the following theorem on the stability of linear delay - differential equations.

**Theorem 2.6 :** Let $A_0, A_1 \in R^{n \times n}$, and consider the linear delay - differential equation (*) $\dot{x}(t) = A_0 x(t) + A_1 x(t-h)$. If $-Z_0 > 0$ and $-Z_0 + A_1^T (Z_0)^{-1} A_1 > 0$ for the matrix $Z_0 = \frac{1}{2}(A_0^T + A_0)$, then the system (*) is asymptotically stable. In this case, for $2\gamma = \lambda_{max}\left(\begin{pmatrix} Z_0 & A_1 \\ A_1^T & Z_0 \end{pmatrix}\right)$ and $\alpha = \inf\{\beta : g_h(s) \text{ has no zeros with } Re(s) \geq \beta\}$, we have $\alpha\left(1 + \lambda_{max}(-Z_0) \cdot \frac{1 - e^{-2\alpha h}}{2\alpha}\right) \leq (1 + e^{-2\alpha h})\gamma$.

**Proof:** This follows directly from Theorems 2.2, 2.3, and 2.4 by setting $K = -Z_0$. Q. E. D.

It is noteworthy that the condition that there exist positive Hermitian K making the matrix $S_K < 0$ does not in any way depend on the number h, i.e. the type of stability given in the above theorems does not depend on the delay duration. This phenomenon is known as stability independent of delay, and has been examined in other contexts by several authors ( [1], [2], [3] ). Considering the inequality $\alpha\left(1 + \lambda_{max}(K) \cdot \frac{1 - e^{-2\alpha h}}{2\alpha}\right) \leq (1 + e^{-2\alpha h})\gamma$ satisfied by the real part of any eigenvalue of the system (*), it is interesting that this expression does involve the delay duration. In fact, if $h = 0$, we learn that $\alpha \leq 2\gamma$, which limits the system spectrum precisely by the maximum eigenvalue of the matrix $S_K$. For nonzero values of h, the information in the inequality depends on the relation between $\lambda_{max}(K)$ and $\gamma$.

## Section 3

In this section the author reviews, and to some extent improves upon, the main theorems resulting from his investigations ( [ 6 ] ) into the stability of nonlinear delay - differential systems in terms of the stability of their linearizations. The stability of these linearizations will be viewed in light of the functional of Theorem 2.2. We will consider nonlinear delay - differential equations of the form (†) $\dot{x}(t) = f(x(t), x(t-h))$, where $f : R^n \times R^n \to R^n$ is continuously differentiable throughout $R^{2n}$. We define the nonlinear delay transformation $F : C[-h,0] \to R^n$ by $F(\phi) = f(\phi(0), \phi(-h))$, and with this the system (†) is written as (†) $\dot{x}(t) = F(x_t)$. Given any such system, it is known that for each $\phi \in C[-h,0]$, there exists $\beta > 0$ and a unique $x(\cdot) = x(\phi,\cdot)$ with $x(u) = \phi(u)$ for $-h \leq u \leq 0$, which satisfies (†) over $[0,\beta)$. It is known ( [ 2 ] ) that if there exists $\beta' \geq \beta$ such that $x(\cdot) = x(\phi,\cdot)$ is a noncontinuable solution over the interval $[0,\beta')$, then the solution $x(\cdot)$ is unbounded in $R^n$ over $[0,\beta')$. For $x,y \in R^n$, we will write the derivative of f at (x,y) as $f'(x,y) = (A_0(x,y), A_1(x,y))$, or merely as $f'(x,y) = (A_0, A_1)$, where $A_0, A_1 \in R^{n \times n}$. The following facts from linear algebra will be valuable in the stability analysis of the nonlinear system (†). The proof of Lemma 3.1 is contained in the author's paper mentioned above.

**Lemma 3.1 :** Let $J = \begin{pmatrix} A & B \\ B^T & 0 \end{pmatrix}$, where $A,B \in R^{nxn}$, and $A = A^T$. Then $J$ is not strictly sign definite.

**Lemma 3.2 :** Let $J_K = \begin{pmatrix} A + K & B \\ B^T & -K \end{pmatrix}$, where $A,B \in R^{nxn}$, $A = A^T$, and $K$ is symmetric and positive. Then for $\lambda_0 = \lambda_{max}(J_K)$, we have $\lambda_0 I + K \geq 0$.

**Proof:** For each $z = (x,y) \in R^n \times R^n$, we have $z^T(J_K)z \leq \lambda_0 |z|^2$. Writing this out in terms of $x$ and $y$, we have $x^T(A + K)x + 2x^TBy - y^TKy \leq \lambda_0(x^Tx + y^Ty)$. Setting $z = (0,y)$, we have $-y^TKy \leq \lambda_0 y^Ty$ for each $y \in R^n$. Thus $0 \leq y^T(\lambda_0 I + K)y$ for all $y \in R^n$, i.e. $\lambda_0 I + K \geq 0$. Q.E.D.

The next three lemmas provide a basis for comparing the solutions of a type of differential inequality with delay to the solutions of a corresponding delay - differential equation. Before giving the first of these lemmas, we introduce the notation $C_-[-h,0]$ to denote the set of all functions $\psi : [-h,0] \to R$ for which both i) $\psi$ is continuous over the half - open interval $[-h,0)$, and ii) $\lim_{u \to 0-} \psi(u)$ exists and is finite. Here it is not required that $\psi(0) = \lim_{u \to 0-} \psi(u)$. It will be useful to consider the norm $\|\cdot\|_-$ defined on this space by $\|\psi\|_- = ( |\psi(0)|^2 + \int_{-h}^{0} |\psi(u)|^2 du )^{1/2}$, i.e. $(\|\psi\|_-)^2 = |\psi(0)|^2 + (\|\psi\|_2)^2$, where $\|\psi\|_2$ is the norm of $\psi$ in $L^2(-h,0)$. The proofs of these lemmas are, as before, found in the paper ( [ 6 ] ) mentioned above.

**Lemma 3.3 :** Let $h > 0$, $d > 0$, and let $m : [-h,d) \to R$ have $m_0 \in C_-[-h,0]$, i.e. suppose that $m$ is continuous over $[-h,0)$, and that $\lim_{u \to 0-} m(u)$ exists and is finite. Let $a < 0$, $|a| > b \geq 0$, and suppose that $m(\cdot)$ is a solution over $[0,d)$ to the delay - differential inequality $m'(t) \leq am(t) + bm(t - h)$. Now let $n(\cdot)$ be the solution over $[0,\infty)$ to the delay - differential *equation* $n'(t) = an(t) + bn(t - h)$ having initial data $n_0 = m_0 \in C_-[-h,0]$. Then $m(t) \leq n(t)$ for $0 \leq t < d$.

**Lemma 3.4 :** Let $a < 0$, $|a| > |b|$, and consider the delay - differential equation (§) $n'(t) = an(t) + bn(t - h)$. Then there exist constants $\alpha \in (a + |b|, 0)$, $C > 0$ such that the following holds :

For each $\psi \in C_-[-h,0]$, the solution $n(\psi,\cdot)$ of (§) satisfies $|n(\psi,t)| \leq C(\|\psi\|_-)e^{\alpha t}$ for all $t \geq 0$.

**Lemma 3.5 :** Let $h > 0$, and let $a < 0$, $b \geq 0$, with $|a| > b$. Then there exist constants $C > 0$, $\alpha < 0$ such that the following holds :

If $d > 0$ and $m : [-h,d) \to [0,\infty)$ is any function having $m_0 \in C_-[-h,0]$ and satisfying the delay - differential inequality $m'(t) \leq am(t) + bm(t - h)$ for $0 \leq t < d$, then $0 \leq m(t) \leq C(\|m_0\|_-)e^{\alpha t}$ for $0 \leq t < d$.

We are now in a position to give a theorem on the global stability of nonlinear delay - differential systems in terms of the stability of their linearizations. This theorem will be a somewhat broader version of one given in the author's paper on the subject ( [ 6 ] ), with a proof which, although similar, should be of interest.

**Theorem 3.1:** Consider the delay - differential equation (†) $\dot{x}(t) = f(x(t),x(t-h))$, where f is continuously differentiable throughout $R^{2n}$. For $x,y \in R^n$, write $f'(x,y) = [A_0 \; A_1]$, and suppose that there exist a symmetric matrix $K > 0$ and real $\gamma < 0$ such that for each $(x,y) \in R^n \times R^n$, the matrix $S_K = \begin{pmatrix} A_0^T + A_0 + K & A_1 \\ A_1^T & -K \end{pmatrix}$ satisfies $\lambda_{max}(S_K) \le 2\gamma$. If $2\gamma + \lambda_{max}(K) < -2\gamma + \lambda_{min}(K)$, then for each $\phi \in C^1[-h,0]$, the solution $x(\phi,\cdot)$ is defined over $[0,\infty)$, and in fact there exists $\bar{x} \in R^n$ such that $x(\phi,t)$ converges at an exponential rate to $\bar{x}$ as $t \to \infty$. Furthermore, $f(\bar{x},\bar{x}) = 0$.

**Proof:** Take any $\phi \in C^1[-h,0]$, and let $\beta$ be any positive real number with the solution $x(\cdot) = x(\phi,\cdot)$ defined over $[0,\beta)$. Now form a real function of t by setting $V(x_t) = F(x_t)^T F(x_t) + \int_{t-h}^{t} \dot{x}^T(u)K\dot{x}(u)du$. Noting for $0+ \le t < \beta$ that $F(x_t) = \dot{x}(t)$, and calculating the right derivative of this function $V(x_t)$, we find for $0+ \le t < \beta$ that

$$\dot{V}(x_t) = 2\langle\!\langle f'(x(t),x(t-h)) \cdot \begin{pmatrix} \dot{x}(t) \\ \dot{x}(t-h) \end{pmatrix}, \dot{x}(t) \rangle\!\rangle + \dot{x}^T(t)K\dot{x}(t) - \dot{x}^T(t-h)K\dot{x}(t-h)$$

$$= 2\langle\!\langle [A_0 \; A_1] \cdot \begin{pmatrix} \dot{x}(t) \\ \dot{x}(t-h) \end{pmatrix}, \dot{x}(t) \rangle\!\rangle + \dot{x}^T(t)K\dot{x}(t) - \dot{x}^T(t-h)K\dot{x}(t-h)$$

$$= (\dot{x}^T(t)A_0^T + \dot{x}^T(t-h)A_1^T)\dot{x}(t) + \dot{x}^T(t)(A_0\dot{x}(t) + A_1\dot{x}(t-h)) + \dot{x}^T(t)K\dot{x}(t) - \dot{x}^T(t-h)K\dot{x}(t-h),$$

where the derivatives $\dot{x}(t)$ and $\dot{x}(t-h)$ are taken on the right. Writing $S_K = S_K(x,y)$ with $x(t) = x$, $x(t-h) = y$, we now see that

$$\dot{V}(x_t) = (\dot{x}^T(t) \;\; \dot{x}^T(t-h)) S_K \begin{pmatrix} \dot{x}(t) \\ \dot{x}(t-h) \end{pmatrix}.$$

Noting the Cauchy - Schwarz inequality, we have $\dot{V}(x_t) \le 2\gamma \cdot |\dot{x}(t) \;\; \dot{x}(t-h)|^2$. Recalling the definition of V, we obtain $\frac{d\, \dot{x}^T(t)\dot{x}(t)}{dt} + \dot{x}^T(t)K\dot{x}(t) - \dot{x}^T(t-h)K\dot{x}(t-h) \le 2\gamma \cdot |\dot{x}(t)|^2 + 2\gamma \cdot |\dot{x}(t-h)|^2$. Thus we have $\frac{d\, \dot{x}^T(t)\dot{x}(t)}{dt} \le \dot{x}^T(t)(2\gamma I - K)\dot{x}(t) + \dot{x}^T(t-h)(2\gamma I + K)\dot{x}(t-h)$ for $0+ \le t < \beta$, and we see that $\frac{d\, |\dot{x}(t)|^2}{dt} \le a|\dot{x}(t)|^2 + b|\dot{x}(t-h)|^2$, where $a = 2\gamma - \lambda_{min}(K)$, $b = 2\gamma + \lambda_{max}(K)$. Now set $m(t) = |\dot{x}(t)|^2$

for $-h \le t < \beta$, with $\dot{x}(0)$ taken as usual on the right, and note that $m'(t) \le am(t) + bm(t-h)$ for $0 \le t < \beta$. Since $\gamma < 0$, we know that $a < 0$, and since $2\gamma = \lambda_{max}(S_K)$, one can note Lemma 3.2, and see that $2\gamma I + K \ge 0$. Thus $0 \le \lambda_{max}(2\gamma I + K) = 2\gamma + \lambda_{max}(K) = b$, and we have $a < 0 \le b < |a|$. For $0 \le t < \beta$, we know that $m(\cdot)$ satisfies the delay - differential inequality $m'(t) \le am(t) + bm(t-h)$ having initial data $m_0 \in C_-[-h,0]$ given by $m_0(u) = |\dot{\phi}(u)|^2$ for $-h \le u < 0$, and $m_0(0) = |F(\phi)|^2 = |\dot{x}(0+)|^2$. Since $a < 0, 0 \le b < |a|$, we note Lemma 3.5, and immediately see that there exist constants $C_0 > 0, \alpha_0 < 0$ with $m(t) \le C_0(\|m_0\|_-)e^{\alpha_0 t}$ for $0 \le t < \beta$. Setting $C = (C_0 \cdot \|m_0\|_-)^{1/2}, \alpha = \alpha_0/2$, and noting $m(t) = |\dot{x}(t)|^2$, we have $|\dot{x}(t)| \le Ce^{\alpha t}$ for $0 \le t < \beta$.

From this inequality for $|\dot{x}|$, we see that for $0 \le t \le \tau < \beta$, one has

$$|x(\tau) - x(t)| = \left|\int_t^\tau \dot{x}\right| \le \int_t^\tau |\dot{x}| \le C\frac{e^{\alpha\tau} - e^{\alpha t}}{\alpha}, \text{ i.e. } |x(\tau) - x(t)| \le \left(\frac{C}{|\alpha|}\right)e^{\alpha t} \text{ for } 0 \le t \le \tau < \beta.$$

Thus $|x(t) - x(0)| \le \frac{C}{|\alpha|}$ over $[0,\beta)$. Since $\beta$ was an arbitrary positive real number for which the solution $x(\cdot)$ is defined over $[0,\beta)$, we see that there is a fixed bounded subset of $R^n$ which $x(\cdot)$ does not escape over its interval of existence, and conclude that the solution $x(t)$ is defined over $[0,\infty)$.

Using the Cauchy criterion with the bound $|x(\tau) - x(t)| \le C(|\alpha|)^{-1}e^{\alpha t}$, valid for $\tau \ge t \ge 0$, one can easily now see that $\lim_{t\to\infty} x(t)$ exists, and we denote this limit by $\bar{x}$. Writing $\bar{x} = \lim_{\tau\to\infty} x(\tau)$ and noting this bound, we have $|\bar{x} - x(t)| \le C(|\alpha|)^{-1}e^{\alpha t}$ for $t \ge 0$, and thus the rate of convergence of $x(t)$ to $\bar{x}$ is exponential, with exponent less than or equal to $\alpha t$. Since $x(t) \to \bar{x}$ and $x(t-h) \to \bar{x}$ as $t \to \infty$, we have $f(x(t),x(t-h)) \to f(\bar{x},\bar{x})$ as $t \to \infty$. Recalling the inequality $|\dot{x}(t)| \le Ce^{\alpha t}$, we see that $\dot{x}(t) \to 0$ as $t \to \infty$, and since $\dot{x}(t) = f(x(t),x(t-h))$, we know also that $\dot{x}(t) \to f(\bar{x},\bar{x})$ as $t \to \infty$, and thus $f(\bar{x},\bar{x}) = 0$. Q. E. D.

**Corollary 3.1:** Consider the delay - differential equation (†) $\dot{x}(t) = f(x(t),x(t-h))$, with the hypotheses of Theorem 3.1 . For each $\phi \in C^1[-h,0]$, define $m_\phi \in C_-[-h,0]$ by $m_\phi(u) = |\dot{\phi}(u)|^2$ for $-h \le u < 0$, and $m_\phi(0) = |F(\phi)|^2$. Then for each $\varepsilon > 0$, there exists $\delta > 0$ such that if $\|m_\phi\|_- < \delta$, then $|x(\phi,t) - \phi(0)| < \varepsilon$ for all $t \ge 0$.

**Proof:** In the bound given in the theorem, we have $|x(\phi,t) - \phi(0)| \le C(|\alpha|)^{-1}$ for $0 \le t < \infty$, i.e. $|x(\phi,t) - \phi(0)| \le (C_0 \cdot \|m_\phi\|_-)^{1/2}(|\alpha|)^{-1}$ for $t \ge 0$. For any $\delta$ with $0 < \delta < (\frac{\alpha^2}{C_0})\varepsilon^2$ and $t \ge 0$, we thus have $\|m_\phi\|_- < \delta \Rightarrow |x(\phi,t) - \phi(0)| \le (C_0)^{1/2}(|\alpha|)^{-1}(\|m_\phi\|_-)^{1/2} \Rightarrow |x(\phi,t) - \phi(0)| < \varepsilon$, i.e. if $\|m_\phi\|_- < \delta$, then $|x(\phi,t) - \phi(0)| < \varepsilon$ for all $t \ge 0$. Q. E. D.

The next two lemmas are given here in order to provide a proof of the uniqueness of the equilibrium in the above theorem. Although the proof of Lemma 3.7 is not given here, it is by no means a routine matter. For the interested reader, the proof of this lemma is once again contained in the author's paper mentioned above ([6]).

**Lemma 3.6 :** Let $f(x,y)$ be continuously differentiable on an open subset of $\mathbb{R}^n \times \mathbb{R}^n$, and let $K$ be symmetric with $K > 0$. Again consider the matrix $S_K = \begin{pmatrix} A_0^T + A_0 + K & A_1 \\ A_1^T & -K \end{pmatrix}$, where $f'(x,y) = [\ A_0\ A_1\ ]$. Set $g(x) = f(x,x)$. If $g(\bar{x}) = 0$ and $S_K$ is sign definite at $(x,y) = (\bar{x},\bar{x})$, then $\bar{x}$ is an isolated zero of $g(x)$.

**Proof :** We begin by noting that if $A_0 + A_1$ were a singular matrix, then we would have nonzero $v \in \mathbb{R}^n$ with $(A_0 + A_1)v = 0$, and thus by a simple calculation one would have $[\ v^T\ v^T\ ]\, S_K \begin{pmatrix} v \\ v \end{pmatrix} = 0$. Since $S_K$ is sign definite at $(x,y) = (\bar{x},\bar{x})$, we see that $A_0(\bar{x},\bar{x}) + A_1(\bar{x},\bar{x})$ is nonsingular, and now noting that $g'(x) = f'(x,x) \cdot \begin{pmatrix} I \\ I \end{pmatrix} = A_0(x,x) + A_1(x,x)$, we see that $g'(\bar{x})$ is nonsingular. From the inverse function theorem, we conclude that $\bar{x}$ is an isolated zero of $g(x)$. Q. E. D.

**Lemma 3.7 :** Again consider the delay - differential equation $(\dagger)$ $\dot{x}(t) = f(x(t),x(t-h))$, with the hypotheses of Theorem 3.1 . For each $\phi \in C^1[-h,0]$, there exists $r > 0$ such that if $\tilde{\phi} \in C[-h,0]$ and $\| \tilde{\phi} - \phi \|_s < r$, then the solution $x(\tilde{\phi},t)$ is defined over $[0,\infty)$, and in fact one has $\lim_{t\to\infty} x(\tilde{\phi},t) = \lim_{t\to\infty} x(\phi,t)$.

We can now give a simple proof of the uniqueness of the equilibrium for which existence is guaranteed by Theorem 3.1 .

**Theorem 3.2 :** Again consider the delay - differential equation $(\dagger)$ $\dot{x}(t) = f(x(t),x(t-h))$, with the same hypotheses as in Theorem 3.1 . Then there is a unique point $\bar{x} \in \mathbb{R}^n$ having $f(\bar{x},\bar{x}) = 0$. For each $\phi \in C^1[-h,0]$, one has $x(\phi,t) \to \bar{x}$ at an exponential rate as $t \to \infty$.

**Proof:** Noting Theorem 3.1 , one need only show the uniqueness of the zero of the function $g(x) = f(x,x)$. Let $C^1[-h,0]$ have the supremum norm $\| \phi \|_s = \sup \{\ |\phi(u)| : -h \le u \le 0\ \}$, and let $\mathbb{R}^n$ have the standard norm. Define the map $T : C^1[-h,0] \to \mathbb{R}^n$ by $T\phi = \lim_{t\to\infty} x(\phi,t)$. Noting that $C^1[-h,0]$ is connected , and noting from Lemma 3.7 that $T$ is continuous, we see that the image of $T$ is connected . From Lemma 3.6 , we know that the zeros of the function $g(x)$ are isolated , and since the image of $T$ is contained in the set of zeros of $g(x)$, we see that the image of $T$ is a single point.

Now let $\{ \bar{x} \}$ be the image of $T$. For any $x' \in \mathbb{R}^n$ having $g(x') = 0$, let $\phi(u) = x'$ for $-h \le u \le 0$. Since $x(\phi,t) = x'$ for all $t \ge 0$, we have $x' = \lim_{t\to\infty} x(\phi,t) = T\phi = \bar{x}$, i.e. $x' = \bar{x}$. Thus we see that the zero of $g(x)$ in $\mathbb{R}^n$ is unique. Q. E. D.

### *Bibliography*

1. Brauer, F. ( 1987 ) , *Absolute Stability in Delay Equations,* Journal of Differential Equations, Vol. 69, pp. 185 - 191.

2. Hale, J.K. ( 1977 ) , *Theory of Functional Differential Equations,* Springer - Verlag, New York.

3. Hale, J.K., Infante, E.F., and Tsen, F. ( 1985 ) , *Stability in Linear Delay Equations,* Journal of Mathematical Analysis and Applications, Vol. 105, No. 2, pp. 533 - 555.

4. Hill, D. and Mareels, M. ( 1990 ) , *Stability Theory for Differential / Algebraic Systems with Application to Power Systems,* IEEE Transactions on Circuits and Systems, Vol. 37, No. 11, pp. 1416 - 1423.

5. Krasovskii, N. ( 1963 ) , *Stability of Motion,* Moscow, 1959. Translation, Stanford University Press.

6. Louisell, J. , *A Velocity Functional for an Analysis of Stability in Delay - Differential Equations* , submitted to SIAM Journal of Mathematical Analysis, ( 1992 ) .

7. Louisell, J. ( 1992 ) , *Growth Estimates and Asymptotic Stability for a Class of Differential - Delay Equation Having Time - Varying Delay,* Journal of Mathematical Analysis and Applications, Vol. 164, No. 2, pp. 453 - 479.

8. Markus, L. and Yamabe, H. ( 1960 ) , *Global Stability Criteria for Differential Systems,* Osaka Math. J., Vol.12, pp. 305 - 317.

# 12 Approximating Piston-Driven Flow of a Non-Newtonian Fluid

**David S. Malkus** University of Wisconsin, Madison, Wisconsin

**John A. Nohel** ETH-Zentrum, Zürich, Switzerland, and University of Wisconsin, Madison, Wisconsin

**Bradley J. Plohr** State University of New York at Stony Brook, Stony Brook, New York

ABSTRACT. As observed in recent experiments of Lim and Schowalter, the piston-driven channel flow of a highly elastic and very viscous non-Newtonian fluid exhibits persistent oscillations. We suggest an explanation for these oscillations in terms of a constitutive model that has a non-monotonic relationship between steady shear stress and strain rate. We describe the reduction of the three-dimensional equations of motion and stress to approximating systems that can be solved by a combination of numerical and analytic methods. One such approximation results in a quadratic system of first-order functional differential equations in which the volumetric flow rate is imposed by a feedback control. Numerical solution of this system exhibits a transition to a regime with persistent oscillations that compare favorably to the Lim–Schowalter observations. To better understand this behavior, we make a further approximation suggested by the numerical solution of the feedback system. The resulting quadratic system of four ordinary differential equations provides precise predictions, through eigenvalue analysis of its linearization, of the transitions of the feedback system.

## 1. INTRODUCTION

Non-Newtonian fluids, such as polymers, suspensions, and emulsions, are often used in advanced materials engineering and process design. Applications include injection molding of high-strength polymers, spinning of synthetic fibers, and the flow of lubricants improved with additives. Intriguing phenomena have been observed experimentally in highly elastic and very viscous fluids, such as polymer solutions and melts. Such non-Newtonian materials can be modeled mathematically as viscoelastic fluids with fading memory. In certain flow regimes, these fluids exhibit instabilities that severely disrupt polymer processing [2].

Laboratory observations have found "spurt" instabilities in pressure-gradient driven flows [17], persistent oscillations in flow at fixed volumetric flow rate [7], and anomalies in step shear strain experiments [11]. Many researchers attribute the observations to "slip" or

1980 *Mathematics Subject Classification* (1985 *Revision*). 34C35, 34K35, 65N30, 76A05, 76A10

*Key words and phrases.* Shear flow, piston-driven, non-Newtonian, non-monotone, oscillations, spurt, feedback control, quadratic dynamical systems, Hopf bifurcation, limit cycles.

This work was supported in part by: the National Science Foundation under Grant DMS-8907264.

"apparent slip," i.e., loss of adhesion of the fluid to the wall. For instance, oscillations are thought to reflect periodic detachment and reattachment of the fluid; the regular temporal variation of normal stresses leads to spatially periodic distortion of the extrudate [2].

We have been investigating an alternative explanation for these phenomena. Our hypothesis is that all three have a common origin in bulk material properties, rather than adhesive properties [2]. To test this hypothesis, we have examined the corresponding one-dimensional shear flows: pressure-driven and piston-driven flow in a slit die, and Couette flow. The crucial feature of the fluid models we employ is a non-monotone relation between steady shear stress and strain rate. Analysis and numerical simulations show that the polymer system changes state in a thin layer near the wall, giving the appearance of a slip layer. Furthermore, the layer has complicated dynamics.

The sharp change of behavior is a consequence of the non-monotone stress/strain-rate relation. Although several constitutive models based on molecular dynamics exhibit such non-monotonicity, this feature has been regarded as a defect. We believe that this assessment should be reconsidered. Indeed, properly conceived non-monotonic constitutive equations violate no known physical law, and they lead to mathematically well-posed initial/boundary-value problems. Moreover, the solutions of these problems bear remarkable similarity to observed anomalies and instabilities in shear flows. Our conclusion is that the key to explaining process-disrupting instabilities is nonlinear dynamics.

In this paper, our focus is the mathematical modeling of the Lim–Schowalter experiment, in which the volumetric flow rate is prescribed by driving the fluid with a piston moving at fixed speed. Experimental data suggest that far away from the piston, the flow is essentially one-dimensional. Furthermore, the data are consistent with choosing a constitutive law that exhibits a non-monotone relationship between steady shear stress and strain rate. Consequently, we model (unsteady) shear flow of a highly elastic and very viscous non-Newtonian fluid in a channel by a singularly perturbed system of quasilinear partial differential equations obeying the Johnson-Segalman–Oldroyd differential constitutive law. To model the Lim and Schowalter experiment, the (unknown) driving pressure gradient is constrained to maintain a prescribed volumetric flow rate. Mathematically, the governing system is a feedback control problem.

Our earlier investigation of shear flows driven by a fixed pressure gradient has identified the key feature of the constitutive relation as the non-monotonicity of the steady stress-strain rate curve [8,9]. Driving pressure gradients implying a wall shear stress exceeding the local shear stress maximum lead to a rich dynamic structure that explains spurt, along with related phenomena that occur under cyclic loading and unloading. For sufficiently large volumetric flow rates, the corresponding pressure gradient exceeds the critical value, and interesting behavior is to be expected. These considerations determine a range of parameters on which the present study is focused. A deeper analytical understanding of these solutions is possible because the system contains a small parameter, the ratio of Reynolds number to Deborah number. Setting this parameter to zero yields a one-parameter family of quadratic, planar dynamical systems. The qualitative behavior of these systems is completely understood through phase plane analysis [8,9]. Moreover, as has recently been shown in Ref. [13], the full governing system of quasilinear PDEs exhibits essentially the same dynamics as the reduced system, provided that the the ratio of Reynolds number to Deborah number is sufficiently small.

In the present case of piston-driven flow, the dynamical systems are coupled by the control equation, which is an integral constraint. This complicates the analysis significantly, but leads to systems of mathematical interest and challenge. Moreover, the control apparently destabilizes the system in a way that is essential to the modeling of the instabilities observed by Lim and Schowalter.

The outline of this paper is as follows. In Sec. 2, we summarize the development the mathematical model for piston-driven channel flow of a highly elastic and viscous non-Newtonian fluid. The model is based on balance laws and the Johnson-Segalman–Oldroyd constitutive law in three dimensions. When the volumetric flow-rate is prescribed, there results a semilinear system of reaction-diffusion equations with a feedback control. We also deduce the inertialess approximation for such a flow, which yields a quadratic system of functional differential equations. In Sec. 3, we establish the global well-posedness in time of the initial/boundary-value problem governing piston-driven flows (Theorem 3.1), and we exhibit how the governing system arises as a singular perturbation (Theorem 3.2 and Remark 3.3). In Sec. 4, we formulate numerical algorithms and use them to simulate piston-driven flows, particularly inertialess flows, numerically. In Sec. 5, we use results of numerical simulations to approximate inertialess flows further using a quadratic system of four first-order ODEs. Numerical simulation of this system leads to a natural conjecture about its global dynamics that is currently under investigation. Finally, in Sec. 6, we use the results of Secs. 4 and 5 to offer an alternative explanation of the mechanism underlying the Lim–Schowalter experiment.

## 2. A Mathematical Model for Piston-Driven Flow

**2.1 Channel Flow of a Johnson-Segalman–Oldroyd Fluid.** To model the experiments of Lim and Schowalter, we consider the planar Poiseuille flow of a Johnson-Segalman–Oldroyd fluid within a channel. We summarize the derivation of the system of partial differential equations that governs such a flow, one-dimensional flow, starting from balance laws and constitutive equations in three dimensions; for more details, see Refs. [8,9].

The channel is aligned along the $y$-axis, and extends between $x = -h/2$ and $x = h/2$ (see Fig. 1). The flow is assumed to be symmetric about the centerline $x = 0$ of the channel, and we restrict attention to the interval $x \in [-h/2, 0]$. Since the fluid undergoes simple shearing, the velocity and stress variables are independent of $y$. In particular, the velocity field is $\mathbf{v} = (0, v(x,t))$, so that the flow is incompressible and the conservation of mass equation is automatically satisfied.

The total stress on the fluid is the sum of three contributions, an isotropic pressure, a Newtonian stress, and an extra stress $\boldsymbol{\pi}$. The conservation of momentum in the $x$-direction implies that the pressure takes the form $p = p_0(x,t) - f(t)\,y$, with $f$ being the pressure gradient. We adopt the Johnson-Segalman–Oldroyd differential constitutive law [14,3] to determine the extra stress. In shear flow, the extra stress is expressible in terms of two variables, the shear stress $\sigma(x,t) := \boldsymbol{\pi}^{xy}$ and the principal normal stress difference $Z(x,t) :=$

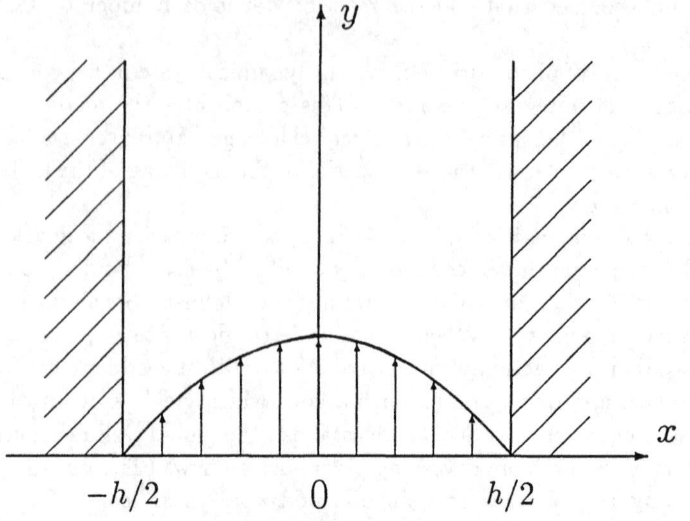


**Fig. 1:** Shear flow through a slit-die.

$-\frac{1}{2}(1-a^2)(\pi^{yy}-\pi^{xx})$; here $a \in (-1,1)$ is the slip parameter of the fluid. The fluid variables $v$, $\sigma$, and $Z$ are governed by the $y$-momentum equation and constitutive differential equations. We introduce two dimensionless parameters that characterize the flow: $\alpha$, the ratio of Reynolds number to Deborah number; and $\varepsilon$, the ratio of Newtonian viscosity to shear viscosity. After suitable scaling, the governing equations become

$$\begin{aligned} \alpha v_t - \sigma_x &= \varepsilon v_{xx} + f \,, \\ \sigma_t - (Z+1)v_x &= -\sigma \,, \\ Z_t + \sigma v_x &= -Z \,. \end{aligned} \tag{JSO}$$

on the interval $[-1/2, 0]$. The boundary conditions are

$$v(-1/2,t) = 0 \quad \text{and} \quad v_x(0,t) = 0 \,, \tag{BC}$$

and the initial conditions have the form

$$v(x,0) = v_0(x) \,, \quad \sigma(x,0) = \sigma_0(x) \,, \quad \text{and} \quad Z(x,0) = Z_0(x) \,. \tag{IC}$$

For consistency of $(IC)$ with $(BC)$, we assume that $v_0(-1/2) = 0$ and $v_0'(0) = 0$; to maintain continuity at $x = 0$, we require that $\sigma_0(0) = 0$.

For the purpose of analysis, it is useful to rewrite the governing equations in the equivalent form of the following degenerate system of reaction diffusion equations. Define the quantities

$$T(x,t) := -f(t)x \tag{2.1}$$

and

$$S := \varepsilon v_x + \sigma - T \,; \tag{2.2}$$

also let $S_0(x) := \varepsilon v_0'(x) + \sigma_0(x) - T(x,0)$. The initial/boundary-value problem $(JSO)$, $(BC)$, $(IC)$ for the quantities $v$, $\sigma$, and $Z$ is equivalent to the following one with unknown functions $S$, $\sigma$, and $Z$: the equations

$$\begin{aligned} \alpha S_t &= \varepsilon S_{xx} + \alpha \left[ (Z+1)\left(\frac{S+T-\sigma}{\varepsilon}\right) - \sigma - T_t \right] , \\ \sigma_t &= (Z+1)\left(\frac{S+T-\sigma}{\varepsilon}\right) - \sigma , \\ Z_t &= -\sigma\left(\frac{S+T-\sigma}{\varepsilon}\right) - Z \end{aligned} \qquad (JSO')$$

on the interval $[-1/2, 0]$, together with the boundary conditions

$$S_x(-1/2, t) = 0 \quad \text{and} \quad S(0,t) = 0 , \qquad (BC')$$

and the initial conditions

$$S(x,0) = S_0(x) , \quad \sigma(x,0) = \sigma_0(x) , \quad \text{and} \quad Z(x,0) = Z_0(x) . \qquad (IC')$$

The compatibility conditions $S_0'(-1/2) = 0$ and $S_0(0) = 0$ and the continuity condition $\sigma_0(0) = 0$ are assumed.

For future refence, it is useful to note that solutions of systems $(JSO)$ or $(JSO')$ satisfy a Lyapunov equation,

$$\frac{d}{dt}\left\{\sigma^2 + (Z+1)^2\right\} = -2\left[\sigma^2 + (Z+\tfrac{1}{2})^2 - \tfrac{1}{4}\right] . \qquad (L)$$

Observe that Eq. $(L)$ is independent of $\alpha$, $\varepsilon$, and $T$. It follows easily from Eq. $(L)$ that $\sigma, Z$ remain bounded for as long as the solution $(S, \sigma, Z)$ of system $(JSO')$ exists.

Steady-state solutions play an important role in the study of $(JSO)$ and $(JSO')$. We use an overbar to indicate such a solution, which satisfies the following relations (see Ref. [9]):

$$\overline{\sigma} = \frac{\overline{v}_x}{1+\overline{v}_x^2} , \qquad \overline{Z}+1 = \frac{1}{1+\overline{v}_x^2} , \qquad (2.3)$$

$$\overline{S} = \varepsilon\overline{v}_x + \frac{\overline{v}_x}{1+\overline{v}_x^2} - \overline{T} = 0 . \qquad (2.4)$$

Thus the steady strain rate $\overline{v}_x$ satisfies the steady stress/strain-rate relation

$$w(\overline{v}_x) = \overline{T} , \qquad (2.5)$$

where $w$ is the function defined by

$$w(\xi) := \frac{\xi}{1+\xi^2} + \varepsilon\xi . \qquad (2.6)$$

When $\varepsilon < 1/8$, the function $w$ is not monotone, as illustrated in Fig. 2. In this case, there are multiple values for $\overline{v}_x(x)$ satisfying Eq. (2.5) when $\overline{T}$ lies in the range between $\overline{T}_m$ and $\overline{T}_M$. This allows for steady-state solutions in which the strain rate $\overline{v}_x$ is discontinuous and the velocity profile $\overline{v}$ has kinks. An example of such a solution is shown in Fig. 3.

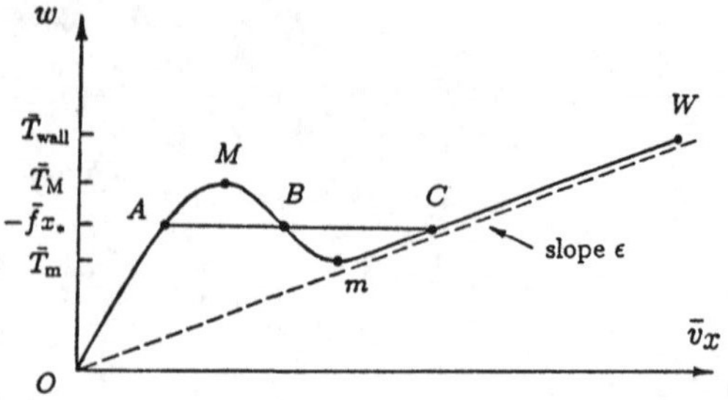


Fig. 2: The steady shear stress $\overline{T} = w(\overline{v}_x)$ plotted *vs.* shear strain rate $\overline{v}_x$.

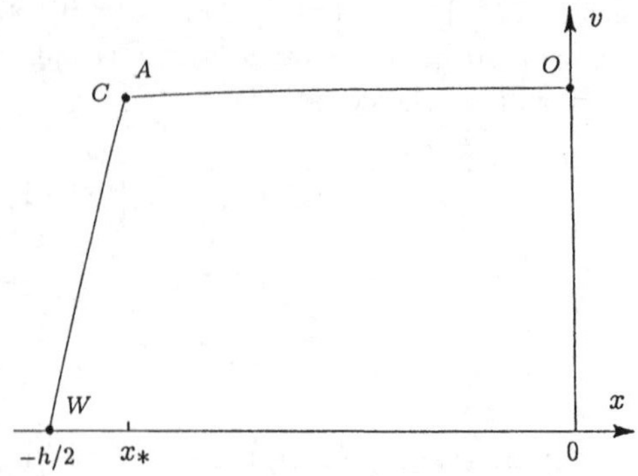


Fig. 3: The velocity profile in steady flow for the path $OACW$ shown in Fig. 2.

**2.2 Piston-Driven Flow.** The distinction between piston-driven and pressure-driven channel flow lies in how the pressure gradient $f$ is determined. If $f$ is prescribed, then the initial/boundary-value problem $(JSO)$, $(BC)$, $(IC)$ describes pressure-driven flow. This problem has been analyzed in Refs. [15,4,8,12,9,13]. For piston-driven flow, by contrast, the volumetric flow rate $Q_0$ is prescribed, and the pressure gradient is adjusted to maintain this flow rate. Thus piston-driven flow is modeled as an instantaneous feedback-control problem in which $f$, or equivalently $T$, is the feedback, and $Q_0$ is the control.

To determine $T$ from $Q_0$, notice that the flow rate is given as

$$Q_0(t) = 2\int_{-1/2}^{0} v(x,t)\,dx \tag{2.7}$$

in terms of the velocity, or as

$$Q_0(t) = 2\int_{-1/2}^{0}\int_{-1/2}^{x} v_x(\hat{x},t)\,d\hat{x}\,dx \tag{2.8}$$

in terms of the velocity gradient. After substituting for $v_x$ using Eq. (2.2), we obtain the relationship

$$Q_0(t) = \frac{2}{\varepsilon}\int_{-1/2}^{0}\int_{-1/2}^{x} [S(\hat{x},t) + T(\hat{x},t) - \sigma(\hat{x},t)]\,d\hat{x}\,dx\ , \tag{2.9}$$

and a routine simplification, using integration by parts, leads to the feedback equation

$$T(x,t) = -12\varepsilon Q_0(t)x + 24x\int_{-1/2}^{0} [\sigma(\hat{x},t) - S(\hat{x},t)]\,\hat{x}\,d\hat{x}\ . \tag{$FB$}$$

Therefore, the model for piston-driven channel flow reduces to the following feedback-control problem: for a prescribed volumetric flow rate $Q_0$, find the solution $S$, $\sigma$, $Z$, of the initial/boundary-value problem $(JSO')$, $(BC')$, $(IC')$, where $T$ is determined by the feedback equation $(FB)$.

**2.3 Inertialess Flows.** In the case of highly elastic polymer systems, $\alpha$ is observed to be very small, 7 to 10 orders of magnitude smaller than $\varepsilon$, which is of the order of $10^{-3}$. Therefore it seems natural to make the inertialess approximation $\alpha = 0$. This approximation has been used successfully in studies of pressure-driven flow in Refs. [9,13].

If $\alpha = 0$, the first equation in $(JSO')$, combined with the boundary conditions $(BC')$, as well as Eq. (2.2), implies that

$$S(x,t) \equiv 0 \ , \quad i.e., \quad v_x = \frac{T-\sigma}{\varepsilon} \ . \tag{2.10}$$

As a result, the system $(JSO')$ simplifies to be a family of quadratic differential equations, parameterized by $x$; adding the feedback equation $(FB)$ (with $S(x,t) \equiv 0$) couples the resulting system. To write this system in compact form, note that the feedback equation $(FB)$ can be regarded as a functional equation involving the operator $P$ defined by

$$[P\sigma](x,t) := 24x \int_{-1/2}^{0} \sigma(\hat{x},t)\hat{x}\, d\hat{x} \ ; \tag{2.11}$$

D. L. Russell [16] observed that $P$ is an orthogonal projection onto the subspace of $L^2[-1/2,0]$ spanned by the function $\sqrt{24}x$. In other words, inertialess piston-driven flow is governed by the one-parameter quadratic system of functional differential equations,

$$\begin{aligned} \sigma_t &= (Z+1)\left(\frac{T-\sigma}{\varepsilon}\right) - \sigma \ , \\ Z_t &= -\sigma\left(\frac{T-\sigma}{\varepsilon}\right) - Z \ , \\ T &= T_0 + P\sigma \ , \end{aligned} \tag{QFDE}$$

parametrized by $x \in [-1/2, 0]$, where $T_0(x,t) := -12\varepsilon Q_0(t)x$ is prescribed, the non-local projection operator $P$ is defined Eq. (2.11), and where $\sigma, Z$ satisfy the initial conditions $(IC')$. Since the Lyapunov equation $(L)$ is satisfied by solutions of $(QFDE)$, all of its solutions are globally bounded on $0 \le t < \infty$.

To our knowledge, the extensive literature on quadratic differential equations and on functional differential equations does include any results on the global qualitative structure of solutions of $(QFDE)$. Determining the complete dynamics of system $(QFDE)$ and then showing that the full control problem $(JSO')$, $(BC'')$, $(IC'')$, and$(FB)$ has similar dynamics for $\alpha$ sufficiently small, is the ultimate objective of this study.

## 3. Mathematical Results

**3.1 Global Well-Posedness.** When $\alpha > 0$, solutions of the initial/boundary-value problem $(JSO')$, $(BC'')$, $(IC'')$ satisfying the feed-back condition $(FB)$ exist for all time and for initial data of arbitrary size. This result can be proved following the same approach as used by Nohel, Pego, and Tzavaras [12] for the case of pressure-driven flow. In the following statements, the function spaces $C^s$ and $W^{s,p}$ refer to the spatial interval $[-1/2, 0]$. For simplicity, the prescribed volumetric flow rate $Q_0$ is taken to be independent of time here and throughout the rest of the paper.

**Theorem 3.1:**
*(a) If $S_0 \in H^s$, where $s > 3/2$, and $\sigma_0, Z_0 \in C^1$, then the unique solution is classical, meaning that $S \in C([0,\infty); C^1) \cap C((0,\infty); H^2)$, $S_t \in C((0,\infty); C^1)$, and $\sigma, Z \in C^1([0,\infty); C^1)$. Furthermore, $v \in C([0,\infty); C^2)$ and $v_t \in C((0,\infty); C^2)$.*

*(b) If $S_0 \in H^1$ and $\sigma_0, Z_0 \in L^\infty$, then there exists a unique solution of $(JSO')$, $(BC')$, $(IC')$, $(FB)$ that is semi-classical, in the sense that $S \in C([0,\infty); H^1) \cap C((0,\infty); W^{2,\infty})$, $S_t \in C((0,\infty); H^s)$ for all $s < 2$, and $\sigma, Z \in C^1([0,\infty); L^\infty)$. Furthermore, $v \in C([0,\infty); W^{1,\infty})$ and $v_t \in C((0,\infty); W^{1,\infty})$.*

The existence of solutions admitting discontinuities in the stress components and in the velocity gradients in (b) was crucial in our explanation of spurt and related phenomena in Refs. [9,13] as a material bulk property rather than an adhesive property in pressure-driven flow; it is equally important here.

**Proof:** We rewrite $(JSO')$, $(BC')$, $(IC')$, $(FB)$ in a form such that the proof given in Ref. [12] applies. If we introduce $\tilde{S} := (I-P)S$ and $\tilde{T} := T_0 + P\sigma$, so that $S + T = \tilde{S} + \tilde{T}$, where $P$ is the projection operator defined in Eq. (2.11), then system $(JSO')$ in conjunction with $(FB)$ becomes

$$\begin{aligned} \alpha\tilde{S}_t &= \varepsilon\tilde{S}_{xx} + \alpha(I-P)\left[(Z+1)\left(\frac{\tilde{S}+\tilde{T}-\sigma}{\varepsilon}\right) - \sigma\right], \\ \sigma_t &= (Z+1)\left(\frac{\tilde{S}+\tilde{T}-\sigma}{\varepsilon}\right) - \sigma, \\ Z_t &= -\sigma\left(\frac{\tilde{S}+\tilde{T}-\sigma}{\varepsilon}\right) - Z. \end{aligned} \qquad (JSO'')$$

Furthermore, one of the boundary conditions for $S$ becomes $\tilde{S}(0,t) = 0$, whereas the other, $(PS)_x(-1/2,t) = -\tilde{S}_x(-1/2,t)$, determines $PS$ once $\tilde{S}$ is known. Applying the operator $P$ to the first equation in $(JSO'')$ shows that $P\tilde{S} = 0$ if and only if $\tilde{S}_x(-1/2,t) = 0$, assuming that the initial value for $\tilde{S}$ is taken to be $\tilde{S}_0 := PS_0$. Therefore, the appropriate boundary and initial conditions are as in $(BC')$ and $(IC')$, with $S$ replaced by $\tilde{S}$. The existence of solutions of this equivalent initial/boundary-value problem is easily deduced using the approach of Ref. [12], generalized slightly to account for the bounded projection operator $P$. □

Identity $(L)$, which implies the global boundedness of $\sigma$ and $Z$, is used in an essential way to establish global existence. Although solutions to $(JSO')$ exist globally, it does not follow that $S$ remains bounded. A proof that $S$ is, in fact, bounded under the assumption that $\alpha$ is sufficiently small (of the order of $\varepsilon^2$) is given in the following result proved as in Ref. [13].

**Theorem 3.2:** *Assume the hypotheses of Theorem 3.1, part (b). For sufficiently small $\alpha > 0$, there is a time $t_0(\alpha) > 0$, which depends only on $\alpha$, and a constant $C$, which depends only on the initial data, such that*

$$||S(\cdot,t)||_{H^1} \le C\alpha, \qquad (3.1)$$

*for all* $t \geq t_0(\alpha)$.

**3.2 Inertialess Flow.** The initial-value problem for the functional differential equations obtained for the inertialess approximation $\alpha = 0$ is also well-posed, globally in time. To see this, we regard the system $(QFDE)$ as an ordinary differential equation for $\sigma(\cdot,t), Z(\cdot,t) \in L^\infty$. It is easily verified that the flux for the differential equation is $C^1$, so that, locally in time, a solution exists and is unique. Furthermore, the Lyapunov equation $(L)$ shows that the local solution remains bounded, so that it can be extended to the interval $0 \leq t < \infty$. A similar conclusion holds for $\sigma(\cdot,t), Z(\cdot,t) \in C^0$.

**Remark** 3.3: For small $\alpha > 0$, the initial/boundary-value problem governing the piston-driven flow can be viewed as a singular perturbation of the inertialess flow problem, in which $\alpha = 0$. Using the estimate in Eq. (3.1) and a simple argument involving the Gronwall inequality, it is shown as in Ref. [13] that for $\alpha > 0$ sufficiently small, the solution of the inertialess piston-driven flow problem approximates the solution $(S, \sigma, Z)$ of the full piston-driven system to within terms of order $\alpha$, provided that $t$ is restricted to a compact time interval $[t_0, t_{max}]$, where $t_0 = t_0(\alpha) > 0$ and $t_{max} > t_0$ is arbitrary.

## 4. Numerical Simulation of Piston-Driven Flows

**4.1 Basic Numerical Method.** The numerical method for the piston-driven shear flows is based on our method for pressure gradient driven flow [8]. We first consider the general case in which inertial effects are included. A Galerkin equation for momentum transfer with test/trial functions, $\phi$, can be written in time discrete form at time level $n+1$ in terms of $\sigma_{n+1}$, which can in turn be calculated from $\sigma_n$, $Z_n$, $(v_x)_n$, and $(v_x)_{n+1}$ by the scheme:

$$\begin{cases} \int_{-1/2}^{0} \{\alpha\phi(v_t)_{n+1} + \phi_x\sigma_{n+1} + \varepsilon\phi_x(v_x)_{n+1} - \phi f_{n+1}\}\, dx = 0 \\ \sigma_{n+1} = (1-\Delta t)\sigma_n + \Delta t Z_n (v_x)_n + \Delta t (v_x)_{n+1} \\ v_{n+1} = v_n + \Delta t\{(1-\beta)(v_t)_n + \beta(v_t)_{n+1}\} \end{cases} \qquad (GAL)$$

Regarding $(GAL)$ as a predictor for $v_x$, the stress equations for $\sigma, Z$ are differenced and advanced at each time step after the solution of the momentum equations using the latest value available, $(v_x)_{n+1}$ by the "corrector" scheme that also advances $Z$:

$$\begin{cases} \sigma_{n+1} = (1-\Delta t)\sigma_n + \Delta t (Z_n + 1)(v_x)_{n+1} \\ Z_{n+1} = (1-\Delta t) Z_n - \Delta t \sigma_{n+1}(v_x)_{n+1} \end{cases} \qquad (4.1)$$

In $(GAL)$, linear line elements were employed to discretize the momentum equation; the second and third equations apply pointwise on $(-1/2, 0)$, while the last is a standard trapezoidal approximation [1]. Observe that there are no nonlinear equations to solve at each time step, and the method is termed "semi-implicit." The implicit treatment of the $\varepsilon$ term in the momentum equation is aimed at avoiding a $\Delta t \sim (\Delta x)^2$ diffusion CFL condition, and the implicit treatment of the last term in the second equation of Eqs. $(GAL)$ is aimed at avoiding a $\Delta t \sim \Delta x$ CFL condition based on elastic wave transit time. A detailed analysis of the stability of the algorithm described above is carried out in Ref. [10]. It was found that stability and accuracy were relatively insensitive to $\beta$ in most cases, and for the present purpose, $\beta = 0.6$ was used throughout.

**4.2 Numerical Approximation of Inertialess, Piston-Driven Flows.** In the inertialess case, Eqs. $(GAL)$ are bypassed in favor of a force balance determined by the driving mechanism. This leads to a prediction of $(v_x)_{n+1}$ that is put directly into Eq. (4.1). For example, in the case of pressure-gradient driven flows,

$$(v_x)_{n+1} = \frac{T^k - \sigma_n^k}{\varepsilon} \tag{4.2}$$

where

$$T^k = -\frac{(x_k + x_{k+1})}{2}\overline{f} \tag{4.3}$$

The modification of the basic numerical method to handle flows modeled by Eqs. $(QFDE)$ and (2.11) is straight-forward: we add the equation

$$(v_x)_{n+1} = \frac{T_n^k - \sigma_n^k}{\varepsilon} \tag{4.4}$$

as the first equation in system (4.1) for the $k^{th}$ element in the mesh, where $k$ runs from 1 to the number of elements $N$, and $\sigma_n^k$ is the value of the shear stress in the $k^{th}$ element at time level $n$. The value for $T_n^k$ is computed by evaluating $P\sigma$ in Eq. $(QFDE)$ by the composite midpoint rule:

$$T_n^k = \frac{(x_k + x_{k+1})}{2}\left(-12\varepsilon Q_0(t) + 24\sum_{i=1}^{N-1}\sigma_n^i\,(x_{i+1} - x_i)\,(x_i + x_{i+1})\,/2\right). \tag{4.5}$$

The evaluation of the volumetric constraint is thus treated explicitly.

**4.3 Preliminary Results of Simulation of systems $(QFDE)$.** In numerical simulations of $(QFDE)$, we have observed four distinct flow regimes corresponding to different ranges for the volumetric flow rate $Q_0$: the classical, spurt I, oscillatory, and spurt II regimes. Typical plots of $f(t)$ *vs.* $t$ for the four regimes are shown in Figs. 4a–d.

In the classical regime, steady solutions that have no discontinuities are achieved. Steady solutions are also achieved in the spurt I and II regimes; these solutions are, however, discontinuous, having the appearance of the spurt solutions achieved in pressure-gradient driven flows [8,9]. As $Q_0$ is varied and the transition to the oscillatory regime is approached, the time to settle to a steady solution increases, and the settling is accompanied by damped oscillations in $f(t)$ and in the velocity near the wall. Finally, at a critical value for $Q_0$, the oscillations appear to become undamped and fail to settle down; this is the oscillatory regime. Figures 5a and 5b depict the shear stress profiles across the channel in the oscillatory regime; the two profiles represent the extremes of the oscillations. If $Q_0$ is increased further, a second transition is observed, and steady, spurt-like solutions are again achieved; this is the spurt II regime. (For technical reasons, Fig. 4d terminates before the oscillations have died away to graphical accuracy; however, the simulation has been carried out for much longer time, and there is no doubt that the oscillations die out.)

We wish to emphasize that because we rely on numerical simulation, and because it is difficult to accurately distinguish slightly damped systems from undamped systems numerically, the existence of a true oscillatory regime for the full system $(QFDE)$ can only be conjectured at this point.

We have studied the transitions between these regimes in detail, as a function of the imposed volumetric flow rate $Q_0$, for Vinogradov's PI-7 sample polyisoprene [17] (*i.e.*, $\varepsilon = 0.001417$). As $Q_0$ is increased from zero, the transition from the classical regime to the spurt I regime occurs at approximately $Q_0 = 0.1$. The transition between the steady spurt I regime and the oscillatory regime occurs at a critical $Q_0$ of about 0.3, whereas the transition from the oscillatory regime back to the steady spurt II regime occurs at a critical $Q_0$ of about 1.4.

We note that Lim [6] has experimentally observed four separate flow regimes as well, the third of which is oscillatory and the fourth of which is a relatively steady regime at high shear rate (inferred from $Q_0$). This corresponds exactly to the results of our numerical simulation of $(QFDE)$.

## 5. Approximation of Inertialess Piston-Driven Flows

In order to analyze the transition from the steady spurt regime to the oscillatory regime and back again, we develop a simpler, finite-dimensional quadratic system of differential equations with solutions that qualitatively resemble those of the infinite-dimensional system $(QFDE)$. The basis for this approximation is the profile of shear stress $\sigma$ across the channel observed when the numerical solution has settled into seemingly periodic oscillations. As seen in Figs. 5a and 5b, $\sigma$ is very nearly piecewise linear, with a single jump discontinuity at a point $x_*$. This suggests determining $x_*$ *a priori* from the numerical simulations of system $(QFDE)$ and then replacing $\sigma$ in the feedback equation for $T$ in system $(QFDE)$ by a piecewise linear approximant. In this manner we obtain a finite-dimensional system that should mimic the late-time dynamics of $(QFDE)$.

**5.1 Approximation by Quadratic Systems of ODEs.** A piecewise linear approximant of the form in Fig. 5a is determined by three values: $\sigma_-(t) := \sigma(x_*^-, t)$ on the layer side of the discontinuity, $\sigma_+(t) := \sigma(x_*^+, t)$ on the core side of the discontinuity, and $\sigma_w(t) := \sigma(-1/2, t)$ at the channel wall. We define the approximation $\sigma_{\text{approx}}$ of $\sigma$ by

$$\sigma_{\text{approx}}(x,t) = \begin{cases} \sigma_w(t)\left(1 - \frac{x+1/2}{x_*+1/2}\right) + \sigma_-(t)\frac{x+1/2}{x_*+1/2} & \text{if } -1/2 \le x < x_*, \\ \sigma_+(t)\, x/x_* & \text{if } x_* < x \le 0. \end{cases} \tag{5.1}$$

The approximate total stress is then $T_{\text{approx}} := T_0 + P\sigma_{\text{approx}}$, where $T_0 = T_0(x) = -12\varepsilon Q_0 x$ is independent of $t$, and where $P$ is the projection operator defined in Eq. (2.11). To be more explicit, let $\gamma_- = -x_*(1+2x_*)(1-4x_*)$, $\gamma_+ = -8x_*^3$, and $\gamma_w = -2x_*(1+2x_*)(1-x_*)$; notice that $\gamma_- + \gamma_+ + \gamma_w/(-2x_*) = 1$. In these terms,

$$T_{\text{approx}}(x,t) = T_0(x) + \frac{x}{x_*}\left[\gamma_-\sigma_-(t) + \gamma_+\sigma_+(t) + \gamma_w\sigma_w(t)\right] . \tag{5.2}$$

The key point to observe is that $T_{\text{approx}}$ depends only on $\sigma_-$, $\sigma_+$, and $\sigma_w$.

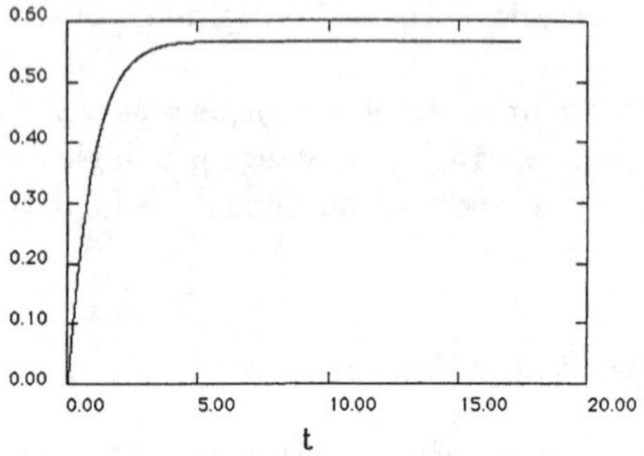


**Fig. 4a**: $f(t)$ *vs.* $t$ in the classical regime of system ($QFDE$).

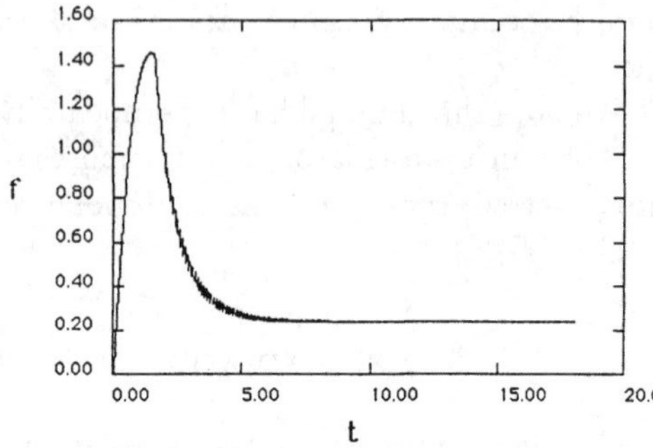


**Fig. 4b**: $f(t)$ *vs.* $t$ in the spurt I regime of system ($QFDE$).

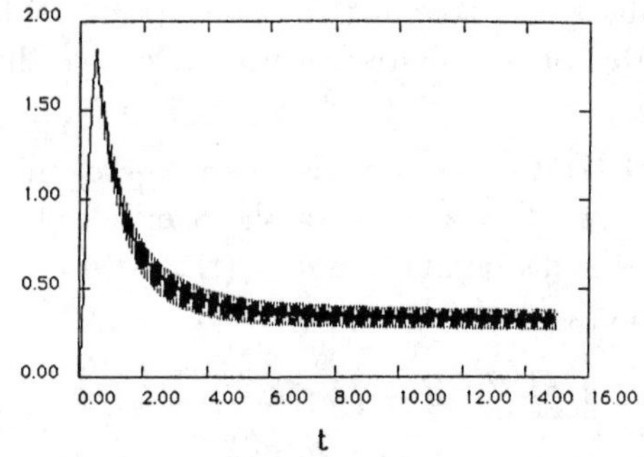


**Fig. 4c**: $f(t)$ *vs.* $t$ in the oscillatory regime of system ($QFDE$).

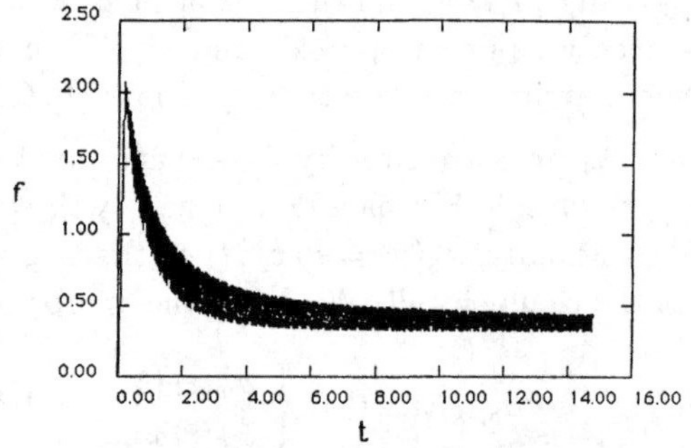


**Fig. 4d**: $f(t)$ *vs.* $t$ in the spurt II regime of system ($QFDE$).

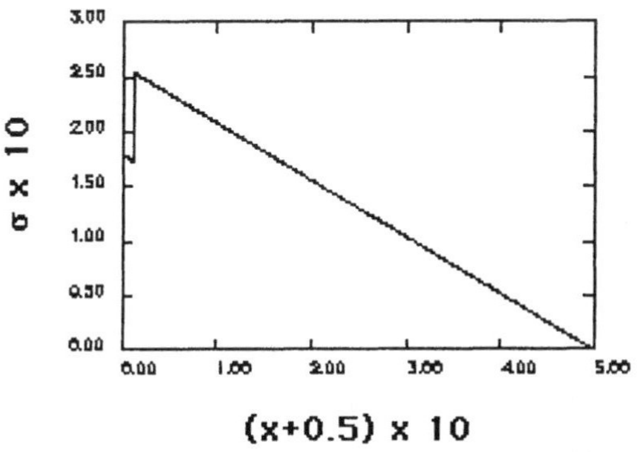


**Fig. 5a**: The shear stress profile during the oscillation stage. The point of minimum wall stress in the cycle is shown.

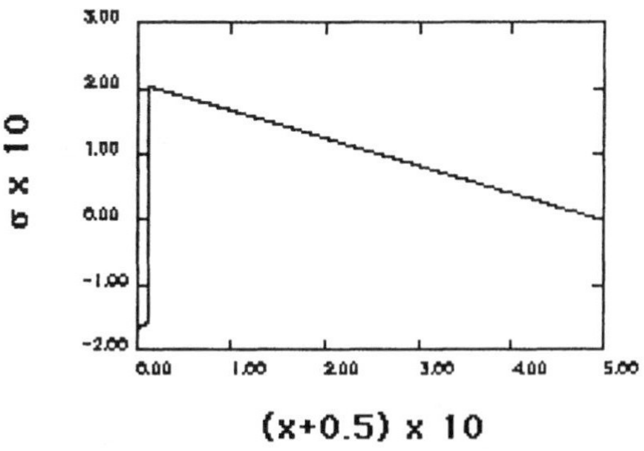


**Fig. 5b**: The shear stress profile during the oscillation stage. The point of maximum wall stress in the cycle is shown.

The approximation to inertialess, piston-driven flow corresponding to Figs. 5 is obtained by replacing $T$ by $T_{\text{approx}}$ in system $(QFDE)$. The resulting system can be reduced to a quadratic system of six ordinary differential equations, as follows. Notice that the first two equations in system $(QFDE)$ represent, for each fixed $x$, a pair of ordinary differential equations governing $\sigma(x,\cdot)$ and $Z(x,\cdot)$. Coupling between pairs corresponding to distinct values of $x$ occurs because of the feedback $T = T_0 + P\sigma$. If, however, $T$ is replaced by $T_{\text{approx}}$, then the pairs corresponding to $x = x_*^-$, $x = x_*^+$, and $x = -1/2$ decouple from the rest. Therefore the approximation to $(QFDE)$ reduces to a system of six quadratic ordinary differential equations for the variables $\sigma_-$, $Z_- := Z(x_*^-,\cdot)$, $\sigma_+$, $Z_+ := Z(x_*^+,\cdot)$, $\sigma_w$, and $Z_w := Z(-1/2,\cdot)$.

A cruder, but still effective, approximation to system $(QFDE)$ is obtained by by assuming that $\sigma$ has the same slope in the layer as in the core. In this case, $\sigma_w = \sigma_- - [1 + 1/(2x_*)]\sigma_+$, and we can dispense with the two equations at $x = -1/2$. Defining $\gamma := 3(-x_*)(1 - 4x_*^2)$, the resulting approximating system of four coupled quadratic ODEs can be written

$$\begin{aligned}
\dot{\sigma}_- &= (Z_- + 1)\left(\frac{T^* - \sigma_-}{\varepsilon}\right) - \sigma_-\,, \\
\dot{Z}_- &= -\sigma_-\left(\frac{T^* - \sigma_-}{\varepsilon}\right) - Z_-\,, \\
\dot{\sigma}_+ &= (Z_+ + 1)\left(\frac{T^* - \sigma_+}{\varepsilon}\right) - \sigma_+\,, \\
\dot{Z}_+ &= -\sigma_+\left(\frac{T^* - \sigma_+}{\varepsilon}\right) - Z_+\,,
\end{aligned} \qquad (Q_4)$$

where the coupling term $T^*$ is

$$T^*(t) := T_0(x_*) + \gamma\sigma_-(t) + [1-\gamma]\sigma_+(t) \ . \tag{5.3}$$

System $(Q_4)$ is endowed with the identities

$$\frac{d}{dt}\left\{\sigma_i^2 + (Z_i+1)^2\right\} = -2\left[\sigma_i^2 + (Z_i+\tfrac{1}{2})^2 - \tfrac{1}{4}\right], \quad i = -, + \ . \tag{5.4}$$

Since these identities are independent of any parameters, all solutions are globally bounded on $0 \le t < \infty$.

**5.2 Numerical Simulation, Analysis, and Conjectures about system $(Q_4)$.** To utilize system $(Q_4)$, we first obtain a sequence of transient solutions to $(QFDE)$, using the numerical algorithm described in Sec. 4, for various values of $Q_0$. Corresponding values of $x_*$ are estimated from these solutions and used to calculate the parameters $\gamma$ and $T_0(x_*)$ that appear in Eq. (5.3). System $(Q_4)$ is then solved using a stiff ODE solver. In principle, the starting condition for $(\sigma_-, Z_-, \sigma_+, Z_+)$ should be $(0,0,0,0)$; however, the solution would then maintain $(\sigma_-, Z_-) = (\sigma_+, Z+)$ for all time. To break this symmetry, we take the initial value of $\sigma_-$ to be slightly negative instead of 0.

The numerical simulations of systems $(Q_4)$ revealed four regimes of solutions, as depicted in Fig. 6; the regimes correspond exactly to the four regimes of $(QFDE)$. Furthermore, the transition points are approximately at $Q_0 = 0.3$ and $Q_0 = 1.4$, just as for system $(QFDE)$. (It should be noted that the transition from the classical to the spurt I regime corresponds to having $x_*$ move away from the wall; therefore this transition cannot be tested independently using system $(Q_4)$. Also, there is some difference between the early-time behavior of $(QFDE)$ and $(Q_4)$ in the spurt I regime, which is not understood at present.) These results suggest the following

**Conjecture**: System $(Q_4)$ possesses a periodic orbit for $Q_0$ in the approximate range $0.3 \le Q_0 \le 1.4$ when $\varepsilon = 0.001417$, and this orbit is a stable limit cycle in four-dimensional phase space.

We are currently attempting to resolve this conjecture. The remainder of this section describes some preliminary steps in analyzing system $(Q_4)$. Suppose that $x_*$, and therefore $T_0(x_*)$ and $\gamma$, have been obtained from the numerical simulation of $(QFDE)$ for a given $Q_0$. The following results hold true for the many values of $Q_0$, in a certain range, that we have sampled.

We first locate the critical points of system $(Q_4)$. The $(\sigma_-, \sigma_+)$ coordinates of a critical point satisfy the simultaneous algebraic equations

$$\begin{aligned} Q(\sigma_-, T^*) &= 0 \ , \\ Q(\sigma_+, T^*) &= 0 \ , \end{aligned} \tag{C}$$

where

$$Q(\sigma, T) := \sigma^3 - 2T\sigma^2 + \left[T^2 + \varepsilon(1+\varepsilon)\right]\sigma - \varepsilon T \ . \tag{5.5}$$

To ascertain the number of solutions, the curve in $(\sigma_-, \sigma_+)$-space along which $Q(\sigma_-, T^*) = 0$ is drawn superimposed on the curve where $Q(\sigma_+, T^*) = 0$, as shown in Fig. 7 (which was

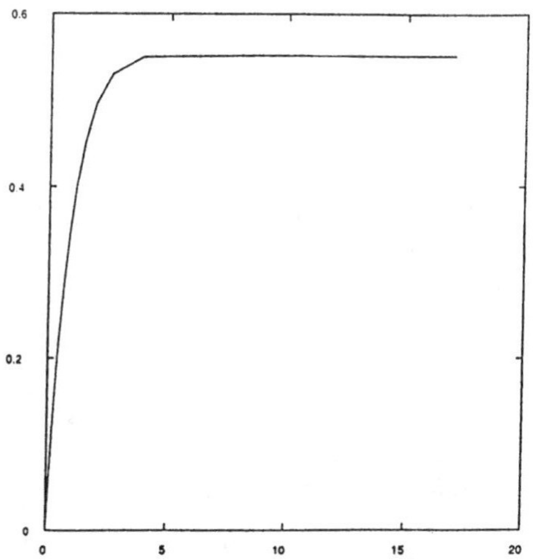


**Fig. 6a**: $f(t)$ *vs.* $t$ in the classical regime of system $(Q_4)$.

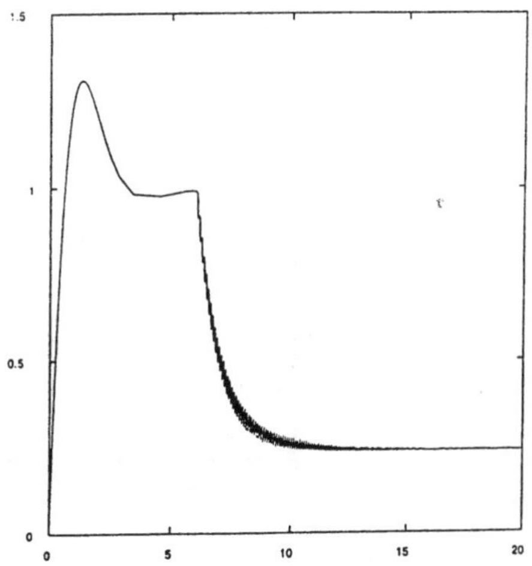


**Fig. 6b**: $f(t)$ *vs.* $t$ in the spurt I regime of system $(Q_4)$.

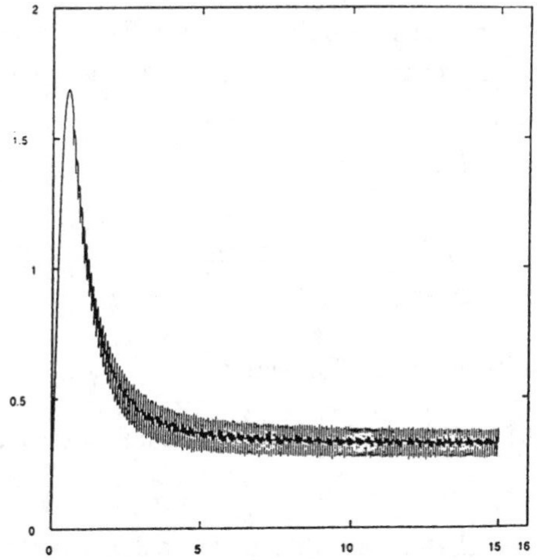


**Fig. 6c**: $f(t)$ *vs.* $t$ in the oscillatory regime of system $(Q_4)$.

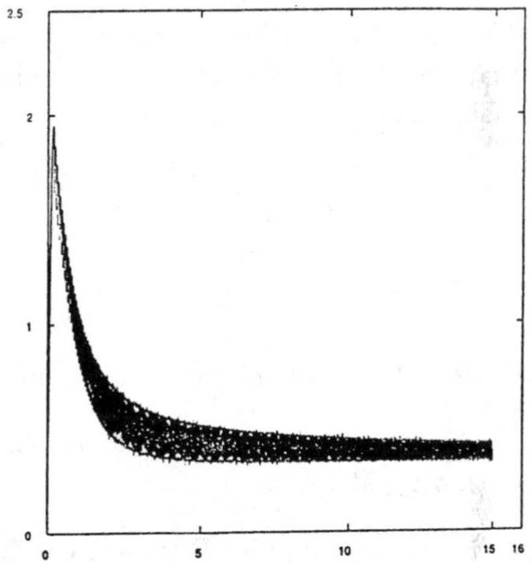


**Fig. 6d**: $f(t)$ *vs.* $t$ in the spurt II regime of system $(Q_4)$.

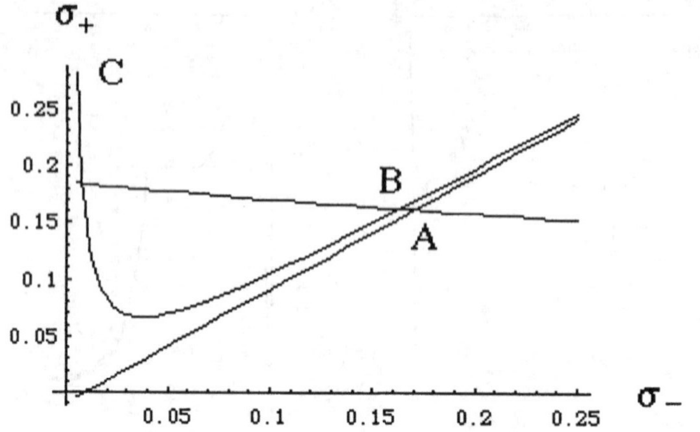


**Fig. 7**: Thee critical points of system $(Q_4)$ for the case $\epsilon = 0.001417$, $Q_0 = 1$, and $\delta = 0.001$.

produced using the program *Mathematica*). This plot indicates that there are three isolated critical points, labeled $A$, $B$, and $C$. The critical points are then found numerically.

To determine the local structure of orbits near the critical points, we studied the linearizations of systems $(Q_4)$ at $A$, $B$, and $C$. Thus we computed the eigenvalues of the $4 \times 4$ Jacobian matrix

$$J = \begin{pmatrix} -\frac{(1-\gamma)(Z_-+1)}{\varepsilon} - 1 & \frac{T_0+(1-\gamma)(\sigma_+-\sigma_-)}{\varepsilon} & \frac{(1-\gamma)(Z_-+1)}{\varepsilon} & 0 \\ \frac{2(1-\gamma)\sigma_- - (1-\gamma)\sigma_+ - T_0}{\varepsilon} & -1 & \frac{(1-\gamma)\sigma_-}{\varepsilon} & 0 \\ \frac{\gamma(Z_++1)}{\varepsilon} & 0 & -\frac{\gamma(Z_++1)}{\varepsilon} - 1 & \frac{T_0+\gamma(\sigma_--\sigma_+)}{\varepsilon} \\ -\frac{\gamma\sigma_+}{\varepsilon} & 0 & \frac{2\gamma\sigma_+ - \gamma\sigma_- - T_0}{\varepsilon} & -1 \end{pmatrix}$$

evaluated at the critical points. The numerical results show the following.

(1) Every critical point $A$ is a local attractor (all four eigenvalues of the Jacobian are real and negative).

(2) Every critical point $B$ is locally saddle-like: two of the eigenvalues of the Jacobian are real, one being positive the other negative; the remaining two eigenvalues have negative real parts. Thus there is a one-dimensional unstable manifold and a three-dimensional stable manifold at $B$.

(3) At the critical point $C$, the local character changes as $Q_0$ is varied; there is a transition at $Q_0{}^* \approx 0.3$ and another transition at $Q_0{}^{**} \approx 1.4$. For $Q_0 < Q_0{}^*$, $C$ is a local attractor (all four eigenvalues of the Jacobian have negative real parts). For $Q_0{}^* < Q_0 < Q_0{}^{**}$, one complex conjugate pair of eigenvalues has a positive real part, while the other pair has a negative real part; also, the eigenvalue pair with real part that changes sign at $Q_0{}^*$ crosses the imaginary axis with a non-zero imaginary part. For $Q_0 > Q_0{}^{**}$, $C$ is again an attractor, and the eigenvalue pair with real part that changes sign at $Q_0{}^{**}$

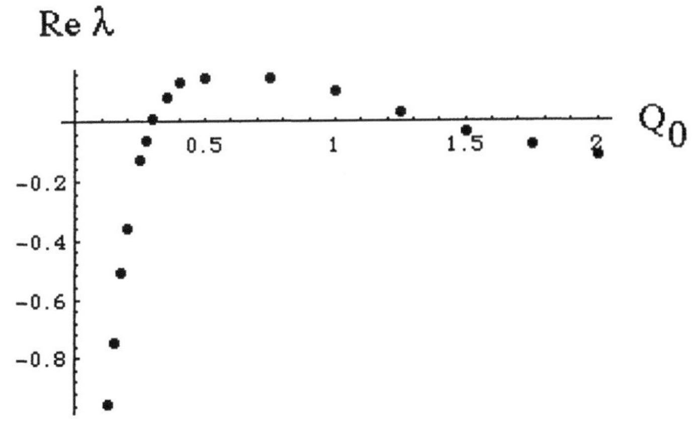


**Fig.** 8: The eigenvalue of the linearization of $(Q_4)$, evaluated at critical point $C$ of Fig. 7, as a function of $Q_0$.

> ${Q_0}^{**}$ crosses the imaginary axis with a non-zero imaginary part. Fig. 8 illustrates this feature, which suggests that system $(Q_4)$ undergoes Hopf bifurcations at ${Q_0}^{*}$ and ${Q_0}^{**}$ that cause the birth and death of a stable limit cycle.

Additionally, from the Lyapunov identity (5.4), we know that the point at infinity is a repeller. These properties suggest that system $(Q_4)$ has an invariant torus in the phase space $\mathbf{R}^4$. An objective of our current research is to determine the precise structure of this invariant set by showing it to be the stable limit cycle, as conjectured on the basis of numerical simulation.

Until the dynamics of system $(QFDE)$ and $(Q_4)$ can be compared rigorously, we note that one quantitative comparison is the frequency of oscillation. Our numerical results indicate that the frequency of solutions of $(QFDE)$, estimated directly from the period of oscillation of the velocity near the wall, agrees, to within 0.5% to 2.5%, with the frequency predicted from the imaginary part of the eigenvalues calculated from the linearization of $(Q_4)$ at $C$. This is illustrated in Fig. 9. The oscillation frequency predicted by $Q_4$ is seen to be proportional to $\varepsilon^{-1}$. That system $(QFDE)$ displays this dependence so clearly amounts to a physical prediction of our model. The dimensional analysis of Refs. [4,9] shows that if the non-dimensional frequency of oscillation at fixed $Q_0$ is proportional to $\varepsilon^{-1}$, then the dimensional frequency is independent of molecular weight of the sample (as long as the molecular weight is high enough to produce oscillations). In Lim's experiments [6], there are two samples that exhibit the oscillations; they have significantly different molecular weight, yet at the same $Q_0$ (i.e., nominal shear rate), they have roughly the same frequency of oscillation.

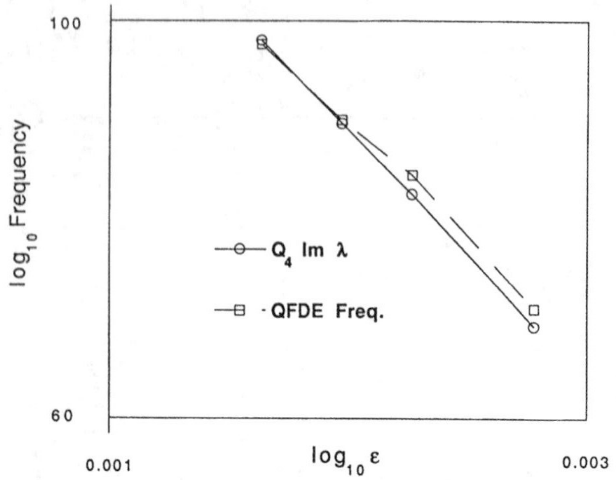


**Fig. 9**: The frequency of oscillation in the oscillatory regime for systems ($Q_4$) and ($QFDE$) with $Q_0 = 1$. The frequency of system ($Q_4$) is determined from the imaginary part of the appropriate eigenvalue; that of system ($QFDE$) is measured from plots of pressure *vs.* time, like Fig. 4c.

## 6. Conclusion

We have presented a mathematical model aimed at explaining the experiments of Lim and Schowalter. Our model of piston-driven channel flow of a highly elastic and very viscous non-Newtonian fluid uses a constitutive model that has a non-monotonic relationship between steady shear stress and strain rate. We have described the reduction of the three-dimensional equations of motion and stress to approximating systems that can be solved by a combination of numerical and analytic methods. One such approximation results in an infinite-dimensional system in which the volumetric flow rate is imposed by a feedback control. Numerical solution of this system exhibits four flow regimes, as do the experiments of Lim and Schowalter. The third of these regimes exhibits persistent oscillations that compare favorably to the Lim–Schowalter observations.

If our model does correctly model the observations of Lim and Schowalter, then the details of the flow are different from what these experimentalists assumed. Rather than a stick-slip flow, the flow we predict in the oscillatory regime is has a thin “apparent slip” layer that exists for all time. The layer is unstable, in the sense that there are large and persistent time variations in the apparent slip velocity; these are associated with a persistent cycle in the controlling pressure gradient. Our earlier analysis of the “spurt phenomenon” establishes that the control is the source of the instability: the flow would become steady if the control were removed by fixing the value of the pressure gradient at any time (though, of course, the desired flow rate might not be achieved).

To better understand the transitions between flow regimes and the nature of the oscillatory behavior in the third flow regime, we made a further approximation suggested by the numerical solution of the feedback system. The resulting four-dimensional quadratic system exhibits solutions that are qualitatively and quantitatively similar to those of the infinite-dimensional feedback system, and it provides precise predictions of the transitions of the feedback system through eigenvalue analysis of its linearization. Moreover, the four-dimensional system allows a precise determination of the frequencies of oscillations in the oscillatory flow regime, and this frequency agrees with observations of Lim. Our current work holds out the promise that deeper global analysis can reveal the nature of these transitions more completely.

## References

[1] R. Cook, D. Malkus, and M. Plesha, *Concepts and Applications of Finite Element Analysis*, John Wiley and Sons, New York, 1989.

[2] M. Denn, "Issues in viscoelastic fluid dynamics," *Ann. Rev. Fluid Mech.* **22** (1989), 13–34.

[3] M. Johnson and D. Segalman, "A Model for Viscoelastic Fluid Behavior which Allows Non-Affine Deformation," *J. Non-Newtonian Fluid Mech.* **2** (1977), 255–270.

[4] R. W. Kolkka, D. S. Malkus, M. G. Hansen, G. R. Ierley, and R. A. Worthing, "Spurt Phenomena of the Johnson-Segalman Fluid and Related Models," *J. Non-Newtonian Fluid Mech.* **29** (1989), 303–325.

[5] R. W. Kolkka, D. S. Malkus, J. A. Nohel, "Singularly Perturbed Non-Newtonian Flows," in preparation, 1992.

[6] F. J. Lim, "Wall slip in narrow molecular weight distribution polybutadienes," Ph.D. Thesis, Princeton Univ., 1988.

[7] F. J. Lim and W. R. Schowalter, "Wall Slip of Narrow Molecular Weight Distribution Polybutadienes," *J. Rheology* **33** (1989), 1359.

[8] D. S. Malkus, J. A. Nohel, and B. J. Plohr, "Dynamics of Shear Flow of a Non-Newtonian Fluid," *J. Comp. Phys.* **87** (1990), 464–487.

[9] D. S. Malkus, J. A. Nohel, and B. J. Plohr, "Analysis of New Phenomena in Shear Flow of Non-Newtonian Fluids," *SIAM J. Appl. Math.* **51** (1991), 899–929.

[10] D. S. Malkus, Y.-C. Tsai, and R. W. Kolkka, "New transient algorithms for non-Newtonian flows," *Finite Elements in Fluids* **8** (1992), 401–424.

[11] F. Morrison and R. Larson, "A study of stress relaxation anomalies in binary mixtures of monodisperse polystyrenes," *J. Polymer Sci.* **30** (1992), 943–950.

[12] J. Nohel, R. Pego, and A. Tzavaras, "Stability of Discontinuous Steady States in Shearing Motions of a Non-Newtonian Fluid," *Proc. Royal Soc. Edinburgh, Series A* **115** (1990), 34.

[13] J. Nohel and R. Pego, "Nonlinear Stability and Asymptotic Behavior of Shearing Motions of a Non-Newtonian Fluid," preprint, (1992); *SIAM J. Math. Anal.*, accepted.

[14] J. Oldroyd, "Non-Newtonian Effects in Steady Motion of Some Idealized Elastico-Viscous Liquids," *Proc. Royal Soc. London, Series A* **245** (1958), 278–297.

[15] B. Plohr, "Instabilities in Shear Flow of Viscoelastic Fluids with Fading Memory," in *PDEs and Continuum Models of Phase Transitions*, eds. M. Rascle, D. Serre, and M. Slemrod, Lect. Notes in Phys. **344**, 113–127, Springer-Verlag, New York, 1989.

[16] D. L. Russell, private communication, 1990.

[17] G. V. Vinogradov, A. Ya. Malkin, Yu. G. Yanovskii, E. K. Borisenkova, B. V. Yarlykov, and G. V. Berezhnaya, "Viscoelastic Properties and Flow of Narrow Distribution Polybutadienes and Polyisoprenes," *J. Polymer Sci., Part A-2* **10** (1972), 1061–1084.

# 13 A Necessary and Sufficient Condition for Exponential Stability of Large-Scale Stochastic Delay Systems in Hierarchical Form

**Xuerong Mao** University of Strathclyde, Glasgow, Scotland

ABSTRACT: In this paper we investigate the exponential stability of a large-scale stochastic delay system in hierarchical form, i.e., the system consists of several subsystems and each subsystem interacts only with "lower" subsystems but not with "higher" subsystems. It is shown under a condition, which is weaker than Lipschtiz continuity, that the large-scale system is exponential stable in $L^2$ if and only if each of the corresponding isolated subsystems is exponential stable in $L^2$. The stability of stochastic systems with constants delay is discussed in more details.

KEY WORDS: Exponential stability, large-scale stochastic delay systems, isolated subsystems, hierarchical form.

## 1. INTRODUCTION

Numerous problems in control theory and dynamical systems as well as applications to engineering, economy, biology lead to the study of exponential stability of a large-scale system described, for instance, by an ordinary differential equation

$$\dot{x} = f(x, t) \tag{1.1}$$

($x \in R^d$ and d is large). In theory, there exist several methods in dealing with the exponential stability. However, enormous calculations involved in the large-scale case make the methods difficult to use in practice. For example, it is known that the exponential stability of a linear differential equation $\dot{x} = Ax$ is equivalent to that all the eigenvalues of the $d \times d$ matrix A have negative real parts. Unfortunately it takes ages to calculate all the eigenvalues when d is large even with the help of modern computers.

One of the effective methods developed for large-scale systems is the decomposition technique which has been studied by many authors, for example, Michel, Miller and Tang [9], Siljak [10], Sontag [11], Vidyasagar [12]. The idea is as follows: Decompose the system into several subsystems (with a necessary rearrangement of the order of the variables)

$$\dot{x}_i = f_i(x_1, ..., x_k, t), \quad i = 1, ..., k, \tag{1.2}$$

where $x_i \in R^{d_i}$ and $\sum d_i = d$, and introduce the isolated subsystems

$$\dot{y}_i = f_i(0, y_i, 0, t). \tag{1.3}$$

Under certain hypotheses added on the interconnected terms

$$f_i(x_1, ..., x_k, t) - f_i(0, y_i, 0, t)$$

it is shown that the large-scale system (1.1) is exponentially stable if each of the isolated subsystems (1.3) is exponentially stable.

So far there is little known on exponential stability of large-scale stochastic delay systems, although there exists an extensive literature on the stability of stochastic delay equations (cf. [1-7]). The aim of this paper is to extend the above decomposition technique to a large-scale stochastic delay system

$$dx(t) = f(\hat{x}(t), t)\, dt + g(\hat{x}(t), t)\, dw(t), \tag{1.4}$$

where $\hat{x}(t) = \{x(t+s): -\tau \le s \le 0\}$ and $\tau$ is a positive constant. In the way shown by Vidyasagar [12], one can rewrite the system as a hierarchical form (if necessary rearrange the order of the variables):

$$dx_i(t) = f_i(\hat{x}_1, ..., \hat{x}_i, t)\, dt + g_i(\hat{x}_1, ..., \hat{x}_i, t)\, dw(t), \quad i = 1, ..., k, \tag{1.5}$$

and then introduce the isolated subsystems

$$dy_i(t) = f_i(0, \hat{y}_i, t)\, dt + g_i(0, \hat{y}_i, t)\, dw(t). \tag{1.6}$$

Assume the interconnected terms satisfy a condition: There exists a $K > 0$ such that

$$|f_i(u_1, ..., u_i, t) - f_i(0, ..., 0, v_i, t)|^2 \le K\Big(\sum_{j<i} \|u_j\|_\tau^2 + \|u_i - v_i\|_\tau^2\Big)$$

and

$$\operatorname{trace}\Big( (g_i(u_1, ..., u_i, t) - g_i(0, ..., 0, v_i, t))(g_i(u_1, ..., u_i, t) - g_i(0, ..., 0, v_i, t))^T \Big)$$

$$\le K\Big(\sum_{j<i} \|u_j\|_\tau^2 + \|u_i - v_i\|_\tau^2\Big)$$

for all $u_j, v_j \in C([-\tau, 0], R^{d_j})$ and $i = 2, ..., k$. Obviously this is a weaker condition than Lipschtiz continuity. The main aim of this paper is to show that the large-scale system is exponentially stable in $L^2$ if and only if each of the corresponding isolated subsystems is exponentially stable in $L^2$.

Let us first give the definition of the exponential stability in $L^2$ precisely and make a few comments in Section 2. We next state and prove our main result in Section 3, and then discuss the exponential stability for nonlinear and linear large-scale stochastic systems with constants delay in Sections 4 and 5.

## 2. DEFINITIONS AND PROPOSITIONS

Let $(\Omega, \mathcal{F}, \{\mathcal{F}_t\}_{t\geq 0}, P)$ be a complete probability space with the filtration $\{\mathcal{F}_t\}$ which is right continuous and contains all P-null sets. Let $w(t) = (w_1, \ldots w_m)^T$, $t \geq 0$ be an m-dimensional Brownian motion. Let $\tau > 0$ and denote $C([-\tau, 0], R^d)$ the space of all continuous functions defined on $[-\tau, 0]$ with values in $R^d$. We introduce a norm in this space:

$$\|u\|_\tau = \max\{|u(s)| : -\tau\leq s\leq 0\} \quad \text{if } u \in C([-\tau, 0], R^d).$$

Throughout this paper $|.|$ denotes the Euclidean norm of a vector, and $\|v\|^2 = \text{trace}(vv^T)$ for a matrix v. Let $L^2(\Omega, \mathcal{F}_{t_0}, C([-\tau, 0], R^d))$ denote all $\mathcal{F}_{t_0}$-measurable $C([-\tau, 0], R^d)$-valued random variables $\xi$ with $E\|\xi\|_\tau^2 < \infty$. Write $L^2(\Omega, C([-\tau, 0], R^d))$ for $L^2(\Omega, \mathcal{F}, C([-\tau, 0], R^d))$.

Let $f : C([-\tau, 0], R^d)\times R_+ \to R^d$ and $g : C([-\tau, 0], R^d)\times R_+ \to R^{d\times m}$ be two functionals. Consider a stochastic delay equation

$$\begin{aligned} dx(t) &= f(\hat{x}(t), t)\, dt + g(\hat{x}(t), t)\, dw(t), \quad t \geq t_0(\geq 0), \\ \hat{x}(t_0) &= \xi. \end{aligned} \tag{2.1}$$

Here $\hat{x}(t) = \{x(t+s): -\tau\leq s\leq 0\}$ and $\xi \in L^2(\Omega, \mathcal{F}_{t_0}, C([-\tau, 0], R^d))$. We assume that the equation has a unique solution which is denoted by $x(t, t_0, \xi)$. We also assume that $f(0, t) \equiv 0$ and $g(0, t) \equiv 0$ so the equation has a trivial solution $x(t, t_0, 0) \equiv 0$.

**Definition 2.1** The trivial solution of the equation (or simply, the equation) is said to be exponentially stable in $L^2(\Omega, C([-\tau, 0], R^d))$ if there exist two positive constants $\lambda$ and M such that

$$E\|\hat{x}(t, t_0, \xi)\|_\tau^2 \leq ME\|\xi\|_\tau^2\, e^{-\lambda(t-t_0)} \quad \text{on } t \geq t_0 \tag{2.2}$$

for all $t_0 \geq 0$ and $\xi \in L^2(\Omega, \mathcal{F}_{t_0}, C([-\tau, 0], R^d))$.

Let us make a few remarks on the definition. First of all, we would like to point out that the exponential stability in $L^2(\Omega, C([-\tau, 0], R^d))$ implies the almost surely exponential stability. As a matter of fact we have the following proposition.

**Proposition 2.2** If equation (2.1) is exponentially stable in $L^2(\Omega, C([-\tau, 0], R^d))$, it is almost surely exponentially stable. In other words, (2.2) implies

$$\limsup_{t\to\infty} \frac{1}{t} \log|x(t, t_0, \xi)| \le -\lambda/2 \quad \text{a.s.} \tag{2.3}$$

*Proof.* Assume (2.2) hold. Let $\varepsilon \in (0, \lambda)$ arbitrarily and let $k = 1, 2, \dots$ . Then (2.2) gives

$$P\Big(\omega: \|\hat{x}(k\tau+t_0, t_0, \xi)\|_\tau > e^{-(\lambda-\varepsilon)k\tau/2}\Big) \le e^{(\lambda-\varepsilon)k\tau} E\|\hat{x}(k\tau+t_0, t_0, \xi)\|_\tau^2 \le ME\|\xi\|_\tau^2 e^{-\varepsilon k\tau}.$$

From the Borel-Cantelli lemma, we can find that for almost all $\omega \in \Omega$,

$$\|\hat{x}(k\tau+t_0, t_0, \xi)\|_\tau \le e^{-(\lambda-\varepsilon)k\tau/2} \tag{2.4}$$

holds for all but finitely many k. Hence there exists $k_o(\omega)$, for all $\omega \in \Omega$ excluding a P-null set, for which (2.4) holds when $k \ge k_o(\omega)$. Consequently,

$$\frac{1}{t} \log|x(t, t_0, \xi)| \le \frac{1}{(k-1)\tau+t_0} \Big(-(\lambda-\varepsilon)k\tau/2\Big)$$

if $(k-1)\tau+t_0 \le t \le k\tau+t_0$, $k \ge k_o$ almost surely. Therefore

$$\limsup_{t\to\infty} \frac{1}{t} \log|x(t, t_0, \xi)| \le -(\lambda-\varepsilon)/2 \quad \text{a.s.}$$

and (2.3) follows since $\varepsilon$ is arbitrary. The proof is complete.

Our second remark is to compare the exponential stability in $L^2(\Omega, C([-\tau, 0], R^d))$ with the exponential stability in mean square (i.e., in $L^2(\Omega, R^d)$). In order to be precise, let us recall the definition of the exponential stability in mean square (cf. [1-7]).

**Definition 2.3** The trivial solution of the equation (or simply, the equation) is said to be exponentially stable in mean square if there exist two positive constants $\lambda$ and M such that

$$E|x(t, t_0, \xi)|^2 \le ME\|\xi\|_\tau^2 e^{-\lambda(t-t_0)} \quad \text{on } t \ge t_0 \tag{2.5}$$

for all $t_0 \geq 0$ and $\xi \in L^2(\Omega, \mathcal{F}_{t_0}, C([-\tau, 0], R^d))$.

Obviously the exponential stability in $L^2(\Omega, C([-\tau, 0], R^d))$ implies the exponential stability in mean square. Although we do not know whether the exponential stability in mean square implies the exponential stability in $L^2(\Omega, C([-\tau, 0], R^d))$ in general, we shall show it does in many important and interesting cases. For instance, if the coefficients of equation (2.1) are Lipschitz (in fact we only need a weaker condition as shown below) and $\tau$ is small enough, then both of the stabilities are equivalent. This is described as follows:

**Proposition 2.4** Assume there exist a $K > 0$ such that

$$|f(u, t)|^2 \vee \|g(u, t))\|^2 \leq K\|u\|_\tau^2 \tag{2.6}$$

for all $u \in C([-\tau, 0], R^d)$ and $t \geq 0$. If $\tau$ is so small that

$$3K(4+\tau)\tau < 1, \tag{2.7}$$

then both of the exponential stability in $L^2(\Omega, C([-\tau, 0], R^d))$ and the exponential stability in mean square are equivalent.

*Proof.* We need only prove that the exponential stability in mean square implies the exponential stability in $L^2(\Omega, C([-\tau, 0], R^d))$ so we assume (2.5) holds. From (2.7) one can find an $\varepsilon \in (0, \lambda)$ such that

$$3K(4+\tau)\tau\, e^{\varepsilon\tau} < 1. \tag{2.8}$$

We shall show that there exists a $C > 0$ such that

$$E\|\hat{x}(t, t_0, \xi)\|_\tau^2 \leq CE\|\xi\|_\tau^2\, e^{-\varepsilon(t-t_0)} \quad \text{on } t \geq t_0. \tag{2.9}$$

We fix $t_0$ and $\xi$ arbitrarily and write $x(t, t_0, \xi) = x(t)$ simply. Using Doob's martingale inequality and hypotheses (2.5) and (2.6) we can easily derive that

$$E\|\hat{x}(t)\|_\tau^2 \leq 3E|x(t-\tau)|^2 + 3K(4+\tau)\int_{t-\tau}^{t} E\|\hat{x}(r)\|_\tau^2\, dr$$

$$\leq 3ME\|\xi\|_\tau^2\, e^{-\lambda(t-\tau-t_0)} + 3K(4+\tau)\int_{t-\tau}^{t} E\|\hat{x}(r)\|_\tau^2\, dr \quad \text{if } t \geq t_0+\tau. \tag{2.10}$$

On the other hand, if $t_0 \le t \le t_0+\tau$,

$$E\|\hat{x}(t)\|_\tau^2 \le E\|\xi\|_\tau^2 + E\Big(\sup_{t_0\le r\le t} |x(r)|^2\Big) \le 4ME\|\xi\|_\tau^2 + 3K(4+\tau)\int_{t_0}^{t} E\|\hat{x}(r)\|_\tau^2\,dr. \tag{2.11}$$

Combining (2.10) and (2.11) we see that for all $t \ge t_0$,

$$E\|\hat{x}(t)\|_\tau^2 \le 4ME\|\xi\|_\tau^2 e^{-\lambda(t-\tau-t_0)} + 3K(4+\tau)\int_{(t-\tau)\vee t_0}^{t} E\|\hat{x}(r)\|_\tau^2\,dr. \tag{2.12}$$

Therefore, for any $T \ge t_0$,

$$\int_{t_0}^{T} e^{\varepsilon t} E\|\hat{x}(t)\|_\tau^2\,dt \le 4ME\|\xi\|_\tau^2 e^{\lambda(\tau+t_0)}\int_{t_0}^{T} e^{-(\lambda-\varepsilon)t}\,dt + 3K(4+\tau)\int_{t_0}^{T} e^{\varepsilon t}\int_{(t-\tau)\vee t_0}^{t} E\|\hat{x}(r)\|_\tau^2\,dr\,dt$$

$$\le \frac{4M}{\lambda-\varepsilon} E\|\xi\|_\tau^2 e^{\lambda\tau+\varepsilon t_0} + 3K(4+\tau)\int_{t_0}^{T}\Big(\int_{r}^{(r+\tau)\wedge T} e^{\varepsilon t}dt\Big) E\|\hat{x}(r)\|_\tau^2\,dr$$

$$\le \frac{4M}{\lambda-\varepsilon} E\|\xi\|_\tau^2 e^{\lambda\tau+\varepsilon t_0} + 3K(4+\tau)\tau e^{\varepsilon\tau}\int_{t_0}^{T} e^{\varepsilon r} E\|\hat{x}(r)\|_\tau^2\,dr.$$

This implies

$$\int_{t_0}^{T} e^{\varepsilon t} E\|\hat{x}(t)\|_\tau^2\,dt \le C_1 E\|\xi\|_\tau^2 e^{\varepsilon t_0} \quad \text{for all } T \ge t_0, \tag{2.13}$$

where $C_1 = 4Me^{\lambda\tau}\big[(\lambda-\varepsilon)(1-3K(4+\tau)\tau e^{\varepsilon\tau})\big]^{-1}$. Finally (2.12) and (2.13) yield that for $t \ge t_0$,

$$e^{\varepsilon t} E\|\hat{x}(t)\|_\tau^2 \le 4ME\|\xi\|_\tau^2 e^{\lambda\tau+\varepsilon t_0} + 3K(4+\tau)e^{\varepsilon t}\int_{(t-\tau)\vee t_0}^{t} E\|\hat{x}(r)\|_\tau^2\,dr$$

$$\le 4ME\|\xi\|_\tau^2 e^{\lambda\tau+\varepsilon t_0} + 3K(4+\tau)e^{\varepsilon\tau}\int_{(t-\tau)\vee t_0}^{t} e^{\varepsilon r} E\|\hat{x}(r)\|_\tau^2\,dr$$

$$\le \Big(4Me^{\lambda\tau} + 3K(4+\tau)e^{\varepsilon\tau}C_1\Big) E\|\xi\|_\tau^2 e^{\varepsilon t_0}. \tag{2.14}$$

This is required (2.9) by putting $C = 4Me^{\lambda\tau} + 3K(4+\tau)e^{\varepsilon\tau}C_1$. The proof is complete.

Another important case for both of the stabilities to be equivalent is a stochastic differential equation with variable time lags of the form:

$$dx(t) = F(x(\tau_1(t)), ..., x(\tau_n(t)), t)\,dt + G(x(\tau_1(t)), ..., x(\tau_n(t)), t)\,dw(t), \quad t \geq t_0,$$
$$\hat{x}(t_0) = \xi. \tag{2.15}$$

Here $\tau_i$, $1 \leq i \leq n$ are all real functions defined on $R_+$ such that $t-\tau \leq \tau_i(t) \leq t$ for all $t \geq 0$, and $F : R^{d\times n}\times R_+ \to R^d$ and $G : R^{d\times n}\times R_+ \to R^{d\times m}$. For this equation we have the following proposition which gives a weaker condition than that of Proposition 2.4.

**Proposition 2.5** Assume there exists a $K > 0$ such that

$$|F(v, t)|^2 \vee \|G(v, t))\|^2 \leq K\|v\|^2 \tag{2.16}$$

for all $v \in R^{d\times n}$ and $t \geq 0$. Then, for equation (2.15), both of the exponential stability in $L^2(\Omega, C([-\tau, 0], R^d))$ and the exponential stability in mean square are equivalent.

*Proof.* Again we need only to prove that the exponential stability in mean square implies the exponential stability in $L^2(\Omega, C([-\tau, 0], R^d))$ so we assume (2.5) holds. Fix any $t_0$ and $\xi$, and write $x(t, t_0, \xi) = x(t)$ simply. Then we can derive that, for any $t \geq t_0+2\tau$,

$$E\|\hat{x}(t)\|_\tau^2 \leq 3E|x(t-\tau)|^2 + 15K \int_{t-\tau}^{t} \Big(\sum_{i=1}^{n} |x(\tau_i(s)|^2\Big)\,ds$$

$$\leq 3ME\|\xi\|_\tau^2\, e^{-\lambda(t-\tau-t_0)} + 15K \int_{t-\tau}^{t} \Big(\sum_{i=1}^{n} ME\|\xi\|_\tau^2\, e^{-\lambda(\tau_i(s)-t_0)}\Big)\,ds$$

$$\leq (\,3Me^{\lambda\tau} + 15KMn\tau e^{2\tau}\,)\,E\|\xi\|_\tau^2\, e^{-\lambda(t-t_0)}.$$

On the other hand, it is easy to show that there exists a $C_1 > 0$ such that

$$E\|\hat{x}(t)\|_\tau^2 \leq C_1E\|\xi\|_\tau^2 \quad \text{whenever } t_0 \leq t \leq t_0+2\tau.$$

Setting $C = C_1e^{2\tau}\vee(3Me^{\lambda\tau}+15KMn\tau e^{2\tau})$ we therefore have

$$E\|\hat{x}(t)\|_\tau^2 \leq CE\|\xi\|_\tau^2\, e^{-\lambda(t-t_0)} \quad \text{for all } t \geq t_0$$

as required. The proof is complete.

Let us now start to discuss the main topic of this paper - the exponential stability in $L^2(\Omega, C([-\tau, 0], R^d))$ for a large-scale stochastic delay system in hierarchical form.

## 3 LARGE-SCALE STOCHASTIC DELAY SYSTEMS IN HIERARCHICAL FORM

In this section we shall consider a large-scale stochastic delay system in hierarchical form, i.e., the system which consists of several subsystems and each subsystem interacts only with "lower" subsystems but not with "higher" subsystems. In other words we shall study a large-scale stochastic delay system of the form

$$dx_i(t) = f_i(\hat{x}_1(t), \ldots, \hat{x}_i(t), t)\, dt + g_i(\hat{x}_1(t), \ldots, \hat{x}_i(t), t)\, dw(t) \quad \text{on } t \geq t_0,$$
$$\hat{x}_i(t_0) = \xi_i, \quad i = 1, \ldots, k. \tag{3.1}$$

Here $x_i \in R^{d_i}$, $k \geq 2$, $\sum d_i = d$, $f_i : C([-\tau, 0], R^{d_1+\ldots d_i}) \times R_+ \to R^{d_i}$, $g_i : C([-\tau, 0], R^{d_1+\ldots d_i}) \times R_+ \to R^{d_i \times m}$, and $\xi_i \in L^2(\Omega, \mathcal{F}_{t_0}, C([-\tau, 0], R^{d_i}))$. Set $x = (x_1, \ldots, x_k)$ and $\xi = (\xi_1, \ldots, \xi_k)$. Assume the equation has a unique global solution and we denote the solution by $x(t, t_0, \xi)$. Assume also that $f_i(0, t) \equiv 0$ and $g_i(0, t) \equiv 0$ for all i. Therefore the equation admits a trivial solution $x(t, t_0, 0) \equiv 0$. Introduce the isolated subsystems

$$dy_i(t) = f_i(0, \hat{y}_i(t), t)\, dt + g_i(0, \hat{y}_i(t), t)\, dw(t) \quad \text{on } t \geq t_0,$$
$$\hat{y}_i(t_0) = \xi_i. \tag{3.2}$$

The solution of each subsystem is denoted by $y_i(t, t_0, \xi_i)$. We make a standing hypothesis:

(H) There exists a $K > 0$ such that

$$|f_i(u_1, \ldots, u_i, t) - f_i(0, \ldots, 0, v_i, t)|^2 \leq K\Big(\sum_{j<i} \|u_j\|_\tau^2 + \|u_i - v_i\|_\tau^2\Big)$$

and

$$\|(g_i(u_1, \ldots, u_i, t) - g_i(0, \ldots, 0, v_i, t))\|^2 \leq K\Big(\sum_{j<i} \|u_j\|_\tau^2 + \|u_i - v_i\|_\tau^2\Big)$$

for all $u_j, v_j \in C([-\tau, 0], R^{d_j})$, $1 \leq j \leq k$ and $2 \leq i \leq k$.

**Theorem 3.1** Under hypothesis (H), the large-scale delay system (3.1) is exponentially stable in $L^2(\Omega, C([-\tau, 0], R^d))$ if and only if each of the isolated delay systems (3.2) is exponentially stable in $L^2(\Omega, C([-\tau, 0], R^{d_i}))$.

*Proof.* "only if": Assume (3.1) is exponentially stable in $L^2(\Omega, C([-\tau, 0], R^d))$. By definition, there exist two positive constants M and $\lambda$ such that

$$E\|x(t, t_0, \xi)\|_\tau^2 \le ME\|\xi\|_\tau^2 e^{-\lambda(t-t_0)} \quad \text{on } t \ge t_0$$

for all $t_0 \ge 0$ and $\xi \in L^2(\Omega, \mathcal{F}_{t_0}, C([-\tau, 0], R^d))$. Let $t_0 \ge 0$ and $\xi_i \in L^2(\Omega, \mathcal{F}_{t_0}, C([-\tau, 0], R^{d_i})$ be arbitrary. Set $\xi = (0, \xi_i, 0)$ and it is easy to see that

$$y_i(t, t_0, \xi_i) = x_i(t, t_0, \xi).$$

Therefore

$$E\|\hat{y}_i(t, t_0, \xi_i)\|_\tau^2 \le E\|\hat{x}(t, t_0, \xi)\|_\tau^2 \le ME\|\xi\|_\tau^2 e^{-\lambda(t-t_0)} = ME\|\xi_i\|_\tau^2 e^{-\lambda(t-t_0)}.$$

In other words, each of the isolated subsystems (3.2) is exponentially stable in $L^2(\Omega, C([-\tau, 0], R^{d_i}))$.

"if": We give the proof for the case $k = 2$; the general case follows by induction. The hypotheses in this case are that both of the equations

$$dy_1(t) = f_1(\hat{y}_1, t)\,dt + g_1(\hat{y}_1, t)\,dw(t)$$

and

$$dy_2(t) = f_2(0, \hat{y}_2, t)\,dt + g_2(0, \hat{y}_2, t)\,dw(t)$$

are exponentially stable in $L^2(\Omega, C([-\tau, 0], R^{d_1}))$ and $L^2(\Omega, C([-\tau, 0], R^{d_2}))$ respectively. Hence one can find a pair of positive constants $M$ and $\lambda$ such that

$$E\|\hat{y}_i(t, t_0, \xi_i)\|_\tau^2 \le ME\|\xi_i\|_\tau^2 e^{-\lambda(t-t_0)} \quad \text{on } t \ge t_0,\ i = 1, 2, \tag{3.3}$$

for all $t_0 \ge 0$ and $\xi_i \in L^2(\Omega, \mathcal{F}_{t_0}, C([-\tau, 0], R^{d_i}))$. We need to show that the equation

$$\begin{aligned} dx_1(t) &= f_1(\hat{x}_1, t)\,dt + g_1(\hat{x}_1, t)\,dw(t), \\ dx_2(t) &= f_2(\hat{x}_1, \hat{x}_2, t)\,dt + g_2(\hat{x}_1, \hat{x}_2, t)\,dw(t), \\ (\hat{x}_1(t_0), \hat{x}_2(t_0)) &= \xi = (\xi_1, \xi_2) \end{aligned}$$

is exponentially stable in $L^2(\Omega, C([-\tau, 0], R^d))$. Note $x_1(t, t_0, \xi) = y_1(t, t_0, \xi_1)$. We therefore need only to show that there exist two positive constants $M_1$ and $\lambda_1$ such that

$$E\|\hat{x}_2(t, t_0, \xi)\|_\tau^2 \le M_1 E\|\xi\|_\tau^2 e^{-\lambda_1(t-t_0)} \quad \text{on } t \ge t_0.$$

In fact we shall show that for any $\varepsilon \in (0, \lambda/2)$ there exists $C > 0$ such that

$$E\|\hat{x}_2(t, t_0, \xi)\|_\tau^2 \le C\, E\|\xi\|_\tau^2\, e^{-(\lambda-2\varepsilon)(t-t_0)} \quad \text{on } t \ge t_0 \tag{3.4}$$

for all $t_0 \ge 0$ and $\xi \in L^2(\Omega, \mathcal{F}_{t_0}, C([-\tau, 0], R^d))$. Fix $t_0$ and $\xi$ arbitrarily and write $x(t, t_0, \xi) = x(t) = (x_1(t), x_2(t))$ simply. Let $\varepsilon \in (0, \lambda/2)$ and define

$$\delta = \tau \vee \frac{1}{\varepsilon}\log(2M) \quad \text{and} \quad \rho = \delta \vee \frac{1}{\varepsilon}\Big(4K\delta(4+\delta) - \log\big[\lambda/4KM(4+\delta)\big]\Big). \tag{3.5}$$

We first claim that for any $\sigma \ge \rho$,

$$E\|\hat{x}_2(t)\|_\tau^2 \le E\|\hat{x}_2(t_0+\sigma)\|_\tau^2\, e^{-(\lambda-\varepsilon)\delta} + E\|\xi_1\|_\tau^2\, e^{-(\lambda-\varepsilon)\sigma} \quad \text{if } t_0+\sigma+\delta \le t \le t_0+\sigma+2\delta. \tag{3.6}$$

In fact, if $t_0+\sigma \le t \le t_0+\sigma+2\delta$, we define $y_2(t) = y_2(t, t_0+\sigma, \hat{x}_2(t_0+\sigma))$. Then by Doob's martingale inequality, the standing hypothesis (H) and inequality (3.3) we can derive that

$$E\Big(\sup_{t_0+\sigma\le s\le t} |x_2(s)-y_2(s)|^2\Big)$$

$$\le 2K(4+\delta)\int_{t_0+\sigma}^{t} E\|\hat{x}_1(s)\|_\tau^2\, ds + 2K(4+\delta)\int_{t_0+\sigma}^{t} E\|\hat{x}_2(s)-\hat{y}_2(s)\|_\tau^2\, ds$$

$$\le 2KM(4+\delta)E\|\xi_1\|_\tau^2 \int_{t_0+\sigma}^{t} e^{-\lambda(s-t_0)}\, ds + 2K(4+\delta)\int_{t_0+\sigma}^{t} E\Big(\sup_{t_0+\sigma\le s\le r}|x_2(s)-\bar{x}_2(s)|^2\Big)\, dr$$

$$\le \big[2KM(4+\delta)/\lambda\big]\, E\|\xi_1\|_\tau^2\, e^{-\lambda\sigma} + 2K(4+\delta)\int_{t_0+\sigma}^{t} E\Big(\sup_{t_0+\sigma\le s\le r}|x_2(s)-\bar{x}_2(s)|^2\Big)\, dr.$$

The well-known Gronwall inequality gives

$$E\Big(\sup_{t_0+\sigma\le s\le t}|x_2(s)-y_2(s)|^2\Big) \le \big[2KM(4+\delta)/\lambda\big]\, E\|\xi_1\|_\tau^2\, e^{-\lambda\sigma+2K(4+\delta)(t-t_0-\sigma)}.$$

In particular, we have

$$E\Big(\sup_{t_0+\sigma\le s\le t_0+\sigma+2\delta}|x_2(s)-y_2(s)|^2\Big) \le \big[2KM(4+\delta)/\lambda\big]\, E\|\xi_1\|_\tau^2\, e^{-\lambda\sigma+4K\delta(4+\delta)}$$

$$\leq \frac{1}{2}\, E\|\xi_1\|_\tau^2\, e^{-(\lambda-\varepsilon)\sigma},$$

where (3.5) has been used. Hence, if $t_0+\sigma+\delta \leq t \leq t_0+\sigma+2\delta$,

$$\begin{aligned} E\|\hat{x}_2(t)\|_\tau^2 &\leq 2E\|\hat{y}_2(t)\|_\tau^2 + 2E\|\hat{x}_2(t)-\hat{y}_2(t)\|_\tau^2 \\ &\leq 2ME\|\hat{x}_2(t_0+\sigma)\|_\tau^2\, e^{-\lambda(t-t_0-\sigma)} + E\|\xi_1\|_\tau^2\, e^{-(\lambda-\varepsilon)\sigma} \\ &\leq 2ME\|\hat{x}_2(t_0+\sigma)\|_\tau^2\, e^{-\lambda\delta} + E\|\xi_1\|_\tau^2\, e^{-(\lambda-\varepsilon)\sigma} \\ &\leq E\|\hat{x}_2(t_0+\sigma)\|_\tau^2\, e^{-(\lambda-\varepsilon)\delta} + E\|\xi_1\|_\tau^2\, e^{-(\lambda-\varepsilon)\sigma} \end{aligned}$$

which is the required (3.6), where (3.5) has been used again. We now let n = 1, 2, .... Take $\sigma = \rho+(n-1)\delta$ in (3.6) we obtain that, if $t_0+\rho+n\delta \leq t \leq t_0+\rho+(n+1)\delta$,

$$E\|\hat{x}_2(t)\|_\tau^2 \leq E\|\hat{x}_2(t_0+\rho+(n-1)\delta)\|_\tau^2\, e^{-(\lambda-\varepsilon)\delta} + E\|\xi_1\|_\tau^2\, e^{-(\lambda-\varepsilon)(\rho+(n-1)\delta)},$$

and then by induction

$$\begin{aligned} E\|\hat{x}_2(t)\|_\tau^2 &\leq E\|\hat{x}_2(t_0+\rho)\|_\tau^2\, e^{-(\lambda-\varepsilon)n\delta} + nE\|\xi_1\|_\tau^2\, e^{-(\lambda-\varepsilon)(\rho+(n-1)\delta)} \\ &\leq \left( E\|x_2(t_0+\rho)\|_\tau^2 + \frac{1}{\varepsilon\delta e}\, E\|\xi_1\|_\tau^2 \right) e^{-(\lambda-2\varepsilon)n\delta}, \end{aligned} \tag{3.7}$$

where we have used $\rho \geq \delta$ and an elementary inequality

$$n\, e^{-\varepsilon n\delta} \leq \frac{1}{\varepsilon\delta e} \quad \text{for all } n \geq 1.$$

On the other hand, it is easy to show that there exists a positive constant $C_1$ which depends only on K, $\rho$ and $\delta$ such that

$$E\|\hat{x}_2(t)\|_\tau^2 \leq C_1 E\|\xi\|_\tau^2 \quad \text{on } t_0 \leq t \leq t_0+\rho+\delta. \tag{3.8}$$

Therefore, if $t > t_0+\rho+\delta$, we choose n such that $t_0+\rho+n\delta \leq t \leq t_0+\rho+(n+1)\delta$, and then (3.7) and (3.8) gives

$$E\|\hat{x}_2(t)\|_\tau^2 \le \left(C_1+\frac{1}{\varepsilon\delta e}\right) E\|\xi\|_\tau^2\, e^{-(\lambda-2\varepsilon)n\delta}$$

$$\le \left(C_1+\frac{1}{\varepsilon\delta e}\right) E\|\xi\|_\tau^2\, e^{-(\lambda-2\varepsilon)(t-t_o-\rho-\delta)}$$

$$\le CE\|\xi\|_\tau^2\, e^{-(\lambda-2\varepsilon)(t-t_o)}, \tag{3.9}$$

where

$$C = \left(C_1+\frac{1}{\varepsilon\delta e}\right) e^{(\lambda-2\varepsilon)(\rho+\delta)}.$$

However, (3.9) also holds for $t_o \le t \le t_o+\rho+\delta$ obviously. In other words, (3.4) has been proved and the proof of the theorem is complete.

## 4. NONLINEAR STOCHASTIC SYSTEMS WITH CONSTANTS DELAY

In this section we first establish a theorem on the exponential stability in $L^2(\Omega, C([-\tau, 0], R^d))$ for a stochastic differential equation with constant time lags and then extend it to a large-scale case. Consider a delay equation

$$dx(t) = f(x(t), t) + F(x(t-\tau_1), ..., x(t-\tau_n), t)\, dt + G(x(t-\tau_1), ..., x(t-\tau_n), t)\, dw(t) \tag{4.1}$$

on $t \ge t_o$ with initial data $\hat{x}(t_o) = \xi \in L^2(\Omega, \mathcal{F}_{t_o}, C([-\tau, 0], R^d))$, where $0\le\tau_1\le...\le\tau_n=\tau$ are all constants, $f : R^d{\times}R_+ \to R^d$, $F : R^{d\times n}{\times}R_+ \to R^d$ and $G : R^{d\times n}{\times}R_+ \to R^{d\times m}$. We assume the equation has a unique solution denoted by $x(t, t_o, \xi)$ as before.

**Theorem 4.1** Assume there exist constants $\lambda>0$, $\alpha_i\ge 0$ and $\beta_i\ge 0$ $(1\le i\le n)$ such that

$$x^T f(x, t) \le -\frac{\lambda}{2} |x|^2, \tag{4.2}$$

$$|F(v_1, ..., v_n, t)|^2 \le \sum_{i=1}^n \alpha_i |v_i|^2 \quad \text{and} \quad \|G(v_1, ..., v_n, t)\|^2 \le \sum_{i=1}^n \beta_i |v_i|^2, \tag{4.3}$$

$$2\left(\sum_{i=1}^n \alpha_i\right)^{1/2} + \sum_{i=1}^n \beta_i < \lambda, \tag{4.4}$$

for all $x$, $v_i \in R^d$ and $t \ge 0$. Then equation (4.1) is exponentially stable in $L^2(\Omega, C([-\tau, 0], R^d))$.

*Proof.* By condition (4.4) we can find $\gamma \in (0, \lambda)$ such that

$$2\Big(\sum_{i=1}^{n} \alpha_i e^{\gamma\tau_i}\Big)^{1/2} + \sum_{i=1}^{n} \beta_i e^{\gamma\tau_i} < \lambda - \gamma. \tag{4.5}$$

We first claim that there exists a $C > 0$ such that

$$\int_{t_0}^{\infty} e^{\gamma s} E|x(s, t_0, \xi)|^2 \le C e^{\gamma t_0} E\|\xi\|_\tau^2 \tag{4.6}$$

for all $t_0 \ge 0$ and $\xi \in L^2(\Omega, \mathcal{F}_{t_0}, C([-\tau, 0], R^d))$. Fix $t_0$ and $\xi$ and write $x(t, t_0, \xi) = x(t)$ as before. Then Itô's formula and condition (4.2) yield

$$e^{\lambda t} E|x(t)|^2 \le e^{\lambda t_0} E|x(t_0)|^2 + 2E\int_{t_0}^{t} e^{\lambda s} |x(s)|\, |F(x(s-\tau_1), \ldots, x(s-\tau_n), s)|\, ds$$

$$+ E\int_{t_0}^{t} e^{\lambda s} \|G(x(s-\tau_1), \ldots, x(s-\tau_n), s)\|^2\, ds. \tag{4.7}$$

Hence, for any $\theta > 0$,

$$E|x(t)|^2 \le e^{-\lambda(t-t_0)} E\|\xi\|_\tau^2 + \frac{1}{\theta}\int_{t_0}^{t} e^{-\lambda(t-s)} E|x(s)|^2\, ds$$

$$+ \sum_{i=1}^{n} (\theta\alpha_i + \beta_i) \int_{t_0}^{t} e^{-\lambda(t-s)} E|x(s-\tau_i)|^2\, ds.$$

Thus for any $T > t_0$,

$$\int_{t_0}^{T} e^{\gamma t} E|x(t)|^2\, dt \le \int_{t_0}^{T} e^{\gamma t - \lambda(t-t_0)} E\|\xi\|_\tau^2\, dt + \frac{1}{\theta}\int_{t_0}^{T} e^{\gamma t}\int_{t_0}^{t} e^{-\lambda(t-s)} E|x(s)|^2\, ds\, dt$$

$$+ \sum_{i=1}^{n} (\theta\alpha_i + \beta_i) \int_{t_0}^{T} e^{\gamma t}\int_{t_0}^{t} e^{-\lambda(t-s)} E|x(s-\tau_i)|^2\, ds\, dt.$$

$$\le \frac{1}{\lambda-\gamma} e^{\gamma t_0} E\|\xi\|_\tau^2 + \frac{1}{\theta(\lambda-\gamma)} \int_{t_0}^{T} e^{\gamma s} E|x(s)|^2 ds$$

$$+ \frac{1}{\lambda-\gamma} \sum_{i=1}^{n} (\theta\alpha_i+\beta_i) \int_{t_0}^{T} e^{\gamma s} E|x(s-\tau_i)|^2 ds$$

$$\le \frac{1}{\lambda-\gamma} \Big( 1+\sum_{i=1}^{n} (\theta\alpha_i+\beta_i)\tau_i e^{\gamma\tau_i} \Big) e^{\gamma t_0} E\|\xi\|_\tau^2$$

$$+ \frac{1}{\lambda-\gamma} \Big( \frac{1}{\theta} + \sum_{i=1}^{n} (\theta\alpha_i+\beta_i) e^{\gamma\tau_i} \Big) \int_{t_0}^{T} e^{\gamma s} E|x(s)|^2 ds. \tag{4.8}$$

In particular, if we choose

$$\theta = \Big( \sum_{i=1}^{n} \alpha_i e^{\gamma\tau_i} \Big)^{-1/2},$$

then, by (4.6)

$$\frac{1}{\lambda-\gamma} \Big( \frac{1}{\theta} + \sum_{i=1}^{n} (\theta\alpha_i+\beta_i) e^{\gamma\tau_i} \Big) = \frac{1}{\lambda-\gamma} \Big\{ 2\Big( \sum_{i=1}^{n} \alpha_i e^{\gamma\tau_i} \Big)^{1/2} + \sum_{i=1}^{n} \beta_i e^{\gamma\tau_i} \Big\} < 1.$$

Applying this to inequality (4.8) we see that there exists a $C > 0$ such that

$$\int_{t_0}^{T} e^{\gamma t} E|x(t)|^2 dt \le C e^{\gamma t_0} E\|\xi\|_\tau^2.$$

Letting $T \to \infty$ we obtain the required inequality (4.6). In order to complete our proof, we now apply Itô's formula along with conditions (4.2) and (4.3) to obtain that, for $r \ge t_0$,

$$e^{\gamma r} |x(r)|^2 \le e^{\gamma t_0} \|\xi\|_\tau^2 + \int_{t_0}^{r} e^{\gamma s} |x(s)|^2 ds + \sum_{i=1}^{n} (\alpha_i+\beta_i) \int_{t_0}^{r} e^{\gamma s} |x(s-\tau_i)|^2 ds$$

$$+ 2 \int_{t_0}^{r} e^{\gamma s} x^T(s) G(x(s-\tau_1), \ldots, x(s-\tau_n), s) dw(s).$$

Hence, if $t \ge t_0+\tau$,

$$e^{\gamma(t-\tau)} E\|\hat{x}(t)\|_\tau^2 \le E\Big( \sup_{t-\tau \le r \le t} e^{\gamma r} |x(r)|^2 \Big)$$

$$\leq e^{\gamma t_o} \|\xi\|_\tau^2 + \int_{t_o}^{t} e^{\gamma s} E|x(s)|^2 ds + \sum_{i=1}^{n} (\alpha_i+\beta_i) \int_{t_o}^{t} e^{\gamma s} E|x(s-\tau_i)|^2 ds$$

$$+ 2E\Big\{ \int_{t_o}^{t-\tau} e^{\gamma s} x^T(s)\, G(x(s-\tau_1), ..., x(s-\tau_n), s)\, dw(s)$$

$$+ \sup_{t-\tau\leq r\leq t} \int_{t-\tau}^{r} e^{\gamma s} x^T(s)\, G(x(s-\tau_1), ..., x(s-\tau_n), s)\, dw(s) \Big\}$$

$$\leq \Big( 1 + C + \sum_{i=1}^{n} (\alpha_i+\beta_i)(\tau_i+C)e^{\gamma\tau_i} \Big) e^{\gamma t_o} E\|\xi\|_\tau^2$$

$$+ 2E\Big( \sup_{t-\tau\leq r\leq t} \int_{t-\tau}^{r} e^{\gamma s} x^T(s)\, G(x(s-\tau_1), ..., x(s-\tau_n), s)\, dw(s) \Big). \tag{4.9}$$

But, by the well-known Burkholder-Davis-Gundy inequality we have

$$2E\Big( \sup_{t-\tau\leq r\leq t} \int_{t-\tau}^{r} e^{\gamma s} x^T(s)\, G(x(s-\tau_1), ..., x(s-\tau_n), s) dw(s) \Big)$$

$$\leq 6E\Big( \int_{t-\tau}^{t} e^{2\gamma s} |x(s)|^2\, \|G(x(s-\tau_1), ..., x(s-\tau_n), s)\|^2 ds \Big)^{1/2}$$

$$\leq 6E\Big\{ \|\hat{x}(t)\|_\tau \Big( \sum_{i=1}^{n} \int_{t-\tau}^{t} \beta_i e^{2\gamma s} |x(s-\tau_i)|^2 ds \Big)^{1/2} \Big\}$$

$$\leq \frac{1}{2} e^{\gamma(t-\tau)} E\|\hat{x}(t)\|_\tau^2 + 18e^{-\gamma(t-\tau)} \sum_{i=1}^{n} \int_{t-\tau}^{t} \beta_i e^{2\gamma s} E|x(s-\tau_i)|^2 ds$$

$$\leq \frac{1}{2} e^{\gamma(t-\tau)} E\|\hat{x}(t)\|_\tau^2 + 18e^{\gamma\tau} \sum_{i=1}^{n} \int_{t_o-\tau}^{\infty} \beta_i e^{\gamma(s+\tau_i)} E|x(s)|^2 ds$$

$$\leq \frac{1}{2} e^{\gamma(t-\tau)} E\|\hat{x}(t)\|_\tau^2 + 18e^{\gamma\tau} \sum_{i=1}^{n} \beta_i(\tau_i+C)e^{\gamma\tau_i+\gamma t_o} E\|\xi\|_\tau^2. \tag{4.10}$$

Substituting (4.10) into (4.9) gives

$$e^{\gamma t} E|x(t)|^2 \leq M_1 e^{\gamma t_o} E\|\xi\|_\tau^2 \qquad \text{if } t \geq t_o+\tau, \tag{4.11}$$

where

$$M_1 = 2\Big(1 + C + \sum_{i=1}^{n}(\alpha_i+\beta_i)(\tau_i+C)e^{\gamma\tau_i}\Big) + 36e^{\gamma\tau}\sum_{i=1}^{n}\beta_i(\tau_i+C)e^{\gamma\tau_i}.$$

Similarly we can show that

$$e^{\gamma t}\, E|x(t)|^2 \le M_2 e^{\gamma t_0}\, E\|\xi\|_\tau^2 \qquad \text{if } t_0 \le t \le t_0+\tau \tag{4.12}$$

for some positive constant $M_2$. In other words, we have proved that equation (4.1) is exponentially stable in $L^2(\Omega, C([-\tau, 0], R^d))$. The proof is complete.

Let us now consider a large-scale stochastic differential equation with the form

$$\begin{aligned} dx_i(t) &= f_i(x_i(t), t)\,dt + F_i(\tilde{x}_1(t), \ldots, \tilde{x}_i(t), t)\,dt + G_i(\tilde{x}_1(t), \ldots, \tilde{x}_i(t), t)\,dw(t) \quad \text{on } t \ge t_0, \\ \hat{x}_i(t_0) &= \xi_i, \quad i = 1, \ldots, k, \end{aligned} \tag{4.13}$$

where $\tilde{x}_i(t) = (x_i(t-\tau_1), \ldots, x_i(t-\tau_n))$, $f_i : R^{d_i}\times R_+ \to R^{d_i}$, $F_i : R^{(d_1+\ldots+d_i)\times n}\times R_+ \to R^{d_i}$, $G_i : R^{(d_1+\ldots+d_i)\times n}\times R_+ \to R^{d_i\times m}$ and $\xi_i \in L^2(\Omega, \mathcal{F}_{t_0}, C([-\tau, 0], R^{d_i}))$. As usual, we assume that the equation has a unique solution and $f_i(0, t) \equiv 0$, $F_i(0, t) \equiv 0$, $G_i(0, t) \equiv 0$. From Theorems 3.1 and 4.1 we obtain the following useful theorem immediately.

**Theorem 4.2** Assume there exist positive constants $\lambda$, $\alpha$, $\beta$, K such that

(i) $x_i^T f_i(x_i, t) \le -\frac{\lambda}{2}|x_i|^2$,

(ii) $|F_i(u_1, \ldots, u_i, t) - F_i(0, \ldots 0, v_i, t)|^2 \le K\sum_{j<i}\|u_j\|^2 + \alpha\|u_i - v_i\|^2$,

(iii) $\|G_i(u_1, \ldots, u_i, t) - G_i(0, \ldots 0, v_i, t)\|^2 \le K\sum_{j<i}\|u_j\|^2 + \beta\|u_i - v_i\|^2$,

(iv) $2\sqrt{\alpha n} + \beta n < \lambda$,

for all $x_i \in R^{d_i}$, $u_i, v_i \in R^{d_i\times n}\times R_+$ and $i = 1, \ldots, k$. Then the large-scale delay equation (4.13) is exponentially stable in $L^2(\Omega, C([-\tau, 0], R^d))$.

## 5. LINEAR STOCHASTIC SYSTEMS WITH CONSTANTS DELAY

In this section we shall discuss the exponential stability of linear large-scale systems with constants delay. In order to make our theory more understandable and to reduce the complication of notations, we shall only consider a system of the form

$$dx_i(t) = A_i\, x_i(t)\, dt + \sum_{j=1}^{i} \sum_{q=1}^{m} \Big( B_{ijq}x_j(t) + C_{ijq}x_j(t-\tau) \Big)\, dw_q(t) \quad \text{on } t \ge t_0,$$

$$\hat{x}_i(t_0) = \xi_i \in L^2(\Omega, \mathcal{F}_{t_0}, C([-\tau, 0], R^{d_i})), \tag{5.1}$$

where $A_i \in R^{d_i \times d_i}$, $B_{ijq} \in R^{d_i \times d_j}$ and $C_{ijq} \in R^{d_i \times d_j}$. But our theory can obviously be extended to more general cases, for instance, a system with several constant time lags and delay shift terms. Note the isolated subsystems are as follows:

$$dy_i(t) = A_i\, y_i(t)\, dt + \sum_{q=1}^{m} \Big( B_{iiq}y_i(t) + C_{iiq}y_i(t-\tau) \Big)\, dw_q(t) \quad \text{on } t \ge t_0,$$

$$\hat{y}_i(t_0) = \xi_i \in L^2(\Omega, \mathcal{F}_{t_0}, C([-\tau, 0], R^{d_i})). \tag{5.2}$$

We see from Theorem 3.1 that equation (5.1) is exponentially stable in $L^2(\Omega, C([-\tau, 0], R^d))$ if and only if each of subsystems (5.2) is exponentially stable in $L^2(\Omega, C([-\tau, 0], R^{d_i}))$. We therefore can apply Theorem 4.1 to obtain the following theorem.

**Theorem 5.1** Assume that for each $i = 1, \ldots, k$, there exists a constant $\lambda_i$ such that

$$x_i^T A_i x_i \le -\frac{\lambda_i}{2} |x_i|^2 \quad \text{for all } x_i \in R^{d_i} \tag{5.3}$$

and

$$2 \sum_{q=1}^{m} \Big( \|B_{iiq}\|^2 + \|C_{iiq}\|^2 \Big) < \lambda_i. \tag{5.4}$$

Then each of subsystems (5.2) is exponentially stable in $L^2(\Omega, C([-\tau, 0], R^{d_i}))$, and therefore the large-scale delay system (5.1) is also exponentially stable in $L^2(\Omega, C([-\tau, 0], R^d))$.

We now observe that equation (5.2) can be regarded as a stochastically perturbed system of a linear ordinary differential equation

$$\dot{y}_i(t) = A_i\, y_i(t), \tag{5.5}$$

and condition (5.3) means equation (5.5) is exponentially stable. Theorem 5.1 shows that the stochastically perturbed system (5.2) remains exponentially stable provided the stochastic perturbation is small enough; the perturbation is measured by (5.4) and is independent of $\tau$.

Let us now consider a special case of system (5.1) with $B_{iiq} = -C_{iiq}$ but the norms of $B_{iiq}$ may be large so that (5.4) may not hold. In other words we may not be able to apply Theorem 5.1 in this case. However, note that the isolated subsystems become

$$dy_i(t) = A_i\, y_i(t)\, dt + \sum_{q=1}^{m} B_{iiq}(y_i(t)-y_i(t-\tau))\, dw_q(t) \quad \text{on } t \geq t_0,$$
$$\hat{y}_i(t_0) = \xi_i \in L^2(\Omega, \mathcal{F}_{t_0}, C([-\tau, 0], R^{d_i})). \tag{5.6}$$

and observe the stochastic perturbation is dependent of the difference $y_i(t)-y_i(t-\tau)$ which can be nearly zero if $\tau$ is very small. Hence the stochastic perturbation could be so small that it could absorbed by the stable system (5.5) so that system (5.6) remains stable. In other words, we might expect that system (5.6) is still exponentially stable provided $\tau$ small enough. This result has been obtained by Mao [4] and we recall it as a lemma.

**Lemma 5.2** (Mao [4], Theorem 5.3.1) Assume there exist positive constants $M_i$ and $\lambda_i$ such that

$$\|e^{A_i t}\|^2 \leq M_i e^{-\lambda_i t} \quad \text{for all } t \geq 0. \tag{5.7}$$

Assume furthermore that $\tau$ is so small that

$$3\tau M_i \sum_{q=1}^{m} \|B_{iiq}\|^2 < 1 \tag{5.8}$$

and

$$3M_i\, \|I-e^{-A_i\tau}\|^2 \left(1-3\tau M_i \sum_{q=1}^{m} \|B_{iiq}\|^2\right)^{-1} \sum_{q=1}^{m} \|B_{iiq}\|^2 < \lambda_i. \tag{5.9}$$

Then system (5.6) is exponentially stable in mean square.

The following theorem is immediate now.

**Theorem 5.3** Let $B_{iiq} = - C_{iiq}$ for all $1 \leq i \leq k$ in the large-scale stochastic delay system (5.1). Assume for each i there exist $M_i$ and $\lambda_i$ such that (5.7) holds. Assume also both (5.8) and (5.9) hold for all i. Then system (5.1) is exponentially stable in $L^2(\Omega, C([-\tau, 0], R^d))$.

ACKNOWLEDGEMENT

The author wishes to thank the referee for his suggestions.

REFERENCES

[1] Ladde, G. S., Stochastic stability analysis of model ecosystems with time-delay, in "Differential Equations and Applications in Biology, Epidemics, and Population

Problems" edited by S. N. Busenberg and K. Cook, Academic Press (1981), 215-228.

[2] Mao, X., Exponential stability for stochastic differential delay equations in Hilbert space, Quarterly J. Math. Oxford (2), 42 (1991), 77-85.

[3] Mao, X., Almost sure exponential stability for delay stochastic differential equations with respect to semimartingales, Stochastic Analysis and Applications 9(2) (1991), 177-194.

[4] Mao, X., Stability of Stochastic Differential Equations with Respect to Semimartingales, Pitman Research Notes in Mathematics Series 251, Longman Scientific and Technical, 1991.

[5] Mao, X. and Markus, L., Energy bounds for nonlinear dissipative stochastic differential equations with respect to semimartingales, Stochastics and Stochastics Reports, 37 (1991), 1-14.

[6] Mizel, V. J. and Trutzer, V., Stochastic hereditary equations: existence and asymptotic stability, J. Integral Equations 7 (1984), 1-72.

[7] Mohammed, S-E. A., Stability of linear delay equations under a small noise, Proceedings of Edinburgh Mathematical Society 29 (1986), 233-254.

[8] Metivier, M., Semimartingales, Walter de Gruyter, Berlin, New York, 1982.

[9] Michel, A. N., Miller, R. K. and Tang, W., Lyapunov stability of interconnected systems: decomposition into strongly connected subsystems, IEEE Trans. Circuits Syst., CAS-25 (1978), 799-809.

[10] Siljak, D. D., Large-Scale Dynamic Systems: Stability and Structure, Amsterdam, the Netherlands: North-Holland, 1977.

[11] Sontag, E. D., Further facts about input to state stabilization, IEEE Trans. Automat. Contr., AC-35(4) (1980), 473-476.

[12] Vidyasagar, M., Decomposition techniques for large-scale systems with nonadditive interactions: stability and stabilizablity, IEEE Trans. Automat. Contr., AC-25(4) (1980), 773-779.

# 14 The Averaging Method and the Problem on Separation of Variables

**Yu. A. Mitropolsky** Institute of Mathematics, Ukrainian Academy of Sciences, Kiev, Ukraine

The averaging method originally appeared in celestial mechanics and, at the beginning, it was mainly connected with the problems of celestial mechanics. Various averaging schemes were applied to the solution of such problems. The idea of the averaging method is the replacement, in differential equations, of the right-hand sides representing oscillations or rotations by "smoothed," or averaged, functions which do not depend upon the time $t$ and rapidly changing parameters.

Later this method was modified and widely applied. Its application is connected with names of many outstanding mathematicians and engineers.

Significant results in the development of the averaging method were obtained by N. M. Krylov and N. N. Bogoliubov. It was Bogoliubov who created the mathematical theory of the averaging method and showed that it is naturally connected with the existence of the change of variables which permits the elimination of the time $t$ from the right-hand sides of equations, with an arbitrary degree of exactness with respect to a small parameter $\epsilon$. He elaborated the method of construction, not only in the system of the first approximation (averaged system), but in the averaged system of higher orders of approximation. The solutions of such systems approximate solutions of the initial (exact) system with arbitrary preassigned accuracy.

Here we shall describe further development of the averaging method and the extension of areas of its application.

The starting point of investigation of the averaging method is a system in a standard form[†]

$$\frac{dx}{dt} = \epsilon X(x, t, \epsilon), \tag{1}$$

where $x = \text{col}\,\|x_1, \ldots, x_{\bar{n}}\|$; $X(x, t, \epsilon)$ is a $\bar{n}$-dimensional vector satisfying all necessary conditions; $\epsilon$ is a small positive parameter.

The system (1) upon averaging

$$M_t\{X(\xi, t, \epsilon)\} = \lim_{T\to\infty} \frac{1}{T} \int_0^T X(\xi, t, \epsilon)dt = X_0(\xi, \epsilon), \tag{2}$$

[†]By the system in standard form we mean the system of equations with right-hand sides which are proportional to a small $\epsilon$ parameter. Many actual problems can be reduced to a system of the form (1).

and with a special change of variables

$$x = \bar{x} + \epsilon F_1(t, \bar{x}) + \epsilon^2 F_2(t, \bar{x}) + \cdots \epsilon^s F_s(t, \bar{x}) + \cdots, \tag{3}$$

is reduced to the averaged system

$$\frac{d\bar{x}}{dt} = \epsilon X_0^{(1)}(\bar{x}) + \epsilon^2 X_0^{(2)}(\bar{x}) + \cdots + \epsilon^s X_0^{(s)}(\bar{x}) + \cdots, \tag{4}$$

which does not contain the argument $t$ explicitly.

Let us write the initial system (1) in the equivalent form

$$\frac{dx}{dt} = \epsilon X(x, y, \epsilon), \quad \frac{dy}{dt} = 1; \tag{5}$$

also the averaged system (4) of (5) correspondingly in the form

$$\frac{d\bar{x}}{dt} = \epsilon X_0(\bar{x}, \epsilon), \quad \frac{d\bar{y}}{dt} = 1, \tag{6}$$

where $X_0(\bar{x}, \epsilon) = X_0^{(1)}(\bar{x}) + \epsilon X_0^{(2)}(\bar{x}) + \cdots$, and $y = \bar{y}$. Integration of the averaged system (6) is simpler than that of the initial system (5) or of the equivalent system (1) since the variables are separated: the system for slow variables $\bar{x}$ does not contain fast variables $\bar{y}$ and it is integrated independently.

As stated above, this allows one to interpret the averaging method in the following way: the averaging method transforms the system (5) with nonseparated variables into the system (6) with separated slow and fast variables.

As known from the averaging method of N. N. Bogoliubov, functions $F_i(t, \bar{x})$ and $X_0^{(i)}(\bar{x})(i = 1, 2 \ldots)$ in the right hand sides of (3) and (4) can be calculated for known right-hand sides of (1), and functions $X_1^{(1)}(\xi), X_0^{(2)}(\xi), \cdots X_0^{(s)}(\xi), \ldots$ are obtained by averaging right hand side of (1) after the substitution (3). For example (compare notation of (38) below),

$$X_0^{(1)}(\xi) = M_t \{X(\xi, t, 0)\}, \quad X_0^{(2)}(\xi) = M_t \left\{ \left( \tilde{X} \frac{\partial}{\partial \xi} \right) X(\xi, t, 0) \right\}.$$

But generally the convergence of these series for $s \to \infty$ cannot be proved because of the appearance of small divisors in the expressions for functions $F_i(t, \bar{x})$ for $i > s$.

Thus we must consider (3) as formal series which are necessary for the construction of the approximate asymptotic solution

$$x = \bar{x} + \epsilon F_1(t, \bar{x}) + \epsilon^2 F_2(t, \bar{x}) + \cdots + \epsilon^s F_s(t, \bar{x}), \tag{7}$$

reducing system (1) to the following form

$$\frac{d\bar{x}}{dt} = \epsilon X_0^{(1)}(\bar{x}) + \epsilon^2 X_0^{(2)}(x) + \cdots \epsilon^s X_0^{(s)}(x) + \epsilon^{s+1} \mathcal{R}(t, \bar{x}, \epsilon). \tag{8}$$

Omitting the terms $\epsilon^{s+1} \mathcal{R}(t, \bar{x}, \epsilon)$ in the system (8) we obtain the averaged system of $s$-th approximation and correspondingly the separation of variables to within the estimates of $s$-th order of smallness. Under the change of variables (7) in the equations (1) the variable $t$ passes to terms of $(s+1)$-th order of smallness. If we find the exact solution of the system of equations (8) and put it in the right-hand side of the series (7), then the obtained expression will be the exact solution of the system of equations (1).

Let us consider a system of nonlinear equations appearing in control theory and vibrotechnics. Let the oscillating system be described by the system of differential equations

$$\frac{dx}{dt} = X(x, \phi, \epsilon), \qquad \frac{d\phi}{dt} = \lambda\omega(x) + \mathcal{A}(x, \phi, \epsilon), \tag{8$^*$}$$

where $\lambda$ is the large parameter $(\frac{1}{\lambda} = \epsilon)$, $\omega = (\omega_1, \omega_2, \ldots, \omega_m), x = (x_1, x_2, \ldots, x_n), \phi = (\phi_1, \phi_2, \ldots, \phi_m); X = (X_1, X_2, \ldots, X_n)$ and $\mathcal{A} = (\mathcal{A}_1, \mathcal{A}_2, \ldots, \mathcal{A}_m)$ are periodic functions in $\phi$ with a period $2\pi$. The problem on the averaging of the system (8)* is stated as follows: it is necessary to change the system of equations (8)* to the more simple averaged system where the variable $\phi$ is separated from variable $x$. Moreover the phase $\phi$ is to be excluded from the right-hand sides of the averaged equations. Using this method for the solution of such problem it is necessary to find the change of variables of the form

$$x = \bar{x} + \sum_{s=1}^{\infty} \epsilon^s u^{(s)}(\bar{x}, \bar{\phi}). \qquad \phi = \bar{\phi} + \sum_{s=1}^{\infty} \epsilon^s v^{(s)}(\bar{x}, \bar{\phi}). \tag{9}$$

With the help of this change of variables the system (8)* can be easily reduced to the averaged system (analogous to (6))

$$\frac{d\bar{x}}{dt} = \sum_{s=1}^{\infty} \epsilon^s X^{(s)}(\bar{x}), \qquad \frac{d\bar{\phi}}{dt} = \lambda\omega(\bar{x}) + \sum_{s=0}^{\infty} \epsilon^s \mathcal{A}^{(s)}(\bar{x}), \tag{10}$$

where $u^{(s)}(\bar{x}, \bar{\phi})$ are $n$-dimensional vector functions, $v^{(s)}(\bar{x}, \bar{\phi})$ are $m$- dimensional vector functions.

The averaged system (10) is much simpler than the initial system (8)* since the equations for $\bar{x}$ are integrated independently of $\phi$. In this case the series in formulae (9) and right-hand sides of equations (10), which are necessary for the construction of the averaged system of the $s$-th approximation, are considered as formal, i.e., we do not discuss their convergence.

From the physical point of view the transformation (9) consists in expanding the real motion described by the variables $x$, $\phi$ into the averaged motion with coordinates $\bar{x}$ and "vibration" described by angle $\bar{\phi}$, and functions $u^{(s)}(\bar{x}, \bar{\phi})$ and $v^{(s)}(\bar{x}, \bar{\phi})$.

The functions in the right-hand sides of the expressions (9) are not defined uniquely since the different terms of this expansion can be arbitrarily associated with the main or highest terms of the series (9).

As known, some additional conditions are necessary for obtaining unique expressions for terms of series (9). For example $u^{(s)}(\bar{x}, \bar{\phi})$ and $v^{(s)}(\bar{x}, \bar{\phi})$ should not contain the zero harmonics in $\bar{\phi}$. This corresponds to the inclusion of all the averaged motions into $\bar{x}$ and $\bar{\phi}$.

As was previously mentioned the series (9) in the general case are not convergent. Moreover small denominators can appear in the terms $u^{(s)}(\bar{x}, \bar{\phi})$ and $v^{(s)}(\bar{x}, \bar{\phi})$ for some $s$. The smallness of these denominators cannot be compensated by the smallness of the corresponding small coefficient $\epsilon^s$.

In this connection we shall describe further developments of the averaging method which allow the possibility to construct the change of variables so that the small denominators would be suppressed in each successive step of approximation and the obtained series would then be convergent.

In many practically important cases the system of equations of the form (1), with some restrictions, is reduced to the system of the form (to compare development of (10), see [2])

$$\begin{aligned} \frac{dh}{dt} &= Hh + \epsilon F(h, \phi, \epsilon), \\ \frac{d\phi}{dt} &= \omega + \epsilon f(h, \phi, \epsilon), \end{aligned} \tag{11}$$

where $h = (h_1, h_2, \dots, h_n)$, $F = (F_1, F_2, \dots, F_n)$, $\phi = (\phi_1, \phi_2, \dots \phi_m)$, $f = (f_1, f_2, \dots, f_m)$, $(n+m = \bar{n})$; all eigenvalues of the square matrix $H$ have non-zero real parts; $F(h, \phi, \epsilon)$ and $f(h, \phi, \epsilon)$ are regular functions in the domain

$$|h| \leq \eta \qquad |\mathrm{Im}(\phi)| \leq \rho$$

which are periodic in $\phi$ with the period $2\pi$; $\epsilon$ is a small positive parameter.

Equations reduced to the form (11) are very convenient for the solution of various problems dealing with the qualitative investigation of the behavior of solutions. In particular they are useful for the construction of the integral manifolds, for the investigation of solutions on manifolds, and for the construction of algorithms for obtaining the approximate solutions.

It should be noted that the change of variables (9) will contain multiple Fourier sums with a small denominator of the form

$$\sum_{|\kappa| \neq 0} \frac{\mathcal{P}_\kappa e^{i(\kappa\phi)}}{(\kappa\omega)}, \tag{12}$$

where $(\kappa\phi) = \kappa_1\phi_1 + \dots + \kappa_m\phi_m$, $|\kappa| = |\kappa_1| + |\kappa_2| + \dots + |\kappa_m|$, $(\kappa\omega) = \kappa_1\omega_1 + \kappa_2\omega_2 + \dots + \kappa_m\omega_m$ and all $\omega_i$ are assumed to be real and linearly independent over the integers.

To avoid the existence of small divisors it would be appropriate, as was proposed by N. M. Krylov and N. N. Bogoliubov, to introduce the exact frequencies $\bar{\omega}$ and deviations $\Delta = \omega - \bar{\omega}$. Then determine the frequencies $\bar{\omega}$ not by $\omega$ and $\epsilon f$, but find the frequency of zero approximation $\bar{\omega}$ by exact $\omega$. Thus we introduce deviation $\Delta$ in the equations (11), and then instead of equation (11) we consider the following system

$$\begin{aligned} \frac{dh}{dt} &= Hh + \epsilon F(h, \phi, \Delta, \epsilon), \\ \frac{d\phi}{dt} &= \bar{\omega} + \Delta + \epsilon f(h, \phi, \Delta, \epsilon). \end{aligned} \tag{13}$$

Here $h = (h_1, h_2, \dots, h_n)$, $\phi = (\phi_1, \phi_2, \dots, \phi_m)$; $\bar{\omega} = (\bar{\omega}_1, \bar{\omega}_2, \dots, \bar{\omega}_m)$ are linearly independent over the integers; $\epsilon F(h, \phi, \Delta, \epsilon)$, $\epsilon f(h, \phi, \Delta, \epsilon)$ are analytic functions of complex variables $h, \phi, \Delta$ which are sufficiently small in the domain

$$|h| \leq \eta, \qquad |\mathrm{Im}(\phi)| \leq \rho, \quad |\Delta| < \sigma \tag{14}$$

and satisfy certain additional conditions. Further we shall assume that all real parts of the characteristic numbers of the matrix $H$ are negative.

Under these assumptions for the system of equations (13) we find an analytic transformation

$$\phi = \theta^{(1)} + u(h, \theta^{(1)}, \Delta^{(1)}), \qquad \Delta = \Delta(\Delta^{(1)}), \tag{15}$$

which would reduce equations (13) to the form

$$\frac{dh}{dt} = Hh + \epsilon F_1(h, \theta^{(1)}, \Delta^{(1)}, \epsilon); \; \frac{d\theta^{(1)}}{dt} = \bar{\omega} + \Delta^{(1)} + \epsilon^2 f_1(h, \theta^{(1)}, \Delta^{(1)}, \epsilon), \tag{16}$$

where the functions $F_1(h, \theta^{(1)}, \Delta^{(1)}, \epsilon)$ and $f_1(h, \theta^{(1)}, \Delta^{(1)}, \epsilon)$ are analytic with respect to $h, \theta^{(1)}, \Delta^{(1)}$; moreover $\epsilon^2 f_1(h, \theta^{(1)}, \Delta^{(1)}, \epsilon)$ has the magnitude of $\epsilon^2$ order of smallness.

Using the same transformation for the system (16) we obtain the equation of the form

$$\begin{aligned} \frac{dh}{dt} &= Hh + \epsilon F_2(h, \theta^{(2)}, \Delta^{(2)}, \epsilon), \\ \frac{d\theta^{(2)}}{dt} &= \bar{\omega} + \Delta^{(2)} + \epsilon^4 f_2(h, \theta^{(2)}, \Delta^{(2)}, \epsilon), \end{aligned} \tag{17}$$

and so on. The orders of smallness of the functions $\epsilon^{2^s} f_s(h, \theta^{(s)}, \Delta^{(s)}, \epsilon)$ and $\Delta^{(s)}(s = 2, 3, 4 \ldots)$ will be correspondingly $\epsilon^4, \epsilon^8, \epsilon^{16}, \ldots$. We can always express $\phi$ and $\Delta$ by $\theta^{(s)}$ and $\Delta^{(s)}$

$$\phi = \theta^{(s)} + \Phi^{(s)}(h, \theta^{(s)}, \Delta^{(s)}, \epsilon), \Delta = \mathcal{D}^{(s)}(\Delta^{(s)}). \tag{18}$$

Note, that the transformations of the form (15) are the familiar transformations which are well known in nonlinear mechanics and which generalize the classical averaging method (the averaging method of N. N. Bogoliubov).

However in the described process of successive transformations, the ideas of A. N. Kolmogorov [3] and of V. I. Arnold [1] on "accelerated convergence" are used essentially. Here a type of Newton's method was applied to the transformation of equations containing small parameters.

Owing to this process a new aspect appears in our investigation. That is, to obtain the approximated process, we do not improve the transformation (15) by adding terms of higher order of smallness in $\epsilon$ (as in (9)) but instead by the repeated application of the same transformation to the new equation (16), and so on. Thus we obtain very fast convergence in the formulae (18) and very fast decrease of the functions $\epsilon^{2^s} f_s(h, \theta^{(s)}, \Delta^{(s)}, \epsilon)$ as $s$ increases.

For $s \to \infty$ in formulae (18) the system of equations (13), upon the change of variables

$$\begin{aligned} \phi &= \theta + \Phi^{(\infty)}(h, \theta, 0, \epsilon), \\ \Delta &= \mathcal{D}^{(\infty)}_{(0)}, \end{aligned} \tag{19}$$

is transformed to the system of the form

$$\frac{dh}{dt} = Hh + \epsilon F_\infty(h, \theta, 0, \epsilon); \qquad \frac{d\theta}{dt} = \bar{\omega}. \tag{20}$$

Integration of this system (20) is much simpler than integration of the initial system (13) (see for example [2]).

The proposed method allows one to solve many interesting and important problems of nonlinear mechanics.

Note consequently that applying the transformations of type (15), we can make the nonlinear term in the second group of equations for $\phi$ tend to zero, but at the same time we can make nonlinear terms in the first group of equations for $h$ tend to zero (despite the fact that $h$ and $F_s(h,\theta,\Delta,\epsilon)$ do not depend on $h$ periodically).

Thus, summarizing our conclusions we obtain the following theorem (see [2], where $H$ in (13) is diagonalised).

**Theorem** [2]. *Assume that for the system of equations*

$$\frac{dh}{dt} = (\alpha+\xi)h + F(h,\phi,\Delta,\xi),$$
$$\frac{d\phi}{dt} = \omega + \Delta + f(h,\phi,\Delta,\xi), \tag{21}$$

*the conditions described above hold.*

*Then by the change of variables*

$$h = g + U^{(\infty)}(g,\theta,0),$$
$$\phi = \theta + \Phi^{(\infty)}(g + U^{(\infty)}(g,\theta,0),\theta,0) \tag{22}$$

*and with the help of a special choice of* $\Delta = \mathcal{D}^{(\infty)}_{(0)}$ *and* $\xi = \xi^{(\infty)}_{(0)}$, *this system is reduced to the linear system with constant coefficients*

$$\frac{dg}{dt} = \alpha g, \quad \frac{d\theta}{dt} = \omega. \tag{23}$$

Integration of (23) gives the general solution of the initial system (21) of the following form

$$h_t = Ce^{\alpha t} + U^{(\infty)}(Ce^{\alpha t}, \omega t + \theta_0, 0),$$
$$\phi_t = \omega t + \theta_0 + \Phi^{(\infty)}(Ce^{\alpha t} + U^{(\infty)}(Ce^{\alpha t}, \omega t + \theta_0, 0), \omega t + \theta_0, 0), \tag{24}$$

where the $n+m$ arbitrary constants $C, \theta_0$ belong to the domain

$$|C| \le \frac{\eta}{2}, \quad |\operatorname{Im}(\theta_0)| \le \frac{\rho}{2}\left(1 - \frac{1}{2m}\right). \tag{25}$$

With the increase of time the solution of the system (21) will become close to the stationary quasi-periodic solution

$$h(\omega t) = U^{(\infty)}(0, \omega t + \theta_0, 0),$$
$$\phi(\omega t) = \omega t + \theta_0 + \Phi^{(\infty)}(U^{(\infty)}(0, \omega t + \theta_0, 0), \omega t + \theta_0, 0), \tag{26}$$

where $|\operatorname{Im}(\theta_0)| \le \frac{\rho}{2}(1 - \frac{1}{2m})$, according to the estimates

$$|h_t - h(\omega t)| \le \frac{3}{2}|C|e^{-\bar{\alpha}t},$$
$$|\phi_t - \phi(\omega t)| \le \frac{3}{2}C_1(\tau_0)|C|e^{-\bar{\alpha}t}. \tag{27}$$

Consider the system of differential equations

$$\frac{dx}{dt} = \mathcal{A}x + \mathcal{P}(\phi)x, \quad \frac{d\phi}{dt} = \omega, \tag{28}$$

where $\mathcal{A}$ is constant, $\mathcal{P}(\phi)$ is periodic with respect to $\phi = (\phi_1,\ldots,\phi_m)$ with a period $2\pi$, and both are $(n\times n)$-dimensional real matrices for real $\phi$; $\omega = (\omega_1,\ldots,\omega_m)$ are the frequencies of the $\mathcal{P}(\omega t)$ matrix; $x = (x_1,\ldots,x_n)$ is a $n$- dimensional vector; $t$ is the time variable.

The system (28) generally is a particular case of the system (13). However, the systems of equations with quasi-periodic coefficients play an important role in the theory of differential equations and stability theory. Thus it is expedient to consider this equation.

The problem on reducibility of the system (28) to a system with constant coefficients has been considered by many authors, but it still remains unsolved in full generality. Although by asymptotic methods of nonlinear mechanics the asymptotic expression for the solutions constructed and their stability have been studied, the expansions obtained are divergent because of the presence of small divisors. Thus the reducibility of the system (28) to a system with constant coefficients could only be considered through asymptotic analysis.

Now we shall give the formulation of the theorem on reducibility of the system of the type (28), as in [2].

**Theorem** [2]. *Let the right hand side of the system of equations (28) satisfy the following conditions:*

1. *The matrix $\mathcal{P}(\phi)$ is periodic in $\phi$ with a period $2\pi$ and analytic in the domain*

$$|\operatorname{Im}(\phi)| = \sup_{\alpha} |\operatorname{Im}(\phi\alpha)| \leq \rho_0 \quad (\rho_0 > 0), \tag{29}$$

*and real for real $\phi$.*

2. *For some positive $\epsilon$ and $d$ the inequality*

$$|(\kappa, \omega)| \geq \epsilon |\kappa|^{-d} \quad (|\kappa| \neq 0) \tag{30}$$

*holds for all integral $\kappa = (\kappa_1, \ldots, \kappa_m)$.*

3. *The eigenvalues $\lambda = (\lambda_1, \ldots, \lambda_n)$ of the $\mathcal{A}$ matrix have distinct real parts.*

*Then we can find a sufficiently small positive constant $\mu_0$ such that for*

$$|\mathcal{P}(\phi)| = \sum_{i,j=1}^{n} |\mathcal{P}_{ij}(\phi)| \leq \mu_0 \tag{31}$$

*the system of equations* (28), *under a nondegenerated change of variables*

$$x = \Phi(\phi)y \tag{32}$$

*with matrix $\Phi(\phi)$ periodic in $\phi$, with period $2\pi$, analytic, analytically invertible in the domain*

$$|\operatorname{Im}(\phi)| \leq \frac{\rho_0}{2}, \tag{33}$$

*and real for* $\operatorname{Im}(\phi) = 0$, *is reduced to the form:*

$$\frac{dy}{dt} = \mathcal{A}_0 y, \qquad \frac{d\phi}{dt} = \omega \tag{34}$$

*where $\mathcal{A}_0$ is a constant matrix.*

By this theorem the fundamental matrix of the solutions of the system (28) has the form

$$X = \Phi(\omega t + \phi_0)e^{\mathcal{A}_0 t}, \qquad \phi = \omega t + \phi_0, \tag{35}$$

where $\Phi(\omega t + \phi_0)$ is a matrix which is nondegenerated quasi-periodic in $t$, real for real $\phi_0$, and with frequency basis $\omega = (\omega_1, \ldots, \omega_m)$.

By the previously stated method of accelerated convergence many actual problems in nonlinear mechanics can be easily solved. Thus, combining the method of integral manifolds with the presented method, we can obtain some important theorems on the behavior of solutions in the neighborhood of quasi-periodic manifolds for various equations containing both "small" and "large" parameters. Important results can be obtained while considering quasi-periodic regimes for the systems with fast and slow variables, and while considering various nonstationary and transient regimes.

Now we pass to the new approach in the investigation of nonlinear differential equations with a small parameter. The idea of this new approach originates from the averaging method itself.

Consider again the exact perturbed system (5)

$$\frac{dx}{dt} = \epsilon X(x, y, \epsilon), \qquad \frac{dy}{dt} = 1, \tag{5}$$

and the averaged system corresponding to it

$$\frac{d\bar{x}}{dt} = \epsilon X_0(\bar{x}, \epsilon), \qquad \frac{d\bar{y}}{dt} = 1. \tag{6}$$

The described property of separation of motions in the averaging method has group-theoretic characteristics. Put, in the systems (5) and (6), $\epsilon = 0$ and write the initial unperturbed systems (the systems of zero approximation) in the form

$$\frac{dx}{dt} = 0, \qquad \frac{dy}{dt} = 1, \tag{36}$$

and correspondingly

$$\frac{d\bar{x}}{dt} = 0, \qquad \frac{d\bar{y}}{dt} = 1. \tag{37}$$

The systems (36) and (37) coincide except for the notations.

Let vectors $X$ and $X_0$ in the systems (5) and (6) have components

$$X = \text{col}\,\|X_1, X_2, \ldots, X_n\|, \quad X_0 = \text{col}\,\|X_{01}, X_{02}, \ldots, X_{0n}\|$$

Put, in accord with system (5), the linear differential operator of the first order

$$W_0 = W + \epsilon\tilde{W}, \tag{38}$$

where

$$W = \frac{\partial}{\partial y}, \qquad \tilde{W} = X_1 \frac{\partial}{\partial x_1} + \cdots + X_n \frac{\partial}{\partial x_n}.$$

Also put, in accord with system (6), the operator

$$U_0 = U + \epsilon\tilde{U}, \tag{39}$$

where

$$U = \frac{\partial}{\partial y}, \quad \tilde{U} = X_{01} \frac{\partial}{\partial \bar{x}_1} + \cdots + X_{0n} \frac{\partial}{\partial \bar{x}_n}.$$

In our terminology the operators (38) and (39) are called operators associated with the systems (5) and (6). If we put $\epsilon = 0$ in the formulae (38) and (39), then the operator (38) turns into the operator

$$W_0' = W \equiv \frac{\partial}{\partial y}, \tag{40}$$

associated with the system of zero approximation (36) and the operator (39) turns into the operator

$$U_0' = U \equiv \frac{\partial}{\partial y}, \tag{41}$$

associated with the system of zero approximation (37). It is easy to show that the Poisson bracket of the operators $U$ and $\tilde{U}$ is identically equal to zero, i.e.,

$$[U, \tilde{U}] = U\tilde{U} - \tilde{U}U \equiv 0. \tag{42}$$

Consider the one-parameter transformation group determined by operator $U$ and which is given by the Lie series

$$\bar{x} = \exp(sU(\bar{x}_0, \bar{y}_0))\bar{x}_{10}, \ldots, \bar{x}_n = \exp(sU(\bar{x}_0, \bar{y}_0))\bar{x}_{n0}, \quad \bar{y} = \exp(sU(\bar{x}, \bar{y}_0))\bar{y}_0 \tag{43}$$

where $\bar{x}_{10}, \ldots, \bar{x}_{n0}, \bar{y}_0$ are new variables, $U(\bar{x}_0, \bar{y}_0) = \partial/\partial\bar{y}_0$ with a parameter $s$ characterizing the group. The identity (42), as known from the theory of continuous transformation groups, denotes that the system of differential equations (6) is invariant with respect to the group (43), i.e., it is turned into the system (44) after the above change of variables, where

$$\frac{d\bar{x}_0}{dt} = \epsilon X_0(\bar{x}_0), \qquad \frac{d\bar{y}_0}{dt} = 1, \tag{44}$$

which coincides, within notations, with the initial system (6).

The invariance of the system (6) under the transformations (43), in the considered case, can be easily checked since the relations (43) in the finite form are given by formulae

$$\bar{x}_1 = \bar{x}_{10}, \ldots, \bar{x}_n = \bar{x}_{n0}, \qquad \bar{y} = \bar{y}_0 + s.$$

At the same time it is easily checked that in the general case the identity analogous to (42) does not hold for the operators $W, \tilde{W}$ of the perturbed system:

$$[W, \tilde{W}] = W\tilde{W} - \tilde{W}W \not\equiv 0,$$

Hence the system (5) is not invariant under the one-parameter group:

$$x_1 = \exp(sW(x_0, y_0))x_{10}, \ldots, x_n = \exp(sW(x_0, y_0))x_{n0}, \quad y = \exp(sW(x_0, y_0))y_0, \tag{45}$$

where $x_{10}, \ldots, x_{n0}, y_0$ are new variables,

$$W(x_0, y_0) = \frac{\partial}{\partial y_0},$$

and $s$ is a parameter characterizing a group which is generated by the operator $W$ associated with a system of zero approximation (36). Indeed, as is mentioned before, the relations(45) can be easily represented in the form

$$x_1 = x_{10}, \ldots, \quad x_n = x_{n0}, \quad y = y_0 + s \tag{46}$$

Under the transformation (45) the system (5) (exact initial system) turns into the system

$$\frac{dx}{dt} = \epsilon X(x_0, y_0 + s, \epsilon), \qquad \frac{dy_0}{dt} = 1$$

which does not coincide, within notations, with the initial system (5).

Now we have the possibility of giving the following group- theoretic interpretation of the averaging method: the averaging method transforms the system (5) (initial exact system) which is not invariant under the one-parameter transformation group generated by operator $W$, see (40), associated with a system of zero approximation (36), into the averaged system (6). The latter system is invariant under the one-parameter transformation group generated by the operator $U$, see (41), associated with a system of zero approximation (37).

This group-theoretic interpretation of the averaging method is taken as the basis of the asymptotic decomposition method that widens the sphere of application of the averaging method.

As an illustration of this approach consider the system of ordinary differential equations

$$\frac{dx}{dt} = \omega(x) + \epsilon\tilde{\omega}(x), \tag{47}$$

where

$$\omega(x)\,\mathrm{col}\,\|\omega_1(x),\ldots,\omega_n(x)\|, \tilde{\omega}(x) = \mathrm{col}\,\|\tilde{\omega}_1(x),\ldots,\tilde{\omega}_n(x)\|.$$

The differential operator associated with the perturbed system (47) will have the form

$$U_0 = U + \epsilon\tilde{U},$$

where

$$U = \omega_1\frac{\partial}{\partial x} + \cdots + \omega_n\frac{\partial}{\partial x_n}, \quad \tilde{U} = \tilde{\omega}_1\frac{\partial}{\partial x_1} + \cdots + \tilde{\omega}_n\frac{\partial}{\partial x_n}.$$

By a change of variables having the form

$$x = \phi(\bar{x}, \epsilon), \tag{48}$$

the system (47) is transformed to the system

$$\frac{d\bar{x}}{dt} = \omega(\bar{x}) + \sum_{\nu=1}^{\infty}\epsilon^{\nu}b^{(\nu)}(\bar{x}), \tag{49}$$

which we call the centralized system.

The operator for this system is $\bar{U}_0 = \bar{U} + \epsilon\bar{\bar{U}}$, where $\bar{U} = \omega_1(\bar{x})\frac{\partial}{\partial\bar{x}_1} + \cdots + \omega_n(\bar{x})\frac{\partial}{\partial\bar{x}_n}$,

$$\bar{\bar{U}} = \sum_{\nu=1}^{\infty}\epsilon^{\nu}N_{\nu}, \qquad N_{\nu} = b^{(\nu)}(\bar{x})\frac{\partial}{\partial\bar{x}_1} + \cdots + b^{(\nu)}(\bar{x})\frac{\partial}{\partial\bar{x}_n}.$$

The choice of the transformation (48) depends on the condition under which the centralized system (49) is invariant under the one-parameter transformations group,

$$\bar{x} = e^{s\bar{U}(\bar{x}_0)}\bar{x}_0, \tag{50}$$

where $\bar{x}_0$ is the vector of new variables.

Hence after the change of variables (50) the system (49) turns into the system

$$\frac{d\bar{x}_0}{dt} = \omega(x_0) + \sum_{\nu=1}^{\infty}\epsilon^{\nu}b^{(\nu)}(\bar{x}_0), \tag{51}$$

which coincides, to within notations, with the initial system. This means that the identity $[\bar{U}, N_{\nu}] \equiv 0$ holds for the operators $\bar{U}, N_{\nu}, (\nu = 1, 2\ldots)$. It should be noted that, for the application of the above presented scheme of asymptotic decomposition, the transformation (48) is taken in the form of a Lie series.

## REFERENCES


1. Arnold V.I., The small denominators and the problems on stability of motion in classical and celestial mechanics. *Successes of Math. Science.*, 1963, 18, 6, 91-92.
2. Bogoliubov N. N., Mitropolsky Yu. A., and Samoilenko A. M. The method of accelerated convergence in nonlinear mechanics. Naukova Dumka, Kiev, 1969, page 245.
3. Kolmogorov A.N. The general theory of dynamical systems and classical mechanics. The Intern. Mathem. Congress in Amsterdam. *Physmatgys. N.*, 1961, 187-208.
4. Krylov N. M., and Bogoliubov N. N. The application of methods of nonlinear mechanics in the theory of stationary oscillations. The Ukr. Academy of Sciences, Kiev, 1934.
5. Mitropolsky Yu.A. The averaging method in nonlinear mechanics. Naukova Dumka, Kiev, 1971, page 440.
6. Mitropolsky Yu.A., Lopatin A.K. Group-theoretic approach in the asymptotic methods of nonlinear mechanics. Naukova Dumka, Kiev, 1988, page 272.

# 15 Construction of Lyapunov Functions Using Integration by Parts

**Patrick C. Parks** Royal Military College of Science, Shrivenham, England

## 1. Historical introduction

I first met Larry Markus at the highly successful symposium on "Differential Equations and Dynamical Systems" held at the University of Puerto Rico, Mayaguez, Puerto Rico, 27-30 December 1965. This symposium was dedicated to Solomon Lefschetz for the many contributions to the field over some 20 years, and Professor Lefschetz himself gave the first presentation at the symposium. The proceedings appeared as a book [1], edited by J.K. Hale and J.P. La Salle. Subsequently I spent the summer of 1966 at the University of Minnesota in the Electrical Engineering Department, seeing a lot of Larry Markus and E. Bruce Lee at this time.

After returning from the USA to a new post at the University of Warwick I was to meet Larry Markus on many occasions in his capacity as Nuffield Visiting Professor at Warwick and as Director of the newly formed Warwick "Control Theory Centre". This Centre was founded to stimulate an interest in Control Theory and related topics amongst British applied mathematicians. A number of successful symposia were held at this time, for example, another symposium on "Differential Equations and Dynamical Systems" held in July 1969, producing also a book [2], edited by D. Chillingworth.

The present author contributed to both volumes [1] and [2], and the present topic is related to both of these earlier contributions. Lyapunov functions are a particularly timely topic as 1992 is the centenary of A.M. Lyapunov's famous doctoral thesis on "The General Problem of the Stability of Motion" (which has recently appeared for the first time in an English translation by A.T. Fuller [3]). Moreover a contributor to the topic of this paper [4] and [5] has been A.J. Pritchard, now the present Director of the Control Theory Centre at Warwick.

## 2. Construction of Lyapunov functions

Construction of Lyapunov functions for particular stability problems continues to be something of an art, although a number of techniques have been developed in the last 30 years, such as R.W. Brockett's scheme for linear differential equations involving positive real functions [6], and graphical techniques related to it such as the Popov criterion and the circle criterion [6, 7]. The present method using integration by parts was devised by the present author and A.J. Pritchard in the mid-1960s and was itself based on a paper by G. Peyser [8] and was thus named by us as the "P-method". It works well in a number of problems in the sense that rather "sharp" sufficient conditions for stability are obtained - in one case these sufficient conditions turn out also to be necessary. It is not obvious why the "P-method" is as good as it appears to be.

## 3. The "P-method" of constructing Lyapunov functions using integration by parts

The technique was originally devised for dynamical systems described by partial differential equations and will first be described in such terms. Suppose the dynamical system is described by a differential operator in space and time

$$Mz = 0 \tag{3.1}$$

where M is of order m in $\partial/\partial t$. Another operator N of order m-1 is $\partial/\partial t$ is generated from M by differentiation of N partially with respect to the symbol $\partial/\partial t$. The product of Mz and Nz is then integrated with respect over space $\Omega$ and time t to give

$$0 = \iint_G Mz\, Nz\, d\Omega\, dt = \int_\Omega q d\Omega + \iint_G Q d\Omega dt \tag{3.2}$$

where $G = [\Omega \times (0,t)]$. The Lyapunov function (or "functional") is taken as

$$V = \int_\Omega q\, d\Omega \quad \text{for which } dV/dt = - \int_\Omega Q d\Omega. \tag{3.3}$$

q and Q can almost be organised as positive - definite functionals in terms of an appropriate function space metric.

If $Mz = 0$ is an ordinary differential equation then M and N are operators in d/dt and (3.1) becomes

$$0 = \int^{t} Mz\ Nz\ dt = q + \int^{t} Q\ dt \qquad (3.4)$$

and here $V = q$ and $dV/dt = -Q$. (3.5)

Here q and Q can usually be organised as positive-definite quadratic forms.

## 4. First example of the technique: panel flutter

Consider the panel-flutter problem taken from [1] where $Mz = 0$ is given in non-dimensional form by

$$d\frac{\partial^4 z}{\partial x^4} + \mu\frac{\partial^2 z}{\partial t^2} - f\frac{\partial^2 z}{\partial x^2} + M_o\frac{\partial z}{\partial x} + \frac{\partial z}{\partial t} = 0 \qquad (4.1)$$

Hence Nz is given by

$$Nz \equiv 2\mu\frac{\partial z}{\partial t} + z \qquad (4.2)$$

and the boundary conditions are that z and $\partial^2 z/\partial x^2$ are zero at the panel chord extremities $x=0$ and $x=1$. d, $\mu$ and $M_o$ are positive numbers. The product Mz Nz contributes the following terms to V and dV/dt after integration by parts with respect to x and integration with respect to t have taken place (making use of the given boundary conditions):

$$V = \int_0^1 \mu^2 \left(\frac{\partial z}{\partial t}\right)^2 + \mu z \frac{\partial z}{\partial t} + \frac{1}{2} z^2 + \mu f \left(\frac{\partial z}{\partial x}\right)^2$$

$$+ \mu d \left(\frac{\partial^2 z}{\partial x^2}\right)^2 dx \tag{4.3}$$

$$dV/dt = -\int_0^1 \left( f \left(\frac{\partial z}{\partial x}\right)^2 + 2\mu M_o \frac{\partial z}{\partial x} \frac{\partial z}{\partial t} + \mu \left(\frac{\partial z}{\partial t}\right)^2 + d \left(\frac{\partial^2 z}{\partial x^2}\right)^2 \right) dx \tag{4.4}$$

where dV/dt is found by differentiating (4.3) partially with respect to t and then substituting for $\mu\partial^2 z/\partial t^2$ from (4.1). Some integration by parts and use of boundary conditions leads to (4.4).

Now considering a functional space norm ρ given by

$$\rho^2 = \int_0^1 \left( \left(\frac{\partial z}{\partial x}\right)^2 + \left(\frac{\partial z}{\partial t}\right)^2 \right) dx$$

asymptotic stability in terms of ρ may be proved (i.e. $\rho \to 0$ as $t \to \infty$) provided

$$f + \pi^2 d > 0 \text{ and}$$

$$M_o^2 < (f + \pi^2 d)/\mu \tag{4.5}$$

These conditions are sufficient (but not necessary) for stability. However, they are obtained *immediately* from the Lyapunov functional (4.3) whereas in [1] these conditions were obtained *only after an optimisation process* combining two candidate Lyapunov functions. The two conditions in (4.5) also have significant physical interpretations, and, for panels of infinite chord length, they approximate *necessary* conditions for stability. The condition $f > -\pi d^2$ states that the non-dimensional tension f must not be a compression greater than the buckling load $\pi d^2$ of the panel; and the condition $M_o^2 < (f + \pi^2 d)/\mu$ reduces to $M_o^2 < f/\mu$ for a panel of infinite chord length. Now $\surd(f/\mu)$ is the non-dimensional speed of flexural waves, along such a membrane-like panel and $M_o$ is the Mach number of the air flow, also along the panel. This second stability criterion, that the speed of the air flow must be less than the speed of flexural waves in a membrane, is "well known" and usually attributed to J.W. Miles [9].

## 5. A second example - the Hermite-Routh-Hurwitz problem

As a second example the technique is applied to an ordinary differential equation: in fact to the nth order linear differential equation with real constant coefficients given by

$$M(z) \equiv \frac{d^n z}{dt^n} + a_1 \frac{d^{n-1} z}{dt^{n-1}} + a_2 \frac{d^{n-2}}{dt^{n-2}} + \dots + a_{n-1} \frac{dz}{dt} + a_n z = 0. \tag{5.1}$$

This example is taken from the author's paper [13].

The classical stability condition in terms of the coefficients $a_i$ is usually known as the "Routh-Hurwitz criterion", derived from the work of E.J. Routh (1877) [10] and A. Hurwitz (1895) [11]. It is not so well known that C. Hermite had effectively solved this problem some years earlier in 1856 [12]. In addition, it was discovered by the present author that Hermite's stability criterion could be proved by a Lyapunov function argument [14] (but *not* by the argument which follows).

In the present context we form a new Lyapunov function using the "P-method" described in section 3. Here N(z) is given by

$$N(z) \equiv n \frac{d^{n-1}z}{dt^{n-1}} + (n-1)a_1 \frac{d^{n-2}z}{dt^{n-2}} + \ldots + 2a_{n-2} \frac{dz}{dt} + a_{n-1}z \tag{5.2}$$

The product M(z) N(z) may best be displayed in an array

| | $z$ | $\frac{dz}{dt}$ | $\frac{d^2z}{dt^2}$ | ... | $\frac{d^nz}{dt^n}$ |
|---|---|---|---|---|---|
| $z$ | $a_{n-1}a_n$ | $a^2_{n-1}$ | $a_{n-1}a_{n-2}$ | ... | $a_{n-1}$ |
| $\frac{dz}{dt}$ | $2a_{n-2}a_n$ | $2a_{n-2}a_{n-1}$ | $2a^2_{n-2}$ | ... | $2a_{n-2}$ |
| $\frac{d^2z}{dt^2}$ | $3a_{n-3}a_n$ | $3a_{n-3}a_{n-1}$ | $3a_{n-3}a_{n-2}$ | ... | $3a_{n-3}$ |
| | ... | ... | ... | ... | ... |
| . | ... | ... | ... | ... | $(n-1)a_1$ |
| $\frac{d^{n-1}z}{dt^{n-1}}$ | $na_n$ | $na_{n-1}$ | $na_{n-2}$ | $na_1$ | $n$ |

(5.3)

We now note that the terms on the alternate even diagonals may be integrated explicitly with respect to time either immediately or after integration by parts. On the alternative odd diagonals some integration by parts is possible but we obtain also terms such as

$$\int \left(\frac{dz}{dt}\right)^2 dt$$

which *cannot* be integrated explicitly. A systematic procedure of integration by parts on these odd diagonals leads to the array of terms still to be integrated as in (5.4) below.

| | $z$ | $\frac{dz}{dt}$ | $\frac{d^2z}{dt^2}$ | ... | $\frac{d^{n-1}z}{dt^{n-1}}$ |
|---|---|---|---|---|---|
| $z$ | $a_{n-1}a_n$ | 0 | $a_{n-3}a_n$ | | ... |
| $\frac{dz}{dt}$ | 0 | $a_{n-2}a_{n-1}$ $-a_{n-3}a_n$ | 0 | ... | |
| $\frac{d^2z}{dt^2}$ | $a_{n-3}a_n$ | 0 | | | |
| | | | ... | 0 | $a_3$ |
| | | ... | 0 | $a_1a_2-a_3$ | 0 |
| $\frac{d^{n-1}z}{dt}$ | | ... | $a_3$ | 0 | $a_1$ |

(5.4)

It will be noticed that this array is in fact the Hermite matrix, ***H***. In fact here

$$\frac{dV}{dt} = -Q = -\boldsymbol{x}^{\mathrm{T}} \boldsymbol{H} \boldsymbol{x} \tag{5.5}$$

$$\text{where } \boldsymbol{x} = \left( z, \frac{dz}{dt}, \frac{d^2z}{dt^2}, \ldots, \frac{d^{n-1}z}{dt^{n-1}} \right)$$

the phase-space vector corresponding to the linear equation (5.1). We have thus found a Lyapunov function V with a derivative given by (5.5). If we now apply conditions for dV/dt to be negative definite these are formally *sufficient* conditions for stability.

However, it is well known that the conditions for **H** to be positive-definite are *necessary and sufficient* stability conditions. So the "P-method" applied to this problem has found sufficient conditions which are "ideally sharp".

## 6. Conclusions

These two examples suggest that the "P-method" generates Lyapunov functions yielding sufficient stability conditions that are close in some sense to being necessary also. The precise reason for this apparent efficiency of the method is a topic worthy of a more detailed investigation.

## References

[1] Hale, J.K., La Salle J.P. (Eds), "Differential Equations and Dynamical Systems", Academic Press, New York, 1967.

[2] Chillingworth, D. (Ed), "Symposium on Differential Equations and Dynamical Systems, University of Warwick, 1968-1969", Springer Verlag, Berlin, 1971.

[3] Lyapunov, A.M., "The General Problem of the Stability of Motion" (Trans. A.T. Fuller), Int. J. Control 55, 531-773, 1992, also published as a book: Taylor & Francis, London, 1992.

[4] Parks, P.C., Pritchard, A.J., "Stability Analysis in Structural Dynamics using Lyapunov Functionals", J. Sound Vibration, 25, 609-621, 1972.

[5] Pritchard, A.J., "A Study of two of the Classical Problems of Hydrodynamic Stability by the Lyapunov Method", J. Inst. Math. Applic. 4, 78-93, 1968.

[6] Brockett, R.W., "Finite Dimensional Linear Systems", Wiley, New York, 1970.

[7] Willems, J.L., "Stability Theory of Dynamical Systems", Nelson, London, 1970.

[8] Peyser, G., "Energy Integrals for the Mixed Problem in Hyperbolic Partial Differential Equations of Higher Order", J. Math. and Mech. 6, 641-633, 1957.

[9] Miles, J.W., "On the Aerodynamic Stability of Thin Panels", J. Aerospace Sci., 23, 771-780, 1956.

[10] Routh, E.J., "A Treatise on the Stability of a Given State of Motion, Particularly Steady Motion", Adams Prize Essay, University of Cambridge, Macmillan, London, 1877, reprinted in A.T. Fuller (Ed), "Stability of Motion", Taylor and Francis, London, 1976.

[11] Hurwitz, A., "On the Conditions under which an Equation has only Roots with Negative Real Parts", Math. Ann. 46, 273-284, (1895).

[12] Hermite, C., "On the Number of Roots of an Algebraic Equation Contained Between Given Limits", (Transl. P.C. Parks), Int. J. Control, 26, 183-195, 1977, French original J. reine angew. Math. 52, 39-51, (1856).

[13] Parks, P.C., "A Lyapunov Function having the Stability Matrix of Hermite in its Time Derivative", Electronic Letters, 5, 608-609, 1969.

[14] Parks, P.C., "A new look at the Routh-Hurwitz problem using Lyapunov's Second Method", Bull. de l'Acad. Polon. des Sciences, Serie des Sci. Techn. 12, 19-21, 1964.

# 16 Remarks on Williams' Problem

**W. Parry** University of Warwick, Coventry, England

The purpose of these remarks is not, unfortunately, to solve Williams' problem but rather to clarify it. In [4], Williams introduced two equivalence relations for non-negative, integral, irreducible matrices. The first, *strong shift equivalence*, was shown to be a necessary and sufficient condition for the associated shifts of finite type to be topologically conjugate. The second, *shift equivalence*, is implied by strong shift equivalence and has the advantage of being much more tractable - indeed, Kim and Roush [1] have shown that the relation is decidable. However, no one has succeeded in proving (or disproving) that shift equivalence implies strong shift equivalence.

Our aim is to provide a solution of a much simpler problem. Given two topologically conjugate shifts of finite type and an associated strong shift equivalence, by taking products of the matrices defining this strong shift equivalence we arrive at a shift equivalence. One may, therefore, say that the shift equivalence is *resolvable*. Our problem is in the converse direction. How can we decide when a *given* shift equivalence has a resolution? To make this clear it may be useful to note that it is possible to have a shift equivalence between a matrix and itself which is *not* resolvable. However, there is clearly a topological conjugacy between the associated shift and itself, namely the identity map. We are concerned only with the problem of deciding when a shift equivalence arises from a strong shift equivalence.

## §1. Preliminaries

Two (square) 0-1 irreducible matrices P,Q are said to be *strong shift equivalent* if there exist (rectangular) non-negative integral matrices $U_1,\ldots U_\ell$, $V_1,\ldots V_\ell$ such that $P = U_1V_1$, $V_1U_1 = U_2V_2, \ldots V_\ell U_\ell = Q$. Williams [4] proved that P and Q are strong shift equivalent if and only if their associated shifts of finite type $\sigma_P, \sigma_Q$ are topologically conjugate. (For background material on shifts of finite type cf. [2], [3], [4].)

The matrices $P,Q$ are said to be *shift equivalent* (of lag $\ell$) if there exist non-negative integral $U,V$ such that $PU = UQ$, $VP = QV$, $UV = P^\ell$, $VU = Q^\ell$.

It is easy to see that strong shift equivalence implies shift equivalence. For $U,V$ one need only define

$$U = U_1 \dots U_\ell\,, \quad V = V_\ell V_{\ell-1} \dots V_1\,.$$

The (difficult) problem posed by Williams is to prove or disprove the converse.

If we are given a shift equivalence between $P$ and $Q$ which arises in the above way from a strong shift equivalence, we shall say that the shift equivalence is *resolvable*. how can we decide resolvability?

REMARK. It is not hard to see that resolvability amounts to the following:

There should be a specific one-one corresesopondence between the $P$ paths $x_0,\dots x_\ell$ and the $UV$ paths $x_0 \xrightarrow{U} y_0 \xrightarrow{V} x_\ell$ and another between the $Q$ paths $y_0 \dots y_\ell$ and the $VU$ paths $y_0 \xrightarrow{V} x_\ell \xrightarrow{U} y_\ell$ with the property that when we define $\varphi(x_0,\dots x_\ell) = y_0$ and $\psi(y_0 \dots y_\ell) = x_\ell$ then $Q(\varphi(x_0,x_1 \dots x_\ell),$ $\varphi(x_1,\dots x_{\ell+1})) = 1$ and $P(\psi(y_0,y_1 \dots y_\ell), \psi(y_1,\dots y_{\ell+1})) = 1$

and moreover

$$\psi(\varphi(x_0,\dots x_\ell), \varphi(x_1,\dots x_{\ell+1}),\dots \varphi(x_\ell,\dots x_{2\ell})) = x_\ell$$
$$\varphi(\psi(y_0,\dots y_\ell), \psi(y_1,\dots y_{\ell+1}),\dots \psi(y_\ell,\dots y_{2\ell})) = y_\ell\,.$$

We can then define

$$\varphi(x) = \{\varphi(x_n,\dots x_{n+\ell})\},\ \psi(y) = \{\psi(y_n,\dots y_{n+\ell})\}$$

so that $\varphi\sigma_P = \sigma_Q\varphi$, $\psi\sigma_Q = \sigma_P\psi$ and $\psi \circ \varphi = \sigma_Q^\ell$, $\varphi \circ \psi = \sigma_P^\ell$.

If $R$ is a non-negative integral matrix (with no trivial rows or columns) it can be written as a product

$$R = DA$$

where $D$ is a *division* (a 0–1 matrix with unit vector columns and no trivial rows) and $A$ is an *amalgamation* (the transpose of a division). Moreover this decomposition is essentially unique in that any other such decomposition takes the form

$$R = DPP^{-1}A$$

where $P$ is a permutation matrix. Furthermore, the product of two divisions is a division and the product of two amalgamations is an amalgamation.

If $P$ is a $k \times k$ 0–1 matrix defining the shift of finite type $\sigma_P$ (with states $1,2,\ldots k$) we can define a matrix $P_1$ with states $(i,j)$ such that $P(i,j) = 1$ by requiring $P_1((i,j), (j,k)) = 1$ and $P_1((i,j), (j',k)) = 0$ when $j \neq j'$. The shift $\sigma_{P_1}$ is essentially the same as the shift $\sigma_P$ – one concentrates on 'words' of length two rather than on single states. More generally we can define a matrix $P_n$ with states $(x_0,\ldots,x_n)$ such that $P(x_0,x_1) = P(x_1,x_2) = \cdots = P(x_{n-1},x_n) = 1$ (an allowable $n+1$ word) by requiring $P((x_0,\ldots x_n), (y_0,\ldots y_n)) = 1$ precisely when $x_1 = y_0, x_2 = y_1,\ldots x_n = y_{n-1}$. And again the shift $\sigma_{P_n}$ is essentially the shift $\sigma_P$. The process here described is sometimes referred to as 'going to a higher block system'.

Another way of defining $P_1,\ldots P_n$ is as follows: let $P = d_1a_1$ where $d_1,a_1$ are the division and amalgamation defined by transitions

$$i \xrightarrow{d_1} (i,j) \xrightarrow{a_1} j$$

and let $P_1 = a_1d_1 = d_2a_2$

where $d_2,a_2$ are the division and amalgamation

$$(i,j) \xrightarrow{d_2} (i,j,k) \xrightarrow{a_2} (j,k)$$

and so on. Thus we have a commutative diagram (of graphs):

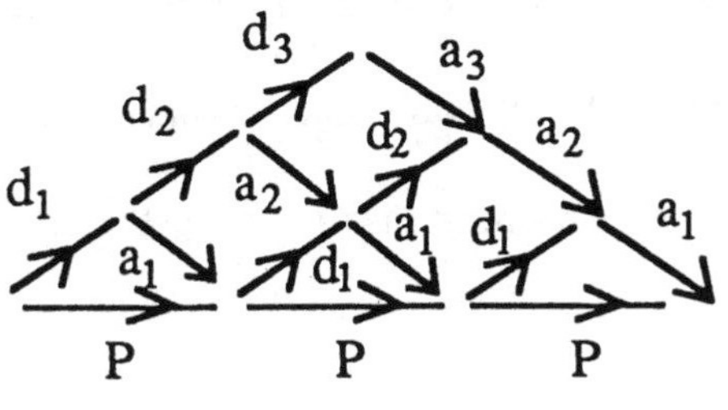


(Here $P_1 = a_1d_1 = d_2a_2$, $P_2 = a_2d_2 = d_3a_3$.)

Finally, we shall need a slightly stronger form of strong shift equivalence which can be derived from any given strong shift equivalence by manipulations of matrices using the DA decomposition observation mentioned above (cf. [2]).

LEMMA 1. Any strong shift equivalence can be displayed in the form of commutative diagrams:

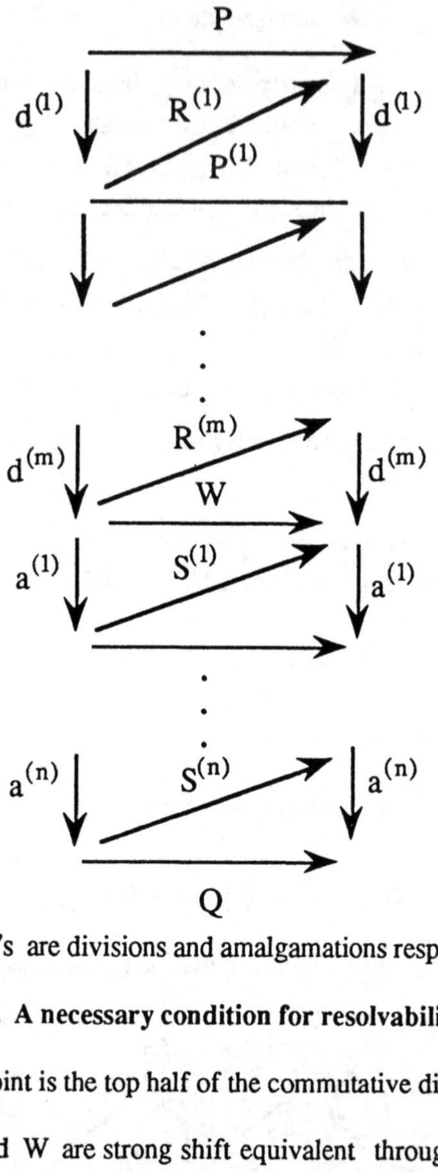


where the d's and a's are divisions and amalgamations respectively.

## §2. A necessary condition for resolvability

Our starting point is the top half of the commutative diagram in Lemma 1.

LEMMA 2. If $P$ and $W$ are strong shift equivalent through a sequence of $m$ divisions

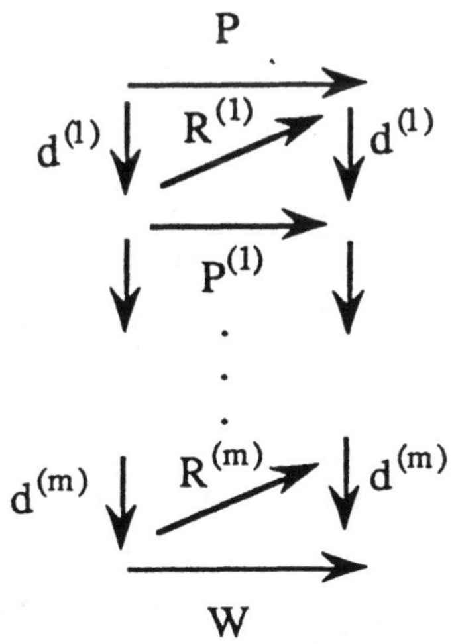


then we have a commutative diagram

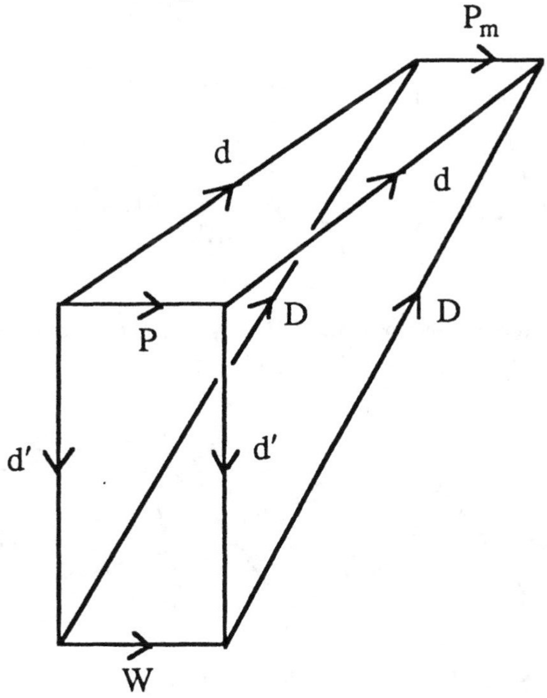


where $d' = d^{(1)} \ldots d^{(m)}$ $d = d_1 \ldots d_m$ and D is a division. (Here all rectangles and triangles are commutative.)

PROOF. We shall prove the lemma for the case $m = 2$. The general case is similar. Suppose we have the commutative diagram

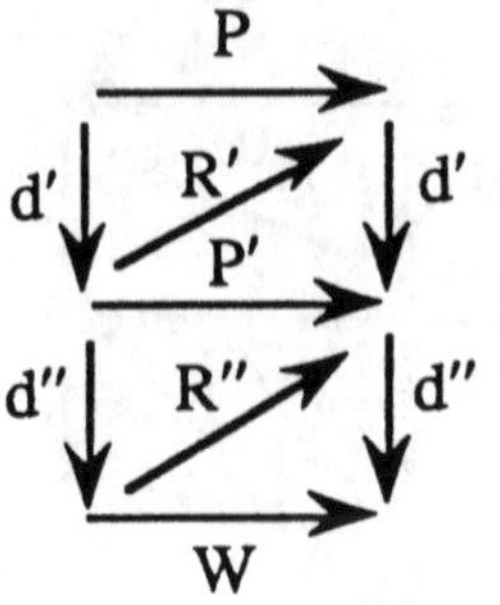


We first use d–a decompositions for the top commutative diagram to get

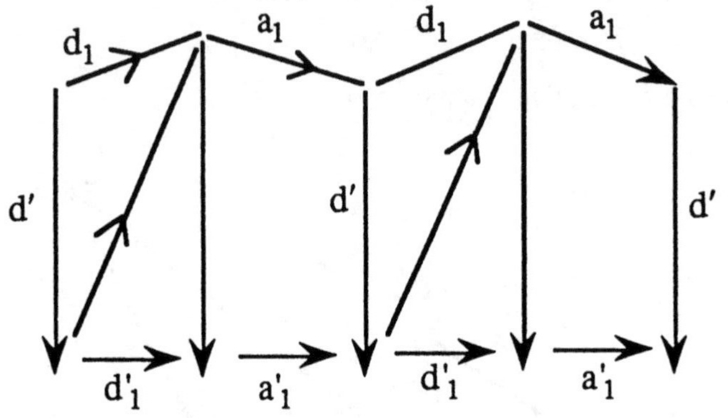


where the unmarked arrows are divisions. We can do the same with the next commutative diagram so that together we obtain

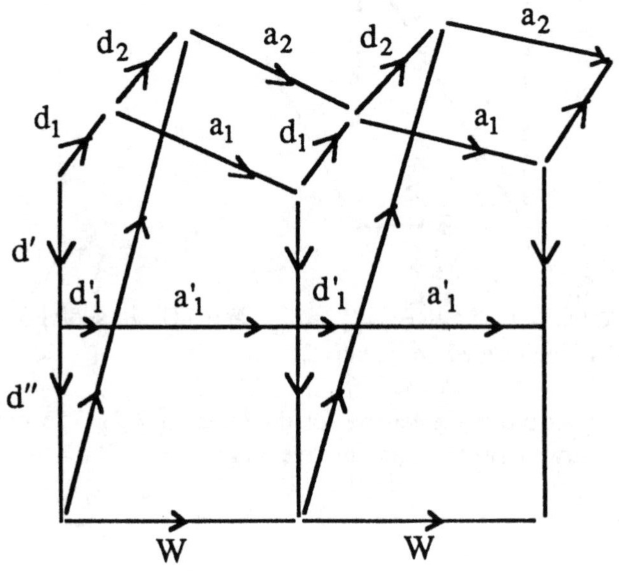

and again the upward pointing unmarked arrows compose to make a division D so that

$$d'd''D = d_1d_2 \text{ and } DP_2 = WD.$$

REMARK. With the obvious modifications a similar result holds for the lower half of the diagram in lemma 1 so that we obtain the following:

COROLLARY. Given the commutative diagram of Lemma 1 we have commutative diagrams:

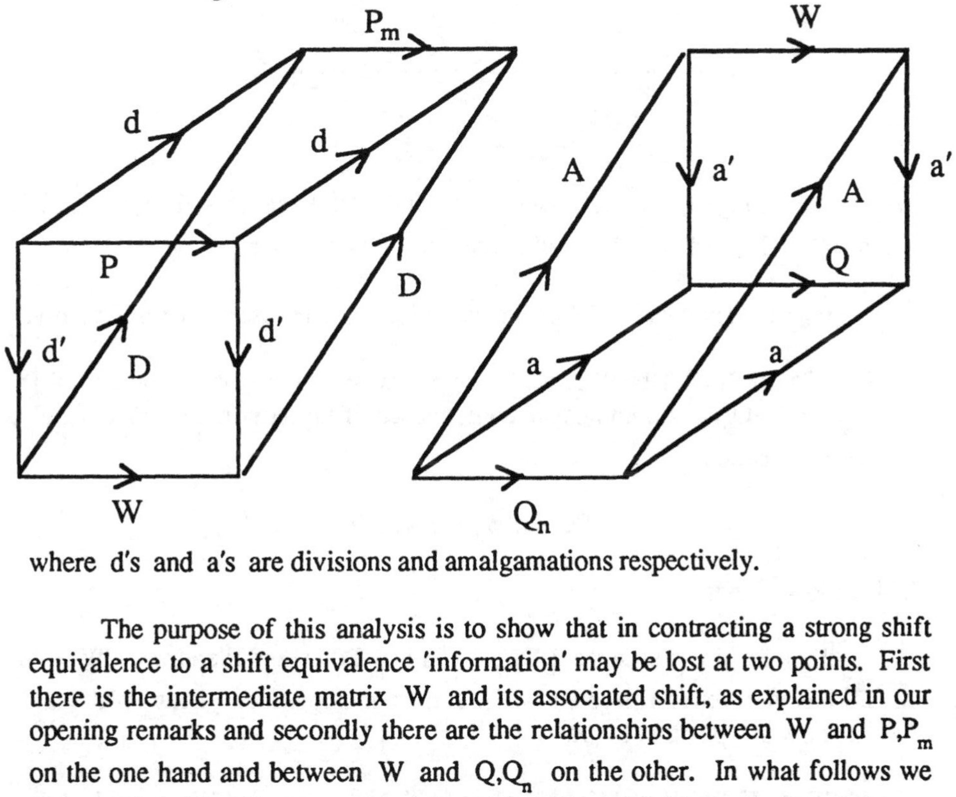


where d's and a's are divisions and amalgamations respectively.

The purpose of this analysis is to show that in contracting a strong shift equivalence to a shift equivalence 'information' may be lost at two points. First there is the intermediate matrix W and its associated shift, as explained in our opening remarks and secondly there are the relationships between W and $P, P_m$ on the one hand and between W and $Q, Q_n$ on the other. In what follows we shall retain these relationships and show that *with them* shift equivalence implies strong shift equivalence. Put another way, these relationships guarantee that a shift equivalence is resolvable.

## §3. Sufficiency of the condition

For simplicity of presentation we suppose we have a shift equivalence between P and Q of lag 2:

$$PU = UQ, VP = QV, UV = P^2, VU = Q^2.$$

by decomposing $U = d'a'$ and then decomposing $a'V$ and $Vd'$ we arrive at a commutative diagram:

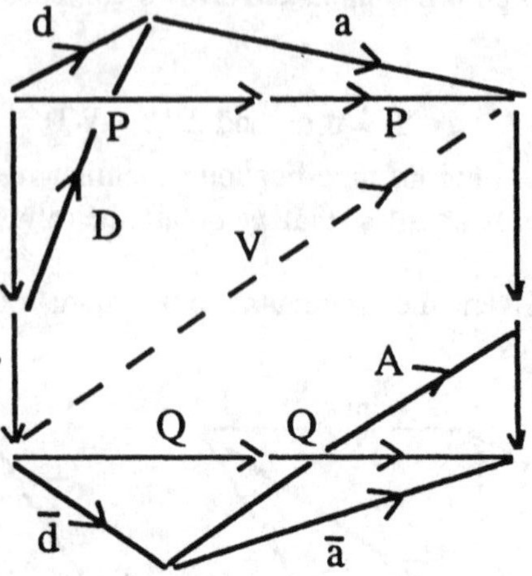


with the usual significance attached to d's and a's. Here $P^2 = da$, $Q^2 = \bar{d}\bar{a}$, $U = d'a'$, $a'V = Da$, $Vd' = \bar{d}A$. Furthermore $d = d'D$, $\bar{a} = Aa'$.

In §2 we saw that the following condition was necessary for resolvability:

RESOLVABILITY CONDITION. There exists $W$ solving the equations $WD = DP_2$, $AW = Q_2A$. An immediate consequence of this condition is that $W$ also solves the equations

$$Pd' = d'W, \; Wa' = a'Q.$$

This follows from:

$dP_2 = Pd$, $d = d'D$ since $d'DP_2 = d'WD = Pd'D$ and therefore $d'W = Pd'$ as $D$ has unitvectors for columns. A similar manipulation yields $Wa' = a'Q$. We are now in a position to prove

THEOREM. If $P,Q$ are shift equivalent (with lag 2, say) and if the resolvability condition is satisfied then $P,Q$ are strong shift equivalent, i.e. the shift equivalence is resolvable.

PROOF. From our assumptions we have the commutative diagram:

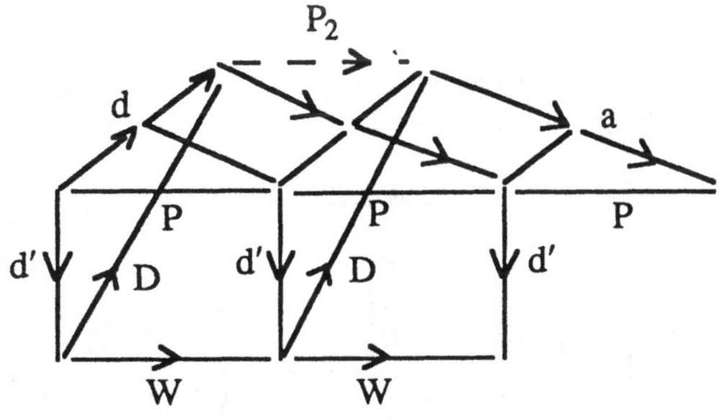


We let $\sigma$ with an appropriate suffix denote the indicated shift (of finite type) and let $\varphi$ with an appropriate suffix (of a division matrix) denote a one block map, e.g. since d' is a division the map $\varphi_{d'}(y_0) = x_0$ when $d'(x_0,y_0) = 1$ is well defined. Moreover $\varphi_d = \varphi_{d'} \circ \varphi_D$ and $\varphi_D \circ \sigma_{P_2} = \sigma_W \circ \varphi_D$ and $\varphi_{d'} \circ \sigma_W = \sigma_P \circ \varphi_{d'}$. Since $\sigma_{P_2}, \sigma_P$ are obviously topologically conjugate ($\varphi_d(x_0x_1x_2) = x_0$) we see that $\sigma_W$ is also topologically conjugate to $\sigma_P$. Another way of seeing this is to note that if $\alpha$ is the partition of the space $X_P$ (on which $\sigma_P$ acts) into states then the corresponding partition associated with $\sigma_{P_2}$ is just $\alpha \vee \sigma_P^{-1}\alpha \vee \sigma_P^{-2}\alpha = \alpha^2$ and the partition $\beta$ associated with $\sigma_W$ satisfies $\alpha \leq \beta \leq \alpha^2$.

We see then that $\sigma_W$ is topologically conjugate to $\sigma_P$ and a similar argument shows that $\sigma_W$ is topologically conjugate to $\sigma_Q$ and this completes the proof.

REMARK. We can see the topological conjugacy given by the theorem in either of two ways. The first way is to view it from the point of view of the intermediate shift $\sigma_W$ and the one block (homeomorphisms) $\varphi_{d'}$, $\psi_{a'}$ given by $\varphi_{d'}(y_0) = x_0$ when $d'(x_0,y_0) = 1$ and $\psi_{a'}(y_0) = z_0$ when $a'(y_0,z_0) = 1$. Alternatively we can construct 3 block maps as follows: define $\varphi_D(x_0,x_1,x_2) = y_0$ when $D(y_0(x_0,x_1,x_2)) = 1$ and compose with $\psi_{a'}$ to obtain the 3 block map $f(x_0,x_1,x_2) = \psi_{a'}(\varphi_D(x_0,x_1,x_2))$. This provides a homomorphism f from $\sigma_P$ to $\sigma_Q$. Similarly we can define a homomorphism g from $\sigma_Q$ to $\sigma_P$ (given by a 3 block map) and it is easy to see that $g \circ f = \sigma_P^2$, $f \circ g = \sigma_Q^2$. Put another way we have 3 block maps with the following property:

If $f(x_0,x_1,x_2) = z_0$, $f(x_1,x_2,x_3) = z_1$ then $Q(z_0,z_1) = 1$ and if $g(z_0,z_1,z_2) = x_0$, $g(x_1,x_2,x_3) = x_1$ then $P(x_0,x_1 = 1$ furthermore $g(f(x_0,x_1,x_2), f(x_1,x_2,x_3), f(x_2,x_3,x_4)) = x_2$ and $f(g(z_0,z_1,z_2), g(z_1,z_2,z_3), g(z_2,z_3,z_4)) = z_2$.

EXAMPLES.

We conclude with two examples of shift equivalences (of lag 2) which are not resolvable. Presenting resolvable shift equivalences is a straight forward matter.

1. This was first brought to my attention by B. Marcus.

$$\text{Let } P = \begin{pmatrix} 1 & 1 \\ 1 & 1 \end{pmatrix}, Q = \begin{pmatrix} 1 & 1 & 0 & 0 \\ 0 & 0 & 1 & 1 \\ 1 & 1 & 0 & 0 \\ 0 & 0 & 1 & 1 \end{pmatrix}, U = d' = \begin{pmatrix} 1 & 0 & 0 & 1 \\ 0 & 1 & 1 & 0 \end{pmatrix},$$

$$V = \begin{pmatrix} 11 \\ 11 \\ 11 \\ 11 \end{pmatrix} \text{ so that } PU = UQ = \begin{pmatrix} 1 & 1 & 1 & 1 \\ 1 & 1 & 1 & 1 \end{pmatrix}$$

$$VP = QV = \begin{pmatrix} 2\,2 \\ 2\,2 \\ 2\,2 \\ 2\,2 \end{pmatrix}, UV = \begin{pmatrix} 2 & 2 \\ 2 & 2 \end{pmatrix} = P^2, VU = Q^2 = \begin{pmatrix} 1 & 1 & 1 & 1 \\ 1 & 1 & 1 & 1 \\ 1 & 1 & 1 & 1 \\ 1 & 1 & 1 & 1 \end{pmatrix}.$$

It is clear that d' does not implement a topological conjugacy by an examination of the fixed points of $\sigma_Q$ which are *both* mapped to a fixed point of $\sigma_P$ . This can also be seen as follows:

$$P_2 = \begin{pmatrix} 1 & 1 & 1 & 1 & 0 & 0 & 0 & 0 \\ 0 & 0 & 0 & 0 & 1 & 1 & 1 & 1 \end{pmatrix} \begin{pmatrix} 1 & 0 \\ 0 & 1 \\ 1 & 0 \\ 0 & 1 \\ 1 & 0 \\ 0 & 1 \\ 1 & 0 \\ 0 & 1 \end{pmatrix} = d_1 d_2 a_2 a_1 = da .$$

$$V = \begin{pmatrix} 1 & 1 & 0 & 0 & 0 & 0 & 0 & 0 \\ 0 & 0 & 0 & 0 & 1 & 1 & 0 & 0 \\ 0 & 0 & 0 & 0 & 0 & 0 & 1 & 1 \\ 0 & 0 & 1 & 1 & 0 & 0 & 0 & 0 \end{pmatrix} \begin{pmatrix} 1 & 0 \\ 0 & 1 \\ 1 & 0 \\ 0 & 1 \\ 1 & 0 \\ 0 & 1 \\ 1 & 0 \\ 0 & 1 \end{pmatrix} = Da$$

and $\begin{pmatrix} 1 & 1 & 1 & 1 & 0 & 0 & 0 & 0 \\ 0 & 0 & 0 & 0 & 1 & 1 & 1 & 1 \end{pmatrix} =$

$$\begin{pmatrix} 1 & 0 & 0 & 1 \\ 0 & 1 & 1 & 0 \end{pmatrix} \begin{pmatrix} 1 & 1 & 0 & 0 & 0 & 0 & 0 & 0 \\ 0 & 0 & 0 & 0 & 1 & 1 & 0 & 0 \\ 0 & 0 & 0 & 0 & 0 & 0 & 1 & 1 \\ 0 & 0 & 1 & 1 & 0 & 0 & 0 & 0 \end{pmatrix}$$

i.e. $d = d'D$.

However it is *not* true that $DP_2 = QD$ nor can we achieve this equation by 'adjusting' with permissible permutations. Hence the shift equivalence is not resolvable. If $d'$ were replaced by $\begin{pmatrix} 1 & 0 & 1 & 0 \\ 0 & 1 & 0 & 1 \end{pmatrix}$ the resulting shift equivalence would be resolvable.

2. This is an example where $P = Q$ so that obviously $\sigma_P, \sigma_Q$ are topologically conjugate.

However the shift equivalence we present is not resolvable.

$$\text{Let } P = Q = \begin{pmatrix} 1 & 0 & 0 & 1 \\ 0 & 1 & 1 & 0 \\ 1 & 0 & 0 & 1 \\ 0 & 1 & 1 & 0 \end{pmatrix} = U$$

and let $V = \begin{pmatrix} 1 & 1 & 0 & 0 \\ 0 & 0 & 1 & 1 \\ 1 & 1 & 0 & 0 \\ 0 & 0 & 1 & 1 \end{pmatrix}$

then $PU = UP = \begin{pmatrix} 1 & 1 & 1 & 1 \\ 1 & 1 & 1 & 1 \\ 1 & 1 & 1 & 1 \\ 1 & 1 & 1 & 1 \end{pmatrix} = VP = PV = P^2$.

The allowable P 3-blocks are

$$\begin{array}{llll} 1\,1\,1 \to 1, & 2\,2\,2 \to 3, & 3\,1\,1 \to 1, & 4\,2\,2 \to 3 \\ 1\,1\,4 \to 4, & 2\,2\,3 \to 2, & 3\,1\,4 \to 4, & 4\,2\,3 \to 2 \\ 1\,4\,2 \to 1, & 2\,3\,1 \to 3, & 3\,4\,2 \to 1, & 4\,3\,1 \to 3 \\ 1\,4\,3 \to 4, & 2\,3\,4 \to 2, & 3\,4\,3 \to 4, & 4\,3\,4 \to 2 \end{array}$$

where we have indicated the f values in accordance with the table of UV blocks:

$$\begin{array}{llll} 1\,1\,1, & 2\,2\,3, & 3\,1\,1, & 4\,2\,3 \\ 1\,1\,2, & 2\,2\,4, & 3\,1\,2, & 4\,2\,4 \\ 1\,4\,3, & 2\,3\,1, & 3\,4\,3, & 4\,3\,1 \\ 1\,4\,4, & 2\,3\,2, & 3\,4\,4, & 4\,3\,2. \end{array}$$

However, note that $f(114) = 4$ and $f(142) = 1$ and $P_2((114), (142)) = 1$. Resolvability would require that $P(4,1) = 1$, which is not the case.

References

1. K.H. Kim and F.W. Roush, *Decidability of shift equivalence*, Springer Lecture Notes 1342, Springer, Berlin, 1988, 374-424.

2. W. Parry. Notes on coding problems for finite state processes, Bull. London. Math. Soc., **23** (1991), 1-33.

3. W. Parry and S. Tuncel. *Classification problems in ergodic theory*, London Math. Soc. lecture note series 67, Cambridge, Cambridge University Press, 1982.

4. R.F. Williams. Classification of sub-shifts of finite type, Ann. of Math. **98** (1973), 120-153; Errata, Ann. of Math. **99** (1974), 380-381.

Mathematics Institute, University of Warwick, Coventry CV4 7AL, U.K.

# 17 Approximations of the Long-Time Dynamics of the Navier-Stokes Equations

**Victor A. Pliss** University of St. Petersburg, St. Petersburg, Russia

**George R. Sell** University of Minnesota, Minneapolis, Minnesota

**Abstract.** In this paper we present a rigorous foundation for comparing the long-time dynamics of two Navier-Stokes problems: the first is a 3D problem on a thin domain, and the second is the reduced 2D problem one obtains when the aspect ratio $\epsilon$ goes to 0. For this theory we use recent results on the existence of global attractors and inertial forms for both the 2D and 3D problems. A rigorous comparison of the dynamics residing in certain hyperbolic attractors is made possible by using the new theory of Approximation Dynamics.

**Key words.** approximation dynamics, attractors, hyperbolic sets, inertial forms, inertial manifolds, Navier-Stokes equations

**1. Introduction.** About 30 years ago, researchers began to discover salient finite dimensional structures in the dynamical theory of infinite dimensional systems arising in theory of nonlinear partial differential equations. These finite dimensional structures include the phenomena of a finite number of determining modes (see Foias and Prodi (1967)), the existence of (compact) global attractors (see Billotti and LaSalle (1971), and Ladyzhenskaya (1972)), the finite dimensionality of these attractors (see Mallet-Paret (1976), Mañé (1981), and Foias and Temam (1979)), as well as the theory of inertial manifolds and inertial forms (see Foias, Sell, and Temam (1988), Hale (1988), Henry (1981), Kwak (1991, 1992), Mallet-Paret and Sell (1988), Mora (1983), and Temam (1988)).[1]

In this paper our interest centers around the long-time dynamics of the Navier-Stokes equations, in two and three dimensions. Recall that these equations form the basic model for an incompressible, viscous fluid flow. We assume that this fluid flow occurs in an open, bounded region $\Omega$ in $R^2$ or $R^3$.

The theory of the approximations of the Navier-Stokes equations we present here is motivated by the the desire to find meaningful methods for the study of the long-time dynamics of these equations. The issues one encounters in the approximation of the long-time dynamics are quite different from those arising in the classical approximation theories. As

This research was supported in part by a grant from the Army Research Office. Both authors express appreciation to the Institute for Mathematics and its Applications and the AHPCRC, in Minneapolis, and the Faculty of Mathematics and Mechanics, in St. Petersburg, for their help in sponsoring this project.

[1] Additional references are available in the AHPCRC Preprint on References in Dynamical Systems, by G. R. Sell.

is well-known, the latter theories usually lead to useful information about the solutions, but on finite time intervals only. It turns out that the recent theory of Approximation Dynamics plays a central role in the analysis presented below. In particular, by putting together two recent developments in Navier-Stokes dynamics (see Raugel and Sell (1992a, 1992b, 1992c), Kwak (1991), and Kwak, Sell, and Shao (1993)) with the theory of hyperbolic attractors presented in Pliss and Sell (1991), we are able to present the foundations for an overall theory of approximations of the long-time dynamics of these equations.

We are especially interested in the connections between the 2 dimensional Navier-Stokes equations (2DNS) and the 3 dimensional Navier-Stokes equations (3DNS) on thin 3D domains. From a physical point of view, one may expect that the long-time dynamics of the full 3DNS on thin domains is well-approximated by the dynamics of some imbedded 2DNS. One of our goals in this paper is to present a rigorous mathematical foundation underlying this physical intuition.

It turns out that the 3DNS on a thin domain is a perturbation of a suitable 2D problem. However this perturbation, which we will describe in Section 3, is a singular perturbation of the 2D problem. The main objective of Section 3 is to describe the recent theory of the 3DNS on thin domains, see Raugel and Sell (1992a, 1992b, 1992c). The first step in this theory is to show that, under suitable boundary behavior, the singular perturbation can be ***regularized***. In this case, the long-time dynamics of the limiting system, which is referred to as the ***reduced*** 3DNS, is precisely the dynamics of the 2DNS. More importantly, we will see that on the thin domain, the full 3DNS has a global attractor $\mathfrak{A}_\epsilon$, when the aspect ratio $\epsilon$ is small, and that, under reasonable assumptions, this attractor is upper semicontinuous at $\epsilon = 0$. Thus the long-time dynamics of the 3DNS, which is entirely contained in $\mathfrak{A}_\epsilon$, lies in a small neighborhood of the global attractor $\mathfrak{A}_0$ of the 2DNS.

The existence of this global attractor for the 3DNS on thin domains is the first of the two recent developments in the Navier-Stokes dynamics mentioned above. The second development is the Kwak theory of inertial forms for the Navier-Stokes equations. The main objective of the latter theory, which is presented in Section 4, is to show that, under suitable conditions, the long-time dynamics of the Navier-Stokes equations is completely described by the dynamics of an inertial form, i.e., a finite dimensional system of ordinary differential equations, see Kwak (1991). It turns out that the theory of Kwak is applicable both to the 2DNS, with periodic boundary conditions, and to the 3DNS on thin domains, also with periodic boundary conditions. Of particular interest here is the behavior of the associated inertial forms, as $\epsilon \to 0$, see Kwak, Sell, and Shao (1993).

This bring us to the main topic of this paper, *viz.* the Approximation Dynamics of the Navier-Stokes equations. The need to find new theories and new algorithms for the study of the long-time dynamical behavior of very high dimensional systems of differential equations is the basic goal of the relatively new theory of Approximation Dynamics. What underlies this new theory is the principle that

***a good approximate solution of a given system of differential equations can be represented as an actual solution of a small perturbation of the given system of equations.*** For example, in Sell (1992) it is shown that, under reasonable conditions, every approximate inertial manifold for the Navier-Stokes equations is an actual inertial manifold for a small perturbation of these equations. This fact, which is in principle valid for other systems of differential equations, offers a convenient framework for the study of the long-time dynamics of these equations.

The basic issue one faces in Approximation Dynamics is how well one can approximate the long-time dynamics of the solutions of a given differential equation

$$u' = F(u) \tag{1.1}$$

by those of an approximate equation

$$v' = F(v) + E(v), \tag{1.2}$$

where $E$ is small in a suitable norm, see Pliss and Sell (1991). Approximation problems of this form are ubiquitous. For example, in the numerical study of turbulence in fluid flows, equation (1.1) might represent a model for a calculation involving both low modes and high modes, and the error term $E(v)$ in equation (1.2) arises when one restricts the calculation to the low modes only. Another example is a multigrid calculation, where equation (1.1) represents the fine grid model and the error term $E(v)$ in equation (1.2) arises when one restricts to the coarse grid.

The principal dynamical objects of interest in this paper are compact, invariant sets, and especially, compact attractors, with suitable hyperbolic structures. In Section 5 we shall give precise definitions of a **weakly hyperbolic** and a **weakly, normally hyperbolic** invariant set. As explained below, these concepts generalize and include the notion of an Anosov flow, see Markus (1961, 1971), Smale (1967), and Arnold (1983), as well as a normally hyperbolic manifold, see Sacker (1969), Fenichel (1971), Hirsch, Pugh and Shub (1977), and Pliss (1977).

We assume that both (1.1) and (1.2) are given on an open set $O$ in the Euclidean space $R^n$ and that both $F$ and $E$ are $C^1$-functions from $O$ to $R^n$. Let $\|\cdot\|$ denote the Euclidean norm on $R^n$. In addition we shall assume that

$$\|E\|_{C^1} \overset{\text{def}}{=} \sup_{u\in O} \max\left(\|E(u)\|, \|DE(u)\|_{\text{op}}\right) .$$

is finite, where $DE$ is the spatial derivative of $E$, and for any bounded linear operator $L$, the operator norm $\|L\|$ is defined by $\|L\| = \|L\|_{op} \overset{\text{def}}{=} \sup\{\|Lv\| : \|v\| \leq 1\}$. The following two theorems, which address the issue of the robustness, or stability, of certain hyperbolic attractors, are proved in Pliss and Sell (1991).

THEOREM A. *Let $\mathcal{K}$ be a given weakly, normally hyperbolic attractor for (1.1) that satisfies the Lipschitz property. Then for every $\epsilon > 0$ there exist $\delta_i$, $i = 1, 2, 3$, such that $0 < \delta_3 \leq \delta_2 \leq \delta_1$ and the following three properties hold:*

(1) *If $\|E\|_{C^1} \leq \delta_1$, then there is a continuous mapping $h : \mathcal{K} \to R^n$ such that the image $\mathcal{K}^E \stackrel{\text{def}}{=} h(\mathcal{K})$ is a compact, invariant set for (1.2) and $\|h(x) - x\| \leq \epsilon$ for all $x \in \mathcal{K}$.*

(2) *If, in addition, $\|E\|_{C^1} \leq \delta_2$, then $\mathcal{K}^E$ is a weakly, normally hyperbolic attractor for (1.2).*

(3) *Moreover, if $\|E\|_{C^1} \leq \delta_3$, then the mapping $h$ is a homeomorphism.*

Furthermore, the associated dynamics $S_2(t)$ on $\mathcal{K}^E$ is essentially a faithful representation of the unperturbed dynamics $S_1(t)$ on $\mathcal{K}$, in a sense we now make precise. Let $\mathcal{K}$ be a compact, invariant set for (1.1). A continuous mapping $h : \mathcal{K} \to R^n$ is said to be a **homomorphism** if there is an induced flow $S_1^E(t)$ on $\mathcal{K}$ such that

$$S_2(t)h(u_0) = h(S_1^E(t)u_0), \qquad \text{for all } u_0 \in \mathcal{K}, t \geq 0. \tag{1.3}$$

The flow $S_1^E(t)$ is induced by the perturbation $E$ in (1.2). We refer to $S_1^E(t)$ as the **shadow flow** on $\mathcal{K}$. We now have the following result:

THEOREM B. *Let the hypotheses and conclusions of Theorem 1 be satisfied. Then for every $E$ with $\|E\|_{C^1} \leq \delta_1$, there is a shadow flow[2] $S_1^E(t)$ on $\mathcal{K}$ satisfying (1.3). In addition, for any $\mu > 0$ and $T > 0$, there is a $\delta(\mu, T) > 0$ such that if $\|E\|_{C^1} \leq \delta(\mu, T)$, then one has*

$$\begin{cases} \|S_1^E(t)u_0 - S_1(t)u_0\| \leq \mu, & \text{for all } u_0 \in \mathcal{K},\ 0 \leq t \leq T, \\ \|DS_1^E(t) - DS_1(t)\|_{op} \leq \mu, & \text{for all } 0 \leq t \leq T, \end{cases} \tag{1.4}$$

*where $D$ denotes the derivative with respect to $u_0$.*

Inequality (1.4) is simply a statement of the continuity of the mapping $E \to S_1^E(t)$. This is a classical result, and it is of limited interest for the study of the long-time dynamics of (1.1) and (1.2). The commutivity relationship (1.3) contains more dynamical information. For example, if $\mathcal{K}$ is a stable periodic orbit, then its homeomorphic image $\mathcal{K}^E$ is also a cycle, and the perturbed flow $S_1^E(t)$ differs from the unperturbed dynamics $S_1(t)$ by a change in the speed along the cycle $\mathcal{K}$.

The main objective of this paper is to show that Theorems A and B can be applied to the infinite dimensional dynamics generated by the 2DNS and by the 3DNS on thin domains. The facts that both of these problems have global attractors and inertial forms lead to a rigorous reduction of the infinite dimensional problem to a finite system of ordinary differential equations. We begin with a brief review of the classical theory of the Navier-Stokes equations in both 2D and 3D.

---

[2]The shadow flow $S_1^E(t)$ is described in more detail in Section 5.

**2. Classical Theory of the Navier-Stokes Equations.** The Navier-Stokes equations on an open, bounded region $\Omega \subset R^d$, $d = 2, 3$, are given by

$$\begin{aligned} U_t - \nu \Delta U + (U \cdot \nabla) U + \nabla P &= F, \\ \nabla \cdot U &= 0, \end{aligned} \tag{2.1}$$

with various boundary conditions, where $\nabla$ is the gradient operator, $\Delta$ is the Laplacian, and $U_t = \frac{d}{dt} U$ is the time derivative. In this paper we examine the case where $\Omega = Q_2$ is a 2 dimensional domain and where $\Omega = \Omega_\epsilon$ is a thin 3-dimensional domain with

$$Q_2 = (0, \ell_1) \times (0, \ell_2), \qquad \Omega_\epsilon = Q_2 \times (0, \epsilon),$$

$\ell_1$ and $\ell_2$ are positive, and $\epsilon$ is a small positive parameter. In order to simplify the treatment in this paper, we will consider equation (2.1) with periodic boundary conditions[3]

$$U(y + \ell_i e_i, t) = U(y, t), i = 1, 2, \qquad U(y + \epsilon e_3, t) = U(y, t), \tag{2.2}$$

where $\{e_1, e_2, e_3\}$ is the natural basis in $R^3$, and

$$\int_{\Omega_\epsilon} U \, dy = 0. \tag{2.3}$$

We require, of course, that the initial data $U_0$ and the forcing function $F$ satisfy

$$\int_{\Omega_\epsilon} U_0 \, dy = \int_{\Omega_\epsilon} F \, dy = 0, \tag{2.4}$$

see Temam (1983). If (2.4) holds, then any solution $U(t)$ of (2.1) will satisfy (2.3) for all $t \geq 0$.

As is customary, see Constantin and Foias (1988) or Temam (1983), we let $\mathbf{V}$ denote the divergence-free vector fields $U$ in $H^1(\Omega) = H^1(\Omega, R^d)$ that satisfy (2.2) and (2.3), and we let $\mathbf{H}$ denote the closure of $\mathbf{V}$ in $L^2(\Omega) = L^2(\Omega, R^d)$, where $\Omega = Q_2$, or $\Omega_\epsilon$. Also we let $\mathbf{P}_d$ denote the orthogonal projection of $L^2(\Omega)$ onto $\mathbf{H}$. Then the Navier-Stokes equations (2.1) on $\Omega$ can be written in the abstract form

$$U_t + \nu A U + B(U, U) = \mathbf{P}_d F, \tag{2.5}$$

where $AU = -\mathbf{P}_d \Delta U$, and $B(U, V) = \mathbf{P}_d (U \cdot \nabla) V$. We will be interested in solutions of (2.5) under the assumption that the initial data $U_0$ satisfy

$$U_0 \in D(A^{\frac{1}{2}}) = \mathbf{V},$$

[3]To some extent, generalizations of the theory described herein to other boundary conditions are possible, see Raugel and Sell (1992a) for details.

where $D(L)$ denotes the domain of the linear operator $L$. Recall that the two norms $\|U\|_{\mathbf{V}}$ and $\|A^{\frac{1}{2}}U\|_{L^2(\Omega)}$ are equivalent on $\mathbf{V}$.

While the theory we describe here remains valid, to a great extent, when the forcing function $F = F(t)$ satisfies $F(\cdot) \in L^\infty(0,\infty;L^2(\Omega))$, we will restrict to the case where $F \in L^2(\Omega)$ is independent of time $t$. See Raugel and Sell (1992a) for more details.

Let $U_0 \in V$. A **regular**, or **strong**, solution of (2.5) on an time interval $[0,T)$, where $0 < T \le \infty$, is defined to be a solution $U(\cdot)$ satisfying $U(0) = U_0$ and

$$U(\cdot) \in L^2(0,\tau;D(A)) \cap L^\infty(0,\tau;\mathbf{V}) \cap C^0([0,\tau];\mathbf{V}), \qquad \text{for every } \tau, 0 < \tau < T.$$

We say that a function

$$U(\cdot) \in L^2(0,T;\mathbf{V}) \cap L^\infty(0,T;\mathbf{H}) \cap C^0_w([0,T);\mathbf{H})$$

is a weak (Leray) solution of (2.5) if

$$\langle U', W\rangle + \nu\langle AU, W\rangle + \langle B(U,U), W\rangle \overset{\text{a.e.}}{=} \langle \mathbf{P}F, W\rangle, \qquad \text{for all } W \in \mathbf{V},$$

and $U(0) = U_0$, while

$$\frac{1}{2}\|U(t)\|^2 + \nu \int_s^t \|A^{1/2}U(\sigma)\|^2\, d\sigma \le \frac{1}{2}\|U(s)\|^2 + \int_s^t |\langle \mathbf{P}F, U(\sigma)\rangle|\, d\sigma$$

for $s = 0$ and for almost all $0 \le s \le t$ in $[0,T)$, where $\langle\cdot,\cdot\rangle$ is the duality pairing between $\mathbf{V}$ and $\mathbf{V}'$, see Constantin and Foias (1988). We recall that $C^0_w([0,T);\mathbf{H})$ is the subspace of $L^\infty(0,T;\mathbf{H})$ consisting of functions which are weakly continuous, that is, the mapping $t \to \langle U(t), W\rangle$ is a continuous function in $t$, for all $W \in \mathbf{H}$. In particular, the initial condition $U(0) = U_0$ is taken in this sense. The study of the regularity of solutions, both in 2D and 3D, has attracted widespread interest beginning with Leray (1933, 1934a, 1934b). Let us review some results on the existence and uniqueness of strong and weak solutions.

For 2DNS it is known that, for any $F \in L^2(\Omega)$ and any $U_0 \in \mathbf{H}$, there exists a unique weak solution $U(\cdot)$ on $[0,\infty)$ satisfying $U(0) = U_0$,

$$U(\cdot) \in L^2(0,\tau;\mathbf{V}) \cap L^\infty(0,\infty;\mathbf{H}) \cap C^0_w([0,\infty);\mathbf{H}), \qquad \text{for every } \tau > 0, \tag{2.6}$$

and for each $t_0 > 0$ the function $\hat{U}(t) = U(t + t_0)$ is a strong solution of (2.5) satisfying $\hat{U}(0) = U(t_0)$. Moreover, if $U_0 \in \mathbf{V}$, then the weak solution $U(\cdot)$ is a strong solution and satisfies

$$U(\cdot) \in L^2(0,\tau;D(A)) \cap L^\infty(0,\infty;\mathbf{V}) \cap C^0([0,\infty);\mathbf{V}), \qquad \text{for every } \tau > 0.$$

In addition, one has $U(\cdot) \in C^0((0,\infty); D(A))$ and there exist positive constants $L_1$ and $L_2$, which are independent of the initial data $U_0$, such that

$$\limsup_{t\to\infty} \|U(t)\|_{\mathbf{V}} \le L_1 \qquad \text{and} \qquad \limsup_{t\to\infty} \|AU(t)\|_{\mathbf{H}} \le L_2. \tag{2.7}$$

These classical results can be found in Constantin and Foias (1988) and Temam (1977, 1983, 1988), for example.

As a consequence of (2.7), one can show that the nonlinear semigroup $S(t)$ acting on $\mathbf{H}$ (or also on $\mathbf{V}$), and defined by the weak solutions (or the strong solutions) of the 2DNS, has a global attractor $\mathfrak{A}$ in $\mathbf{H}$ (or in $\mathbf{V}$). That is to say, $\mathfrak{A}$ is a nonempty, compact, invariant set, and $\mathfrak{A}$ attracts any bounded set $B$ in $\mathbf{H}$ (or in $\mathbf{V}$). Recall that $\mathfrak{A}$ is invariant if $S(t)\mathfrak{A} = \mathfrak{A}$, for all $t \ge 0$, and that $\mathfrak{A}$ attracts a bounded set $B$ in $\mathbf{H}$ (or in $\mathbf{V}$), if for every $\delta > 0$ there is a time $\tau = \tau(B,\delta) \ge 0$ such that

$$S(t)B \subset \mathcal{N}_X(\mathfrak{A},\delta), \qquad \text{for all } t \ge \tau,$$

where $\mathcal{N}_X(\mathfrak{A},\delta)$ denotes the $\delta$-neighborhood of $\mathfrak{A}$ in $X$ and $X = \mathbf{H}$ (or $X = \mathbf{V}$). In addition, the global attractor $\mathfrak{A}$ is uniformly asymptotically stable in $\mathbf{H}$ (or in $\mathbf{V}$), see Sell and You (1993). Notice that (2.7) implies that $\mathfrak{A}$ is a bounded set in $D(A)$. Actually the global attractor $\mathfrak{A}$ of $S(t)$ on $\mathbf{V}$ is a compact set in $H^2(\Omega)$, and it is also the global attractor of the restriction of $S(t)$ to $H^2(\Omega) \cap \mathbf{V}$. Furthermore, this global attractor $\mathfrak{A}$ has finite dimension, see Mallet-Paret (1976), Mañé (1981), and Foias and Temam (1979).

The situation is quite different for the 3DNS. It is known that, for any $F \in L^2(\Omega)$ and any $U_0 \in \mathbf{H}$, there exists a weak solution $U(\cdot)$ on $[0,\infty)$ satisfying $U(0) = U_0$ and (2.6). (However, unlike the 2DNS, it is not known whether this weak solution is uniquely determined by the data, $U_0$ and $F$.) Furthermore, if $U_0 \in \mathbf{V}$, then there exists a time $T = T(F, U_0)$, $0 < T \le \infty$, such that (2.5) has a unique strong, or regular, solution $U(\cdot)$ on $[0,T)$ satisfying $U(0) = U_0$,

$$U(\cdot) \in L^2(0,\tau; D(A)) \cap L^\infty(0,\tau;\mathbf{V}) \cap C^0([0,\tau];\mathbf{V}), \quad \text{and} \quad D_tU(\cdot) \in L^2(0,\tau;\mathbf{H}),$$

for every $\tau$, $0 < \tau < T$. Moreover, if both the initial data $U_0$ in $\mathbf{V}$ and the forcing function $F$ are small, then one has $T = \infty$, i.e., $U(\cdot)$ is a globally regular solution of (2.5). See the references cited above for the details.

**3. Navier-Stokes Equations on Thin 3D Domains.** Our objective in this section is to describe the recent theory of Raugel and Sell (1992a, 1992b, 1992c), which guarantees the existence of a global attractor for the Navier-Stokes equations on the thin 3 dimensional domain considered here. The fact that $\Omega_\epsilon$ is close to $Q_2$ suggests that the 3DNS on $\Omega_\epsilon$ is a perturbation of the 2DNS on $Q_2$. This is the case, however this perturbation is singular. The regularization of this singular perturbation is done in two steps, and it follows the methods of Hale and Raugel (1989a, 1989b). First one maps $\Omega_\epsilon$ onto $Q_3 = Q_2 \times (0,1)$ by means of dilation $(y_1, y_2, y_3) \longrightarrow (x_1, x_2, x_3)$, where $x_1 = y_1$, $x_2 = y_2$, and $x_3 = \epsilon^{-1} y_3$. The Navier-Stokes equations (2.1) on $\Omega_\epsilon$ are then transformed to the dilated Navier-Stokes equations on $Q_3$, see below. This dilation alone does not remove the singular perturbation because some of the differential operators appearing in the dilated equations contain coefficients with $\epsilon^{-1}$, or $\epsilon^{-2}$, and $\epsilon$ is small. The second step is accomplished by introducing the orthogonal projection $v = Mu$, for $u$ defined on $Q_3$, where

$$v(x_1, x_2) = \int_0^1 u(x_1, x_2, s)\, ds.$$

By applying $M$ and $(I - M)$ to the dilated Navier-Stokes evolutionary equations, one finds an equivalent system in terms of $v$ and $w$, where $w = u - v$. Then, by using some improved Sobolev inequalities given in Raugel and Sell (1992a), we show that this leads to a regularization of the singular perturbation of the 3DNS on $\Omega_\epsilon$, for small $\epsilon$.

Let us now turn to some of the details. The linear operator $J_\epsilon$ given by $U = J_\epsilon u$, where $U(y_1, y_2, y_3) = u(y_1, y_2, \epsilon^{-1} y_3)$, sets up a one-to-one correspondence between the measurable functions on $\Omega_\epsilon$ and the measurable functions on $Q_3$. (Throughout this section, we shall use capital Roman letters to denote functions on $\Omega_\epsilon$ and lower case Roman letters for functions on $Q_3$.) By using the operator $J_\epsilon$, one transforms the Navier-Stokes equations (2.1) to the following system on $Q_3$:

$$\begin{aligned} u_t - \nu \Delta_\epsilon u + (u \cdot \nabla_\epsilon) u + \nabla_\epsilon p &= f, \\ \nabla_\epsilon \cdot u &= 0, \end{aligned} \tag{3.1}$$

where $\nabla_\epsilon = (\partial_1, \partial_2, \epsilon^{-1}\partial_3)$, $\Delta_\epsilon = \partial_1^2 + \partial_2^2 + \epsilon^{-2}\partial_3^2$, $\partial_i = \frac{\partial}{\partial x_i}$, $i = 1, 2, 3$, $u = J_\epsilon^{-1} U$, $p = J_\epsilon^{-1} P$, and $f = J_\epsilon^{-1} F$. Of course (3.1) is supplemented with the appropriate analogues of (2.2) and (2.4). We will refer to (3.1) as the **dilated** 3DNS on $Q_3$.

Let $L^2(Q_3) = L^2(Q_3, \mathbf{R}^3)$ and let $\|\cdot\|$ denote the usual $L^2$-norm. For $m = 0, 1, 2, \ldots$ we let $H^m(Q_3)$ denote the Sobolev space $H^m(Q_3, \mathbf{R}^3)$, and let $H_p^m(Q_3) = H_p^m(Q_3, \mathbf{R}^3)$ be the closure in $H^m(Q_3, \mathbf{R}^3)$ of those smooth functions that are periodic in space, i.e., $u(x + \ell_i e_i) = u(x), i = 1, 2, 3$, where $\ell_3 = 1$. One then has $H_p^0(Q_3) = L^2(Q_3)$. Finally, we denote by $\mathbf{H}_\epsilon = \mathbf{H}_\epsilon(Q_3)$ (respectively by $\mathbf{V}_\epsilon = \mathbf{V}_\epsilon(Q_3)$), the closure in $L^2(Q_3)$ (respectively $H^1(Q_3)$) of those smooth functions $u$ that are periodic on $Q_3$ and satisfy

$$\int_{Q_3} u\, dx = 0 \quad \text{and} \quad \operatorname{div}_\epsilon u = \nabla_\epsilon \cdot u = \partial_1 u_1 + \partial_2 u_2 + \epsilon^{-1}\partial_3 u_3 = 0.$$

We let $\mathbf{H}_0 = \mathbf{H}_0(Q_3)$ (respectively $\mathbf{V}_0 = \mathbf{V}_0(Q_3)$) denote those functions $u = u(x_1, x_2)$ in $\mathbf{H}_\epsilon$ (respectively $\mathbf{V}_\epsilon$) that do not depend on $x_3$.

We let $\mathbf{P}_\epsilon$ denote the orthogonal projection of $L^2(Q_3)$ onto $\mathbf{H}_\epsilon$, and we introduce the operators

$$A_\epsilon u = -\mathbf{P}_\epsilon \Delta_\epsilon u, \quad B_\epsilon(u^1, u^2) = \mathbf{P}_\epsilon(u^1 \cdot \nabla_\epsilon) u^2.$$

By applying the operator $\mathbf{P}_\epsilon$ to (3.1), we obtain the following abstract nonlinear evolutionary equation on $\mathbf{H}_\epsilon$, called the **dilated Navier-Stokes evolutionary equation** (or dilated 3DNS for short),

$$u_t + \nu A_\epsilon u + B_\epsilon(u,u) = \mathbf{P}_\epsilon f. \tag{3.2}$$

We define $\mathbf{V}_\epsilon^m$, for $m = 1, 2, \ldots$, by $\mathbf{V}_\epsilon^m = \mathbf{V}_\epsilon \cap H_p^m(Q_3)$. One then has:

$$\mathbf{V}_\epsilon = \mathbf{V}_\epsilon^1 = D(A_\epsilon^{1/2}) \quad \text{and} \quad \mathbf{V}_\epsilon^2 = D(A_\epsilon).$$

In the purely periodic case, which we study here, one has $A_\epsilon u = -\Delta_\epsilon u$, $u \in D(A_\epsilon)$, and $A_\epsilon u$ can be represented in terms of the Fourier series expansion of $u$. We will study (3.2) with initial conditions $u_0 \in \mathbf{V}_\epsilon$.

We define an orthogonal projection $M$ on $L^2(Q_3)$ by

$$(Mu)(x_1, x_2) = v(x_1, x_2) \stackrel{\text{def}}{=} \int_0^1 u(x_1, x_2, s)\, ds$$

and set $w = (I - M)u$. One then has

$$MA_\epsilon u = A_\epsilon M u, \quad u \in D(A_\epsilon), \tag{3.3}$$

and the following Poincaré inequality is valid

$$\|w\| \le C_5 \epsilon \|A_\epsilon^{1/2} w\|, \quad w \in \mathbf{V}_\epsilon,\ Mw = 0. \tag{3.4}$$

We now apply the projections $M$ and $(I - M)$ to the equation (3.2) where $v = Mu$ and $w = (I - M)u$. Since one has $MB_\epsilon(v, v) = B_\epsilon(v, v)$, it follows from (3.3) that one obtains the system:

$$\begin{cases} v' + \nu A_\epsilon v + B_\epsilon(v,v) = M\mathbf{P}_\epsilon f - M(B_\epsilon(w,w) + B_\epsilon(w,v) + B_\epsilon(v,w)), \\ w' + \nu A_\epsilon w = (I - M)\mathbf{P}_\epsilon f - (I - M)(B_\epsilon(v,w) + B_\epsilon(w,v) + B_\epsilon(w,w)). \end{cases} \tag{3.5}$$

The initial condition $u(0) = v_0 + w_0$ also splits into $v$ and $w$ components. We are going to study the solutions $(v(t), w(t))$ of (3.5) that satisfy $v(0) = v_0 = Mu_0, w(0) = w_0 = (I - M)u_0$. Note that $A_\epsilon v = D_1^2 v + D_2^2 v$, i.e., $A_\epsilon v$ is independent of $\epsilon$, as is $B_\epsilon(v, v)$. Also $M$ maps $\mathbf{H}_\epsilon$ (respectively $\mathbf{V}_\epsilon$) onto $\mathbf{H}_0$ (respectively $\mathbf{V}_0$).

When $(I - M)\mathbf{P}_\epsilon f = 0$, then the set $\{w = 0\}$ is positively invariant, that is, if $u_0$ depends only on $x_1$ and $x_2$, or equivalently, if $w_0 = 0$, then $w(t) \equiv 0$, for $t \ge 0$, and $\bar{v} = v(t)$ satisfies the equation

$$\bar{v}' + \nu A_\epsilon \bar{v} + B_\epsilon(\bar{v}, \bar{v}) = M\mathbf{P}_\epsilon f, \tag{3.6}$$

with $\bar{v}(0) = v_0$. We refer to (3.6) as the **reduced** 3DNS on $Q_3$.

The reduced 3DNS on $Q_3$ incorporates the 2DNS on $Q_2$. In particular, if we set $\bar{v} = (m_1, m_2, \bar{v}_3)$, where $m = (m_1, m_2) = (\bar{v}_1, \bar{v}_2)$, then $\bar{v}$ is a solution of the reduced 3DNS (3.6) if and only if $m = (\bar{v}_1, \bar{v}_2)$ is a solution of the 2DNS

$$m' - \nu \mathbf{P}_2(\partial_1^2 + \partial_2^2)m + \mathbf{P}_2(m \cdot \nabla)m = (g_1, g_2), \tag{3.7}$$

and $\bar{v}_3$ is a solution of the linear equation

$$\bar{v}_3' - \nu(\partial_1^2 + \partial_2^2)\bar{v}_3 + (\bar{v}_1\partial_1 + \bar{v}_2\partial_2)\bar{v}_3 = g_3, \tag{3.8}$$

where $g = (g_1, g_2, g_3) = M\mathbf{P}_\epsilon f$. Note that, since $\bar{v}_3$ is a slave variable, the long-time dynamics of (3.6) is completely determined by the 2DNS.

The following two theorems are proved in Raugel and Sell (1992a, 1992b, 1992c).

THEOREM C. *Consider the dilated 3DNS on $Q_3$ given by (3.2), with purely periodic boundary conditions. Then there exists a positive number $\epsilon_0 = \epsilon_0(\nu)$ such that, for every $\epsilon$, $0 < \epsilon \leq \epsilon_0$, there exist large sets $\mathcal{R}(\epsilon)$ and $\mathcal{S}(\epsilon)$ with*

$$\mathcal{R}(\epsilon) \subset \mathbf{V}_\epsilon \qquad \text{and} \qquad \mathcal{S}(\epsilon) \subset L^2(Q_3)$$

*such that, if $u_0 \in \mathcal{R}(\epsilon)$ and $f \in \mathcal{S}(\epsilon)$, then (3.2) has a unique, globally regular solution $u(\cdot)$ satisfying $u(0) = u_0$ and*

$$u(\cdot) \in L^2_{\text{loc}}(0, \infty; D(A_\epsilon)) \cap L^\infty(0, \infty; \mathbf{V}_\epsilon) \cap C^0([0, \infty); \mathbf{V}_\epsilon).$$

*Furthermore, there exist positive constants $L_1$ and $L_2$, which do not depend on $u_0$, such that*

$$\limsup_{t \to \infty} \|u(t)\|_{\mathbf{V}_\epsilon} \leq L_1 \qquad \text{and} \qquad \limsup_{t \to \infty} \|A_\epsilon u(t)\|_{\mathbf{H}_\epsilon} \leq L_2.$$

*Moreover, the set of strong solutions of the dilated 3DNS has a local attractor $\mathfrak{A}_\epsilon = \mathfrak{A}_\epsilon(f)$, which attracts every bounded set in $\cup_{t \geq 0} S_\epsilon(t)\mathcal{R}(\epsilon)$, and this attractor is a compact set in $D(A_\epsilon) = \mathbf{V}_\epsilon^2$.*

More than being a local attractor for the strong solutions, $\mathfrak{A}_\epsilon(f)$ is, in fact, a global attractor for the weak solutions, i.e., one has the following result.

THEOREM D. *Let the hypotheses of Theorem C be satisfied. Then there is an $\epsilon_1 > 0$ such that, for $0 < \epsilon \leq \min(\epsilon_0, \epsilon_1)$, the local attractor $\mathfrak{A}_\epsilon(f)$ is the global attractor for the weak (Leray) solutions of the 3DNS, provided $f$ belongs to $\mathcal{S}(\epsilon)$.*

Next we turn to the precise description[4] of the sets $\mathcal{R}(\epsilon)$ and $\mathcal{S}(\epsilon)$ referred to in Theorem C. We shall say that **Hypothesis H** is satisfied if one has

$$p = -\frac{29}{24} + \delta_1, \quad q_1 = -\frac{5}{12} + \delta_2, \quad q_2 = -\frac{5}{12} + \delta_2, \quad r = -2 + \delta_4,$$

[4] A somewhat more general description is given in Raugel and Sell (1992b).

where $\delta_i$ are arbitrarily small positive numbers, and

$$\eta_i^{-1}(\epsilon) = (-\log \epsilon)^{\alpha_i}, \qquad \alpha_i > 0,\ i = 1, 2, 3, 4.$$

Next we define $\mathcal{R}(\epsilon)$ and $\mathcal{S}(\epsilon)$ as follows:

$$\mathcal{R}(\epsilon) \stackrel{\text{def}}{=} \{u_0 \in \mathbf{V} : \|A_\epsilon^{1/2} v_0\|^2 \le \epsilon^{q_1}\eta_1^{-2}, \|A_\epsilon^{1/2} w_0\|^2 \le \epsilon^p \eta_3^{-2}\}$$

and

$$\mathcal{S}(\epsilon) \stackrel{\text{def}}{=} \{f \in L^2(Q_3) : \|M\mathbf{P}_\epsilon f\|^2 \le \epsilon^{q_2}\eta_2^{-2}, \|(I - M)\mathbf{P}_\epsilon f\|^2 \le \epsilon^r \eta_4^{-2}\}.$$

The proof of Theorem C begins by verifying that when Hypothesis H is satisfied then there exist a positive number $\epsilon_0$, a positive constant $C$, which is independent of $\epsilon$ and $t$, and, for all $\epsilon$, $0 < \epsilon \le \epsilon_0$, there is a time $T_1 = T_1(\epsilon) > 0$, such that, whenever $u_0 \in \mathcal{R}(\epsilon)$ and $f \in \mathcal{S}(\epsilon)$, then (3.2) has a (unique) strong solution $u$ that belongs to $C^0([0, \infty); \mathbf{V}_\epsilon) \cap L^\infty((0, \infty); \mathbf{V}_\epsilon)$, i.e., one has

$$\|A_\epsilon^{1/2} u(t)\|^2 \le K_1^2, \qquad t \ge 0,$$

where $K_1$ is independent of $t$. Moreover the components of $u = v + w$ satisfy

$$\|A_\epsilon^{1/2} v(t)\|^2 \le C\left(1 + \epsilon^{q_2}\eta_2^{-2} + \epsilon^{2+r}\eta_4^{-2} + \epsilon^{4+2r}\eta_4^{-4}\right)^3, \qquad t \ge T_1, \tag{3.9}$$

and

$$\|A_\epsilon^{1/2} w(t)\|^2 \le k_2^2 \epsilon^{2+r}\eta_4^{-2}, \qquad t \ge T_1, \tag{3.10}$$

where $k_2$ is a positive constant depending only on $\nu$ and $Q_2$. Notice that the bounds in inequalities (3.9) and (3.10) do not depend on the initial data $u_0$. Next we turn to the description of the local attractor $\mathfrak{A}_\epsilon = \mathfrak{A}_\epsilon(f)$ generated in Theorem C.

We denote by $u(t) = S_\epsilon(t)u_0$ the strong solution of (3.2) with initial data $u_0 \in \mathbf{V}_\epsilon$, and we assume that Hypothesis H is satisfied. Let $\epsilon_0 > 0$ be given as in Theorem C. For $0 < \epsilon \le \epsilon_0$, we fix $f \in \mathcal{S}(\epsilon)$, and define

$$\mathcal{B}_\epsilon \stackrel{\text{def}}{=} \bigcup_{t \ge 0} S_\epsilon(t)\mathcal{R}(\epsilon).$$

Due to the above argument, the set $\mathcal{B}_\epsilon$ is a bounded set in $V_\epsilon$. Also $S_\epsilon(t)$ is a $C^0$-semigroup, or semiflow, on $\mathcal{B}_\epsilon$, for $t \ge 0$. Furthermore, one can show that $S_\epsilon(\cdot)u_0$ belongs to $C^0([0, \infty); \mathbf{V}_\epsilon^1)$ whenever $u_0$ is in $\mathcal{B}_\epsilon$. Moreover, for $\tau \ge T_1 + 1$, the set

$$\bigcup_{t \ge \tau} S_\epsilon(t)\mathcal{B}_\epsilon$$

lies in the following bounded set in $\mathbf{V}_\epsilon^2$,

$$\{u \in D(A_\epsilon) : \|A_\epsilon u\|^2 \leq k_2^2 \epsilon^{2+r} \eta_4^{-2} + C(1 + G_2(\epsilon))^3\}, \tag{3.11}$$

where $G_2(\epsilon)$ depends on $f$, but not on the initial data $u_0$. Since the set in (3.11) is bounded in $\mathbf{V}_\epsilon^2$, it lies in a compact set in $\mathbf{V}_\epsilon$. Therefore, $\mathfrak{A}_\epsilon = \omega(\mathcal{B}_\epsilon)$, the $\omega$-limit set of $\mathcal{B}_\epsilon$ in $\mathbf{V}_\epsilon$, is a nonempty, compact, invariant set in $\mathbf{V}_\epsilon$, and it attracts $\mathcal{B}_\epsilon$ in $\mathbf{V}_\epsilon$. Since

$$S_\epsilon(t)\mathcal{B}_\epsilon \subset \mathcal{B}_\epsilon, \qquad \text{for all } t \geq 0,$$

$\mathfrak{A}_\epsilon$ is a local attractor for the strong solutions of (3.2) in $\mathbf{V}_\epsilon$ and the basin of attraction contains $\mathcal{B}_\epsilon$. (Recall that $\mathfrak{A}_\epsilon$ is a **local** attractor of $S_\epsilon(t)$ if there is a neighborhood $\mathcal{N}_\epsilon$ of $\mathfrak{A}_\epsilon$ in $\mathbf{V}_\epsilon$ such that $\mathfrak{A}_\epsilon$ attracts every bounded set of $\mathcal{N}_\epsilon$.) To put it another way, $\mathfrak{A}_\epsilon$ is the global attractor of the restriction of $S_\epsilon(t)$ to $\mathcal{B}_\epsilon$, see Hale (1988) and Sell and You (1993).

One can verify that, for any $\tau > 0$, there is a compact subset $\mathcal{K}_\epsilon(\tau)$ of $\mathbf{V}_\epsilon^2$ such that

$$S_\epsilon(\mathbf{P}_\epsilon f, t)\mathcal{B}_\epsilon \subset \mathcal{K}_\epsilon(\tau), \qquad t > \tau. \tag{3.12}$$

From (3.12) we deduce that $\mathfrak{A}_\epsilon$ is a compact set in $\mathbf{V}_\epsilon^2$, and it attracts the set $\mathcal{B}_\epsilon \cap \mathbf{V}_\epsilon^2$ in the space $\mathbf{V}_\epsilon^2$. We thus have completed the proof of Theorem C.

In order to show that the local attractor $\mathfrak{A}_\epsilon$ is the global attractor for the weak (Leray) solutions of the dilated 3DNS given by (3.2), we first fix $f \in \mathcal{S}(\epsilon)$. Then one shows that, for any $r > 0$, there exists a time $T_\epsilon = T_\epsilon(r)$, such that for any weak (Leray) solution $u(t)$ of (3.2) with $\|u(0)\| \leq r$, there is a positive time $t_0 = t_0(r)$, $0 < t_0 \leq T_\epsilon(r)$ such that $u(t_0) \in \mathcal{S}(\epsilon)$. As a result, $u(t)$ becomes a strong solution of (3.2) for $t \geq t_0$, and the (local) attractor $\mathfrak{A}_\epsilon$ given in Theorem C is the global attractor for the weak (Leray) solutions of (3.2), provided $\epsilon$ is sufficiently small, see Raugel and Sell (1992a, 1992b, 1992c).

**Behavior of Attractors as $\epsilon \to 0$.** Let us now consider the reduced 3DNS given by (3.6) and let us denote by $S_0(g,t)\bar{v}_0 = \bar{v}(t)$ the strong solution of (3.6) with initial data $\bar{v}_0$ in $\mathbf{V}_0$, where $g = M\mathbf{P}_\epsilon f$. The theory of the 2DNS implies that $S_0(g,t)$ is a $C^0$-semigroup, or semiflow, from $\mathbf{V}_0$ into $\mathbf{V}_0$ and has a global attractor $\mathfrak{A}_0 \equiv \mathfrak{A}_0(g)$ in $\mathbf{V}_0$, see Raugel and Sell (1992a). It is not difficult to see that if $(I - M)\mathbf{P}_\epsilon f = 0$, then the attractors $\mathfrak{A}_\epsilon$ and $\mathfrak{A}_0(g)$ coincide.

If $(I - M)\mathbf{P}_\epsilon f \neq 0$, the comparison of the two attractors $\mathfrak{A}_\epsilon$ and $\mathfrak{A}_0(g)$ is more complicated. However, by using the uniform asymptotic stability of $\mathfrak{A}_0(g)$ and the fact that the reduced 3DNS regularizes the full 3DNS, we can show that $\mathfrak{A}_\epsilon$ is upper semicontinuous at $\epsilon = 0$, as is stated in the following result, see Raugel and Sell (1992a) for the proof.

THEOREM E. *Let Hypothesis H hold, and let $\epsilon_2 = \min(\epsilon_0, \epsilon_1)$ be given by Theorems C and D. Let $0 < \epsilon_n \leq \epsilon_2$ and let $f_n$ be a sequence in $L^2(Q_3)$ satisfying $\epsilon_n \to 0$, as $n \to \infty$, and*

$$\lim_{n \to \infty} \|M\mathbf{P}_{\epsilon_n} f_n - g_0\| = 0,$$

*for some function $g_0$ in $\mathbf{H}_0 \equiv M\mathbf{H}_\epsilon$. Assume further that $f_n \in \mathcal{S}(\epsilon_n)$. Then the attractors $\mathfrak{A}_{\epsilon_n}$ of (3.2), with forcing term $\mathbf{P}_{\epsilon_n} f_n$, are upper semicontinuous at $\epsilon = 0$ in $\mathbf{V}_{\epsilon_n}$, i.e., for every $\delta > 0$ there is a $N \geq 1$ such that*

$$\mathfrak{A}_{\epsilon_n} \subset \left\{ u \in \mathbf{V}_{\epsilon_n} : \inf_{v \in \mathfrak{A}_0(g_0)} \|A_{\epsilon_n}^{\frac{1}{2}}(u - v)\| \leq \delta \right\}, \qquad \text{for } n \geq N,$$

*where $\mathfrak{A}_0(g_0)$ is the global attractor of (3.6) with $g_0 = M\mathbf{P}_\epsilon f$. If in addition, one has*

$$\lim_{n \to \infty} \|f_n - f_0\| = 0,$$

*where $f_0 \in L^2(Q_3)$, then the attractors $\mathfrak{A}_{\epsilon_n}$ are upper semicontinuous at $\epsilon = 0$ in $\mathbf{V}_{\epsilon_n}^2$, i.e., for every $\delta > 0$ there is a $N \geq 1$ such that*

$$\mathfrak{A}_{\epsilon_n} \subset \left\{ u \in \mathbf{V}_{\epsilon_n}^2 : \inf_{v \in \mathfrak{A}_0(g_0)} \|A_{\epsilon_n}(u - v)\| \leq \delta \right\}, \qquad \text{for } n \geq N. \tag{3.13}$$

This result on the upper semicontinuity of $\mathfrak{A}_\epsilon$ at $\epsilon = 0$ contains some important information for comparing the long-time dynamics on $\mathfrak{A}_\epsilon$ with that on $\mathfrak{A}_0(g_0)$, when $\epsilon$ is small. However, as we will show in the next section, the Kwak Transformation and the associated theory of inertial forms will enable us to extract a much deeper insight into these comparisons of the long-time dynamics.

**4. Navier-Stokes Equations: The Kwak Transformation.** As above we consider the Navier-Stokes equations (2.5) on $\Omega$, with periodic boundary conditions. In this section we shall impose an additional spatial smoothness assumption on the forcing term $F$ and assume that $F \in D(A) \subset H^2(Q_2)$. One of the reasons for this added smoothness assumption is to insure that the global attractor for the 2DNS lies in the space $\mathbf{V}^3 = \mathbf{V} \cap \mathbf{H}^3(Q_2, R^2)$. See Kwak (1991) for more details.

Next we will describe the Kwak Transformation. As we shall see, this transformation is a nonlinear change of variables in either the 2DNS or the 3DNS, which has the property that it imbeds the Navier-Stokes equations on $\Omega$ into a system of reactions diffusion equations on $\Omega$ of the form

$$\frac{\partial}{\partial t}\mathsf{U} + \mathsf{A}\mathsf{U} = \mathsf{F}(\mathsf{U}). \tag{4.1}$$

The significance of this transformation, as we shall see, is that the nonlinear term $(U \cdot \nabla)U$ is decomposed into simpler pieces which are easier to analyze.

The first step is to write the nonlinear term $B(U, U) = \mathsf{P}_d(U \cdot \nabla)U$, where $d = 2$, or 3, in a different form. We will assume that the vector field $U$ is a column vector and that the gradient $\nabla$ is also a column vector. In the 3D case, one has

$$U = \begin{pmatrix} U_1 \\ U_2 \\ U_3 \end{pmatrix}, \qquad \nabla = \begin{pmatrix} D_1 \\ D_2 \\ D_3 \end{pmatrix},$$

and we define

$$U \otimes U \stackrel{\text{def}}{=} UU^T = \begin{pmatrix} U_1^2 & U_1U_2 & U_1U_3 \\ U_1U_2 & U_2^2 & U_2U_3 \\ U_1U_3 & U_2U_3 & U_3^2 \end{pmatrix},$$

where $^T$ denotes the matrix transpose operation. Let $\Sigma^d$ denote the collection of all $(d \times d)$ real symmetric matrices. Then one has $U \otimes U \in \Sigma^d$, for $d = 2$, or 3, and

$$(U \cdot \nabla)U = \nabla^T UU^T = \nabla^T U \otimes U.$$

Next define $R$ by $R \stackrel{\text{def}}{=} U \otimes U = UU^T$. Then (2.5) can be rewritten as

$$U_t + \nu AU + \mathbf{P}_d \nabla^T R = \mathbf{P}_d F. \tag{4.2}$$

The next step in the construction of the Kwak Transformation is to calculate the equation that the new variable $R \in \Sigma^d$ must satisfy, given that $U$ satisfies (2.5), or (4.2). As an intermediate step in this calculation, we define

$$Z = (Z^{(1)}, Z^{(2)}), \qquad \text{or} \qquad Z = (Z^{(1)}, Z^{(2)}, Z^{(3)}),$$

for $d = 2$, or 3, where

$$Z^{(i)} = D_i U = \frac{\partial}{\partial y_i} U, \qquad 1 \le i \le d.$$

It is convenient to define the operation $\nabla * U$ by

$$\nabla * U \stackrel{\text{def}}{=} Z.$$

Notice that since $\nabla \cdot U = 0$, one has $\nabla \cdot Z^{(i)} = 0$, for $1 \le i \le d$. Next we apply $D_i$ to (4.2) to obtain

$$Z_t^{(i)} + \nu AZ^{(i)} + \mathbf{P}_d D_i \nabla^T R = \mathbf{P}_d D_i F, \qquad 1 \le i \le d.$$

(Note that with the periodic boundary conditions one has $D_i \mathbf{P}_d U = \mathbf{P}_d D_i U$, for all $U \in \mathbf{V}$.) The equation for $Z$ then becomes

$$Z_t + \nu AZ + \mathbf{P}_d \nabla * (\nabla^T R) = \mathbf{P}_d \nabla * F. \tag{4.3}$$

By using the identity $Z_1^{(1)} + Z_2^{(2)} + Z_3^{(3)} = \nabla \cdot U = 0$, it is easily verified that $\nabla^T R = M(U, Z)$, where

$$M(U, Z) = \begin{pmatrix} U_1 Z_1^{(1)} + U_2 Z_1^{(2)} + U_3 Z_1^{(3)} \\ U_1 Z_2^{(1)} + U_2 Z_2^{(2)} + U_3 Z_2^{(3)} \\ U_1 Z_3^{(1)} + U_2 Z_3^{(2)} + U_3 Z_3^{(3)} \end{pmatrix}.$$

Finally the equation for $R$ becomes

$$R_t - \nu\Delta R = -N(U, Z) + (\mathsf{P}_d F \otimes U + U \otimes \mathsf{P}_d F), \tag{4.4}$$

where

$$N(U, Z) = -2\nu \sum_{i=1}^{d} Z^{(i)} \otimes Z^{(i)} - (\mathsf{P}_d M(U, Z) \otimes U + U \otimes \mathsf{P}_d M(U, Z)).$$

By combining (4.2), (4.3), and (4.4), we obtain the following system

$$\begin{cases} U_t + \nu AU + \mathsf{P}_d \nabla^T R & = \mathsf{P}_d F \\ Z_t + \nu AZ + \mathsf{P}_d \nabla * (\nabla^T R) & = \mathsf{P}_d \nabla * F \\ R_t - \nu \Delta R & = -N(U, Z) + (\mathsf{P}_d F \otimes U + U \otimes \mathsf{P}_d F). \end{cases} \tag{4.5}$$

The **Kwak Transformation** is then defined by

$$\mathsf{U} = G(U) = (G_0, G_1, G_2), \qquad U \in \mathbf{V}, \tag{4.6}$$

where $G_0(U) = U$, $G_1(U) = \nabla * U$, and $G_2(U) = U \otimes U = UU^T$.

The system of equations (4.5) for $U$, $Z$, and $R$ describes a candidate for the system of reaction diffusion equations (4.1), where $\mathsf{U} = (U, Z, R)^T$ In fact, the left side of (4.5) defines a candidate for the linear operator $\frac{\partial}{\partial t}\mathsf{U} + \mathsf{A}\mathsf{U}$, while the right side of this system is a candidate for the nonlinearity $\mathsf{F}(\mathsf{U})$. Note that if equation (4.1) is defined in this way, it remains meaningful whether or not the vector $\mathsf{U}$ is in the range of the Kwak Transformation $G$. As a result, one has considerable latitude in the construction of the reaction diffusion equation (4.1). For example, any nonlinearity $\mathsf{F}(\mathsf{U})$ that agrees with the right side of (4.5) in a neighborhood of the range of $G$ is another candidate for realizing the Kwak imbedding of the Navier-Stokes equations into a system of reaction diffusion equations.

This observation opens up the possibility that one might be able to select the nonlinearity $\mathsf{F}(\mathsf{U})$ so that the equation (4.1) agrees with (4.5) in a neighborhood of the range of $G$ ***and*** so that other desirable properties hold. At the same time one can add linear stabilizing terms of the form $k_1^2(Z - G_1(U))$ to the left side of the $Z$ equation in (4.5) without changing the dynamics of (4.5) on the range of $G$. Similarly one can add $k_1^2 U$ and $k_1^2 R$ to both sides of the $U$ and $R$ equations without altering these dynamics. Since we are primarily interested in the long-time dynamics, it suffices to insure that all such changes do not alter the equation (4.5) on $G(\mathfrak{A}_{\mathrm{NS}})$, where $\mathfrak{A}_{\mathrm{NS}}$ is the global attractor of the 2DNS on $Q_2$, or the 3DNS on $\Omega_\epsilon$. One objective then is to make such alterations so that the new system is dissipative, and as a result, equation (4.1) will have a global attractor $\mathfrak{A}_{RD}$. Moreover, $\mathfrak{A}_{RD}$ will necessarily satisfy the relationship $\mathfrak{A}_{RD} \supset G(\mathfrak{A}_{NS})$. The revised system we seek will then have the form

$$\text{(4.7)} \qquad \begin{cases} U_t + \nu AU + \mathsf{P}_d\nabla^T R + k_1^2 U & = \mathsf{F}_0(\mathsf{U}) + k_1^2 U \\ Z_t + \nu AZ + \mathsf{P}_d\nabla * (\nabla^T R) + k_1^2(Z - G_1(U)) & = \mathsf{F}_1(\mathsf{U}) \\ R_t - \nu\Delta R + k_1^2 R & = \mathsf{F}_2(\mathsf{U}) + k_1^2 R, \end{cases}$$

where $\mathsf{F} = (\mathsf{F}_0, \mathsf{F}_1, \mathsf{F}_2)$ agrees with the right side of (4.5) in a neighborhood of $G(\mathfrak{A}_{NS})$.

Let us restrict our attention for the moment to the 2DNS on $Q_2$. In this case we consider (4.7) on the function space

$$\mathcal{H} = \mathcal{H}(Q_2) \stackrel{\text{def}}{=} \mathbf{H}(Q_2)^3 \times L^2(Q_2, \Sigma^2).$$

The following result, for the 2DNS on $Q_2$, is proved in Kwak (1991).

THEOREM F. *Let the 2DNS be given on $Q_2$ with periodic boundary conditions. Then there is a system of reaction diffusion equations of the form (4.7) on $\mathcal{H} = \mathcal{H}(Q_2)$ with the following properties:*

(1) *Equation (4.7) agrees with (4.5) in a neighborhood of $G(\mathfrak{A}_{NS})$, where $G$ is the Kwak Transformation (4.6), and $\mathfrak{A}_{NS}$ is the global attractor of the 2DNS.*

(2) *The linear operator $\mathsf{A}$ is a sectorial operator with compact resolvent on $\mathcal{H}$. Furthermore, the fractional powers of $\mathsf{A}$ are well-defined, and we set $\mathcal{V} = D(\mathsf{A}^{\frac{1}{2}})$.*

(3) *The nonlinearity $\mathsf{F} = \mathsf{F}(\mathsf{U})$ is a locally Lipschitz continuous mapping of $\mathcal{V}$ into $\mathcal{H}$. Furthermore $\mathsf{F}$ is a Lipschitz continuous mapping of the domain $D(\mathsf{A})$ into itself, and there exist constants $K_0$ and $K_1$ such that one has $\|\mathsf{AF}(\mathsf{U})\| \le K_0$, for all $\mathsf{U} \in D(\mathsf{A})$ and*

$$\|\mathsf{A}(\mathsf{F}(\mathsf{U}_1) - \mathsf{F}(\mathsf{U}_2))\| \le K_1\|\mathsf{A}(\mathsf{U}_1 - \mathsf{U}_2)\|, \qquad \textit{for all } \mathsf{U}_1, \mathsf{U}_2 \in D(\mathsf{A}).$$

(4) *Equation (4.7) is dissipative on the space $\mathcal{V}$, and the global attractor $\mathfrak{A}_{RD}$ is a bounded set in $D(\mathsf{A})$ that satisfies*

$$\mathfrak{A}_{RD} = G(\mathfrak{A}_{NS}).$$

*Moreover, the restriction $G : \mathfrak{A}_{NS} \to \mathfrak{A}_{RD}$ is a homeomorphism which preserves all the long-time dynamics of the 2DNS.*

Notice that, in both the 2D and 3D problems, the linear operator defined by the left side of (4.5) can be written in block form as

$$\mathsf{A} = \begin{pmatrix} \nu A + k_1^2 & 0 & * \\ * & \nu A + k_1^2 & * \\ 0 & 0 & -\nu\Delta + k_1^2 \end{pmatrix}$$

where * denotes various partial differential operators. This operator $\mathsf{A}$ is not self-adjoint. However, because of the block triangular form for $\mathsf{A}$, it is not difficult to show that $\mathsf{A}$ has exactly the same spectrum as $(A + k_1^2 I)$, where $A$ is the Stokes operator. (Of course, the multiplicities, or the dimensions of the generalized eigenspaces, for these two operators will differ.) Hence, the spectrum $\sigma(\mathsf{A})$ consists of real eigenvalues only. Returning to the 2DNS on $Q_2$, we let $\lambda_1 < \lambda_2 < \lambda_3 < \ldots$ denote the distinct eigenvalues of $\mathsf{A}$. For the dilated 3DNS on $Q_3$, we will denote the distinct eigenvalues of $\mathsf{A}$ by $\Lambda_1 < \Lambda_2 < \Lambda_3 < \ldots$.

The theory of inertial manifolds for problems that are not self-adjoint, see Sell and You (1992), can be applied to the 2DNS on $Q_2$. In particular, because of the boundedness and Lipschitz properties of $\mathsf{F}$, one can show that there exist constants $C_0 > 0$ and $C_1 > 0$ such that if for some integer $m \geq 1$ the spectral gap condition

$$\lambda_{m+1} - \lambda_m \geq C_1 \quad \text{and} \quad \lambda_{m+1} \geq C_0 \tag{4.8}$$

holds, then there exists a finite dimensional inertial manifold $\mathfrak{M}$ for (4.7). For this problem, the inertial manifold $\mathfrak{M}$ can be realized as the graph of a suitable function

$$\Phi : P\mathcal{H} \to Q\mathcal{H} \cap D(\mathsf{A}),$$

where $P$ is the orthogonal projection of $\mathcal{H}$ onto the sum of the generalized eigenspaces of $\mathsf{A}$ corresponding to eigenvalues $\lambda$ satisfying $\lambda \leq \lambda_m$ and $Q = I - P$. As usual, the restriction of (4.7) to $\mathfrak{M}$ gives rise to the inertial form, or system of ordinary differential equations[5]

$$p' + \mathsf{A}p = P\mathsf{F}(p + \Phi(p)). \tag{4.9}$$

If $p(t)$ is a solution of (4.9), then $\mathsf{U}(t) = p(t) + \Phi(p(t))$ is a solution of (4.7) and $\mathsf{U}(t) \in \mathfrak{M}$, for all $t \geq 0$. Since $\mathfrak{A}_{\mathrm{RD}} \subset \mathfrak{M}$, it follows from Theorem F that all the long-time dynamics of the 2DNS is described completely by (4.9), provided that (4.8) is satisfied.

This theory can be applied in the rectangular region $Q_2 = (0, \ell_1) \times (0, \ell_2)$ provided $(\ell_1 \ell_2^{-1})^2$ is rational, because in this case it is known that the eigenvalues of the Stokes operator $A$ satisfies the asymptotic relationship

$$\limsup_{n \to \infty} (\lambda_{m+1} - \lambda_m) = \infty, \tag{4.10}$$

see Richards (1982) and Mallet-Paret and Sell (1988). As a result, (4.8) is satisfied for infinitely many choices of $m$.

One of the steps used in the proof of Theorem F in Kwak (1991) is a modification of the 2DNS given by (2.5), prior to the application of the Kwak Transformation $G$. This modification has the effect of leaving the original equation (2.5) unchanged in a set $B_0$ and

[5]This is an example, in the infinite dimensional setting, of the Reduction Principle, see Pliss (1964).

forcing the nonlinear terms to vanish outside another set $B_1$, where $B_0 \subset B_1$ and both of these sets are open, bounded sets in $D(A)$. In Kwak (1991) it is shown that these bounded sets can be chosen so that

(1) $\mathfrak{A}_{NS} \subset B_0$, and

(2) the semiflow generated by the modified Navier-Stokes equations is dissipative and has the same global attractor $\mathfrak{A}_{NS}$.

Let us now turn to the 3DNS on a thin domain $\Omega_\epsilon$. As in Section 3, we apply the dilation operator $J_\epsilon^{-1}$ to obtain the dilated 3DNS on $Q_3$ given by (3.1). Next we apply $\mathbf{P}_\epsilon$ to (3.1) to obtain (3.2). By using Theorems C and D we see that once again the global attractor exists and is a bounded set in $D(A_\epsilon)$, see (3.11) and (3.12).

The Kwak Transformation associated with the dilated 3DNS on $Q_3$ now takes on the form

$$\mathsf{U} = G_\epsilon(u) = (G_0, G_1, G_2), \qquad u \in \mathbf{V}_\epsilon, \tag{4.11}$$

where $G_0(u) = u$, $G_1(u) = z = \nabla_\epsilon * u$, and $G_2(u) = r = u \otimes u$. For $\epsilon > 0$ we define $\mathcal{H}_\epsilon$ by

$$\mathcal{H}_\epsilon = \mathcal{H}_\epsilon(Q_3) \overset{\text{def}}{=} \mathbf{H}_\epsilon(Q_3)^4 \times L^2(Q_3, \Sigma^3).$$

Let $\mathcal{H}_0$ denote those functions $\mathsf{U} = \mathsf{U}(x_1, x_2)$ in $\mathcal{H}_\epsilon$ that do not depend on $x_3$. Thus one has

$$\mathcal{H}_0 = \mathcal{H}_0(Q_3) \overset{\text{def}}{=} \mathbf{H}_0(Q_3)^4 \times L^2(Q_2, \Sigma^3).$$

For the 3D problem, the Kwak Transformation is a mapping of $\mathbf{V}_\epsilon(Q_3)$ in $\mathcal{H}_\epsilon$.

Let $u = v \in \mathbf{V}_0(Q_3)$, i.e., $Mu = u = v$. Then the image $\mathsf{U} = G_\epsilon(u)$ under the Kwak Transformation does not depend on $\epsilon$ and is a point in $\mathcal{H}_0$. As noted in Section 3, the space $\mathbf{V}_0(Q_3)$ is an invariant set under the dynamics $S_\epsilon(t)$ of the 3DNS, provided $(I-M)\mathbf{P}_\epsilon f = 0$, see (3.6) and the discussion preceeding the statement of Theorem E. Furthermore from (3.7) and (3.8), we see that if $(I-M)\mathbf{P}_\epsilon f = 0$ and $u_0 = \bar{v}_0 \in \mathbf{V}_0$, then the solution $S_\epsilon(t)\bar{v}_0$, which does not depend on $\epsilon$, satisfies $\bar{v} = (m_1, m_2, \bar{v}_3)$, where $m = (m_1, m_2)$ is a solution of the 2DNS and $\bar{v}_3$ is a slave variable. In this way, the dynamics of the 2DNS is imbedded in the reduce 3DNS on $Q_3$, which in turn is imbedded in the dynamics of the dilated 3DNS on $Q_3$. Furthermore, this imbedding of the 2DNS into the 3DNS is preserved by the Kwak Transformation G, which as we have seen, maps $\mathbf{V}_0(Q_3)$ into $\mathcal{H}_0(Q_3)$. We now have the following result concerning the reduced 3DNS on $Q_3$, see Kwak, Sell, and Shao (1993).

THEOREM G. *The conclusions of Theorem F remain valid for the reduced 3DNS on $Q_3$, where (4.7) is replaced by*

$$\frac{d}{dt}\mathsf{U} + \mathsf{A}_0\mathsf{U} = \mathsf{F}_0(\mathsf{U}), \qquad \mathsf{U} \in \mathcal{H}_0. \tag{4.12}$$

*In particular, if the eigenvalues of the 2DNS on* $Q_2$ *satisfy (4.8), then there exists an inertial manifold* $\mathfrak{M}_0$ *for the reduced 3DNS on* $Q_3$. *Furthermore,* $\mathfrak{M}_0$ *can be realized be realized as the graph of a suitable function*

$$\Phi_0 : P\mathcal{H}_0 \to Q\mathcal{H}_0 \cap D(A_0),$$

*where* $P$ *is the orthogonal projection of* $\mathcal{H}_0$ *onto the sum of the generalized eigenspaces of* $A_0$ *corresponding to eigenvalues* $\lambda$ *satisfying* $\lambda \le \lambda_m$ and $Q = I - P$. *Moreover, the restriction of (4.12) to* $\mathfrak{M}_0$ *gives rise to the inertial form*

$$p' + A_0 p = P\mathsf{F}_0(p + \Phi_0(p)).$$

For the dilated 3DNS on $Q_3$, we obtain the following result, see Kwak, Sell, and Shao (1993).

THEOREM H. *Let the dilated 3DNS be given on* $Q_3$ *by (3.2), with periodic boundary conditions. Then there is an* $\epsilon_3 > 0$ *such that* $\epsilon_3 \le \min(\epsilon_0, \epsilon_1)$, *where* $\epsilon_0$ *and* $\epsilon_1$ *are given by Theorems C and D, and for every* $\epsilon$, $0 < \epsilon \le \epsilon_3$, *there is a system of reaction diffusion equations of the form*

$$\frac{d}{dt}\mathsf{U} + A_\epsilon \mathsf{U} = \mathsf{F}_\epsilon(\mathsf{U}), \qquad \mathsf{U} \in \mathcal{H}_\epsilon, \tag{4.13}$$

*with the following properties:*

(1) *Equation (4.13) agrees with the counterpart of (4.7) in a neighborhood of* $G(\mathfrak{A}_{\epsilon;\mathrm{NS}})$, *where* $G_\epsilon$ *is the Kwak Transformation (4.11), and* $\mathfrak{A}_{\epsilon;\mathrm{NS}}$ *is the global attractor of the dilated 3DNS on* $Q_3$.

(2) *The linear operator* $A_\epsilon$ *is a sectorial operator with compact resolvent on* $\mathcal{H}_\epsilon$. *Furthermore, the fractional powers of* $A_\epsilon$ *are well-defined, and we set* $\mathcal{V}_\epsilon = D(A_\epsilon^{\frac{1}{2}})$.

(3) *There exist constants* $K_0 > 0$ *and* $K_1 > 0$ *such that for all* $\epsilon$, $0 < \epsilon \le \epsilon_3$, *the nonlinearity* $\mathsf{F}_\epsilon = \mathsf{F}_\epsilon(\mathsf{U})$ *is a locally Lipschitz continuous mapping of* $\mathcal{V}_\epsilon$ *into* $\mathcal{H}_\epsilon$. *Furthermore* $\mathsf{F}_\epsilon$ *is a Lipschitz continuous mapping of the domain* $D(A_\epsilon)$ *into itself with* $\|A_\epsilon \mathsf{F}_\epsilon(\mathsf{U})\| \le K_0$, *for all* $\mathsf{U} \in D(A_\epsilon)$ *and*

$$\|A_\epsilon(\mathsf{F}_\epsilon(\mathsf{U}_1) - \mathsf{F}_\epsilon(\mathsf{U}_2))\| \le K_1 \|A_\epsilon(\mathsf{U}_1 - \mathsf{U}_2)\|, \qquad \text{for all } \mathsf{U}_1, \mathsf{U}_2 \in D(A_\epsilon).$$

(4) *Equation (4.13) is dissipative on the space* $\mathcal{V}_\epsilon$, *and the global attractor* $\mathfrak{A}_{\epsilon;\mathrm{RD}}$ *is a bounded set in* $D(A_\epsilon)$ *that satisfies*

$$\mathfrak{A}_{\epsilon;\mathrm{RD}} = G_\epsilon(\mathfrak{A}_{\epsilon;\mathrm{NS}}).$$

*Moreover, the restriction* $G_\epsilon : \mathfrak{A}_{\epsilon;\mathrm{NS}} \to \mathfrak{A}_{\epsilon;\mathrm{RD}}$ *is a homeomorphism which preserves all the long-time dynamics of the dilated 3DNS on* $Q_3$.

In order to compare the eigenvalues of the 2D and 3D problems, it is convenient to use the $v$ and $w$ equations (3.5). Because of the commutivity relationship (3.3) we see that for the dilated 3DNS on $Q_3$ one has

$$A_\epsilon u = \lambda u \iff A_\epsilon v = \lambda v \quad \text{and} \quad A_\epsilon w = \lambda w,$$

where $v = Mu$ and $w = (I - M)u$. In other words, the spectrum of $A_\epsilon$ satisfies

$$\sigma(A_\epsilon) = \sigma(A_\epsilon M) \cup \sigma(A_\epsilon(I - M)),$$

which implies that

$$\sigma(A_\epsilon + k_1^2 I) = \sigma(A_\epsilon M + k_1^2 I) \cup \sigma(A_\epsilon(I - M) + k_1^2 I).$$

Now the spectrum of $A_\epsilon M$ does not depend on $\epsilon$ and is precisely the spectrum of the Stokes operator for the 2DNS. Using the convention described above for the 2DNS, we denote the distinct eigenvalues of $(A_\epsilon M + k_1^2 I)$ by $\lambda_1 < \lambda_2 < \lambda_3 < \dots$, and the distinct eigenvalues of $(A_\epsilon + k_1^2 I)$ are denoted by $\Lambda_1 < \Lambda_2 < \Lambda_3 < \dots$.

For the eigenvalues of $A_\epsilon(I - M)$, we use (3.4) twice to obtain

$$\|w\| \le C_5^2 \epsilon^2 \|A_\epsilon w\|, \qquad w \in \mathbf{V}_\epsilon^2,\ Mw = 0. \tag{4.14}$$

Now (4.14) implies that $A_\epsilon(I - M)$ is a positive operator. If one has

$$A_\epsilon(I - M)w = A_\epsilon w = \mu w, \qquad \text{for } \mu \ge 0,$$

where $\|w\| = 1$, then (4.14) implies that $1 \le C_5^2 \epsilon^2 \|\mu w\| = C_5^2 \epsilon^2 \mu$, or $\mu \ge C_5^{-2} \epsilon^{-2}$.

Using the facts that the bound $K_0$ and the Lipschitz coefficient $K_1$ for $\mathsf{F}_\epsilon$ do not depend on $\epsilon$, for $\epsilon$ small, one can show that the constants $C_0$ and $C_1$ appearing in the spectral gap condition (4.8) can be chosen independent of $\epsilon$ as well, see Sell and You (1992). As a result, we obtain the following result.

COROLLARY OF THEOREM H. *Let the dilated 3DNS be given on $Q_3$ by (3.2), with periodic boundary conditions. Use (4.10) to select and fix $\lambda_m$ and $\lambda_{m+1}$ so that (4.8) holds. Next fix $\epsilon$ to satisfy $0 \le \epsilon \le \epsilon_4$, where $\epsilon_4 = \min(\epsilon_3, C_5^{-1} \lambda_{m+1}^{-\frac{1}{2}})$. Then the eigenvalues for the 3D problem satisfy:*

(1) *$\Lambda_i = \lambda_i$, and these eigenvalues have the same multiplicities for $1 \le i \le m + 1$;*

(2) *the generalized eigenspaces of $\mathsf{A}_\epsilon$ corresponding to eigenvalues $\lambda$ with $\lambda = \Lambda_i$, for some $i$ with $1 \le i \le m+1$, consist entirely of functions $u = u(x_1, x_2)$ which do not depend on $x_3$; and*

(3) *one has*

$$\Lambda_{m+1} - \Lambda_m \ge C_1 \quad \text{and} \quad \Lambda_{m+1} \ge C_0.$$

Now let $P$ denote the orthogonal projection on $\mathcal{H}_\epsilon$ into the spaces spanned by the generalized eigenspaces of $\mathsf{A}_\epsilon$ corresponding to eigenvalues $\lambda$ with $\lambda = \Lambda_i$, for some $i$ with $1 \leq i \leq m+1$. Because of property (2) above, it follows that $P\mathcal{H}_\epsilon$ is a subspace of $\mathcal{H}_\epsilon$ consisting of functions which do not depend on $x_3$. In other words, one has $P\mathcal{H}_\epsilon = P\mathcal{H}_0$, for all $\epsilon$ with $0 < \epsilon \leq \epsilon_4$. By using the theory of Sell and You (1992), it is shown in Kwak, Sell, and Shao (1993) that (4.12) has an inertial manifold $\mathfrak{M}_\epsilon$ which can be realized as the graph of a function

$$\Phi_\epsilon : P\mathcal{H}_\epsilon = P\mathcal{H}_0 \to Q\mathcal{H}_\epsilon \cap D(\mathsf{A}_\epsilon), \qquad 0 \leq \epsilon \leq \epsilon_4, \tag{4.15}$$

where $Q = I - P$. (Notice that we have included $\epsilon = 0$ in (4.15) to include the inertial manifold $\mathfrak{M}_0$ for the reduced 3DNS on $Q_3$.) The associated inertial forms for (4.12) and (4.13) are then given by

$$p' + \mathsf{A}_\epsilon p = P\mathsf{F}_\epsilon(p + \Phi_\epsilon(p)), \qquad 0 \leq \epsilon \leq \epsilon_4. \tag{4.16}$$

Let us examine the various terms in (4.16) with some care. First of all, since $P\mathcal{H}_\epsilon = P\mathcal{H}_0$, for $0 \leq \epsilon \leq \epsilon_4$, we see that each of the three terms in (4.16) describe functions of $x_1$ and $x_2$, which do not depend on $x_3$. Since $\mathsf{A}_\epsilon p$ does not depend on $x_3$, one has $\mathsf{A}_\epsilon p = \mathsf{A}_0 p$, for $0 \leq \epsilon \leq \epsilon_4$. At $\epsilon = 0$ the mapping $\Phi_0$ assumes values in $Q\mathcal{H}_0 \cap D(\mathsf{A}_0)$, and as a result $\Phi_0(p)$ is also a function which does not depend on $x_3$. The only place where $x_3$-dependence can occur in (4.16) is in the term $\Phi_\epsilon(p)$, for $\epsilon > 0$. Recall that if $p(t)$ is a solution of (4.16), for $t \geq 0$, then $\mathsf{U}(t) = p(t) + \Phi_\epsilon(p(t))$ is a solution of (4.12) or (4.13) and $\mathsf{U}(t) \in \mathfrak{M}_\epsilon$, for all $t \geq 0$ and $0 \leq \epsilon \leq \epsilon_4$.

As a result, we see that equation (4.16) can be written as

$$p' + \mathsf{A}_0 p = P\mathsf{F}_0(p + \Phi_0(p)) + E_\epsilon(p), \qquad 0 \leq \epsilon \leq \epsilon_4, \tag{4.17}$$

where $E_\epsilon(p) = P\mathsf{F}_\epsilon(p + \Phi_\epsilon(p)) - P\mathsf{F}_0(p + \Phi_0(p))$. The following result is proved in Kwak, Sell, and Shao (1993).

THEOREM I. *Let the dilated 3DNS be given on $Q_3$ by (3.2), with periodic boundary conditions. Let $\Lambda_m$ and $\Lambda_{m+1}$ be chosen by the Corollary to Theorem H, and let $\mathfrak{M}_\epsilon =$ Graph $\Phi_\epsilon$ denote the associated inertial manifolds and (4.17) related inertial forms, for $0 \leq \epsilon \leq \epsilon_4$. Assume that the forcing function $f$ in (3.2) satisfies*

$$\mathsf{P}_{\epsilon_n} f \in \mathcal{S}(\epsilon_n) \cap \mathbf{V}^3_{\epsilon_n}, \qquad \text{for } 0 < \epsilon_n \leq \epsilon_4, \tag{4.18}$$

*where $\epsilon_n$ is a sequence satisfying $\epsilon_n \to 0$, as $n \to \infty$. Assume further that there is a function $g \in \mathbf{V}^3_0$ such that*

$$\|A_{\epsilon_n}\mathsf{P}_{\epsilon_n} f - A_{\epsilon_n} g\| + \|A_{\epsilon_n}\mathsf{P}_{\epsilon_n}\nabla_{\epsilon_n} * f - A_{\epsilon_n}\nabla_{\epsilon_n} * g\| \to 0, \qquad \text{as } n \to \infty. \tag{4.19}$$

*Then one has $\|E_{\epsilon_n}\|_{C^1} \to 0$, as $n \to \infty$, where the norm is defined by*

$$\|E_\epsilon\|_{C^1} = \sup_{p \in P\mathcal{H}_0} \|E_\epsilon(p)\| + \sup_{p \in P\mathcal{H}_0} \|DE_\epsilon(p)\|_{op}.$$

Note that if $\mathbf{P}_{\epsilon_n} f$ satisfies (4.18) and (4.19), then all the hypotheses of Theorem E are satisfied. Consequently, the global attractors $\mathfrak{A}_{\epsilon_n}$ for the dilated 3DNS on $Q_3$ are upper semicontinuous in $\mathbf{V}^2_{\epsilon_n}$ at $\epsilon = 0$, and (3.13) is valid.

The proof of Theorem I consists in showing that the nonlinearity $\mathsf{F}_\epsilon$ is a $C^1$-mapping from $D(\mathsf{A}_\epsilon)$ into itself and has bounded support. Furthermore, one has

$$\|\mathsf{A}_\epsilon(\mathsf{F}_\epsilon - \mathsf{F}_0)\|_{C^1} \to 0, \qquad \text{as } \epsilon \to 0^+.$$

Next one shows that the mappings $\Phi_\epsilon$ are also $C^1$-functions with bounded support and that

$$\|\mathsf{A}_\epsilon(\Phi_\epsilon - \Phi_0)\|_{C^1} \to 0, \qquad \text{as } \epsilon \to 0^+.$$

See Kwak, Sell, and Shao (1993) for the details.

**5. Approximation Dynamics for the Navier-Stokes Equations.** Because of the convergence $\|E_{\epsilon_n}\|_{C^1} \to 0$, as $n \to \infty$, shown in the last section, we see that Theorems A and B are applicable to the dilated 3DNS on $Q_3$. In this section we wish to give a precise description of the concept of a hyperbolic attractor referred to in Theorems A and B, along with some comments on the dynamical features one can expect to see in such an attractor. It should be emphasized that we are not assuming that the global attractor itself have a hyperbolic structure. Instead, we are studying some distinguished compact, invariant sets $\mathcal{K}$, which are necessarily contained in the global attractor, and which are themselves local attractors. For example, $\mathcal{K}$ may represent the orbit of a stable, periodic pattern, such as a Taylor cell occurring in a fluid flow. Another example is a stable torus, as is seen after a Hopf bifurcation of a periodic orbit.

Let $\mathcal{K}$ denote a given compact invariant set for (1.1). For $u_0 \in \mathcal{K}$, we let $u(t, u_0) = S_1(t)u_0$ denote the solution of (1.1) that satisfies $u(0, u_0) = u_0$. For each $u_0 \in \mathcal{K}$ we let $\Phi(t, u_0)$ denote the fundamental operator solution of the linear system

$$\frac{d}{dt}v = \frac{\partial F(u(t, u_0))}{\partial u} v \tag{5.1}$$

that satisfies $\Phi(0, u_0) = I$, where $I$ is the identity operator on $R^n$. We shall say that the linear system (5.1) is **weakly hyperbolic along** $u(t, u_0)$ **with constants** $a$, $\lambda_1$, and $\lambda_2$ provided $\lambda_2 < \lambda_1$, $\lambda_1 > 0$, $a \geq 1$ and there exist complementary linear spaces $\mathfrak{U}^n(t, u_0)$ and $\mathfrak{U}^s(t, u_0)$, $\dim \mathfrak{U}^n = k$, $\dim \mathfrak{U}^s = n - k$, and

$$\Phi(t, u_0)\mathfrak{U}^s(0, u_0) = \mathfrak{U}^s(t, u_0), \qquad \Phi(t, u_0)\mathfrak{U}^n(0, u_0) = \mathfrak{U}^n(t, u_0)$$

for all $t \in R$. Furthermore if $v \in \mathcal{U}^s(\tau, u_0)$, then

$$|\Phi(t, u_0)\Phi^{-1}(\tau, u_0)v| \leq a|v|e^{-\lambda_1(t-\tau)}, \qquad \text{for } t \geq \tau, \quad t, \tau \in R,$$

and if $v \in \mathcal{U}^n(\tau, u_0)$, then

$$|\Phi(t, u_0)\Phi^{-1}(\tau, u_0)v| \leq a|v|e^{-\lambda_2(t-\tau)}, \qquad \text{for } t \leq \tau, \quad t, \tau \in R.$$

See Pliss (1977), Sacker and Sell (1978, 1980), and Sell (1978) for related concepts.

We shall say that a compact, invariant set $\mathcal{K}$ for (1.1) is a **weakly, normally hyperbolic attractor** if $\mathcal{K}$ satisfies the following two properties:

(1) The linear system (5.1) is hyperbolic along $u(t, u_0)$ with constants $a$, $\lambda_1$, and $\lambda_2$ for every $u_0 \in \mathcal{K}$.

(2) There exists an $r > 0$ such that for each $u_0 \in \mathcal{K}$ there exists a $k$-dimensional locally invariant disk $\mathcal{D}(u_0) \subset \mathcal{K}$ with the center at the point $u_0$ and radius $r$ such that if $u \in \mathcal{D}(u_0)$, then at the point $u$, the disk $\mathcal{D}(u_0)$ is tangent to the linear space $\mathcal{U}^n(0, u)$.

A weakly, normally hyperbolic attractor $\mathcal{K}$ is said to satisfy the **Lipschitz property** if the mapping $u_0 \to \mathcal{U}^n(0, u_0)$ is a Lipschitz continuous mapping of $\mathcal{K}$ into the Grassmanian of $k$-dimensional subspaces of $R^n$. The notion of hyperbolicity introduced here has some antecedents in the literature, see for example, Fenichel (1971), Hirsch, Pugh and Shub (1977), Meyer and Sell (1989), Pliss (1977), Sacker (1969), Sell (1978), and Smale (1967). In the sequel we shall let $\mathcal{K}$ denote a given hyperbolic attractor for (1.1) that satisfies the Lipschitz property.

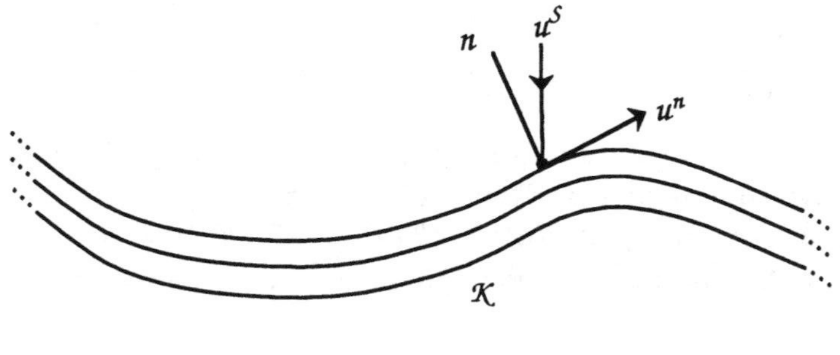


Figure: Schematic of a Hyperbolic Set $\mathcal{K}$.

Let for $u_0 \in \mathcal{K}$ we define the sets $\mathcal{S}_1(u_0)$, $\mathcal{S}_2(u_0)$, $\mathcal{S}_3(u_0)$, ..., $\mathcal{S}(u_0)$ by

$$\mathcal{S}_1(u_0) = \bigcup_{u \in \mathcal{D}(u_0)} \mathcal{D}(u), \qquad \mathcal{S}_{i+1}(u_0) = \bigcup_{u \in \mathcal{S}_i(u_0)} \mathcal{D}(u), \qquad \text{for } i \geq 1,$$

and

$$\mathcal{S}(u_0) = \bigcup_{i=1}^{\infty} \mathcal{S}_i(u_0).$$

Note that the set $\mathcal{S}(u_0)$ is invariant and $\mathcal{S}(u_0) \subset \mathcal{K}$, for every $u_0 \in \mathcal{K}$. The Lipschitz property insures that the leaf through each point $u_0$ is uniquely determined by $u_0$, that is, if $u_1 \in \mathcal{S}(u_0)$, then $\mathcal{S}(u_1) = \mathcal{S}(u_0)$, see Pliss and Sell (1991). This implies that if $\mathcal{S}(u_0) \cap \mathcal{S}(u_1)$ is nonempty, then $\mathcal{S}(u_0) = \mathcal{S}(u_1)$. The set $\mathcal{S}(u_0)$ is referred to as the **leaf** of $\mathcal{K}$ through $u_0$.

Let $\mathcal{U}^s(u_0) = \mathcal{U}^s(0, s_0)$, $\mathcal{U}^n(u_0) = \mathcal{U}^n(0, u_0)$. It is clear that $\mathcal{U}^s(t, u_0) = \mathcal{U}^s(u(t, u_0))$ and $\mathcal{U}^n(t, u_0) = \mathcal{U}^n(u(t, u_0))$ for all $t$. Let $\mathcal{N}(u)$, $u \in \mathcal{K}$ be the $(n-k)$-dimensional hyperplane perpendicular to $\mathcal{U}^n(u)$ at the point $u$.

There are several points to note concerning this definition of a weakly, normally hyperbolic attractor $\mathcal{K}$ and the dynamics on $\mathcal{K}$. These are:

(1) the set $\mathcal{K}$ may have many leafs, even infinitely many;

(2) in the case of infinitely many leafs in $\mathcal{K}$, these leafs may be dense in $\mathcal{K}$; and

(3) the set $\mathcal{K}$ may contain fixed points of (1.1), even though $\mathcal{K}$ itself is not a fixed point.

The basic approach in the proof of Theorem A begins by fixing a point $u_0 \in \mathcal{K}$ and then seeking another vector $\psi(u_0) \in R^n$ which satisfies $\psi(u_0) \in \mathcal{N}(u_0)$ and so that

$$h(u_0) \stackrel{\text{def}}{=} u_0 + \psi(u_0)$$

is the desired mapping of $\mathcal{K}$ onto $\mathcal{K}^E$. Among other things, we want $\|h(u_0) - u_0\| = \|\psi(u_0)\| < \epsilon$. The construction of $\psi$ is based on the Graph Transform Method, and ultimately, the proof reduces to using a fixed point theorem on a suitable function space, see Pliss and Sell (1991). It is shown that $\psi : \mathcal{K} \to R^n$ is a continuous mapping, and for each leaf $\mathcal{S}$ in $\mathcal{K}$, the mapping $\psi : \mathcal{S} \to R^n$ is locally Lipschitz continuous.

Let us now turn to the shadow flow $S_1^E(t)u_0$ mentioned in Theorem B. For each point $u \in \mathcal{K}$, there is an orthonormal coordinate system generated by $\mathcal{U}^n(u)$ and $\mathcal{N}(u)$. Let $P(u)$ denote the orthogonal projection of $R^n$ onto $\mathcal{U}^n(u)$, and set $Q(u) = I - P(u)$. The ranges of $P(u)$ and $Q(u)$ then satisfy $\mathcal{R}(P(u)) = \mathcal{U}^n(u)$ and $\mathcal{R}(Q(u)) = \mathcal{N}(u)$. Since $\mathcal{U}^n(u)$ is tangent to the leaf $\mathcal{S}(u)$ passing through $u$, at each $u \in \mathcal{K}$, the vector field $F$ in (1.1) must satisfy

$$P(u)F(u) = F(u), \qquad \text{for all } u \in \mathcal{K}.$$

Furthermore, the Lipschitz property of the mapping $u \to \mathcal{U}^n(u)$ implies that the mapping $u \to P(u)$ is also Lipschitz continuous. Now let $\psi$ be the fixed point described above. Then the shadow flow $S_1^E(t)u_0$ on $\mathcal{K}$ is generated by the initial value problem

$$\sigma' = P(\sigma)\left[F(\sigma + \psi(\sigma) + F(\sigma + \psi(\sigma))\right], \qquad \sigma(0) = u_0,\ u_0 \in \mathcal{K}. \tag{5.2}$$

Since all the terms in (5.2) are locally Lipschitz continuous functions of $\sigma$ on each leaf $\mathcal{S}$ in $\mathcal{K}$, and since these leafs are uniquely determined, equation (5.2) has a unique solution. Furthermore, since $\mathcal{R}(P(\sigma)) = \mathcal{U}^n(\sigma)$ is tangent to the leaf $\mathcal{S}(\sigma)$ for each $\sigma \in \mathcal{K}$, it follows that the solution $\sigma(t) = S_1^E(t)u_0$ of (5.2) lies in $\mathcal{S}(u_0) \subset \mathcal{K}$, for all $t \in R$.

**Examples.** A stable fixed point and a stable periodic orbit are hyperbolic attractors that satisfy the Lipschitz property. More generally, a smooth, stable, normally hyperbolic, invariant manifold is a weakly, normally hyperbolic attractor that satisfies the Lipschitz property. The fact that a normally hyperbolic, invariant manifold persists under a small $C^1$-perturbation is, of course, well-known. See for example, Fenichel (1971), Hirsch, Pugh, and Shub (1977), and Sacker (1969). Other examples of weakly, normally hyperbolic attractors are given in Pliss and Sell (1991). These include

(1) attractors which are not manifolds, and

(2) invariant manifolds which are not normally hyperbolic.

Let us examine one of these examples in more detail.

The $k$-dimensional torus $T^k$ can be imbedded into an open set $O$ in $R^{k+\ell}$ by using the coordinate system

$$(\theta;\sigma) = (\theta_1,\dots,\theta_k;\sigma_1,\dots,\sigma_\ell), \qquad 0 \le \theta_i < 2\pi, \quad \sigma = (\sigma_1,\dots,\sigma_\ell) \in \mathcal{D}^\ell,$$

where $\mathcal{D}^\ell$ denotes an $\ell$-dimensional disk. Let $\mathcal{K}_0$ denote a given hyperbolic attractor with constants $a$, $\lambda_1$, and $\lambda_2$ for an ordinary differential equation $x' = X(x)$ on $R^n$. Then $\mathcal{K}_1 = \mathcal{K}_0 \times T^k$ is a hyperbolic attractor for the product flow

$$\text{(5.3)} \qquad \begin{cases} x' = F(x,\theta,\sigma) \\ \theta' = \Theta(x,\theta,\sigma) \\ \sigma' = -A\sigma + G(x,\theta,\sigma) \end{cases}$$

on $R^n \times O$, provided $A$ is an $(\ell \times \ell)$ matrix whose eigenvalues $\mu_i$ satisfy Re $\mu_i > \max(0,\lambda_2)$ and $F$, $\Theta$, and $G$ are $C^1$-functions that satisfy $F(x,\theta,0) = X(x)$, $\Theta(x,\theta,0) = \omega$, $G(x,\theta,0) = 0$, and $D_3G(x,\theta,0) = 0$, where $D_3 = \frac{\partial}{\partial z}$ and $\omega = (\omega_1,\dots,\omega_k)$ is a constant vector. If $\mathcal{K}_0$ satisfies the Lipschitz property, then so does $\mathcal{K}_1$.

Let $\mathcal{D}_i(u_0)$, $u_0 \in \mathcal{K}_0$, denote the disks in $\mathcal{K}_i$, for $i = 0, 1$, as prescribed in the definition of a hyperbolic attractor. Then $\dim \mathcal{D}_0 = k_0$ for some $k_0$, and one has $\dim \mathcal{D}_1 = k_0 + k$. If there is an eigenvalue $\mu_i$ that satisfies $\max(0,\lambda_2) <$ Re $\mu_i \le \lambda_1$, then the attractor $\mathcal{K}_1$ is not normally hyperbolic, see Hirsch, Pugh and Shub (1977). Even though $\mathcal{K}_1$ need not be normally hyperbolic, Theorem A is applicable to perturbations of the underlying equation.

It is instructive to compare this theory of Approximation Dynamics, as it applies to (5.3), with the examples constructed in Fenichel (1971) of perturbations of invariant manifolds which are not normally hyperbolic. Theorem A asserts that if one adds any sufficiently small $C^1$-perturbation term $Y$ to (5.3), then there is a homeomorphism $h$ of $\mathcal{K}_1$ onto a hyperbolic attractor $\mathcal{K}^Y$ for the perturbed equation. The Fenichel construction suggests that, since $\mathcal{K}_1$ is not normally hyperbolic, there exists a small perturbation such that the perturbed equation does not have a *smooth* invariant manifold near $\mathcal{K}_1$. In such a case, we conclude that, while $h : \mathcal{K}_1 \to \mathcal{K}^Y$ and $h^{-1} : \mathcal{K}^Y \to \mathcal{K}_1$ are both continuous, at least one of them is not differentiable. As is shown in Pliss and Sell (1991), for each leaf $\mathcal{S} \subset \mathcal{K}_1$, the restriction $h\mid_{\mathcal{S}}$ and its inverse are locally Lipschitz continuous. It follows that a breakdown in smoothness can only occur in the direction normal to $\mathcal{S}$. Wrinkles can only arise in the stable direction!

**6. Concluding Remarks.** Each of the three theories described above, namely

(1) the existence of global attractors for the weak (Leray) solutions of the Navier-Stokes equations on thin 3D domains,

(2) the existence of inertial forms for the 2D and 3D Navier-Stokes equations, and

(3) the theory of Approximation Dynamics,

are presented to illustrate new and salient finite dimensional structures in the theory of the long-time dynamical behavior of the Navier-Stokes equations. By putting these three theories together, as we have done in this paper, we were able to derive a rigorous foundation for the study of all the long-time dynamics of the dilated 3DNS on $Q_3$, in terms of finite dimensional systems of ordinary differential equations. We also showed the robustness, or stability, of the dynamics contained in certain hyperbolic attractors, when comparing the thin 3D domain problem with the limiting 2D problem.

It is widely believed, based primarily on physical intuition, that there is very little difference in the behavior of turbulence in 2D fluid flows and in 3D flows on thin domains. We have presented in this paper the basic mathematical tools for addressing this issue. Furthermore, we have given the first rigorous proof of this physical intuition concerning turbulence, at least in the context of the hyperbolic attractors described above.

One issue that cannot be overemphasized is the importance of the $C^1$-convergence of the error terms in (1.2). The importance of this issue, which is well understood in the dynamics literature, seems to have been overlooked by some researchers. For example, some theories of approximate inertial manifolds are based on error estimates in a $C^0$-norm. While a theory of $C^0$-approximations for dynamical issues is under development, see Pilyugin (1992), it is known that the $C^0$-approximations are too crude to accurately detect such delicate dynamical features as is seen in bifurcation theory. There is a need to develop approximate inertial manifold methods based on $C^1$-convergence, see Sell (1992).

The shadow flow described in Theorem B contains special meaning for the approximations of the long-time dynamics of the Navier-Stokes equations. Take a simple case, where $\mathcal{K}$ is a stable, periodic orbit. In this case, the shadow flow on $\mathcal{K}$ describes (1) a change in the speed of moving along the perturbed periodic orbit and (2) a change in the phase of the perturbed orbit. Phase changes are also evident in more complicated dynamical behavior. For example, what one actually sees in some videos of different approximations of the global attractor of the Navier-Stokes equations is that the phase changes are spatially dependent. A report on this phenomenon is now under preparation by Sell and Yin Yan.

Theorems A and B are important contributions to the theory of Approximation Dynamics, but there is more to be done. The Lorenz attractor, for example, does not seem to satisfy the definition of a weakly, normally hyperbolic attractor used above. Nevertheless, the Lorenz attractor seems to be stable, in some sense. It would be good to be able to give a rigorous explanation of this phenomenon. We also need new methods to treat chaotic behavior and turbulence seen in many fluid flows.

Finally there is a very important problem underlying all modeling of all 3 dimensional, incompressible, viscous, fluid flows. This is the problem of the global regularity of the solutions of the 3DNS. In the theory of the dilated 3DNS on $Q_3$, Raugel and Sell (1992a, 1992b, 1992c) made a significant contribution in helping to resolve this problem, because they were able to show that every weak (Leray) solution of these equations becomes a strong solution at some time $t_0 > 0$ and remains a strong (or regular) solution for all $t \geq t_0$. Even though the theory of Raugel and Sell is a major inroad into the global regularity issue, this issue remains one of the major problems in the theory of the Navier-Stokes equations in 3 dimensions.

## REFERENCES


V. I. ARNOLD (1983), *Geometrical Methods in the Theory of Ordinary Differential Equations*, Springer Verlag, New York.

J. E. BILLOTTI, J. P. LASALLE (1971), *Dissipative periodic processes*, Bull. Amer. Math. Soc., 77, pp. 1082-1088.

P. CONSTANTIN AND C. FOIAS (1988), *Navier-Stokes Equations*, Univ. Chicago Press, Chicago.

N. FENICHEL (1971), *Persistence and smoothness of invariant manifolds for flows*, Indiana Univ. Math. J., 21, pp. 193-226.

C. FOIAS AND G. PRODI (1967), *Sur le comportement global des solutions non stationnaires des équations de Navier-Stokes en dimension 2*, Rend. Sem. Univ. Padova, 39.

C. FOIAS, G. R. SELL AND R. TEMAM (1988), *Inertial manifolds for nonlinear evolutionary equations*, J. Differential Equations, 73, pp. 309-353.

C. FOIAS AND R. TEMAM (1979), *Some analytic and geometric properties of the solutions of the Navier-Stokes equations*, J. Math. Pures Appl., 58, pp. 339-368.

J. K. HALE (1988), *Asymptotic Behavior of Dissipative Systems*, Math. Surveys and Monographs, Vol. 25, Amer. Math. Soc., Providence, R. I..

J. K. HALE AND G. RAUGEL (1988), *Upper semicontinuity of the attractor for a singularly perturbed hyperbolic equation*, J. Differential Equations, 73, pp. 197-214.

J. K. HALE AND G. RAUGEL (1989), *Lower semicontinuity of the attractor for gradient systems*, Annali Mat. Pura Appl., 154, pp. 281-326.

D. B. HENRY (1981), *Geometric Theory of Semilinear Parabolic Equations*, Lecture Notes in Mathematics, No. 840, Springer Verlag, New York.

M. W. HIRSCH, C. C. PUGH AND M. SHUB (1977), *Invariant Manifolds*, Lecture Notes In Math, No. 583, Springer Verlag, New York.

M. KWAK (1991), *Finite dimensional inertial forms for the 2D Navier-Stokes equations*, University of Minnesota PhD Thesis, AHPCRC Preprint 91-30, Indiana J. Math. (to appear).

M. Kwak (1992), *Finite dimensional description of convective reaction diffusion equations*, J. Dynamics and Differential Equations, 4, pp. 515-543.

M. Kwak, G. R. Sell, and Z. Shao (1993), *Finite dimensional structures for the Navier-Stokes equations on thin 3D domains*, AHPCRC Preprint.

O. A. Ladyzhenskaya (1972), *On the dynamical system generated by the Navier-Stokes equations*, English translation, J. Soviet Math., 3, pp. 458-479.

J. Leray (1933), *Etude de diverses équations intégrales nonlinéaires et de quelques problèmes que pose l'hydrodynamique*, J. Math. Pures Appl., 12, pp. 1-82.

J. Leray (1934a), *Essai sur les mouvements plans d'un liquide visqueux que limitent des parois*, J. Math. Pures Appl., 13, pp. 331-418.

J. Leray (1934b), *Sur le mouvement d'un liquide visqueux emplissant l'espace*, Acta Math., 63, pp. 193-248.

J. Mallet-Paret (1976), *Negatively invariant sets of compact maps and an extension of a theorem of Cartwright*, J. Differential Equations, 22, pp. 331-348.

J. Mallet-Paret and G. R. Sell (1988), *Inertial manifolds for reaction diffusion equations in higher space dimensions*, J. Amer. Math. Soc., 1, pp. 805-866.

R. Mañé (1981), *On the dimension of the compact invariant sets of certain nonlinear maps*, in *Lecture Notes in Math., Vol. 898*, Springer Verlag, New York, pp. 230-242.

L. Markus (1961), *Structurally stable differential systems*, Ann. Math., 73, pp. 1-19.

L. Markus (1971), *Lectures in Differentiable Dynamics*, CBMS Lecture Notes in Math., No. 3, Amer. Math. Soc., Providence.

K. R. Meyer and G. R. Sell (1989), *Melnikov transforms, Bernoulli bundles and almost periodic perturbations*, Trans. Amer. Math. Soc., 314, pp. 63-105.

X. Mora (1983), *Finite dimensional attracting manifolds in reaction diffusion equations*, Contemporary Math., 17, pp. 353-360.

S. Yu. Pilyugin (1992), *The Space of Dynamical Systems with the $C^0$-Topology*, Preprint.

V. A. Pliss (1964), *A reduction principle in the theory of the stability of motion*, Izv. Akad. Nauk SSSR, Mat. Ser, 28, pp. 1297-1324.

V. A. Pliss (1977), *Integral Sets of Periodic Systems of Differential Equations*, Russian, Izdat. Nauka, Moscow.

V. A. Pliss and G. R. Sell (1991), *Perturbations of attractors of differential equations*, J. Differential Equations, 92, pp. 100-124.

G. Raugel and G. R. Sell (1992a), *Navier-Stokes equations on thin 3D domains I: Global attractors and global regularity of solutions*, AHPCRC Preprint 90-04, J. Amer. Math. Soc. (to appear).

G. Raugel and G. R. Sell (1992b), *Navier-Stokes equations on thin 3D domains II: Global regularity of spatially periodic solutions*, AHPCRC Preprint 92-062, Proc. College de France (to appear).

G. Raugel and G. R. Sell (1992c), *Navier-Stokes equations on thin 3D domains III: Global and local attractors*, IMA Proceedings (to appear).

I. Richards (1982), *On the gaps between numbers which are the sum of two squares*, Adv. Math, 46, pp. 1-2.

R. J. Sacker (1969), *A perturbation theorem for invariant manifolds and Hölder continuity*, J. Math. Mech., 18, pp. 705-762.

R. J. Sacker and G. R. Sell (1978), *A spectral theory for linear differential systems*, J. Differential Equations, 27, pp. 320-358.

R. J. Sacker and G. R. Sell (1980), *The spectrum of an invariant submanifold*, J. Differential Equations, 38, pp. 135-160.

G. R. Sell (1978), *The structure of a flow in the vicinity of an almost periodic motion*, J. Differential Equations, 27, pp. 359-393.

G. R. Sell (1992), *An optimality condition for approximate inertial manifolds*, in *IMA Proceedings on Dynamical Systems Approaches to Turbulence*, (to appear), Springer Verlag, New York.

G. R. Sell and Y. You (1992a), *Inertial manifolds: The non-self adjoint case*, J. Differential Equations, 96, pp. 203-255.

G. R. Sell and Y. You (1993), *Semiflows and global attractors*, Preprint.

S. Smale (1967), *Differentiable dynamical systems*, Bull. Amer. Math. Soc., 73, pp. 747-817.

R. TEMAM (1977), *Navier-Stokes Equations*, North-Holland, Amsterdam.

R. TEMAM (1983), *Navier-Stokes Equations and Nonlinear Functional Analysis*, CBMS Regional Conference Series, No. 41, SIAM, Philadelphia.

R. TEMAM (1988B), *Infinite Dimensional Dynamical Systems in Mechanics and Physics*, Springer Verlag, New York.

# 18 About the Solution of Some Inverse Problems in Differential Galois Theory by Hamburger Equations

**J. P. Ramis** Université de Louis Pasteur, Strasbourg, France

## 1. Introduction.

In differential Galois theory [1] (like in classical Galois theory) there are two natural questions :

*(i)* To compute the differential Galois group of a given differential equation.

*(ii)* If a linear algebraic group $G$ is given, to find a differential equation $Dy = 0$ whose differential Galois group $Gal_K(D)$ is $G$.

A lot of recent works are devoted to the solution of Problem (i) (cf. [Be 1], [B], [BBH], [Kat 2, 3, 4], [KP], [MR 1], [Mi 1, 2, 3, 4, 5], [Ra 3]), but the general case remains largely open.

Problem (ii) is **the inverse problem**. We will limit ourselves to **linear** differential equations (Picard-Vessiot theory). In that case a **necessary** condition is that the given group $G$ is a **linear algebraic group**. This condition is also sufficient and we can even choose $D$ as an algebraic differential operator on the Riemann sphere. The inverse problem was solved by C. and M. TRETKOFF [TT] (cf. also M. SINGER [Si 2,3], and [BB], [Kov 1], [Kov 2]) :

**Theorem 1.1.** —

*Let $G$ be a complex linear algebraic group. Then there exists a linear algebraic differential equation $Dy = 0$ on the Riemann sphere whose differential Galois group is $G$.*

The main tools of the proof of this result are the **Riemann-Hilbert correspondence** ([R], [D 1], [Kat 1]), a Schlesinger's theorem [Sc] (the differential Galois group is (Zariski) topologically generated by the monodromy group) and a property of linear algebraic groups (there exists a finitely generated Zariski dense subgroup $H$). A consequence of this proof is that one obtains a fuchsian (regular singular) differential equation having "many" singular points on the Riemann sphere (the number

---

(1) In all this paper the field of constants of all the differential fields is the field of complex numbers, and all the differential equations are complex analytic.

of generators of the subgroup $H$ plus one). The aim of the present paper is to give a partial solution of the inverse problem : we will solve it for **all semi-simple** complex linear algebraic groups using **only** differential equations with a "small" number of singular points, more precisely one or two. With such a constraint the class of fuchsian equations is too poor, so it is necessary to suppose that one singularity is irregular. Of course, up to an homography on the Riemann sphere, we can then suppose that $\infty$ is an irregular singular point and that zero is regular or regular singular for the differential equation $Dy = 0$. For such equations called (1) **Hamburger equations** the solution of the inverse problem is equivalent to the solution of a **local** inverse problem. The main tools (2) of the solution of this local inverse problem are an extension to irregular connexions of the Riemann-Hilbert correspondence (the *wild* Riemann-Hilbert correspondence [MR 2]), a generalization of Schlesinger's theorem (the differential Galois group is topologically generated by the **wild** monodromy group [Ra 3], [MR 1], [MR 2]), and the description of a semi-simple Lie algebra in terms of a **Chevalley basis** using the Cartan matrix [Se 1,2]. This proof, in some sense, mimics C. and M. Tretkoff's and is inspired by the idea of J. ECALLE of building a great number of "natural" Lie algebras by **representations of resurgence algebras** [E 1, 2, 3, 4].

In order to solve the inverse problem there is a completely different approach (3) : we can use the (partial) solution of the direct problem. We limit ourselves to Hamburger equations, so we have to check the known computations of differential Galois groups for such equations. Among these equations there are the **generalized confluent hypergeometric differential equations** [Er]. For this family a weak version of the direct problem (4) is known for a large class : [Kat 2, 3, 4], [KP] (computation of the identity component of the Galois differential group for "non degenerate equations"). A. DUVAL and, C. MITSCHI solved the direct problem (i. e. the computation of the differential Galois group itself), [DM], [Mi 1, 2, 3 , 4, 5] for some smaller classes. We will recall the main results in part 2. The advantage of this approach is to give an **explicit** answer (and in some sense to get the "simplest solution" for a given group). Unfortunately some complex semi-simple algebraic groups are **not** obtained by this method. More precisely, if we limit ourselves to simple complex Lie groups, it is

---

(1) This definition gives a slightly bigger class than Ince's, cf. infra...

(2) These tools are described in part 3.

(3) The corresponding partial solution to our problem is described in part 2. This part is independant of the rest of the paper.

(4) The problem is to find a differential equation, the identity component of its differential Galois group being given.

possible [5] by this method to get all the groups belonging to the "big classes" corresponding to the simple Lie algebras $A_n = sl(n+1)$ $(n \geq 1)$, $B_n = so(2n+1)$ $(n \geq 2)$, $C_n = sp(2n)$ $(n \geq 3)$, $D_n = so(2n)$ $(n \geq 4)$, and to the exceptional algebra $G_2$. It is however **impossible** to get the exceptionnal cases $F_4$, $E_6$, $E_7$, $E_8$.

We will now recall some basic definitions and give some references.

The best introduction to "classical" differential algebra is certainly Kaplansky's book [Kap]. For a more recent point of view see [Be 1], [B], [De 2], [MR 1]. Complete proofs are detailed in [Ko].

A complex linear algebraic group is a subgroup $G \subset GL(n; \mathbb{C})$ with the property that there exist polynomials $P_r \in \mathbb{C}[X_{11}, ..., X_{nn}]$ in the $n^2$ variables $X_{ij}$ such that $G = \{g_{ij} \in GL(n; \mathbb{C}) / \ P_r(g_{ij}) = 0\}$.

By definition a Lie algebra of positive dimension is semi-simple if it has no non zero commutative ideal. A (Zariski-)connected algebraic group is semi-simple if it has no non zero commutative invariant subgroup. A connected algebraic group is semi-simple if and only if its Lie algebra is semi-simple [Hu] (13.5). On a complex semi-simple group there exists a unique structure of complex algebraic group [Se 1].

For the proof of our main theorem we will make an essential use of a Chevalley basis. Let $\mathcal{L}$ be a semi-simple Lie algebra. A Chevalley basis for $\mathcal{L}$ is a set $\{H_i, X_i, Y_i\}_{i=1,...,n}$ of generators satisfying the relations :

(1.1) *(i)* $[H_i, H_j] = 0$

*(ii)* $[H_i, X_j] = n(i,j)X_j$, $[H_i, Y_j] = -n(i,j)Y_j$

*(iii)* $[X_i, Y_i] = H_i$

*(vi)* $ad(X_i)^{-n(i,j)+1}(X_j) = 0$, $ad(Y_i)^{-n(i,j)+1}(Y_j) = 0$ $(i \neq j)$.

The $n(i,j)$'s are integers and given by the **Cartan matrix** ($n(i,j) \leq 0$ if $i \neq j$) of the Lie algebra $\mathcal{L}$. Each semi-simple complex Lie algebra admits a Chevalley basis [Se 1], VI.

**Lemma 1.2.** —

*Let $G$ be a complex linear semi-simple algebraic group, $G \subset GL(n; \mathbb{C})$. Let $\mathcal{L}$ be its Lie algebra. Let $\{H_i, X_i, Y_i\}_{i=1,...,n}$ be a Chevalley basis of $\mathcal{L}$. Then the Zariski closure in $GL(n; \mathbb{C})$ of the subgroup generated by the exp $X_i$'s and the exp $Y_i$'s is the group $G$ itself.*

We denote by $G_1$ the Zariski closure of the subgroup generated by the *exp* $X_i$'s and the *exp* $Y_i$'s and by $\mathcal{L}_1$ its Lie algebra. It is a subgroup of $G$, and its Lie algebra is a subalgebra of $\mathcal{L}$. But the Lie algebra contains the $X_i$'s and the $Y_i$'s, so it contains the $H_i$'s (by (1.2) *(iii)*). Then $\mathcal{L}_1 = \mathcal{L}$, and $G_1 = G$.

---

(5) For orthogonal groups detailed proofs are written only for small values of the rank $n$ [Mi 1], cf. infra.

We will now define the **Hamburger differential equations**. Our definition is slightly more general than Ince's.

**Definition 1.3.** —

*A linear algebraic homogeneous differential equation (resp. operator) on the Riemann sphere is a Hamburger equation (resp. operator) if*

*(i) There are at most two singular points, namely* $0$ *and* $\infty$ *;*

*(ii) The origin is regular or a regular singular singularity*

*(iii) The point at infinity is an irregular singularity*

INCE adds the following conditions : the origin is **always** a singularity and the point at infinity is an essential singularity for **every** solution ([I] 17.6 p. 430, [H] p. 238).

A linear algebraic homogeneous differential equation is a Hamburger equation (in our sense) if and only if it may be written in the form :

$$(1.2) \quad z^n \frac{d^n y}{dz^n} + z^{n-1} p_{n-1} \frac{d^{n-1} y}{dz^{n-1}} + \ldots + p_0 y = 0,$$

where $p_0, \ldots, p_{n-1}$ are polynomials, or equivalently in the form

$$(1.3) \quad (z\frac{d}{dz})^n y + q_{n-1}(z\frac{d}{dz})^{n-1} + \ldots + q_0 y = 0,$$

where $q_0, \ldots, q_{n-1}$ are polynomials.

**Examples 1.4.** —

The Kummer equation ($a,\ c \in \mathbb{C}$)

$$zy'' + (c-z)y' - ay = 0,$$

the Whittaker equation ($\chi,\ m \in \mathbb{C}$)

$$y'' + (-\frac{1}{4} + \chi z^{-1} + (\frac{1}{4} - m^2)z^{-2})y = 0,$$

and the Bessel equation ( $\nu \in \mathbb{C}$)

$$z^2 y'' + zy' + (z^2 - \nu^2) = 0$$

are clearly Hamburger equations.

More generally the generalized hypergeometric operators

$$(1.4) \quad D_{q,p} = (-1)^{q-p} z \prod_{j=1,\ldots,p} (\partial + \mu_j) - \prod_{j=1,\ldots,q} (\partial + \nu_j - 1),$$

where $\partial = z\frac{d}{dz}$ and $p, q$ are positive integers with $1 \leq p < q$, are Hamburger operators.

Equations which may be written in the form

$$(1.5) \quad y^{(n)} - \sum_{k=0,\ldots,p} a_k z^k y^{(k)} = 0,$$

where $1 \leq p < n$, and $a_0, \ldots, a_p \in \mathbb{C}$, are also Hamburger equations and are easily "reducible" to confluent hypergeometric equations [PW] 3, p. 57. The differential equations for Ecalle's accelerating and decelerating functions (with a rational parameter) are of this type [MR 2], p.353.

The case $p = 2$

$$(1.6) \quad y^{(n)} - \alpha z y' - \beta = 0$$

seems particularly interesting [Kat 3], [Mi 4] (cf. infra).

Equations (1.5) are **regular** at the origin. So their monodromy is **trivial**. Then their Galois differential group is Zariski-connected [Kat], [MR 2].

## 2. Some explicit solutions of the inverse Galois problem by generalized confluent hypergeometric differential equations.

For differential equations of order two and three there are algorithms for the computation of the differential Galois groups and it is easy to get explicit examples for all the semi-simple cases [Du], [DM], [DL], [SU]. For orders $n \geq 4$ the only known explicit computations are related to generalized hypergeometric differential equations.

The generalized hypergeometric differential operators are

$$(2.1) \quad D_{q,p} = (-1)^{q-p} z \prod_{j=1,\ldots,p} (\partial + \mu_j) - \prod_{j=1,\ldots,q} (\partial + \nu_j - 1),$$

where $\partial = z\frac{d}{dz}$ and $p, q$ are positive integers. For $p = q$ these operators are Fuchsian (with singularities at 0, 1, $\infty$). For $p > q \geq 1$, these operators are irregular (confluent) : they are Hamburger operators.

We briefly recall some fundamental properties of hypergeometric differential equations (for more details, proofs and references see [B])

**Lemma 2.1.** —

*The operator (2.1) is irreducible (that is it does not factor) over $\mathbb{C}(x)$ if and only if $\mu_i \neq \nu_j$ (mod $\mathbb{Z}$) for all $i, j$.*

**Lemma 2.2.** —

*If $D_{q,p}y = 0$ and $D'_{q',p'}y = 0$ are two irreducible hypergeometric differential equations with parameters $\mu_1, \ldots, \mu_p; \lambda_1, \ldots, \lambda_q$ and $\mu'_1, \ldots, \mu'_{p'}; \lambda'_1,$*

*$\ldots, \lambda'_{q'}$, then they are equivalent over $\mathbb{C}(x)$ if and only if $p = p'$, $q = q'$ and, after renumbering if necessary $\mu_i = \mu'_i$ (mod $\mathbb{Z}$), $\nu_j = \nu'_j$ (mod $\mathbb{Z}$), for $i = 1, \ldots, p$; $j = 1, \ldots, q$.*

Let $V$ be a vector space and $G$ a group acting on $V$. We shall say that $G$ is imprimitive on $V$ if there exists a direct sum decomposition $V = V_1 \oplus \ldots \oplus V_m$ $(m > 1)$ such that $G$ permutes non trivially the $V_i$'s. If it is not the case we will call $G$ **primitive** on $V$.

**Theorem 2.3.** —

*Let G be the differential Galois group of $D_{q,p}$. We suppose that $D_{q,p}$ is irreducible. Then G (acting on the space of solutions) is imprimitive if and only if the equation $D_{q,p}$ is Kummer induced, or Belyi induced or inverse Belyi induced.*

For definitions of *Kummer induced, Belyi induced* and *inverse Belyi induced* cf. [B]. The corresponding conditions are exceptional relations among parameters.

We will denote by $G^0$ the identity component of an algebraic group G.

**Theorem 2.4.** —

*Let G be the differential Galois group of $D_{q,p}$. We suppose $D_{q,p}$ irreducible and confluent ($p > q \geq 1$, $D_{q,p}$ is a Hamburger equation) and G primitive. Then $G^0$ acts irreducibly (on the space of solutions).*

We are only interested here in Hamburger equations so we will not recall anything about the case $p = q$. We refer the interested reader to [B], [BBH], and for the reducible case to [Bou 1, 2].

If $V$ is a finite dimensional complex vector space, $G \subset GL(V)$ is an algebraic group and $C \subset \mathbb{C}^*$ is an algebraic group of scalars, we set

$$C.G = \{\lambda g / \lambda \in C,\ g \in G\}.$$

We introduce a "weak inverse problem" :

*(ii')* If a Zariski-connected linear algebraic group $G'$ is given, find a differential equation $Dy = 0$ such that the identity component $G^0$ of its differential Galois group $Gal_K(D)$ is $G'$.

If we limit ourselves to generalized confluent hypergeometric equations, we can **explicitely** solve the weak inverse problem for some set of semi-simple algebraic groups (to be precised later). In order to do that we will use a very important result of N. Katz [Kat 4] ($C \subset \mathbb{C}^*$ is an algebraic group of scalars depending on some conditions) :

**Theorem 2.5.** —

*Let G be the differential Galois group of the hypergeometric differential $D_{q,p}$. (More precisely we will consider its representation as a group acting* *on the space of solutions V.) We suppose that $D_{q,p}$ is irreducible and confluent ($p > q \geq 1$, $D_{q,p}$ is a Hamburger equation) and G primitive. Then we have the following possiblities for G (and only these possibilities) :*

*(i) If $q - p = 1$ then $G = GL\ (V)$;*

*(ii) If $q - p$ is odd then $G = C.SL\ (V)$;*

*(iii) If $q - p$ is even, then $G = C.SL\ (V)$ or $G = C.SO\ (V)$;, or $G = C.Sp\ (V)$ or, in addition (if $q - p = 6$) :*

*if $q = 7$, $p = 1$, $G = C.G_2$;*

*if $q = 8$, $p = 2$, $G = C.Spin(7)$, Spin (7) in the eight dimensional spin representation;*

*if $q = 8$, $p = 2$, $G = C.SL(3)$, SL(3) in adjoint representation;*

*if $q = 8$, $p = 2$, $G = C.SL(2) \times SL(2) \times SL(2)$, $SL(2) \times SL(2) \times SL(2)$ in tensor product of standard representation;*

*if $q = 8$, $p = 2$, $G = C.SL(2) \times Sp(4)$, $SL(2) \times Sp(4)$ in tensor product of standard representation;*

*if $q = 8$, $p = 2$, $G = C.SL(2) \times SL(4)$, $SL(2) \times SL(4)$ in tensor product of standard representation;*

*if $q = 9$, $p = 3$, $G = C.SL(3) \times SL(3)$ $SL(3) \times SL(3)$ in tensor product of standard representation.*

Looking carefully at Katz's results [Kat 4] and limiting ourselves to the case of a generalized confluent hypergeometric equation whose Wronskian is in $\mathbb{C}(x)$ (i. e. the differential Galois group is a subgroup of $SL\ (V)$), we get

**Theorem 2.6.**

*The weak inverse problem in differential Galois theory can be solved for the following semi-simple algebraic groups (more precisely in the representations of Theorem 2.4) : $SL\ (n;\mathbb{C})$, $SO\ (n;\mathbb{C})$, $Sp\ (2n;\mathbb{C})$ ($n \in \mathbb{N}^*$), $G_2$, $Spin\ (7;\mathbb{C})$, $SL\ (2;\mathbb{C}) \times SL\ (2;\mathbb{C}) \times SL\ (2;\mathbb{C})$, $SL\ (2;\mathbb{C}) \times Sp\ (4;\mathbb{C})$, $SL\ (2;\mathbb{C}) \times SL\ (4;\mathbb{C})$, $SL\ (3;\mathbb{C}) \times SL\ (3;\mathbb{C})$ with generalized confluent hypergeometric differential equations (Hamburger equations). Moreover it is possible to give families of explicit solutions for each case.*

For an arbitrary generalized confluent hypergeometric equations it is possible to compute explicitely the formal monodromy, the exponential torus and the Stokes multipliers (as well as the wild monodromy representation, cf. part 3 below). Then it is possible to compute explicitely the differential Galois group itself (and not only its identity component). Some of these computations were done by A. Duval and C. Mitschi [DM], [Mi 1,2,3,4,5].

Looking at the results we get

**Theorem 2.7.** —

*The inverse problem in differential Galois theory can be solved for the*

*following semi-simple algebraic groups $SL\ (n;\mathbb{C})$, $Sp\ (2n;\mathbb{C})$ $(n \in \mathbb{N}^*)$, $SO\ (n;\mathbb{C})$ $(n = 3,4,5,7)$, $G_2$, with generalized confluent hypergeometric differential equations (Hamburger equations). Moreover it is possible to give families of explicit solutions for each case.*

**Examples 2.8.** —

The differential Galois group of the confluent hypergeometric operator

$$D_{7,1} = z(\partial + \frac{1}{2}) - \partial^3(\partial^2 - \frac{1}{16})^2$$

is $G_2$ [Mi 3].

The differential Galois group of the differential equation of type (1.6)

$$D_K y = y^{(7)} + \frac{z}{7}y' + \frac{1}{14} = 0$$

is also $G_2$ [Kat 3,4], [Mi 4].

## 3. Wild fundamental groups, meromorphic classification and Galois differential groups.

We recall some fundamental definitions and results of [Ra 4] and [MR 2].

Let $\Delta = d/dx - A$, with $A \in End(n;\mathbb{C}\{x\}[x^{-1}])$, be a germ of meromorphic differential operator at the origin of the complex plane $\mathbb{C}$. It is well known that $\Delta$ admits a formal fundamental solution :

$$\hat{F}(x) = \hat{H}(u)u^{\nu L}e^{Q(u^{-1})},$$

where $u^\nu = x$ (for some $\nu \in \mathbb{N}^*$), $L \in End(n;\mathbb{C})$, $\hat{H} \in GL(n;\mathbb{C}[[u]][u^{-1}])$, and $Q$ is a diagonal matrix with entries in $u^{-1}\mathbb{C}[u^{-1}]$, which is invariant, up to permutations of the diagonal entries, under the transformation corresponding to $u \mapsto e^{\frac{2i\pi}{\nu}}u$ $(x \mapsto e^{2i\pi}x)$ and satisfies $[e^{2i\pi\nu L}, Q] = 0$. (If $\nu = 1$ $[L,Q] = 0$, and $L$ can be supposed in Jordan form.). If $Q = Diag\ \{q_1, q_2, ..., q_n\}$, then the set $\{q_1, q_2, ..., q_n\}$ is a subset of $u^{-1}\mathbb{C}[u^{-1}]$ which is independent of the choice of the fundamental solution $\hat{F}$ ($\nu$ is chosen minimal).

We set $K = \mathbb{C}\{x\}[x^{-1}]$ and we denote by $K < \hat{F} >$ the differential field generated on $K$ by the entries of $\hat{F}$.

Replacing $x$ by $e^{2i\pi}x$ in $\hat{F}(x)$, we get a new fundamental solution of the operator $\Delta$. We have $\hat{F}(e^{2i\pi}x) = \hat{F}(x)\hat{M}$, with $\hat{M} \in GL\ (n;\mathbb{C})$; $\hat{M}$ is the **formal monodromy**. The corresponding formal monodromy operator belongs to $Aut_K\ K < \hat{F} > \approx Gal_K\ (\Delta)$.

The matrix $\hat{H}(u)$ is **multisummable** [MR 2], [MaR] in every direction but perhaps a finite set of singular directions $\Sigma(\hat{F})$.

For every $d \notin \Sigma(\hat{F})$ ($d$ is drawn onto the Riemann surface of the Logarithm), the summation operator in the direction $d$ extends uniquely in a summation operator

$$S_d : K < \hat{F} > \to \mathcal{A}_d,$$

where $\mathcal{A}_d$ is the differential algebra of functions analytic on germs of sectors bisected by $d$. This map $S_d$ is an injective homomorphism of differential algebras [MR 2].

If $d$ is a singular direction, we can sum "just before $d$" by $S_d^-$ and "just after $d$" by $S_d^+$. The map $(S_d^+)^{-1}S_d^-$ is an automorphism of differential fields of $K < \hat{F} >$. It is the **Stokes multiplier** associated to the singular direction $d$. It defines an element of $Aut_K\ K < \hat{F} > \approx Gal_K\ (\Delta)$.

We will now define the **exponential torus**. Let $\hat{\mathbb{K}} = \hat{K}_\nu < u^L, e^Q >$ be the differential field generated on $\hat{K}_\nu = \mathbb{C}[[u]][u^{-1}]$ by the entries of $u^{\ \nu L}$ and $e^Q$. We set $K_\nu = \mathbb{C}\{u\}[u^{-1}]$, $\hat{\mathbb{L}}_\nu = \hat{K}_\nu < e^Q >= \hat{K}_\nu < e^{q_1}, \ldots, e^{q_n} > \subset \mathbb{K}$, $\mathbb{L}_\nu = K_\nu < e^Q >$. If $\mu$ is the dimension of the free $\mathbb{Z}$-module $\mathcal{E}_\Delta \subset u^{-1}\mathbb{C}[u^{-1}]$ generated by $q_1, \ldots, q_n$, the Galois differential group $Aut_{\hat{K}_n}\ \hat{\mathbb{L}}_\nu = Aut_{K_n}\ \mathbb{L}_\nu$ is a **torus** $\mathcal{T}_\Delta = \mathcal{T}(q_1, \ldots, q_n)$ which is isomorphic to $(\mathbb{C}^*)^\nu$. It can be identified with a subgroup of $Aut_K\ K < \hat{F} > \approx Gal_K\ (\Delta)$ which is independant of the choice of $\hat{F}$. We call it the **exponential torus.**

The following extension of the theorem of Schlesinger [Sch] to the **irregular** case is the **principal tool** of this paper (and of the solutions of the direct problem by A. Duval and C. Mitschi used in part 2).

**Theorem 3.1**

*Let $K = \mathbb{C}\{x\}[x^{-1}]$. Let $\Delta = d/dx - A$, with $A \in End(n; K)$, be a germ of meromorphic differential operator at the origin of the complex plane $\mathbb{C}$. Let $\hat{F}$ be a formal fundamental solution of $\Delta$. Let $\mathcal{H}$ be the subgroup of $GL(n; \mathbb{C})$ generated by the formal monodromy matrix $\hat{M}$, the exponential torus $\mathcal{T}$, and the Stokes matrices of $\Delta$ associated to the given formal fundamental solution $\hat{F}$. Then the representation of the differential Galois group $Gal_K(\Delta)$ of $\Delta$ in $GL(n; \mathbb{C})$, given by $\hat{F}$, is the Zariski closure of $\mathcal{H}$ in $GL(n; \mathbb{C})$.*

We recall very briefly some basic definitions. For more details see [MR 2].

We set $K = \mathbb{C}\{x\}[x^{-1}]$ and, for $\nu \in \mathbb{N}^*$, $K = \mathbb{C}\{x^{1/\nu}\}[x^{-1/\nu}]$ ($K \subset K_\nu$). Let $\mathcal{E}_\nu = x^{-1/\nu}\mathbb{C}[x^{-1/\nu}]$. We set $\mathcal{E} = \bigcup_{\nu \in \mathbb{N}^*} \mathcal{E}_\nu$. For $q = \{q_1, \ldots, q_m\} \subset \mathcal{E}_\nu$, *globally invariant* by the monodromy transform $x \mapsto e^{2i\pi}x$, we set

$\mathcal{E}(q) = \mathbb{Z}\, q_1 + \ldots + \mathbb{Z}\, q_m$. Then $\mathcal{E}$ is a direct limit of finitely generated free $\mathbb{Z}$-modules (lattices) : $\mathcal{E} = \varinjlim_q \mathcal{E}(q)$.

We define an action of a "formal loop" $\hat{\gamma}_0$ at the origin on $\mathcal{E}$ by

$$\hat{\gamma}_0 : q(x) \mapsto q(e^{2i\pi}x) = q\hat{\gamma}_0(x).$$

We suppose that $q = \{q_1, \ldots, q_m\} \subset \mathcal{E}_\nu$ is globally invariant by $\hat{\gamma}_0$. Then the differential Galois group $\mathcal{T}(q) = Aut_{K_\nu}\ K_\nu < e^q >= Aut_{K_\nu}\ K_\nu < e^{q_1}, \ldots, e^{q_m} >$ is a torus. We identify it with a torus of $Aut_K\ K < e^q >$. The lattice $\mathcal{E}(q)$ is identified with the lattice of **weights** of the Lie algebra $Lie\ \mathcal{T}(q)$ of the torus $\mathcal{T}(q)$ (cf. [MR 2], p. 361) :

We identify $\mathcal{E}(q)$ with the lattice of characters (that is the dual group) of the torus $\mathcal{T}(q)$ as follows : to $p \in \mathcal{E}(q)$ corresponds

$$p : \mathcal{T}(q) \to \mathbb{C}^*$$

$$p : \tau \to p(\tau),$$

where $\tau(e^p) = p(\tau)e^p =< p, \tau > e^p$. We also denote by $p$ the infinitesimal map (weight)

$$p : Lie\ \mathcal{T}(q) \to \mathbb{C}$$

associated to

$$p : \mathcal{T}(q) \to \mathbb{C}^*.$$

We define the exponential torus by $\mathcal{T} = \varprojlim_q \mathcal{T}(q)$. For the wild fundamental group this infinite dimensional torus plays the role of a Cartan torus. We denote by $Lie\ \mathcal{T} = \varprojlim_q\ Lie\ \mathcal{T}(q)$ the Lie algebra of $\mathcal{T}$. We define an action of $\hat{\gamma}_0$ on the exponential torus by

$$q(\hat{\gamma}_0\tau) =< q, \hat{\gamma}_0\tau >=< q\hat{\gamma}_0^{-1}, \tau >= q\hat{\gamma}_0(\tau)$$

($\tau \in \mathcal{T}$ and $q \in \mathcal{E}$).

The differential Galois group $Aut_K\ K < e^q >$ is the semi-direct product of the torus $\mathcal{T}$ and of a finite cyclic group defined by the adjoint action of $Aut_K\ K_\nu$ on $\mathcal{T}(q) \subset Aut_K\ K < e^q >$. (The formal loop $\hat{\gamma}_0$ corresponds to the generator $x^{1/\nu} \mapsto e^{2i\pi/\nu}x^{1/\nu}$ of $Aut_K\ K_\nu$.)

Then we denote by $(\hat{\gamma}_0)$ the free group generated (multiplicatively) by the "formal loop" $\hat{\gamma}_0$, and by $\pi_{1,sf}(\mathbb{C}^*)$ the semi-direct product defined by the action of $(\hat{\gamma}_0)$ on $\mathcal{T}$ :

$$\pi_{1,sf}(\mathbb{C}^*) = (\hat{\gamma}_0) \ltimes \mathcal{T},$$

$\pi_{1sf}(\mathbb{C}^*)$ is the wild formal fundamental group at the origin.

We have clearly

$$\pi_{1,sf}(\mathbb{C}^*) = (\hat{\gamma}_0) \ltimes \mathcal{T} = \varprojlim_q Aut_K\ K < e^q > .$$

To $q \in \mathcal{E}$, and a line of **maximal decay** $d$ of $e^q$ (drawn onto the Riemann surface of Logarithm), we associate a symbol $\dot{\Delta}_{q,d}$. All these symbols generate a complex infinite dimensional **free** Lie algebra *Lit* $\mathcal{R}$, the **resurgence algebra**. We define an action of the wild formal fundamental group $\pi_{1,sf}(\mathbb{C}^*)$ on *Lie* $\mathcal{R}$ by (cf. [MR 2], p. 386)

$$\hat{\gamma}_0 \dot{\Delta}_{q,d} \hat{\gamma}_0^{-1} = \dot{\Delta}_{q\gamma_0, e^{2i\pi} d}$$

$$\tau \dot{\Delta}_{q,d} \tau^{-1} = q(\tau) \dot{\Delta}_{q,d}.$$

Using a version of the local wild Riemann-Hilbert correspondence [MR 2] (Theorem 17, p. 387) we can represent a germ of meromorphic connection $\nabla$ at the origin by a pair of representations $(\hat{\rho}_f, L\rho_r)$, where

$$\hat{\rho}_f :\ \pi_{1,sf}(\mathbb{C}^*, 0)) \to\ GL\,(n; \mathbb{C}),$$

is a finite dimensional representation of the wild formal fundamental group, and

$$L\rho_r :\ Lie\ \mathcal{R} \to\ End\,(n; \mathbb{C})$$

is a finite dimensional representation of the resurgent algebra. This pair of representations is compatible with the action of $\pi_{1,sf}(\mathbb{C}^*, 0))$ on *Lie* $\mathcal{R}$, corresponding to the semi-direct product

$$\pi_{1,s}(\mathbb{C}^*, 0)) = \pi_{1,sf}(\mathbb{C}^*, 0)) \triangleright \mathcal{R}.$$

As $\pi_{1,sf}(\mathbb{C}^*, 0))$ is also a semi-direct product

$$\pi_{1,sf}(\mathbb{C}^*, 0)) = (\hat{\gamma}_0) \ltimes \mathcal{T},$$

we can write the representation $\hat{\rho}_f$ as a pair of representations $(\hat{\rho}_m, \rho_e)$, and $\hat{\rho}_f$ is finally defined by the image $\hat{\rho}_m(\hat{\gamma}_0)$ (the "formal monodromy"), and the finite dimensional representation of the exponential torus $\rho_e$ : $\mathcal{T} \to\ GL\,(n; \mathbb{C})$. Finally the version of the wild Riemann-Hilbert correspondence which we will use below is a natural bijection between germs of meromorphic connexions at the origin and triples of representations $(\hat{\rho}_m, \rho_e, L\rho_r)$ satisfying the preceding conditions.

In a preceding version of this paper [Ra 2] we used representations of the wild fundamental group $\pi_{1,s}(\mathbb{C}^*, 0))$ defined in [Ra 1], [MR 2]. Here we slightly change our presentation, avoiding any deep use of the wild fundamental group.

Usually, via the Riemann-Hilbert correspondence, we interpret a fuchsian connection $\nabla$ on the Riemann sphere by its monodromy representation, that is by a finite dimensional linear representation of the fundamental group $\pi_1(P^1(\mathbb{C}) - S)$, where $S$ is the finite set of singularities of $\nabla$. So we use the finitely generated free group $\pi_1(P^1(\mathbb{C}) - S)$ in place of the enormous universal group $\pi_1(P^1(\mathbb{C}) - \ldots) = \varprojlim \pi_1(P^1(\mathbb{C}) - S)$. We can do the same in the irregular case : for a **fixed** connection $\nabla$, we can replace the wild fundamental group $\pi_{1,s}$ by a "smaller" quotient $\pi'_{1,s}$.

**Definition 3.2.** —

*Let $\nabla$ be a germ of meromorphic connection at the origin. We denote by*

$$\rho : \pi_{1,s}(\mathbb{C}^*, 0)) \rightarrow GL(n; \mathbb{C}),$$

*the corresponding representation.*

*An acting wild fundamental group for $\nabla$ (or any corresponding differential operator or equation) is a quotient $\pi'_{1,s}$, of $\pi_{1,s}(\mathbb{C}^*, 0))$ where Lie $\pi'_{1,s} =$ Lie $\pi'_{1,sf} \ltimes$ Lie $\mathcal{R}'$ and $\pi'_{1,sf} = (\hat{\gamma}_0) \ltimes T'$, such that :*

*(i) There exists a representation*

$$\rho' : \pi'_{1,s} \rightarrow GL(n; \mathbb{C}),$$

*such that $\rho = \rho' p'$, where $p' : \pi_{1,s}(\mathbb{C}^*, 0)) \rightarrow \pi'_{1,s}$ is the natural projection;*

*(ii) The acting exponential torus $T'$ is a finite dimensional quotient of the exponential torus $T$, such that the representation $\rho'_e : T' \rightarrow GL(n; \mathbb{C})$, is faithfull;*

*(iii) The acting resurgence algebra Lie $\mathcal{R}'$ is a free algebra generated by symbols $\dot{\Delta}_{q,d}$ whose image by Lie $\rho'$ is non trivial. It is a quotient of the resurgent Lie algebra $\mathcal{R}$.*

**Remarks 3.3.** —

The notion of acting algebra (*algèbre agissante*) is due to J. Ecalle, who introduced it in a more general context.

For practical purposes it is sufficient to know a **finitely** generated subalgebra of the acting algebra : we can pick only one symbol $\dot{\Delta}_{q,d}$ for each orbit under the action of $\hat{\gamma}_0$. (This action corresponds to a permutation of the lines $d$ by a turn, and, in the image by the representation, to the conjugation by the formal monodromy.)

Using the descriptions of [MR 2] it is easy to check that the images of the generators $\dot{\Delta}_{q,d}$ of the acting algebra *Lie* $\mathcal{R}'$ are exactly **all** the non-trivial Stokes multipliers of the given connection $\nabla$.

Finally we can associate to the germ of meromorphic connection $\nabla$ a pair $(\pi'_{1,sf} = (\hat{\gamma}_0) \ltimes \mathcal{T}'; \, Lie\ \mathcal{R}')$, with an action of $\pi'_{1,sf}$ on $Lie\ \mathcal{R}'$, and a pair of representations

$$\hat{\rho}'_f : \ \pi'_{1,sf} \to \ GL\,(n;\mathbb{C}),$$

$$L\rho'_r : \ Lie\ \mathcal{R}' \to \ End\,(n;\mathbb{C});$$

which are compatible with the action.

The important point is now to notice that it is possible conversely to "build" germs of meromorphic connections, starting from a finite dimensional representation of an "abstract" acting wild fundamental group : we associate the representation $\rho = \rho' p'$ to the representation $\rho'$ , and $\nabla$ to $\rho$ via the inverse wild Riemann-Hilbert correspondance. Equivalently we can associate the pair of representations $(\hat{\rho}_f, L\rho_r)$ to the pair of representations $(\hat{\rho}'_f, L\rho'_r)$, and we get the connection $\nabla$ by a local version of the wild Riemann-Hilbert correspondence.

In the next part we will build a germ of meromorphic connection $\nabla$ with a given semi-simple differential Galois group $G$. Starting from a Chevalley basis of the semi-simple Lie algebra $Lie\ G$, we will first build an "abstract" acting exponential torus $\mathcal{T}'$ and an "abstract" acting resurgence algebra $Lie\ \mathcal{R}'$, with an action of $\mathcal{T}'$ on $Lie\ \mathcal{R}'$ (the acting exponential torus is isomorphic to the Cartan torus whose Lie algebra is generated by the $H_i$'s, and the acting resurgent Lie algebra is generated by symbols corresponding to the $X_i$'s and the $Y_i$'s, **"forgetting"** some relations). Finally we will get the connection from a pair of representations $(\hat{\rho}'_f, L\rho'_r)$. This construction is essentially **trivial** : key points are the local wild Riemann-Hilbert correspondence [MR 2], and the theory of Chevalley basis for semi-simple Lie algebras. (Compare with [TT].)

## 4. The main theorem.

**Theorem 4.1.** —

*Let $G$ be a semi-simple complex algebraic group. Then*

*(i) There exists a Hamburger linear differential equation $Dy = 0$ on the Riemann sphere such that its global differential Galois group (Picard-Vessiot group) $Gal_{\mathbb{C}(x)}\,(D)$ is isomorphic to $G$ ;*

*(ii) There exists a germ of meromorphic linear differential equation $Dy = 0$ at the origin on the Riemann sphere such that its differential Galois group $Gal_{\mathbb{C}\{x\}[x^{-1}]}\,(D)$ is isomorphic to $G$.*

*Moreover, if the group $G$ has a faithful linear representation of dimension $n$, then we can get an equation $Dy = 0$ of order $n$ in (i) and (ii).*

Assertions (i) and (ii) are equivalent. For a Hamburger operator $D$ the differential Galois group is the Zariski closure of the group generated by the two local Galois groups at zero and infinity : $Gal_{\mathbb{C}\{z\}[z^{-1}]}$ $(D)$ and $Gal_{\mathbb{C}\{x\}[x^{-1}]}$ $(D)$ $(x = z^{-1})$. The point 0 is a regular singularity for $D$ and the local differential Galois group is the Zariski closure of the group $\mathcal{M}$ generated by the operator of monodromy [Sc]. But the monodromy around $\infty$ is the same (up to orientation) as the monodromy around the origin. Then $\mathcal{M}$ is contained in the local differential Galois group at $\infty$, and this group is equal to the global group $Gal_{\mathbb{C}(x)}$ $(D)$. So (i) $\Rightarrow$ (ii). Conversely, if (ii) is true, we can replace the germ of meromorphic differential equation $D$ at infinity by a meromorphically equivalent Hamburger equation (using the Birkhoff's algebraization theorem [Bi]). This operation does not change the local Galois group at infinity, so (ii) $\Rightarrow$ (i).

We will now prove assertion (ii). We will work with the coordinate $x$ ($\infty$ corresponds to $x = 0$). We set $K = \mathbb{C}\{x\}[x^{-1}]$.

We will denote by $\mathfrak{g}$ the Lie algebra of the group $G$. We fix a faithful finite dimensional representation of $G$ and we identify $G$ and $\mathfrak{g}$ to their images in $GL\ (n; \mathbb{C})$ and $End\ (n; \mathbb{C})$ respectively. We choose a Chevalley basis $\{H_i, X_i, Y_i\}_{i=1,\dots,m}$ for $\mathfrak{g}$. Next we choose an **unramified** sublattice $\mathcal{E}'$ of dimension $m$ of the exponential lattice $\mathcal{E}$ $(\mathcal{E}' \subset \frac{1}{x}\mathbb{C}[\frac{1}{x}])$. The quotient $\mathcal{T}' = Aut_K\ K < \mathcal{E}' >$ of the exponential torus $\mathcal{T}$ will be our acting exponential torus. We choose a basis $p = (p_1, \dots p_m)$ of $\mathcal{E}'$ : $\mathcal{E}' = \mathbb{Z}p_1 \oplus \dots \oplus \mathbb{Z}p_m$. We identify $\mathcal{E}'$ with the lattice of characters of the torus $\mathcal{T}'$ by $(q \in \mathcal{E}')$ :

$$q : \mathcal{T}' \to \mathbb{C}^*$$

$$q : \theta \to q(\theta),$$

where $(e^q)\theta = q(\theta)e^q$. We denote also by $q$ the infnitesimal map

$$q : Lie\ \mathcal{T}' \to \mathbb{C}$$

associated to

$$q : \mathcal{T}' \to \mathbb{C}^*.$$

So we identify $\mathcal{E}'$ with the lattice of weights of the commutative Lie algebra $Lie\ \mathcal{T}'$.

We denote by $\{E_1, ..., E_m\}$ the $\mathbb{C}$-basis of the Lie algebra $Lie\ \mathcal{T}$ of $\mathcal{T}$ which is dual to the basis $p = \{p_1, \dots p_m\} : p_j(Ei) = \delta_{i,j};\ i, j = 1, \dots, m$.

Let $C = (n(i,j))$ be a fixed square $(n, n)$ matrix with entries in $\mathbb{Z}$. We define weights $q_j \in \mathcal{E}'$ on $Lie\ \mathcal{T}'$ by the formulae :

$$(4,1) \quad q_j(Ei) = n(i,j);\ i, j = 1, \dots, m.$$

Explicitely we have $q_j = \bigoplus_{i=1,\dots,m} n(i,j)p_i$.

The $q_j$'s are polynomials in $x^{-1}$. For each $q_j$ we choose a line of maximal decay $\alpha_j$ and a line of maximal growth $\beta_j$, such that the $(q_j, \alpha_j)$'s (resp. the $(q_j, \alpha_j)$'s) are distinct elements of the set $\mathcal{E}' \times \mathbb{R}$. Some $q_j$'s can be equal for different values of $j$ ($q_j = q'_j$ if and only if $n(i,j) = n(i,j')$ for $i = 1, \ldots m$), in that case it can be necessary to change our initial choice of the $p_i$'s : the degree of the new $p_i$'s being larger than the maximal number of values of $j$ corresponding to a same subset $\{n(i,j)/i = 1, \ldots m\}$ of $\mathbb{N}$. Then we denote by $Lie\ \mathcal{R}'$ the **free** complex Lie algebra generated by the $2m$ symbols $\dot{\Delta}_{q_j,\alpha_j}$ and $\dot{\Delta}_{-q_j,\beta_j}$ ($j = 1, \ldots, m$). We define an action of the abelian Lie algebra $Lie\ \mathcal{T}'$ on $Lie\ \mathcal{R}'$ by

$$[E_i, \dot{\Delta}_{q_j,\alpha_j}] = q_j(E_i)\dot{\Delta}_{q_j,\alpha_j} = n(i,j)\dot{\Delta}_{q_j,\alpha_j}$$

$$[E_i, \dot{\Delta}_{-q_j,\beta_j}] = -q_j(E_i)\dot{\Delta}_{-q_j,\beta_j} = -n(i,j)\dot{\Delta}_{-q_j,\beta_j}.$$

The free Lie algebra $\mathcal{R}'$ is a quotient of the resurgent Lie algebra $\mathcal{R}$, and the action which just defined of $Lie\ \mathcal{T}'$ on $Lie\ \mathcal{R}'$ corresponds evidently to the action of $Lie\ \mathcal{T}$ on $Lie\ \mathcal{R}$ by taking quotients. The corresponding semi-direct product $Lie\ \mathcal{T}' \ltimes Lie\mathcal{R}'$ is a quotient of the semi-direct product $Lie\ \pi_{1,s}(\mathbb{C}^*, 0) = Lie\ \mathcal{T} \ltimes Lie\ \mathcal{R}$ : The natural map

$$Lie\ p' : Lie\ \pi_{1,s}(\mathbb{C}^*, 0) \to Lie\ \pi'_{1,s}$$

is surjective.

Now if we choose $C$ to be the Cartan matrix of the Chevalley basis $\{H_i, X_i, Y_i\}_{i=1,\ldots,m}$ of $\mathfrak{g}$, we can define a unique homomorphism of Lie-algebras

$$\psi : Lie\ \mathcal{T}' \ltimes Lie\ \mathcal{R}' \to \mathfrak{g}$$

by $\psi(E_i) = H_i, \psi(\Delta_{q_j,\alpha_j}) = X_i, \psi(\Delta_{-q_j,\beta_j}) = Y_i$. This homomorphism is clearly **surjective**. It is easy to check that the restriction of $\psi$ to the acting resurgent algebra $Lie\ \mathcal{R}'$ is already surjective.

We set $\hat{\rho}(\hat{\gamma}_0) = id$ (i. e. we decide that the formal monodromy is trivial). Then, using $Lie\ p'$ and $\psi$, we get a pair of representations $(\hat{\rho}_f, L\rho_r)$, where

$$\hat{\rho}_f : \ \pi_{1,sf}(\mathbb{C}^*, 0)) \to \ GL\ (n; \mathbb{C}),$$

$$L\rho_r : \ Lie\ \mathcal{R} \to \ End\ (n; \mathbb{C}).$$

This pair of representations is clearly compatible with the action of $\pi_{1,sf}(\mathbb{C}^*, 0))$ on $Lie\ \mathcal{R}$ corresponding to the semi-direct product.

By the inverse wild Riemann-Hilbert correspondance the pair $(\hat{\rho}_f, L\rho_r)$ defines a unique germ of meromorphic connection $\nabla$. Let $D$ be a corresponding germ of meromorphic system of order one and rank $n$.

We denote by $\mathcal{H}$ the Cartan subgroup of $G$ corresponding to the subalgebra $\mathfrak{h}$ of $\mathfrak{g}$ generated by the $H_i$'s. This Cartan subgroup is clearly the exponential torus of $D$. The $exp\ X_i$'s and the $exp\ Y_i$'s are **all** the Stokes multipliers of $D$ [MR 2].

Using Theorem 3.1, we see that the differential Galois group of $D$ is the Zariski closure of the group generated by the exponential torus $\mathfrak{H}$ and the Stokes multipliers $exp\ X_i$ and $exp\ Y_i$ $(i = 1, \dots m)$. This group is also the Zariski closure of the group generated by the Stokes multipliers $exp\ X_i$, $exp\ Y_i$ $(i = 1, \dots m)$. Using Lemma 1.2, we see that this group is the given semi-simple group $G$. This ends the proof of Theorem 4.1.

**Example 4.2.** —

We will do explicitely the preceding construction for the group $SL\ (2; \mathbb{C})$. We have $\mathfrak{g} = sl(2; \mathbb{C})$.

We choose as a Chevalley basis of $\mathfrak{g}$ the usual $sl_2$ triple :

$$H = \begin{pmatrix} 1 & 0 \\ 0 & -1 \end{pmatrix}$$

$$X = \begin{pmatrix} 0 & 1 \\ 0 & 0 \end{pmatrix}$$

$$Y = \begin{pmatrix} 0 & 0 \\ 1 & 0 \end{pmatrix}.$$

We have $m = 1$, so we do not write the index $i$. We have the relations $[H, X] = 2X$, $[H, Y] = -2Y$, $[X, Y] = H$, $n(i, j) = 2$.

We choose $\mathcal{E}' = \mathbb{Z}\ e^{-\frac{1}{z}}$ as an acting exponential lattice $(p = e^{-\frac{1}{z}})$. The corresponding acting exponential torus is $\mathcal{T}' = \mathbb{C}^* : \mu \in \mathbb{C}^*$ transforms $e^{-\frac{1}{z}}$ in $\mu e^{-\frac{1}{z}}$. We have $q = e^{-\frac{2}{z}}$, $\alpha = \mathbb{R}^+$, $\beta = \mathbb{R}^-$. Then $\dot{\Delta}_{q,\alpha} = \dot{\Delta}_{-\frac{2}{x}, \mathbb{R}^+}$, and $\dot{\Delta}_{-q,\beta} = \dot{\Delta}_{\frac{2}{x}, \mathbb{R}^-}$.

The representation

$$\rho'_e : T' = \mathbb{C}^* \to GL\ (2; \mathbb{C})$$

is given by

$$\mu \to \begin{pmatrix} \mu & 0 \\ 0 & \mu^{-1} \end{pmatrix}.$$

We set $L\rho'_e(E) = H$, $L\rho'_r(\dot{\Delta}_{-\frac{2}{x}, \mathbb{R}^+}) = X$, $L\rho'_r(\dot{\Delta}_{\frac{2}{x}, \mathbb{R}^-}) = Y$, and $\hat{\rho}_m(\hat{\gamma}_0) = id$.

We will now prove that the connexion defined by the wild monodromy representation $(\hat{\rho}_m, \rho_e, L\rho_r)$, can be represented by a **Whittaker equation.**

We set $z = x^{-1}$. We look at the general Whittaker differential equation :

$$y'' + (-\frac{1}{4} + \chi z^{-1} + (\frac{1}{4} - m^2)z^{-2})y = 0.$$

We work in a given basis of formal solutions at infinity ($x = 0$) $\mathcal{B} = \{\hat{W}_{\kappa,m}, \hat{W}_{-\kappa,m}\}$ (cf. [MR 1], 3.3.7 p. 192).

The formal monodromy is (cf. [MR 1], 3.3.8 p. 192)

$$\hat{M} = \begin{pmatrix} e^{2i\pi\kappa} & 0 \\ 0 & e^{-2i\pi\kappa} \end{pmatrix}.$$

So we choose $\kappa = 0$. We get $\hat{M} = id$.

The Stokes multipliers are (cf. [MR 1], Théorème 3.2.2 p. 193)

$$Sto_0 = \begin{pmatrix} 0 & \rho_2(m) \\ 0 & 0 \end{pmatrix},$$

and

$$Sto_\pi = \begin{pmatrix} 0 & 0 \\ \rho_1(m) & 0 \end{pmatrix},$$

with $\rho_1(m) = \rho_2(m) = -2i\pi \frac{1}{\Gamma(m+\frac{1}{2})\Gamma(-m+\frac{1}{2})}$. We have

$$\Gamma(m + \frac{1}{2})\Gamma(-m + \frac{1}{2}) = \frac{\pi}{sin\ \pi(m + \frac{1}{2})},$$

and

$$\rho_1(m)\rho_2(m) = -4sin^2\ \pi(m + \frac{1}{2}).$$

We want $\rho_1(m)\rho_2(m) = 1$. So we choose $m = -\frac{1}{2} + \frac{i}{\pi}Log\ \frac{1+\sqrt{5}}{2}$. We transform the basis $\mathcal{B}$ of formal solutions by the transformation

$$\begin{pmatrix} 1 & 0 \\ 0 & i \end{pmatrix}.$$

In the new basis the Stokes multipliers are

$$Sto_0 = \begin{pmatrix} 0 & 1 \\ 0 & 0 \end{pmatrix} = X$$

and

$$Sto_\pi = \begin{pmatrix} 0 & 0 \\ 1 & 0 \end{pmatrix} = Y.$$

So the connexion defined by the wild monodromy representation $(\hat{\rho}_m, \rho_e, L\rho_r)$, can be represented by the Whittaker equation

$$\mathcal{W}_{0,-\frac{1}{2}+\frac{i}{\pi}Log\ \frac{1+\sqrt{5}}{2}} : y''+(-\frac{1}{4}+\frac{i}{\pi}Log\ \frac{1+\sqrt{5}}{2}(1-\frac{i}{\pi}Log\ \frac{1+\sqrt{5}}{2})z^{-2})y = 0.$$

**Remark 4.3.** —

In the proof of Theorem 4.1, there are many choices for the acting exponential torus $\mathcal{T}'$. The choice for the torus corresponds to the choice for the lattice $\mathcal{E}'$. We can for example take $\mathcal{E}' = \mathbb{Z}\ \frac{\lambda_1}{x} \oplus \ldots \oplus \mathbb{Z}\ \frac{\lambda_m}{x}$ (where the $\lambda_i$'s are free over $\mathbb{Q}$), and we get a "one-level" differential equation $D$, or $\mathcal{E}' = \mathbb{Z}\ \frac{1}{x} \oplus \ldots \oplus \mathbb{Z}\ \frac{1}{x^m}$, and we get a "multi-level" differential equation. If there are symetries in the Cartan matrix it is natural to try to reflect these symetries in the different choices in our construction. Then it can be interesting to generalize this construction (cf. the following remark).

**Remark 4.4.** —

In the proof of Theorem 4.1, we can build more general classes of differential equations corresponding to a given differential Galois group $G$. We can get equations with a non trivial formal monodromy. We start from the preceding construction. We have built an acting exponential torus $\mathcal{T}'$, an acting resurgent Lie algebra $Lie\ \mathcal{R}'$, and surjective homomorphisms $p'_e : \mathcal{T} \to \mathcal{T}'$ and $Lp'_r : Lie\ \mathcal{R} \to Lie\ \mathcal{R}'$, as well as pairs of representations $\rho'_e : \mathcal{T}' \to GL\ (n;\mathbb{C})$, $L\rho'_r : Lie\ \mathcal{R}' \to End\ (n;\mathbb{C})$, and $\rho_e = \rho'_e p'_e : \mathcal{T} \to GL\ (n;\mathbb{C})$, $L\rho_r = L\rho'_r Lp'_r : Lie\ \mathcal{R} \to End\ (n;\mathbb{C})$. We choose a matrix $\hat{M}$ in $G \subset GL\ (n;\mathbb{C})$ such that, if we set $\hat{\rho}_m(\hat{\gamma}_0) = \hat{M}$, then the action of $\hat{M}$ on $\mathfrak{g}$ corresponds by $(L\rho_e, L\rho_r)$ to the action of $\hat{\gamma}_0$ on $Lie\ \pi_{1,s}(\mathbb{C}^*)$.

**Remark 4.5.** —

In the construction of the acting torus and of the acting resurgence algebra we only used the rank $m$ and the matrix $C$. It is not necessary that this matrix satisfies the conditions for a Cartan matrix. So it is possible to do this construction for example with a generalized Cartan matrix corresponding to a Kac-Moody algebra [K]. We get so a surjective representation of $Lie\ \pi_{1,s}(\mathbb{C}^*, 0)$ on a Kac-Moody algebra. I do not know if this representation is related to some dynamical system.

## 5. Some open problems.

We have partially solved the **local** inverse problem :

*(L)* If a linear algebraic group $G$ is given, to find a germ of linear meromorphic differential equation $Dy = 0$ (at the origin) whose differential Galois group $Gal_{\mathbb{C}\{x\}[x^{-1}]}(D)$ is $G$;

and the inverse problem **limited to Hamburger equations** :

*(H)* If a linear algebraic group $G$ is given, to find a linear Hamburger differential equation $Dy = 0$ whose global differential Galois group $Gal_{\mathbf{C}(x)}(D)$ is $G$.

It is not possible to solve these problems without a condition on the group $G$ :

**Proposition 5.1.** —

*(i) Let $Dy = 0$ be a germ of linear meromorphic differential equation at the origin. We denote by $G$ its differential Galois group and by $G^0$ the identity component of $G$. Then the finite group $G/G^0$ is cyclic. More precisely this group is generated by the image of the (actual) monodromy.*

*(ii) Let $Dy = 0$ be a linear Hamburger differential equation. We denote by $G$ its global differential Galois group and by $G^0$ the identity component of $G$. Then the finite group $G/G^0$ is cyclic. More precisely this group is generated by the image of the (actual) monodromy.*

*(iii) Let $Dy = 0$ be a linear algrbraic differential equation with $m + 1$ singular points on the Riemann sphere. Then the finite group $G/G^0$ can be generated by $m$ elements. More precisely this group is generated by the image of the monodromy group (or Poincaré group) of $D$.*

Assertion *(ii)* follows easily from assertion *(i)*, and *(iii)* is a generalization of *(i)* and is proved along the same lines. We will prove *(i)*. Let $Dy = 0$ be a germ of meromorphic differential equation at the origin. Its differential Galois group $G$ is the Zariski closure of the group generated by the exponential torus, the Stokes multipliers and the **actual** monodromy $M$ [MR 2] (1). We set $C = G/G^0$. The exponential torus and the Stokes multipliers are contained in $G^0$ [MR 2], so the group generated in $C$ by the image $M'$ of the monodromy $M$ is Zariski dense in $C$. But the group $C$ is finite, then $C$ is a finite cyclic group generated by $M'$.

Finally the following inverse problems remain open for a non semisimple algebraic group $G$ (2) :

*(L')* If a linear algebraic group $G$ is given, with $G/G^0$ cyclic, to find a germ of linear meromorphic differential equation $Dy = 0$ (at the origin) whose differential Galois group $Gal_{\mathbf{C}\{x\}[x^{-1}]}(D)$ is $G$;

*(H')* If a linear algebraic group $G$ is given, with $G/G^0$ cyclic, to find a

---

(1) We work with a fixed representation of $G$ as a subgroup of the linear group of a space of actual solutions on a germ of sector at the origin.

(2) If there exists a Zariski dense subgroup of $G$ generated by one element, then it is possible to solve problem (L') with a germ of regular singular linear differential equation [TT].

linear Hamburger differential equation $Dy = 0$ whose differential Galois group $Gal_{\mathbb{C}(x)}(D)$ is $G$;

*(M)* If a linear algebraic group $G$ is given, with $G/G^0$ generated by $m$ elements, to find a linear algebraic differential equation $Dy = 0$, with $m+1$ singular points on the Riemann sphere, whose differential Galois group $Gal_{\mathbb{C}(x)}(D)$ is $G$. More generally, if a linear algebraic group $G$ is given, to find a linear algebraic differential equation $Dy = 0$, with a minimal number of singular points on the Riemann sphere, whose differential Galois group is $G$.

It is not possible to solve problems *(L')* and *(H')* for an arbitrary group $G$. For example there exists no solution if $G$ is the commutative additive group $\mathbb{C}^2$. The problem (L') will be solved in [Ra 5].

A first version of this paper was published as a Strasbourg I.R.M.A. preprint [Ra 2].

## REFERENCES

[Be 1] D. BERTRAND, *Groupes algébriques et équations différentielles linéaires*, Sém. Bourbaki, 44ème année, 1991-92, num. 750.

[Be 2] D. BERTRAND, *Groupes algébriques linéaires et théorie de galois différentielle*, Cours de troisième cycle 1985-86, Université Paris VI, notes rédigées par R. Lardon.

[B] F. BEUKERS, *Differential Galois Theory*, in *Théorie des Nombres et Physique*, Springer Lecture Notes, Springer Verlag

[BBH] F. BEUKERS, F. BROWNAWELL, W. D. HECKMANN, *Monodromy for the Hypergeometric Function $_nF_{n-1}$*, Inv. Math. 95 (1989), p. 325–354.

[BB] A. BIALYNICKI-BIRULA, *On the inverse problem of Galois theory of differential fields*, Bull. Amer. Math. Soc. 16 (1963), p. 960–964.

[Bi] G. D. BIRKHOFF, *Equinalent singular points of ordinary differential equations*, Math. Ann. 74 (1915), p. 134–139.

[Bou 1] K. BOUSSEL, *Groupes de Galois des équations hypergéométriques réductibles*, Thèse Paris 1992.

[Bou 2] K. BOUSSEL, *Groupes de Galois des équations hypergéométriques...*, CRAS Paris 309 (1989), p. 587–589.

[D 1] P. DELIGNE, *Equations différentielles à points singuliers réguliers*, Springer Lectures Notes in math. 163, Springer Verlag (1970).

[D 2] P. DELIGNE, *Catégories tannakiennes*, Grothendieck Festshrift 1991.

[Du] A. DUVAL, *Triconfluent Heun Equations*, in a book in preparation

(Ronveaux ed.) (1989).

[D LR] A. DUVAL, M. LODAY-RICHAUD, *Kovacic's algorithm and its application to some families of special functions*, (1992).

[D M] A. DUVAL, C. MITSCHI, *Matrices de Stokes et groupes de Galois des équations hypergéométriques confluentes généralisées*, Pacific Journal of Mathematics, vol. 138, 1 (1989), p. 25–56.

[E 1] J. ECALLE, *Les Fonctions Résurgentes*, t. I, Publications Mathématiques d'Orsay (1981).

[E 2] J. ECALLE, *Les Fonctions Résurgentes*, t. II, Publications Mathématiques d'Orsay (1981).

[E 3] J. ECALLE, *Les Fonctions Résurgentes*, t. III, Publications Mathématiques d'Orsay (1985).

[Er] A. ERDÉLYI and..., *Higher Transcendantal Functions* (Bateman manuscript project) vol. II, Ch. 7, Mc Graw Hill, New York 1953.

[H] HAMBURGER, J. für Math. 103 (1888).

[Ho] G. P. HOCHSCHILD, *Basic Theory of Algebraic Groups and Lie Algebras*, Springer Verlag New York 1981.

[Hu] J. E. HUMPHREYS, *Linear Algebraic Groups*, Springer Verlag, New York 1975

[I] E. L. INCE, *Ordinary Differential equations*, Dover Publications.

[K] V. KAC, *Infinite dimensional Lie Algebras*, Birkhäuser 1983.

[Kap] I. KAPLANSKY, *An introduction to differential algebra*, Hermann, Paris 1957.

[Kat 1] N.M. KATZ, *An Overview of Deligne's Work on Hilbert's Twenty-First Problem*, Mathematical Developments Arising From Hilbert Problems, American Mathematical Society, Providence (1976).

[Kat 2] N.M. KATZ, *On the calculation of some differential Galois Groups*, Inventiones Math. 87 (1987).

[Kat 3] N.M. KATZ, *Exponential sums and differential equations*, (1987).

[Kat 4] N.M. KATZ, *Exponential sums over finite fields and differential equations over the complex numbers : some interactions*, A.M.S. Winter meeting (1989).

[KP] N.M. KATZ, R. PINK, *A note on pseudo CM-representations and differential galois groups* Duke Math. J. 54 (1987), p. 57–65.

[Ko] E.R. KOLCHIN, *Differential Algebra and Algebraic Groups*, Academic

Press, New-York (1973).

[Kov 1] J. Kovacic, *The inverse problem in the Galois theory of differential fields*, Ann. of Math. 89 (1969), p. 583–608.

[Kov 2] J. Kovacic, *On the inverse problem in the Galois theory of differential fields*, Ann. of Math. 93 (1971), p. 269–284.

[MR 1] J. Martinet, J.P. Ramis, *Théorie de Galois Différentielle et Resommation*, in *Computer Algebra and Differential equations* (E. Tournier ed.), Academic Press (1989), p. 117–214.

[MR 2] J. Martinet, J.P. Ramis, *Elementary acceleration and multisummability*, Ann. Inst. Henri Poincaré, Physique Théorique, 54, 4 (1991), p.331-401

[Mi 1] C. Mitschi, *Groupes de Galois différentiels et G-fonctions*, Thèse, Strasbourg (1989).

[Mi 2] C. Mitschi, *Groupes de Galois différentiels de certaines équations hypergéométriques généralisées d'ordre 7*, preprint I.R.M.A. Strasbourg (1989).

[Mi 3] C. Mitschi, *Groupe de Galois différentiel des équations hypergéométriques confluentes généralisées*, C. R. Acad. Sci. Paris, t. 309, Série I (1989), p. 217–220.

[Mi 4] C. Mitschi, *Differential Galois groups and G-functions*, Proceedings CADE 2, 1990, Academic Press (1991).

[Mi 5] C. Mitschi, *Differential Galois groups of confluent generalized hypergeometric equations*, in preparation (1993).

[PW] R. B. Paris, A. D. Wood, *Asymptotics of high order differential equations*, Pitman Research Notes in Mathematics series 129, Longman Scientific and Technical 1986.

[Ra 1] J. P. Ramis, *Irregular connections, wild $\pi_1$, and confluence,* Proceedings of a Conference at Katata, Japan 1987, Taniguchi Fundation.

[Ra 2] J. P. Ramis, *Un problème inverse en Théorie de Galois différentielle*, preprint IRMA Strasbourg (1989).

[Ra 3] J. P. Ramis, *About my theorem on Galois differential theory : a "mode d'emploi"*. Preprint IRMA Strasbourg (1991).

[Ra 4] J.P. Ramis, *Filtration Gevrey sur le groupe de Picard-Vessiot d'une équation différentielle irrégulière*,Preprint Instituto de Matematica Pura e Aplicada (IMPA), Rio de Janeiro, 45 (1985), p.1-38.

[Ra 5] J.P. Ramis, *About the inverse problem in differential Galois theory,* in preparation, IRMA Strasbourg (1993).

[R] H. RÖHRL, *Das Riemmannsch-Hilbertsche Problem der Theorie der linearen Differentielgleichungen*, Math. Annalen, 133 (1957), p. 1–25.

[Sc] L. SCHLESINGER, *Handbuch der Theorie der Linearen Differentialgleichungen*, Teubner, Leipzig 1897.

[Se 1] J. P. SERRE, *Lie Algebras and Lie groups*, Benjamin, New York 1965.

[Se 2] J. P. SERRE, *Algèbres de Lie semi-simples complexes*, Benjamin New York 1966.

[Si 1] M. SINGER, *An outline of differential galois theory*, in *Computer Algebra and Differential equations* (E. Tournier ed.), Academic Press (1989).

[Si 2] M. SINGER, *On the inverse Problem in differential Galois theory*, preprint (1989).

[Si 3] M. SINGER, *Moduli of linear differential equations on the Riemann sphere with fixed Galois group*, to appear in the Pacific Journal of Mathematics.

[SU] M. SINGER, F. ULMER, *Galois groups of second and thirs order differential equations*, preprint (1992).

[TT] M. TRETKOFF, C. TRETKOFF, *Solution of the inverse problem of differential Galois theory in the classical case*, Am. J. Math. 101 (1979).

# 19 A Remark on Bessel Functions

**Yasutaka Sibuya** University of Minnesota, Minneapolis, Minnesota

§.0. **Abstract**: On September 24, 1990, Lawrence Markus gave the following problem to the author.

**Problem.** *Consider the differential field $F < J_0, J_1, J_2, \cdots >$, where $F$ is the differential field of all elementary functions over $C(x)$ . Is $\int^x J_1(t)^3\, dt$ in this field $F < J_0, J_1, J_2, \cdots >$ ? It seems unlikely but I have no ideas.*

The author has not found any answer to this question. But, in this paper, an approach to this problem will be explained.

§1. **Introduction**: Let $C(x)$ be the ordinary differential field of rational functions of a complex variable $x$. Also, for each nonnegative integer $n$, let us denote by $J_n(x)$ the Bessel function of the first kind of order $n$, i.e.

$$J_n(x) = \left(\frac{x}{2}\right)^n \sum_{\ell=0}^{+\infty} \frac{(-1)^\ell}{\ell!\,(n+\ell)!} \left(\frac{x}{2}\right)^{2\ell} . \tag{1.1}$$

We consider the ring $C(x)[J_0, \delta J_0]$ of polynomials in $J_0$ and $\delta J_0$ with coefficients in $C(x)$, where $\delta = x\frac{d}{dx}$. The ring $C(x)[J_0, \delta J_0]$ is a differential algebra over the field $C$ of complex numbers, since $y = J_0$ is a solution of the differential equation:

$$\delta^2 y + x^2 y = 0 . \tag{1.2}$$

**Observation 1.1**: By utilizing recurrence relations:

$$x J_{n+1}(x) = n J_n(x) - \delta J_n(x) \qquad (n = 0, 1, \cdots) \tag{1.3}$$

(cf. G.N.Watson[2 ; p.45]), we can show that

$$J_n(x) \in C(x)[J_0, \delta J_0] \qquad (n = 0, 1, \cdots). \tag{1.4}$$

In particular, note that

$$J_1(x) = -\frac{dJ_0(x)}{dx} . \tag{1.5}$$

To illustrate our main idea, we shall first prove the following theorem.

**Theorem 1.2.** *We have*

$$\int^x J_n(t)\,dt \begin{cases} \in C(x)[J_0, \delta J_0] & \text{if } n \text{ is odd}, \\ \notin C(x)[J_0, \delta J_0] & \text{if } n \text{ is even}, \end{cases} \tag{1.6}$$

*where $\int^x J_n(t)\,dt$ denotes the indefinite integral of $J_n(x)$.*

By utilizing a similar idea, we shall also show that

$$\int^x J_1(t)^3\,dt \notin C(x)[J_0, \delta J_0]\,. \tag{1.7}$$

§2. **Proof of Theorem 1.2**: In case when $n$ is an odd integer, we can derive formular (1.6) by utilizing recurrence formulas

$$J_{n+1}(x) = J_{n-1}(x) - 2\frac{dJ_n(x)}{dx} \qquad (n = 1, 2, \cdots) \tag{2.1}$$

(cf. G.N.Watson[2 ; p.45]) and

$$J_1(x) = -\frac{dJ_0(x)}{dx}\,. \tag{1.5}$$

In case when $n$ is an even integer, we can reduce Theorem 1.2 to the case $n = 0$ by utilizing recurrence formulas (2.1). In fact, it is easily seen that, if

$$\int^x J_0(t)\,dt \notin C(x)[J_0, \delta J_0]\,, \tag{2.2}$$

then

$$\int^x J_n(t)\,dt \notin C(x)[J_0, \delta J_0] \qquad \text{for} \quad n = 2, 4, 6, \cdots\,. \tag{2.3}$$

We shall prove (2.2) by deriving a contradiction from the assumption that

$$\int^x J_0(t)\,dt \in C(x)[J_0, \delta J_0]\,. \tag{2.4}$$

Assumption (2.4) means that there exists a polynomial $F(x, \xi, \eta)$ in two variables $\xi$ and $\eta$ whose coefficients are rational functions of $x$ (i.e. $F \in C(x)[\xi, \eta]$) such that

$$\int^x J_0(t)\,dt = F(x, J_0(x), \delta J_0(x))\,. \tag{2.5}$$

From (2.5) we derive

$$\begin{aligned} x J_0(x) &= x\frac{\partial F}{\partial x}(x, J_0(x), \delta J_0(x)) + \frac{\partial F}{\partial \xi}(x, J_0(x), \delta J_0(x))\delta J_0(x) \\ &\quad + \frac{\partial F}{\partial \eta}(x, J_0(x), \delta J_0(x))\delta^2 J_0(x) \\ &= x\frac{\partial F}{\partial x}(x, J_0(x), \delta J_0(x)) + \frac{\partial F}{\partial \xi}(x, J_0(x), \delta J_0(x))\delta J_0(x) \\ &\quad - \frac{\partial F}{\partial \eta}(x, J_0(x), \delta J_0(x))x^2 J_0(x). \end{aligned} \tag{2.6}$$

Since $J_0$, $\delta J_0$ and $x$ are algebraically independent over the field $C$ (cf. C.L.Siegel [1 ; Satz 3 on p.216]), the polynomial $F(x,\xi,\eta)$ must satisfy the equation:

$$x\xi \;=\; x\frac{\partial F}{\partial x}(x,\xi,\eta) \;+\; \eta\frac{\partial F}{\partial \xi}(x,\xi,\eta) \;-\; x^2\xi\frac{\partial F}{\partial \eta}(x,\xi,\eta). \tag{2.7}$$

Write $F$ in the form:

$$F(x,\xi,\eta) \;=\; F_0(x) \;+\; F_1(x)\xi \;+\; F_2(x)\eta \;+\; \cdots ,$$

where $F_0$, $F_1$, $F_2$ are rational functions of $x$, and $\cdots$ denotes those terms that are nonlinear in $\xi$ and $\eta$. Then (2.7) implies that

$$x \;=\; \delta F_1 \;-\; x^2 F_2 , \qquad 0 \;=\; \delta F_2 \;+\; F_1 ,$$

and hence

$$\delta^2 F_2 + \; x^2 F_2 \;=\; -x . \tag{2.8}$$

At $x = \infty$, let us write $F_2$ in the form:

$$F_2(x) \;=\; \sum_{m=m_0}^{+\infty} \; a_m \, x^{-m} ,$$

where $m_0$ is an integer, $a_m \in C$ for $m \geq m_0$, and $a_{m_0} \neq 0$. Since $F_2$ is a rational function of $x$, the series $\sum_{m\geq m_0} a_m \, x^{-m}$ is convergent for $|x| > R$ if $R$ is a sufficiently large positive number. From (2.8) we derive

$$m_0 = 1, \;\; a_1 = -1, \;\; a_2 = 0, \;\; m^2 a_m + a_{m+2} = 0 \;\; \textit{for} \;\; m \geq 3 . \tag{2.9}$$

Utilizing (2.9), we can prove that the series $\sum_{m\geq m_0} a_m \, x^{-m}$ is divergent. Thus we derived a contradiction from assumption (2.4). **q.e.d.**

§3. **Proof of (1.7)**: Assuming that

$$\int^x J_1(t)^3 \, dt \in C(x)[J_0, \delta J_0] , \tag{3.1}$$

we shall derive a contradiction. Assumption (3.1) implies that there exists a polynomial $F(x,\xi,\eta) \in C(x)[\xi,\eta]$ such that

$$-\eta^3 = x^3 \frac{\partial F}{\partial x} + x^2 \eta \frac{\partial F}{\partial \xi} - x^4 \xi \frac{\partial F}{\partial \eta} . \tag{3.2}$$

Set

$$F = F_0(x) + \sum_{N=1}^{+\infty} \sum_{p=0}^{N} F_{N,p}(x) \xi^p \eta^{N-p} , \tag{3.3}$$

where $F_0$ and the $F_{N,p}$ are rational functions of $x$ . Set also

$$H_p(x) = F_{3,p}(x) \qquad (p = 0, 1, 2, 3) .$$

Then (3.2) yields

$$\begin{cases} -1 = x^2 \delta H_0 + x^2 H_1 , \\ 0 = x^2 \delta H_1 + 2x^2 H_2 - 3x^4 H_0 , \\ 0 = x^2 \delta H_2 + 3x^2 H_3 - 2x^4 H_1 , \\ 0 = x^2 \delta H_3 - x^4 H_2 . \end{cases} \tag{3.4}$$

If we set

$$\begin{cases} A_0 = \begin{bmatrix} 0 & 1 & 0 & 0 \\ 0 & 0 & 2 & 0 \\ 0 & 0 & 0 & 3 \\ 0 & 0 & 0 & 0 \end{bmatrix} , \quad A_1 = \begin{bmatrix} 0 & 0 & 0 & 0 \\ 3 & 0 & 0 & 0 \\ 0 & 2 & 0 & 0 \\ 0 & 0 & 1 & 0 \end{bmatrix} , \\ \vec{H}(x) = \begin{bmatrix} H_0 \\ H_1 \\ H_2 \\ H_3 \end{bmatrix} , \quad \vec{e} = \begin{bmatrix} 1 \\ 0 \\ 0 \\ 0 \end{bmatrix} , \end{cases} \tag{3.5}$$

we can write (3.4) in the form:

$$\delta \vec{H}(x) + [\, A_0 - x^2 A_1 \,] \vec{H}(x) = -x^{-2} \vec{e} . \tag{3.6}$$

We claim that Equation (3.6) can not admit any rational solutions. In fact, any rational solution $\vec{H}$ of (3.6) must have a form:

$$\vec{H}(x) = \frac{1}{2x^2}\,\vec{e} + \sum_{m=0}^{M} x^{2m}\vec{B}_m \,,$$

where $M$ is a nonnegative integer, $\vec{B}_m \in C^4$ and $\vec{B}_M \neq \vec{0}$. This means that

$$\begin{cases} A_0\vec{B}_0 - \frac{1}{2}A_1\vec{e} = \vec{0}\,, \\ [\,(2m)I + A_0\,]\vec{B}_m - A_1\vec{B}_{m-1} = \vec{0} \qquad \textit{for} \quad m \geq 1\,, \\ \vec{B}_M = \begin{bmatrix} 0 \\ 0 \\ 0 \\ \gamma_1 \end{bmatrix}, \end{cases}$$

where $I$ is the $4 \times 4$ identity matrix and $\gamma_1$ is a nonzero complex number.

It is easily seen that

$$\vec{B}_0 = \begin{bmatrix} \gamma_2 \\ 0 \\ \frac{3}{4} \\ 0 \end{bmatrix} \qquad \textit{and} \qquad \vec{B}_{M-1} = \gamma_1 \begin{bmatrix} 0 \\ \frac{3}{2} \\ 2M \\ \gamma_3 \end{bmatrix},$$

where $\gamma_2$ and $\gamma_3$ are certain complex numbers. This means that $M > 1$ and that

$$[\,2(M-1)I + A_0\,]\vec{B}_{M-1} - A_1\vec{B}_{M-2} = \begin{bmatrix} \frac{3}{2}\gamma_1 \\ \cdots \\ \cdots \\ \cdots \end{bmatrix} \neq \vec{0}\,.$$

This is a contradiction. **q.e.d.**

**Acknowledgements**: The main part of this paper was prepared at University of Groningen, the Netherlands while the author was a guest researcher supported partially by the Netherlands Organisation for Scientific Research and the Department of Mathematics, University of Groningen. The author wishes to thank these institutes for the invitation and the hopspitality. In particular, he expresses his appreciation to Professor M. van der Put for useful advises. The author was also partially supported by a grant from National Science Foundation.

## References:

[1] C.L.Siegel, *Über* einige Anwendungen diophantischer Approximationen, Abhandlungen der Preussischen Akademie der Wissenschaften, Physikalisch-mathematische Klasse 1929, Nr 1; C.L.Siegel Gesammelte Abhandlungen, Band 1, Springer-Verlag, 1966, 209-266.

[2] G.N.Watson, A Treaties on the Theory of Bessel Functions, second edition, Cambridge , 1962.

# 20 The Broadwell System, Self-Similar Ordinary Differential Equations, and Fluid Dynamical Limits

**Marshall Slemrod**[1] University of Wisconsin, Madison, Wisconsin

**Athanasios E. Tzavaras**[2] University of Wisconsin, Madison, Wisconsin

[1]This research was supported in part by NSF Grants INT-8914473 and DMS-9006945.

[2]This research supported in part by the U.S. Army Research Office under Grant No. DAAL03-88-K-0185.

We begin by considering the Broadwell model[1],[4] of a gas of identical particles each of mass $m$ contained in $\mathbb{R}^3$. A point in $\mathbb{R}^3$ is identified with its coordinates $(x, y, z)$. Particles move with six fixed velocities $\underline{v}_1 = c(1,0,0)$, $\underline{v}_2 = c(-1,0,0)$, $\underline{v}_3 = c(0,1,0)$, $\underline{v}_4 = c(0,-1,0)$, $\underline{v}_5 = c(0,0,1)$, $\underline{v}_6 = c(0,0,-1)$. As the particles move they may collide in a binary fashion with only elastic collisions permitted. Elastic collisions conserve mass, momentum, and energy.

Let us consider for example the collision of two particles traveling toward each other with velocities $\underline{v}_1$ and $\underline{v}_2$. Their total mass is $2m$, total momentum is zero, and total kinetic energy is $mc^2$. An elastic collision within the restriction of our six velocity rule which preserves mass, momentum, and energy (which is in fact kinetic energy since attraction between particles is not considered) can only yield that the particles collide and head off in opposite directions with speed $c$ along axes parallel to the $x, y, z$ axes. In other words a pair of particles with velocities $\underline{v}_1$ and $\underline{v}_2$ becomes after collision a pair of particles with velocities $\underline{v}_2$ and $\underline{v}_1$ or $\underline{v}_3$ and $\underline{v}_4$ or $\underline{v}_5$ and $\underline{v}_6$. Since each of these possibilities is equally likely each one can occur with probability $\frac{1}{3}$. We thus see the effect of $\underline{v}_1$, $\underline{v}_2$ collision: There is a $\frac{1}{3}$ probability of retaining the $\underline{v}_1$, $\underline{v}_2$ particles but a $\frac{2}{3}$ probability that they may become some other pair of particles. Of course similar statements can be made for the other two types of "head on" collisions. What about collisions of particles coming at right angles to one another, e.g. particles with velocities $\underline{v}_1$, $\underline{v}_3$. In this their total momentum is $m(\underline{v}_1 + \underline{v}_3) = mc(1,1,0)$ and the only way to conserve this momentum after a collision is that the $\underline{v}_1$, $\underline{v}_2$ particles preserve their velocities i.e. the particles must pass through each other as if there was no collision. Hence particles coming at right angles do not interact in the model.

Next we denote by $f_j(x, y, z, t)$ the number of particles per unit volume with velocity $\underline{v}_j$, ; $j = 1, ..., 6$. If particles did not collide we could have $f_j$ satisfying the simple advection equation

$$\frac{\partial f_j}{\partial t} + v_j \cdot \nabla f_j = 0$$

e.g. with $j = 1$

$$\frac{\partial f_1}{\partial t} + c\frac{\partial f_1}{\partial x} = 0 \; .$$

But particles do collide and for example according to our above discussion we can lose $\underline{v}_1$ particles in a collision with $\underline{v}_2$ particles with probability $\frac{2}{3}$ yet will gain $\underline{v}_1$ particles due to $\underline{v}_3$, $\underline{v}_4$ and $\underline{v}_5$, $\underline{v}_6$ collisions, each with probability $\frac{1}{3}$. Hence if we assume that rate of change of $\underline{v}_1$ particles is proportional to the product of the densities of particles which will yield increase or decrease of $\underline{v}_1$ particles via collisions weighted by the probability of occurence of such collisions we would find

$$\frac{\partial f_1}{\partial t} + c\frac{\partial f_1}{\partial x} = \frac{1}{\varepsilon}(\frac{1}{3}f_3 f_4 + \frac{1}{3}f_5 f_6 - \frac{2}{3}f_1 f_2) \; .$$

The proportional factor $\frac{1}{\varepsilon} > 0$ scales the rate at which collisions occur. Small $\varepsilon > 0$ means rapid collisions, large $\varepsilon > 0$ means long periods between collisions. For convenience we set

$3\varepsilon$ to $\varepsilon$ and now write the full set of transport equations

$$\begin{aligned}
\frac{\partial f_1}{\partial t} + c\frac{\partial f_1}{\partial x} &= \frac{1}{\varepsilon}(f_3 f_4 + f_5 f_6 - 2 f_1 f_2)\,, \\
\frac{\partial f_2}{\partial t} - c\frac{\partial f_2}{\partial x} &= \frac{1}{\varepsilon}(f_3 f_4 + f_5 f_6 - 2 f_1 f_2)\,, \\
\frac{\partial f_3}{\partial t} + c\frac{\partial f_3}{\partial y} &= \frac{1}{\varepsilon}(f_1 f_2 + f_5 f_6 - 2 f_3 f_4)\,, \\
\frac{\partial f_4}{\partial t} - c\frac{\partial f_4}{\partial y} &= \frac{1}{\varepsilon}(f_1 f_2 + f_5 f_6 - 2 f_3 f_4)\,, \\
\frac{\partial f_5}{\partial t} + c\frac{\partial f_5}{\partial z} &= \frac{1}{\varepsilon}(f_1 f_2 + f_3 f_4 - 2 f_5 f_6)\,, \\
\frac{\partial f_6}{\partial t} - c\frac{\partial f_6}{\partial z} &= \frac{1}{\varepsilon}(f_1 f_2 + f_3 f_4 - 2 f_5 f_6)\,.
\end{aligned} \tag{1}$$

The parameter $\varepsilon > 0$ is called the Knudsen number.

If we sum the equations we recover the identity

$$\begin{aligned}
\frac{\partial}{\partial t}(f_1 + f_2 + f_3 + f_4 + f_5 + f_6) + c\frac{\partial}{\partial x}(f_1 - f_2) + c\frac{\partial}{\partial y}(f_3 - f_4) \\
+ c\frac{\partial}{\partial z}(f_5 - f_6) = 0
\end{aligned} \tag{2}$$

which simply expresses the conservation of mass, i.e. the density of particles $f \doteq m(f_1 + f_2 + f_3 + f_4 + f_5 + f_6)$, and momentum of particles $\rho u \doteq mc(f_1 - f_2)$, $\rho v \doteq mc(f_3 - f_4)$, $\rho w \doteq mc(f_5 - f_6)$ satisfy the usual mass balance equation

$$\frac{\partial \rho}{\partial t} + \frac{\partial}{\partial x}(\rho u) + \frac{\partial}{\partial y}(\rho v) + \frac{\partial}{\partial z}(\rho w) = 0\,. \tag{3}$$

Similarly if we subtract the second equation from the first, fourth from the third, and sixth from the fifth we find

$$\begin{aligned}
\frac{\partial}{\partial t}(f_1 - f_2) + c\frac{\partial}{\partial x}(f_1 + f_2) = 0\,, \\
\frac{\partial}{\partial t}(f_3 - f_4) + c\frac{\partial}{\partial y}(f_3 + f_4) = 0\,, \\
\frac{\partial}{\partial t}(f_5 - f_6) + c\frac{\partial}{\partial z}(f_5 + f_6) = 0\,,
\end{aligned} \tag{4}$$

Multiplying these equations by $c$ and defining

$$\Pi_{xx} = -\rho u^2 + c^2 m(f_1 + f_2),\ \Pi_{yy} = -\rho v^2 + c^2 m(f_3 + f_4)\ ,$$

$$\Pi_{zz} = -\rho w^2 + c^2 m(f_5 + f_6),\ \Pi_{xy} = -\rho uv,\ \Pi_{xy} = -\rho uw\ ,$$

$$\Pi_{yz} = -\rho vw,\ \Pi_{xy} = \Pi_{yx},\ \Pi_{xz} = \Pi_{zx},\ \Pi_{yz} = \Pi_{zy}\ ,$$

we find the usual balance of linear momentum equations are satisfied

$$\begin{aligned}
&\frac{\partial}{\partial t}(\rho u) + \frac{\partial}{\partial x}(\rho u^2) + \frac{\partial}{\partial y}(\rho uv) + \frac{\partial}{\partial z}(\rho uw) \\
&\qquad + \frac{\partial}{\partial x}\Pi_{xx} + \frac{\partial}{\partial y}\Pi_{xy} + \frac{\partial}{\partial z}\Pi_{xz} = 0\ , \\
&\frac{\partial}{\partial t}(\rho v) + \frac{\partial}{\partial x}(\rho vu) + \frac{\partial}{\partial y}(\rho v^2) + \frac{\partial}{\partial z}(\rho vw) \\
&\qquad + \frac{\partial}{\partial x}\Pi_{yx} + \frac{\partial}{\partial y}\Pi_{yy} + \frac{\partial}{\partial z}\Pi_{yz} = 0\ , \\
&\frac{\partial}{\partial t}(\rho w) + \frac{\partial}{\partial x}(\rho wu) + \frac{\partial}{\partial y}(\rho wv) + \frac{\partial}{\partial z}(\rho w^2) \\
&\qquad + \frac{\partial}{\partial x}\Pi_{zx} + \frac{\partial}{\partial y}\Pi_{zy} + \frac{\partial}{\partial z}\Pi_{zz} = 0\ .
\end{aligned} \tag{5}$$

$\underline{\Pi}$ is the stress tensor; its symmetry guarantees balance of moment of momentum.

We thus see that our simple model which had six equations (1) in the six *microscopic* unknowns $f_1,\ f_2,\ f_3,\ f_4,\ f_5,\ f_6$ has associated with it four conservation laws (3, 4) or equivalently (3, 5). These four conservation laws are not a closed system: there are only four equations in six unknowns. The object of the fluid dynamic limit problem is to rigorously close the system in the limit as $\varepsilon \to 0$ producing a closed system of four balance laws in the four macroscopic dependent variables $\rho,\ u,\ v,\ w$. Success in this pursuit will rigorously show the connection of a microscopic kinetic view of gas dynamics with a macroscopic kinetic picture.

The formal computation of the closed macroscopic system is easy. We can simply attempt a formal asymptotic expansion in powers of the (assumed) small parameter $\varepsilon > 0$

of the form

$$f_j(x,y,z,t) = f_j^0(x,y,z,t) + \varepsilon f_j^1(x,y,z,t) + \dots \tag{6}$$

and substitute this ansatz into (1). Equating powers of $\varepsilon$ we find the order $\varepsilon^{-1}$ terms on the right sides of (1) must equate with zero and hence we see

$$0 = f_3^0 f_4^0 + f_5^0 f_0^0 - 2 f_1^0 f_2^0 = f_1^0 f_2^0 + f_5^0 f_6^0 - 2 f_3^0 f_4^0 = f_1^0 f_2^0 + f_3^0 f_4^0 - 2 f_5^0 f_6^0 \,. \tag{7}$$

The expansion (6) is the *Hilbert expansion*[3] for the Broadwell system (1) and the solutions of (7) are the *Maxwellians* for the Broadwell system. Solutions of (7) are given by the four parameter family of solutions

$$(f_1^0, f_2^0, f_3^0, f_4^0, f_5^0, f_6^0) = C(e^{c_1}, e^{-c_1}, e^{c_2}, e^{-c_2}, e^{c_3}, e^{-c_3}) \,. \tag{8}$$

Next substitution of the Hilbert expansion into our conservation laws of mass and momentum (2), (4) and again equating powers of $\varepsilon$ shows that the $f_j^0$ must satisfy (2), (4), i.e. the four parameter family of Maxwellians given by (8) should satisfy the four conservation laws of mass and momentum. Since there is a one-to-one map of $C$, $c_1$, $c_2$, $c_3$ this is equivalent to saying that four equations (3), (5) can be expressed in terms of the four macroscopic variables $\rho$, $u$, $v$, $w$. For computational simplicity we for the moment retain the variables $C$, $c_1$, $c_2$, $c_3$ and record the result that the formal $\varepsilon \to 0$ limit of (1) is the hyperbolic system of conservation laws

$$\frac{\partial}{\partial t}(C(e^{c_1}+e^{-c_1}+e^{c_2}+e^{-c_2}+e^{c_3}+e^{-c_3}))$$
$$+c\frac{\partial}{\partial x}(C(e^{c_1}-e^{-c_1}))+c\frac{\partial}{\partial y}(C(e^{c_2}-e^{-c_2}))+c\frac{\partial}{\partial z}(C(e^{c_3}-e^{-c_3}))=0$$

(conservation of mass) ,

$$\frac{\partial}{\partial t}(C(e^{c_1}-e^{-c_1}))+c\frac{\partial}{\partial x}(C(e^{c_1}+c^{-c_1}))=0\,, \tag{9}$$
$$\frac{\partial}{\partial t}(C(e^{c_2}-e^{-c_2}))+c\frac{\partial}{\partial y}(C(e^{c_2}+c^{-c_2}))=0\,,$$
$$\frac{\partial}{\partial t}(C(e^{c_3}-e^{-c_3}))+c\frac{\partial}{\partial z}(C(e^{c_3}+c^{-c_3}))=0\,,$$

(conservation of linear momentum) .

This is our closed system of four equations in the four macroscopic variables $C$, $c_1$, $c_2$, $c_3$ which will be functions of the independent variables $x$, $y$, $z$, $t$.

This issue of interest now is not what is the form of the limiting macroscopic equations but can the limiting procedure be rigorously justified. First to simplify the computations let us consider the special case of (1) when $f_3 = f_4 = f_5 = f_6$ and all the $f_j$ depend only on $x$, $t$ with no $y$, $z$ dependence allowed. Then (1) becomes the system

$$\begin{aligned}\frac{\partial f_1}{\partial t}+c\frac{\partial f_1}{\partial x}&=\frac{1}{\varepsilon}(f_3^2-f_1f_2)\,,\\ \frac{\partial f_2}{\partial t}-c\frac{\partial f_2}{\partial x}&=\frac{1}{\varepsilon}(f_3^2-f_1f_2)\,,\\ \frac{\partial f_3}{\partial t}&=\frac{1}{2\varepsilon}(f_1f_2-f_3^2)\,,\end{aligned} \tag{10}$$

where we have trivially redefined $\varepsilon$.

In this case the relevant conservation laws are

$$\frac{\partial}{\partial t}(f_1+f_2+4f_3)+c\frac{\partial}{\partial x}(f_1-f_2)=0 \tag{11}$$

(conservation of mass) ,

and

$$\frac{\partial}{\partial t}(f_1 - f_2) + c\frac{\partial}{\partial x}(f_1 + f_2) = 0 \tag{12}$$

(conservation of linear momentum) ,

and goal is to show the limit of solution of (10) as $\varepsilon \to 0+$ is a solution of fluid dynamic $2 \times 2$ system of hyperbolic conservation laws

$$\frac{\partial}{\partial t}(f_1 + f_2 + 4(f_1 f_2)^{1/2}) + c\frac{\partial}{\partial x}(f_1 - f_2) = 0\,, \tag{13}$$

$$\frac{\partial}{\partial t}(f_1 - f_2) + c\frac{\partial}{\partial x}(f_1 + f_2) = 0\,. \tag{14}$$

(Of course the above system can be equivalent expressed in terms of our earlier exposited system (9) with two dependent variables $C$, $c_1$ and two independent variables $x$, $t$.)

In his original paper Broadwell derived an explicit construction of travelling wave solutions to (11), (12)

$$(f_1, f_2, f_3) = (\hat{f}_1(\theta), \hat{f}_2(\theta), \hat{f}_3(\theta)),\ \theta = \frac{x - st}{\varepsilon}\,,$$

with Maxwellian data $\hat{f}_j = \begin{cases} f_j^+ \\ f_j^- \end{cases}$ at $\theta = \pm\infty$ , i.e. $f_1^{\pm} f_2^{\pm} = (f_3^{\pm})^2$, where $s$ and the data satisfy the consistency condition

$$\begin{aligned} -s(f_1^+ + f_2^+ + 4f_3^+) + (f_1^+ - f_2^+) &= \\ -\,s(f_1^- + f_2^- + 4f_3^-) + (f_1^- - f_2^-)\,, \end{aligned} \tag{15}$$

$$-s(f_1^+ - f_2^+) + (f_1^+ + f_2^+) = -s(f_1^- - f_2^-) + (f_1^- + f_2^-)\,,$$

where by virtue of (15) the data is consistent with the Rankine-Hugoniot jump conditions

$$\begin{aligned} -s(f_1^+ + f_2^+ + 4(f_1^+ f_2^+)^{1/2}) + (f_1^+ - f_2^+) &= \\ -s(f_1^- + f_2^- + 4(f_1^- f_2^-)^{1/2}) + (f_1^- - f_2^-) \end{aligned} \tag{16}$$

$$-s(f_1^+ - f_2^+) + (f_1^+ + f_2^+) = -s(f_1^- - f_2^-) + (f_1^- + f_2^-)\,.$$

Thus Broadwell has shown for Riemann data satisfying the Rankine-Hugoniot jump conditions (16) passage to the fluid dynamic limit for the Broadwell model can be achieved via an explicit construction of travelling wave solutions. This limit provides a weak solution to the initial Riemann initial value problem for (13), (14) with initial data

$$\hat{f}_j = \begin{cases} f_j^+ & x > 0 \\ \\ f_j^- & x < 0 \end{cases} \quad , \qquad f_1^\pm f_2^\pm = (f_3^\pm)^2 \, .$$

This leaves open the question as to what can be done regarding the fluid dynamic limit for arbitrary Riemann data which is not necessarily consistent with (16).

The approach we take is motivated by the following observation: Any system of conservation laws

$$\frac{\partial F(U)}{\partial t} + \frac{\partial G(U)}{\partial x} = 0 \, , \tag{17}$$

$U : (-\infty, \infty) \times (0, \infty) \to \mathbb{R}^N$; $F, G : R^N \to \mathbb{R}^N$ with Riemann data

$$U(x, 0) = U_0(x) \, ,$$

$$U_0(x) = U^-, \; x < 0; \; U(x) = U^+, \; x > 0 \, , \tag{18}$$

must possess space-time dilational invariance. This means that for any positive constant $\alpha > 0$, the change of variable $(x, t) \to (\alpha x, \alpha t)$ preserves both the equations and the initial data. Hence solutions of Riemann problems should depend only on the similarity variable $\xi = \frac{x}{t}$ i.e. $U(x, t) = U(\xi)$.

As an example of the implications of the above reasoning consider the problem of attempting to solve the Riemann problem for a system of conservation laws as a "viscous" limit of the system

$$\frac{\partial F(U)}{\partial t} + \frac{\partial G(U)}{\partial x} = \varepsilon \frac{\partial^2 U}{\partial x^2} \, , \tag{19}$$

where $\varepsilon > 0$.

Consideration of space-time dilational invariance for (17), (18) might make us consider the ansatz $U(x,t) = U(\xi)$ for (19). But unfortunately (19) unlike (17) does not possess space-time dilational invariance. It was this reason that Dafermos[2] suggested the "viscous" limit problem

$$\frac{\partial F(U)}{\partial t} + \frac{\partial G(U)}{\partial x} = \varepsilon t \frac{\partial^2 U}{\partial x^2} ,$$

$t > 0$, which does possess space-time dilational invariance.

Since the Broadwell system does not possess space-time dilational invariance we are motivated to consider the artificial Broadwell system

$$\begin{aligned} \frac{\partial f_1}{\partial t} + \frac{\partial f_1}{\partial x} &= \frac{1}{\varepsilon t}(f_3^2 - f_1 f_2) , \\ \frac{\partial f_2}{\partial t} - \frac{\partial f_2}{\partial x} &= \frac{1}{\varepsilon t}(f_3^2 - f_1 f_2) , \\ \frac{\partial f_3}{\partial t} &= \frac{1}{2\varepsilon t}(f_1 f_2 - f_3^2) \end{aligned} \tag{20}$$

which does indeed possess the desired space-time dilational invariance. We now make the ansatz $f_1(x,t) = f_1(\xi)$, $f_2(x,t) = f_2(\xi)$, $f_3(x,t) = f_3(\xi)$ and substitute into (20) to obtain the system of non-autonomous ordinary differential equations

$$\begin{aligned} -(\xi - 1) f_1'(\xi) &= (f_3^2 - f_1 f_2)/\varepsilon , \\ -(\xi + 1) f_2'(\xi) &= (f_3^2 - f_1 f_2)/\varepsilon , \\ -\xi f_3'(\xi) &= (f_1 f_2 - f_3^2)/2\varepsilon . \end{aligned} \tag{21}$$

Since we wish $f_j(x,t) \to f_j^{\pm}$ for $x \gtrless 0$ as $t \to 0+$ for $j = 1,2,3$ we impose boundary data

$$f_j(-\infty) = f_j^- , \ f_j(+\infty) = f_j^+ , \ j = 1,2,3 , \tag{22}$$

where $f_1^- f_2^- = (f_3^-)^2$, $f_1^+ f_2^+ = (f_3^+)^2$. We note, however, that since $(f_1, f_2, f_3) = (f_1^{\pm}, f_2^{\pm}, f_3^{\pm})$ are constant solutions of (21), (22) on $1 < \xi < \infty$ and $-\infty < \xi < -1$

respectively the boundary condition (22) is replaced by the boundary condition

$$f_j(-1) = f_j^-,\ f_j(+1) = f_j^+,\ j = 1,2,3 \tag{23}$$

and hence we need only consider (21) on $-1 < \xi < 1$.

We close with a statement of our results on existence of solutions to the boundary value problem (21), (22) and passage to the fluid dynamic limit. Proofs and further details will appear in a longer publication, [5].

**Theorem.** *There is a $\delta > 0$ so that if $|f_1^+ - f_1^-| + |f_2^+ - f_2^-| + |f_3^+ - f_3^-| < \delta$ then the non-linear boundary value problem* (21), (23) *has a solution in $C^1((-1,0) \cup (0,1)) \cap C[-1,1]$. The parameter $\delta$ may depend on $\varepsilon$.*

Since our long term goal is to let $\varepsilon \to 0$ to obtain a resolution to the fluid dynamic limit problem the possible dependence of $\delta$ on $\varepsilon$ is unpleasant. Our next result obtained by a continuation argument shows there is a nontrivial class of boundary data for which (0.13), (0.15) possesses a solution for all $\varepsilon > 0$.

**Theorem.** *For all $\varepsilon > 0$ and for all positive Maxwellian boundary data satisfying $f_1^+ \geq f_1^-$, $f_2^+ \leq f_2^-$, $f_3^+ = f_3^-$ the boundary value problem* (21), (23) *possesses a solution.*

It would be nice to produce a larger class of boundary data then that given by the above theorem but so far if such a set exists, it has eluded us. In any case, the theorem provides us with a non-trivial set of solutions of (21), (23) existing for all $\varepsilon > 0$ for which we attempt to pass to the fluid dynamic limit $\varepsilon \to 0$. This is discussed next.

**Theorem.** *For all positive Maxwellian data satisfying $f_1^+ \geq f_1^-$, $f_2^+ \leq f_2^-$, $f_3^+ = f_3^-$ the modified Broadwell system* (20) *with Riemann data*

$$\lim_{t \to 0+} f_j(x,t) = \begin{Bmatrix} f_j^- & x < 0 \\ f_j^+ & x > 0 \end{Bmatrix},$$

*possess a positive solution* $f_j^\varepsilon(x,t) = f_j^\varepsilon(\xi)$, $\xi = \frac{x}{t},; t > 0$, $j = 1,2,3$. *The sequence* $\{f_j^\varepsilon(x,t)\}$ *possesses a subsequence which converges boundedly a.e. as* $\varepsilon \to 0+$ *to a positive solution* $f_j(x,t)$, $j = 1,2,3$, *of the fluid limit Riemann problem.*

Also we note the papers [6], [7] where the authors have approximated solutions of the fluid dynamic limit equation (13), (14) by solutions of the Broadwell system. This is an approximation program which is different from the compactness program given here.

**Acknowledgement.** This research was supported in part by the National Science Foundation grants INT-8914473 and DMS-9006945 (Slemrod) and by the U.S. Army Research Office under Grant No. DAAL03-88-K-0185 (Tzavaras).

**References**

[1] J. E. Broadwell, *Shock structure in a simple discrete velocity gas, Physics of Fluids* 7 (1964) 1243-1247.

[2] C. M. Dafermos, *Solution of the Riemann problem for a class of hyperbolic systems of conservation laws by the viscosity method, Archive for Rational Mechanics and Analysis* **52** (1973) 1-9.

[3] C. Cercignani, The Boltzmann equation and its applications, Springer-Verlag, New York, 1988.

[4] T. Platkowski and R. Illner, Discrete velocity models of the Boltzmann equation: a survey of the mathematical aspects of the theory, SIAM Review **30** (1988), 213-255.

[5] M. Slemrod and A. E. Tzavaras, Self-similar fluid dynamic limits for the Broadwell system, to appear Archive for Rational Mechanics and Analysis.

[6] R. E. Caflisch and G. C. Papanicolaou, The fluid dynamic limit of a nonlinear model of the Boltzmann equation, Comm. Pure Appl. Math. **32** (1979), 584-616.

[7] Z. Xin, The fluid dynamic limit for the Broadwell model of the nonlinear Boltzmann equation in the presence of shocks, Comm. Pure Appl. Math. **44** (1991), 679-714.

# 21 The Statistical Mechanics of Asset Prices

**Michael Stutzer** University of Minnesota, Minneapolis, Minnesota

It is difficult to think of another scientific concept which has been used as broadly as entropy. The choice of a probability distribution which minimizes relative entropy subject to constraints, dubbed the *maximum entropy formalism* (MEF), is a unifying framework for studying *statistical mechanics*, i.e. the modelling of phenomena consisting of large numbers of interacting components. The MEF has also found hundreds of applications in the social sciences, including economics and econometrics. However, as Zellner [48, p.21] has recently noted, "much more empirical and theoretical work needs to be done to get a 'maximum entropy, thermodynamic model' that performs well in explaining and predicting the behavior of economic systems."

In fact, the amount of work required is less than appears. For this paper shows that the well-established Arrow-Debreu analysis of contingent claims pricing may be conducted in an MEF framework. Utilizing a discrete time approximation of a constant coefficient, vector Ito process to describe the evolution of asset prices, the MEF is used to provide an alternative derivation of the Black-Scholes option pricing formula, and an entropy generalization of the popular variance bound diagnostic for empirical specifications of consumption-based asset pricing models.

The paper is self-contained, intended for systems scientists wishing to explore this connection between information theory, statistical mechanics, stochastic differential equations, and asset pricing.

Helpful comments at early stages of my explorations into this subject were made by Michael Dothan, John Baxter, and S.R.S. Varadhan.

## 1 Introduction

I recently visited Minnesota's iron mining region, and was surprised to learn that Prof. Markus was born there. I searched his curriculum vita for corroborating evidence of his Iron Range roots, expecting to find an early paper or two written in Serbo-Croatian. I wasn't able to unearth any, but it was obvious that Markus has staked his claims in many places, and profitably mined many. But there the similarities end. For unlike the Iron Range, Markus has diversified his interests, plowing new ground in many disciplines.

Josiah Willard Gibbs [1839-1903] was another polymath, contributing to mathematics, physics, and engineering. From Yale in 1863, he received the first doctorate in engineering (only the second doctorate in the sciences) granted in the U.S., and was appointed to its newly created chair in mathematical physics in 1871 [46]. Gibbs contributions to thermodynamics, made relatively early in his career, have influenced the work of economist and Nobel laureate Paul Samuelson, who also notes that Gibbs helped supervise Irving Fisher's doctoral dissertation in 1891 [38].

Just prior to Gibbs' death, his treatise on statistical mechanics was published, detailing his creative use of a family of probability distributions dubbed *canonical distributions.* A canonical density is proportional to $\exp[\gamma'\mathbf{R}]$, where $\gamma$ is a vector of Lagrange multipliers resulting from a convex optimization problem soon to be described, and $\mathbf{R}$ is vector of measurable functions on the sample space. The process of generalizing and applying Gibbs' method is now referred to as the *maximum entropy formalism* (MEF) [28]. In honor of Gibbs, distributions which solve constrained entropy problems are also called *Gibbs states.*

Despite the hundreds of applications of Gibbs states in the physical and social sciences, many motivated by the viewpoint of information theory [7], the MEF has seen limited use in financial economics. Osborne [37] used the Weber-Fechner Law and Gibbs' ensemble method to rationalize the choice of a lognormal distribution for stock prices. Cozzolino and Zechner [9] made an analogous contribution. In addition, Griffeath and Snell [20] used the MEF to revise path probabilities in a finite dimensional Markov chain, and applied this to optimal investment policy. Cover[chap.15] [7] uncovered relationships between entropy and dynamic portfolio strategies of investors with logarithmic utility. Gibbs states also play a key role in understanding the properties of Ising-like, random field models formulated by Föllmer [14], Durlauf [12],[11] and Brock [5], [4], the latter proposing the use of random field models in modeling different types of heterogeneous trader interdependencies. Finally, and most closely related to this paper, Grandy [18] modified the standard single period state pricing model, incorporating expected utility maximizing agents who use the MEF to estimate the probabilities of uncertain states from observations on expected asset prices. And independent work by Föllmer and Schweizer [15] used the MEF to choose a martingale measure used in their characterization of the "intrinsic risk" inherent in hedging non-redundant contingent claims in continuous time.

This paper takes a different tack. Section 2 reviews the concept of Arrow-Debreu (arbitrage-free) state pricing of contingent claims [1, chap.4] in a single period setting [1]. When markets are complete, there is a unique probability measure under which any asset's price is its risklessly discounted, expected future payoff. But analysts may not observe enough asset return data to compute this *risk neutral* probability measure, leaving them to choose from a continuum of measures consistent with observed return data. Alternatively, financial markets may just be incomplete, creating an operationally equivalent selection problem for the analyst. While all admissible measures will

[1]See Dothan [10, chaps.1-2].

produce the same valuation of a state-contingent claim which is duplicated by a portfolio of the observed assets, these valuations will generally differ on a state-contingent claim which cannot be duplicated by a portfolio of these assets. Section 3 uses Bayesian, information theoretic arguments similar to those motivating the econometric applications of entropy in Theil [43], Theil and Fiebig [44], and Zellner [50],[49],[48] to axiomatically rationalize the selection of a particular risk neutral measure under these circumstances. Upon normalizing it by the sum of its components, the resulting risk neutral measure is a Gibbs state, called *canonical risk neutral probabilities.*

To develop interesting applications of canonical risk neutral probabilities, section 4 starts by developing the simplest multiperiod generalization of the model in section 2. To do so, we adopt an incomplete market version of Hua He's [24] discrete time approximation to a continuous time model, in which risky asset prices follow a constant coefficient, vector Ito process. We compute the canonical risk neutral distribution, and the product measure formed from it. The latter, called the *canonical martingale measure*, is one of many martingale measures. Under any of these measures, the prices of state-contingent claims duplicated by dynamic trading strategies involving the observed assets are expected, (risklessly) discounted present values of their random payoffs, called *risk neutral valuations.* [2]

But risk neutral valuations of a state-contingent claim which is *not* duplicated by trading strategies involving the observed assets may not agree. Motivated by arguments in section 3, we investigate the canonical risk neutral valuation of contingent claims. Passing to the continuous time limit of our discrete time model yields a limiting parameter vector $\hat{\gamma}$ of the canonical risk neutral distribution. Normalizing $\hat{\gamma}$ by the sum of its components yields the vector of portfolio weights for the familiar mean-variance efficient *tangency portfolio* of the observed risky assets.

We first use this limit result to provide a very simple calculation of the limiting canonical martingale measure in the simplest case, i.e. when only a single stock's price process is observed in a model with $K > 2$ states. It is shown that the canonical risk neutral valuation of a European call option on this stock is the Black-Scholes price. While there are many other derivations of the Black-Scholes model, this one interprets Black-Scholes as the MEF inference in a dynamically incomplete model with frequent trading, using data only about the underlying stock's price process and the riskless rate of interest.

Moving from the theoretical domain to the empirical, the canonical risk neutral probabilities are shown to be of use in empirically testing consumption-based and/or other asset pricing models which don't admit arbitrage opportunities. The popular variance bound diagnosis of an asset pricing model [22],[21] incorporates the restriction that the expected products of asset returns of interest to the analyst and the model's intertemporal marginal rate of substitution (IMRS) must equal one. Imposing these moment restrictions may alternatively be viewed as the selection of a particular risk neutral distribution computed from the model's IMRS. By construction, the relative entropy of this distribution can not be less than the minimal relative entropy attained by the canonical risk neutral distribution. The aforementioned limit result is then used to show that under the assumed price process, frequently sampled data will produce variance bound diagnostics agreeing with entropy bound diagnostics. In this situation, information theory thus provides an alternative, axiomatically rationalized interpretation of commonly employed diagnostic techniques which utilize the variance bound. In other cases, the relative entropy bound provides an additional diagnostic tool.

Finally, section 5 uses large deviations theory to develop frequentist rationale for our focus on

[2] See Varian [45] and Huang and Litzenberger [26, chap.8] for applications oriented discussions. For a complete and rigorous theoretical treatment, see Harrison and Kreps [23].

canonical risk neutral probabilities. First, it is shown that as the length of the return series grows to infinity, the canonical probabilities may be characterized as approximate conditional probabilities of state occurrence, where conditioning is done on the set of realizations for which the observed sample average returns are approximately the riskless rate. Second, it is shown that the relative entropy bound of section 4.3 is the rate at which the probability of empirically observing near zero risk premia approaches zero as the sample length increases. As such, the relative entropy bound is larger when risk premia are more likely to be observed. The large deviations perspective is then used to exposit the isomorphism mapping canonical risk neutral valuation to the equilibrium statistical mechanics of large mechanical systems. Because of the outstanding success of statistical mechanics methods throughout the sciences, one can hope that future extensions of this interdisciplinary connection may also prove useful in financial economics.

## 2 Review of the Standard Model

We adopt the standard, finite dimensional state preference model with a finite number of states[3]. Traders are endowed with a single consumption good at either the beginning of the period, the end of the period, or both. Uncertain states drawn from a set $\Omega = \{\omega_1, \ldots, \omega_K\}$ determine both end of period endowments (if any) and marketed asset payoffs. There are N + 1 primary assets available for trade at the beginning of the period; asset i pays $X_i(\omega_j)$ of the consumption good to the buyer from the seller when state j occurs at the the period's end.

A trader's feasible consumption by asset trading must then satisfy the following condition.

**Definition 2.1** *The bundle* $(c_0, \mathbf{c})$ *is in the feasible set* $C(c_0, e_0)$ *if and only if* $\theta$ *satisfies*

$$\begin{pmatrix} \mathbf{P}' \\ \mathbf{X} \end{pmatrix} \theta \geq \begin{pmatrix} c_0 - e_0 \\ \mathbf{c} - \mathbf{e} \end{pmatrix} \tag{1}$$

*where*

$$\mathbf{P}' = (P_0, \ldots, P_N)$$

$$\mathbf{X} = (\mathbf{X}_0, \ldots, \mathbf{X}_N)$$

$$\theta = (\theta_0, \ldots, \theta_N)'$$

*and*

$$\mathbf{c} - \mathbf{e} = (c(\omega_1) - e(\omega_1), \ldots, c(\omega_K) - e(\omega_K))'$$

The first inequality in (1) requires that the beginning of period consumption $c_0$ not exceed the beginning of period endowment $e_0$, plus any end of period revenue generated by short sales, net the cost of any long positions. The other inequalities similarly constrain end of period consumption in each state after all positions are closed. A trader's portfolio choice problem would then be to solve

[3]See, e.g. Dothan [10, chaps. 1-2], or Willinger and Taqqu [47]

$$\max_{\theta} U(c_0, \mathbf{c}) \tag{2}$$
$$subject\ to\ (1)$$

where U is assumed to be strictly increasing in its arguments.

Any condition necessary for (2) to have a solution must also be necessary for any equilibrium notion in which traders solve (2). The following no-arbitrage property of **P** and **X** is such a condition.

**Definition 2.2** *There are no arbitrage opportunities when there are no $\theta$ satisfying*

$$\begin{pmatrix} -\mathbf{P}' \\ \mathbf{X} \end{pmatrix} \theta \overset{>}{\neq} \mathbf{0} \tag{3}$$

By Steimke's Theorem (Mangasarian [36, p.32]), there is no arbitrage opportunity, i.e. (3) has no solution, if and only if there exists a strictly positive vector **v** satisfying

$$\begin{pmatrix} -\mathbf{P} & \mathbf{X}' \end{pmatrix} \mathbf{v} = \mathbf{0} \tag{4}$$

In what follows, we maintain the assumption that the zeroth asset is riskless, i.e. that $X_0(\omega_j) = 1$ for all j. Divide each equation in (4) by $v_0 > 0$, and define the strictly positive probabilities $Q_j = \frac{v_j/v_0}{\sum_{j=1}^{K} v_j/v_0} = \frac{v_j/v_0}{P_0}$, $j = 1, \ldots, K$. Finally, define the *gross riskless rate* $r \equiv 1/P_0$. This yields the following well-known result.

**Theorem 2.1** *There are no arbitrage opportunities if and only if there exists a strictly positive K-vector of probabilities* **Q** *satisfying*

$$P_i = \mathcal{E}_{\mathbf{Q}}\left[\mathbf{X}_i/r\right], i = 0, \ldots, N \tag{5}$$

where $\mathcal{E}_{\mathbf{Q}}$ denotes the mathematical expectation with respect to **Q**. Under the interpretation that $X_i(\omega_j)$ is the j th possible end of period price of asset i, (5) states that its current price $P_i$ is the **Q** - expected value of its risklessly discounted future price. Thus, each price process $(P_i, \mathbf{X}_i/r)$ is a **Q**-martingale, so **Q** is referred to as an *equivalent martingale measure.* The multiperiod equivalent martingale measure described in the next section will be a product of these one period measures, so in accord with standard use we will refer to a measure **Q** satisfying (5) as a *risk neutral probability measure.*

If, in addition, financial markets are *complete*, i.e. the matrix **X** is square and has full rank $K$, then **X** is invertible, and any other state-contingent claim's payoff $\mathbf{y} = \mathbf{X}\theta$ with $\theta = \mathbf{X}^{-1}\mathbf{y}$, i.e. it may be *duplicated* by a portfolio of the assets in **X**. To preclude arbitrage opportunities, the price of this claim must be the cost of the portfolio, i.e. $\theta'\mathbf{P} = (\mathbf{X}\theta)'\mathbf{Q}/r = \mathcal{E}_{\mathbf{Q}^*}[\mathbf{y}/r]$, where (5) defines the unique measure $\mathbf{Q}^* \equiv r\mathbf{X}'^{-1}\mathbf{P}$ of *normalized Arrow-Debreu state prices* [1, chap.4].

In what is to follow, it will prove convenient to define the i th asset's *gross rate of return* $R_i(\omega_j) \equiv X_i(\omega_j)/P_i$. The no-arbitrage constraints (5) can then be rewritten as

$$\mathcal{E}_{\mathbf{Q}}[\mathbf{R}_i] = r,\ i = 0, \ldots, N. \tag{6}$$

That is, the **Q**-expected, or *risk adjusted* gross return of each asset is the gross riskless rate $r$. If financial markets are complete, then $N = K - 1$ and the unique vector of *normalized Arrow-Debreu state prices* $\mathbf{Q}^*$ is computed as:

$$\mathbf{Q}^* \equiv \mathbf{R}'^{-1}\mathbf{r} \tag{7}$$

where the matrix

$$\mathbf{R} = (\mathbf{r}, \mathbf{R}_1, \ldots, \mathbf{R}_N)$$

and

$$\mathbf{r} = (r, \ldots, r)'$$

A portfolio of the riskless asset and the $K-1$ risky assets has gross return vector $\mathbf{y} = \sum_{i=0}^{K-1} \theta_i \mathbf{R}_i$, where $\theta_i$ is the fraction of the portfolio's cost spent on asset $i$. If the asset is sold short, this fraction is negative. Again, *when markets are complete*, any state-contingent claim's return $\mathbf{y}$ is duplicated by some weighted average of the columns of $\mathbf{R}$, and the absence of arbitrage implies $\mathcal{E}_{\mathbf{Q}^*}[\mathbf{y}] = r$.

Now suppose additionally that traders' agree about the actual state probabilities, denoted $\pi = (\pi_1, \ldots, \pi_K) > 0$, and maximize (possibly different) strictly concave expected utility $U(c_0, \mathbf{c}) = u(c_0) + \sum_j \pi_j u(c(\omega_j))$ in (2). Then, it is well known [19] that the first order conditions for an interior solution to (2) yield [4]

$$Q_j^* = \frac{m_j}{\mathcal{E}_\pi[m]}\pi_j \tag{8}$$

$$\textit{where } m_j \equiv u'(c(\omega_j))/u'(c_0).$$

In the multiperiod setting of section 4, the random variable $m$ is called the *intertemporal marginal rate of substitution* from that utility function, or the IMRS [26, pp.201-2].

**Example 2.1** *Suppose that in addition to the riskless asset, just one risky asset, a stock, is needed to complete the market. Then $K$ must equal 2. The unique measure satisfying (7) is the normalized Arrow-Debreu state prices $Q_1^* = (r - R_1(\omega_2)/(R_1(\omega_1) - R_1(\omega_2)) > 0$ and $Q_2^* = 1 - Q_1^*$. In this single period setting, a call option on the risky asset with exercise price $S < \max[X_1(\omega_1), X_1(\omega_2)]$ has return vector $\mathbf{y} = (\max[X_1(\omega_1) - S, 0]/p, \max[X_1(\omega_2) - S, 0]/p)'$, where $p$ is the option's price. Because of the complete markets assumption, $\mathbf{y}$ may be written as a weighted average of the columns of $\mathbf{R}$, i.e. the option may be duplicated by a portfolio of the 2 assets. The hypothesis of no arbitrage then implies that $\mathbf{y}$ also satisfies (6), i.e. its risk adjusted return is the riskless rate $r$. This, of course, may differ from its actual expected return $\mathcal{E}_\pi[\mathbf{y}]$. For example, suppose this time period has a length of one day. If $\pi_1 = 1/2$, $R_1(\omega_1) = 1.01611$, $R_1(\omega_2) = .984708$, and $r = 1 + .05/365$, then $Q_1^* = .491338$. The actual expected (gross) return of the stock is $\mathcal{E}_\pi[\mathbf{R}_1] = 1 + .15/365$, rather than its risk adjusted return $\mathcal{E}_{\mathbf{Q}^*}[\mathbf{R}_1] = 1 + .05/365$. The actual variance of the stock's return is .09/365.*

But suppose that in addition to the riskless asset return $r$, analysts only work with returns from $N < K - 1$ risky assets. This may be because some assets' prices are not observed, are unobservable, or the assets just don't exist. Operationally, analysis of any of the above situations is tantamount to assuming that financial markets are *incomplete*. Then the *observed* matrix $\tilde{\mathbf{R}}$ is of

[4] The first order conditions equate the (unnormalized) state price in each state $j$ to $\pi_j m_j$. The sum of these state prices is then $\mathcal{E}_\pi[m]$. Normalizing by the sum produces (8).

rank $N+1 < K$, and (6) defines a convex polytope of measures. Any $\mathbf{Q}$ satisfying (6) can be used to price portfolios of the observed $N$ risky assets and the riskless asset. That is, to preclude arbitrage opportunities, the return from any portfolio $\mathbf{y} = \tilde{\mathbf{R}}\theta$ of the *observed* assets must be $\mathcal{E}_{\mathbf{Q}}[\mathbf{y}] = r$, no matter which $\mathbf{Q}$ satisfying (6) is chosen.

But if a contingent claim *cannot* be duplicated by a portfolio of the $N$ assets and the riskless asset, then different measures satisfying (6) will produce different expected values. *So, without further assumptions, there is no obvious way to price these claims which are not "spanned" by these $N+1$ assets.*

**Example 2.2** *In the previous example, the stock could either go up or down by roughly 1.6 percent. But suppose there is a nonzero probability that the stock could remain unchanged, i.e. there exist $K = 3$ states and $R_1(\omega_3) = 1.0$. Then, unless we observe a third asset linearly independent of this stock and the riskless asset, i.e. unless we observe the third column in $\mathbf{R}$, we cannot use (7) to compute the normalized Arrow-Debreu state prices. System (6) is 2 equations in the 3 unknown risk neutral probabilities. Its solutions include $\mathbf{Q}^1 = (.3, .307089, .392911)$ and $\mathbf{Q}^2 = (.2, .201739, .598261)$. But a state-contingent claim with return $\mathbf{y} = (1.02, .98, 1)$ is not duplicated by a portfolio of the stock and the riskless asset. Consequently, $\mathcal{E}_{\mathbf{Q}^1}[\mathbf{y}] = .999858$, which is not equal to $\mathcal{E}_{\mathbf{Q}^2}[\mathbf{y}] = .999965 \neq 1 + .05/365$. And, as Cox, Ross and Rubinstein state, "From either the hedging or complete markets approaches, it should be clear that three-state or trinomial stock price movements will not lead to an option pricing formula based solely on arbitrage considerations."[8, p.240].*

Roughly speaking, this indeterminacy arises in discrete time whenever the number of shocks affecting asset prices in each period exceeds the number of assets used to construct state prices. The next section proposes a particularly simple way of resolving this indeterminacy.

# 3 Canonical Risk Neutral Probabilities

This section uses the information theoretic approach to density estimation adopted by, among others cited in the introduction, Theil [43],[44], and Zellner [50],[49],[48] to resolve this indeterminacy by statistically inferring a particular measure $\hat{\mathbf{Q}}$ satisfying (6), called *canonical risk neutral probabilities*, which is a Gibbs state. By construction, $\hat{\mathbf{Q}}$ will exactly price all state-contingent claims duplicated by portfolios formed from the observed subset of assets. With specific asset price processes in the multiperiod setting of section 4, we will see that canonical probabilities also facilitate the derivation of powerful asset pricing predictions for contingent claims, as well as the empirical testing of other asset pricing theories.

To implement this approach, we first choose a prior estimate of risk neutral probabilities, consistent solely with our knowledge of the riskless asset return $r$ but no explicit knowledge of any risky assets' returns. More specifically, the prior adopted is that $Q_j = \pi_j$, where $\pi_j$ is the actual probability of state $j = 1, \ldots, K$ occurring. Then, having obtained information about $N$ risky assets' returns, select a particular measure $\hat{\mathbf{Q}}$ satisfying (6), which contains the least additional information other than that there are no arbitrage opportunities involving portfolios formed from the riskless asset and the observed risky assets. To implement the above properties, one needs to quantify the additional information, or *information gain* $I(\mathbf{Q}, \pi)$, in passing from the prior estimate $\pi$ to some other measure $\mathbf{Q}$. Then, the canonical measure will be that measure minimizing $I$ subject to (6). While there are many conceivable ways to quantify the information gain in passing from one measure to another, Hobson [25] and Shore and Johnson [40] provide separate and appealing axiomatic foundations for choosing the *Kullback-Leibler number* [32], also called the *relative entropy* [7]

$$I(\mathbf{Q},\pi) \equiv \sum_{j=1}^{K} Q_j \log(Q_j/\pi_j) \tag{9}$$

as the correct function for this purpose.

Adopt the following definition:

**Definition 3.1** *The canonical risk neutral distribution* $\hat{\mathbf{Q}}$ *solves the constrained relative entropy problem:*

$$\min_{\mathbf{Q}} \sum_{j=1}^{K} Q_j \log(Q_j/\pi_j) \tag{10}$$

*subject to:*

$$\mathcal{E}_{\mathbf{Q}}[\mathbf{R}_i] = r,\ i = 0,\ldots,N \leq K$$

When a uniform prior is adopted, as it will be in the dynamic model of the next section, solving (10) is equivalent to constrained maximization of the *Shannon entropy* $S \equiv -\sum_j Q_j \log Q_j$. This is a measure of the uncertainty, or lack of information, inherent in $\mathbf{Q}$, and has been axiomatically rationalized as such by Khinchin [30]. Under logarithmic utility, Arrow [chap.12][1] has shown that it may also serve as the value of information in a problem of choice under uncertainty. The introduction to this paper cites just a few of the numerous uses this choice has found in analogous situations. In each application, the researcher seeks to produce predictions which incorporate *only* information made explicit by known constraints. As noted by Henri Theil:

> Maximum entropy (ME) is a general principle of estimating the distribution function of one or several random variables. The basic idea is that the statistician knows something (but not everything) about this distribution, and that it is rational from his point of view to estimate this distribution by a fitted distribution which displays maximum ignorance on his part subject to the constraint of what he knows. This is indeed rational, because any other fit would automatically imply the assumption of more knowledge than the statistician actually has. [43, pp.2-3].

The problem a statistician faces in selecting a risk neutral probability distribution $\mathbf{Q}$, subject only to the constraint (6) on its first moment information, is of this form. However appealing the aformentioned axiomatic rationalizations underlying this logic, the value of focusing on the canonical risk neutral probabilities will only begin to be evident after seeing the applications in section 4 and discussion in section 5.

## 3.1 Computation of Canonical Probabilites

If the analyst has information about all $N = K - 1$ risky assets forming a full basis with the riskless asset, the unique measure (7) of normalized Arrow-Debreu state prices solves problem (10) trivially, because it is the only feasible point. Otherwise, straightforward differentiation of its Lagrangian [7, p.294] shows that the solution is:

$$\hat{Q}_j = \pi_j e^{\sum_{i=1}^{N} \hat{\gamma}_i R_i(\omega_j)}/Z, \ j = 1, \ldots, K \tag{11}$$

where the *partition function* $Z$ is the normalizing constant

$$Z \equiv \sum_{j=1}^{K} \pi_j e^{\sum_{i=1}^{N} \hat{\gamma}_i R_i(\omega_j)} = \mathcal{E}_\pi \left[ e^{\sum_i \hat{\gamma}_i \mathbf{R}_i} \right]. \tag{12}$$

The Lagrange multiplier vector $\hat{\gamma}$ may be computed by minimizing the moment generating function of the excess return vector

$$\hat{\gamma} = argmin\ M(\gamma; r) \equiv \mathcal{E}_\pi \left[ e^{\sum_i \hat{\gamma}_i (\mathbf{R}_i - r)} \right] \tag{13}$$

and the minimal relative entropy attained by $\hat{\mathbf{Q}}$ may also be computed from (13) by [5]

$$I(\hat{\mathbf{Q}}, \pi) = -\log M(\hat{\gamma}; r). \tag{14}$$

**Example 3.1** *In the previous example,* $K = 3$ *states existed, and we observed the returns from only* $N = 1$ *stock, and the riskless rate* $r = 1 + .05/365$. *Suppose we adopt a uniform prior, i.e.* $\pi_j = 1/3$, *for j=1,2,3. If* $R_1(\omega_1) = 1.01984$, $R_1(\omega_2) = .981388$, *and* $R_1(\omega_3) = 1$, *then the actual mean gross return on the stock will still be* $\mathcal{E}_\pi[\mathbf{R}_1] = 1 + .15/365$, *while its variance is* $.09/365$. *Numerically minimize (13) to compute the single (because* $N = 1$*) multiplier* $\hat{\gamma} = -1.10526$. *Substituting in (11) yields the canonical risk neutral probabilities* $\hat{Q}_1 = .326202$, $\hat{Q}_2 = .340634$, *and* $\hat{Q}_3 = .333434$. *Substituting into (9) or using (14) yields the minimal relative entropy* $I(\hat{\mathbf{Q}}, \pi) = .000426$.

We now develop a few applications of the canonical risk neutral probabilities (11) in a multiperiod setting.

# 4 A Simple Multiperiod Model

In addition to the existence of a riskless asset paying a constant interest rate in each time period $t = 1, \ldots T$, we assume there are $N$ risky assets which may be traded each period. The one-period model previously described is replicated at each node of the event tree, resulting in *path independence*. That is, if $P_i$ denotes the price of asset $i$ at some time and node in the event tree, its price one period later will be $P_i R_i(\omega_j)$ if state $j$ occurs, which it does with probability $\pi_j$. After yet another period, its price will change to $P_i R_i(\omega_j) R_i(\omega_k)$ if state $k$ occurs then. So, the price two periods later will be the same whether state $j$ occurs before state $k$ or vice versa, and hence occurs with probability $2\pi_j \pi_k$. The resulting event tree is called a multinomial lattice, and financial markets will be complete if there are $K - 1$ independent risky assets [35]. For example, the complete market with $K = 2$ and $N = 1$ risky asset is a binomial lattice used to produce the binomial option model [26, pp.248-54].

[5] To derive (13) and (14), first note that the equivalent problem of minimizing the log of $M(\gamma; r)$ in (13) has first order conditions equivalent to the constraints in (10), when $\mathbf{Q}$ is given by (11) and (12). The second order condition is just that the risk adjusted covariance matrix of returns is positive definite. (14) follows upon substitution of $\hat{\mathbf{Q}}$ into (9) and simplifying.

For $K > 2$, one obtains a K-nomial lattice. A random payoff at period $T$ realized by a dynamic trading strategy involving the $N$ assets may again be valued as an expected discounted present value. But now the expectation must be taken with respect to the $T$-fold product measure $\mathbf{Q}^T$ formed from risk neutral probabilities $\mathbf{Q}$ satisfying (6), which is an equivalent martingale measure. That is, along a particular state realization $(\omega(t),\ t = 1,\ldots,T)$, $\mathbf{Q}^T = \prod_{t=1}^{T} Q_{\omega(t)}$. If financial markets are complete, all state-contingent claims may be priced in this way, but doing so requires all $K^2$ elements of the gross return matrix $\mathbf{R}$ needed to compute (7).

In what is to follow, we assume that the $N + 1 < K$ asset returns used by the analyst are generated by Hua He's [24] discrete time approximation of the following continuous time price process:

$$
\begin{aligned}
dP_i &= \mu_i P_i dt + \sum_{l=1}^{K-1} \sigma_{il} P_i dW_l \\
dP_0 &= \iota P_0 dt
\end{aligned}
\tag{15}
$$

for the $i = 1,\ldots,N$ risky assets. Here, $W_l$ denotes component $l$ of an $K-1$-dimensional standard Wiener process $\mathbf{W}$, $\mu_i$ is a drift parameter and $\sigma_{il}$ is component $(i,l)$ of a matrix $\sigma$ determining the correlation among returns, and their volatilities. The riskless asset follows the deterministic process $dP_0 = \iota P_0 dt$. To approximate this process over the time interval [0,1] by a K-nomial lattice with discrete trading times, He assumes that the actual state distribution is uniform, i.e. $\pi_j \equiv 1/K$. Gross returns corresponding to his price process approximation are then given by:

$$
\begin{aligned}
R_i(\omega_j) &= 1 + \mu_i/n + \sum_{l=1}^{K-1} e_{jl}\sigma_{il}/\sqrt{n} \\
r &= 1 + \iota/n
\end{aligned}
\tag{16}
$$

where $\iota$ is the continuously compounded annual interest rate, $n$ is the number of trading periods of length $1/n$, and $e_{jl}$ is element $(j,l)$ of a $K \times K-1$ matrix $\mathbf{e} = (\mathbf{e}_1,\ldots,\mathbf{e}_{K-1})$, whose columns form an orthonormal basis for the $K-1$ dimensional linear space which is orthogonal to the K-vector of ones. Because of the uniform actual distribution of states, the columns of $\mathbf{e}$ are uncorrelated random variables with zero mean and unit variance. Thus, column $l$ is used to approximate the increment $dW_l$. When financial markets are complete, He [24] proves that the K-nomial lattice price process, with $K-1$ risky asset gross returns (16) and an invertible (covariance) matrix $\mathbf{C} = \sigma\sigma'$, converges weakly to the continuous time price process (15) as the number of trading periods $n \to \infty$. It is reasonable to assume that the number of uncorrelated shocks $K$ which could influence asset returns is quite large, so a large number of assets would be required to complete the discrete time trading market.

Accordingly, it is not unreasonable to assume that the analyst uses only $N < K-1$ of the discrete time, risky asset returns (16), and again there will be a multiplicity of risk neutral probability measures. The canonical risk neutral probabilities are produced by substituting (16) into (13), solving, and calculating (11). The *canonical martingale measure* is the product measure formed from this, which easily simplifies to the following value along a particular state realization $\omega(t)$, $t = 1,\ldots,T$:

$$\hat{\mathbf{Q}}^T = \exp\left[\sum_{i=1}^{N} \hat{\gamma}_i \sum_{t=1}^{T} R_i(\omega(t))\right] / \left[\sum_{j=1}^{K} \exp[\sum_i \hat{\gamma}_i R_i(\omega_j)]\right]^T \tag{17}$$

## 4.1 The Canonical Multinomial Option Model

To illustrate the use of the canonical martingale measure (17), assume that data on only a single asset $i$ is used to price a European call option on it. Denoting the number of times $\omega_j$ occurs in a particular realization by $\nu_j$, use the canonical martingale measure (17) formed from the $N = 1$ asset to predict the canonical option price $P(n)$:

$$P(n) = \mathcal{E}_{\hat{\mathbf{Q}}^T}\left[\max\left(P_i \prod_{t=1}^{T} R_i(\omega(t)) - S, 0\right) / r^T\right] \tag{18}$$

$$= \sum_{\nu:\sum_j \nu_j = T} \left[\nu_1, \overset{T}{\ldots}, \nu_K\right] \prod_{j=1}^{K} \hat{Q}_j^{\nu_j} * \max\left(P_i \prod_{j=1}^{K} R_i(\omega_j)^{\nu_j} - S, 0\right) / r^T \tag{19}$$

where

$T$ = the expiration date, in periods

$\left[\nu_1, \overset{T}{\ldots}, \nu_K\right]$ = a multinomial coefficient

$P_i$ = asset i's current price

$S$ = the exercise price

**Example 4.1** *Consider the special case of $K = 3$ states. Suppose that the stock has a mean annualized net growth rate of $\mu_i = .15$, and that $\sigma_{i1} = \sigma_{i2} = \sqrt{.045}$, so the stock's variance is $.09 = \sum_{l=1}^{2} \sigma_{il}^2$. Suppose the net riskless rate of interest is $\iota = .05$, and that the stock's current price is $P_\iota = 40$.*

*To price an option with an exercise price of $S = 45$ which has 30 days to maturity, first let $n = 365$, so $T = 30$. We randomly generate a suitable orthonormal basis $\mathbf{e}$ twenty five times, each time computing canonical risk neutral probabilities from (11)-(13) and the resulting canonical trinomial model (18). This produced option prices with a mean of 15.8 cents, and a standard deviation of only .008. The multiplier $\hat{\gamma} = -1.11$ exhibited almost no basis-dependent variation. 16 cents is the Black-Scholes value, also attained by the usual binomial model, as reported by Geske and Shastri [16, p.59]. The most computationally costly procedure by far is to compute the $T + K - 1!/T!K - 1!$ multinomial coefficients, one for each terminal node in the lattice. With daily time steps, the binomial model thus needs compute only $T + 1 = 31$ binomial coefficients. But by choosing a step size of 5 days (i.e. $n = 73$ and $T = 6$), the trinomial model still produces 28 terminal nodes. With this longer trading interval, the mean option price is still 15.8 cents, but the basis-dependent standard deviation increases to .021.*

Note that a close approximation to Black-Scholes is attained, despite the fact that the calculation used no data on the second risky asset needed to complete the market, calculate (7), and resolve the indeterminacy described in Example 2.2. We will soon see that the exponential form of the canonical risk neutral probabilities makes it easy to prove that the canonical option price converges to the Black-Scholes price as $n \to \infty$, i.e. as the trading interval shrinks toward continuous trading, regardless of the number of shocks $K$.

## 4.2 The Continuous Time Limit

It is quite easy to evaluate the continuous time limit of the solution to (13) with returns given by (16). To do so, expand (13) in a Taylor series about the risky assets' mean gross returns $1 + \mu_i/n$, take the expectation with respect to $\pi$, and take the logarithm to obtain the following equivalent minimization:

$$\min_{\gamma_1,\ldots,\gamma_N} \sum_{l=1}^{N} \gamma_l(\mu_l - \iota)/n + \log\left[1 + \gamma'\sigma\sigma'\gamma/2n + o(1/n)\right] \tag{20}$$

where $\sigma$ is the matrix of volatility parameters $\sigma_{il}$ in (16). For each $n$, minimization of (20) is equivalent to the minimization of $n$ times (20). Doing so, the familiar series expansion for $e^x$ identifies the second term to be $\gamma'\sigma\sigma'\gamma/2$ as $n \to \infty$, yielding a quadratic minimization having the limit solution:

$$\lim_{n\to\infty} \hat{\gamma} = -\mathbf{C}^{-1}\mathbf{x} \tag{21}$$

where $\mathbf{x}$ is the N-vector of annualized mean excess returns from the annualized riskless rate $\iota$, and $\mathbf{C} = \sigma\sigma'$ is the symmetric matrix of annualized return covariances, assumed to be positive definite[6]. Normalizing (21) by the sum of its components yields the portfolio weights in the familiar mean-variance efficient *tangency portfolio* [29, p.435] of the $N$ risky assets. We will repeatedly make use of (21) in what follows.

**Example 4.2** *Suppose the analyst knows the return process of just the $N = 1$ risky asset of example 4.1. Then $\mathbf{x} = (\mu - \iota) = (.15 - .05) = .10$, while $\mathbf{C} \equiv \sigma\sigma' = .09 = \sigma^2$. The limiting result is $\hat{\gamma} = -(\mu - \iota)/\sigma^2 = -.10/.09 = -1.11$, in agreement to two places with the exact computation done there for $n = 365$.*

We will now show that the limit of the canonical multinomial option price is the Black-Scholes price. Because we defined the parameters from annualized data, the number of time steps $s$ for an option with $T$ days to expiration is $s = [nT/365]$, using brackets to denote the nearest integer function. To calculate the option price, first note that $\mathcal{E}_{\mathbf{Q}^s}(\bullet) \equiv \mathcal{E}_{\pi^s}(\xi_n^s \bullet)$, where the *discrete-time likelihood ratio* $\xi_n^s \equiv \prod_{t=1}^{s} \mathbf{Q}(\omega(t))/\pi(\omega(t))$. The following theorem shows that the canonical discrete time likelihood ratio $\hat{\xi}_n^s$ converges to the density of the continuous time martingale measure used by Dothan in deriving the Black-Scholes model [10, p.210].

[6] Because the state space has finite cardinality, all Taylor expansions and expectations used here exist. Because $\mathbf{C}$ is assumed to be positive definite, the logarithm in (20) exists, and the sufficient conditions for this interior solution are realized.

**Theorem 4.1**

$$\lim_{n\to\infty} \hat{\xi}_n^{[nT/365]} = \exp[-\frac{\mu-\iota}{\sigma}W_{T/365} - 1/2\frac{(\mu-\iota)^2}{\sigma^2}T/365] \quad (22)$$

*where $W_{T/365} \sim N(0, T/365)$. This is the Radon-Nikodym derivative of the martingale measure for the Black-Scholes model [10, p.210].*

**Proof.** Substitute $\hat{\mathbf{Q}}(\omega(t)) \equiv (1/K)\exp[\hat{\gamma}R_i(\omega(t))]/Z$ into the definition and simplify to obtain the canonical likelihood ratio:

$$\hat{\xi}_n^{[nT/365]} = \exp[\hat{\gamma}(n) \sum_{t=1}^{[nT/365]} R_i(\omega(t))]/Z^{[nT/365]} \quad (23)$$

where we now explicitly notate the dependence of $\hat{\gamma}$ on $n$, the number of time steps per year.

Take logs, substitute for the single stock $i$ from (16), and as done earlier, use the Taylor expansion of a moment generating function (12) about the mean asset returns to obtain:

$$\begin{aligned}\log \hat{\xi}_n^{[nT/365]} &= -\left(\sum_{t=1}^{[nT/365]} -\hat{\gamma}(n)\sum_{l=1}^{K-1} \sigma_{il}e_{\omega(t)l}/\sqrt{n}\right) \\ &\quad -1/2\hat{\gamma}(n)^2\sigma^2 - [nT/365]o(1/n)\end{aligned} \quad (24)$$

As $n \to \infty$, the Lindeberg central limit theorem guarantees that the above weighted sum of uniformly bounded discrete random variables $e_{\omega(t)l}$ converges to $N(0, \sigma^2 \lim_{n\to\infty} \hat{\gamma}(n)^2 T/365)$. Substituting (21) computed in the scalar Example 4.2 yields the correct density. ∎

Of course, when financial markets are complete, the continuous time limit of the unique discrete time martingale measure computed in He [24] would also produce the Black Scholes price for an option on a single stock. Here, we see that the same limit is obtained when the discrete time model is incomplete in the above way, i.e. only data on the asset the option is written on is used, despite there being $K$ states in each period. Thus, in this particular discrete time model of incomplete markets, the MEF statistical inference from incomplete information (the canonical risk neutral valuation) produces the Black-Scholes model in the continuous time limit.

### 4.3 The Relative Entropy Bound on the IMRS

In section 2, it was noted that a complete market, consumption-based asset pricing model identifies the unique risk neutral probabilities with a function of its intertemporal marginal rate of substitution (IMRS) $m$, via the change of measure (8). For example, in the representative agent model of Lucas [33] with constant relative risk aversion parameter equal to $\alpha$ and time discount factor parameter equal to $\delta$,

$$m_j(\alpha, \delta) = \delta(c_{t+1}(\omega_j)/c_t)^{-\alpha} \quad (25)$$

where $c_t$ denotes consumption at time $t$.

Hansen and Jagannathan [22] noted that the absence of arbitrage requires $m$ to have a mean $\mathcal{E}_\pi[m] = 1/r$, and a variance $\sigma^2[m]$ at least as large as that attained by a variance minimizing "benchmark" random variable $m^*$ with the same mean $\mathcal{E}_\pi[m^*] = \mathcal{E}_\pi[m] = 1/r$, computed from the returns of $N$ assets the analyst is interested in. Letting $\Theta$ denote a vector of parameters required by an IMRS specification $m(\Theta)$, and using overlines to denote sample means (i.e. time averages), a contrary finding that its sample variance $s^2[m(\Theta)] < s^2[m^*]$ when $r = 1/\overline{m(\Theta)}$ indicates that the specification fails, a finding which may additionally be checked for statistical significance [21].

For example, both Bekaert and Hodrick [2, p.503] and Ferson and Harvey [13, p.519] use the variance bound inequality

$$\sigma^2[m] \geq \tilde{x}'\tilde{C}^{-1}\tilde{x}/r^2 = \sigma^2[m^*] \tag{26}$$

for $m$ satisfying $\mathcal{E}_\pi[m] = 1/r$, where $\tilde{C}$ denotes the covariance matrix of *period* (not annualized) returns, and $\tilde{x}$ denotes the mean excess return vector of period returns. The right hand side of (26) defines the *variance bound frontier* in the $(r, \sigma^2)$ plane. One then checks to see whether the sample point $(1/\overline{m(\Theta)}, s^2[m(\Theta)])$ (hopefully) lies above the estimated variance bound frontier. Using returns data from $N = 5$ assets formed as portfolios of stocks, bonds, and bills, and the specification $m(\alpha, 1)$ in (25) with seasonally adjusted consumption, Ferson and Harvey [13, p.545] showed that the sample point only lay above the estimated variance bound frontier (26) when the coefficient of relative risk aversion parameter $\alpha > 48$. Because this high a value is not in accord with some analysts' prior information about its value, they propose an alternative, time-nonseparable specification of $m$ which does not have this problem.

In addition, it may be intrinsically interesting to measure the marginal effects particular assets have on the frontier. For example, Snow [41] finds that, for some but not all time periods, the addition of an asset formed from a portfolio of small firm stocks significantly raises variance bound frontiers already incorporating stock market proxies. This lends weight to the claim that "small firm effects" are not invariant over time.

Different diagnostics may be produced by using the relative entropy induced by the change of measure (8), rather than the variance of $m$. Substituting (8) into (9), the relative entropy associated with an IMRS $m$ (or indeed any strictly positive $m$) [7] must satisfy

$$I(\mathbf{Q}^*, \pi) = \mathcal{E}_\pi\left[(m/\mathcal{E}_\pi[m])\log(m/\mathcal{E}_\pi[m])\right] \geq I(\hat{\mathbf{Q}}, \pi) \tag{27}$$

at the implied riskless rate $r = 1/\mathcal{E}_\pi[m]$. As with the variance bound, one checks whether the sample point $(1/\overline{m}, \overline{(m/\overline{m})\log(m/\overline{m})})$ in the $(r, I)$ plane lies above the estimated relative entropy bound frontier, computed at each value of $r$ by unconstrained numerical minimization of the sample moment generating function

$$\hat{M}(\gamma; r) \equiv \frac{1}{T}\sum_{t=1}^{T} e^{\sum_{i=1}^{N}\gamma_i(R_i(\omega(t))-r)} \tag{28}$$

and then using (14) to obtain the estimated relative entropy bound $-\log \hat{M}$.

The qualitative properties of the relative entropy bound frontier are readily apparent. Equivalently minimizing the log of (13), the envelope theorem and (14) immediately show that

$$\partial I(\hat{\mathbf{Q}}, \pi)/\partial r = \sum_i \hat{\gamma}_i. \tag{29}$$

[7] Because these diagnostic procedures exploit only the unconditional moment restriction $\mathcal{E}_\pi[R_i m] = 1$, they apply to other asset pricing models as well ( see, e.g. Hansen and Jagannathan [21] ).

That is, the slope of the relative entropy bound frontier is just the sum of the Gibbs state's parameters. The usual comparative statics of unconstrained, convex minimization yields

$$\partial\left(\sum_{i=1}^{N}\hat{\gamma}_i\right)/\partial r > 0. \tag{30}$$

Combining (29) and (30), we see that the relative entropy bound frontier is strictly convex, and has an interior minimum (if any) at a value of $r$ for which the $\sum_i \hat{\gamma}_i = 0$.

The relative entropy bound is the (axiomatically rationalized) information gained by the risk neutral change of measure. The larger the bound, the more Kullback-Leibler "distance" separates risk neutral probabilities from the actual probabilities $\pi$. Failure of (8) to exceed the bound is thus attributed to its failure to differ enough from $\pi$, i.e. to provide enough information to perform the risk adjustment (6) equivalent to the absence of arbitrage among the $N$ assets.

The variance bound in (26) is a quadratic approximation of the relative entropy bound. To derive the quadratic approximation and understand when it is likely to be good, note that the terms in (20) involving higher than second order moments are of $o(1/n)$. When minimizing $n * (20)$, they may be dropped when $n$ is sufficiently large, and the log term may be approximated by the near zero value $\gamma'\sigma\sigma'\gamma/2n$. Substitute the limiting solution (21) into this quadratic approximation, take the log and simplify to obtain the approximation

$$\begin{aligned} I(\hat{\mathbf{Q}},\pi) &= -\log M(\hat{\gamma};r) \\ &\approx \frac{1}{2}\mathbf{x}'\mathbf{C}^{-1}\mathbf{x}/n \\ &= \frac{1}{2}(\mathbf{x}/n)'(\mathbf{C}/n)^{-1}(\mathbf{x}/n)'. \end{aligned} \tag{31}$$

For any sampling frequency $n$, the expected excess *period* return vector $\tilde{x} = \mathbf{x}/n$, and the covariance matrix of the period returns $\tilde{C} = \mathbf{C}/n$. Comparing (26) and (31), we see that the approximate relative entropy bound is just $r^2/2$ times the variance bound. For large $n$, $r = 1 + \iota/n \approx 1$ over the range of reasonable riskless rates, in which case the relative entropy bound is approximately half the variance bound throughout the range. Then, when data is sampled often enough from the returns process assumed herein, use of (26) will produce similar results to use of (27), and will share the information theoretic interpretation of the latter.

In other circumstances, the diagnostics may provide different results. For example, Brainard, et.al.[3] have shown that the consumption CAPM model explains data better at longer horizons (up to three years) than it does at much shorter intervals, and provide intuitive reasons for the failure of consumption-based models to fit monthly data. In addition, it is well known that returns data are often autocorrelated and characterized by moments which are not functions of the first and second. In these circumstances, the relative entropy bound diagnostics may be run in conjunction with the variance bound diagnostics and/or the alternative moment bounds developed in Snow [41].

# 5 Large Deviations and Canonical Probabilities

A simplified special case of the large deviations results in Stroock and Zeitouni [42] is used to provide additional rationale for the canonical measure. Recall that the discrete time approximation used here assumes that the state process is i.i.d., i.e. its realizations are modeled as the product space

$\left(\Omega^T, \pi^T\right)$, where $(\pi_1, \ldots, \pi_K)$ was assumed to be uniform. So by the law of large numbers, for each fixed number of time periods per year $n$, as the total number of time periods $T$ grows to infinity, the relative frequency distribution of the state $\omega$ up to time $T$, denoted $\mathbf{L}_T$, approaches $\pi$, the uniform distribution, for almost all realizations. For fixed $n$, the asymptotic time average of each asset i's gross rate of return (16) is the limiting integral:

$$\begin{aligned}
\lim_{T\to\infty} \overline{R_i} &\equiv \lim_{T\to\infty} 1/T \sum_{t=1}^{T} R_i(\omega(t)) \\
&= \lim_{T\to\infty} \int R_i d\mathbf{L}_T \\
&= \sum_{j=1}^{K} R_i(\omega_j)\pi_j \\
&= \mathcal{E}_\pi[R_i] \\
&= 1 + \mu_i/n, i = 1, \ldots, N < K - 1 \qquad (32)
\end{aligned}$$

Because not all $1 + \mu_i/n = r$, the set of realizations in which these time averages equal the riskless short rate $r$ has measure zero. Of course, if risk neutral probabilities $\mathbf{Q}$ governed the actual process, the observed time averages would equal $r$ for almost all realizations, by construction. But $\pi$ governs the process. Nonetheless, for each $T$ and $\delta > 0$, it can be shown that if we condition on the (vanishingly small) set of realizations for which the time averaged gross rates of return up to $T$ are within $\delta$ of $r$, then it is highly likely that the relative frequency distribution $\mathbf{L}_T$ is close to the canonical distribution $\hat{\mathbf{Q}}$ when $T$ is large and $\delta$ is small. More precisely, the following is proven:

**Theorem 5.1** *Let $G \subset \mathcal{R}^K$ be any open subset in the space of discrete probability measures containing the canonical measure $\hat{\mathbf{Q}}$, computed from (11) - (13). Then, the conditional probability*

$$\overline{\lim_{\delta\downarrow 0}}\, \overline{\lim_{T\to\infty}}\, Prob\left[\mathbf{L}_T \notin G \mid r - \delta \leq \int R_i d\mathbf{L}_T \leq r + \delta, i = 1, \ldots, N\right] = 0 \qquad (33)$$

**Proof.** Because of the i.i.d. process used here, the proof is much simpler than that given for the general result of Stroock and Zeitouni [42], and can be based on a simple version of Sanov's large deviation theorem, given below.[8]

Denote the assets' gross return vector $\mathbf{R} \equiv (R_1, \ldots, R_N)$, the closed set $F_\delta \equiv \{\mathbf{L}_T \ni r - \delta \leq \int \mathbf{R} d\mathbf{L}_T \leq r + \delta\}$, and the open set $G_\delta \equiv \{\mathbf{L}_T \ni r - \delta < \int \mathbf{R} d\mathbf{L}_T < r + \delta\}$. Then the conditional probability of the theorem is written

$$\overline{\lim_{\delta\downarrow 0}}\, \overline{\lim_{T\to\infty}} \frac{\pi^T\left[\mathbf{L}_T \in G^c \bigcap F_\delta\right]}{\pi^T\left[\mathbf{L}_T \in F_\delta\right]} \qquad (34)$$

The proof utilizes Sanov's Theorem below.

**Theorem 5.2** *(Sanov)[6, pp.28-29] For any closed set $F$ in the (Prohorov) metric space of probability measures, $\overline{\lim}_{T\to\infty} 1/T \log\left(\pi^T\left[\mathbf{L}_T \in F\right]\right) \leq -\inf_{\mathbf{Q}\in F} I(\mathbf{Q}, \pi)$, while for any open set $G$,*

[8] Using innovative combinatorial methods applying only in the finite state case, an analogous "conditional limit theorem" is proved in Cover [7, sec.12.6]

$\underline{\lim}_{T\to\infty} 1/T \log\left(\pi^T [\mathbf{L}_T \in G]\right) \geq -\inf_{\mathbf{Q}\in G} I(\mathbf{Q},\pi)$ *where the rate function I is the relative entropy function (9).*

This result will be used to show that the numerator in (34) approaches zero at a faster rate than the denominator. For the latter, Sanov's Theorem has $\underline{\lim}_{T\to\infty} 1/T \log\left(\pi^T [\mathbf{L}_T \in G_\delta]\right) \geq -\inf_{\mathbf{Q}\in G_\delta} I(\mathbf{Q},\pi)$ To compute the right hand side, note that for $\delta$ sufficiently small, the unconstrained optimum $\pi \notin G_\delta$. Then, the Kuhn-Tucker Theorem demonstrates that the constrained infimum occurs on the boundary, so the right hand side is $-\min[I(\hat{\mathbf{Q}}(r-\delta),\pi), I(\hat{\mathbf{Q}}(r+\delta),\pi)] \equiv -M_\delta$. Because $F_\delta \supset G_\delta$, the denominator goes to zero at an exponential rate no faster than $M_\delta$ when $\delta$ is this small. Examining the numerator, note that there also exists $\delta$ sufficiently small so that $\hat{\mathbf{Q}}(r-\delta), \hat{\mathbf{Q}}(r+\delta) \notin G^c$. Thus for $\delta$ sufficiently small, $\inf_{\mathbf{Q}\in G^c\cap F_\delta} I(\mathbf{Q},\pi) > M_\delta$ and the numerator goes to zero at a rate larger than $M_\delta$. To guarantee that $\delta$ is sufficiently small to do this, the theorem takes $\delta \downarrow 0$. ∎

Theorem 5.1 shows that canonical risk neutral probabilities approximate the conditional state probabilities given that the sample risklessly discounted asset price paths appear to be martingales.

It is also easy to use Sanov's Theorem to interpret the relative entropy bound. Note that the denominator of (34) is the probability of observing all sample average returns (weakly) within $\delta$ of $r$. As in the above proof, Sanov's Theorem yields an estimated lower bound on this probability of $e^{-T\inf_{\mathbf{Q}\in F_\delta} I(\mathbf{Q},\pi)}$ for large $T$. The above proof gave an upper bound of $e^{-T\inf_{\mathbf{Q}\in G_\delta} I(\mathbf{Q},\pi)}$ for the probability of observing all sample average returns strictly within $\delta$ of $r$. But the relative entropy bound is just $\min_{\mathbf{Q}\in F_0} I(\mathbf{Q},\pi)$, which must approximate these estimates when $\delta$ is small. So the larger the relative entropy bound, the less likely it is that all assets' sample averages will be near zero, i.e. the more likely it is that at least one asset's sample average return will not be near the riskless rate. As such, the relative entropy bound may be interpreted as a scalar indicator of the importance of risk premia in the data. If an additional asset significantly raises the relative entropy bound at some $r$, it does so because it significantly decreases the probability of observing a near zero vector of expected excess returns (risk premia) at that value of $r$.

## 5.1 Statistical Mechanics and Asset Pricing

The canonical option model with a single observed stock in an arbitrage-free economy is isomorphic to an energy conserving mechanical system, with a finite number of energy states, placed in a heat bath. The presence of the heat bath requires the physical scientist to condition on realizations consistent with the internal energy, as fixed by the heat bath. Because of Theorem 5.1, physical scientists know that the canonical change of measure will approximate the conditional probabilities when the number of components $T$ is large, i.e. in the thermodynamic limit. This rationalizes the statistical mechanics procedure of computing expectations with the canonical measure. In asset pricing, the presence of the no-arbitrage constraint requires analysts to restrict attention to realizations consistent with the risk adjusted mean return, as fixed by the riskless rate of interest. Because of Theorem 5.1, canonical risk neutral probabilities will approximate the conditional probabilities when $n$ is large enough for the number of time steps $T$ to be large as well, i.e. in the continuous time limit.

More generally, arbitrage-free pricing with $N$ observed risky assets and a constant riskless rate of interest is isomorphic to physical systems with $N$ conservation laws, in which "heat baths" fix the same mean value for each quantity conserved in isolation. Under the isomorphism, (13) maps to the free energy function. The mean and volatility parameters governing asset rates of return in (16) are identified with mechanical parameters [27] affecting state-dependent energy and other quantities conserved in isolation. The riskless rate of interest becomes a thermal parameter controlling the fixed internal energy and any other fixed internal quantities that would be conserved in isolation. Comparative statics result (29) describes the thermodynamic relationship between entropy and (generalized) temperature, while (30) describes the relationship between temperature and internal energy.

Finally, critical phenomena, like phase transitions, are often identified with the failure of free energy convexity and associated comparative statics results dependent on it (Kindermann and Snell[p.48] [31]). In our setting, suppose that (10) fails to have a minimum because there is no feasible point. That is, suppose that there is no solution to (6). By theorem 2.1, this is equivalent to the presence of arbitrage opportunities. This failure of the riskless interest rate and the asset mean and volatility parameters to assume values consistent with the absence of arbitrage is a critical phenomenon. Starting from parameter values consistent with the absence of arbitrage, suppose that a change in monetary policy changes the riskless rate to a value inconsistent with a solution to (6). Market forces should cause changes in the assets' mean and volatility parameters to eliminate the arbitrage opportunity, but if these happen slowly enough relative to the trading frequency, transient arbitrage opportunities would be possible. This heuristic discussion suggests that deeper investigation of asset pricing problems using other methods of statistical mechanics may be warranted.

## 6 Summary and Conclusions

It is well-known that prices of contingent claims in complete and arbitrage-free securities markets can be computed using normalized Arrow-Debreu state prices, the unique *risk neutral probability measure*. But in situations operationally equivalent to incomplete markets, there are many risk neutral measures. This paper uses simple mathematics to begin exploring the value of an information theoretic approach to density estimation, called the maximum entropy formalism (MEF), in selecting a particular risk neutral probability measure for use in situations operationally equivalent to incomplete financial markets. The resulting risk neutral probability measure is the solution to a constrained, convex minimization of relative entropy, called a *Gibbs state*, and is from an exponential family called *canonical distributions*. These distributions are at the core of Gibbs' [17] equilibrium statistical mechanics of large systems of interacting components, and numerous other applications of the MEF cited herein. The investigation is conducted within what is perhaps the simplest possible multiperiod setting for asset pricing, i.e. a discrete time approximation to a constant coefficient, vector Ito process for asset returns.

The canonical risk neutral probability distribution has a form which facilitates passing to the continuous time limit. Doing so shows that the distribution's parameters approach a vector $\hat{\gamma}$ which, when normalized by the sum of its components, is the vector of portfolio weights in the familiar mean-variance efficient *tangency portfolio* of the observed risky assets. This limiting result is used to painlessly obtain a few illustrative asset pricing results. First, we produced a simple derivation and alternative interpretation of the Black-Scholes option pricing model, as an MEF statistical inference from analysts' incomplete information. Second, we showed how the minimal relative entropy realized by the canonical risk neutral distribution provides a relative entropy bound diagnostic for the intertemporal marginal rate of substitution (IMRS) in specifications of the consumption-based asset pricing model. The popular variance bound diagnostic [22],[21] is shown to be a quadratic approximation of the relative entropy bound. This provides alternative interpretations of variance bound diagnostic procedures when the approximation is good, and provides an additional diagnostic procedure when a quadratic approximation is inappropriate.

Large deviations theory is then used to provide a frequentist interpretation of the canonical distribution and its uses, augmenting the axiomatic foundation frequently adopted in the econometric literature (for recent surveys, see Maasoumi [34] and Sengupta [39]). Finally, the isomorphism between canonical risk neutral pricing and equilibrium statistical mechanics and thermodynamics is described.

Zellner [48, p.21] recently stated that "much more empirical and theoretical work needs to be done to get a 'maximum entropy, thermodynamic model' that performs well in explaining and predicting the behavior of economic systems." While this paper makes some progress in this regard, it alone does not constitute an entropy-based foundation for financial economics. Perhaps this will never come. As Samuelson notes:

> ...I have come over the years to have some impatience and boredom with those who try to find an analogue of the entropy of Clausius or Boltzmann or Shannon to put into economic theory. It is the *mathematical* structure of *classical* (phenomenological, macroscopic, nonstochastic) *thermodynamics* that has isomorphisms with *theoretical economics.* [38, p.263]

Similarly, this paper shows that the mathematical structure of Gibbsian statistical mechanics is isomorphic to an important part of financial economics. But will this isomorphism ultimately prove as useful as those alluded to by Samuelson for microeconomic theory?

## References

[1] Kenneth J. Arrow. *Essays in the Theory of Risk-Bearing.* North Holland, 1974.

[2] Geert Bekaert and Robert J. Hodrick. Characterizing predictable components in excess returns on equity and foreign exchange markets. *Journal of Finance*, 47(2):467–509, 1992.

[3] William C. Brainard, William R. Nelson, and Matthew D. Shapiro. The consumption beta explains expected returns at long horizons. 1991. Dept. of Economics, University of Michigan.

[4] William Brock. Beyond randomness, or, emergent noise: interactive systems of agents with cross dependent demands. March 1992. Dept. of Economics, University of Wisconsin.

[5] William Brock. Understanding macroeconomic time series using complex systems theory. *Structural Change and Economic Dynamics*, 2(1):119–141, 1991.

[6] James A. Bucklew. *Large Deviation Techniques in Decision, Simulation, and Estimation.* Wiley, 1990.

[7] Thomas M. Cover and Joy A. Thomas. *Elements of Information Theory.* Wiley, 1991.

[8] John C. Cox, Stephen A. Ross, and Mark Rubinstein. Option pricing: a simplified approach. *Journal of Financial Economics*, 7:229–263, 1979.

[9] J.M. Cozzolino and M.J. Zahner. The maximum entropy distribution of the future market price of a stock. *Operations Research*, 1973.

[10] Michael U. Dothan. *Prices in Financial Markets.* Oxford University Press, 1990.

[11] Stephen Durlauf. Locally interacting systems, coordination failure, and the behavior of aggregate activity. 1989. Dept. of Economics, Stanford University.

[12] Stephen Durlauf. *Nonergodic Economic Growth.* Technical Report 7, Stanford Institute for Theoretical Economics, 1991.

[13] Wayne E. Ferson and Campbell R. Harvey. Seasonality and consumption-based asset pricing. *Journal of Finance*, 47(2):511–552, 1992.

[14] Hans Follmer. Random economies with many interacting systems. *Journal of Mathematical Economics*, 1:51–62, 1974.

[15] Hans Follmer and Martin Schweizer. Hedging of contingent claims under incomplete information. In M.H.A. Davis and R.J. Elliott, editors, *Applied Stochastic Analysis*, Gordon and Breach, 1990.

[16] Robert Geske and Kuldeep Shastri. Valuation by approximation: a comparison of alternative option valuation techniques. *Journal of Financial and Quantitative Analysis*, 20(1):45–71, 1985.

[17] Josiah Willard Gibbs. *Elementary Principles of Statistical Mechanics.* Yale Bicentennial Publications, 1902.

[18] Christopher Grandy. The principle of maximum entropy and the difference between risk and uncertainty. In W. T. Grandy and L. H. Schick, editors, *Maximum Entropy and Bayesian Methods*, pages 39–47, Kluwer, 1991.

[19] Richard C. Green and Sanjay Srivastava. Risk aversion and arbitrage. *Journal of Finance*, 40(1):257–268, 1985.

[20] David Griffeath and J. Laurie Snell. Optimal stopping in the stock market. *Annals of Probability*, 2, 1974.

[21] Lars Peter Hansen and Ravi Jagannathan. Assessing specification errors in stochastic discount factor models. 1991. Dept. of Finance, Carlson School of Management, University of Minnesota.

[22] Lars Peter Hansen and Ravi Jagannathan. Implications of security market data for models of dynamic economies. *Journal of Political Economy*, 99(2):225–262, 1991.

[23] M. Harrison and D. Kreps. Martingales and arbitrage in multiperiod securities markets. *Journal of Economic Theory*, 20:381–408, 1979.

[24] Hua He. Convergence from discrete to continuous-time contingent claim prices. *Review of Financial Studies*, 3(4):523–546, 1990.

[25] Arthur Hobson. *Concepts in Statistical Mechanics.* Gordon and Breach, 1971.

[26] Chi-fu Huang and Robert H. Litzenberger. *Foundations for Financial Economics.* North-Holland, 1988.

[27] Edward Jaynes. Information theory and statistical mechanics. *Physics Review*, 106:620–630, 1957.

[28] Edward Jaynes. Where do we stand on maximum entropy ? In Raphael D. Levine and Myron Tribus, editors, *The Maximum Entropy Formalism*, MIT Press, 1979.

[29] J.D. Jobson and Bob Korkie. Potential performance and tests of portfolio efficiency. *Journal of Financial Economics*, 10:433–466, 1982.

[30] A.I. Khinchin. *Mathematical Foundations of Information Theory.* Dover, 1957.

[31] Ross Kindermann and J.Laurie Snell. *Markov Random Fields and Their Applications.* Volume 1 of *Contemporary Mathematics*, American Mathematical Society, 1980.

[32] S. Kullback and R. A. Leibler. On information and sufficiency. *Annals of Mathematical Statistics*, 22:79–86, 1961.

[33] Robert E. Lucas. Asset prices in an exchange economy. *Econometrica*, 46:1429–45, 1978.

[34] Esfandiar Maasoumi. Information theory. In *The New Palgrave: Econometrics*, Norton, 1990.

[35] Dilip B. Madan, Frank Milne, and Hersh Shefrin. The multinomial option pricing model and its brownian and poisson limits. *The Review of Financial Studies*, 2(2):251–265, 1989.

[36] Olvi L. Mangasarian. *Nonlinear Programming.* Mc-Graw Hill, 1969.

[37] M.F.M. Osborne. Brownian motion in the stock market. In Paul Cootner, editor, *The Random Character of the Stock Market*, MIT Press, 1970.

[38] Paul A. Samuelson. Gibbs in economics. In G. Caldi and G.D. Mostow, editors, *Proceedings of the Gibbs Symposium*, pages 255–267, American Mathematical Society, 1990.

[39] Jati K. Sengupta. *Maximum Entropy in Applied Econometric Research.* Technical Report 7-90, Dept. of Economics, University of California at Santa Barbara, 1990.

[40] J.E. Shore and R.W. Johnson. Axiomatic derivation of the principle of maximum entropy and the principle of minimum cross-entropy. *IEEE Transactions on Information Theory*, IT-26(1):26–37, 1980.

[41] Karl N. Snow. Diagnosing asset pricing models using the distribution of asset returns. *Journal of Finance*, 46(3):955–983, 1991.

[42] D.W. Stroock and O. Zeitouni. *Microcanonical Distributions, Gibbs' States, and the Equivalence of Ensembles.* Technical Report P-219, Center For Intelligent Control Systems, Massachusetts Institute of Technology, Room 35-311, Cambridge, MA. 02139, 1990.

[43] Henri Theil. *The Maximum Entropy Distribution: A Progress Report.* Technical Report 8043, Center for Mathematical Studies in Business and Economics, University of Chicago, 1980.

[44] Henri Theil and Denzil Fiebig. *Exploiting Continuity: Maximum Entropy Estimation of Continuous Distributions.* Ballinger, 1984.

[45] Hal R. Varian. The arbitrage principle in financial economics. *Journal of Economic Perspectives*, 1(2):55–72, 1987.

[46] Lynde Phelps Wheeler. *Josiah Willard Gibbs.* Yale University Press, 1952.

[47] Walter Willinger and Murad Taquu. The analysis of finite security markets using martingales. *Advances in Applied Probability*, 19:1–25, 1987.

[48] Arnold Zellner. Bayesian methods and entropy in economics and econometrics. In W. T. Grandy and L. H. Schick, editors, *Maximum Entropy and Bayesian Methods*, pages 17–31, Kluwer, 1991.

[49] Arnold Zellner. Maximal data information prior distributions. In A. Aycac and C. Brumat, editors, *New Developments in the Applications of Bayesian Methods*, North Holland, 1977.

[50] Arnold Zellner and R.A. Highfield. Calculation of the maximum entropy distributions and approximation of marginal posterior distributions. *Journal of Econometrics*, 37:195–209, 1988.

# 22 The Winding Problem for Stochastic Oscillators

**Ananda P. N. Weerasinghe** Iowa State University, Ames, Iowa

## §1. Introduction

Consider the stochastic dynamical system governed by

$$\ddot{Y} + g(Y) = h\dot{W} \tag{1.1}$$

with the initial condition $(Y(0), \dot{Y}(0)) = (y_0, v_0) \in R^2$, where the solution $Y(t) \in R$ and the function $g : R \to R$ is locally Lipschitz-continuous and satisfying $xg(x) > 0$ for all $x \neq 0$. $\dot{W}$ represents the one-dimensional "white noise" where $\{W(t) : t \geq 0\}$ is a one-dimensional Brownian motion. $h > 0$ is a constant. Then, as a special case of the stochastic dynamical systems studied by Markus and Weerasinghe ([3], [4]), the following results are true.

(1.2) We have existence and strong uniqueness of the solution $Y(t)$ for all $t \geq 0$.

(1.3) The solution $Y(t)$ has infinitely many zeros and these zeros are simple and isolated.

Introduce $V(t) = \dot{Y}(t)$. Then (1.1) can be written in the form

$$\begin{aligned} dY(t) &= V(t) \\ dV(t) &= hdW(t) - g(Y(t))dt \end{aligned} \tag{1.4}$$

with the initial condition $(V(0), Y(0)) = (v_0, y_0) \in R^2$. Then (1.4) can be considered as a stochastic differential equation driven by a one-dimensional Brownian motion $(W(t) : t \geq 0)$ on a probability space $(\Omega, F, P)$. The solution $(V(t), Y(t))$ exists and strong uniqueness holds for all $t \geq 0$. Let us assume $(V(0), Y(0)) \neq (0,0) \in R^2$. Then from (1.3), it follows that $(V(t), Y(t)) \neq (0,0)$ with probability one. Therefore, in this case we can define the winding angle process to be the continuous determination $\theta(t)$ of the argument of $(V(t), Y(t))$.

In [2], McKean proved the strong laws for the speed of winding in the case $g(y) \equiv 0$. In this article, we would like to study the asymptotic behavior of the winding angle $\theta(t)$, and the process $(V(t), Y(t))$ in the case $g(y) = ky$ where $k$ is a positive constant. In section 2, we prove a central limit theorem for $\theta(t)$.

In section 3, we use the results in section 2, to derive a central limit theorem for the process $N(t)$, where $N(t)$ is the number of zeros of the process $\{Y(s) : 0 \leq s \leq t\}$.

By letting $h = 0$ we have the corresponding deterministic oscillator. Section 4 consists of limit theorems for the deviations of the stochastic oscillator from the corresponding deterministic oscillator.

We intend to obtain limit theorems for non-linear stochastic oscillators (when $g(y)$ is non-linear) by means of stochastic comparison theorems. This issue will be addressed in a future article.

## §2. The Winding Angle

Let $(V(t), Y(t))$ be the solution to the stochastic oscillator

$$\begin{aligned} dY(t) &= V(t)dt \\ dV(t) &= -kY(t)dt + hdW(t) \end{aligned} \tag{2.1}$$

with the initial condition $(V(0), Y(0)) = (v_0, y_0) \neq (0,0)$ where $k$ and $h$ are positive constants, and $W(t)$ is a one dimensional Brownian motion defined on the probability space $(\Omega, F, P)$. The solution process $(V(t), Y(t))$ take values in $R^2$ and is also defined on the probability space $(\Omega, F, P)$. For our consideration of the winding angle, we begin with the initial condition $(V(0), Y(0)) = (1, 0)$. The case of general initial condition can be reduced to this case by waiting until the process hits the axis $[Y = 0$, and $V > 0]$ for the first time (these hitting times are finite a.s. [3], [4]) and then the problem reduces to the initial condition $(V(0), Y(0)) = (C, 0)$ with $C > 0$. Next the transformation $(V, Y) \to (V/C, Y/C)$ reduces it to $(V(0), Y(0)) = (1, 0)$. Hence the solution to (2.1) with the initial condition $(V(0), Y(0)) = (1, 0)$ can be written in the form

$$Y(t) = \frac{1}{\sqrt{k}} \sin(\sqrt{k}t) + \frac{h}{\sqrt{k}} \int_0^t \sin(\sqrt{k}(t-s))dW(s) \tag{2.2}$$

and

$$V(t) = \cos(\sqrt{k}t) + h \int_0^t \cos(\sqrt{k}(t-s))dW(s). \tag{2.3}$$

Using Itô's formula one can carefully check that (2.2) and (2.3) satisfy (2.1) together with the boundary condition $(V(0), Y(0) = (1, 0))$.

Let $\theta(t)$ be the continuous argument of the winding angle of the process $(V(t), Y(t))$ on the $xy$-plane. We define the complex valued process $Z(t)$ by $Z(t) = V(t) + iY(t)$, where $i^2 = -1$. Then we have the following theorem.

**Theorem 2.1.** *With the notation above,*

(2.4) $\dfrac{2(\theta(t) - \sqrt{k}t)}{\log(t)}$ *converges in distribution to a Cauchy distribution with*

*parameter one.*

**Remark**: If $Z$ is a random variable with a Cauchy distribution with parameter 1, then its probability distribution is given by $P[Z \leq y] = \frac{1}{\pi} \int_{-\infty}^{y} \frac{1}{1+x^2} dx$ for each $y \in R$.

Proof: Notice

$$\begin{aligned} Z(t) &= Z_1(t) + i(1 - \sqrt{k})Y(t) \\ \text{where } \ Z_1(t) &= V(t) + i\sqrt{k}Y(t). \end{aligned} \tag{2.5}$$

Since $Z(t)$ and $Z_1(t)$ are continuous processes with $Z(0) = Z_1(0) = 1$, which differ by an imaginary number, it follows that

$$\phi(t) - \pi \leq \theta(t) \leq \phi(t) + \pi$$

where $\theta(t)$ and $\phi(t)$ are the continuous arguments of the processes $Z(t)$ and $Z_1(t)$ respectively.

From (2.2) and (2.3) it follows that

$$Z_1(t) = e^{i\sqrt{k}t} + h\int_0^t e^{i\sqrt{k}(t-s)}dW(s). \tag{2.6}$$

Therefore

$$Z_1(t)e^{-i\sqrt{k}t} = 1 + Z_2(t) \tag{2.7}$$

where $$Z_2(t) = h\int_0^t e^{-i\sqrt{k}s}dW(s).$$

From (2.7) and from the continuity of the processes $Z_1$ and $Z_2$ on the complex plane it follows that

$$\theta(t) - \sqrt{k}t = \arg(1 + Z_2(t)) + 2n_0\pi$$

where $n_0$ is a constant integer. Now, since

$$\arg(Z_2(t)) - \pi < \arg(1 + Z_2(t)) < \arg(Z_2(t)) + \pi,$$

we have

$$\arg(Z_2(t)) - \alpha < \phi(t) - \sqrt{k}t < \arg(Z_2(t)) + \beta \tag{2.8}$$

where $\alpha$ and $\beta$ are constant angles.

Notice

$$Z_2(t) = M_1(t) + iM_2(t) \tag{2.9}$$

where

$$M_1(t) = h\int_0^t \cos(\sqrt{k}s)dW(s)$$

and $$M_2(t) = -h\int_0^t \sin(\sqrt{k}s)dW(s).$$

$M_1$ and $M_2$ are mean-zero martingales as well as Gaussian processes. Furthermore,

$$E[M_1^2(t)] = \frac{h^2 t}{2} + \frac{h^2}{4\sqrt{k}} \sin(2\sqrt{k}t),$$
$$E[M_2^2(t)] = \frac{h^2 t}{2} - \frac{h^2}{4\sqrt{k}} \sin(2\sqrt{k}t).$$

Now consider the processes $\left\{ M_1\left(\frac{n\pi}{\sqrt{k}}\right) : n = 1,2,3\ldots \right\}$ and $\left\{ M_2\left(\frac{n\pi}{\sqrt{k}}\right) : n = 1,2,3\ldots \right\}$. Both processes are Gaussian and independent of each other, and also each process has Gaussian increments. These follow from the easy computations

$$E[M_1\left(\frac{m\pi}{\sqrt{k}}\right) \cdot M_2\left(\frac{n\pi}{\sqrt{k}}\right)] = 0 \quad \text{for all } m, n = 1,2,3\ldots$$

and for $n > m$;

$$M_1\left(\frac{n\pi}{\sqrt{k}}\right) - M_1\left(\frac{m\pi}{\sqrt{k}}\right) = h \int_{\frac{m\pi}{\sqrt{k}}}^{\frac{n\pi}{\sqrt{k}}} \cos(\sqrt{k}s)dW(s).$$

Therefore $M_1\left(\frac{n\pi}{\sqrt{k}}\right) - M_1\left(\frac{m\pi}{\sqrt{k}}\right)$ is independent of $\left\{ M_1\left(\frac{\pi}{\sqrt{k}}\right), \ldots, M_1\left(\frac{m\pi}{\sqrt{k}}\right) \right\}$ and is Gaussian with mean 0 and variance $\frac{h^2}{2\sqrt{k}}(n-m)\pi$.

Let $\{(B_1(t), B_2(t)) : t \geq 0\}$ be a two-dimensional Brownian motion. Therefore the processes

$$\left\{ \left( M_1\left(\frac{n\pi}{\sqrt{k}}\right), M_2\left(\frac{n\pi}{\sqrt{k}}\right) \right) : n = 1,2,3,\ldots \right\} \quad \text{and}$$

$$\left\{ \left( B_1\left(\frac{h^2 n\pi}{2\sqrt{k}}\right), B_2\left(\frac{h^2 n\pi}{2\sqrt{k}}\right) \right) : n = 1,2,3,\ldots \right\}$$

are both Gaussian with mean zero and with the same variance-convariance matrix. Let $\underline{B}(t) = B_1(t) + iB_2(t)$ where $i^2 = -1$.

From Spitzer's theorem (Ref. [1]), $\frac{2\arg(B(t))}{\log(t)}$ converges in distribution to $C_1$ where $C_1$ is a random variable with Cauchy distribution with parameter 1. Hence it follows from (2.9), that $\dfrac{2\arg\left(Z_2\left(\frac{n\pi}{\sqrt{k}}\right)\right)}{\log\left(\frac{n\pi}{\sqrt{k}}\right)}$ also converges in distribution to $C_1$ (since $\log(\frac{h^2}{2}\frac{n\pi}{\sqrt{k}}) =$

$\log(\frac{n\pi}{\sqrt{k}}) + \log(\frac{h^2}{2})$). Then using (2.6), (2.7) and (2.8) together with the above statement, we see that

$$\frac{2\left(\theta\left(\frac{n\pi}{\sqrt{k}}\right) - \sqrt{k}\left(\frac{n\pi}{\sqrt{k}}\right)\right)}{\log\left(\frac{n\pi}{\sqrt{k}}\right)} \tag{2.10}$$

converges in distribution to a Cauchy distribution with parameter 1. Now we have our conclusion on the lattice $\left\{\frac{n\pi}{\sqrt{k}} : n = 1, 2, 3, \ldots\right\}$, and need to generalize it to continuous time $t$ as $t \to \infty$.

Notice that in our defintion of $\theta(t), V(t)$ is measured along $x$-axis and $Y(t)$ is measured along $y$-axis. From (2.1), $\frac{dY}{dt} = V(t)$, therefore, on the right-half plane ($V \geq 0$), $Y(t)$ is increasing and on the left-half plane ($V \leq 0$), $Y(t)$ is decreasing. Therefore the process cannot go from the first quadrant to the fourth quadrant directly (without passing through the second and third quadrants). Similarly it cannot go from the third quadrant to the second quadrant directly. Therefore, if $t_1 > t_2$, we have

$$\theta(t_1) > \theta(t_2) - \pi.$$

Consequently if $\frac{(n+1)\pi}{\sqrt{k}} \geq t \geq \frac{n\pi}{\sqrt{k}}$ we have

$$\theta\left(\frac{(n+1)\pi}{\sqrt{k}}\right) + \pi \geq \theta(t) \geq \theta\left(\frac{n\pi}{\sqrt{k}}\right) - \pi. \tag{2.11}$$

Hence

$$\theta\left(\frac{(n+1)\pi}{\sqrt{k}}\right) - \sqrt{k}\left(\frac{(n+1)\pi}{\sqrt{k}}\right) + 2\pi \geq \theta(t) - \sqrt{k}t \geq \theta\left(\frac{n\pi}{\sqrt{k}}\right) - \sqrt{k}\left(\frac{n\pi}{\sqrt{k}}\right) - 2\pi. \tag{2.12}$$

Using (2.12) together with the fact that $\frac{\log(n)}{\log(n+1)} \to 1$ as $n \to \infty$ gives the conclusion of the theorem.

## §3. Number of Zeros of $Y$

Let us consider (2.1) with the initial condition $(V(0), Y(0)) = (1, 0)$. Let $N(t)$ be the number of zeros on the time interval $[0, t]$. Then we have the following limit theorem for $N(t)$.

**Theorem 3.1.** $\dfrac{2\pi(N(t) - \frac{\sqrt{k}}{\pi}t)}{\log(t)}$ *converges in distribution to a Cauchy distribution with parameter one.*

Proof: This is an easy consequence of Theorem 2.1 and the following facts. First, if $t_1 < t_2$ and $Y(t_1) = Y(t_2) = 0$ then by the mean value theorem, there is a point $s$, such that $t_1 < s < t_2$ and $V(s) = 0$.

Second, consider the process $(V(t), Y(t))$ on the $xy$-plane. Then as we observed before, $Y(t)$ is increasing on the right-half plane ($V \geq 0$) and $Y(t)$ is decreasing on the left-half plane ($V \leq 0$). Therefore the above facts together with (1.3) imply:

$$N(t) = m \quad \text{if and only if} \quad (m-1)\pi \leq \theta(t) < m\pi \quad \text{for all} \quad t.$$

Hence

$$(N(t)-1)\pi \leq \theta(t) < N(t)\pi \qquad \text{for all} \quad t \geq 0.$$

Now, applying Theorem 2.1, we derive the conclusion of the theorem.

Let $T_2 = inf\{t > 0 : Y(t) = 0\}$. Recall $Y(0) = 0$, but $T_2$ is well-defined by (1.3). Using Theorem 3.1 we can derive the following estimate for the probability $P[T_2 > t]$ for large $t$.

**Corollary 3.2.** $P[T_2 > t] \sim \dfrac{1}{2} + \dfrac{1}{\pi}\arctan\left(\dfrac{4\pi - 2\sqrt{k}t}{\log t}\right)$ *for large* $t > 0$.

Proof:

$$\begin{aligned} P[T_2 > t] &= P[N(t) < 2] \\ &= P\left[\frac{2\pi N(t) - 2\sqrt{k}t}{\log(t)} < \frac{4\pi - 2\sqrt{k}t}{\log(t)}\right]. \end{aligned}$$

Now using the central limit theorem for $N(t)$, we see that this probability is approximately $F\left(\frac{4\pi - 2\sqrt{k}t}{\log(t)}\right)$ as $t \to \infty$, where $F(x)$ is the distribution function of the Cauchy distribution of parameter 1. This yields the proof of the corollary.

## §4. Deviations from the deterministic oscillator

In (2.1), if we make $h = 0$, (2.1) reduces to a simple harmonic oscillator with the solution $Y_0(t) = \frac{1}{\sqrt{k}}\sin(\sqrt{k}t)$ and $V_0(t) = \cos(\sqrt{k}t)$.

Define the complex valued process $Z_3(t)$ by

$$Z_3(t) = (V(t) - \cos(\sqrt{k}t)) + i\left(\sqrt{k}Y(t) - \sin(\sqrt{k}t)\right) \tag{4.1}$$

where the processes $V$ and $Y$ are given by (2.2) and (2.3). Hence

$$Z_3(t) = h\int_0^t e^{i\sqrt{k}(t-s)}dW(s) \tag{4.2}$$

and consequently $Z_3(t)e^{-i\sqrt{k}t} = h\int_0^t e^{-i\sqrt{k}s}dW(s)$. With the above representation, we have the following result.

**Theorem 4.1.** *With the above notation,*

a) $\frac{1}{h}\sqrt{\frac{2}{t}}\left(Z_3(t)e^{-i\sqrt{k}t}\right)$ *converges in distribution to a random variable* $(X_1 + iX_2)$ *where* $X_1$ *and* $X_2$ *are independent and have normal distributions with mean zero and variance one.*

b) $\frac{\sqrt{2}|Z_3(t)|}{h\sqrt{t}}$ *converges in distribution to a random variable* $\sqrt{X_1^2 + X_2^2}$ *where* $X_1$ *and* $X_2$ *have normal distributions with mean zero and variance one.*

Proof: Consider $\frac{1}{h}Z_3(t)e^{-i\sqrt{k}t} = \int_0^t \cos(\sqrt{k}s)dW(s) - i\int_0^t \sin(\sqrt{k}s)dW(s)$ by (4.2). Take $M_1(t) = \int_0^t \cos(\sqrt{k}s)dW(s)$ and $M_2(t) = -\int_0^t \sin(\sqrt{k}s)dW(s)$. Notice the joint process $(M_1(t), M_2(t))$ is Gaussian with mean (0,0) and covariance matrix

$$\begin{pmatrix} \frac{t}{2} + \frac{\sin(2\sqrt{k}t)}{4\sqrt{k}}, & -\frac{\sin^2(\sqrt{k}t)}{2\sqrt{k}} \\ -\frac{\sin^2}{2\sqrt{k}}(\sqrt{k}t), & \frac{t}{2} - \frac{\sin(2\sqrt{k}t)}{4\sqrt{k}} \end{pmatrix}. \quad \text{Now as} \quad t \to +\infty,$$

$$\frac{1}{t}\begin{pmatrix} \frac{t}{2} + \frac{\sin(2\sqrt{k}t}{2\sqrt{k}}, & -\frac{1}{2\sqrt{k}}\sin^2(\sqrt{k}t) \\ -\frac{\sin^2(\sqrt{k}t)}{2\sqrt{k}}, & \frac{t}{2} - \frac{\sin(2\sqrt{k}t)}{4\sqrt{k}} \end{pmatrix} \quad \text{approaches the limit} \quad \begin{pmatrix} \frac{1}{2}, & 0 \\ 0, & \frac{1}{2} \end{pmatrix}.$$

Since $(M_1(t), M_2(t))$ is a mean-zero Gaussian process, we can conclude that $\frac{\sqrt{2}}{\sqrt{t}}(M_1(t), M_2(t))$ converges in distribution to a Gaussian random variable $(N_1, N_2)$ where $N_1$ and $N_2$ are independent with mean 0 and variance 1. This yields part a) of the theorem.

Part b) is an immediate consequence of part a).

**Remark**: In the case $k = 1$, the above theorem implies the weak convergence of the process $[(V(t) - \cos(t))^2 + (Y(t) - \sin(t))^2]^{\frac{1}{2}}$ as $t \to \infty$. The next proposition deals with the deviations of $Y(t)$ and $V(t)$ from the corresponding functions of the deterministic oscillator.

**Proposition 4.2.** : *Let $Y(t)$ and $V(t)$ be as in (2.2) and (2.3) repectively. Then*

a) $\sqrt{2}\dfrac{(\sqrt{k}Y(t) - \sin(\sqrt{k}t))}{h\sqrt{t}}$ *converges in distribution to a normally-distributed random variable with mean zero and variance one.*

b) $\sqrt{2}\dfrac{(V(t) - \cos(\sqrt{k}t))}{h\sqrt{t}}$ *also converges in distribution to a normally-distributed random variable with mean zero and variance one.*

Proof: a) Notice $\frac{(\sqrt{k}Y(t) - \sin(\sqrt{k}t))}{h}$ is a Gaussian process with mean 0 and variance $\frac{t}{2} - \frac{1}{4\sqrt{k}}\sin(2\sqrt{k}t)$, and with continuous sample paths.

Hence $\frac{\sqrt{2}(\sqrt{k}Y(t) - \sin(\sqrt{k}t))}{h\sqrt{t}}$ also Gaussian with mean 0 and variance converging to 1 as $t \to \infty$. This yields part a) of the proposition. Proof of part b) is similar.

**Acknowledgement**: I would like to thank Professor K. B. Athreya for some helpful discussions.

## References

[1] Durrett, R. (1982), A new proof of Spitzer's result on the winding of two-dimensional Brownian motion, Ann. Prob. **10**, 244-246.

[2] McKean, H. P. (1963), A winding problem for a resonator driven by a white noise, J. Math. Kyoto Univ. **2**, 227-235.

[3] Markus, L. and Weerasinghe, A. (1988), Stochastic Oscillators, J. Diff. Eqs. **71** no. 2, 288-314.

[4] Markus, L. and Weerasinghe, A. (1989), Oscillation and energy bounds for non-linear dissipative stochastic differential systems, Univ. of Minnesota, Mathematics report no. 88-119.

# 23 On the Convexity of Carrying Simplices in Competitive Lotka-Volterra Systems

**E. C. Zeeman** University of Oxford, Oxford, England

**M. L. Zeeman** University of Texas at San Antonio, San Antonio, Texas

**Abstract**

In [2] M.W. Hirsch proves that for a competitive Lotka-Volterra system on the positive cone in $\mathbf{R^n}$, there is a globally attracting Lipschitz submanifold that projects radially homeomorphically to the unit simplex $\Delta$ in $\mathbf{R^n}$ (here $\Delta = \{x \in \mathbf{R^n} : x_i \geq 0, \forall i \text{ and } \sum_{i=1}^{n} x_i = 1\}$ ). This hypersurface is called the carrying simplex, and denoted by $\Sigma$. In this paper we prove that for a two-dimensional competitive Lotka-Volterra system $\Sigma$ is convex. We give an example to show that the analogous statement in three-dimensions is false, and announce some results relating the convexity of $\Sigma$ to the dynamics on $\Sigma$ in higher dimensions.

*Partially supported by a grant from the Office of Research Development of the University of Texas at San Antonio

## Introduction.

Consider a community of $n$ mutually competing species modeled by the Lotka-Volterra system

$$\dot{x}_i = x_i(b_i - \Sigma_{j=1}^n a_{ij}x_j), \qquad i = 1, \ldots, n \tag{1}$$

where $x_i$ is the population size of the $i$th species at time $t$, and $\dot{x}_i$ denotes $\frac{dx_i}{dt}$.

Each $k$-dimensional coordinate subspace of $\mathbf{R^n}$ is invariant under system (1), ($k \in \{1 \ldots n\}$), and we adopt the tradition of restricting attention to the closed positive cone $\mathbf{R^n_+}$. We denote the open positive cone by $\mathring{\mathbf{R}}^{\mathbf{n}}_{+}$, and call a vector $x$ *positive* if $x \in \mathbf{R^n_+}$, *strictly* positive if $x \in \mathring{\mathbf{R}}^{\mathbf{n}}_{+}$.

The mutual competition between the species dictates that $a_{ij} > 0$ for all $i \neq j$. In addition we assume throughout that for each $i$, $b_i > 0$ and $a_{ii} > 0$, meaning that each species, in isolation, would exhibit logistic growth. That is, when we consider system (1) restricted to the $i$th coordinate axis, we have

$$\dot{x}_i = x_i(b_i - a_{ii}x_i), \qquad b_i,\ a_{ii} > 0$$

in which the repulsion at 0 (growth of small populations) and the repulsion at $\infty$ (competition within large populations) balance at an attracting fixed point, $R_i$, at the *carrying capacity* $\frac{b_i}{a_{ii}}$. Note that the invariance of the axes ensures that $R_i$ is also fixed by the full $n$-dimensional system. We call $R_i$ the $i$th *axial fixed point* of (1).

These ideas can be generalised to the full system as follows. For all $x \in \mathbf{R^n}$ of sufficiently small magnitude $F(x) \in \mathbf{R^n_+}$ (growth of small populations), and for all $x \in \mathbf{R^n_+}$ of sufficiently large magnitude, $-F(x) \in \mathbf{R^n_+}$ (competition of large populations). M. W. Hirsch proves in [2] that this growth and competition balance on a globally attracting hypersurface, which we call the *carrying simplex* and denote by $\Sigma$.

A certain amount is known about the structure of $\Sigma$. Hirsch [2] proves that it is a Lipschitz submanifold everywhere transverse to all strictly positive vectors, and homeomorphic to the unit simplex $\Delta$ in $\mathbf{R^n_+}$ via radial projection (where $\Delta = \{x \in \mathbf{R^n_+} : \Sigma_{i=1}^n x_i = 1\}$ ), whilst Brunovski [1] and Mierczynski [5] give criteria under which it is at least $C^1$,

In this paper we study the relationship between the convexity of $\Sigma$ and the dynamics of system (1). In this context, we use the following definition of convexity:

**Definition 1** *We call a $C^1$ hypersurface $M$ in $R^n$ convex if, for all $m \in M$, the surface*

*M lies entirely in one of the closed half spaces determined by the hyperplane T tangent to M at m. We call M strictly convex if, for all* $m \in M$, $T \cap M = m$.

## Results.

We describe a simple lemma (the pink lemma [1]) which captures, in a global geometric package, one aspect of the algebraic simplicity of Lotka-Volterra systems. We apply the pink lemma to show that the carrying simplex $\Sigma$ of a two-dimensional competitive Lotka-Volterra system is convex (theorem 6).

We then use the two-dimensional result to show that the analogous statement in higher dimensions is false. That is, we use an example inspired by [6, 7] (system (3)) to illustrate how a three-dimensional competitive Lotka-Volterra system may have a non-convex carrying simplex. To explore the higher dimensional situation, we complete the paper with the announcement of two further results (theorems 9 and 10), and some questions that we believe to be open.

Theorem 9 states that an $n$-dimensional competitive Lotka-Volterra system with a strictly convex carrying simplex has no non-trivial recurrence in $\overset{\circ}{\mathbf{R}}{}^{\mathbf{n}}_{+}$. Thus any three-dimensional competitive Lotka-Volterra system with a non-trivial periodic orbit in $\overset{\circ}{\mathbf{R}}{}^{\mathbf{3}}_{+}$ (of which there are numerous examples [6, 7]) has a carrying simplex which is not strictly convex. Theorem 9 is certainly not sharp. For example, system (3) (see figure 4) has a non-convex carrying simplex, but only trivial recurrence.

Theorem 10 states that, under certain conditions, we can use the pink lemma to relax the convexity condition from the carrying simplex $\Sigma$ to an appropriately defined quadratic function $Q$, and still conclude that the system has no non-trivial recurrence.

## The carrying simplex.

We begin by stating precisely some results about the carrying simplex. As mentioned in the introduction, it is easy to see that 0 is a repelling fixed point of system (1), and that the basin of repulsion of 0 in $\mathbf{R}^{\mathbf{n}}_{+}$ is bounded. The carrying simplex $\Sigma$ is the boundary of that basin. To be precise, we define $B(0) = \{x \in \mathbf{R}^{\mathbf{n}}_{+} : \alpha(x) = 0\}$, then $\Sigma = \partial B(0) \setminus B(0)$, where $\alpha(x)$ denotes the *alpha*-limit set of the trajectory through $x$ and $\partial B(0)$ denotes the boundary of $B(0)$ taken in $\mathbf{R}^{\mathbf{n}}_{+}$. We remove $B(0)$ from $\partial B(0)$ to avoid topological awkwardness at the coordinate subspaces.

[1] So named because it was first proved on a tiny piece of pink paper

Applying a theorem of Hirsch ([2], theorem 1.7), we have:

**Theorem 2 (Hirsch.)** *Given system (1), every trajectory in* $\mathbf{R}^n_+ \setminus \{0\}$ *is asymptotic to one in* $\Sigma$, *and* $\Sigma$ *is a Lipschitz submanifold homeomorphic to the unit simplex* $\Delta$ *via radial projection.* $\Sigma$ *is also homeomorphic to its image (on either axis) under projection along any strictly positive vector.*

## The pink lemma.

In lemma 3 we state the pink lemma in full generality. We then restrict our attention to the special case of quadratic vector fields, restating the pink lemma as corollary 4, and illustrating how this statement suits our current application (corollary 5).

Let $i : \mathbf{R}^m \to \mathbf{R}^n$ and $\pi : \mathbf{R}^n \to \mathbf{R}^k$ be affine maps. (Here *affine* means a linear map followed by a translation.) The proof of the following lemma is immediate.

**Lemma 3 (The Pink Lemma)** *If* $F : \mathbf{R}^n \to \mathbf{R}^n$ *is a polynomial vector field of degree* $d$, *then* $\pi \circ F \circ i : \mathbf{R}^m \to \mathbf{R}^k$ *is also a polynomial of degree* $d$.

Our interest in this paper is in quadratic systems. We shall usually apply the pink lemma by choosing an affine hyperplane $H \subset \mathbf{R}^n$. (Here *affine hyperplane* means the translate of an $(n-1)$-dimensional vector subspace of $\mathbf{R}^n$.) Choose an affine embedding $i : \mathbf{R}^{n-1} \to \mathbf{R}^n$ that maps $\mathbf{R}^{n-1}$ onto $H$. Let $\mathbf{R}$ be the vector subspace of $\mathbf{R}^n$ normal to $H$, and let $\pi$ be the orthogonal projection of $\mathbf{R}^n$ onto $\mathbf{R}$. Then the pink lemma reduces to:

**Corollary 4** *If* $F : \mathbf{R}^n \to \mathbf{R}^n$ *is quadratic, then* $Q = \pi \circ F \circ i : \mathbf{R}^{n-1} \to \mathbf{R}$ *is a quadratic function.*

### Notational remark.

To simplify notation in our discussion of $Q$, we shall henceforth identify $H$ with the domain of $Q$, allowing the same coordinate to denote corresponding points in either space, and using the context to avoid ambiguity.

To see how the pink lemma packages algebraic information into a geometric tool, consider the case when $F$ is a quadratic vector field on $\mathbf{R}^2$. Let $H$ be a line in the plane. Then $Q : \mathbf{R} \to \mathbf{R}$, defined as above, measures the normal component of $F$ on $H$. In particular, the sign of $Q$ indicates to which side of $H$ the vector field points. A zero of $Q$ corresponds to a point on $H$ which is either a zero of $F$, or at which $F$ is tangent to $H$. The pink lemma says that $Q$ is a quadratic function, so that a nontrivial $Q$ has at most two zeros. Thus we can conclude information about the behaviour of trajectories, relative to $H$, as follows.

Given $y \in \mathbf{R}^2$, let $\phi$ denote the orbit of $y$ under the system $\dot{x} = F(x)$.

**Corollary 5** *For each $y \in \mathbf{R}^2$, either $\phi \subseteq H$, or $\phi$ meets $H$ in no more than three points that are consecutive intersections along $\phi$ and in the same order along $H$.*

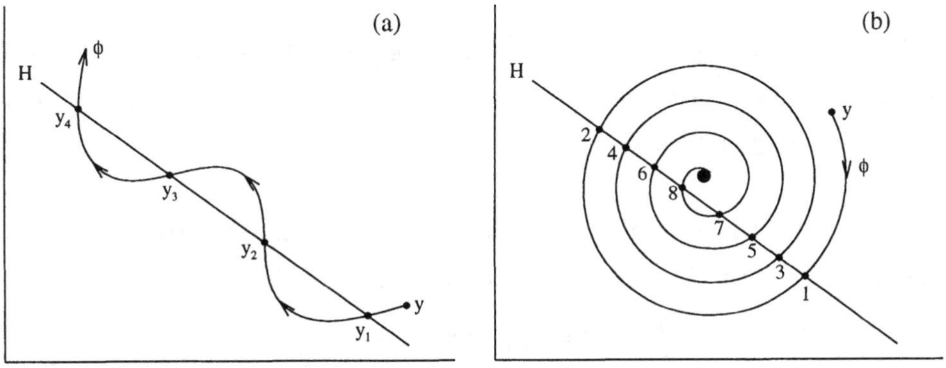


Figure 1: The orbit $\phi$ of $y$ meeting $H$.

Figure 1(a) we shall prove impossible. By contrast, figure 1(b) is possible because [2] quadratic systems in the plane can have spiral attractors. However, in figure 1(b), although 1234 are consecutive intersections along $\phi$ they are not in the same order along $H$, and although 1357 are in the same order along both $\phi$ and $H$, they are not consecutive intersections along $\phi$.

**Proof of corollary 5.**

If $\phi \cap H$ contains an interval then $Q$ vanishes on that interval, and is therefore identically zero. Hence $\phi \subseteq H$. Therefore let us assume that $\phi \cap H$ does not contain an interval.

Suppose, for contradiction, that $y \notin H$ and that $\phi$ meets $H$ in at least four points $y_1, \ldots, y_4$ that are consecutive intersections along $\phi$ and in the same order along $H$. Consider first the case when $\phi$ is transverse to $H$ at each $y_i$, as shown in figure 1(a). As

[2]But not possible in two-dimensional competitive Lotka-Volterra systems because the latter do not have spiral attractors.

mentioned above, the sign of $Q$ at $y_i$ indicates to which side of $H$ the vector field points at $y_i$. The transversality of $\phi$ and $H$ at each $y_i$ thus ensures that $Q(y_i) \neq 0$ and

$$\text{sgn}Q(y_1) = -\text{sgn}Q(y_2) = \text{sgn}Q(y_3) = -\text{sgn}Q(y_4).$$

So $Q$ changes sign three times on $\mathbf{R}$, and hence (by the intermediate value theorem) has three zeros on $\mathbf{R}$. This contradicts the fact that $Q$ is a quadratic on $\mathbf{R}$. For future convenience we shall abbreviate this proof by saying "$y_1y_2y_3y_4$ give a pink contradiction on $H$."

If $\phi$ is tangent to $H$ at any of the $y_i$, we can make a slight translation of $H$ to a nearby, parallel line $H_\epsilon$, to remove the tangency. We can then apply the previous argument to $H_\epsilon$, to reach a contradiction. Q.E.D.

## Two dimensional competitive Lotka-Volterra systems.

When $n = 2$, system (1) reduces to a system $\dot{x} = F(x)$ on $\mathbf{R}^2_+$, where

$$\begin{cases} F_1(x) = x_1(b_1 - a_{11}x_1 - a_{12}x_2) \\ F_2(x) = x_2(b_2 - a_{21}x_1 - a_{22}x_2) \end{cases} \tag{2}$$

If we define $A = (a_{ij})$, and $B = (b_1, b_2)$, then $P = A^{-1}B^T$ is a fixed point of system (2). It is well known that if $P \in \mathring{\mathbf{R}}^2_+$, then $P$ is globally attracting on $\mathring{\mathbf{R}}^2_+$ if $\det A > 0$, and $P$ is a saddle if $\det A < 0$. If $\det A = 0$, the system has a line of degenerate fixed points in $\mathbf{R}^2$.

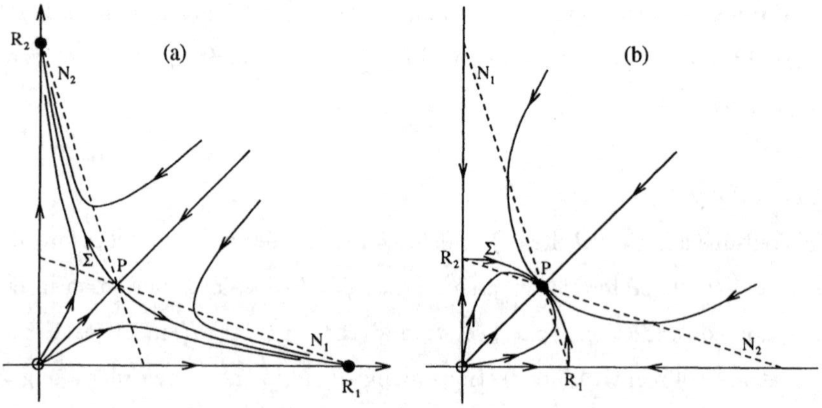


Figure 2: The dynamics of two-dimensional competitive Lotka-Volterra systems with a fixed point $P \in \mathbf{R}^2_+$. (a) $P$ is a saddle when $\det A < 0$, and (b) $P$ is an attractor when $\det A > 0$. An attractor is represented by a solid dot, and a repellor by an open dot.

The classical way to see this is by a geometric analysis of the nullclines of the system: the sets on which one component of the vector field vanishes. The $x_1$ nullcline is given by

$$\dot{x_1} = 0 \Leftrightarrow x_1(b_1 - a_{11}x_1 - a_{12}x_2) = 0 \Leftrightarrow \begin{cases} x_1 = 0 \\ \text{or } a_{11}x_1 + a_{12}x_2 = b_1 \end{cases}$$

and so consists of the $x_2$-axis together with the line $N_1$ which has axial intercepts $R_1 = (\frac{b_1}{a_{11}}, 0)$ and $(0, \frac{b_1}{a_{12}})$. Similarly, the $x_2$ nullcline consists of the $x_1$-axis together with the line $N_2$ with axial intercepts $(\frac{b_2}{a_{21}}, 0)$ and $R_2 = (0, \frac{b_2}{a_{22}})$. See figure 2.

The fixed points of the system lie at the intersections of the two nullclines. Generically, there are four such intersections. They are at 0, $R_1$, $R_2$ and $P = N_1 \cap N_2$. Now, $\det A$ provides information about the geometric configuration of the $N_i$ via the axial intercepts. More precisely, the assumption that $P \in \overset{\circ}{\mathbf{R}}{}^2_+$ ensures that $\text{sgn}(\frac{b_1}{a_{11}} - \frac{b_2}{a_{21}}) = \text{sgn}(\frac{b_2}{a_{22}} - \frac{b_1}{a_{12}})$. So

$$\det A < 0 \Leftrightarrow a_{11}a_{22} < a_{12}a_{21} \Leftrightarrow \frac{b_1}{a_{11}} > \frac{b_2}{a_{21}} \text{ and } \frac{b_2}{a_{22}} > \frac{b_1}{a_{12}},$$

as in figure 2(a). Similarly

$$\det A > 0 \Leftrightarrow \frac{b_1}{a_{11}} < \frac{b_2}{a_{21}} \text{ and } \frac{b_2}{a_{22}} < \frac{b_1}{a_{12}},$$

as in figure 2(b).

When $\det A = 0$, the lines $N_i$ coincide, producing a line of degenerate fixed points.

With this fixed point information there are plenty of elementary arguments with which to verify that the dynamical behaviour is indeed as pictured in figure 2. When $\det A < 0$, $P$ is a saddle point, and when $\det A > 0$, $P$ is an attractor. See May [4], Hofbauer and Sigmund [3], Zeeman [7]. In particular, notice that in either case the carrying simplex is composed of trajectories joining the $R_i$ and $P$, and that when $P$ is a saddle, $P$ repels on $\Sigma$. By theorem 2, $\Sigma$ is everywhere transverse to all strictly positive vectors, and thus trapped in the region between the lines $N_i$, where neither $F(x)$ nor $-F(x)$ is strictly positive. The following theorem verifies that $\Sigma$ is in fact convex, as shown.

**Theorem 6** *Given system (2), $\Sigma$ is convex.*

**Proof.**

When $\det A = 0$, $\Sigma$ is simply a line of fixed points, and hence convex.

Assume that $\det A \neq 0$ and that $P \in \mathring{\mathbf{R}}^2_+$. In this case, $\Sigma$ is not a line, since $\Sigma$ joins the $R_i$ and $P$. Indeed, $\Sigma$ contains no line segments. If it did, that line segment would contain $P$ and one of the $R_i$ (by corollary 5), and hence lie in the nullcline $N_i$. But $N_i$ is not invariant when $\det A \neq 0$. We proceed to show that $\Sigma$ is strictly convex. The idea is simple, but a careful labeling of points is required.

Suppose, for contradiction, that $\Sigma$ is not strictly convex. Then there is a line $H$ in $\mathbf{R}^2$ which meets $\Sigma$ in at least three points. By slightly perturbing $H$ if necessary, we may assume that $H$ is transversal to $\Sigma$, and contains no fixed points. We shall use the three points of $H \cap \Sigma$ together with the added dynamical information classically known about $F$ to construct pink contradictions on $H$.

Notice that $H$ meets the positive quadrant in a finite interval, for the following reason. Since $H$ joins two points of $\Sigma$ it is parallel to the tangent at some point of $\Sigma$ in between (by the mean value theorem). In other words, $H$ is parallel to the vector field at this point, and thus has negative slope because $\Sigma$ is trapped in the region between the nullclines where the vector field has negative slope. Therefore $H$ meets the positive quadrant in a finite interval, $h_1 h_2$ say. From now on we shall assume that $H$ denotes the finite interval, since it suffices to confine attention to the positive quadrant.

Now, $H$ meets $\Sigma$ in a finite number of points, since they are compact and transversal. Also, the points of intersection lie in the same order along $\Sigma$ and $H$ because $\Sigma$ projects radially homeomorphically onto $H$ (by theorem 2).

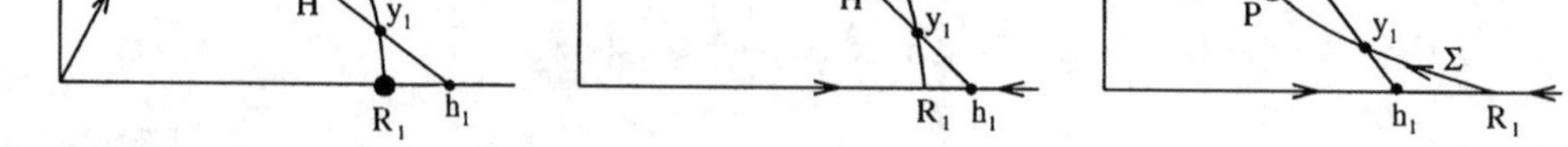

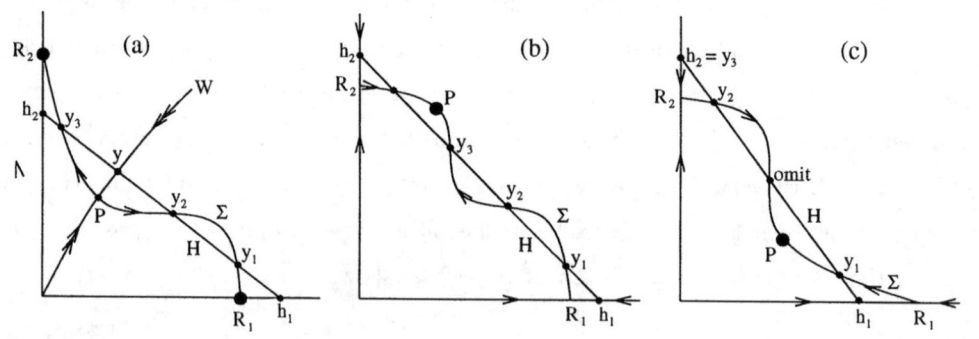

Figure 3: Pink contradictions on $H$ (a) when $P$ repels on $\Sigma$, and (b,c) when $P$ is an attractor. An attractor is represented by a (large) solid dot.

Consider first the case when $P$ is a saddle. Then $P$ repels on $\Sigma$, and $P$ has a one-dimensional stable manifold $W$, which runs from the origin to infinity in the positive quadrant, and therefore crosses $H$, at $y$ say.

Let $y_1, y_2, y_3$ be three consecutive points of $\Sigma \cap H$ such that there are no other points of $\Sigma \cap H$ in between these points and $P$, along $\Sigma$. Then, independent of the position of $P$ relative to the $y_i$, the four points $yy_1y_2y_3$ give a pink contradiction on $H$, as is illustrated by the example in figure 3(a).

Now consider the case when $P$ attracts on $\Sigma$. Starting from the end $h_1$ of $H$ let $y_1, y_2, y_3$ label the first three consecutive points of $\Sigma \cap H$, subject to the following two provisos: (i) Omit the first point of $\Sigma \cap H$ after $P$ along $\Sigma$, if there is one. (ii) If there are only three points in $\Sigma \cap H$ and we have omitted one by (i) then let $y_3 = h_2$. Two examples are illustrated in figures 3(b,c). Then $h_1y_1y_2y_3$ give a pink contradiction on $H$.

Finally, consider the case when $P \notin \overset{\circ}{\mathbf{R}}{}^2_+$. Then $\Sigma$ consists of a single trajectory together with its alpha and omega limit points (one on each axis). We leave it to the reader to verify that if $\Sigma$ were not strictly convex, this would lead to another pink contradiction. Q.E.D.

By combining theorem 6 with the classically known facts about the location of the fixed points on the nullclines (figure 2), and using theorem 2 to view $\Sigma$ as a graph over either axis, we have:

**Corollary 7** *Given system (2) with $P \in \overset{\circ}{\mathbf{R}}{}^2_+$, if $P$ repels on $\Sigma$ then $\Sigma$ is concave up, and if $P$ attracts on $\Sigma$, then $\Sigma$ is concave down.*

## Side note on periodic orbits in the plane.

Applying the pink lemma slightly differently also has simple consequences for the geometric structure of periodic orbits of quadratic systems in the plane.

For example, let $F : \mathbf{R}^2 \to \mathbf{R}^2$ be quadratic, and let $\gamma$ be a hyperbolic periodic orbit of the system $\dot{x} = F(x)$. By Poincaré-Bendixson theory, there must be a fixed point $P$ inside $\gamma$.

**Corollary 8** *If the hyperbolic periodic orbit $\gamma$ is a circle, then the fixed point $P$ is not at the centre of the circle.*

**Proof.**

Suppose that $P$ is at the centre of $\gamma$. Let $H$ be any line through $P$, so that $H$ meets $\gamma$ at antipodal points $\gamma_1$, $\gamma_2$. Let $i : \mathbf{R} \to \mathbf{R}^2$, be inclusion onto $H$ (as before), let $\pi : \mathbf{R}^2 \to \mathbf{R}$

be orthogonal projection onto $H$ (as opposed to onto the normal to $H$ as we had before), and let $Q = \pi \circ F \circ i$, as usual. By the pink lemma, $Q : \mathbf{R} \to \mathbf{R}$ is quadratic.

Now, at each $\gamma_i$, $F$ is tangent to $\gamma$, and hence normal to $H$. Thus $Q = 0$ at each $\gamma_i$. In addition, $Q = 0$ at the fixed point $P$, so $Q$ has three zeros on $\mathbf{R}$, and hence must be identically zero on $\mathbf{R}$. That is: the vector field $F$ is everywhere normal to the radial lines through $P$. Hence the trajectories of the system lie on concentric circles (centre $P$), so that $P, \gamma$ neither attract nor repel neighbourhoods of themselves. Therefore $\gamma$ is not hyperbolic. Q.E.D.

## Three-dimensional competitive Lotka-Volterra systems.

In the previous section we showed that the carrying simplex $\Sigma$ of a two-dimensional competitive Lotka-Volterra system is convex. In this section we show by example that the analogous statement in three dimensions is false. For contrast with the results announced in the next section, we choose an example which has a fixed point $P \in \mathring{\mathbf{R}}^2_+$, but no non-trivial recurrence.

Consider the system

$$\begin{cases} \dot{x_1} = x_1(10 - 3x_1 - 5x_2 - 2x_3) \\ \dot{x_2} = x_2(10 - 4x_1 - 3x_2 - 3x_3) \\ \dot{x_3} = x_3(10 - 2x_1 - 1x_2 - 7x_3) \end{cases} \tag{3}$$

In [7], a partial classification of three dimensional competitive Lotka-Volterra was developed, using a generalisation of the classical nullcline analysis. System (3) lies in class 25 of that classification, and hence has a phase portrait on $\Sigma$ as shown in figure 4(a). Note that we are using theorem 2 to picture the carrying simplex as a flat two-dimensional manifold, homeomorphic to the unit simplex $\Delta \in \mathbf{R}^3_+$.

We shall now verify some of this dynamical information, and use it to look at how $\Sigma$ sits in $\mathbf{R}^3_+$. First note that $P = (1, 1, 1)$ is a fixed point of system (3). Next we consider the restriction of the system to each coordinate plane $x_i = 0$. This restriction is a two-dimensional competitive Lotka-Volterra system whose carrying simplex is precisely $\Sigma \cap \{x_i = 0\}$. Thus the boundary $\partial\Sigma$ of $\Sigma$ is composed of the three carrying simplices of the restricted systems in the coordinate planes.

In the coordinate plane $x_i = 0$, the restricted system has matrix $A_i = \begin{pmatrix} a_{jj} & a_{jk} \\ a_{kj} & a_{kk} \end{pmatrix}$, and $B_i = (b_j, b_k)$, where $j < k$ and $i, j, k$ are distinct. Thus $A_3 = \begin{pmatrix} 3 & 5 \\ 4 & 3 \end{pmatrix}$, $B_3 = (10, 10)$, and there is a fixed point $P_3$ at $\frac{10}{11}(2, 1, 0)$ which repels on $\partial\Sigma$ since $\det A_3 < 0$. Similarly there are fixed points $P_1$, $P_2$ in the coordinate planes $x_1 = 0$, $x_2 = 0$ respectively, both of which attract on $\partial\Sigma$ since $\det A_1, \det A_2 > 0$

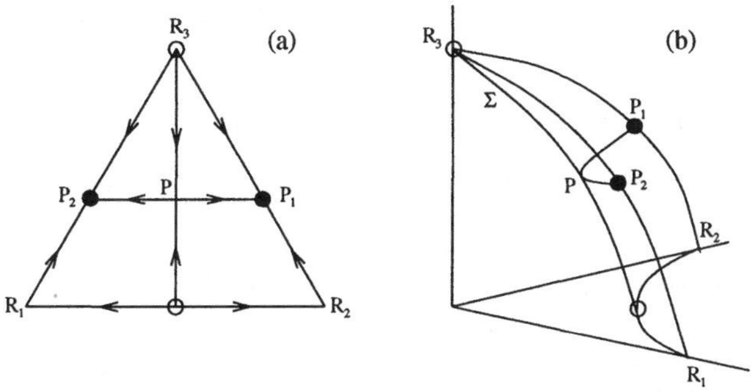


Figure 4: The dynamics (a) and location (b) of the carrying simplex $\Sigma$ of system (3). Fixed point notation as in figure 2.

By corollary 7 the component of $\partial\Sigma$ in the coordinate plane $x_3 = 0$ is concave up as a graph over the $x_1$- or $x_2$-axis, whereas the components of $\partial\Sigma$ in the other two coordinate planes are concave down as graphs over the $x_3$-axis. See figure 4(b). Hence $\Sigma$ is not convex as a surface in $\mathbf{R}^3$ (definition 1).

## Further results.

We now turn to competitive Lotka-Volterra systems of arbitrary dimension $n$. It is well known that if there is no fixed point in $\overset{\circ}{\mathbf{R}}{}^{\mathbf{n}}_{+}$, then there is a Liapunov function, and hence no recurrence in $\overset{\circ}{\mathbf{R}}{}^{\mathbf{n}}_{+}$ (see Hofbauer and Sigmund [3, section 9.2]). The following two theorems apply to competitive Lotka-Volterra systems of arbitrary dimension for which there is a fixed point $P$ in $\overset{\circ}{\mathbf{R}}{}^{\mathbf{n}}_{+}$. For proofs, see Zeeman and Zeeman [8].

**Theorem 9** *Given system (1), if $\Sigma$ is a strictly convex hypersurface in $\mathbf{R}^{\mathbf{n}}$, then there is no non-trivial recurrence in $\overset{\circ}{\mathbf{R}}{}^{\mathbf{n}}_{+}$*

Theorem 10 uses the pink lemma to relax the convexity condition from $\Sigma$ to an appropriately defined quadratic function $Q$. Given competitive Lotka-Volterra system $\dot{x} = F(x)$ with fixed point $P \in \overset{\circ}{\mathbf{R}}{}^{\mathbf{n}}_{+}$, define $H$ to be the hyperplane tangent to $\Sigma$ at $P$, and let $Q : \mathbf{R}^{\mathbf{n-1}} \to \mathbf{R}$ be the corresponding quadratic function.

**Theorem 10** *If $P$ repels on $\Sigma$, the boundary $\partial\Sigma$ attracts on $\Sigma$, and $Q$ is definite, then the basin of repulsion of $P$ is the interior of $\Sigma$.*

**Remarks**

The ecological interpretation of theorem 10 is that under the given hypotheses, there can be no coexistence of all $n$ species. That is, at least one of the species will be driven to extinction.

An analogous result holds for the case when $P$ attracts on $\Sigma$ and $\partial\Sigma$ repels on $\Sigma$, provided that we add a mild hypothesis about the relative strengths of the eigenvalues at $P$. Then we can conclude that $P$ is globally attracting on $\overset{\circ}{\mathbf{R}}{}^n_+$, so that all strictly positive initial conditions lead to a stable coexistence of all $n$ species.

## Open Questions.

Interesting dynamics can occur in the three-dimensional case when there is an attracting or repelling fixed point in $\overset{\circ}{\mathbf{R}}{}^3_+$, and the carrying simplex has negative curvature at $P$ (and hence the corresponding $Q$ is indefinite). We know that generic Hopf bifurcations can occur in this case, spinning off small hyperbolic closed orbits shaped like the seam on a tennis ball, but we do not know what the global dynamics on the carrying simplex are, nor how many closed orbits there can be.

## References

[1] P. Brunovski. (1992) Controlling Non-Uniqueness of Local Invariant Manifolds. *Preprint.*

[2] M. W. Hirsch. (1988) Systems of Differential Equations that are Competitive or Cooperative. III: Competing Species. *Nonlinearity,* **1**: 51–71.

[3] J. Hofbauer and K. Sigmund. (1988) *The Theory of Evolution and Dynamical Systems.* Cambridge University Press, Cambridge.

[4] R. M. May. (1975) *Stability and Complexity in Model Ecosystems.* Princeton University Press, Princeton.

[5] J. Mierczynski (1992) The $C^1$ Property of Carrying Simplices for a Class of Competitive Systems of Ordinary Differential Equations. *Journal of Differential Equations,* To appear.

[6] E. C. Zeeman. (1980) Population dynamics from game theory. *Global Theory of Dynamical Systems.* Springer Lecture Notes in Math, 819. Springer-Verlag, New York, 471–497.

[7] M. L. Zeeman. (1993) Hopf bifurcations in competitive three-dimensional Lotka-Volterra systems. *Dynamics and Stability of Systems,* Vol. 8. (to appear).

[8] E. C. Zeeman and M. L. Zeeman. Ruling out recurrence in competitive Lotka-Volterra systems. *To appear.*

# 24 The Shuffle Product and Symmetric Groups

**A. A. Agrachev** Steklov Mathematics Institute, Moscow, Russia
**R. V. Gamkrelidze** Steklov Mathematics Institute, Moscow, Russia

ABSTRACT. The shuffle product of permutations is investigated. This operation is an algebraic model of the standard triangulation of the Cartesian product of symplexes. A nontrivial connection is established between the shuffle product and the usual composition product of permutations, which leads to interesting algebraic–combinatorial implementations.

## §1. INTRODUCTION

A wide range of evolution equations lead to the investigation of asymptotic series of the form

$$\sum_{n=0}^{\infty} \int\limits_{t\le\tau_n\le\ldots\le\tau_1\le T} \cdots \int \varphi_n(\tau_1,\ldots,\tau_n)d\tau_1,\ldots d\tau_n, \tag{1}$$

where $\tau_i$ are reals and $\varphi_n(\tau_1,\ldots,\tau_n)$ belong to the appropriate associative algebra — to an operator algebra, to a tensor algebra or, in the simplest case, to a (commutative) field of scalars. An expression of the form (1) is of a very general nature and describes an arbitrary evolutionary process on the interval $[t,T]$, when the state of the process at every time instant does not depend on the future. Taking into account this connection, we call the series of the form (1) a *chronological series*.It is natural to presuppose that the product of two chronological series is again a chronological series, though an explicit formula representing a product of two chronological series in the form (1) is not at all trivial and is connected with interesting combinatorial and algebraic structures.

For simplicity we take $t=0, T=1$ and introduce the notation

$$\Delta^n = \{(\tau_1,\ldots,\tau_n)\big|0\le\tau_n\le\ldots\le\tau_1\le 1\}.$$

We have

$$\left.\begin{aligned}&\sum_{n=0}^{\infty}\int\limits_{\Delta^n}\cdots\int\varphi_n(\tau_1,\ldots,\tau_n)d\tau_1\ldots d\tau_n\sum_{n=0}^{\infty}\int\limits_{\Delta^n}\cdots\int\psi_n(\tau_1,\ldots,\tau_n)d\tau_1\ldots d\tau_n=\\&\quad=\sum_{n=0}^{\infty}\sum_{k=0}^{n}\int\limits_{\Delta^k\times\Delta^{n-k}}\cdots\int\varphi_k(\tau_1,\ldots,\tau_k)\psi_{n-k}(\tau_{k+1},\ldots,\tau_n)d\tau_1\ldots d\tau_n.\end{aligned}\right\} \tag{2}$$

The integral over the product of simplexes $\Delta^k\times\Delta^{n-k}$ is easily represented as a sum of integrals over the standard simplex $\Delta^n$ if we use special permutations of variables — the *shuffles*.

*Key words and phrases.* Permutation, shuffle, symmetric group, Volterra series.

Let $\Sigma_n$ be the group of all permutations of the set $\{1, 2, \ldots, n\}$ and $k \in \{1, \ldots, n\}$. Put

$$S(k, n-k) = \{\sigma \in \Sigma_n | \sigma(i) < \sigma(i+1), \forall i \neq k\}.$$

The following identity is valied for an arbitrary summable $f$:

$$\int\limits_{\Delta^k \times \Delta^{n-k}} \cdots \int f(\tau_1, \ldots, \tau_n) d\tau_1 \ldots d\tau_n = \int\limits_{\Delta^n} \cdots \int \sum_{\sigma \in S_n(k)} f(t_{\sigma(1)}, \ldots, t_{\sigma(n)}) dt_1 \ldots d\tau_n. \tag{3}$$

The elements of the set $S(k, n-k)$ are called *the $(k, n-k)$-type shuffles*. Thus, in order to represent the product (2) as a chronological series, we have to reshuffle the variables in the tensor products of the functions $\varphi_k$ and $\psi_{n-k}$ under the integrals and sum up. It seems that the reshuffling procedure reflects some fundamental properties common to all evolutionary processes.

This article is devoted to purely algebraic study of this procedure, regardless to the properties of concrete chronological series. As a central result of the paper we consider a remarkable connection of the shuffle operation with the group algebras of symmetric groups and some combinatorial–algebraic identities, derived from this connection.

A close relation of integral expressions of the form (1) with shuffles was discovered some time ago. In [6] a commutative algebra was introduced with the aid of the shuffle operation and its connection with the "iterated integrals" was established. This theme was further developed in the works of M.Fliess and others, cf. [5]. Among more recent papers we should mention [4], where further references could be found. In contrast to these investigations we introduce here a *noncommutative* shuffle algebra Ш, which accumulates all basic properties of shuffles. The noncommutativity is very essential since the primary role will be plaied not even by the algebra Ш, but rather by a certain Lie algebra for which Ш is the universal enveloping algebra.

In the next section the shuffle product of permutations is defined, and its relations with the Volterra series and the commutative shuffle product are established. In §3 we investigate the structure of the shuffle algebra Ш and of some algebraic and combinatorial objects closely connected with this algebra. In §4 the key object is introduced — the Lie algebra of variations, which turns out to be, together with the corresponding group, a connection bridge between the shuffle algebra and the group algebras of the symmetric groups. This is shown in §5. In the final section we establish some symmetry relations for Lie elements in free associative algebras.

## §2. The shuffle product of permutations

We use the following notations. By $\nu_1 \circ \nu_2$, $\nu_1, \nu_2 \in \Sigma_n$, we denote the composition of permutations: $\nu_1 \circ \nu_2(i) \stackrel{\text{def}}{=} \nu_1(\nu_2(i)), i = 1, \ldots n$. Let $\mu \in \Sigma_m, \nu \in \Sigma_n$, then by $\mu \times \nu$ we denote an element of $\Sigma_{m+n}$ defined by the formula

$$(\mu \times \nu)(i) = \begin{cases} \mu(i), 1 \leq i \leq n \\ \nu(i-n) + n, n < i \leq n+m. \end{cases}$$

By the symbol $\mathbb{R}[\Sigma_n]$ we denote the group algebra of the group $\Sigma_n$ over the reals, i.e. the algebra over $\mathbb{R}$ for which $\Sigma_n$ is a basis with the table of compositions

$$(\nu_1, \nu_2) \mapsto \nu_1 \circ \nu_2 \ \forall \nu_1, \nu_2 \in \Sigma_n.$$

**Definition.** The shuffle product of two permutations $\mu \in \Sigma_m, \nu \in \Sigma_n$ is defined as an element $\mu ш \nu \in \mathbb{R}[\Sigma_{m+n}]$, given by the formula

$$\mu ш \nu = \sum_{\sigma \in S(m,n)} \sigma \circ (\mu \times \nu).$$

If we put $\mathbb{R}[\Sigma] \stackrel{\text{def}}{=} \bigoplus_{n=0}^{\infty} \mathbb{R}[\Sigma_n]$ and extend the ш–operation by linearity over the whole $\mathbb{R}[\Sigma]$ we obtain a graded algebra. Evidently, it is associative and for arbitrary $\nu_i \in \Sigma_{n_i}$, $i = 1, \ldots, k$ we have

$$\nu_1 ш \ldots ш \nu_k = \sum_{\sigma \in S(n_1, \ldots, n_k)} \sigma \circ (\nu_1 \times \ldots \times \nu_k),$$

where

$$S(n_1, \ldots, n_k) = \left\{ \sigma \in \Sigma_{n_1 + \cdots + n_k} \middle| \begin{array}{l} \sigma(i) < \sigma(i+1), \\ \forall i \neq n_1 \cdots + n_j, \; j = 1, \ldots, k-1 \end{array} \right\}.$$

Denote by $\mathbb{I}_n$ the unity in the group $\Sigma_n$, i.e. the identity permutation: $\mathbb{I}_n(i) = i$, $i = 1, \ldots, n$, and put $\mathbb{I}_0 = 1 \in \mathbb{R}$.

**Definition.** The graded ш–subalgebra in $\mathbb{R}[\Sigma]$ generated by the elements $\mathbb{I}_n$, $n = 0, 1, 2, \ldots$, will be called the *shuffle algebra*

$$Ш = \bigoplus_{n=0}^{\infty} Ш_n, \quad Ш_n = Ш \cap \mathbb{R}[\Sigma_n].$$

Let $\mathcal{S}$ be a set, then

$$Ass(\mathcal{S}) \stackrel{\text{def}}{=} \bigoplus_{n=0}^{\infty} Ass_n(\mathcal{S})$$

will denote the free associative algebra over $\mathbb{R}$ with the set $\mathcal{S}$ of free generators, so that $Ass_n(\mathcal{S})$ consists of linear combinations of $n$–letter words of the alphabet $\mathcal{S}$. Denote by $\theta_n$ the right natural action of the group $\Sigma_n$ on the $n$–letter words:

$$\theta_n(\sigma) : s_1 \ldots s_n \mapsto s_{\sigma(1)} \ldots s_{\sigma(n)} \; \forall \sigma \in \Sigma_n, \; s_i \in \mathcal{S}, \; i = 1, \ldots, n,$$

and by $\overline{\theta}_n(\sigma)$ the adjoint left action:

$$\overline{\theta}_n(\sigma) : s_1 \ldots s_n \mapsto s_{\sigma^{-1}(1)} \ldots s_{\sigma^{-1}(n)}.$$

Extending the operators $\theta_n, \overline{\theta}_n$ by linearity, we can suppose that they define right and left actions of the group algebra $\mathbb{R}[\Sigma_n]$ on $Ass_n(\mathcal{S})$ for every $n$.

The commutative shuffle product in $Ass(\mathcal{S})$ mentioned in the Introduction is expressed by the shuffle product of the permutations introduced here and the representation $\overline{\theta}_n$ by the formula

$$x ш y \stackrel{\text{def}}{=} \overline{\theta}_{m+n}(\mathbb{I}_m ш \mathbb{I}_n)(xy) \; \forall x \in Ass_m(\mathcal{S}), y \in Ass_n(\mathcal{S}). \tag{4}$$

The following identity follows directly from the definitions and from the fact that $\overline{\theta}_n$ is a left action on the group algebra:

$$x_1 ш \ldots ш x_k = \overline{\theta}_{n_1 + \cdots + n_k}(\mathbb{I}_{n_1} ш \ldots ш \mathbb{I}_{n_k}) \, x_1 \ldots x_k, \; \forall x_i \in Ass_{n_i}(\mathcal{S}), \; i = 1, \ldots, k.$$

The mapping

$$\mathbb{I}_{n_1} \text{Ш} \dots \text{Ш} \mathbb{I}_{n_k} \mapsto \overline{\theta}_{n_1+\dots+n_k}(\mathbb{I}_{n_1} \text{Ш} \dots \text{Ш} \mathbb{I}_{n_k}) x_{n_1} \dots x_{n_k}$$

defines a homomorphism of the algebra **Ш** into the commutative algebra with the multiplication defined by (4). A much reacher collection of representations of the algebra **Ш** is given by the following construction.

Let $L$ be a real Lie algebra and $U_L$ — its universal enveloping (associative) algebra. Furthermore, let $X_t$, $t \in [0,1]$, be a summable vector function with values in $L$. Define a linear mapping $X_* : \mathbb{R}[\Sigma] \to U_L$ by the formula

$$X_*(\nu) = \int \dots \int_{\Delta^n} \theta_n(\nu) X_{t_1} \dots X_{t_n} dt_1 \dots dt_n, \ \forall \nu \in \Sigma_n, \ n = 1, 2, \dots .$$

The equality (3) implies

$$X_*(\nu_1 \text{Ш} \nu_2) = X_*(\nu_1) X_*(\nu_2), \ \forall \nu_1, \nu_2 \in \mathbb{R}[\Sigma],$$

hence, $X_*$ defines a homomorphism of the algebra **Ш** into $U_L$.

Note that the homomorphism $X_*$ transforms the formal power series in $\varepsilon$ $\sum_{n=0}^{\infty} \varepsilon^n \mathbb{I}_n$ into the Volterra series generated by $X_t$ :

$$\sum_{n=0}^{\infty} \varepsilon^n X_*(\mathbb{I}_n) = 1 + \sum_{n=1}^{\infty} \varepsilon^n \int \dots \int_{\Delta^n} X_{t_1} \dots X_{t_n} dt_1 \dots dt_n.$$

Suppose that an additive basis of $L$ is given, $\mathcal{X} \subset L$. The linear space $Ass(X)$ with the commutative shuffle multiplication (4) is a commutative algebra which we denote by $Sh(\mathcal{X})$. We extend the field of scalars of $L$ by considering the Lie algebra $Sh(\mathcal{X}) \otimes L$ over the commutative ring of scalars $Sh(\mathcal{X})$. In case of a finite set $\mathcal{X} = \{X_1, \dots, X_r\}$ every homomorphism $X_*$ could be factored through a certain universal homomorphism

$$\mathbf{I}_{\mathcal{X}} : \text{Ш} \to Sh(\mathcal{X}) \otimes L.$$

Indeed, we can put

$$\mathbf{I}_{\mathcal{X}}(\mathbb{I}_n) = \sum_{i_1, \dots, i_n = 1}^{r} X_{i_1} \dots X_{i_n} \otimes X_{i_1} \dots X_{i_n}, \ n = 1, 2, \dots . \qquad (5).$$

In the next section we shall show that the algebra **Ш** is freely generated by the elements $\mathbb{I}_n$, $n = 1, 2, \dots$, hence, since $\mathbf{I}_{\mathcal{X}}$ is defined by (5) on the generators, it can be extended to a homomorphism of **Ш** into $Sh(\mathcal{X}) \otimes L$.

Suppose $X(t) = \sum_{i=1}^{r} u_i(t) X_i$, $t \in [0,1]$, where $u_i(t)$ are summable real valued functions. Then

$$X_*(\mathbb{I}_n) = \sum_{i_1, \dots, i_n = 1}^{r} \int \dots \int_{\Delta^n} u_{i_1}(t_1) \dots u_{i_n}(t_n) dt_1 \dots dt_n X_{i_1} \dots X_{i_n}.$$

At the same time, the mapping

$$U : X_{i_1} \dots X_{i_n} \mapsto \int \dots \int_{\Delta^n} u_{i_1}(t_1) \dots u_{i_n}(t_n) dt_1 \dots dt_n$$

defines a homomorphism of the algebra $Sh(\mathcal{X})$ into $\mathbb{R}$, hence, we have $X_* = U \circ \mathbf{I}_{\mathcal{X}}$. The described construction can be easily generalized to the case of an infinite $\mathcal{X}$, if certainly, $span\{X(t) | t \in [0,1]\}$ is finite dimensional.

## §3. The monotonicity type of permutations

**Definition.** Suppose $n > 0$, $\sigma \in \Sigma_n$. The *monotonicity type* of the permutation $\sigma$ is defined as the word $Mon(\sigma) = e_1 \dots e_{n-1}$ of the length $n-1$ in the alphabet of two letters $\{\alpha, \beta\}$, given by the rule

$$e_i = \begin{cases} \alpha, \text{if } \sigma(i) < \sigma(i+1) \\ \beta, \text{if } \sigma(i) > \sigma(i+1), \ i = 1, \dots, n-1. \end{cases}$$

Since there are $2^{n-1}$ words of length $n-1$ the classification according the monotonicity types partitions the group $\Sigma_n$, which consists of $n!$ elements, into $2^{n-1}$ subsets. It is easily seen that none of these subsets is empty. Moreover, for $n > 1$ only two of them contain only one element, namely the sets defined by the monotonicity types $\underbrace{\alpha \dots \alpha}_{n-1} = \alpha^{n-1}$ and $\underbrace{\beta \dots \beta}_{n-1} = \beta^{n-1}$ :

$$Mon(\sigma) = \alpha^{n-1} \iff \sigma = \mathbb{I}_n, \ Mon(\sigma) = \beta^{n-1} \iff \sigma = \mathbb{J}_n,$$

where by $\mathbb{J}_n$ we denote the inversion $\mathbb{J}_n(i) = n - i + 1$. For $n = 1$ we have $\mathbb{J}_1 = \mathbb{I}_1$, $Mon(\mathbb{I}_1)$ is an empty word.

Let $a$ be a monotonicity type, i.e. a word in the alphabet $\{\alpha, \beta\}$. We introduce the notation

$$iaj = \sum_{Mon(\sigma)=a} \sigma.$$

Thus, $iaj \in \mathbb{R}[\Sigma]$. The reasons for such a notation will be clear from our considerations below.

**Proposition 1.** *Let $a = e_1 \dots e_{n_1}$ be a monotonicity type, $e_i \in \{\alpha, \beta\}$, $i = 1, \dots, n-1$. Denote $T_a = \{i | e_i = \beta\}$ and suppose that $\# T_a = m$. Than*

$$iaj = \sum_{k=0}^{m} (-1)^{m-k} \sum_{\{i_1, \dots, i_k\} \subset T_a} \mathbb{I}_{i_1} ш \mathbb{I}_{i_2 - i_1} ш \dots ш \mathbb{I}_{i_k - i_{k-1}} ш \mathbb{I}_{n-i_k}. \tag{6}$$

**Proof.** According to the definition we have

$$\mathbb{I}_{i_1} ш \mathbb{I}_{i_2 - i_1} ш \dots ш \mathbb{I}_{n-i_k} = \sum_{\sigma \in S(i_1, i_2 - i_1, \dots, n - i_k)} \sigma. \tag{7}$$

Hence, the element (7) is a sum of all permutations with monotonicity type containing the letter $\alpha$ at all places not coinciding with $i_1, \dots, i_k$, (and at these places can stand either of the letters $\alpha$ or $\beta$.) Thus, the sum (6) contains every permutation of the monotonicity type $a$ with the coefficient 1, and every permutation of a monotonicity type containing $\beta$ at the places $j_1, \dots, j_r$, where $\{j_1, \dots, j_r\}$ is a proper subset of $T_a$, with the coefficient $\sum_{k=r}^{m} (-1)^{m-k} \binom{m-r}{k-r} = 0$. Indeed, $\{j_1, \dots, j_r\}$ is contained in exactly $\binom{m-r}{k-r}$ subsets of $T_a$ with $k$ elements.

**Proposition 2.** *The shuffle algebra* $\mathrm{Ш}$ *is a free associative graded algebra with one generator in every degree* $n = 1, 2, \ldots$ . *The elements of the form* $iaj$, *where* $a$ *is an arbitrary monotonicity type, constitute an additive basis of the algebra* $\mathrm{Ш}$.

**Proof.** Proposition 3 implies that $iaj \in \mathrm{Ш}_n$ for every word $a = e_1 \ldots e_{n-1}$, $e_i \in \{\alpha, \beta\}$. The elements $iaj$ for different words $a$ are linearly independent. Therefore, $dim\, \mathrm{Ш}_n \geq 2^{n-1}$. On the other hand, the algebra $\mathrm{Ш}$ is generated by the elements $\mathbb{I}_n$, $n = 1, 2, \ldots$ , hence can be represented as a factor–algebra of a free associative graded algebra with one generator in every degree $n = 1, 2, \ldots$ . It remains to remark that the dimension of the $n$–th component of such an algebra is equal to $2^{n-1}$.

Thus, the subspace $\mathrm{Ш} \subset \mathbb{R}[\Sigma]$ consists of all linear combinations of permutations such that permutations with identical monotonicity type enter in every linear combination with equal coefficients.

**Proposition 3.** *For arbitrary monotonicity types* $a, b$ *the following identity holds*

$$(iaj)\mathrm{ш}(ibj) = ia\alpha bj + ia\beta bj.$$

**Proof.** Suppose $iaj \in \mathrm{Ш}_n$, $ibj \in \mathrm{Ш}_m$, then

$$(iaj)\mathrm{ш}(ibj) = \sum_{\sigma \in S(n,m)} \sigma \circ (iaj \times ibj).$$

Since every permutation $\sigma \in S(n, m)$ is monotonously increasing on the segment $1, \ldots , n$ and separately on the segment $n+1, \ldots , n+m$, the monotonicity type $Mon(\sigma \circ (\nu \times \mu))$ of the permutation $\sigma \circ (\nu \times \mu)$ for every $\nu \in \Sigma_n$, $\mu \in \Sigma_m$ is equal to $Mon(\nu)\alpha Mon(\mu)$, or to $Mon(\nu)\beta Mon(\mu)$. Hence,

$$(iaj)\mathrm{ш}(ibj) = k_1(ia\alpha bj) + k_2(ia\beta bj),$$

where $k_1, k_2$ are nonnegative integers.

Consider the element $\overline{\sigma} \in S(n, m)$ defined by

$$\overline{\sigma}(i) = \begin{cases} i + m, & i \leq n \\ i - n, & i > n. \end{cases}$$

We have

$$Mon(\mathbb{I}_{n+m} \circ (\mu \times \nu)) = Mon(\nu \times \mu) = Mon(\nu)\alpha Mon(\mu) : \; Mon(\overline{\sigma} \circ (\nu \times \mu)) = Mon(\nu)\beta Mon(\mu).$$

Hence, $k_1 k_2 \neq 0$.

Furthermore, since $\sigma_1 \circ (\nu_1 \times \mu_1) = \sigma_2 \circ (\nu_2 \times \mu_2)$, we have for arbitrary $\nu_i \in \Sigma_n$, $\mu_i \in \Sigma_m$, $\sigma_i \in S(n, m)\, i = 1, 2$, that $\sigma_1 = \sigma_2$, $\nu_1 = \nu_2$, $\mu_1 = \mu_2$. Indeed, put $\nu = \nu_1 \circ \nu_2^{-1}$, $\mu = \mu_1 \circ \mu_2^{-1}$, then $\sigma_2 = \sigma_1 \circ (\nu \times \mu)$. Consequently,

$$\alpha^{n-1} * \alpha^{m-1} = Mon(\sigma_2) = Mon(\sigma_1 \circ (\nu \times \mu)) = Mon(\nu) * Mon(\mu),$$

where instead of $*$ we can put either $\alpha$ or $\beta$. Thus,

$$Mon(\nu) = \alpha^{n-1},\, Mon(\mu) = \alpha^{m-1} \Longrightarrow \nu = \mathbb{I}_n,\, \mu = \mathbb{I}_m,\, \nu \times \mu = \mathbb{I}_{n+m},\, \sigma_1 = \sigma_2,$$

hence, $k_1 = k_2 = 1$.

Denote by $\mathcal{A}$ the associative algebra with generators $\alpha, \beta, i, j$ and defining relations

$$ji = \alpha + \beta,\ i^2 = j^2 = \alpha i = \beta j = j\alpha = j\beta = 0.$$

We introduce in $\mathcal{A}$ a graduation by halfintegers $\mathcal{A} = \bigoplus_{k=0}^{\infty} \mathcal{A}_{\frac{k}{2}}$, assigning to the variables the following degrees:

$$deg\alpha = \deg\beta = 1,\ degi = degj = \frac{1}{2}.$$

Proposition 3 implies that $\text{Ш} \subset \bigoplus_{n=0}^{\infty} \mathcal{A}_n$, i.e. the graded algebra Ш is naturally identified with a subalgebra in $\mathcal{A}$, consisting of linear combinations of homogeneous elements of integer degrees. We shall constantly use this identification $\text{Ш} \subset \mathcal{A}$ in the sequel, hence, when representing the elements of the shuffle algebra as polynomials of (noncommutative) variables $\alpha, \beta, i, j$, the product symbol ш will be omitted.

## §4. The Lie algebra of variations

Denote by $\overline{\text{Ш}}$ the completion of the graded algebra $\text{Ш} = \bigoplus_{n=0}^{\infty} \text{Ш}_n$ in the topology of termwise convergence, i.e. the space of formal series $\sum_{n=0}^{\infty} x_n$, $x_n \in \text{Ш}_n$, $n = 0, 1, 2, \dots$, with the "Cauchy multiplication"

$$\Big(\sum_{n=0}^{\infty} x_n\Big) ш \Big(\sum_{n=0}^{\infty} y_n\Big) = \sum_{n=0}^{\infty}\Big(\sum_{k=0}^{n} x_k ш\, y_{n-k}\Big).$$

We shall consider one-parameter families of formal series, their differentiation, integration, etc. are carried out termwise. Put for every $\varepsilon \in \mathbb{R}$ : $\mathbb{I}(\varepsilon) = \sum_{n=0}^{\infty} \varepsilon^n \mathbb{I}_n$, $\mathbb{I}(\varepsilon) \in \overline{\text{Ш}}$. We have

$$\mathbb{I}(\varepsilon) = 1 + \varepsilon \sum_{n=0}^{\infty} \varepsilon^n i\alpha^n j = 1 + \varepsilon i(1 - \varepsilon\alpha)^{-1} j, \tag{8}$$

and the element $\mathbb{I}(\varepsilon)$ is, evidently, invertible in $\overline{\text{Ш}}$. Proposition 3 permits to compute explicitly the inverse element:

$$\mathbb{I}(\varepsilon)^{-1} = \sum_{n=0}^{\infty} (-\varepsilon)^n \mathbb{J}_n = 1 - \varepsilon \sum_{n=0}^{\infty} (-\varepsilon)^n i\beta^n j = 1 - \varepsilon i(1 + \varepsilon\beta)^{-1} j. \tag{9}$$

Indeed,

$$(1 + \varepsilon i(1 - \varepsilon\alpha)^{-1} j)(1 - \varepsilon i(1 + \varepsilon\beta)^{-1} j) = 1 + \varepsilon i((1 - \varepsilon\alpha)^{-1} -$$
$$-(1 + \varepsilon\beta)^{-1}) j - \varepsilon i(1 - \varepsilon\alpha)^{-1}(\varepsilon\alpha + \varepsilon\beta)(1 + \varepsilon\beta)^{-1} j = 1.$$

**Definition.** We denote by $\mathcal{V}$ the subgroup of invertible elements of $\overline{\text{Ш}}$, generated by the elements $\mathbb{I}(\varepsilon)$, $\varepsilon \in \mathbb{R}$, and call it the *group of variations*.

Thus, we have

$$\mathcal{V} = \{X_1(\varepsilon_1) ш \dots ш X_k(\varepsilon_k) | X_i \in \{\mathbb{I}, \mathbb{J}\},\ \varepsilon_i \in \mathbb{R},\ i = 1, \dots, k,\ k > 0\}.$$

If the group $\mathcal{V}$ were a Lie group, we could define the Lie algebra of variations as the Lie algebra of $\mathcal{V}$. Since this is not the case, we shall define the Lie algebra of variations by explicitly introducing its generators and show that the relation of the introduced Lie algebra to the group $\mathcal{V}$ is similar to that in the Lie theory.

For $x, y \in \text{Ш}$ we put $[x, y] = x \text{ш} y - y \text{ш} x$ — the "shuffle commutator" of $x, y$, and $(adx)y \overset{\text{def}}{=} [x, y]$. For $\mathcal{X} \subset \text{Ш}$ we have the Lie algebra generated by the elements of $\mathcal{X}$:

$$Lie\mathcal{X} = span\{(adx_1)\dots(adx_{k-1})x_k | x_i \in \mathcal{X},\ i = 1, \dots, k,\ k > 0\}.$$

We introduce three power series, defined through $\mathbb{I}(\varepsilon)$:

$$\lambda(\varepsilon) = ln\mathbb{I}(\varepsilon) = \sum_{n=1}^{\infty} \frac{(-1)^{n-1}}{n} \left( \sum_{k=1}^{\infty} \varepsilon^k \mathbb{I}_k \right)^{\text{Ш}n},$$

$$\omega(\varepsilon) = \left( \frac{d}{d\varepsilon} \mathbb{I}(\varepsilon) \right) \text{ш}\, \mathbb{I}(\varepsilon)^{-1}, \quad \widehat{\omega}(\varepsilon) = \mathbb{I}(\varepsilon)^{-1} \text{ш} \left( \frac{d}{d\varepsilon} \mathbb{I}(\varepsilon) \right).$$

Put

$$\lambda(\varepsilon) = \sum_{n=1}^{\infty} \varepsilon^n \lambda_n, \quad \omega(\varepsilon) = \sum_{n=1}^{\infty} \varepsilon^{n-1} \omega_n, \quad \widehat{\omega}(\varepsilon) = \sum_{n=1}^{\infty} \varepsilon^{n-1} \widehat{\omega}_n.$$

**Proposition 4 and Definition.** *The following relations are valid:*

$$Lie\{\lambda_n | n \geq 1\} = Lie\{\omega_n | n \geq 1\} = Lie\{\widehat{\omega}_n | n \geq 1\}. \tag{10}$$

*The Lie Algebra (10) (with the Lie bracket [,]) is called the Lie algebra of variations and is denoted by* $\mathbf{V}$.

Let $\overline{\mathbf{V}} \subset \overline{\text{Ш}}$ and $\overline{\mathcal{V}} \subset \overline{\text{Ш}}$ be closures of $\mathbf{V}$, and $\mathcal{V}$ in the topology of the termwise convergence.

**Proposition 5.** *The mappings*

$$v \mapsto e^v = \sum_{n=0}^{\infty} \frac{v^{\text{Ш}n}}{n!}, \quad v \in \overline{\mathbf{V}},$$

$$x \mapsto lnx = \sum_{n=1}^{\infty} \frac{(-1)^{n-1}}{n} (x-1)^{\text{Ш}n}, \quad x \in \overline{\mathcal{V}},$$

*are reciprocal bijections of* $\overline{\mathbf{V}}$ *and* $\overline{\mathcal{V}}$.

Propositions 4, 5 can be proved by standard Lie theory techniques and remain valid if Ш is replaced by an arbitrary graded associative algebra, and $\mathbb{I}(\varepsilon)$ — is replaced by an arbitrary power series starting with the unity. We shall omit the proofs.

**Proposition 6.** *The Lie algebra of variations* $\mathbf{V}$ *is a free Lie algebra. Each of the sets*

$$\{\lambda_i | i \geq 1\}, \quad \{\omega_i | \geq 1\}, \quad \{\widehat{\omega}_i | i \geq 1\}$$

*is the set of free generators of the Lie algebra* **V** *and of the associative algebra* **Ш**.

**Proof.** We give the proof for the set $\{\omega_i | i \geq 1\}$, the proofs for other two cases are similar. From the definition of $\omega(\varepsilon)$ we obtain

$$\mathbb{I}(\varepsilon) = 1 + \int_0^\varepsilon \omega(\tau) ш \mathbb{I}(\tau) d\tau,$$

hence,

$$\mathbb{I}_n = \frac{1}{n} \sum_{k=1}^{n} \omega_k ш \mathbb{I}_{n-k},\ n = 1, 2, \ldots .$$

In particular, $\mathbb{I}_1 = \omega_1, \mathbb{I}_2 = \frac{1}{2}(\omega_1 ш \omega_1 + \omega_2)$, etc. . By induction on $n$ we obtain that $\mathbb{I}_n$ is a shuffle polynomial in $\omega_k$, $k \leq n$. Hence, the elements $\omega_n$ generate the algebra **Ш**. Since it is a free algebra with one generator in every homogeneous component $\mathbf{Ш}_n$, the elements $\omega_n$, $n \geq 1$, are free generators of the associative algebra **Ш**. Standard results from the theory of Lie algebras imply that the elements $\omega_n$, $n \geq 1$ constitute free generators of the Lie subalgebra in **Ш**, which is generated by these elements.

**Corollary.** *The imbedding $V \subset$ **Ш** induces an isomorphism $U_V \cong$ **Ш** where $U_V$ is the universal enveloping algebra of the Lie algebra $V$.*

Indeed, the universal enveloping algebra of a free Lie algebra is a free associative algebra with the same set of generators.

An invariant definition of the Lie algebra of variations and its additional properties could be found in [2].

Now we give explicit expressions for the families of generators $\lambda_n$, $\omega_n$, $\widehat{\omega}_n$, $n \geq 1$, through the monotonicity types. For the monotonicity type $a = e_1 \ldots e_{n-1}$, $e_i \in \{\alpha, \beta\}$, we put

$$d_\alpha(a) = \#\{i || e_i = \alpha\},\ d_\beta(a) = \#\{i | e_i = \beta\}. \tag{11}$$

**Proposition 7.** *The following identities are valied:*

$$1)\ \lambda(\varepsilon) = ln(1 + i\varepsilon(1 - \varepsilon\alpha)^{-1} j),\ \lambda_n = \sum_a (-1)^{d_\beta(a)} \frac{d_\alpha(a)! d_\beta(a)!}{n!} iaj,\ n = 1, 2, \ldots ,$$

*where the summation is carried over all words $a$ consisting of $n-1$ letters of the alphabet $\{\alpha, \beta\}$;*

$$2) \omega(\varepsilon) = i(1 - \varepsilon\alpha)^{-1}(1 + \varepsilon\beta)^{-1} j,\ \omega_n = \sum_{k=1}^{n} (-1)^{k-1} i \alpha^{n-k} \beta^{k-1} j,\ n = 1, 2, \ldots ;$$

$$3) \widehat{\omega}(\varepsilon) = i(1 + \varepsilon\beta)^{-1}(1 - \varepsilon\alpha)^{-1} j,\ \widehat{\omega}_n = \sum_{k=1}^{n} (-1)^{k-1} i \beta^{k-1} \alpha^{n-k} j,\ n = 1, 2, \ldots .$$

**Proof.** 1) Formula for $\lambda(\varepsilon)$ follows from (8). An explicit expression for $\lambda_n$ can be obtained by a direct calculation, which will be shortly described here.

First of all, the (numerical) coefficient before $iaj$ in the series

$$\lambda(1) = ln(1 + \sum_{n=0}^{\infty} i\alpha^n j)$$

depends only on $d_\alpha(a)$ and $d_\beta(a)$. Suppose, for $d_\beta(a) = k$, $d_\alpha(a) = l$ it is equal to $\kappa(k,l)$. After reducing the terms we obtain the equality

$$\kappa(k,l) = \sum_{i=0}^{l} \frac{(-1)^{k+i}}{k+i+1}\binom{l}{i}$$

which can be given the form

$$\kappa(k,l) = \int_{-1}^{0} t^k(1+t)^l dt.$$

Integrating by parts, we obtain

$$\kappa(k,l) = -\frac{l}{k+1}\kappa(k+1,l-1),\ \kappa(k,0) = \frac{(-1)^k}{k+1}.$$

Hence,

$$\kappa(k,l) = \frac{(-1)^k}{(k+1)\binom{k+l+1}{l}} = (-1)^k\frac{k!l!}{(k+l+1)!}.$$

2) According to (9) we have $\mathbb{I}(\varepsilon)^{-1} = 1 - i\varepsilon(1+\varepsilon\beta)^{-1}j$. Hence,

$$\omega(\varepsilon) = \left(\frac{d}{d\varepsilon}(i\varepsilon(1-\varepsilon\alpha)^{-1}j)\right)(1 - i\varepsilon(1+\varepsilon\beta)^{-1}j) = i(1-\varepsilon\alpha)^{-2}j(1 - i\varepsilon(1+\varepsilon\beta)^{-1}j) =$$
$$= i(1-\varepsilon\alpha)^{-2}(1-(\varepsilon\alpha+\varepsilon\beta)(1+\varepsilon\beta)^{-1})j = i(1-\varepsilon\alpha)^{-1}(1+\varepsilon\beta)^{-1}j.$$

An explicit expression for $\omega_n$ is directly obtained from the formula for $\omega(\varepsilon)$. The identities 3) are proved similarly.

Let $L$ be a Lie algebra, $U_L$ — its universal enveloping algebra, $X_\tau$, $\tau \in [0,1]$ — a summable curve in $L$. We consider "left" and "right" formal Volterra series with coefficients in $U_L$:

$$\overleftarrow{exp}\int_0^1 \varepsilon X_\tau d\tau \stackrel{\text{def}}{=} 1 + \sum_{n=1}^{\infty} \varepsilon^n \int\cdots\int_{\Delta^n} X_{\tau_1}\ldots X_{\tau_n} d\tau_1\ldots d\tau_n,$$

$$\overrightarrow{exp}\int_0^1 \varepsilon X_\tau d\tau \stackrel{\text{def}}{=} 1 + \sum_{n=1}^{\infty} \varepsilon^n \int\cdots\int_{\Delta^n} X_{\tau_n}\ldots X_{\tau_1} d\tau_1\ldots d\tau_n.$$

In §2 the homomorphism $X_* : \mathbf{III} \longrightarrow U_L$ was introduced. Applying this homomorphism to the objects introduced in this section leads us to useful relations for Volterra series, which initially motivated us to introduce the algebraic structures considered here, cf. [2]. We have

$$X_*(\mathbb{I}(\varepsilon)) = \overleftarrow{exp}\int_0^1 \varepsilon X_\tau d\tau = e^{X_*(\lambda(\varepsilon))} = \overleftarrow{exp}\int_0^\varepsilon X_*(\omega(\theta))d\theta =$$
$$= \overrightarrow{exp}\int_0^\varepsilon X_*(\hat{\omega}(\theta))d\theta;\ X_*(\mathbb{I}(\varepsilon)^{-1}) = \overrightarrow{exp}\int_0^1 -\varepsilon X_\tau d\tau.$$

## §5. The main result — the multiplicativity relation

Here we establish the relation between the shuffle product "ш" and the standard (composition) product "∘" in the group algebras $\mathbb{R}[\Sigma]$ :

$$\mu \circ \nu : i \mapsto \mu(\nu(i)),\ i = 1, \dots, n,\ \forall \mu, \nu \in \Sigma_n. \tag{12}$$

We call the product (12) the *symmetry product* to distinguish it from the shuffle product. We extend the symmetry product to $\overline{\mathbb{R}[\Sigma]}$ by the relation $\mu \circ \nu = 0 \forall \mu \in \Sigma_m, \nu \in \Sigma_n,\ m \neq n$. Thus,

$$\left(\sum_{n=0}^{\infty} x_n\right) \circ \left(\sum_{n=0}^{\infty} y_n\right) \overset{\text{def}}{=} \sum_{n=0}^{\infty} x_n \circ y_n,\ x_n, y_n \in \Sigma_n,\ n = 0, 1, \dots .$$

**Theorem 1.** *For every $x \in \mathcal{V}$ and arbitrary $\mu, \nu \in \overline{\mathbb{R}[\Sigma]}$ the following multiplicativity relation holds*

$$x \circ (\mu ш \nu) = (x \circ \mu) ш (x \circ \nu).$$

Before proving the theorem, we derive some of its consequences. A remarkable consequence is the fact, that the symmetry product of two elements of the shuffle algebra is an element of the shuffle algebra. Moreover, the following Proposition explicitly expresses the symmetry product in the shuffle algebra.

**Proposition 8.** *For arbitrary integers $n, m; i_1, \dots, i_n; j_1, \dots, j_m$, the following identity holds:*

$$(\mathbb{I}_{i_1} ш \dots ш \mathbb{I}_{i_n}) \circ (\mathbb{I}_{j_1} ш \dots ш \mathbb{I}_{j_m}) = \sum_K \mathbb{I}_{k_{11}} ш \dots ш \mathbb{I}_{k_{n1}} ш \mathbb{I}_{k_{12}} ш \dots ш \mathbb{I}_{k_{n2}} ш \dots ш \mathbb{I}_{k_{1m}} ш \dots ш \mathbb{I}_{k_{nm}},$$

*where the summation is over all matrices*

$$K = \begin{pmatrix} k_{11} & \dots & k_{1m} \\ \vdots & \dots & \vdots \\ k_{n1} & \dots & k_{nm} \end{pmatrix},$$

*subject to the conditions*

$$\sum_{i=1}^{n} k_{il} = j_l,\ l = 1, \dots, m;\ \sum_{j=1}^{m} k_{lj} = i_l,\ l = 1, \dots, n.$$

**Proof.** Theorem 1 implies

$$(\mathbb{I}(t_1) ш \dots ш \mathbb{I}(t_n)) \circ (\mathbb{I}_{j_1} ш \dots ш \mathbb{I}_{j_m}) =$$
$$= ((\mathbb{I}(t_1) ш \dots ш \mathbb{I}(t_n)) \circ \mathbb{I}_{j_1}) ш \dots ш ((\mathbb{I}(t_1) ш \dots ш \mathbb{I}(t_n)) \circ \mathbb{I}_{j_m}).$$

Equating coefficients in the left– and right–handsides we obtain the desired equality.

The direct sum $\mathbf{Ш}_+ = \bigoplus_{n=1}^{\infty} \mathbf{Ш}_n$ is a maximal ideal in the shuffle algebra, hence, $\mathbf{Ш} = \mathbb{R} \oplus \mathbf{Ш}_+$. Let $\mathbf{Ш}_+^k = \underbrace{\mathbf{Ш}_+ ш \dots ш \mathbf{Ш}_+}_{k \text{ times}}$ be the linear hull of the elements in $\mathbf{Ш}_+$, which are representable as products of at least $k$ factors. From the obtained table for the symmetry product directly follows the following

**Corollary.** *For every integer $k > 0$ the inclusions*

$$\mathrm{III}_+ \circ \mathrm{III}_+^k \subset \mathrm{III}_+^k, \ \mathrm{III}_+^k \circ \mathrm{III}_+ \subset \mathrm{III}_+^k. \tag{13}$$

*hold.*

Observe that $\forall n > 0$ the space $\mathrm{III}_n$ is an algebra relative to the symmetry product — an associative subalgebra of the group algebra $\mathbb{R}[\Sigma_n]$.

**Definition.** We call the space $\mathrm{III}_n$ with the symmetry product "$\circ$" the *algebra of types of monotonicity* of permutations of order $n$ and denote it by $\mathrm{III}_n^\circ$.

We collect the properties of the monotonicity type algebras, already proved, in the following

**Proposition 9.** *For every integer $n > 0$ the following relations hold:*
*a) $dim\mathrm{III}_n^\circ = 2^{n-1}$;*
*b) $\mathbb{I}_n$ is the unity in the algebra $\mathrm{III}_n^\circ$:*
*c) The subspace $\mathrm{III}_+^k \cap \mathrm{III}_n$ is a two-sided ideal in the algebra $\mathrm{III}_n^\circ$ of dimension $\sum_{i=k}^{n} \binom{n-1}{i-1}$ for $k = 1, 2, \ldots$.*

**Proof.** a) follows from Proposition 2, b) is evident, c) is contained in the corollary of Theorem 1 and in the Proposition 2.

**Proposition 10 and Definition.** *Every element $x \in \mathrm{III}_n$ is uniquely represented as*

$$x = \rho_n(x)\mathbb{I}_n + y, \ \rho_n(x) \in \mathbb{R}, y \in \mathrm{III}_+^2.$$

*The defined mapping $\rho_n : \mathrm{III}_n^\circ \longrightarrow \mathbb{R}$ is a homomorphism of the monotonicity types algebra into $\mathbb{R}$. Additionally, we have $\rho_n(iaj) = (-1)^{d_\beta a}$ for every monotonicity type $a$, where, as above, $d_\beta(a)$ is the number of the letters $\beta$ in the word $a$.*

**Proof.** The homomorphism $\rho_n$ is well defined since, according to the Proposition 9, the subspace $\mathrm{III}_+^2 \cap \mathrm{III}_n$ is an ideal of codimension 1 in $\mathrm{III}_n^\circ$, and $\mathbb{I}_n$ is the unity in $\mathrm{III}_n^\circ$. An explicit dependence of $\rho_n$ from the monotonicity type directly follows from (6).

Remark that the functional $\rho_n$, defined on the monotonicity types of permutations is very similar to the parity of the individual permutations, which defines a unique nontrivial homomorphism of $\mathbb{R}[\Sigma_n]$ into $\mathbb{R}$. We remind that the letter $\beta$ denotes the places where a permutation of the given monotonicity type decreases(i.e. has a disorder). At the same time $\rho_n$ does not coincide with the parity: for example, for an inversion $\mathbb{J}_n = i\beta^{n-1}j$ with parity $(-1)^{\frac{n(n-1)}{2}}$ we have $\rho_n(\mathbb{J}_n) = (-1)^{n-1}$. Indeed, to compute $\rho_n$ one should not take into account all disorders, but only those with neighboring values of the argument. The nature of the homomorphisms $\rho_n$ is clearly seen from the following Proposition, which could be considered as an infinitesimal version of Theorem 1. (We remind that $V$ denotes the Lie algebra of variations).

**Proposition 11.** *For all $v \in V \cap \mathrm{III}_n$, $x \in \mathrm{III}_n$ the relation $v \circ x = \rho_n(x)v$ holds.*

**Proof.** Since $v \circ \mathbb{I}_n = v$ the identity we have to prove is equivalent to the relation $v \circ \mathrm{III}_+^2 = 0$. Proposition 5 implies that for $\forall t \in \mathbb{R}$ the series $e^{tv} = 1 + \sum_{n=1}^{\infty} \frac{t^n}{n} v^{\mathrm{III}n}$ belongs to $\overline{V}$. Theorem 1 implies the identity

$$e^{tv} \circ (x_1 \mathrm{III}\, x_2) = (e^{tv} \circ x_1) \mathrm{III}\, (e^{tv} \circ x_2) \forall x_1, x_2 \in \mathrm{III}_+.$$

Differentiating it by $t$ for $t = 0$ and remembering that the symmetry product of two permutations of different degrees is equal to zero, we obtain

$$v \circ (x_1 ш x_2) = (v \circ x_1) ш (1 \circ x_2) + (1 \circ x_1) ш (v \circ x_2) = 0.$$

It turns out that the last statement can be inverted and we come to the following characterization of $V$ in terms of the symmetry product:

*Let $w \in Ш_+$, then*

$$w \in V \Longleftrightarrow w \circ (x_1 ш x_2) = 0 \, \forall x_1, x_2 \in Ш_+.$$

We shall prove this assertion in a more general setting, as Proposition 12 below.

**Corollary** (to Proposition 11). *We have*

$$v \circ v = \rho_n(v) v \; \forall v \in V \cap Ш_n.$$

*In particular, for $\rho(v) = 1$ $v$ is an idempotent of the algebra $\mathbb{R}[\Sigma]$. For the sequences of generators of the Lie algebra of variations; described in §4, we obtain*

$$\rho_n(\lambda_n) = 1, \; \rho_n(\omega_n) = \rho_n(\widehat{\omega}_n) = n, \; i = 1, 2, \ldots .$$

Put

$$V^n = span\{v_1 ш \ldots ш v_k | v_i \in V, \; i = 1, \ldots, k, \; k \leq n\}, \; n = 1, 2, \ldots .$$

Since Ш is the universal enveloping algebra for the Lie algebra $V$ the subspaces $V^n$ constitute an increasing filtration in $Ш_+$, (whereas the subspaces $Ш_+^k$ form a decreasing filtration).

**Proposition 12.** *1) The following relation holds*

$$V^n = \{x \in Ш_+ | x \circ Ш_+^{n+1} = 0\}.$$

*2) Let $v_1, \ldots, v_n \in V$ be homogeneous elements of the Lie algebra $V$. For every nonempty subset $T = \{i_1, \ldots, i_k\} \subset \{1, \ldots, n\}$, $i_1 < \ldots < i_k$, put $v_T = v_{i_1} ш \ldots ш v_{i_k}$. Then*

$$(v_1 ш \ldots ш v_n) \circ (\mathbb{I}_{i_1} ш \ldots ш \mathbb{I}_{i_k}) = \sum v_{T_1} ш \ldots ш v_{T_k}, \tag{14}$$

*where the summation is over all partitions of $\{1, \ldots, n\}$ into $k$ subsets $T_j$ such that $v_{T_j} \in Ш_{i_j}$, $j = 1, \ldots, k$.*

**Proof.** We start with 2). The relation (14) generalizes Proposition 11. The same could be said about the proof. For arbitrary $t_1, \ldots, t_n \in \mathbb{R}$ the element $e^{t_1 v_1} ш \ldots ш e^{t_n v_n}$ belongs to $\overline{V}$. Hence,

$$\begin{aligned}
&\left(e^{t_1 v_1} ш \ldots ш e^{t_n v_n}\right) \circ (\mathbb{I}_{i_1} ш \ldots ш \mathbb{I}_{i_k}) = \\
&= \left(\left(e^{t_1 v_1} ш \ldots ш e^{t_n v_n}\right) \circ \mathbb{I}_{i_1}\right) ш \ldots ш \left(\left(e^{t_1 v_1} ш \ldots ш e^{t_n v_n}\right) \circ \mathbb{I}_{i_k}\right).
\end{aligned}$$

Equating the coefficients before $t_1 \ldots t_n$ in left– and right–handsides of this relation we obtain (14).

Now we turn to the relation 1). First of all, (14) implies $V^n \circ Ш_+^{n+1} = 0$ We have only to prove that $x \circ Ш_+^{n+1} \neq 0$, if $x \notin V^n$.

Let $v_1, v_2, \ldots$ be an additive basis of $V$, consisting of homogeneous elements, and linearly ordered in such a way that the degrees of the elements are monotonically not decreasing. The Poincaré–Birkhoff–Witt theorem implies that the elements $v_{i_1} ш \ldots ш v_{i_m}$, $i_1 \leq \ldots \leq i_m$, $m \leq k$, constitute a basis of the space $V^k \; \forall k > 0$. Union over all $k$'s yields an additive basis of $\mathbf{Ш}$. For every $x \in \mathbf{Ш}$ we denote by $x_{i_1 \ldots i_m}$ the coefficient at $v_{i_1} ш \ldots ш v_{i_m}$ in the representation according to this basis.

Let $x \in V^k \setminus V^{k-1}$; we call the *principal term in the development of $x$ according to this basis* the element $x_{j_1 \ldots j_k} v_{j_1} ш \ldots ш v_{j_k}$, where the sequence $j_1, \ldots, j_k$ is lexicographically minimal among all sequences $i_1, \ldots, i_k$ such that $x_{i_1 \ldots i_k} \neq 0$.

Let $v_{j_\iota} \in \mathbf{Ш}_{r_\iota}$, $\iota = 1, \ldots, k$, then, according to (14), the principal term in the development of the element $x \circ (\mathbb{I}_{r_1} ш \ldots ш \mathbb{I}_{r_k})$ is equal to

$$x_{j_1 \ldots j_k} l_1! \ldots l_m! v_{j_1} ш \ldots ш v_{j_k} \neq 0,$$

where $l_1, \ldots, l_m$ denote the lengths of the constancy intervals of the non–decreasing integer–valued function $\iota \mapsto r_\iota$, $\iota = 1, \ldots, k$.

**Corollary.** *The subspaces $V^k \cap \mathbf{Ш}_n$ are two–sided ideals in $\mathbf{Ш}_n^\circ \; \forall k, n > 0$:*

$$V^k \circ \mathbf{Ш}_+ = \mathbf{Ш}_+ \circ V^k = V^k.$$

**Proof of Theorem 1.** Let $x = \sum_{k=1}^{\infty} x_k$ be an element of the group $\mathcal{V}$, $x_k \in \mathbf{Ш}_k, \forall k \geq 0, \mu \in \Sigma_m, \nu \in \Sigma_n$. We have to prove the relation

$$x_{m+n} \circ (\mu ш \nu) = (x_m \circ \mu) ш (x_n \circ \nu).$$

It is sufficient to prove this relation for the case $\mu = \mathbb{I}_m, \nu = \mathbb{I}_n$ since

$$x_{n+m} \circ (\mu ш \nu) = x_{n+m} \circ (\mathbb{I}_m ш \mathbb{I}_n) \circ (\mu \times \nu)$$

and

$$(x_m \circ \mu) ш (x_n \circ \nu) = (\mathbb{I}_m ш \mathbb{I}_n) \circ (x_m \circ \mu \times x_n \circ \nu) = (\mathbb{I}_m ш \mathbb{I}_n) \circ (x_m \times x_n) \circ (\mu \times \nu) = (x_m ш x_n) \circ (\mu \times \nu).$$

Thus, Theorem 1 is equivalent to the relations

$$x_{n+m} \circ (\mathbb{I}_m ш \mathbb{I}_n) = x_m ш x_n \; \forall m, n \geq 0, \; \left( \sum_{k=0}^{\infty} x_k \right) \in \mathcal{V} \tag{15}$$

We remind that the group $\mathcal{V}$ is generated by the series $\mathbb{I}(t) = \sum_{n=0}^{\infty} t^n \mathbb{I}_n$, $t \in \mathbb{R}$, where $\mathbb{I}(t)^{-1} = \sum_{n=0}^{\infty} (-t)^n \mathbb{J}_n$, $\mathbb{I}_0 = \mathbb{J}_0 = 1$. Thus, an arbitrary element $x = \left( \sum_{i=0}^{\infty} x_n \right) \in \mathcal{V}$ is represented as

$$x = \left( \sum_{i=0}^{\infty} t_0^i a_i^0 \right) ш \left( \sum_{i=0}^{\infty} t_1^i a_i^1 \right) ш \ldots ш \left( \sum_{i=0}^{\infty} t_k^i a_i^k \right), \tag{16}$$

where for every $j = 0, 1, \dots, k$ the sequence $\{a_j^i\}_{i=0}^{\infty}$ either coincides with the sequence $\{\mathbb{I}_i\}_{i=0}^{\infty}$ or with the sequence $\{\mathbb{J}_i\}_{i=0}^{\infty}$; $t_0, t_1, \dots, t_k$ are arbitrary real numbers, $k$ — an arbitrary nonnegative integer.

The prove of the basic relation (15) given below makes use of an artificial auxiliary construction. In this respect the prove of purely combinatorial relation (15) seems to be unsatisfactory. Let $\mathfrak{A}$ be an associative algebra with $n+m$ generators $s_1, \dots, s_{n+m}$, defined by the following condition: every (noncommutative) monomial of $s_1, \dots, s_{n+m}$ of degree not less than two in at least one variable is zero. The monomials $s_{i_1} \dots s_{i_k}$, $i_j \neq i_{j'}$ for $j \neq j', k \leq n+m$, form an additive basis of $\mathfrak{A}$. (One should not confuse $\mathfrak{A}$ with the exterior algebra!)

As in a free associative algebra, in $\mathfrak{A}$ a right action $\theta$ on the polynomials of degree $k$ in $s_1, \dots, s_{n+m}$ is defined:

$$\theta(\sigma)s_{i_1} \dots s_{i_k} = s_{i_{\sigma(1)}} \dots s_{i_{\sigma(k)}} \,\forall \sigma \in \Sigma_k, k > 0;\; \theta(\sigma \circ \nu) = \theta(\nu) \circ \theta(\sigma).$$

Let $\eta = \{\eta_1, \eta_2, \dots\}$ be a sequence of monomials of first degree of the algebra $\mathfrak{A}$ of which only finite numbers could be nonzero. Put $\Delta_k(\eta) = \sum_{i_1 < \dots < i_k} \eta_{i_1} \dots \eta_{i_k}$. Among all sequences of monomials of first degree we distinguish a "standard" sequence of the form $\mathrm{s} = \{s_1, \dots, s_{n+m}, 0, 0, \dots\}$. We have $\Delta_{n+m}(\mathrm{s}) = s_1 \dots s_{n+m}$.

**Lemma.** *For every admissible sequence $\eta$ and arbitrary $k, l \geq 0$ the following identity is valid*

$$\Delta_k(\eta)\Delta_l(\eta) = \theta(\mathbb{I}_k ш \mathbb{I}_l)\Delta_{k+l}(\eta). \tag{17}$$

**Proof.** Since $\eta_{i_1} \dots \eta_{i_{k+l}} = 0$ for $i_j = i_{j'}$, $j \neq j'$, we obtain

$$\Delta_k(\eta)\Delta_l(\eta) = \sum_{\substack{i_1 < \dots i_k \\ i_{k+1} < \dots < i_{k+l}}} \eta_{i_1} \dots \eta_{i_{k+l}} = \sum_{\sigma \in S(k,l)} \sum_{j_1 < \dots < j_{k+l}} \eta_{j_{\sigma(1)}} \dots \eta_{j_{\sigma(k+l)}} = \theta(\mathbb{I}_k ш \mathbb{I}_l)\Delta_{k+l}(\eta).$$

Let $x \in \mathcal{V}$ be of the form (16), then

$$x_{n+m} = \sum_{i_0 + \dots + i_k = n+m} t_0^{i_0} a_{j_0}^0 ш\, t_1^{i_1} a_{i_1}^1 ш \dots ш\, t_k^{i_k} a_{i_k}^k.$$

By means of the element $x$ we construct a special sequence $\eta = \eta(x)$ according to the relations

$$\eta_{j(n+m)+i} = t_j s_{a_{n+m}^j(i)},\; j = 0, 1, \dots, k,\; i = 1, \dots, n+m;\; \eta_l = 0 \text{ for } l > (k+1)(n+m).$$

The following identity holds

$$\Delta_l(\eta(x)) = \theta(x_l)\Delta_l(\mathrm{s}) \,\forall l > 0. \tag{18}$$

Indeed,

$$\Delta_l(\eta(x)) = \sum_{j_1 < \dots < j_l} \eta_{j_1} \dots \eta_{j_l} = \sum_{i_0 + \dots + i_k = l} (t_0^{i_0} \theta(a_{i_0}^0)\Delta_{i_0}(\mathrm{s})) \dots (t_k^{i_k} \theta(a_{i_k}^k)\Delta_{i_k}(\mathrm{s})) =$$

$$= \sum_{i_0 + \dots + i_k = l} t_0^{i_0} \dots t_k^{i_k} \theta(a_{i_0}^0 \times \dots \times a_{i_k}^k)\Delta_{i_0}(\mathrm{s}) \dots \Delta_{i_k}(\mathrm{s}).$$

At the same time, according to (17), we have

$$\theta(a^0_{i_0}\times\ldots\times a^k_{i_k})\Delta_{i_0}(\mathrm{s})\ldots\Delta_{i_k}(\mathrm{s})=\theta(a^0_{i_0}\times\ldots\times a^k_{i_k})\circ\theta(\mathbb{I}_{i_0}ш\ldotsш\mathbb{I}_{i_k})\Delta_l(\mathrm{s})=$$
$$\theta\left((\mathbb{I}_{i_0}ш\ldotsш\mathbb{I}_{i_k})\circ(a^0_{i_0}\times\ldots\times a^k_{i_k})\right)\Delta_l(\mathrm{s})=\theta(a^0_{i_0}ш\ldotsш a^k_{i_k})\Delta_l(\mathrm{s}).$$

Hence,

$$\Delta_l(\eta(x))=\sum_{i_1+\ldots i_k=l}t_0^{i_0}\ldots t_k^{i_k}\theta(a^0_{i_0}ш\ldotsш a^k_{i_k})\Delta_l(\mathrm{s})=\theta(x_l)\Delta_l(\mathrm{s}).$$

The relations (17) and (18) easily imply (15):

$$\theta(x_{n+m}\circ(\mathbb{I}_mш\mathbb{I}_n))\Delta_{n+m}(\mathrm{s})=\theta(\mathbb{I}_mш\mathbb{I}_n)\circ\theta(x_{n+m})\Delta_{n+m}(\mathrm{s})=\theta(\mathbb{I}_mш\mathbb{I}_n)\Delta_{n+m}(\eta(x))=$$
$$\Delta_m(\eta(x))\Delta_n(\eta(x))=(\theta(x_m)\Delta_m(\mathrm{s}))(\theta(x_n)\Delta_n(\mathrm{s}))=\theta(x_m\times x_n)\Delta_m(\mathrm{s})\Delta_n(\mathrm{s})=$$
$$=\theta(x_m\times x_n)\circ\theta(\mathbb{I}_mш\mathbb{I}_n)\Delta_{n+m}(\mathrm{s})=\theta((\mathbb{I}_mш\mathbb{I}_n)\circ(x_m\times x_n))\Delta_{m+n}(\mathrm{s})=\theta(x_mш x_n)\Delta_{n+m}(\mathrm{s}).$$

Since $\Delta_{n+m}(\mathrm{s})=s_1\ldots s_{n+m}$ we obtain $x_{n+m}\circ(\mathbb{I}_mш\mathbb{I}_n)=x_mш x_n$.

## §6. Symmetries of the Lie polynomials.

Let $S$ be a set, $Ass(S)=\sum_{n=0}^{\infty}Ass_n(S)$ — a free associative algebra with the set $S$ of free generators, and let $Lie(S)=\sum_{n=1}^{\infty}Lie_n(S)$ be a free Lie algebra with the same set of generators, realized as a subspace in $Ass(S)$ and consisting of all commutator (Lie) polynomials in the variables from the set $S$. Thus,

$$Ass_n(S)=span\{s_1\ldots s_n|s_i\in S,\ i=1,\ldots,n\},$$
$$Lie_n(S)=span\{[s_1,[s_2,\ldots,s_n]\ldots]|s_i\in S,\ i=1,\ldots,n\},\ [x,y]=xy-yx\,\forall x,y\in Ass(S).$$

By $\theta_n$ we denote the standard right action of the group algebra $\mathbb{R}[\Sigma_n]$ on the space $Ass_n(S)$, (cf. §2).

Finally, $\rho_n:\mathrm{III}_n^{\circ}\longrightarrow\mathbb{R}$ — is the homomorphism introduced in the previous section,( cf. Proposition 10).

**Theorem 2.** *For every integer $n>0$ the following relations are valid:*

1) $\theta_n(v)Ass_n(S)\subset Lie_n(S)\ \forall v\in V\cap\mathrm{III}_n$;
2) $\theta_n(x)q=\rho_n(x)q\ \forall x\in\mathrm{III}_n, q\in Lie_n(S)$.

Theorem 2 implies that for every $v\in V\cap\mathrm{III}_n$, which satisfies the relation $\rho(v)=1$, the mapping $\theta_n(v)$ is a projector $Ass_n(S)$ onto the subspace $Lie_n(S)$. Before proving this theorem in full generality, we shall show that at least one of the operators $\theta_n(v)$ coincides with a well known projector of this sort.

We remind that $\omega_n=\sum_{k=1}^{n}(-1)^{k-1}i\alpha^{n-k}\beta^{k-1}j$ belongs to $V\cap\mathrm{III}_n$, and $\rho(\omega_n)=n$.

**Proposition 13.** *For arbitrary $s_i \in S$, $i = 1, \dots, n$, the following relation is valid*

$$\theta_n(\omega_n)s_1 s_2 \dots s_n = (ads_1)\dots(ads_{n-1})s_n. \tag{19}$$

**Proof.** The equation (19) can be proved by a direct combinatorial consideration, if an explicit expression for $\omega_n$ through monotonicity types is used. We shall choose a more convenient way exploiting the elements of chronological calculus and working with the whole series $\omega(\epsilon) = \sum_{n=1}^{\infty} \varepsilon^{n-1}\omega_n$.

Let $X_t \in spanS$, $t \in [0,1]$, and suppose that $dimspan\{X_t | t \in [0,1]\} < \infty$ and that the vector–function $t \mapsto X_t$ is summable. The vector–function $X_t$ defines a homomorphism $X_* : \text{Ш} \longrightarrow Ass(S)$ according to the formula, (cf. §2):

$$X_*(\nu) = \int \cdots \int_{\Delta^n} \theta(\nu) X_{t_1} \dots X_{t_n} dt_1 \dots dt_n, \; \forall \nu \in \mathbb{R}[\Sigma_n].$$

We have

$$X_*(\mathbb{I}(\varepsilon)) = \overleftarrow{\exp} \int_0^1 \varepsilon X_t dt,$$

hence,

$$X_*(\omega(\varepsilon)) = X_* \left( \frac{d}{d\varepsilon} \mathbb{I}(\varepsilon) \text{ш} \, \mathbb{I}(\varepsilon)^{-1} \right) = \left( \frac{d}{d\varepsilon} \overleftarrow{exp} \int_0^1 \varepsilon X_t dt \right) \circ \left( \overleftarrow{exp} \int_0^1 \varepsilon X_t dt \right)^{-1}.$$

The differentiation formula of the chronological exponent with respect to a parameter, cf. [1]–[3], gives

$$X_*(\omega(\varepsilon)) = \int_0^1 \overleftarrow{exp} \int_t^1 \varepsilon ad X_\tau d\tau X_t dt.$$

Equating the coefficients at $\varepsilon^{n-1}$, we obtain

$$\int \cdots \int_{\Delta^n} \theta(\omega_n) X_{t_1} \dots X_{t_n} dt_1 \dots dt_n = \int \cdots \int_{\Delta^n} ad X_{t_1} \dots ad X_{t_{n-1}} X_{t_n} dt_1 \dots dt_n. \tag{20}$$

Since (20) is valid for an arbitrary vector–function $X_t$ we easily obtain (19).

**Proof of Theorem 2.** 1) Proposition 13 and standard facts about free Lie algebras yield that $\frac{1}{n}\theta_n(\omega_n)$ is a projector from $Ass_n(S)$ onto $Lie_n(S)$. Furthermore, Proposition 11 implies that for every $v \in V$ we have $\theta_n(v) = \theta_n(v \circ \omega_n \frac{1}{n}) = \frac{1}{n}\theta_n(\omega_n) \circ \theta_n(v)$, hence,

$$\theta_n(v) Ass_n(S) \subset \theta_n(\omega_n) Ass_n(S) = Lie_n(S).$$

2) Let $x \in \text{Ш}_n$, $q \in Lie_n(S)$. Using again Proposition 11, we obtain

$$\theta_n(x)q = \theta_n(x) \circ \theta_n(\frac{1}{n}\omega_n)q = \frac{1}{n}\theta_n(\omega_n \circ x)q = \frac{\rho(x)}{n}\theta_n(\omega_n)q = \rho(x)q.$$

Recalling that $dim\text{Ш}_n = 2^{n-1}$. Thus, assertion 2) of the Theorem describes $2^{n-1}$ independent symmetry relations which should be satisfied by the Lie polynomials of the $n$ degree. Assertion 1) gives those among these relations which are satisfied only by Lie polynomials. Expressing the relations trough the monotonicity types we come to the following

**Corollary.** *Let $q \in Lie_n(S)$ and suppose that $a$ is a monotonicity type of permutations of order $n$. Then*

$$\sum_{\{\sigma | Mon(\sigma)=a\}} \theta(\sigma)q = (-1)^{d_\beta a} q,$$

*where, as above, $d_\beta a$ is the number of letters $\beta$ in the word $a$ or, what is the same, is the number of places where a permutation of the monotonicity type $a$ decreases.*

We conclude the article with an additional characterization of the Lie algebra of variations $V$. Initially it was defined in terms of the shuffle product as the "Lie algebra of the group generated by the series $\mathbb{I}(\varepsilon), \varepsilon \in \mathbb{R}$". Then we obtained that, cf. Proposition 12, for every $n > 0$ the homogeneous component $V \cap \text{Ш}_n$ is a left annulator of the maximal ideal $(\text{Ш}_+ ш \text{Ш}_+) \cap \text{Ш}_n$ of the algebra of monotonicity types $\text{Ш}_n^\circ$ with the symmetry product.

Put

$$Lie_n = \{y \in \mathbb{R}[\Sigma_n] | \theta_n(y) Ass_n(S) \subset Lie_n(S) \forall S\}.$$

Evidently, $Lie_n$ is a left ideal in $\mathbb{R}[\Sigma_n]$.

**Proposition 14.** *For every $n > 0$ the following relations hold*

$$V \cap \text{Ш}_n = Lie_n \cap \text{Ш}_n; \ \mathbb{R}[\Sigma_n] \circ v = Lie_n, \forall v \in V \cap \text{Ш}_n, \rho(v) \neq 0.$$

**Proof.** Suppose $x \in Lie_n \cap \text{Ш}_n$, $v \in V \cap \text{Ш}_n$, $\rho(v) = 1$. Then $\theta_n(x) Ass(S) \subset Lie(S)$ and, according to the assertion 2) of Theorem 2, $\theta_n(v) \circ \theta_n(x) = \theta_n(x)$. At the same time, for $\forall S$ we have $\theta_n(v) \circ \theta_n(x) = \theta_n(x \circ v)$, hence, $\theta_n(x \circ v) = \theta_n(x)$. If $\#S \geq n$, then $x \circ v = x$. Taking into account that $V \cap \text{Ш}_n$ is a two-sided ideal in $\text{Ш}_n^\circ$, cf. Corollary of Proposition 12, we obtain $x \in V \cap \text{Ш}_n$. Furthermore, the assertion 2) of Theorem 2 yields

$$\theta_n(Lie_n \circ v) = \theta_n(v) \circ \theta_n(Lie_n) = \theta_n(Lie_n).$$

For $\#S \geq n$ we obtain $Lie_n \circ v = Lie_n$. Since $Lie_n$ is a left ideal in $\mathbb{R}[\Sigma_n]$, it is generated by the idempotent $v$.

## References


1. Agrachev A.A., Gamkrelidze R.V., *Exponential representation of flows and chronological calculus*, Matem. Sborn. **107**, **#4** (1978), 467–532. (Russian)
2. Agrachev A.A., Gamkrelidze R.V., *Volterra series and groups of substitutions*, Itogi Nauki. Sovremen. Probl Matem. Noveyshie Dost. (1991), VINITI, Moscow, 3–40. (Russian)
3. Agrachev A.A., Gamkrelidze R.V., Sarychev A.V., *Local invariants of smooth control systems*, Acta Appl. Math. **14** (1989), 191–237.
4. Crouch P. E., Lamnabhi–Lagarrigue F., *Algebraic and multiple integral identities*, Acta Appl. Math. **15** (1989), 235–274.
5. Fliess M., *Fonctionnelles causales non linéaires et indéterminées non commutatives*, Bull.Soc.Math. France **109** (1981), 3–40.
6. Ree R., *Lie elements and an algebra associated with shuffles*, Annals of Math. **68**, **#2** (1958), 210–220.



Steklov Inst., ul.Vavilova 42, Moscow GSP-1, 117966, Russia
*E-mail address*: gam@post.mian.su

# 25 Optimal Control of Infinite Dimensional Systems Governed by Integro Differential Equations

**N. U. Ahmed** University of Ottawa, Ottawa, Ontario, Canada

**ABSTRACT.**

In this paper we consider the problem of optimal control for a class of systems governed by integro differential equations on Banach space. We prove the existence of optimal controls and also present necessary conditions of optimality.

## 1 INTRODUCTION.

The dynamics of many physical systems, such as, visco elastic fluid, thermodynamics, electrodynamics, continuum mechanics, population biology are governed by integro differential equations on Banach space. The abstract mathematical model for such systems can be described as follows:

$$\begin{aligned} dy(t)/dt &= \int_0^t da(r)Ay(t-r) + F(y(t)), t \geq 0, \\ y(0) &= y_0, \end{aligned} \tag{1.1}$$

where $A$ is typically a linear unbounded operator in a suitable Banach space and $F$ a nonlinear operator, $a$ is a scalar valued function and $y_0$ is the initial state. A corresponding control system may be described as:

$$\begin{aligned} dy(t)/dt &= \int_0^t da(r)Ay(t-r) + F(y(t)) + B(t)u(t), \\ y(0) &= y_0, \end{aligned} \tag{1.2}$$

where $B$ is the control operator and $u$ is the control.

For the sake of motivation we present an example from the field of visco-elasticity.Consider a three dimensional isotropic and incompressible material. Let $v$ denote the displacement and $u$ the velocity of material points . Conservation of momentum leads to the following expression

$$u_t(t,\xi) = h(t,\xi) + div_\xi \sigma(t,\xi) \tag{1.3}$$

where $h$ represents body force and $\sigma$ the stress tensor which is related to the strain tensor through the constitutive law

$$\begin{aligned} \sigma(t,\xi) &\equiv -p(t,\xi)I + 2\int_0^\infty da(r)E_t(t-r,\xi) \\ E(t,\xi) &\equiv (1/2)((\nabla v) + (\nabla v)') \end{aligned} \tag{1.4}$$

where $p$ denotes the pressure, $a$ the stress relaxation function and $E$ the deformation tensor. Substituting (1.4) into (1.3) we obtain

$$u_t(t,\xi) = f(t,\xi) - \nabla p(t,\xi) + \int_0^t da(r)\, \triangle_\xi\, u(t-r,\xi) \tag{1.5}$$

where $f$, given by

$$f(t,\xi) \equiv h(t,\xi) + \int_t^\infty da(r)\, \triangle_\xi\, u(t-r,\xi),$$

includes the past history of stress to which the material had been subjected to. Defining the operator $A$ as the restriction of the Laplacian on divergence free vector fields , for example, with

$$D(A) \equiv \{\phi \in L_2(\Omega) : \triangle\phi \in L_2(\Omega) \text{ and } div\phi = 0\}$$

we can write this system in the abstract form,

$$du(t)/dt = \int_0^t da(r)Au(t-r) + f(t), t \geq 0,$$

in the Hilbert space $H \equiv L_2^\sigma(\Omega)$ where $L_2^\sigma$ denotes the completion of divergence free $C^\infty$ vector fields in the norm topology of $L_2$. Note that the operator $A$ is called the Stokes operator and it is easily verified that it is self adjoint and dissipative. The stress relaxation kernel $a$ is, in general, of the form

$$a(t) \equiv a_0 + a_1 t + \int_0^t a_2(\tau) d\tau$$

with $a_0 \geq 0$ representing the Newtonian viscosity, $a_1 \geq 0$ representing the elasticity modulus and $a_2(t) \geq 0, t \geq 0$, is a nonincreasing function with $lim_{t \to \infty} a_2(t) = 0$, representing the second law of thermodynamics. In any case this is a mathematical model for linear visco elasticity, a special case of the general model as presented in the introduction. The question of boundary controlabilty of such systems have been studied by Leugering [6] and stabilizability by Littman and Markus [7] under specific assumptions. Optimal control and identification of systems governed by diifferential equations on Banach space have been the subject of continuing studies in recent years.The reader is referred to [2,3] and the reference therein. Optimal control of systems described by integro differential equations in finite dimensional spaces have been extensively studied in the literature in the sixties and seventies [8,9]. To the knowledge of the author, very little is known for infinite dimensional differential systems with delays.
To work with the general model we shall consider a general Banach space $X$ to be the state space and $V$ a separable reflexive Banach to be the space where the controls take their values from. For any Banach space $Y$ and any interval $I \equiv [0, T], T \leq \infty$, $L_p(I, Y), 1 \leq p \leq \infty$ will denote the Banach space of strongly measurable $Y$ valued functions having $p-th$ power summable norms. For any two Banach spaces $X$ and $Y$, $\mathcal{L}(X, Y)$ will denote the space of bounded linear operators from $X$ to $Y$. We shall introduce further notations in the sequel as required.

## 2 EXISTENCE OF SOLUTIONS.

In this section we consider the system (1.2) and discuss the question of existence and uniqueness of solutions and their regularity properties. In this regard the homogeneous equation

$$\begin{aligned} dz(t)/dt &= \int_0^t da(r) A z(t-r) \\ z(0) &= x, \end{aligned} \tag{2.1}$$

plays a central role as does the equation $dz(t)/dt = Az(t), z(0) = x$ for differential equations

$$dz(t)/dt = Az(t) + f(t), z(0) = x. \tag{2.2}$$

It is known that in the later case if the operator $A$ is the generator of a $C_0$–semigroup $S(t), t \geq 0$, in $X$ then by use of the so called variation of constants formula one can write the solution of the nonhomogeneous problem as

$$z(t) = S(t)x + \int_0^t S(t-s)f(s)ds, t \geq 0. \tag{2.3}$$

In recent years there has been growing interest in the development and extension of semigroup theory to cover systems governed by integro differential equations [see Da Prato Ianelli,4; Pruss,5 ]. For convenience we shall quote a result due to Da Prato and Ianelli [4] which extends the Hille-Yosida generation theorem for the problem (2.1). Consider the problem

$$C(A,a) : \begin{cases} dz(t)/dt = \int_0^t da(r)Az(t-r) \\ z(0) = x. \end{cases}.$$

**Definition 1.** A strongly continuous operator valued function $S(t), t \geq 0$ in $X$ is said to be a resolvent or a transition operator or a solution operator of the problem $C(A,a)$ if (i): S(0) = I (identity operator) and (ii): there exist constants $\omega \in R$ and $M \geq 1$ such that

$$\| S(t) \|_{\mathcal{L}(X)} \leq Me^{\omega t} \quad \text{for} \quad t \geq 0,$$

and, for $x \in D(A), t \to S(t)x \in C(I,X)$ and it is once continuously differentiable on the open interval $(0,T)$, $S(t)$ commutes with $A$ on $D(A)$ and it satisfies the Cauchy problem $C(A,a)$ for all $t \in I$.

For convenience and to avoid confusion with the expression "resolvent" generally used for $R(\lambda, A) \equiv (\lambda I - A)^{-1}, \lambda \in \rho(A)$, the resolvent set of $A$, we shall use the term "transition operator". The following generation theorem is due to Da Prato and Ianelli [4].

**Lemma 2.** Suppose

(ai) A is a closed densely defined linear operator with domain and range $D(A), R(A) \subset X$,

(aii) $a \in BV_{loc}(R_+)$ with $\int_0^\infty e^{-\omega t}|da(t)| < \infty$ for some $\omega \in R$.

Then the necessary and sufficient conditions for the existence of a transition operator, $S(t), t \geq 0$, for the problem $C(A, a)$ are

(1): the Laplace transform (of $a$) $\hat{a}(\lambda) \neq 0, (1/\hat{a}(\lambda)) \in \rho(A)$ for $\lambda > \omega$,

(2): $R(\lambda) \equiv (1/\lambda)(I - \hat{a}(\lambda)A)^{-1}$ exists for all $\lambda > \omega$ and

$$\| R^{(n)}(\lambda)/n! \| \leq M/(\lambda - \omega)^{n+1} \quad \text{for all} \quad \lambda > \omega, n \in N_0 \equiv \{0, 1, 2\}.$$

Thus under the assumptions of Lemma 2, the Cauchy problem $C(A, a)$ has a unique classical solution for each $x \in D(A)$ and it is given by $z(t) = S(t)x, t \geq 0$.

Now let us consider the nonhomogenous problem :

$$C(A, a, f) : \begin{cases} dz(t)/dt = \displaystyle\int_0^t da(r)Az(t - r) + f(t), t \geq 0, \\ z(0) = x \end{cases} \tag{2.4}$$

**Definition 3.** A function $z$ is said to be a mild solution of the problem $C(A, a, f)$ if

(M1): $z \in C(I, X)$

(M2) $z(t) = S(t)x + \int_0^t S(t - r)f(r)dr, t \geq 0$, where $S(t), t \geq 0$ is the transition operator of the problem $C(A, a)$. In contrast, a function $z$ is said to be a classical solution if

(C1): $z \in C(I, X) \cap C^1((0, T), X)$

(C2): $z(t) \in D(A), t \in I$

(C3): $z$ satisfies the equation (2.4) for all $t \in I$.

Only under very strong regularity conditions on the data, such as $x \in D(A), f \in C^1(I, X)$, one can prove the existence of classical solutions. Certainly for control problems this is a severe restriction and hence the concept of mild solutions is more appropriate. In general a mild solution does not satisfy equation (2.4) but one can find a sequence of classical solutions that converges to the mild solution. Throughout the paper we are interested only in the mild solutions. Now we shall consider the semilinear control system (1.2) and prove the existence and uniqueness of solutions.

**Theorem 4.** Suppose
(b1): assumptions of lemma 2 hold, $I \equiv [0,T], T < \infty$.
(b2): $B \in L_q(I, \mathcal{L}(V,X)), 1 < q \leq \infty$.
(b3): $\mathcal{U}_{ad} \subseteq L_p(I,V), 1 < p < \infty, (1/p) + (1/q) = 1$.
(b4): $F : X \longrightarrow X$ is locally Lipschitz and satisfies the growth condition,

$$\| F(x) \|_X \leq \alpha(1 + \| x \|_X) \quad \text{for some } \alpha > 0.$$

Then for each $y_0 \in X$ the semilinear system (1.2) has a unique mild solution $y \in C(I,X)$.
*Proof.* By using the transition operator, $S(t), t \geq 0$, corresponding to the problem $C(A,a)$, whose existence is assured by Lemma 2, and using the variation of constants formula we obtain the following integral equation,

$$y(t) = \phi(t,u) + \int_0^t S(t-r)F(y(r))dr, t \in I, \tag{2.5}$$

where

$$\phi(t,u) \equiv S(t)y_0 + \int_0^t S(t-r)B(r)u(r)dr, t \in I.$$

By virtue of assumptions (b2) and (b3) and the strong continuity of $S(t), t \geq 0$, we have for each $u \in \mathcal{U}_{ad}$, $\phi(.,u) \in C(I,X)$. Then by virtue of assumption (b4) and Banach fixed point theorem one can easily verify that the integral equation (2.5) has unique solution $y \in C(I,X)$. Hence by definition, this is also the unique mild solution of system (1.2). □

## 3 EXISTENCE OF OPTIMAL CONTROLS.

We consider the Lagrange problem
(P): find $u^0 \in \mathcal{U}_{ad}$ such that

$$J(u^0) \equiv \eta(u^0, y^0) \leq \eta(u, y^u) \equiv J(u), \quad \text{for all } u \in \mathcal{U}_{ad}$$

$$\text{where } \eta(u, y^u) \equiv \int_I \ell(t, y^u(t), u(t))dt.$$

Here $y^u$ denotes the solution of the system (1.2) corresponding to the control $u \in \mathcal{U}_{ad}$.

For the existence problem we shall introduce the following assumptions. First, let $\tilde{B}$ denote the Nemytski operator: $(\tilde{B}u)(t) = B(t)u(t), t \in I$. Clearly under the assumption (b2) of theorem 4, $\tilde{B}$ maps $L_p(I, V)$ to $L_1(I, X)$.

**ASSUMPTIONS:**

(A1): $\mathcal{U}_{ad} = L_p(I, V)$

(A2): Either $\tilde{B}$ is strongly continuous or for each $t \in I$, the operator $u \longrightarrow L_t(u) \equiv \int_0^t S(t-\theta)B(\theta)u(\theta)d\theta$ is strongly continuous from $L_p(I, V)$ to $X$.

(A3): $\ell : I \times X \times V \longrightarrow R \cup \{+\infty\}$ is Borel measurable satisfying the following conditions:

(i) $\ell(t, ., .)$ is sequentially lower semicontinuous on $X \times V$ for almost all $t \in I$,

(ii) $\ell(t, x, .)$ is convex on $V$ for each $x \in X$ and almost all $t \in I$,

(iii): there exist $b \geq 0, c > 0$ and $h \in L_1(I, R)$ such that

$$\ell(t, x, v) \geq h(t) + b \parallel x \parallel_X + c \parallel v \parallel_V^p .$$

**REMARK 5.** It would be interesting to find necessary and sufficient conditions for compactness of the transition operator $S(t)$ for $t > 0$. For semigroups this is well known. However we may mention that if $S(t)$ is a compact solution operator then the corresponding $R(\lambda)$ is necessarily compact.The converse is not necessarily true even in the semigroup case.

For convenience we shall denote by $\mathcal{A}_{ad}$ the set of admissible control-state pairs given by $\{(u, y) \in \mathcal{U}_{ad} \times C(I, X) : y = y^u\}$ where $y^u$ is the solution of (1.2) corresponding to the control $u$.

**Theorem 6.** Under the assumptions (A1),(A2) and (A3), the optimal control problem (P) has a solution, that is there exists an admissible state-control pair $\{u^0, y^0\}$ such that

$$J(u^0) \equiv \eta(u^0, y^0) \leq \eta(u, y^u) = J(u) \quad \text{for all} \quad u \in \mathcal{U}_{ad}.$$

*Proof.* If $Inf\{J(u), u \in \mathcal{U}_{ad}\} = +\infty$ there is nothing to prove. So we assume $Inf\{J(u), u \in \mathcal{U}_{ad}\} = m < \infty$. Let $\{u^n, y^n\} \in \mathcal{A}_{ad}$ be a minimizing sequence. By virtue of assumption (A3)-(iii) $\{u^n\}$ is contained in a bounded subset of $L_p(I,V)$. Since $L_p(I,V)$ is a separable reflexive Banach space, it has a subsequence relabeled as $\{u^n\}$ and an element $u^0 \in \mathcal{U}_{ad}$ such that $u^n \xrightarrow{w} u^0$. Let $\{y^n\} \in C(I,X)$ denote the corresponding sequence of solutions of the integral equation,

$$y^n(t) = \phi(t,u^n) + \int_0^t S(t-\theta)F(y^n(\theta))d\theta.$$

Since the controls are contained in a bounded set and $F$ satisfies the growth condition (see theorem 4), using Gronwall inequality and the integral equation above one can easily verify that there exists an $r > 0$ such that

$$Sup\{\| y^n(t) \|_X \; n \in N, t \in I\} \leq r < \infty.$$

Let $B_r(X)$ denote the closed ball of radius $r$ in $X$. Clearly $y^n(t) \in B_r(X)$ for all $t \in I$ and for all $n \in N$. Since $F$ is locally Lipschitz there exists a constant $K_r$ such that

$$\| F(x) - F(z) \|_X \leq K_r \| x - z \|_X,$$

for all $x, z \in B_r(X)$. Let $y^0$ denote the solution corresponding to $u^0$, that is

$$y^0(t) = \phi(t,u^0) + \int_0^t S(t-\theta)F(y^0(\theta))d\theta, t \geq 0.$$

Since $\{y^0(t), y^n(t), t \in I\} \subset B_r(X)$, by the local Lipschitz property we have

$$\| y^n(t) - y^0(t) \|_X \leq \| \phi(t,u^n) - \phi(t,u^0) \|_X + MK_r \int_0^t e^{w(t-s)} \| y^n(s) - y^0(s) \|_X \, ds. \tag{3.1}$$

Under the assumptions of theorem 4, specifically (b2) and (b3), it is clear that for every $u \in L_p(I,V)$, $\phi(.,u) \in C(I,X)$. Hence under the assumption (A2), for each $t \in I$, $\| \phi(t,u^n) - \phi(t,u^0) \|_X \longrightarrow 0$ as $n \to \infty$. Using this fact,the general Gronwall inequality and the Lebesgue dominated convergence theorem, it follows from (3.1) that

$$y^n(t) \xrightarrow{s} y^0(t) \text{ in } X \text{ for each } t \in I.$$

Using this and once more applying dominated convergence theorem one can easily verify that $y^n \xrightarrow{s} y^0$ also in $L_1(I, X)$. Note that our assumption (A3)-(iii) implies the assumption (2.3) of Balder [ Theorem 2.1, p1400]. Hence by theorem 2.1 of Balder we conclude that $(u, y) \longrightarrow \eta(u, y)$ is sequentially lower semi continuous in the weak topology of $L_p(I, V) \subset L_1(I, V)$ and strong topology of $L_1(I, X)$. Hence $J$ is weakly lower semi continuous on $L_p(I, V)$ and since by (A3)-(iii), $J > -\infty$, $J$ attains its infimum in $\mathcal{U}_{ad}$. This proves that the optimal control problem (P) has a solution. □

**Remark 7**. Theorem 6 gives sufficient conditions for the existence of optimal controls in case $\mathcal{U}_{ad} = L_p(I, V)$. In case $\mathcal{U}_{ad}$ is a closed bounded convex subset of $L_p(I, V)$, the condition (A3)-(iii) can be relaxed. This is stated in the following theorem.

Let $\mathcal{U}$ be any weakly compact subset of $V$ and let $t \longrightarrow U(t)$ be a measurable set valued map with values $U(t) \in cc(\mathcal{U})$ where $cc(\mathcal{U})$ denotes the class of nonempty, closed, convex subsets of $\mathcal{U}$.

**Theorem 8.** Suppose the following assumptions hold:

(d1): $\mathcal{U}_{ad} \equiv \{u : I \longrightarrow V \text{ strongly measurable } : u(t) \in U(t), t \in I\}$

(d2): same as (A2)

(d3): same as (A3) with (A3)-(iii) replaced by $\ell(t, x, v) \geq h(t) - b(\| x \|_X + \| v \|_V)$ for some $b > 0$. Then there exists an optimal control for the problem (P).

*Proof.* The proof is similar to that of theorem 6 with the only difference being in the argument leading to the boundedness of the minimizing sequence of controls $\{u^n\}$. Under the present hypotheses this follows from (d1). □

## 4 NECESSARY CONDTIONS OF OPTIMALITY.

In this section we present some necessary conditions of optimality. For this we shall encounter the following adjoint system :

$$\begin{aligned} d\psi(t)/dt &= -\int_0^{T-t} da(r) A^* \psi(t+r) + g(t) \\ \psi(T) &= \xi. \end{aligned} \tag{4.1}$$

Let $X^*$ denote the dual of $X$ and $X^*_w$ the space $X^*$ endowed with the $w^*$ topology and $C(I, X^*_w)$ the topological space of $w^*$-continuous $X^*$-valued functions defined on the interval $I = [0, T]$. Let $<,>$ denote the duality pairing between $X^*$ and $X$.

**Lemma 9.** For each $\xi \in X^*$ and $g \in L_1(I, X^*)$, the adjoint problem (4.1) has a unique solution $\psi \in C(I, X^*_w)$ given by

$$\psi(t) = S^*(T-t)\xi - \int_t^T S^*(\theta - t)g(\theta)d\theta \tag{4.2}$$

which satisfies equation (4.1) in the weak sense, that is, for each $\eta \in D(A)$,

$$\begin{aligned} d/dt < \psi(t), \eta > &= -\int_0^{T-t} da(r) < \psi(t+r), A\eta > + < g(t), \eta > \\ Lim_{t \to T} < \psi(t), \eta > &=< \xi, \eta >, \end{aligned} \tag{4.3}$$

for all $\eta \in D(A)$.

*Proof.* We present an outline of the proof. For each $\eta \in D(A)$, we have

$$d/dt < \psi(t), \eta >= (d/dt)\{< \xi, S(T-t)\eta > - \int_t^T < g(\theta), S(\theta - t)\eta > d\theta\}.$$

Now note that for $\eta \in D(A)$, $S(t)\eta$ is differentiable and that $S(t)$ commutes with $A$ on $D(A)$ and that

$$(d/dt)S(t)\eta = \int_0^t da(\tau)AS(t-\tau)\eta, t \in I.$$

Using these facts and Fubini's theorem one can easily conclude that

$$d/dt < \psi(t), \eta >= -\int_0^{T-t} da(\tau) < \psi(t+\tau), A\eta > + < g(t), \eta >,$$

for all $\eta \in D(A)$. □

The main result of this section is the necessary conditions of optimality for the problem (P). In what follows we shall assume that an optimal control exists [ see theorems 6 and 8]. For the necessary conditions of optimality we shall need some additional conditions on $F$ and $\ell$.

**Theorem 10.** Suppose the following conditions hold:

(e1): $\mathcal{U}_{ad} \equiv \{u \in L_p(I,V) : u(t) \in \mathcal{U} \ a.e \}$ where $\mathcal{U}$ is a closed convex subset of $V$;

(e2): $F$ is Frechet differentiable on $X$ with the Frechet derivative $F'$ continuous and uniformly bounded on $X$;

(e3): $\ell(t,x,v)$ is continuously Frechet differentiable in both $x$ and $v$ on $X$ and $V$ respectively so that $\ell_y \in L_1(I,X^*)$ and $\ell_u \in L_q(I,V^*)$ in the neighbourhood of optimal trajectories.

Then in order that $\{u^0, y^0\}$ be the optimal pair it is necessary that there exists a $\psi \in C(I, X_w^*)$ such that the following equations and inequalities hold:

$(1) : dy^0/dt = \int_0^t da(r)Ay^0(t-r) + F(y^0(t)) + B(t)u^0(t), y(0) = y_0.$

$(2) : -d\psi(t)/dt = \int_0^{T-t} da(r)A^*\psi(t+r) + (F'(y^0(t)))^*\psi(t) + \ell_y^0(t), \psi(T) = 0.$

$(3) : \int_I < B^*\psi + \ell_u^0, u - u^0 >_{V^*,V} dt \geq 0$ for all $u \in \mathcal{U}_{ad}$,

where $\ell_u^0(t) \equiv \ell_u(t, y^0(t), u^0(t))$, $\ell_y^0(t) \equiv \ell_y(t, y^0(t), u^0(t))$.

*Proof.* Let $\{u^0, y^0\}$ be the optimal pair (existence assured by theorems 6 and 8) and $u \in \mathcal{U}_{ad}$ an arbitrary control.Clearly, by convexity of $\mathcal{U}_{ad}$, $u^\epsilon \equiv u^o + \epsilon(u - u^0) \in \mathcal{U}_{ad}$ for $0 \leq \epsilon \leq 1$. By theorem 4, the state equation (1.2) has a unique solution $y^\epsilon \in C(I,X)$ corresponding to the control $u^\epsilon$. Define $w^\epsilon \equiv (1/\epsilon)(y^\epsilon - y^0)$. Note that $w^\epsilon$ satisfies the integral equation

$$\begin{aligned} w^\epsilon(t) = \int_0^t S(t-\tau)\left(\int_0^1 F'(y^0(\tau) + s(y^\epsilon(\tau) - y^0(\tau)))ds\right)w^\epsilon(\tau)d\tau \\ + \int_0^t S(t-\tau)B(\tau)(u(\tau) - u^0(\tau))d\tau. \end{aligned} \tag{4.4}$$

By virtue of assumption (e2) and continuous dependence of solutions on the controls (see theorem 4), one can justify taking $\epsilon$ to zero in the above equation to obtain

$$w^0(t) = \int_0^t S(t-\tau)F'(y^0(\tau))w^0(\tau)d\tau + \int_0^t S(t-\tau)B(u - u^0)d\tau. \tag{4.5}$$

Indeed, by assumption (e2), it is clear that

$$C^{\epsilon}(t) \equiv \int_0^1 F'(y^0(t) + s(y^{\epsilon}(t) - y^0(t)))ds \longrightarrow F'(y^0(t)) \equiv C(t)$$

in the uniform operator topology of $\mathcal{L}(X)$ as $\epsilon \to 0$. Further, it is clear that the operator valued function $t \longrightarrow C(t)$ is continuous taking values from $\mathcal{L}(X)$ and that there exists a constant $K$ such that $\| C(t) \|_{\mathcal{L}(X)} \leq K$. Hence for the given $\{u, u^0\}$, by virtue of Banach fixed point theorem, the linear Volterra integral equation,

$$w(t) = \int_0^t S(t-\tau)C(\tau)w(\tau)d\tau + \int_0^t S(t-\tau)B(\tau)(u(\tau) - u^0(\tau))d\tau,$$

has a unique solution $w \in C(I, X)$, and that it coincides with $w^0$. Hence $w$ is a mild solution of the equation

$$\begin{aligned} dw/dt &= \int_0^t da(r)Aw(t-r) + C(t)w + B(t)(u - u^0) \\ w(0) &= 0. \end{aligned} \tag{4.6}$$

Note that $w$ is the Gateaux differential of $y$ in the direction $u - u^0$.

By definition of optimality

$$J(u^0) = \eta(u^0, y^0) \leq \eta(u^{\epsilon}, y^{\epsilon}) = J(u^{\epsilon}).$$

Hence by use of hypothesis (e3), after some elementary computations, one obtains

$$\int_I < l_y^0, w >_{X^*,X} dt + \int_I < l_u^0, u - u^0 >_{V^*,V} dt \geq 0 \tag{4.7}$$

for all $u \in \mathcal{U}_{ad}$. Now we can eliminate $l_y^0$ by invoking the adjoint equation (2). Since by (e3) $l_y^0 \in L_1(I, X^*)$, by use of Lemma 9 one can verify the existence of a unique weak solution of equation (2) in the sense defined earlier. Using (4.2) we write equation (2) as a Volterra integral equation,

$$\psi(t) = h(t) - \int_t^T S^*(\theta - t)C^*(\theta)\psi(\theta)d\theta$$

$$h(t) \equiv - \int_t^T S^*(\theta - t)l_y^0(\theta)d\theta.$$

Defining $G$ as the operator for the right hand expression, we show that , for sufficiently large $n$, the $n$-th power of G is a contraction in the Banach space $L_\infty(I, X^*)$ proving that the integral equation has a unique solution there. Since $h \in C(I, X_w^*)$, it follows from the first equation that actually $\psi$ belongs to $C(I, X_w^*)$. Returning to the expression (4.7), the first term is given by

$$\begin{aligned}
&\int_I < l_y^0, w > dt \\
&= -\int_I \{< \dot{\psi}(t) + \int_0^{T-t} da(r) A^* \psi(t+r) + C^*(t)\psi(t), w(t) >_{X^*,X}\} dt \\
&= \int_I \{< \psi(t), \dot{w}(t) - \int_0^t da(r) Aw(t-r) - C(t)w(t) >_{X^*,X}\} dt \\
&= \int_I < \psi(t), B(t)(u(t) - u^0(t)) >_{X^*,X} dt.
\end{aligned} \tag{4.8}$$

The third equality follows from the second by virtue of equation (4.6). Since the mild solution $w(t)$ need not belong to $D(A)$, the step from the first equality to the second is crucial. In order to justify this step we use the Yosida approximation of the identity, $I_n \equiv nR(n, A)$ where $R(\lambda, A)$ is the resolvent of $A$ corresponding to $\lambda \in \rho(A)$. It is easy to verify that $I_n \longrightarrow I$(identity operator) in the strong operator topology in $\mathcal{L}(X)$, and for any $x \in X, I_n x \in D(A)$ for $n \in \rho(A)$. In equation (4.6) we replace the operators $C, B$ by $I_nC, I_nB$ and replace $w$ in the first expression of (4.8) by $w_n$ which is the solution of equation

$$\begin{aligned}
dw_n/dt &= \int_0^t da(r) Aw_n(t-r) + (I_nC(t))w_n + (I_nB(t))(u - u^0) \\
w_n(0) &= 0.
\end{aligned} \tag{4.9}$$

Equation (4.9) has a unique strong solution $w_n$ with $w_n(t) \in D(A)$ for almost all $t \in I$, provided $n \in \rho(A)$. Note that a strong solution is obviously a mild solution. Comparing the integral equations for the mild solutions of equations (4.6) and (4.9), it follows from Gronwall inequality that

$$\| w_n(t) - w(t) \|_X \leq \alpha_n \; Exp\{Me^{\gamma T}. \int_0^t \| I_nC(\tau) \|_{\mathcal{L}(X)} d\tau\}, t \in I, \tag{4.10}$$

where $\gamma \equiv |\omega|$ and

$$\alpha_n \equiv (Me^{\gamma T}) \int_0^T \{\| I_n(B(u-u^0))-B(u-u^0)) \|_X + \| (I_nCw-Cw) \|_X\}dt.$$

Since $I_n \rightarrow I$ in the strong operator topology, $B(u - u^0) \in L_1(I, X)$ and $Cw \in C(I, X)$, it follows from dominated convergence theorem that $Lim_n \alpha_n \longrightarrow 0$. Further, note that, for any given number $M_1 > M$, $\| I_nC(t) \|_{\mathcal{L}(X)} \leq (M_1K)$ for all $n \geq n_1$ for a suitable $n_1 \in N$. Using these facts it follows from (4.10) that $w_n \longrightarrow w$ in the usual topology of $C(I, X)$. Now returning to the expression (4.8) with $w$ replaced by $w_n$ we have

$$\begin{aligned}
&-\int_I < \dot{\psi}(t) + \int_0^{T-t} da(r)A^*\psi(t+r) + C^*(t)\psi(t), w_n(t) >_{X^*,X} dt \\
&= \int_I < \psi(t), \dot{w}_n(t) - \int_0^t da(r)Aw_n(t-r) - C(t)w_n(t) >_{X^*,X} dt \\
&= \int_I < \psi(t), I_nCw_n - Cw_n + I_nB(t)(u(t) - u^0(t)) >_{X^*,X} dt.
\end{aligned} \tag{4.11}$$

Hence

$$\begin{aligned}
&-\int_I < \dot{\psi}(t) + \int_0^{T-t} da(r)A^*\psi(t+r) + C^*(t)\psi(t), w(t) >_{X^*,X} dt \\
&= Lim_n \int_I < \psi(t), I_nCw_n - Cw_n + I_nB(t)(u(t) - u^0(t)) >_{X^*,X} dt.
\end{aligned} \tag{4.12}$$

For $n > n_1$, we have

$$\| I_nCw_n - Cw_n \|_X \leq K(1 + M_1)(\| w_n - w \|) + \| I_n(Cw) - Cw \| .$$

This estimate and strong convergence of $I_n$ to $I$ and uniform convergence of $w_n$ to $w$ imply that

$$Lim_n \| I_n(Cw_n)(t) - (Cw_n)(t) \|_X = 0 \quad \text{uniformly in} \quad t \in I$$
$$Lim_n I_n(B(t)(u(t) - u^0(t))) \longrightarrow B(t)(u(t) - u^0(t)) \ \text{a.e} \ t \in I.$$

Replacing $w$ of equation (4.7) by $w - w_n + w_n$ and then taking the limit, the inequality (3) follows by virtue of (4.12) and the analysis immediately following it. This ends the justification of equation (4.8) and also completes proof. □

In the above result we assumed that $\ell$ is Frechet differentiable in the control variable. In case $\ell(t,x,v)$ is merely continuous in $v$ and Frechet differentiable in $x$ and $\mathcal{U} \subset V$ is a closed bounded convex set we can prove a Pontryagin type minimum principle. Define

$$M \equiv \{u : I \longrightarrow V, \text{ strongly measurable } : u(t) \in \mathcal{U} \ a.e;\}$$

with the topology induced by the metric

$$\rho(u,v) \equiv \mu\{t \in I : u(t) \neq v(t)\},$$

where $\mu$ denotes the Lebesgue measure. Since $\mathcal{U}$ is a closed subset of a Banach space, the set $M$, with the metric $\rho$ as defined above, is a complete metric space. We shall need the following lemma.

**Lemma 11.** Suppose the assumptions (b1) and (b4) of theorem 4 hold and $B \in L_\infty(I, \mathcal{L}(V,X))$ and $\mathcal{U}_{ad} = M$. Then for the semilinear system (1.2), the mapping

$$u \longrightarrow y^u$$

is continuous from $M$ to $C(I,X)$ in the respective topologies and further there exists a constant $\beta = \beta(T, M, \omega, K, \| B \|_{L_\infty(I,\mathcal{L}(V,X))})$ such that

$$\| y^u - y^v \|_{C(I,X)} \leq \beta\rho(u,v)$$

for all $u, v \in \mathcal{U}_{ad}$.

*Proof.* Using the integral equation (2.5) for controls $u$ and $v$, the proof follows from Gronwall inequality. □

**Theorem 12.** Consider the problem (P) and suppose $B \in L_\infty(I, \mathcal{L}(V,X))$ and $\mathcal{U}_{ad} = M$ and $v \longrightarrow \ell(t,x,v)$ is merely continuous, and $x \longrightarrow \ell(t,x,v)$ is continuously Frechet differentiable with $\ell_x \in L_1(I, X^*)$ along admissible state-control pairs.Then the optimality conditions (1),(2) of theorem 10 hold and (3) is replaced by (3)* :

$$\ell(t, y^0(t), u^0(t)) + < \psi(t), B(t)u^0(t) > \leq \ell(t, y^0(t), v) + < \psi(t), B(t)v > \tag{3)*}$$

for all $v \in \mathcal{U}$ and almost all $t \in I$.

*Proof.* Let $u^0$ be the optimal control and $\sigma \subset I$ any measurable set and $v \in \mathcal{U}$. Define

$$u^\sigma(t) = \begin{cases} u^0(t), & t \in I \setminus \sigma\,; \\ v, & t \in \sigma. \end{cases}$$

Since $u^0$ is optimal we have

$$\int_\sigma \ell(t, y^0, u^0)dt \leq \int_\sigma \ell(t, y^\sigma, v)dt + \int_{I\setminus\sigma} (\ell(t, y^\sigma, u^0) - \ell(t, y^0, u^0))dt. \tag{4.13}$$

By virtue of Frechet differentiability of $\ell(t, x, v)$ in $x$, it follows from Lagrange formula that

$$\ell(t, y^\sigma, u^0) - \ell(t, y^0, u^0) = \\ < \ell_y(t, y^0, u^0), y^\sigma - y^0 > + \int_0^1 < \ell_y^{\theta,\sigma} - \ell_y^0, y^\sigma - y^0 > d\theta, \tag{4.14}$$

where $\ell_y^{\theta,\sigma} \equiv \ell_y(t, y^0 + \theta(y^\sigma - y^0), u^0)$ and $\ell_y^0 \equiv \ell_y(t, y^0, u^0)$. By continuity of the Frechet differential of $\ell$ on $X$ and the fact that $y^\sigma \longrightarrow y^0$ as $\mu(\sigma) \longrightarrow 0$ (see Lemma 11), we have

$$\int_{I\setminus\sigma} \{ \int_0^1 < \ell_y^{\theta,\sigma} - \ell_y^0, y^\sigma - y^0 > d\theta \} dt = o(\mu(\sigma)).$$

Here o(.) stands for small order of approximation. Hence the expression (4.13) can be written as

$$\int_\sigma \ell(t, y^0, u^0)dt \\ \leq \int_\sigma \ell(t, y^\sigma, v)dt + \int_{I\setminus\sigma} < \ell_y(t, y^0, u^0), y^\sigma - y^0 > dt + o(\mu(\sigma)). \tag{4.15}$$

The second term in the above equation can be split into two terms; one involving integration over $I$ and the other involving integration over the set $\sigma$. Since $\ell_y^0 \in L_1(I, X^*)$, it follows from Lemma 11 that

$$\int_\sigma < \ell_y^0, y^\sigma - y^0 >_{X^*,X} dt = o(\mu(\sigma)).$$

Hence expression (4.15) reduces to

$$\int_\sigma \ell(t,y^0, u^0)dt$$
$$\leq \int_\sigma \ell(t, y^\sigma, v)dt + \int_I < \ell_y(t, y^0, u^0), y^\sigma - y^0 > dt + o(\mu(\sigma)). \tag{4.16}$$

Using the adjoint equation (2) of theorem 10 and following similar arguments as in that theorem, one can verify that

$$\int_I < \ell_y(t, y^0, u^0), y^\sigma - y^0 > dt$$
$$= \int_I < \psi, F(y^\sigma) - F(y^0) - C(t)(y^\sigma - y^0) + \chi_\sigma(t)B(v - u^0) > dt,$$

where $\chi_\sigma$ is the indicator function of the set $\sigma$.
Substituting this in equation (4.16) and using Lemma 11 along with the assumption on $\ell$, we obtain

$$\int_\sigma \{\ell(t, y^0, u^0) + < B^*\psi, u^0 >_{X^*,X}\}dt \leq$$
$$\int_\sigma \{\ell(t, y^0, v) + < B^*\psi, v >_{X^*,X}\}dt$$
$$+ \int_I < \psi, F(y^\sigma) - F(y^0) - C(t)(y^\sigma - y^0)dt + o(\mu(\sigma)). \tag{4.17}$$

Since $C(t) = F'(y^0(t))$ and $y^\sigma \to y^0$ in $C(I, X)$ as $\mu(\sigma) \to 0$, and the Frechet differential $F'$ of $F$ is continuous in the uniform operator topology of $\mathcal{L}(X)$, it is easy to verify that

$$\int_I < \psi, \int_0^1 (F'(y^0 + \theta(y^\sigma - y^0)) - C)(y^\sigma - y^0)d\theta > dt = o(\mu(\sigma)).$$

Hence (4.17) reduces to

$$\int_\sigma \{\ell(t, y^0, u^0) + \langle B^*\psi, u^0 \rangle_{X^*,X}\} dt \leq \int_\sigma \{\ell(t, y^0, v) + \langle B^*\psi, v \rangle_{X^*,X}\} dt + o(\mu(\sigma)). \tag{4.18}$$

Let $t$ be any Lebesgue density point of $u^0$ and $\sigma$ any measurable set containing $t$ and shrinking to the one point set $\{t\}$ as $\mu(\sigma) \to 0$. Dividing (4.17) by $\mu(\sigma)$ and letting it converge to zero, we obtain the inequality (3)*. This ends the proof. □

**Corollary 13.** The pointwise minimum principle, as given by Theorem 12, is equivalent to the following integral minimum principle: :

$$\int_I \{\ell(t, y^0(t), u^0(t)) + \langle \psi(t), B(t)u^0(t) \rangle\} dt \leq \int_I \{\ell(t, y^0(t), v(t)) + \langle \psi(t), B(t)v(t) \rangle\} dt.$$

for all $v \in \mathcal{U}_{ad}$.

*Proof.* The proof is standard. □

Another more popular model [1] for systems governed by integro differential equations on Banach space is given by the following:

$$\begin{aligned} dy(t)/dt &= -Ay(t) + f(y(t)) + \int_{-a}^{t} h(t-s)g(y(s))ds, t \in [0, b] \\ y(t) &= \phi(t), t \in [-a, 0]. \end{aligned} \tag{4.19}$$

Here $-A$ is the infinitesimal generator of an analytic semigroup $\{T(t), t \geq 0\}$ in $X$. Given the past history $\phi$, one can rewrite this equation as

$$\begin{aligned} dy(t)/dt &= -Ay(t) + f(y(t)) + \int_0^t h(t-s)g(y(s))ds + B(t)u(t), t \in [0, b] \\ y(0) &= y_0, \end{aligned} \tag{4.20}$$

where now the history term is absorbed in the operator $f$ and a control term has been added. For the existence of solutions of equation (4.19) see [1,3]. Here we shall present the necessary conditions of optimality for the Lagrange problem as discussed before.
We shall introduce the following assumptions

($\alpha$): $-A$ is the generator of an analytic semigroup $\{T(t), t \geq 0\}$ in $X$.

($\beta$): For $0 \leq \alpha < 1, X_\alpha \equiv [D(A^\alpha)]$ is the Banach space with respect to the graph topology induced by the graph norm $\| \zeta \|_\alpha \equiv \| A^\alpha \zeta \|$ for $\zeta \in D(A^\alpha)$.

($\gamma$): $f, g$ map $X_\alpha$ to $X$ and there exists a constant $C > 0$ such that for $q \equiv f, g$

$$\| q(\xi) - q(\zeta) \|_X \leq C \| \xi - \zeta \|_\alpha$$
$$\| q(\zeta) \| \leq C(1 + \| \zeta \|_\alpha) \quad \text{for all} \quad \xi, \zeta \in X_\alpha;$$

and their Frechet derivatives are continuous in the uniform operator topology of $\mathcal{L}(X_\alpha, X)$.

($\delta$): $h \in L_1(I, R)$ where $I \equiv [0, b]$.

($\vartheta$): The operator $B$, the cost integrand $\ell$ and the admissible controls satisfy the assumptions of theorem 12.

**Theorem 14.** Suppose the assumptions $(\alpha), (\beta), (\gamma), (\delta), (\vartheta)$ hold. Let $\{u^0, y^0\}$ be an admissible control-state pair and let $F(t), G(t)$ denote the Frechet differentials of $f, g$ along the trajectory $y^0$, and $\ell_y^0$ the Frechet differential of $\ell$ along the pair $\{u^0, y^0\}$. Then, in order that the pair $\{u^0, y^0\}$ be optimal for the Lagrange problem P subject to the dynamic constraint (4.20), it is necessary that the following relations hold :

$$(\aleph): \begin{array}{l} dy^0/dt + Ay^0(t) = \int_0^t h(t-s) g(y^0(s)) ds + f(y^0(t)) + B(t) u^0(t), \\ y(0) = y_0, t \in I, \end{array}.$$

$$(\Re): \begin{array}{l} -d\psi/dt + A^* \psi(t) = F^*(t)\psi(t) + G^*(t) \int_t^b h(s-t)\psi(s) ds + \ell_y^0(t), \\ \psi(b) = 0 \end{array}.$$

$$(\Im): \int_I \{\ell(t, y^0(t), u^0(t)) + \langle \psi(t), B(t)u^0(t) \rangle\} dt \leq \int_I \{\ell(t, y^0(t), v(t)) + \langle \psi(t), B(t)v(t) \rangle\} dt,$$

for all $v \in \mathcal{U}_{ad}$.

*Proof.* The proof is similar to that of Theorem 12. □

**Remark 15.** Note that all the results presented here do also hold for time varying $f, g$.

## REFERENCES.


[1] N.U.Ahmed, Nonlinear Evolution Equations on Banach Space; Journal of Applied Mathematics and Stochastic Analysis, 4, 3, (1991),pp187-202.

[2] N.U.Ahmed, Optimization and Identification of Systems Governed by Evolution Equations on Banach Space, Pitman Research Notes in Math. Series, 184, (1988).

[3] N.U.Ahmed, Semigroup Theory with Applications to Systems and Control, Pitman Research Notes in Math. Series, 246, (1991).

[4] G.Da Prato, M.Iannelli, Linear Integro Differential Equations in Banach Space, Rend. Sem. Math. Padova 62, (1980), 207-219.

[5] J.Pruss, Regularity and Integrability of Resolvents of Linear Volterra Equations. Pitman Research Notes in Math Series, vol 190,(1989), pp339-367.

[6] G. Leugering, Boundary Controllability of a Viscoelastic String, Pitman Research Notes in Math Series, vol 190,(1989) pp 258-270.

[7] W.Littman, L.Markus, Stabilization of a Hybrid System of Elasticity by Feedback Boundary Damping, Annali di Matematica pura ed applicata (IV), vol. CLII, (1988) pp 281-330.

[8] M.N.Oguztörelli, Time Lag Control Systems, Academic Press, New York, (1966).

[9] J. Warga, Optimal Control of Differential and Functional Equations, Academic Press, New York, London, (1972).

[10] E.Balder, Necessary and sufficient conditions for $L_1$-strong-weak lower semicontinuity of integral functionals, Nonlinear Analysis,11,(1987), pp1399-1404.

# 26 Data Analysis of a Lumped System

**Louis Auslander** City University of New York, New York, New York

## 1 Introduction

I live in an old house with an ancient heating system, that has been converted from coal to oil, and is controlled by a single thermostat. Several times a winter, we have a visitor who usually sits in the same chair and on occasion complains that our house is "cold". Whenever this happens, my wife jumps up and raises the setting on the thermostat. Soon, I feel hot. I cannot insult my guest by getting out a thermometer and placing it near her chair. So Dear Abby, is there any way of doing some measurements when nobody is here that will enable me to see how valid the complaints are?

The idea would be to keep track of the temperature at the thermostat and the on-off condition of the furnace as a function of time while my guest is here. From this data I would like to be able, with high probability, to estimate the temperature at the chair. Naturally, when no one is here, I can take measurements at the chair, at the thermostat and track the on-off condition of the furnace.

This can be formalized as follows: I can collect a vector function of time $V(t)$ whose first component $H(t)$ is 0 or 1, depending on the furnace being on or off, whose second component, $T(t)$, is the temperature at the thermostat, and whose third component is $C(t)$, the temperature at the chair. Let

$$V(t) = \begin{pmatrix} H(t) \\ T(t) \\ C(t) \end{pmatrix} \quad \text{and} \quad O(t) = \begin{pmatrix} H(t) \\ T(t) \end{pmatrix}.$$

Of course, when my visitor is present, I will only know $O(t)$ and would like to infer $C(t)$.

After looking at this problem for a while, I thought that the hidden Markov models, HMMs, used in isolated word recognition in speech would be a good tool. Let me remind the reader that L. R. Rabiner in [2] has presented an excellent tutorial on this subject. However, on reviewing Rabiner's paper, I found that I could not use HMMs directly, but I had to introduce a slightly weaker structure that I call hidden stochastic models. Part of the justification for introducing hidden stochastic models can already be found from the following quote from page 284 of Rabiner's work [2].

"A. Limitations of HMMs.

Although use of HMM technology has contributed greatly to recent advances in speech recogntion there are some inherent limitations to this type of statistical model for speech. A major limitation is the assumption that successive observations (frames of speech) are independent and therefore the probability of a sequence of observations $P(O_1, O_2, ..., O_T)$ can be written as a product of probabilites of individual observations, i.e.

$$P(O_1, ..., O_T) = \prod_{i=1}^{T} P(O_i)$$

...Finally the Markov assumption itself... is clearly inappropriate for speech sounds where dependence often extends through several states. However, in spite of these limitations, this type of statistical model has worked extremely well for certain types of speech recognition problems."

## 2 Hidden Stochastic Models

In this section we will introduce statistical models that I call hidden stochastic models, HSMs, that are more general than hidden Markov models, HMMs. We will see how to specialize HSMs to HMMs and relate, at a very superficial level, these two statistical models. We have followed Rabiner's notation and order of presentation to make as transparent as possible the relation of HSMs to HMMs.

Assume we are given a set $S$ of states $\{S_1, ..., S_N\}$. Let $S^T$ denote the paths of time length $T$ formed from the states in $S$. In Rabiner's notation

$$Q = q_1, ..., q_T$$

is an element of $S^T$. Assume we have a probability measure $\mu$ on $S^T$. For HMMs this measure is obtained from a Markov process denoted by $(\pi, A)$.

Denote by $V = \{V_1, ..., V_m\}$ the observation symbols and let $V^T$ be the paths of length $T$ formed from $V$. An element in $V^T$ is denoted in [2] by

$$O = O_1...O_T \qquad O_t \in V^T$$

and is called an observation sequence. Assume for each $Q \in S^T$ we have a conditional probability defined

$$P(O|Q), \;\; O \in V^T$$

and define the probability of $O$ by

$$P(O) = \sum_Q P(O|Q)P(Q).$$

A discussion of $P(O|Q)$ and $P(O)$ for HMMs is given on page 262 of [2] and is directly determined by Rabiner's matrix $B$ and the assumption of statistical independence of observations. In our setting, we will call $\lambda = \{\mu, P(O|Q)\}$ the hidden stochastic model.

The standard treatment of HMMs centers on the following ideas (see [2] p.261).

"C. The Three Basic Problems for HMMs.

Given the form of HMM of the previous section, there are three basic problems of interest that must be solved for the model to be useful in real-world applications. These problems are the following:

Problem 1. Given the observation sequence $O = O_1O_2...O_T$ and a model $\lambda = \{A, B, \pi\}$, how do we efficiently compute $P(O|\lambda)$, the probability of the observation sequence, given the model?

Problem 3. How do we adjust the model parameters $\lambda = \{A, B, \pi\}$ to maximize $P(O|\lambda)$?"

We see that HSMs really amount to the assumption that we have a solution to Problem 1 after dropping the words "efficiently compute." However, since our definition of observation is in the whole patch, we have no way of solving problem 3 by pulling the structure apart state by state as for HMMs. Finally, for HSMs, since we have a less structured model, we must ask for more information to solve "real-world applications."

We define a word $W$ as a subset of $S^T$. One statistical solution to relating HSMs to real-world problems is the following.

Assume we have performed $K$ experiments with outcomes $Q \in W$ and the observables $O(1), ..., O(K)$. Define

$$P(Q|W) = \frac{\text{frequency of } Q \text{ occuring}}{K}$$

and

$$P(O|Q,W) = \frac{\text{frequency of } O \text{ when } Q \text{ occured}}{\text{frequency of } Q \text{ occuring}}.$$

The conditional probability $P(O|W)$ is then defined in the obvious way by

$$P(O|W) = \sum_{Q \in W} P(O|Q,W)P(Q|W).$$

Of course, this minimal approach may involve a great deal of computation and may not be satisfactory for reasons we will soon see.

# 3 Monotone Functions and Markov Processes

The minimal solution for defining a HSM presented in section 1 is the most naive and assumes the least structure. It suffers from the difficulty that two paths of observations which may contain the same long subpath are considered as entirely different. If we asssume more structure on the problem, we will see that we can use Markov processes to refine the method of assigning probability to paths in order to overcome this difficulty. Before carrying this out, let us present the following quote from Leo Breiman [1], that should help motivate our construction.

"The basic property characterizing Markov chains is a probalistic analogue of a familiar property of dynamical systems. If one has a system of particles and the position and velocities of all particles are given at time t, the equations of motion can be completely solved for the future development of the system. Therefore, any other information given concerning the past of the process up to time t is superfluous as far as future development is concerned. The present state of the system contains all relevant information concerning the future. Probabilistically, we formalize this by defining a Markov chain..."

In our original problem $T(t)$ and $C(t)$ were numerically valued functions. It is clear that if we know that $f(t)$ is monotone on an interval, knowing

$f(t_0)$ gives information about $f(t)$ for $t > t_0$. It is natural to think of relating monotone functions and Markov chains. We will now carry this out.

We are interested in states that are quantized numerical values. More precisely, consider an interval $[a, b]$ and partition $[a, b]$ into M equal pieces by letting $a = a_0 < a_1 < \cdots < a_M = b$ where

$$a_i - a_{i-1} = \frac{b-a}{M}, \quad i = 1, ..., M$$

and $S_i = [a_{i-1}, a_i)$ is called the $i^{th}$ state. We will say that a path of states $q_1, q_2, ..., q_T$ is monotone increasing if $q_k = S_\alpha$ and $q_{k+1} = S_\beta$ implies $\beta \geq \alpha$ for $k = 1, ..., M$. Similarly, if $\alpha \leq \beta$ we will say the path is monotone decreasing.

A path of states will be called monotone if it is monotone increasing or decreasing. It is important to note that given a collection of states, with repetition allowed, it determines precisely one monotone increasing or monotone decreasing path.

Recall that a subset $W \subset S^T$ is called a word. We say a word is monotone increasing (decreasing) if all the paths in $W$ are monotone increasing (decreasing). A word is called monotone if it is either monotone increasing or decreasing.

If $W$ is a monotone word we assign a Markov process $(\pi, A)$ to $W$ as follows:

$$\begin{aligned} \pi_i &= \text{expected frequency of state } S_i \text{ at } t = 1 \\ a_{ij} &= \frac{\text{expected number of transitions from state } S_i \text{ to } S_j}{\text{expected number of transitions from state } S_i}. \end{aligned}$$

Notice that if two paths have a subpath in common, this subpath will make the same contribution to the probability of each path.

Remark: For a monotone word the Markov process assigned above is a left-right model in the language of [2].

Now let $f(t)$, $c \leq t \leq d$, be a function whose range is contained in $[a, b]$ and assume $f(t)$ is sampled at $t_1 < t_2 < \cdots < t_T$. We assign to the sampled function $f(t_i)$, $i = 1, ..., T$ an element of $S^T$ by

$$f(t_i) = S_{\alpha(i)}$$

if $a_{\alpha(i)-1} \leq f(t_i) < a_{\alpha(i)}$. If $S_{\alpha(1)} \cdots S_{\alpha(T)}$ is a monotone sequence, we will call the sampled function monotone.

If the sampled function is not monotone we may break the sampled function into a sequence of monotone increasing followed by monotone decreasing sampled functions or vice versa.

Let us now look at the problem of defining $P(O|Q,W)$. The naive conditional probability defined by

$$P(O|Q,W) = \frac{\text{number of times } O \text{ occurs for } Q \in W}{\text{number of times } Q \text{ occurs}}$$

during some training runs has a serious flaw. Just because an outcome has not occurred in any of our training runs, we should not assign it probability 0. We may just not have met it yet. One of the virtues of HMMs is that every outcome has a probability assigned to it, even if it has not been met in a training run. We will now define $P(O|Q,W)$ so as to meet this objective.

Let $W$ be a monotone word, let $Q \in W$ and let $O$ be the observations in $K$ training runs that realize $Q$. Assume further that $O$ is monotone. Define a Markov process $(\gamma, B)$ as follows:

$$\gamma_i = \frac{\text{number of times } f(t_1) \in S_i}{K}$$

$$b_{ij} = \frac{\text{expected number of times } f(t_\alpha) \in S_i \text{ and } f(t_{\alpha+1}) \in S_j}{\text{expected number of times } f(t_\alpha) \in S_i}$$

where the expectation is taken over the $K$ training runs that realize $Q$. Use this Markov process to define a probability distribution on the set of observations and make it a conditional probability by dividing by $P(Q|W)$.

We see that if we deal with monotone words and assume observation paths are monotone, we can construct a hidden stochastic model that can assign probabilities to observation sequences that have not occurred in training runs. However, it is important to observe that even with these restrictions our training runs <u>must</u> explore all the state paths in a word $W$.

We are finally in a position to close the argument. Let $W_1, \ldots, W_L$ be a collection of monotone words and let $f(t_i)$ be a sampled measurement that is monotone for each of the $L$ words. Assume we cannot see the state path but only the outcome sequence $O = O_1 \cdots O_T$, $O_i = f(t_i)$. We compute $P(O|W_\alpha)$, $\alpha = 1, \ldots, L$ and say $W_\beta$ has occurred if $P(O|W_\beta) = \max P(O|W_\alpha)$.

This is analogous to the scheme given on page 261 of [2].

# 4 My Living Room Revisited

To use the hidden stochastic models discussed in §2, we must break $C(t)$ up into words. To be useful, this breakup must be recognizable from $O(t)$. Thus if $t_1 < t_2 < \cdots$ is a sequence of times during which $H(t)$ is successively 1 or 0 we define a piece $f_k(t)$ by $f_k(t) = C(t)$, $t_k \leq t < t_{k+1}$. Since $H(t)$ is recognizable from $O(t)$ this breakup is observable. We next observe that if we can quantize our problem to ignore overshoot or undershoot, a sampled version of $f_k(t)$ will be monotone.

We next come to the crucial definition. We say that two sampled pieces $f_k(t)$ and $f_l(t)$ are in the same word if their corresponding state paths have the same initial and terminal states. We have now defined our collection of words.

We next notice that $T(t)$, with the same caveat about over and undershoot, when sampled will be monotone when $f(t)$ is. We are thus in a position to apply the theory indicated in the previous section.

Of course now I must really do the job and see how it works. Maybe I should apply for some sort of contract or grant and try it out.

# References


[1] Leo Breiman, *Probability*, Addison-Wesley Pub. Co., 1968.

[2] Lawrence R. Rabiner, "A Tutorial on Hidden Markov Models and Selected Applications in Speech Recognition," Proc. IEEE, **77** (1989), 257-286.

# 27 Ergodic Bellman Systems for Stochastic Games

**Alain Bensoussan** Institut National de Recherche en Informatique et en Automatique, Le Chesnay, France

**Jens Frehse*** Institüt für Angewandte Mathematik der Universität Bonn, Bonn, Germany

## 1 Introduction

The object of this article is to generalize our previous results obtained for ergodic control to ergodic differential games.The main tool is the Bellman equation which leads to a system of equations in the case of games. Let us briefly recall here the results published in [2]. Consider the second order differential operator

(1. 1)
$$A = -\sum_{i,j=1}^{n} \frac{\partial}{\partial x_i} a_{ij} \frac{\partial}{\partial x_j}$$

We assume that

(1. 2) $a_{ij}(x)$ is measurable,periodic and bounded

with period 1 in all components,(for convenience but not necessary),($x \in R^n$). We denote $Y = (0,1)^n$.

(1. 3) $a_{ij}(x)\xi_i\xi_j \geq \beta|\xi|^2,\ \forall \xi \in R^n, \beta > 0$

Let now $H(x,p)$ be the Hamiltonian defined on $R^n \times R^n$ such that

(1. 4) $H$ is Caratheodory, periodic in $x$

(1. 5)
$$-k_1(1+|p|^2) \leq H(x,p) \leq -k_0|p|^2 + K.$$

In ([2]) we proved the following result

**Theorem 1.1** *We assume (1. 2)to (1. 5). Then there exists a pair $z, \rho$ ,where $z \in W^{2,p}(Y)$ periodic and $\rho$ is a scalar ,such that*

(1. 6)
$$Az + \rho = H(x, Dz)$$

To equation (1. 6) is associated in a natural way the sequence of equations

(1. 7)
$$Au_\epsilon + \epsilon u_\epsilon = H(x, Du_\epsilon)$$

and we have the following convergence results

(1. 8)
$$\epsilon u_\epsilon(x) \to \rho\ \forall x$$

*The 2nd author made this research during a stay at INRIA supported by the Von Humboldt price

(1. 9) $$u_\epsilon(x) - u_\epsilon(x_0) \to z(x) \ \in W^{2,p}(Y) \text{ weakly} .$$

Equation (1. 6) is the Bellman equation corresponding to an ergodic control problem,whenever we take

(1. 10) $$H(x,p) = \inf_v (l(x,v) + p.g(x,v))$$

and $A$ is given in the non divergence form

(1. 11) $$A = - \sum_{i,j=1}^{n} a_{i,j} \frac{\partial^2}{\partial x_j \partial x_i}$$

Write the matrix $a$ as

$$a = \frac{1}{2}\sigma\sigma*;$$

then consider the stochastic differential equation

(1. 12) $$\begin{aligned} dy &= g(y,v)dt + \sigma(y)dw \\ y(0) &= x \end{aligned}$$

and the cost function

(1. 13) $$J_{\epsilon,x}(v) = E \int_0^\infty e^{-\epsilon t} l(y,v)\, dt$$

We have

(1. 14) $$u_\epsilon(x) = \inf_v J_{\epsilon,x}(v)$$

We refer to [1] for a thorough discussion of ergodic control. Our purpose here is more to investigate the situation of games.

# 2 Presentation of the problem and main results

## 2.1 Notation and Assumptions

We shall keep the operator $A$ as defined in (1. 1),but we shall introduce $N$ functions (analogous to $H$), namely $H^\nu(x,p^1,\dots,p^N), \nu = 1,\dots,N$ . We call them Hamiltonians for convenience.We shall denote by $p = (p^1,\dots,p^N) \in R^{nN}$.We make the following assumptions

(2. 1) $$\begin{gathered} H^\nu(x,p) \text{is Caratheodory ,periodic in} x \\ H^\nu(x,p) \le k_1(1 + |p||p^\nu|) \end{gathered}$$

and the global condition

(2. 2) $$\sum_{\nu=1}^{N} H^\nu(x,p) \ge k_0|p|^2 - K$$

**Remark 2.1** *When $N = 1$ the assumptions (2. 1),(2. 2) do not reduce to (1. 5). We could give assumptions which generalize (1. 5),but curiously we have not been able to construct examples arising from stochastic games which could fit such assumptions. On the contrary, we shall see in the next section that there exist stochastic games which yield Hamiltonians satisfying (2. 1),(2. 2)*

**Remark 2.2** *It is easy to check that*

$$|H^\nu(x,p)| \le K(1 + |p|^2)$$

**Remark 2.3** *For applications in stochastic analysis, the ellipticity constant $\beta$ may be "small". Thus a smallness condition for $K$ in terms of $\beta$ is not acceptable,and not required here.*

## 2.2 Verification of the assumptions

The Hamiltonians will be related to game problems, as in our previous paper [3]. We consider $N$ controls

$$v = (v_1, \cdots, v_N)$$

and functions

$$l^\nu(x, v) \quad , g(x, v) .$$

We set

$$L^\nu(x, p^\nu, v) = l^\nu(x, v) + p^\nu . g(x, v)$$

and we look for a Nash point in $(v_1, \cdots, v_N)$ of the functionals $L^\nu(x, p^\nu, v)$ . Assuming that such a Nash point exists denoted by

$$\hat{V}(x, p) = \left(\hat{V}_1(x, p), \cdots, \hat{V}_N(x, p)\right)$$

we write

$$H^\nu(x, p) = L^\nu(x, p^\nu, \hat{V}(x, p))$$

Such Hamiltonians are said to derive from Nash games.We recall the definition of a Nash point .We say that $\hat{V}$ is a Nash point if for any $\nu = 1, cdots, N$ ,when we freeze in $L^\nu$ all the components of $v$ except $v_\nu$ to the corresponding components of $\hat{V}$,then the minimum in $v_\nu$ is attained for $\hat{V}_\nu$.

Let us show with an example that we can construct Hamiltonians deriving from Nash games,which satisfy the required assumptions (2. 1),(2. 2). We take ,with $N = 2$,

$$\begin{aligned} l^1(x, v_1, v_2) &= 1/2 v_1^2 + \theta v_1 v_2 + f^1(x) \\ l^2(x, v_1, v_2) &= 1/2 v_2^2 + \theta v_1 v_2 + f^2(x) \end{aligned}$$

where $f^1(x), f^2(x)$ are bounded periodic functions. We define also

$$g(x, v_1, v_2) = v_1 + v_2$$

We thus have

$$\begin{aligned} L^1(x, p^1, v) &= 1/2 v_1^2 + \theta v_1 v_2 + f^1(x) + p^1(v_1 + v_2) \\ L^2(x, p^2, v) &= 1/2 v_2^2 + \theta v_1 v_2 + f^2(x) + p^2(v_1 + v_2) \end{aligned}$$

We can compute the Nash point

$$\hat{V}_1(x, p) = \frac{1}{(1 - \theta^2)}(-p^1 + \theta p^2)$$
$$\hat{V}_2(x, p) = \frac{1}{(1 - \theta^2)}(-p^2 + \theta p^1)$$

and the Hamiltonians

$$H^1(x, p) = -\tfrac{1}{2}(\hat{V}_1)^2 + p^1 . \hat{V}_2 + f^1(x)$$
$$H^2(x, p) = -\tfrac{1}{2}(\hat{V}_2)^2 + p^2 . \hat{V}_1 + f^2(x)$$

The assumption (2. 1) is easily satisfied. To check (2. 2) we first notice that

$$(1 - \theta^2)^2(H^1 + H^2 - f^1 - f^2) = \left((p^1)^2 + (p^2)^2\right)\left(\frac{-1}{2} + \theta - \frac{\theta^2}{2} - \theta^3\right) + 2p^1 p^2(-1 + \theta + \theta^2)$$

Therefore ,in order that assumption (2. 2) be satisfied,it is necessary and sufficient that

$$(2.\ 3) \qquad |-1 + \theta + \theta^2| < \frac{-1}{2} + \theta - \frac{\theta^2}{2} - \theta^3$$

We leave it as an exercise to check the following

**Lemma 2.1** *The property (2. 3) is satisfied if and only if* $\theta < -\frac{3}{2}$

### 2.3 Main results

We consider the following system of equations(called the Bellman system )

$$
\begin{aligned}
Az^\nu + \rho^\nu &= H^\nu(x, Dz) \\
z^\nu \in W^{2,p}(Y), \text{ periodic} \quad & \quad \rho^\nu \text{ constant}
\end{aligned}
\tag{2. 4}
$$

Our main result is the following

**Theorem 2.1** *We make the assumptions (1. 1),(1. 2),(1. 3),(2. 1), (2. 2),and dimension $n = 2$ then there exists a solution $z, \rho$ of (2. 4)*

**Remark 2.4** *The assumption $n = 2$ comes from the technique. It seems very likely that Theorem 2.1 can be improved. Nevertheless it is the only result which is available for this problem,to the best of our knowledge.*

**Remark 2.5** *Theorem 2.1 can be used to give the solution of a Nash stochastic differential game.This is done as in the control case described above with the necessary adaptations as in [3]*

**Remark 2.6** *For $\epsilon$ fixed,the $L^\infty$ estimate can be obtained up to $\theta \leq \frac{1}{2}$.Corresponding existence theorems for stochastic games can be derived*

## 3 Approximation

### 3.1 Description of the approximated problem

The approximated problem is defined by

$$
\begin{aligned}
Au^\nu_\epsilon + \epsilon u^\nu_\epsilon &= H^\nu(x, Du_\epsilon) \\
u^\nu_\epsilon \in W^{2,p}(Y), \text{ periodic} &
\end{aligned}
\tag{3. 1}
$$

We have the following

**Theorem 3.1** *We make the assumptions (1. 1),(1. 2),(1. 3),(2. 1), (2. 2); then there exists a solution ;$u_\epsilon$ of (3. 1)*

The proof of Theorem 3.1 uses another approximation ,as follows

$$Au^\nu_{\epsilon,\delta} + \epsilon u^\nu_{\epsilon,\delta} = \frac{H^\nu(x, Du_{\epsilon,\delta})}{1 + \delta |Du_{\epsilon,\delta}|^2}$$

for which we know ,since the right hand side is bounded,that there exists a solution in $W^{2,p}(Y)$. The method then consists of obtaining a priori estimates,uniform with respect to $\delta$,and allowing to let $\delta$ tend to 0.
To simplify a little the presentation,we shall perform the calculations leading to the a priori estimates directly on solutions of (3. 1).

### 3.2 First estimates

We begin with the following lemma,(Maximum Principle)

**Lemma 3.1**

$$|\epsilon u^\nu_\epsilon(x)| \leq C \quad \textit{where } C \textit{ is independant of } \epsilon$$

PROOF:
Using the assumption (2. 1) we can write

$$Au_\epsilon^\nu + \epsilon u_\epsilon^\nu \leq k_1(1 + |Du_\epsilon||Du_\epsilon^\nu|)$$

Using the weak Maximum Principle, it is straightforward from the above inequality to prove the upper estimate

(3. 2) $$\epsilon u_\epsilon^\nu \leq k_1$$

Next adding the equations (3. 1) with $\nu$ running from 1 to $N$ we obtain,using the assumption(2. 2),

$$A\sum_{\nu=1}^{N} u_\epsilon^\nu + \epsilon\sum_{\nu=1}^{N} u_\epsilon^\nu \geq -K$$

and from the weak Maximum Principle again we deduce

(3. 3) $$\epsilon\sum_{\nu=1}^{N} u_\epsilon^\nu \geq -K$$

and from (3. 2),(3. 3)the desired result follows.
♠

**Lemma 3.2**

$$\int_Y |Du_\epsilon|^2\, dx \leq C$$

PROOF:
We have from the assumption(2. 2)

$$A\sum_{\nu=1}^{N} u_\epsilon^\nu + \epsilon\sum_{\nu=1}^{N} u_\epsilon^\nu \geq k_0|Du_\epsilon|^2 - K$$

and integrating over $Y$ we obtain, thanks to Lemma 3.1,

$$k_0\int_Y |Du_\epsilon|^2\, dx \leq K + C$$

the result follows.
♠

### 3.3 Important inequalities

We shall consider for any point $x_0 \in Y$,the ball $B_R(x_0)$,with $R < R_0$. The number $R_0$ is sufficiently large so that

$$B_{R_0}(x_0) \supset Y, \forall\, x_0 \in Y$$

We consider cut off functions

$$\tau_R \text{ Lipschitz } \tau = 1 \text{ on } B_R,\ \text{supp } \tau \subset B_{2R}, |D\tau_R| \leq \frac{1}{R}$$

We begin with "Cacciopoli's inequality"

**Lemma 3.3**

$$\int_{B_R} |Du_\epsilon|^2\,dx \le K_1 \int_{B_{2R}-B_R} |Du_\epsilon|^2\,dx + K_2 \int_{B_{2R}} |Du_\epsilon|^2 |u_\epsilon - \overline{u_{\epsilon,2R}}|\,dx + K_3 R^{n+2}$$

*where*

$$\overline{u_{\epsilon,2R}} = \frac{1}{|B_{2R} - B_R|} \int_{B_{2R}-B_R} u_\epsilon\,dx$$

PROOF:
We test equation(3. 1) with

$$(u_\epsilon^\nu - \overline{u_{\epsilon,2R}^\nu})\tau_R^2.$$

Using the quadratic growth of $H^\nu$ and Hölder's inequality,we obtain easily the estimate ,(we use a generic notation for the constants):

$$\beta \int_{B_{2R}} |Du_\epsilon^\nu|^2 \tau_R^2\,dx \le K_1 \int_{B_{2R}} |u_\epsilon^\nu - \overline{u_{\epsilon,2R}^\nu}|^2 |D\tau_R|^2\,dx + K_2 \int_{B_{2R}} |Du_\epsilon|^2 |u_\epsilon^\nu - \overline{u_{\epsilon,2R}^\nu}| \tau_R^2\,dx + K_3 \int_{B_{2R}} |u_\epsilon^\nu - \overline{u_{\epsilon,2R}^\nu}| \tau_R^2\,dx$$

We notice that

$$\int_{B_{2R}} |u_\epsilon^\nu - \overline{u_{\epsilon,2R}^\nu}|^2 |D\tau_R|^2\,dx \le \frac{1}{R^2} \int_{B_{2R}-B_R} |u_\epsilon^\nu - \overline{u_{\epsilon,2R}^\nu}|^2\,dx$$

$$\le C \int_{B_{2R}-B_R} |Du_\epsilon^\nu|^2\,dx$$

where we have used Poincaré's inequality.
Now ,using again Poincaré's inequality, we have

$$\int_{B_{2R}} |u_\epsilon^\nu - \overline{u_{\epsilon,2R}^\nu}| \tau_R^2\,dx \le C R \int_{B_{2R}} |Du_\epsilon^\nu|\,dx$$
$$\le \delta \int_{B_{2R}} |Du_\epsilon^\nu|^2\,dx + \frac{C}{\delta} R^{n+2}$$

where $\delta$ is arbitrarily small. We split the integral over $B_{2R}$ into an integral over $B_R$ and an integral over $B_{2R} - B_R$.Since $\delta$ is small the integral over $B_R$ can be absorbed by the left hand side. Collecting results , "Cacciopoli's inequality" has been proven.
♠ We shall also need the following inequality ,called the "inhomogeneous hole filling inequality "

**Lemma 3.4**

$$\int_{B_R} |Du_\epsilon|^2\,dx \le K \left( \int_{B_{2R}-B_R} |Du_\epsilon|^2\,dx \right)^{\frac{1}{2}} + K R^n$$

PROOF:
We add up all equations (3. 1) and test with $\tau_R$. Using the assumption (2. 2) and Hölder inequality for the left hand side,the desired result is easily obtained.
♠

## 4 Hölder estimates

In this section key estimates will be proved ,namely that the sequence $u_\epsilon(x) - u_\epsilon(x_0)$ remains in a bounded subset of a space $C^\delta(\overline{Y})$, for some convenient $\delta > 0$. We begin with an interesting property,which is the consequence of the "inhomogeneous hole filling inequality."

## 4.1 Proof of the smallness condition

The following technique has been introduced by the second author, [4]. We shall prove the

**Proposition 4.1**

$$\int_{B_R} |Du_\epsilon|^2\, dx \le C_0 \frac{1+\log \log_2 \frac{R_0}{2R}}{1+\log_2 \frac{R_0}{2R}} \quad \forall R \le \frac{R_0}{4}$$

PROOF:
We first notice that

$$R^n \le k_n \frac{1}{\log_2 \frac{R_0}{R}} \forall R < R_0$$

with

$$k_n = \frac{R_0^n\, e^{-1}}{n \log_2}$$

hence the "inhomogeneous hole filling inequality " Lemma 3.4 implies also

$$\int_{B_R} |Du_\epsilon|^2\, dx \le K \left( \int_{B_{2R}-B_R} |Du_\epsilon|^2\, dx \right)^{\frac{1}{2}} + K k_n \frac{1}{\log_2 \frac{R_0}{R}} \forall R < R_0 \tag{4. 1}$$

We next state the following algebraic result

**Lemma 4.1** *Let $s_i$ be a sequence of positive real numbers such that*

$$\begin{array}{lll} s_{i+1} & \le & s_i \\ s_i & \le & a\sqrt{s_{i-1}-s_i} + bi^{-1} \end{array} \tag{4. 2}$$

*where $a, b > 0, \ i \ge 1$.*
*Then one has the property*

$$s_i \le \frac{1}{i+1} s_0 + (b + \frac{a^2}{4}) \frac{1}{i+1}(1 + \log\, i), \quad i \ge 1 \tag{4. 3}$$

PROOF:
Note that (4. 4)

$$s_i \le i(s_{i-1} - s_i) + (b + \frac{a^2}{4}) i^{-1}$$

hence,setting $b_0 = b + \frac{a^2}{4}$

$$s_i \le \frac{i}{i+1} s_{i-1} + b_0 \frac{1}{i(i+1)}$$

By induction,one checks that

$$s_{i+j} \le \frac{i}{i+j+1} s_{i-1} + b_0 \frac{1}{i+j+1} \sum_{l=i}^{j+i} \frac{1}{l}$$

In the previous relation we let $i = 1, j = j-1$ we obtain

$$s_j \le \frac{1}{j+1} s_0 + b_0 \frac{1}{j+1} \sum_{l=1}^{j} \frac{1}{l}$$

with $j \geq 1$.
Now we use the property

$$1 + \log j \geq \sum_{l=1}^{j} \frac{1}{l} \; \forall j \geq 1$$

to obtain (4. 5).
♠
PROOF OF PROPOSITION 4.1:
We set for $j \geq 1$

$$R_j = R_0 2^{-j}, \; s_j = \int_{B_{R_j}} |Du_\epsilon|^2 \, dx$$

By definition

$$j = \log_2 \frac{R_0}{R_j}$$

and the condition $j \geq 1$ implies $R_j \leq \frac{R_0}{2}$. Applying (4. 1) we deduce

$$s_j \leq \sqrt{s_{j-1} - s_j} + K k_n \frac{1}{j}$$

which coincides with (4. 4). Therefore applying Lemma 4.1 we obtain

$$\int_{B_{R_j}} |Du_\epsilon|^2 \, dx \leq C_0 \frac{1 + \log \log_2 \frac{R_0}{R_j}}{1 + \log_2 \frac{R_0}{R_j}}$$

where $C_0$ is a suitable constant. Using the fact that the function $\frac{1 + \log \log_2 x}{1 + \log_2 x}$ is decreasing for $x \geq 4$, it follows that for any $R \leq \frac{R_0}{4}$ we can find $j \geq 2$ such that

$$R_{j+1} \leq R \leq R_j;$$

hence $2R \leq R_j$. The desired result follows easily.
♠

## 4.2 Hölder and Morrey's norm

We introduce the Hölder semi-norm

$$[u_\epsilon]_\alpha = \sup\{\frac{|u_\epsilon(x) - u_\epsilon(y)|}{|x - y|^\alpha} \mid x, y \in Y\}$$

and Morrey's semi norm

$$|||Du_\epsilon|||_{2,\alpha}^2 = \sup\{\frac{1}{R^{n-2+2\alpha}} \int_{B_R(x_0)} |Du_\epsilon|^2 \, dx \, | x_0 \in Y, \; R < R_0\}$$

Then one has Morrey's Lemma

$$[u_\epsilon]_\alpha \leq C|||Du_\epsilon|||_{2,\alpha}$$

where the constant does not depend on the function,hence on $\epsilon$. We shall prove the following result

**Proposition 4.2** *Assume $n = 2$, then one has*

$$|||Du_\epsilon|||_{2,\alpha} \leq C_0$$

*where*

$$\alpha = \frac{1}{2}\log_4(1 + K_1^{-1})$$

PROOF:
We start with "Cacciopoli's inequality",from which we deduce ,[5]

$$\int_{B_R} |Du_\epsilon|^2\, dx \leq \frac{K_1}{K_1+1}\int_{B_{2R}} |Du_\epsilon|^2\, dx + k_2\int_{B_{2R}} |Du_\epsilon|^2 |u_\epsilon - \overline{u_{\epsilon,2R}}|\, dx + k_3 R^{n+2}.$$

Multiplying by $R^{-n+2-2\alpha}$, with $\alpha$ to be defined later, yields

$$R^{-n+2-2\alpha}\int_{B_R} |Du_\epsilon|^2\, dx \leq \frac{K_1}{K_1+1}R^{-n+2-2\alpha}\int_{B_{2R}} |Du_\epsilon|^2\, dx + k_2 R^{-n+2-2\alpha}\int_{B_{2R}} |Du_\epsilon|^2 |u_\epsilon - \overline{u_{\epsilon,2R}}|\, dx + k_3 R^{4-2\alpha}$$

(4. 4)
Now we use

$$R^{-n+2-2\alpha}\int_{B_{2R}} |Du_\epsilon|^2\, dx \leq 2^{n-2+2\alpha}|||Du_\epsilon|||_{2,\alpha}^2$$

and

$$|u_\epsilon(x) - \overline{u_{\epsilon,2R}}| \leq (4R)^\alpha [u_\epsilon]_\alpha.$$

Therefore also (taking account of Morrey's Lemma),

$$R^{-n+2-2\alpha}\int_{B_{2R}} |Du_\epsilon|^2 |u_\epsilon - \overline{u_{\epsilon,2R}}|\, dx \leq C R^{\frac{-n+2}{2}} 2^{\frac{n-2}{2}+3\alpha}\left(\int_{B_{2R}} |Du_\epsilon|^2\, dx\right)^{\frac{1}{2}} |||Du_\epsilon|||_{2,\alpha}^2$$

and from (4. 4) we deduce, collecting results

$$R^{-n+2-2\alpha}\int_{B_R} |Du_\epsilon|^2\, dx \leq \left(\frac{K_1}{K_1+1}2^{n-2+2\alpha} + k_2 C R^{\frac{-n+2}{2}} 2^{\frac{n-2}{2}+3\alpha}\left(\int_{B_{2R}} |Du_\epsilon|^2\, dx\right)^{\frac{1}{2}}\right) |||Du_\epsilon|||_{2,\alpha}^2 + k_3 R^{4-2\alpha}$$

(4. 5)
and if $n = 2$ ,(4. 5) yields

$$(4.\ 6)\quad R^{-2\alpha}\int_{B_R} |Du_\epsilon|^2\, dx \leq \left(\frac{K_1}{K_1+1}2^{2\alpha} + k_2 C 2^{3\alpha}\left(\int_{B_{2R}} |Du_\epsilon|^2\, dx\right)^{\frac{1}{2}}\right) |||Du_\epsilon|||_{2,\alpha}^2 + k_3 R^{4-2\alpha}$$

We now choose $\alpha > 0$ so small such that $\frac{K_1}{K_1+1}2^{2\alpha} < 1$. For instance, we take

$$\alpha = \frac{1}{2}\log_4(1 + \frac{1}{K_1})$$

which means

$$2^{2\alpha} = (\frac{1+K_1}{K_1})^{1/2}.$$

Now from Proposition 4.1 ,it follows that for any $\delta$, there exists $R_\delta$ independant of $\epsilon$ such that

$$R^{-2\alpha}\int_{B_R} |Du_\epsilon|^2\, dx \leq \delta,\ \forall R \leq R_\delta.$$

Choose then $\delta$ such that (recall the choice of $\alpha$))

$$\sqrt{\delta} = \frac{1}{2k_2 C}\frac{K_1}{K_1+1}^{3/4}(1 - \frac{K_1}{K_1+1})^{1/2}$$

We deduce

$$R^{-2\alpha}\int_{B_R}|Du_\epsilon|^2\,dx \le \frac{1}{2}(1+\frac{K_1}{1+K_1}^{1/2})|||Du_\epsilon|||^2_{2,\alpha} + k_3R^{4-2\alpha}_{\delta/2}\,\forall R < R_{\delta/2}.$$

Note also

$$R^{-2\alpha}\int_{B_R}|Du_\epsilon|^2\,dx \le R^{-2\alpha}_{\delta/2}\int_Y|Du_\epsilon|^2\,dx\,\forall R > R_{\delta/2}$$

Since either the sup in $R$ of the left hand side is attained for $R < R_{\delta/2}$ ,or it is attained for $R \geq R_{\delta/2}$ ,we can assert that either

$$1/2(1-\sqrt{\frac{K_1}{1+K_1}})|||Du_\epsilon|||^2_{2,\alpha} \le k_3R^{4-2\alpha}_{\delta/2}$$

or

$$|||Du_\epsilon|||^2_{2,\alpha} \le R^{-2\alpha}_{\delta/2}\int_Y|Du_\epsilon|^2\,dx$$

Hence $|||Du_\epsilon|||^2_{2,\alpha}$ is estimated by a constant independant of $\epsilon$ ,and the desired result follows. ♠

## 5 Proof of the main result

We can now proceed with the proof of Theorem 2.1. We set

$$z_\epsilon(x) = u_\epsilon(x) - u_\epsilon(x_0)$$

then,using Proposition 4.2 and Morrey's Lemma we deduce

$$||z_\epsilon||_{(C^\alpha(\overline{Y}))^N} \le C_1$$

and ,of course, from Lemma 3.2 we also have

$$||z_\epsilon||_{(H^1(Y))^N} \le C_1\,.$$

Recalling Lemma 3.1 we can extract a subsequence such that

$$\epsilon\, u_\epsilon \to \rho \text{ constant } \in (L^\infty(Y))^N$$

$$z_\epsilon \to z \text{ in } (L^\infty(Y))^N \text{ strongly}$$

from the compactness of the injection of $(C^\alpha(\overline{Y}))^N$ into $(L^\infty(Y))^N$. Also

$$z_\epsilon \to z \text{ in } (H^1(Y))^N \text{ weakly}$$

In fact,from the equations,the weak convergence in $H^1$ and the strong convergence in $L^\infty$ it also follows,by classical arguments that

$$z_\epsilon \to z \text{ in } (H^1(Y))^N \text{ strongly}\,.$$

This allows to pass to the limit in equations (2. 4) and we obtain equations (3. 1).The passage from $C^\alpha$ to $W^{2,p}$ is standard .This concludes the proof of Theorem 2.1.

## References

[1] Alain Bensoussan ,*Perturbation Methods in Optimal Control*,Wiley,Gauthier-Villars Series in Modern Applied Mathematics .

[2] Alain Bensoussan , Jens Frehse ,On Bellman equation of ergodic type with quadratic growth Hamiltonian,*Contributions to Modern Calculus of Variations*, Pitman Res.Notes Math.Series,ed. L.Cesari,Longman,vol 148(1987),13-26.

[3] Alain Bensoussan ,Jens Frehse ,Nonlinear elliptic systems in stochastic game theory ,Journal für die reine und angewandte Mathematik ,Band 350(1983),23-67.

[4] Jens Frehse ,On two-dimensional quasilinear elliptic systems, Manuscripta Mathematica 28 ,(1979) 21-49.

[5] Kjell-O. Widman, Hölder continuity of solutions of elliptic systems, Manuscripta Mathematica 5, (1971) (299-308).

# 28 Some Results on Feedback Stabilizability of Nonlinear Systems in Dimension Three or Higher

**William M. Boothby**[1] Washington University, St. Louis, Missouri
**Riccardo Marino**[2] University of Rome "Tor Vergata," Rome, Italy

## 1. INTRODUCTION.

In this note we consider nonlinear control systems of the form

$$(*) \qquad \dot{z} = f(z) + u_1 g_1(z) + \ldots + u_m g_m(z), z \in R^n$$

defined on a neighborhood $U_0$ of the origin $z = 0$ in $R^n$ which have 0 as an equilibrium point, i.e. $f(0) = 0$. It is no loss of generality to assume that $g_1(z), \ldots, g_m(z)$ are linearly independent for all $z \in U$, and we shall do so. We will also assume for simplicity of exposition that $f, g_1, \ldots, g_m \in C^\infty(U_0)$, i.e. smooth on $U_0$ , and that the system is strongly accessible at $z = 0$, i.e. that the strong accessibility Lie algebra $L_0$ has rank n at $z = 0$. These assumptions are more a matter of convenience than necessity: $C^2$ differentiability is usually sufficient for our purposes and we could, in general, restrict to a lower dimensional submanifold of $U_0$ on which strong accessibility holds. The basic concepts of nonlinear control theory which we use, and many general references may be found in Isidori [7]; more particularly the ideas applied here may be found in the authors' papers (see References). An interesting and important question which arises for such systems is the following:

*Does there exist a smooth (local) static, state feedback $u(z)$ such that $\dot{z} = f(z) + g(z)u(z)$ is asymptotically stable at $z = 0$. In brief: is $(*)$ feedback stabilizable ?*

[We have used the notation: $g(z) = (g_{ij}(z))$ is an nxm matrix whose columns are the components of $g_1, \ldots, g_m$ and that $u(z) = (u_1(z), \ldots, u_m(z))^t$.] Two systems of the form $(*)$ are said to be *feedback equivalent* if they are related by a change of variables and static state feedback of the form $u(z) = \alpha(z) + \beta(z)v$, $\alpha(0) = 0$, $\alpha(z)$ and $v$ are m-vectors and $B(z)$ is a nonsingular mxm matrix (all $C^\infty$). All of

[1] Support from the NSF (during early stages of this work) and from the University of Rome II, "Tor Vergata" are gratefully acknowledged

[2] Partially supported by the Ministero della Universita e della Ricerca Scientifica

our conditions and results are invariant under such equivalence. There is a fairly extensive literature on this question, especially for the two dimensional single input case. See, for example [5],[9], and [13] and the extensive references they cite. One of the principal differences between our work and that of [5] and [9] is that we are only interested in conditions for smooth (as opposed to continuous) feedback stabilizability. In [5], for example, there is a very complete analysis of the continuous case in dimension two. The techniques for the smooth case are often quite different from those used in the continuous case. Our purpose here is to try to extend, at least to $n = 3$, some of the results we obtained for smooth stabilizability by applying the Center Manifold Theorem to the planar single input case ($n = 2, m = 1$) in our papers [1]-[3]. Although in the present note we primarily consider three dimensional systems, in several cases this restriction is unnecessary as we shall point out.

This problem is somewhat narrower in scope than it would appear. Recall that the answer to the question posed is well known except in the critical case, in which the associated linear system

$$\dot{z} = Az + Bu,\ A = (\partial f/\partial z)_0,\ B = (g_1(0), ..., g_m(0))$$

is not controllable and the set of eigenvalues corresponding to the uncontrollable modes contains no eigenvalues with positive real part and at least one with real part zero. It is the latter condition that determines the criticality. If there are any eigenvalues in this set which have positive real part, then it is impossible to make the origin asymptotically stable by smooth (even $C^1$) feedback. On the other hand, if all the eigenvalues in this set have negative real part, then it is easy to show that the system can be stabilized asymptotically at the origin by linear feedback.

To apply the techniques of nonlinear geometric control theory, one must make assumptions about dimension and involutivity of certain naturally defined distributions. To state the particular assumptions made in this note, we use the notation of [10] as follows: Let $G(z) = G^0(z) = span\{g_1(z), ..., g_m(z)\}$ in $T_z(R^n)$ and let $G = G^0$ denote the corresponding $C^r(U_0)$ module. Then define $G^j = span\{G^{j-1}, [f + G, G^{j-1}]\}$ as a $C^r(U_0)$ module and let $G^j(z)$ be the corresponding subspace of $T_z(R^n)$. Recall that according to [10, Theorem 2] there exists $p > 0$ such that $G^p = G^{p+1} = ... = L_0$ and $G^{p-1} \not\subseteq G^p$; we have assumed $\dim L_0(z) = n$ on $U^0$. We next make the following two definitions:

**Definition 1.** We say that dim $G^j = k$ if dimension (or rank) of $G^j(z) = k$ (as a linear subspace of $T_z(R^n)$ on an open set U such that $0 \in \overline{U}$. Noting that dim $G^j(0) \leq dim G^j$, we say that $G^j$ is *regular* at 0 if equality holds; otherwise it is said to be *singular*.

Note that in case the data are real analytic the regular points of $G^j$ form an open, dense set.

**Definition 2.** The integer $\nu$, if it exists, is the largest positive integer such that $G^0, G^1, ...G^{\nu-1}$ are involutive and of constant rank on some neighborhood U of 0. (Recall that $G^j$ is involutive if $[G^j, G^j] \subseteq G^j$).

**Remarks.**

(1). $\nu$ exists and is at least 1 if $G = G^0$ is involutive since we have assumed $\{g_1(0), ..., g_m(0)\}$ is a linearly independent set, i.e. that $G^0$ is regular at 0.

(2). $G^j$ can be both singular at 0 and involutive, but it is of constant rank on a neighborhood of 0 if and only if it is regular at 0.

(3). It is quite possible for $G^j, G^{j+1}, ..., G^{l-1}$ to be singular and $G^l$ to be regular at 0.

In all that follows, we shall assume that $\nu$ exists. This is equivalent to requiring that $G^0$ be regular at 0 **and** involutive. When $m = 1$, this assumption simply means that $g(0) \neq 0$ since a one dimensional distribution is automatically involutive. When $m > 1$, it is more restrictive. It then means that $(*)$ is feedback equivalent to a system in which m of the variables can be controlled directly, i.e. we can introduce coordinates around 0 such that $g_j$, for $j = 1, ..., m$, has components which are all 0 except for the $(r + j) - th$ which is 1 ($r = n - m$ the codimension of the state space $R^n$).

In summary then, we assume :

*(i) The system $(*)$ is critical: all eigenvalues of the uncontrollable modes have nonpositive real parts and for least one of them the real part is zero.*
*(ii) $\nu$ is defined, i.e. G is both regular at 0 (of dimension m) and involutive.*

We have also assumed strong accessibility: dim $L_0 = n$, although, as noted, this assumption is one of convenience. Of course, all of these conditions are completely independent of the choice of local coordinates. The conditions (i) and (ii) reduce to $df = 0$ and $g(0) \neq 0$ at $z = 0$ in the planar, single input case, which, although already very complicated, presents many interesting problems and has been extensively studied, as we have noted.

We may begin our development by noting the following rather self-evident proposition which shows that we need only consider cases with $m < n$. We remark that the proposition does not use (i) or (ii), that strong accessibility is implied by the hypotheses and moreover that $C^2$ differentiability is sufficient.

**Proposition.** *If $m \geq n$ and dimension of $G^0 = m$, then $(*)$ is stabilizable at $z = 0$ by $C^\infty$ static state feedback.*

*Proof.* It is enough to consider the case $m = n$. Let $U$ be a connected neighborhood of 0 on which $\{g_1(z), ..., g_n(z)\}$ are everywhere independent. Then, since $f(0) = 0$, there is a unique nxn matrix F(z) defined on a neighborhood of 0 inside U with the properties that $f(z) = F(z)z$ (matrix multiplication) and $F(0) = (\partial f/\partial z)_0$, the Jacobean matrix of f(z) at $z = 0$. Using the assumption that the $g_i$'s form a basis on $U$, there exist unique $(u_1(z), ..., u_n(z))$ on $U$ such that $\sum u_i(z)g_i(z) = -F(0) - z$ With these $u_i(z)$ we have

$$\dot{z} = f(z) + \sum u_j(z)g_j(z) = F(z)z - F(0)z - z$$

Since the linear part of the right side is $-z$, $z = 0$ is a stable equilibrium point of this system.

We now classify the various possibilities for systems $(*)$ on $R^3$ satisfying our assumptions (i) and (ii) according to the possible values of $m \leq 2$ and $\nu \geq 1$ and try to use the special features in each instance to reduce the difficulties of the problem. In fact, although we are mainly interested in $n = 3$, some of our results are valid for $n > 3$ as well, but the classification is exhaustive only for $n = 3$ .

The procedure is to first use (smooth) nonlinear feedback as defined above to obtain an equivalent system in canonical form which is more transparent for the smooth stabilization problem than $(*)$. We reduce the problem to five cases. Two of these, (3) and (4), turn out to be feedback linearizable and thus stabilizable and (1) can be reduced in a sense to a (difficult) two dimensional problem. The remaining two cases (2) and (5) are the most interesting and also can be generalized to $n > 3$. The technique in cases (1), (2) and (5) involves the Center Manifold Theorem (in the form given in the Appendix) to reduce the problem to one of lower dimension somewhat as we did in [1]-[3]. The results we obtain certainly do not give a complete answer to the question, even with our two assumptions, but they do simplify the problem and allow us to state some sufficient conditions for smooth feedback stabilizability.

## 2. SINGLE INPUT CASES.

**Case 1.** $m = 1, \nu = 1$.

We are assuming $g(0) \neq 0$ so that $g$ determines a distribution of dimension one, which is automatically involutive on U. Since $\nu = 1, G^1$ is either singular at 0 or not involutive. Let $\phi_2$ and $\phi_3$ be functions such that $\langle g, d\phi_2\rangle = 0 = \langle g, d\phi_3\rangle$ and $d\phi_2 \wedge d\phi_3 \neq 0$ on a neighborhood of 0, and choose a third function $\phi_1$ such that

$d\phi_1, d\phi_2, d\phi_3$ are independent around 0. Taking new coordinates $z' = (z_1', z_2', z_3') = (\phi_1(z), \phi_2(z), \phi_3(z))$, we have a system which is feedback equivalent to $(*)$:

$$\begin{aligned} \dot{z}_1' &= L_f\phi_1 + L_g\phi_1 u \\ \dot{z}_2' &= L_f\phi_2(z) \\ \dot{z}_3' &= L_f\phi_3(z) \end{aligned}$$

Now let $x_1, x_2$ denote $z_2', z_3'$ and $y$ denote $z_1'$ and using this new notation for $L_f\phi_i, i = 2, 3$ together with feedback: $v = L_f\phi_1 + L_g\phi_1 u$, we have, finally

$$\begin{aligned} \dot{x}_1 &= f_1(x_1, x_2; y) \\ \dot{x}_2 &= f_2(x_1, x_2; y) \\ \dot{y} &= v \end{aligned}$$

Thus the original system is feedback stabilizable if and only if this one is. From the version of the Center Manifold Theorem given in the Lemma of the Appendix, we see that if there is a $C^\infty$ surface through the origin of the form: $y = h(x_1, x_2)$, $h(0,0) = 0$, for which the projection $f^S$ of the vector field $f$, corresponding to $v = 0$, defined by this canonical form, has the origin as an asymptotically stable equilibrium point, this system, hence $(*)$, is smoothly feedback stabilizable. In other words, if the two dimensional system

$$\begin{aligned} \dot{x}_1 &= f_1(x_1, x_2; h(x_1, x_2)) \\ \dot{x}_2 &= f_2(x_1, x_2; h(x_1, x_2)) \end{aligned}$$

for some choice of $h(x_1, x_2), h(0,0) = 0$, has the origin as an asymptotically stable equilibrium point, then so too does $(*)$.

**Case 2.** $m = 1, \nu = 2$.

Then $G^0, G^1$ are involutive of constant rank 1 and 2 respectively and $G^2$ must be singular at 0. This latter fact follows from our assumption that rank $L_0(0) = 3$ for, if we suppose that $G^2$ is regular at 0, then we must have either (a) rank $(G^2) = 3$ so $G^2$ is involutive and $\nu = 3$ or (b) rank $(G^2) =$ rank $G^1$ so that rank $L_0 < 3$ (see the definition of $G^j$ above). In either case our assumptions are contradicted. We now may choose functions $\phi_1(z), \phi_2(z), \phi_3(z)$ as follows. First take $\phi_1$ so that $\langle g, d\phi_1\rangle = 0$ but $\langle ad_f g, d\phi_1\rangle$ is never zero on a neighborhood of 0, let $\phi_2 = L_f\phi_1$, and then choose $\phi_3$ such that $\langle g, d\phi_3\rangle = 0 = \langle ad_f g, d\phi_3\rangle$. Then $d\phi_1 \wedge d\phi_2 \wedge d\phi_2 \neq 0$ and we may define new coordinates $z' = (\phi_1(z), \phi_2(z), \phi_3(z))$. Note that $\{dz_3 = 0\} = span\{\partial/\partial z_1', \partial/\partial z_2'\} = span\{g, ad_f g\}$. On the other hand $\langle g, dz_1'\rangle \equiv 0$ so $g = \alpha\, \partial/\partial z_2'$ for some $\alpha(z') \neq 0$.

The system of equations $(*)$ in the $z'$ coordinates is

$$\begin{aligned}\dot z_1' &= L_f\phi_1 = z_2'\\ \dot z_2' &= L_f\phi_1 + L_gL_f\phi_1 u\\ \dot z_3' &= L_f\phi_3(z_1', z_2', z_3')\end{aligned}$$

Using feedback we take $L_f\phi_1 + L_gL_f\phi_1 u = v$ and we further note that

$$\langle g, dL_f\phi_3\rangle = \alpha\,\langle \partial/\partial z_2', dL_f\phi_3\rangle = \alpha[L_f\,\langle g, d\phi_3\rangle + \langle ad_f g, d\phi_3\rangle] \equiv 0$$

so $L_f\phi_3$ is independent of $z_2'$. Hence the system, with the $z'$ written as $z$, is equivalent to the canonical one

$$\begin{aligned}\dot z_1 &= z_2\\ \dot z_2 &= v\\ \dot z_3 &= f_3(z_1, z_3)\end{aligned}$$

Now suppose that we can find a feedback function $u(z_1, z_3)$, $u(0,0) = 0$, which stabilizes the two dimensional system

$$\begin{aligned}\dot z_3 &= f_3(z_1, z_3)\\ \dot z_1 &= u\end{aligned}$$

Then we shall see that $(*)$ can be stabilized. For returning to the canonical three dimensional system above, which we now reorder to

$$\begin{aligned}\dot z_1 &= z_2\\ \dot z_3 &= f_3(z_1, z_3)\\ \dot z_2 &= v\end{aligned}$$

we see that if we take S to be defined by $z_2 = u(z_1, z_3)$, then an application of the Lemma of the Appendix shows that this system can be stabilized by a suitable choice of feedback (given explicitly in the Appendix). This reduces this case to the much better understood two dimensional one. In fact in [1] it was shown that generically there exists a smooth curve $z_1 = w(z_3)$ with $w(0) = 0$ such that $t = 0$ is an asymptotically stable equilibrium point of the scalar equation $\dot z_3 = f_3(w(t), t)$. The Center Manifold Theorem was then applied to give an explicit expression for smooth feedback $u(z_1, z_3)$, which makes the origin an asymptotically stable equilibrium point of the resulting closed loop system. The proof involved a simplified version of the Lemma of the Appendix. We remark that this procedure could be extended to the case $n > 3, m = 1, \nu = n - 1$, reducing it to stabilizing a two dimensional system.

**Case 3.** $m = 1, \nu = 3$.

In this case $G^0, G^1, G^2$ are each involutive of constant rank 1, 2, and 3 respectively (using the fact that rank $L_0 = 3$). This is exactly the condition that the system is feedback equivalent to a controllable linear system (see [10] where this is discussed in relation to the criteria proved in [6] and [8]). Thus it has the canonical form:

$$\begin{aligned} \dot{z}_1 &= z_2 \\ \dot{z}_2 &= z_3 \\ \dot{z}_3 &= v \end{aligned}$$

Such a system is, of course, equivalent by linear state feedback to a stable system. The same is also true for $n > 3, m = 1, \nu = n - 1$

## 3. THE MULTI-INPUT CASES.

We now consider cases of two (or, when $n > 3$, more) inputs.

**Case 4.** $n = 3, m = 2, \nu = 2$.

We have $G^0$ involutive of constant rank 2 and $G^1$ involutive and of constant rank 3; in fact $G^1 = L_0$. This is again a system which is feedback equivalent to a controllable linear system. By feedback as defined above (i.e. including change of variables) it can be written in the form:

$$\begin{aligned} \dot{z}_1 &= z_2 \\ \dot{z}_2 &= v_1 \\ \dot{z}_3 &= v_2 \end{aligned}$$

Again, as is known from linear control theory, it is stabilizable by linear state feedback and the same holds for $n > 3, m = n - 1, \nu = 2$, (see [10]).

**Case 5.** $m = 2, \nu = 1$.

$G^0$ is of constant rank 2 and is involutive (by our basic assumptions), but $G^1$ is either singular at 0 or not involutive. This is in some ways the most interesting case, and is just as easily treated for arbitrary n, with $m = n - 1$ and $G^0$ involutive and not singular at the origin, i.e. dim $G^0(0) = m = n - 1$. Using the same techniques as above we introduce coordinates such that $\langle G^0, dz_3 \rangle = 0$ and use feedback to write the system in the form

$$\begin{aligned} \dot{z}_1 &= v_1 \\ \dot{z}_2 &= v_2 \\ \dot{z}_3 &= \phi(z_1, z_2, z_3) \end{aligned}$$

According to the Lemma of the Appendix, the system will be stabilized if we can find a curve $z = (\tilde{z}_1(z_3), \tilde{z}_2(z_3), z_3)$ passing through the origin which is such that the scalar differential equation $\dot{z}_3 = \phi(\tilde{z}_1(z_3), \tilde{z}_2(z_3), z_3)$ has $z_3 = 0$ as a stable singular point (cf. [1]-[3]). Sufficient conditions for such a curve would be difficult to establish in general; however we can find some cases in which the question can be answered. As remarked, it is just as simple to do so for arbitrary n, in which case the system has the general form:

$$\dot{z}_1 = v_1, ..., \dot{z}_{n-1} = v_{n-1}, \dot{z}_n = \phi(z_1, ..., z_n)$$

and we seek a curve $\tilde{z}(z_n) = (\tilde{z}_1(z_n), ..., \tilde{z}_{n-1}(z_n), z_n)$ through $(0, ..., 0)$ and defined on some $(-\epsilon, \epsilon)$ interval, such that $\dot{z}_n = \phi(\tilde{z}_1(z_n), ..., \tilde{z}_{n-1}(z_n), z_n)$ has $z_n = 0$ as a stable equilibrium point. Equivalently, we look for a $C^2$ curve $\tilde{z}(t) = (\tilde{z}_1(t), ..., \tilde{z}_n(t))$ defined on some open interval $(-\epsilon, \epsilon)$ with $\tilde{z}(0) = 0$, $\dot{\tilde{z}}(0) \neq 0$ such that $\dot{z} = \phi(\tilde{z}_1(t), ..., \tilde{z}_n(t))$ has $t = 0$ as a stable equilibrium point, i.e. for $0 < t < \epsilon$, $\phi(\tilde{z}(t)) < 0$ and for $,-\epsilon < t < 0, \phi(\tilde{z}(t)) > 0$. In other words $t\phi(\tilde{z}(t)) < 0$ for $t \neq 0$ on $(-\epsilon, \epsilon)$. Now write the Taylor expansion of $\phi(z)$ at $z = 0$,

$$\phi(z) = \phi^{(k)}(z) + \phi^{(k+1)}(z) + ...$$

where $\phi^{(j)}(z)$ denotes a homogeneous polynomial of degree j in $z_1, ..., z_n$ and $\phi^{(k)}(z) \neq 0$ is the leading term of the expansion. By considering the associated linear system in the light of (i), it is easy to see that $f$ has no linear terms, i.e. that $k \geq 2$. In order that a curve $\tilde{z}(t)$ with $\tilde{z}(0) = 0$, satisfy the condition $t\phi(\tilde{z}(t)) < 0$ for $t \neq 0$ on some $(-\epsilon, \epsilon)$, it is sufficient that $t\phi^{(k)}(\tilde{z}((t)) < 0$ on some $(-\epsilon, \epsilon)$. On the other hand, if $\phi^{(k)}(z)$ does not change sign on some deleted neighborhood $U \setminus \{0\}$ of the origin, then there can be no curve with the required properties.

We first consider the case $k = 2$, which is "generic" given our assumptions. We then have the non-vanishing form $\phi^{(2)}(z) = \sum a_{ij} z_i z_j$ as leading term. Either $\phi^{(2)}(z) \geq 0$ for all z, in which case the equilibrium point $z = 0$ is unstable or $\phi^{(2)}(z)$ must have both positive and negative values arbitrarily close to $z = 0$. We claim that in this case there is indeed a curve $z = \tilde{z}(t)$ through $z = 0$ with $\dot{\tilde{z}} \neq 0$ which satisfies $t\phi(\tilde{z}(t)) < 0$ for all $t \neq 0$ on some interval $(-\epsilon, \epsilon)$. For convenience we restrict ourselves to the case of quadratic forms of maximum rank. Thus let $\phi^{(2)}(z)$ be of maximum rank and take both positive and negative values. Then there is an orthonormal change of coordinates $z = z(x)$ such that $\phi^{(2)}(z(x)) = \lambda_1 x_1^2 + ...\lambda_p x_p^2 - \lambda_{p+1} x_{p+1}^2 ... - \lambda_n x_n^2$, where $0 < p < n$ and $\lambda_i > 0, i = 1, ..., n$. For simplicity we denote the form $\phi^{(2)}$ in the x-coordinates simply by "$\sigma$", and note that

$$\frac{1}{2} \operatorname{grad} \sigma = \lambda_1 x_1, ..., \lambda_p x_p, \lambda_{p+1} x_{p+1}, ..., \lambda_n x_n$$

Hence $\|\text{grad}\,\sigma\| \geq 0$ and $= 0$ if and only if $x = 0$. It follows that on $R^n\setminus\{0\}, \sigma = 0$ is a nonsingular orientable hypersurface which consists of a cone of lines through the origin. We note that if $\sigma(a) > 0$, then $\sigma(ta) > 0$ for all $t \neq 0$ and that each component of $\sigma = 0$ separates $R^n\setminus\{0\}$ (see, e.g. [12]) with positive values on one side of the hypersurface and negative values on the other. We can see this directly as follows: let $a \neq 0$ be a point of the hypersurface $\sigma = 0$ and (grad $\sigma)_a$ be the gradient at $a$. It will be orthogonal to the hypersurface. Now $p(s) = a + s(\text{grad}\,\sigma)_a$ is the line through $a$ along the gradient vector; consider the values of $\sigma$ along this curve, i.e. consider $\sigma(p(s))$. We have that $d\sigma/ds = \langle(\text{grad}\,\sigma)_{p(s)}, (\text{grad}\,\sigma)_a\rangle$, hence $(d\sigma/ds)_{s=0} = \|(\text{grad}\,\sigma)_a\|^2 > 0$. Since $\sigma(p(a)) = 0$, it follows that for some interval $-\epsilon_a < s < \epsilon_a$ the values of $\sigma$ increase with s. However $\sigma(p(0)) = 0$ so we see that $\sigma$ changes from negative to positive as we cross through the hypersurface $\sigma = 0$ at $a = p(0)$. In particular $\sigma(\epsilon_a) > 0$ and $\sigma(-\epsilon_a) < 0$.

We can now construct the desired curve $z(t)$ through the origin, namely $z(t) = ta - \epsilon_a(\text{grad}\,\sigma)_a$, $-1 \leq t \leq 1$. Note that $z(1) = a - \epsilon_a(\text{grad}\,\sigma)_a$ and $z(-1) = -a - \epsilon_a(\text{grad}\,\sigma)_a = -(a + \epsilon_a(\text{grad}\,\sigma)_a) = p(-\epsilon_a)$. Thus $\sigma(-p(\epsilon_a)) = \sigma(p(\epsilon_a)) > 0$ and $\sigma(z(1)) = \sigma(p(-\epsilon_a)) < 0$. In fact, $\sigma(z(t)) = \sigma(ta - t^2\epsilon_a(\text{grad}\,\sigma)_a) = t^2\sigma(a + \epsilon_a(-t)(\text{grad}\,\sigma)_a)$ and we note that for $0 < t < 1$ and for $-1 < t < 0, \sigma(z(t)) > 0$. In other words, $t\sigma(z(t)) < 0, t \neq 0$ on $(-\epsilon_a, \epsilon_a)$, as required for $t = 0$ to be a stable equilibrium point of $\dot{z}_n = \sigma(z(t)) = \sigma(\dot{z}_1(z_n), ..., \dot{z}_{n-1}(z_n), z_n)$.

Finally we have required that, when written in terms of the z-coordinates, the curve $z(t) = (z_1(t), ..., z_n(t))$ satisfy $\dot{z}_n(0) \neq 0$, thus allowing us to use $z_n$ as parameter on the curve. Equivalently, we must be able to choose the curve $z(t)$ so that its tangent vector at the origin, $\dot{z}(0)$, is not orthogonal to the vector $e_n = (0, 0, ..., 1)$ . If we write $z(t)$ in the x-coordinates, $z(t) = ta - t^2\epsilon_a(\text{grad}\,\sigma)_a$, we see that $\dot{z}(0) = a$, where $a$ is any point other than the origin on the hypersurface $\sigma = 0$. But we can find an $a$ with $\sigma(a) = 0$ which is not orthogonal to $e_n$ unless $\sigma = 0$ lies on the hyperplane of vectors orthogonal to $e_n$. This could happen if and only if $\sigma$ has rank one, contrary to our assumption. In summary, we have proved:

*Suppose that $\dot{z}_1 = v_1, ..., \dot{z}_{n-1}, = v_{n-1}, \dot{z}_n = \phi(z_1, ..., z_n)$ satisfies (i) and (ii) and is not stabilizable by linear feedback. Then $\phi(z)$ has as the leading term in its Taylor expansion a form of rank $\geq 2$. Assume that it is a quadratic form $\sigma(z) = \sum a_{ij}z_iz_j$ which is of maximum rank (the generic case). If $\sigma(z) \geq 0$ for all z, then the origin is an unstable equilibrium point for any choice of controls. Otherwise we may choose the controls as functions of z so that the origin is an asymptotically stable equilibrium point.*

We may illustrate this in the case $n = 3$. Suppose that $\sigma(x) =$

$x_1^2 + x_2^2 - x_3^2$. Then $\sigma(x) = 0$ is a cone, and choosing $a = (1,0,1)$ we see that $\sigma(a) = 0$, $\sigma < 0$ inside the cone, and $\sigma > 0$ outside. We have, with $\epsilon_a = 1/2$,

$$z(t) = t(1,0,1) - t^2\epsilon_a(-1,0,-1) = (t - t^2\epsilon_a, 0, t + t^2\epsilon_a).$$

This curve is easily seen to be a parabola in the $x_1x_3$ plane which passes from outside the cone ($x_3 < 0$) thru the origin to inside the cone ($x_3 > 0$) as t goes through $t = 0$. We have $\dot{x}_3(0) = a$. A similar parabola can be obtained taking any $a \neq 0$ on the cone. (Such form a basis of $R^3$ in fact). We remark, parenthetically, that it is clear that maximum rank is stronger than necessary for the result above.

Finally, we consider the case in which the first non-vanishing term $\phi^{(k)}$ in the Taylor expansion of $\phi$ is a homogeneous polynomial of odd degree. We show that in this case there is always a straight line $z(t) = t(c_1, c_2, ..., c_n)$ through the origin such that $t\phi^{(k)}(z(t)) < 0$ for all $t \neq 0$. Thus we prove that:

*If the leading term $\phi^{(k)}$ in the Taylor expansion of $\phi$ is of odd degree $k \geq 3$, then the system $\dot{z}_1 = v_1, ..., \dot{z}_{n-1} = v_{n-1}, \dot{z}_n = \phi(z_1, ..., z_n)$ is locally feedback stabilizable at $z = 0$.*

**Proof.** First suppose that there is a point $c = (c_1, c_2, ..., c_n) \neq 0$ at which $\phi(c) < 0$. Then there would be an open set of such points and we could with no loss of generality suppose c is such that $c_1 \neq 0$. Define $z(t) = tc = (tc_1, tc_2, ..., tc_n)$ and note that $dz_1/dt = c_1 \neq 0$. Further, consider $t\phi(z(t)) = t\phi^{(k)}(tc) = t^{k+1}\phi^{(k)}(c)$. Since $k+1$ is even and $\phi(c) < 0$, this expression would always be negative. Hence if there exists $c \neq 0$ such that $\phi(c) < 0$, the result is proved.

To see that such a c exists, we write $\phi^{(k)}(z) = \sum a_{i_1 \cdots i_k} z_{i_1} ... z_{i_k}$ summed over all k-tuples chosen from $(1, ..., n)$. If there is a nonzero term of the form $az_i^k$, then we may choose $c = (0, ..., \pm 1, 0, ..., 0)$ with $+1$ in the ith place if $a < 0$ and $-1$ if $a > 0$, and all other entries zero, yielding $\phi(c) < 0$ as desired. If there is no such term in the expression for $\phi^{(k)}(z)$ , we can easily change coordinates so that there is such. Suppose, for example that $a_{1...k} \neq 0$. Introduce new coordinates $z_1 = x_1, z_2 = x_1 + x_2, z_3 = x_1 + x_3, ..., z_k = x_1 + x_k, z_{k+1} = x_{k+1}, ..., z_n = x_n$. Then in the new cordinates $\phi$ contains the term $a_{1...k}x_1^k$. This completes the proof.

## APPENDIX

Consider a $C^r (r \geq 2)$ system

$$(*) \qquad \dot{z} = f(z) + u_1 g_1(z) + \ldots + u_m g_m(z), z \in U_0 \subset R^n$$

where $f(0) = 0$ and $G(z) = span\{g_1(z), \dots g_m(z)\}$ has dimension $m$ at each point of the neighborhood $U_0$ of the origin $z = 0$. Let $S$ denote a $C^2$ submanifold of $U_0$ of dimension $r = n - m$ passing through the origin and transversal to $G(z)$ at each point $z \in S$. Then the tangent space to $R^n$ at each point $z$ of $S$ splits into the direct sum $G(z)$ and the subspace tangent to $S$. Hence there is a well defined projection of the vector field $f(z)$ at each $z \in S$ which defines a $C^2$ tangent vector field $f^S$ on $S$. It is important to note that $f^S$ is uniquely determined by $f, G$ and $S$. It is independent of the coordinates used and of the values of $u = (u_1, \dots, u_m)$ and thus is the same for any system feedback equivalent to $(*)$. We first note the following.

**PROPOSITION.** *Let $S$ and $f^S$ be as described above and suppose that $G$ is involutive. Then there exists $C^2$ feedback $u(z) = (u_1(z), ..., u_m(z))$ such that $S$ is an invariant manifold (locally) of the resulting closed loop flow $f(z) + u_1(z)g_1(z) + \dots + u_m(z)g_m(z)$ , i.e. this vector is tangent to $S$ for $z \in S$; indeed, this vector is exactly $f^S(z)$.*

**Proof.** To prove this we introduce new coordinates in a neighborhood $U \subset U_0$ of $z = 0$ , which we do in two steps. First, choose coordinates $(x; y) = (x_1, ..., x_r; y_1, ..., y_m)$ such that (i) $z = 0$ corresponds to $(x; y) = (0; 0)$ and (ii) $G = span\{\partial/\partial y_1, ..., \partial/\partial y_m\}$. This is possible since we have assumed $G$ to be involutive. Now since $G$ is transverse to $S$ at each of its points, in these coordinates the equations of the submanifold $S$ in $U$ takes the form $y_j = h_j(x_1, ..., x_r)$, $j = 1, ..., m$, or $y = h(x)$ and $(*)$ takes the following "rectified" form

$$(*')\qquad \begin{aligned} \dot{x} &= f^{(1)}(x; y) \\ \dot{y} &= f^{(2)}(x; y) + g(x; y)u \end{aligned}$$

where g is a nonsingular mxm matrix and $u = (u_1, ..., u_m)^t$.

Secondly, we make a further coordinate change: $(\tilde{x}; \tilde{y}) = (x; y - h(x))$. In these coordinates the equations $(*')$ become

$$(\tilde{*})\qquad \begin{aligned} \dot{\tilde{x}} &= f^{(1)}(\tilde{x}; \tilde{y}) = f^{(1)}(x; y - h(x)) \\ \dot{\tilde{y}} &= v + (\partial h/\partial x)\tilde{f}^{(1)}(\tilde{x}; \tilde{y}) + \tilde{f}^{(2)}(\tilde{x}; \tilde{y}) \end{aligned}$$

with $\partial h/\partial x$ denoting the Jacobian and $\tilde{f}^{(2)}(\tilde{x}; \tilde{y}) = f^{(2)}(x; y - h(x))$. Finally taking $v = \tilde{v} - (\partial h/\partial x)\tilde{f}^{(1)}(\tilde{x}; \tilde{y}) - \tilde{f}^{(2)}(\tilde{x}; \tilde{y})$, we have

$$(\tilde{*})\qquad \begin{aligned} \dot{\tilde{x}} &= \tilde{f}^{(1)}(\tilde{x}; \tilde{y}) \\ \dot{\tilde{y}} &= \tilde{v} \end{aligned}$$

Note that $S$ now has equations $\tilde{y} = 0$, corresponding to $y = h(x)$ in the previous coordinate system. This is the main reason for introducing these new coordinates, since we see at once that the proposition holds with feedback $\tilde{v} = 0$. Indeed, $S$ now coincides with the r-coordinate plane $\tilde{y} = 0$ and the vector field given by the closed loop with $f(z)$ whose components in the $(\tilde{x}; \tilde{y})$ coordinates are $(\tilde{f}^{(1)}(\tilde{x}; 0); 0)$ is tangent to this coordinate plane and is exactly the projection $f^S$ of f onto this plane.

These preliminaries allow us to state the version of the Center Manifold Theorem which we have used above at several points. In [1]-[3] we were able to use this theorem precisely as stated in Carr [4] and Appendix B of Isidori [7], but here we have prepared the ground for employing a useful corollary of it which was proved by Isidori (Lemma, page 442, *op. cit.*). We restate this lemma using the Proposition above.

**LEMMA.** *Assume that the system* $(*)$ *satisfies the assumptions (i) and (ii) of this note and that* $S$ *is a submanifold as in the Proposition above. Then if the origin is an asymptotically stable equilibrium point of the flow* $f^S$ *on* $S$*, the system* $(*)$ *is* $C^2$ *stabilizable by feedback.*

This follows from the Proposition and the Lemma of Isidori applied to the system $(\tilde{*})$ with the feedback $\tilde{v} = A\tilde{y}$, where $A$ is an mxm matrix all of whose eigenvalues have negative real part, or, equivalently with the feedback $u = g^{-1}(x; y)(Ay - Ah - f^{(2)}(x; y) - (\partial h/\partial x)f^{(1)}(x; y - h(x))$ applied to $(*')$. Isidori's Lemma states that given a system of the form

$$\begin{aligned} \dot{z} &= f(z; y) \\ \dot{y} &= Ay + p(z, y) \end{aligned}$$

where (a) all eigenvalues of $A$ have negative real parts, (b) $p(z, 0) = 0$ for all $z$ near $z = 0$, and (c) $\partial p/\partial y = 0$ at $(0, 0)$; then, if $\dot{z} = f(z, 0)$ has an asymptotically stable equilibrium at $z = 0$, it follows that the full system has an asymptotically stable equilibrium at $(0, 0)$. In our case $p(z, y) = 0$ and $A$ is determined by our choice of feedback.

## REFERENCES

[1] W.M. Boothby and R. Marino, Feedback stabilization of planar nonlinear systems, *Systems and Control Letters,* **12** (1989), 87-92

[2] W.M. Boothby and R. Marino, Feedback stabilization of planar nonlinear sytems II, *28th IEEE-CDC Conference, Tampa Florida, December* (1989), 1970-1974

[3] W.M. Boothby and R. Marino, The Center Manifold Theorem in feedback stabilization of planar, single input systems, *Control Theory and Adv. Technol.,* **6** (1990), 517-532

[4] J. Carr, Applications of center manifold theory, *Springer, New York* (1981)

[5] W.P. Dayawansa, C.F Martin, and G. Knowles, Asymptotic stabilization of a class of smooth two dimensional systems, *SIAM J. Control and Opt.*,**28** (1990), 1321-1349

[6] L.R. Hunt, R. Su and G. Meyer, Design for multi-input nonlinear systems, in *Differential geometric control theory, Brockett, Millman, and Sussman, editors, Birkhaeuser, Boston* (1983)

[7] A. Isidori, Nonlinear control systems, 2nd Edition, *Springer Verlag, New York* (1989)

[8] B. Jakubczyk and W. Respondek, On linearization of control systems, *Bull. Acad. Polon. Sci.* **28** (1980), 517-522

[9] M. Kawski, Stabilization of nonlinear systems in the plane, *Systems and Control Letters,* **12** (1989), 169-175

[10] R. Marino, W.M. Boothby and D.L. Elliott, Geometric properties of linearizable control systems, *Math. Systems Theory,* **18** (1985), 97-123

[11] R. Marino, On the largest feedback linearizable subsystem, *Systems and Control Letters* **6** (1986), 345-351

[12] H. Samelson, Orientability of hypersurfaces in $R^n$, *Proc. Amer. Math. Soc.,* **22** (1969), 301-2

[13] E.D. Sontag, Feedback stabilization of nonlinear systems, *Proceedings of MTNS '89, Birkhaeuser, Boston* (1991)

[14] W.M. Wonham, Linear multivariable control: a geometric approach, *Springer-Verlag, New York* (1971)

# 29 Robust Stabilization of Infinite-Dimensional Systems with Respect to Coprime Factor Perturbations

**Ruth F. Curtain** University of Groningen, Groningen, The Netherlands

**A. J. Pritchard** University of Warwick, Coventry, England

**Abstract**

This paper solves the problem of robust stabilization of a class of infinite-dimensional systems with respect to stable factor perturbations of a fixed normalized left-coprime factorization. The class of infinite-dimensional systems is that described by the state-space operators $(A, B, C, D)$ where $A$ generates a strongly continuous semigroup on Hilbert spaces $V$ and $W$, $D(A^V) \hookrightarrow W \hookrightarrow V$ and the system has a bounded observability map from $V$ and a bounded controllability map onto $W$ (the Pritchard-Salamon class). $C \in \mathcal{L}(W, Y)$ and $B \in \mathcal{L}(V, U)$, where $U$ and $Y$ are separable Hilbert spaces and in addition, we assume that $(A, B)$ is admissibly exponentially stabilizable, $(A, C)$ is admissibly exponentially detectable.

**Key words** : robust stabilization, Riccati equations, coprime factorization, Hankel operators, Nehari theorem.

# 1 Introduction

There are many different theories for designing stabilizing controllers (both finite and infinite dimensional) for infinite dimensional systems (see [5] for a survey). Since in practice one only usually has an approximate model of the system, it is important to know that the controller one designs for a nominal model also stabilizes other systems which are close to it; this is the robustness issue. The robustness definition we use here is based on a frequency domain description of the perturbed systems: they are stable factor perturbations in a normalized left-coprime factorization of the nominal plant. Recently, in [17] a complete, explicit solution was obtained for the finite-dimensional case in terms of the solution to two Riccati equations. Later an abstract solution to this problem for a general class of infinite-dimensional was obtained in [12]. However, the next step of obtaining a parametrization of robustly stabilizing controllers is more difficult and only partial results are available. These include results for single input-single output systems with a feedthrough delay in [13], for infinite-dimensional systems of positive real type in [6], and for strictly proper Pritchard-Salamon systems $\Sigma(A, B, C)$ in [3]. In this paper we extend the results of [3] in two directions. First we allow for proper, but not necessarily, strictly proper Pritchard-Salamon systems $\Sigma(A, B, C, D)$. This seemingly simple generalization is mathematically much more complicated due to the fact that the perturbed operators are of the form $A - BKC$, where both $B$ and $C$ are unbounded. This type of perturbation has recently been studied in [7] and it has enabled a theory for Riccati equations with these type of operators to be developed. The second extension is to allow for infinite-dimensional input and output spaces and the difference here is that the Hankel operator need not be compact as was the case in [3]. For this extension we have used the new results in [10].

In Section 2 we formulate the robustness stabilization problem, introducing the relevant mathematical background, and we state the known results which are to be used in the sequel. In Section 3 we introduce the Pritchard-Salamon class of infinite-dimensional systems and state their relevant properties. We also prove a key technical result in Lemma 3.11. In Section 4 explicit formulae for a normalized doubly coprime factorization of an admissibly exponentially stabilizable and detectable Pritchard-Salamon sytem $\Sigma(A, B, C.D)$ are obtained. These were derived in [3] for the case $D = 0$ but for the case $D \neq 0$ and infinite-dimensional input and output spaces this is a new result. Finally, in Section 5 we solve the robust stabilization problem, namely the maximal robustness margin is determined and we derive explicit formulae parametrizing all robustly stabilizing controllers which achieve a given robustness margin.

# 2 The robust stabilization problem

We consider here the problem of robust stabilization with respect to coprime factor perturbations which has been studied for finite-dimensional systems in [17], for Pritchard-Salamon systems in [3] and for a general class of infinite-dimensional systems in [12]. In our approach we use results on Hardy spaces of vector-valued functions which can be found in [20].

**Definition 2.1** For the Banach space $X$ and the separable Hilbert space $Z$ we define the following Hardy spaces

$$\mathbb{H}_\infty(X) = \{x : \mathbb{C}_0^+ \to X \mid x \text{ is holomorphic and } \sup_{\mathrm{Re}\, s>0} \|x(s)\| < \infty\},$$

and

$$H_2(Z) = \left\{ \begin{array}{ll} z : \mathbb{C}_0^+ \to Z \mid z & \text{is holomorphic and} \\ & \|z\|_2^2 = \sup_{\xi>0}((2\pi)^{-1} \int_{-\infty}^{\infty} \|z(\xi + j\omega)\|^2 d\omega) < \infty \end{array} \right\}.$$

We recall that $\mathbb{H}_\infty(X)$ is a Banach algebra under the norm

$$\|G\|_\infty = \sup_{\mathrm{Re}\, s>0} \|G(s)\|_X$$

and $H_2(Z)$ is a Hilbert space under the equivalent norm

$$\|z\|^2 = (2\pi)^{-1} \int_{-\infty}^{\infty} \|\tilde{z}(j\omega)\|_Z^2 d\omega$$

where $\tilde{z}(j\omega)$ is the boundary function in $L_2((-j\infty, j\infty); Z)$ which we identify with $z \in H_2(Z)$.

Our class of systems will be in terms of transfer operators which will be elements of $\mathbb{H}_\infty(\mathcal{L}(U,Y))$ where $U$ and $Y$ are separable Hilbert spaces and $\mathcal{L}(U,Y)$ denotes the space of bounded linear operators from $U$ to $Y$. The unstable systems $\mathbb{UT}(\mathcal{L}(U,Y))$ will be represented by the algebra

$$\mathbb{UT}(\mathcal{L}(U,Y)) = \left\{ \begin{array}{l} G : D(G) \subset \mathbb{C}_0^+ \to \mathcal{L}(U,Y), \text{where D(G) differs from} \\ \mathbb{C}_0^+ \text{by at most countably many points} \end{array} \right\}.$$

$\mathbb{UT}(\mathcal{L}(U,Y))$ is an extremely large class of unstable systems and in most applications it is too large a class to work with. In our work we will concentrate our attention on a small subset: those elements $G$ which have left- and right-coprime factorizations.

**Definition 2.2** Suppose that $M \in \mathbb{H}_\infty(\mathcal{L}(U), N \in \mathbb{H}_\infty(\mathcal{L}(U,Y))$. Then $M$ and $N$ are *right-coprime* if there exist $\tilde{X} \in \mathbb{H}_\infty(\mathcal{L}(U)), \tilde{Y} \in \mathbb{H}_\infty(\mathcal{L}(Y,U))$ such that the following *Bezout identity* holds on $\mathbb{C}_0^+$

$$\tilde{X}M - \tilde{Y}N = I. \tag{1}$$

*Left-coprimeness* is defined analogously.

$G \in \mathbb{UT}(\mathcal{L}(U,Y))$ has a *right-coprime factorization* if there exist $M \in \mathbb{H}_\infty(\mathcal{L}(U)), N \in \mathbb{H}_\infty(\mathcal{L}(U,Y))$ which are right-coprime and $G = NM^{-1}$. *Left-coprime factorizations* are defined analogously.

$G \in \mathbb{UT}(\mathcal{L}(U,Y))$ has a *doubly coprime factorization* if there exist $\tilde{N}, N \in \mathbb{H}_\infty(\mathcal{L}(U,Y))$, $X, \tilde{M} \in \mathbb{H}_\infty(\mathcal{L}(Y))$, $Y, \tilde{Y} \in \mathbb{H}_\infty(\mathcal{L}(Y,U))$, $\tilde{X}, M \in \mathbb{H}_\infty(\mathcal{L}(U))$ such that

$$\begin{bmatrix} \tilde{X} & -\tilde{Y} \\ -\tilde{N} & \tilde{M} \end{bmatrix} \begin{bmatrix} M & Y \\ N & X \end{bmatrix} = \begin{bmatrix} I & 0 \\ 0 & I \end{bmatrix} \quad \text{on } \mathbb{C}_0^+. \tag{2}$$

The right-coprime factorization $G = NM^{-1}$ is *normalized* if

$$N(j\omega)^* N(j\omega) + M(j\omega)^* M(j\omega) = I, \quad \omega \in \mathbb{R} \tag{3}$$

(i.e. $\begin{bmatrix} N \\ M \end{bmatrix}$ is inner).

The left-coprime factorization $G = \tilde{M}^{-1}\tilde{N}$ is *normalized* if

$$\tilde{N}(j\omega)\tilde{N}(j\omega)^* + \tilde{M}(j\omega)\tilde{M}(j\omega)^* = I, \quad \omega \in \mathbb{R}. \tag{4}$$

(i.e. $[\tilde{N} \;\; \tilde{M}]$ is co-inner) □

We emphasize that elements of $\mathbb{UT}(\mathcal{L}(U,Y))$ need not possess right- or left-coprime factorizations, in general, but in Section 4 we shall consider a class for which we can always construct a normalized doubly coprime factorization. Our aim is to stabilize unstable systems $G \in \mathbb{UT}(\mathcal{L}(U,Y))$ in the following sense.

**Definition 2.3** Let us suppose that the system $G \in \mathbb{UT}(\mathcal{L}(U,Y))$. Then $K \in \mathbb{UT}(\mathcal{L}(Y,U))$ *stabilizes* $G$ if $S = (I - GK)^{-1}$, $KS$, $SG$ and $I + KSG \in \mathbb{H}_\infty(X)$ for $X$ equal to one of $\mathcal{L}(Y), \mathcal{L}(Y,U), \mathcal{L}(U), \mathcal{L}(U,Y)$. □

In our applications $G$ always possesses a left-coprime factorization. As in Vidyasagar [21] p 363 it is easy to show that if $G \in \mathbb{UT}(\mathcal{L}(U,Y))$ possesses a left-coprime factorization and $K \in \mathbb{UT}(\mathcal{L}(Y,U))$ stabilizes $G$ then $K$ possesses a right-coprime factorization. We are interested in stabilizing not just one system, but a whole family of perturbations of the system.

**Definition 2.4** *(Robust stabilization with respect to normalized coprime factor perturbations)*
Given a system $G \in \mathbb{UT}(\mathcal{L}(U,Y))$ with a normalized left-coprime factorization $G = \tilde{M}^{-1}\tilde{N}$, find a controller $K \in \mathbb{UT}(\mathcal{L}(Y,U))$ which stabilizes $G$ and all perturbed systems $G_\Delta \in \mathcal{G}_\varepsilon$, where

$$\mathcal{G}_\varepsilon = \left\{ \begin{array}{l} G_\Delta = (\tilde{M} + \Delta_M)^{-1}(\tilde{N} + \Delta_N) \text{ such that } [\Delta_M, \Delta_N] \in \mathbb{H}_\infty(\mathcal{L}(Y \oplus U, Y)), \\ \tilde{M} + \Delta_M, \tilde{N} + \Delta_N \text{ are left-coprime and } \|[\Delta_M, \Delta_N]\|_\infty < \varepsilon \end{array} \right\}.$$

A controller $K$ which achieves this is said to *robustly stabilize* $G$ with *robustness margin* $\varepsilon$. □

Note that although $\Delta_M$ and $\Delta_N$ are stable perturbations, they introduce perturbed plants $G_\Delta$ with very different unstable poles through the denominator term $\tilde{M} + \Delta_M$. This means that this is a very general class of perturbations. The following necessary and sufficient conditions for robust stabilization with respect to normalized coprime factorizations was first shown in [17] for the rational case using a small gain argument. This proof was extended to a class of infinte-dimensional systems in Chapter 9 of [11] and it is not difficult to see that the same reasoning holds here in our situation.

**Theorem 2.5** *Suppose that $G \in \mathbb{UT}(\mathcal{L}(U,Y))$ has a normalized doubly coprime factorization $G = NM^{-1} = \tilde{M}^{-1}\tilde{N}$ satisfying (2)–(4). Then there exists a controller $K \in \mathbb{UT}(\mathcal{L}(Y,U))$ which robustly stabilizes $G$ with robustness margin $\varepsilon$ if and only if $K$ stabilizes $G$ and*

$$\left\| \begin{pmatrix} K \\ I \end{pmatrix} (I - GK)^{-1}\tilde{M}^{-1} \right\|_\infty \le \varepsilon^{-1}. \tag{5}$$

The next step is to convert the problem (5) to a Nehari-extension problem as in [17] and [11].

**Theorem 2.6** *Suppose that $G \in \mathbb{UT}(\mathcal{L}(U,Y)))$ satisfies the assumptions in Theorem 2.5. Then $K \in \mathbb{UT}(\mathcal{L}(Y,U))$ robustly stabilizes $G$ with robustness margin $\varepsilon$ if and only if $K$ has a right-coprime factorization $K = UV^{-1}$ satisfying*

$$\left\| \begin{bmatrix} -\tilde{N}^* \\ \tilde{M}^* \end{bmatrix} + \begin{bmatrix} U \\ V \end{bmatrix} \right\|_\infty \le (1 - \varepsilon^2)^{1/2}. \tag{6}$$

From [1] it is known that for $G \in \mathbb{H}_\infty(\mathcal{L}(U,Y))$

$$\inf_{K(-s)\in\mathbb{H}_\infty(\mathcal{L}(U,Y))} \|G+K\|_\infty = \|\Gamma\|, \tag{7}$$

where $\Gamma$ is the Hankel operator associated with the symbol $G$. (7) is known as the Nehari extension problem and in this paper we use the results from [10]) where they prove (7) and obtain explicit solutions to the suboptimal Nehari extension problem for a special class of systems; the Pritchard-Salamon class (see Section 3).
The *Hankel operator* $\Gamma$ of a stable system $M \in \mathbb{H}_\infty$ can be defined as the orthogonal projection onto $\mathbb{H}_2^\perp$ of the multiplication operator induced by $M^*$ restricted to the Hardy space $\mathbb{H}_2$.. Applying (7) to $G = [-\tilde{N}, \tilde{M}]$ yields the following corollary.

**Corollary 2.7** *Suppose that $G \in \mathbb{UT}(\mathcal{L}(U,Y))$ has a normalized doubly coprime factorization $G = NM^{-1} = \tilde{M}^{-1}\tilde{N}$ satisfying (2)-(4). Then there exists a controller $K \in \mathbb{UT}(\mathcal{L}(Y,U))$ which robustly stabilizes $G$ with robustness margin $\varepsilon$ if and only if $K$ stabilizes $G$ and*

$$\varepsilon \leq \varepsilon_{max} = (1-\|\Gamma\|^2)^{-1/2}, \tag{8}$$

*where $\Gamma$ is the* Hankel operator *associated with the system $[\tilde{N}, \tilde{M}] \in \mathbb{H}_\infty(\mathcal{L}(Y\oplus U, Y))$, and $\varepsilon_{max}$ is the* maximal robustness margin.

# 3 The Pritchard-Salamon class

The Pritchard-Salamon class (denoted by $\mathbb{PS}$) of infinite-dimensional linear systems was first introduced in [18], [19] in connection with the optimal control problem. Later, it was seen to describe a particularly nice class of systems with attractive systems theoretic properties and a simple structure (see [2] and [4]). The class has a well-defined transfer function and it is closed under feedback connections, in particular, "$BKC$" connections (see [7]). Here we summarize known properties of this class and prove some new ones needed to solve our problem.

**Definition 3.1** *(Pritchard-Salamon class)*
Let $V$, and $W$ are separable Hilbert spaces with continuous, dense injections satisfy

$$W \hookrightarrow Z \hookrightarrow V. \tag{9}$$

Suppose that $A$ is the infinitesimal generator of a consistent strongly continuous semigroup $S(t)$ on $W$, $Z$ and $V$, and $U$ and $Y$ are separable Hilbert spaces, the input and output spaces, respectively.
$B \in \mathcal{L}(U,V)$ is an *admissible input operator* for $S(t)$ if there exists a constant $\beta > 0$ such that

$$\left\| \int_0^{t_1} S(t_1-s)Bu(s)ds \right\|_W \leq \beta \|u\|_{L^2(0,t_1;U)} \tag{10}$$

for all finite $t_1 > 0$.
$C \in \mathcal{L}(W,Y)$ is an *admissible output operator* for $S(t)$ if there exists a constant $\gamma > 0$ such that

$$\left( \int_0^{t_1} \|CS(t)x\|_Y^2 dt \right)^{1/2} \leq \gamma \|x\|_V \tag{11}$$

for all $x \in W$ and all finite $t_1 > 0$.
Under the above assumptions $\Sigma(A, B, C, D)$ is called a *Pritchard-Salamon system* for any $D \in \mathcal{L}(U, Y)$. □

Where confusion may arise we use superscripts, e.g. $A^X$, $S^X(t)$, to denote the operators on $X = V$, $Z$ or $W$. We denote the growth bound on $X$ by $\omega^X$. These are not the same, in general, (see [7]).
We remark that (11) shows that $CS(\cdot)$ extends to a bounded operator from $V$ to $L^2(0, t_1; Y)$ and we use the same symbol for this extension. We now collect several properties of $\mathbb{PS}$ systems which have been proved in [7], [2] and the references therein.

**Property 3.2** Let $F \in \mathcal{L}(W, U)$ be an admissible output operator for $S(t)$ and suppose that $\Sigma(A, B, C, D)$ is a $\mathbb{PS}$ system. Then $\Sigma(A_{BF}, B, C, D)$ is also a $\mathbb{PS}$ system, where $A_{BF}$ is the infinitesimal generator of $S_{BF}(t)$ which is the unique solution of

$$S_{BF}(t)x = S(t)x + \int_0^t S(t-\tau)BFS_{BF}(\tau)x d\tau\,. \tag{12}$$

Furthermore, if

$$\mathcal{D}(A^V) \hookrightarrow W\,, \tag{13}$$

then $\mathcal{D}(A^V_{BF}) = \mathcal{D}(A^V)$ and

$$A^V_{BF}x = Ax + BFx \quad \text{for all} \quad x \in \mathcal{D}(A^V). \tag{14}$$

Note that any $F \in \mathcal{L}(V, U)$ is an admissible output operator.

**Property 3.3** Let $H \in \mathcal{L}(Y, V)$ be an admissible input operator for $S(t)$ and suppose that $\Sigma(A, B, C, D)$ is a $\mathbb{PS}$ system. Then $\Sigma(A_{HC}, B, C, D)$ is also a $\mathbb{PS}$ system, where $A_{HC}$ is the infinitesimal generator of $S_{HC}(t)$ which is the unique solution of

$$S_{HC}(t)x = S(t)x + \int_0^t S_{HC}(t-\tau)HCS(\tau)x d\tau\,. \tag{15}$$

Furthermore, if

$$\mathcal{D}(A^V) \hookrightarrow W\,, \tag{16}$$

then $\mathcal{D}(A^V_{HC}) = \mathcal{D}(A^V)$ and

$$A^V_{HC}x = Ax + HCx \quad \text{for all} \quad x \in \mathcal{D}(A^V). \tag{17}$$

Note that any $H \in \mathcal{L}(Y, W)$ is an admissible input operator.

**Property 3.4** A $\mathbb{PS}$ system $\Sigma(A, B, C, D)$ has a well-defined transfer operator $G(s)$ given by

$$\hat{y}(s) = G(s)\hat{u}(s) \quad \text{for} \quad s \in \mathbb{C}^+_\alpha\,, \tag{18}$$

where $\alpha > \max(\omega^V, \omega^W)$, and

$$G(s) = D + C(sI - A^V)^{-1}B = D + C(sI - A^W)^{-1}B\,. \tag{19}$$

**Property 3.5** A $\mathbb{PS}$ system $\Sigma(A, B, C, D)$ satisfies

$$C \int_0^t S(\tau)Bud\tau = \int_0^t CS(\tau)Bud\tau \quad \text{for all} \quad u \in U\,. \tag{20}$$

Although conditions for the growth bounds of $S(t)$ on $V$ and $W$ to be equal are given in [7], they do not have nice perturbations properties. To avoid such problems we use the form of stabilizability and detectibility introduced in [7].

**Definition 3.6** Let $\Sigma(A, B, C, D)$ be a $\mathbb{PS}$ system. It is *admissibly exponentially stabilizable* if there exists an admissible output operator $F \in \mathcal{L}(W, U)$ such that $S_{BF}(\cdot)$ defined by (12) is exponentially stable on $V$ and $W$. Similarly, it is *admissibly exponentially detectable* if there exists an admissible input operator $H \in \mathcal{L}(Y, V)$ such that $S_{HC}(\cdot)$ defined by (15) is exponentially stable on $V$ and $W$. □

We recall that $F \in \mathcal{L}(V, U)$ is automatically an admissible output operator. The following lemma is therefore rather surprising, since it shows that we could have restricted Definition 3.6 to bounded feedbacks $F$.

**Lemma 3.7** *Suppose that $\Sigma(A, B, C, D)$ is a $\mathbb{PS}$ system. Then it is admisssibly exponentially stabilizable if and only if there exists a bounded output operator $F \in \mathcal{L}(V, U)$ such that $S_{BF}(\cdot)$ is exponentially stable on $V$ and $W$.*
*Furthermore, if $\Sigma(A, B, C, D)$ is admissibly exponentially stabilizable and there exists $F \in \mathcal{L}(V, U)$ such that $S_{BF}(\cdot)$ is exponentially stable on $V$, then it is also exponentially stable on $W$.* □

We also need the concept of a dual system. In order to describe this we assume that the Hilbert spaces $Z, U, Y$ are identified with their duals. Then $B' \in \mathcal{L}(V', U)$, $C' \in \mathcal{L}(Y, W')$, $D^* \in \mathcal{L}(Y, U)$ and

$$V' \hookrightarrow Z \hookrightarrow W'\,. \tag{21}$$

**Lemma 3.8** *Suppose that $\Sigma(A, B, C, D)$ is an admissibly exponentially stabilizable and detectable $\mathbb{PS}$ system, then $\Sigma(A', C', B', D^*)$ is also an admissibly exponentially stabilizable and detectable $\mathbb{PS}$ system with respect to (21). Moreover if (13) holds then*

$$\mathcal{D}((A')^{W'}) \hookrightarrow V', \tag{22}$$

**Proof:** The fact that $\Sigma(A', C', B', D^*)$ is a $\mathbb{PS}$ system is immediate by duality. And the stabilizability and detectability results are a consequence of Lemma 3.7 and the fact that $S_{BF}(\cdot)$ is exponentially stable on $V$ and $W$ if and only if $S'_{BF}(\cdot)$ is exponentially stable on $V'$ and $W'$. The final statement follows from the facts that $A' \in \mathcal{L}(V', (\mathcal{D}(A^V))')$ where $\mathcal{D}(A^V)$ is endowed with the graph norm and $V' \hookrightarrow W' \hookrightarrow (\mathcal{D}(A^V))'$ see [15]. □

A $\mathbb{PS}$ system which satisfies (13) will be called a *regular* $\mathbb{PS}$ system. We are now in a position to state results on Riccati equations

**Theorem 3.9** *Suppose that $\Sigma(A,B,C,D)$ is an admissibly exponentially stabilizable and detectable regular $\mathbb{PS}$ system. If*

$$R = I + DD^*\,, \quad S = I + D^*D\,, \quad A_{-BS^{-1}D^*C} = A - BS^{-1}D^*C \tag{23}$$

*then for every $z \in \mathcal{D}(A^V)$ the following* control Riccati equation *with values in $(\mathcal{D}(A^V))'$ has a unique, nonnegative, self-adjoint solution $Q \in \mathcal{L}(V, V')$*

$$A'_{-BS^{-1}D^*C}Qz + QA_{-BS^{-1}D^*C}z - QBS^{-1}B'Qz + C'R^{-1}Cz = 0. \tag{24}$$

*Moreover, $Q \in \mathcal{L}(\mathcal{D}(A^V), \mathcal{D}((A')^{W'}))$ and if $F = -S^{-1}(D^*C + B'Q)$, then $A_{BF}$ is the infinitesimal generator of a strongly continuous semigroup $S_{BF}(\cdot)$ given by (12) which is exponentially stable on $V$ and $W$.*
*If*

$$A_{-BD^*R^{-1}C} = A - BD^*R^{-1}C \tag{25}$$

*then for all $z \in \mathcal{D}((A')^{W'})$ the following* filter Riccati equation *with values in $(\mathcal{D}((A')^{W'}))'$ has a unique, nonnegative, self-adjoint solution $P \in \mathcal{L}(W', W)$*

$$A_{-BD^*R^{-1}C}Pz + PA'_{-BD^*R^{-1}C}z - PC'R^{-1}CPz + BS^{-1}B'z = 0. \tag{26}$$

*Moreover, $P \in \mathcal{L}(\mathcal{D}((A')^{W'}), \mathcal{D}(A^V))$ and if $H = -(BD^* + PC')R^{-1}$, then $A_{HC}$ is the infinitesimal generator of a strongly continuous semigroup $S_{HC}(\cdot)$ given by (15) which is exponentially stable on $V$ and $W$.*

**Proof:** First note that $-S^{-1}D^*C$ is an admissible output operator for $\Sigma(A,B,C,D)$ and hence by Property 3.2 $\Sigma(A_{-BS^{-1}D^*C}, B, C, D)$ is a $\mathbb{PS}$ system. Moreover, it is easy to see that $\Sigma(A_{-BS^{-1}D^*C}, B, C, D)$ is admissibly exponentially stabilizable and detectable. So by Lemma 3.7 it is admissibly exponentially stabilizable by a bounded feedback in $\mathcal{L}(V,U)$. We may therefore apply the main theorem in [18] to conclude the first part of the theorem (see also [14]). Now (24) may be written as the following equation with values in $(\mathcal{D}(A^V))'$

$$A'_{-BF}Qz = -QA_{-BS^{-1}D^*C}z - C'R^{-1}Cz$$

for all $z \in \mathcal{D}(A^V)$. But note that the RHS lies in $W'$ and $0 \in \rho^{W'}(A'_{-BF})$ and hence $Q \in \mathcal{L}(\mathcal{D}(A^V), \mathcal{D}((A'_{-BF})^{W'})) = \mathcal{L}(\mathcal{D}(A^V), \mathcal{D}((A')^{W'}))$.
To prove the second part note that by applying the first part to the dual system we obtain that for all $z \in \mathcal{D}((A')^{W'})$ there exists a unique, nonnegative, self-adjoint solution $P \in \mathcal{L}(W', W)$ for the equation

$$A''_{-BD^*R^{-1}C}Pz + PA'_{-BD^*R^{-1}C}z - PC'R^{-1}CPz + BS^{-1}B'z = 0. \tag{27}$$

Moreover, $P \in \mathcal{L}(\mathcal{D}((A')^{W'}), \mathcal{D}((A'')^V))$. But $\mathcal{D}((A'')^V) = \mathcal{D}(A^V)$ and $A'' = A$ on $\mathcal{D}(A^V)$ (see for example [15]). So the proof is complete. □

We now examine some properties of $\mathbb{PS}$ systems $\Sigma(A,B,C,D)$ for which $A$ generates an exponentially stable $C_0$-semigroup on both $V$ and $W$.

**Definition 3.10** If $A$ generates an exponentially stable $C_0$-semigroup on $V$ and $W$, then we say the $\mathbb{PS}$ system $\Sigma(A, B, C, D)$ is *exponentially stable.*
Its *controllability operator* $\mathcal{B} \in \mathcal{L}(L^2(0,\infty;U), W)$ is defined by

$$\mathcal{B}u = \int_0^\infty S(s)Bu(s)ds \tag{28}$$

and its *observability operator* $\mathcal{C} \in \mathcal{L}(V, L^2(0,\infty;Y))$ is defined by

$$\mathcal{C}v = CS(t)v \quad \text{for} \quad v \in W\,. \tag{29}$$

Its *controllability gramian* $L_B \in \mathcal{L}(W', W)$ and its *observability gramian* $L_C \in \mathcal{L}(V, V')$ are defined by

$$L_B = \mathcal{B}\mathcal{B}'\,, \quad L_C = \mathcal{C}'\mathcal{C}\,. \tag{30}$$

The Hankel operator $\Gamma \in \mathcal{L}(L^2(0,\infty;U), L^2(0,\infty;Y))$ is defined by $\Gamma = \mathcal{C}\mathcal{B}$ and $\|\Gamma\|$ will denote the induced norm. □

That $\mathcal{C}$ and $\mathcal{B}$ are well-defined was shown in [2], [9]. The above definition of the Hankel operator differs from that given at the end of Section 2, but it is shown in [10] that the two definitions are equivalent for the $\mathbb{PS}$ class of transfer operators. The following results were established in [2], [9], [7] and [11].

**Lemma 3.11** *Suppose that $\Sigma(A, B, C, D)$ is an exponentially stable regular $\mathbb{PS}$ system, then we have the following:*
*(a) For every $z \in \mathcal{D}(A^V)$, $L_C$ is the unique solution in $\mathcal{L}(V, V')$ of the following Liapunov equation with values in $(\mathcal{D}(A^V))'$*

$$A'L_Cz + L_CAz + C'Cz = 0. \tag{31}$$

*And for every $z \in (\mathcal{D}((A')^{W'})$, $L_B$ is the unique solution in $\mathcal{L}(W', W)$ of the following Liapunov equation with values in $(\mathcal{D}((A')^{W'})'$*

$$AL_Bz + L_BA'z + BB'z = 0. \tag{32}$$

*Moreover,*

$$L_B \in \mathcal{L}(\mathcal{D}((A')^{W'}), \mathcal{D}(A^V))\,, \quad L_C \in \mathcal{L}(\mathcal{D}(A^V), \mathcal{D}((A')^{W'})). \tag{33}$$

*(b) $(I + \alpha L_BL_C)^{-1} \in \mathcal{L}(\mathcal{D}(A^V)) \cap \mathcal{L}(V)$, $(I + \alpha L_CL_B)^{-1} \in \mathcal{L}(\mathcal{D}((A')^{W'})) \cap \mathcal{L}(W')$ for $-1/\alpha \in \rho^V(L_BL_C) \cap \rho^W(L_BL_C)$.*
*(c)$\sigma(L_BL_C) = \sigma(\Gamma^*\Gamma)$, and the spectral radius of $L_BL_C$ equals $\|\Gamma\|^2$.*

**Proof:** (a) (32) may be viewed as a special case of the Riccati equation (26) with $C = D = 0$. Since $\Sigma(A, B, C, D)$ is an exponentially stable $\mathbb{PS}$ system it is necessarily admissibly exponentially stabilizable and detectable. We may therefore apply Theorem 3.9 to conclude the uniqueness statement and (33).
(b) Clearly $L_BL_C \in \mathcal{L}(V)$ and hence for $-1/\alpha \in \rho^V(L_BL_C)$, $(I+\alpha L_BL_C)^{-1} \in \mathcal{L}(V)$.

By setting $z = L_C x$, $x \in \mathcal{D}(A^V)$ in (32) and mutiplying (31) by $L_B$ and setting $z = x$ we obtain the following two equations with values in $(\mathcal{D}((A')^{W'}))'$

$$\begin{aligned} AL_BL_Cx + L_BA'L_Cx + BB'L_Cx &= 0 \\ L_BA'L_Cx + L_BL_CAx + L_BC'Cx &= 0, \end{aligned}$$

Multiplying these equations by $\alpha$ and subtracting yields

$$A(I + \alpha L_BL_C)x = (I + \alpha L_BL_C)Ax + \alpha L_BC'Cx - \alpha BB'L_Cx.$$

Note that for $x \in \mathcal{D}(A^V)$ the RHS is in $V$, and so for $-1/\alpha \in \rho^V(L_BL_C)$ we may reformulate it as an equation with values in $V$ of the form

$$\begin{aligned} A(I + \alpha L_BL_C)x &= (I + \alpha L_BL_C)\tilde{A}x && (34) \\ \tilde{A} &= A + (I + \alpha L_BL_C)^{-1}(\alpha L_BC'C - \alpha BB'L_C), && (35) \end{aligned}$$

$x \in \mathcal{D}(A^V)$. Now $\alpha(I + \alpha L_BL_C)^{-1}BB'L_C \in \mathcal{L}(V)$ and since $-1/\alpha \in \rho^W(L_BL_C)$, $\alpha(I + \alpha L_BL_C)^{-1}L_BC' \in \mathcal{L}(Y, W)$ is an admissible input operator. It follows that $\tilde{A}$ is a generator on $V$ with $\mathcal{D}(A^V) = \mathcal{D}(\tilde{A}^V)$ and there exists $\lambda \in \rho^V(A) \cap \rho^V(\tilde{A})$. For such $\lambda$ we have by (34)

$$(I + \alpha L_BL_C)(\lambda I - \tilde{A})^{-1} = (\lambda I - A)^{-1}(I + \alpha L_BL_C)$$

on $V$. Setting $E = (I + \alpha L_BL_C)^{-1}$, we have for $x \in \mathcal{D}(A^V)$,

$$\begin{aligned} Ex &= E(\lambda I - A)^{-1}E^{-1}E(\lambda I - A)^{-1}x \\ &= EE^{-1}(\lambda I - \tilde{A})^{-1}E(\lambda I - A)^{-1}x \\ &= (\lambda I - \tilde{A})^{-1}E(\lambda I - A)^{-1}x \\ &\in \mathcal{D}(A^V). \end{aligned}$$

Thus $E \in \mathcal{L}(\mathcal{D}(A^V))$. The proof of the other statement in (b) is similar.
(c) See [10]. □

# 4 A normalized doubly coprime factorization

In this section we obtain a normalized doubly coprime factorization for our system.

**Theorem 4.1** *Suppose that $\Sigma(A, B, C, D)$ is an admissibly exponentially stabilizable and detectable regular $\mathbb{PS}$ system.*
*A normalized doubly coprime factorization of $G(s) = D + C(sI - A)^{-1}B$ is given by*

$$G(s) = N(s)S^{-1/2}(M(s)S^{-1/2})^{-1} = (R^{-1/2}\tilde{M}(s))^{-1}R^{-1/2}\tilde{N}(s),$$

*where*

$$\begin{aligned} M(s) &= I + F(sI - A_{BF})^{-1}B & \tilde{M}(s) &= I + C(sI - A_{HC})^{-1}H, \\ N(s) &= I + C_F(sI - A_{BF})^{-1}B & \tilde{N}(s) &= D + (sI - A_{HC})^{-1}B_H, \\ X(s) &= I - C_F(sI - A_{BF})^{-1}H & \tilde{X}(s) &= I - F(sI - A_{HC})^{-1}B_H, \\ Y(s) &= -F(sI - A_{BF})^{-1}H & \tilde{Y}(s) &= -F(sI - A_{HC})^{-1}H, \end{aligned}$$

$$\begin{array}{rclrcl} C_F & = & C + DF & B_H & = & B + HD\,, \\ F & = & -S^{-1}(D^*C + B'Q) & H & = & -(BD^* + PC')R^{-1}\,, \\ R & = & I + DD^* & S & = & I + D^*D\,. \end{array}$$

*$Q$ and $P$ are the unique nonnegative solutions of the Riccati equations (24) and (26) respectively, and $A_{BF}$ and $A_{HC}$ are the infinitesimal generators of the semigroups defined by (12) and (15) respectively.*

**Proof:** (a) Since $C$ is admissible and $B'Q \in \mathcal{L}(V,U)$ it follows that $F \in \mathcal{L}(W,U)$ is an admissible output operator. Similarly, $H$ is an admissible input operator. So by Property 3.4 all the transfer matrices are well defined. Moreover $A_{BF}$ and $A_{HC}$ both generate semigroups which are exponentially stable on both $V$ and $W$ and hence all the transfer functions are holomorphic and bounded on Re $s > \varepsilon$.
(b) We prove that $G = NM^{-1} = \tilde{M}^{-1}\tilde{N}$ is a doubly coprime factorization in five steps.
(i) First we show that $G = NM^{-1}$ on Re $s > \max(\omega^V, \omega^W, 0)$.
To do this we use the following identities on $V$

$$(sI - A^V_{BF})^{-1} = (sI - A^V)^{-1} + (sI - A^V)^{-1}BF(sI - A^V_{BF})^{-1}\,, \tag{36}$$
$$(sI - A^V_{BF})^{-1} = (sI - A^V)^{-1} + (sI - A^V_{BF})^{-1}BF(sI - A^V)^{-1} \tag{37}$$

noting that $B \in \mathcal{L}(U,V)$, $F \in \mathcal{L}(W,U)$ is admissible and $A_{BF} = A + BF$ by (14). These identities hold for $s \in \rho(A^V) \cap \rho(A^V_{BF}) \supset$ Re $s > \max(\omega^V, \omega^W, 0)$ since $A^V_{BF}$ generates an exponentially stable semigroup.
From (36) we obtain

$$\begin{aligned} M(s) - I &= F(sI - A^V_{BF})^{-1}B \\ &= F(sI - A^V)^{-1}B + F(sI - A^V)^{-1}BF(sI - A^V_{BF})^{-1}B \\ &= F(sI - A^V)^{-1}BM(s) \end{aligned}$$

and so

$$M(s)[I - F(sI - A^V)^{-1}B] = I\,.$$

Similarly, using (37), it follows that

$$[I - F(sI - A^V)^{-1}B]M(s) = I$$

and so $M(s)$ is invertible on $\rho(A^V) \cap \rho(A^V_{BF})$ and

$$M^{-1}(s) = I - F(sI - A^V)^{-1}B. \tag{38}$$

Now for Re $s > \max(\omega^V, \omega^W, 0)$

$$\begin{aligned} N(s)M^{-1}(s) &= [D + (C + DF)(sI - A^V_{BF})^{-1}B][I - F(sI - A^V)^{-1}B] \\ &= D - DF(sI - A^V)^{-1}B + (C + DF)(sI - A^V_{BF})^{-1}B \\ &- (C + DF)(sI - A^V_{BF})^{-1}BF(sI - A^V)^{-1}B. \end{aligned}$$

Using (37), we have

$$\begin{aligned} N(s)M^{-1}(s) &= D + C(sI - A^V)^{-1}B \\ &= D + C(sI - A)^{-1}B \end{aligned}$$

since by (19) we can take $A$ to be $A^V$ or $A^W$.
(ii) Now we show

$$\tilde{X}M - \tilde{Y}N = I. \tag{39}$$

First we multiply it by $M^{-1}$ from the left to yield

$$\tilde{X} - \tilde{Y}G = M^{-1}. \tag{40}$$

And we establish (40) using the resolvent identity

$$(sI - A^W_{HC})^{-1} = (sI - A^W)^{-1} + (sI - A^W_{HC})^{-1}HC(sI - A^W)^{-1} \tag{41}$$

on $\rho(A^W) \cap \rho(A^W_{HC})$, noting that $H$ is an input admissible operator and so $(sI - A^W_{HC})^{-1}H \in \mathbb{H}_\infty$ by Property 3.4. For Re $s > \max(\omega^V, \omega^W, 0)$, we have

$$\begin{aligned}\tilde{X}(s) - \tilde{Y}(s)G(s) &= I - F(sI - A^W_{HC})^{-1}(B + HD) \\ &\quad + F(sI - A^W_{HC})^{-1}H(D + C(sI - A^W)^{-1}B) \\ &= I - F(sI - A^W_{HC})^{-1}B + F(sI - A^W_{HC})^{-1}HC(sI - A^W)^{-1}B.\end{aligned}$$

And using (41)

$$\begin{aligned}\tilde{X}(s) - \tilde{Y}(s)G(s) &= I - F(sI - A^W)^{-1}B \\ &= I - F(sI - A^V)^{-1}B \\ &= M^{-1}(s)\end{aligned}$$

by (38). Thus (39) holds on Re $s \geq \max(\omega^V, \omega^W, 0)$. But for $\varepsilon > 0$ sufficiently small all terms on the left are holomorphic on Re $s \geq -\varepsilon$ and so the equality may be extended to $\overline{\mathbb{C}}_0^+$.
(iii) The proof that $G = \tilde{M}^{-1}\tilde{N}$ and $-\tilde{N}Y + \tilde{M}X = I$ is essentially the same as in (i), (ii).
(iv) Next we show that

$$-\tilde{N}M + \tilde{M}N = 0. \tag{42}$$

Now $\tilde{M}^{-1}(-\tilde{N}M + \tilde{M}N)M^{-1} = -G + G = 0$ on Re $s \geq \max(\omega^V, \omega^W, 0)$. So we have established (42) on Re $s \geq \max(\omega^V, \omega^W, 0)$ and it extends to $\overline{\mathbb{C}}_0^+$, since $M$, $N$, $\tilde{M}$, $\tilde{N}$ are holomorphic on Re $s \geq -\varepsilon$ for some $\varepsilon > 0$.
(v) Now we prove the last condition for a doubly coprime factorization, namely

$$\tilde{X}Y - \tilde{Y}X = 0. \tag{43}$$

For Re $s > \max(\omega^V, \omega^W, 0)$, we have

$$\begin{aligned}\tilde{X}Y - \tilde{Y}X &= -(I - F(sI - A_{HC})^{-1}(B + HD))F(sI - A_{BF})^{-1}H \\ &\quad + F(sI - A_{HC})^{-1}H(I - (C + DF)(sI - A_{BF})^{-1}H) \\ &= -F(sI - A_{BF})^{-1}H + F(sI - A_{HC})^{-1}(B + HD)F(sI - A_{BF})^{-1}H \\ &\quad + F(sI - A_{HC})^{-1}H - F(sI - A_{HC})^{-1}H(C + DF)(sI - A_{BF})^{-1}H \\ &= -F(sI - A^V_{BF})^{-1}H + F(sI - A^V_{HC})^{-1}BF(sI - A^V_{BF})^{-1}H \\ &\quad + F(sI - A^V_{HC})^{-1}H - F(sI - A^V_{HC})^{-1}HC(sI - A^V_{BF})^{-1}H \\ &= F(sI - A^V_{HC})^{-1}(-sI + A^V_{HC} + BF)(sI - A^V_{BF})^{-1}H \\ &\quad + F(sI - A^V_{HC})^{-1}(sI - A^V_{BF} - HC)(sI - A^V_{BF})^{-1}H \\ &= 0\end{aligned}$$

since by (14) $A^V_{BF}x = (A+BF)x$ and similarly $A^V_{HC}x = (A+HC)x$ for $x \in \mathcal{D}(A^V)$.
(c) Finally, we prove

$$N(\jmath\omega)^*N(\jmath\omega) + M(\jmath\omega)M(\jmath\omega)^* = S\,. \tag{44}$$

First note that $I - DS^{-1}D^* = R^{-1}$, and so

$$C_F = C + DF = R^{-1}C - DS^{-1}B'Q\ , D^*(C+DF) + F = -BQ\,, \tag{45}$$

$$C'_F C_F + F'F = (F'D^* + C')(DF + C) + F'F = C'R^{-1}C + QBR^{-1}B'Q \tag{46}$$

Hence

$$\begin{aligned}
&N(\jmath\omega)^*N(\jmath\omega) + M(\jmath\omega)M(\jmath\omega)^* \\
&\quad = (D^* + B'(-\jmath\omega I - A'_{BF})^{-1}C'_F)(D + C_F(\jmath\omega I - A_{BF})^{-1}B) \\
&\qquad +(I + B'(-\jmath\omega I - A'_{BF})^{-1}F')(I + F(\jmath\omega I - A_{BF})^{-1}B) \\
&\quad = D^*D + D^*C_F(\jmath\omega I - A_{BF})^{-1}B + B'(-\jmath\omega I - A_{BF})^{-1}C'_F D \\
&\qquad\quad +B'(-\jmath\omega I - A'_{BF})^{-1}C'_F C_F(\jmath\omega I - A_{BF})^{-1}B + I \\
&\qquad\quad +B'(-\jmath\omega I - A'_{BF})^{-1}F' + F(\jmath\omega I - A_{BF})^{-1}B \\
&\qquad\quad +B'(-\jmath\omega I - A'_{BF})^{-1}F'F(\jmath\omega I - A_{BF})^{-1}B.
\end{aligned}$$

And using (45) and (46)

$$\begin{aligned}
&N(\jmath\omega)^*N(\jmath\omega) + M(\jmath\omega)M(\jmath\omega)^* \\
&\quad = I + D^*D - B'(-\jmath\omega I - A'_{BF})^{-1}QB - B'Q(\jmath\omega I - A_{BF})^{-1}B \\
&\qquad +B'(-\jmath\omega I - A'_{BF})^{-1}(C'R^{-1}C + QBS^{-1}B'Q)(\jmath\omega I - A_{BF})^{-1}B \\
&\quad = I + D^*D + B'(-\jmath\omega I - A'_{BF})^{-1}(-Q(\jmath\omega I - A_{BF}) + (\jmath\omega I + A'_{BF})Q \\
&\qquad\qquad +C'R^{-1}C + QBS^{-1}B'Q)(\jmath\omega I - A_{BF})^{-1}B\,.
\end{aligned}$$

Now $(\jmath\omega I - A_{BF})^{-1}Bu \in \mathcal{D}(A^V)$ for any $u \in U$ and so using the Riccati equation (24) we obtain

$$N(\jmath\omega)^*N(\jmath\omega) + M(\jmath\omega)M(\jmath\omega)^* = I + D^*D = S.$$

This shows that $N(s)S^{-1/2}$, $M(s)S^{-1/2}$ satisfy the normalization condition (3) and in a similar manner one can verify that $R^{-1/2}\tilde{N}(s)$, $R^{-1/2}\tilde{M}(s)$ satisfy the normalization condition (4). □

# 5 Solution to the robust stabilization problem

In Section 2 we gave an abstract condition for the existence of robust stabilizing controllers in terms of the norm of the Hankel operator of $[R^{-1/2}\tilde{N}, R^{-1/2}\tilde{M}]$. Here we derive explicit formulae for these controllers in terms of the parameters $A, B, C, D, L_B$ and $L_C$ of the system. First we need some extra information concerning the Hankel operators associated with exponentially stable regular $\mathbb{PS}$ systems.

**Lemma 5.1** *Suppose that $G$ is the transfer operator of the exponentially stable regular $\mathbb{PS}$ system $\Sigma(A, B, C, D)$ and let $\Gamma$ be the associated Hankel operator and $L_B, L_C$ the controllability and observability gramians. Then:*
*(a) There exist $\{w_n\} \in W, \{v_n\} \in V'$, $\|w_n\|_W = 1$, such that*

$$\|L_B v_n - \|\Gamma\| w_n\|_X \to 0 \text{ as } n \to \infty \text{ for } X = V \text{ or } W \tag{47}$$

*and*

$$L_C w_n = \|\Gamma\| v_n. \tag{48}$$

*(b) If $K(-s) \in \mathbb{H}_\infty(\mathcal{L}(U, Y))$ satisfies*

$$\|G + K\|_\infty = \|\Gamma\|, \tag{49}$$

*then*

$$\|(G + K)g_n - \|\Gamma\| f_n\|_2 \to 0 \text{ as } n \to \infty, \tag{50}$$

*where*

$$\begin{aligned} g_n(s) &= B'(sI - A')^{-1} v_n & (51)\\ f_n(s) &= C(sI - A)^{-1} w_n. & (52) \end{aligned}$$

**Proof:** (a) First recall that since $\Gamma^*\Gamma$ is self-adjoint, the spectral radius $r(\Gamma^*\Gamma) = \|\Gamma\|^2$ and $\|\Gamma\|^2 \in \sigma(\Gamma^*\Gamma)$. In [10] it is shown that $\sigma(\Gamma^*\Gamma) = \sigma(L_B L_C)$ and there is no residual spectrum. Thus $\|\Gamma\|^2 \in \sigma(L_B L_C)$ and there exists $\{w_n\}$ such that $\|w_n\|_W = 1$ and $\|L_B L_C w_n - \|\Gamma\|^2 w_n\|_X \to 0$ as $n \to \infty$ for $X = V$ or $W$. Setting $\|\Gamma\| v_n = L_C w_n$ yields (47).
(b) Since $A$ generates an exponentially stable $C_0$-semigroup (11) extends to $t_1 = \infty$ and so $\int_0^\infty \|CS(t)w_n\|_Y^2 dt < \infty$. By the Paley-Wiener theorem $f_n \in H_2(Y)$ and similarly $g_n \in H_2(U)$. We now calculate their norms.

$$\begin{aligned} \|g_n(j\omega)\|_2^2 &= \langle(-j\omega I - A)^{-1} BB'(j\omega I - A')^{-1} v_n, v_n\rangle_{W,W'} \\ &= \langle(j\omega I + A)^{-1}[AL_B + L_B A'](j\omega I - A')^{-1} v_n, v_n\rangle_{W,W'} \\ &= \langle(j\omega I + A)^{-1}[(j\omega I + A)L_B + L_B(A' - j\omega I)](j\omega I - A')^{-1} v_n, v_n\rangle_{W,W'} \\ &= \langle L_B(j\omega I - A')^{-1} v_n, v_n\rangle_{W,W'} - \langle(j\omega I + A)^{-1} L_B v_n, v_n\rangle_{W,W'} \\ &= -2\text{Re}\,\langle(j\omega I + A)^{-1} L_B v_n, v_n\rangle_{W,W'}. \end{aligned}$$

Similarly

$$\|f_n(j\omega)\|_2^2 = 2\text{Re}\,\langle L_C(j\omega I - A)^{-1} w_n, w_n\rangle = 2\|\Gamma\|\text{Re}\,\langle(j\omega I - A)^{-1} w_n, v_n\rangle.$$

Hence by (47),(48) we see that $\|f_n(j\omega)\|^2 \to \|g_n(-j\omega)\|^2$ as $n \to \infty$, and since $\|w_n\|_W = 1, \|v_n\|_{V'} \leq$ const, there holds

$$\|f_n\|_{H_2(Y)}^2 \to \|g_n\|_{H_2(U)}^2 \text{ as } n \to \infty. \tag{53}$$

In the following we denote $g_n(-s)$ by $g_n^-$ and the inner product on $L_2((-j\infty, j\infty); Y)$ by $\langle\cdot,\cdot\rangle_2$ and the norm by $\|\cdot\|_2$. Consider

$$\begin{aligned} \|(G + K)g_n^- - \|\Gamma\| f_n\|_2^2 &= \|(G + K)g_n^-\|_2^2 + \|\Gamma\|^2\|f_n\|_2^2 - 2\text{Re}\,\langle(G + K)g_n^-, \|\Gamma\| f_n\rangle_2 \\ &= \|(G + K)g_n^-\|_2^2 + \|\Gamma\|^2\|f_n\|_2^2 - 2\|\Gamma\|\text{Re}\,\langle Gg_n^-, f_n\rangle_2, \end{aligned}$$

since $K(s)g_n(-s) \in H_2(Y)^\perp$.
Now

$$\begin{aligned} G(j\omega)g_n(-j\omega) &= C(j\omega I - A)^{-1}BB'(-j\omega I - A')^{-1}v_n \\ &= -C(j\omega I - A)^{-1}[AL_B + L_B A'](-j\omega I - A')^{-1}v_n \\ &= C(j\omega I - A)^{-1}[(j\omega I - A)L_B - L_B(j\omega I + A')](-j\omega I - A')^{-1}v_n \\ &= -CL_B(j\omega I + A')^{-1}v_n + C(j\omega I - A)^{-1}L_B v_n. \end{aligned}$$

But $CL_B(sI + A')^{-1}v_n \in H_2(U)^\perp$ and so

$$\begin{aligned} \langle Gg_n^-, f_n\rangle_2 &= \langle C(j\omega I - A)^{-1}L_B v_n, f_n\rangle_2 \\ &= \langle C(j\omega I - A)^{-1}L_B v_n, C(j\omega I - A')^{-1}w_n\rangle_2 \\ &\to \|\Gamma\|\|f_n\|^2 \text{ as } n \to \infty \end{aligned}$$

by (47), since $\|v_n\|_{V'} \leq$ const.
Thus

$$\begin{aligned} &\|(G+K)g_n^- - \|\Gamma\|f_n\|_2^2 \\ &= \|(G+K)g_n^-\|_2^2 + \|\Gamma\|^2\|f_n\|_2^2 - 2\|\Gamma\|\text{Re }\langle C(j\omega I - A)^{-1}L_B v_n, f_n\rangle_2 \\ &\leq \|G+K\|_\infty^2\|g_n^-\|_2^2 + \|\Gamma\|^2\|f_n\|_2^2 - 2\|\Gamma\|\text{Re }\langle C(j\omega I - A)^{-1}L_B v_n, f_n\rangle_2 \\ &\to \|G+K\|_\infty^2\|f_n\|_2^2 - \|\Gamma\|^2\|f_n\|_2^2 \end{aligned}$$

by the above and (53).
But $K$ satisfies (49) and so (50) holds. □

In Section 2 we saw that the key to the solution of the robust stabilization problem lies in the Hankel operator associated with $[\tilde{N}_1, \tilde{M}_1]$, where $G = \tilde{M}_1^{-1}\tilde{N}_1$ is a left-normalized coprime factorisation of $G$.

**Lemma 5.2** *Suppose that $\Sigma(A, B, C, D)$ is an admissibly stabilizable and detectable regular $\mathbb{PS}$ system and $G(s) = \tilde{M}_1(s)^{-1}\tilde{N}_1(s)$ is the normalized left-coprime factorisation constructed in Theorem 4.1. If $\Gamma_1$ is the Hankel operator associated with $[\tilde{N}_1, \tilde{M}_1]$, then $\|\Gamma_1\| < 1$.*

**Proof:** First we recall from [10], [1] that the Nehari problem has the solution

$$\inf_{K(-s)\in\mathbb{H}_\infty(\mathcal{L}(U,Y))} \|[\tilde{N}_1, \tilde{M}_1] + K\|_\infty = \|\Gamma_1\| \tag{54}$$

and inserting $K = 0$, we obtain $\|\Gamma_1\| \leq \|[\tilde{N}_1, \tilde{M}_1]\|_\infty = 1$.
Suppose now that $\|\Gamma_1\| = 1$. Then $K = 0$ is a solution to (54) and from Lemma 5.1 we know that there exist $g_n \in H_2(U \oplus Y)$ and $f_n \in H_2(Y)$ such that

$$\|[\tilde{N}_1, \tilde{M}_1]g_n^- - f_n\|_2 \to 0 \text{ as } n \to \infty.$$

From Theorem 4.1 there exist $X \in \mathbb{H}_\infty(\mathcal{L}(Y))$ and $Y \in \mathbb{H}_\infty(\mathcal{L}(U,Y))$ such that

$$[\tilde{N}_1, \tilde{M}_1]\begin{bmatrix} -Y \\ X \end{bmatrix} = I$$

on Re $s \geq -\varepsilon$. So

$$\begin{aligned}
\|[\tilde{N}_1, \tilde{M}_1]g_n^- - f_n\|_2 &= \|[\tilde{N}_1, \tilde{M}_1]g_n^- - [\tilde{N}_1, \tilde{M}_1]\begin{bmatrix} -Yf_n \\ Xf_n \end{bmatrix}\|_2 \\
&= \|[\tilde{N}_1, \tilde{M}_1]\|_\infty \|g_n^- - \begin{bmatrix} -Yf_n \\ Xf_n \end{bmatrix}\|_2 \text{ since } [\tilde{N}_1, \tilde{M}_1] \text{ is co-inner,} \\
&= \|g_n^- - \begin{bmatrix} -Yf_n \\ Xf_n \end{bmatrix}\|_2
\end{aligned}$$

But the LHS tends to zero as $n \to \infty$, whereas $Xf_n$ and $-Yf_n$ are both $H_2$-functions and $g_n^- \in H_2(U \oplus Y)^\perp$ and so RHS cannot converge to zero. This contradiction proves the lemma. □

**Theorem 5.3** *Suppose that $G$ is the transfer operator of the admissibly exponentially stabilizable and detectable regular $\mathbb{PS}$ system $\Sigma(A, B, C, D)$ and let $G(s) = (R^{-1/2}\tilde{M}(s))^{-1}R^{-1/2}\tilde{N}(s)$ be the normalized left-coprime factorization constructed in Theorem 4.1. Then the controllability and observability gramians of $[R^{-1/2}\tilde{N}, R^{-1/2}\tilde{M}]$ are given by*

$$L_B = P\,, \quad L_C = Q(I + PQ)^{-1} \tag{55}$$

*respectively, where $P$ and $Q$ are the unique, self-adjoint nonnegative solutions to the filter and control algebraic Riccati equations (26) and (24), respectively. Moreover,*

$$\|\Gamma_{[\tilde{N},\tilde{M}]}\|^2 = r(L_B L_C) = r(PQ)(1 + r(PQ))^{-1} \tag{56}$$

*where $r(PQ)$ is the spectral radius of $PQ$.*

**Proof:** By Theorem 4.1, $[R^{-1/2}\tilde{N}, R^{-1/2}\tilde{M}]$ has a state-space realization $\Sigma(A_{HC}, B_P, C_P, D_P)$, where $A_{HC} = A_H C$, $B_P = [B + HD, H]$, $C_P = R^{-1/2}C$, $D_P = [R^{-1/2}D, R^{-1/2}]$, $H = -(PC' + BD^*)R^{-1}$. $R = I + DD^*$, $S = I + D^*D$ and $P$ is the unique solution of the filter algebraic Riccati equation (26). $A_{HC}$ generates an exponentially stable semigroup and so by Lemma 3.11, the controllability gramian $L_B$ of $[R^{-1/2}\tilde{N}, R^{-1/2}\tilde{M}] = \Sigma(A_{HC}, B_P, C_P, D_P)$ is the unique solution of the Liapunov equation

$$A_{HC}L_B x + L_B A'_{HC}x = -B_P B'_P x \quad \text{for} \quad x \in \mathcal{D}((A')^{W'})\,. \tag{57}$$

To show that $L_B = P$ we first rewrite the Riccati equation (26) for $P$ in the form

$$A_{HC}Px + PA'_{HC}x = -PC'R^{-1}CPx - BS^{-1}B'x\,,$$

where $A_{HC} = A - PC'R^{-1}C - BD^*R^{-1}C$. Now

$$\begin{aligned}
B_P B'_P &= (B + HD)(B + HD)' + HH' \\
&= BB' + BD^*H' + HDB' + HRH' \\
&= BB' - BD^*R^{-1}(PC' + BD^*)' - (PC' + BD^*)R^{-1}DB' \\
&\quad +(PC' + BD^*)R^{-1}(CP + DB') \\
&= BS^{-1}B' + PC'R^{-1}CP
\end{aligned} \tag{58}$$

and from the uniqueness it is clear that $L_B = P$.
The observability gramian $L_C$ of $[R^{-1/2}\tilde{N}, R^{-1/2}\tilde{M}]$ is the unique solution of the Liapunov equation

$$A'_{HC}L_C x + L_C A_{HC} x = -C'R^{-1}Cx \quad \text{for} \quad x \in \mathcal{D}(A^V). \tag{59}$$

From Lemma 5.2 $\|\Gamma\| < 1$ and so by Lemma 3.11 $(I - L_B L_C)^{-1} \in \mathcal{L}(\mathcal{D}(A^V)) \cap \mathcal{L}(V)$. Then by (33) we see that $X := L_C(I - L_B L_C)^{-1} \in \mathcal{L}(\mathcal{D}(A^V)), \mathcal{D}((A')^{W'}) \cap \mathcal{L}(V, V')$. We will verify that $X$ satisfies the control Riccati equation (24). First we reformulate this equation into the following equivalent form

$$QA_{HC}x + A'_{HC}Qx - QBS^{-1}B'Qx + C'R^{-1}Cx = -C'R^{-1}CL_BQx - QL_BC'R^{-1}Cx \tag{60}$$

for $x \in \mathcal{D}(A^V)$.
For $x \in \mathcal{D}(A^V)$, we consider

$$XA_{HC}x + A'_{HC}Xx \quad = \quad L_C(I - L_BL_C)^{-1}A_{HC}x + A'_{HC}L_C(I - L_BL_C)^{-1}x.$$

But for $x \in V$

$$L_C(I - L_BL_C)^{-1}x = (I - L_CL_B)^{-1}L_Cx.$$

Hence

$$\begin{aligned} &XA_{HC}x + A'_{HC}Xx \\ &= (I - L_CL_B)^{-1}L_CA_{HC}x + A'_{HC}L_C(I - L_BL_C)^{-1}x \\ &= (I - L_CL_B)^{-1}(L_CA_{HC}(I - L_BL_C) + (I - L_CL_B)A'_{HC}L_C)(I - L_BL_C)^{-1}x \\ &= (I - L_CL_B)^{-1}(L_CA_{HC} + A'_{HC}L_C - L_C(A_{HC}L_B + L_BA'_{HC})L_C)(I - L_BL_C)^{-1}x. \end{aligned}$$

Now $(I - L_BL_C)^{-1}x \in \mathcal{D}(A^V)$ and $L_C(I - L_BL_C)^{-1}x \in \mathcal{D}((A')^{W'})$, so using the Liapunov equations (57) and (59) we obtain

$$XA_{HC}x + A'_{HC}Xx = (I - L_CL_B)^{-1}(-C'R^{-1}C + L_CB_PB'_PL_C)(I - L_BL_C)^{-1}x.$$

Substituting for $B_PB'_P$ from (58) and using the expression for $X$ and the fact that $P = L_B$, yields

$$\begin{aligned} &XA_{HC}x + A'_{HC}Xx \\ &= (I - L_CL_B)^{-1}(L_CL_BC'R^{-1}CL_BL_C - C'R^{-1}C)(I - L_BL_C)^{-1}x + XBS^{-1}B'Xx \\ &= -C'R^{-1}C - XL_BC'R^{-1}Cx - C'R^{-1}CL_BXx + XBS^{-1}B'Xx. \end{aligned}$$

So we have verified that $X$ satisfies the Riccati equation (60) and by the uniqueness

$$Q = (I - L_CL_B)^{-1}L_C = (I - L_CP)^{-1}L_C.$$

Hence

$$PQ = L_B(I - L_CL_B)^{-1}L_C = (I - L_BL_C)^{-1}L_BL_C$$

shows that

$$(I + PQ) = (I - L_BL_C)^{-1},$$

and (56) follows readily. □

As a consequence of the above lemma and Corollary 2.7 we have the following result on robust stabilization.

**Theorem 5.4** *Suppose that* $\Sigma(A,B,C,D)$ *is an admissibly exponentially stabilizable and detectable regular* $\mathbb{PS}$ *system. Then there exists a controller* $K \in \mathbb{UT}(\mathcal{L}(Y,U))$ *which robustly stabilizes* $G$ *with robustness margin* $\varepsilon$ *if and only if*

$$\varepsilon \le \varepsilon_{max} = (1 + r(PQ))^{1/2},$$

*where* $P$ *and* $Q$ *are the unique self-adjoint solutions of the Riccati equations (26) and (24) respectively.* □

Finally, explicit formulas for robustly stabilizing controllers with an apriori robustness margin $\epsilon < \varepsilon_{max}$ can be derived as in Chapter 8 in [11]. The derivation there is for bounded $B$ and $C$ operators ($V = W$) and uses the solution to the suboptimal Nehari problem (6) in Theorem 2.6 provided by [9]. However, the solution to the suboptimal Nehari problem in [10] applies equally well to exponentially stable $\mathbb{PS}$ systems with infinite-dimensional input and output spaces which is exactly our situation; in Theorem 4.1 we give a realization of $[R^{-1/2}\tilde{N}, R^{-1/2}\tilde{M}]$ as an exponentially stable $\mathbb{PS}$ system. Moreover, all the realizations of the controllers become $\mathbb{PS}$ systems if we choose the parameter to be the transfer matrix of a $\mathbb{PS}$ system. Consequently, all the formulae for robustly stabilizing controllers for $G$ remain valid when interpreted in this sense. This is due to the fact that $\mathbb{PS}$ systems are closed under feedback connections. Since this extension to regular $\mathbb{PS}$ is straightforward we do not give the details here, but just quote the result.

**Theorem 5.5** *Suppose that* $\Sigma(A,B,C,D)$ *is an admissibly exponentially stable regular* $\mathbb{PS}$ *system. Robustly stabilizing controls* $K \in \mathbb{UT}(\mathcal{L}(Y,U))$ *with right-coprime factorizations for* $G(s) = D + C(sI - A)^{-1}B$ *which achieve a robustness margin* $\epsilon$, $0 < \epsilon < \epsilon_{max}$ *are all given by*

$$K(s) = (\theta_{11}(s)H(s) + \theta_{12}(s))(\theta_{21}(s)H(s) + \theta_{22}(s))^{-1}, \tag{61}$$

*where* $H \in \mathbb{H}_\infty(\mathcal{L}(Y))$, $\|H\|_\infty \le 1$ *and*

$$\begin{aligned}
\theta_{11}(s) &= (\sigma\epsilon)^{-1}(I + \sigma^{-2}F(sI + A_{BF})^{-1}E^*B)S^{-1/2}\\
\theta_{12}(s) &= (\sigma^{-1}D^* - \sigma^{-3}F(sI + A_{BF})^{-1}E^*PC')R^{-1/2}\\
\theta_{21}(s) &= (\sigma\epsilon)^{-1}(D + \sigma^{-2}(C + DF)(sI + A_{BF})^{-1}E^*B)S^{-1/2}\\
\theta_{22}(s) &= -(\sigma^{-1}I + \sigma^{-3}(C + DF)(sI + A_{BF})^{-1}E^*B)S^{-1/2}
\end{aligned}$$

*where* $E = (I + PQ - \sigma^{-2}PQ)^{-1}$ *and* $P, Q, R, S, F, A_{BF}$ *are as in Theorem 4.1 and* $\sigma = (1 - \epsilon^2)^{1/2}$.

This theorem demonstrates the overwhelming choice of robustly stabilizing controllers. The most popular one is the so called *central* controller.

**Corollary 5.6** *Suppose that* $\Sigma(A,B,C,D)$ *is an admissibly exponentially stable regular* $\mathbb{PS}$ *system. Then the following controller* $K_0 \in \mathbb{UT}(\mathcal{L}(Y,U))$ *stabilizes* $G$ *with a robustness margin* $\epsilon, 0 < \epsilon < \epsilon_{max}$.

$$K_0(s) = -D^* - \sigma^{-2}B'Q(sI - \hat{A}_{BF})^{-1}E^*PC' \tag{62}$$

*where* $E, F, P, Q, \sigma$ *are as in the above theorem and* $\hat{A}_{BF}$ *is the infinitesimal generator of the* $C_0$*-semigroup generated by the perturbation* $BF - \sigma^{-2}E^*PC'(C + DF)$.

# References

[1] V.M. Adamajan, D.Z. Arov, M.G. Krein : Infinite Hankel Block Matrices and Related Extension Problems, *Amer. Math. Soc. Transl.*, Vol 111, 1978, pp. 133-156.

[2] R.F. Curtain : Equivalence of input-output stability and exponential stability for infinite-dimensional systems, *J. Math. Systems Theory*, **21**, 1988, pp. 19-48.

[3] R.F. Curtain : Robust stabilizability of normalized coprime factors; the infinite-dimensional case, *Int. J. Control*, **51**, 1990, pp. 1173-1190.

[4] R.F. Curtain : A synthesis of time and frequency domain methods for the control of infinite-dimensional systems: a system theoretic approach, pp 171-224 in "Control and Estimation in Distributed Parameter Sytems", *Frontiers in Applied Mathematics*, ed H.T. Banks, SIAM, Philadelphia, 1992.

[5] R.F. Curtain : A comparison of finite and infinte dimensional contoller designs for distributed parameter systems, Rapport de Recherche, No. 1647, 1992, INRIA, France.

[6] R.F. Curtain, B.A.M. Van Keulen : Robust control with respect to coprime factors of infinite-dimensional positive real systems, *IEEE Trans. Aut. Control*, **37**, 1992, pp. 868-871.

[7] R.F. Curtain, H. Logemann, S. Townley, H.J. Zwart : Well-posedness, stabilizability and admissibility for Pritchard-Salamon systems, Report No. 260, 1992, Institut für Dynamische Systeme, Universität Bremen, Germany.

[8] R.F. Curtain, A.J. Pritchard : *Infinite Dimensional Linear Systems Theory*, LNCIS No. 8, Springer-Verlag, Berlin, 1978.

[9] R.F. Curtain, A. Ran : Explicit formulas for Hankel norm approximations of infinite-dimensional systems, *J. Integral Equations and Operator Theory*, **13**, 1989, pp. 455-469.

[10] R.F. Curtain, H.J. Zwart : The Nehari problem for the Pritchard-Salamon class of infinite-dimensional linear systems, Report W-9216, Institute of Mathematics, University of Groningen, The Netherlands, 1992

[11] R.F. Curtain, H.J. Zwart : Lecture Notes on Infinite-Dimensional Systems Theory, Faculty of Applied Mathematics, University of Twente, The Netherlands,

[12] T.T. Georgiou, M.C. Smith : Optimal robustness in the gap metric, *IEEE Trans. Aut. Control*, 1990, pp. 673-686.

[13] T.T. Georgiou, M.C. Smith : Robust stabilization in the gap metric: controller design for distributed plants, *IEEE Trans. Aut. Control* (to appear).

[14] B.M.A. Van Keulen : Equivalent conditions for the solvability of the infinite-dimensional $LQ$-problem with unbounded input and output operators, (manuscript).

[15] B.M.A. Van Keulen : Doctorate thesis, Groningen, in preparation.

[16] H. Logemann : Circle criteria, small-gain conditions and internal stability for infinite-dimensional systems, *Automatica*, **27**, 1991, pp. 677-690.

[17] D. MacFarlane, K. Glover : Robust controller design using normalized coprime factor plant description, LNCIS No. 138, Spinger-Verlag, Berlin, 1990.

[18] A.J. Pritchard, D. Salamon : The linear quadratic control problem for infinite-dimensional systems, SIAM J. Control and Optimiz., **25**, 1987, pp. 121-144.

[19] A.J. Pritchard, D. Salamon : The linear quadratic control problem for retarded systems with delays in control and observation, *IMA J. Control and Information*, **2**, 1985, pp. 335-362.

[20] M. Rosenblum, R. Rovnyak : Hardy Classes and Operator Theory, Oxford University Press, 1985.

[21] M. Vidyasagar : Control System Synthesis: A Factorization Approach, MIT Press, Cambridge, 1985

# 30 Time-Delayed Perturbations and Robust Stability

**Richard F. Datko** Georgetown University, Washington, D.C.


## Abstract

We present examples of systems which lose stability under perturbations induced by small time delays. We also describe a general class of infinite-dimensional dynamical systems which are stably robust with respect to time delays.


## Introduction

Over a period of years we have observed that certain infinite-dimensional dynamical systems which are uniformly exponentially stable become either stable or unstable when arbitrary small time delays are introduced into their dynamics. Another pseudo-anomaly, which should be well know, but which appears from time to time, is to consider a system whose spectrum lies in $Re\ \lambda < 0$ asymptotically stable when in fact it is unstable (see e.g. [1] or [3]). For instance in [3] two systems are described which have the same point spectrum, but one is asymptotically stable and the other is unstable.

How does one explain away such apparent pathologies? In our opinion one answer is that they fail a robustness litmus test in that some of the poles of the transfer function of the unperturbed or nominal system are shifted in or out of the left half plane by seemingly small time delays. Since it is well know that the Laplace transform, although linear, is not continuous this should surprise no one. And yet real-time dynamical systems which possess anomalous stability behavior with respect to time-delayed perturbations continue to be studied.

Our purpose is to offer some examples showing precisely the manner in which this poles shifting occurs and to present a general class of dynamical systems for which it does not happen. Loosely speaking the problem is traceable to the structure of the infinitesimal generators of certain $C_0$-semigroups. If these generate analytic semigroups, robustness of stability with respect to small time delays may be expected, otherwise care must be taken.

# 1 Examples of Nonrobustness and a General Result Guaranteeing Its Presence

**Notation:** Since the terms uniform exponential stability and asymptotic stability we occur several times in the exposition we shall use the respective abbreviations *u.e.s.* and *a.s.* for these conditions.

**Example 1.** Consider the two systems of scalar neutral functional-differential equations

$$P_1(\alpha): \quad \frac{d}{dt}[x(t) - \alpha x(t-1)] + x(t) = 0$$

$$P_2(\alpha): \quad \frac{d^2}{dt^2}[x(t) - 2\alpha x(t-1) + \alpha^2 x(t-2)] + 2\frac{d}{dt}[x(t) - \alpha x(t-1)] + x(t) = 0,$$

where $0 < \alpha \leq 1$ is a parameter.

Using the work of D. Henry [12] we can assert that if the functions

$$g(\lambda, \alpha) = 1 - \alpha e^{-\lambda} \tag{2}$$

and

$$f(\lambda, \alpha) = \lambda(1 - \alpha e^{-\lambda}) + 1 \tag{3}$$

have all their zeros in $Re\,\lambda < 0$, then the systems $P_1(\alpha)$ and $P_2(\alpha)$ are *u.e.s.* Let $0 < \alpha < 1$. Clearly $g(\lambda, \alpha)$ has all its zeros in $Re\,\lambda < 0$. To see that

$f(\lambda, \alpha)$ does also we look at the functional defined along real solutions of $P_1(\alpha)$ given by

$$V(x(t), \alpha) = \frac{1}{2}[x(t) - \alpha x(t-1)]^2 + \frac{1}{2}\int_{t-1}^{t} (x(\sigma))^2 dv_1. \tag{5}$$

Since

$$\begin{aligned} \frac{dV}{dt}(x(t), \alpha) &= -\frac{1}{2}(x(t) - \alpha x(t-1))^2 - \frac{1}{2}(1-\alpha^2)(x(t-1))^2 \\ &\leq 0 \end{aligned} \tag{6}$$

it is easy to deduce in the usual way that all the zeros of (3) lie in $Re\ \lambda < 0$ for **all** $\alpha \in (0, 1]$. Thus $P_1(\alpha)$ and $P_2(\alpha)$ are *u.e.s.* for $\alpha \in (0, 1)$. However since the zeros of $g(\lambda, 1)$ are $\lambda = 2n\pi i, n = 0, \pm 1, \ldots,$ Henry's criterion fails for $\alpha = 1$. Again using the work of Henry we know there exists a sequence of zero of $\{\lambda_n\}$, such that

$$\left[\begin{array}{l} Re\ \lambda_n = -\delta_n, \delta_n > 0, \delta_n \to 0, \\ Im \lambda_n = w_n, w_n \to \infty \text{ and} \\ f(\lambda_n, 1) = 0. \end{array}\right. \tag{7}$$

However, since (6) is satisfied for all $\alpha \in [0, 1], P_1(1)$ can be shown to be *a.s.* The case for $P_2(1)$ is quite different.

Because of the structure of the zeros of $f(\lambda_n, 1)$ described by (7) and the fact that the characteristic equation of $P_2(\alpha)$ has double roots we can find a sequence of solutions of $P_2(1)$ of the form

$$\begin{aligned} x_n(t) &= \tfrac{t}{\lambda_n} e^{\lambda_n t} \\ \dot{x}_n(t) &= \tfrac{e^{\lambda_n t}}{\lambda_n}[1 + \lambda_n t]. \end{aligned} \tag{8}$$

In the Banach space $C^1[-2, 0]$ of all continuously differentiable functions on $[-2, 0]$ with the norm

$$|\phi| = \sup[|\phi(\sigma)| + |\phi(\sigma)| : -2 \leq \sigma \leq 0] \tag{9}$$

the solutions of $P_2(1)$ generate a $C_0$-semigroup $T(t)(\phi)$ (see e.g.[11]). A simple calculations shows that the solutions of (8) satisfy the estimates

$$|x_n(-\frac{1}{\delta_n})| + |\dot{x}_n(-\frac{1}{\delta_n})| \geq \frac{1}{2}(\frac{1}{\delta_n}); \tag{9}$$

$$|x_n(\sigma)| + |\dot{x}_n(\sigma)| \leq 4, \quad \sigma \in [-2,0] \tag{10}$$

for $n$ sufficiently large. Thus the Principle of Uniform Boundedness is violated (see e.g. [15]) in that a uniformly bound set of functions $|\phi_n| \leq 4$ in $C^1[-2,0]$ has the property that $|T(t_n)\phi_n| \to \infty$ for some sequence $\{t_n\} \subset R^+$. This implies that there exists a $\phi \in C^1[-2,0]$ such that $|T(t)\phi| \to \infty$ as $t \to \infty$. Hence $P_2(1)$ is unstable.

**Discussion 1.** Example 1 shows that $P_1(\alpha)$ and $P_2(\alpha)$ are *u.e.s* for $0 < \alpha < 1$, that $P_1(1)$ is *a.s.* and that $P_2(1)$ is unstable. Misguided intuition might have indicated that at least $P_2(1)$ would be stable because the eigenvalues of $P_2(1)$ lie in $Re\,\lambda < 0$. This is another example of the all too common habit of basing stability criteria for linear dynamical systems on their eigenvalues alone. (Reference for this could be supplied but our object in this article is to be constructive.)

The next example is one of a robustness theory which when viewed in a certain light lacks robustness.

**Example 2.** Consider the scalar functional-differential equation

$$\dot{x}(t) = -x(t) - x(t-h) - \frac{1}{2}x(t-2h), \tag{11}$$

where $h \geq 0$ is a constant. The system (11) is *u.e.s.* for all $h \geq 0$. To see this observe that when $h = 0$ the system is *u.e.s.* If this property is lost for some $h > 0$, then by a standard procedure (see e.g. [2] ) the equation

$$iw = -1 - e^{-i\alpha} - \frac{1}{2}e^{-2i\alpha}, \alpha = wh \tag{12}$$

must have a solution for some $w > 0$ and $h > 0$. Since it does not,all solutions of (11) are *u.e.s.* Thus (11) is robust with respect to stability for all $h \geq 0$.

However, is a system of this type a realistic model for all $h \geq 0$? Suppose the delays are of the form

$$\dot{x}(t) = -x(t) - x(t-h) - \frac{1}{2}x(t-(2+\epsilon)h) \tag{13}$$

where $\epsilon > 0$ is fixed. Or suppose the system is

$$\dot{x}(t) = -x(t) - x(t-h_1) - \frac{1}{2}x(t-h_2), \tag{14}$$

where $h_1 \geq 0$ and $h_2 \geq 0$ are independent of each other. It is clear from a simple computation that although (14) is *u.e.s* for $h_1 = h_2 = 0$ it is not for $h_1 \geq 0$ and $h_2 \geq 0$ arbitrary. But even in the case of (13) there exist $\epsilon_n \to 0$ such that (13) is not *u.e.s.* for all $h$. For example if

$$\epsilon_n = \frac{1}{2}\frac{1}{2n+1}, \qquad n = 1, 2, ..., \tag{15}$$

it is shown in [4] that when $h > 2(2n+1)\pi$ and $\epsilon_n$ satisfies (15) the system (13) is unstable. Hence robustness with respect to stability for all $h$ in the case of the system (11) is only in the eye of the beholder, since for a system with two time delays both delays cannot be assumed to be known with absolute precision.

**Discussion 2.** If we let $Z = e^{-\lambda h}$ then the characteristic equation associated with (11) is

$$\lambda = -1 - Z - \frac{Z^2}{2} \tag{16}$$

The one for (13) is

$$\lambda = -1 - Z - \frac{Z^{2+\epsilon}}{2}. \tag{17}$$

If $Z_1 = e^{-\lambda h_1}$ and $Z_2 = e^{-\lambda h_2}$ the characteristic equation for (14) is

$$\lambda = -1 - Z_1 - \frac{Z_2}{2}. \tag{18}$$

The determination of lose of *u.e.s.* for some finite delay in the case of (11) reduces to the existence of a solution to the equation

$$iw = -1 - Z - \frac{Z^2}{2}, w > 0, |Z| = 1, \tag{19}$$

and in the case of (13) the equivalent equations is

$$iw = -1 - Z - Z^{2+\epsilon}, w > 0, |Z| = 1. \tag{20}$$

The right hand side of (19) is a polynomial, and hence an entire function and for $0 < \epsilon < 1$ the right hand side of (20) is a transcendental function. Thus it is not too surprising that (11) and (13) have anomalous stability behavior.

The next example is that of a damped vibrating system set in an infinite dimensional Hilbert space. The abstract model is

$$\ddot{x} + B\dot{x} + Ax = 0, \tag{21}$$

where $A$ is positive definite and unbounded and $B$ is nonnegative and unbounded. A model such as (21) in a finite-dimensional space can tolerate small time delays in the damping term. That is, if in (21) we replace $B\dot{x}(t)$ by $B\dot{x}(t-h)$, $h > 0$, the resulting system will remain *u.e.s.* for small $h$ provided the original system was. It has been shown in the infinite dimensional setting [5] that, if $B$ is symmetric and $A^{-\frac{1}{2}}BA^{-\frac{1}{2}}$ is a compact operator, arbitrary time delays in the damping make the system unstable. Example 3 below conforms to this structure, but an even more dramatic result occurs in this case.

**Example 3.** Consider the nominal system

$$w_{tt} = w_{xx}, 0 < x < 1, t > 0 \tag{22}$$

$$w(0,t) = 0, w_x(1,t) = -Kw_t(1,t), \tag{23}$$

where $K > 1$ is constant.

The term on the right side of (23) is sometimes termed a stabilizing feedback introduced through a Neumann boundary condition (see e.g. [14]). It is known (again see [14]) that (22), (23) is *u.e.s.* and its point spectrum is easily seen to satisfy the equation

$$-\frac{1}{K} = \tanh \lambda, \tag{24}$$

*i.e.*

$$\lambda = \frac{1}{2}\log\frac{(1-K)}{(1+K)}. \tag{25}$$

Thus the point spectrum lies along the vertical line

$$Re\,\lambda = \frac{1}{2}\log\frac{K-1}{1+K} \tag{26}$$

in the left half complex plane.

If we now assume a time delay in the feedback term $-Kw_t(1,t)$ in (23), *i.e.*

$$w_x(1,t) = -Kw_t(1,t-h), w(0,t) = 0 \tag{27}$$

the resulting system (22),(27) has its spectrum given by solutions of the equation

$$-\frac{1}{K} = e^{-\lambda h}\tanh\lambda. \tag{28}$$

Observe that as $Re\,\lambda \to \infty, \tanh\lambda \to 1$. Thus from a naive viewpoint one might assume (28) has some solutions asymptotic to the equation

$$-\frac{1}{K} = e^{\lambda h}, \tag{29}$$

that is

$$\lambda = \frac{1}{h}\log(-K). \tag{30}$$

While this is not quite correct, it is shown in [6] that if

$$h \cong \frac{(\log K)^2}{(2k+1)\pi}, \tag{31}$$

then (28) has solutions whose real parts satisfy

$$Re\,\lambda \cong \frac{(2k+1)\pi}{\log K}, \tag{32}$$

where $k$ is an arbitrary large positive integer. Thus as $h \to 0^+$ the system (22), (27) has solutions which diverge with respect to time at an arbitrary exponential rate.

**Discussion 3.** If $K$ in (27) satisfies $0 < K \leq 1$ the above phenomenon cannot occur, but the systems (22), (27) nevertheless can be shown to be unstable for arbitrary small time delays. While one might make a case for the lack of robustness in Examples 1 and 2, we believe that a nominal model such as that given by (22), (23), where $K > 1$, simply cannot be justified as representing some physical reality when small time delays cause the havoc demonstrated in Example 3.

The three examples given above have a common denominator, which is that small parametric changes in the real-time systems result in transfer functions whose singularities are radically different than those of the nominal systems. This is a form of nonrobustness which we believe deserves more attention.

In order not to be too negative we shall now present a robustness result which overcomes this difficulty. But we shall first need some notation.

**Notation.**

1) $X$ will denote a Banach space and $L(X)$ the corresponding space of bounded linear operators. $I$ will stand for the identity element in $L(X)$. The norms in $X$ and $L(X)$ will be denoted by $|\cdot|$.

2) $A_0$ is an unbounded linear operator which generates an analytic semigroup, $T(t)$, on $X$. We assume $(\lambda I - A_0)^{-1}$ is compact wherever it exists and that there exists $\alpha > 0, \beta > 0$ and $M_1 > 0$ such that

$$|(\lambda I - A_0)^{-1}| \leq \frac{M_1}{|\lambda + \alpha|} \tag{33}$$

whenever $Re\,\lambda > -\beta$. This condition guarantees that

$$|T(t)| \leq M_2 e^{-\frac{\alpha}{2}t}$$

for some $M_2 \geq 1$ and all $t \geq 0$ (see e.g. [10] or [13]).

3) The operators $\{A_j\}, 1 \leq j \leq m$, are in $L(X)$ and $h_j \geq 0, 1 \leq j \leq m$, are constants. We let

$$h = \sup_{1 \leq j \leq m} h_j, \tag{34}$$

and define the Banach space $C(h)$ of continuous mappings from $[-h, 0]$ into $X$ whose norm is

$$|\phi| = \sup\{\phi(t)| : -h \leq t \leq 0\}. \tag{35}$$

4) We let

$$R_m^+ = \{(h_1, ...., h_m) : h_j \geq 0, 1 \leq j \leq m\}. \tag{36}$$

The vectors in $R_m^+$ are denoted by $\hat{h}$ and their norms $|\hat{h}|$ are given by (34). We also define

$$[-2\pi, 2\pi]^m = \{(\alpha_1, ..., \alpha_m) : -2\pi \leq \alpha_j \leq 2\pi, 1 \leq j \leq m\}. \tag{37}$$

In terms of the above notation we consider the abstract functional-differential equation defined on $C(h)$ by

$$\frac{d}{dt}(x(t)) = A_o x(t) + \sum_{j=1}^{m} A_j x(t - h_j) \text{ if } t \geq 0 \tag{38}$$

$$x(t) = \phi(t), \quad \phi \in C(h), \text{ if } -h \leq t \leq 0. \tag{39}$$

The system (38), (39) is a differential equation in the sense that its formal Laplace transform is actually the Laplace transform of a continuous function $x_t(\phi) \in C(h)$ for each $\phi \in C(h)$. Moreover the solutions of the system when understood in this sense generate a $C_0$-semigroup in $C(h)$ (see e.g. [7] for proofs of these statements). In fact the solutions of (38), (39) can for each $\phi \in C(h)$ be described by the equation

$$x(t, \phi) = S(t)\phi(0) + \sum_{j=1}^{m} \int_{-h_j}^{0} S(t - \sigma - h_j) A_j \phi(\sigma) d\sigma, \tag{40}$$

where $S : [0, \infty) \to L(X)$ is the inverse Laplace transform of the holomonphic mapping

$$\hat{S}(\lambda) = (\lambda I - A_0 - \sum_{j=1}^{m} A_j e^{\lambda h_j})^{-1}, \tag{41}$$

again see [7] for proof of this statement. The following is a robustness theorem.

**Theorem** Suppose $A_0$ and $\{A_j\}$ have the properties described in the Notation and assume when $\hat{h} = 0$ in $R_m^+$ the system (38), (39) is *u.e.s.* Then a necessary and sufficient condition for (38), (39) to be *u.e.s.* for all $\hat{h}$ in $R_m^+$ is that

$$[I - (iwI - A_0)^{-1} \sum_{j=1}^{m} A_j e^{i\alpha_j}]^{-1}$$

exist for all $w \in (-\infty, \infty) - \{0\}$ and $-2\pi \leq \alpha_j \leq 2\pi, 1 \leq j \leq m$.

**Sketch of the proof.** (The complete proof is given in [9]). Because of the representation of solutions (40) it is clear that the system (38), (39) is *u.e.s.* if and only if for a given $\hat{h} \in R_m^+$ the operator function $S : [0, \infty) \to L(X)$ has an exponential decay rate, *i.e.* there exists $\propto_0 > 0$ and $M_3 \geq 1$ such that

$$|S(t)| \leq M_3 e^{-\propto_0 t}, t \geq 0. \tag{42}$$

This in turn is a function of the poles of $\hat{S}(\lambda)$ since, because $(\lambda I - A_0)^{-1}$ is compact, $\hat{S}(\lambda)$ is compact wherever its exists. Hence any singularities $\hat{S}(\lambda)$ possesses are isolated poles of finite order (see e.g. [10] , p.592). Thus the problem reduces to the location of the poles of $\hat{S}(\lambda)$. It is also not difficult to see that $\hat{S}(\lambda)$ has a pole $Re\,\lambda_0 > -\beta$ if and only the operator-valued function

$$N(\lambda, \hat{h}) = [I - (\lambda I - A_0)^{-1} \sum_{j=1}^{m} A_j e^{-\lambda h_j}], \tag{43}$$

where $\hat{h} = (h_1, ..., h_m)$, has the eigenvalue one at $\lambda = \lambda_0$. Using the methods described in [8] we can show that the function $\tau : R_m^+ \to R$ defined by

$$\tau(\hat{h}) = \sup\{Re\,\lambda : N(\lambda, \hat{h}) \text{ has one as an eigenvalue }\} \tag{44}$$

is continuous on $R_m^+$.

Thus, since by hypothesis $\tau(0) < 0$, if (38), (39) looses the property of being *u.e.s.* if at some $\hat{h}_0 \in R_m^+, \tau(\hat{h}_0) = 0$. This means (38), (39) is *u.e.s.* for all $\hat{h} \in R_m^+$ if and only if

$$\tau(\hat{h}) < 0 \text{ for all } \hat{h} \in R_m^+.$$

Moreover if $\tau(\hat{h}_0) = 0$ then there exists $w \in (-\infty, \infty) - \{0\}$ such that $N(iw, \hat{h}_0)$ has one as an eigenvalue. Assume

$$\hat{h}_0 = (h_1, ..., h_m) \tag{45}$$

and $\tau(\hat{h}_0) = 0$ then, if

$$\hat{\alpha} = (wh_1, ...., wh_m) = (\alpha_1, ..., \alpha_m), \tag{46}$$

the operator

$$I - (iwI - A_0)^{-1} \sum_{j=1}^{m} A_j e^{-i\alpha_j} \tag{47}$$

also has one as an eigenvalue. But for each $\hat{\alpha}$ of the form (46) we can find

$$\hat{\alpha}_0 = (\alpha_1^0, ..., \alpha_m^0), -2\pi \leq \alpha_j^0 \leq 2\pi \tag{48}$$

which satisfies (47) which proves the theorem.

## References

[1] W.E. Brumley, On the asymptotic behavior of solutions of differential-difference equations of neutral type, J. Differential Equations **7** (1970), 175-188.

[2] K.L. Cooke and J.M. Ferreira, Stability conditions for linear retarded functional differential equations, J. Math. Anal. Appl. **96** (1983), 480-504.

[3] R. Datko, An example of an unstable neutral equation, Int. J. Control **38** (1983), 263-267.

[4] R. Datko, Remarks concerning the asymptotic stability and stabilization of linear delay differential equations, J. Math. Anal. Appl. **111** (1985), 571-584.

[5] R. Datko and Y.C. You, Some second-order vibrating systems cannot tolerate small time delays in their damping, J. of Optimization Theory and Appl. **70**, (1991), 521-536.

[6] R. Datko, Two examples of illposedness with respect to small time delays in stablilized elastic systems, to appear IEEE Trans. on Auto. Control.

[7] R. Datko, Linear autonomous neutral differential equations in a Banach space, J. Differential Equations **25** (1977), 258-274.

[8] R. Datko, The uniform exponential stability of a class of linear differential-difference equations in a Hilbert space, Proceedings of the Royal Soc. of Edinburgh **89A** (1981), 201-215.

[9] R. Datko, The robustness of uniform exponential stability of generalized neutral functional-differential equations in Banach spaces, to appear.

[10] N. Dunford and J.T. Schwartz, *Linear Operators, Part I*, Wiley, New York, 1958.

11] J. Hale, *Theory of Functional Differential Equations*, Spring-Verlag, New York, 1977.

[12] D. Henry, Linear autonomous neutral functional differential equations, J. Differential Equations, **15** (1974), 106-128.

[13] E. Hille and R.S. Phillips, *Functional Analysis and Semi-groups*, A.M.S. Colloquium Publications, **31**, Providence, R.I., 1957.

[14] J.-L . Lions, Exact contrallability, stabilization and perturbations for distributed parameter systems, SIAM Rev. **30** (1988), 1-68.

[15] A.E. Taylor and D.C. Lay, *Introduction to Functional Analysis*, Second Edition, Robert E. Krieger Publ. Co, Malabar, Florida, 1980.

# 31 Positive Controllability of Linear Systems with Delay

**Mohamed A. El-Hodiri** University of Kansas, Lawrence, Kansas

**F. S. Van Vleck** University of Kansas, Lawrence, Kansas

## I. INTRODUCTION

Recently T. Schanbacher [Sch-2] posed the problem of determining which positive states of a positivity preserving linear control system can be reached using nonnegative controls. He considered both finite and infinite dimensional systems of the form

$$(1.1) \qquad \dot{x}(t) = Ax(t) + Bu(t),\ x(0) = x_0,$$

where x is the state vector and and u is the control vector. In an appropriate sense, solutions of (1.1) can be written as

$$(1.2) \qquad x(t) = e^{At}x_0 + \int_0^t e^{A(t-s)}Bu(s)\ ds$$

where u is an admissible controller. Schanbacher assumed $e^{At} \geq 0$ for each t and $B \geq 0$ (precise definitions are given in Section 2) and investigated, for $x_0 = 0$, the question of determining which states in the nonnegative orthant of the range space are attained as u ranges over all nonnegative admissible controls.

Here we will restrict our attention to linear control systems in finite dimensional spaces but with delay in the control. So we will be considering

$$(1.3) \qquad x(t) = Ax(t) + Bu(t) + Cu(t - h),$$

where $h > 0$ is a real number, x is a real n-vector, u is a real m-vector, A is an $n \times n$ real constant matrix, and B and C are $n \times m$ real constant matrices. For convenience and simplicity, we have assumed that there is only one delay term and we will also take $h = 1$.

We were originally lead to a related problem by the discussion, in the political economy of development, centered around the question of feasibility of certain targets of economic growth. For instance, if an economy starts at time zero at a level $y_0$ of per capita gross national product (GNP), is it possible for this economy to reach a level of $2y_0$ given the technological and resource limitations to which the country's economy is constrained. If we let p(t) represent the path of prices, assumed to be perfectly known, over the time period under consideration and if x(t) is the n-vector of final goods and services per capita, then the per capita GNP y(t), at time t, is given by: $y(t) = p(t)x(t)$. Let x(t) be produced according to the production functions:

$$x_i(t) = f^i(k_i(t), \ell_i(t), R_i(t)),$$

where $k_i$, $\ell_i$, and $R_i$ are respectively the capital, labor and resource inputs to industry i. We will assume that resource and labor availability are determined outside the model at quantities R(t) and L(t) respectively. The like resource and labor constraints are given by:

$$\Sigma\, \ell_i(t) \leqq L(t), \qquad \Sigma\, R_i(t) \leqq R(t).$$

Capital of type i grows by the amount of new investment and declines by the amount of equipment that becomes obsolete. Thus

$$\dot{k}_j = u_j(t) - u_j(t - \tau_j),\ k_j(0) = k_j^0,$$

where $u_i(t)$ represents newly installed machinery at time t and where $\tau_i$ is the lifetime of machines of type i. The output of good i is allocated between investments, $J_i$, in new machines, and consumption $C_i$. The investment requirements in terms of good i are given by the functions:

$$J_i = \sum_{j=1}^{n} h_{ij}(u_j)$$

we may thus write the allocation constraint as:

$$x_i = \Sigma\, h_{ij}(u_j) + C_i.$$

Assume, further that $C_i$ and $\ell_i$ are determined in advance, $\tau_j = \tau$ for all j, and that the functions f and h are linear. Then we can reduce our problem to a controllability problem for a system:

$$\dot{k} = Ak(t) + Bu(t) + Cu(t - \tau),$$

with contraints on u of the form:

$$0 \leqq u \leqq \bar{u}.$$

This problem will not be explored further here, but rather we will show how Schanbacher's results for the finite dimensional nondelay case can be combined with older results for linear control systems with delays to give results about positive cone controllability using nonnegative controllers.

In Section 2 we give the notation, definitions, and background results that we will be using in the later sections. In Section 3 we will focus on "Euclidean" type controllability, while in Section 4, "set-point" type controllability will be treated.

## 2. NOTATION, DEFINITIONS AND BACKGROUND RESULTS

As usual $\mathbb{R}^n$ denotes real n-dimensional Euclidean space. If $x, y \in \mathbb{R}^n$, $x \geq y$ iff $x_i \geq y_i$ for each i, i = 1, 2, ..., n, and $x > y$ iff $x \geq y$ and $x \neq y$. So $x \geq 0$ iff $x_i \geq 0$ for i = 1, 2, ..., n and $x > 0$ iff $x \geq 0$ and $x \neq 0$. The nonnegative orthant of $\mathbb{R}^n$, denoted by $\mathbb{R}^n_+$, is just the set of all $x \in \mathbb{R}^n$ such that $x \geq 0$. A linear transformation $S : \mathbb{R}^p \to \mathbb{R}^q$ is nonnegative, written $S \geq 0$, iff $Sx \geq 0$ for all $x \in \mathbb{R}^p_+$.

For a subset U of $\mathbb{R}^p$, cl U, co U, cone U and cocone U denote the closure of U, the smallest convex set containing U, the smallest cone containing U and 0, and the smallest convex cone containing U and 0, respectively.

The class of admissible controllers u will be all measurable functions $u : \mathbb{R}_+ \to \mathbb{R}^m_+$ such that u is in $L^1$ on [0, t] for each $t > 0$. A, B and C will always denote real constant matrices of sizes $n \times n$, $n \times m$ and $n \times m$, respectively.

Consider the system

$$\dot{x}(t) = Ax(t) + Bu(t), \quad x(0) = x_0. \tag{2.1}$$

Assume $e^{At} \geq 0$ and $B \geq 0$. Since solutions of (2.1) are given by

$$x(t) = e^{At}x_0 + \int_0^t e^{A(t-s)}Bu(s)\ ds \tag{2.2}$$

for any admissible u, it is clear that $x_0 \geq 0$ implies $x(t) \geq 0$ for $t \geq 0$ and so the system preserves positivity.

**Remark.** For a given nonempty subset U of $\mathbb{R}^m$, the admissible controllers can be taken to be the functions u in $L^1$ on [0, t] for each $t > 0$ with $u(t) \in$ cone U. In this case the following lemma [Sch-2, Proposition 4.3] indicates what conditions $e^{At}$ and B should satisfy.

**Lemma 2.1.** The following are equivalent for the system (2.1).

(a) $x(t) \geq 0$ for all $x_0 \geq 0$ and all admissible u with $u(t) \in$ cone U for $t \geq 0$

(b) $e^{At} \geq 0$ for all $t \geq 0$ and $BU \geq 0$.

We henceforth assume $U = \mathbb{R}^m_+$ and $x_0 = 0$, and define

$$K^+_t = \{x(t) = \int_0^t e^{A(t-s)}Bu(s)\, ds \mid u \text{ admissible}\},\ t \geq 0$$

and

$$K^+ = \bigcup_{t>0} K^+_t.$$

Then, the system (2.1) (with $x_0 = 0$) is said to be positively controllable in time t, $(PC)_t$, iff $K^+_t = \mathbb{R}^n_+$. Positive controllability (PC), for (2.1) means that $K^+ = \mathbb{R}^n_+$. Analogously approximate positive controllability in time t, $(APC)_t$, or approximate positive controllability, (APC), occur when cl $K^+_t = \mathbb{R}^n_+$, or cl $K^+ = \mathbb{R}^n_+$, respectively.

Next, we state some results due to Schanbacher [Sch-2] that will be used in Section's 3 and 4. Let e(1), ..., e(m) denote the standard unit coordinate vectors in $\mathbb{R}^m$ and let e[1], ..., e[n] denote those of $\mathbb{R}^n$.

**Proposition 2.2.** [Sch-2, Proposition 4.7]. With the above assumptions:

(a) $\text{cl } K^+_t = \text{cl}\{\text{co}\{e^{As} Bu \mid 0 \leq s \leq t, u \in \mathbb{R}^m_+\}\}$

$\text{cl } K^+ = \text{cl}\{\text{co}\{e^{As} Bu \mid 0 \leq s, u \in \mathbb{R}^m_+\}\}$

(b) $\text{cl } K^+_t = \text{cl}\{\text{cocone}\{e^{As} Be(j) \mid 0 \leq s \leq t, j = 1, \ldots, m\}\}$

$\text{cl } K^+ = \text{cl}\{\text{cocone}\{e^{As}Be(j) \mid 0 \leq s, j = 1, \ldots, m\}\}$.

**Theorem 2.3.** [Sch-2, Theorem 4.9]. With the above assumptions and $x_0 = 0$, the system (2.1) is

(a) $(APC)_t$ iff for each i, $1 \leq i \leq n$, there exist j, $1 \leq j \leq m$, and $\mu > 0$ such that $e[i] = \mu Be(j)$, and

(b) (APC) iff for each i, $1 \leq i \leq n$, there exist j, $1 \leq j \leq m$, and $\mu > 0$ such that $e[i] = \mu Be(j)$ or there exists j, $1 \leq j \leq m$, such that $e[i] = \lim_{t\to\infty} \dfrac{e^{At}Be(j)}{\|e^{At}Be(j)\|}$.

**Remark.** The above limit exists under the assumptions given above and is an eigenvector of A; see[Sch-1].

As a corollary to Proposition 2.2 and Theorem 2.3 we have the following statements.

**Corollary 2.4.** [Sch-2, Corollary 4.10]. With the above assumptions and $x_0 = 0$,

(a) the system (2.1) is $(APC)_t$ iff $B\mathbb{R}^m_+ = \mathbb{R}^n_+$ (and hence $m \geq n$);

(b) if (2.1) is (APC), then $2m \geq n$ and for each i, $1 \leq i \leq n$, either e[i] is an eigenvector of A or $e[i] = \mu Be(j)$ for some j, $1 \leq j \leq m$, and some $\mu > 0$;

(c) if (2.1) is (PC), then (2.1) is $(APC)_t$ for each $t > 0$.

**Remark.** The statement in Corollary 2.4(c), while true, is not obvious. From the above and the definitions the following implications are true

$$\begin{array}{ccccc} & & (PC)_t & & \\ & \Downarrow & & \Downarrow & \\ (PC) & \Rightarrow & (APC)_t & \Rightarrow & (APC) \end{array}$$

**Remark.** P. Monson, C. Shannon and F. Van Vleck [MSV] have shown that (PC) implies $(PC)_t$ for each $t > 0$ for (2.1) and so the diagram can be extended to

$$\begin{array}{ccccc} & & (PC)_t & & \\ & \Updownarrow & & \Downarrow & \\ (PC) & \Rightarrow & (APC)_t & \Rightarrow & (APC) \end{array}$$

Examples given in [Sch-2] and [MSV] show that all other possible implications are false.

We now recall some controllability results for the system

$$\dot{x}(t) = Ax(t) + Bu(t) + Cu(t - 1), \; x(0) = x_0, \tag{2.3}$$

with A, B, C as in (1.3).

The system (2.3) is Euclidean controllable (EuC) on $[0, t_1]$ if for every $x_0, x_1 \in \mathbb{R}^n$ there is a control function $u : [-1, t_1] \to \mathbb{R}^m$ with u specified on [-1, 0] steering the system from $x_0$ at time 0 to $x_1$ at time $t_1$. The system is setpoint controllable (SpC) on $[0, t_1]$ if $x_0$ can be steered to $x_1$ at time $t_1$ with u given on both

$[-1, 0]$ and $[t_1 - 1, t_1]$. Chyung [C], Sebakhy [Se], and Banks, Jacobs and Latina [BJL] proved Euclidean controllability results for (2.3) that are analogous to the standard controllability result for (2.1). Banks, Jacobs and Latina also gave setpoint controllability results for (2.3). We summarize these results below. We assume $t_1 > 1$.

**Theorem 2.5.** [Se, Theorem; BJL, Theorems 3.1 and 3.2]. The system (2.3) is (EuC) on any $[0, t_1]$ iff rank $[B, AB, \ldots, A^{n-1} B, C, AC, \ldots, A^{n-1} C] = n$.

We henceforth denote the system (2.3) by the triple (A, B, C). This makes (A, B, 0) the system (2.1).

**Theorem 2.6.** [BJL, Theorem 3.3]. If (A, B + C, 0) is controllable, then (A, B, C) is (EuC).

**Theorem 2.7.** [BJL, Theorems 3.6 and 3.7]. (A, B, C) is (SpC) on any $[0, t_1]$, $t_1 > 1$, iff $(A, B + e^{-A}C, 0)$ is controllable, i.e. rank $[B + e^{-A}C, A(B + e^{-A}C), \ldots, A^{n-1}(B + e^{-A}C)] = n$.

**Remark.** [BJL, Remark 3.4]. Rank $[B + e^{-A}C, A(B + e^{-A}C), \ldots, A^{n-1}(B + e^{-A}C)] = n$ implies rank $[B, AB, \ldots, A^{n-1}B, C, AC, \ldots, A^{n-1}C] = n$. The converse is not true.

## 3. EUCLIDEAN POSITIVE CONTROLLABILITY USING POSITIVE CONTROLLERS.

We consider the system

$$\dot{x}(t) = Ax(t) + Bu(t) + Cu(t - 1),\ x(0) = x_0, \tag{2.3}$$

as in Section 2. We make no specific assumptions initially except $t_0 = 0$ and $t > 1$. Then, for any admissible controller, the solution x(t) satisfies the variation of parameter formula

$$x(t) = e^{At}x_0 + e^{At} \int_0^t e^{-As}[Bu(s) + Cu(s - 1) \, ds \tag{3.1}$$

$$= e^{At}[x_0 + \int_{-1}^{0} e^{-A(s+1)}Cu(s) \, ds] + \int_0^{t-1} e^{A(t-1-s)}(e^{A}B + C) \, u(s) \, ds$$

$$+ \int_{t-1}^{t} e^{A(t-s)}Bu(s) \, ds,$$

where the second formula is obtained from the first using a simple substitution. Letting $\hat{x}_0 = x_0 + \int_{-1}^{0} e^{-A(s+1)}Cu(s)\, ds$ we see that there is a "drift" term $e^{At}\hat{x}_0$, with $\hat{x}_0$ fixed in both Euclidean and setpoint type controllability, and an "ending" term $\int_{t-1}^{t} e^{A(t-s)}Bu(s)\, ds$, which is specified in setpoint controllability.

Note that

$$(3.2) \qquad \{\int_{t-1}^{t} e^{A(t-s)}Bu(s)\, ds \mid u \text{ admissible}\} = \{\int_{0}^{1} e^{A(1-s)}B\, \tilde{u}(s)\, ds \mid \tilde{u} \text{ admissible}\}$$

where $\tilde{u}(s) = u(t + s - 1)$.

We next introduce the analogues of Schanbacher's set of attainability in time t, $K_t$, as follows: Let

$$K_t(A, B, C) = \{x(t) - e^{At}\hat{x}_0 \mid u \text{ admissible}\}$$

$$= \{x(t) \mid u \text{ admissible}, x_0 = 0, u(s) = u_0(s) = 0 \text{ for } s \in [-1, 0]\}$$

and

$$K_t^+(A, B, C) = \{x(t) \in K_t(A, B, C) \mid u(s) \geq 0 \text{ for } 0 \leq s \leq t\}.$$

Note that $K_t^+(A, B, 0)$ is just Schanbacher's $K_t^+$.

By the variation parameter formula (3.1) and the note (3.2), we see that

$$(3.3) \qquad K_t(A, B, C) = K_{t-1}(A, e^AB + C, 0) + K_1(A, B, 0)$$

and

$$(3.4) \qquad K_t^+(A, B, C) = K_{t-1}^+(A, e^AB + C, 0) + K_1^+(A, B, 0)$$

This last relation, (3.4) is central to all of our later arguments. (3.4) also suggests that we replace Schanbacher's $e^{At} \geq 0$ and $B \geq 0$ with $e^{At} \geq 0$, $B \geq 0$ and $e^AB + C \geq 0$. These last three conditions are weaker than the more obvious conditions $e^{At} \geq 0$, $B \geq 0$, $C \geq 0$ and are what we will usually assume in the remainder of this paper.

Next, two more definitions. We say that (2.3), with $x_0 = 0$, $u(s) = u_0(s) = 0$ on

[-1, 0] is Euclidean postively controllable in time t, $(\text{EuPC})_t$, if $K_t^+(A, B, C) = \mathbb{R}_+^n$; Euclidean positively controllable, (EuPC), if $K^+(A, B, C) = \bigcup_{t>0} K_t^+(A, B, C) = \mathbb{R}_+^n$; Euclidean approximately positively controllable in time t, $(\text{EuAPC})_t$, if $\text{cl } K_t^+(A, B, C) = \mathbb{R}_+^n$; and Euclidean approximately positively controllable, (EuAPC), if $\text{cl } K^+(A, B, C) = \mathbb{R}_+^n$.

**Theorem 3.1.** Assume $e^{At} \geq 0$, $B \geq 0$ and $e^A B + C \geq 0$. Then the system (2.3) is $(\text{EuAPC})_t$ iff for each i, $1 \leq i \leq n$, there is a j, $1 \leq j \leq m$, and $u > 0$ such that $e[i] = \mu Be(j)$ or $e[i] = \mu(e^A B + C)\, e(j)$.

**Proof.** If $e[i] = \mu Be(j)$ or $e[i] = \mu(e^A B + C)\, e(j)$, then by Proposition 2.2(b), $e[i] \in \text{cl } K_1^+(A, B, 0)$ or $e[i] \in \text{cl } K_{t-1}^+(A, e^A B + C, 0)$ and hence by (3.4) in $K_t^+(A, B, C)$. Thus $\text{cl } K_t^+(A, B, C) = \mathbb{R}_+^n$.

Conversely, if $\text{cl } K_t^+(A, B, C) = \mathbb{R}_+^n$, then for each i, $1 \leq i \leq n$, $e[i] \in \text{cl } K_t^+(A, B, C)$. So there is a sequence $(x_p) \to e[i]$ with $e[i] \in \text{cl } K_t^+(A, B, C)$. Write $x_p$ as $x_p = y_p + z_p$ with $y_p \in K_{t-1}^+(A, C^A B + C, 0)$ and $z_p \in K_1^+(A, B, 0)$. Since the sequences $(y_p)$ and $(z_p)$ are composed of nonnegative terms and their sum converges, each must be bounded. Hence there are subsequences $(y_{pk})$ and $(z_{pk})$ that converge to, say, y and z, respectively. Then $y + z = e[i]$ and so $y = \lambda e[i]$ and $z = (1 - \lambda)e[i]$ for some $\lambda$, $0 \leq \lambda \leq 1$. Since either $\lambda$ and $1 - \lambda$ must be nonzero, either $e[i] \in \text{cl } K_{t-1}^+(A, e^A B + C, 0)$ or $e[i] \in \text{cl } K_1^+(A, B, 0)$. By [Sch-2, proof of Theorem 4.9(6)], either $e[i] = \mu(e^A B + C)e(j)$ or $e[i] = \mu Be(j)$ for some j and some $\mu > 0$.

**Corollary 3.2.** Suppose $A = 0$, $B \geq 0$, and $C \geq 0$. Then $\dot{x} = Bu(t) + Cu(t - 1)$, $x_0 = 0$, $u_0(s) = 0$, $-1 \leq s \leq 0$, is $(\text{EuAPC})_t$ iff for each i, $1 \leq i \leq n$, there is a j, $1 \leq j \leq m$, and $\mu > 0$ such that $e[i] = \mu Be(j)$ or $e[i] = \mu Ce(j)$.

**Proof.** By Theorem 3.1, either $e[i] = \mu Be(j)$ or $e[i] = \mu(B + C)\, e(j)$. If the latter is true, then since both $B \geq 0$ and $C \geq 0$, we must have $\mu Be(j) = \lambda e[i]$ and $\mu Ce(j) = (1 - \lambda)\, e[i]$ for some $\lambda$, $0 \leq \lambda \leq 1$. The result now follows by considering cases.

**Remark.** The statement $(\text{EuAPC})_t$ in Corollary 3.2 can be replaced by any of $(\text{EuPC})_t$, (EuPC) and (EuAPC) since, for $A = 0$,

$$K_t^+(0, B, 0) = \text{cl } K_t^+(0, B, 0) = K^+(0, B, 0) = \text{cl } K^+(0, B, 0)$$

$$= \text{cocone } \{Be(j) \mid j = 1, 2, \ldots, m\}.$$

**Corollary 3.3.** If $A \geq 0$, $B \geq 0$ and $C = -B$, then (2.3) is $(\text{EuAPC})_t$ iff, for each i, $1 \leq i \leq n$, there is a j, $1 \leq j \leq m$, and $\mu > 0$ such that $e[i] = \mu Be(j)$.

**Proof.** Note that $e^A B + C = e^A B - B = (e^A - I) B \geq B \geq 0$, since $A \geq 0$. If the system is $(EuAPC)_t$, then either $e[i] = \mu Be(j)$, and we are done, or $e[i] = \mu(e^A - I) Be(j)$. In the latter case, $e[i] = \mu(e^A - I) Be(j) \geq \mu Be(j)$ and so there is a $v$, $0 \leq v \leq 1$, such that $v\, e[i] = \mu Be(j)$. If $v = 0$, then $Be(j) = 0$, and we have a contradiction. Hence $v \neq 0$ and $e[i] = \frac{\mu}{v} Be(j)$. The converse is immediate.

The next results are related to Theorem 2.6 [BJL, Theorem 3.3].

**Theorem 3.4.** Assume $e^{At} \geq 0$, $B \geq 0$ and $C \geq 0$. If $\dot{x}(t) = Ax(t) + (B + C)u(t)$, $x_0 = 0$, is $(APC)_t$, then (2.3) is $(EuAPC)_t$.

Before we prove this result, we give results for systems without delay.

**Proposition 3.5.** Assume $e^{At} \geq 0$, $B \geq 0$ and $C \geq 0$. If $\dot{x}(t) = Ax(t) + (B + C)u(t)$, $x_0 = 0$, is $(APC)_t$, then $\dot{x}(t) = Ax(t) + [B, C]\begin{bmatrix} u_1(t) \\ u_2(t) \end{bmatrix}$, $x_0 = 0$ is $(APC)_t$.

(Here $[B, C]$ is the partitioned $n \times 2m$ matrix whose first $m$ columns are those of $B$ and whose next $m$ columns are those of $C$.)

**Proof.** By Theorem 2.3(a), for each $i$ there are $j$ and $\mu > 0$ such that $\mu(B + C)\, e(j) = e[i]$. Since $B, C \geq 0$, we see that $\mu Be(j) = \lambda e[i]$ and $\mu Ce(j) = (i - \lambda)\, e[i]$, where $0 \leq \lambda \leq 1$. Thus, for any $\lambda$, $0 \leq \lambda \leq 1$, there exist $j$ and $n > 0$ such that $e[i] = \eta\, [B, C]\begin{bmatrix} e(j) \\ 0 \end{bmatrix}$ or $e[i] = \eta\, [B, C]\begin{bmatrix} 0 \\ e(j) \end{bmatrix}$ and hence $\dot{x}(t) = Ax(t) + [B, C]\begin{bmatrix} u_1 \\ u_2 \end{bmatrix}$ is $(APC)_t$.

To see that the converse need not be true, let $A = \begin{bmatrix} 0 & 0 \\ 0 & 0 \end{bmatrix}$, $B = \begin{bmatrix} 1 & 1 \\ 0 & 0 \end{bmatrix}$ and $C = \begin{bmatrix} 0 & 0 \\ 1 & 1 \end{bmatrix}$. Then $e[1] = Be(j)$, $e[2] = Ce(j)$, but $(C + B)e(j) = \begin{bmatrix} 1 \\ 1 \end{bmatrix}$, for $j = 1, 2$.
The conclusion then follows from the following proposition, which merely says that $\dot{x} = Ax + Bu_1 + Cu_2$ can be controlled using either $B$ or $C$.

**Proposition 3.6.** Assume $e^{At} \geq 0$, $B \geq 0$ and $C \geq 0$. The system $\dot{x}(t) = Ax(t) + [B, C]\begin{bmatrix} u_1(t) \\ u_2(t) \end{bmatrix}$, $x_0 = 0$, is $(APC)_t$ iff for each $i$, $1 \leq i \leq n$, there exist $j$, $1 \leq j \leq m$, and $\mu > 0$ such that $e[i] = \mu Be(j)$ or $e[i] = \mu Ce(j)$.

**Proof of Theorem 3.4.** As in the proof of Proposition 3.5, for each $i$, there exist $j$,

$\mu > 0$ and $\lambda$, $0 \le \lambda \le 1$, such that $\mu Be(j) = \lambda e[i]$ and $\mu Ce(j) = (1 - \lambda)e[i]$. If $\lambda = 0$, $Be(j) = 0$ and $\mu Ce(j) = e[i]$, so $\mu(e^A B + C)e(j) = e[i]$. If $\lambda \ne 0$, then $\frac{\mu}{\lambda} Be(j) = e[i]$. Thus, by Theorem 3.1, (2.3) is $(EuAPC)_t$.

The next result is the extension of Schanbacher's result for approximate positive controllability to the situation being considered here.

**Theorem 3.7.** Assume $e^{At} \ge 0$, $B \ge 0$ and $e^A B + C \ge 0$. Then the system (2.3) is (EuAPC) iff, for each i, $1 \le i \le n$, there is j, $1 \le j \le m$, such that either

(a) there is $\mu > 0$ such that $e[i] = \mu Be(j)$ or $e[i] = \mu(e^A B + C)e(j)$

or

(b) $e[i] = \lim_{t \to \infty} \dfrac{e^{At}(e^A B + C)e(j)}{\|e^{At}(e^A B + C)e(j)\|}$

**Proof.** This follows from formula (3.4), Theorem 3.1, Proposition 2.2(b) and Schanbacher's proof of Theorem 2.3(b).

## 4. SET-POINT POSITIVE CONTROLLABILITY USING POSITIVE CONTROLLERS

As before, we consider equation

$$\dot{x}(t) = Ax(t) + Bu(t) + Cu(t - 1),\ x(0) = x_0, \tag{2.3}$$

on $[0, t]$ where $t > 1$ and $u(s) = u_0(s)$, $-1 \le s \le 0$, and $u(s) = u_1(s)$, $t - 1 \le s \le t$, are given functions. In this case the variation of parameters formula (3.1) becomes

$$x(t) = e^{At}\tilde{x}_0 + \int_0^{t-1} e^{A(t-1-s)}(e^A B + C)u(s)\, ds, \tag{4.1}$$

where $\tilde{x}_0 = x_0 + \int_{-1}^{0} e^{-A(s+1)} Cu_0(s)\, ds + \int_{t-1}^{t} e^{-As} Bu_1(s)\, ds$. When we take $x_0 = 0$, $u_0(s) = 0$, $-1 \le s \le 0$, and $u_1(s) = 0$, $t - 1 \le s \le t$, $\tilde{x}_0 = 0$ and we have the analogue of (3.4):

$$K_t^+(A, B, C) = K_{t-1}^+(A, e^A B + C, 0) \tag{4.2}$$

The following two theorems are immediate consequences of (4.2).

**Theorem 4.1.** Assume $e^{At} \ge 0$ and $e^A B + C \ge 0$. Then (2.3), with $x_0 = 0$, $u_0 = 0$, $u_1$

= 0, is $(SpAPC)_t$ iff the non-delay system $\dot{x}(t) = Ax(t) + (e^A B + C)u(t)$ is $(APC)_{t-1}$, which is true iff for each i, $1 \le i \le n$, there is j, $1 \le j \le m$, and $\mu > 0$ such that $e[i] = \mu(e^A B + C)e(j)$.

**Theorem 4.2.** With the assumption of Theorem 4.1, (2.3) is (SpAPC) iff the non-delay system $\dot{x}(t) = Ax(t) + (e^A B + C)u(t)$ is (APC).

We have given here several analogues to Schanbacher's results on approximate positive controllability of linear systems using nonnegative controllers. These results parallel the earlier extensions of results for controllability of linear systems to linear systems with a single delay. Results for systems with multiple delays or nonconstant delays, though more complicated, follow along these same lines.

## REFERENCES

[BJL] H. T. Banks, M. Q. Jacobs and M. R. Latina. The Synthesis of Optimal Controls for Linear, Time-Optimal Problems with Retarded Controls. Journal of Optimization Theory and Applications 8 (1971), 319-366.

[C] D. H. Chyung. On the Controllability of Linear Systems with Delays in Control. IEEE Transactions on Automatic Control AC-15 (1970), 255-257.

[MSV] P. Monson, C. Shannon and F. S. Van Vleck. Aspects of Positivity in Controllers and Controllability. M. A. Research Project Report, University of Kansas 1988.

[Sch-1] T. Schanbacher. Asymptotic Behavior of Positive Semigroups. Mathematische Zeitschrift 195 (1987), 481-485.

[Sch-2] T. Schanbacher. Aspects of Positivity in Control Theory. SIAM Journal on Control and Optimization 27 (1989), 457-475.

[Se] O. Sebakhy. A Simplified Criterion for the Controllability of Linear Systems with Delay in Control. IEEE Transactions on Automatic Control AC-16 (1971), 364-365.

# 32 Discrete Time Partially Observed Control

**Robert J. Elliott** University of Alberta, Edmonton, Alberta, Canada

**John B. Moore** The Australian National University, Canberra, Australia

**Acknowledgments:** Research partially supported by NSERC Grant A7964 and Boeing BCAC. The hospitality of the Department of Systems Engineering at the Australian National University is gratefully acknowledged.

**Key Words:** Stochastic control, reference probability, unnormalized density, separated control, dynamic programming, minimum principle, adjoint process, dual control.

**Abstract:** A discrete time, partially observed control problem is discussed by explicitly constructing a reference probability under which the observations are independent. Using the unnormalized conditional probabilities as information states the problem is treated in separated form. Dynamic programming and minimum principle results are obtained.

## 1. Introduction.

Much effort has been expended on discussing the optimal control of partially observed diffusions in continuous time, but the results are still not entirely satisfactory. Discrete time control problems are treated in the books of Kumar and Varaiya [5] and Caines [1]. In this paper we discuss the discrete time, partially observed control problem using the reference probability. This idea is described in references [3] and [4]; the reference probability is constructed explicitly, and the role of the dynamics in the separated problem is clarified. The unnormalized conditional probabilities, which describe the state of the process given the observations, play the role of 'information states', and the control problem can be re-cast as a fully observed optimal control problem. A dynamic programming result and minimum principle are obtained, in terms of separated controls, and an adjoint process is described. Finally, when some of the parameters of the model are unknown it is shown how the methods extend to dual control problems.

## 2. Dynamics.

We shall consider a finite time horizon control problem and, for simplicity, we suppose noise is additive in the state and observation processes. All processes are defined initially on a probability space $(\Omega, \mathcal{F}, P)$.

The state process $\{x_k\}$, $k = 0, 1, \ldots, M$, take values in $R^d$ and has dynamics

$$x_{k+1} = F_k(x_k, u_k) + w_{k+1}. \tag{2.1}$$

We suppose the initial density $\pi_0(z)$ of $x$ is known.

The observation process $\{y_k\}$, $k = 0, 1, \ldots, M$ takes values in $R^m$ and has dynamics

$$y_{k+1} = H_k(x_k) + b_{k+1}. \tag{2.2}$$

We suppose $y_0 = 0 \in R^m$. For $0 \leq k \leq M$ write $y^k = \{y_0, y_1, \ldots, y_k\}$. $\{G_k\}$ is the complete filtration generated by $x$ and $y$. $\{\mathcal{Y}_k\}$ is the complete filtraton generated by $y$.

The noise in the state process is a sequence $\{w_k\}$, $1 \leq k \leq M$, of independent $R^d$ valued random variables having densities $\psi_k$.

The noise in the observation process is a sequence $\{b_k\}$, $1 \leq k \leq M$, of independent $R^m$ valued random variables having positive densities $\phi_k$, $\phi_k(b) > 0$ for all $b \in R^m$. The parameter $u_k$ in (2.1) represents the control variable, and takes values in a set $U \subset R^p$. At time $k$, $u_k$ is $\mathcal{Y}_k$ measurable, that is, $u_k$ is a function of $y^k$. For $0 \leq k < M$ write $\underline{U}(k)$ for the set of such control functions and

$$\underline{U}(k, k+\ell) = \underline{U}(k) \cup \underline{U}(k+1) \cup \cdots \cup \underline{U}(k+\ell).$$

For $u \in \underline{U}(0, M-1)$, $x^u$ will denote the trajectory $(x_0, x_1^u, x_2^u, \ldots, x_M^u)$ determined by (2.1).

## 3. Unnormalized Densities.

We review the recurrence relation for the unnormalized conditional density of the state given the observations. For details see [3] and [4].

Suppose we have an equivalent probability measure $\overline{P}$ on $(\Omega, G_M)$ such that under $\overline{P}$:

1) $\{y_k\}$ is a sequence of independent random variables having positive densities $\phi_k$,
2) for any $u \in \underline{U}(0, M-1)$, $x_{k+1}^u \in R^d$ satisfies the dynamics

$$x_{k+1}^u = F_k(x_k^u, u_k) + w_k, \qquad k \in Z^+,$$

where $w_k$ is a sequence of independent random variables having densities $\psi_k$.

Suppose $u \in \underline{U}(0, M-1)$. Define $\bar{\gamma}_\ell^u = \phi_\ell(y_\ell - H_{\ell-1}(x_{\ell-1}^u))/\phi_\ell(y_\ell)$ and $\overline{\Lambda}_n^u = \prod_{\ell=1}^n \bar{\gamma}_\ell^u$. Then a probability $P^u$ can be defined by setting the restriction of $\frac{dP^u}{d\overline{P}}$ to $G_M$ equal to $\overline{\Lambda}_M^u$. It is under $P^u$ that the state and observation processes have the form (2.1) and (2.2).

Suppose $\Phi_\ell$ is any $G$-adapted process. Then a version of Bayes' theorem states that

$$E^u[\Phi_n \mid \mathcal{Y}_n] = \overline{E}[\overline{\Lambda}_n^u \Phi_n \mid \mathcal{Y}_n] / \overline{E}[\overline{\Lambda}_n^u \mid \mathcal{Y}_n]. \tag{3.1}$$

Write $q_n^u(z)$ for the unnormalized conditional density such that

$$\overline{E}[\overline{\Lambda}_n^u I(x_n^u \in dz) \mid \mathcal{Y}_n] = q_n^u(z)dz.$$

The equation (3.1) indicates why we consider $q_n^u(z)$, because the normalized conditional density $p_n^u(z)$ is then given by:

$$p_n^u(z) = q_n^u(z) / \int_{R^d} q_n^u(\xi)d\xi,$$

and for any Borel test function $f$

$$E^u[f(x_n^u) \mid \mathcal{Y}_n] = \int_{R^d} f(z)p_n^u(z)dz.$$

Now for $u \in \underline{U}(0, M-1)$ and any Borel test function $f$ consider

$$\overline{E}[f(x_n^u)\overline{\Lambda}_n^u \mid \mathcal{Y}_n] = \int_{R^d} f(z)q_n^u(z)dz = \langle f(z) q_n^u(z) \rangle \tag{3.2}$$

$$= \overline{E}[f(F_{n-1}(x_{n-1}^u, u_{n-1}) + w_n)\overline{\Lambda}_{n-1}^u \phi_n(y_n - H_{n-1}(x_{n-1}^u)) \mid \mathcal{Y}_n] / \phi_n(y_n)$$

$$= \phi_n(y_n)^{-1}\overline{E}\Big[\int_{R^d} f(F_{n-1}(x_{n-1}^u, u_{n-1}) + w)\psi_n(w)dw\overline{\Lambda}_{n-1}^u \phi_n(y_n - H_{n-1}(x_{n-1}^u)) \mid \mathcal{Y}_n\Big]$$

$$= \phi_n(y_n)^{-1} \int\int f(F_{n-1}(\xi, u_{n-1}) + w)\psi_n(w)\phi_n(y_n - H_{n-1}(\xi))q_{n-1}^u(\xi)dwd\xi.$$

Substituting $z = F_{n-1}(\xi, u_{n-1}) + w$ this is

$$= \phi_n(y_n)^{-1} \int\int f(z)\psi_n(z - F_{n-1}(\xi, u_{n-1}))\phi_n(y_n - H_{n-1}(\xi))q^u_{n-1}(\xi)d\xi dz. \qquad (3.3)$$

The equality of (3.2) and (3.3) holds for all Borel test functions $f$, so we have the following recurrence relation for $q^u_n$:

THEOREM 3.1.

$$q^u_n(z) = \phi_n(y_n)^{-1} \int_{R^d} \psi_n(z - F_{n-1}(\xi, u_{n-1}))\phi_n(y_n - H_{n-1}(\xi))q^u_{n-1}(\xi)d\xi. \qquad (3.4)$$

REMARKS 3.2 This equation describes the observable dynamics of a separated problem. $q^u_n(\cdot)$ is an "information state" in the sense of Kumar and Varaiya [5]. That is, if we know $q^u_{n-1}(\cdot)$, $y^n$ and $u_{n-1}$, equation (3.4) enables us to determine $q^u_n(\cdot)$.

The initial information state $q_0$ is just $\pi_0$, the (normalized) density of $x_0$. Note that, even if $\pi_0$ is a unit mass at a particular $x_0$, $q^u_1(z) = \phi_1(y_1)^{-1}\psi_1(z - F_0(x_0, u_0))\phi_1(y_1 - H_0(x_0))$, and the consequent terms $q^u_2, q^u_3, \ldots$ follow from equation (3.4).

## 4. Cost.

Suppose, given $x_0$ and $u \in \underline{U}(0, M-1)$, the cost function associated with the problem is of the form

$$C(x_0, u) = \sum_{k=0}^{M-1} c_k(x^u_k, u_k) + c_M(x^u_M).$$

Then the expected cost, if control $u$ is used and the density of $x_0$ is $\pi_0(\cdot)$, is

$$V_0(\pi_0, u) = E[C(x_0, u)].$$

This can be expressed

$$V_0(\pi_0, u) = \overline{E}\Big[\overline{\Lambda}^u_M\Big(\sum_{k=0}^{M-1} c_k(x^u_k, u_k) + c_M(x^u_M)\Big)\Big]$$

$$= \sum_{k=0}^{M-1} \overline{E}[\overline{\Lambda}^u_k c_k(x^u_k, u_k)] + \overline{E}[\overline{\Lambda}^u_M c_M(x^u_M)]$$

$$= \overline{E}\Big[\sum_{k=0}^{M-1} \langle c_k(z, u_k), q^u_k(z)\rangle + \langle c_M(z), q^u_M(z)\rangle\Big]$$

$$= \overline{E}\Big[\overline{E}\Big[\sum_{k=0}^{M-1} \langle c_k(z, u_k), q^u_k(z)\rangle + \langle c_M(z), q^u_M(z)\rangle \mid \mathcal{Y}_M\Big]\Big]$$

where, for example, we write

$$\langle c_k(z,u_k), q_k^u(z)\rangle = \int_{R^d} c_k(z,u_k) q_k^u(z)dz$$

$$= \overline{E}[\overline{\Lambda}_k^u c_k(x_k^u, u_k) \mid \mathcal{Y}_k].$$

REMARKS 4.1. We have seen the information state at time $k$ belongs to the set $S$ of positive measures $q(\cdot)$ on $R^d$. Note the probability measures are a subset of $S$.

$S$ is an infinite dimensional space. A metric can be defined on $S$ using the $L^1$ norm, so that for $q^1(\cdot), q^2(\cdot) \in S$

$$d(q^1,q^2) = \|q^1 - q^2\| = \int_{R^d} |q^1(z) - q^2(z)|dz.$$

Any $q \in S$ can be normalized to give a probability measure $\pi(q) = q(\cdot)/\|q\|$.

Consider the process starting from some intermediate time $k$, $0 \le k \le M$, from some state $q(\cdot) \in S$. Then, for $u \in \underline{U}(k, M-1)$

$$q_{k+1}^u(z) = \phi_{k+1}(y_{k+1})^{-1} \int_{R^d} \psi_{k+1}(z - F_k(\xi,u_k))\phi_{k+1}(y_{k+1} - H_k(\xi))q(\xi)d\xi. \tag{4.1}$$

The remaining information states $q_n^u(\cdot)$, $k+1 < n \le M$, are similarly obtained from (3.4).

The expected cost accumulated, starting from state $q(\cdot) \in S$ and using control $u \in \underline{U}(k, M-1)$ is, therefore

$$V_k(q,u) = \overline{E}\Big[\sum_{j=k}^{M-1} \langle c_j(z,u_j), q_j^u(z)\rangle + \langle c_M(z), q_M^u(z)\rangle \mid q_k = q\Big]. \tag{4.2}$$

REMARKS 4.2. The problem is now in a separated form. The filtering recursively determines the unnormalized, conditional probabilities which are the information states, $q_k^u(\cdot)$. These evolve according to the dynamics (3.4), and the cost is expressed in terms of these information states.

DEFINITION 4.3. A control $u \in \underline{U}(0, M-1)$ is said to be separated if $u_k$ depends on $y^k$ only through the information state $q_k^u(\cdot)$. Write $\underline{U}_S(0, M-1)$ for the set of separated controls.

DEFINITION 4.4. For $0 \le k \le M-1$ the cost process is defined by:

$$V(k,q) = \bigwedge_{u \in \underline{U}(k,M-1)} V_k(q,u).$$

Here $V_k(q,u)$ is given by (4.2). Also, set $V(M,q) = \langle c_M(z), q(z)\rangle$. We now establish the dynamic programming identity.

THEOREM 4.5. *For* $0 \leq k \leq M-1$ *and* $q \in S$

$$V(k,q) = \bigwedge_{u \in \underline{U}(k)} \overline{E}[\langle c_k(z,u_k), q(z)\rangle + V(k+1, q^u_{k+1}) \mid q_k = q]. \tag{4.3}$$

Proof.

$$\begin{aligned} V(k,q) &= \bigwedge_{u \in \underline{U}(k,M-1)} V_k(q,u) = \bigwedge_{u \in \underline{U}(k)} \bigwedge_{v \in \underline{U}(k+1,M-1)} V_k(q,u) \\ &= \bigwedge_{u \in \underline{U}(k)} \bigwedge_{v \in \underline{U}(k+1,M-1)} \overline{E}\Big[\overline{E}\Big[\langle c_k(z,u_k), q(z)\rangle \\ &\quad + \sum_{j=k+1}^{M-1} \langle c_j(z,v_j), q^v_j(z)\rangle + \langle c_M(z), q^v_M(z)\rangle \mid \mathcal{Y}_{k+1}\Big] q_k = q\Big] \\ &= \bigwedge_{u \in \underline{U}(k)} \Big\{\overline{E}[\langle c_k(z,u_k), q(z)\rangle \mid q_k = q] \\ &\quad + \bigwedge_{v \in \underline{U}(k+1,M-1)} \overline{E}\Big[\overline{E}\Big[\sum_{j=k+1}^{M-1} \langle c_j(z,v_j), q^v_j(z)\rangle + \langle c_M(z), q^v_M(z)\rangle \mid \mathcal{Y}_{k+1}\Big] q_k = q\Big]\Big\}. \end{aligned}$$

Using the Lattice property for the controls, (see Lemma 16.14 of [2]), the inner minimization and first expectation can be interchanged, so this is

$$\begin{aligned} &= \bigwedge_{u \in \underline{U}(k)} \Big\{\overline{E}[\langle c_k(z,u_k), q(z)\rangle \mid q_k = q] \\ &\quad + \overline{E}\Big[\bigwedge_{v \in \underline{U}(k+1,M-1)} \overline{E}\Big[\sum_{j=k+1}^{M-1} \langle c_j(z,v_j), q^v_j(z)\rangle + \langle c_M(z), q^v_M(z)\rangle \mid \mathcal{Y}_{k+1}\Big] q_k = q\Big\} \\ &= \bigwedge_{u \in \underline{U}(k)} \overline{E}[\langle c_k(z,u_k), q(z)\rangle + V(k+1, q^u_{k+1}) \mid q_k = q], \end{aligned}$$

and the result follows.

COROLLARY 4.6. *Write* $\underline{U}_S(k, M-1)$ *for the set of separated controls on* $\{k, k+1, \ldots, M-1\}$. *Then for* $q \in S$

$$V(k,q) = \bigwedge_{u \in \underline{U}(k,M-1)} V_k(q,u) = \bigwedge_{u \in \underline{U}_S(k,M-1)} V_k(q,u).$$

Proof. The proof will use backward induction in $k$. Clearly $V(M,q) = V_M(q) = \langle c_M(z), q(z)\rangle$ and the result holds for $k = M$. Suppose the result is true for $k+1$, $k+2$, $\ldots$, $M$. Then from Theorem 4.5

$$V(k,q) = \bigwedge_{u\in\underline{U}(k)} \overline{E}\big[\langle c_k(z,u_k), q(z)\rangle + V(k+1, q^u_{k+1}) \mid q_k = q\big].$$

It is clear that a minimizing $u_k$, (or a sequence of minimizing $u_k$), depend only on the information state $q_k = q$. Therefore,

$$\begin{aligned} V(k,q) &= \bigwedge_{u\in\underline{U}_S(k)} \overline{E}\Big[\langle c_k(z,u_k), q(z)\rangle + \bigwedge_{v\in\underline{U}_S(k+1,M-1)} V_{k+1}(q^u_{k+1}, v) \mid q_k = q\Big] \\ &= \bigwedge_{u\in\underline{U}_S(k,M-1)} V_k(q,u). \end{aligned} \tag{4.4}$$

THEOREM 4.7. *Suppose $u^*$ is a separated control such that, for each $q \in S$, $u^*_k(q)$ achieves the minimum in (4.3). Then $V_k(q,u^*) = V(k,q)$, and $u^*$ is an optimal control.*

Proof. We shall again prove the result by backward induction in $k$. Clearly

$$\begin{aligned} V_M(q,u^*) &= \langle c_M(z), q(z)\rangle \\ &= V(M,q). \end{aligned}$$

Suppose the result holds for $k+1, k+2, \ldots, M$. Then

$$\begin{aligned} V_k(q,u^*_k) &= \overline{E}[\langle c_k(z,u^*), q(z)\rangle + V_{k+1}(q^{u^*}_{k+1}, u^*) \mid q_k = q] \\ &= \overline{E}[\langle c_k(z,u^*), q(z)\rangle + V(k+1, q^{u^*}_{k+1}) \mid q_k = q] \\ &= V(k,q). \end{aligned}$$

Now for any other $u \in \underline{U}(0, M-1)$

$$V_k(q,u^*) = V(k,q) \le V_k(q,u),$$

and, in particular, $V_0(q,u^*) \le V_0(q,u)$, so $u^*$ is optimal.

## 5. The Adjoint Process.

Consider any control $u \in \underline{U}(0, M-1)$. We shall suppose for simplicity of notation that the cost is purely terminal at the final time $M$, so

$$C(x_0, u) = c_M(x_M^u).$$

Then

$$\begin{aligned} V(\pi_0, u) &= E[c_M(x_M^u)] \\ &= \overline{E}[\langle c_M(z), q_M^u(z)\rangle]. \end{aligned}$$

THEOREM 5.1. *There is a process $\kappa_k^u(z, y^k)$, adapted to $\mathcal{Y}_k$, such that for $0 \le k \le M$*

$$\overline{E}[\langle c_M(z), q_M^u(z)\rangle \mid \mathcal{Y}_k] = \langle \kappa_k^u(z, y^k), q_k^u(z)\rangle.$$

*Further, $\kappa_k^u$ evolves in reverse time so that*

$$\kappa_k^u(\xi, y^k) = \int_{R^d}\int_{R^m} \kappa_{k+1}^u(z, y^k, y_{k+1})\phi_{k+1}(y_{k+1} - H_k(\xi))\psi_{k+1}(z - F_k(\xi, u_k))dz dy_{k+1}.$$

Proof. Again we use backward induction. Define $\kappa_M^u(z, y^M) = c_M(z)$ so

$$\begin{aligned} \overline{E}[\langle c_M(z), q_M^u(z)\rangle \mid \mathcal{Y}_M] &= \langle c_M(z), q_M^u(z)\rangle \\ &= \langle \kappa_M^u(z, y^M), q_M^u(z)\rangle. \end{aligned}$$

Suppose $\kappa_{k+1}^u(z, y^{k+1})$ has been defined. Then

$$\langle \kappa_{k+1}^u(z, y^{k+1}), q_{k+1}^u(z)\rangle = \int_{R^d} \kappa_{k+1}^u(z, y^{k+1}) q_{k+1}^u(z) dz$$

and

$$\begin{aligned} &\overline{E}\big[\langle \kappa_{k+1}^u(z, y^{k+1}), q_{k+1}^u(z)\rangle \mid \mathcal{Y}_k\big] \\ &= \int_{R^d}\int_{R^d}\int_{R^m} \kappa_{k+1}^u(z, y^k, y_{k+1})\phi_{k+1}(y_{k+1})^{-1}\phi_{k+1}(y_{k+1} - H_k(\xi)) \\ &\quad \times \psi_{k+1}(z - F_k(\xi, u_k)) q_k^u(\xi)\phi_{k+1}(y_{k+1}) dz d\xi dy_{k+1} \\ &= \langle \kappa_k^u(\xi, y^k), q_k^u(\xi)\rangle \end{aligned}$$

where

$$\kappa_k^u(\xi, y^k) = \int_{R^d}\int_{R^m} \kappa_{k+1}^u(z, y^k, y_{k+1})\phi_{k+1}(y_{k+1} - H_k(\xi))\psi_{k+1}(z - F_k(\xi, u_k))dz dy_{k+1}.$$

REMARKS 5.2. Note in particular

$$\begin{aligned} V(\pi_0, u) &= \overline{E}[\langle c_M(z), q_M^u(z)\rangle] \\ &= \overline{E}[\langle \kappa_0^u(\xi, y_0), \pi_0(\xi)\rangle] \\ &= \overline{E}[\langle \kappa_k^u(\xi, y^k), q_k^u(\xi)\rangle]. \end{aligned}$$

## 6. Parameter Estimation and Dual Control.

Suppose we have a situation where the model contains unknown parameters $\theta^1, \theta^2, \theta^3$, ..., which we also wish to estimate. That is, suppose the state dynamics and observation processes are of the form:

$$x_{k+1} = F_k(x_k, u_k, \theta^1, \theta^2,) + w_k$$

$$y_{k+1} = H_k(x_k, \theta^3) + b_k, \quad 0 \le k \le M.$$

Here $\theta^i$ takes values in some measure space ($\Theta^i$, $\beta^i$, $\lambda^i$), with $\lambda^i$ a probability measure. $\Theta^i$ could be a (subset of a) Euclidean space.

For example, see [4], a simple case would be (one dimensional) linear dynamics and observations of the form

$$x_{k+1} = \theta^1 x_k + \theta^2 u_k + w_k$$

$$y_{k+1} = \theta^3 x_k + b_k.$$

The analysis of the previous sections goes through, taking the $\theta^i$ to be additional state variables. The unnormalized conditional density $q_n^u(z, \lambda^1, \lambda^2, \lambda^3)$ is defined by

$$\overline{E}[\overline{\Lambda}_n^u I(x_n^u \in dz) I(\theta^1 \in d\lambda^1) I(\theta^2 \in d\lambda^2) I(\theta^3 \in d\lambda^3) \mid \mathcal{Y}_n] = q_n^u(z, \lambda^1, \lambda^2, \lambda^3) dz d\lambda^1 d\lambda^2 d\lambda^3,$$

and the recursive equations (3.4) and dynamic programming results are exactly as before.

## 7. Conclusion.

A discrete time, partially observed control problem is discussed in separated form. Under a reference probability, which is explicitly constructed, the dynamics are given by recursive equations for the unnormalized, conditional probabilities. Dynamic programming and minimum principle results are obtained, and the extension to dual control, parameter estimation problems indicated.

## References.

1. P.E. Caines, 1988. *Linear Stochastic Systems.* John Wiley, New York-Toronto.
2. R.J. Elliott, 1982. *Stochastic Calculus and Applications.* Springer–Verlag, New York-Heidelberg-Berlin.
3. R.J. Elliott. A general recursive discrete time filter. Submitted.
4. R.J. Elliott and J.B. Moore. State and parameter estimation for linear systems.
5. P.R. Kumar and P. Varaiya, 1986. *Stochastic Systems.* Prentice Hall, New Jersey.

# 33 Symmetries of Differential Systems

**Fabio Fagnani** Scuola Normale Superiore, Pisa, Italy

**Jan C. Willems** University of Groningen, Groningen, The Netherlands

## 1 Introduction.

In the classical theory of dynamical systems, symmetry has always played a central role. However it is only in the last 15 years it has also emerged as an issue in systems and control theory. Using the setting established in the behavioral approach to systems [9], [10], [11], a systematic analysis of symmetries for linear systems has been undertaken by the authors in [1], [2], [3] where the reader is also referred for a list of earlier system theoretic references on the subject.

In this paper we will focus on finite group symmetries of linear differential systems and present a rather complete theory on the subject. Together with new results, we also present, in a new light, a unified picture which encompasses many of the results in [2] and [3].

In section 1 we briefly recall some basic facts on the theory of linear differential behaviors which are defined as kernels in $C^\infty(\mathbf{R}, W)$ of linear constant coefficient ordinary differential operators where $W$ is a $k$-vector space. In section 2 we will introduce our definition of symmetry, as a representation of a finite group $G$ on the infinite-dimensional space $C^\infty(\mathbf{R}, W)$, which preserves the class $\mathsf{D}[W]$ of linear differential behaviors. Our main goal is to study the structure and the representation of behaviors in $\mathsf{D}[W]$ which are symmetric, namely, those which are fixed by the action of $G$. In section 4, we prove a result (Theorem 1) on the classification of the symmetries on $\mathsf{D}[W]$. This result is then used in section 5 to show (Theorem 3) how the action of $G$ on $C^\infty(\mathbf{R}, W)$ can be lifted to a dual action on the free $k[z]$-module $W^*[z]$. These can be thought as the set of the differential operators from $C^\infty(\mathbf{R}, W)$ to $C^\infty(\mathbf{R}, \mathbf{R})$. This permits to shift the analysis to a pure algebraic level and in section 6 we prove Theorems 4 and 5 which exploit the structure of finite group actions on free $k[z]$-modules. Finally, in section 7 the algebraic results are used to establish canonical differential representations of symmetric systems. We close with a number of examples involving static symmetrics and time-reversibility

The important interplay between symmetries and control problems will be studied in a later paper.

It is a pleasure to dedicate this paper to Larry Markus on his 70-th birthday. His remarkable style of exposition and his talent for combining advanced mathematical ideas

---

Research supported by EEC Contract SC1-0433-C(A).

with engineering relevance which I (JCW) was privileged to experience as a beginning researcher had a great influence on my later scientific work.

## 2 Differential behaviors. Preliminary facts.

Let $k$ be equal $\mathbf{R}$ or $\mathbf{C}$ and denote by $k^*$ the multiplicative group $k \setminus \{0\}$. Throughout this paper $W$ and $E$ will always denote finite dimensional vector spaces over $k$. Denote by $C_W^\infty = C^\infty(\mathbf{R}, W)$ the $k$-vector space of infinitely differentiable functions from $\mathbf{R}$ to $W$ equipped with the canonical Frechet topology of uniform convergence on compact subsets of $\mathbf{R}$. Denote $R = k[z]$ and $E[z] = R \otimes_k E$ the $R$-module of polynomials with coefficients in $E$.

If $D \in \mathrm{Hom}_k(\mathrm{W}, \mathrm{E})[\mathrm{z}]$ we can define the differential operator

$$D\left(\frac{d}{dt}\right) : C_W^\infty \to C_E^\infty$$

as follows: if $D = \sum_{i=0}^n D_i z^i$ where $D_i \in \mathrm{Hom}(\mathrm{W}, \mathrm{E})$ and $w \in C_W^\infty$ then,

$$\left(D\left(\frac{d}{dt}\right) w\right)(t) := \sum_{i=0}^{n} D_i \left(\frac{d^i w}{dt^i}(t)\right)$$

Following [10] we define a *differential behavior* over $W$ (called the *signal space*), a subspace $\mathcal{B}$ of $C_W^\infty$ which is the kernel of a differential operator $D\left(\frac{d}{dt}\right)$. $D$ is called a *polynomial matrix representation* of $\mathcal{B}$ and $E$ is called the *equation space* of $D$. The class of differential behaviors over $W$ is denoted by $\mathbf{D}[W]$.

Let $\mathcal{B} \in \mathbf{D}[W]$. Consider the annihilators of $\mathcal{B}$, defined by

$$\mathcal{B}^\perp := \left\{p \in W^*[z] \mid p\left(\frac{d}{dt}\right) w = 0 \ \forall w \in \mathcal{B}\right\}$$

where $W^* := \mathrm{Hom}_k(\mathrm{W}, k)$. Clearly $\mathcal{B}^\perp$ is an $R$-submodule of the $R$-free module $W^*[z]$, hence it is finitely generated and free. On the other hand, if $M$ is an $R$-submodule of $W^*[z]$, one can consider

$${}^\perp M := \left\{f \in C_W^\infty \mid p\left(\frac{d}{dt}\right) f = 0 \ \forall p \in M\right\}$$

It is a standard fact [6], [10] that

$${}^\perp(\mathcal{B}^\perp) = \mathcal{B} \quad , \quad ({}^\perp M)^\perp = M$$

for all $\mathcal{B} \in \mathbf{D}[W]$ and for all submodules $M$ of $W^*[z]$. This yields a bijection between $\mathbf{D}[W]$ and $\mathbf{S}[W^*[z]]$, the set of all submodules of $W^*[z]$. For $\mathcal{B} \in \mathbf{D}[W]$, denote by $p(\mathcal{B})$ the rank of the free module $\mathcal{B}^{\perp}$. Clearly $p(\mathcal{B}) \leq \dim_k W$ and we can find $D \in \mathrm{Hom}(\mathrm{W}, \mathrm{k}^{\mathrm{p}(\mathcal{B})})[\mathrm{z}]$ such that $\ker D\left(\frac{d}{dt}\right) = \mathcal{B}$. This simply shows that $\mathcal{B}$ can be described by $p(\mathcal{B})$ differential equations and no less. Any polynomial matrix representation of $\mathcal{B}$ with equation space of dimension $p(\mathcal{B})$ will therefore be called *minimal.*

Consider $M$, a submodule of $E[z]$, and

$$(1) \qquad M_i := \{m \in M \mid \deg(m) \leq i\}$$

where $\deg(m)$ denotes the degree of the polynomial $m \in E[z]$. Clearly the $M_i$'s are $k$-vector spaces and there holds

$$(2) \qquad M_i + zM_i \subseteq M_{i+1}.$$

It is a standard fact (see [9] for more details) that equality holds in (2) except for at most finitely many $i$'s. Denote by $h$ the largest of them. We can then construct an $R$-basis of $M$ in the following inductive way. Let $B_0 = \{r_1^0, \ldots, r_{\gamma_0}^0\}$ be a $k$-basis of $M_0$ and let $B_i = \{r_1^i, \ldots, r_{\gamma_i}^i\}$ be a $k$-basis of a complementary subspace of $M_{i-1} + zM_{i-1}$ inside $M_i$. It can be proven that $\cup_i B_i$ is an $R$-basis for $M$. It will be called a *canonical basis.* For $\mathcal{B} \in \mathrm{D}[W]$, let $\{r_j^i \mid j = 1, \ldots, \gamma_i, i = 1, \ldots, h\}$ be such a canonical basis for $\mathcal{B}^{\perp}$. Consider

$$D(z) = \begin{bmatrix} r_1^0(z) \\ \vdots \\ r_{\gamma_h}^h(z) \end{bmatrix} \in \mathrm{Hom}[\mathrm{W}, \mathrm{k}^{\mathrm{h}}][\mathrm{z}].$$

$D$ is evidently a minimal representation of $\mathcal{B}$ and it will be called *canonical.* Define $n(\mathcal{B}) = \sum_{i=0}^{h} i\gamma_i$. These two integers $p(\mathcal{B})$ and $n(\mathcal{B})$ each have an important system theoretic interpretation: $p(\mathcal{B})$ is the number of output variables in any input/output representation of $\mathcal{B}$ while $n(\mathcal{B})$ is the Mc-Millan degree of $\mathcal{B}$, namely, the dimension of the state space in any minimal state space representation of $\mathcal{B}$. See [9], [10], and [11] for precise statements and details.

The choice of working in $C^{\infty}$ is mostly done for the matter of simplicity. More general settings can indeed be chosen, see [11] and [1]. In particular, all the results we present in this paper are still true if we replace $C^{\infty}$ with the space of distributions $\mathcal{D}'$.

## 3 Symmetries of differential behaviors.

Denote by $GL_k(C_W^{\infty})$ the group of all topological vector space isomorphisms of $C_W^{\infty}$. Let $G$ be a group and let $T : G \to GL_k(C_W^{\infty})$ be a representation of $G$. Clearly, $T$ induces an action of $G$ on the class of all (closed) subspaces of $C_W^{\infty}$:

$$g \cdot \mathcal{B} := \{T_g w \mid w \in \mathcal{B}\}$$

We will say that $(G,T)$ is a *symmetry on* $\mathbf{D}[W]$ if the set $\mathbf{D}[W]$ is invariant by the action of $G$. A behavior $\mathcal{B} \in \mathbf{D}[W]$ is said to be *symmetric* if it is fixed by the action of $G$ (i.e., $T_g w \in \mathcal{B}$ for every $w \in \mathcal{B}$ and $g \in G$). The subset of all symmetric behaviors in $\mathbf{D}[W]$ will be denoted by $\mathbf{D}[W]^G$.

Because of the bijective correspondence between $\mathbf{D}[W]$ and $\mathbf{S}[W^*[z]]$, $G$ also acts in a natural way on $\mathbf{S}[W^*[z]]$. If $M \in \mathbf{S}[W^*[z]]$ and $g \in G$, define

$$g \cdot M := (g(^{\perp}M))^{\perp}.$$

Denote by $\mathbf{S}[W^*[z]]^G$ the subset of the submodules in $\mathbf{S}[W^*[z]]$ which are fixed by this action of $G$. It follows that $\mathbf{D}[W]^G$ and $\mathbf{S}[W^*[z]]^G$ are also in bijective correspondence through the bijection $\mathcal{B} \leftrightarrow \mathcal{B}^{\perp}$. It is not a priori evident that the action of $G$ on $\mathbf{S}[W^*[z]]$ is induced by some action of $G$ on $W^*[z]$. This is in fact true and will be proven in this section. First let us present some examples.

A symmetry $(G,T)$ on $\mathbf{D}[W]$ is called *static* if there exists a representation $\rho$ of $G$ on $W$ such that $(T_g w)(t) = \rho_g(w(t))$ for all $t \in \mathbf{R}$ and $g \in G$. In the sequel we will identify $T$ and $\rho$. A typical example of a static symmetry is as follows. Let $W = \mathbf{R}^q$, $G = S_q$ the permutation group on $q$ elements and $T$ the permutation representation on $\mathbf{R}^q$ defined by

$$T_\sigma\, {}^t(w_1,\dots,w_q) := {}^t(w_{\sigma(1)},\dots,w_{\sigma(q)})$$

where $w_i \in \mathbf{R}$ and where $^t$ denotes transpose. $\mathcal{B} \in \mathbf{D}[\mathbf{R}^q]$ is symmetric if and only if

$${}^t(w_1,\dots,w_q) \in \mathcal{B} \Rightarrow {}^t(w_{\sigma(1)},\dots,w_{\sigma(q)}) \in \mathcal{B} \quad \forall \sigma \in S_q$$

where $w_i \in C^\infty$. We can think of this static symmetry as occurring when we model the position of $q$ identical particles in $\mathbf{R}^q$.

If $(G,T)$ is a static symmetry on $\mathbf{D}[W]$, it is easy to see that the associated action of $G$ on $\mathbf{S}[W^*[z]]$ is given by

$$g \cdot M = \{T_g^* \cdot m \mid m \in M\}$$

where $T^*$ is the dual representation of $T$ and where if $m = \sum m_i z^i$, $T_g^* m := \sum (T_g m_i) z^i$ Thus, in this case, the action of $G$ on $\mathbf{S}[W^*[z]]$ is induced by the $R$-linear action of $G$ on $W^*[z]$ given by $T^*$. Static symmetries have been studied in much detail in [1] and [3].

An important example of non-static symmetry is *time-reversibility*. In this case $G = \mathbf{Z}_2 = (\{-1,+1\},\cdot)$, $(T_{-1}w)(t) = w(-t)$. Differential behaviors which are symmetric with respect to this symmetry are called *time-reversible*. In this case the $G$-action on $W^*[z]$ which induces the $G$-action on $\mathbf{S}[W^*[z]]$ is as follows. If $p(z) \in W^*[z]$ and $g \in G$, then $(gp)(z) := p(gz)$. See [2] for more details about time-reversibility and some generalizations.

## 4 A classification result.

We will now set up some more notation and then prove a preliminary result of independent interest. Aside from the differential operators, there are other maps acting between $C^\infty$-spaces which will be of interest to us. For $x \in \mathbf{R}$ denote by $\sigma^x : C^\infty \to C^\infty$ the shift operator $\sigma^x w(t) := w(t+x)$. If $\xi \in k$ denote by $M_\xi : C^\infty \to C^\infty$ the multiplicative operator given by $M_\xi w(t) := e^{\xi t} w(t)$. If $\eta \in \mathbf{R}$ denote by $S_\eta : C^\infty \to C^\infty$ the scaling operator given by $S_\eta w(t) := w(\eta t)$. Clearly, $\sigma^x$, $M_\xi$ and $S_\eta$ can be extended to operators on $C^\infty_W$ by taking their tensor product with the identity on $W$. We will use the same symbol to denote both. If $\lambda \in k$ define $e_\lambda : \mathbf{R} \to k$ by $e_\lambda(t) = e^{\lambda t}$.

The following commutation rules are easily verified:

$$
\begin{aligned}
M_\xi \circ \sigma^x &= e^{-\xi x} \cdot \sigma^x \circ M_\xi \\
S_\eta \circ \sigma^x &= \sigma^{\eta^{-1} x} \circ S_\eta \\
M_\xi \circ R\left(\frac{d}{dt}\right) &= R\left(\frac{d}{dt} - \xi\right) \circ M_\xi \\
S_\eta \circ R\left(\frac{d}{dt}\right) &= R\left(\eta^{-1}\frac{d}{dt}\right) \circ S_\eta \\
M_\xi \circ S_\eta &= S_\eta \circ M_{\xi \eta^{-1}}
\end{aligned} \tag{3}
$$

for all $x \in \mathbf{R}$, $\xi \in k$, $\eta \in \mathbf{R}^*$, $R \in GL_R(W[z])$.

**Theorem 1:** *Let $\Lambda \in GL(C^\infty_W)$. The following conditions are equivalent*

(i) *$\Lambda$ leaves the class $\mathsf{D}[W]$ invariant.*

(ii) *There exist (unique) $x \in \mathbf{R}$, $\eta \in \mathbf{R}^*$, $\xi \in k$, and $R \in GL_R(W[z])$ such that*

$$\Lambda = \sigma^x \circ R\left(\frac{d}{dt}\right) \circ \mathcal{M}_\xi \circ S_\eta .$$

**Proof:** The implication (ii)⇒(i) can be straightforwardly deduced by applying the above commutation rules.

(i)⇒(ii). Notice that the only one dimensional subspaces in $\mathsf{D}[W]$ are $\text{span}_k\{e_\lambda w\}$ where $\lambda \in k$ and $w \in W$. Therefore, there exist maps

$$\alpha : k \to GL_k(W) \quad \tilde{\alpha} : k \to GL_k(W)$$

$$\beta : k \to k \quad \tilde{\beta} : k \to k$$

which satisfy

$$\alpha(\lambda)\tilde{\alpha}(\lambda) = \tilde{\alpha}(\lambda)\alpha(\lambda) = I_q \quad \forall \lambda \in k \tag{4}$$

$$\tilde{\beta} \circ \beta = \beta \circ \tilde{\beta} = Id \tag{5}$$

such that

$$\begin{aligned} \Lambda(e_\lambda w) &= e_{\beta(\lambda)}\alpha(\lambda)w \quad \forall \lambda \in k \\ \Lambda^{-1}(e_\lambda w) &= e_{\tilde{\beta}(\lambda)}\tilde{\alpha}(\lambda)w \quad \forall \lambda \in k. \end{aligned} \tag{6}$$

The map $\lambda \mapsto e_\lambda w$ from $k$ to $C^\infty_W$ is entire (in the sense that it admits a global power series expansion) and, since $\Lambda$ is continuous, also the map from $k$ to $C^\infty_W$ given by $\lambda \mapsto e_{\beta(\lambda)}\alpha(\lambda)w$ is entire for all $w \in W$. This implies that $\alpha$ and $\beta\alpha$ are entire and consequently also $\det \alpha$ and $\beta \det \alpha$ are so. For the same reason $\det \tilde{\alpha}$ and $\tilde{\beta} \det \tilde{\alpha}$ are entire. It follows from (4) that $\beta$ and $\tilde{\beta}$ are also. From (5) it immediately follows that the extension of $\beta$ to $\mathbf{C}$ is an entire automorphism of $\mathbf{C}$. Consequently, there exist $\eta \in k^*$ and $\xi \in k$ such that $\beta(\lambda) = \eta\lambda - \xi$. A simple argument based on the continuity of $\Lambda$ shows that we must have $\eta \in \mathbf{R}^*$.

Consider now $\Omega := \Lambda \circ S_{\eta^{-1}} \circ M_{-\xi}$. The result now follows if we prove that $\Omega$ is the composition of a differential operator and a shift. It immediately follows from (6) that

$$(\Omega \circ \sigma^x)e_\lambda w = (\sigma^x \circ \Omega)e_\lambda w \quad \forall \lambda \in k \;\; \forall w \in W \;\; \forall x \in \mathbf{R}.$$

Since $\{e_\lambda w \mid \lambda \in k,\ w \in W\}$ is dense in $C^\infty_W$ it follows that

$$\Omega \circ \sigma^x = \sigma^x \circ \Omega \quad \forall x \in \mathbf{R}. \tag{7}$$

This equality implies that there exists a compact support distribution $\Gamma \in (C^\infty)' \otimes_k \mathrm{Hom}_k[W, W]$ such that

$$\Omega w = \Gamma * w.$$

Now fix a basis of $W$ and write $\Gamma$ in matrix form $(\Gamma_{ij})$ where $\Gamma_{ij} \in (C^\infty)'$ for $i, j = 1, \ldots, q$. Consider the differential behavior

$$\mathcal{B}_j = \{0\} \times \cdots \times \{0\} \times C^\infty \times \{0\} \times \cdots \times \{0\}$$

with $C^\infty$ in the $j$-th place.

Then $\mathcal{A} = \Omega(\mathcal{B}_j)$ has to be a differential behavior, as well as all its projections [11]

$$\mathcal{A}_{hk} = \left\{ \begin{pmatrix} \Gamma_{hj} * w \\ \Gamma_{kj} * w \end{pmatrix} \mid w \in C^\infty \right\}$$

where $h \neq k$. We claim that $\mathcal{A}_{hk} \neq C^\infty \oplus C^\infty$ for all pairs $(h, k)$. Indeed, assume to the contrary that $\mathcal{A}_{hk} = C^\infty \oplus C^\infty$ for some pair $(h, k)$. Consider then the differential behavior

$\mathcal{A}' := \{w \in \mathcal{A} \mid w_h = 0\}$. Clearly $\{0\} \subsetneq \mathcal{A}' \subsetneq \mathcal{A}$ which yields $\{0\} \subsetneq \Omega^{-1}\mathcal{A}' \subsetneq \mathcal{B}_i$. From standard results on differential systems it follows that $\Omega^{-1}\mathcal{A}'$ has to be finite dimensional. Then $\mathcal{A}'$ also has to be finite dimensional, but this is impossible since it contains a subspace isomorphic to $C^\infty$. Therefore $\mathcal{A}_{hk} \neq C^\infty \oplus C^\infty$ for all pairs $(h, k)$. Consider now a pair $(h, k)$ such that $\Gamma_{hj}$ and $\Gamma_{kj}$ are both not zero. Then there exist distributions $r_1$ and $r_2$ with support in 0, both different from zero, such that

$$r_1 * \Gamma_{hj} + r_2 * \Gamma_{kj} = 0.$$

Passing to Fourier transforms, we obtain

$$\hat{r}_1 \cdot \hat{\Gamma}_{hj} + \hat{r}_2 \cdot \hat{\Gamma}_{kj} = 0. \tag{8}$$

All the Fourier transforms we are considering are in the algebra $\mathrm{Hol}_{\exp}(\mathbf{C})$ of the exponential holomorphic functions slowly increasing along the real axis. Distributions with support in zero correspond to the subalgebra of polynomials $\mathbf{C}[z]$. We can evidently assume that $\hat{r}_1$ and $\hat{r}_2$ are coprime in $\mathbf{C}[z]$. From (8) and from standard results on holomorphic exponential functions [8], it immediately follows that $\hat{\Gamma}_{kj} = \hat{Y}\hat{r}_2$ and $\hat{\Gamma}_{hj} = -\hat{Y}\hat{r}_1$ for a suitable $\hat{Y} \in \mathrm{Hol}_{\exp}(\mathbf{C})$. By examining now all the $\hat{\Gamma}_{hj}$ for $h = 1, \ldots, q$ we see that there exist $\hat{Y}_j \in \mathrm{Hol}_{\exp}(\mathbf{C})$ and $\hat{p}_{hj} \in \mathbf{C}[z]$ such that $\hat{\Gamma}_{hj} = \hat{Y}_j \hat{p}_{hj}$. We now return to the time domain and obtain $\Gamma_{hj} = Y_j * p_{hj}$ where the $p_{ij}$ are distributions with support in 0 and $Y_1, \ldots, Y_q$ are distributions with compact support. Applying now $\Omega$ to the differential behaviors

$$\mathcal{B}_{ij} = \{{}^t(0, \ldots, 0, w, 0, \ldots, 0, w, 0, \ldots, 0) \mid w \in C^\infty\}$$

with $w$ in the $i$-th and $j$-th place, and arguing as before, we finally deduce that we can assume that $Y_1 = Y_2 = \cdots = Y_q = Y$. Since $Y$ is necessarily invertible in the convolution algebra $(C^\infty)'$ we obtain, by a classic result of Lions [5], that $Y = \alpha\delta_x$ for some $\alpha \in k^*$ and $x \in \mathbf{R}$. This concludes the proof of existence.

Finally, uniqueness is straightforward.

□

## 5 Group actions on polynomial modules.

Throughout the remainder of this paper we will assume that $G$ is a finite group. Consider a representation $T : G \to GL_k(C^\infty_W)$ such that $(G, T)$ is a symmetry on $\mathbf{D}[W]$. Then, by virtue of Theorem 1, there exist maps

$$x : G \to \mathbf{R}, \quad \xi : G \to k, \quad \eta : G \to \mathbf{R}^*$$

$$R : G \to GL_R(W[z])$$

such that

$$T_g = \sigma^{x_g} \circ R_g\left(\frac{d}{dt}\right) \circ M_{\xi_g} \circ S_{\eta_g}. \tag{9}$$

By using the commutation rules (3), we easily obtain the following relations

$$\begin{aligned} x_{gg'} &= x_g + \eta_g^{-1} x_{g'} \\ \xi_{gg'} &= \xi_g + \eta_g \xi_{g'} \\ \eta_{gg'} &= \eta_g \eta_{g'} \\ R_{gg'}(z) &= \exp(-\eta_{g^{-1}}\xi_g x_{g'}) R_g(z) R_{g'}(\eta_{g^{-1}} z - \eta_{g^{-1}}\xi_g) \end{aligned} \tag{10}$$

for all $g$ and $g'$ in $G$. In particular, $\eta$ is a homomorphism and, since $G$ is finite, we have that $\eta(G) \subseteq \{-1, +1\}$. Let $N = \ker \eta$. It is clear from (10) that $x_{|N} : N \to \mathbf{R}$ and $\xi_{|N} : N \to \mathbf{R}$ are homomorphism, and, since $G$ is finite, it follows that they are both identically zero. This yields

$$\begin{aligned} 0 = \xi_e &= \xi_{g^{-1}} + \eta_{g^{-1}}\xi_g \\ 0 = x_e &= x_{g^{-1}} + \eta_g x_g \end{aligned} \qquad \forall g \in G \tag{11}$$

where $e$ is the identity of the group $G$. Denote by $\mathrm{Aut}(R)$ the group of ring automorphisms of $R = k[z]$. Define the group homomorphism

$$\tau : G \to \mathrm{Aut}(R) \tag{12}$$

by

$$(\tau_g p)(z) = (p^g)(z) := p(\eta_{g^{-1}} z - \eta_{g^{-1}}\xi_g). \tag{13}$$

The action of $G$ on $R$ above introduced, can be extended to any tensor product of the type $R \otimes_k W$. It will still be denoted by $\tau$ or by right superscript. Using (11) and (12), the last equation of (10) can be rewritten as

$$R_{gg'} = \exp(\xi_{g^{-1}} x_{g'}) R_g R_{g'}^g. \tag{14}$$

Consider now $D \in \mathrm{Hom_k(W, E)[z]}$. Then we see, using (3) and (11), that

$$D\left(\frac{d}{dt}\right) \circ T_{g^{-1}} = \sigma^{x_{g^{-1}}} \circ M_{\xi_{g^{-1}}} \circ S_{\eta_{g^{-1}}} \circ (D^g R_{g^{-1}}^g)\left(\frac{d}{dt}\right)$$

which shows that if $\mathcal{B} = \ker D\left(\frac{d}{dt}\right)$, then

$$g \cdot \mathcal{B} = \ker(D^g R^g_{g^{-1}})\left(\frac{d}{dt}\right). \tag{15}$$

Equation (15) indicates how $G$ should act on $W^*[z]$. Define

$$U : G \to GL_k(W^*[z]) \tag{16}$$

by

$$U_g = \alpha_g(\tau_g \circ R^*_g) \tag{17}$$

where $\alpha : G \to \mathbf{R}^*$ is a normalizing factor which we will determine later and where $R^*$ is defined as follows. If $R_g = \sum R_{i,g} z^i$, then $R^*_g := \sum R^*_{i,g^{-1}} z^i$. We have the following

**Proposition 2:** $U$ is a homomorphism if and only if $\alpha$ is of type

$$\alpha_g = \exp\left(\frac{\xi_{g^{-1}} x_{g^{-1}}}{2}\right)\beta_g$$

where $\beta : G \to k^*$ is a homomorphism.

**Proof:** From (14) it easily follows that

$$R^*_{gg'} = \exp(\xi_{g'} x_{g^{-1}})(R^*_g)^{g'^{-1}} R^*_{g'}.$$

Therefore,

$$U_{gg'} = \alpha_{gg'} \cdot \tau_{gg'} \circ R^*_{gg'} = \alpha_{gg'} \cdot \exp(\xi_{g'} x_{g^{-1}}) \cdot \tau_{gg'} \circ (R^*_g)^{g'^{-1}} R^*_{g'}.$$

On the other hand,

$$U_g U_{g'} = \alpha_g \alpha_{g'} \cdot \tau_g \circ R^*_g \circ \tau_{g'} \circ R^*_{g'} = \alpha_g \alpha_{g'} \cdot \tau_{gg'} \circ (R^*_g)^{g'^{-1}} R^*_{g'},$$

This shows that $U$ is a homomorphism if and only if

$$\alpha_{gg'} = \exp(-\xi_{g'} x_{g^{-1}})\alpha_g \alpha_{g'} \quad \forall g, g' \in G. \tag{18}$$

We claim that if $\beta : G \to k^*$ is a homomorphism then $\alpha_g = \exp\left[\frac{1}{2}\xi_{g^{-1}} x_{g^{-1}}\right]\beta_g$ satisfies (18). Indeed, with this choice,

$$\alpha_{gg'} = \exp\left[\frac{1}{2}(\xi_{g'^{-1}} + \eta_{g'^{-1}}\xi_{g^{-1}})(x_{g'^{-1}} + \eta_{g'} x_{g^{-1}})\right]\beta_{gg'} = \alpha_g \alpha_{g'} \exp\left[\frac{1}{2}(-\xi_{g^{-1}} x_{g'} - \xi_{g'} x_{g^{-1}})\right].$$

In order to prove the claim it clearly suffices to prove that

$$\xi_{g'} x_g = \xi_g x_{g'} \quad \forall g, g' \in G. \tag{19}$$

Equation (19) is true if $g$ or $g'$ is in $N$. Since $N$ is a subgroup at $G$ of index 2, we can assume that $g' = gg_0$ for some $g_0 \in N$. It then follows from (10) that $x'_g = x_g$ and $\xi'_g = \xi_g$, and so (19) is proven. On the other hand, if $\alpha$ and $\alpha'$ are two solutions of (18) it immediately follows that $\alpha' = \alpha\beta$ where $\beta : G \to k^*$ is a homomorphism. This completes the proof.

□

Consider a homomorphism $\tau : G \to \mathrm{Aut}(R)$. Notice that it will necessarily be of the form (13). Let $M$ be an $R$-free module of finite rank $q$. Consider a representation $U : G \to GL_k(M)$. $U$ is called a *τ-linear representation* (or also a *quasi-linear representation*) if

$$U_g(pm) = (\tau_g p) U_g(m) \quad \forall p \in R,\ m \in M,\ g \in G.$$

If $M$ is an $R$-graded module (e.g. $E[z]$) then $U$ is said to be *degree-preserving* if $\deg(U_g m) = \deg(m)$ for all $m \in M$ and $g \in G$.

We have thus proven the following result.

**Theorem 3:** *Let $(G, T)$ be a symmetry on $\mathbb{D}[W]$ with $G$ finite group. Then there exists a quasi-linear representation $U : G \to GL_k(W^*[z])$ such that*

$$(T_g \mathcal{B})^\perp = U_g(\mathcal{B}^\perp) \quad \forall g \in G \text{ and } \forall \mathcal{B} \in \mathbb{D}[W]. \tag{20}$$

*Moreover, $U$ is unique up to the multiplication by a homomorphisms from $G$ to $k^*$.*

**Proof:** If we choose $\alpha_g = \exp\left(\frac{1}{2}\xi_{g^{-1}} x_{g^{-1}}\right)$, Proposition 2 shows that the corresponding $U$ in (17) is a $\tau$-linear representation where $\tau$ is given in (13). So, (20) follows from (15). Finally, uniqueness follows from (15) and from Proposition 2.

□

We will refer to $U$ as the quasi-linear representation associated with the symmetry $(G, T)$.

**Remark:** Notice that in the case of static symmetries and time-reversibility discussed in section 3, the corresponding $U$ is degree-preserving. Moreover, $U$ is linear ($\tau = 1$) if and only if for all $g \in G$, $T_g$ is a differential operator followed by a shift.

## 6 Quasilinear representation on polynomial modules.

A simple and, as we will see, canonical way to construct a quasi-linear representation is the following. Let $E$ be a $k$-vector space, $\rho : G \to GL_k(E)$ a linear representation, and $\tau : G \to \mathrm{Aut}(R)$ a homomorphism. Consider the $R$-module $E[z]$ and the representation $U : G \to GL_k(E[z])$ given by $U_g(\sum e_i z^i) = \sum \rho_g(e_i)(\tau_g z)^i$. It is evident that $U$ is a $\tau$-linear representation on $E[z]$. $U$ will be called a *quasi-linear split representation.* Notice that in this case $U$ is always degree-preserving. Let us now prove the following result.

**Theorem 4:** *Let $E$ be a $k$-vector space and assume that $E[z]$ is equipped with a $\tau$-linear split degree-preserving representation $U$. Let $M \subseteq E[z]$ be a $R$-submodule which is $G$-invariant. Then there exists a $k$-vector space $F$ equipped with a filtration $F_i$, a $\tau$-linear split representation on $F[z]$ such that the $F_i$'s are $G$-invariant, and a $G$-equivariant $R$-isomorphism $\psi : F[z] \to M$, such that $\psi(F_i) = M_i$.*

**Proof:** Consider the filtration of $k$-vector spaces $M_i$ associated with $M$ as in (1). Clearly, the $M_i$'s are $G$-invariant. Let $N_0 = M_0$ and let $N_i$ be a complementary $G$-invariant $k$-subspace of $M_{i-1} + zM_{i-1}$ inside $M_i$. Let $F = N_0 \oplus N_1 \oplus \cdots \oplus N_h$ where $h$ is the largest integer for which $M_{i-1} + zM_{i-1} \neq M_i$. Clearly $F$ is equipped, in a natural way, with a $k$-linear representation $\rho$ of $G$. Consider $F[z]$ with the quasi-linear split representation induced by $\tau$ and $\rho$. Define $F_i = N_0 \oplus N_1 \oplus \cdots \oplus N_i$. Define $\psi : F[z] \to M$ as the $R$-linear extension of the natural inclusion $F \hookrightarrow M$. It is straightforward to check that $\psi$ satisfies all the required properties.

□

We now prove the following general fact.

**Theorem 5:** *Let $M$ be a free $R$-module of rank $p$ and let $U : G \to GL_k(M)$ be a $\tau$-linear representation. Then there exist a $k$-vector space $E$ of dimension $p$, a $\tau$-linear split representation on $E[z]$, and a $R$-isomorphism $\psi : M \to E[z]$ which is $G$-equivariant.*

**Proof:** First consider the case $U$ is an $R$-linear representation. In this case the theorem had already been proven in [2]. For the sake of completeness though we briefly show how to do it. Consider the specialization of $M$ at 0, namely $M_0 = M/zM$. $M_0$ is a finite dimensional $k$-vector space on which $G$ acts $k$-linearly. Consider the following two $k(z)$-vector spaces: $M \otimes_{k[z]} k(z)$ and $M_0 \otimes_k k(z)$. It is easy to check that the representations of $G$ naturally induced by them have the same character. Therefore there exists a $G$-equivariant $k(z)$-isomorphism between them. By multiplying this isomorphism by a suitable $p \in k[z]$ we then get an injective $G$-equivariant $R$-map $\psi : M \to M_0 \otimes_k k[z]$. The theorem now follows from Theorem 4.

Consider now the general case. Let $M$ be a free $R$-module equipped with a $\tau$-linear representation $U$ where $\tau : G \to \mathrm{Aut}(R)$ is a homomorphism. Let $N = \ker \tau$. Since $U_{|N}$ is an $R$-linear representation we can assume, from the previous case, that $M = E[z]$ for some $k$-vector space $E$ equipped with an $N$-representation $\rho$ and that $U_{|N} = \rho$. It is easy to see, using (3), that there exists a $t_0 \in \mathbf{R}$ such that $(\tau_g p)(t_0) = p(t_0)$ for all $g \in G$ and for all $p \in R$. Consider $M_0 := M/(z - t_0)M$ equipped with the induced $k$-linear representation $U_0$. Let $j : M \to M_0[z]$ be given by $j(\sum_{i=0}^{l} m_i z^i) := \sum_{i=0}^{l} \tilde{m}_i z^i$ where $\tilde{m}_i$ is the projection of $m_i$ in $M_0$. Consider

$$R_h : M \to M_0[z]$$

given by

$$R_h = j \circ \tau_h \circ U_h + U_{0h} \circ j.$$

Using the fact that $h^2 \in N$ for all $h \in G$, it is easy to see that every $R_h$ is a $R$-linear injective map and

$$R_h \circ U_h = (\tau_h \circ U_{0h}) \circ R_h \quad \forall h \in G. \tag{21}$$

Fix any $\tilde{h} \in G \setminus N$. If $g_0 \in N$, it follows that

$$R_{g_0 \tilde{h}} = U_{0g_0} \circ R_{\tilde{h}} \quad R_{\tilde{h} g_0} = R_{\tilde{h}} \circ U_{g_0}. \tag{22}$$

Now, if $g \in G \setminus N$, there exists a $g_0 \in N$ such that $g = g_0 \tilde{h}$. By (21) and (22) it follows that

$$R_{\tilde{h}} \circ U_g = U_{0g_0^{-1}} \circ \tau_g \circ U_{0g} \circ R_g = (\tau_g \circ U_{0\tilde{h}g\tilde{h}^{-1}}) \circ R_{\tilde{h}}. \tag{23}$$

On the other hand, for $g \in N$, it follows from (22) that

$$R_{\tilde{h}} \circ U_g = R_{\tilde{h}g} = R_{(\tilde{h}g\tilde{h}^{-1})\tilde{h}} = U_{0\tilde{h}g\tilde{h}^{-1}} \circ R_{\tilde{h}}. \tag{24}$$

Equations (23) and (24) show that if we consider, on $M_0[z]$, the quasi-linear split representation induced by $\tilde{\rho}_g = U_{0\tilde{h}g\tilde{h}^{-1}}$ and $\tau$, then the $R$-map $R_{\tilde{h}}$ is $G$-equivariant. We can now again apply Theorem 4.

□

**Remark:** In [3] the case when $U$ is $R$-linear is treated in slightly greater generality considering also the case of compact $G$. Extensions to reductive groups have also been considered (see [4]). In the case $G = \mathbf{Z}_2$, Theorem 5 had already been proven in [1] and, partially, in [2].

## 7 Representations of symmetric differential behaviors.

Theorems 4 and 5 are very useful in constructing nice canonical differential representations of symmetric systems. Let $(G, T)$ be a symmetry on $\mathbb{D}[W]$ with $G$ a finite group. Let $\mathcal{B} \in \mathbb{D}[W]^G$ and apply Theorem 5 to $\mathcal{B}^\perp \in \mathbb{S}[W^*[z]]^G$. Then there exists a vector space $E$, a quasi-linear split representation on $E[z]$ induced by some representation $\rho$ on $E$, and, by the associated homomorphism $\tau : G \to \mathrm{Aut}(R)$, a $G$-equivariant injective homomorphism

$$\psi : E[z] \to W^*[z]$$

such that $\mathrm{Im}(\psi) = \mathcal{B}^\perp$. Consider the dual map

$$D = \psi^* : W[z] \to E^*[z].$$

Then, clearly

$$\mathcal{B} = \ker D\left(\frac{d}{dt}\right)$$

and it is a minimal representation. $D$ has a very special structure reflecting the symmetry on $\mathcal{B}$. Namely,

$$\alpha_{g^{-1}} D R_g = \rho_g^* D^g \quad \forall g \in G. \tag{25}$$

Moreover, it follows from Theorem 4, that in the case $U$ is degree-preserving, $D$ can be chosen to be canonical.

Let us now see what (25) becomes when affixed to some particular cases.

**Example 1:** *Static symmetries.* In the notation established in paragraph 5, we have: $x = \xi = 0, \eta = 1, R$ constant (in $z$). In this case (25) becomes

$$D(z) R_g = \rho_g^* D(z) \quad \forall g \in G. \tag{26}$$

By choosing suitable bases in $W$ and in $E^*$, relative to a decomposition of $R$ and $\rho^*$ into irreducible components, (26) can be further structured ending up with a diagonal block structure for $D(z)$. For more details and concrete examples, see [3].

**Example 2:** *Time-reversibility.* In this case $T_g = S_g$ for $g \in \{-1, +1\}$, and (25) becomes

$$D(-z) = AD(z)$$

where $A \in GL_k(E)$ is such that $A^2 = I$. By choosing a suitable basis of $E$ we can assume that $A$ is a signature matrix and obtain the following block structure for $D$.

$$D(z) = \begin{bmatrix} D_1(z^2) \\ zD_2(z^2) \end{bmatrix}$$

**Example 3:** *Involutive symmetries.* A class of involutive symmetries (i.e $G = \mathbf{Z}_2$) are given by the following: $T_{-1} = R\left(\frac{d}{dt}\right) \circ M_\xi \circ S_{-1}$ where $\xi$ is any element of $k$ and where

$$P(z)P(\eta^{-1}z - \xi) = I.$$

This is clearly a generalization of Example 2. In this case (25) is equivalent to

$$D(-z - \xi)R(-z - \xi) = AD(z)$$

where $A$ is a signature matrix.

**References.**


[1] F. FAGNANI *Symmetries of linear dynamical systems*, Ph.D. dissertation, Department of Mathematics, University of Groningen, the Netherlands, February 1992.

[2] F. FAGNANI, J.C. WILLEMS *Representations of time-reversible systems*, Journal of Mathematical Systems, Estimation and Control, Vol. 1, pp. 5-28, 1991.

[3] F. FAGNANI, J.C. WILLEMS *Representations of symmetric linear dynamical systems*, to appear in SIAM Journal on Control and Optimization, 1993.

[4] H. KRAFT *G-vector bundles and the linearization problem*, Proceedings of the Conference on Group Actions and Invariant Theory, Montreal 1988, CMS Conference Proceedings 10, 1988, 111-134.

[5] B. MALGRANGE *Existence et approximation des solutions des équations aux dérivées partielles et des équations des convolution*, Annales de l'Institut Fourier, 6, pp. 271-355, 1955-56.

[6] J.M. SCHUMACHER *Transformations of linear systems under external equivalence*, Linear Algebra and its Applications, Vol. 102, pp. 1-34, 1988.

[7] J.-P. SERRE *Linear representations of finite groups*, Springer-Verlag, 1977.

[8] F. TREVES *Topological vector spaces, distributions and kernels*, Academic Press, 1967.

[9] J.C. WILLEMS *From time series to linear systems, Part I: Finite dimensional linear time-invariant systems*, Automatica, Vol. 22, 1986, pp. 561-580.

[10] J.C. WILLEMS *Models for dynamics*, Dynamics Reported, Vol. 2, pp. 171-269, 1989.

[11] J.C. WILLEMS *Paradigms and puzzles in the theory of dynamical systems*, IEEE Transactions on Automatic Control, Vol. 36, pp. 259-294, 1991.

# 34 Relaxation in Semilinear Infinite Dimensional Control Systems

**H. O. Fattorini** University of California–Los Angeles, Los Angeles, California

Abstract. We show that trajectories of semilinear infinite dimensional control systems where the control appears linearly can be approximated by trajectories driven by extremal controls (that is, controls taking values at extremal points of the control set). This is done under two sets of hypotheses. The results are relevant to systems driven either by ordinary or relaxed controls. There are two applications to nonlinear distributed parameter systems, one hyperbolic, the other parabolic.

Keywords: relaxed controls, optimal controls, relaxation.

1980 Mathematics subject classifications: 93E20, 93E25.

§**1. Introduction.** We consider semilinear control systems

$$y'(t) = Ay(t) + f(t, y(t)) + F(t, y(t))u(t), \qquad y(0) = \zeta. \tag{1.1}$$

where $A$ generates a strongly continuous semigroup $S(t)$ in a Banach space $E$; $f(t, y)$ (resp. $F(t, y)$) is defined in $[0, T] \times E$, and takes values in $E$ (resp. in the space $L(X^*, E)$ of linear bounded operators from the dual $X^*$ of a Banach space $X$ into $E$). The control $u(t)$ takes values in a convex subset $\mathbf{U}$ of $X^*$. The main results (Theorems 5.2 and 6.5) state that, under suitable assumptions on (1.1), trajectories corresponding to controls

This work was supported in part by the NSF under grant PDQ-9801234

$u(t)$ taking values in $\mathbf{U}$ can be uniformly approximated by trajectories corresponding to controls taking values in the extremal points of $\mathbf{U}$. In the first result, $A$ generates a compact semigroup; in the second, there are no restrictions on the semigroup but the conditions on $F(t, y)$ are more demanding. As a particular case we obtain theorems on approximation of relaxed trajectories by ordinary trajectories. There are two applications to distributed parameter systems.

§2. **The control system.** Let $X$ be an arbitrary Banach space, $X^*$ its dual. A $X^*$-valued function $g(\cdot)$ is $X$-**weakly measurable** if $\langle g(\cdot), f\rangle$ is measurable for each $f \in X$. The space $L^\infty_w(0, T; X^*)$ consists of all $X$-weakly measurable $X^*$-valued functions $g(\cdot)$ such that there exists $C > 0$ with

$$|\langle g(t), f\rangle| \le C\|f\| \qquad \text{a.e. in} \quad 0 \le t \le T \tag{2.1}$$

for every $f \in E$ ("a.e" depending on $f$). We equip $L^\infty_w(0, T; X^*)$ with the norm $\|g\| =$ infimum of all $C$ such that (2.1) holds. As a consequence of the Dunford - Pettis theorem [1], [8] this space can be identified with the dual of $L^1(0, T; X)$ with the pairing

$$\langle g(\cdot), f(\cdot)\rangle = \int_0^T \langle g(\sigma), f(\sigma)\rangle d\sigma \,. \tag{2.2}$$

If $X$ is separable then the norm $\|g(\cdot)\|$ of any $g(\cdot) \in L^\infty_w(0, T; X^*)$ is measurable, and $\|g\|$ is the essential supremum norm. In the general case, this is not true but it can be proved [8] that there exists a linear **lifting operator** $S : L^\infty_w(0, T; X^*) \to L^\infty_w(0, T; X^*)$ such that (a) $Sg$ belongs to the equivalence class $[g]$ of $g$. (b) The function $t \to \|(Sg)(t)\|$ is measurable in $0 \le t \le T$. (c) $\sup_{0\le t\le T}\|(Sg)(t)\| = \|g\|$. For more information on these spaces, see [8].

The space of admissible controls is

$$\mathfrak{C}_{ad}(0, T; \mathbf{U}) = \{u(\cdot) \in L^\infty_w(0, T; X^*); u(t) \in \mathbf{U}\}\,.$$

where $\mathbf{U} \subseteq X^*$ and "$u(t) \in \mathbf{U}$" means "there exists $\tilde{u}(\cdot)$ in the equivalence class $[u(\cdot)]$ of $u(\cdot)$ in $L^\infty_w(0, T; X^*)$ such that $\tilde{u}(t) \in \mathbf{U}$ in $0 \le t \le T$." In most of the results we assume that $\mathbf{U}$ is convex and $X$-weakly compact in $X^*$. By Alaoglu's Theorem, this will be satisfied if $\mathbf{U}$ is convex, bounded and $X$-weakly closed.

Except in §9 and unless otherwise stated, the space $E$ is **reflexive** and **separable.** The functions $f : [0, T] \times E \to E$ and $F : [0, T] \times E \to L(X^*, E)$ satisfy Hypothesis $(I) = (I_1) \cup (I_2)$ below.

$(I_1)$ $f(t, y)$ is strongly measurable for $y$ fixed and (a) for every $c > 0$ there exists $\alpha_1(\cdot, c) \in L^1(0, T)$ with

$$\|f(t, y)\| \le \alpha_1(t, c) \qquad (0 \le t \le T, \|y\| \le c) \tag{2.3}$$

(b) For every $c > 0$ there exists $\beta_1(\cdot\,, c) \in L^1(0, T)$ such that

$$\|f(t, y') - f(t, y)\| \le \beta_1(t, c)\|y' - y\| \quad (0 \le t \le T, \|y\|, \ \|y'\| \le c) \tag{2.4}$$

$(I_2)$ $F(t, y)^* E^* \subseteq X$ and for each $y \in E$ and $y^* \in E^*$ fixed $F(t, y)^* y^*$ is strongly measurable in $X$. (a) for every $c > 0$ there exists $\alpha_2(\cdot\,, c) \in L^1(0, T)$ such that

$$\|F(t, y)\| \le \alpha_2(t, c) \qquad (0 \le t \le T, \|y\| \le c)\,. \tag{2.5}$$

(b) For every $c > 0$ there exists $\beta_2(\cdot\,, c) \in L^1(0, T)$ such that

$$\|F(t, y') - F(t, y)\| \le \beta_2(t, c)\|y' - y\| \quad (0 \le t \le T, \|y\|, \ \|y'\| \le c)\,. \tag{2.6}$$

Obviously, (b) of $(I_1)$(resp. (b) of $(I_2)$) implies continuity of $f(t, y)$ in $E$ (resp. of $F(t, y)$ in $L(X^*, E)$) for $t$ fixed. By definition, solutions of (1.1) are solutions of the integral equation

$$y(t) = S(t)\zeta + \int_0^t S(t - \sigma) f(\sigma, y(\sigma)) d\sigma \\ + \int_0^t S(t - \sigma) F(\sigma, y(\sigma)) u(\sigma) d\sigma\,. \tag{2.7}$$

If $y(\cdot) \in C(0, T; E)$ then we see (approximating $y(\cdot)$ by a piecewise constant function) that $t \to f(t, y(t))$ is strongly measurable, thus it belongs to $L^1(0, T; E)$ by (2.3); hence the first integral in (2.7) makes sense. For the second, we argue with $(I_2)$ in the same way and obtain that $t \to F(t, y(t))^* y^*$ is strongly measurable for any $y^* \in E^*$ and $y(\cdot) \in C(0, T; E)$. Hence $t \to \langle y^*, F(t, y(t)) u(t)\rangle = \langle F(t, y(t))^* y^*, u(t)\rangle$ is measurable. It follows that $t \to F(t, y(t)) u(t)$ is weakly measurable, hence strongly measurable by separability of $E$ [7]. The bound (2.5) implies that $\|F(\cdot\,, y(\cdot))\| \in L^1(0, T; E)$.

Under Hypothesis (I), the equation (2.7) can be locally solved by successive approximations for any $u(\cdot) \in L^\infty_w(0, T; X^*)$. Solutions exist and are unique in $0 \le t \le T$ or in a maximal interval of existence $[0, \tilde{t})$, $\tilde{t} < T$ (depending on $u(\cdot)$) and $\limsup_{t \to \tilde{t}} \|y(t)\| = +\infty$.

§3. **Compactness results**. Below, $E$ is an arbitrary Banach space, $S(\cdot)$ a strongly continuous semigroup in $E$.

**Theorem 3.1.** *The operator*

$$\Lambda(\cdot) = \int_0^t S(t-\sigma)g(\sigma)d\sigma\,, \tag{3.1}$$

*is bounded from $L^1(0,T;E)$ into $C(0,T;E)$. If $S(t)$ is compact for $t>0$ and $\{g_\kappa(\cdot);\kappa\in K\}$ is a generalized sequence in $L^1(0,T;E)$ such that the integrals of the $\|g_\kappa(\cdot)\|$ are equicontinuous in $0\le t\le T$, then $\{\Lambda g_\kappa(\cdot)\}$ has a generalized subsequence convergent in $C(0,T;U)$.*

**Proof.** Let $\chi_t(\sigma)$ be the characteristic function of $[0,t]$. For $g(\cdot)\in L^1(0,T;E)$, $t<t'$ we have

$$\Lambda g(t') - \Lambda g(t) = \int_0^T \{\chi_{t'}(\sigma)S(t'-\sigma) - \chi_t(\sigma)S(t-\sigma)\}g(\sigma)d\sigma$$

so that by the dominated convergence theorem $\Lambda g(t')\to\Lambda g(t)$ as $t'\to t$; this shows that $\Lambda g(\cdot)\in C(0,T;E)$. Obviously,

$$\|\Lambda g(\cdot)\|_{C(0,T;E)} \le M\|g(\cdot)\|_{L^1(0,T;E)} \tag{3.2}$$

($M$ a bound for $\|S(\sigma)\|$ in $0\le t\le T$), so that (a) is proved. Note that (b) states that the closure of $\{\Lambda g_\kappa(\cdot);\kappa\in K\}$ is compact in $C(0,T;E)$. Since, in a metric space, compactness and sequential compactness are equivalent [1, p.21], it is enough to prove (b) for a sequence $\{g_n(\cdot)\}$. Let $\delta>0$,

$$\Lambda_\delta g(\cdot) = \int_0^{t-\delta} S(t-\sigma)g(\sigma)d\sigma = S(\delta)\int_0^{t-\delta} S(t-\delta-\sigma)g(\sigma)d\sigma \tag{3.3}$$

for $t\ge\delta$, $\Lambda_\delta y(t)=0$ for $t<\delta$. We have

$$\|(\Lambda-\Lambda_\delta)g_n\|_{C(0,T;E)} \le M\sup_{\delta\le t\le T}\int_{t-\delta}^t \|g_n(\sigma)\|d\sigma \tag{3.4}$$

where the right hand side tends to zero independently of $n$. Assume we can prove that for each $\delta>0$ $\{\Lambda_\delta g_n\}$ has a convergent subsequence in $C(0,T;E)$. Then, if $\{\delta_m\}$ is a sequence of positive numbers tending to zero, we can use the diagonal sequence trick to construct a sequence (call it also $\{g_n\}$) such that $\{\Lambda_{\delta_m}g_n\}$ is convergent for all $m$. If $\varepsilon>0$, using (3.4) we may insure $\|\Lambda g_n - \Lambda_{\delta_m}g_n\|\le\varepsilon/3$ for $m$ large enough. Accordingly, $\|\Lambda g_n - \Lambda g_k\| \le 2\varepsilon/3 + \|\Lambda_{\delta_m}g_n - \Lambda_{\delta_m}g_k\| \le \varepsilon$ for $m$, $k\ge n_0$ and it follows that $\{\Lambda g_n\}$ is Cauchy, hence convergent.

We show then that for each $\delta>0$ $\{\Lambda_\delta g_n\}$ has a convergent subsequence in $C(0,T;E)$. It follows from the second expression (3.3) for $\Lambda_\delta$ that for every $t$ fixed $\{\Lambda_\delta g_n(t)\}$ has a convergent subsequence. Applying this to every rational $t \in [0,T]$ and using again the diagonal sequence trick, we come up with a subsequence of $\{g_n(\cdot)\}$ (denoted with the same symbol) such that $\{\Lambda_\delta g_n(t)\}$ is convergent for $t$ rational. To show that $\{\Lambda_\delta g_n(\cdot)\}$ is convergent in $C(0,T;U)$ we only have to show that $\{\Lambda_\delta g_n(t)\}$ is equicontinuous. If this is not true, there exist sequences $\{s_n\}$, $\{t_n\} \subseteq [0,T]$ such that $s_n < t_n$, $t_n - s_n \to 0$ and $\|\Lambda g_n(t_n) - \Lambda g_n(s_n)\| \geq 2\varepsilon > 0$. We may assume that $\{s_n\}$, $\{t_n\}$ are convergent to $\bar{t} \in [0,T]$. For each $n$, select $y_n^* \in E^*$ with $\|y_n^*\| = 1$,

$$\langle y_n^*, \Lambda g_n(t_n) - \Lambda g_n(s_n)\rangle \geq \varepsilon > 0 . \tag{3.5}$$

If $\bar{t} > \delta$, then $t_n$, $s_n > \delta$ for $n$ large enough and we rewrite this inequality as

$$\int_{s_n-\delta}^{t_n-\delta} \langle S(t_n-\sigma)^* y_n^*,\ g_n(\sigma)\rangle d\sigma + \int_0^{s_n-\delta} \langle (S(t_n-\sigma)^* - S(s_n-\sigma)^*) y_n^*,\ g_n(\sigma)\rangle d\sigma \geq \varepsilon . \tag{3.6}$$

The first integral tends to zero by the equicontinuity condition. Due to continuity of $S(t)$ (thus of $S(t)^*$) in $t > 0$ in the norm of $L(E,E)$, $(S(t_n-\sigma)^* - S(s_n-\sigma)^*)y_n^* \to 0$ uniformly in $0 \leq t \leq T$, thus the second integral also tends to zero, a contradiction with (3.5). The case $\bar{t} < \delta$ is excluded, since (3.5) becomes impossible for sufficiently large $n$. Finally, if $\bar{t} = \delta$ we use (3.4) if $s_n \geq \delta$; if $s_n < \delta$, the first integral is from 0 to $t_n - \delta$ and the second disappears.

It is easy to see that $\Lambda$ is not a compact operator from $L^1(0,T;E)$ into $C(0,T;E)$. To do this take a "$\delta$-sequence" $\{\phi_m(\cdot)\}$ where $\phi_m(t) = m\phi(mt)$, $\phi(\cdot)$ a continuous nonnegative function with support in $0 \leq t \leq 1$ and integral 1. Pick a sequence $\{y_n\} \subset E$ with $\|y_n\| = 1$, $\|y_n - y_k\| \geq 5\varepsilon > 0$ for $n \neq k$ and a sequence $\{t_n\}$, $t_n > 0$ such that $\|S(t_n)y_n - y_n\| \leq \varepsilon$, so that $\|S(t_n)y_n - S(t_k)y_k\| \geq 3\varepsilon$ for $n \neq k$. Define $g_n(t) = \phi_{m(n)}(t)y_n$, where $m(n)$ is chosen for each $n$ so large that $\|\Lambda g_n(t_n) - S(t_n)y_n\| \leq \varepsilon$; then $\{\Lambda g_n(\cdot)\}$ cannot be convergent in $C(0,T;E)$. However, we have

**Corollary 3.2.** *If $S(t)$ is compact for $t > 0$ then* (a) *$\Lambda$ is compact from $L^p(0,T;E)$ into $C(0,T;E)$ for $p > 1$,* (b) *$\Lambda$ is compact from $L^1(0,T;E)$ into $L^q(0,T;E)$ for $q > 1$.*

To show (a) it is enough to use Theorem 3.1 and note that if $\{g_n(\cdot)\}$ is a bounded set in $L^p(0,T;E)$ then, by Hölder's inequality,

$$\int_e \|g_n(\sigma)\| d\sigma \leq |e|^{1-1/p} \|g_n\|_{L^p(0,T;E)}$$

for any measurable $e$, where $|\cdot|$ denotes Lebesgue measure. To prove (b), notice that $\Lambda$ can be written as a convolution $(\chi_T S) * (\chi_T g)$, where $\chi_T(t)$ is the characteristic function of $[0,T]$. Let $\phi_n(\cdot)$ be a $\delta$-sequence. Consider the operator $\Lambda_n = (\chi_\delta S) * \phi_n * (\chi_T g) = \phi_{n*}(\chi_\delta S) * (\chi_T y) = \phi_{n*}(\Lambda y)$. By Young's inequality, the operator $\phi_{n*}(\chi_T y)$ is bounded from $L^1(0,T;E)$ into $L^q(0,T;E)$, thus $\Lambda_n$ is compact from $L^1(0,T;E)$ into $C(0,T;U)$, a fortiori from $L^1(0,T;E)$ into $L^q(0,T;E)$. Finally,

$$\|\Lambda_n g(t) - \Lambda g(t)\| \le \int_0^T \|(\phi_{n*} S)(t-\sigma) - S(t-\sigma)\| \, \|g(\sigma)\| d\sigma .$$

Using continuity of $S(t)$ in $t > 0$ in the norm of $L(E,E)$ we show that

$$\int_0^T \|(\phi_{n*} S)(t-\sigma) - S(t-\sigma)\|^q d\sigma \to 0$$

as $n \to \infty$; then, by Young's inequality, $\Lambda_n \to \Lambda$ in the norm of $L(L^1, L^q)$, thus $\Lambda$ is compact. This ends the proof.

§4. **Weak approximations of admissible controls.** Let $X$ be a Banach space, $\mathbf{U} \subseteq X^*$ convex and $X$-weakly closed. A set $D \subseteq \mathbf{U}$ is **total** in $\mathbf{U}$ if $\overline{\mathrm{conv}}(D) = \mathbf{U}$, where $\overline{\mathrm{conv}}$ is closed convex hull in the $X$-weak topology of $X^*$. A point $u \in \mathbf{U}$ is **extremal** if it is not a proper convex combination of two different points in $\mathbf{U}$; in other words, if $u = \alpha v + (1-\alpha)w$ with $v, w \in \mathbf{U}$, $0 < \alpha < 1$, then $v = w = u$. If $\mathbf{U}$ is $X$-weakly compact, then it has extremal points [1 p. 439]; moreover, by the Krein-Milman theorem [1 p. 440] the set $D$ of its extremal points is total.

Denote by $PC(0,T;D)$ the set of all functions defined in $0 \le t \le T$ with $u(t) \in D$, constant in each of a finite number of disjoint subintervals $\subseteq [0,T]$ covering $[0,T]$.

**Theorem 4.1.** *Let $\mathbf{U}$ be $X$-weakly compact, $D \subseteq \mathbf{U}$ total in $\mathbf{U}$. Then $PC(0,T;D)$ is $L^1(0,T;X)$-weakly dense in $\mathfrak{C}_{ad}(0,T;\mathbf{U})$.*

The proof is essentially the same as that of Theorem 3.2 in [6], thus we only sketch the various steps. In the first we show that the space $C(0,T;\mathbf{U})$ of (strongly) continuous, $\mathbf{U}$-valued functions $u(\cdot)$ defined in $0 \le t \le T$ is $L^1(0,T;X)$-weakly sequentially dense in $\mathfrak{C}_{ad}(0,T;\mathbf{U})$. To see this we take $u(\cdot) \in \mathfrak{C}_{ad}(0,T;\mathbf{U})$, a $\delta$-sequence $\{\phi_n\}$ (see §3), extend $u(\cdot)$ to $t < 0$ and $t > T$ setting $u(t) = u \in \mathbf{U}$ there and define

$$\langle u_n(t), f\rangle = \int_{-\infty}^{\infty} \phi_n(t-\sigma)\langle u(\sigma), f\rangle d\sigma \qquad (f \in X).$$

Since $\mathbf{U}$ is $X$-weakly closed, $u_n(t) \in \mathbf{U}$. That $u_n(\cdot)$ is continuous for each $n$ follows from the estimate

$$|\langle u_n(t') - u_n(t), f\rangle| \leq \int_{-\infty}^{\infty} \{\phi_n(t' - \sigma) - \phi_n(t - \sigma)\} \|u(\sigma)\| \; \|f\| d\sigma$$

($f \in X$). Finally, the convergence statement is $\langle u_n(\cdot), f(\cdot)\rangle \to \langle u(\cdot), f(\cdot)\rangle$ in $L^1(0, T)$ for every $f(\cdot) \in L^1(0, T; X)$: this is

$$\langle u_n(\cdot), f(\cdot)\rangle = \int_{-\infty}^{\infty} \phi_n(\cdot - \sigma)\langle u(\sigma), f(\cdot)\rangle d\sigma \to \langle u(\cdot), f(\cdot)\rangle$$

in $L^1(0, T)$, which follows from standard results on mollifiers. Approximation by continuous functions thus proved, it is clear that the space $PC(0, T; \mathbf{U})$ is as well $L^1(0, T; X)$-weakly sequentially dense in $\mathfrak{C}_{ad}(0, T; \mathbf{U})$. Then, we only have to prove that $PC(0, T; D)$ is dense in $PC(0, T; \mathbf{U})$. Since each $u(\cdot) \in PC(0, T; \mathbf{U})$ is constant in each of a finite number of intervals, we can construct the approximations in each interval and piece them together, thus we may assume that $u(\cdot) \equiv u$ is constant. Let $\{u_\kappa\}$ be a generalized sequence in $X^*$ such that $u_\kappa \to u$ $X^*$-weakly. Define $u_\kappa(t) \equiv u_\kappa$; then if $f(t) \in L^1(0, T; X)$ we have

$$\langle u_\kappa(\cdot), f(\cdot)\rangle = \int_0^T \langle u_\kappa, f(t)\rangle dt \to \int_0^T \langle u, f(t)\rangle dt$$

thus, since $D$ is total in $\mathbf{U}$, we only have to show the approximation for $u(\cdot) \in \mathfrak{C}_{ad}(0, T; \mathbf{U})$ of the form

$$u(t) = u = \sum_{k=1}^{m} \alpha_k u^k \qquad (\alpha_k \geq 0, \Sigma\alpha_k = 1, u^k \in D).$$

Take a partition $\{t_j\}$, $t_j = jh = jT/n$ ($j = 0,\ 1, \ldots, n$) of the interval $[0, T]$ and divide each interval $[t_{j-1}, t_j]$ into $m$ subintervals $I_{jk}$ of length $|I_{jk}| = \alpha_\kappa h$. Define an element of $PC(0, T; D)$ by $u_n(t) = u^k$ ($t \in I_{jk}, 1 \leq j \leq n, 1 \leq k \leq m$). Then, if $f(\cdot) \in C(0, T; X)$

$$\begin{aligned} \langle u_n(\cdot), f(\cdot)\rangle &= \sum_{j=1}^{n} \sum_{k=1}^{m} \int_{I_{jk}} \langle u^k, f(t)\rangle dt \\ &\to \sum_{k=1}^{m} \alpha_k \int_0^T \langle u^k, f(t)\rangle dt = \langle u(\cdot), f(\cdot)\rangle . \end{aligned}$$

Since $C(0, T; X)$ is dense in $L^1(0, T; X)$ and the $u_n(\cdot)$ are bounded in $L^\infty_w(0, T; X)$, we deduce that $u_n(\cdot) \to u$ $L^1(0, T; X)$-weakly in $L^\infty_w(0, T; X^*)$. This ends the proof of Theorem 4.1.

## §5. The first relaxation theorem

**Lemma 5.1.** *Assume that $S(t)$ is compact for $t > 0$ and that Hypothesis $(I)$ holds. Let $u(\cdot) \in L^\infty_w(0,T;X^*)$ be such that the solution $y(\cdot,u)$ of (2.7) exists in $0 \le t \le \bar{t}$, and let $\{u_\kappa(\cdot);\kappa \in K\}$ be a bounded generalized sequence in $L^\infty_w(0,\bar{t};X^*)$ such that $u_\kappa(\cdot) \to u(\cdot)$ $L^1(0,\bar{t};X)$-weakly. Then there exists $\kappa_0 \in K$ such that $y(\cdot,u_\kappa)$ exists in $0 \le t \le \bar{t}$ if $\kappa \ge \kappa_0$ ("$\ge$" the ordering in $K$) and*

$$y(t,u_\kappa) \to y(t,u)$$

*uniformly in $0 \le t \le \bar{t}$.*

**Proof.** Let $[0,\bar{t}_\kappa] \subseteq [0,\bar{t}]$ be the maximal interval where $y(t,u_\kappa)$ exists and satisfies $\|y(t,u_\kappa) - y(t,u)\| \le 1$. For any $\kappa$ we have

$$\begin{aligned} y(t,u_\kappa) - y(t,u) &= \int_0^t S(t-\sigma)\{f(\sigma,y(\sigma,u_\kappa)) - f(\sigma,y(\sigma,u))\}d\sigma \\ &+ \int_0^t S(t-\sigma)\{F(\sigma,y(\sigma,u_\kappa))u_\kappa(\sigma) - F(\sigma,y(\sigma,u))u_\kappa(\sigma)\}d\sigma \\ &+ \int_0^t S(t-\sigma)F(\sigma,y(\sigma,u))(u_\kappa(\sigma)) - u(\sigma))d\sigma \end{aligned} \tag{5.1}$$

in $0 \le t \le \bar{t}_\kappa$. Estimating and using Gronwall's inequality,

$$\|y(t,\mu_\kappa) - y(t,\mu\| \le C\|\rho_\kappa(t)\|,$$

where

$$\rho_\kappa(t) = \Lambda\{F(\cdot,y(\cdot,u))(u_\kappa(\cdot)) - u(\cdot))\},$$

$\Lambda$ the operator in Theorem 3.1. In view of (2.5) the (generalized) sequence $\{g_\kappa(\cdot)\} = \{F(\cdot,y(\cdot,u))(u_\kappa(\cdot)) - u(\cdot))\}$ has equicontinuous integrals in $0 \le t \le \bar{t}$, so that by Theorem 3.1 there exists a generalized subsequence (denoted with the same symbols) with $\rho_\kappa(t) \to \rho(t) \in C(0,\bar{t};U)$. Now, $u_\kappa(\cdot) \to u(\cdot)$ $L^1(0,\bar{t};X)$-weakly in $L^\infty_w(0,\bar{t};X^*)$ and

$$\langle y^*, \rho_\kappa(t)\rangle = \int_0^{\bar{t}} \langle \chi_t(\sigma)F(\sigma,y(\sigma,u))^* S(t-\sigma)y^*, u_\kappa(\sigma) - u(\sigma)\rangle d\sigma\,. \tag{5.2}$$

Using $(I_2)$ for $F(t,y)$ and approximating $S(t-\cdot)y^*$ by a piecewise constant function we deduce that $\chi_t(\cdot)F(\cdot,y(\cdot,u))^*S(t-\cdot)y^* \in L^1(0,\bar{t};X)$, thus it follows that $\langle y^*,\rho(t)\rangle = \lim_\kappa\langle y^*\rho_\kappa(t)\rangle = 0$, thus $\rho(t) = 0$. It follows that for $\kappa \ge$ a certain $\kappa_0$ we have $t_\kappa = \bar{t}$ and that the uniform convergence statement holds. This ends the proof.

**Relaxation Theorem 5.2.** *Assume that* $\mathbf{U}$ *is* $X$*-weakly compact, that* $D$ *is total in* $\mathbf{U}$*, that* $S(t)$ *is compact and that Hypothesis* $(I)$ *holds. Let* $u(\cdot) \in \mathfrak{C}_{ad}(0, \bar{t}; \mathbf{U})$ *be such that the solution* $y(\cdot, u)$ *of* (1.1) *exists in* $0 \le t \le \bar{t}$ *and* $\varepsilon > 0$*. Then there exists* $\nu(\cdot) \in PC(0, T; D)$ *such that* $y(t, \nu)$ *exists in* $0 \le t \le \bar{t}$ *and*

$$\|y(t,u) - y(t,\nu)\| \le \varepsilon \quad (0 \le t \le \bar{t}).$$

**Proof.** Combine Lemma 5.1 with Theorem 4.1.

**§6. The second relaxation theorem.** In the first two results, $E$ is reflexive but need not be separable; $\mathbf{U} \subseteq X^*$ is $X$-weakly compact and $D \subseteq \mathbf{U}$ is total. In Lemma 6.3 $Y$ is an arbitrary Banach space. The space $CV(0, T; V)$ ($V$ a subset of $X^*$) consists of all countably valued functions $u(\cdot)$ such that $u(t) \in V$ a.e.

**Lemma 6.1.** *Let* $\mathbf{U} \subseteq X^*$ *be* $X$*-weakly compact,* $F \in L(X^*, E)$ *such that* $F^* : E^* \to X$*. Then* $F(\mathbf{U}) = \{Fu; u \in \mathbf{U}\}$ *is convex and closed in* $E$ *and we have* $\overline{\text{conv}}(F(D)) = F(\mathbf{U})$ *(closure taken in the strong topology).*

**Proof.** Obviously, $F(\mathbf{U})$ is convex. Let $\{y_n\} \subset F(\mathbf{U})$. Then $y_n = Fu_n$, $\{u_n\} \subseteq \mathbf{U}$. Select a generalized subsequence of $\{u_n\}$ (same name) $X$-weakly convergent to $u \in \mathbf{U}$, and let $y^* \in E^*$. Then $\langle y^*, y_n\rangle = \langle y^*, Fu_n\rangle = \langle F^*y^*, u_n\rangle \to \langle F^*y^*, u\rangle = \langle y^*, Fu\rangle$, so that $Fu_n \to Fu$ $E^*$-weakly in $E$. It follows that $F(\mathbf{U})$ is $E^*$-weakly compact in $E$, hence weakly closed; since $E$ is reflexive, $F(\mathbf{U})$ is as well strongly closed. The same argument shows that, if $u \in \mathbf{U}$ and $\{u_\kappa\} \subseteq \text{conv}(D)$ is a generalized sequence with $u_\kappa \to u$ $X$-weakly then $Fu_\kappa \to Fu$ $E$-weakly. It follows that the weak closure of $\text{conv}(F(D))$ equals $F(\mathbf{U})$. Since $\text{conv}(F(D))$ is convex, its strong closure equals its weak closure.

**Lemma 6.2.** *Let* $t \to g(t) \in E$ *be strongly measurable and* $t \to F(t) \in L(X^*, E)$ *be strongly measurable in the sense of the norm of* $L(X^*, E)$*. Assume that* $D$ *is total in* $\mathbf{U}$ *and*

$$g(t) \in F(t)\mathbf{U} \qquad \text{a.e. in} \quad 0 \le t \le T.$$

*Then, given* $\varepsilon > 0$ *there exists* $v(\cdot) \in CV(0, T; \text{conv}(D))$ *such that*

$$\|F(t)v(t) - g(t)\| \le \varepsilon \qquad (0 \le t \le T). \tag{6.1}$$

**Proof.** Choose a countably valued $L(X^*, E)$-valued function $F_\varepsilon(t)$ and a countably $E$-valued function $g_\varepsilon(t)$ such that

$$\|F_\varepsilon(t) - F(t)\| \le \varepsilon \qquad \|g_\varepsilon(t) - g(t)\| \le \varepsilon$$

a.e. in $0 \le t \le T$. Let $\Xi_\varepsilon$ be a countable collection of pairwise disjoint measurable sets where $F_\varepsilon(t) = F_\varepsilon$ and $g_\varepsilon(t) = g_\varepsilon$ are constant. Since $g_\varepsilon \in F(\mathbf{U})_\varepsilon = \{v; \mathrm{dist}(v, F(\mathbf{U})) \le \varepsilon\}$, then there exists $v(e) \in \mathrm{conv}(D)$ such that $\|F_\varepsilon v(e) - g_\varepsilon\| \le 2\varepsilon$. Defining $v(t) = v(e)$ in each $e \in \Xi_\varepsilon$, we have $\|F(t)v(t) - g(t)\| \le 4\varepsilon$.

**Lemma 6.3.** *Let $E$ be reflexive and separable, $t \to F(t) \in L(X^*, E)$ strongly measurable in $0 \le t \le T$ in the sense of the norm of $L(X^*, E)$ with $\|F(\cdot)\| \in L^1(0,T)$, and let $t \to S(t) \in L(E, Y)$ ($Y$ a Banach space) be strongly continuous. Then, if $u(\cdot) \in \mathfrak{C}_{ad}(0,T;\mathbf{U})$ there exists a countably valued $\nu(\cdot) \in CV(0,T;D)$ such that*

$$\left\| \int_0^T \mathfrak{S}(t)\{F(t)\nu(t) - F(t)u(t)\}dt \right\|_Y \le \varepsilon. \tag{6.2}$$

**Proof.** The function $g(\cdot) = F(\cdot)u(\cdot)$ is strongly measurable in $E$ (see §1). Using Lemma 6.2 select $v(\cdot) \in CV(0,T;\mathrm{conv}(D))$ such that $\|F(t)v(t) - F(t)u(t)\| \le \varepsilon$. Then select a countably valued $L(X^*, E)$-valued function $F_\varepsilon(\cdot)$ such that $\|F_\varepsilon(t) - F(t)\| \le \varepsilon$, so that

$$\|F_\varepsilon(t)v(t) - F_\varepsilon(t)u(t)\| \le (2M+1)\varepsilon,$$

$M$ a bound for $\mathbf{U}$. Let $\Xi_\varepsilon$ be a collection of pairwise disjoint measurable sets covering $[0,T]$ and such that both $F_\varepsilon(\cdot) = F_\varepsilon$ and $v(\cdot) = v \in \mathrm{conv}(D)$ are constant in each $e \in \Xi_\varepsilon$. Let $v = \Sigma\, \alpha_j(e)d_j(e)$, $X^*(e)$ the subspace of $X^*$ generated by $d_1(e), \ldots, d_n(e)$. Considering $\mathfrak{S}(\cdot)$ as a $L(X^*(e), Y)$-valued function $\mathfrak{S}(\cdot)$ is strongly continuous, thus (since $X^*(e)$ is finite dimensional) continuous in the uniform topology of operators and can be approximated uniformly in $t \in e$ in the $L(X^*(e), Y)$-norm by a piecewise constant function $\mathfrak{S}_\varepsilon(t)$,

$$\|\mathfrak{S}_\varepsilon(t) - \mathfrak{S}(t)\|_{L(X^*(e);Y)} \le \varepsilon \qquad (t \in e).$$

Let, finally $d \subseteq e$ be a set where $\mathfrak{S}_\varepsilon(t) = \mathfrak{S}_\varepsilon$ is constant. Divide $d$ into $n$ disjoint measurable sets with $\mathrm{meas}(d_j) = \alpha_j(e)\mathrm{meas}(d)$, and define $\nu(t) = d_j(e)$ for $t \in d_j$. Then

$$\int_e \mathfrak{S}_\varepsilon(t)F_\varepsilon(t)\nu(t)dt = \int_e \mathfrak{S}_\varepsilon(t)F_\varepsilon(t)v(t)dt.$$

The control $\nu(\cdot) \in CV(0,T;D)$ obtained piecing together all the $\nu(\cdot)$ constructed in each $d \subseteq e$ for all $e \in \Xi_\varepsilon$ satisfies (6.2) with a different $\varepsilon$. This ends the proof.

**Lemma 6.4.** *Same assumptions as in Lemma 6.3. Let the $L(E,E)$-valued function $S(t,s)$ be strongly continuous in $0 \le s \le t \le T$. Then if $u(\cdot) \in \mathfrak{C}_{ad}(0,T;\mathbf{U})$ and $\varepsilon > 0$ there exists $\nu(\cdot) \in CV(0,T;D)$ such that*

$$\left\| \int_0^t S(t,\sigma)F(\sigma)\nu(\sigma) - F(\sigma)u(\sigma)\}d\sigma \right\|_E \le \varepsilon \quad (0 \le t \le T). \tag{6.3}$$

**Proof.** Extend $S(t,s)$ to the square $0 \le s, t \le T$ setting $S(t,s) = 0$ for $s > t$. Let $M$ be the maximum of $\|S(t,s)\|$ in the square. Then construct a strongly continuous operator valued function $S_\varepsilon(t,s)$ with the same $M$ and such that $S_\varepsilon(t,s) = S(t,s)$ in $s \le t$, $S(t,s) = 0$ in $s \ge t+\varepsilon$; for instance, $S_\varepsilon(t,s) = (1/\varepsilon)(t+\varepsilon-s)S((t+s)/2, (t+s)/2))$ in $t \le s \le t+\varepsilon$. Applying Lemma 6.3 with $Y = C(0,T;E)$, $\mathfrak{S}(t) = S_\varepsilon(\cdot\,,t)$, (6.3) results for $S_\varepsilon(t,s)$; since the $L^1$-norm of $F(\cdot)\nu(\cdot)$ is bounded independent of $\nu(\cdot)$, (6.3) follows as well for $S(t,s)$.

**Relaxation Theorem 6.5.** *Let the space $E$ be reflexive and separable and* $\mathbf{U}$ *$X$-weakly compact in $X^*$. Let the control system* (1.1) *satisfy Assumption (I); further assume that for every $y$ fixed the function $t \to F(t,y)$ is strongly measurable in the sense of the norm of $L(X^*,E)$. Let $u(\cdot) \in \mathfrak{C}_{ad}(0,T;\mathbf{U})$ be such that $y(t,u)$ exists in $0 \le t \le \bar{t}$ and $\varepsilon > 0$. Then there exists $\nu(\cdot) \in CV(0,T;D)$ such that $y(t,\nu)$ exists in $0 \le t \le \bar{t}$ and*

$$\|y(t,u) - y(t,\nu)\| \le \varepsilon \qquad (0 \le t \le \bar{t}).$$

**Proof.** Let $\nu(\cdot) \in CV(0,T;D)$ be the control that performs (6.3) for $S(t,s) = S(t-s)$, $F(t) = F(t, y(t,u))$. We have

$$\begin{aligned} &y(t,\nu) - y(t,u) \\ &\quad= \int_0^t S(t-\sigma)\{f(y(\sigma,\nu)) - f(y(\sigma,u))\}d\sigma \\ &\quad+ \int_0^t S(t-\sigma)\{F(\sigma,y(\sigma,\nu))\nu(\sigma) - F(t,y(\sigma,u))\nu(\sigma)\}d\sigma \\ &\quad+ \int_0^t S(t-\sigma)\{F(\sigma,y(\sigma,u))\nu(\sigma) - F(t,y(\sigma,u))u(\sigma)\}d\sigma\,. \end{aligned}$$

Estimating and using Gronwall's inequality, we obtain

$$\|y(t,\nu) - y(t,\mu)\| \le C\varepsilon$$

in the maximal interval $0 \le t \le t_\nu$ where $y(t,\nu)$ exists and satisfies $\|y(t,\nu) - y(t,\mu)\| \le 1$. The proof ends like that of Lemma 5.1.

Relaxation results that partly include Theorem 6.5 in the linear case are in [2].

§7. **Relaxed controls.** We shall use three spaces of relaxed controls with increasing degree of generality. A **probability measure** in a set $U$ is a measure defined in a field of subsets of $U$ with $\mu \ge 0$, $\mu(U) = 1$; $\mu$ need not be countably additive.

**Class 1.** $U$ = compact metric space, $X = C(U) = \{$continuous functions in $U$ with supremum norm$\}$, $X^* = \sum_{rca}(U) = \{$regular countably additive measures defined in the Borel field of $U\}$ (the $\sigma$-field generated by the closed sets of $U$), $\mathbf{U} = \prod_{rca}(U) = \{$probability measures in $\sum_{rca}(U)\}$.

**Class 2.** $U$ = normal topological space, $X = BC(U) = \{$bounded continuous functions in $U$ with supremum norm$\}$, $X^* = \sum_{rba}(U) = \{$regular finitely additive measures defined in the field generated by the closed sets of $U\}$, $\mathbf{U} = \prod_{rba}(U) = \{$probability measures in $\sum_{rba}(U)\}$.

**Class 3.** $U$ = arbitrary set, $X = B(U) = \{$bounded functions in $U$ with supremum norm$\}$, $X^* = \sum_{ba}(U) = \{$finitely additive measures defined in the field of all subsets of $U\}$, $\mathbf{U} = \prod_{ba}(U) = \{$probability measures in $\sum_{ba}(U)\}$.

If $D$ is the set of all Dirac measures $\delta(\cdot - u)du$, we have $\overline{\text{conv}}(D) = \mathbf{U}$ in all three cases (see [6]); moreover, $D$ is the set of extremal points of $\mathbf{U}$. We note that the control space $\mathfrak{C}_{ad}(0, T; \mathbf{U})$ is $L^1(0, T; E)$-weakly compact in $L^\infty_w(0, T; X^*)$ [5], [6]; this is the basic ingredient in existence theorems for optimal control problems [5].

The control system

$$y'(t) = Ay(t) + f(t, y(t), u(t)), \qquad u(t) \in U \tag{7.1}$$

where $A$ is the infinitesimal generator of a strongly continuous semigroup $S(t)$ and $U$ is an arbitrary set is fitted with one of the three classes of relaxed control depending on the available properties of the control set $U$. To fix ideas, assume $U$ is a normal topological space; then we can use class 2. The relaxed system is

$$y'(t) = Ay(t) + F(t, y(t))\mu(t) \tag{7.2}$$

where we call $\mu(\cdot)$ the elements of $\mathfrak{C}_{ad}(0, T; \mathbf{U})$; $F(t, y)$ is defined as the unique element of $E$ satisfying

$$\langle y^*, F(t, y)\mu\rangle = \int_U \langle y^*, f(t, y, u)\rangle \mu(du).$$

To give sense to this definition we assume that $f(t, y, u)$ satisfies:

$f(t, y, \cdot) \in BC(U)$ for $t, y$ fixed and, for every $y^* \in E^*$ the function $t \to \langle y^*, f(t, y, \cdot)\rangle$ is a strongly measurable $BC(U)$-valued function. Moreover, (a) for every $c > 0$ there exists $\alpha(\cdot, c) \in L^1(0, T)$ such that

$$\|f(t, y, u)\| \le \alpha(t, c) \qquad (0 \le t \le T,\ \|y\| \le c,\ u \in U). \tag{7.3}$$

(b) For every $c > 0$ there exists $\beta(\cdot, c) \in L^1(0,T)$ such that

$$\|f(t,y',u) - f(t,y,u)\| \le \beta(t,c) \qquad (7.4)$$
$$(0 \le t \le T,\ \|y\|,\ \|y'\| \le c,\ u \in U).$$

Under these conditions, we check easily that $F(t,y) \in L(\sum_{rba}(U), E)$ satisfies assumption $(I_2)$ in §2; note that $F(t,y)^* \in L(E^*, BC(U))$ is given by $F(t,y)^*y^* = \langle y^*, f(t,y,\cdot)\rangle$. The additional assumption in Lemma 6.3, Lemma 6.4 and Theorem 6.5 is: the function $t \to f(t,y,\cdot)$ is a strongly measurable $BC(U;E)$-valued function, where $BC(U;E)$ is the space of all bounded continuous $E$-valued functions with the supremum norm.

If $\mu(\cdot)$ takes values in $D$ (that is, if $\mu(t) = \delta(\cdot - u(t))$ with $u(t) \in U$) then $F(t,y)\mu(t) = f(t,y,u(t))$ and the relaxed system (7.2) reduces to the original system (7.1). Thus, the extremal trajectories in Theorem 5.2 and 6.5 are correspond to the original system (7.1). Both theorems then state that trajectories of the relaxed system (7.2) can be approximated by trajectories of (7.1).

§8. **A semilinear hyperbolic system.** Consider the controlled wave equation

$$y_{tt}(t,x) = \sum_{j=1}^{m}\sum_{k=1}^{m} \partial^j(a_{jk}(x)\partial^k y(t,x)) - \phi(y(t,x)) + u(t,x) \qquad (8.1)$$

in a bounded domain $\Omega$ of class $C^{(1)}$ with boundary $\Gamma$ in $m$-dimensional Euclidean space $\mathbf{R}^m$; $x = (x_1, \ldots, x_m)$, $a_{jk}(x) = a_{kj}(x)$ and $A = \Sigma\Sigma\,\partial^j(a_{jk}(x))\partial^k y$ is uniformly elliptic. The equation is combined with a boundary condition $\beta$ on $\Gamma$, either of Dirichlet type $y = 0$ or of variational type $\partial_\nu y = \alpha(x)y$, where $\partial_\nu$ denotes the derivative in the direction of the **conormal** vector $(\nu_1(x), \ldots, \nu_m(x))$, with components $\nu_j(x) = \Sigma a_{jk}(x)\eta_k(x)$, $(\eta_1(x), \ldots, \eta_m(x))$ the outer normal vector on $\Gamma$. We take $X = L^1(\Omega)$, so that $X^* = L^\infty(\Omega)$; the control set $\mathbf{U}$ is the unit ball of $L^\infty(\Omega)$. Accordingly, $\mathfrak{C}_{ad}(0,T;\mathbf{U})$ is the unit ball of $L^\infty_w(0,T;L^\infty(\Omega))$, that is, the unit ball of $L^\infty((0,T)\times\Omega)$; admissible controls are measurable functions $u(t,x)$ satisfying

$$|u(t,x)| \le 1 \quad \text{a.e. in} \quad (t,x) \in (0,T)\times\Omega\,. \qquad (8.2)$$

Assuming for simplicity that $\beta$ is the Dirichlet boundary condition we reduce (8.1) to a system of the form (1.1) for a 2-dimensional vector function $\mathbf{y} = (y, y_t) = (y, y_1)$,

$$\mathbf{y}'(t) = \mathbf{A}(\beta)\mathbf{y}(t) + \mathbf{f}(\mathbf{y}(t)) + \mathbf{F}u(t) \qquad (8.3)$$

in the space $E = H_0^1(\Omega) \times L^2(\Omega)$, where

$$\mathbf{A}(\beta) = \begin{bmatrix} 0 & I \\ \sum\sum \partial^j(a_{jk}(x)\partial^k) & 0 \end{bmatrix},$$
$$\mathbf{f}((y,y_1)) = \begin{bmatrix} 0 \\ \phi(y) \end{bmatrix}, \qquad \mathbf{F}u = \begin{bmatrix} 0 \\ u \end{bmatrix}.$$

If the $a_{jk}(x)$ are continuously differentiable in $\overline{\Omega}$, the operator $\mathbf{A}(\beta)$ generates a strongly continuous group $\mathbf{S}(\cdot)$ in $E = H_0^1(\Omega)\times L^2(\Omega)$ (see [6] for a precise description of the domain $D(\mathbf{A}(\beta))$ and more details). The operator $\mathbf{F}$ maps $X^* = L^\infty(\Omega)$ into $\mathbf{E}$; the adjoint is $\mathbf{F}^*\mathbf{y} = F^*((y, y_1)) = y_1$, a bounded operator from $\mathbf{E}$ into $L^2(\Omega)$, thus into $L^1(\Omega)$. It follows that $\mathbf{F}$ satisfies Assumption $(I_2)$; moreover, since $\mathbf{F}$ is independent of time, it also satisfies automatically the assumption on strong $L(X^*, E)$-measurability needed in Theorem 6.5.

The assumptions on $\phi$ depend on the dimension $m$ of the space (see [6]):

**Dimension.** $m > 2$. Let $\alpha = m/(m-2)$. Then

$$|\phi(y)| \le C(1 + |y|^\alpha) \qquad (y \in \mathbf{R}, u \in U). \tag{8.4}$$

$$|\phi(y') - \phi(y)| \le K(1 + |y|^{\alpha-1} + |y'|^{\alpha-1})|y' - y|. \tag{8.5}$$

**Dimension.** $m = 2$. There exists some $\alpha > 0$ such that (8.4) and (8.5) hold.

**Dimension.** $m = 1$. $\phi$ is Lipschitz continuous.

The set $D$ of extremal points of the unit ball of $X^* = L^\infty(\Omega)$ consists of the functions $u(x)$ with $|u(x)| = 1$ a.e. Accordingly we obtain directly from Theorem 6.5:

**Theorem 8.1.** *Let $y(t, \cdot, u)$ be a trajectory of* (8.1) *in $0 \le t \le \bar{t}$ corresponding to a control $u(t, x)$ satisfying* (8.2). *Then, given $\varepsilon > 0$ there exists a control $\nu(t, x)$ with*

$$|\nu(t,x)| = 1 \quad a.e.\ in \quad (t,x) \in (0,\bar{t}) \times \Omega,$$
$$\|y(t,\cdot,u) - y(t,\cdot,\nu)\|^2_{H^1(\Omega)} + \|y_t(t,\cdot,u) - y_t(t,\cdot,\nu\|^2_{L^2(\Omega)} \le \varepsilon^2 \quad (0 \le t \le \bar{t}),$$

*where $y(t, \cdot, \nu)$ is the trajectory of* (8.1) *corresponding to $\nu$ with the same initial conditions.*

**§9. The $^\odot$-theory.** Let $E$ be a Banach space, $S(t)$ a strongly continuous semigroup in $E$. Denote by $E^\odot \subseteq E^*$ the closure of the domain of $D(A^*)$ in $E^*$, and by $S^\odot(t)$ the **Phillips adjoint** of $S(t)$, that is, the restriction of $S(t)^*$ to $E^\odot$. The subspace $E^\odot$ is **determining** in $E^*$, that is, the norm $\|y\|_0 = \sup|\langle y^*, y\rangle|$ $(y^* \in E^\odot, \|y^*\| \le 1)$ in $E$ is equivalent to the original norm $\|y\|$ of $E$. It follows that the canonical pairing of $E^\odot$ and $E$ produces a bicontinuous linear imbedding of $E$ into $(E^\odot)^*$. In fact, the imbedding is into $(E^\odot)^\odot = E^{\odot\odot}$ and we have $A \subseteq (A^\odot)^\odot = A^{\odot\odot}$ [7 p. 430]. We assume in this section that $S(t)E \subseteq D(A)$ and $AS(t)$ is continuous in the operator norm in $t > 0$, that $E$ and $E^\odot$ are separable and that $E$ is $^\odot$**-reflexive** with respect to $S(\cdot)$, that is,

$$E^{\odot\odot} = E. \tag{9.1}$$

This implies $S^{\odot\odot}(t) = S(t), A^{\odot\odot} = A$. Also,

$$S(t)^* E^* \subseteq E^{\odot}, \qquad S(t)^*|_{E^{\odot}} = S^{\odot}(t). \tag{9.2}$$

$$S^{\odot}(t)^*(E^{\odot})^* \subseteq E \qquad S^{\odot}(t)^*|_E = S(t). \tag{9.3}$$

In fact, since $S(t)E \subseteq D(A)$ and $AS(t)$ is bounded for $t > 0$ we obtain taking adjoints and using commutativity of $A$ and $S(t)$ that $S(t)^*E^* \subseteq D(A^*) \subseteq E^{\odot}$, which is the first relation (9.2); the second is just the definition of $S^{\odot}(t)$. The same argument, this time applied in $E^{\odot}$ to the semigroup $S^{\odot}(\cdot)$ proves that $S^{\odot}(t)^*(E^{\odot})^* \subseteq E^{\odot\odot} = E$. By definition of $S^{\odot\odot}(t)$, we have $S^{\odot}(t)^*|_E = S^{\odot\odot}(t)$, and by $^{\odot}$-reflexivity, the second relation (9.3) follows. We note that (9.2) and (9.3) imply

$$S(t+\varepsilon)^* = S(t)^{\odot}S(\varepsilon)^*, \; S(t+\varepsilon) = S(t)S^{\odot}(\varepsilon)^* \tag{9.4}$$

for $t \geq 0$, $\varepsilon > 0$. In particular, (9.4) implies that $S(t)^*$ and $S^{\odot}(t)^*$ are continuous in the operator norm in $t > 0$.

This sort of abstract setup is justified by one important example, elliptic differential operators $A(\beta)$, where $\beta$ is a boundary condition of Dirichlet or variational type. For a variational boundary condition we use the space $E = C(\overline{\Omega})$ of continuous functions in the closure $\overline{\Omega}$ of a domain $\Omega$ in Euclidean space; $E^*$ is the space $\Sigma(\overline{\Omega})$ of regular Borel measures in $\overline{\Omega}$ with the total variation norm and $E^{\odot} = L^1(\Omega)$, $(E^{\odot})^* = L^{\infty}(\Omega)$. If $\beta$ is the Dirichlet boundary condition then $C(\overline{\Omega})$ is replaced by the subspace $C_0(\overline{\Omega})$ defined by $u = 0$ on $\Gamma$, $\Sigma(\overline{\Omega})$ by the subspace $\Sigma_0(\overline{\Omega})$ consisting of measures vanishing on $\Gamma$. The same scheme applies to operators in $L^1(\Omega)$; here $L^1(\Omega)^* = L^{\infty}(\Omega)$, $L^1(\Omega)^{\odot} = C(\overline{\Omega})$ ($C_0(\overline{\Omega})$ for the Dirichlet boundary condition). For further details on these spaces and operators, and on the Dirichlet boundary condition see [6, Section §7]; we note that the semigroups $S(t)^*$ and $S^{\odot}(t)^*$ are not strongly continuous at $t = 0$.

We consider the system (2.1) under hypothesis $(I^{\odot})$ below, consisting again of two parts. The hypotheses on $f(t, y)$ are the same as in $(I_1)$. For $F(t, y)$, the only difference is that $F : [0, T] \times E \to L(X^*, (E^{\odot})^*)$, $F(t, y)^*E^{\odot} \subseteq X$ and for each $y^* \in E^{\odot} F(t, y)^*y^*$ is strongly measurable. Assumptions (a) and (b) are the same. We have $\langle y^*, F(t, y(t))u(t)\rangle = \langle F(t, y(t))^*y^*, u(t)\rangle$ so that $t \to F(t, y(t))u(t)$ is merely $E^{\odot}$-weakly measurable. This does not imply strong measurability, but since $E^{\odot}$ is separable, if $g(\cdot)$ is an arbitrary $E^{\odot}$-weakly measurable $(E^{\odot})^*$-valued function then $\|g(\cdot)\|$ is measurable. This gives sense to the definition of such spaces as $L^1_w(0, T; (E^{\odot})^*)$. The corresponding integral equation is

$$\begin{aligned} y(t) = S(t)\zeta &+ \int_0^t S(t-\sigma)f(\sigma, y(\sigma))d\sigma \\ &+ \int_0^t S^{\odot}(t-\sigma)^*F(\sigma, y(\sigma))u(\sigma)d\sigma \end{aligned} \tag{9.5}$$

if $\zeta \in E$; by (9.3), the integrand takes values in $E$. We may also take the initial condition $\zeta$ in $(E^{\odot})^*$, in which case $S(t)\zeta$ is replaced by $S^{\odot}(t)^*\zeta$. Equation (9.5) is slightly nonstandard, since solutions take values in $E^{\odot\odot} = E$ but the right hand side $F(t, y(t))\mu(t)$ takes values in $(E^{\odot})^* \supseteq E^{\odot\odot} = E$. It is interpreted using

**Theorem 9.1.** (a) *Let $g(\cdot)$ be a $E^{\odot}$-weakly measurable $(E^{\odot})^*$-valued function defined in $0 \le t \le T$. Then*

$$\sigma \to S^{\odot}(t-\sigma)^* g(\sigma)$$

*is strongly measurable in $0 \le \sigma \le t$.* (b) *The operator*

$$\Lambda g(t) = \int_0^t S^{\odot}(t-\sigma)^* g(\sigma) d\sigma$$

*is bounded from $L^1_w(0, T; (E^{\odot})^*)$ into $C(0, T; E)$. If $S(t)$ is compact for $t > 0$ and $\{g_\kappa(\cdot); \kappa \in K\}$ is a generalized sequence in $L^1_w(0, T; (E^{\odot})^*)$ such that the integrals of the $\|g_\kappa(\cdot)\|$ are equicontinuous in $0 \le t \le T$, then $\{\Lambda g_\kappa(\cdot)\}$ has a generalized subsequence convergent in $C(0, T; E)$.*

**Proof.** For a proof of (a) see [6, Lemma 6.1]. The proof of (b) is exactly the same as that of Theorem 3.1 and is omitted.

**Lemma 9.2.** *Assume that $S(t)$ is compact for $t > 0$ and that Hypothesis $(I^{\odot})$ holds. Let $u(\cdot) \in L^\infty_w(0, T; X^*)$ be such that the solution $y(\cdot, u)$ of (2.7) exists in $0 \le t \le \bar{t}$, and let $\{u_\kappa(\cdot); \kappa \in K\}$ be a bounded generalized sequence in $L^\infty_w(0, \bar{t}; X^*)$ such that $u_\kappa(\cdot) \to u(\cdot)$ $L^1(0, \bar{t}; X)$-weakly. Then there exists $\kappa_0 \in K$ such that $y(\cdot, u_\kappa)$ exists in $0 \le t \le \bar{t}$ if $\kappa \ge \kappa_0$ ("$\ge$" the ordering in $K$) and*

$$y(t, u_\kappa) \to y(t, u)$$

*uniformly in $0 \le t \le \bar{t}$.*

The proof is essentially the same as that of Lemma 5.1. The only difference is that in (5.2) we pick $y^* \in E^{\odot}$, taking advantage of the fact that $E^{\odot}$ is determining.

**Theorem 9.3.** *Assume that $\mathbf{U}$ is $X$-weakly compact, that $D$ is total in $\mathbf{U}$, that $S(t)$ is compact and that Hypothesis $(I^{\odot})$ holds. Let $u(\cdot) \in \mathfrak{C}_{ad}(0, \bar{t}; \mathbf{U})$ be such that the solution $y(\cdot, u)$ of (1.1) exists in $0 \le t \le t$. Then there exists $\nu(\cdot) \in PC(0, T; D)$ such that $y(t, \nu)$ exists in $0 \le t \le \bar{t}$ and*

$$\|y(t, u) - y(t, \nu)\| \le \varepsilon \qquad (0 \le t \le \bar{t}).$$

The proof results combining Lemma 9.2 with Theorem 4.1. The results are applied to relaxation as follows. The system (7.1) is relaxed to (7.2), but $F(t,y)\mu$ is the unique element of $(E^{\odot})^*$ that satisfies

$$\langle y^*, F(t,y,\mu)\rangle = \int_U \langle y^*, f(t,y,u)\rangle \mu(du)$$

for every $y^* \in E^{\odot}$. We check easily that $F(t,y)$ satisfies Assumption $(I^{\odot})$. Theorem 9.3 then shows that trajectories of the relaxed system (7.2) can be approximated by trajectories of (7.1) (see the end of §7).

**§10. A semilinear parabolic system.** We consider a controlled parabolic equation

$$y_t(t,x) = \sum_{j=1}^{m}\sum_{k=1}^{m} \partial^j(a_{jk}(x)\partial^k y(t,x)) + f(y(t,x)) + u(t,x) \tag{10.1}$$

in a bounded domain $\Omega$ of class $C^{(2)}$ with boundary $\Gamma$ in $m$-dimensional Euclidean space $\mathbf{R}^m$. Notations are the same in §8; $a_{jk}(x) = a_{kj}(x)$, $A = \Sigma\Sigma\partial^j(a_{jk}(x)\partial^k y$ is uniformly elliptic and the coefficients are twice continuously differentiable in $\overline{\Omega}$. The equation is combined with a boundary condition $\beta$ on $\Gamma$. The generation properties of the operator $A(\beta)$ have been described in §9. Assuming that $\beta$ is a variational boundary condition we apply the $^{\odot}$-theory in §9 with $E = C(\overline{\Omega})$; we have $E^{\odot} = L^1(\Omega)$, $(E^{\odot})^* = L^1(\Omega)^* = L^\infty(\Omega)$. We take $X = L^1(\Omega)$, so that $X^* = L^\infty(\Omega)$; the control set $\mathbf{U}$ is the unit ball of $X^*$, so that (see §8) admissible controls are measurable functions $u(t,x)$ satisfying

$$|u(t,x)| \le 1 \quad \text{a.e. in} \quad (t,x) \in (0,T)\times\Omega\,. \tag{10.2}$$

In abstract form, the equation is

$$y'(t) = A(\beta)y(t) + f(y(t)) + Fu(t) \tag{10.3}$$

where $F : L^\infty(\Omega) \to L^\infty(\Omega)$ is the identity operator. Assumption $(I_1)$ for the nonlinear term is satisfied if $f(y)$ is Lipschitz continuous; that $F$ satisfies $(I_2)$ is obvious. We apply this time Theorem 5.2:

**Theorem 10.1.** *Let $y(t,\cdot,u)$ be a trajectory of* (10.1) *in $0 \le t \le \bar t$ corresponding to a control $u(t,x)$ satisfying* (10.2). *Then, given $\varepsilon > 0$ there exists a control $\nu(t,x)$ with*

$$|\nu(t,x)| = 1 \quad a.e. \quad in \quad (0,\bar t)\times\Omega\,,$$
$$\|y(t,\cdot,u) - y(t,\cdot,\nu)\|_{C(\bar\Omega)} \le \varepsilon \qquad (0 \le t \le \bar t)\,,$$

*where* $y(t,\cdot,\nu)$ *is the trajectory of* (10.1) *corresponding to* $\nu$ *with the same initial condition.*

For the Dirichlet boundary condition, we replace $C(\overline{\Omega})$ by $C_0(\overline{\Omega})$.

**References**


[1] N. DUNFORD and J. T. SCHWARTZ, **Linear operators**, part 1, Interscience, New York 1958.

[2] H. O. FATTORINI, A remark on the "bang-bang" principle for linear control systems in infinite dimensional spaces, SIAM J. Control 6 (1968) 109-113.

[3] H. O. FATTORINI, The time optimal control problem in Banach spaces, Appl. Math. Optimization 1 (1974/75) 163-188.

[4] H. O. FATTORINI, Relaxed controls in infinite dimensional systems, Int. Series Numer. Math. vol. 100, Birkhäuser, Basel (1991), 115-128.

[5] H. O. FATTORINI, Existence theory and the maximum principle for relaxed infinite dimensional optimal control problems, to appear in SIAM J. Control Optimization.

[6] H. O. FATTORINI, Relaxation theorems, differential inclusions and Filippov's theorem for relaxed controls in semilinear infinite dimensional systems, to appear in J. Differential Equations.

[7] E. HILLE and R. S. PHILLIPS, **Functional Analysis and Semi-Groups**, Amer. Math. Soc., Providence, 1957.

[8] A. IONESCU TULCEA and C. IONESCU TULCEA, **Topics in the Theory of Lifting**, Springer, Berlin, 1969.

[9] J. WARGA, **Optimal control of differential and functional equations**, Academic Press, New York, 1971.

[10] L. C. YOUNG, **Lectures on the Calculus of Variations and Optimal Control Theory**, W. B. Saunders, Philadelphia 1969.

# 35 An Algebraic Approach to Hankel Norm Approximation Problems

**P. A. Fuhrmann*** Ben-Gurion University of the Negev, Beer Sheva, Israel

**Abstract**

The polynomial approach introduced in Fuhrmann [1991] is extended to cover the crucial area of AAK theory, namely the characterization of zero location of the Schmidt vectors of the Hankel operators. This is done using the duality theory developed in that paper but with a twist. First we get the standard, lower bound, estimates on the number of unstable zeroes of the minimal degree Schmidt vectors of the Hankel operator. In the case of the Schmidt vector corresponding to the smallest singular the lower bound is in fact achieved. This leads to a solution of a Bezout equation. We use this Bezout equation to introduce another Hankel operator which has singular values that are the inverse of the singular values of the original Hankel operator. Moreover the singular vectors are closely related to the original singular vectors. The lower bound estimates on the number of antistable zeroes of the new singular vectors lead to an upper bound estimate on the number of antistable zeroes of the original singular vectors. These two estimates turn out to be tight and give the correct number of antistable zeroes. From here the standard results on Hankel norm approximation and Nehari complementation follow easily.

*Earl Katz Family Chair in Algebraic System Theory

Partially supported by the Israeli Academy of Sciences under Grant No. 249/50

# 1 INTRODUCTION

In Fuhrmann [1991] a polynomial approach to AAK theory, see Adamjan, Arov and Krein [1968a,1968b,1971,1978], was given. While many, one might dare to say even interesting, results were given in that paper, there was also a fundamental underlying weakness. At a crucial point the paper used the original AAK results concerning the number of antistable zeroes of the minimal degree singular vectors of the given Hamkel operator. Thus that paper was not self contained. In point of fact, the determination of the zero location of Hankel singular values seems to be the bottleneck of AAK theory. No simple method was found to this problem, and even efforts at making a reasonably elementary exposition of AAK theory, e.g. Young [1988] and Partington [1988], have failed in this respect. Even the case of singular vectors corresponding to the largest singular vector, the only relatively easy case in AAK theory, which is disposed in a one line proof, is not really elementary inasmuch as it uses inner/outer factorizations.

The object of this paper is to address itself to this problem and to provide an elementary solution. It should be clarified at the outset that the context in which the problem is solved is that of rational functions. In fact the proof uses rationality in a crucial way. Thus the method, at least as presented in this paper, is not as general as others. However it has the advantage of simplicity. It can be truly said that this method brings AAK theory to a level that can be safely presented to the undergraduate student.

To achieve our goal we redevelop the theory with a little twist. The twist in our approach is that we focus first on the zeroes of the singular vector corresponding to the smallest singular vector. This turns out to be rather trivial to figure out. Once this is achieved a natural Bezout equation presents itself and leads to a related, one should really say a dual, Hankel operator. For both Hankel operators we have lower bound estimates on the number of antistable zeroes of the corresponding singular vectors. However the dual estimates translate into upper bound estimates of the original singular vectors. Moreover the estimates are tight, i.e. they determine the number of antistable zeroes of the minimal degree singular values. From this point the results on Hankel norm approximation follow as in Fuhrmann [1991].

A natural question presents itself. How was this approach been overlooked so long. It seems that the explanation lies in the tremendous authority of M.G. Krein. Once he put the limelight on the largest singular value, everything else remained in the dark.

In writing this paper a basic decision had to be made. Most of the development presented in the current approach is based on the results in Fuhrmann [1991]. It could be presented via a long list of pointers to the relevant parts of that paper. That would mean a short, but also unreadable, presentation. The other alternative, the one eventually adopted, was to make this paper self contained. This means that there is substantial duplication of results, but the order of the development is different.

The paper is structured as follows. In section 2 we collect basic information on Hankel operators, invariant subspaces and their representation via Beurling's theorem. Next we introduce model intertwining operators. We do this using the frequency domain representation of the right translation semigroup. We study the basic properties of intertwining maps and in particular their invertibility properties. The important point here is the connection of invertibility to the solvability of an $H_+^\infty$ Bezout equation. We follow this by defining Hankel operators. For the case of a rational, antistable function we give specific, Beurling type, representations for the cokernel and the image of the corresponding Hankel operator. Of importance is the connection between Hankel operators and intertwining maps. This connection, coupled with invertibility properties of intertwining maps are the key to duality theory.

In section 3 we do a detailed analysis of Schmidt pairs of a Hankel operator with scalar, rational symbol. Some important lemmas, due to Adamjan, Arov and Krein [1971], are rederived in this setting from an algebraic point of view. These lemmas lead to a polynomial formulation of the singular value singular vector equation of the Hankel operator. This equation, we refer to it as the Fundamental Polynomial Equation, is easily reduced, using the theory of polynomial models, to a standard eigenvalue problem.

Using nothing more than the polynomial division algorithm, the subspace of all singular vectors corresponding to a given singular value, is parametrized via the minimal degree solution of the FPE. We obtain a connection between the minimal degree solution and the multiplicity of the singular value.

The FPE can be transformed, using a simple algebraic manipulation to a form that leads immediately to lower bound estimates on the the number of antistable zeroes of $p_k$, the minimal degree solution corresponding to the $k$-th Hankel singular value. This lower bound is shown to actually coincide with the degree of the minimal degree solution for the special case of the smallest singular value. Thus this polynomial turns out to be antistable. Another algebraic manipulation of the FPE leads to a Bezout equation over $H_+^\infty$. This provides the key to duality.

Section 4 has duality theory is its main theme. Using the previously obtained Bezout equation, we invert the intertwining map corresponding to the initial Hankel operator. The inverse intertwining map is related to a new Hankel operator which has inverse singular values to those of the original one. Moreover we can compute the Schmidt pairs corresponding to this Hankel operator in terms of the original Schmidt pairs.

Section 5 applies the previous information. The same estimates on the number of antistable zeroes of the minimum degree solutions of the FPE that were obtained for the original Hankel operator Schmidt vectors are applied now to the new Hankel operator Schmidt vectors. Thus we obtain a second set of inequalities. The two sets of inequalities, taken together, lead to precise information on the number of antistable zeroes of the minimal degree solutions corresponding to all singular values. Utilizing this information leads to the solution of the Nehari problem as well as that of the general Hankel norm approximation problem.

It is fitting that the new insight into this problem came while preparing for a seminar at the Department of Mathematics of the University Kaiserslautern, where much of the research on the previous paper has been done. For providing this intellectually stimulating atmosphere I would like to thank D. Prätzel-Wolters. This particular piece of research was done while working on a large joint project with R. Ober. It is a pleasure to acknowledge his creative criticism and the endless conversations that no doubt helped in getting this work done.

# 2 PRELIMINARIES

Hankel operators are generally defined in the time domain and via the Fourier transform their frequency domain representation is obtained. We will skip this part and introduce Hankel operators directly as frequency domain objects. Our choice is to develop the theory of continuous time systems. This means that the relevant frequency domain spaces are the Hardy spaces of the left and right half planes. Thus we will study Hankel operators defined on half plane Hardy spaces rather than on those of the unit disc as was done by Adamjan, Arov and Krein [1971]. In this we follow the choice of Glover [1984]. This choice seems to be a very convenient ones as all results on duality simplify significantly, due to the greater symmetry between the two half planes in comparison to the unit disc and its exterior.

## 2.1 HARDY SPACES

Our setting will be that of Hardy spaces. Thus $H^2_+$ is the Hilbert space of all analytic functions in the open right half plane with

$$||f||^2 = \sup_{x>0} \frac{1}{\pi} \int_{-\infty}^{\infty} |f(x+iy)|^2 dy.$$

The space $H^2_-$ is similarly defined in the open left half plane. It is a theorem of Fatou that guarrantees the existence of boundary values of $H^2_\pm$-functions on the imaginary axis. Thus the spaces $H^2_\pm$ can be considered as closed subspaces of $L^2(i\mathbf{R})$, the space of Lebesgue square integrable functions on the imaginary axis. It follows from the Fourier-Plancherel and Paley-Wiener theorems that

$$L^2(i\mathbf{R}) = H^2_+ \oplus H^2_-,$$

with $H^2_+$ and $H^2_-$ the Fourier-Plancherel transforms of $L^2(0,\infty)$ and $L^2(-\infty,0)$ respectively. Also $H^\infty_+$ and $H^\infty_-$ will denote the spaces of bounded analytic functions on the open right and left half planes respectively. These spaces can be considered as subspaces of $L^\infty(i\mathbf{R})$, the space of Lebesgue measurable and essentially bounded functions on the imaginary axis. An extensive discussion of these spaces can be found in Hoffman [1962], Duren [1970] and Garnett [1981].

We will define $f^*(s) = f(-\bar{s})^*$.

## 2.2 INVARIANT SUBSPACES

Before the introduction of Hankel operators we digress a bit on invariant subspaces of $H^2_+$. Since we are using the half planes for our definition of the $H^2$ spaces, we do not have the shift operators coveniently at our disposal. This forces us to a slight departure from the usual convention.

The algebra $H_+^\infty$ can be made an algebra of operators on $H_+^2$ by letting, for $\psi \in H^\infty$, induce a map $T_\psi : H_+^2 \longrightarrow H_+^2$ which is defined by

$$T_\psi f = \psi f, \qquad f \in H_+^2. \tag{1}$$

The next proposition characterizes the adjoints of this class of operators.

**Proposition 2.1** *Let $\psi \in H^\infty$ and $T_\psi$ be defined by (1). The adjoint of $T_\psi$ is given by*

$$T_\psi^* f = P_+ \psi^* f, \qquad f \in H_+^2.$$

Both $T_\psi$ and $T_\psi^*$ are special cases of Toeplitz operators.

**Definition 2.1**

- *A subspace $M \subset of H_+^2$ is called an* **invariant subspace** *if, for each $\psi \in H_+^\infty$ we have*

$$T_\psi M \subset M.$$

- *A subspace $M \subset of H_+^2$ is callaed a* **backward invariant subspace** *if, for each $\psi \in H_+^\infty$ we have*

$$T_\psi^* M \subset M.$$

Clearly backward invariant subspaces are just orthogonal complements of invariant subspaces.

Invariant subspaces have been characterized by Beurling [1949]. For this we need the notion of an inner function.

**Definition 2.2** *A function $m \in H_+^\infty$ is called* **inner** *if $\|m\|_\infty \leq 1$ and its boundary values on the imaginary axis are unitary a.e.*

Thus on the imaginary axis we have $m^* m = 1$. The next result, Beurling's theorem, is central. We quote it to put some results in the right perspective. We do not actually use it as in our setup we can directly calculate the relevant invariant subspaces and identify the corresponding inner functions. Thus we will not give a prooof of this theorem.

**Theorem 2.1 (Beurling)** *A nontrivial subspace $M \subset H_+^2$ is an invariant subspace if and only if*

$$M = m H_+^2$$

*for some inner function $m$.*

## 2.3 MODEL OPERATORS AND INTERTWINING MAPS

Given an inner function $m \in H_+^\infty$ we consider the left invariant subspace

$$H(m) = \{mH_+^2\}^\perp = H_+^2 \ominus mH_+^2.$$

The algebra $H_+^\infty$, or equivalently the algebra of analytic Toeplitz operators, induces an algebra of bounded operators in $\{mH_+^2\}^\perp$. Thus for $\Theta \in H_+^\infty$ the maps $T_\Theta : H(m) \longrightarrow H(m)$ are defined by

$$T_\Theta f = P_{H(m)}\Theta f, \quad \textit{for} \;\; f \in H(m). \tag{2}$$

Clearly, if $\Theta \in H_+^\infty$, we have $||T_\Theta|| \le ||\Theta||_\infty$.

We note that, for $\tau \le 0$, the functions $exp_\tau(s) = e^{-\tau s}$ are all in $H_+^\infty$. The operators $T_{exp_\tau}$ form a strongly continuous semigroup of operators on $\{mH_+^2\}^\perp$. The following is a continuous time version of the Sarason [1968] commutant lifting theorem.

**Theorem 2.2** *A bounded operator $X$ on $\{mH_+^2\}^\perp$ satisfies*

$$XT_{exp_\tau} = T_{exp_\tau}X$$

*for all $\tau \le 0$ if and only if there exists a $\Theta \in H_+^\infty$ such that*

$$X = T_\Theta.$$

The next theorem sums up duality properties of operators commuting with shifts.

**Theorem 2.3** *Let $\Theta, m \in H_+^\infty$ with $m$ an inner function, and let $T_\Theta$ be defined by*

$$T_\Theta f := P_{H(m)}\Theta f, \quad \textit{for} \;\; f \in H(m).$$

*Then*

1. *Its adjoint $T_\Theta^*$ is given by*

$$T_\Theta^* f = P_+\Theta^* f, \quad \textit{for} \;\; f \in H(m).$$

2. *The operator $\tau_m : H(m) \longrightarrow H(m)$ defined by*

$$\tau_m f := mf^*$$

   *is unitary.*

3. *The operators $T_{\Theta^*}$ and $T_\Theta^*$ are unitarily equivalent. More specifically we have*

$$T_\Theta \tau_m = \tau_m T_\Theta^*$$

Proof:

1. Let $f, g \in H(m)$. Then
$$\begin{aligned}(T_\Theta f, g) &= (P_{H(m)}\Theta f, g) = (mP_- m^*\Theta f, g)\\ &= (P_- m^*\Theta f, m^* g) = (m^*\Theta f, P_- m^* g)\\ &= (m^*\Theta f, m^* g) = (\Theta f, g) = (f, \Theta^* g)\\ &= (P_+ f, \Theta^* g) = (f, P_+\Theta^* g) = (f, T_\Theta^* g).\end{aligned}$$
Here we used the fact that $g \in H(m)$ if and only if $m^* g \in H_-^2$.

2. Clearly the map $\tau_m$, as a map in $L^2$, is unitary. From the orthogonal direct sum decomposition
$$L^2 = H_-^2 \oplus H(m) \oplus mH_+^2$$
it follows, by conjugation, that
$$L^2 = m^* H_-^2 \oplus \{H_-^2 \ominus m^* H_-^2\} \oplus H_+^2.$$
Hence $m\{H_-^2 \ominus m^* H_-^2\} = H(m)$.

3. We compute
$$T_\Theta \tau_m f = T_\Theta m f^* = P_{H(m)}\Theta m f^* = mP_- m^*\Theta m f^* = mP_-\Theta f^*$$
Now
$$\tau_m T_\Theta^* = \tau_m(P_+\Theta^* f) = m(P_+\Theta^* f)^* = mP_-\Theta f^*.$$

∎

The following spectral mapping theorem has been proved in Fuhrmann [1968a]. A vectorial generalization is given in Fuhrmann [1968b]. This will be instrumental in the analysis of Hankel operators restricted to their cokernels.

**Theorem 2.4 (Fuhrmann)** *Let $\Theta, m \in H_+^\infty$ with $m$ an inner function. The following statements are equivalent.*

1. *The operator $T_\Theta$ defined in (2) is invertible.*

2. *There exists a $\delta > 0$ such that*
$$|\Theta(s)| + |m(s)| \geq \delta, \quad \textit{for all } s \textit{ with } \mathrm{Re}\, s > 0. \tag{3}$$

3. *There exist $\xi$, $\eta \in H_+^\infty$ that solve the Bezout equation*
$$\xi\Theta + \eta m = 1. \tag{4}$$

*In this case we have*
$$T_\Theta^{-1} = T_\xi.$$

**Proof:** We will not give a proof which can be found in Fuhrmann [1968a,1981]. We remark only that by the Carleson corona theorem, Carleson [1962], the strong coprimeness condition of (3) is equivalent to the solvability of the Bezout equation (4) over $H^\infty$. ∎

## 2.4 HANKEL OPERATORS

We proceed to define Hankel operators and we do this directly in the frequency domain. Readers interested in the time domain definition and the details of the transformation into frequency domain are refered to Fuhrmann [1981], Glover [1984].

**Definition 2.3** *Given a function* $\phi \in L^\infty(i\mathbf{R})$ *the* Hankel operator $H_\phi : H^2_+ \longrightarrow H^2_-$ *is defined by*

$$H_\phi f = P_-(\phi f), \quad for\ f \in H^2_+. \tag{5}$$

*The adjoint operator* $(H_\phi)^* : H^2_- \longrightarrow H^2_+$ *is given by*

$$(H_\phi)^* f = P_+(\phi^* f), \quad for\ f \in H^2_-.$$

Here $\phi^*(z) = \overline{\phi(-\overline{z})}$.

In the algebraic theory of Hankel operators the kernel and image of a Hankel operator are directly related to the coprime factorization of the symbol over the ring of polynomials. The details can be found for example in Fuhrmann [1983]. In the same way the kernel and image of a large class of Hankel operators are related to a coprime facorization over $H^\infty$. This theme, originating in the work of Douglas, Shapiro and Shields [1971] and that of D. N. Clark, see Helton [1974], is developed extensively in Fuhrmann [1981]. Of course if the symbol of the Hankel operator is rational and in $H^\infty_-$ these two coprime factorizations are easily related.

Thus assume $\phi = \frac{n}{d} \in H^\infty_-$ and $n \wedge d = 1$. So our assumption is that $d$ is antistable. In spite of the slight ambiguity we will write $n = \deg d$. It will always be clear from the context what $n$ means. This leads to

$$\phi = \frac{n}{d} = \frac{n}{d^*}\frac{d^*}{d}.$$

Thus

$$\phi = m^*\eta$$

with

$$\eta = \frac{n}{d^*}\ , \quad m = \frac{d}{d^*}$$

is a coprime factorization in $H^\infty_+$.

The next theorem discusses the functional equation of Hankel operators. It can be shown, quite easily using the commutant lifting theorem of Sz.-Nagy and Foias [1970], that the Hankel operators are the only solutions of this functional equation. For more information one can consult Nikolskii [1985].

**Theorem 2.5** *1. For every* $\psi \in H^\infty_+$ *the Hankel operator* $H_\phi$ *satisfies the functional equation*

$$P_-\psi H_\phi f = H_\phi \psi f, \quad f \in H^2_+.$$

*2. $KerH_\phi$ is an invariant subspace, i.e. for $f \in KerH_\phi$ and $\psi \in H_+^\infty$ we have $\psi f \in KerH_\phi$ .*

It follows from a theorem of Beurling [1949] that $KerH_\phi = mH_+^2$ for some inner function $m \in H_+^\infty$. Since we are dealing with the rational case the next theorem can make this more specific and characterizes the kernel and image of a Hankel operator and also clarifies the connection between them and polynomial and rational models. A closely related derivation can be found in Young [1983] and Lindquist and Picci [1985].

**Theorem 2.6** *Let $\phi = \frac{n}{d} \in H_-^\infty$ and $n \wedge d = 1$ Then*

*1. $KerH_\phi = \frac{d}{d^*}H_+^2$*

*2. $\{KerH_\phi\}^\perp = \{\frac{d}{d^*}H_+^2\}^\perp = X^{d^*}$*

*3. $ImH_\phi = H_-^2 \ominus \frac{d^*}{d}H_-^2 = X^d$*

Proof: $\{KerH_\phi\}^\perp$ contains only rational functions. Let $f = \frac{p}{q} \in \{\frac{d}{d^*}H_+^2\}^\perp$, then $\frac{d^*p}{dq} \in H_-^2$. So $q \mid d^*p$. But, as $p \wedge q = 1$ it follows that $q \mid d^*$, i.e. $d^* = qr$. Hence $f = \frac{rp}{d^*} \in X^{d^*}$.

Conversely, let $\frac{p}{d^*} \in X^{d^*}$ then, $\frac{p}{d^*} = \frac{p}{d}\frac{d}{d^*}$ or $\frac{d^*}{d}\frac{p}{d^*} \in H_-^2$. So we have $\frac{p}{d^*} \in \{\frac{d}{d^*}H_+^2\}^\perp$. ∎

The previous theorem, though of an elementary nature, is central to all further development as it provides the direct link between the infinite dimensional object, namely the Hankel operator, and the well developed theory of polynomial and rational models. This link will be continuously exploited.

There is a very close connection between a wide class of Hankel operators and intertwining maps. This is summarized in the following.

**Theorem 2.7** *A map $H : H_+^2 \longrightarrow H_-^2$ is a Hankel operator with a nontrivial kernel $mH_+^2$, with $m$ inner, if and only if we have*

$$H = m^* X P_{H(m)}$$

*where $X : H(m) \longrightarrow H(m)$ is an intertwining map, i.e. of the form*

$$X = T_\Theta$$

*for some $\Theta \in H_+^\infty$.*

**Proof:** Assume $X = T_\Theta$ with $\Theta \in H_+^\infty$. We define $H : H_+^2 \longrightarrow H_-^2$ by

$$Hf = m^* X P_{H(m)}.$$

Then, since $m$ is inner and $P_{H(m)}$ an orthogonal projection, the operator $H$ is bounded. Moreover

$$\begin{aligned} H exp_\tau f &= m^* X P_{H(m)} exp_\tau f \\ &= m^* P_{H(m)} \Theta P_{H(m)} exp_\tau P_{H(m)} f \\ &= m^* P_{H(m)} \Theta exp_\tau P_{H(m)} f \\ &= m^* P_{H(m)} exp_\tau \Theta P_{H(m)} f \\ &= m^* P_{H(m)} exp_\tau P_{H(m)} \Theta P_{H(m)} f \\ &= m^* m P_- m^* exp_\tau P_{H(m)} \Theta P_{H(m)} f \\ &= P_- m^* exp_\tau P_{H(m)} \Theta P_{H(m)} f \\ &= P_- exp_\tau m^* m P_- m^* \Theta P_{H(m)} f \\ &= P_- exp_\tau m^* X P_{H(m)} f \\ &= P_- exp_\tau H f \end{aligned}$$

Thus $H$ satisfies the functional equation of a Hankel operator.

Conversely, assume $H = H_\phi$ is a Hankel operator with a nontrivial kernel given by $mH_+^2$, for some inner function $m \in H^\infty$. Then we define a map $T : H(m) \longrightarrow H(m)$ by

$$Tf = mHf, \qquad f \in H(m).$$

Then we compute

$$\begin{aligned} T P_{H(m)} exp_\tau f &= m H P_{H(m)} exp_\tau f \\ &= m H exp_\tau f \\ &= m P_- exp_\tau H f \\ &= m P_- m^* m exp_\tau H f \\ &= P_{H(m)} exp_\tau m H f \\ &= P_{H(m)} exp_\tau T f \end{aligned}$$

So $T$ is an intertwining map. ∎

The previous theorem opens the way to prove Nehari's theorem from Sarason's lifting theorem as well as prove Sarason's theorem from Nehari's. This equivalence is known for a long time and can be found in Page [1970], Nikolskii [1985] and Adamjan, Arov and Krein [1968], to cite a few references.

Hankel operators in general and those with rational symbol in particular are never invertible. Still we may want to invert the Hankel operator as a map from its cokernel, i.e. the orthogonal complement of its kernel, to its image. We saw that such a restriction of a Hankel operator is of considerable interest because of its connection to intertwining maps of model operators. Now theorem 2.4 gave a full characterization of invertibility properties of intertwining maps. These can be applied now to the inversion of the restricted Hankel operators. This will turn out to be of great importance in the development of duality theory.

# 3 SCHMIDT PAIRS OF RATIONAL HANKEL OPERATORS

It is quite well known, see Gohberg and Krein [1969], that singular values of operators are closely related to the problem of best approximation by operators of finite rank. That this basic method could be applied to the approximation of Hankel operators by Hankel operators of lower ranks through the detailed analysis of singular values and the corresponding Schmidt pairs is a fundamental contribution of Adamjan, Arov and Krein [1971].

We recall that, given a bounded operator $A$ on a Hilbert space, $\mu$ is a *singular value* of $A$ if there exists a nonzero vector $f$ such that

$$A^*Af = \mu^2 f.$$

Rather than solve the previous equation we let $g = \frac{1}{\mu}Af$ and go over to the equivalent system

$$\begin{cases} Af &= \mu g \\ A^*g &= \mu f \end{cases},$$

i.e. $\mu$ is a singular value of both $A$ and $A^*$.

The analysis of Schmidt pairs of Hankel operators goes back to Adamjan, Arov and Krein [1971]. Here, for the rational case we present an algebraic derivation of some of their results.

We proceed to compute the singular vectors of the Hankel operator $H_\phi$. In view of the preceeding remarks we have to solve

$$H_\phi f = \mu g$$

$$H_\phi^* g = \mu f$$

or

$$P_- \frac{n}{d}\frac{p}{d^*} = \mu \frac{\hat{p}}{d}$$

$$P_+ \frac{n^*}{d^*}\frac{\hat{p}}{d} = \mu \frac{p}{d^*}$$

This means there exist polynomials $\pi$ and $\xi$ such that

$$\frac{n}{d}\frac{p}{d^*} = \mu \frac{\hat{p}}{d} + \frac{\pi}{d^*}$$

$$\frac{n^*}{d^*}\frac{\hat{p}}{d} = \mu \frac{p}{d^*} + \frac{\xi}{d}.$$

These equations can be rewritten as polynomial equations

$$np = \mu d^* \hat{p} + d\pi \qquad (6)$$

$$n^*\hat{p} = \mu dp + d^*\xi. \tag{7}$$

Equation (6), considered as an equation modulo the polynomial $d$, is not an eigenvalue equation as there are too many unknowns. More specifically, we have to find the coefficients of both $p$ and $\hat{p}$. To overcome this difficulty we study in more detail the structure of Schmidt pairs of Hankel operators.

**Lemma 3.1** *Let* $\{\frac{p}{d^*}, \frac{\hat{p}}{d}\}$ *and* $\{\frac{q}{d^*}, \frac{\hat{q}}{d}\}$ *be two Schmidt pairs of the Hankel operator* $H_{\frac{n}{d}}$, *corresponding to the same singular value* $\mu$. *Then*

$$\frac{p}{\hat{p}} = \frac{q}{\hat{q}},$$

*i.e. this ratio is independent of the Schmidt pair.*

Proof: The polynomials $p$, $\hat{p}$ correspond to one Schmidt pair and let the polynomials $q$, $\hat{q}$ correspond to another Schmidt pair, i.e.

$$nq = \mu d^*\hat{q} + d\rho \tag{8}$$

$$n^*\hat{q} = \mu dq + d^*\eta. \tag{9}$$

Now, from equations (6) and (9) we get

$$0 = \mu d(p\hat{q} - q\hat{p}) + d^*(\xi\hat{q} - \eta\hat{p}).$$

Since $d$ and $d^*$ are coprime it follows that $d^* \mid p\hat{q} - q\hat{p}$. On the other hand, from equations (6) and (8), we get

$$0 = \mu d^*(\hat{p}q - \hat{q}p) + d(\pi q - \rho p),$$

and hence that $d \mid \hat{p}q - \hat{q}p$. Now both $d$ and $d^*$ divide $\hat{p}q - \hat{q}p$, and, since $\deg(\hat{p}q - \hat{q}p) < \deg d + \deg d^*$ it follows that

$$\hat{p}q - \hat{q}p = 0.$$

Equivalently

$$\frac{p}{\hat{p}} = \frac{q}{\hat{q}}$$

i.e. $\frac{p}{\hat{p}}$ is independent of the particular Schmidt pair associated to the singular value $\mu$. ∎

**Lemma 3.2** *Let* $\{\frac{p}{d^*}, \frac{\hat{p}}{d}\}$ *be a Schmidt pair associated with the singular value* $\mu$. *Then* $\frac{p}{\hat{p}}$ *is unimodular or all pass.*

Proof: Going back to equation (7) and the dual of (6) we have

$$n^*\hat{p} = \mu dp + d^*\xi$$

$$n^*p^* = \mu d(\hat{p})^* + d^*\pi^*$$

It follows that

$$0 = \mu d(pp^* - \hat{p}(\hat{p})^*) + d^*(\xi p^* - \pi^*\hat{p})$$

and hence $d^* \mid (pp^* - \hat{p}(\hat{p})^*)$. By symmetry also $d \mid (pp^* - \hat{p}(\hat{p})^*)$, and so necessarily

$$pp^* - \hat{p}(\hat{p})^* = 0.$$

This can be rewritten as

$$\frac{p}{\hat{p}}\frac{p^*}{(\hat{p})^*} = 1,$$

i.e. $\frac{p}{\hat{p}}$ is all pass. ∎

We will say that a pair of polynomials $(p, \hat{p})$, with $\deg p, \deg \hat{p} < \deg d$, is a *solution pair* if there exist polynomials $\pi$ and $\xi$ such that equations (6) and (7) are satisfied.

The next lemma characterizes all solution pairs.

**Lemma 3.3** *Let $\mu$ be a singular value of the Hankel operator $H_{\frac{n}{d}}$. Then there exists a unique, up to a constant factor, solution pair $(p, \hat{p})$, of minimal degree. The set of all solutions pairs is given by*

$$\{(q, \hat{q}) \mid q = pa,\ \hat{q} = \hat{p}a \ \ , \deg a < \deg q - \deg p\}.$$

Proof: Clearly, if $\mu$ is a singular value of the Hankel operator, then a nonzero solution pair $(p, \hat{p})$, of minimal degree exists. Let $(q, \hat{q})$ be any other solution, pair with $\deg q$, $\deg \hat{q} < \deg d$. By the division rule for polynomials $q = ap + r$ with $\deg r < \deg p$. Similarly, $\hat{q} = \hat{a}\hat{p} + \hat{r}$ with $\deg \hat{r} < \deg \hat{p}$. From equation (6) we get

$$n(ap) = \mu d^*(a\hat{p}) + d(a\pi) \tag{10}$$

whereas equation (8) yields

$$n(ap + r) = \mu d^*(\hat{a}\hat{p} + \hat{r}) + d(\tau) \tag{11}$$

By subtraction we obtain

$$nr = \mu d^*((\hat{a} - a)\hat{p} + \hat{r}) + d(\tau - a\pi) \tag{12}$$

Similarly from Equation (9) we get

$$n^*(\hat{a}\hat{p} + \hat{r}) = \mu d(ap + r) + d^*\xi. \tag{13}$$

whereas equation (7) yields

$$n^*(a\hat{p}) = \mu d(ap) + d^*(a\xi). \tag{14}$$

Subtracting the two gives

$$n^*((\hat{a}-a)\hat{p}+\hat{r}) = \mu dr + d^*(\eta - a\xi). \tag{15}$$

Equations (12) and (15) imply that $\{\frac{r}{d^*}, \frac{(\hat{a}-a)\hat{p}+\hat{r}}{d}\}$ is a $\mu$ Schmidt pair. Since necessarily $\deg r = \deg(\hat{a}-a)\hat{p}+\hat{r})$ we get $\hat{a}=a$. Finally, since we assumed $(p,\hat{p})$ to be of minimal degree we must have $r=\hat{r}=0$.

Conversely, if $a$ is any polynomial with $\deg a < \deg d - \deg p$ then from Equations (6) and (7) it follows by multiplication that $(pa, \hat{p}a)$ is also a solution pair. ∎

**Lemma 3.4** *Let $p$, $q$ be coprime polynomials with real coefficients such that $\frac{p}{q}$ is all pass. Then $q = \pm p^*$.*

Proof: Since $\frac{p}{q}$ is all pass, it follows that

$$\frac{p}{q}\frac{p^*}{q^*} = 1$$

or $pp^* = qq^*$. As the polynomials $p$ and $q$ are coprime it follows that $p \mid q^*$ and hence $q^* = \pm p$. ∎

In the general case we have the following.

**Lemma 3.5** *Let $p$, $q$ be polynomials with real coefficients such that $p \wedge q = 1$ and $\frac{p}{q}$ is all pass. Then, with $r = p \wedge \hat{p}$, we have*

$$p = rs$$

$$\hat{p} = \pm rs^*$$

Proof: Write $p = rs$, $\hat{p} = r\hat{s}$. Then $s \wedge \hat{s} = 1$ and $\frac{s}{\hat{s}}$ is all pass. The result follows by applying the previous lemma. ∎

The next theorem is of central importance due to the fact that it reduces the analysis to one polynomial. Thus we get an equation which is easily reduced to an eigenvalue problem.

**Theorem 3.1** *Let $\mu$ be a singular value of $H_\phi$ and let $(p,\hat{p})$ be a nonzero, minimal degree solution pair of equations (6) and (7). Then $p$ is a solution of*

$$np = \lambda d^* p^* + d\pi, \tag{16}$$

*with $\lambda$ real and $|\lambda| = \mu$.*

**Proof:** Let $(p, \hat{p})$ be a nonzero, minimal degree solution pair of equations (6) and (7). By taking their adjoints we can easily see that $(\hat{p}^*, p^*)$ is also a nonzero, minimal degree solution pair. By uniqueness of such a solution, i.e. by Lemma 3.3, we have

$$\hat{p}^* = \epsilon p. \tag{17}$$

Since $\frac{\hat{p}}{p}$ is all pass and both polynomials are real we have $\epsilon = \pm 1$. Let us put $\lambda = \epsilon\mu$, then (17) can be rewritten as

$$\hat{p} = \epsilon p^*$$

and so (16) follows from (6). ∎

We will refer to equation (16) as the **fundamental polynomial equation**. It will be the source of all future derivations.

**Corollary 3.1** *Let $\mu_i$ be a singular value of $H_\phi$ and let $p_i$ be the minimal degree solution of the fundamental polynomial equation, i.e.*

$$np_i = \lambda_i d^* p_i^* + d\pi_i.$$

*Then*

1. $$\deg p_i = \deg p_i^* = \deg \pi_i.$$

2. *Putting $p_i(z) = \sum_{j=0}^{n-1} p_{i,j} z^j$ and $\pi_i(z) = \sum_{j=0}^{n-1} \pi_{i,j} z^j$ we have the equality*

$$\pi_{i,n-1} = \lambda_i p_{i,n-1}. \tag{18}$$

**Corollary 3.2** *Let $p$ be a minimal degree solution of equation (16). Then*

1. *The set of all singular vectors of the Hankel operator $H_{\frac{n}{d}}$, corresponding to the singular value $\mu$, is given by*

$$Ker(H_{\frac{n}{d}}^* H_{\frac{n}{d}} - \mu^2 I) = \{\frac{pa}{d^*} \mid a \in R[z], \ \deg a < \deg d - \deg p\}$$

2. *The multiplicity of $\mu = ||H_\phi||$ as a singular value of $H_\phi$ is equal to $m = \deg d - \deg p$ where $p$ is the minimum degree solution of (16).*

3. *There exists a constant $c$ such that $c + \frac{n}{d}$ is a constant multiple of an antistable all-pass function if and only if $\mu_1 = \cdots = \mu_n$.*

Proof: We will prove (3) only. Assume all singular values are equal to $\mu$. Thus the multiplicity of $\mu$ is $\deg d$. Hence the minimal degree solution $p$ of (16) is a constant and so is $\pi$. Putting $c = -\frac{\pi}{p}$ then (16) can be rewritten as

$$\frac{n}{d} + c = \lambda \frac{d^* p^*}{dp},$$

and this is a multiple of an antistable all-pass function.

Conversely assume, without loss of generality, that $\frac{n}{d} + c$ is antistable all-pass. Then the induced Hankel operator is isometric and all its singular values are equal to 1. ∎

Part 3 of the corollary is due to Glover [1984].

The following simple proposition is important in the study of zeroes of singular vectors.

**Proposition 3.1** *Let $\mu_k$ be asingular value of $H_\phi$ and let $p_k$ be the minimal degree solution of*

$$np_k = \lambda_k d^* p_k^* + d\pi_k$$

*Then*

- *The polynomials $p_k$ and $p_k^*$ are coprime.*
- *The polynomial $p_k$ has no imaginary axis zeroes.*

**Proof:**

- Let $e = p_k \wedge p_k^*$. Without loss of generality we may assume that $e = e^*$. The polynomial $e$ has no imaginary axis zeroes, for that would imply that $e$ and $\pi_k$ have a nontrivial common divisor. Thus the fundamental polynomial equation could be divided by a suitable polynomial factor. This in contradiction to the assumption that $p_k$ is a minimal degree solution.
- This clearly follows from the first part.

∎

The fundamental polynomial equation is easily reduced to either a generalized eigenvalue equation or to a regular eigenvalue equation. There are several reductions of this kind in the literature, e.g. Kung [1980], Harshavardhana, Jonckheere and Silverman [1984], to cite a few. The one proposed here is simple and uses polynomial models.

## 3.1 Zeroes of singular vectors

We begin now the study of the zero location of the numerator polynomials of singular vectors. This is of course the same as the study of the zeroes of minimal degree solutions of equation (16). The following proposition provides a lower bound on the number of zeroes the minimal degree solutions of (16) can have in the open left half plane. However the lower bound is sharp in one special case. This is enough to lead us eventually to a full characterization, given originally by Adamjan, Arov and Krein [1968], and this will be given in Theorem 5.1.

**Proposition 3.2** *Let $\phi = \frac{n}{d} \in H_-^\infty$. Let $\mu_k$ be a singular value of $H_\phi$ satisfying $\mu_1 \geq \cdots \geq \mu_{k-1} > \mu_k = \cdots = \mu_{k+\nu-1} > \mu_{k+\nu} \geq \cdots \geq \mu_n$ i.e. $\mu_k$ is a singular value of multiplicity $\nu$. Let $p_k$ be the minimum degree solution of (16) corresponding to $\mu_k$. Then the number of antistable zeroes of $p_k$ are $\geq k-1$.*

*If $\mu_n$ is the smallest singular value of $H_\phi$ and is of multiplicity $\nu$, i.e. $\mu_1 \geq \cdots \geq \mu_{n-\nu} > \mu_{n-\nu+1} = \cdots = \mu_n$, and $p_{n-\nu+1}$ is the corresponding minimum degree solution of (16), then all the zeroes of $p_{n-\nu+1}$ are antistable.*

**Proof:** From equation (16), i.e.

$$np_k = \lambda_k d^* p_k^* + d\pi_k,$$

we get, dividing by $dp_k$,

$$\frac{n}{d} - \frac{\pi_k}{p_k} = \lambda_k \frac{d^* p_k^*}{dp_k}$$

which implies of course that

$$\|H_{\frac{n}{d}} - H_{\frac{\pi_k}{p_k}}\| \leq \|\frac{n}{d} - \frac{\pi_k}{p_k}\|_\infty = \mu_k \|\frac{d^* p_k^*}{dp_k}\|_\infty = \mu_k.$$

This means, by the definition of singular values, that $rank H_{\frac{\pi_k}{p_k}} \geq k-1$. But this implies, by Kronecker's theorem, that the number of antistables poles of $\frac{\pi_k}{p_k}$ which is the same as the number of antistable zeroes of $p_k$ is $\geq k-1$.

If $\mu_n$ is the smallest singular value and has multiplicity $\nu$, and $p_{n-\nu+1}$ is the minimal degree solution of equation (16), then it has degree $n-\nu$. But by the previous part it must have at least $n-\nu$ antistable zeroes. So this implies that all the zeroes of $p_{n-\nu+1}$ are antistable.

∎

The previous result is extremely important from our point of view. It shifts the focus from the largest singular value, the starting point in all derivations sofar, to the smallest singular value. Certainly the derivation is elementary, inasmuch as we use only the definition of singular values and Kronecker's theorem. The great advantage is that at this stage we can solve an important Bezout equation which is the key to duality theory.

We have now at hand all that is needed to obtain the optimal Hankel norm approximant corresponding to the smallest singular value. We shall delay this analysis to a later stage and develop duality theory first.

From equation (16) we obtain, dividing by $\lambda_n d^* p_n^*$, the Bezout equation

$$\frac{n}{d^*}\left(\frac{1}{\lambda_n}\frac{p_n}{p_n^*}\right) - \frac{d}{d^*}\left(\frac{1}{\lambda_n}\frac{\pi_n}{p_n^*}\right) = 1. \tag{19}$$

Since the polynomials $p_n$ and $d$ are antistable all four functions appearing in the Bezout equation are in $\in H_+^\infty$. We shall discuss next the implications of this Bezout equation.

# 4 DUALITY

In this section we develop a duality theory in the context of Hankel norm approximation problems. There are three operations applied to a given, antistable, transfer function. Namely, inversion of the restricted Hankel operator, taking the adjoint map and finally one sided multiplication by unitary operators. The last two operations do not change the singular values, whereas the first operation inverts them.

We will say that two Hilbert space operators $T : H_1 \longrightarrow H_2$ and $T' : H_3 \longrightarrow H_4$ are *equivalent* if there exist unitary operators $U : H_1 \longrightarrow H_3$ and $V : H_2 \longrightarrow H_4$ such that

$$VT = T'U.$$

**Lemma 4.1** *Let $T : H_1 \longrightarrow H_2$ and $T' : H_3 \longrightarrow H_4$ be equivalent. Then $T$ and $T'$ have the same singular values.*

Proof: Let $T^*Tx = \mu^2 x$. Since $VT = T'U$ it follows that

$$U^*T'^*T'Ux = T^*V^*VTx = T^*Tx = \mu^2 x,$$

or

$$T'^*T'(Ux) = \mu^2(Ux).$$

∎

The following proposition is bordering on the trivial and no proof need be given. However, when applied to Hankel operators it has far reaching implications. In fact it provides a key to duality theory and leads eventually to the proof of the AAK results.

**Proposition 4.1** *Let $T$ be an invertible linear transformation. Then, if $x$ is a singular vector of the operator $T$ corresponding to the singular value $\mu$, i.e. $T^*Tx = \mu^2 x$ then*

$$T^{-1}(T^{-1})^*x = \mu^{-2}x$$

*i.e. $x$ is also a singular vector for $(T^{-1})^*$ corresponding to the singular value $\mu^{-1}$.*

In view of this proposition, it is of interest to compute $[(H_\phi|H(m))^{-1}]^*$. Before proceeding with this we compute the inverse of a related operator. This is a special case of Theorem 2.4 for the rational case. Note that, since $||T_\Theta^{-1}|| = \mu_n^{-1}$, there exists, by Sarason's theorem, a $\xi \in H_+^\infty$ such that $T_\Theta^{-1} = T_\xi$ and $||\xi||_\infty = \mu_n^{-1}$. The next theorem provides this $\xi$. For an algebraic analogue of the next two theorems we refer to Helmke and Fuhrmann [1989].

**Theorem 4.1** *Let $\phi = \frac{n}{d} \in H_-^\infty$. Then $\theta = \frac{n}{d^*} \in H_+^\infty$. The operator $T_\theta$ defined by equation (2) is invertible and its inverse given by $T_{\frac{1}{\lambda_n}\frac{p_n}{p_n^*}}$ where $\lambda_n$ is the last signed singular value of $H_\phi$ and $p_n$ is the minimal degree solution of*

$$np_n = \lambda_n d^* p_n^* + d\pi_n.$$

<u>Proof:</u> From the previous equation we obtain the Bezout equation

$$\frac{n}{d^*}\left(\frac{1}{\lambda_n}\frac{p_n}{p_n^*}\right) - \frac{d}{d^*}\left(\frac{\pi_n}{\lambda_n p_n^*}\right) = 1.$$

By Theorem 3.2 the polynomial $p_n$ is antistable so $\frac{p_n}{p_n^*} \in H_+^\infty$. This, by Theorem 2.4 implies the result. ∎

It is well known that stabilizing controllers are related to solutions of Bezout equations over $H^\infty$. Thus we expect equation (19) to lead to a stabilizing controller. The next corollary is a result of this type.

**Corollary 4.1** *Let $\phi = \frac{n}{d} \in H_-^\infty$. The controller $k = \frac{p_n}{\pi_n}$ stabilizes $\phi$. If the multiplicity of $\mu_n$ is $m$ there exists a stabilizing controller of degree $n - m$.*

<u>Proof:</u> Since $p_n$ is antistable, we get from (16) that $np_n - d\pi_n = \lambda_n d^* p_n^*$ is stable. We compute

$$\frac{\phi}{1-k\phi} = \frac{\frac{n}{d}}{1 - \frac{p_n}{\pi_n}\frac{n}{d}} = \frac{-n\pi_n}{np_n - d\pi_n} = \frac{-n\pi_n}{\lambda_n d^* p_n^*} \in H_+^\infty.$$

∎

This corollary is related to questions of robust control. For more on this see Glover [1986].

**Theorem 4.2** *Let $\phi = \frac{n}{d} \in H_-^\infty$. Let $H : X^{d^*} \longrightarrow X^d$ be defined by $H = H_\phi | X^{d^*}$. Then*

1. *$H_\phi^{-1} : X^d \longrightarrow X^{d^*}$ is given by*

$$H_\phi^{-1} h = \frac{1}{\lambda_n}\frac{d}{d^*} P_- \frac{p_n}{p_n^*} h \tag{20}$$

*2.* $(H_\phi^{-1})^* : X^{d^*} \longrightarrow X^d$ *is given by*

$$(H_\phi^{-1})^* f = \frac{1}{\lambda_n}\frac{d^*}{d}P_+\frac{p_n^*}{p_n}f \tag{21}$$

Proof:

1. Let $m = \dfrac{d}{d^*}$ and let $T$ be the map given by $T = mH_{\frac{n}{d}}$. Thus we have the following commutative diagram

$$\begin{array}{ccc} X^{d^*} & \xrightarrow{H_\phi} & X^d \\ T & \searrow & \downarrow\; m \\ & & X^{d^*} \end{array}$$

Now

$$\begin{aligned} Tf &= \frac{d}{d^*}P_-\frac{n}{d}f = \frac{d}{d^*}P_-\frac{d^*}{d}\frac{n}{d^*}f \\ &= P_{H(\frac{d}{d^*})}\frac{n}{d^*}f = P_{X^{d^*}}\frac{n}{d^*}f \end{aligned}$$

i.e. $T = T_\theta$ where $\theta = \frac{n}{d^*}$. Now, from $T_\theta = mH_{\frac{n}{d}}$ we have, by Theorem 4.1,

$$T_\theta^{-1} = T_{\frac{1}{\lambda_n}\frac{p_n}{p_n^*}}.$$

So, for $h \in X^d$,

$$\begin{aligned} H_{\frac{n}{d}}^{-1}h &= \frac{1}{\lambda_n}P_{H(\frac{d}{d^*})}\frac{p_n}{p_n^*}\frac{d}{d^*}h \\ &= \frac{1}{\lambda_n}\frac{d}{d^*}P_-\frac{d^*}{d}\frac{p_n}{p_n^*}\frac{d}{d^*}h \qquad (22) \\ &= \frac{1}{\lambda_n}\frac{d}{d^*}P_-\frac{p_n}{p_n^*}h. \end{aligned}$$

2. equation (22) can be written also as

$$H_{\frac{n}{d}}^{-1}h = T_{\frac{1}{\lambda_n}\frac{p_n}{p_n^*}}mh.$$

Therefore, using Theorem 2.3, we have, for $f \in X^{d^*}$,

$$(H_\phi^{-1})^*f = m^*(T_{\frac{1}{\lambda_n}\frac{p_n}{p_n^*}})^*f = \frac{d^*}{d}P_+\frac{1}{\lambda_n}\frac{p_n^*}{p_n}f = \frac{1}{\lambda_n}\frac{d^*}{d}P_+\frac{p_n}{p_n^*}f.$$

∎

**Corollary 4.2** *There exist polynomials $\alpha_i$, of degree $\le n-2$, such that*

$$\lambda_i p_n^* p_i - \lambda_n p_n p_i^* = \lambda_i d^* \alpha_i, \quad i = 1, \ldots, n-1.$$

*This holds also formally for $i = n$ with $\alpha_n = 0$.*

Proof: Since

$$H_{\frac{n}{d}} \frac{p_i}{d^*} = \lambda_i \frac{p_i^*}{d}$$

it follows that

$$(H_{\frac{n}{d}}^{-1})^* \frac{p_i}{d^*} = \lambda_i^{-1} \frac{p_i^*}{d}.$$

So, using equation (21), we have

$$\frac{1}{\lambda_n} \frac{d^*}{d} P_+ \frac{p_n^*}{p_n} \frac{p_i}{d^*} = \frac{1}{\lambda_i} \frac{p_i^*}{d},$$

i.e.

$$\frac{\lambda_n}{\lambda_i} \frac{p_i^*}{d^*} = P_+ \frac{p_n^*}{p_n} \frac{p_i}{d^*}.$$

This implies, by partial fraction decomposition, the existence of polynomials $\alpha_i$, $i = 1, \ldots, n$ such that $\deg \alpha_i < \deg p_n = n-1$, and

$$\frac{p_n^*}{p_n} \frac{p_i}{d^*} = \frac{\lambda_n}{\lambda_i} \frac{p_i^*}{d^*} + \frac{\alpha_i}{p_n},$$

i.e.

$$\lambda_i p_n^* p_i - \lambda_n p_n p_i^* = \lambda_i d^* \alpha_i. \tag{23}$$

∎

We saw, in Theorem 4.2, that for the Hankel operator $H_\phi$ the map $(H_\phi^{-1})^*$ is not a Hankel map. However there is an equivalent Hankel map. We sum this up in the following.

**Theorem 4.3** *Let $\phi = \frac{n}{d} \in H_-^\infty$. Let $H : X^{d^*} \longrightarrow X^d$ be defined by $H = H_\phi | X^{d^*}$. Then*

1. *The operator $(H_\phi^{-1})^*$ is equivalent to the Hankel operator $H_{\frac{1}{\lambda_n}\frac{d^* p_n}{d p_n^*}}$.*

2. *The Hankel operator $H_{\frac{1}{\lambda_n}\frac{d^* p_n}{d p_n^*}}$ has singular values $\mu_1^{-1} < \cdots < \mu_n^{-1}$.*

3. *The Schmidt pairs of $H_{\frac{1}{\lambda_n}\frac{d^* p_n}{d p_n^*}}$ are $\{\frac{p_i^*}{d^*}, \frac{p_i}{d}\}$.*

Proof: We saw that

$$(H_{\frac{n}{d}}^{-1})^* = \frac{d^*}{d} T^*_{\frac{1}{\lambda_n}\frac{p_n}{p_n^*}}.$$

Since multiplication by $\frac{d^*}{d}$ is a unitary map of $X^{d^*}$ onto $X^d$, the operator $(H_{\frac{n}{d}}^{-1})^*$ has, by Lemma 4.1, the same singular values as $T^*_{\frac{1}{\lambda_n}\frac{p_n}{p_n^*}}$. These are the same as those of the adjoint operator $T_{\frac{1}{\lambda_n}\frac{p_n}{p_n^*}}$. However the last operator is equivalent to the Hankel operator $H_{\frac{1}{\lambda_n}\frac{d^* p_n}{dp_n^*}}$.

Indeed,

$$\frac{d^*}{d} T_{\frac{1}{\lambda_n}\frac{p_n}{p_n^*}} f = \frac{d^*}{d} P_{H(\frac{d}{d^*})} \frac{1}{\lambda_n}\frac{p_n}{p_n^*} f = \frac{d^*}{d}\frac{d}{d^*} P_- \frac{d^*}{d}\frac{1}{\lambda_n}\frac{p_n}{p_n^*} f = H_{\frac{1}{\lambda_n}\frac{d^* p_n}{dp_n^*}} f.$$

This Hankel operator has singular values $\mu_1^{-1} < \cdots < \mu_n^{-1}$. and its Schmidt pairs are $\{\frac{p_i^*}{d^*}, \frac{p_i}{d}\}$. Indeed

$$H_{\frac{d^* p_n}{dp_n^*}} \frac{p_i^*}{d^*} = P_- \frac{d^* p_n}{dp_n^*}\frac{p_i^*}{d^*} = P_- \frac{p_n p_i^*}{dp_n^*}.$$

Now, from equation (23) we get

$$p_n p_i^* = \frac{\lambda_i}{\lambda_n} p_n^* p_i - \frac{\lambda_i}{\lambda_n} d^* \alpha_i,$$

or taking the dual of that equation

$$p_n p_i^* = \frac{\lambda_n}{\lambda_i} p_n^* p_i + d\alpha_i^*,$$

So

$$\frac{p_n p_i^*}{dp_n^*} = \frac{\lambda_n}{\lambda_i}\frac{p_n^* p_i}{dp_n^*} + \frac{d\alpha_i^*}{dp_n^*} = \frac{\lambda_n}{\lambda_i}\frac{p_i}{d} + \frac{\alpha_i^*}{p_n^*}.$$

Hence

$$P_- \frac{p_n p_i^*}{dp_n^*} = \frac{\lambda_n}{\lambda_i}\frac{p_i}{d}.$$

Therefore

$$\frac{1}{\lambda_n} H_{\frac{d^* p_n}{dp_n^*}} \frac{p_i^*}{d^*} = \frac{1}{\lambda_i}\frac{p_i}{d}.$$

∎

# 5 HANKEL NORM APPROXIMATION

The duality results obtained before allow us now to complete our study of the zero structure of minimal degree solutions of the fundamental polynomial equation (16). This in turn leads to an elementary proof of the central theorem in the AAK theory.

**Theorem 5.1 (Adamjan, Arov and Krein)** *Let $\phi = \frac{n}{d} \in H_{-}^{\infty}$. Let $\mu_k$ be a singular value of $H_\phi$ satisfying $\mu_1 \geq \cdots \geq \mu_{k-1} > \mu_k = \cdots = \mu_{k+\nu-1} > \mu_{k+\nu} \geq \cdots \geq \mu_n$ i.e. $\mu_k$ is a singular value of multiplicity $\nu$. Let $p_k$ be the minimum degree solution of (16) corresponding to $\mu_k$. Then the number of antistable zeroes of $p_k$ is exactly $k-1$.*

*If $\mu_1$ is the largest singular value of $H_\phi$ and is of multiplicity $\nu$, i.e. $\mu_1 = \cdots = \mu_\nu > \mu_{\nu+1} \geq \cdots \geq \mu_n$, and $p_1$ is the corresponding minimum degree solution of (16), then all the zeroes of $p_1$ are stable, this is equivalent to saying that $p_1$ is outer.*

**Proof:** We saw, in the proof of Proposition 3.2, that the number of antistable zeroes of $p_k$ is $\geq k-1$. Now, by Theorem 4.3, $p_k^*$ is the minimum degree solution of the fundamental equation corresponding to the transfer function $\frac{1}{\lambda_n}\frac{d^* p_n}{dp_n^*}$ and the singular value $\mu_{k+\nu-1}^{-1} = \cdots = \mu_k^{-1}$. Clearly we have $\mu_n^{-1} \geq \cdots \geq \mu_{k+\nu}^{-1} > \mu_{k+\nu-1}^{-1} = \cdots = \mu_k^{-1} > \mu_{k-1}^{-1} \geq \cdots \geq \mu_1^{-1}$. In particular, applying Proposition 3.2, the number of antisatble zeroes of $p_k^*$ is $\geq n-k-\nu+1$. Since the degree of $p_k^*$ is $n-\nu$ it follows that the number of stable zeroes of $p_k^*$ is $\leq k-1$. However this is the same as saying the number of antistable zeroes of $p_k$ is $\leq k-1$. Combining the two inequalities, it follows that the number of antistable zeroes of $p_k$ is exactly $k-1$.

The first part implies that the minimum degree solution of (16) has only stable zeroes, i.e. it is an outer function.

∎

We now come to apply some results of the previous section to the case of Hankel norm approximation. We use here the characterization of singular values $\mu_1 \geq \mu_2 \geq \cdots$ of a linear transformation $A : V_1 \longrightarrow V_2$ as approximation numbers, namely

$$\mu_k = \inf\{\|A - A_k\| \mid rank A_k leq k-1\}.$$

See Gohberg and Krein [1969] for an extensive treatment of this topic.

**Theorem 5.2 (Adamjan, Arov and Krein)** *Let $\phi = \frac{n}{d} \in H_{-}^{\infty}$ be a scalar, strictly proper, transfer function, with $n$ and $d$ coprime polynomials and $d$ is monic of degree $n$. Assume that $\mu_1 \geq \cdots \geq \mu_{k-1} > \mu_k = \cdots = \mu_{k+\nu-1} > \mu_{k+\nu} \geq \cdots \geq \mu_n > 0$ are the singular values of $H_\phi$. Then*

$$\begin{aligned}\mu_k &= \inf\{\|H_\phi - A\| | rank A \leq k-1\} \\ &= \inf\{\|H_\phi - H_\psi\| | rank H_\psi \leq k-1\} \\ &= \inf\left\{\|\phi - \psi\|_\infty | \psi \in H_{[k-1]}^{\infty}\right\}\end{aligned}$$

*Moreover, the infimum is attained on a unique function $\psi_k = \phi - \frac{H_\phi f_k}{f_k} = \phi - \mu\frac{g}{f}$, where $(f_k, g_k)$ is an arbitrary Schmidt pair of $H_\phi$ that corresponds to $\mu_k$.*

Proof: Given $\psi \in H^\infty_{[k-1]}$, we have, by Kronecker's theorem, that $rank H_\psi = k-1$. Therefore we clearly have

$$\begin{aligned} \mu_k &= \inf\{\|H_\phi - A\| | rank A \leq k-1\} \\ &\leq \inf\{\|H_\phi - H_\psi\| | rank H_\psi \leq k-1\} \\ &\leq \inf\left\{\|\phi - \psi\|_\infty | \psi \in H^\infty_{[k-1]}\right\} \end{aligned}$$

so the proof will be complete if we can exhibit a function $\psi_k \in H^\infty_{[k-1]}$ for which the equality $\mu_k = \|\phi - \psi\|_\infty$ holds. To this end let $p_k$ be the minimal degree solution of (16), and define $\psi_k = \frac{\pi_k}{p_k}$. From the equation

$$np_k = \lambda_k d^* p_k^* + d\pi_k$$

we get, dividing by $dp_k$, that

$$\frac{n}{d} - \frac{\pi_k}{p_k} = \lambda_k \frac{d^* p_k^*}{dp_k}.$$

This is of course equivalent to

$$\psi_k = \frac{\pi_k}{p_k} = \frac{n}{d} - \lambda_k \frac{d^* p_k^*}{dp_k} = \phi - \frac{H_\phi f_k}{f_k},$$

as for $f_k = \frac{p_k^*}{d}$ we have $H_\phi f_k = \lambda_k \frac{p_k}{d^*}$ and, by Lemma 3.1, the ratio $\frac{H_\phi f_k}{f_k}$ is independent of the particular Schmidt pair. So

$$\|\phi - \psi\|_\infty = \|\frac{n}{d} - \frac{\pi_k}{p_k}\|_\infty = \|\lambda_k \frac{d^* p_k^*}{dp_k}\|_\infty = \mu_k$$

Moreover $\frac{\pi_k}{p_k} \in H^\infty_{[k-1]}$, as $p_k$ has exactly $k-1$ antistable zeroes. ∎

**Corollary 5.1** *The polynomials $\pi_k$ and $p_k$ have no common antistable zeroes.*

**Proof:** Follows from the fact that $rank H_{\frac{\pi_k}{p_k}} \geq k-1$.

∎

# References

[1968a] V. M. Adamjan, D. Z. Arov and M. G. Krein, "Infinite Hankel matrices and generalized problems of Caratheodory-Fejer and F. Riesz", *Funct. Anal. Appl.* 2, 1-18.

[1968b] V. M. Adamjan, D. Z. Arov and M. G. Krein, "Infinite Hankel matrices and generalized problems of Caratheodory-Fejer and I. Schur", *Funct. Anal. Appl.* 2, 269-281.

[1971] V. M. Adamjan, D. Z. Arov and M. G. Krein, "Analytic properties of Schmidt pairs for a Hankel operator and the generalized Schur-Takagi problem", *Math. USSR Sbornik* 15 (1971), 31-73.

[1978] V. M. Adamjan, D. Z. Arov and M. G. Krein, "Infinite Hankel block matrices and related extension problems", Amer. Math. Soc. Transl., series 2, Vol. 111, 133-156.

[1949] A. Beurling, "On two problems concerning linear transformations in Hilbert space", *Acta Math.*, 81, pp. 239-255.

[1962] L. Carleson, "Interpolation by bounded analytic functions and the corona problem", *Ann. of Math.* 76, 547-559.

[1971] R.G. Douglas, H.S. Shapiro & A.L. Shields, "Cyclic vectors and invariant subspaces for the backward shift", *Ann. Inst. Fourier, Grenoble* 20,1, 37-76.

[1970] P. Duren, *Theory of $H^p$ Spaces*, Academic Press, New York.

[1968a] P. A. Fuhrmann, "On the corona problem and its application to spectral problems in Hilbert space", *Trans. Amer. Math. Soc.* 132(1968), 55-67.

[1968b] P. A. Fuhrmann, "A functional calculus in Hilbert space based on operator valued analytic functions", *Israel J. Math.* 6, 267-278.

[1981] P. A. Fuhrmann, *Linear Systems and Operators in Hilbert Space*, McGraw-Hill, New York.

[1983] P. A. Fuhrmann, "On symmetric rational transfer functions", *Linear Algebra and Appl.*, 50,167-250.

[1991] P. A. Fuhrmann, "A polynomial approach to Hankel norm and balanced approximations", *Linear Algebra and Appl.*, 146, 133-220.

[1981] J.B. Garnett, *Bounded Analytic Functions*, Academic Press, New York.

[1984] K. Glover, "All optimal Hankel-norm approximations and their $L^\infty$-error bounds", *Int. J. Contr.* 39, 1115-1193.

[1986] K. Glover, "Robust stabilization of linear multivariable systems, relations to approximation", *Int. J. Contr.* 43, 741-766.

[1969] I. Gohberg and M. G. Krein *Introduction to the Theory of Nonselfadjoint Operators*, Amer. Math. Soc., Providence.

[1984] P. Harshavardhana, E. A. Jonckheere and L. M. Silverman, "Eigenvalue and generalized eigenvalue formulations for Hankel norm reduction directly from polynomial data", $23^{rd}$ *IEEE Conf. on Decision and Control*, Las Vegas, Nevada, December 1984, 111-119.

[1989] U. Helmke and P. A. Fuhrmann, "Bezoutians", *Lin. Alg. Appl.*, v. 122-124, 1039-1097.

[1974] J. W. Helton, "Discrete time systems, operator models and scattering theory", *J. Funct. Anal.*, 16, 15-38.

[1962] K. Hoffman, *Banach Spaces of Analytic Functions*, Prentice Hall.

[1980] S. Kung, "Optimal Hankel-norm reductions: scalar systems", *1980 Proc. Joint Automat. Contr. Conf.*, San Francisco.

[1985] A. Lindquist and G. Picci, "Realization theory for multivariate stationary Gaussian processes", *SIAM J. Contr. & Optim.*, Vol. 23, 809-857.

[1985] N. K. Nikolskii, *Treatise on the Shift Operator*, Springer Verlag, Berlin.

[1970] L.B. Page, "Applications of the Sz.-Nagy and Foias lifting theorem", *Indiana Univ. Math. J.*, 20, 135-145.

[1988] J. Partington, *An Introduction to Hankel Operators*, Cambridge University Press, Cambridge.

[1967] D. Sarason, "Generalized interpolation in $H^\infty$", *Trans. Amer. Math. Soc.* 127, 179-203.

[1970] B. Sz.-Nagy and C. Foias, *Harmonic analysis of Operators on Hilbert Space*, North Holland, Amsterdam.

[1983] N. Young, "The singular value decomposition of an infinite Hankel matrix", *Linear Algebra and Appl.*, 50,639-656.

[1988] N. Young, *An Introduction to Hilbert Space*, Cambridge University Press, Cambridge.

# 36 Stabilizing Solutions to Riccati Inequalities and Stabilizing Compensators with Disturbance Attenuation

**Aristide Halanay** University of Bucharest, Bucharest, Romania

ABSTRACT. Necessary and sufficient conditions for existence of stabilizing compensators with disturbance attenuation are obtained for linear systems with time-varying coefficients in terms of Riccati inequalities. A generalized Popov-Yakubovich theory is used.

## 1. The problem. Main results.

We shall describe here the problem and the main results, without insisting upon the general situations corresponding to abstract evolution operators, where these results at least partially hold.

Consider a linear system with time-varying coefficients

$$(\Sigma) \qquad \begin{aligned} x' &= A(t)x + B_1(t)u_1 + B_2(t)u_2 \\ y_1 &= C_1(t)x + D_{12}(t)u_2 \\ y_2 &= C_2(t)x + D_{21}(t)u_1. \end{aligned}$$

Here $u_1$ stands for the disturbance, $u_2$ is the control, $y_2$ is the measured output to be used in the compensator, and $y_1$ is a quality or regulated output. A compensator (controller) for $(\Sigma)$ is a linear system

$$(\Sigma_c) \qquad \begin{aligned} x_c' &= A_c(t)x_c + B_c(t)u_c \\ y_c &= C_c(t)x_c + D_c(t)u_c. \end{aligned}$$

The compensator is strictly proper, if $D_c = 0$.

The compensator $(\Sigma_c)$ is coupled to $(\Sigma)$ by taking $u_c = y_2$ and $u_2 = y_c$. After such coupling, assuming $D_c = 0$, one obtains the resulting closed loop system

$$(\Sigma_R) \qquad \begin{aligned} x' &= A(t)x + B_2(t)C_c(t)x_c + B_1(t)u_1 \\ x_c' &= B_c(t)C_2(t)x + A_c(t)x_c + B_c(t)D_{21}(t)u_1 \\ y_1 &= C_1(t)x + D_{12}(t)C_c(t)x_c. \end{aligned}$$

The compensator is <u>stabilizing</u> if the evolution of $(\Sigma_R)$ corresponding to $u_1 = 0$ is exponentially stable. After stabilization, for every disturbance $u_1 \in L^2(\mathbb{R}, U^1)$ there is a unique solution $\binom{x}{x_c} \in L^2(\mathbb{R}, \Psi)$, and to this solution there corresponds a unique output $y_1$. A linear operator $T_{y_1 u_1} : L^2(\mathbb{R}, U^1) \to L^2(R, Y^1)$ is defined in this way and the compensator is <u>$\gamma$-disturbance attenuating</u> if $\|T_{y_1 u_1}\| < \gamma$.

To simplify, we shall assume that the usual structure assumptions are satisfied, that is $D_{12}^* D_{12} = I$, $D_{12}^* C_1 = 0$, $D_{21} D_{21}^* = I$, $D_{21} B_1^* = 0$, $\gamma = 1$. The final result is given below.

**Theorem 1.** *If the system $(\Sigma)$ admits a strictly proper stabilizing controller with disturbance attenuation, there exist $R > 0$, $S > 0$, bounded on $\mathbb{R}$, and $\alpha > 0$, $\beta > 0$, $\gamma > 0$ satisfying*

$$R' + RA(t) + A^*(t)R + R(t)[B_1(t)B_1^*(t) - B_2(t)B_2^*(t)]R(t) + C_1^*(t)C_1(t) \leq -\alpha I,$$

$$S' - \{A(t)S + SA^*(t) + S[C_1^*(t)C_1(t) - C_2^*(t)C_2(t)]S + B_1(t)B_1^*(t)\} \geq \beta I,$$

$$s^{-1}(t) - R(t) \geq \gamma I,$$

*and there exist $F$, $H$ bounded on $\mathbb{R}$ and $\hat{\alpha} > 0$, $\hat{\beta} > 0$ such that $A + B_2F$, $A + HC_2$ define exponentially stable evolutions, and*

$$\begin{aligned} &R'(t) + R(t)[A(t) + B_2(t)F(t)] + [A(t) + B_2(t)F(t)]^* R(t) + C_1^*(t)C_1(t) \\ &\quad + F^*(t)F(t) + R(t)B_1(t)B_1^*(t)R(t) \leq -\hat{\alpha} I, \end{aligned}$$

$$\begin{aligned} &S'(t) - \{A(t) + H(t)C_2(t)]S(t) + S(t)[A(t) + H(t)C_2(t)]^* \\ &\quad + S(t)C_1^*(t)C_1(t)S(t) + H(t)H^*(t) + B_1(t)B_1^*(t)\} \geq \hat{\beta} I. \end{aligned}$$

Conversely if $R, S, F, H$ exist with the properties above, then by taking

$$C_c = F, \qquad B_c = -(I - SR)^{-1}H,$$

$$P_1 = -[R' + R(A + B_2F) + (A + B_2F)^*R + C_1^*C_1 + F^*F + RB_1B_1^*R],$$

$$K = F^*B_2^*R + F^*F - (S^{-1} - R)B_1B_1^*R + P_1,$$

$$A_c = A + B_2F - B_cC_2 - (S^{-1} - R)^{-1}K,$$

a stabilizing compensator with disturbance attenuation is obtained.

This theorem extends to the case where the controlled system is as in Sampei et.al. (1990) [6]. The proof reduces to the state feedback situation as considered by Zhou and Khargonekar (1988) [8]. As an important step we shall discuss a generalized Popov-Yakubovich theory.

## 2. Generalized Popov-Yakubovich theory.

**Definition 1.** Let $\mathcal{H}$ be a Hilbert space. An evolutionary process $X$ on $\mathcal{H}$ is a function $X : \Delta \to \mathcal{L}^2(\mathcal{H}, \mathcal{H})$, with $\Delta = \{(t, s) \subset \mathbb{R} \times \mathbb{R} : t \geq s\}$, $X(s, s) = I$, $X(t, s)X(s, \sigma) = X(t, \sigma)$ for $t \geq s \geq \sigma$, and with $X(\cdot, \cdot)$-strongly continuous.

*Remark 2.* In a recent report Pandolfi (1992) [5], following Curtain & Pritchard (1978) [1], considers a weaker continuity property for $X$, namely $t \longmapsto X(t, T)x_o$ is continuous on $t \geq T$ for each $x_o \in \mathcal{H}$.

**Definition 3.** The evolutionary process $X$ is exponentially stable if there exist $\beta \geq 1$, $\alpha > 0$ such that

$$|X(t, s)x_o|_{\mathcal{H}} \leq \beta e^{-\alpha(t-s)}|x_o|_{\mathcal{H}} \text{ for all } x_o \in \mathcal{H}.$$

The following theorem is an extension of the result in Yakubovich (1974) [7].

**Theorem 4.** *Consider an exponentially stable evolutionary process $X$, and operator-valued functions $B, K, L, M$, each continuous and bounded on $\mathbb{R}$, with*

$B(t) : \tilde{\mathcal{H}} \to \mathcal{H}$, $K(t) : \tilde{\mathcal{H}} \to \tilde{\mathcal{H}}$, $L(t) : \tilde{\mathcal{H}} \to \mathcal{H}$, $M(t) : \mathcal{H} \to \mathcal{H}$, $K(t)$, $M(t)$ *selfadjoint,* $K(t)$ *invertible and* $K^{-1}(t)$ *uniformly bounded with respect to* $t \in \mathbb{R}$. *Define*

$$\phi_{t_o} : \mathcal{H} \to L^2(t_o, \infty; \mathcal{H}) \text{ by } (\phi_{t_o} x_o)(t) = X(t, t_o)x_o, \text{ and}$$

$$\mathcal{L}_{t_o} : L^2(t_o, \infty; \tilde{\mathcal{H}}) \to L^2(t_o, \infty; \mathcal{H}) \text{ by } (\mathcal{L}_{t_o} U)(t) = \int_{t_o}^{t} X(t,s)B(s)u(s)\, ds.$$

*Let* $R_{t_o} : L^2(t_o, \infty; \tilde{\mathcal{H}}) \to L^2(t_o, \infty; \tilde{\mathcal{H}})$ *be defined by* $R_{t_o} = K + L^*\mathcal{L}_{t_o} + \mathcal{L}_{t_o}^* L + \mathcal{L}_{t_o}^* M \mathcal{L}_{t_o}$, *where* $K, L, M$ *denote here the multiplication operators associated to the operator-valued functions* $K, L, M$. *Assume that* $\mathcal{R}_{t_o}^{-1}$ *is defined for all* $t_o \in \mathbb{R}$ *and that there exists* $c > o$ *such that for all* $t_o \in \mathbb{R}$, $\|\mathcal{R}_{t_o}^{-1}\| \le c < \infty$. *Then the operator-valued functions* $R : \mathbb{R} \to \mathcal{L}(\mathcal{H}, \mathcal{H})$, $F : \mathbb{R} \to \mathcal{L}(\mathcal{H}, \tilde{\mathcal{H}})$ *defined by* $R(t) = \mathcal{P}_t^o - \mathcal{P}_t \mathcal{R}_t^{-1} \mathcal{P}_t^*$, $F(t) = -K^{-1}(t)[L^*(t) + B^*(t)R(t)]$ *with* $\mathcal{P}_t^o = \phi_t^* M \phi_t$, $\mathcal{P}_t = \phi_t^*(L + M\mathcal{L}_t)$ *have the properties*

(i) *The evolutionary process* $X_F$ *defined by*

$$X_F(t, t_o)x_o = [(\phi_{t_o} - \mathcal{L}_{t_o} \mathcal{R}_{t_o}^{-1} \mathcal{P}_{t_o}^*)x_o](t)$$

*is exponentially stable and satisfies*

$$X_F(t, t_o)x_o = X(t, t_o)x_o + \int_{t_o}^{t} X(t,s)B(s)F(s)X_F(s, t_o)x_o\, ds$$

(ii) $R(t)x_o = \int_t^\infty X^*(s,t)[M(s) - F^*(s)K(s)F(s)]X(s,t)x_o\, ds$.

*Proof.* a) Consider the Popov-Yakubovich functional

$$J(t_o, x_o, u) = \int_{t_o}^{\infty} \{(K(t)u(t), u(t))_{\tilde{H}} + 2(L(t)u(t), x(t))_{\mathcal{H}} + (M(t)x(t), x(t))_{\mathcal{H}}\}\, dt$$

with $x(t) = X(t, t_o)x_o + \int_{t_o}^t X(t,s)B(s)u(s)\, ds = (\phi_{t_o} x)(t) + (\mathcal{L}_{t_o} u)(t)$. A direct calculation leads to $J(t_o, x_o, u) = \langle u, \mathcal{R}_{t_o} u\rangle_{t_o} + 2(x_o, \mathcal{P}_{t_o} u)_{\mathcal{H}} + (\mathcal{P}_{t_o}^o x_o, x_o)_{\mathcal{H}}$ where $\langle u, v\rangle_{t_o} = \int_{t_o}^\infty (u(t), v(t))_{\tilde{\mathcal{H}}}\, dt$.

b) For $t \geq s$ we have

$$\mathcal{P}_t^* x(t) + \mathcal{R}_t u = P_{t,s}(\mathcal{P}_s^* x(s) + \mathcal{R}_s u),$$

where $P_{t,s}$ is the projection of $L^2(s,\infty;\tilde{\mathcal{H}})$ onto $L^2(t,\infty;\tilde{\mathcal{H}})$ defined by restriction. We have indeed, from the definitions,

$$\begin{aligned}\mathcal{P}_t^* x(t) + \mathcal{R}_t u &= L^* \phi_t x(t) + \mathcal{L}_t M \phi_t x(t) + Ku + L^* \mathcal{L}_t u + \mathcal{L}_t^* L u + \mathcal{L}_t^* M \mathcal{L}_t u \\ &= (L^* + \mathcal{L}_t^* M)[\phi_t x(t) + \mathcal{L}_t u] + \mathcal{L}_t^* L u + Ku,\end{aligned}$$

and a direct calculation shows that $[\phi_t x(t) + \mathcal{L}_t u](r) = [\phi_s x(s)](r) + (\mathcal{L}_s u)(r)$, $r > t \geq s$. Since $(\mathcal{L}_{t_o}^* x)(r) = B^*(r)\int_r^\infty X^*(t,r)x(t)dt$, $r \geq t_o$, we have $\mathcal{L}_t^* = P_{t,s}\mathcal{L}_s^*$, for $t > s$, and the claim is proved.

c) Let $u^{(t_o,x_o)} = -\mathcal{R}_{t_o}^{-1}\mathcal{P}_{t_o}^* x_o$.

Replace $\mathcal{P}_{t_o}^* x_o = -\mathcal{R}_{t_o} u^{(t_o,x_o)}$ in the expression for $J(t_o, x_o, u)$ to get

$$J(t_o, x_o, u) = (x_o, R(t_o)x_o)_{\mathcal{H}} + \langle \mathcal{R}_{t_o}(u - u^{(t_o,x_o)}), u - u^{(t_o,x_o)} \rangle_{t_o}$$

d) For $x \in L^2(t_o, \infty : \mathcal{H})$, $u \in L^2(t_o, \infty : \tilde{\mathcal{H}})$, define $\lambda(t) = \int_t^\infty X^*(s,t)[M(s)x(s) + L(s)u(s)]ds$. Then for arbitrary $u \in L^2(t_o, \infty, \tilde{\mathcal{H}})$, $v \in L^2(t_o, \infty; \tilde{\mathcal{H}})$, we have

$$J(t_o, x_o, u + v) = J(t_o, x_o, u) + 2\langle v, L^* x + Ku + B\lambda \rangle_{t_o} + J(t_o, 0, v).$$

To prove this claim, denote $z(t) = (\mathcal{L}_{t_o} v)(t)$ and start with

$$\begin{aligned}J(t_o, x_o, u + v) = {} & J(t_o, x_o, u) + J(t_o, 0, v) \\ & + 2\int_{t_o}^\infty (v(t),\ K(t)u(t))_{\tilde{\mathcal{H}}}\, dt + 2\int_{t_o}^\infty (L(t)v(t), x(t))_{\mathcal{H}}\, dt \\ & + 2\int_{t_o}^\infty (L(t)u(t), z(t))_{\mathcal{H}}\, dt + 2\int_{t_o}^\infty (z(t), M(t)x(t))_{\mathcal{H}}\, dt.\end{aligned}$$

The result follows from

$$\int_{t_o}^{\infty} (z(t), L(t)u(t) + M(t)x(t))_{\mathcal{H}}\, dt$$
$$= \int_{t_o}^{\infty} (v(s), \int_{s}^{\infty} B^*(s)X^*(t,s)[L(t)u(t) + M(t)x(t)]\, dt)_{\tilde{\mathcal{H}}}\, ds$$
$$= \int_{t_o}^{\infty} (v(s), B^*(s)\lambda(s))_{\tilde{\mathcal{H}}}\, ds.$$

e) Let

$$x^{(t_o,x_o)} = \phi_{t_o} x_o + \mathcal{L}_{t_o} u^{(t_o,x_o)} \lambda^{(t_o,x_o)}(t)$$
$$= \int_{t}^{\infty} x^*(s,t)[M(s)x^{(t_o,x_o)}(s) + L(s)u^{(t_o,x_o)}(s)]ds.$$

Then $L^*(t)x^{(t_o,x_o)}(t) + K(t)u^{(t_o,x_o)}(t) + B^*(t)\lambda^{(t_o,x_o)}(t) = 0$ a.e. In the formula in d), take $u = u^{(t_o,x_o)}$, $v = w - u^{(t_o,x_o)}$ to get $J(t_o, x_o, w) = J(t_o, x_o, u^{(t_o,x_o)}) + J(t_o, 0, w - u^{(t_o,x_o)}) + 2\langle w - u^{(t_o,x_o)}, L^*x^{(t_o,x_o)} + Ku^{(t_o,x_o)} + B^*\lambda^{(t_o,x_o)}\rangle_{t_o}$, and we obtain the conclusion since $w$ is arbitrary. Since $B, K, L$ are assumed to be continuous and $K$ is uniformly invertible, we may deduce that $u^{(t_o,x_o)}$ is continuous.

f) Take $u = u^{(t_o,x_o)}$, $s = t_o$ in b); then $\mathcal{P}_t^* x^{(t_o,x_o)}(t) + \mathcal{R}_t u^{(t_o,x_o)} = 0$, hence $u^{(t_o,x_o)} = -\mathcal{R}_t^{-1}\mathcal{P}_t^* x^{(t_o,x_o)}(t)$.

Define $F(t)$ by $F(t)x_o = -(\mathcal{R}_t^{-1}\mathcal{P}_t^* x_o)(t)$, $x_o \in \mathcal{H}$. Then $u^{(t_o,x_o)}(t) = F(t)x^{(t_o,x_o)}(t)$, hence

$$x^{(t_o,x_o)}(t) = X(t,t_o)x_o + \int_{t_o}^{t} X(t,s)B(s)F(s)x^{(t_o,x_o)}(s)\, ds.$$

Since we have also

$$x^{(t_o,x_o)}(t) = X(t,t_o)x_o - \int_{t_o}^{t} X(t,s)B(s)(\mathcal{R}_{t_o}^{-1}\mathcal{P}_{t_o}^* x_o)(s)\, ds,$$

we may write $x^{(t_o,x_o)}(t) = \hat{X}(t,t_o)x_o$, where $\hat{X}(t,t_o)$ is a linear operator and $\hat{X}(\cdot,\cdot)x_o$ is continuous. We shall see that $\hat{X}$ is an evolutionary process. Let $t \geq T \geq t_o$ and write

$$\hat{X}(t,t_o)x_o = X(t,T)[X(T,t_o)x_o - \int_{t_o}^{t} X(T,s)B(s)(\mathcal{R}_{t_o}^{-1}\mathcal{P}_{t_o}^* x_o)(s)\, ds].$$

On the other hand

$$\begin{aligned}\hat{X}(t,T)\hat{X}(T,t_o)x_o &= X(t,T)[\hat{X}(T,t_o) - \int_T^t X(T,s)B(s)(\mathcal{R}_T^{-1}\mathcal{P}_T^*\hat{X}(T,t_o)x_o)(s)\,ds] \\ &= X(t,T)[X(T,t_o)x_o - \int_{t_o}^T X(T,s)B(s)(\mathcal{R}_{t_o}^{-1}\mathcal{P}_{t_o}^*)(s)\,ds \\ &\quad - \int_T^t X(T,s)B(s)(\mathcal{R}_T^{-1}\mathcal{P}_T^*\hat{X}(T,t_o)x_o)(s)\,ds].\end{aligned}$$

But we know that $\mathcal{R}_T^{-1}\mathcal{P}_T^*\hat{X}(T,t_o)x_o = \mathcal{R}_T^{-1}\mathcal{P}_T^*x^{(t_o,x_o)}(T) = -u^{(t_o,x_o)} = \mathcal{R}_{t_o}^{-1}\mathcal{P}_{t_o}^*x_o$ and we deduce that $\hat{X}(t,T)\hat{X}(T,t_o)x_o = \hat{X}(t,t_o)x_o$, hence $\hat{X}$ is indeed an evolutionary process. We shall denote it by $X_F$. It can be seen that $X_F$ satisfies the integral relation in part i) of the theorem statement. Since the norms of the operation $\phi_{t_o}$, $\mathcal{L}_{t_o}$, $\mathcal{R}_{t_o}^{-1}$, $\mathcal{P}_{t_o}^*$ are uniformly bounded with respect to $t_o$, we deduce that $\langle x^{(t_o,x_o)}, x^{(t_o,x_o)}\rangle_{t_o} \leq c^2|x_o|_{\mathcal{H}}^2$, and by a Datko-type argument (Datko 1970, 1973) [2], [3], we deduce that $X_F$ is exponentially stable.

g) Let us now remark that

$$\lambda^{(t_o,x_o)}(t) = \phi_t^*[Mx^{(t_o,x_o)} + Lu^{(t_o,x_o)}],$$

and a simple calculation gives

$$\lambda^{(t_o,x_o)}(t) = \mathcal{P}_t^0 x^{(t_o,x_o)}(t) + \mathcal{P}_t u^{(t_o,x_o)} = R(t)x^{(t_o,x_o)}(t) \qquad t \geq t_o.$$

We use this formula in the relation in e) to get

$$L^*(t)x^{(t_o,x_o)}(t) + K(t)F(t)x^{(t_o,x_o)}(t) + B^*(t)R(t)x^{(t_o,x_o)}(t) = 0,$$

and for $t = t_o$ we obtain the formula for $F$ in the theorem statement.

We see also that

$$R(t)x_o = \int_t^\infty X^*(s,t)[M(s) - F^*(s)K(s)F(s) - R(s)B(s)F(s)]X_F(s,t)x_o\,ds.$$

and we will deduce the formula for $R$ in the theorem statement from this. In the above formula, replace $X_F(s,t_o)x_o$ by the corresponding expression in the theorem statement. Denote $\tilde{M}(s) = M(s) - F^*(s)K(s)F(s)$ and $\tilde{R}(t)x_o = R(t)x_o - \int_t^\infty X^*(s,t)\tilde{M}(s)X(s,t)x_o\,ds$.

A direct calculation gives $\tilde{R}(t)x_o = -\int_t^\infty X^*(\sigma,t)\tilde{R}(\sigma)B(\sigma)F(\sigma)X_F(\sigma,t)x_o\,d\sigma$. Next write $\tilde{S}(t)x_o = X^*(t,t_o)\tilde{R}(t)x_o$, $\hat{S}(t)x_o = \tilde{S}(t)x_o - \tilde{S}(T)X(T,t)x_o$. A calculation shows that $\hat{S}(t)x_o = -\int_t^T \hat{S}(\theta)B(\theta)F(\theta)X_F(\theta,t)x_o\,d\theta$. A standard argument gives $\hat{S}(t)x_o = 0$ for $t \le T$. Hence $\tilde{S}(t)x_o = \tilde{S}(T)X(T,t)x_o, t \le T$. Hence $\tilde{S}(t)x_o = \tilde{S}(T)X(T,t)x_o,\ t \le T$ and $X^*(t,t_o)\tilde{R}(t)x_o = X^*(T,t_o)\tilde{R}(T)X(T,t_o)x_o, T \ge t \ge t_o$. For $t = t_o$ we deduce $\tilde{R}(t)x_o = X^*(T,t_o)\tilde{R}(T)X(T,t_o)x_o$. Using exponential stability of $X$ and boundedness of $\tilde{R}$ we deduce that $\tilde{R}(t)x_o = 0$, and we obtain the formula for $R$ in the statement of the theorem. The theorem is completely proved. □

Let us remark that the theorem states essentially that the assumption concerning invertibility of $\mathcal{R}_{t_o}$ implies existence of $R$ and $F$ satisfying (i), (ii) and $F = -K^{-1}(L^* + B^*R)$.

Let us assume that $X$ is generated by $A$, that is

$$\frac{d}{dt}X(t,s) = A(t)X(t,s), \qquad X(s,s) = I.$$

Then $\frac{d}{dt}X(s,t) = -X(s,t)A(t)$ and

$$R'(t) = -M(t) + F^*(t)K(t)F(t) - A^*(t)R(t) - R(t)A(t),$$

that is,

$$R(t) + A^*(t)R(t) + R(t)A(t) - M(t) - [L(t) + R(t)B(t)]K^{-1}(t)[L^*(t) + B^*(t)R(t)] = 0.$$

Hence $R$ solves a Riccati equation and this solution is bounded on $\mathbb{R}$.

We see further that

$$\frac{d}{dt}X_F(t,t_o) = A(t)X_F(t,t_o) + B(t)F(t)X_F(t,t_o),$$

hence $X_F$ is generated by $A + BF$, and since $X_F$ is exponentially stable we see that $R$ is a stabilizing solution of the Riccati equation. This solution is unique, as we shall prove in the general situation. Such uniqueness allows us to deduce that in the almost periodic case it is an almost periodic solution to the Riccati equation, in the periodic case it is periodic and if all coefficients are constant it is a solution to the "algebraic Riccati equation".

Let us end this section by proving uniqueness.

**Proposition 5.** *Let $B, C$ be continuous, operator-valued functions, bounded on $\mathbb{R}$, with $B(t) = B^*(t)$, $C(t) = C^*(t)$; let $R_1$, $R_2$ be solutions to*

$$R(t) = X^*(T,t)R(T)X(T,t) + \int_t^T X^*(s,t)[R(s)B(s)R(s) + C(s)]X(s,t)\,ds,$$

*where $X$ is an evolutionary process.*

Let $X_{R_1}$, $X_{R_2}$ be defined by

$$X_{R_1}(t,s) = X(t,s) + \int_s^t X(t,\tau)B(\tau)R_1(\tau)X_{R_1}(\tau,s)\,d\tau$$

$$X_{R_2}(t,s) = X(t,s) + \int_s^t X(t,\tau)B(\tau)R_2(\tau)X_{R_2}(\tau,s)\,d\tau.$$

Then

(i) $R_1(t) - R_2(t) = X^*_{R_2}(T,t)[R_1(T) - R_2(T)]X_{R_1}(T,t)$,

(ii) If $R_1, R_2$ are bounded and $X_{R_1}$, $X_{R_2}$ are exponentially stable, then $R_1 = R_2$.

*Proof.* Denote $P = R_1 - R_2$; then

$$P(t) = X^*(T,t)P(T)X(T,t) + \int_t^T X^*(s,t)[R_1(s)B(s)R_1(s) - R_2(s)B(s)R_2(s)]X(s,t)\,ds.$$

Use

$$X(T,t) = X_{R_1}(T,t) - \int_t^T X_{R_1}(T,t)B(\tau)R_1(T)X(\tau,t)\,d\tau$$

$$= X_{R_2}(T,t) - \int_t^T X_{R_2}(T,\tau)B(\tau)R_2(\tau)X(\tau,t)\,d\tau$$

to obtain

$$
\begin{aligned}
X^*(T,t)P(T)X(T,t) &= X^*_{R_2}(T,t)P(T)X_{R_1}(T,t) \\
&\quad - \int_t^T X^*(\tau,t)R_2(\tau)B(\tau)X_{R_2}(T,\tau)P(T)X_{R_1}(T,t)\,d\tau \\
&\quad - \int_t^T X^*_{R_2}(T,t)P(T)X_{R_1}(T,\tau)B(\tau)R_1(\tau)X(\tau,t)\,d\tau \\
&\quad + \int_t^T \int_t^T X^*(\tau,t)R_2(\tau)B(\tau)X^*_{R_2}(T,\tau)P(T)X_{R_1}(T,\theta)B(\theta)R_1(\theta)X(\theta,t)d\tau\,d\theta.
\end{aligned}
$$

Write $X^*_{R_2}(T,t)P(T)X_{R_1}(T,t) = \tilde{P}(t)$, Direct calculations lead to

$$P(t)-\tilde{P}(t) = \int_t^T X^*(\tau,t)\{R_2(\tau)B(\tau)[P(\tau)-\tilde{P}(\tau)]+[P(\tau)-\tilde{P}(\tau)]B(\tau)R_1(\tau)\}X(\tau,t)\,d\tau.$$

By a standard argument it follows that $P(t) = \tilde{P}(t)$, and thus (i) is proved. Uniqueness follows by letting $t \to \infty$. We obtain from here also the formula

$$P(t) = X^*(T,t)P(T)X(T,t) + \int_t^T X^*(s,t)[P(s)B(s)R_1(s) + R_2(s)B(s)P(s)]X(s,t)\,ds.$$

## 3. A Riccati inequality via input-output operators.

**Theorem 1.** *Let the evolutionary process $X$ and continuous and bounded operator-valued functions $B, C, D$ be given, $B(t) : \mathcal{U} \to \mathcal{X}$, $C(t) : \mathcal{X} \to \mathcal{Y}$, $D(t) : \mathcal{U} \to Y$, where $\mathcal{X}$, $\mathcal{U}$, $\mathcal{Y}$ are Hilbert spaces. Assume $\delta > o$ exists such that $\gamma^2 I - D^*(t)D(t) \geq \delta^2 I$ for all $t \in \mathbb{R}$. Let $V$ be such that $\gamma^2 I - D^*(t)D(t) = V^*(t)V(t)$ with $V(t)$ investible and $V^{-1}(t)$ bounded uniformly with respect to $t \in \mathbb{R}$.*

Assume there exist operator valued functions $P, R, W$ satisfying:

a) $P(t) \geq \alpha I$, $\alpha > 0$, $t \longmapsto |R(t)|$ is bounded,

b) $C^*(t)D(t) + R(t)B(t) + W^*(t)V(t) = 0$,

c) $R(t)x_o = \int_t^\infty X^*(s,t)[C^*(s)C(s) + W^*(s)W(s) + P(s)]X(s,t)x_o\,ds.$

Then:

(i) $X$ is exponentially stable.

(ii) The input-output operator $T : L^2(\mathbb{R}, \mathfrak{U}) \to L^2(\mathbb{R}, \mathfrak{Y})$ defined by

$$(Tu)(t) = D(t)u(t) + \int_{-\infty}^{t} C(t)X(t,s)B(s)u(s)\,ds, \text{ satisfies } \|T\| < \gamma.$$

Conversely, assume that $X$ is exponentially stable and $\|T\| < \gamma$. Then $\gamma^2 I - D^*(t)D(t) \geq \delta^2 I$ and, with $V$ as above, there exist $\epsilon > 0$ and bounded $R$ and $W$ satisfying

$$D^*C + B^*R + V^*W = 0,$$

$$R(t)x_o = \int_{t}^{\infty} X^*(s,t)[C^*(s)C(s) + W^*(s)W(s) + \epsilon^2 I]X(s,t)x_o\,ds.$$

*Proof.* a) The exponential stability follows directly by a Datko type argument.

b) We have $(T^*y)(s) = D^*(s)y(s) + \int_s^\infty B^*(s)X^*(t,s)C^*(t)y(t)\,dt$ and a direct calculation using relations for $V, W, R$ leads to

$$\begin{aligned}[(\gamma^2 I - T^*T)u](s) &= [\gamma^2 I - D^*(s)D(s)]u(s) \\ &+ \int_s^\infty B^*(s)X^*(\sigma,s)W^*(\sigma)V(\sigma)u(\sigma)d\sigma + \int_{-\infty}^{s} V^*(s)W(s)X(s,\sigma)B(\sigma)u(\sigma)\,d\sigma \\ &+ \int_s^\infty \left\{\int_{-\infty}^{\tau} B^*(s)X^*(\tau,s)[W^*(\tau)W(\tau) + P(\tau)]X(\tau,\sigma)B(\sigma)u(\sigma)d\sigma\right\} d\tau\end{aligned}$$

Consider the input-output operator $G$ associated to $X$, $B$, $(W^*W + P)^{1/2}$, $(W^*W + P)^{1/2}W^*V$.

A calculation leads to $\langle(\gamma^2 I - T^*T)u, u\rangle = \langle G^*Gu, u\rangle + \langle(I + W^*PW)^{-1}Vu, Vu\rangle \geq \delta^2\langle u, u\rangle$, and (ii) is proved.

c) Assume now that $X$ is exponentially stable and $\|T\| < \gamma$. Then, by a suitable choice of the input $u$ we deduce that $\gamma^2 I - D^*(t)D(t) \geq \delta^2 I$. Since from exponential stability of $X$ it follows that $\|x\|^2 \leq c^2\|u\|^2$, we deduce that $\epsilon > o$ will exist such that for every $t_o$, we have $\int_{t_o}^\infty [\gamma^2(u,u)_{\mathfrak{U}} - (y,y)_{\mathfrak{Y}} - \epsilon^2(x,x)_{\mathfrak{X}}]\,dt \geq \frac{\delta^2}{2}\int_{t_o}^\infty (u,u)_{\mathfrak{U}}\,dt$, where we have denoted

$y = Cx + Du$, $x(t) = \int_{t_o}^{t} X(t,s)B(s)u(s)\,ds$. Denote $K(t) = \gamma^2 I - D^*(t)D(t)$, $L(t) = -C^*(t)D(t)$, $M(t) = -C^*(t)C(t) - \varepsilon^2 I$, and use Theorem 4 in Section 2 to deduce existence of $N$, $F$ with $F = -K^{-1}(L^* + B^*N)$, $X_F$ exponentially stable, and

$$N(t)x_o = \int_t^\infty X^*(s,t)[-C^*(s)C(s) - \varepsilon^2 I - F^*(s)K(s)F(s)].$$

Introduce $V, W$ as $\gamma^2 I - D^*D = V^*V$, $L^* + B^*N = V^*W$, take $R = -N$ and deduce the formula for $R$ in the theorem statement.

## 4. A Riccati inequality via stabilizing state feedback with disturbance attenuation.

**Theorem 1.** *Consider the evolutionary process $X$ and operator-valued functions $B_1$, $B_2$, $C_1$, $D_{12}$ with $D_{12}^*D_{12} = I$, $D_{12}^*C_1 = 0$. Assume there exists an operator-valued function $F_2$ and a corresponding exponentially stable evolutionary process $X_{F_2}$ satisfying*

$$X_{F_2}(t,t_o)x_o = X(t,t_o)x_o + \int_{t_o}^{t} X(t,s)B_2(s)F_2(s)X_{F_2}(s,t_o)x_o\,ds,$$

*such that the operator $T$ defined for $u_1 \in L^2(\mathbb{R}, U^1)$ by*

$$(Tu_1)(t) = \int_{-\infty}^{t} [C_1(t) + D_{12}(t)F_2(t)]X_{F_2}(t,s)B_1(s)u_1(s)\,ds$$

*satisfies $\|T\| < 1$. Then there exists an operator-valued function $R$ satisfying the Riccati inequality,*

$$R(t) - \int_t^\infty X_{F_2}^*(s,t)\{C_1^*(s)C_1(s) + F_2(s)F_2(s)$$
$$+ R(s)B_1(s)B_1^*(s)R(s)\}X_{F_2}(s,t)\,ds \geq \alpha I,\ \alpha > 0.$$

*Conversely, if $R \geq 0$ satisfies*

$$R(t) = \int_t^\infty X^*(s,t)\{C_1^*(s)C_1(s) + R(s)[B_1(s)B_1^*(s) - B_2(s)B_2^*(s)R(S) + P(s)\}$$
$$X(s,t)\,ds, \quad P(s) \geq \alpha I,\ \alpha > 0,$$

*and if we set* $F_2(s) = B_2^*(s)R(s)$*, then* $X_{F_2}$ *defined by*

$$X_{F_2}(t,t_o)x_o = X(t,t_o)x_o + \int_{t_o}^{t} X(t,s)B_2(s)F_2(s)X_{F_2}(s,t_o)\,ds$$

*is exponentially stable and* $T$ *above satisfies* $\|T\| < 1$.

*Proof.*

a) We shall use Theorem 1 in Section 3 for $(X_{F_2}, B_1, C_1 + D_{12}F_2, 0)$ and $\gamma = 1$. Then $V + I$, $W = -RB_1$ and $R(t)x_o = \int_t^\infty X_{F_2}^*(s,t)[C_1^*(s)C_1(s) + F_2^*(s)F_2(s) + R(s)B_1(s)B_1^*(s)R(s) + \epsilon^2 I]X_{F_2}(s,t)x_o\,ds$, and the first claim in the statement is obtained.

b) We shall use now some complementary formulae. As is proved in Curtain & Pritchard [1], if $X_F$ satisfies

$$X_F(t,t_o)x_o = X(t,t_o)x_o + \int_{t_o}^{t} X(t,s)B(s)F(s)X_F(s,t_o)x_o\,ds,$$

then $X(t,t_o)x_o = X_F(t,t_o)x_o - \int_{t_o}^{t} X_F(t,s)B(s)F(s)X(s,t_o)x_o\,ds$ (see also Pandolfi [5]).

c) If $R(t)x_o = \int_t^\infty X^*(s,t)[M(s) - F^*(s)K(s)F(s)]X(s,t)x_o\,ds$,
$F(s) = K^{-1}(s)[L^*(s) + B^*(s)R(s)]$, then
$R(t)x_o = \int_t^\infty X_F^*(s,t)[M(s) - L(s)K^{-1}(s)L^*(s) + R(s)B(s)K^{-1}(s)B^*(s)R(s)]X_F(s,t)x_o\,ds$.
To check this formula, recall the formulae in the proof of Theorem 4, Section 2, to see

$$\begin{aligned} R(t)x_o &= \int_t^\infty X^*(s,t)[M(s) - F^*(s)K(s)F(s) - R(s)B(s)F(s)]X_F(s,t)x_o\,ds \\ &= \int_t^\infty X^*(s,t)[M(s) + L(s)F(s)]X_F(s,t)x_o\,ds \\ &= \int_t^\infty X^*(s,t)[M(s) - L(s)K^{-1}(s)L^*(s) + R(s)B(s)B^*(s)R(s) \\ &\qquad + F^*(s)B(s)R(s)]X_F(s,t)x_o\,ds. \end{aligned}$$

Set $\hat{M}(s) = M(s) - L(s)K^{-1}(s)L^*(s) + R(s)B(s)B^*(s)R(s)$, write

$$R(t)x_o = \int_t^\infty X^*(s,t)\hat{M}(s)X_F(s,t)x_o\,ds + \int_t^\infty X^*(s,t)F^*(s)B^*(s)R(s)X_F(s,t)x_o\,ds,$$

and use the formula in b) to get

$$R(t)x_o = \int_t^\infty X_F^*(s,t)\hat{M}(s)X_F(s,t)x_o\,ds - \int_t^\infty \left[\int_t^s X^*(\sigma,t)F^*(\sigma)B^*(\sigma)X_F^*(s,\sigma)d\sigma\right]\hat{M}(s)X_F(s,t)x_o\,ds + \int_t^\infty X^*(\sigma,t)F^*(\sigma)B^*(\sigma)R(\sigma)X_F(\sigma,t)x_o\,d\sigma.$$

Denote $\hat{R}(t) = R(t) - \int_t^\infty X_F^*(s,t)\hat{M}(s)X_F(s,t)\,ds$ and deduce $\hat{R}(t)x_o = \int_t^\infty X^*(\sigma,t)F^*(\sigma)B^*(\sigma)\hat{R}(\sigma)X_F(\sigma,t)x_o\,d\sigma$. Further write $\hat{S}(t) = \hat{R}(t)X_F(t,t_o)$ and deduce by direct calculation that

$$[\hat{S}(t) - X_F^*(T,t)\hat{S}(T)]x_o = \int_t^T X^*(s,t)F^*(s)B^*(s)[\hat{S}(s) - X_F^*(T,s)\hat{S}(T)]x_o\,ds.$$

From here, by standard arguments we obtain $\hat{S}(t) - X_F^*(T,t)\hat{S}(T) = 0$, $R(t)X_F(t,t_o) = X_F^*(T,t)\hat{R}(T)X_F(T,t_o)$, $\hat{R}(t) = X_F^*(T,t)\hat{R}(T)X_F(T,t)$, $T > t$, and since $X_F$ is exponentially stable and $\hat{R}$ is bounded, we deduce that $\hat{R} \equiv 0$ and the claim in c) is proved.

d) If $R$ satisfies the Riccati inequality in the statement and if we set $F_2(s) = -B_2^*(s)R(s)$, we deduce from the above formulae with $K = I$, $L = 0$, $M = RB_1B_1^*R + C_1^*C_1 + P$, that

$$R(t)x_o = \int_t^\infty X_{F_2}(s,t)[C_1^*(s)C_1(s) + R(s)B_1(s)B_1^*(s)R(s) + F_2^*(s)F_2(s)]X_{F_2}(s,t)x_o\,ds + \int_t^\infty X_{F_2}^*(s,t)P(s)X_{F_2}(s,t)x_o\,ds.$$

In fact, if we do not know that $X_{F_2}$ is exponentially stable from the proof in c), we have only to write

$$R(t)x_o = X_{F_2}^*(T,t)R(T)X_{F_2}(T,t)x_o + \int_t^T X_{F_2}(s,t)\hat{P}(s)X_{F_2}(s,t)x_o\,ds,$$
$$\hat{P}(s) \geq \alpha I,\ \alpha > 0;$$

we deduce from this that

$$\alpha\int_t^T |X_{F_2}(s,t)x_o|^2 ds \leq (R(t)x_o, x_o) - (R(T)X_{F_2}(T,t)x_o,\ X_{F_2}(T,t)x_o),$$

and since $R(T) \geq 0$, it follows that

$$\int_t^{\infty} |X_{F_2}(s,t)x_o|^2 ds \leq \frac{\rho}{\alpha}|x_o|^2, \quad R(t)(x_o, x_o) \leq \rho |x_o|^2.$$

A Datko-type argument shows that $X_{F_2}$ is exponentially stable, and reasoning as in the proof of Theorem 1 in Section 3 shows that $T$ is a contraction. In this way the theorem is proved.

The result in Theorem 1 is an extension of the one in van Keulen et.al. (1991), [4], since by an argument in Pandolfi [5], existence of $R$ satisfying the Riccati inequalilty implies existence of a solution to the Riccati equation.

## 5. Stabilizing compensators with disturbance altenuation.

We are now in position to sketch the proof of Theorem 1 in Section 1. The system $(\Sigma_R)$ may be written

$$\begin{pmatrix} x' \\ x'_c \end{pmatrix} = \mathcal{A}\begin{pmatrix} x \\ x_c \end{pmatrix} + \mathcal{B}u_1, \quad y_1 = \mathcal{C}\begin{pmatrix} x \\ x_c \end{pmatrix},$$

$$\mathcal{A} = \begin{pmatrix} A & B_2C_c \\ B_cC_2 & A_c \end{pmatrix}, \quad \mathcal{B} = \begin{pmatrix} B_1 \\ B_cD_{21} \end{pmatrix}, \quad \mathcal{C} = (\, C_1 \quad D_{12}C_c \,).$$

For this system the input-output operator is strictly contracting and we may use Theorem 1 in Section 3 to deduce existence of $\mathcal{R}$, $\mathcal{W}$, $\mathcal{P}$ with $\mathcal{R}(t) \geq \rho I$, $\quad \mathcal{P}(t) \geq \alpha I$, $\quad \rho > o$, $\alpha > o$,

$$\mathcal{R}\mathcal{B} + \mathcal{W}^* = o, \quad \mathcal{R}' + \mathcal{A}^*\mathcal{R} + \mathcal{R}\mathcal{A} + \mathcal{C}^*\mathcal{C} + \mathcal{W}^*\mathcal{W} + \mathcal{P} = 0.$$

Write $\mathcal{R} = \begin{pmatrix} R_{11} & R_{12} \\ R_{12}^* & R_{22} \end{pmatrix}$, $\quad \mathcal{W} = (\, W_1 \quad W_2 \,)$, $\quad \mathcal{P} = \begin{pmatrix} P_{11} & P_{12} \\ P_{12} & P_{22} \end{pmatrix}$ to obtain equations for $R_{11}$, $R_{12}$ $R_{22}$. If we denote $R_1 = R_{11} - R_{12}R_{22}^{-1}R_{12}^*$, we obtain

$$R_1' + R_1A + A^*R_1 + C_1^*C_1 + R_1(B_1B_1^* - B_2B_2^*)R_1 + P_1$$
$$+ (C_cR_{22}^{-1}R_{12}^* - B_2^*R_1)^*(C_cR_{22}^{-1}R_{12}^* - B_2^*R_1) = 0,$$

$$P_1 = P_{11} - R_{12}R_{22}^{-1}P_{12}^* - P_{12}R_{22}^{-1}R_{12}^* + R_{12} + R_{22}^{-1}P_{22}R_{22}^{-1}R_{12}^*,$$

and $P_1 \geq \bar{\alpha} I$. If we define $R = (1 - \varepsilon)R_1$ we see that for $\varepsilon > 0$ small enough, $R$ satisfies the first inequality in the statement.

We take next $S = R_{11}^{-1}$, and the equation for $R_1$ leads to

$$S' = SA^* + AS + S(C_1^*C_1 - C_2^*C_2)S + B_1B_1^* + SP_{11}S + (SC_2^* + SR_{12}B_c)(C_2S + B_c^*R_{12}^*S),$$

that is, $S$ satisfies the second Riccati inequality. We see also that $S^{-1} - R = R_{11} + \epsilon R_{11} - (R_{11} - R_{12}R_{22}^{-1}R_{12}^*) \geq \gamma I$. From the Riccati equation for $R_1$, with $F = -C_cR_{22}^{-1}R_{12}^*$, we see by a Liapunov argument that $A + B_2F$ defines an exponentially stable evolution, and in the same way from the equation for $R_{11}$ with $H = R_{11}^{-1}R_{12}B_c = SR_{12}B_c$, we see that $A + HC_2$ defines an exponentially stable evolution. The last two Riccati inequalities are then checked by direct inspection. To prove that the compensator in the statement has the required properties, we take $\mathcal{R} = \begin{pmatrix} S^{-1} & -S^{-1} + R \\ R - S^{-1} & S^{-1} - R \end{pmatrix}$, check that $\mathcal{R} > 0$, define $\mathcal{W}$ by $\mathcal{R}\mathcal{B} + \mathcal{W} = 0$, and compute $\mathcal{R}' + \mathcal{A}^*\mathcal{R} + \mathcal{R}\mathcal{A} + \mathcal{C}^*\mathcal{C} + \mathcal{W}^*\mathcal{W}^*$. After some manipulations, it is seen by a Riccati inequality argument that the compensator is stabilizing with disturbance attenuation.

We last remark that we preferred for this result to work with standard situations instead of considering general evolutinoary processes.

**Acknowledgement.** This paper is dedicated to Lawrence Markus. Almost periodic Riccati equations represented the first topics on which our mathematical trajectories did intersect; Popov ideas were a subject of our discussions during his visit to Bucharest and also at Warwick. The author also wishes to thank the reviewer for helping with the final preparation of the manuscript.

## REFERENCES

1. Curtain, R. F., Pritchard, A. J., *Infinite dimensional systems theory*, Lecture Notes in Control and Information Sciences, 8, Springer Verlag, 1978.

2. Datko, R., *Extending a theorem of A. M. Liapunov to Hilbert space*, Journal of Math. Anal. Appl. **52** (1970), 610–616.

3. Datko, R., *Uniform asymptotic stability of evolutionary processes in Banach space*, SIAM J. Math. Anal. **3** (1973), 428–445.

4. van Keulen B., Peters, M., Curtain, R., *$H_\infty$- control with state feedback: The infinite dimensional case*, Paper W-9015, MTNS 91, Kobe, Japan, June 17–21 (1991).

5. Pandolfi, L., *Existence of solutions to the Riccati equation for time-varying distributed systems*, Raporte Interno **1** Dipartimento di Matematica, Politecnico di Torino, 1992.

6. Sampei, M., Mita, T., Kakamichi, M., *An algebraic approach to output feedback control problems*, Systems and Control Letters **14** (1990), 13–24.

7. Yakubovich, V. A., *The frequency theorem for the case where the state and control spaces are Hilbert spaces*, Sib. Mat. Zh. T. **15 3** (1974), 639–669. (Russian)

8. Zhou, K., and Khargonekar, P. P., *An algebraic approach to $H^\infty$-optimization*, Systems and Control Letters **11** (1988), 85–92.

# 37 Vector Field Approximations Preserving Structural Properties

**Henry Hermes** University of Colorado, Boulder, Colorado

## 1. Introduction

Our goal is to study approximations of a vector field, or control system, which preserve structural properties. For example, if the structural property is asymptotic stability of the rest solution $x = 0$ for the equation $\dot{x} = X(x)$ with $X(0) = 0$, one might attempt to write $X(x) = A(x) + R(x)$ where $A$ would be an approximating vector field and $R$ a "higher order" remainder vector field. If $x = 0$ is an asymptotically-stable solution of $\dot{x} = A(x)$ when is this also true for $\dot{x} = X(x)$? We will deal with the situation where the linear approximation gives no definitive information concerning the desired property (and this aspect is coordinate free), i.e., in the above example $A$ would be nonlinear in any choice of local coordinates.

Let $Z$ be a real analytic vector field on $\mathbb{R}^n$ with $Z(0) = 0$ and $x = 0$ an asymptotically stable (either local or global) solution of $\dot{x} = Z(x)$. We first describe what will be meant by $X^{(m)}$ is a $Z$-homogeneous vector field of degree $m$. As motivation, if in local coordinates $x = (x_1, \ldots, x_n)$ we have $Z(x) = \sum_{i=1}^{n} - x_i \frac{\partial}{\partial x_i}$ and $X^{(m)}(x) = \sum_{i=1}^{n} a_i(x) \frac{\partial}{\partial x_i}$, then $X^{(m)}$ being $Z$ homogeneous of degree $m$ has the classical meaning, i.e., for $\varepsilon > 0$, $a_i(\varepsilon x) = \varepsilon^m a_i(x)$, $i = 1, \ldots, n$. In this case we say the $a_i$ are homogeneous of degree $m$ with respect to the standard dilation, denoted $\delta_\varepsilon^1 : \mathbb{R}^n \to \mathbb{R}^n$, where $\delta_\varepsilon^1 x = (\varepsilon x_1, \ldots, \varepsilon x_n)$. A more general dilation $\delta_\varepsilon^r x = (\varepsilon^{r_1} x_1, \ldots, \varepsilon^{r_n} x_n)$, $1 \le r_1 \le \cdots \le r_n$ integers, is associated to the vector field $Z(x) = \sum_{i=1}^{n} - r_i x_i \frac{\partial}{\partial x_i}$. Specifically a function $h : \mathbb{R}^n \to \mathbb{R}^1$ is homogeneous of degree $m$ with respect to $\delta_\varepsilon^r$ if $h(\delta_\varepsilon^r x) = \varepsilon^m h(x)$; we denote this by $h \in H_m$. A vector field

This research was supported by NSF grant DMS 9100439.

$X^{(m)}(x) = \sum_{i=1}^{n} a_i(x) \frac{\partial}{\partial x_i}$ is homogeneous of degree $m$ with respect to $\delta_\varepsilon^r$ if $a_i \in H_{r_i+m-1}$, $i = 1, \ldots, n$. Massera, [M], showed that if a smooth vector field $X$ with $X(0) = 0$ could be written as $X(x) = X^{(m)}(x) + R(x)$ where $X^{(m)}$ is homogeneous of degree $m$ with respect to $\delta_\varepsilon^1$ while the remainder $R$ is the sum of homogeneous vector fields of degree greater than $m$, then the zero solution of $\dot{x} = X^{(m)}(x)$ being asymptotically stable (and this is global by homogeneity) implies the same is true for the zero solution of $\dot{x} = X(x)$. In [H1] it is shown that this result holds with $\delta_\varepsilon^1$ replaced by $\delta_\varepsilon^r$. Here $X^{(m)}$ is considered the approximation of $X$.

One may easily show, e.g. [H2], that $X^{(m)}$ is homogeneous of degree $m$ with respect to a dilation $\delta_\varepsilon^r$ having $r = (r_1, \ldots, r_n)$ if and only if when $Z(x) = \sum_{i=1}^{n} -r_i x_i \frac{\partial}{\partial x_i}$ one has the Lie product $[Z, X^{(m)}] = (m-1)X^{(m)}$. This motivates the definition that if $Z$ is any real analytic vector field with $Z(0) = 0$ and $x = 0$ an asymptotically stable solution of $\dot{x} = Z(x)$, then *a vector field $X^{(m)}$ is $Z$-homogeneous of degree $m$ if $[Z, X^{(m)}] = (m-1)X^{(m)}$*. The analog of Massera's theorem for $X^{(m)}$ being $Z$-homogeneous of degree $m$ and the remainder $R$ a sum of vector fields all $Z$-homogeneous of degree greater than $m$ is shown in [H2] and stated as Theorem 1, below. A better feeling for what is required of the remainder $R$ can be obtained as follows. Let $(\exp tZ)(y)$ denote the solution, at time $t$, of the equation $\dot{x} = Z(x)$, $x(0) = y$ and $(\exp tZ)_*$ the induced tangent space isomorphism of the map $y \to (\exp tZ)(y)$. A critical result in [H2] is

**Proposition 1.** *Let $Z$ be a real analytic vector field with $Z(0) = 0$ and the rest solution $x = 0$ of $\dot{x} = Z(x)$ asymptotically stable. Then a continuous vector field $X$, with $X(0) = 0$, which is real analytic in a deleted neighborhood of zero is $Z$-homogeneous of degree $m$ if and only if*

$$(\exp -sZ)_* X((\exp sZ)(y)) = e^{-(m-1)s} X(y). \tag{1}$$

If $R$ is a sum of vector fields $Z$-homogeneous of degree greater than $m$, (1) gives a strong bound on the growth of $R$. Indeed, the main theorem of [H2] may be stated as follows.

**Theorem 1.** *Let $Z$ be a real analytic vector field with $Z(0) = 0$ and $x = 0$ an asymptotically stable solution of $\dot{x} = Z(x)$. Let $X$, with $X(0) = 0$, be a continuous vector field which is real analytic in a deleted neighborhood of the origin and assume we can write*

$$X(x) = X^{(m)}(x) + R(x) \tag{2}$$

*where $X^{(m)}$ is $Z$-homogeneous of degree $m \geq 1$ while the remainder vector field $R$ satisfies*

$$|(\exp -sZ)_* R((\exp sZ)(y))| = o(e^{-(m-1)s}) \quad \text{as} \quad s \to \infty \tag{3}$$

*for $y$ in a neighborhood of zero. Then if $x = 0$ is an asymptotically stable solution of $\dot{x} = X^{(m)}(x)$ the same is true for $\dot{x} = X(x)$. Furthermore, if $R$ is the sum of vector fields $Z$-homogeneous of degree greater than $m$ property (3) holds.*

Homogeneous approximating vector fields relative to a dilation $\delta_\varepsilon^r$ are basic tools in the proofs of theorems on small time local controllability (STLC), see [Su1], [H3], [St1]. Specifically, a control system

$$\dot{x} = X_0(x) + u(t)X_1(x), \quad x(0) = 0, \quad X_0(0) = 0, \quad X_1(0) \neq 0 \tag{4}$$

with $u$ the control and zero in the interior of the (otherwise arbitrary) set $U$ of admissible control values, is STLC at zero if for any $t_1 > 0$ the set of points which can be reached in time $t_1$ by solutions of (4) corresponding to all admissible controls (i.e., measurable $u : [0, t_1] \to U$) contains a full neighborhood of zero. The property STLC is a structural property, i.e., persists under sufficiently small ($C^1$) perturbations of the vector fields $X_0, X_1$. (For sharp results on questions of this type see [Su2].) Let $H_0$ consist of constant functions so $X^{(0)}$ is defined. Write

$$X_0(x) = X_0^{(m)}(x) + X_0^{(m+1)}(x) + \cdots, \quad X_1(x) = X_1^{(0)}(x) + X_1^{(1)}(x) + \cdots$$

(if the first component of $X_1$ does not vanish at zero and $r_1 = 1$, for example, $X_1^{(0)}(0) \neq 0$ and consider the "approximating system"

$$\dot{x} = X_0^{(m)}(x) + u(t)X_1^{(0)}(x), \qquad x(0) = 0. \tag{5}$$

One may show, [Br], [St1], [H4] that if system (5) is STLC at zero, the same is true for the original system (4). This reduces the problem to the study of the (usually) much simpler system (5). We prove, here, an analogous result, Theorem 2, for $Z$-homogeneous (rather than necessarily homogeneous with respect to a dilation) approximating vector fields. In particular, when dealing with homogeneity relative to a dilation, as above, a crucial subalgebra of the Lie algebra generated by $X_0^{(m)}$ and $X_1^{(0)}$ will be nilpotent which leads to "polynomial" calculations, see [H4].

Examples are given to illustrate the need and use of these types of approximations as are specific characterizations of $Z$-homogeneous vector fields for several interesting vector fields $Z$.

## 2. Examples.

If the linear control system formed by replacing the vector fields $X_0, X_1$ in (4) by their linear approximations is completely controllable, then it is well known, [LM, Theorem 1, pg. 366], that system (4) is locally controllable (STLC) at zero. If the linear approximation gives no information, it is natural to consider expanding all components of $X_0, X_1$ in a Taylor series and form an approximating system by retaining the lowest order, nonvanishing, terms in each component. Our first example will show that this crude attempt can result in an approximating system which is STLC at zero when the original system is not. Our purpose is to stress that approximations in the nonlinear case must be more carefully chosen if they are to "preserve" the structural property being considered.

*Example 2.1.*

Consider a control system of the form (4) on $\mathbb{R}^3$ with

$$X_0(x) = x_1^3 \frac{\partial}{\partial x_2} + (x_2^3 + x_1^4) \frac{\partial}{\partial x_3}, \qquad X_1 = \frac{\partial}{\partial x_1}.$$

The approximating system obtained by retaining the lowest order, nonvanishing, terms of a Taylor series expansion about zero of each component is

$$\dot{x} = Y_0(x) + uY_1, \quad Y_0(x) = x_1^3 \frac{\partial}{\partial x_2} + x_2^3 \frac{\partial}{\partial x_3}, \quad Y_1 = \frac{\partial}{\partial x_1}. \tag{6}$$

Let $(\operatorname{ad} V_1, V_2)$ denote the Lie product $[V_1, V_2]$ and inductively $(\operatorname{ad}^{k+1} V_1, V_2) = [V_1, (\operatorname{ad}^k V_1, V_2)]$. The Lie products in (6) which do not vanish at zero are $Y_1(0), (\operatorname{ad}^3 Y_1, Y_0)(0)$ and $(\operatorname{ad}^3(\operatorname{ad}^3 Y_1, Y_0), Y_0)(0)$, all of which have an odd number of factors $Y_1$. A theorem on STLC, [Su], [H3], states: Let $\mathcal{S}^j$ denote the Lie products of $Y_0, Y_1$ containing at most $j$ factors $Y_1$ and $\mathcal{S}^j(0)$ the elements of $\mathcal{S}^j$ evaluated at zero. If dim span $\mathcal{S}^k(0) = n$ for some $k$ and the values of $j$ at which dim span $\mathcal{S}^j(0)$ increases are all odd, then the system is STLC at zero. Here the increases occur at $j = 1, 3, 9$ hence system (6) is STLC at zero. For the original system, now letting $\mathcal{S}^j$ denote the Lie products of $X_0, X_1$ containing at most $j$-factors $X_1$, we have dim span $\mathcal{S}^3(0) = 2$ while $(\operatorname{ad}^4 X_1, X_0)(0) \in \mathcal{S}^4(0)$ is linearly independent of span $\mathcal{S}^3(0)$. This is an obstruction to STLC at zero, see [St2], and hence our original system is not STLC at zero. □

For asymptotic stability, the proof of Theorem 1 (as given in [H2]) can be utilized to show the following. Let $X$ be a smooth vector field on $\mathbb{R}^n$, with $X(0) = 0$, which can be written as $X(x) = A(x) + R(x)$ with $x = 0$ an asymptotically stable solution of $\dot{x} = A(x)$. If

$$(\exp -sA)_* R((\exp sA))(y)) \to 0 \tag{7}$$

as $s \to \infty$ for $y$ in a neighborhood of zero, then $x = 0$ is also an asymptotically stable solution of $\dot{x} = X(x)$. Condition (7) is very stringent! Indeed, if for example $A(x) = -x_1^3 \frac{\partial}{\partial x_1} - x_2^5 \frac{\partial}{\partial x_2}$ one finds $(\exp -sA)_*$ has singularities for finite values of $s > 0$. This, again, illustrates the need for "homogeneous" approximations. We next illustrate the use of Theorem 1 with

*Example 2.2.*

Let $Z(x) = -x_1 \frac{\partial}{\partial x_1} - 3x_2 \frac{\partial}{\partial x_2}$ which leads to homogeneity relative to the dilation $\delta_\varepsilon^r x = (\varepsilon x_1; \varepsilon^3 x_2)$, i.e., $r = (1, 3)$. Consider the question of asymptotic stability of $x = 0$

for $\dot{x} = X(x)$ with $X(x) = (-x_1^3 + x_2^2)\frac{\partial}{\partial x_1} - x_2^{5/3}\frac{\partial}{\partial x_2}$. Then $-x_1^3 \in H_3 = H_{r_1+2}$, $x_2^2 \in H_6 = H_{r_1+5}$, $-x_2^{5/3} \in H_5 = H_{r_2+2}$ and hence $X(x) = X^{(3)}(x) + X^{(6)}(x)$ where $X^{(3)}(x) = -x_1^3\frac{\partial}{\partial x_1} - x_2^{5/3}\frac{\partial}{\partial x_2}$, $X^{(6)}(x) = x_2^2\frac{\partial}{\partial x_1}$. Clearly the rest solution of $\dot{x} = X^{(3)}(x)$ is asymptotically stable and by Theorem 1, this is also true for $\dot{x} = X(x)$. The main point is the term $-x_1^3\frac{\partial}{\partial x_1}$ has "more influence" than $x_2^2\frac{\partial}{\partial x_1}$ relative to this dilation. □

Our next examples illustrate the construction of $Z$-homogeneous approximations for several reference vector fields $Z$. The well known characterization of a $Z$-homogeneous vector field $X^{(m)}$ of order $m$ when $Z(x) = \sum_{i=1}^{n} -r_i x_i \frac{\partial}{\partial x_i}$, $1 \le r_1 \le \cdots \le r_n$ integers, was given in the introduction.

*Example 2.3.*

On $\mathbb{R}^2$ let $Z(x) = Bx$ where $B = \begin{pmatrix} -2 & 1 \\ -1 & -2 \end{pmatrix}$. The eigenvalues of $B$ are $(-2 \pm i)$ so the rest solution $x = 0$ of $\dot{x} = Z(x)$ is asymptotically stable; solutions spiral to the origin. To compute what it means for a vector field $X^{(m)}$ to be $Z$-homogeneous of degree $m$ one must solve $[Z, X] = (m-1)X$ for $X$. If we let $X(x) = a^1(x)\frac{\partial}{\partial x_1} + a^2(x)\frac{\partial}{\partial x_2}$ and subscripts $a^i_{x_j}$ denote partial derivatives, this means we must solve the linear, partial differential equations

(i) $$(-2x_1 + x_2)a^1_{x_1} - (x_1 + 2x_2)a^1_{x_2} = -(1+m)a^1 + a^2$$

(ii) $$(-2x_1 + x_2)a^2_{x_1} - (x_1 + 2x_2)a^2_{x_2} = -a^1 - (1+m)a^2$$

Proceeding via the method of characteristics, let $x(t) = (x_1(t), x_2(t))$. The characteristic equations $\dot{x}_1 = -2x_1 + x_2$, $x_1(0) = c_1$; $\dot{x}_2 = -x_1 - 2x_2$, $x_2(0) = c_2$ have as solution

(iii) $$x(t) = c_1 e^{-2t}\begin{pmatrix} \cos t \\ -\sin t \end{pmatrix} + c_2 e^{-2t}\begin{pmatrix} \sin t \\ \cos t \end{pmatrix}.$$

With $x(t)$ given by (iii) we then must solve

(iv) $$\frac{d}{dt}\begin{pmatrix} a^1(x(t)) \\ a^2(x(t)) \end{pmatrix} = \begin{pmatrix} -(1+m) & 1 \\ -1 & -(1+m) \end{pmatrix}\begin{pmatrix} a^1(x(t)) \\ a^2(x(t)) \end{pmatrix}.$$

The solution of (iv) is (letting $a = \begin{pmatrix} a^1 \\ a^2 \end{pmatrix}$)

$$\text{(v)} \qquad a(x(t)) = k_1 e^{-(1+m)t} \begin{pmatrix} \cos t \\ -\sin t \end{pmatrix} + k_2 e^{-(1+m)t} \begin{pmatrix} \sin t \\ \cos t \end{pmatrix}.$$

From (iii), $e^{-2t}\cos t = \dfrac{c_1x_1 + c_2x_2}{c_1^2 + c_2^2}$, $e^{-2t}\sin t = \dfrac{c_2x_1 - c_1x_2}{c_1^2 + c_2^2}$.

Now write (v) as

$$\begin{aligned} a(x(t)) &= e^{-(m-1)t} \begin{pmatrix} k_1 e^{-2t}\cos t + k_2 e^{-2t}\sin t \\ -k_1 e^{-2t}\sin t + k_2 e^{-2t}\cos t \end{pmatrix} \\ &= \frac{(x_1^2 + x_2^2)^{(m-1)/4}}{(c_1^2 + c_2^2)^{(m+3)/4}} \begin{pmatrix} k_1c_1 + k_2c_2 & k_1c_2 - k_2c_1 \\ -k_1c_2 + k_2c_1 & k_1c_1 - k_2c_2 \end{pmatrix} \begin{pmatrix} x_1 \\ x_2 \end{pmatrix}. \end{aligned}$$

Combining constants gives the final result, i.e., the general form of a vector field $X^{(m)}$ which is $Z$-homogeneous of degree $m$ is

$$X^{(m)}(x) = (x_1^2 + x_2^2)^{(m-1)/4} \begin{pmatrix} \alpha & \beta \\ -\beta & \alpha \end{pmatrix} \begin{pmatrix} x_1 \\ x_2 \end{pmatrix}.$$

Note that, as expected, $Z$ itself is $Z$-homogeneous of degree one, i.e., $[Z, Z] = 0$. □

*Example 2.4.*

Let $Z(x) = Bx$ with $B = \begin{pmatrix} -2 & 1 \\ 0 & -2 \end{pmatrix}$. Then the zero solution of $\dot{x} = Z(x)$ is asymptotically stable. Proceeding as in Example 2.3, one finds that the general form of a vector field $X^{(m)}$ which is $Z$-homogeneous of degree $m$ is

$$X^{(m)}(x) = \begin{pmatrix} c_1x_1|x_2|^{(m-1)/2} + c_2|x_2|^{(m+1)/2} \\ c_1|x_2|^{(m+1)/2} \end{pmatrix}.$$

Again, $Z$ itself is homogeneous of degree one. Here $X^{(m)}$ is real analytic off the plane $x_2 = 0$, i.e., not in a deleted neighborhood of $x = 0$. □

## 3. $Z$-Homogeneous Approximations Preserve STLC.

When dealing with a dilation $\delta_\varepsilon^r$ the functions $h$ homogeneous of degree zero, i.e., $h(\delta_\varepsilon^r x) = h(x)$ are the constant functions, denoted $H_0$. If $r = (r_1, \dots, r_n)$ with $r_1 = 1$

the vector field $X = \frac{\partial}{\partial x_1}$ is homogeneous of degree zero with respect to $\delta_\varepsilon^r$, i.e., we can approximate a vector field $X_1$, as in (4), which has $X_1(0) \neq 0$, as $X_1^{(0)} + X_1^{(1)} + \cdots$. The results for asymptotic stability, as discussed in the introduction, dealt with vector fields $X$ having $X(0) = 0$ and $Z$-homogeneous approximations of degree $m \geq 1$. Examples 2.3, 2.4 show that, in general, $Z$-homogeneous vector fields of degree zero may have singularities at the origin hence in what follows we will not form homogeneous approximations of vector fields which do not vanish at the origin. For a single input system such as (4) with $X_1(0) \neq 0$ one can always choose local coordinates $x = (x_1, \ldots, x_n)$ such that $X_1 = \frac{\partial}{\partial x_1}$. We assume this has been done and consider our system, on $\mathbb{R}^n$, as

$$\dot{x} = X_0(x) + u(t)X_1, \qquad X_0(0) = 0,\ X_1 = \frac{\partial}{\partial x_1}. \tag{8}$$

**Theorem 2.** *Let $Z$ be a real analytic vector field with $Z(0) = 0$ and $x = 0$ an asymptotically stable solution of $\dot{x} = Z(x)$. Let $X_0$, as in (8), be continuous, real analytic in a deleted neighborhood of zero, and expandable as $X_0(x) = X_0^{(m)}(x) + R(x)$ where $X_0^{(m)}$ is $Z$-homogeneous of degree $m \geq 1$ while $R$ satisfies the remainder estimate (3). If the approximating system*

$$\dot{x} = X_0^{(m)}(x) + u(t)X_1 \tag{9}$$

*is STLC at zero, then system (8) is also STLC at zero.*

**Lemma.** *(a) Let $t \to x(t,u)$ denote a solution of (8) for some initial data and admissible control $u$. Consider $s > 0$ fixed (for the moment) and rescale time as $t = t(\tau) = e^{(m-1)s}\tau$. Define $x^s(\tau,u) = (\exp -sZ) \circ x(t(\tau),u)$. Then $\tau \to x^s(\tau,u)$ satisfies*

$$\begin{aligned} \frac{d}{d\tau}\, x^s(\tau,u) = {} & (X_0^{(m)}(x^s(\tau,u)) + e^{(m-1)s} \exp(-sZ)_* R((\exp sZ) \circ x^s(\tau,u))) \\ & + u(t(\tau))e^{(m-1)s} X_1. \end{aligned} \tag{10}$$

*(b) Conversely if $\tau \to x^s(\tau,u)$ is a solution of (10) and $\tau(t) = e^{(1-m)s}t$, then $x(t,u) = (\exp sZ) \circ x^s(\tau(t),u)$ is a solution of (8).*

*Proof.* (a) Using Proposition 1, we compute

$$\frac{d}{d\tau}\, x^s(\tau,u) = (\exp -sZ)_*\dot{x}(t(\tau),u)t'(\tau) =$$

$$(\exp -sZ)_*(X_0^{(m)}((\exp sZ)\circ x^s(\tau,u)) + R((\exp sZ)\circ x^s(\tau,u)) + u(t(\tau))X_1)e^{(m-1)s} =$$

$$(X_0^{(m)}(x^s(\tau,u) + e^{(m-1)s}(\exp -sZ)_*R((\exp sZ)\circ x^s(\tau,u))) + e^{(m-1)s}u(t(\tau))X_1$$

as was desired.

(b) Conversely, assuming $x^s(\tau,u)$ satisfies (10) we compute

$$\begin{aligned}\dot{x}(t,u) &= (\exp sZ)_*(X_0^{(m)}((\exp -sZ)\circ x(t,u)) \\ &\quad + e^{(m-1)s}(\exp -sZ)_*R(x(t,u)))e^{(1-m)s} + u(t)X_1 \\ &= (X_0^{(m)}(x(t,u)) + R(x(t,u)) + u(t)X_1. \quad \square\end{aligned}$$

*Proof of Theorem 2.* (This now follows closely the proof given in [St1] for homogeneity relative to a dilation.) Let $A_1(t)$ denote the attainable set at time $t \geq 0$ for system (8), i.e., the set of points which can be reached in time $t$ by solutions of (8), initiating from the origin at $t = 0$, using all admissible controls. Let $A_2(t)$ denote the attainable set for system (9) and $A_3(\tau)$ the attainable set for system (10). Recall that STLC at zero requires that zero be in the interior of the attainable set at any time $t_1 > 0$, where admissible controls are measurable with values in an arbitrary set $U \subset I\!R^1$ having $0 \in \text{int.}\,U$.

We assume that system (9) is STLC at zero, i.e., $0 \in \text{int.}\,A_2(t_1)$ for any $t_1 > 0$. Let $B_\rho$ be an origin centered ball of radius $\rho > 0$ contained in $A_2(t_1)$ and $\mathcal{U} = \{u \in \mathcal{L}_1[0,t_1] : u(t) \in U\}$ denote the set of admissible controls. For $p \in B_\rho$ let $\mathcal{U}_p$ denote those controls in $\mathcal{U}$ which have corresponding solutions of (9) with value $p$ at time $t_1$. Then $\mathcal{U}_p \neq \emptyset$ for $p \in B_\rho$ and for $\rho > 0$ sufficiently small, the map $p \to u_p$ admits a continuous selection, [St1, Lemma 2.1] which we denote $p \to u_p$. Then the map $p \to x^s(\tau(t_1), u_p)$ is continuous. Since $R$ satisfies the remainder estimate (3), $|e^{(m-1)}(\exp -sZ)_*R((\exp sZ)(y))| = o(1)$, i.e., goes to zero as $s \to \infty$, uniformly for $y$ in a neighborhood of zero. This means, see eq. (10), that for $s > 0$ sufficiently large we can assure that $|p - x^s(\tau(t_1,u_p))| < \rho/2$ for all $p$

in the boundary of $B_\rho$. By the corollary [LM, pg. 252] the map $p \to x^s(\tau(t_1), u_p)$, $p \in B_\rho$, then covers a $\rho/2$ ball about zero, i.e.,

$$\text{(i)} \qquad 0 \in \text{int.}\, A_3(\tau(t_1)) \quad \text{and} \quad s > 0 \text{ sufficiently large.}$$

Next, from the lemma, we have:

$$q = x(t(\tau), u) \in A_1(t(\tau)) \Longleftrightarrow (\exp -sZ)(q) \in A_3(\tau)$$

or equivalently

$$\text{(ii)} \qquad A_1(t(\tau)) = (\exp sZ)A_3(\tau).$$

Pick $t_1 > 0$, let $s > 0$ be such that (i) holds and choose $\tau$ so that $t(\tau) = e^{(m-1)s}\tau = t_1$. Then $0 \in \text{int.}\, A_3(\tau(t_1))$ and (ii) shows $0 \in \text{int.}\, A_1(t_1)$. □

## References


[Br] A. Bressan; Local asymptotic approximations of nonlinear control systems, *Internat. J. Control* **41** (1985), 1331–1336.

[H1] H. Hermes; Homogeneous coordinates and continuous asymptotically stabilizing feedback controls, *Differential Equations, Stability and Control* (S. Elaydi, ed.), Lecture Notes in Pure and Applied Math #127, Marcel Dekker, Inc., NY (1991), 249–260.

[H2] H. Hermes; Vector field approximations; flow homogeneity (to appear) *Ordinary Differential Equations and Delay Equations* (J. Wiener & J. Hale, eds.), Longman Press.

[H3] H. Hermes; Control systems which generate decomposable lie algebras, *J. Differential Eqs.* **44** (1982), 166–187.

[H4] H. Hermes; Nilpotent and high order approximation of vector field systems, *SIAM Review* **33** (1991), 238–264.

[LM] E. B. Lee and L. Markus; *Foundations of Optimal Control Theory*, John Wiley, NY (1967).

[M] J. L. Massera; Contributions to stability theory, *Ann. Math.* **64** (1956), 182–206.

[St1] G. Stefani; Polynomial approximation to control systems and local controllability, Proc. 24th IEEE Conference on Decision and Control, I, (1985), 33–38.

[St2] G. Stefani; Local properties of nonlinear control systems, *Geometric Theory of Nonlinear Control Systems*, (B. Jakubuzyk, W. Respondek and K. Tchon, eds.), Tech. University Wroclaw (1984), 219–226.

[Su1] H. Sussmann; A general theorem on local controllability, *SIAM J. Control & Opt.* **25** (1987), 158–194.

[SU2] H. Sussmann; Some properties of vector field systems that are not altered by small perturbations, *J. Diff. Eqs.* **20** (1976), 292–315.

# 38 Nonlinear Boundary Stabilization of a von Kármán Plate Equation

**Mary Ann Horn*** University of Minnesota, Minneapolis, Minnesota

**Irena Lasiecka**† University of Virginia, Charlottesville, Virginia

## 1 Introduction

### 1.1 Statement of the Problem

Let $\Omega$ be an open bounded domain in $R^2$ with a sufficiently smooth (e.g., $C^\infty$) boundary, $\Gamma$. In $\Omega$, we consider the following von Kármán system in the variables $w(t,x)$ and $\chi(w(t,x))$ with nonlinear feedback controls, $g$, $f_1$, and $f_2$:

$$w_{tt} - \gamma^2 \Delta w_{tt} + \Delta^2 w + b(x) w_t = [w, \chi(w)] \quad \text{in } Q_\infty = (0,\infty) \times \Omega \tag{1.1.a}$$

$$\left.\begin{aligned} w(0,\cdot) &= w_0 \\ w_t(0,\cdot) &= w_1 \end{aligned}\right\} \quad \text{in } \Omega \tag{1.1.b}$$

*This material is based upon work partially supported under a National Science Foundation Mathematical Sciences Postdoctoral Research Fellowship.

†Research partially supported by the National Science Foundation Grant NSF DMS 8902811.

$$\Delta w + (1-\mu)B_1 w = -f_1(\tfrac{\partial}{\partial\nu} w_t) \qquad \text{on } \Sigma_\infty = (0,\infty)\times\Gamma \tag{1.1.c}$$

$$\tfrac{\partial}{\partial\nu}\Delta w + (1-\mu)B_2 w - \gamma^2 \tfrac{\partial}{\partial\nu} w_{tt} - w = g(w_t) - \tfrac{\partial}{\partial\tau} f_2(\tfrac{\partial}{\partial\tau} w_t) \quad \text{on } \Sigma_\infty = (0,\infty)\times\Gamma, \tag{1.1.d}$$

where $b(x) \in L^\infty(\Omega)$ satisfies $b(x) > 0$ *a.e.* in $\Omega$, $0 < \mu < \frac{1}{2}$ is Poisson's ratio, the operators $B_1$ and $B_2$ are given by

$$\begin{aligned} B_1 w &= 2n_1 n_2 w_{xy} - n_1^2 w_{yy} - n_2^2 w_{xx} \\ B_2 w &= \tfrac{\partial}{\partial\tau}[(n_1^2 - n_2^2)w_{xy} + n_1 n_2 (w_{yy} - w_{xx})], \end{aligned} \tag{1.2}$$

and the controls, $g$ and $f_i$ are continuous, monotone functions and are subject to the following constraints:

$$\left.\begin{aligned} g(s)s &> 0 && \text{for } s \neq 0 \\ f_i(s)s &> 0 && \text{for } s \neq 0 \\ m|s| \le |f_i(s)| &\le M|s| && \text{for } |s| > 1,\ i = 1,2 \\ m|s|^2 \le g(s)s &\le M|s|^{r+1} && \text{for } |s| > 1, \end{aligned}\right\} (H-1)$$

where $r$ is any positive constant.

*Remark 1.1:* No assumptions are made on the behavior of $g$ and $f_i$, $i = 1,2$, at the origin.

In (1.1), $\chi(w)$ satisfies the system of equations

$$\left.\begin{aligned} \Delta^2 \chi &= -[w,w] \\ \chi &= \tfrac{\partial}{\partial\nu}\chi = 0 \quad \text{on } \Sigma_T, \end{aligned}\right\} \tag{1.3}$$

where

$$[\phi,\psi] = \frac{\partial^2\phi}{\partial x^2}\frac{\partial^2\psi}{\partial y^2} + \frac{\partial^2\phi}{\partial y^2}\frac{\partial^2\psi}{\partial x^2} - 2\frac{\partial^2\phi}{\partial x\partial y}\frac{\partial^2\psi}{\partial x\partial y}. \tag{1.4}$$

Define the bilinear form

$$a(w,v) = \int_\Omega (\Delta w\Delta v + (1-\mu)(2w_{xy}v_{xy} - w_{xx}v_{yy} - w_{yy}v_{xx}))d\Omega. \tag{1.5}$$

We define the energy functional by

$$\begin{aligned} E_w(t) &= \tfrac{1}{2}\int_\Omega\{|w_t|^2+\gamma^2|\nabla w_t|^2+|\Delta\chi(w)|^2\}d\Omega+\tfrac{1}{2}\int_\Gamma w^2 d\Gamma+\tfrac{1}{2}a(w,w) \\ &\equiv E_{w,1}(t)+E_{w,2}(t), \end{aligned} \tag{1.6}$$

where $E_{w,2}(t)$ is defined by

$$E_{w,2}(t) \equiv \frac{1}{2}\int_\Omega |\Delta\chi(w)|^2 d\Omega. \tag{1.7}$$

In view of this, the associated space of finite energy is $\mathcal{H} \equiv H^2(\Omega)\times H^1(\Omega)$, with the norm

$$\|(w,w_t)\|^2_{\mathcal{H}} \equiv \|w\|^2_{H^2(\Omega)}+\|w_t\|^2_{L_2(\Omega)}+\gamma^2\|\nabla w_t\|^2_{L_2(\Omega)}. \tag{1.8}$$

The following well-posedness theorem for problem (1.1)-(1.3) is a very special case of the result in [2].

**Theorem 1.1** *(See [2].) For any $w_0 \in H^2(\Omega)$, $w_1 \in H^1(\Omega)$, and $T>0$, there exists a unique solution to (1.1), $w \in C(0,T;H^2(\Omega))\cap C^1(0,T;H^1(\Omega))$, such that*

$$\nabla w_t|_\Gamma \in L_2(0,T;L_2(\Gamma)). \tag{1.9}$$

*Remark 1.2:* We note that the light internal damping alone (i.e., with out boundary dissipation), which is represented by $bw_t$, is not enough to cause uniform decay for $E_w(t)$.

*Remark 1.3:* Notice that the regularity property in (1.9) does not follow from a priori interior regularity of $w$ ($w_t \in H^1(\Omega)$). It is an independent regularity result.

Our goal is to show that the boundary controls, $f_1$, $f_2$, and $g$, cause the energy of our system, (1.6), to decay uniformly with respect to the initial energy as time increases.

## 1.2 Literature

The problem of boundary stabilization has attracted considerable attention in recent years (see [3], [10], [11], [8] and references therein). We shall concentrate on the results pertinent to model (1.1).

In the context of control theory and, in particular, stabilization theory, the von Kármán model was introduced for the first time in [3]. In fact, in [3], the exponential decay rates for the solutions to (1.1) with $\gamma = 0$ (rotational forces neglected) and with *linear* feedbacks, $f_i$ and $g$, were established. This result of [3] was derived under the geometric conditions on $\Gamma$ which required that $\Omega$ be "star-shaped." Subsequently, in [1], the results of [3] were extended to the case when:

*(i.)* $\gamma \neq 0$, i.e., the rotational forces are taken into account;

*(ii.)* no geometric conditions are imposed on $\Gamma$.

These generalizations required techniques different from those in [3] and were based on microlocal estimates combined with a nonlinear compactness/uniqueness argument (see [1]) (rather than the Liapunov function techniques used in [3]). Other works relevant to the stabilization of the von Kármán system are: [5], where the one dimensional problem was treated; and [4] and [12], where a system of equations different from (1.1) is considered for two dimensional "star-shaped" domains. The main goal of this paper is to treat a fully nonlinear case. This is to say that in addition to the nonlinearity appearing in the equation, the feedback controls, $f_i$ and $g$, are also nonlinear. Additionally, we do not assume any growth conditions hold at the origin, which is in contrast with most of the literature related to the subject (see [3], [5], etc.). The presence of nonlinear feedbacks introduces genuine new difficulties to the problem. For example, in this fully nonlinear case, we do not have, in general, smooth solutions (even if the initial data are assumed to be very regular). On the other hand, a rigorous derivation of partial differential equation estimates (needed for the solution to the stabilization problem), including the most fundamental "dissipation energy" estimate (see Lemma 3.1) requires a certain amount of regularity of the solutions which, typically, is not guaranteed by the existence theorem. To cope with this difficulty, we shall introduce a certain regularization/approximation procedure which will lead to an "approximating" problem for which partial differential equation calculus can be rigorously justified (similar ideas were used in the context of the wave equation in [6]). Passage to the limit on the approximation will reconstruct the needed estimates for the original nonlinear problem. It should be noted that this procedure requires an existence theorem (see Theorem 1.1) which supplies certain "a priori" regularity of the traces of the solutions (not only the usual interior regularity).

## 1.3 Statement of Main Results

To state our stability result, we will need the following notation. Let the function $h(x)$ be defined by:

$$h(x) \equiv h_0(x) + h_1(x) + h_2(x), \tag{1.10}$$

where $h_i(x)$ are concave, strictly increasing functions with $h_i(0) = 0$ such that

$$\begin{aligned} &h_0(sg(s)) \geq s^2 + g^2(s) \quad |s| \leq 1 \\ &h_i(sf_i(s)) \geq s^2 + f_i^2(s) \quad |s| \leq 1 \quad i = 1, 2. \end{aligned} \tag{1.11}$$

(Such functions can be easily constructed. See [7].) Then $h(x)$ enjoys the same properties, i.e., it is concave, strictly increasing, and $h(0) = 0$. Define

$$\tilde{h}(x) \equiv h(\frac{x}{mes\ \Sigma_T}). \tag{1.12}$$

Since $\tilde{h}$ is monotone increasing, for every $c \geq 0$, $cI + \tilde{h}$ is invertible. Setting

$$p(x) \equiv (cI + \tilde{h})^{-1}(Kx), \tag{1.13}$$

where $K$ is a positive constant, we see that $p$ is a positive, continuous, strictly increasing function with $p(0) = 0$.

We are now in a position to state our result.

**Theorem 1.2** *Assume hypothesis (H-1) holds. Let $w$ be the solution to system (1.1). Then for some $T_0 > 0$,*

$$E_w(t) \leq \mathcal{S}(\frac{t}{T_0} - 1) \; for \; t > T_0, \tag{1.14}$$

*where $\mathcal{S}(t) \to 0$ when $t \to \infty$ and $\mathcal{S}(t)$ is the solution (contraction semigroup) of the differential equation*

$$\begin{cases} \frac{d}{dt}\mathcal{S}(t) + q(\mathcal{S}(t)) = 0 \\ \mathcal{S}(0) = E_w(0), \end{cases} \tag{1.15}$$

*and $q(x)$ is given by*

$$q(x) \equiv x - (I + p)^{-1}(x) \; for \; x > 0. \tag{1.16}$$

*In this case, the constant $K$ will generally depend on $E_w(0)$ and the constant $c = \frac{1}{mes\Sigma_T}(m^{-1} + M)$.*

*Remark 1.4:* One could also consider feedback controls acting on a portion of the boundary only. In this case, the appropriate geometric conditions imposed on the *uncontrolled* part of the boundary are needed (see [1]).

## 2 Preliminary Energy Estimate

Our goal is to prove energy decay rates for problem (1.1). In order to do this, one needs to perform certain partial differential equation calculations on the problem. These calculations require regularity of the solutions higher than is available from Theorem 1.1. Since our nonlinear problem may not have a sufficiently regular solution (even if the initial data are smooth), we resort to an approximation argument (this argument was used in the context of wave equations in [7]). In fact, the idea here is to approximate solutions to the nonlinear problem (1.1) by solutions to different (linear) problems. Since this linear problem admits regular solutions for smooth initial data, the partial differential equation calculations can be performed on this problem. Final passage to the limit on the approximation problem allows us to obtain needed energy identities for the original nonlinear problem.

To follow our program, we start by defining the following approximations. To do this, we need the following corollary of Theorem 1.1.

**Corollary 2.1** *Let $w$ be a solution to (1.1). Then*

$$f_1(\frac{\partial}{\partial \nu} w_t) \in L_2(0,T;L_2(\Gamma)) \tag{2.1}$$

*and*

$$g(w_t) - \frac{\partial}{\partial \tau} f_2(\frac{\partial}{\partial \tau} w_t) \in L_2(0,T;H^{-1}(\Gamma)). \tag{2.2}$$

**Proof of Corollary 2.1:** Hypothesis (H-1) together with (1.9) of Theorem 1.1 imply

$$\begin{aligned} &f_1(\tfrac{\partial}{\partial \nu} w_t) \in L_2(\Sigma_T) \\ &f_2(\tfrac{\partial}{\partial \tau} w_t) \in L_2(\Sigma_T). \end{aligned} \tag{2.3}$$

Hence

$$\frac{\partial}{\partial \tau} f_2(\frac{\partial}{\partial \tau} w_t) \in L_2(0,T;H^{-1}(\Gamma)). \tag{2.4}$$

On the other hand, with $\phi \in L_2(0,T;H^1(\Gamma))$,

$$\begin{aligned}\int_0^T \int_\Gamma |g(w_t(t,x))\phi(t,x)|dxdt &\le \int_0^T \int_\Gamma |w_t(t,x)|^r|\phi(t,x)|dxdt \\ &\le C\int_0^T \|\phi(t)\|_{H^1(\Gamma)} \int_\Gamma |w_t(t,x)|^r dxdt \\ &\le C\int_0^T \|\phi(t)\|_{H^1(\Gamma)}\|w_t(t)\|^r_{L_p(\Gamma)}dt \\ &\le C\int_0^T \|\phi(t)\|_{H^1(\Gamma)}\|w_t(t)\|^r_{H^{1/2}(\Gamma)}dt \qquad (2.5) \\ &\le C\int_0^T \|\phi(t)\|_{H^1(\Gamma)}\|w_t(t)\|^r_{H^1(\Omega)}dt \\ &\le C\int_0^T \|\phi(t)\|_{H^1(\Gamma)}E_w(t)^{r/2}dt \\ &\le CE_w(0)^{r/2}\int_0^T \|\phi(t)\|_{H^1(\Gamma)}dt,\end{aligned}$$

where the first inequality follows from hypothesis (H-1), the second and third follow from Sobolev Imbeddings and the boundedness of $\Gamma$, and the fourth from trace theory. Hence,

$$g(w_t) \in L_2(0,T;H^{-1}(\Gamma)), \tag{2.6}$$

which, together with (2.3) and (2.4) prove (2.1) and (2.2). □

Let $w$ be the solution of the original problem (1.1). By using the regularity properties in (1.9), (2.1), and (2.2), along with density of approximate (see below) Sobolev spaces, we are in a position to define

$$f_n \in H^{1,1}(Q_T); \qquad \|f_n - [w,\chi(w)]\|_{L_2(0,T;H^{-1}(\Omega))} \longrightarrow 0 \tag{2.7}$$

$$f_{1n} \in H^{1,1}(\Sigma_T); \qquad \|f_{1n} - f_1(\tfrac{\partial}{\partial\nu}w_t)\|_{L_2(\Sigma_T)} \longrightarrow 0 \tag{2.8}$$

$$f_{2n} \in H^{1,1}(\Sigma_T); \quad \|f_{2n} - [g(w_t) - \tfrac{\partial}{\partial\tau}f_2(\tfrac{\partial}{\partial\tau}w_t)]\|_{L_2(0,T;H^{-1}(\Gamma))} \longrightarrow 0 \tag{2.9}$$

$$\alpha_n \in H^{1,1}(\Sigma_T); \qquad \|\alpha_n - \tfrac{\partial}{\partial\nu}w_t\|_{L_2(\Sigma_T)} \longrightarrow 0 \tag{2.10}$$

$$\beta_n \in H^{1,1}(\Sigma_T); \qquad \|\beta_n - (w_t - \tfrac{\partial^2}{\partial\tau^2}w_t)\|_{L_2(0,T;H^{-1}(\Gamma))} \longrightarrow 0, \tag{2.11}$$

where $Q_T \equiv \Omega \times (0,T)$ and $\Sigma_T \equiv \Gamma \times (0,T)$. We consider the following approximating problem:

$$\begin{cases} w_{n,tt} - \gamma^2 \Delta w_{n,tt} + \Delta^2 w_n + b w_{n,t} = f_n \\ w_n(0) = w_{n,0}; \quad w_{n,t}(0) = w_{n,1} \\ \Delta w_n + (1-\mu) B_1 w_n + \frac{\partial}{\partial \nu} w_{n,t}|_\Gamma = -f_{1n} + \alpha_n \\ \frac{\partial}{\partial \nu}\Delta w_n + (1-\mu) B_2 w_n - \gamma^2 \frac{\partial}{\partial \nu} w_{n,tt} - w_n - w_{n,t} + \frac{\partial^2}{\partial \tau^2} w_{n,t}|_\Gamma = f_{2n} - \beta_n, \end{cases} \tag{2.12}$$

where

$$\|w_{n,0} - w_0\|_{H^2(\Omega)} \to 0; \quad \|w_{n,1} - w_1\|_{H^1(\Omega)} \to 0, \tag{2.13}$$

and $(w_{n,0}, w_{n,1}) \in \mathcal{D}$, where $\mathcal{D}$, as dense set of $\mathcal{H}$, consists of $w_{n,0} \in H^4(\Omega)$, $w_{n,1} \in H^3(\Omega)$, where $w_{n,0}, w_{n,1}$ satisfy the appropriate compatability conditions on the boundary. By standard linear semigroup methods, one easily shows that the linear problem, (2.12), admits a classical solution,

$$w_n \in C(0,T; H^4(\Omega)) \cap C^1(0,T; H^3(\Omega)). \tag{2.14}$$

The following proposition plays a critical role in our development.

**Proposition 2.1** *Let $w_n$ (respectively, $w$) be a solution of (2.12) (respectively, (1.1)). Then as $n \to \infty$, the following convergence holds.*

$$w_n \to w \text{ in } C(0,T; H^2(\Omega)) \cap C^1(0,T; H^1(\Omega)) \tag{2.15}$$

$$\nabla w_{n,t}|_\Gamma \to \nabla w_t \text{ in } L_2(\Sigma_T). \tag{2.16}$$

**Proof:** Consider the equation satisfied by the difference $w_n - w_m$. Multiplying this equation by $w_{n,t} - w_{m,t}$ and integrating the result from 0 to $T$ yields

$$\begin{aligned} E_{w_n - w_m,1}(T) &+ \int_0^T \int_\Gamma [\hat{w}_t^2 + |\nabla \hat{w}_t|^2] d\Gamma dt + \int_0^T \int_\Omega b \hat{w}_t^2 d\Omega dt \\ &= \int_0^T \int_\Omega (f_n - f_m) \hat{w}_t d\Omega dt + \int_0^T \int_\Gamma (f_{1n} + \alpha_n - f_{1m} - \alpha_m) \frac{\partial}{\partial \nu} \hat{w}_t d\Gamma dt \\ &+ \int_0^T \int_\Gamma (f_{2n} + \beta_n - f_{2m} - \beta_m) \hat{w}_t d\Gamma dt + E_{w_n - w_m,1}(0), \end{aligned} \tag{2.17}$$

where $\hat{w} \equiv w_n - w_m$. Hence,

$$\begin{aligned} C_0\|\hat{w}(T)\|^2_{H^2(\Omega)} &+\|\hat{w}_t(T)\|^2_{H^1(\Omega)} + \|\nabla \hat{w}_t\|^2_{L_2(\Sigma_T)} + \|\hat{w}_t\|^2_{L_2(\Sigma_T)} \\ &\leq \tfrac{1}{2}\|f_n - f_m\|^2_{L_2(0,T;H^{-1}(\Omega))} + \tfrac{1}{2}\int_0^T \|\hat{w}_t(t)\|^2_{H^1(\Omega)} dt \\ &+\tfrac{1}{2}\|f_{1n} + \alpha_n - f_{1m} - \alpha_m\|^2_{L_2(\Sigma_T)} + \tfrac{1}{2}\|\tfrac{\partial}{\partial \nu}\hat{w}_t\|^2_{L_2(\Sigma_T)} \\ &+\tfrac{1}{2}\|f_{2n} + \beta_n - f_{2m} - \beta_m\|^2_{L_2(0,T;H^{-1}(\Omega))} \\ &+\tfrac{1}{2}\|\hat{w}_t\|^2_{L_2(0,T;H^1(\Omega))} + E_{w_n - w_m,1}(0), \end{aligned} \tag{2.18}$$

and

$$\begin{aligned} \|\hat{w}(T)\|^2_{H^2(\Omega)} &+\|\hat{w}_t(T)\|^2_{H^1(\Omega)} + \|\nabla \hat{w}_t\|^2_{L_2(\Sigma_T)} \\ &\leq C[\|f_n - f_m\|^2_{L_2(0,T;H^{-1}(\Omega))} + \|f_{1n} - f_{1m}\|^2_{L_2(\Sigma_T)} + \|\alpha_n - \alpha_m\|^2_{L_2(\Sigma_T)} \\ &+\|f_{2n} - f_{2m}\|^2_{L_2(0,T;H^{-1}(\Omega))} + \|\beta_n - \beta_m\|^2_{L_2(0,T;H^{-1}(\Omega))} + E_{w_n - w_m,1}(0)] \\ &\longrightarrow 0 \ as \ n \to \infty, \end{aligned} \tag{2.19}$$

where the limit follows by using (2.7)-(2.11). Thus, by (2.13) and Corollary 2.1,

$$\begin{aligned} w_n &\to w^* \ in \ C(0,T;H^2(\Omega)) \cap C^1(0,T;H^1(\Omega)) \\ \nabla w_n|_\Gamma &\to \nabla w^*|_\Gamma \ in \ L_2(\Sigma_T). \end{aligned} \tag{2.20}$$

This allows us to pass with the limit on the linear equation, (2.12). We obtain

$$\begin{cases} w^*_{tt} - \gamma^2 \Delta w^*_{tt} + \Delta^2 w^* + b w^*_t = f = [w, \chi(w)] \\ w^*(0) = w_0; \quad w^*_t(0) = w_1 \in H^2(\Omega) \\ \Delta w^* + (1-\mu)B_1 w^* + \frac{\partial}{\partial \nu} w^*_t|_\Gamma = -f_1(\frac{\partial}{\partial \nu} w_t) + \frac{\partial}{\partial \nu} w_t \\ \frac{\partial}{\partial \nu}\Delta w^* + (1-\mu)B_2 w^* - \gamma^2 \frac{\partial}{\partial \nu} w^*_{tt} - w^* - w^*_t + \frac{\partial^2}{\partial \tau^2} w^*_t|_\Gamma \\ \qquad = g(w_t) - \frac{\partial}{\partial \tau} f_2(\frac{\partial}{\partial \tau} w_t) - w_t + \frac{\partial^2}{\partial \tau^2} w_t. \end{cases} \tag{2.21}$$

Since $w$ satisfies (2.21) and the solution to (2.21) is unique, we infer that $w \equiv w^*$ and

$$\begin{aligned} w_n &\to w \ in \ C(0,T;H^2(\Omega)) \cap C^1(0,T;H^1(\Omega)) \\ \nabla w_n|_\Gamma &\to \nabla w|_\Gamma \ in \ L_2(\Sigma_T), \end{aligned} \tag{2.22}$$

as desired. □

Now we are in a position to prove the fundamental energy relation for problem (1.1).

**Lemma 2.1** *(Energy Identity) Let $w$ be the solution to (1.1). Then the following energy identity holds.*

$$E_w(T) - E_w(0) + \int_{\Sigma_T} [g(w_t)w_t + f_1(\frac{\partial}{\partial\nu}w_t)\frac{\partial}{\partial\nu}w_t + f_2(\frac{\partial}{\partial\tau}w_t)\frac{\partial}{\partial\tau}w_t]d\Gamma dt = 0. \tag{2.23}$$

**Proof:** We first prove this energy identity for the solution, $w_n$, of the approximation problem, (2.12). Indeed, by applying a standard energy argument to (2.12), we obtain

$$\begin{aligned} E_{w_n,1}(T) - E_{w_n,1}(0) \quad &+ \textstyle\int_{\Sigma_T} |\frac{\partial}{\partial\nu}w_{n,t}|^2 d\Gamma dt + \int_{\Sigma_T}(w_{n,t})^2 d\Gamma dt + \int_{\Sigma_T} |\frac{\partial}{\partial\tau}w_{n,t}|^2 d\Gamma dt \\ &= \textstyle\int_0^T \int_\Omega f_n w_{n,t} d\Omega dt + \int_{\Sigma_T}(f_{1n} + \alpha_n)\frac{\partial}{\partial\nu}w_{n,t} d\Gamma dt \\ &\quad - \textstyle\int_{\Sigma_T}(f_{2n} + \beta_n)w_{n,t} d\Gamma dt. \end{aligned} \tag{2.24}$$

Using convergence properties (2.7)-(2.11) and the result of Proposition 2.1, we obtain

$$\begin{aligned} E_{w,1}(T) - E_{w,1}(0) \quad &+ \textstyle\int_{\Sigma_T} |\frac{\partial}{\partial\nu}w_t|^2 d\Gamma dt + \int_{\Sigma_T}(w_t)^2 d\Gamma dt + \int_{\Sigma_T} |\frac{\partial}{\partial\tau}w_t|^2 d\Gamma dt \\ &= \textstyle\int_0^T \int_\Omega [w, \chi(w)]w_t d\Omega dt + \int_{\Sigma_T}[-f_1(\frac{\partial}{\partial\nu}w_t) + \frac{\partial}{\partial\nu}w_t]\frac{\partial}{\partial\nu}w_t d\Gamma dt \\ &\quad - \textstyle\int_{\Sigma_T}[g(w_t) - \frac{\partial}{\partial\tau}f_2(\frac{\partial}{\partial\tau}w_t)]w_t d\Gamma dt + \int_{\Sigma_T}(w_t^2 + |\frac{\partial}{\partial\tau}w_t|^2) d\Gamma dt. \end{aligned} \tag{2.25}$$

After canceling boundary terms, we have

$$\begin{aligned} E_{w,1}(T) - E_{w,1}(0) \quad &= \textstyle\int_0^T \int_\Omega [w, \chi(w)]w_t d\Omega dt - \int_{\Sigma_T} f_1(\frac{\partial}{\partial\nu}w_t)\frac{\partial}{\partial\nu}w_t d\Gamma dt \\ &\quad - \textstyle\int_{\Sigma_T}[g(w_t) - \frac{\partial}{\partial\tau}f_2(\frac{\partial}{\partial\tau}w_t)]w_t d\Gamma dt. \end{aligned} \tag{2.26}$$

Since

$$\begin{aligned} &\textstyle\int_0^T \int_\Omega [w, \chi(w)]w_t d\Omega dt = \int_\Omega [w, \chi(w)]w d\Omega|_0^T - \int_0^T \int_\Omega [w_t, \chi(w)]w d\Omega dt \\ &\Longrightarrow \textstyle\int_0^T \int_\Omega [w, \chi(w)]w_t d\Omega dt = E_{w,2}(0) - E_{w,2}(T), \end{aligned} \tag{2.27}$$

by the symmetricity of the trilinear form ([3], Lemma 5.2.1), our desired result follows directly from (2.26) and (2.27). □

# 3 A Priori Estimates

To proof Theorem 1.2, we first show the following inequality holds.

**Lemma 3.1** *Let $w$ be the solution to (1.1), $0 < \alpha < T/2$ and $\epsilon > 0$ be arbitrary. Then there exist constants, $C$, $C_{T,\alpha,\epsilon}$, and $C(E_w(0))$ such that the following inequality holds:*

$$\begin{aligned}\int_\alpha^{T-\alpha} E_w(t)dt - CE_w(0) \leq\ & C_{T,\alpha,\epsilon}\{\int_{\Sigma_T}(|w_t|^2 + \gamma^2|\nabla w_t|^2)d\Gamma dt + \|f_1(\tfrac{\partial}{\partial\nu}w_t)\|^2_{L_2(\Sigma_T)} \\ & + \|f_2(\tfrac{\partial}{\partial\tau}w_t)\|^2_{L_2(\Sigma_T)} + C(E_w(0))\int_{\Sigma_T} g(w_t)w_t d\Gamma dt \\ & + \int_{\Sigma_A}|g(w_t)|^2 d\Gamma dt + \int_{Q_T} b(x)w_t^2 d\Omega dt\},\end{aligned} \tag{3.1}$$

*where $C(E_w(0))$ is an increasing function of $E_w(0)$ and $\Sigma_A \equiv \{(t,x) \in \Sigma_T : |w_t(t,x)| < 1\}$.*

## 3.1 Multiplier Methods

To facilitate the proof of Lemma 3.1, we begin by using a multiplier method on the approximation problem (2.12) to prove the following preliminary estimate.

**Proposition 3.1** *Let $(w_0, w_1) \in \mathcal{D}$. Then the energy of system (2.12) as given by (1.6) satisfies the following estimate:*

$$\begin{aligned}\tfrac{1}{2}\int_0^T E_{w_n,1}(t)dt + \tfrac{1}{2}\int_{Q_T} f_n w_n d\Omega dt\ & -C_1E_{w_n,1}(T) - C_2E_{w_n,1}(0) \\ & \leq |(\tfrac{\partial}{\partial\nu}w_n, \tilde{f}_{1n})_{L_2(\Sigma_T)}| + |(w_n, \tilde{f}_{2n})_{L_2(\Sigma_T)}| \\ & \quad + |(\tfrac{\partial}{\partial\nu}(\vec{h}\cdot\nabla w_n), \tilde{f}_{1n})_{L_2(\Sigma_T)}| + |(\vec{h}\cdot\nabla w_n, \tilde{f}_{2n})_{L_2(\Sigma_T)}| \\ & \quad \int_{Q_T} f_n\vec{h}\cdot\nabla w_n d\Omega dt \\ & \quad + C_3\int_{\Sigma_T}\vec{h}\cdot\nu(w_{n,t}^2 + \gamma^2|\nabla w_{n,t}|^2)d\Gamma dt \\ & \quad + C_4\int_{\Sigma_T}(|\tfrac{\partial^2 w_n}{\partial\tau^2}|^2 + (|\tfrac{\partial^2 w_n}{\partial\nu^2}|^2 + (|\tfrac{\partial^2 w_n}{\partial\nu\partial\tau}|^2)d\Gamma dt \\ & \quad + C_5\int_{Q_T} b(x)w_{n,t}^2 d\Omega dt + C_7 l.o.(w_n),\end{aligned} \tag{3.2}$$

*where*

$$\begin{aligned} l.o.(w_n) &\equiv \|w_n\|^2_{L_2(0,T;H^{2-\epsilon}(\Omega))}, \\ \tilde{f}_{1n} &\equiv f_{1n} - \alpha_n + \tfrac{\partial}{\partial\nu} w_{n,t}, \\ \tilde{f}_{2n} &\equiv f_{2n} - \beta_n + w_{n,t} - \tfrac{\partial^2}{\partial\tau^2} w_{n,t}, \end{aligned} \tag{3.3}$$

$0 < \epsilon < 1/2$, *and* $h \equiv x - x_0$ *for some* $x_0 \in R^2$.

**Proof of Proposition 3.1:** *Step 1: Identities.* From [3] (p. 84, (4.5.17)), with adjustments to take both the nonhomogeneous right-hand side of (2.12) and the boundary conditions into account, we have

$$\begin{aligned} \int_0^T E_{w,1}(t)dt \quad &+ \tfrac{1}{2}\int_{Q_T} f_n w_n d\Omega dt \\ &\le -\tfrac{1}{2}[(w_{n,t}, \vec{h}\cdot\nabla w_n)_{L_2(\Omega)} + \gamma^2(\nabla w_{n,t}, \nabla(\vec{h}\cdot\nabla w_n))_{L_2(\Omega)}]_0^T \\ &\quad + \tfrac{1}{2}[(w_{n,t}, w_n)_{L_2(\Omega)} + \gamma^2(\nabla w_{n,t}, \nabla w_n)_{L_2(\Omega)}]_0^T \\ &\quad - \tfrac{1}{2}(\tfrac{\partial}{\partial\nu} w_n, \tilde{f}_{1n})_{L_2(\Sigma_T)} - \tfrac{1}{2}(w_n, \tilde{f}_{2n})_{L_2(\Sigma_T)} \\ &\quad - \int_{\Sigma_T}[\tfrac{\partial}{\partial\nu}(\vec{h}\cdot\nabla w_n)\tilde{f}_{1n} - (\vec{h}\cdot\nabla w_n)\tilde{f}_{2n}]d\Gamma dt - \int_{\Sigma_T}(\vec{h}\cdot\nabla w_n)w_n d\Gamma dt \\ &\quad + \tfrac{1}{2}\int_{\Sigma_T}\vec{h}\cdot\nu(w_{n,t}^2 + \gamma^2|\nabla w_{n,t}|^2)d\Gamma dt \\ &\quad - \tfrac{1}{2}\int_{\Sigma_T}\vec{h}\cdot\nu[w_{n,xx}^2 + w_{n,yy}^2 + 2\mu w_{n,xx}w_{n,yy} + 2(1-\mu)w_{n,xy}^2]d\Gamma dt \\ &\quad - \int_{Q_T} b(x)w_{n,t}\vec{h}\cdot\nabla w_n d\Omega dt + \int_{Q_T} f_n\vec{h}\cdot\nabla w_n d\Omega dt \\ &\quad - \tfrac{1}{2}\int_{Q_T} f_n w_n d\Omega dt + \tfrac{1}{2}(b(x)w_n, w_n)_{L_2(\Omega)}|_0^T. \end{aligned} \tag{3.4}$$

Notice that the regularity of the solution given by (2.14) allows us to justify the calculations in [3].

*Step 2: Bounding Linear Terms.* To bound the first term on the second to last line of (3.4), we use equation (2.7) from [1] which gives us

$$-\int_{Q_T} b(x)w_{n,t}\vec{h}\cdot\nabla w_n d\Omega dt \le C_1\int_{Q_T} b(x)w_{n,t}^2 d\Omega dt + C_2 l.o.(w_n). \tag{3.5}$$

All terms which need to be evaluated at 0 and $T$, including the first and second lines and the last term on the right-hand side of (3.4), can be bounded by

$$C_1 E_{w_n}(T) + C_2 E_{w_n}(0). \tag{3.6}$$

Finally, note that the sixth line and the last term on the fourth line of the right-hand side of (3.4) can be bounded by second order traces of the solution $w_n$ and $l.o.(w_n)$.

With the above three estimates, we obtain our desired result, (3.2). □

Next, we take the limit of (3.2) as $n \to \infty$ to obtain a similar inequality for the solution to (1.1), which we state in the following proposition.

**Proposition 3.2** *Let $(w_0, w_1) \in \mathcal{H}$. Then the energy of system (1.1) as given by (1.6) satisfies the following estimate:*

$$\begin{aligned}
\int_0^T E(t)dt \quad & -C_1 E(T) - C_2 E(0) \\
& \leq C_3 \int_{\Sigma_T} (|w_t|^2 + \gamma^2 |\nabla w_t|^2) d\Gamma dt + C_4 E^2(0) \int_{\Sigma_T} |\Delta \chi(w)| d\Gamma dt \\
& + C_5 \int_{\Sigma_T} (|\tfrac{\partial^2 w}{\partial \tau^2}|^2 + |\tfrac{\partial^2 w}{\partial \nu^2}|^2 + |\tfrac{\partial^2 w}{\partial \nu \partial \tau}|^2) d\Gamma dt \\
& + C(E(0)) \int_{\Sigma_T} g(w_t) w_t d\Gamma dt + C_6 \int_{\Sigma_A} |g(w_t)|^2 d\Gamma dt \\
& + C_7 \int_{Q_T} b(x) w_t^2 d\Omega dt + C_8 l.o.(w),
\end{aligned} \tag{3.7}$$

*where $\Sigma_A \equiv \{(t, x) \in \Sigma_T : |w_t(t, x)| < 1\}$.*

**Proof:** *Step 1: Approximation Results.* Taking the limit as $n \to \infty$ in (3.2), by virtue of (2.7)-(2.11) and Proposition 2.1, we find

$$
\begin{aligned}
\int_0^T E_{w,1}(t)dt \quad & +\tfrac{1}{2}\int_{\Sigma_T}[w,\chi(w)]w d\Omega dt - C_1 E_{w,1}(T) - C_2 E_{w,1}(0) \\
& \leq |(\tfrac{\partial}{\partial\nu}w, \mathcal{F}_1(w_t))_{L_2(\Sigma_T)}| + |(w, \mathcal{F}_2(w_t))_{L_2(\Sigma_T)}| \\
& \quad +|(\tfrac{\partial}{\partial\nu}(\vec{h}\cdot\nabla w), \mathcal{F}_1(w_t))_{L_2(\Sigma_T)}| + |(\vec{h}\cdot\nabla w, \mathcal{F}_2(w_t))_{L_2(\Sigma_T)}| \\
& \quad +C_3\int_{\Sigma_T}\vec{h}\cdot\nu(w_t^2+\gamma^2|\nabla w_t|^2)d\Gamma dt \\
& \quad +C_4\int_{\Sigma_T}(|\tfrac{\partial^2 w}{\partial\tau^2}|^2 + (|\tfrac{\partial^2 w}{\partial\nu^2}|^2 + (|\tfrac{\partial^2 w}{\partial\nu\partial\tau}|^2)d\Gamma dt \\
& \quad +\int_{Q_T}[w,\chi(w)]\vec{h}\cdot\nabla w d\Omega dt \\
& \quad +C_5\int_{Q_T}b(x)w_t^2 d\Omega dt + C_6 l.o.(w),
\end{aligned}
\tag{3.8}
$$

where $\mathcal{F}_i$ are the control terms, defined as follows:

$$
\begin{aligned}
\mathcal{F}_1(w_t) &\equiv -f_1(\tfrac{\partial}{\partial\nu}w_t) \\
\mathcal{F}_2(w_t) &\equiv g(w_t) - \tfrac{\partial}{\partial\tau}f_2(\tfrac{\partial}{\partial\tau}w_t).
\end{aligned}
\tag{3.9}
$$

*Step 2: Bounding Nonlinear Terms.* To rewrite the terms in (3.8) involving the von Kármán nonlinearity, we recall results from [3]:

$$(\Delta^2\chi, \chi)_{L_2(\Omega)} = \|\Delta\chi\|^2_{L_2(\Omega)}, \tag{3.10}$$

$$([w,\chi(w)], \vec{h}\cdot\nabla w)_{L_2(\Omega)} = -\frac{1}{2}\|\Delta\chi\|^2_{L_2(\Omega)} - \frac{1}{2}\int_\Gamma \vec{h}\cdot\nu(\Delta\chi)^2 d\Gamma. \tag{3.11}$$

From [1], we find

$$\int_{\Sigma_T}|\Delta\chi(w)|^2 d\Gamma dt \leq \frac{1}{4C_4}\int_0^T E_w(t)dt + CE_w^2(0)\int_{\Sigma_T}|\Delta\chi(w)|d\Gamma dt. \tag{3.12}$$

Recalling the definitions of $\mathcal{F}_i$, we bound the remaining nonlinear terms in (3.2) as follows:

$$
\begin{aligned}
|(\tfrac{\partial}{\partial\nu}w, \mathcal{F}_1(w_t))_{L_2(\Sigma_T)}| &\leq C(\|\tfrac{\partial}{\partial\nu}w\|^2_{L_2(\Sigma_T)} + \|f_1(\tfrac{\partial}{\partial\nu}w_t)\|^2_{L_2(\Sigma_T)}) \\
&\leq C(l.o.(w) + \|f_1(\tfrac{\partial}{\partial\nu}w_t)\|^2_{L_2(\Sigma_T)}),
\end{aligned}
\tag{3.13}
$$

$$|(w,\mathcal{F}_2(w_t))_{L_2(\Sigma_T)}| \le C(l.o.(w) + \|f_1(\frac{\partial}{\partial\tau}w_t)\|^2_{L_2(\Sigma_T)} + |(w,g(w_t))_{L_2(\Sigma_T)}|), \tag{3.14}$$

$$|(\frac{\partial}{\partial\nu}(\vec{h}\cdot\nabla w),\mathcal{F}_1(w_t))_{L_2(\Sigma_T)}| \le C(\|\frac{\partial^2 w}{\partial\nu^2}\|^2_{L_2(\Sigma_T)} + \|\frac{\partial^2 w}{\partial\nu\partial\tau}\|^2_{L_2(\Sigma_T)} + \|f_1(\frac{\partial}{\partial\nu}w_t)\|^2_{L_2(\Sigma_T)}), \tag{3.15}$$

$$\begin{aligned} |(\vec{h}\cdot\nabla w,\mathcal{F}_2(w_t))_{L_2(\Sigma_T)}| &\le C(\|\tfrac{\partial^2 w}{\partial\tau^2}\|^2_{L_2(\Sigma_T)} + \|\tfrac{\partial^2 w}{\partial\nu\partial\tau}\|^2_{L_2(\Sigma_T)} \\ &\quad + \|f_2(\tfrac{\partial}{\partial\tau}w_t)\|^2_{L_2(\Sigma_T)} + |(\vec{h}\cdot\nabla w, g(w_t))_{L_2(\Sigma_T)}|). \end{aligned} \tag{3.16}$$

To bound the terms involving $g(w_t)$, we proceed as follows. Define $\mathcal{B}_i$ to be

$$\begin{aligned} \mathcal{B}_1 &\equiv |(w,g(w_t))_{L_2(\Sigma_T)}| \\ \mathcal{B}_2 &\equiv |(\vec{h}\cdot\nabla w, g(w_t))_{L_2(\Sigma_T)}|. \end{aligned} \tag{3.17}$$

*Estimates for* $\mathcal{B}_2$. Denote $\Sigma_A \equiv \{x\in\Sigma_T; |w_t(t,x)|<1\}$ and $\Sigma_B \equiv \Sigma_T\backslash\Sigma_A$. Then

$$\int_{\Sigma_A} |g(w_t)\vec{h}\cdot\nabla w| d\Gamma dt \le C(\int_{\Sigma_A} |g(w_t)|^2 d\Gamma dt + \int_{\Sigma_A} |\nabla w|^2 d\Gamma dt) \tag{3.18}$$

and

$$\int_{\Sigma_B} |g(w_t)\vec{h}\cdot\nabla w| d\Gamma dt \le \frac{C_1}{\epsilon}\int_{\Sigma_B} |g(w_t)|^\beta d\Gamma dt + \epsilon C_2 \int_{\Sigma_A} |\nabla w|^\alpha d\Gamma dt, \tag{3.19}$$

where $\frac{1}{\alpha}+\frac{1}{\beta}=1$. Recall hypothesis (H-1). Selecting $r(\beta-1)=1$, i.e., $\beta=\frac{1+r}{r}>1$, $\alpha=r+1$, and using (H-1), we find

$$\begin{aligned} \textstyle\int_{\Sigma_B} |g(w_t)|^\beta d\Gamma dt &= \textstyle\int_{\Sigma_B} |g(w_t)|^{\beta-1}|g(w_t)| d\Gamma dt \\ &\le C \textstyle\int_{\Sigma_B} |w_t^{r(\beta-1)} g(w_t)| d\Gamma dt = C\int_{\Sigma_B} |w_t g(w_t)| d\Gamma dt. \end{aligned} \tag{3.20}$$

Collecting (3.19) and (3.20), using trace theory, Sobolev's imbeddings, and the following inequality,

$$\|\nabla w(t)\|^\alpha_\Gamma \le E_w(t)^{\alpha/2} = E_w(t)E_w(t)^{\alpha/2-1} \le E_w(0)^{\alpha/2-1}E_w(t), \tag{3.21}$$

yields

$$\int_{\Sigma_B} |g(w_t)\vec{h}\cdot\nabla w| d\Gamma dt \le \frac{C}{\epsilon}\int_{\Sigma_B} |w_t g(w_t)| d\Gamma dt + \epsilon C(E_w(0))\int_0^T E_w(t)dt. \tag{3.22}$$

*Estimates for $\mathcal{B}_1$.* Splitting the integral as before, we find

$$\int_{\Sigma_A} |g(w_t)w| d\Gamma dt \leq C\left(\int_{\Sigma_A} |g(w_t)|^2 d\Gamma dt + \int_{\Sigma_A} |w|^2 d\Gamma dt\right) \tag{3.23}$$

and

$$\int_{\Sigma_B} |g(w_t)w| d\Gamma dt \leq \frac{C}{\epsilon} \int_{\Sigma_B} |w_t g(w_t)| d\Gamma dt + \epsilon C(E_w(0)) \int_0^T E_w(t) dt. \tag{3.24}$$

Taking $\epsilon$ to be appropriately small, using assumption (H-1), then (3.18)-(3.24), completes the proof of Proposition 3.2. □

Next, we bound the second-order traces by using the following proposition.

**Proposition 3.3** *Let $w$ be the solution to (1.1). Then for any $\alpha > 0$ and $\epsilon > 0$, $w$ satisfies the following inequality:*

$$\begin{aligned} \int_\alpha^{T-\alpha} \int_\Gamma \left(\left|\frac{\partial^2 w}{\partial \tau^2}\right|^2 \right. & \left. + \left|\frac{\partial^2 w}{\partial \nu^2}\right|^2 + \left|\frac{\partial^2 w}{\partial \nu \partial \tau}\right|^2\right) d\Gamma dt \\ & \leq C_{T,\alpha,\epsilon}\{\|w_t\|^2_{L_2(\Sigma_T)} + \|\nabla w_t\|^2_{L_2(\Sigma_T)} + E_w^2(0)\|\chi(w)\|_{L_1(0,T;H^{3-\epsilon}(\Omega))} + l.o.(w) \\ & + \|f_1(\tfrac{\partial}{\partial \nu} w_t)\|^2_{L_2(\Sigma_T)} + \|f_2(\tfrac{\partial}{\partial \tau} w_t)\|^2_{L_2(\Sigma_T)} \\ & + C(E_w(0)) \int_{\Sigma_T} g(w_t) w_t d\Gamma dt + \int_{\Sigma_A} |g(w_t)|^2 d\Gamma dt\}. \end{aligned} \tag{3.25}$$

To prove this proposition, we require the following result which was proven in [9] using microlocal analysis methods.

**Proposition 3.4 ([9], Theorem 1.1)** *Let $p(t,x)$ be a solution to the following linear problem (in the sense of distributions)*

$$\left.\begin{array}{ll} p_{tt} - \gamma^2 \Delta p_{tt} + \Delta^2 p = f & in\ Q_T \\ p(0,\cdot) = p_0;\ p_t(0,\cdot) = p_1 & in\ \Omega \\ \Delta p + (1-\mu) B_1 p = g_1 & on\ \Sigma_T \\ \frac{\partial}{\partial \nu} \Delta p + (1-\mu) B_2 p - \gamma^2 \frac{\partial}{\partial \nu} p_{tt} - p = g_2 & on\ \Sigma_T. \end{array}\right\} \tag{3.26}$$

*For every $T > \alpha > 0$ and $\frac{1}{2} > \epsilon > 0$, the following estimate holds:*

$$\int_{\alpha}^{T-\alpha} \int_{\Gamma}(|\frac{\partial^2 w}{\partial \tau^2}|^2 + |\frac{\partial^2 w}{\partial \nu^2}|^2 + |\frac{\partial^2 w}{\partial \nu \partial \tau}|^2) d\Gamma dt$$
$$\leq C_{T,\alpha}\{\|f\|^2_{L_2([0,T];H^{-\epsilon}(\Omega))} + \|g_1\|^2_{L_2(\Sigma_T)} + \|g_2\|^2_{L_2([0,T];H^{-1}(\Omega))} \tag{3.27}$$
$$+\|p_t\|^2_{L_2(\Sigma_T)} + \|\nabla p_t\|^2_{L_2(\Sigma_T)} + \|p\|^2_{L_2([0,T];H^{3/2+\epsilon}(\Omega))}\}.$$

**Proof of Proposition 3.3:** We apply the result of Proposition 3.4 to system (1.1) with

$$f \equiv -b(x)w_t + [\chi(w), w]$$
$$g_1 \equiv -f_1(\frac{\partial}{\partial \nu} w_t) \tag{3.28}$$
$$g_2 \equiv g(w_t) - \frac{\partial}{\partial \tau} f_2(\frac{\partial}{\partial \tau} w_t).$$

From [1], we know that

$$\int_0^T \|f\|^2_{H^{-\epsilon_0}(\Omega)} dt \leq C\{l.o.(w) + E_w^2(0) \int_0^T \|\chi(w)\|_{H^{3-\epsilon_1}(\Omega)} dt\}. \tag{3.29}$$

Splitting the integral of the $g(w_t)$ term as in the third step of the proof of Proposition 3.2 and using (3.20) with $\beta = 1$, we find

$$\|g_2\|^2_{L_2([0,T];H^{-1}(\Omega))} \leq C(E_w(0)) \int_{\Sigma_T} g(w_t) w_t d\Gamma dt + C_1 \int_{\Sigma_A} |g(w_t)|^2 d\Gamma dt + C_2 \|f_2(\frac{\partial}{\partial \tau} w_t)\|^2_{L_2(\Sigma_T)}. \tag{3.30}$$

Combining (3.28)-(3.30) with (3.27) and choosing $\epsilon = \epsilon_0$, we arrive at our desired result. □

At this point, we are able to prove the following energy estimate.

**Lemma 3.2** *Let $w$ be the solution to (1.1), $0 < \alpha < T$ and $\epsilon > 0$ be arbitrary. Then*

$$\int_{\alpha}^{T-\alpha} E_w(t) dt - CE_w(0) \leq C_{T,\alpha,\epsilon}\{\int_{\Sigma_T}(|w_t|^2 + \gamma^2|\nabla w_t|^2) d\Gamma dt + E_w^2(0) \int_{\Sigma_T} |\Delta \chi(w)| d\Gamma dt$$
$$+E_w^2(0) \int_{\Sigma_T} \|\chi(w)\|_{H^{3-\epsilon}(\Omega)} dt + \|f_1(\frac{\partial}{\partial \nu} w_t)\|^2_{L_2(\Sigma_T)}$$
$$+\|f_2(\frac{\partial}{\partial \tau} w_t)\|^2_{L_2(\Sigma_T)} + C(E_w(0)) \int_{\Sigma_T} g(w_t) w_t d\Gamma dt$$
$$+ \int_{\Sigma_A} |g(w_t)|^2 d\Gamma dt + \int_{Q_T} b(x) w_t^2 d\Omega dt + l.o.(w)\}. \tag{3.31}$$

**Proof of Lemma 3.2:** Applying the result of Proposition 3.2 on the interval $[\alpha, T-\alpha]$ yields

$$\begin{aligned}\int_{\alpha}^{T-\alpha} E_w(t)dt \quad & -C_1E_w(T-\alpha) - C_2E_w(\alpha) \\ & \leq C_3\int_{\Sigma_T}(|w_t|^2 + \gamma^2|\nabla w_t|^2)d\Gamma dt + C_4E_w^2(\alpha)\int_{\Sigma_T}|\Delta\chi(w)|d\Gamma dt \\ & +C_5\int_{\Sigma_{T_\alpha}}(|\tfrac{\partial^2 w}{\partial\tau^2}|^2 + |\tfrac{\partial^2 w}{\partial\nu^2}|^2 + |\tfrac{\partial^2 w}{\partial\nu\partial\tau}|^2)d\Gamma dt \\ & +C(E_w(\alpha))\int_{\Sigma_T}g(w_t)w_t d\Gamma dt + C_6\int_{\Sigma_A}|g(w_t)|^2 d\Gamma dt \\ & +C_7\int_{Q_T}b(x)w_t^2 d\Omega dt + C_8 l.o.(w),\end{aligned} \tag{3.32}$$

where $\Sigma_{T_\alpha} \equiv \Gamma \times [\alpha, T-\alpha]$. Since the energy of the system is nonincreasing, $E_w(T) \leq E_w(T-\alpha) \leq E_w(\alpha) \leq E_w(0)$. Also, $E_w^2(\alpha) \leq E_w^2(0)$. Applying Proposition 3.3 to the second-order traces on the right-hand side of (3.32) and recalling Lemma 2.1 then gives us our desired result. □

## 3.2 Completion of the Proof of Lemma 3.1

From [1], we have

$$\int_{\Sigma_T}|\Delta\chi(w)|d\Gamma dt \leq C\int_0^T \|\chi(w)\|_{H^{3-\epsilon}(\Omega)}dt. \tag{3.33}$$

Thus, to remove terms involving $\chi(w)$ and lower-order terms, we need the following result.

**Proposition 3.5** *Let $w$ be the solution to (1.1). Then $w$ satisfies the following inequality:*

$$\begin{aligned}\int_0^T \|\chi(w)\|_{H^{3-\epsilon}(\Omega)}dt + l.o.(w) \quad & \leq C(E(0))\{\|w_t\|^2_{L_2(\Sigma_T)} + \|\nabla w_t\|^2_{L_2(\Sigma_T)} + \int_{Q_T}b(x)w_t^2 d\Omega dt \\ & +\|f_1(\tfrac{\partial}{\partial\nu}w_t)\|^2_{L_2(\Sigma_T)} + \|f_2(\tfrac{\partial}{\partial\tau}w_t)\|^2_{L_2(\Sigma_T)} \\ & + \int_{\Sigma_T}g(w_t)w_t d\Gamma dt + \int_{\Sigma_A}|g(w_t)|^2 d\Gamma dt\}.\end{aligned} \tag{3.34}$$

**Proof of Proposition 3.5:** Since this proof follows along similar lines as the proof in [1], for the sake of brevity, we omit it. □

**Completion of the Proof of Lemma 3.1:** By combining the results of Lemma 3.2 and Proposition 3.5, we obtain the desired result of Lemma 3.1, (3.1). □

# 4 Final Estimates

Let the functions $h(x)$, $h_i(x)$, $i = 0, 1, 2$, and $\tilde{h}(x)$ be defined as in (1.10), (1.11), and (1.12), respectively.

$$h(x) \equiv h_0(x) + h_1(x) + h_2(x), \tag{4.1}$$

where $h_i(x)$ are concave, strictly increasing functions with $h_i(0) = 0$. Then $h(x)$ enjoys the same properties. Moreover, we assume that

$$\begin{aligned} &h_0(sg(s)) \geq s^2 + g^2(s) \quad |s| \leq 1 \\ &h_i(sf_i(s)) \geq s^2 + f_i^2(s) \quad |s| \leq 1 \quad i = 1, 2. \end{aligned} \tag{4.2}$$

By the hypotheses imposed on functions $h_i(x)$, we obtain

$$\int_{\Sigma_T} |f_1(\frac{\partial}{\partial\nu} w_t)|^2 d\Gamma dt = \int_{\Sigma_{A_1}} |f_1(\frac{\partial}{\partial\nu} w_t)|^2 d\Gamma dt + \int_{\Sigma_{B_1}} |f_1(\frac{\partial}{\partial\nu} w_t)|^2 d\Gamma dt, \tag{4.3}$$

where $\Sigma_{A_1} \equiv \{(t, x) \in \Sigma_T : |\frac{\partial}{\partial\nu} w_t| \leq 1\}$ and $\Sigma_{B_1} \equiv \Sigma_T \backslash \Sigma_{A_1}$. Hence, using hypothesis (H-1) on $\Sigma_{B_1}$, we find

$$\begin{aligned} &\int_{\Sigma_T} |\tfrac{\partial}{\partial\nu} w_t|^2 d\Gamma dt \quad + \int_{\Sigma_T} |f_1(\tfrac{\partial}{\partial\nu} w_t)|^2 d\Gamma dt \\ &\qquad \leq \int_{\Sigma_{A_1}} [|\tfrac{\partial}{\partial\nu} w_t|^2 + |f_1(\tfrac{\partial}{\partial\nu} w_t)|^2] d\Gamma dt + (M + \tfrac{1}{m}) \int_{\Sigma_{B_1}} f_1(\tfrac{\partial}{\partial\nu} w_t) \tfrac{\partial}{\partial\nu} w_t d\Gamma dt \\ &\qquad \leq \int_{\Sigma_{A_1}} h_1(\tfrac{\partial}{\partial\nu} w_t f_1(\tfrac{\partial}{\partial\nu} w_t)) d\Gamma dt + (M + \tfrac{1}{m}) \int_{\Sigma_T} f_1(\tfrac{\partial}{\partial\nu} w_t) \tfrac{\partial}{\partial\nu} w_t d\Gamma dt. \end{aligned} \tag{4.4}$$

Similarly, the same argument applied to $f_2$ yields

$$\begin{aligned} &\int_{\Sigma_T} |\tfrac{\partial}{\partial\tau} w_t|^2 d\Gamma dt \quad + \int_{\Sigma_T} |f_2(\tfrac{\partial}{\partial\tau} w_t)|^2 d\Gamma dt \\ &\qquad \leq \int_{\Sigma_{A_2}} h_2(\tfrac{\partial}{\partial\tau} w_t f_2(\tfrac{\partial}{\partial\tau} w_t)) d\Gamma dt + (M + \tfrac{1}{m}) \int_{\Sigma_T} f_2(\tfrac{\partial}{\partial\tau} w_t) \tfrac{\partial}{\partial\tau} w_t d\Gamma dt \end{aligned} \tag{4.5}$$

and, finally, for $g$, we have

$$\begin{aligned}\int_{\Sigma_A}|g(w_t)|^2d\Gamma dt+\int_{\Sigma_T}|w_t|^2d\Gamma dt &\le \int_{\Sigma_A}[|g(w_t)|^2+|w_t|^2d\Gamma dt+\int_{\Sigma_B}|w_t|^2d\Gamma dt\\ &\le \int_{\Sigma_A}h_0(w_tg(w_t))d\Gamma dt+\frac{1}{m}\int_{\Sigma_T}g(w_t)w_td\Gamma dt.\end{aligned} \tag{4.6}$$

Define

$$\tilde{h}_i(x)\equiv h_i(\frac{x}{mes\ \Sigma_T}). \tag{4.7}$$

Then, by Jensen's inequality,

$$\begin{aligned}\int_{\Sigma_T}\{|w_t|^2 &+|\nabla w_t|^2+|f_1(\tfrac{\partial}{\partial\nu}w_t)|^2+|f_2(\tfrac{\partial}{\partial\tau}w_t)|^2+g(w_t)w_t\}d\Gamma dt+\int_{\Sigma_A}|g(w_t)|^2d\Gamma dt\\ &\le C_1\int_{\Sigma_T}\{g(w_t)w_t+f_1(\tfrac{\partial}{\partial\nu}w_t)\tfrac{\partial}{\partial\nu}w_t+f_2(\tfrac{\partial}{\partial\tau}w_t)\tfrac{\partial}{\partial\tau}w_t\}d\Gamma dt\\ &\quad +C_2\left[\tilde{h}_0(\int_{\Sigma_T}g(w_t)w_td\Gamma dt)+\tilde{h}_1(\int_{\Sigma_T}f_1(\tfrac{\partial}{\partial\nu}w_t)\tfrac{\partial}{\partial\nu}w_td\Gamma dt)\right.\\ &\quad \left.+\tilde{h}_2(\int_{\Sigma_T}f_2(\tfrac{\partial}{\partial\tau}w_t)\tfrac{\partial}{\partial\tau}w_td\Gamma dt)\right]\\ &\le C_1\int_{\Sigma_T}\{g(w_t)w_t+f_1(\tfrac{\partial}{\partial\nu}w_t)\tfrac{\partial}{\partial\nu}w_t+f_2(\tfrac{\partial}{\partial\tau}w_t)\tfrac{\partial}{\partial\tau}w_t\}d\Gamma dt\\ &\quad +C_2\sum_{i=0}^{2}\tilde{h}_i(\int_{\Sigma_T}\{g(w_t)w_t+f_1(\tfrac{\partial}{\partial\nu}w_t)\tfrac{\partial}{\partial\nu}w_t+f_2(\tfrac{\partial}{\partial\tau}w_t)\tfrac{\partial}{\partial\tau}w_t\}d\Gamma dt\\ &\quad +\int_{Q_T}b(x)w_t^2d\Omega dt),\end{aligned} \tag{4.8}$$

where the last inequality follows from the monotonicity of the functions $\tilde{h}_i$.

Denoting $\mathcal{F}\equiv\int_{\Sigma_T}\{g(w_t)w_t+f_1(\frac{\partial}{\partial\nu}w_t)\frac{\partial}{\partial\nu}w_t+f_2(\frac{\partial}{\partial\tau}w_t)\frac{\partial}{\partial\tau}w_t\}d\Gamma dt+\int_{Q_T}b(x)w_t^2d\Omega dt$, we obtain from Lemma 3.1 and (4.8), using the monotonicity of $\tilde{h}$ once more to include $\int_{Q_T}b(x)w_t^2d\Omega dt$,

$$\int_{\alpha}^{T-\alpha}E_w(t)dt-C_1E_w(0)\le C_{T,\alpha,\epsilon}(E_w(0))[\mathcal{F}+\tilde{h}(\mathcal{F})]. \tag{4.9}$$

Since

$$\int_0^{\alpha}E_w(t)dt+\int_{T-\alpha}^{T}E_w(t)dt\le 2\alpha E_w(0), \tag{4.10}$$

we find

$$\int_0^{T}E_w(t)dt-C_{1,\alpha}E_w(0)\le C_{T,\alpha,\epsilon}(E_w(0))[\mathcal{F}+\tilde{h}\mathcal{F}], \tag{4.11}$$

and by Lemma 2.1,

$$\begin{aligned} &\int_0^T E_w(t)dt \le C_{T,\alpha,\epsilon}(E_w(0))[\mathcal{F} + \tilde{h}\mathcal{F}] + C_{1,\alpha}E_w(0) \\ &\Longrightarrow TE_w(T) \le C_{T,\alpha,\epsilon}(E_w(0))[\mathcal{F} + \tilde{h}(\mathcal{F})] \qquad (4.12) \\ &\Longrightarrow E_w(T) \le C_T(E_w(0))[\mathcal{F} + \tilde{h}(\mathcal{F})]. \end{aligned}$$

Hence, recalling (2.1),

$$(I + \tilde{h})^{-1}(\frac{E_w(T)}{C_T(E_w(0))}) \le \mathcal{F} = E_w(0) - E_w(T). \qquad (4.13)$$

Setting

$$p(s) \equiv (I + \tilde{h})^{-1}(\frac{s}{C_T(E_w(0))}), \qquad (4.14)$$

we have proven the following proposition.

**Proposition 4.1** *Let $w$ be the solution to (1.1) and $E_w(t)$ be the corresponding energy at time $t$. If $T$ is sufficiently large, then there exists a monotone increasing function, $p$, such that*

$$p(E_w(T)) + E_w(T) \le E_w(0). \qquad (4.15)$$

To arrive at the conclusion of Theorem 1.2, we need to apply the result of Lemma 3.3 in [7].

**Lemma 4.1 ([7], Lemma 3.3)** *Let $p$ be a positive, increasing function such that $p(0) = 0$. Since $p$ is increasing, we can define a function $q$ such that $q(x) = x - (I + p)^{-1}(x)$. Notice that $q$ is also an increasing function. Consider a sequence $s_n$ of positive numbers which satisfy:*

$$s_{m+1} + p(s_{m+1}) \le s_m. \qquad (4.16)$$

*Then $s_m \le \mathcal{S}(m)$, where $\mathcal{S}(t)$ is a solution of a differential equation*

$$\begin{cases} \frac{d}{dt}\mathcal{S}(t) + q(\mathcal{S}(t)) = 0 \\ \mathcal{S}(0) = s_0. \end{cases} \qquad (4.17)$$

*Moreover, if* $p(x) > 0$ *for* $x > 0$, *then* $\lim_{t\to\infty} \mathcal{S}(t) = 0$.

Applying the result of Proposition 4.1, we obtain

$$E_w(m(T+1)) + p(E_w(m(T+1))) \leq E_w(mT), \tag{4.18}$$

for $m = 0, 1, ...$ Thus, applying Lemma 4.1 with

$$s_m \equiv E_w(mT), \tag{4.19}$$

yields

$$E_w(mT) \leq \mathcal{S}(m),\ m = 0, 1, ... \tag{4.20}$$

Setting $t = mT + \tau$, $0 \leq \tau < T$, and recalling the evolution property gives

$$E_w(t) \leq E_w(mT) \leq \mathcal{S}(m) \leq \mathcal{S}(\frac{t-\tau}{T}) \leq \mathcal{S}(\frac{t}{T} - 1)\ for\ t > T, \tag{4.21}$$

which completes the proof of Theorem 1.2. □

# References


[1] M. E. Bradley and I. Lasiecka. Global decay rates for the solutions to a von Kármán plate without geometric conditions. *Journal of Mathematical Analysis and Applications.* To appear.

[2] A. Favini and I. Lasiecka. Wll-posedness and regularity of second-order abstract non-linear evolution equations.

[3] J. E. Lagnese. *Boundary Stabilization of Thin Plates.* Society for Industrial and Applied Mathematics, Philadelphia, 1989.

[4] J. E. Lagnese. Uniform asymptotic energy estimates for solutions of the equation of dy-namical plane elasticity with nonlinear dissipation at the boundary. *Nonlinear Analysis*, 16(1):35–54, 1991.

[5] J. E. Lagnese and G. Leugering. Uniform stabilization of a nonlinear beam by nonlinear boundary feedback. *Journal of Differential Equations*, 91(2):355–388, 1991.

[6] I. Lasiecka. Global uniform decay rates for the solutions to wave equations with non-linear boundary conditions. *Applicable Analysis.* To appear.

[7] I. Lasiecka and D. Tataru. Uniform boundary stabilization of semilinear wave equations with nonlinear boundary damping. *Differential and Integral Equations.* To appear.

[8] I. Lasiecka and R. Triggiani. *Differential and Algebraic Riccati Equations with Application to Boundary/Point Control Problems: Continuous Theory and Approximation Theory.* Springer-Verlag, New York, 1991.

[9] I. Lasiecka and R. Triggiani. Sharp trace estimates of solutions to Kirchoff and Euler-Bernoulli equations. In *Proceedings of the International Conference on Abstract Evolution Equations*, University of Bologne, Italy, July 1991. Mercel Dekker.

[10] J. L. Lions. *Contrôlabilité exacte, perturbations et stabilization de systèmes distribués*, volume 1. Masson, Paris, 1989.

[11] W. Littman. Boundary control theory for beams and plates. *Proceedings of 24th Conference on Decision and Control*, pages 2007–2009, 1985.

[12] J. Puel and M. Tucsnak. Boundary stabilization for the von Kármán equations. *C. R. Acad. Sc.*, 314:609–612, 1992.

# 39 Global Null Controllability of Linear Control Processes with Positive Lyapunov Exponents

**Russell Johnson*** Universitá di Firenze, Florence, Italy

**Mahesh Nerurkar#** Rutgers University, Camden, New Jersey

## Abstract

We give conditions under which a locally null controllable linear process may admit positive Lyapunov exponents and yet be globally null controllable.

## Key Words

Local and global null controllability, Lyapunov exponents, exponential dichotomy.

---

* The research of this author is supported by N.S.F. Grant DMS-9001483.

# The research of this author is supported by a Research Council Grant from the Rutgers University.

## Section (1) Introduction

The purpose of this paper is to respond to a question of [9] concerning constrained control processes

$$x' = A(t)x + B(t)u \qquad (x \in \mathbf{R}^n, u \in \Omega \subset \mathbf{R}^k) \quad (1.1)$$

where $\Omega$ is a compact subset of $\mathbf{R}^k$ which contains the origin, and $A(\circ),B(\circ)$ are recurrent (e.g. Bohr almost periodic) functions into the set of matrices of appropriate sizes. Namely, suppose that (1.1) is locally null controllable. Suppose further that the linear system

$$x' = A(t)x \qquad (1.2)$$

admits solutions with positive exponential growth; i.e., has at least one positive Lyapunov exponent. The question is: when is (1.1) globally null controllable?

One's first response might well be: "never"; global null controllability cannot hold because of the presence of positive Lyapunov exponents in (1.2). If $A$ is a constant matrix, then the presence of a positive Lyapunov exponent means that some eigenvalue of $A$ has positive real part. In this case, it is well-known that (1.1) cannot be globally null controllable. Similarly, if $A(\circ)$ is periodic, then (1.2) admits a positive Lyapunov exponent if and only if a Floquet exponent of (1.2) has positive real part, and in this case it is also easy to see that (1.1) is not globally null controllable.

We will see, however, that in the case of non-periodic but recurrent coefficient matrices $A$, it is rather often the case that (1.1) is globally null controllable even when (1.2) admits a positive Lyapunov exponent. In fact, if (1.2) satisfies an "irreducibility" condition,and if the dichotomy spectrum (the analoue of the set of real parts of eigenvalues) is an interval containing the origin, then most (residually many) processes in the hull of (1.1) are globally null controllable.

The paper is organized as follows. In the rest of Section(1), we introduce the definitions and terminology needed for our discussion. In particular we define the terms underlined above. Our notion of irreducibility for recurrent linear systems (1.2) is introduced in Section(2). Then in Section(3), we discuss the global controllability of locally controllable linear processes (1.1) whose homogeneous parts (1.2) admit positive Lyapunov exponents.

We begin our review of terminology by recalling that the process (1.1) is locally null controllable if there exists a neighbourhood U of the origin in $\mathbf{R}^n$ and a time $T > 0$ such that, for each $x_0 \in U$, there exists an integrable function $u:[0,T] \to \Omega$ with the following property: if $x(t)$ is the solution of

$$x' = A(t)x + B(t)u$$
$$x(0) = x_0,$$

then $x(T) = 0$. We say that $x_0$ can be steered to zero in time T.

We assume throughout that $\Omega$ is compact in $\mathbf{R}^k$ and contains the origin. In this case, there is a simple necessary and sufficient condition for local null controllability of (1.1), [1]. To formulate it, let $H_\Omega:\mathbf{R}^k \to \mathbf{R}$ be the support function of $\Omega$ defined by setting,

$$H_\Omega(\alpha) = \mathrm{Sup}\{\langle\alpha,\omega\rangle \mid \omega\in\Omega\}.$$

Here, $\langle\ ,\ \rangle$ denotes the Euclidean inner product on $\mathbf{R}^k$. One has the following lemma.

**Lemma(1.3)** [1] The process (1.1) is locally null controllable if and only if there exists $\varepsilon > 0$ such that

$$\int_0^\infty H_\Omega(B^*(s)\Phi^*(s)^{-1}x_0)ds \geq \varepsilon \quad (x_0 \in \mathbf{R}^n, \| x_0 \| = \langle x_0,x_0\rangle^{1/2} = 1)$$

where $\Phi(t)$ is the fundamental matrix solution of (1.2) satisfying $\Phi(0) = I$ and * denotes adjoint.

Next recall that global null controllability of (1.1) means that every vector $x_0 \in \mathbf{R}^n$ can be steered to the origin in finite time (which of course may depend on $x_0$). The relation between local and global null controllability for recurrent linear processes is discussed in [9].

It is now time to discuss the concept of recurrence. Briefly, a matrix valued function $f(t)$ is called recurrent if, given $\varepsilon > 0$, there exists $T > 0$ such that each interval $[t,t+T] \subset \mathbf{R}$ of length T contains a point $s$ with the property that $d(f_s,f) < \varepsilon$. Here $f_s(t) = f(s+t)$ is the $s$-translate of the matrix function $f$ and $d$ is a metric on the set of such functions.

More precisely, for each pair of integers $n,k > 0$, let $M(n,k)$ be the set of real $n \times k$ matrices. Let $C_{n,k}$ be the set of all uniformly bounded, uniformly continuous functions from $\mathbf{R}$ to $M(n,k)$, equipped with the topology of uniform convergence on compact sets (compact-open topology).

For each $s \in \mathbf{R}$, define the translation $\tau_s$: $C_{n,k} \to C_{n,k}$ by

$$(\tau_s f)(t) = f(t+s) \quad (t \in \mathbf{R};\ f \in C_{n,k}).$$

For fixed positive integers $n$ and $k$, let $A \in C_{n,n}$, $B \in C_{n,k}$ be matrix functions. Say that the pair $(A,B) \in C_{n,n} \times C_{n,k}$ is <u>recurrent</u> if for each neighbourhood $V$ of $(A,B)$ in $C_{n,n} \times C_{n,k}$, there exists $T > 0$ such that each interval $[t,t+T] \subset \mathbf{R}$ of lenght $T$ contains a number $s$ such that $\tau_s(A,B) \in V$. Here of course $\tau_s(A,B)$ is the pair $(A(s + \circ),B(s + \circ))$. This definition agrees with that given previously if we fix a metric $d$ on $C_{n,n} \times C_{n,k}$ which is compatible with the compact-open topology.

It is easy to see that, if $(A,B)$ is recurrent, then the <u>hull</u>

$$E = \mathrm{cls}\{(\tau_t A,\tau_t B) \mid t \in \mathbf{R}\} \subseteq C_{n,n} \times C_{n,k}$$

is compact. Also, the set of translations $\{\tau_t{:}E \to E \mid t \in \mathbf{R}\}$ defines a <u>flow</u> on $E$; that is, $\tau{:}E \times \mathbf{R} \to E{:}(\xi,t) \to \tau_t(\xi)$ is jointly continuous and

$$\tau_t \circ \tau_s = \tau_{t+s}, \quad (t,s \in \mathbf{R})$$
$$\tau_0 = \text{identity map.}$$

Moreover, the flow $(E,\{\tau_t\})$ is <u>minimal</u> in the sense that each orbit $\{\tau_t(\xi) \mid t \in \mathbf{R}\}$ is dense in $E$. See, e.g. [2].

We Remark that, if $A$ and $B$ are Bohr almost periodic functions ([4]), then $E$ is minimal. If $\xi = (\widetilde{A},\widetilde{B}) \in E$, define

$$a(\xi) = \widetilde{A}(0),\ b(\xi) = \widetilde{B}(0).$$

Observe that $\widetilde{A}(t) = a(\tau_t(\xi))$, $\widetilde{B}(t) = b(\tau_t(\xi))$. We will study the <u>collection</u> of control processes

$$x' = a(\tau_t(\xi))x + b(\tau_t(\xi))u \quad (\xi \in E). \qquad (1.1)_\xi$$

Let $\Phi(\xi,t)$ be the fundamental matrix solution of

$$x' = a(\tau_t(\xi))x \qquad (1.2)_\xi$$

which satisfies $\Phi(\xi,0) = I$. Then $\Phi$ is jointly continuous in $\xi$ and $t$ and satisfies the following cocycle identity.

$$\Phi(\xi,t + s) = \Phi(\tau_t(\xi),s)\Phi(\xi,t) \quad (\xi \in E;\ t,s \in \mathbf{R}). \qquad (1.4)$$

Next, we consider the topic of Lyapunov exponents. Let $\xi \in E$ and consider the numbers

$$\beta(x_0,\xi) = \overline{\lim_{t \to \infty}} \frac{1}{t} \ln \| \Phi(\xi,t)x_0 \|,$$

where $x_0$ ranges over the set of non-zero vectors $x_0 \in \mathbf{R}^n$. Each such number is called a Lyapunov exponent of equation $(1.2)_\xi$. It is easily seen that, for fixed $\xi$, there are only finitely many distinct Lyapunov exponents (e.g., [11]).

In order to discuss the dependence of the set of Lyapunov exponents on $\xi$, we need to discuss ergodic measures on E.

**Definition (1.5)** Let E be a compact metric space with a continuous flow $\{\tau_t \mid t \in \mathbf{R}\}$. A Radon probability measure $\mu$ on E is said to be invariant if, for each Borel set $B \subset E$, one has $\mu(\tau_t(B)) = \mu(B)$ for all $t \in \mathbf{R}$. An invariant measure $\mu$ is ergodic if in addition the following indecomposability criterion is satisfied: if $B \subset E$ is a Borel set satisfying $\mu(\tau_t(B)\Delta B) = 0$ for each $t \in \mathbf{R}$ ( $\Delta$ denotes the symmetric difference), then $\mu(B) = 0$ or $\mu(B) = 1$.

It is well-known that, if $(E,\{\tau_t\})$ is a continuous flow with E a compact metric space, then E admits at least one ergodic measure [15]. Furthermore, if $\mu$ is ergodic, then for $\mu$-a.a. $\xi \in E$, the set of Lyapunov exponents of equations $(1.2)_\xi$ is independent of $\xi$ [16]. We write $\{\beta_i(\mu) \mid 1 \leq i \leq s\}$ for the set of distinct Lyapunov exponents which are independent of $\xi$ for $\mu$-a.a. $\xi \in E$.

We turn now to the definition of exponential dichotomy. We say that equations $(1.2)_\xi$ have an exponential dichotomy (ED for short) if there exists a continuous family of projections $P = P(\xi)\colon \mathbf{R}^n \to \mathbf{R}^n$ and constants $K > 0$, $\alpha > 0$ such that

$$\| \Phi(\xi,t)P(\xi)\Phi(\xi,s)^{-1} \| \leq Ke^{-\alpha(t-s)} \quad (t \geq s)$$

$$||\Phi(\xi,t)(I - P(\xi))\Phi(\xi,s)^{-1}|| \le Ke^{\alpha(t-s)} \quad (t \le s).$$

For each $\lambda \in \mathbf{R}$, consider the translated equations

$$x' = (-\lambda I + a(\tau_t(\xi))x \tag{1.6}$$

where $I$ is the $n \times n$ identity matrix. The book of Massera and Schaeffer [12] gives a detailed description of circumstances under which (1.6) has an ED if and only if the linear operator $L = -\lambda I + a(\tau_t(\xi))$ is invertible. This motivates the following definition [17].

**Definition(1.7)** The dichotomy spectrum $\Lambda$ is the set $\lambda \in \mathbf{R}$ for which the translated equations (1.6) do not admit an exponential dichotomy.

The dichotomy spectrum is a finite union of intervals: $\Lambda = [c_1,d_1] \cup \cdots \cup [c_p,d_p]$ ($p \le n$; see [18],[19]). It is known [18] that corresponding to each interval $[c_j,d_j]$ in $\Lambda$, there is a closed vector subbundle $W_j \subset E \times \mathbf{R}^n$ which is invariant in the sense that, if $(\xi,x_0) \in W_j$, then $(\tau_t(\xi),\Phi(\xi,t)x_0) \in W_j$ for all $t \in \mathbf{R}$. If $(\xi,x_0) \in W_j$, then

$$c_j \le \liminf_{t \to \pm\infty} \frac{1}{t} \log ||\Phi(\xi,t)x_0|| \le \limsup_{t \to \pm\infty} \frac{1}{t} \log ||\Phi(\xi,t)x_0|| \le d_j.$$

Finally, $E \times R^n = \bigoplus_{j=1}^{p} W_j$ (direct sum decomposition).

The following facts are proved in [10].

**Proposition(1.8)** (a) If $\mu$ is an ergodic measure on $E$, then $\beta_i(\mu) \in \Lambda$ for each Lyapunov exponent $\beta_i(\mu)$ of equations $(1.2)_\xi$ which is constant $\mu$-a.e.

(b) If $\beta = c_j,d_j$ is an end point of an interval in $\Lambda$, then there exists an ergodic measure $\mu$ on $E$ and a corresponding Lyapunov exponent $\beta_i(\mu)$ such that $\beta = \beta_i(\mu)$.

## Section(2) Irreducibility

In this section, we discuss our irreducibility criterion for equations $(1.2)_\xi$. This criterion is stated in term of a "recurrent triangularization" for equations $(1.2)_\xi$. As we will see, it is not a straightforward generalization of any natural "irreducibility" concept for constant coefficient systems $x' = Ax$. However, it seems rather natural in a control-theoretic context.

Let us first review a basic construction [10]. Let $Y = E \times O(n)$ where $O(n)$ is the orthogonal group on $\mathbf{R}^n$. We define a flow on $Y$ by using the Gram-Schmidt orthogonalization process, as follows. As above let $\Phi(\xi,t)$ be the fundamental matrix solution of $(1.2)_\xi$ satisfying $\Phi(\xi,0) = I$. Let $\xi \in E$, $g_0 \in O(n)$, and write $y = (\xi,g_0) \in Y$. By the Gram-Schmidt process, we obtain

$$\Phi(\xi,t)g_0 = \widehat{\Gamma}(y,t)R(y,t) = \widehat{\Gamma}(\xi,g_0,t)R(\xi,g_0,t) \tag{2.1}$$

where $\widehat{\Gamma}(y,t) \in O(n)$ and $R(y,t)$ is a triangular matrix with zeroes above the main diagonal and positive diagonal entries. Using the cocycle identity (1.4), we have

$$\begin{aligned}\widehat{\Gamma}(y,t+s)R(y,t+s) &= \Phi(\xi,t+s)g_0 = \Phi(\tau_t(\xi),s)\Phi(\xi,t)g_0 \\ &= \Phi(\tau_t(\xi),s)\widehat{\Gamma}(y,t)R(y,t) \\ &= \widehat{\Gamma}(\widehat{\Gamma}(y,t),s)R(\tau_t(\xi)),\widehat{\Gamma}((y,t),s)R(y,t).\end{aligned}$$

By uniqueness in the Gram-Schmidt decomposition, we have

$$\widehat{\Gamma}(y,t+s) = \widehat{\Gamma}(\widehat{\Gamma}(y,t),s) \quad (t,s \in \mathbf{R}).$$

Hence the mappings $\hat{\tau}_t\colon Y \to Y$ defined by

$$\hat{\tau}_t(\xi,g_0) = (\tau_t(\xi),\widehat{\Gamma}(y,t))$$

define a flow on $Y$. Furthermore the map $\Gamma(y,t) = \widehat{\Gamma}(y,t)g_0^{-1}$ satisfies

$$\Gamma(y,t+s) = \Gamma(\hat{\tau}_t(y),s)\Gamma(y,t).$$

Finally,

$$R(y,t+s) = R(\hat{\tau}_t(y),s)R(y,t) \quad (t,s \in \mathbf{R}).$$

That is, $\Gamma$ and $R$ satisfy the cocycle identity (1.4) with respect to the flow $(Y,\{\hat{\tau}_t\})$.

Let $M$ be a minimal set of $Y$. It is easily seen that such an $M$ exists; see, e.g., [2]. Consider the restriction of the cocycle $R$ to $M \times \mathbf{R}$. We write

$$r(m) = \frac{d}{dt} R(\hat{\tau}_t(m))\Big|_{t=o},$$

so that $R(m,t)$ is the fundamental matrix solution of the differential equation

$$w' = r(\hat{\tau}_t(m))w \qquad (2.2)_m$$

which satisfies $R(m,0) = I$. We recall that $r(\hat{\tau}_t(m)) = [r_{i,j}(\hat{\tau}_t(m))]$ is a matrix with all the entries above the main diagonal equal to zero.

There is a close relationship between equations $(1.2)_\xi$ and equations $(2.2)_m$. Namely, we can lift equations $(1.2)_\xi$ to $M$ by introducing the natural projection $\pi: M \to E$, $\pi(\xi,g) = \xi$ and defining

$$\hat{a}(m) = a \circ \pi(m) \quad (m \in M).$$

The fundamental matrix solution $\hat{\Phi}(m,t)$ of the lifted equation $x' = \hat{a}(\hat{\tau}_t(m))x$ satisfies $\hat{\Phi}(m,t) = \hat{\Phi}(\pi(m),t) = \Phi(\xi,t)$, where $\xi = \pi(m)$.

Let $y = (\xi,g) \in Y$, and define $F: Y \to O(n)$ by

$$F(y) = g. \qquad (2.3)$$

Then the orthogonal change of variables $x = F(\hat{\tau}_t(m))w$ transforms equations $(1.2)_\xi$ (lifted to $M$ via $\pi$) into equations $(2.2)_m$. We have

$$R(m,t) = F(\hat{\tau}_t(m))\Phi(\xi,t)F(m)^{-1} \quad (m \in M). \qquad (2.4)$$

We can now define what we mean by irreducibility of equations $(1.2)_\xi$. Introduce the quantities,

$$\rho^-(i) = \inf_{m \in M} \liminf_{t \to \infty} \frac{1}{t} \int_0^t r_{i,i}(\hat{\tau}_s(m))ds$$

$$\rho^+(i) = \sup_{m \in M} \limsup_{t \to \infty} \frac{1}{t} \int_0^t r_{i,i}(\hat{\tau}_s(m))ds.$$

where $1 \le i \le n$. Let $J_i$ be the interval $[\rho^-(i),\rho^+(i)] \subset \mathbf{R}$, $(1 \le i \le n)$.

**Definition(2.5)** Equations $(1.2)_\xi$ are called <u>irreducible</u> if there exists a minimal subset $M \subset Y = E \times O(n)$ such that the intervals $J_i$ have a point in common; i.e., $\bigcap_{i=1}^{n} J_i \neq \phi$.

To clarify the significance of irreducibility, let $\mu$ be an ergodic measure on $E$, and let $\hat{\mu}$ be a probability measure on $M$ which is ergodic with respect to $\{\hat{\tau}_t \mid t \in \mathbf{R}\}$ and which is a lift of $\mu$; i.e., $\hat{\mu}(\pi^{-1}(B)) = \mu(B)$ for each Borel set $B \subset E$. Such a measure $\hat{\mu}$ can always be found [10]. Then for $\hat{\mu}$-a.a. $m \in M$, the set of Lyapunov exponents of equations $(2.2)_m$ equals the set of mean values

$$\int_M r_{i,i}(m)d\hat{\mu}(m) \qquad (1 \le i \le n). \tag{2.6}$$

See, e.g., [16] for a proof. Hence for $\mu$-a.a. $\xi \in E$, the set of Lyapunov exponents of $(1.2)_\xi$ equals the set of mean values (2.6).

Now by the Birkhoff ergodic theorem:

$$\lim_{t \to \infty} \frac{1}{t} \int_0^t r_{i,i}(\hat{\tau}_s(m))ds = \int_M r_{i,i}(m)d\hat{\mu}(m)$$

for $\hat{\mu}$-a.a. m. Thus, roughly speaking, each interval $J_i$ is bounded by "lower" and "upper" Lyapunov exponents. More precisely, the following result can be proved [10].

**Proposition(2.7)** For each $1 \le i \le n$, there exists ergodic measures $\hat{\mu}_i, \hat{v}_i$ on M such that

$$\rho^-(i) = \int_M r_{i,i}(m)d\hat{\mu}_i(m)$$

$$\rho^+(i) = \int_M r_{i,i}(m)d\hat{v}_i(m).$$

Letting $\mu_i, v_i$ be the projections of $\hat{\mu}_i, \hat{v}_i$ to E, we see that $\rho^\pm(i)$ are the true almost everywhere exponents of equations $(1.2)_\xi$, though in general with respect to distinct ergodic measures on E. If E admits a <u>unique</u> ergodic measure $\mu$ (this is the case if, for example, A(t) and B(t) are Bohr almost periodic), then each value $\rho^\pm(i)$ is one of the exponents $\{\beta_i(\mu) \mid 1 \le i \le s\}$ corresponding to $\mu$. It should be emphasized that M need not be uniquely ergodic even if E is uniquely ergodic.

We give an example to show that a family of equations $(1.2)_\xi$ which is irreducible in the sense of Definition(2.5) may be reducible in a natural topological sense. For this, let $A_1(t)$ be the $2 \times 2$ almost periodic matrix valued function constructed by Millionschikov [13].

The dichotomy spectrum of the equation $x' = A_1(t)x$ is a non-degenerate interval $[-\beta,\beta]$, see [13], [7]. Define

$$A(t) = \begin{pmatrix} A_1(t) & 0 \\ 0 & a \end{pmatrix}$$

where a is a real number in $[-\beta,\beta]$. It can be shown that, after triangularization, two of the three intervals $J_i$ are equal to $[-\beta,\beta]$ and the third reduces to $\{a\}$. On the other hand, the block form of $A(t)$ clearly induces a reduction of the space of solutions of $x' = A(t)x$ to a sum of a two-dimensional bundle and a one-dimensional bundle.

It is natural to ask about the relation between irreducibility and properties of the dichotomy spectrum $\Lambda$. We prove a priliminary result.

**Proposition(2.8)** The dichotomy spectrum $\Lambda$ equals $\bigcup_{i=1}^{n} J_i$.

**Proof** Perhaps surprisingly, the proof is not trivial. Note first that, by (2.4) equations $(2.2)_m$ have the same dichotomy spectrum as equations $(1.2)_\xi$.

Suppose that $\beta \notin \bigcup_{i=1}^{n} J_i$. We claim that $\beta \notin \Lambda$, i.e. that the translated equations

$$w' = [-\beta I + r(\widehat{\tau}_t(m))]w$$

have an ED. We suppose WLOG that $\beta = 0$ (replace r by $-\beta I + r$).

We proceed by induction on the dimension n of equations $(2.2)_m$. If $n = 1$, one can check directly that, if $0 \notin J_1$, then equations $(2.2)_m$ admit an ED. So, suppose that for all recurrent lower triangular linear systems of dimension $d = 1,2,\cdots,n-1$, the condition $0 \notin \bigcup_{i=1}^{d} J_i$. implies that the system has ED.

We will use the following criterion of Selgrade [19]: equations $(2.2)_m$ have ED if and only if for each $m \in M$, the only solution $w(t)$ of $(2.2)_m$ which is bounded, both as $t \to \infty$ and $t \to -\infty$ is $w(t) \equiv 0$. Suppose then that $w(t)$ is a solution of some equation $(2.2)_m$ which is bounded as $t \to \pm\infty$. Write $w(t) = (w_1(t), \cdots, w_n(t))^T$. Since $0 \notin J_1$, and since $r(\circ)$ is lower triangular, we see that $w_1(t) \equiv 0$. On the other hand, the set $\{w_1 \in \mathbf{R}^n \mid w_1 = 0\}$ is invariant under solutions of equations $(2.2)_m$. By the induction assumption and by Selgrade's criterion, no solution $w(t) = (0, w_2(t), \cdots, w_n(t))^T$ can be bounded unless it is identically zero. This proves that $\Lambda \subset \bigcup_{i=1}^{n} J_i$.

Now we prove that $\bigcup_{i=1}^{n} J_i \subset \Lambda$. Let $\beta \in \bigcup_{i=1}^{n} J_i$, and suppose WLOG that $\beta = 0$. We want to show that equations $(2.2)_m$ do not have an ED.

Again we proceed by induction on the dimension $n$ of the recurrent lower triangular equations $(2.2)_m$. If $n = 1$, one shows immediately that, if $0 \in J_1$, then the equations $(2.1)_m$ does not admit an ED.

Suppose now that $n \geq 2$ and that $0 \in J_n$. Then the one-dimensional equation $w_n' = r_{n,n}(\hat{\tau}_t(m))w_n$ admits a non-zero bounded solution for some $m \in M$ ( by Selgrade's criterion). However, $w(t) = (0, 0, \cdots, w_n(t))^T$ is a solution of the full equation $(2.2)_m$. So again by Selgrade's criterion, $0 \in \Lambda$.

If $0 \notin J_n$, then $0 \in J_i$ for some $1 \leq i < n$. By the induction assumption, there is a non-trivial bounded solution $\tilde{w}(t)$ of the truncated equations

$$\begin{pmatrix} w_1 \\ \vdots \\ w_{n-1} \end{pmatrix}' = \begin{pmatrix} r_{1,1}(\hat{\tau}_t(m)) & \cdots & 0 \\ \vdots & & \vdots \\ r_{n-1,1}(\hat{\tau}_t(m)) & \cdots & r_{n-1,n-1}(\hat{\tau}_t(m)) \end{pmatrix} \begin{pmatrix} w_1 \\ \vdots \\ w_{n-1} \end{pmatrix}.$$

Now since $0 \notin J_n$, the equation $w_n' - r_{n,n}(\hat{\tau}_t(m))w_n = \sum_{i=1}^{n-1} r_{n,i}(\hat{\tau}_t(m))\tilde{w}_i(t)$ admits a unique bounded solution $\tilde{w}_n(t)$. Then $w(t) = (\tilde{w}(t), \tilde{w}_n(t))$ is a non-trivial bounded solution of $(2.2)_m$, hence $0 \in \Lambda$. This completes the proof of Proposition(2.8).

**Corollary(2.9)** If equations $(1.2)_\xi$ are irreducible, then the dichotomy spectrum $\Lambda$ is a single interval.

We finish the section by considering systems of dimension $n = 2$ or $3$. We assume in (2.10) and (2.11) that the flow $(E,\{\tau_t\})$ admits a unique ergodic measure $\mu$. This condition holds if, for example, $A(t)$ and $B(t)$ are Bohr almost periodic.

**Proposition(2.10)** If $n = 2$ and if equations $(1.2)_\xi$ do not have an exponential dichotomy, then $J_1$ and $J_2$ are both equal to the dichotomy spectrum $\Lambda$.

**Proof** This follows from the structure theory for two-dimensional systems presented in [7]. The results given there are stated for almost periodic systems, but the arguments needed for Proposition(2.10) carry over to the case when $(E,\{\tau_t\})$ is uniquely ergodic.

**Proposition(2.11)** If $n = 3$, then equations $(1.2)_\xi$ are irreducible if and only if $\Lambda$ is a single interval.

This proposition is not true if $E$ is allowed to carry more than one ergodic measure.

**Proof** We need only prove the "if" statement. Let

$$C = \{\beta_1(\mu),\beta_2(\mu),\beta_3(\mu)\}$$

be the almost everywhere Lyapunov exponents of equations $(1.2)_\xi$ with respect to $\mu$ (see Section(1)). By Proposition(2.7) and the discussion following it, we see that $\rho^{\pm}(i) \in C$ for each choice of $\pm$ and $i = 1,2,3$. We can assume that $\beta_1(\mu) \leq \beta_2(\mu) \leq \beta_3(\mu)$. Now, we know that the union of the intervals $J_i$ equals $\Lambda = [\beta_1(\mu),\beta_3(\mu)]$. It follows that the intervals $J_i$ must have $\beta_2(\mu)$ in common.

## Section(3) Irreducibility and Controllability

We return to the family of control processes $(1.1)_\xi$, where $\xi$ ranges over the minimal set $E$ and the control $u$ takes values in a fixed compact subset $\Omega \subset \mathbf{R}^k$ which contains the origin. We assume from now on that, for some $\xi_0 \in E$, the process $(1.1)_{\xi_c}$ is locally controllable (see Section(1)). Then we have the following, [9, Theorem(2.10)].

**Proposition(3.1)** The family of control processes $(1.1)_\xi$ $(\xi \in E)$ is uniformly null controllable if for some $\xi_0 \in E$, the process $(1.1)_{\xi_0}$ is locally controllable. Uniform controllability means there exists a neighbourhood U of the origin in $\mathbf{R}^n$ and a time $T > 0$, both of which are independent of $\xi$, such that the following property holds. If $\xi \in E$ and $x_0 \in U$, then there is a measurable control $u:[0,T] \to \Omega$ such that the solution $x(t)$ of $(1.1)_\xi$ with $x(0) = x_0$ satisfies $x(T) = 0$.

We are interested in conditions assuring that the above notion of local controllability implies global controllability, especially when equations $(1.2)_\xi$ admit a positive Lyapunov exponent. Suppose that the dichotomy spectrum of equations $(1.2)_\xi$ is $\Lambda = [c_1,d_1] \cup \cdots \cup [c_p,d_p]$ where $c_1 \leq d_1 < c_2 \leq \cdots c_p \leq d_p$. Suppose further that $0 \in [a_r,b_r]$ where $1 \leq r \leq p$. Let $W_1, \cdots, W_p$ be the corresponding subbundles of $E \times \mathbf{R}^n$ (see Section(1)).

If $(\xi, x_0) \in W_i$ for $i < r$, then $x_0$ can easily be steered to zero: let $u = 0$ until $x_0$ enters the neighbourhood U of Proposition(3.1), then steer to zero using local controllability. On the other hand, if $i > r$, then we leave as an exercise to show that, if $\| x_0 \|$ is large enough, then $x_0$ cannot be steered to zero.

It is natural, then, to restrict attention to the case when $\Lambda$ is a single interval $[c,d]$, i.e. to restrict attention to the subbundle $W_r$. The reader should be warned, however, that there are problems with this process. Especially, $W_r \subset E \times \mathbf{R}^n$ need not be topologically trivial (e.g. Palmer [17]), and thus it is not clear how to "restrict" equations $(1.2)_\xi$ to $W_r$. However, one can trivialize $W_r$ with a recurrent change of variables $x = C(t)w$ ([17], also [3]), and this is usually sufficient to treat solutions in $W_r$ as if they arose from a family $(1.2)_\xi$ with $\Lambda = [c_r, d_r]$.

For the rest of the paper, then, we suppose that $\Lambda = [c,d]$ is an interval. We are particularly interested in global controllability of $(1.1)_\xi$ when $c \leq 0 < d$. In this case, equations $(1.2)_\xi$ admit a positive Lyapunov exponent $\beta = d$. We are going to show that, if equations $(1.2)_\xi$ are irreducible and if $\bigcap_{i=1}^{n} J_i$ intersects $(-\infty, 0]$, then there is a residual set $E_1 \subset E$ such that the process $(1.1)_\xi$ is globally null controllable for each $\xi \in E_1$.

**Theorem(3.2)** Suppose that the processes $(1.1)_\xi$ are locally null controllable, and that $\Lambda = [c,d]$ is a single interval. Suppose there exists $\beta \leq 0$ such that $\beta \in \bigcap_{i=1}^{n} J_i$. Then there is a residual set $E_1 \subset E$ such that, if $\xi \in E_1$, then $(1.1)_\xi$ is globally null controllable.

Recall that a subset $E_1$ of a metric space $E$ is <u>residual</u> if it contains a countable intersection of open dense sets.

**Proof** The proof uses a technique applied in proving [9, Theorem(3.2)]. Let $M \subseteq Y = E \times O(n)$, be a minimal set, and let $\pi: M \to E$, $\pi(\xi, g) = \xi$ be the projection. Note that the orthogonal change of variables $x = F(\hat{\tau}_t(m))w$ which transforms $(1.2)_\xi$ to $(2.2)_m$ takes the control process $(1.1)_\xi$ to

$$w' = r(\hat{\tau}_t(m))w + F^{-1}(\hat{\tau}_t(m))b(\tau_t(\xi))u. \qquad (3.3)_m$$

Clearly $(1.1)_\xi$ is locally (respectively globally) null controllable if and only if $(3.3)_m$ is locally (respectively globally) null controllable.

Next write $w$ in the component form: $w = (w_1, \cdots, w_n)^T$. Observe that the first component $w_1$ satisfies

$$w_1' = r_{1,1}(\hat{\tau}_t(m))w_1 \quad (m \in M). \qquad (3.4)_m$$

By hypotheses, the interval $J_1$ contains $\beta \leq 0$. Now, it is easy to show (e.g., [5],[8]) that one of the following two possibilities holds.

(i) There is a residual subset $M_1 \subseteq M$ such that

$$\liminf_{t \to +\infty} \int_0^t r_{1,1}(\hat{\tau}_s(m))ds = -\infty \qquad (3.5)$$

for each $m \in M_1$.

(ii) There is a continuous function $R_{1,1}: M \to \mathbf{R}$ such that

$$R_{1,1}(\hat{\tau}_t(m)) - R_{1,1}(m) = \int_0^t r_{1,1}(\hat{\tau}_s(m))ds \qquad (3.6)$$

for all $m \in M$. (Condition (ii) implies that $J_1 = \{0\}$.)

Let $U \subset \mathbf{R}^n$ and $T > 0$ be as in Proposition(3.1). Let $\overline{w} = (\overline{w}_1, \cdots, \overline{w}_n)^T$ be an element of $\mathbf{R}^n$, and suppose that alternative (i) holds. Let $m \in M_1$. There is a time $\eta > 0$ such that, if $w_1(t)$ is the solution of $(3.4)_m$ with $w_1(0) = \overline{w}_1$, then

$$(w_1(\eta), 0, \cdots, 0) \in U.$$

There is a measurable control function $u^1:[\eta, \eta + T] \to \Omega$ such that, if w(t) is a solution of $(3.3)_m$ satisfying $w(o) = \overline{w}$, then $w_1(\eta + T) = 0$. Set $T_1 = \eta + T$,

$$u_1(t) = \begin{cases} 0 & \text{if } t \notin [\eta, \eta+T] \\ u^1(t) & \text{if } t \in [\eta, \eta+T]. \end{cases}$$

Then $u_1$ steers $\overline{w}$ in time $T_1$ to the vector $\overline{w}^1$ whose first component is zero.

Now suppose (ii) holds. Let $\delta > 0$ be so small that the ball of radius $\delta$ and center $w = 0$ is contained in U. By (3.6) and minimality of M, we can find a time $\eta^1$ such that

$$| w_1(\eta^1) - \overline{w}_1 | < \delta/2.$$

Here $w_1(t)$ satisfies $(3.4)_m$ with $w_1(0) = \overline{w}_1$. Hence there is a control u with values in $\Omega$ defined on $[0, \eta^1 + T]$ which steers $\overline{w}$ to a vector $(\omega_1, \cdots, \omega_n)$ satisfying

$$| \omega_1 | < (1 - \delta/2) | w_1 |.$$

Repeating this construction at most $\frac{2|\overline{w}_1|}{\delta}$ times, we obtain a time $T_1$ and a control $u_1$ which steers $\overline{w}$ to a vector $\overline{w}^1$ whose first component at time $T_1$ is zero.

We can now apply the above argument to the second component $\overline{w}_2^1$ of the vector $\overline{w}^1$. Using the fact that $\{w \mid w_1 = 0\}$ is invariant under solutions of the homogeneous system $(2.2)_m$, and using the analogue of (3.5) or (3.6) for $r_{2,2}(m)$, we see that $\overline{w}^1$ can be steered in time $T_2$ via a control $u_2$, supported on $(T_1,\infty)$, to a vector $\overline{w}^2$ whose first two components are zero.

Repeating n times the above arguments, we obtain a control $u = u_1 + \cdots + u_n$ which steers $\overline{w}$ to zero in finite time $T = T_1 + \cdots + T_n$. Thus there is a residual set $M_1 \subseteq M$ of points such that $(3.3)_m$ is globally controllable. Now, the projection $E_1 = \pi(M_1)$ is residual since M is minimal [6]. This completes the proof of Theorem(3.2).

**Corollary(3.7)** Suppose $(E,\{\tau_t\})$ admits a unique invariant measure. Suppose further that $n = 2$ and that $0 \in \Lambda = [c,d]$. If equations $(1.1)_\xi$ are locally controllable, then there is a residual set $E_1 \subseteq E$ such that $(1.1)_\xi$ is globally null controllable for all $\xi \in E_1$.

**Proof** The corollary follows from Proposition(2.10) and Theorem(3.2).

We also have a specific result when $n = 3$. Suppose that $(E,\{\tau_t\})$ admits a unique invariant measure $\mu$. Suppose that there is only one spectral interval and let $\beta_1(\mu) \le \beta_2(\mu) \le \beta_3(\mu)$ be the Lyapunov exponents. As in the proof of Proposition(2.11), we have

$\beta_2(\mu) \in \bigcap_{i=1}^{3} J_i$. Hence we have the following.

**Corollary(3.8)** Under the above condition : if $\beta_2(\mu) \leq 0$, then equations $(1.1)_\xi$ are globally null controllable for a residual set of $\xi \in E$.

We state a last result, this time in arbitrary dimension $n$. Suppose again that $\Lambda = [c,d]$ and that $(E,\{\tau_t\})$ admits a unique ergodic measure $\mu$. Suppose that there are $n$ distinct Lyapunov exponents $\beta_1 < \cdots < \beta_n$. (Millionschikov [14] shows that distinctness of the exponents is generic for almost periodic systems.) Using Proposition(2.8), one can show that each interval $J_i$ must contain $\beta_r$ where $1 \leq r \leq n - 1$. Hence if $\beta_{n-1} \leq 0$, the proof of Theorem(3.2) applies, and one concludes that $(1.1)_\xi$ is globally null controllable for a residual set of $\xi \in E$. Observe that irreducibility is not required for this argument.

Finally, we observe that, if $(E,\{\tau_t\})$ is uniquely ergodic and if $d > 0$, then the residual set $E_1 \subseteq E$ satisfies $\mu(E_1) = 0$. This follows from [9, Theorem(3.7)].

## References

[1] **B.Barmish and W.Schmittendorf**, Null controllability of linear systems with constrained controls, SIAM Journal of Control and Optimization, 18 (1980).

[2] **R.Ellis**, Lectures on Topological Dynamics, Benjamin, New York, (1967).

[3] **R.Ellis and R.Johnson**, Topological dynamics and linear differential systems, Journal of Differential Equations 44 (1982), 21-39.

[4] **A.Fink**, Almost Periodic Differential Equations, Lecture notes in mathematics #377, Springer-Verlag, Heidelberg (1974).

[5] **H.Furstenberg**, Strict ergodicity and transformations of the torus, American Journal of Mathematics, 85 (1963), 477-515.

[6] **W.Gottschalk and G.Hedlund**, Topological Dynamics, A.M.S. Colloquim Publications, Providence, R.I. (1955).

[7] **R.Johnson**, Ergodic theory and linear differential equations, Journal of Differential Equations, 28 (1978), 23-34.

[8] **R.Johnson**, Minimal functions with unbounded integral, Israel Journal of Mathematics, 31 (1978), 133-141.

[9] **R.Johnson and M.Nerurkar**, On null controllability of linear systems with recurrent coefficients and constrained controls, Journal of Dynamics and Differential Equations, Vol.4, no 2 (1992), 259-273.

[10] **R.Johnson, K.Palmer and G.Sell**, Ergodic theory of linear dynamical systems, SIAM Journal of Mathematical Analysis 18 (1987),1-33.

[11] **A.Lyapunov**, Probléme général de la stabilité du mouvement (1892), in Ann. Math. Studies 17 (1947).

[12] **J.Massera and J.Schaeffer**, Linear Differential Equations and Function Spaces, Academic Press, New York/London, (1966).

[13] **V.Millionschikov**, Proof of the existence .....almost periodic coefficients, Differential Equations, 4 (1968), 203-205.

[14] **V.Millionschikov**, Typicality of almost reducible systems with almost periodic coefficients, Differential Equations, 14 (1978), 448-450.

[15] **V.Nemytskii and V.Stepanov**, Qualitative Theory of Ordinary Differential Equations, Princeton University Press (1960).

[16] **V.Oseledec**, A multiplicative ergodic theorem, Lyapunov characteristic numbers for dynamical systems, Trans. Moscow Math. Soc. 19 (1968), 197-231.

[17] **K.Palmer,** On reducibility of almost periodic systems of linear differential equations, Journal of Differential Equations, Volume 35 (1980) 374-390.

[18] **R.Sacker and G.Sell**, A spectral theory for linear differential systems, Journal of Differential Equations, Volume 27 (1978) 320-358.

[19] **J.Selgrade,** Isolated invariant sets for flows on vector bundles, Trans. A.M.S. Volume 203, (1975) 359-390.

# 40 Linear Two-Dimensional Systems with Deviating Arguments

**Tadeusz Kaczorek** Department of Electrical Engineering, Warsaw University of Technology, Warsaw, Poland

**Abstract.** Models of linear 2-D systems with deviating arguments are defined and some relationships between them are given. The general response formula for general model of 2-D linear systems with deviating arguments is derived. The Cayley-Hamilton theorem is extended for the general model. The concepts of local reachability and local controllability are also extended for the systems with deviating arguments. Necessary and sufficient conditions for local reachability and local controllability are established. The minimum energy control problem for the general model of 2-D linea systems with deviating arguments is fomulated and solved.

## 1. Introduction

Recently the classical 2-D state-space Roesser model [19], th Fornasini-Marchesini models [2,3] and the Kurek model [16] hav been extended for the singular case [1,4-9,18]. A survey of recen results in singular 2-D linear systems has been given in [1,13] In [5] the general response formula has been derived and the minimum energy control problem has been solved for singular general model of 2-D linear systems. A new concept of complete controllability of singular 2-D linear systems with variable coefficients has been introduced by Klamka in [15]. The minimum energy control problem for singular 2-D linear systems with variable coefficients has been formulated and solved also in [15]. In this paper singular and regular models of 2-D linear systems with deviating arguments are introduced. The general response formula for general regular model of 2-D linear systems with deviating arguments is derived. The well-known concepts of local reachability and local controllability [14,6,13] are extended for the general model of 2-D linear systems with deviating arguments. The minimum energy control problem for the general model is formulated and solved.

## 2. Models of 2-D linear systems with deviating arguments

Consider the generalized model of 2-D linear systems with deviating arguments

$$Ex_{i+1,j+1} = A_o x_{ij} + A_1 x_{i+1,j} + A_2 x_{i,j+1} + B_o x_{i-d_1,j-d_2} + B_1 x_{i-d_1+1,j-d_2} + B_2 x_{i-d_1,j-d_2+1} + C_o u_{ij} + C_1 u_{i+1,j} + C_2 u_{i,j+1} \quad i,j \in Z \tag{1a}$$

$$y_{ij} = D_o x_{ij} + D_1 u_{ij} \tag{1b}$$

where $x_{ij} \in R^n$ is the semistate vector at the point $(i,j)$, $u_{ij} \in R^m$ is the input vector, $y_{ij} \in R^p$ is the output vector and $E, A_k, B_k \in R^{qxn}$, $C_k \in R^{qxm}$, $D_o \in R^{pxn}$, $D_1 \in R^{pxm}$, $d_1, d_2$ are real numbers, $R^{qxp}$ is the set of real matrices with q rows and p columns, Z is the set of nonnegative integers, $k = 0,1,2$.
If $q \neq n$ or $\det E = 0$ when $q = n$ the model (1) is called singular.
If $q = n$ and $\det E \neq 0$ the model (1) is called regular. In this case premultiplying (1a) by $E^{-1}$ we obtain a model with $E = I$ (the identity matrix).
When all or some entries of the matrices E, $A_k$, $B_k$, $C_k$, $D_o$ and $D_1$ depend on i and j then (1) is called a model with variable coefficients.
If $d_1>0$ and $d_2>0$ then (1) is called a model with delays (retarded arguments) and if $d_1<0$ and $d_2<0$ then it is called a model with advanced arguments.
From (1) for $C_1 = 0$ and $C_2 = 0$ we obtain the first generalized Fornasini-Merchesini model of 2-D linear systems with deviating arguments. Similarily, from (1) for $A_o = 0$, $B_o = 0$ and $C_o = 0$ we obtain the second generalized Fornasini-Marchesini model of 2-D linear systems with deviating arguments.
The generalized Roesser model of 2-D linear systems with deviating arguments has the form

$$E\begin{bmatrix} x^h_{i+1,j} \\ x^v_{i,j+1} \end{bmatrix} = \begin{bmatrix} A_{11} & A_{12} \\ A_{21} & A_{22} \end{bmatrix}\begin{bmatrix} x^h_{ij} \\ x^v_{ij} \end{bmatrix} + \begin{bmatrix} B_{11} & B_{12} \\ B_{21} & B_{22} \end{bmatrix}\begin{bmatrix} x^h_{i-d_1,j-d_2} \\ x^v_{i-d_1,j-d_2} \end{bmatrix} + \begin{bmatrix} C_{10} \\ C_{20} \end{bmatrix} u_{ij} \quad i,j \in Z \tag{2a}$$

$$y_{ij} = \begin{bmatrix} D_{01}/D_{02} \end{bmatrix} \begin{bmatrix} x^h_{ij} \\ x^v_{ij} \end{bmatrix} + D_1 u_{ij} \tag{2b}$$

where $x^h_{ij} \in R^{n_1}$ is the horizontal semistate vector, $x^v_{ij} \in R^{n_2}$ is the vertical semistate vector, $u_{ij} \in R^m$ is the input, $y_{ij} \in R^p$ is the output vector and $A_{11}$, $B_{11} \in R^{q_1 x n_1}$, $A_{12}$ $B_{12} \in R^{q_1 x n_2}$, $A_{21}$, $B_{21} \in R^{q_2 x n_1}$, $A_{22}$, $B_{22} \in R^{q_2 x n_2}$, $C_{10} \in R^{q_1 x m}$, $C_{20} \in R^{q_2 x m}$, $D_{01} \in R^{p x n_1}$, $D_{02} \in R^{p x n_2}$, $D_{10} \in R^{p x n}$.

If the matrix E is square and det E ≠ 0 the model (2) is called regular.

Defining

$$x_{ij} := \begin{bmatrix} x^h_{ij} \\ x^v_{ij} \end{bmatrix}$$

we may write (2a) in the form

$$[E_1 \; 0] x_{i+1,j} + [0 \; E_2] x_{i,j+1} = \begin{bmatrix} A_{11} & A_{12} \\ A_{21} & A_{22} \end{bmatrix} x_{ij} + \begin{bmatrix} B_{11} & B_{12} \\ B_{21} & B_{22} \end{bmatrix} x_{i-d_1, j-d_2} + \begin{bmatrix} C_{10} \\ C_{20} \end{bmatrix} u_{ij}$$

where

$$E = [E_1/E_2], \; E_1 \in R^{q x n_1}, \; E_2 \in R^{q x n_2}$$

Therefore, the generalized Roesser model (2) is a special case of the model (1).

Defining

$$x^h_{ij} := \begin{bmatrix} x_{ij} \\ x_{i,j+1} \end{bmatrix}, \; x^v_{ij} := x_{i+1,j}$$

we may write (1a) in the form

$$[0 \; 0 \; E] \begin{bmatrix} x_{i+1,j} \\ x_{i+1,j+1} \\ x_{i+1,j+1} \end{bmatrix} = [A_0 A_2 A_1] \begin{bmatrix} x_{ij} \\ x_{i,j+1} \\ x_{i+1,j} \end{bmatrix} + [B_0 B_2 B_1] \begin{bmatrix} x_{i-d_1, j-d_2} \\ x_{i-d_1, j-d_2+1} \\ x_{i-d_1+1, j-d_2} \end{bmatrix} +$$

$$+ [C_0 C_1 C_2] \begin{bmatrix} u_{ij} \\ u_{i+1,j} \\ u_{i,j+1} \end{bmatrix}$$

Therefore the model (1) is a special case of the model (2).
In particular case when $B_2 = 0$ defining

$$x^h_{ij} := Ex_{i,j+1} - A_1 x_{ij} - B_1 x_{i-d_1,j-d_2} \text{ and } x^v_{ij} := x_{ij}$$

we may write (1a) in the form

$$\begin{bmatrix} I & -A_2 \\ 0 & E \end{bmatrix} \begin{bmatrix} x^h_{i+1,j} \\ x^v_{i,j+1} \end{bmatrix} = \begin{bmatrix} 0 & A_0 \\ I & A_1 \end{bmatrix} \begin{bmatrix} x^h_{ij} \\ x^v_{ij} \end{bmatrix} + \begin{bmatrix} 0 & B_0 \\ 0 & B_1 \end{bmatrix} \begin{bmatrix} x^h_{i-d_1,j-d_2} \\ x^v_{i-d_1,j-d_2} \end{bmatrix} + \begin{bmatrix} C_0 & C_1 & C_2 \\ 0 & 0 & 0 \end{bmatrix} \begin{bmatrix} u_{ij} \\ u_{i+1,j} \\ u_{i,j+1} \end{bmatrix}$$

In what follows it is assumed that

$$d_1 := \frac{f}{g}, \quad d_2 := \frac{r}{t} \tag{3}$$

where f, g, r and t are positive or negative integers.
Defining

$$h := \frac{1}{g}, \quad k := \frac{1}{t} \quad \text{and} \quad i := h\alpha, \quad j := k\beta \tag{4}$$

we may write (1a) in the form

$$Ex_{(\alpha+g)h,(\beta+t)k} = A_0 x_{h\alpha,k\beta} + A_1 x_{(\alpha+g)h,k\beta} + A_2 x_{h\alpha,(\beta+t)k} + B_0 x_{(\alpha-f)h,(\beta-r)k} +$$

$$+ B_1 x_{(\alpha+g-f)h,(\beta-r)k} + B_2 x_{(\alpha-f)h,(\beta+t-r)k} + C_0 u_{h\alpha,k\beta} + C_1 u_{(\alpha+g)h,k\beta} +$$

$$+ C_2 u_{h\alpha,(\beta+t)k} \tag{5}$$

Let the matrices $E, A_k, B_k, C_k$, $k=0,1,2$ of (1a) and $u_{ij}$ for $i,j \in Z$ be given. What are the boundary conditions which we have to know to be able to compute from (5) $x_{ij}$ for all $i,j>0$.
First we shall consider (5) for $E = I$. In this case the desired boundary conditions are given by

$$x_{ij} \text{ for } -fh \le i \le (g-1)h,\ j \ge -rk \quad \text{and} \quad -rk \le j \le (t-1)k,\ i > (g-1)h \tag{6}$$

Similarily, assuming (4) we may write (2a) in the form

$$E\begin{bmatrix} x^h_{(\alpha+g)h,\,k\beta} \\ x^v_{h\alpha,\,(\beta+t)k} \end{bmatrix} = \begin{bmatrix} A_{11} & A_{12} \\ A_{21} & A_{22} \end{bmatrix}\begin{bmatrix} x^h_{h\alpha,\,k\beta} \\ x^v_{h\alpha,\,k\beta} \end{bmatrix} + \begin{bmatrix} B_{11} & B_{12} \\ B_{21} & B_{22} \end{bmatrix}\begin{bmatrix} x^h_{(\alpha-f)h,\,(\beta-r)k} \\ x^v_{(\alpha-f)h,\,(\beta-r)k} \end{bmatrix} + \begin{bmatrix} C_{10} \\ C_{20} \end{bmatrix} u_{h\alpha,\,k\beta} \tag{7}$$

For (7) with E = I the boundary conditions are given by

$x^h_{ij}$ for $-fh\le i$, $-rk\le j<0$ and $-fh\le i\le(g-1)h$, $j\ge0$ (8a)

and

$x^v_{ij}$ for $-rk\le j$, $-fh\le i<0$ and $-rk\le j\le(t-1)k$, $i\ge0$ (8b)

Let $U_{ad}$ be a given set of admissible input vectors of (5). Boundary conditions (6) of (5) are called admissible (acceptable) if for given $u_{ij}$ there exists a solution $x_{h\alpha,k\beta}$ for $\alpha,\beta\ge0$ of (5). In general case when E is singular not all boundary conditions (6) are admissible (5). The set of admissible boundary conditions (6), denoted by $X_{ad}$, depends on the matrices $E, A_k, B_k, C_k$, k=0,1,2 but also on $U_{ad}$. The sets $X_{ad}$ and $U_{ad}$ can be found in a similar way as for singular 2-D linear systems without delays [11,12].

## 3. General response formula for 2-D linear systems with deviating arguments.

To simplify the notation without loss of generality we may assume $h = k = 1/\alpha = i$, $\beta = j$ / and write (5) with E = I in the form

$$x_{i+g,j+t}=A_0x_{ij}+A_1x_{i+g,j}+A_2x_{i,j+t}+B_0x_{i-f,j-r}+B_1x_{i+g-f,j-r}+$$
$$+B_2x_{i-f,j+t-r}+C_0u_{ij}+C_1u_{i+g,j}+C_2u_{i,j+t} \tag{9}$$

Let $X(z_1,z_2)$ be the 2-D Z transform of $x_{ij}$ defined by

$$X(z_1,z_2) = Z(x_{ij}) = \sum_{i=0}^{\infty}\sum_{j=0}^{\infty} x_{ij}z_1^{-i}z_2^{-j}$$

Taking into account that [10]

$$Z(x_{i+g,j+t}) = z_1^g z_2^t X(z_1,z_2) - \sum_{i=0}^{g-1}\sum_{j=0}^{t-1} x_{ij}z_1^{g-i}z_2^{t-j} +$$

$$-\sum_{i=0}^{g-1}\sum_{j=t}^{\infty} x_{ij}z_1^{g-i}z_2^{t-j} - \sum_{i=g}^{\infty}\sum_{j=0}^{t-1} x_{ij}z_1^{g-i}z_2^{t-j}$$

from (9) we obtain

$$G(z_1,z_2)X(z_1,z_2)=\sum_{i=0}^{g-1}\sum_{j=0}^{t-1} x_{ij}z_1^{g-i}z_2^{t-j}+\sum_{i=0}^{g-1}\sum_{j=t}^{\infty} x_{ij}z_1^{g-i}z_2^{t-j}+$$

$$+\sum_{i=g}^{\infty}\sum_{j=0}^{t-1} x_{ij}z_1^{g-i}z_2^{t-j}+\sum_{i=0}^{g-1}\sum_{j=0}^{\infty} A_1x_{ij}z_1^{g-i}z_2^{-j}-\sum_{i=0}^{\infty}\sum_{j=0}^{t-1} A_2x_{ij}z_1^{-i}z_2^{t-j} -$$

$$+\sum_{i=0}^{-f-1}\sum_{j=0}^{-r-1} B_0x_{ij}z_1^{-f-i}z_2^{-r-j}+\sum_{i=0}^{-f-1}\sum_{j=-r}^{\infty} B_0x_{ij}z_1^{-f-i}z_2^{-r-j} -$$

$$+\sum_{i=-r}^{\infty}\sum_{j=0}^{-r-1} B_0x_{ij}z_1^{-f-i}z_2^{-r-j} + \sum_{i=0}^{g-f-1}\sum_{j=0}^{-r-1} B_1x_{ij}z_1^{g-f-i}z_2^{-r-j} -$$

$$+\sum_{i=0}^{g-f-1}\sum_{j=-r}^{\infty} B_1x_{ij}z_1^{g-f-i}z_2^{-r-j} + \sum_{i=g-f}^{\infty}\sum_{j=0}^{-r-1} B_1x_{ij}z_1^{g-f-i}z_2^{-r-j} -$$

$$+\sum_{i=0}^{-f-1}\sum_{j=0}^{t-r-1} B_2x_{ij}z_1^{-f-i}z_2^{t-r-j} + \sum_{i=0}^{-f-1}\sum_{j=t-r}^{\infty} B_2x_{ij}z_1^{-f-i}z_2^{t-r-j} -$$

$$+\sum_{i=-f}^{\infty}\sum_{j=0}^{t-r-1} B_2x_{ij}z_1^{-f-i}z_2^{t-r-j}+U'(z_1,z_2) \tag{10}$$

where

$$G(z_1,z_2)=\left[Iz_1^gz_2^t-A_0-A_1z_1^g-A_2z_2^t-B_0z_1^{-f}z_2^{-r}-B_1z_1^{g-f}z_2^{-r}-B_2z_1^{-f}z_2^{t-r}\right] \tag{11}$$

$$U'(z_1,z_2)=C_0Z(u_{ij})+C_1Z(u_{i+g,j})+C_2Z(u_{i,j+t})$$

From (11) it follows that $G(z_1,z_2)$ is nonsingular for some $(z_1,z_2)$. Let

$$G^{-1}(z_1,z_2) = \sum_{i=0}^{\infty}\sum_{j=0}^{\infty} T_{ij}z_1^{-(g+i)}z_2^{-(t+j)} \tag{12}$$

where $T_{ij}\in R^{nxn}$ is the transition matrix of (9) defined by
$T_{00}:= I$ (the identity matrix)

$$T_{ij} = A_0T_{i-g,j-t}+A_1T_{i,j-t}+A_2T_{i-g,j}+B_0T_{i-g-f,j-t-r}+ \\ + B_1T_{i-f,j-t-r}+B_2T_{i-f-g,j-r} \qquad \text{for } i,j\geq 0 \tag{13}$$

$T_{ij} := 0$ (the zero matrix) for $i<0$ or/and $j<0$

From the equality $G(z_1,z_2)G^{-1}(z_1,z_2)=G^{-1}(z_1,z_2)G(z_1,z_2)$ and (11), (12) it follows that

$$A_0T_{i-g,j-t}+A_1T_{i,j-t}+A_2T_{i-g,j}+B_0T_{i-g-f,j-t-r}+B_1T_{i-f,j-t-r}+B_2T_{i-f-g,j-r}=$$

$$T_{i-g,j-t}A_0+T_{i,j-t}A_1+T_{i-g,j}A_2+T_{i-g-f,j-t-r}B_0+T_{i-f,j-t-r}B_1+T_{i-f-g,j-r}B_2$$

for all $i,j\geq 0$.

Substitution of (12) into (10) yields

$$X(z_1,z_2)=\sum_{i=0}^{\infty}\sum_{j=0}^{\infty}T_{ij}z_1^{-(g+i)}z_2^{-(t+j)}\left[\sum_{i=0}^{g-1}\sum_{j=0}^{t-1}x_{ij}z_1^{g-i}z_2^{t-j}+\right.$$

$$+\sum_{i=0}^{g-1}\sum_{j=t}^{\infty}x_{ij}z_1^{g-i}z_2^{t-j}+$$

$$+\sum_{i=g}^{\infty}\sum_{j=0}^{t-1}x_{ij}z_1^{g-i}z_2^{t-j}-\sum_{i=0}^{g-1}\sum_{j=0}^{\infty}A_1z_1^{g-i}z_2^{-j}-\sum_{i=0}^{\infty}\sum_{j=0}^{t-1}A_2z_1^{-i}z_2^{t-j}+$$

$$-\sum_{i=0}^{-f-1}\sum_{j=0}^{-r-1}B_0x_{ij}z_1^{-f-i}z_2^{-r-j}-\sum_{i=0}^{-f-1}\sum_{j=-r}^{\infty}B_0x_{ij}z_1^{-f-i}z_2^{-r-j}+$$

$$-\sum_{i=-r}^{\infty}\sum_{j=0}^{-r-1}B_0x_{ij}z_1^{-f-i}z_2^{-r-j}-\sum_{i=0}^{g-f-1}\sum_{j=0}^{-r-1}B_1x_{ij}z_1^{g-f-i}z_2^{-r-j}+ \tag{14}$$

$$-\sum_{i=0}^{g-f-1}\sum_{j=-r}^{\infty}B_1x_{ij}z_1^{g-f-i}z_2^{-r-j}-\sum_{i=g-f}^{\infty}\sum_{j=0}^{-r-1}B_1x_{ij}z_1^{g-f-i}z_2^{-r-j}+$$

$$-\sum_{i=0}^{-f-1}\sum_{j=0}^{t-r-1}B_2x_{ij}z_1^{-f-i}z_2^{t-r-j}-\sum_{i=0}^{-f-1}\sum_{j=t-r}^{\infty}B_2x_{ij}z_1^{-f-i}z_2^{t-r-j}+$$

$$\left.-\sum_{i=-f}^{\infty}\sum_{j=0}^{t-r-1}B_2x_{ij}z_1^{-f-i}z_2^{t-r-j}+U'(z_1,z_2)\right]$$

Using the inverse 2-D Z transformation for (14) and taking into account that

$$U'(z_1,z_2)=\left[C_0+C_1z_1^g+C_2z_2^t\right]U(z_1,z_2)-\sum_{i=0}^{g-1}\sum_{j=0}^{\infty}C_1u_{ij}z_1^{g-i}z_2^{-j}+$$

$$-\sum_{i=0}^{\infty}\sum_{j=0}^{t-1}C_2u_{ij}z_1^{-i}z_2^{t-j}$$

and (13) we obtain

$$x_{ij}=\sum_{\alpha=0}^{i-g}\sum_{\beta=0}^{j-t}T_{i-g-\alpha,\,j-t-\beta}C_0u_{\alpha\beta}+\sum_{\alpha=0}^{i}\sum_{\beta=0}^{j-t}T_{i-\alpha,\,j-t-\beta}C_1u_{\alpha\beta}+$$

$$+\sum_{\alpha=0}^{i-g}\sum_{\beta=0}^{j}T_{i-g-\alpha,\,j-\beta}C_2u_{\alpha\beta}+\sum_{\alpha=0}^{g-1}\sum_{\beta=0}^{t-1}T_{i-\alpha,\,j-\beta}x_{\alpha\beta}+$$

$$+\sum_{\alpha=0}^{g-1}\sum_{\beta=t}^{j}T_{i-\alpha,\,j-\beta}x_{\alpha\beta}+\sum_{\alpha=g}^{i}\sum_{\beta=0}^{t-1}T_{i-\alpha,\,j-\beta}x_{\alpha\beta}+ \tag{15}$$

$$-\sum_{\alpha=0}^{g-1}\sum_{\beta=0}^{j-t}T_{i-\alpha,\,j-t-\beta}[A_1,C_1]\begin{bmatrix}x_{\alpha\beta}\\u_{\alpha\beta}\end{bmatrix}-\sum_{\alpha=0}^{i-g}\sum_{\beta=o}^{t-1}T_{i-g-\alpha,\,j-\beta}[A_2,C_2]\begin{bmatrix}x_{\alpha\beta}\\u_{\alpha\beta}\end{bmatrix}+$$

$$-\sum_{\alpha=0}^{-f-1}\sum_{\beta=0}^{-r-1}T_{i-g-f-\alpha,\,j-t-r-\beta}B_0x_{\alpha\beta}-\sum_{\alpha=0}^{-f-1}\sum_{\beta=-r}^{j-t-r}T_{i-g-f-\alpha,\,j-t-r-\beta}B_0x_{\alpha\beta}+$$

$$-\sum_{\alpha=-r}^{i-g-f}\sum_{\beta=0}^{-r-1}T_{i-g-f-\alpha,\,j-t-r-\beta}B_0x_{\alpha\beta}-\sum_{\alpha=0}^{g-f-1}\sum_{\beta=0}^{-r-1}T_{i-f-\alpha,\,j-t-r-\beta}B_1x_{\alpha\beta}+$$

$$-\sum_{\alpha=0}^{g-f-1}\sum_{\beta=-r}^{j-t-r}T_{i-f-\alpha,\,j-t-r-\beta}B_1x_{\alpha\beta}-\sum_{\alpha=g-f}^{i-f}\sum_{\beta=0}^{-r-1}T_{i-f-\alpha,\,j-t-r-\beta}B_1x_{\alpha\beta}+$$

$$-\sum_{\alpha=0}^{-f-1}\sum_{\beta=0}^{t-r-1}T_{i-g-f-\alpha,\,j-r-\beta}B_2x_{\alpha\beta}-\sum_{\alpha=0}^{-f-1}\sum_{\beta=t-r}^{j-r}T_{i-g-f-\alpha,\,j-r-\beta}B_2x_{\alpha\beta}+$$

$$-\sum_{\alpha=-f}^{i-g-f}\sum_{\beta=0}^{t-r-1}T_{i-g-f-\alpha,\,j-r-\beta}B_2x_{\alpha\beta}\qquad\text{for } i,j\geq 0$$

The desired general response formula can be obtained by substitution of (15) into (1b).

## 4. Extension of Cayley-Hamilton theorem for 2-D linear systems with deviating arguments

The well-known Cayley-Hamilton theorem can be extended for 2-D linear systems with deviating arguments.

Let

$$\det G(z_1,z_2) = \sum_{i=0}^{n_1'} \sum_{j=0}^{n_2'} d_{ij} z_1^i z_2^j \tag{16a}$$

and

$$\mathrm{Adj}\, G(z_1,z_2) = \sum_{i=0}^{m_1} \sum_{j=0}^{m_2} H_{ij} z_1^i z_2^j \tag{16b}$$

Theorem 1

The matrices $T_{ij}$, defined by (13), satisfy the equations

$$\sum_{i=0}^{n_1'} \sum_{j=0}^{n_2'} d_{ij} T_{i-g-k,\, j-t-l} = 0 \quad \text{for} \begin{cases} k<0 \text{ and } m_1<k\le n_1'-g \\ l<0 \text{ and } m_2<l\le n_2'-t \end{cases} \tag{17}$$

Proof.

From (12) and (16) we have

$$\sum_{i=0}^{m_1} \sum_{j=0}^{m_2} H_{ij} z_1^i z_2^j = \left[\sum_{i=0}^{n_1'} \sum_{j=0}^{n_2'} d_{ij} z_1^i z_2^j\right]\left[\sum_{k=0}^{\infty} \sum_{l=0}^{\infty} T_{kl} z_1^{-(g+k)} z_2^{-(t+l)}\right] \tag{18}$$

Equating the coefficient matrices at the same powers of (18) for inddices defined by (17) and taking into account that $T_{ij} = 0$ for $i<0$ and/or $j<0$ we obtain the desired equations.■

## 5. Local reachability and local controllability

Consider the model (9) in the rectangle

$$[0,M]\times[0,N] := \{(i,j): 0\le i\le M,\ 0\le j\le N\} \tag{19}$$

with boundary conditions

$$x_{ij} \text{ for } -f\le i\le g-1,\ -r\le j\le N \text{ and } -r\le j\le t-1,\ g-1<i\le M \tag{20}$$

Definition 1

The model (9) is locally reachable in the rectangle (19) if for any boundary conditions (20) and every vector $x_f \in R^n$ there exists a sequence of input vectors

$$\{u_{00}, u_{01}, \ldots, u_{0N}, u_{10}, \ldots, u_{1N}, u_{20}, \ldots, u_{M-1,N} U_{M0}, \ldots, u_{M,N-1}\} \tag{21}$$

such that $x_{MN} = x_f$.

Theorem 2

The model (9) is locally reachable in the rectangle (19) iff

$$\text{Rank } R_{MN} = n \tag{22}$$

where

$$R_{MN} := [P_{00}/P_{01}/\ldots/P_{0N}/P_{10}/\ldots/P_{1N}/P_{20}/\ldots/P_{MN}],\quad P_{ij} := P^C_{ij} + P^{C_1}_{ij} + P^{C_2}_{ij}$$

$$P^C_{ij} := T_{M-g-i,N-t-j}C_0 + T_{M-i,N-t-j}C_1 + T_{M-g-i,N-j}C_2 \text{ for } 0\le i\le M,\ 0\le j\le N$$

$$P^{C_1}_{ij} := \begin{cases} T_{M-i,N-t-j}C_1 & \text{for } 0\le i\le g-1,\ 0\le j\le N \\ 0 & \text{for } g\le i\le M,\ \ 0\le j\le N \end{cases} \tag{23}$$

$$P^{C_2}_{ij} := \begin{cases} T_{M-g-i,N-j}C_2 & \text{for } 0\le i\le M,\ \ 0\le j\le t-1 \\ 0 & \text{for } 0\le i\le M,\ \ t\le j\le N \end{cases}$$

and $T_{ij}$ are defined by (13).

Proof.

Using (15), (23) and taking into account that $x_{MN} = x_f$ we obtain

$$x_f - \sum_{\alpha=0}^{g-1}\sum_{\beta=0}^{t-1} T_{M-\alpha,N-\beta}x_{\alpha\beta} - \sum_{\alpha=0}^{g-1}\sum_{\beta=t}^{N} T_{M-\alpha,N-\beta}x_{\alpha\beta} - \sum_{\alpha=g}^{M}\sum_{\beta=0}^{t-1} T_{M-\alpha,N-\beta}x_{\alpha\beta} +$$

$$+\sum_{\alpha=0}^{g-1}\sum_{\beta=0}^{N-t} T_{M-\alpha,N-t-\beta}A_1x_{\alpha\beta}+\sum_{\alpha=0}^{M-g}\sum_{\beta=0}^{t-1} T_{M-g-\alpha,N-\beta}A_2x_{\alpha\beta}+$$

$$+\sum_{\alpha=0}^{-f-1}\sum_{\beta=0}^{-r-1} T_{M-g-f-\alpha,N-t-r-\beta}B_0x_{\alpha\beta}+\sum_{\alpha=0}^{-f-1}\sum_{\beta=-r}^{N-t-r} T_{M-g-f-\alpha,N-t-r-\beta}B_0x_{\alpha\beta}+$$

$$+\sum_{\alpha=-r}^{M-g-f}\sum_{\beta=0}^{-r-1} T_{M-g-f-\alpha,N-t-r-\beta}B_0x_{\alpha\beta}+\sum_{\alpha=0}^{g-f-1}\sum_{\beta=0}^{-r-1} T_{M-f-\alpha,N-t-r-\beta}B_1x_{\alpha\beta}+$$

$$+\sum_{\alpha=0}^{g-f-1}\sum_{\beta=-r}^{N-t-r} T_{M-f-\alpha,N-t-r-\beta}B_1x_{\alpha\beta}+\sum_{\alpha=g-f}^{M-f}\sum_{\beta=0}^{-r-1} T_{M-f-\alpha,N-t-r-\beta}B_1x_{\alpha\beta}+$$

$$+\sum_{\alpha=0}^{-f-1}\sum_{\beta=0}^{t-r-1} T_{M-g-f-\alpha,N-r-\beta}B_2x_{\alpha\beta}+\sum_{\alpha=0}^{-f-1}\sum_{\beta=t-r}^{N-r} T_{M-g-f-\alpha,N-r-\beta}B_2x_{\alpha\beta}+$$

$$+\sum_{\alpha=-f}^{M-g-f}\sum_{\beta=0}^{t-r-1} T_{M-g-f-\alpha,N-r-\beta}B_2x_{\alpha\beta} = R_{MN}u \tag{24}$$

where

$$u^T= \left[u_{00}^T/u_{01}^T/\dots/u_{0N}^T/u_{10}^T/\dots/u_{1N}^T/u_{20}^T/\dots/u_{MN}^T\right]$$

the upper index T denotes the transposition.

From (24) it follows that the model (9) is locally reachable in the rectangle (19) iff (22) holds.■

Definition 2

The model (9) is locally controllable in the rectangle (19) if for any boundary conditions (20) there exists a sequence of input vectors (21) such that $x_{MN} = 0$.

Theorem 3

The model (9) is locally controllable in the rectangle (19) iff

$$\text{rank } R_{MN} = \text{rank } [R_{MN}/Q'_{MN}] \tag{25a}$$

or equivalently

$$\text{Im } Q'_{MN} \subset \text{Im } R_{MN} \tag{25b}$$

where $R_{MN}$ is defined by (23) and

$$Q'_{MN}:= [Q_{00}/Q_{01}/\dots/Q_{0N}/Q_{10}/\dots/Q_{1N}/Q_{20}/\dots/Q_{MN}] \tag{26a}$$

$$Q_{ij} := Q^{x}_{ij} + Q^{A_1}_{ij} + Q^{A_2}_{ij} + Q^{B_0}_{ij} + Q^{B_1}_{ij} + Q^{B_2}_{ij} \quad (26b)$$

$$Q^{x}_{ij} := \begin{cases} -T_{M-i,N-j} & \text{for } 0\le i\le g-1,\ 0\le j\le N \text{ and } g\le i\le M,\ 0\le j\le t-1 \\ 0 & \text{for } g\le i\le M,\ t\le j\le N \end{cases} \quad (26c)$$

$$Q^{A_1}_{ij} := \begin{cases} T_{M-i,N-t-j}A_1 & \text{for } 0\le i\le g-1,\ 0\le j\le N \\ 0 & \text{for } g\le i\le M,\ 0\le j\le N \end{cases} \quad (26d)$$

$$Q^{A_2}_{ij} := \begin{cases} T_{M-g-i,N-j}A_2 & \text{for } 0\le i\le M,\ 0\le j\le t-1 \\ 0 & \text{for } 0\le i\le M,\ t\le j\le N \end{cases} \quad (26e)$$

$$Q^{B_0}_{ij} := \begin{cases} T_{M-g-j-i,N-t-r-j}B_0 & \text{for } 0\le i\le -f-1, 0\le j\le N \text{ and } -f\le i\le M, 0\le j\le -r-1 \\ 0 & \text{for } -f\le i\le M,\ -r\le j\le N \end{cases} \quad (26f)$$

$$Q^{B_1}_{ij} := \begin{cases} T_{M-f-i,N-t-r-j}B_1 & \text{for } 0\le i\le g-f-1, 0\le j\le N \text{ and } g-f\le i\le M, 0\le j\le -r-1 \\ 0 & \text{for } g-f\le i\le M,\ -r\le j\le N \end{cases} \quad (26h)$$

$$Q^{B_2}_{ij} := \begin{cases} T_{M-f-g-i,N-r-j}B_2 & \text{for } 0\le i\le -f-1, 0\le j\le N \text{ and } -f\le i\le M, 0\le j\le t-r-1 \\ 0 & \text{for } -f\le i\le M,\ t-r\le j\le N \end{cases} \quad (26i)$$

and Im stands for the image.

Proof. Using (15),(23),(26) and setting $x_{MN} = 0$ we obtain

$$R_{MN}u = Q'_{MN}x \quad (27)$$

where

$$x^T = \left[x^T_{00}/x^T_{01}/\dots/x^T_{0N}/x^T_{10}/\dots/x^T_{1N}/x^T_{20}/\dots/x^T_{MN}\right]$$

From (27) it follows that for any boundary conditions (20) the model (9) is locally controllable in the rectangle (19) iff (25a) holds. The equivalence of (25a) and (25b) is obvious.■

Remark. From (22) and (25) it follows that the local reachability implies always the local controllability but the local controllability does not imply the local reachability of the model (9).

## 6. Minimum energy control

Consider the model (9) with boundary conditions (20) and the performance index

$$I(u) := \sum_{i=0}^{M} \sum_{j=0}^{N} u_{ij}^T Q u_{ij} \tag{28}$$

where Q is an mxm symmetric and positive definite weighting matrix.

The minimum energy control problem for the model (9) can be stated as follows. Given $A_k, B_k, C_k$, k=0,1,2, the boundary conditions (20), the matrix Q and M,N, find a sequence of input vectors (21) which transfer the model from the boundary conditions to the desired final state $x_f, x_{MN} = x_f$, and minimizes the performance index (28).

To solve the problem we define the matrix

$$W_{MN} := R_{MN} Q_d R_{MN}^T = \sum_{i=0}^{M} \sum_{j=0}^{N} P_{ij} Q^{-1} P_{ij}^T \tag{29}$$

where $R_{MN}$ is defined by (23) and

$$Q_d := \text{diag}\,[Q^{-1}, \ldots, Q^{-1}]$$

It is easy to show that $W_{MN}$ is nonsingular (positive definite) iff the model (9) is reachable in the rectangle (19).

Let us define

$$\hat{u}_{ij} := Q^{-1} P_{ij}^T W_{MN}^{-1} (x_f - x_{bc}) \text{ for } 0 \le i \le M,\ 0 \le j \le N \tag{30}$$

where

$$x_{bc} := \sum_{i=0}^{g-1} \sum_{j=0}^{t-1} T_{M-i,N-j} x_{ij} + \sum_{i=0}^{g-1} \sum_{j=t}^{N} T_{M-i,N-j} x_{ij} + \sum_{i=g}^{M} \sum_{j=0}^{t-1} T_{M-i,N-j} x_{ij} +$$

$$- \sum_{i=0}^{g-1} \sum_{j=0}^{N-t} T_{M-i,N-t-j} A_1 x_{ij} - \sum_{i=0}^{M-g} \sum_{j=0}^{t-1} T_{M-g-i,N-j} A_2 x_{ij} +$$

$$- \sum_{i=0}^{-f-1} \sum_{j=0}^{-r-1} T_{M-g-f-i,N-t-r-j} B_0 x_{ij} - \sum_{i=0}^{-f-1} \sum_{j=-r}^{N-t-r} T_{M-g-f-i,N-t-r-j} B_0 x_{ij} +$$

$$- \sum_{i=-r}^{M-g-f} \sum_{j=0}^{-r-1} T_{M-g-f-i,N-t-r-j} B_0 x_{ij} - \sum_{i=0}^{g-f-1} \sum_{j=0}^{-r-1} T_{M-f-i,N-t-r-j} B_1 x_{ij} +$$

$$- \sum_{i=0}^{g-f-1} \sum_{j=-r}^{N-t-r} T_{M-f-i,N-t-r-j} B_1 x_{ij} - \sum_{i=g-f}^{M-f} \sum_{j=0}^{-r-1} T_{M-f-i,N-t-r-j} B_1 x_{ij} +$$

$$-\sum_{i=0}^{-f-1}\sum_{j=0}^{t-r-1} T_{M-g-f-i,N-r-j}B_2x_{ij}-\sum_{i=0}^{-f-1}\sum_{j=t-r}^{N-r} T_{M-g-f-i,N-r-j}B_2x_{ij}+$$

$$-\sum_{i=-f}^{M-g-f}\sum_{j=0}^{t-r-1} T_{M-g-f-i,N-r-j}B_2x_{ij}$$

Theorem 4

Let us assume that:

i) the model (9) is locally reachable in the rectangle (19),

ii) $\bar{u}_{ij}$ is any sequence of the form (21) which transfers the model from boundary conditions (20) to $x_f$.

Then the sequence (30) accomplies the same task and

$$I(\bar{u}) \geq I(\hat{u}) \tag{31}$$

The minimum value of (28) is given by

$$I(\hat{u}) = (x_f-x_{bc})^T W_{MN}^{-1}(x_f-x_{bc}) \tag{32}$$

Proof. First we shall show that (30) transfers the model from boundary conditions (20) to $x_f$. Using (15) for $i = M$, $j = N$, (23), (30) and (29) we obtain

$$x_{MN} = \sum_{i=0}^{M}\sum_{j=0}^{N} P_{ij}Q^{-1}P_{ij}^T W_{MN}^{-1}(x_f-x_{bc})+x_{bc} = x_f$$

Since $\bar{u}_{ij}$ and $\hat{u}_{ij}$ transfer the model from boundary conditions (20) to $x_f$, the

$$\sum_{i=0}^{M}\sum_{j=0}^{N} P_{ij}(\bar{u}_{ij}-\hat{u}_{ij}) = 0 \tag{33}$$

Using (33) it is easy to show that

$$\sum_{i=0}^{M}\sum_{j=0}^{N} \bar{u}_{ij}^T Q\bar{u}_{ij}=\sum_{i=0}^{M}\sum_{j=0}^{N} \hat{u}_{ij}^T Q\hat{u}_{ij}+\sum_{i=0}^{M}\sum_{j=0}^{N} (\bar{u}_{ij}-\hat{u}_{ij})^T Q(\bar{u}_{ij}-\hat{u}_{ij}) \tag{34}$$

Note that the inequality (31) holds since the last term in (34) is always nonnegative.

To obtain the minimum value of (28) we substitute (30) into (28)

$$I(\hat{u})=\sum_{i=0}^{M}\sum_{j=0}^{N}\hat{u}_{ij}^{T}Q\hat{u}_{ij}=\sum_{i=0}^{M}\sum_{j=0}^{N}\left[Q^{-1}P_{ij}^{T}W_{MN}^{-1}(x_f-x_{bc})\right]^{T}Q\left[Q^{-1}P_{ij}^{T}W_{MN}^{-1}(x_f-x_{bc})\right]=$$

$$=(x_f-x_{bc})^{T}W_{MN}^{-1}\left[\sum_{i=0}^{M}\sum_{j=0}^{N}P_{ij}Q^{-1}P_{ij}^{T}\right]W_{MN}^{-1}(x_f-x_{bc})=(x_f-x_{bc})^{T}W_{MN}^{-1}(x_f-x_{bc}).\blacksquare$$

In a similar way as in [9] the linear - quadratic optimal regulator problem can be also extended for singular 2-D linear systems with deviating arguments.

## 7. Concluding remarks

Regular and singular models of 2-D linear systems with deviating arguments have been introduced and some relationships between them are established. The general response formula for general regular model of 2-D linear systems with deviating arguments has been derived. In a similar way as in [6,7] the formula may be extended for singular general model of 2-D linear systems with constant and variable coefficients. Necessary and sufficient conditions for local reachability and local controllability for the general regular model have been established. With slight modifications the conditions can be also extended for singular general model of 2-D linear systems with constant and variable coefficients [6,7]. Following Klamka [15] we may also introduce the notion of complete controllability of the 2-D linear systems with deviating arguments. The concepts of strong observability and strong reconstructibility introduced by Kurek in [17] may be extended for the 2-D linear systems with deviating arguments. The minimum energy control problem for the general regular model of 2-D linear systems with deviating arguments has been formulated and solved. An extension of this problem for singular general model of 2-D linear systems with deviating arguments is easy.

With slight modifications the above considerations can be extended for n-D, n>2, linear systems with deviating arguments.

References


[1] G.Beauchamp, F.L.Lewis and B.G.Mertzios, Recent results in 2-D singular systems, Proc. IFAC Workshop on System Structure and Control, Prague, Sept. 1989, pp. 253-256.

[2] E.Fornasini and G.Marchesini, State space realization of two-dimensional filters, IEEE Trans. Automat.Contr. AC-21, 1976, pp. 484-491.

[3] E.Fornasini and G.Marchesini, Doubly indexed dynamical systems: State space models and structural properties. Math. System Theory vol. 12, No.1, 1978.

[4] T.Kaczorek, Singular general model of 2-D systems and its solution, IEEE Trans.Automat.Contr. AC-33,1988,pp. 1060-1061.

[5] T. Kaczorek, Singular multidimensional linear discrete systems, Proc.IEEE Inter.Symp.Circuits and Systems, Helsinki, June 7-9, 1988, pp. 105-108.

[6] T.Kaczorek, General response formula and minimum energy control for general singular model of 2-D systems, IEEE Trans.Automat.Contr. AC-34, 1989, pp. 433-436.

[7] T.Kaczorek, General response formula, controllability and observability for singular 2-D linear systems with variable coefficients, Proc. IMACS-IFAC Inter.Symp.Math. and Intel. Models in System Simulation, Sept. 3-7, Brussels 1990, pp. VII. A.2-1 - A.2-6.

[8] T.Kaczorek, Observability and reconstructibility of singular 2-D systems, Bull.Pol.Acad.Sci.Techn. vol.37,1989,pp.531-535.

[9] T. Kaczorek, The linear-quadratic optimal regulator for singular 2-D systems with variable coefficients, IEEE Trans. Automat.Contr. AC-34, 1989, pp. 565-566.

[10] T. Kaczorek, Two-Dimensional Linear Systems, New York, Springer-Verlag 1985.

[11] T.Kaczorek, Determination of boundary conditions for singular general model of 2-D linear systems, Submitted to IEEE Trans. Autom. Contr. 1992.

[12] T.Kaczorek, Acceptable input sequences for singular 2-D linear systems, IEEE Trans.Automat.Contr. vol. 38, June 1993 (in press).

[13] T.Kaczorek, Some recent results in singular 2-D systems theory, Kybernetika, vol.27, 1991, No 3, pp. 253-262.

[14] J.Klamka, Controllability of M-dimensional systems, Found. Contr.Engin. vol. 8, No 2, 1983, pp. 65-74.

[15] J.Klamka, Optimal control problems for singular 2-D systems, Proc. IMACS-IFAC Inter.Symp.Math. and Intel. Models in System Simul. Sept. 3-7,1990, Brussels, pp. VII. A.3-1 - A.3-4.

[16] J.Kurek, The general state-space model for a two-dimensional linear digital systems, IEEE Trans.Automat.Contr. AC-30, 1985, pp. 600-602.

[17] J.Kurek, Strong observability and strong reconstructibility of a system described by the 2-D Roesser model, Inter. J. Control, vol. 47, 1988, pp. 633-641.

[18] F.L. Lewis and B.G. Mertzios, On the analysis of two - dimensional discrete singular systems. Math. Control Signal Systems 1992 (in press).

[19] R.P.Roesser, A discrete state-space model for linear image processing, IEEE Trans.Automat.Contr. AC-21, 1975, pp. 1-10.

# 41 Boundary Controllability in Transmission Problems for Thin Plates

**John E. Lagnese** Georgetown University, Washington, D.C.

## 1 Problem formulation

The purpose of this paper is to study the question of exact controllability via boundary controls of a thin, elastic plate with discontinuous but piecewise constant elastic parameters.

Let $\Omega$, $\Omega_1$ be bounded, simply connected open sets in $\Re^2$ with Lipschitz boundaries consisting of a finite number of smooth arcs and suppose that $\overline{\Omega}_1 \subset \Omega$. Let

$$\Omega_2 = \Omega/\overline{\Omega}_1, \;\; \Gamma = \partial\Omega, \;\; \Gamma_1 = \partial\Omega_1, \;\; \Gamma_2 = \partial\Omega_2 = \Gamma \cup \Gamma_1.$$

Let $\boldsymbol{\nu}_i$ and $\boldsymbol{\nu}$ denote the *exterior* unit normals to $\Gamma_i$ and $\Gamma$, respectively, wherever the normals exist. Thus $\boldsymbol{\nu}_2 = \boldsymbol{\nu}$ on $\Gamma$ and $\boldsymbol{\nu}_2 = -\boldsymbol{\nu}_1$ on $\Gamma_1$. We imagine that $\Omega$ represents the mid-plane of a thin plate of uniform thickness $h$ when the plate is in equilibrium. Thus the equilibrium position of the plate is

$$\mathcal{P} = \Omega \times (-h/2, h/2).$$

Let

$$\mathcal{P}_1 = \Omega_1 \times (-h/2, h/2), \;\; \mathcal{P}_2 = \Omega_2 \times (-h/2, h/2).$$

It is assumed that each part $\mathcal{P}_i$ is elastically isotropic and homogeneous, but that the material parameters of the two parts of $\mathcal{P}$ may differ. Let $E_i$ and $\mu_i$ be Young's modulus and Poisson's ratio, respectively, and $D_i = E_i h^3/12(1-\mu_i^2)$ be the flexural rigidity of $\mathcal{P}_i$.

Let $\mathcal{S}$ denote the set of second order symmetric tensors and $C_i : \mathcal{A} \mapsto \mathcal{A}$ be defined by

$$C_i[\varepsilon] = D_i[\mu_i(\varepsilon_{11} + \varepsilon_{22})\mathcal{I} + (1-\mu_i)\varepsilon], \;\; \varepsilon \in \mathcal{A}, \tag{1.1}$$

where $\mathcal{I}$ is the identity in $\mathcal{A}$. For a function $\mathbf{u} : \Re^2 \mapsto \Re^2$ set

$$\varepsilon(\mathbf{u}) = \frac{1}{2}(\nabla\mathbf{u} + (\nabla\mathbf{u})^*) = \frac{1}{2}\left(\frac{\partial u_j}{\partial x_k} + \frac{\partial u_k}{\partial x_j}\right).$$

Research supported by the Air Force Office of Scientific Research through grant F49620-92-J-0031.

We shall consider the following system of dynamic equations, transmission conditions and boundary conditions (where $\dot{} = \partial/\partial t$):

$$\begin{cases} I_{\rho_i}\ddot{\boldsymbol{\phi}}_i - \operatorname{div} C_i[\varepsilon(\boldsymbol{\phi}_i)] + K_i(\nabla W_i + \boldsymbol{\phi}_i) = 0, \\ \rho_i \ddot{W}_i - K_i \operatorname{div}(\nabla W_i + \boldsymbol{\phi}_i) = 0 \quad \text{in } \Omega_i,\, t > 0, \;\; i = 1,2; \end{cases} \tag{1.2}$$

$$\begin{cases} W_1 = W_2, \;\; \boldsymbol{\phi}_1 = \boldsymbol{\phi}_2, \\ C_1[\varepsilon(\boldsymbol{\phi}_1)]\boldsymbol{\nu}_1 + C_2[\varepsilon(\boldsymbol{\phi}_2)]\boldsymbol{\nu}_2 = 0, \\ K_1(\nabla W_1 + \boldsymbol{\phi}_1)\cdot \boldsymbol{\nu}_1 + K_2(\nabla W_2 + \boldsymbol{\phi}_2)\cdot\boldsymbol{\nu}_2 = 0 \quad \text{on } \Gamma_1,\, t > 0; \end{cases} \tag{1.3}$$

$$W_2 = f, \;\; \boldsymbol{\phi}_2 = \mathbf{m} \quad \text{on } \Gamma_2,\, t > 0. \tag{1.4}$$

Equations (1.2) are those associated with *Reissner-Mindlin* plate theory. In these equations $\rho_i/h$ is the mass density of $\mathcal{P}_i$ per unit of reference volume, $K_i$ is its shear modulus and $I_{\rho_i} = \rho_i h^2/12$. The components of the vector "angle" $\boldsymbol{\phi}_i(x_0, y_0)$ represent the total rotation angles of the cross-sections $x = x_0$, $y = y_0$, respectively, due to bending and shearing; $\nabla W_i + \boldsymbol{\phi}_i$ represents the rotation due to shearing alone of these cross-sections.

The first two conditions in (1.3) are continuity requirements for the transverse displacements and cross-sectional rotations along the "junction curve" $\Gamma_1$; the last two are balance laws for the transmission of moments and transverse forces, respectively, along the junction. Equations (1.3) together comprise a special case of the junction conditions for a general three-dimensional network of Reissner-Mindlin plates derived in [3, Section 3].

In (1.4), $f$ and $\mathbf{m} = (m_1, m_2)$ are *control inputs.* One may, of course, consider other than Dirichlet boundary conditions on $\Gamma$ and obtain results similar to those presented below, but we shall not do so here.

To complete the description of the system, initial conditions are specified:

$$\begin{cases} \boldsymbol{\phi}_i(0) = \boldsymbol{\phi}_i^0, \;\; \dot{\boldsymbol{\phi}}_i(0) = \boldsymbol{\phi}_i^1, \\ W_i(0) = W_i^0, \;\; \dot{W}_i(0) = W_i^1 \quad \text{in } \Omega_i, \;\; i = 1,2. \end{cases} \tag{1.5}$$

Naturally, the initial conditions will have to satisfy appropriate transmission conditions along $\Gamma_1$.

## 2 Formulation of main results

For $s \geq 0$ define

$$\mathcal{H}^s(\Omega) = H^s(\Omega;[\mathbf{i}]) \bigoplus H^s(\Omega;[\mathbf{j}]) \bigoplus H^s(\Omega;[\mathbf{k}]),$$

where $\mathbf{i}, \mathbf{j}, \mathbf{k}$ is the natural basis for $\Re^3$ and $H^s(\Omega;[\mathbf{e}])$ consists of functions $u(\cdot)\mathbf{e}$ with $u \in H^s(\Omega)$, the usual (real) Sobolev space of order $s$. Let $\mathcal{H}_0^s(\Omega)$ denote the closed subspace of $\mathcal{H}^s(\Omega)$ consisting of those functions which vanish on $\Gamma$ together with their derivatives to order $s-1$. We set

$$H = \mathcal{H}^0(\Omega), \quad V = \mathcal{H}_0^1(\Omega)$$

with norms defined as follows. Let

$$\mathbf{\Phi} = \phi_1 \mathbf{i} + \phi_2 \mathbf{j} + W\mathbf{k} := \boldsymbol{\phi} + W\mathbf{k}$$

belong to $H$ or to $V$. We set

$$\|\mathbf{\Phi}\|_H = \left\{ \sum_{i=1}^{2} \int_{\Omega_i} (\rho_i |W|^2 + I_{\rho_i} |\boldsymbol{\phi}|^2) d\Omega_i \right\}^{1/2},$$

$$\|\mathbf{\Phi}\|_V = \left\{ \sum_{i=1}^{2} \int_{\Omega_i} [(C_i[\varepsilon(\boldsymbol{\phi})], \varepsilon(\boldsymbol{\phi})) + K_i |\nabla W + \boldsymbol{\phi}|^2] d\Omega_i \right\}^{1/2},$$

where for $\xi, \eta \in \mathcal{A}$ the matrix inner product $(\xi, \eta) = \sum \xi_{ij} \eta_{ij}$. One may show that $\|\cdot\|_V$ defines a norm on $V$ equivalent to that obtained by using the standard Sobolev norm of order one. The Hilbert space $V$ is dense in $H$ with compact injection so if $H$ is identified with its dual space one has the compact embeddings $V \subset H \subset V'$, where $V'$ denotes the dual of $V$ with respect to $H$.

We denote by $\mathbf{\Phi}^0, \mathbf{\Phi}^1$ the initial data (1.5), i.e.,

$$\mathbf{\Phi}^0 = \begin{cases} \mathbf{\Phi}_1^0 \text{ in } \Omega_1, \\ \mathbf{\Phi}_2^0 \text{ in } \Omega_2, \end{cases} \quad \mathbf{\Phi}^1 = \begin{cases} \mathbf{\Phi}_1^1 \text{ in } \Omega_1, \\ \mathbf{\Phi}_2^1 \text{ in } \Omega_2, \end{cases},$$

where

$$\mathbf{\Phi}_i^0 = \boldsymbol{\phi}_i^0 + W_i^0 \mathbf{k}, \quad \mathbf{\Phi}_i^1 = \boldsymbol{\phi}_i^1 + W_i^1 \mathbf{k}.$$

We also set

$$\mathbf{F} = m_1 \mathbf{i} + m_2 \mathbf{j} + f\mathbf{k} := \mathbf{m} + f\mathbf{k},$$

$$U = L^2(\Gamma; [\mathbf{i}]) \bigoplus L^2(\Gamma; [\mathbf{j}]) \bigoplus L^2(\Gamma; [\mathbf{k}]).$$

**Theorem 2.1** *(Well-posedness of the control problem.) Assume that*

$$\mathbf{\Phi}^0 \in H, \quad \mathbf{\Phi}^1 \in V', \quad \mathbf{F} \in L^2(0,T;U).$$

*There is a function $\mathbf{\Phi} \in C([0,T];H) \cap C^1([0,T];V')$ such that $\mathbf{\Phi}_i := \mathbf{\Phi}|_{\Omega_i}$, $i = 1,2$ is the unique solution of* (1.2)–(1.5).

**Remark 2.1** The definition of a solution of (1.2)–(1.4) will be made precise in the proof of Theorem 2.1.

Suppose that $\mathbf{\Phi}^0 = \mathbf{\Phi}^1 = 0$, let $T > 0$ and $\mathbf{\Phi}$ be the function of Theorem 2.1. The *reachability problem* (equivalent to the exact controllability problem for the system under consideration) is to determine the set

$$R_T = \{(\mathbf{\Phi}(T), \dot{\mathbf{\Phi}}(T)) | \, \mathbf{F} \in L^2(0,T;U)\}.$$

By Theorem 2.1, $R_T \subset H \times V'$ for each $T > 0$. Under some restrictions on the elastic parameters and on the geometry of $\Gamma_1$ we can prove the opposite inclusion.

**Theorem 2.2** *Assume that $\Gamma_1$ is star-shaped with respect to a point $x_1 \in \Omega_1$:*

$$(\mathbf{x} - \mathbf{x}_0) \cdot \boldsymbol{\nu} \geq 0, \;\; x \in \Gamma_1. \tag{2.1}$$

*Assume further that*

$$\rho_1 \leq \rho_2, \;\; K_1 \geq K_2, \;\; D_1(1 \pm \mu_1) \geq D_2(1 \pm \mu_2). \tag{2.2}$$

*Then there is a time $T_0 > 0$ such that $R_T = H \times V'$ for $T > T_0$.*

**Remark 2.2** Results analogous to Theorem 2.2 have been obtained by J.-L. Lions for a wave equation in a general $n$-dimensional region [7, Chapter VI], and by Chen, Delfour, Krall and Payre for a collection of "serially connected" Euler-Bernoulli beams [1]. In particular, both results require certain monotonicity conditions on the parameters of the problems. On the other hand, exact controllability of a general connected planar network of Timoshenko beams has been established in [4] without any such monotonicity assumptions. See also [5], where this result is extended to networks of Timoshenko beams in $\Re^3$. It is probably the case that conditions (2.2) are artifacts of our method of proof of Theorem 2.2 and are not essential to the validity of its conclusion.

**Remark 2.3** Theorem 2.2 easily extends to the following situation involving $\Omega$ and $n$ open sets $\omega_1, \ldots, \omega_n$, $n \geq 2$, with

$$\overline{\omega}_i \subset \omega_{i+1}, \;\; i = 1, \ldots, n-1, \;\; \omega_n = \Omega.$$

Set

$$\Omega_1 = \omega_1, \;\; \Omega_i = \omega_i / \overline{\omega}_{i-1}, \;\; i = 2, \ldots, n.$$

One should assume that $\Gamma_1$ is star-shaped with respect to a point in $\Omega_1$ and that the elastic parameters associated with $\Omega_i$ satisfy appropriate monotonicity conditions with respect to $i$.

On the other hand, there are geometries of interest for which the multiplier method employed below seems inadequate to establish exact controllability, for example, if $\Omega$ contains two *disjoint* parts $\overline{\Omega}_1$ and $\overline{\Omega}_2$ whose associated material parameters are different than those associated with $\Omega_3 := \Omega/(\overline{\Omega}_1 \cup \overline{\Omega}_2)$. Another type of geometry for which the multiplier method formally works but which leads to technical difficulties regarding *regularity* of the solution is one in which $\Omega_1 \subset \Omega$ but $\Gamma_1 \cap \Gamma \neq \emptyset$, for example, if $\Omega$ is a circle and $\Omega_1$ as an angular sector of that circle.

# 3 Proof of Theorem 2.1

Let $A$ denote the Riesz isomorphism of $V$ onto $V'$, set

$$D_A = \{\Phi |\, A\Phi \in H\}, \quad \|\Phi\|_{D_A} = \|A\Phi\|_H,$$

and $D'_A$ be the dual of $D_A$ with respect to $H$. Then $A$ is an isometric isomorphism of $D_A$ onto $H$, of $V$ onto $V'$ and may be extended to an isometric isomorphism of $H$ onto $D'_A$ through the formula

$$\langle A\Phi, \Psi\rangle_V = (\Phi, A\Psi)_H, \quad \forall \Phi \in V, \Psi \in D_A,$$

where $\langle \cdot, \cdot \rangle_V$ denotes the inner produce in the $V'$–$V$ duality pairing.

**Remark 3.1** Let $\Phi \in D_A$ and $\Phi_i = \Phi|_{\Omega_i}$. If $\Gamma$ and $\Gamma_1$ are sufficiently smooth (say $C^2$), one may use standard techniques of elliptic regularity to show that $\Phi_i \in \mathcal{H}^2(\Omega_i)$, $i = 1, 2$. Under the milder regularity requirements on $\Gamma$ and $\Gamma_1$ assumed above, it follows from results of Nicaise [8] that $\Phi_i \in \mathcal{H}^{3/2+\epsilon}(\Omega_i)$ for some $\epsilon \in (0, 1/2]$. Therefore, all of the transmission conditions in (1.3) hold in the sense of traces. Note that $\Phi$ is *not* in $\mathcal{H}^{3/2+\epsilon}(\Omega)$ in general since $\partial\Phi_1/\partial\boldsymbol{\nu} \neq \partial\Phi_2/\partial\boldsymbol{\nu}$ on $\Gamma_1$ unless $D_1 = D_2$, $\mu_1 = \mu_2$ and $K_1 = K_2$. For $i = 1, 2$ one has the continuous embeddings

$$\|\Psi_i\|_{\mathcal{H}^{3/2+\epsilon}(\Omega_i)} \leq C\|\Psi\|_{D_A}.$$

We next write a variational equation of the system (1.2)–(1.4). Let $\hat{\Phi} \in D_A$, $\hat{\Phi} = \hat{\phi} + \hat{W}\mathbf{k}$, and form

$$0 = \sum_{i=1}^{2} \int_{\Omega_i} \{[\rho_i \ddot{W}_i - K_i \operatorname{div}(\nabla W_i + \phi_i)]\hat{W} + [I_{\rho_i}\ddot{\phi}_i - \operatorname{div} C_i[\varepsilon(\phi_i)] + K_i(\nabla W_i + \phi_i)] \cdot \hat{\phi}\} d\Omega_i. \quad (3.1)$$

After some integrations by parts in (3.1), in which we utilize the Green's formula

$$\int_{\Omega_i} (C_i[\varepsilon(\phi)], \varepsilon(\psi)) d\Omega_i = -\int_{\Omega_i} \psi \cdot (\operatorname{div} C_i[\varepsilon(\psi)]) d\Omega_i + \int_{\Gamma_i} \psi \cdot (C_i[\varepsilon(\phi)]\boldsymbol{\nu}) d\Gamma_i, \quad (3.2)$$

we obtain (the integrals over $\Gamma_1$ vanish because of (1.3))

$$(\ddot{\Phi}, \hat{\Phi})_H + (\Phi, A\hat{\Phi})_H = \int_\Gamma \{K_2 f \frac{\partial \hat{W}}{\partial \boldsymbol{\nu}} + \mathbf{m} \cdot C_2[\varepsilon(\hat{\phi})]\boldsymbol{\nu}\} d\Gamma, \quad (3.3)$$

where

$$\Phi = \begin{cases} \Phi_1 & \text{in } \Omega_1, \\ \Phi_2 & \text{in } \Omega_2 \end{cases}, \quad \Phi_i = \phi_i + W_i\mathbf{k}.$$

Write $\hat{\mathbf{\Phi}} = A^{-1}\tilde{\mathbf{\Phi}}$ where $\tilde{\mathbf{\Phi}} \in H$. Then (3.3) takes the form

$$(\ddot{\mathbf{\Phi}}, A^{-1}\tilde{\mathbf{\Phi}})_H + (\mathbf{\Phi}, \tilde{\mathbf{\Phi}})_H = \int_\Gamma \{K_2 f \frac{\partial \hat{W}}{\partial \boldsymbol{\nu}} + \mathbf{m} \cdot C_2[\varepsilon(\hat{\phi})]\boldsymbol{\nu}\} d\Gamma. \tag{3.4}$$

For an appropriate $s > 3/2$ one has

$$\begin{aligned} \| \int_\Gamma \{K_2 f \frac{\partial \hat{W}}{\partial \boldsymbol{\nu}} + \mathbf{m} \cdot C_2[\varepsilon(\hat{\phi})]\boldsymbol{\nu}\} d\Gamma \| &\le C\|\mathbf{F}\|_U \|\hat{\mathbf{\Phi}}_2\|_{\mathcal{H}^s(\Omega_2)} \\ &\le C\|\mathbf{F}\|_U \|\hat{\mathbf{\Phi}}\|_{D_A} \\ &= C\|\mathbf{F}\|_U \|\tilde{\mathbf{\Phi}}\|_H. \end{aligned}$$

Therefore, there is an operator $B \in \mathcal{L}(U, H)$ such that

$$(B\mathbf{F}, \tilde{\mathbf{\Phi}})_H = \int_\Gamma \{K_2 f \frac{\partial \hat{W}}{\partial \boldsymbol{\nu}} + \mathbf{m} \cdot C_2[\varepsilon(\hat{\phi})]\boldsymbol{\nu}\} d\Gamma, \ \forall \tilde{\mathbf{\Phi}} \in H, \ \hat{\mathbf{\Phi}} = A^{-1}\tilde{\mathbf{\Phi}}. \tag{3.5}$$

Equation (3.4) may then be written

$$(\ddot{\mathbf{\Phi}}, A^{-1}\tilde{\mathbf{\Phi}})_H + (\mathbf{\Phi}, \tilde{\mathbf{\Phi}})_H = (B\mathbf{F}, \tilde{\mathbf{\Phi}})_H,$$

which is the same as

$$\ddot{\mathbf{\Phi}} + A\mathbf{\Phi} = AB\mathbf{F} \quad \text{in } D'_A$$

or, by introducing

$$X = \begin{pmatrix} \mathbf{\Phi} \\ \dot{\mathbf{\Phi}} \end{pmatrix}, \quad \mathcal{A} = \begin{pmatrix} 0 & I \\ -A & 0 \end{pmatrix}, \quad \mathcal{B} = \begin{pmatrix} 0 \\ AB \end{pmatrix},$$

as the first order equation

$$\dot{X} = \mathcal{A}X + \mathcal{B}\mathbf{F} \quad \text{in } V' \times D'_A. \tag{3.6}$$

One has $\mathcal{A} \in \mathcal{L}(H \times V', V' \times D'_A)$ and $\mathcal{B} \in \mathcal{L}(U, V' \times D'_A)$. Further, it is well-known that $\mathcal{A}$, considered as an unbounded operator in $V' \times D'_A$ with domain $H \times V'$, is the generator of a $C_0$-group of unitary operators on $V' \times D'_A$. It follows immediately that for

$$\mathbf{F} \in L^2(0, T; U), \quad X^0 = \begin{pmatrix} \mathbf{\Phi}^0 \\ \mathbf{\Phi}^1 \end{pmatrix} \in V' \times D'_A,$$

(3.6) has a unique mild solution $X \in C([0,T]; V' \times D'_A)$ satisfying $X(0) = X^0$. We then define $\mathbf{\Phi}_i = \mathbf{\Phi}|_{\Omega_i}$ to obtain, *by definition*, the corresponding solution of (1.2)–(1.5).

To complete the proof of Theorem 2.1, we need to show that $X \in C([0,T]; H \times V')$ if $X^0 \in H \times V'$. To do so we consider the *homogeneous* system

$$\begin{cases} I_{\rho_i}\ddot{\boldsymbol{\psi}}_i - \operatorname{div} C_i[\varepsilon(\boldsymbol{\psi}_i)] + K_i(\nabla Z_i + \boldsymbol{\psi}_i) = 0, \\ \rho_i \ddot{Z}_i - K_i \operatorname{div}(\nabla Z_i + \boldsymbol{\psi}_i) = 0 \quad \text{in } \Omega_i, t > 0, \ i = 1, 2; \end{cases} \tag{3.7}$$

$$\begin{cases} Z_1 = Z_2, \ \psi_1 = \psi_2, \\ C_1[\varepsilon(\psi_1)]\nu_1 + C_2[\varepsilon(\psi_2)]\nu_2 = 0, \\ K_1(\nabla Z_1 + \psi_1)\cdot \nu_1 + K_2(\nabla Z_2 + \psi_2)\cdot \nu_2 = 0 \quad \text{on } \Gamma_1,\ t > 0; \end{cases} \tag{3.8}$$

$$Z_2 = 0, \ \psi_2 = 0 \quad \text{on } \Gamma_2,\ t > 0; \tag{3.9}$$

$$\begin{cases} \psi_i(0) = \psi_i^0, \ \dot{\psi}_i(0) = \psi_i^1, \\ Z_i(0) = Z_i^0, \ \dot{Z}_i(0) = Z_i^1 \quad \text{in } \Omega_i, \ i = 1, 2. \end{cases} \tag{3.10}$$

As above, (3.7)–(3.10) may be written as

$$\dot{Y} = \mathcal{A}Y, \ Y(0) = Y^0, \tag{3.11}$$

where

$$Y = \begin{pmatrix} \Psi \\ \dot{\Psi} \end{pmatrix}, \quad Y^0 = \begin{pmatrix} \Psi^0 \\ \Psi^1 \end{pmatrix},$$

$$\Psi = \begin{cases} \Psi_1 \ \text{in } \Omega_1, \\ \Psi_2 \ \text{in } \Omega_2 \end{cases}, \quad \Psi_i = \psi_i + Z_i\mathbf{k}.$$

Since $\mathcal{A}$, considered as an unbounded operator in $V\times H$ with domain $D_{\mathcal{A}}\times V$, generates a $C_0$-group of unitary operators on $V\times H$, (3.11) has a unique mild solution $Y \in C([0,T]; V\times H)$ whenever $Y^0 \in V \times H$. Furthermore, one has the following *regularity result* for this solution.

**Lemma 3.1** *Let $Y^0 \in V \times H$ and $Y = \begin{pmatrix} \Psi \\ \dot{\Psi} \end{pmatrix} \in C([0,T]; V\times H)$ be the solution of* (3.11). *Let $\Psi|_{\Omega_i} = \Psi_i = \psi_i + Z_i\mathbf{k}$. Then*

$$C_2[\varepsilon(\psi_2)]\nu|_\Gamma \in L^2(\Gamma\times(0,T)), \quad \left.\frac{\partial Z_2}{\partial \nu}\right|_\Gamma \in L^2(\Gamma\times(0,T))$$

*and*

$$\int_0^T\int_\Gamma \left[ |C_2[\varepsilon(\psi_2)]\nu|^2 + K_2\left(\frac{\partial Z_2}{\partial \nu}\right)^2\right] d\Gamma dt \le C(T+1)\|(\Psi^0,\Psi^1)\|^2_{V\times H}. \tag{3.12}$$

One may conclude from (3.12), upon applying a lifting theorem of Lasiecka and Triggiani [6], that $X \in C([0,T]; H\times V')$. (In fact, (3.12) shows that condition (1.6) of [6] is satisfied with $p = q = 2$ and $X = H \times V'$. More precisely, it shows that the adjoint operator $L_T^*$, in the notation of [6], is continuous from $L^2(0,T; H\times V)$ into $L^2(0,T;U)$.) However, in order to keep the presentation self-contained, we shall provide a proof, based on the idea of transposition. The mild solution of (3.6) having initial data $X(0) = X^0$ is given by

$$X(t) = \exp(t\mathcal{A})X^0 + \int_0^t \exp((t-s)\mathcal{A})\mathcal{B}\mathbf{F}(s)\,ds, \quad 0 \le t \le T,$$

where $\exp(t\mathcal{A})$ is the unitary group on $V' \times D'_A$ generated by $\mathcal{A}$. Suppose that $X^0 \in H \times V'$, let $Y^0 \in V \times D_A$ and $\mathcal{B}' \in \mathcal{L}(V \times D_A; U)$ be the dual of $\mathcal{B}$, defined by

$$\langle \mathcal{B}\mathbf{F}, Y^0 \rangle_{V \times D_A} = (\mathbf{F}, \mathcal{B}' Y^0)_U, \quad \forall \mathbf{F} \in U, \ Y^0 \in V \times D_A.$$

Let $\tau \in (0, T]$ be fixed. We have

$$\langle X(\tau), Y^0 \rangle_{V \times D_A} = \langle X^0, \exp(\tau \mathcal{A}') Y^0 \rangle_{H \times V} + \int_0^\tau (\mathbf{F}(s), \mathcal{B}' \exp((\tau - s)\mathcal{A}') Y^0)_U \, ds. \tag{3.13}$$

Here $\mathcal{A}'$ is the dual of $\mathcal{A}$, defined by

$$\langle \mathcal{A} X^0, Y^0 \rangle_{V \times D_A} = \langle X^0, \mathcal{A}' Y^0 \rangle_{H \times V}, \quad \forall X^0 \in H \times V', \ Y^0 \in V \times D_A.$$

One has

$$\mathcal{A}' = \begin{pmatrix} 0 & A \\ -I & 0 \end{pmatrix}, \quad D(\mathcal{A}') = V \times D_A.$$

As is well-known, $\mathcal{A}'$ generates a unitary group $\exp(t\mathcal{A}')$ on $H \times V$ and $\exp(t\mathcal{A}')$ is the dual of the restriction of $\exp(t\mathcal{A})$ to $H \times V'$. Therefore (3.13) is the same as

$$\langle X(\tau), Y^0 \rangle_{V \times D_A} = \langle X^0, Y(\tau) \rangle_{H \times V} + \int_0^\tau (\mathbf{F}(s), \mathcal{B}' Y(s))_U \, ds, \tag{3.14}$$

where $Y(t) := \exp((\tau - t)\mathcal{A}') Y^0$, $0 \le t \le \tau$, satisfies

$$\dot{Y} = -\mathcal{A}' Y, \quad 0 \le t \le \tau, \quad Y(\tau) = Y^0. \tag{3.15}$$

If we write

$$Y = (\Psi_1, \Psi), \quad Y^0 = (\Psi^1, \Psi^0),$$

(3.15) signifies that

$$\begin{cases} \ddot{\Psi} + A\Psi = 0, \quad \Psi_1 = \dot{\Psi}, \\ \Psi(\tau) = \Psi^0, \quad \dot{\Psi}(\tau) = \Psi^1. \end{cases} \tag{3.16}$$

In addition,

$$\langle \mathcal{B}\mathbf{F}, Y \rangle_{V \times D_A} = \langle AB\mathbf{F}, \Psi \rangle_{D_A} = (B\mathbf{F}, A\Psi)_H.$$

From (3.5) we have

$$(B\mathbf{F}, A\Psi)_H = \int_\Gamma [\mathbf{m} \cdot (C_2[\varepsilon(\boldsymbol{\psi})]\boldsymbol{\nu}) + K_2 f \frac{\partial Z}{\partial \nu}] \, d\Gamma,$$

where $\Psi = \boldsymbol{\psi} + Z\mathbf{k}$. It follows that

$$\mathcal{B}' Y = C_2[\varepsilon(\boldsymbol{\psi})]\boldsymbol{\nu} + K_2 \frac{\partial Z}{\partial \nu} \mathbf{k} \Big|_\Sigma, \tag{3.17}$$

where $\Sigma = \Gamma \times (0,T)$. We insert this expression into (3.14) to obtain the estimate

$$|\langle X(\tau), Y^0\rangle_{V\times D_A}| \le \|X^0\|_{H\times V'}\|Y^0\|_{H\times V} + \|\mathbf{F}\|_{L^2(0,T;U)}\|C_2[\varepsilon(\boldsymbol{\psi})]\boldsymbol{\nu} + K_2\frac{\partial Z}{\partial \boldsymbol{\nu}}\mathbf{k}\|_{L^2(0,T;U)}$$
$$\le C_0(T+1)\{\|X^0\|_{H\times V'} + \|\mathbf{F}\|_{L^2(0,T;U)}\}\|Y^0\|_{H\times V}, \quad 0\le\tau\le T,$$

in view of Lemma 3.1. It follows that $X\in L^\infty(0,T;H\times V')$. One may pass from $L^\infty$ to $C$ by a standard approximation argument. This completes the proof of Theorem 2.1.

**Proof of Lemma 3.1.** It suffices to prove (3.12) for initial data in $D_A\times V$.

The method of the proof is standard (c.f. [7], for example). One multiplies (3.7) with i=2 by $(\nabla\boldsymbol{\psi}_2)\mathbf{h}$ and $\mathbf{h}\cdot\nabla Z_2$ respectively, where $\mathbf{h}$ is a $W^{1,\infty}$ vector field in $\Omega$ such that $\mathbf{h}=\boldsymbol{\nu}$ on $\Gamma$ and $\mathbf{h}=0$ in a neighborhood of $\Gamma_1$, adds the products and integrates the sum over $\Omega_2\times(0,T)$. One thereby obtains

$$\int_0^T\int_{\Omega_2}\{[\rho_2\ddot Z_2 - K_2\operatorname{div}(\nabla Z_2+\boldsymbol{\psi}_2)]\mathbf{h}\cdot\nabla Z_2$$
$$+[I_{\rho_i}\ddot{\boldsymbol{\psi}}_2 - \operatorname{div}C_2[\varepsilon(\boldsymbol{\psi}_2)] + K_2(\nabla Z_2+\boldsymbol{\psi}_2)]\cdot((\nabla\boldsymbol{\psi}_2)\mathbf{h})\}\,d\Omega dt = 0. \quad (3.18)$$

One has

$$\int_0^T\int_{\Omega_2}(\mathbf{h}\cdot\nabla Z_2)\ddot Z_2\,d\Omega dt = \int_{\Omega_2}(\mathbf{h}\cdot\nabla Z_2)\dot Z_2\,d\Omega|_0^T + \frac{1}{2}\int_0^T\int_{\Omega_2}(\operatorname{div}\mathbf{h})|\dot Z_2|^2 d\Omega dt,$$

$$\int_0^T\int_{\Omega_2}(\mathbf{h}\cdot\nabla Z_2)\operatorname{div}(\nabla Z_2+\boldsymbol{\psi}_2)\,d\Omega dt = \int_0^T\int_{\Omega_2}\{\operatorname{div}[(\mathbf{h}\cdot\nabla Z_2)\nabla Z_2] - (\nabla\mathbf{h}\nabla Z_2)\cdot\nabla Z_2$$
$$-\frac{1}{2}\operatorname{div}(\mathbf{h}|\nabla Z_2|)^2 + \frac{1}{2}(\operatorname{div}\mathbf{h})|\nabla Z_2|^2 + (\operatorname{div}\boldsymbol{\psi}_2)(\mathbf{h}\cdot\nabla Z_2)\}d\Omega dt$$
$$=\int_0^T\int_{\Omega_2}\{\frac{1}{2}(\operatorname{div}\mathbf{h})|\nabla Z_2|^2 - (\nabla h\nabla Z_2)\cdot\nabla Z_2 + (\operatorname{div}\boldsymbol{\psi}_2)(\mathbf{h}\cdot\nabla Z_2)\}d\Omega dt + \frac{1}{2}\int_\Sigma\left(\frac{\partial Z_2}{\partial\boldsymbol{\nu}}\right)^2 d\Sigma,$$

$$\int_0^T\int_{\Omega_2}\ddot{\boldsymbol{\psi}}_2\cdot(\nabla\boldsymbol{\psi}_2\mathbf{h})d\Omega dt = \int_{\Omega_2}\dot{\boldsymbol{\psi}}_2\cdot(\nabla\boldsymbol{\psi}_2\mathbf{h})d\Omega\Big|_0^T + \frac{1}{2}\int_0^T\int_{\Omega_2}(\operatorname{div}\mathbf{h})|\dot{\boldsymbol{\psi}}_2|^2 d\Omega dt,$$

$$\int_0^T\int_{\Omega_2}(\nabla\boldsymbol{\psi}_2\mathbf{h})\cdot\operatorname{div}C_2[\varepsilon(\boldsymbol{\psi}_2)]\,d\Omega dt = -\int_0^T\int_{\Omega_2}(C_2[\varepsilon(\boldsymbol{\psi}_2)],\nabla\boldsymbol{\psi}_2\mathbf{h})\,d\Omega dt$$
$$+\int_\Sigma(\nabla\boldsymbol{\psi}_2\mathbf{h})\cdot(C_2[\varepsilon(\boldsymbol{\psi}_2)]\boldsymbol{\nu})\,d\Sigma.$$

When the last four formulas are inserted into (3.18) the result is

$$\int_{\Omega_2}\{\rho_2\dot Z_2(\mathbf{h}\cdot\nabla Z_2) + I_{\rho_i}\dot{\boldsymbol{\psi}}_2\cdot(\nabla\boldsymbol{\psi}_2\mathbf{h})\}d\Omega\Big|_0^T + \frac{1}{2}\int_0^T\int_{\Omega_2}(\operatorname{div}\mathbf{h})\{\rho_2|\dot Z_2|^2 + I_{\rho_i}|\dot{\boldsymbol{\psi}}_2|^2\}d\Omega dt$$
$$+\int_0^T\int_{\Omega_2}\{(C_2[\varepsilon(\boldsymbol{\psi}_2)],\nabla\boldsymbol{\psi}_2\mathbf{h}) + K_2[-\frac{1}{2}(\operatorname{div}\mathbf{h})|\nabla Z_2|^2 + (\nabla\mathbf{h}\nabla Z_2)\cdot\nabla Z_2$$
$$-(\operatorname{div}\boldsymbol{\psi}_2)(\mathbf{h}\cdot\nabla Z_2) + (\nabla Z_2+\boldsymbol{\psi}_2)]\cdot((\nabla\boldsymbol{\psi}_2)\mathbf{h})]\}d\Omega dt$$
$$-\int_\Sigma\{\frac{1}{2}\left(\frac{\partial Z_2}{\partial\boldsymbol{\nu}}\right)^2 + (\nabla\boldsymbol{\psi}_2\mathbf{h})\cdot(C_2[\varepsilon(\boldsymbol{\psi}_2)]\boldsymbol{\nu})\}d\Sigma = 0.$$

On $\Sigma$ on has (see (4.7) and (4.8) below)

$$\nabla\boldsymbol{\psi}_2\mathbf{h} = \frac{\partial\boldsymbol{\psi}_2}{\partial\boldsymbol{\nu}}, \quad C_2[\varepsilon(\boldsymbol{\psi}_2)]\boldsymbol{\nu} = D_2\left[\left(\boldsymbol{\nu}\cdot\frac{\partial\boldsymbol{\psi}_2}{\partial\boldsymbol{\nu}}\right)\boldsymbol{\nu} + \frac{1-\mu_2}{2}\left(\boldsymbol{\tau}\cdot\frac{\partial\boldsymbol{\psi}_2}{\partial\boldsymbol{\nu}}\right)\boldsymbol{\tau}\right]$$

where $\boldsymbol{\tau}$ is the positively oriented unit tangent vector to $\Gamma$. Therefore, on $\Sigma$

$$\begin{aligned}(\nabla\boldsymbol{\psi}_2\mathbf{h})\cdot(C_2[\varepsilon(\boldsymbol{\psi}_2)]\boldsymbol{\nu}) &= D_2\left[\left(\boldsymbol{\nu}\cdot\frac{\partial\boldsymbol{\psi}_2}{\partial\boldsymbol{\nu}}\right)^2 + \frac{1-\mu_2}{2}\left(\boldsymbol{\tau}\cdot\frac{\partial\boldsymbol{\psi}_2}{\partial\boldsymbol{\nu}}\right)^2\right]\\ &\geq (D_2)^{-1}|C_2[\varepsilon(\boldsymbol{\psi}_2)]\boldsymbol{\nu}|^2.\end{aligned}$$

In addition,

$$\int_{\Omega_2}\{\rho_2\dot Z_2(\mathbf{h}\cdot\nabla Z_2) + I_{\rho_i}\dot{\boldsymbol{\psi}}_2\cdot(\nabla\boldsymbol{\psi}_2\mathbf{h})\}d\Omega\Big|_0^T \leq C\|(\boldsymbol{\Psi},\dot{\boldsymbol{\Psi}})\|^2_{L^\infty(0,T;V\times H)},$$

$$\left|\int_0^T\int_{\Omega_2}(\operatorname{div}\mathbf{h})\{\rho_2|\dot Z_2|^2 + I_{\rho_i}|\dot{\boldsymbol{\psi}}_2|^2\}d\Omega dt\right| \leq \|\mathbf{h}\|_{W^{1,\infty}(\Omega_2)}\int_0^T \|\dot{\boldsymbol{\Psi}}\|^2_H dt,$$

$$\begin{aligned}\Big|\int_0^T\int_{\Omega_2}&\{(C_2[\varepsilon(\boldsymbol{\psi}_2)], \nabla\boldsymbol{\psi}_2\mathbf{h}) + K_2[-\frac{1}{2}(\operatorname{div}\mathbf{h})|\nabla Z_2|^2 + (\nabla\mathbf{h}\nabla Z_2)\cdot\nabla Z_2\\ &- (\operatorname{div}\boldsymbol{\psi}_2)(\mathbf{h}\cdot\nabla Z_2) + (\nabla Z_2 + \boldsymbol{\psi}_2)]\cdot((\nabla\boldsymbol{\psi}_2)\mathbf{h})]\}d\Omega dt| \leq C\int_0^T \|\boldsymbol{\Psi}\|^2_V\, dt.\end{aligned}$$

It follows from the above estimates that

$$\begin{aligned}\int_\Sigma\left[|C_2[\varepsilon(\boldsymbol{\psi}_2)]\boldsymbol{\nu}|^2 + K_2\left(\frac{\partial Z_2}{\partial\boldsymbol{\nu}}\right)^2\right]d\Sigma &\leq C\left[\int_0^T \|(\boldsymbol{\Psi},\dot{\boldsymbol{\Psi}})\|^2_{V\times H}dt + \|(\boldsymbol{\Psi},\dot{\boldsymbol{\Psi}})\|^2_{L^\infty(0,T;V\times H)}\right]\\ &= C(T+1)\|(\boldsymbol{\Psi}^0,\boldsymbol{\Psi}^1)\|^2_{V\times H}.\end{aligned}$$

## 4 Proof of Theorem 2.2

As is well-known (using either the Hilbert Uniqueness Method of J.-L. Lions or, equivalently, proceeding via the control-to-state map $F \mapsto (\boldsymbol{\Phi}(T), \dot{\boldsymbol{\Phi}}(T)) : L^2(0,T;U) \mapsto H\times V')$, the conclusion of Theorem 2.2 is equivalent to continuous observability of the adjoint system, i.e.,

$$\|\mathcal{B}'Y\|_{L^2(0,T;U)} \geq c_0(T)\|Y^0\|_{H\times V}, \quad \forall Y^0 \in H\times V, \tag{4.1}$$

for some $c_0(T) > 0$ and for $T$ large enough, where $Y$ is the solution of

$$\dot Y = -\mathcal{A}'Y,\ 0 < t < T, \quad Y(T) = Y^0.$$

If one writes $Y = (\Psi, \dot{\Psi})$, $Y^0 = (\Psi^1, \Psi^0)$, in view of (3.17) it is seen that (4.1) is equivalent to the inequality

$$\int_0^T \int_\Gamma \left\{ |C_2[\varepsilon(\psi_2)]\nu|^2 + K_2 \left| \frac{\partial Z_2}{\partial \nu} \right|^2 \right\} d\Gamma dt \geq c_0 \|(\Psi^0, \Psi^1)\|^2_{V \times H}, \tag{4.2}$$

where $\Psi_i := \Psi|_{\Omega_i} = \psi_i + Z_i \mathbf{k}$ satisfies (3.7)–(3.9) and

$$\Psi_i(T) = \Psi_i^0 = \psi_i^0 + Z_i^0 \mathbf{k}, \quad \dot{\Psi}_i(T) = \Psi_i^1 = \psi_i^1 + Z_i^1 \mathbf{k}. \tag{4.3}$$

Because of the time reversibility of (3.7)–(3.9), it is equivalent to consider initial conditions (3.10) rather than (4.3). So we shall establish (4.2) for the solution of (3.7)–(3.10) with initial data $(\Psi^0, \Psi^1) \in V \times H$. In fact, in view of Lemma 3.1, it is sufficient to prove it for initial data in $D_A \times V$. For such data the solution has regularity $\Psi_i \in C([0,T]; \mathcal{H}^s(\Omega_i)) \cap C^1([0,T]; \mathcal{H}^{s-1}(\Omega_i))$ for some $s > 3/2$ and the transmission conditions (3.8) hold in the ordinary sense of traces. (See Remark 3.1.) This degree of regularity is needed to justify the multiplier method employed below.

Our starting point for the proof is an identity from [2, Lemma 3.3.1] that may be stated as follows.

**Lemma 4.1** *Let $\Omega$ be a bounded, open set in $\Re^2$ with a Lipschitz and piecewise smooth boundary. Let $\Psi = \psi + Z\mathbf{k}$ and assume that $\Psi \in \mathcal{H}^s(\Omega)$ for some $s > 3/2$. Then*

$$\begin{aligned} &\int_\Omega \{(\nabla \psi \mathbf{r}) \cdot [\operatorname{div} C[\varepsilon(\psi)] - K(\nabla Z + \psi)] + K(\mathbf{r} \cdot \nabla Z) \operatorname{div}(\nabla Z + \psi)\} \, d\Omega \\ &\quad = K \int_\Omega (\nabla Z + \psi) \cdot \psi \, d\Omega - \frac{1}{2} \int_\Gamma (\mathbf{r} \cdot \boldsymbol{\nu}) \{(C[\varepsilon(\psi)], \varepsilon(\psi)) + K|\nabla Z + \psi|^2\} \, d\Gamma \\ &\qquad + \int_\Gamma \{(\nabla \psi \mathbf{r}) \cdot (C[\varepsilon(\psi)]\boldsymbol{\nu}) + K(\mathbf{r} \cdot \nabla Z)(\nabla Z + \psi) \cdot \boldsymbol{\nu}\} \, d\Gamma, \end{aligned} \tag{4.4}$$

*where $\mathbf{r}$ denotes any radial vector field $\mathbf{x} - \mathbf{x}_0$.*

In (4.4) the constant $K$ is arbitrary and $C$ is any tensor of the form (1.1):

$$C[\varepsilon] = D[\mu(\varepsilon_{11} + \varepsilon_{22})\mathcal{I} + (1 - \mu)\varepsilon], \quad \varepsilon \in \mathcal{A}.$$

We need to rewrite the integrals over $\Gamma$ in (4.4) in forms more suitable for proving Theorem 2.2. We have on $\Gamma$

$$(\nabla \psi \mathbf{r}) \cdot (C[\varepsilon(\psi)]\boldsymbol{\nu}) = (\mathbf{r} \cdot \boldsymbol{\nu}) \frac{\partial \psi}{\partial \boldsymbol{\nu}} \cdot (C[\varepsilon(\psi)]\boldsymbol{\nu}) + (\mathbf{r} \cdot \boldsymbol{\tau}) \frac{\partial \psi}{\partial \boldsymbol{\tau}} \cdot (C[\varepsilon(\psi)]\boldsymbol{\nu}) \tag{4.5}$$

and

$$\frac{\partial \psi}{\partial x_j} = \nu_j \frac{\partial \psi}{\partial \boldsymbol{\nu}} + \tau_j \frac{\partial \psi}{\partial \boldsymbol{\tau}}, \tag{4.6}$$

where $\boldsymbol{\nu} = (\nu_1, \nu_2)$, $\boldsymbol{\tau} = (\tau_1, \tau_2) = (-\nu_2, \nu_1)$. Write $C[\varepsilon(\boldsymbol{\psi})]\boldsymbol{\nu}$ in terms of normal and tangential components:

$$C[\varepsilon(\boldsymbol{\psi})]\boldsymbol{\nu} = C_\nu(\boldsymbol{\psi})\boldsymbol{\nu} + C_\tau(\boldsymbol{\psi})\boldsymbol{\tau}.$$

It may be verified from (4.6), after some calculation, that

$$C_\nu(\boldsymbol{\psi}) = D\left[\boldsymbol{\nu}\cdot\frac{\partial\boldsymbol{\psi}}{\partial\boldsymbol{\nu}} + \mu\boldsymbol{\tau}\cdot\frac{\partial\boldsymbol{\psi}}{\partial\boldsymbol{\tau}}\right], \tag{4.7}$$

$$C_\tau(\boldsymbol{\psi}) = \frac{(1-\mu)D}{2}\left[\boldsymbol{\tau}\cdot\frac{\partial\boldsymbol{\psi}}{\partial\boldsymbol{\nu}} + \boldsymbol{\nu}\cdot\frac{\partial\boldsymbol{\psi}}{\partial\boldsymbol{\tau}}\right]. \tag{4.8}$$

Therefore

$$\frac{\partial\boldsymbol{\psi}}{\partial\boldsymbol{\nu}} = \left(\boldsymbol{\nu}\cdot\frac{\partial\boldsymbol{\psi}}{\partial\boldsymbol{\nu}}\right)\boldsymbol{\nu} + \left(\boldsymbol{\tau}\cdot\frac{\partial\boldsymbol{\psi}}{\partial\boldsymbol{\nu}}\right)\boldsymbol{\tau}$$
$$= \frac{1}{D}\left[C_\nu(\boldsymbol{\psi})\boldsymbol{\nu} + \frac{2}{1-\mu}C_\tau(\boldsymbol{\psi})\boldsymbol{\tau}\right] - \mu\boldsymbol{\tau}\cdot\frac{\partial\boldsymbol{\psi}}{\partial\boldsymbol{\tau}}\boldsymbol{\nu} - \boldsymbol{\nu}\cdot\frac{\partial\boldsymbol{\psi}}{\partial\boldsymbol{\tau}}\boldsymbol{\tau}.$$

Insertion of this expression into (4.5) yields

$$(\nabla\boldsymbol{\psi}\mathbf{r})\cdot(C[\varepsilon(\boldsymbol{\psi})]\boldsymbol{\nu}) = (\mathbf{r}\cdot\boldsymbol{\nu})\left\{\frac{1}{D}\left[|C_\nu(\boldsymbol{\psi})|^2 + \frac{2}{1-\mu}|C_\tau(\boldsymbol{\psi})|^2\right]\right.$$
$$\left. - \left(\mu\boldsymbol{\tau}\cdot\frac{\partial\boldsymbol{\psi}}{\partial\boldsymbol{\tau}}\boldsymbol{\nu} + \boldsymbol{\nu}\cdot\frac{\partial\boldsymbol{\psi}}{\partial\boldsymbol{\tau}}\boldsymbol{\tau}\right)\cdot(C[\varepsilon(\boldsymbol{\psi})]\boldsymbol{\nu})\right\} + (\mathbf{r}\cdot\boldsymbol{\tau})\frac{\partial\boldsymbol{\psi}}{\partial\boldsymbol{\tau}}\cdot(C[\varepsilon(\boldsymbol{\psi})]\boldsymbol{\nu}).$$

Another rather lengthy calculation show that on $\Gamma$

$$(C[\varepsilon(\boldsymbol{\psi})], \varepsilon(\boldsymbol{\psi})) = \frac{1}{D}\left[|C_\nu(\boldsymbol{\psi})|^2 + \frac{2}{1-\mu}|C_\tau(\boldsymbol{\psi})|^2\right] + D(1-\mu^2)\left|\boldsymbol{\tau}\cdot\frac{\partial\boldsymbol{\psi}}{\partial\boldsymbol{\tau}}\right|^2.$$

In addition,

$$(\mathbf{r}\cdot\nabla Z)(\nabla Z + \boldsymbol{\psi})\cdot\boldsymbol{\nu} = (\mathbf{r}\cdot\boldsymbol{\nu})[|(\nabla Z + \boldsymbol{\psi})\cdot\boldsymbol{\nu}|^2 - \boldsymbol{\nu}\cdot\boldsymbol{\psi}(\nabla Z + \boldsymbol{\psi})\cdot\boldsymbol{\nu}]$$
$$+ (\mathbf{r}\cdot\boldsymbol{\tau})\frac{\partial Z}{\partial\boldsymbol{\tau}}(\nabla Z + \boldsymbol{\psi})\cdot\boldsymbol{\nu},$$

$$|\nabla Z + \boldsymbol{\psi}|^2 = |(\nabla Z + \boldsymbol{\psi})\cdot\boldsymbol{\nu}|^2 + |(\nabla Z + \boldsymbol{\psi})\cdot\boldsymbol{\tau}|^2.$$

Therefore

$$
\begin{aligned}
-\frac{1}{2}\int_\Gamma &(\mathbf{r}\cdot\boldsymbol{\nu})\{(C[\varepsilon(\boldsymbol{\psi})],\varepsilon(\boldsymbol{\psi}))+K|\nabla Z+\boldsymbol{\psi}|^2\}\,d\Gamma \\
&+\int_\Gamma\{(\nabla\boldsymbol{\psi}\mathbf{r})\cdot(C[\varepsilon(\boldsymbol{\psi})]\boldsymbol{\nu})+K(\mathbf{r}\cdot\nabla Z)(\nabla Z+\boldsymbol{\psi})\cdot\boldsymbol{\nu}\}\,d\Gamma \\
=\frac{1}{2}\int_\Gamma&(\mathbf{r}\cdot\boldsymbol{\nu})\left\{\frac{1}{D}\left[|C_\nu(\boldsymbol{\psi})|^2+\frac{2}{1-\mu}|C_\tau(\boldsymbol{\psi})|^2\right]+K|(\nabla Z+\boldsymbol{\psi})\cdot\boldsymbol{\nu}|^2\right\}d\Gamma \\
&-\int_\Gamma(\mathbf{r}\cdot\boldsymbol{\nu})\left\{\frac{D(1-\mu^2)}{2}\left|\boldsymbol{\tau}\cdot\frac{\partial\boldsymbol{\psi}}{\partial\tau}\right|^2+\frac{K}{2}\left|\frac{\partial Z}{\partial\tau}+\boldsymbol{\psi}\cdot\boldsymbol{\tau}\right|^2\right. \\
&+K(\boldsymbol{\psi}\cdot\boldsymbol{\nu})(\nabla Z+\boldsymbol{\psi})\cdot\boldsymbol{\nu}+\left.\left(\mu\boldsymbol{\tau}\cdot\frac{\partial\boldsymbol{\psi}}{\partial\tau}\boldsymbol{\nu}+\boldsymbol{\nu}\cdot\frac{\partial\boldsymbol{\psi}}{\partial\tau}\boldsymbol{\tau}\right)\cdot(C[\varepsilon(\boldsymbol{\psi})]\boldsymbol{\nu})\right\}d\Gamma \\
&+\int_\Gamma(\mathbf{r}\cdot\boldsymbol{\tau})\left\{\frac{\partial\boldsymbol{\psi}}{\partial\tau}\cdot(C[\varepsilon(\boldsymbol{\psi})]\boldsymbol{\nu})+K\frac{\partial Z}{\partial\tau}(\nabla Z+\boldsymbol{\psi})\cdot\boldsymbol{\nu}\right\}d\Gamma. \qquad (4.9)
\end{aligned}
$$

This last expression will eventually be used in the right side of (4.4)

Two other simple identities that will be needed are

$$
\begin{aligned}
\int_\Omega &\boldsymbol{\psi}\cdot[\operatorname{div}C[\varepsilon(\boldsymbol{\psi})]-K(\nabla Z+\boldsymbol{\psi})]\,d\Gamma \\
&=-\int_\Omega[(C[\varepsilon(\boldsymbol{\psi})],\varepsilon(\boldsymbol{\psi}))+K\boldsymbol{\psi}\cdot(\nabla Z+\boldsymbol{\psi})]\,d\Gamma+\int_\Gamma\boldsymbol{\psi}\cdot(C[\varepsilon(\boldsymbol{\psi})]\boldsymbol{\nu})\,d\Gamma, \qquad (4.10)
\end{aligned}
$$

and

$$\int_\Omega Z\operatorname{div}(\nabla Z+\boldsymbol{\psi})\,d\Omega=-\int_\Omega\nabla Z\cdot(\nabla Z+\boldsymbol{\psi})\,d\Omega+\int_\Gamma Z(\nabla Z+\boldsymbol{\psi})\cdot\boldsymbol{\nu}\,d\Gamma. \qquad (4.11)$$

Now suppose that $\boldsymbol{\Psi}_1,\boldsymbol{\Psi}_2$ is a solution of (3.7) with regularity

$$\boldsymbol{\Psi}_i\in C([0,T];\mathcal{H}^s(\Omega_i))\cap C^1([0,T];\mathcal{H}^{s-1}(\Omega_i)),\quad s>3/2. \qquad (4.12)$$

We apply (4.4) to $\boldsymbol{\Psi}_i$ in the region $\Omega_i$ with $K=K_i$ and $C=C_i$, using the radial field $\mathbf{r}_i=\mathbf{x}-\mathbf{x}_i$, and then integrate (4.4) with respect to $t$ from 0 to $T$. Thus

$$
\begin{aligned}
\int_0^T\int_{\Omega_i}&(\nabla\boldsymbol{\psi}_i\mathbf{r}_i)\cdot\{\operatorname{div}C_i[\varepsilon(\boldsymbol{\psi}_i)]-K_i(\nabla Z_i+\boldsymbol{\psi}_i)+K_i(\mathbf{r}_i\cdot\nabla Z_i)\operatorname{div}(\nabla Z_i+\boldsymbol{\psi}_i)\}\,d\Omega dt \\
&=\int_0^T\int_{\Omega_i}\left\{I_{\rho_i}(\nabla\boldsymbol{\psi}_i\mathbf{r}_i)\cdot\ddot{\boldsymbol{\psi}}_i+\rho_i(\mathbf{r}_i\cdot\nabla Z_i)\ddot{Z}_i\right\}d\Omega dt \\
&=\alpha_{i1}(t)|_0^T-\int_0^T\int_{\Omega_i}\left\{I_{\rho_i}(\nabla\dot{\boldsymbol{\psi}}_i\mathbf{r}_i)\cdot\dot{\boldsymbol{\psi}}_i+\rho_i(\mathbf{r}_i\cdot\nabla\dot{Z}_i)\dot{Z}_i\right\}d\Omega dt,
\end{aligned}
$$

where

$$\alpha_{i1}(t)=\int_\Omega[I_{\rho_i}(\nabla\boldsymbol{\psi}_i\mathbf{r}_i)\cdot\dot{\boldsymbol{\psi}}_i+\rho_i(\mathbf{r}_i\cdot\nabla Z_i)\dot{Z}_i]\,d\Omega.$$

One has

$$\int_{\Omega_i} (\nabla\boldsymbol{\psi}_i\mathbf{r}_i)\cdot\dot{\boldsymbol{\psi}}_i\,d\Omega = \frac{1}{2}\int_{\Omega_i}[\operatorname{div}(\mathbf{r}_i|\dot{\boldsymbol{\psi}}_i|^2) - 2|\dot{\boldsymbol{\psi}}_i|^2]\,d\Omega$$
$$= \frac{1}{2}\int_{\Gamma_i}(\mathbf{r}_i\cdot\boldsymbol{\nu}_i)|\dot{\boldsymbol{\psi}}_i|^2 d\Gamma - \int_{\Omega_i}|\dot{\boldsymbol{\psi}}_i|^2 d\Omega$$

and, similarly,

$$\int_{\Omega_i}(\mathbf{r}_i\cdot\nabla\dot{Z}_i)\dot{Z}_i\,d\Omega = \frac{1}{2}\int_{\Gamma_i}(\mathbf{r}_i\cdot\boldsymbol{\nu}_i)|\dot{Z}_i|^2 d\Gamma - \int_{\Omega_i}|\dot{Z}_i|^2 d\Omega.$$

Therefore

$$\int_{Q_i}(\nabla\boldsymbol{\psi}_i\mathbf{r}_i)\cdot\{\operatorname{div} C_i[\varepsilon(\boldsymbol{\psi}_i)] - K_i(\nabla Z_i + \boldsymbol{\psi}_i) + K_i(\mathbf{r}_i\cdot\nabla Z_i)\operatorname{div}(\nabla Z_i + \boldsymbol{\psi}_i)\}\,dQ$$
$$= \alpha_{i1}(t)|_0^T + \int_{Q_i}\left[\rho_i|\dot{Z}_i|^2 + I_{\rho_i}|\dot{\boldsymbol{\psi}}_i|^2\right]dQ$$
$$- \frac{1}{2}\int_{\Sigma_i}(\mathbf{r}_i\cdot\boldsymbol{\nu}_i)\left[\rho_i|\dot{Z}_i|^2 + I_{\rho_i}|\dot{\boldsymbol{\psi}}_i|^2\right]d\Sigma, \quad (4.13)$$

where $Q_i = \Omega_i\times(0,T)$ and $\Sigma_i = \Gamma_i\times(0,T)$. Substitute (4.13) into the left side of (4.4) and (4.9) into the right side of (4.4) (after integrating (4.4) in $t$). We obtain

$$\alpha_{i1}(t)|_0^T + \int_{Q_i}\left[\rho_i|\dot{Z}_i|^2 + I_{\rho_i}|\dot{\boldsymbol{\psi}}_i|^2\right]dQ - K_i\int_{Q_i}(\nabla Z_i + \boldsymbol{\psi}_i)\cdot\boldsymbol{\psi}_i\,dQ$$
$$= \frac{1}{2}\int_{\Sigma_i}(\mathbf{r}_i\cdot\boldsymbol{\nu}_i)\left[\rho_i|\dot{Z}_i|^2 + I_{\rho_i}|\dot{\boldsymbol{\psi}}_i|^2\right]d\Sigma$$
$$+ \frac{1}{2}\int_{\Sigma_i}(\mathbf{r}_i\cdot\boldsymbol{\nu}_i)\left\{\frac{1}{D_i}\left[|C_{\nu_i}(\boldsymbol{\psi}_i)|^2 + \frac{2}{1-\mu_i}|C_{\tau_i}(\boldsymbol{\psi}_i)|^2\right] + K_i|(\nabla Z_i + \boldsymbol{\psi}_i)\cdot\boldsymbol{\nu}_i|^2\right\}d\Sigma$$
$$- \int_{\Sigma_i}(\mathbf{r}_i\cdot\boldsymbol{\nu}_i)\left\{\frac{D_i(1-\mu_i^2)}{2}\left|\boldsymbol{\tau}_i\cdot\frac{\partial\boldsymbol{\psi}_i}{\partial\boldsymbol{\tau}_i}\right|^2 + \frac{K_i}{2}\left|\frac{\partial Z_i}{\partial\boldsymbol{\tau}_i} + \boldsymbol{\psi}_i\cdot\boldsymbol{\tau}_i\right|^2\right.$$
$$\left. + K_i(\boldsymbol{\psi}_i\cdot\boldsymbol{\nu}_i)(\nabla Z_i + \boldsymbol{\psi}_i)\cdot\boldsymbol{\nu}_i + \left(\mu_i\boldsymbol{\tau}_i\cdot\frac{\partial\boldsymbol{\psi}_i}{\partial\boldsymbol{\tau}_i}\boldsymbol{\nu}_i + \boldsymbol{\nu}_i\cdot\frac{\partial\boldsymbol{\psi}_i}{\partial\boldsymbol{\tau}_i}\boldsymbol{\tau}_i\right)\cdot(C_i[\varepsilon(\boldsymbol{\psi}_i)]\boldsymbol{\nu}_i)\right\}d\Sigma$$
$$+ \int_{\Sigma_i}(\mathbf{r}_i\cdot\boldsymbol{\tau}_i)\left\{\frac{\partial\boldsymbol{\psi}_i}{\partial\boldsymbol{\tau}_i}\cdot(C_i[\varepsilon(\boldsymbol{\psi}_i)]\boldsymbol{\nu}_i) + K_i\frac{\partial Z_i}{\partial\boldsymbol{\tau}_i}(\nabla Z_i + \boldsymbol{\psi}_i)\cdot\boldsymbol{\nu}_i\right\}d\Sigma, \quad (4.14)$$

where $C_{\nu_i}(\boldsymbol{\psi}_i), C_{\tau_i}(\boldsymbol{\psi}_i)$ are the components of $C_i[\varepsilon(\boldsymbol{\psi}_i)]$ in the directions of $\boldsymbol{\nu}_i$ and $\boldsymbol{\tau}_i$, respectively.

One also has

$$\int_{Q_i}\boldsymbol{\psi}_i\cdot[\operatorname{div}(C_i[\varepsilon(\boldsymbol{\psi}_i)]) - K_i(\nabla Z_i + \boldsymbol{\psi}_i)]\,dQ = \alpha_{i2}(T)|_0^T - I_{\rho_i}\int_{Q_i}|\dot{\boldsymbol{\psi}}_i|^2 dQ,$$

$$K_i\int_{Q_i}Z_i\operatorname{div}(\nabla Z_i + \boldsymbol{\psi}_i)\,dQ = \alpha_{i3}(t)|_0^T - \rho_i\int_{Q_i}|\dot{Z}_i|^2 dQ,$$

where

$$\alpha_{i2}(t) = \int_{\Omega_i} \boldsymbol{\psi}_i \cdot \dot{\boldsymbol{\psi}}_i \, d\Omega, \quad \alpha_{i3}(t) = \int_{\Omega_i} Z_i \dot{Z}_i \, d\Omega.$$

Use of the last two relations in (4.10) and (4.11) yields

$$\alpha_{i2}(t)|_0^T + \int_{Q_i} [(C_i[\varepsilon(\boldsymbol{\psi}_i)], \varepsilon(\boldsymbol{\psi}_i)) - I_{\rho_i}|\dot{\boldsymbol{\psi}}_i|^2] \, dQ + K_i \int_{Q_i} \boldsymbol{\psi}_i \cdot (\nabla Z_i + \boldsymbol{\psi}_i) \, dQ = 0, \tag{4.15}$$

and

$$\alpha_{i3}(t)|_0^T - \rho_i \int_{Q_i} |\dot{Z}_i|^2 dQ + K_i \int_{Q_i} \nabla Z_i \cdot (\nabla Z_i + \boldsymbol{\psi}_i) \, dQ = K_i \int_{\Sigma_i} Z_i (\nabla Z_i + \boldsymbol{\psi}_i) \cdot \boldsymbol{\nu}_i \, d\Sigma, \tag{4.16}$$

respectively. Let $0 < \delta < 1$. Multiply (4.15) and (4.16) by $1 - \delta$ and by $\delta$, respectively, and then add the product to (4.14). We obtain the following expression, where $\alpha_i = \alpha_{i1} + \delta\alpha_{i2} + (1 - \delta)\alpha_{i3}$:

$$\begin{aligned}
&\alpha_i(t)|_0^T + \int_{Q_i} \{(1-\delta)\rho_i |\dot{Z}_i|^2 + \delta I_{\rho_i} |\dot{\boldsymbol{\psi}}_i|^2 + (1-\delta)(C_i[\varepsilon(\boldsymbol{\psi}_i)], \varepsilon(\boldsymbol{\psi}_i)) + \delta K_i (|\nabla Z_i|^2 - |\boldsymbol{\psi}_i|^2)\} dQ \\
&= \frac{1}{2} \int_{\Sigma_i} (\mathbf{r}_i \cdot \boldsymbol{\nu}_i) \left[\rho_i |\dot{Z}_i|^2 + I_{\rho_i} |\dot{\boldsymbol{\psi}}_i|^2\right] d\Sigma \\
&+ \frac{1}{2} \int_{\Sigma_i} (\mathbf{r}_i \cdot \boldsymbol{\nu}_i) \left\{ \frac{1}{D_i} \left[ |C_{\nu_i}(\boldsymbol{\psi}_i)|^2 + \frac{2}{1-\mu_i} |C_{\tau_i}(\boldsymbol{\psi}_i)|^2 \right] + K_i |(\nabla Z_i + \boldsymbol{\psi}_i) \cdot \boldsymbol{\nu}_i|^2 \right\} d\Sigma \\
&- \int_{\Sigma_i} (\mathbf{r}_i \cdot \boldsymbol{\nu}_i) \left\{ \frac{D_i(1-\mu_i^2)}{2} \left| \boldsymbol{\tau}_i \cdot \frac{\partial \boldsymbol{\psi}_i}{\partial \boldsymbol{\tau}_i} \right|^2 + \frac{K_i}{2} \left| \frac{\partial Z_i}{\partial \boldsymbol{\tau}_i} + \boldsymbol{\psi}_i \cdot \boldsymbol{\tau}_i \right|^2 \right. \\
&\left. + K_i (\boldsymbol{\psi}_i \cdot \boldsymbol{\nu}_i)(\nabla Z_i + \boldsymbol{\psi}_i) \cdot \boldsymbol{\nu}_i + \left( \mu_i \boldsymbol{\tau}_i \cdot \frac{\partial \boldsymbol{\psi}_i}{\partial \boldsymbol{\tau}_i} \boldsymbol{\nu}_i + \boldsymbol{\nu}_i \cdot \frac{\partial \boldsymbol{\psi}_i}{\partial \boldsymbol{\tau}_i} \boldsymbol{\tau}_i \right) \cdot (C_i[\varepsilon(\boldsymbol{\psi}_i)]\boldsymbol{\nu}_i) \right\} d\Sigma \\
&+ \int_{\Sigma_i} (\mathbf{r}_i \cdot \boldsymbol{\tau}_i) \left\{ \frac{\partial \boldsymbol{\psi}_i}{\partial \boldsymbol{\tau}_i} \cdot (C_i[\varepsilon(\boldsymbol{\psi}_i)]\boldsymbol{\nu}_i) + K_i \frac{\partial Z_i}{\partial \boldsymbol{\tau}_i} (\nabla Z_i + \boldsymbol{\psi}_i) \cdot \boldsymbol{\nu}_i \right\} d\Sigma \\
&+ \delta K_i \int_{\Sigma_i} Z_i (\nabla Z_i + \boldsymbol{\psi}_i) \cdot \boldsymbol{\nu}_i \, d\Sigma.
\end{aligned} \tag{4.17}$$

Suppose that, in addition to the above conditions, $\boldsymbol{\Psi}_1, \boldsymbol{\Psi}_2$ satisfy (3.8) and (3.9). Use of the boundary conditions (3.9) in the last identity yields

$$\begin{aligned}
&\alpha(t)|_0^T + \sum_{i=1}^{2} \int_{Q_i} \{(1-\delta)\rho_i |\dot{Z}_i|^2 + \delta I_{\rho_i} |\dot{\boldsymbol{\psi}}_i|^2 + (1-\delta)(C_i[\varepsilon(\boldsymbol{\psi}_i)], \varepsilon(\boldsymbol{\psi}_i)) + \delta K_i (|\nabla Z_i|^2 - |\boldsymbol{\psi}_i|^2)\} dQ \\
&= \frac{1}{2} \int_{\Sigma} (\mathbf{r}_2 \cdot \boldsymbol{\nu}_2) \left\{ \frac{1}{D_2} \left[ |C_{\nu_2}(\boldsymbol{\psi}_2)|^2 + \frac{2}{1-\mu_2} |C_{\tau_2}(\boldsymbol{\psi}_2)|^2 \right] + K_2 \left| \frac{\partial Z_2}{\partial \nu} \right|^2 \right\} d\Sigma + \mathcal{I}_1(\Sigma_1) + \mathcal{I}_2(\Sigma_1),
\end{aligned} \tag{4.18}$$

where $\alpha = \alpha_1 + \alpha_2$ and $\mathcal{I}_i(\Sigma_1)$ denotes the sum of the integrals on the right side of (4.17) over $\Sigma_1$.

One has

$$|\nabla Z_i|^2 - |\psi_i|^2 \geq \frac{1}{2}|\nabla Z_i + \psi_i|^2 - 2|\psi_i|^2$$

and, by Korn's Lemma

$$\sum_{i=1}^{2} \int_{\Omega_i} |\psi_i|^2 d\Omega \leq C(\Omega) \sum_{i=1}^{2} \int_{\Omega_i} (C_i[\varepsilon(\psi_i)], \varepsilon(\psi_i))\, d\Omega.$$

Therefore, if $\delta$ is small enough we obtain the estimate

$$\begin{aligned}
\sum_{i=1}^{2} \int_{Q_i} &\{(1-\delta)\rho_i|\dot{Z}_i|^2 + \delta I_{\rho_i}|\dot{\psi}_i|^2 + (1-\delta)(C_i[\varepsilon(\psi_i)], \varepsilon(\psi_i)) + \delta K_i(|\nabla Z_i|^2 - |\psi_i|^2)\}dQ \\
&\geq c_0 \sum_{i=1}^{2} \int_{Q_i} \{\rho_i|\dot{Z}_i|^2 + I_{\rho_i}|\dot{\psi}_i|^2 + (C_i[\varepsilon(\psi_i)], \varepsilon(\psi_i)) + K_i|\nabla Z_i + \psi_i|^2\}dQ \\
&= c_0 \int_0^T (\|\dot{\Psi}(t)\|_H^2 + \|\Psi(t)\|_V^2)\, dt \qquad (4.19) \\
&= c_0 T \|(\Psi^0, \Psi^1)\|_{V\times H}^2
\end{aligned}$$

for some constant $c_0 > 0$. Insertion of (4.19) into (4.18) yields

$$\begin{aligned}
\alpha(t)|_0^T + c_0 T\|(\Psi^0, \Psi^1)\|_{V\times H}^2 \leq \frac{1}{2}\int_\Sigma (\mathbf{r}_2 \cdot \boldsymbol{\nu}_2) \Big\{ &\frac{1}{D_2}\left[|C_{\nu_2}(\psi_2)|^2 + \frac{2}{1-\mu_2}|C_{\tau_2}(\psi_2)|^2\right] \\
&+ K_2 \left|\frac{\partial Z_2}{\partial \nu}\right|^2 \Big\} d\Sigma + \mathcal{I}_1(\Sigma_1) + \mathcal{I}_2(\Sigma_1). \qquad (4.20)
\end{aligned}$$

We now show that if one chooses $\mathbf{r}_1 = \mathbf{r}_2 = \mathbf{x} - \mathbf{x}_0$, then

$$\mathcal{I}_1(\Sigma_1) + \mathcal{I}_2(\Sigma_1) \leq 0 \qquad (4.21)$$

under hypotheses (2.1) and (2.2) of Theorem 2.2. Since one also has

$$|\alpha(t)|_{L^\infty(0,T)} \leq C(\Omega)\|(\Psi^0, \Psi^1)\|_{V\times H}^2,$$

the conclusion of Theorem 2.2 will follow from (4.20) and (4.21).

On $\Gamma_1$ one has $\boldsymbol{\nu}_1 = -\boldsymbol{\nu}_2$, $\boldsymbol{\tau}_1 = -\boldsymbol{\tau}_2$. The transmission conditions (3.8) imply, in particular, that $C_{\nu_1}(\psi_1) = C_{\nu_2}(\psi_2)$, $C_{\tau_1}(\psi_1) = C_{\tau_2}(\psi_2)$. These conditions also imply, in a straightforward manner, that

$$\begin{aligned}
\mathcal{I}_1(\Sigma_1) + \mathcal{I}_2(\Sigma_1) = &\frac{1}{2}\int_{\Sigma_1} (\mathbf{r}\cdot\boldsymbol{\nu}_1)[(\rho_1 - \rho_2)|\dot{Z}_1|^2 + (I_{\rho_1} - I_{\rho_2})|\psi_1|^2]d\Sigma \\
&+ \frac{1}{2}\int_{\Sigma_1} (\mathbf{r}\cdot\boldsymbol{\nu}_1)\left[\left(\frac{2}{D_1(1-\mu_1)} - \frac{2}{D_2(1-\mu_2)}\right)|C_{\tau_1}(\psi_1)|^2\right. \\
&\left. + \left(\frac{1}{K_1} - \frac{1}{K_2}\right)|(\nabla Z_1 + \psi_1)\cdot\boldsymbol{\nu}_1|^2 - (K_1 - K_2)\left|\frac{\partial Z_1}{\partial \tau_1} + \psi_1\cdot\boldsymbol{\tau}_1\right|^2\right] d\Sigma \\
+ \frac{1}{2}\sum_{i=1}^{2}\int_{\Sigma_1} (\mathbf{r}\cdot\boldsymbol{\nu}_i) &\left[\frac{1}{D_i}|C_{\nu_i}(\psi_i)|^2 - D_i(1-\mu_i^2)\left|\boldsymbol{\tau}_i\cdot\frac{\partial\psi_i}{\partial\tau_i}\right|^2 - 2\mu_i\boldsymbol{\tau}_i\cdot\frac{\partial\psi_i}{\partial\tau_i}C_{\nu_i}(\psi_i)\right] d\Sigma.
\end{aligned}$$

Therefore

$$\mathcal{I}_1(\Sigma_1)+\mathcal{I}_2(\Sigma_1)\le \frac{1}{2}\sum_{i=1}^{2}\int_{\Sigma_1}(\mathbf{r}\cdot\boldsymbol{\nu}_1)\left[\frac{1}{D_i}|C_{\nu_i}(\boldsymbol{\psi}_i)|^2 - D_i(1-\mu_i^2)\left|\boldsymbol{\tau}_i\cdot\frac{\partial\boldsymbol{\psi}_i}{\partial\boldsymbol{\tau}_i}\right|^2 - 2\mu_i\boldsymbol{\tau}_i\cdot\frac{\partial\boldsymbol{\psi}_i}{\partial\boldsymbol{\tau}_i}C_{\nu_i}(\boldsymbol{\psi}_i)\right]d\Sigma \quad (4.22)$$

if

$$\rho_1\le\rho_2,\quad K_1\ge K_2,\quad D_1(1-\mu_1)\ge D_2(1-\mu_2).$$

The bracketed quantity in the last integral may be written

$$\frac{1}{D_i}\left[C_{\nu_i}(\boldsymbol{\psi}_i)-D_i(1+\mu_i)\boldsymbol{\tau}_i\cdot\frac{\partial\boldsymbol{\psi}_i}{\partial\boldsymbol{\tau}_i}\right]\left[C_{\nu_i}(\boldsymbol{\psi}_i)+D_i(1-\mu_i)\boldsymbol{\tau}_i\cdot\frac{\partial\boldsymbol{\psi}_i}{\partial\boldsymbol{\tau}_i}\right],$$

which equals, by virtue of (4.7),

$$D_i\left[\left(\boldsymbol{\nu}_i\cdot\frac{\partial\boldsymbol{\psi}_i}{\partial\boldsymbol{\nu}_i}\right)^2-\left(\boldsymbol{\tau}_i\cdot\frac{\partial\boldsymbol{\psi}_i}{\partial\boldsymbol{\tau}_i}\right)^2\right].$$

It follows from (3.8) and (4.22) that

$$\mathcal{I}_1(\Sigma_1)+\mathcal{I}_2(\Sigma_1)\le \frac{1}{2}\int_{\Sigma_1}(\mathbf{r}\cdot\boldsymbol{\nu}_1)\left[D_1\left(\boldsymbol{\nu}_1\cdot\frac{\partial\boldsymbol{\psi}_1}{\partial\boldsymbol{\nu}_1}\right)^2-D_2\left(\boldsymbol{\nu}_2\cdot\frac{\partial\boldsymbol{\psi}_2}{\partial\boldsymbol{\nu}_2}\right)^2 -(D_1-D_2)\left(\boldsymbol{\tau}_1\cdot\frac{\partial\boldsymbol{\psi}_1}{\partial\boldsymbol{\tau}_1}\right)^2\right]d\Sigma. \quad (4.23)$$

Since $C_{\nu_1}(\boldsymbol{\psi}_1)=C_{\nu_2}(\boldsymbol{\psi}_2)$ on $\Gamma_1$, we obtain from (4.7)

$$D_2\boldsymbol{\nu}_2\cdot\frac{\partial\boldsymbol{\psi}_2}{\partial\boldsymbol{\nu}_2}=D_1\boldsymbol{\nu}_1\cdot\frac{\partial\boldsymbol{\psi}_1}{\partial\boldsymbol{\nu}_1}+(\mu_1D_1-\mu_2D_2)\boldsymbol{\tau}_1\cdot\frac{\partial\boldsymbol{\psi}_1}{\partial\boldsymbol{\tau}_1}. \quad (4.24)$$

Use of (4.24) in (4.23) yields

$$\mathcal{I}_1(\Sigma_1)+\mathcal{I}_2(\Sigma_1)\le \frac{1}{2}\int_{\Sigma_1}(\mathbf{r}\cdot\boldsymbol{\nu}_1)\left[\frac{D_1}{D_2}(D_2-D_1)\left(\boldsymbol{\nu}_1\cdot\frac{\partial\boldsymbol{\psi}_1}{\partial\boldsymbol{\nu}_1}\right)^2 -\frac{2D_1}{D_2}(\mu_1D_1-\mu_2D_2)\left(\boldsymbol{\nu}_1\cdot\frac{\partial\boldsymbol{\psi}_1}{\partial\boldsymbol{\nu}_1}\right)\left(\boldsymbol{\tau}_1\cdot\frac{\partial\boldsymbol{\psi}_1}{\partial\boldsymbol{\tau}_1}\right) -\left(D_1-D_2+\frac{(\mu_1D_1-\mu_2D_2)^2}{D_2}\right)\left(\boldsymbol{\tau}_1\cdot\frac{\partial\boldsymbol{\psi}_1}{\partial\boldsymbol{\tau}_1}\right)^2\right]d\Sigma. \quad (4.25)$$

It is easy to check that the quantity in brackets in (4.25) is a negative semi-definite quadratic form exactly when

$$D_1\ge D_2\ \text{ and }\ (D_1-D_2)^2\ge(\mu_1D_1-\mu_2D_2)^2.$$

This pair of inequalities is the same as

$$D_1(1\pm\mu_1)\ge D_2(1\pm\mu_2).$$

## References


[1] G. Chen, M.C. Delfour, A.M. Krall and G. Payre, "Modeling, stabilization and control of serially connected beams," *SIAM J. Control Opt.*, **25** (1987), pp. 526–546.

[2] Lagnese, J. E., *Boundary Stabilization of Thin Plates*, Studies in Appl. Math., Vol.10, SIAM Publications, Philadelphia, 1989.

[3] Lagnese, J.E. and G. Leugering, "Modelling of dynamic networks of thin elastic plates," *Math. Methods Appl. Sci.*, to appear.

[4] Lagnese, J.E., Leugering, G. and Schmidt, E.J.P.G., "Controllability of planar network of Timoshenko beams," *SIAM J. Control and Opt.*, to appear.

[5] Lagnese, J.E., Leugering, G. and Schmidt, E.J.P.G., "On the analysis and control of hyperbolic systems associated with vibrating networks," to appear.

[6] Lasiecka, I. and Triggiani, R., "A lifting theorem for the time regularity of solutions to abstract equations with unbounded operators and applications to hyperbolic equations," *Proc. AMS*, **104**, 1988, pp.745–755.

[7] Lions, J.-L., *Contrôlabilité Exacte, Perturbations et Stabilisation de Systèmes Distribués; Tome 1, Contrôlabilité Exacte*, Collection RMA, Vol. 8, Masson, Paris, 1988.

[8] Nicaise, S., "About the Lamé system in a polygonal or a polyhedral domain and a coupled problem between the Lamé system and the plate equation I: regularity of the solutions," *Pub. IRMA*, Lille, **22**, No. 5, (1990).

# 42 Approximation of Linear Input/Output Delay-Differential Systems

**E. Bruce Lee** University of Minnesota, Minneapolis, Minnesota

Abstract Two approaches for finding reduced order models for linear delay-differential type systems are described. The first is a balanced approximation approach based on generalized reachability and observability gramians. The second is based on frequency response data in which a rational function (that is, low pass or band pass filter) approximation is sought with a good match to the frequency response data. The second approach is exploited as a way to find root chains of exponential polynomials, that is, roots of the characteristic function of the linear delay-differential type systems.

## I. Introduction

Approximation of input/output delay-differential equation models by simpler models is frequently necessary in order to simplify the analysis and synthesis of control systems involving time delays. For example, the design of a quadratic regulator or an H-infinity optimal robust compensator for linear delay-differential type systems lead to computationally complex results (Alekal, Brunovsky, Chyung and Lee [1] or Lee and Wu [2]). Even if one takes a sampled data approach by selecting the control action to involve piecewise-constant inputs the computational complexity is not significately decreased (Alford and Lee [3]) and further approximations must be admitted without a clear indication of their adequacy.

On the other hand, some constrained optimization approaches to feedback compensator design allow a certain amount of imperfection in the model to be used for controller synthesis (Curtain and Glover [4]). Thus if the imperfect model is sufficiently close to the true system model in an (uniform) H-infinity norm sense and has same number of unstable modes, then a compensator which is stabilizing for the approximant is also stabilizing for the true system model.

Research supported by the National Science Foundation under Grant DMS 9002919

There have been a number of recent studies related to the approximation of the delay-differential type systems from this control/system engineering point of view. These have involved in many cases finding coefficients of a Laguerre-Fourier series expansion with truncation after a certain number of terms based on the error estimates being used or on finding other rational approximates (Partington, Glover, Zwart and Curtain [5]; Partington [6]; Glader, Hognas, Makila and Toivonen [7], Wahlberg [8], Makila [9] and references therein). The results which are documented in these recent papers are most closely related to the results given in section III.

Two basic methods of approximation of the delay-differential systems will be described. The first method is a direct generalization of the balanced approximation method of model reduction initiated by Moore [10] for finite dimensional linear system models. Using this method for infinite dimensional linear delay-differential systems one can retain the infinite dimensional form of the model by eliminating subsystems which are weakly coupled to the input and output variables. This extension of the balanced approximation is possible through use of complex integral representations of the Gramians of the linear delay-differential systems as introduced by Lu, Lee and Zhang [11] and using balancing transformations in a 2-D sense. Some details of this approach, will be given in the next section II. An example will illustrate the procedure and the extent to which the appproximation retains properties of the original model. This approach has also been used in the balanced approximation of linear 2-D discrete-time systems in [11] and further refined by Premaratne, Jury and Mansour [12]. Using this method one can find reduced order models which are often adequate for frequency response and transient response (step response) comparisons.

The second method is based directly on frequency response analysis of the linear delay-differential system. Here we seek an approximation among the proper rational functions (that is, linear low pass or band pass filters) with a good match of the frequency response signatures. Gu, Khargonekar and Lee [13] recently developed this approach to find finite dimensional approximants of certain infinite dimensional systems which can lead to models that are adequate for the design of stabilizing controllers. Error estimates in the transfer function uniform norm on $[-j\infty,j\infty]$, (H-infinity norm), have also been found allowing one to determine the order of the filter truncation which is adequate for the stabilization task, for example. Also the extension of this approximation method to handle the unstable systems has recently been done under the assumption that there are only a finite number of unstable poles (Gu, Khargonekar, Lee and Misra [14]). More recent work has included obtaining user friendly MATLAB based software to convert the frequency response data into the approximate model. In section III this Fourier-Laguerre series approximation method will be outlined and an example will show full details of how to find a good approximant using the software based on MATLAB. Section IV will further exploit this software to find characteristic roots near the imaginary axis of the linear delay-differential systems. Here an example will show how to do this in detail. Further background on the methods being used can be found in papers of Wu and Lee [15], Partington [16] and Kamen, Khargonekar and Tannenbaum [17].

To close this introductory section an example of approximation by the usual dominant mode technique will be given.

Consider a system with input/output delay-differential equation (DDE) model

$$y'(t) + y(t) + 2y(t - 0.3) = 2u(t - 0.3), \qquad (' = \frac{d}{dt}).$$

If we assume zero initial data for a Cauchy initial value problem and assume the input function is square integrable (in $L_2(-\infty,\infty)$) then we could seek a continuous function y(t) which statisfies DDE almost everyhwere to the right of the initial data.

Let F represent the input-output operator of the system, that is, y(t) = (Fu)(t). In some cases we seek an approximation of F, called $F_a$, such that maximum of $|(Fu)(t) - (F_a u)(t)|$ for $0 \le t \le \infty$, is small for restricted classes of inputs such as pulse functions or step functions.

Using Laplace transform techniques the input/output operator in transfer function form is

$$T(s) = \frac{Y(s)}{U(s)} = \frac{2e^{-0.3s}}{1 + s + 2e^{-0.3s}} = \frac{2}{e^{0.3s}(1 + s) + 2} = \left. \frac{2}{z(1 + s) + 2} \right|_{z=e^{0.3s}}$$

Now we may seek an approximation of T called $T_a$ such that the amplification of sinusoidal inputs in going through the linear system is the same as for the approximant at all frequencies, that is, $|T(j\omega) - T_a(j\omega)|$ is uniformly small in $\omega$. The exponential polynomial (characteristic function) $1 + s + 2e^{-0.3s}$ has a pair of dominant (modes) zeros at approximately $s = -2.5 \pm j3.9$ (next pair of zeros ordered in terms of imaginary part is at approximately $-8 \pm j26$). Thus we might take as our first approximation to T(s) the transfer function

$$T_a(s) = \frac{14.3\, e^{-0.3s}}{s^2 + 5s + 21.46}.$$

A comparison of frequency response data, figure 1, shows that the graphs of $T(j\omega)$ and $T_a(j\omega)$ versus $\omega$ are close. Figure 2 shows the magnitude difference versus radian frequency $\omega$ and figure 3 shows the difference $|T(j\omega) - T_a(j\omega)|$ versus $\omega$. The H-infinity norm difference can be determined from figure 3 to be approximately 0.31; about 35% maximum error. Pulse and step responses of F and $F_a$ are also close.

If we seek a truly finite dimensional approximation, we might try replacing the numerator term $e^{-0.3s}$ by its Taylor series (or Pade) approximant. Using a Taylor series does not lead to a good frequency response comparison.

In fact the function $e^{-\tau s}$ for $\tau>0$ is hard to approximate by a (strictly proper stable) rational function in $H_\infty$-norm [18]. To show this let $\sigma$ denote delay (right shift) operator by $\tau$ seconds in $L_2(-\infty,\infty)$; so $(\sigma x)(t) = x(t - \tau)$ for all $x(t) \in L_2(-\infty,\infty)$. In terms of Fourier transform $\mathcal{F}$ we have

$$\mathcal{F}\{\sigma x(t)\} = e^{-j\omega\tau} X(j\omega) \qquad \text{where } \mathcal{F}\{x(t)\} = X(j\omega) \text{ and } j=\sqrt{-1}.$$

Next let $\Sigma$ denote the mapping of $L_2(-j\infty,j\infty)$ into $L_2(-j\infty,j\infty)$ given by $(\Sigma X)(j\omega) = e^{-j\omega\tau} X(j\omega)$ (meaning that $\mathcal{F}\{\sigma x(t)\} = \Sigma\mathcal{F}\{x(t)\}$. Next consider the possibility of

$$\Sigma_n = \frac{b_{n-1}(j\omega)^{n-1} + \dots + b_1(j\omega) + b_o}{(j\omega)^n + a_{n-1}(j\omega)^{n-1} + \dots + a_1(j\omega) + a_o}$$

being an approximant of $\Sigma$ for $n = 1, 2, \dots$ . Here $a_i$'s and $b_i$'s, $i=0, 1, 2, \dots, n-1$ are real numbers to be selected.

Note in the uniform norm on the imaginary axis $j\Re$, $\|\Sigma - \Sigma_n\| \geq 1$, for $n = 1, 2, \dots$; since $|e^{-j\omega\tau}| = 1$ and $\lim_{\omega\to\infty} \Sigma_n = 0$. Thus there is no sequence of approximants of the form $\{\Sigma_n\}$ which converges uniformly to $\Sigma$. However, the product $e^{-s\tau}P(s)$ can be uniformly approximated if $P(s)$ is a stricuy proper rational function (see [7] and [13]).

In the remainder of this report we will be considering linear input/output delay-differential type systems with mathematical models of the following form (in vector-matrix notation with commensurate delay duration $\tau>0$)

$$\frac{dx}{dt}(t) = \sum_{i=0}^{\ell} A_i x(t - i\tau) + B_i\, u(t - i\tau) \tag{1}$$

with output equation

$$y(t) = \sum_{i=0}^{\ell} C_i\, x(t - i\tau) \tag{2}$$

Here $x(t)$ is an n-vector, the input $u(t)$ is an m-vector function and the output $y(t)$ is an p-vector function. $A_i \in R^{n\times n}$, $B_i \in R^{n\times m}$ and $C_i \in R^{p\times n}$ for $i=0, 1, 2, \dots, \ell$. The input/output operator of such a system can be shown to be a convolution operator involving the inpulse response. The transfer function (obtained using the Laplace transform) is

$$T(e^{\tau s},s) = \Big(\sum_{i=0}^{\ell} C_i e^{-i\tau s}\Big)\Big(sI - \sum_{i=0}^{\ell} A_i e^{-i\tau s}\Big)^{-1}\Big(\sum_{i=0}^{\ell} B_i e^{-i\tau s}\Big) \tag{3}$$

Note T is a pxm matrix each element of which is a ratio of exponential polynomials of the form $\dfrac{b_{n-1}(e^{-\tau s})s^{n-1} + \dots + b_1(e^{-\tau s})s + b_o(e^{-\tau s})}{s^n + a_{n-1}(e^{-\tau s})s^{n-1} + \dots + a_1(e^{-\tau s})s + a_o(e^{-\tau s})}$ where $b_i(e^{-\tau s})$ and $a_i(e^{-\tau s})$ for $i=0, 1, 2, \dots, n-1$, are polynomials in $e^{-\tau s}$.

In the next section II we began considering how to approximate the delay-differential equation model having transfer function $T(e^{\tau s}, s) = T(z,s)|_{z=e^{\tau s}}$ by a lower order model in a more systematic way. Here we use a generalization of the balanced-truncation method originally developed for linear finite dimensional systems [10]. Section III is consideration of the same approximation question but restricting the approximant to be finite dimensional (rational transfer function). Section IV shows how to exploit the algorithm (software) as described in section III to obtain roots of the exponential polynomials.

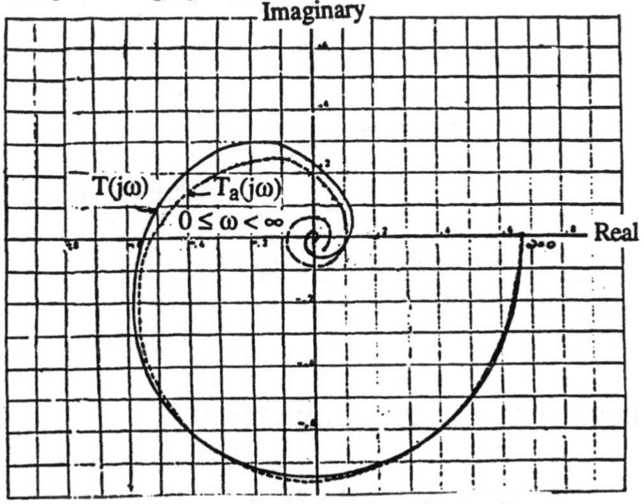


Figure 1 Frequency Response Comparison

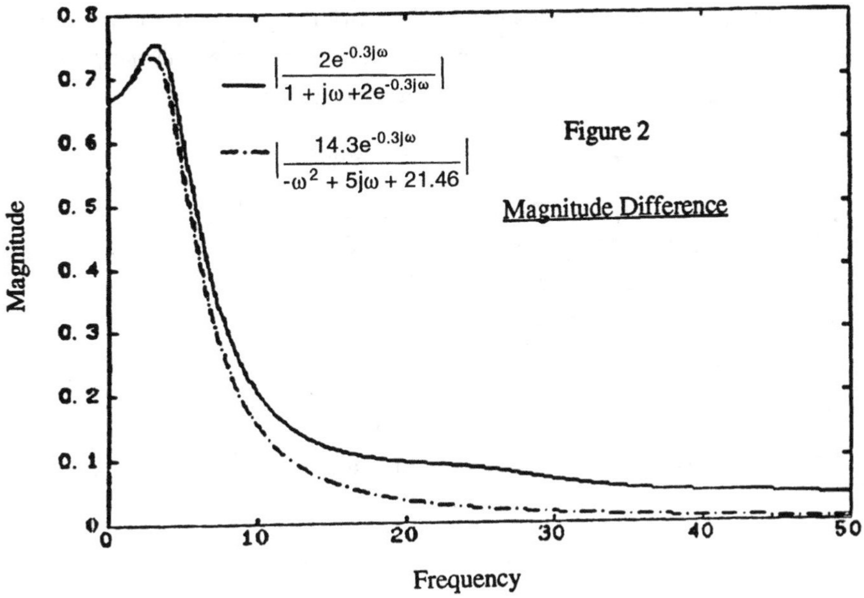

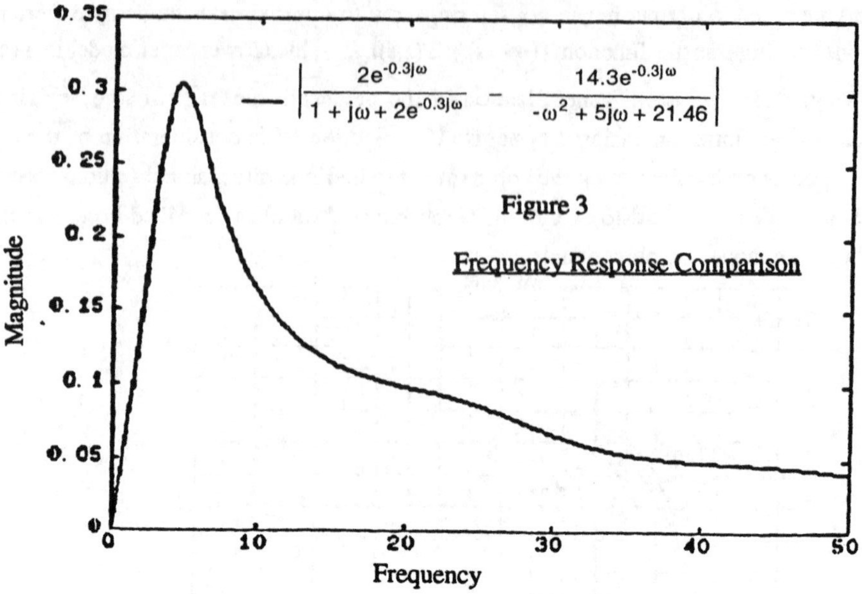


Figure 3

Frequency Response Comparison

## II. Balanced Approximation of delay-differential system models

In this section we consider the task of finding a reduced order model $T_r$ of a given model T of a dynamical system such that $\|T - T_r\|$ is suitably small using a generalization of balanced truncation. Given therefore a transfer function T=T(z,s) with commensurate delays (for a delay-differential system as described at the end of section I), there is a realization (Sontag [19]; Zak, Lee and Lu [20]) in partitioned form as a differential-difference equation

$$\begin{bmatrix} w(t+\tau) \\ \dot{x}(t) \end{bmatrix} = \begin{bmatrix} A_1 & A_2 \\ A_3 & A_4 \end{bmatrix} \begin{bmatrix} w(t) \\ x(t) \end{bmatrix} + \begin{bmatrix} B_1 \\ B_2 \end{bmatrix} u(t) \equiv A \begin{bmatrix} w(t) \\ x(t) \end{bmatrix} + Bu(t)$$

$$y(t) = \begin{bmatrix} C_1 & C_2 \end{bmatrix} \begin{bmatrix} w(t) \\ x(t) \end{bmatrix} \equiv C \begin{bmatrix} w(t) \\ x(t) \end{bmatrix} \tag{4}$$

where $A \in R^{(n_1+n_2)x(n_1+n_2)}$, $B \in R^{(n_1+n_2)xm}$, $C \in R^{px(n_1+n_2)}$.

Here the partitioning is into an infinite dimensional (continuous-time) pure difference equation part and a finite dimensional differential equation part with coupling terms between the two parts given by $A_2$ and $A_3$.

Let $I(z,s) = zI_{n1} \oplus sI_{n2}$ and note $T(z,s) = C[I(z,s) - A]^{-1}B$. Now set $F(z,s) = [I(z,s) - A]^{-1}B$ and $G(z,s) = C[I(z,s) - A]^{-1}$. Then we can define as in [11] generalized reachability and observability gramians in terms of integrals

$$K = \frac{1}{2\pi} \int_{-j\infty}^{j\infty} F(e^{s\tau},s)\, F^*(e^{s\tau},s)\, ds$$

$$W = \frac{1}{2\pi} \int_{-j\infty}^{j\infty} G^*(e^{s\tau},s)\, G(e^{s\tau},s)\, ds \tag{5}$$

respectively.

To do balancing and retain the above partitioned form we consider balancing (similarity) transformations of the form $P=P_1 \oplus P_2$ where nonsingular $P_1 \in R^{n_1 x n_1}$ and nonsingular $P_2 \in R^{n_2 x n_2}$. Thus the linearly equivalent system under the similarity transformation P will become $(\hat{A},\hat{B},\hat{C})=(P^{-1}AP,\ P^{-1}B,\ CP)$, and the Gramians of the tranformed realization become $\hat{K} = P^{-1}KP^{-T}$, and $\hat{W} = P^TWP$ respectively.

To proceed with the balancing in the two subparts partition K, W, $\hat{K}$, $\hat{W}$ as

$$K = \begin{bmatrix} K_{11} & K_{12} \\ K_{21} & K_{22} \end{bmatrix},\ W = \begin{bmatrix} W_{11} & W_{12} \\ W_{21} & W_{22} \end{bmatrix},\ \hat{K} = \begin{bmatrix} \hat{K}_{11} & \hat{K}_{12} \\ \hat{K}_{21} & \hat{K}_{22} \end{bmatrix},\ \text{and } \hat{W} = \begin{bmatrix} \hat{W}_{11} & \hat{W}_{12} \\ \hat{W}_{21} & \hat{W}_{22} \end{bmatrix}.$$

Here $K_{ii}$, $W_{ii}$, $\hat{K}_{ii}$, and $\hat{W}_{ii}$ are square matrices of size $n_i x n_i$ for i = 1, 2.

It is easy to obtain a balancing transformation P using the 1-D algorithm of Laub [21] on each part when $K_{ii}$ and $W_{ii}$ for i=1, 2 are positive definite. A sufficient condition for positive definiteness is that local controllability and local observability hold [11].

Thus in the positive definite case there is a balancing transformation in terms of the subpartitioning. And one can do model reduction to get $T_r$ of lower order by elimination of the least controllable and observable part of the system. An example will illustrate the steps involved in this approximation (reduction).

Example Consider a single-input/single-output system as represented by the differential-difference equation

$$4y''(t) + 4y'(t) + 2y'(t - 1/2) + y(t) = 2u(t) \tag{6}$$

with input u(t) and output y(t).

Using Laplace transform technique the transfer function from input u(t) to output y(t) is

$$T(z,s) = \frac{\frac{z}{2}}{z(s^2 + s + \frac{1}{4}) + \frac{s}{2}} = \frac{\frac{1}{2}}{s^2 + s + \frac{1}{4} + \frac{se^{-s/2}}{2}}, \qquad \text{where } z = e^{\frac{s}{2}}.$$

Such a system takes inputs in $L_2(0,\infty)$ into outputs in $L_2(0,\infty)$ since

$\lambda^2 + \lambda + 1/4 + \frac{\lambda}{2} e^{-\lambda/2} \neq 0$ for Re $\lambda \geq 0$, (See Driver [22] page 327).

Thus we seek a similar stability property for the reduced model and also seek a model which closely duplicates responses to sinusoidal inputs (frequency response).

A realization of T(z,s) in the above partitioned form is

$$\begin{bmatrix} w(t + 1/2) \\ \dot{x}_1(t) \\ \dot{x}_2(t) \end{bmatrix} = \left[\begin{array}{c|cc} 0 & 0 & -1/2 \\ \hline 0 & 0 & 1/2 \\ 1 & -1/2 & -1 \end{array}\right] \begin{bmatrix} w(t) \\ x_1(t) \\ x_2(t) \end{bmatrix} + \left[\begin{array}{c} 0 \\ \hline -1 \\ 0 \end{array}\right] u(t) \tag{7}$$

$$y(t) = [0 \mid 0 \quad 1] \begin{bmatrix} w(t) \\ x_1(t) \\ x_2(t) \end{bmatrix} \tag{8}$$

Using numerical integration compute according to equation 5 in partitioned form

$$K_{22} = \frac{1}{2\pi} \int_{-\infty}^{\infty} \frac{\begin{bmatrix} e^{(\frac{j\omega}{2})} (1 + j\omega) + 1/2 \\ -1/2\, e^{(\frac{j\omega}{2})} \end{bmatrix} \left[ e^{-(\frac{j\omega}{2})}(1 - j\omega) + 1/2 \mid -\frac{1}{2} e^{-(\frac{j\omega}{2})} \right]}{\left| e^{(\frac{j\omega}{2})} \left[\frac{1}{4} - \omega^2 + j\omega\right] + \frac{j\omega}{2} \right|^2} d\omega$$

$$= \begin{bmatrix} 10.2057 & -3.1715 \\ -3.1715 & 1.0680 \end{bmatrix}, \tag{9}$$

and in a similar fashion compute

$$W_{22} = \begin{bmatrix} 0.3399 & 0 \\ 0 & 0.4076 \end{bmatrix}. \tag{10}$$

Using balancing [21] on $K_{22}$ and $W_{22}$ leads to a balancing transformation

$$P_2 = \begin{bmatrix} 1.71454 & 0.17776 \\ -0.53276 & 0.47704 \end{bmatrix}. \tag{11}$$

Thus for similarity transformation $P = I \oplus P_2$ an equivalent realization balanced in the s-direction is with

$$\hat{A} = \left[\begin{array}{c|cc} 0 & 0.26638 & -0.23852 \\ \hline -0.19478 & -0.07603 & 0.23491 \\ 1.87875 & -0.76518 & -0.9237 \end{array}\right], \quad \hat{b} = \left[\begin{array}{c} 0 \\ \hline -0.52272 \\ -0.58379 \end{array}\right], \quad \hat{c} = \left[\begin{array}{c|cc} 0 & -0.53276 & 0.47704 \end{array}\right] \tag{12}$$

When doing the balancing we order the elements of the matrices according to their input/output significance as represent by the singular values, see [11]. Thus truncation by elimination of last row and last column leads to a reduced model

$$\hat{A}_r = \left[\begin{array}{c|c} 0 & 0.26638 \\ \hline -0.19478 & -0.07603 \end{array}\right], \quad \hat{b}_r = \left[\begin{array}{c} 0 \\ \hline -0.52272 \end{array}\right], \quad \hat{c}_r = \left[\begin{array}{c|c} 0 & -0.53276 \end{array}\right] \tag{13}$$

The transfer function of the reduced model is then

$$T_r(z,s) = \hat{c}_r \left[zI \oplus sI - \hat{A}_r\right]^{-1} \hat{b}_r$$

$$= \frac{z\,(0.53276)\,(0.52272)}{z\,(s + 0.07603) + (0.26638)\,(0.19478)}\,. \tag{14}$$

This system has same input/output stability property as original since

$$\lambda + 0.07603 + 0.05189 e^{\left(\frac{-\lambda}{2}\right)} \neq 0 \quad \text{for } \operatorname{Re} \lambda \geq 0.$$

A standard frequency response comparison is to consider the magnitude differences between T and $T_r$ on $j\Re$, namely $|T(e^{\frac{j\omega}{2}}, j\omega)| - |T_r(e^{\frac{j\omega}{2}}, j\omega)|$ for $0 \leq \omega < \infty$. Figure 4 shows this comparison. We have also previously considered this example in reference [11] where the signatures of $T(z,s)$, and $T_r(z,s)$ on $j\Re$ were plotted in the complex plane, from which $|T - T_r|$ can be determined; In any case the error is approximately 10% of the maximum magnitude.

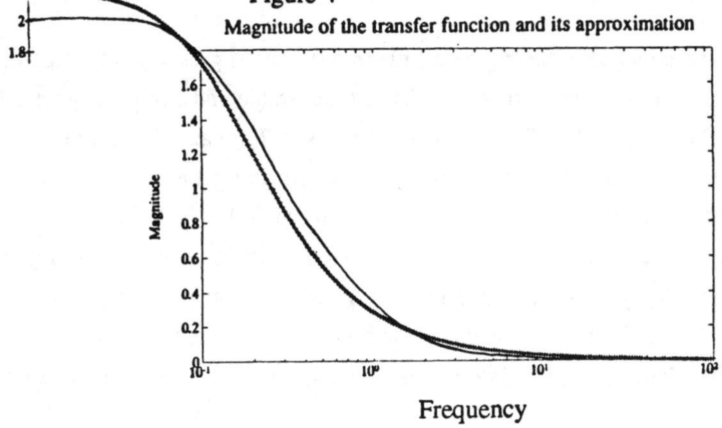


Figure 4

Magnitude of the transfer function and its approximation

## III. Rational approximation of delay-differential system models

Here we again consider the task of finding a reduced order model $T_r$ of a given model T such that $\|T - T_r\|$ is suitably small; but now we require $T_r$ to be finite dimensional even if T is not.

Consider then a transfer function representation T of the input/output behavior of a commensurate delay-differential system. This transfer function is an mxp matrix of elements each of which is a ratio of exponential polynomials of the form

$$\frac{b_{n-1}(e^{-s\tau})s^{n-1} + \ldots + b_0(e^{-s\tau})}{s^n + a_{n-1}(e^{-s\tau})s^{n-1} + \ldots + a_0(e^{-s\tau})} \tag{15}$$

where $b_i(e^{-s\tau})$ and $a_i(e^{-s\tau})$ for i=0, 1, 2, ..., n-1 are real coefficient polynomials in $e^{-s\tau}$ (see section I). Here $\tau>0$ is the basic delay duration; and all other delay terms involve an integer multiple of it in this commensurate delay case. The non-commensurate delay case involving two or more distinct delay durations, say $\tau_1>0$ and $\tau_2>0$, where $\tau_1$ and $\tau_2$ are not rationally related can be approximated by the Fourier-Laguerre series procedure to be outlined since in such a case the models will be linear and we only use frequency response data during the approximation. See also comments and results related to T(s) given by equation 2.33 or equation 4.10 in the paper [13].

Frequency response data (signature of T on $j\Re$) will be the basis for the approximation procedure (set $s=j\omega$, $0\leq\omega<\infty$ in above transfer function). Also the frequency response data could be obtained from the physical system by spectrum analysis procedures. Since it is somewhat easier to do the analysis and comparisons on the unit circle of the complex plane rather than the imaginary axis we convert our frequency response data on $j\Re$ in s-plane to corresponding data on the unit circle in z-plane using a bilinear transformation $s:=(\lambda\frac{1-z}{1+z})$. The bilinear transformation provides a conformal mapping of the right half plane, Res>0, into the unit disc, $|z|\leq 1$ (Complete details in [2] and [13]).

Since there can be poles in the right half plane one way to proceed is to decompose by a partial fraction expansion the transfer function T(s) into its stable and completely unstable parts, namely $T_s$ and $T_u$. So $T = T_s + T_u = T_s(s) + T_u(s)$ where both $T_s(s)$ and $T_u(-s)$ are analytic in the open right half plane. To retain the same stability properties and get a reduced order model we could then just approximate $T_s$ as $\tilde{T}_s$. The reduced model would be $\tilde{T}_s + T_u$. Such a procedure requires knowledge of the unstable poles, which is a computationally demanding task. The approach proposed in [13] by passes this step by treating the frequency response data directly and extracting the unstable data from the gain and phase relationships. From previous results, see for example Bellman and Cooke [23], it is known that there are only a finite number of poles to the right of any vertical line in the complex plane for the above linear delay-differential type system. The details of how to obtain convergence in the uniform norm of the Fourier-Laguerre series representation of $T_s(s)$ with error estimates in term of singular values are contained in the paper [13] and elsewhere in the control literature [7] [16]. To summarize the basis for these results notice that the Stone-Weierstrass theorem asserts that any continuous function of the unit circle $\{|z|=1\}$, can be uniformly approximated by the trigonometric polynomials, (polynomials in $e^{j\theta}$). For T(s) continuous on $j\Re$ the function F defined by

$$F(z) \equiv T(\lambda \frac{1 - z}{1 + z}), \quad \lambda = \text{positive constant}$$

can be approximated on the unit circle as Fourier series

$$F_s(z) + F_u(z) = F(z) \approx \sum_{k=-N}^{N-1} f_k z^k = \sum_{k=0}^{N-1} f_k z^k + \sum_{k=-N}^{-1} f_k z^k \tag{16}$$

where $F_s(z)$ and $F_u(z)$ are stable and unstable parts of F(z). So $F_s(z) = \sum_{k=0}^{\infty} f_k z^k$ and $F_u(z) = \sum_{k=-\infty}^{-1} f_k z^k$.

Hence $F_u(z)$ and $F_s(z)$ can be uniformly approximated on the unit circle as a polynomial in $z=e^{j\theta}$, if a finite number of coefficients $\{f_k\}_{k=-N}^{N-1}$ are known.

According to previous results, [13], or Wu [24] these can be approximately found using the discrete Fourier transform. That is, instead of using $f_k$ we use

$$f_M(k) = \frac{1}{2M} \sum_{r=-M}^{M-1} F(W_{2M}^{\,r})\, W_{2M}^{-rk} \quad \text{for } k = -M, -M+1, \ldots, M-1. \tag{17}$$

where $W_{2M} = e^{j\pi/M}$; this is easily computed using fast Fourier transform (FFT).

A standard result in Harmonic analysis is that if $F_u(z)$ has finite McMillan degree and if $\frac{dF}{dz}$ localized to $z=e^{j\omega}$ is in $L_2[0,2\pi]$ then the sequence $\{\|f_k\|\}$ is absolutely summable and

$$f_M(k) = \sum_{L=\infty}^{\infty} f_{2LM+k} \,.$$

In the paper [14] a constructive procedure for finding the Fourier coefficients is outlined and for the unstable part the following convergence result is established.

<u>Theorem</u> Let $F(z) = T(\lambda \frac{1 - z}{1 + z})$ be given as above with unstable part $F_u(z)$ having McMillan degree n. Define $S_N^M(z) \equiv \sum_{k=1}^{N} f_M(-k) z^{-k}$ with $f_M(k)$ as given above and M>N. Suppose dF/dz localized to |z|=1 is in $L_2[0,2\pi]$ then

$$\lim_{\sqrt{M} \geq N \to \infty} \|F_u - F_{u;n}^{M;N}\|_\infty = 0$$

where $F_{u;n}^{M;N}$ is an Nth-order approximant of $S_N^M(z)$.

Previously [13] reported on this FFT based method for approximating the stable possibly infinite dimensional part of the model, namely $T_s(s)$, giving uniform norm convergence as in the above theorem. They also gave some associated error bounds in terms of Hankel singular values, which are useful in terms of feedback stabilization [4].

This FFT and singular value decomposition (SVD) theory with further truncation by balancing is the basis for an algorithm for taking frequency response data to a good low order approximant.

The algorithm for finding the approximant $\tilde{T}_s(s)$ involves finding $\tilde{T}_u(s)$ and $\tilde{T}_s(s)$, the approximants for the unstable and stable parts respectively. The steps of the algorithm are as follows:

<u>Step 1</u> For given frequency response data $T(j\omega)$ verify if $(\lambda - j\omega)\dfrac{dT(j\omega)}{dj\omega} \in L_2(-\infty,\infty)$ and choose $\lambda>0$ to find frequency response data on unit circle i.e. $F(z) \equiv T(\lambda\dfrac{1 - z}{1 + z})$;

<u>Step 2</u> Use 2M point IFFT to compute $f_M(k)$ as defined above. Select truncation numbers $N_1$ and $N_2$ for stable and unstable Fourier-Laguerre series using plots of $f_M(k)$ versus k;

<u>Step 3</u> Compute Hankel singular values of $F_{N_1} \equiv \sum_{k=0}^{N_1} f_k z^k$ and $S_{N_2}^M \equiv \sum_{k=1}^{N_2} f_M(-k)\, z^{-k}$ where $N_1^2 \leq M$ and $N_2^2 \leq M$. Determine $n_1$, the number of stable poles of T(s) or the error involved in selecting a certain number and determine $n_2$, the number of unstable poles of T(s);

<u>Step 4</u> Use balanced truncation and bilinear tranformation to obtain stable part approximation of order $n_1$ and unstable part approximation of order $n_2$; and

<u>Step 5</u> Add stable part approximation $\tilde{T}_s(s)$ to unstable part approximation $\tilde{T}_u(s)$ to get total approximation $\tilde{T}_s(s)$.

In the next section we exploit this algorithm and its realization in terms of software based on MATLAB, to look at structural properties of delay-differential systems. Especially we consider how to find the finite number of poles to the right of various vertical lines in the complex plane. Using the frequency response data we can find characteristic roots near the imaginary axis $s=j\omega$, $-\infty<\omega<\infty$, for example.

To conclude this section we offer one example of model reduction (approximation) by the above algorithm which shows details of each step. The model is one obtained by Fiagbedzi and

Pearson [25], see also [14], for control of recirculation combustion in a monopropellant rocket engine - their transfer function model is

$$T(s) = \frac{s^3 + (1 + e^{-s})s^2 + (1 + 2e^{-s})s + e^{-s}}{s^4 + (1 + e^{-s})s^3 + 2(1 + e^{-s})s^2 + (1 + 2e^{-s})s + 2e^{-s}} \tag{18}$$

A graph of $T(j\omega)$, $0.1 \leq \omega \leq 100$ is shown as figure 5; there the frequency response data is given in polar coordinates, as magnitude and phase. It is quite clear from the magnitude plot that there are two poles near $j\Re^+$ at about $\omega$=.92 and $\omega$=1.5 radians per second.

Using above algorithm it is easy to verify the conditions of step 1 and to find the bandwidth of the data to be about 0.4. We typically use one to ten times the bandwidth as the parameter $\lambda$ of the bilinear transformation. This trade-off question is discussed in the paper of Cai and Lee [26].

Selecting $\lambda$=4=(10)(0.4) we can then use a 2M point FFT algorithm to compute $f_M(k)$ of step 2. By plotting the numbers $f_M(k)$ versus k we can easily determine how many Fourier-Laguerre terms to include for the stable, ($N_1$) terms, and unstable parts, ($N_2$) terms. A 1024 point FFT algorithm was used to compute $f_M(k)$ . The numbers $f_M(k)$ are then plotted for $0 \leq k \leq 200$ and $825 \leq k \leq 1024$, see figures 6 and 7; this allows one to select how many terms to retain in the Fourier-Laguerre series expansion for the stable part (figure 6) and unstable part (figure 7). Note because of periodicity of the FFT the unstable terms $f_M(-1) = f_M(1024)$, $f_M(-2) = f_M(1023)$ and so on. In this case after about 75 terms in figure 6 the data is nearly zero; so select $N_1$=75 to be number of terms in approximation of stable part. And just before the 950th term in figure 7 the data is approximately zero; so select $N_2$=75 to be number of terms to use in approximating the unstable part. This completes step 2 of the algorithm.

For step 3 compute Hankel singular values as an aid to find how far to reduce the Fourier-Laguerre expansion found in step 2. A plot of the stable case Hankel singular values versus k is given as figure 8; from this we see big gap between 3 and 4 so we could select $n_1$=3, but we select $n_1$=6 (just for comparison) as order of reduced stable model to be found by balanced truncation algorithm. A plot of unstable part Hankel singular values versus k (figure 9) shows big gap between k=2 and k=3; suggesting that there are two unstable poles and a reduced unstable model of order $n_2$=2 will be obtained by balanced truncation. In section IV we show how the algorithm can be further exploited to find more conclusively the unstable poles. This completes step 3 of the algorithm. In step 4 the balancing and bilinear transformation are applied to find the reduced stable part approximant and the unstable part approximant. Figures 10 and 11 show the magnitude and phase of the stable part approximant, while figures 12 and 13 show the magnitude and phase of the unstable part approximant. Finally in step 5 we combine the stable and unstable part approximants, getting thereby Figures 14 and 15 for the total magnitude and total phase. Figure 16 is the error comparison, difference $|T(j\omega)| - |\hat{T}(j\omega)|$. From figure 16, since the phase approximation is good, one can see that the uniform norm error is about 0.003 (about 0.1% error in terms of the peak magnitude of 2.5). In table 1 we display pole and zero locations of the approximate model. There are indeed poles near $j\Re^+$ at about .11330 + j1.52 and at -.1858 + j.9179. The pole zero cancellation possibilities mean that we could have selected $n_1$=3 without much sacrifice in quality.

In the next section IV we will show how this algorithm can be exploited to study properties of the input/output delay-differential system by considering the calculation of roots of the characteristic exponential polynomial of such systems by finding poles near $j\Re$. In particular one can more carefully investigate how many poles are in the right half plane and their locations.

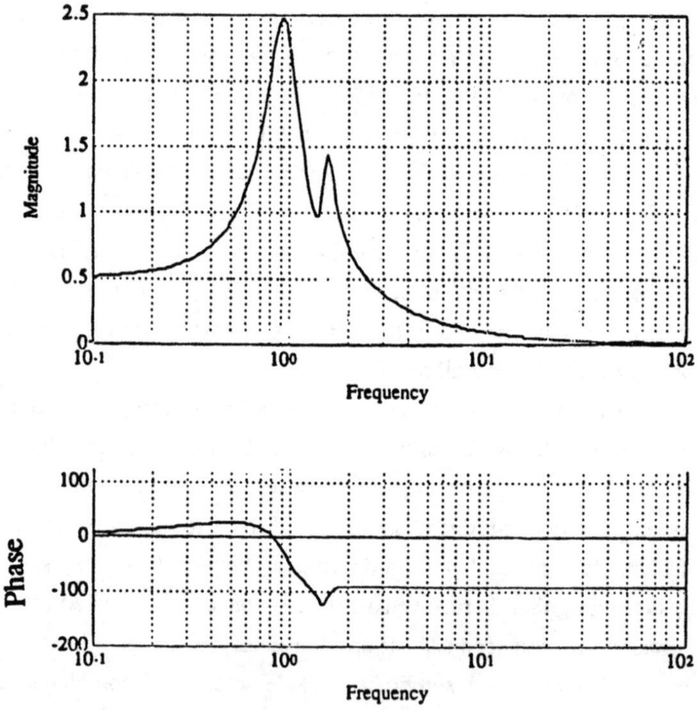


Figure 5 Frequency Response Data

Figure 6

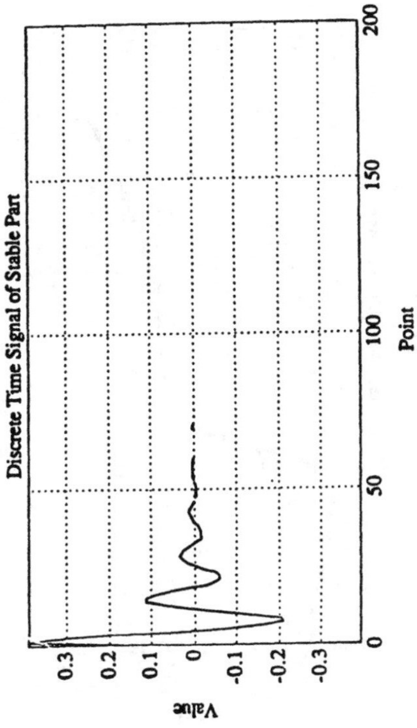


Figure 8

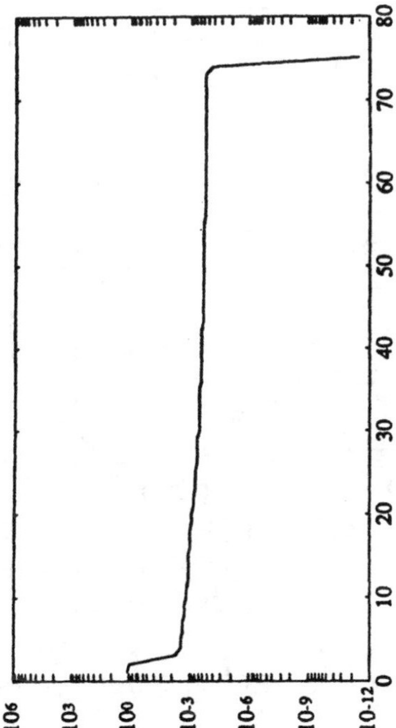


Stable

Figure 7

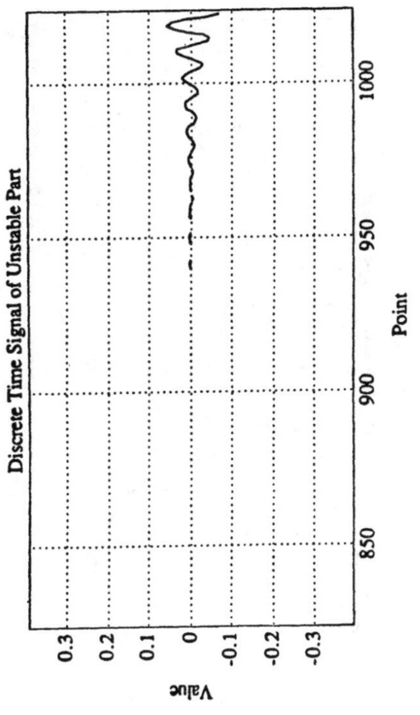


Figure 9

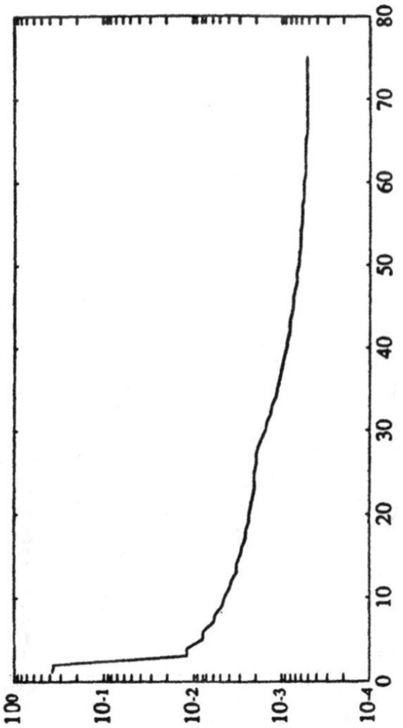


Unstable

**Hankel singular values**

Figure 10

Total magnitude and stable part magnitude

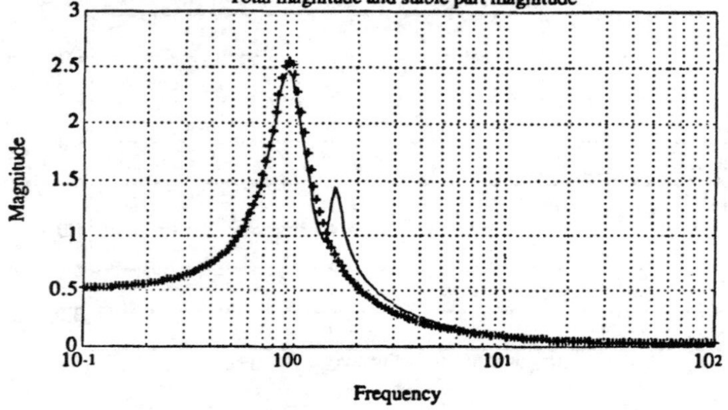


Figure 11

Total phase and stable part phase

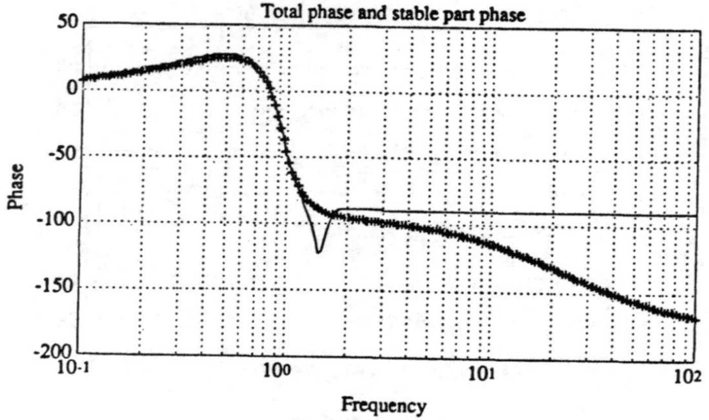


Figure 12

Total magnitude and unstable part magnitude

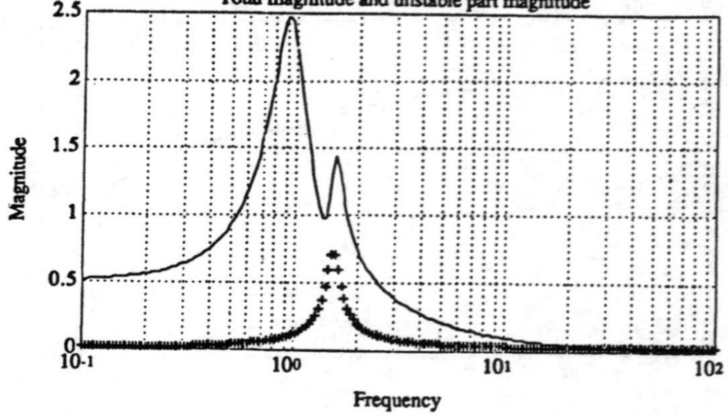


Figure 13

Total phase and unstable part phase

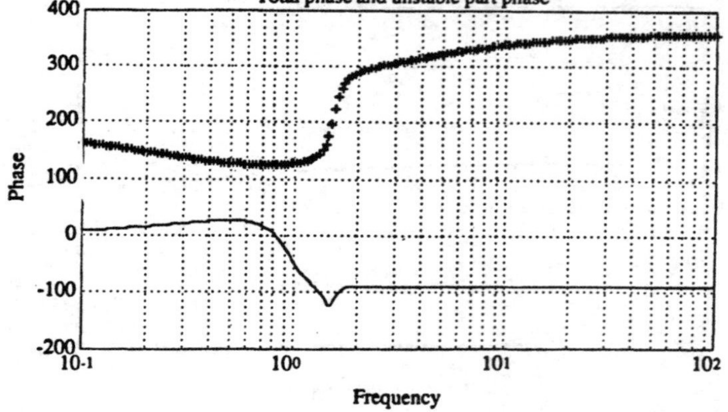

Figure 14

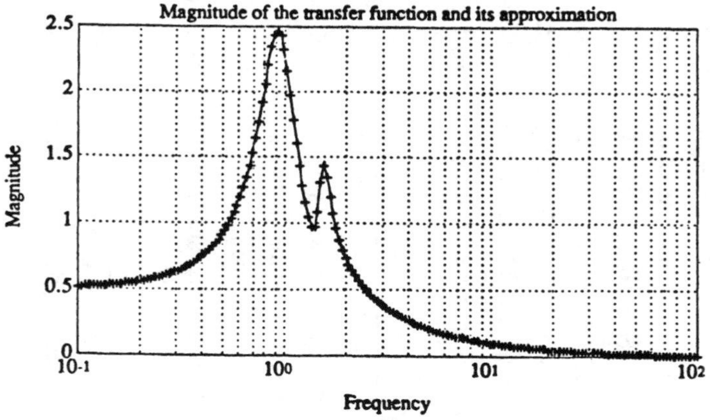


Figure 16

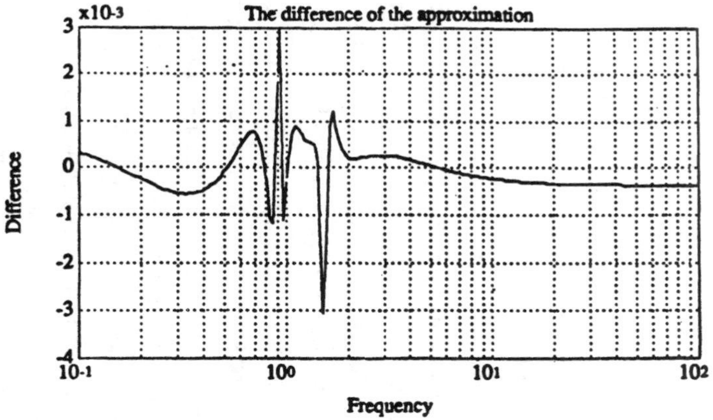


Figure 15

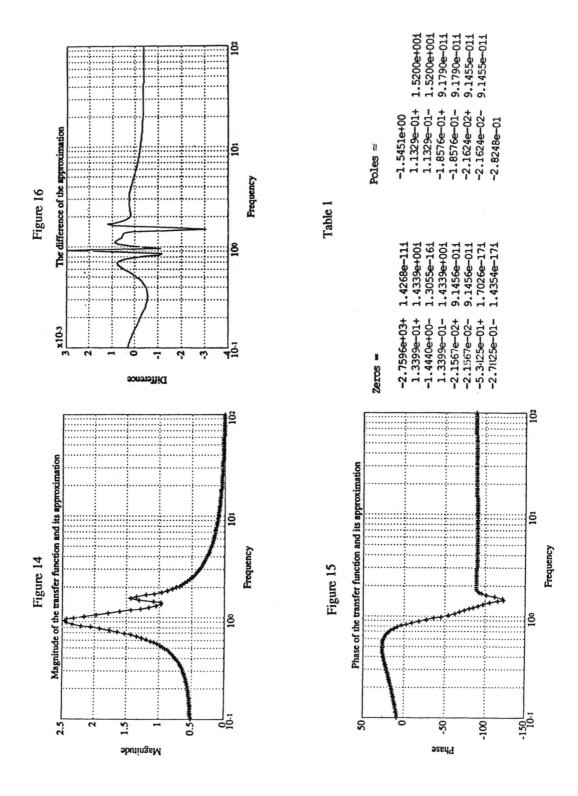

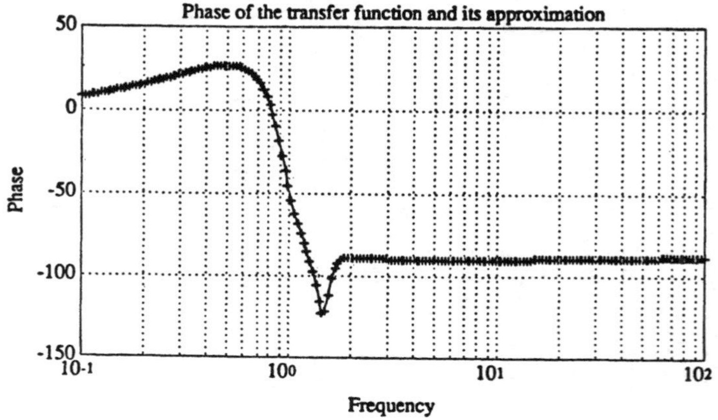


Table 1

Zeros =

| | |
|---|---|
| -2.7596e+03+ | 1.4268e-11i |
| 1.3399e-01+ | 1.4339e+00i |
| -1.4440e+00- | 1.3055e-16i |
| 1.3399e-01- | 1.4339e+00i |
| -2.1567e-02+ | 9.1456e-01i |
| -2.1567e-02- | 9.1456e-01i |
| -5.3125e-01+ | 1.7026e-17i |
| -2.7825e-01- | 1.4354e-17i |

Poles =

| | |
|---|---|
| -1.5451e+00 | |
| 1.1329e-01+ | 1.5200e+00i |
| 1.1329e-01- | 1.5200e+00i |
| -1.8576e-01+ | 9.1790e-01i |
| -1.8576e-01- | 9.1790e-01i |
| -2.1624e-02+ | 9.1455e-01i |
| -2.1624e-02- | 9.1455e-01i |
| -2.8248e-01 | |

## IV Calculating roots of exponential polynomials

In this section it is shown how to exploit the algorithm described in section III by considering the task of finding poles of the transfer function $T(s)=1/(\text{exponential polynomial})$ near $j\Re$. Thus we can extract roots of exponential polynomials which lie in the right half plane or in the left half plane near the imaginary axis. The root chains of exponential polynomials are described in the book [23]. In particular the analysis there shows that to the right of any vertical line in the complex plane one will find only a finite number of roots and roots occur in chains with bounding estimates. If one uses a Padé approximation of $e^{-s}$ or other approximates which lead to rational functions and hence to finding roots of polynomials, usually many false roots are found and some real roots may be missed. Manitius and Tran in 1985 [27] showed how a result of Kuhn on saturated triangles could be applied to effectively find the roots near $j\Re$. Brierley [28] used their algorithm to study dependency of the root locations (root chains) of the exponential polynomials on the delay duration $\tau>0$. Other results on computing dependency on delay duration can be found in paper of Chiasson [29]. Now consider a specific example considered by Brierley namely the exponential polynomial $s^3 + s^2 + 2s + e^{-\tau s} + 1$ with root chains as a function of delay duration $\tau$ shown as figure 11 of Lee and Lu [30] (figure 11 was mislabed with $s^2 + s + se^{-hs} + 1$). Using the algorithm of section III we will look for poles of the corresponding transfer function

$$T(s) = 1/(s^3 + s^2 + 2s + e^{-s} + 1)$$

near $j\Re$; we will study just the case when $\tau=1$. The algorithm uses only FFT and SVD type calculations and it is easy to extract the poles in the right half plane; in this case at approximately $.0392\pm j1.1485$. The figure 17 shows the prominance of these poles in the frequency response data. Table 2 is a tabulation of the zero and pole locations for the approximant $\tilde{T}(s)$.

The substitution $s:=s+1$ in the transfer function $T(s)$ gives rise to a new transfer function $T_1(s)$ which should have no poles in the right half plane; the above right half plane pair is shifted to poles of $T_1(s)$ (computed using the algorithm of section III) at $-0.96087\pm j1.1485$, and the discrete time signal of the unstable part shows no nonzero Fourier-Laguerre coefficients. That is, we consider

$$T_1(s) = \left[\frac{1}{s^3 + s^2 + 2s + e^{-s} + 1}\right]_{s:=s+1} = \left[\frac{1}{s^3 + 4s^2 + 7s + e^{-(s+1)} + 5}\right] \qquad (19)$$

using the algorithm of section III. The frequency response data is shown as figures 18 with pole zero locations as tabulated in Table 3 using an 11th order approximant for the stable part.

To more carefully obtain pole locations near $j\Re$ we can move $j\Re$ using the substitution procedure. We can move past the real pole at $-2.1396$ in the transfer function $T(s)$ by the substitution $s:=(s-3)$; so we study the transfer function

$$T_{-3} = \left.\frac{1}{s^3 + s^2 + 2s + e^{-s} + 1}\right|_{s:=s-3} = \frac{1}{s^3 - 8s^2 + 23s + e^{-(s-3)} - 23}\,. \qquad (20)$$

The pole-zero configuration of table 4 was calculated for $T_{-3}$ by the algorithm of section III using a seventh order approximant for the stable part and a third order approximant for unstable part ($N_1$=50, $N_2$=50).

To move one more pole to right half plane and get j$\Re$ closer to other left half plane poles use the substitution s:=(s-5); Thus,we study the transfer function

$$T_{-5} = \frac{1}{s^3 - 14s^2 + 67s + e^{-(s-5)} - 109} \tag{21}$$

Table 5 is the listing of the approximation data and pole-zero locations for $T_{-5}$ computed using algorithm of section III with $N_1$=50, $N_2$=50, $n_1$=6 and $n_2$=4.

To continue our movement of j$\Re$ we see that by the substitution s:=s-8 in T(s) we can move two more poles to R.H.P. and move several more near j$\Re$. Thus a study of

$$T_{-8} = \frac{1000}{s^3 - 23s^2 + 178s + e^{-(s-8)} - 463} \tag{22}$$

develops a pair of poles at 0.7217 ± j9.001 to be added to the unstable group for $T_{-8}$. Note we have changed the coefficient in the numerator from 1 to 1000 to get realistic numbers for $T_{-8}(j\omega)$, namely $T_{-8}(0)$ is then close to 0.5. Figure 19 shows the original frequency response data and that of the approximant. Table 6 is a listing of pole zero locations.

A study of

$$T_{-9} = \frac{10000}{s^3 - 26s^2 + 227s + e^{-(s-9)} - 665} \tag{23}$$

uncovers other poles near j$\Re$; see frequency response data of figure20 and tabulation of problem data in table 7.

We will not explore for further roots; but using the best data from each proximity of j$\Re$ the root locations covering the right half plane with Re s ≥-10 are shown in figure 21.

To ensure that roots near j$\Re$ are not missed in this frequency response based analysis (since-for example, poles far out in right half plane may not have a significant influence on the frequency response data) we can bound the domain over which to search using estimates for the terms that can occur in the exponential polynomial. This can be illustrated by considering roots in right half plane, Re s>0, of the exponential polynomial $s^2 + 2se^{-s} + 1 = 0$. First write this as $s^2 = -2se^{-s}-1$ and let $s=\alpha+j\omega$ with $\alpha\geq 0$. Consider when $|s^2| > |2se^{-s} + 1|$ to get an estimate of where roots cannot lie as a function of increasing $\omega$ for each fixed $\alpha\geq 0$.

Now $|s^2| > \omega^2$ and $|2se^{-s} + 1| \leq |2se^{-s}| + |1| \leq 1 + 2|s|e^{-\alpha} = 1 + \sqrt{4(\alpha^2 + \omega^2)e^{-2\alpha}}$ $\leq 1 + 2\sqrt{\alpha^2 e^{-2\alpha}} + 2\sqrt{\omega^2 e^{-2\alpha}} = 1 + (2\omega + 2\alpha)e^{-\alpha} \leq 1 + 2\omega + \beta$ where $\beta = \max_{0\leq\alpha} 2\alpha e^{-\alpha}$. Thus for $\omega^2 > 1 + 2\omega + \beta$ we have no roots. Or $\omega > \frac{1}{\omega} + 2 + \frac{\beta}{\omega}$, which for $\omega > 1$ means that $\omega > 1 + 2 + \beta > \frac{1}{\omega} + 1 + \frac{\beta}{\omega}$. Thus if $|\omega| > 3 + \beta$ we have no roots in Re s ≥ 0.

Next consider when $|s^2| > |-2se^{-s} - 1|$ all s with Re s $\geq$ 0 and $|s| > R$. To find R note that $|s| > |2e^{-s} + \frac{1}{s}|$ and $2|e^{-s}| + \frac{1}{|s|} \leq 3$ if $|s| > 1$; so take R=3.

Hence for this example all roots in Re s $\geq$ 0 lie in the rectangle $|\text{Im } s| = |\omega| \leq 3 + ß$, $0 \leq \alpha \leq 3$.

Using such estimates is one way to ensure that roots far out in right half plane are not over looked in the frequency response based calculations as described above.

To get such estimates is routine and in the single delay case are easily available. One standard result based on matrix norms which recently appeared is the following (based on previous work of Mori and Kokane [31] for the vector-matrix DDE model $\dot{x} = Ax(t) + Bx(t-\tau)$ where A, B $\in \Re^{n \times n}$.)

Lemma [32] All right half plane zeros of det[sI - A - Be$^{-s}$] lie in rectangle $|\omega| \leq \ell_2$, $0 \leq \sigma \leq \alpha < \ell_1$, where $\ell_1 = \mu(A) + \|B\|$ and $\ell_2 = \mu(-jA) + \|B\|$ and $\alpha \leq \ell_1$, is any positive value satisfying

$$\mu(A) - \alpha \pm \|B\| e^{-\tau\alpha} < 0.$$

Here $\|X\|$ denotes matrix norm and $\mu(X)$ is matrix measure for matrix X.

Applying this result to the example $s^3 + s^2 + 2s + e^{-s} + 1 = 0$ considered above we find that all right half plane zeros lie in rectangle $|\omega| \leq 2.2247$, $0 \leq \alpha \leq 0.96575$ where $s = \alpha + j\omega$. These estimates assure us that the only roots in right half plane for $s^3 + s^2 + 2s + e^{-s} + 1$ are at approximately: 0.0392±j1.1485, as calculated above and shown in figure 21.

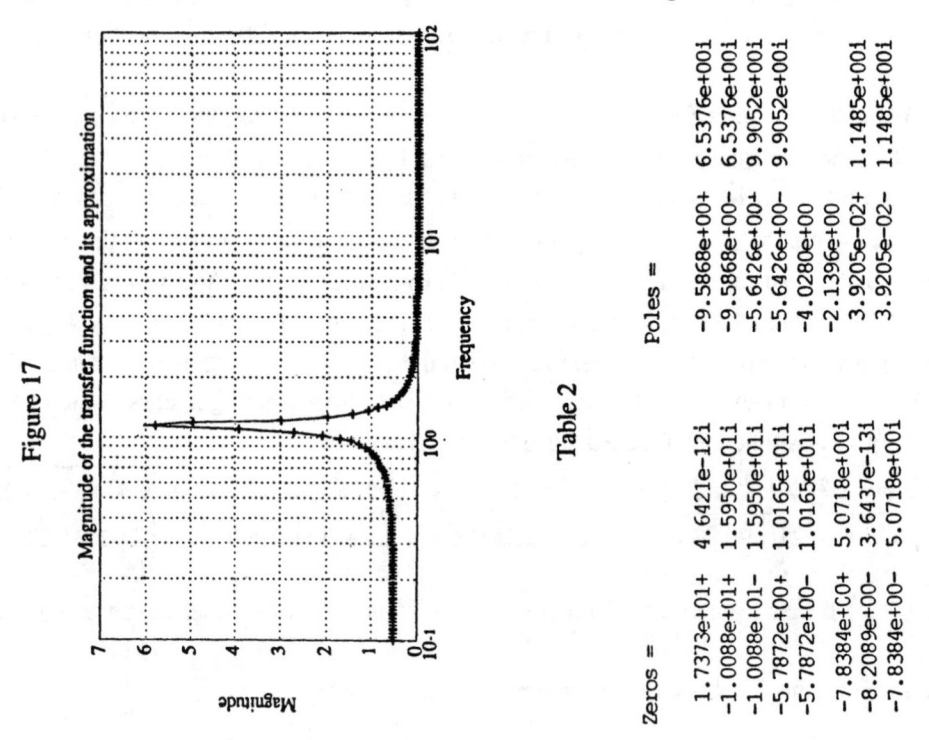

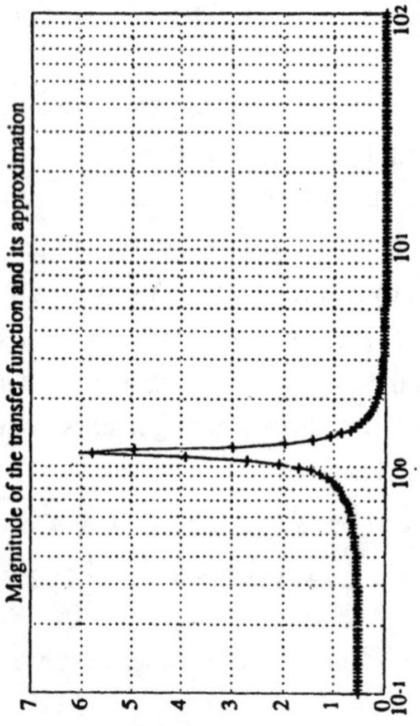


Figure 17

Table 2

| Zeros = | | Poles = | |
|---|---|---|---|
| 1.7373e+01+ | 4.6421e−12i | −9.5868e+00+ | 6.5376e+00i |
| −1.0088e+01+ | 1.5950e+01i | −9.5868e+00− | 6.5376e+00i |
| −1.0088e+01− | 1.5950e+01i | −5.6426e+00+ | 9.9052e+00i |
| −5.7872e+00+ | 1.0165e+01i | −5.6426e+00− | 9.9052e+00i |
| −5.7872e+00− | 1.0165e+01i | −4.0280e+00 | |
| −7.8384e+C0+ | 5.0718e+00i | −2.1396e+00 | |
| −8.2089e+00− | 3.6437e−13i | 3.9205e−02+ | 1.1485e+00i |
| −7.8384e+00− | 5.0718e+00i | 3.9205e−02− | 1.1485e+00i |

Figure 18

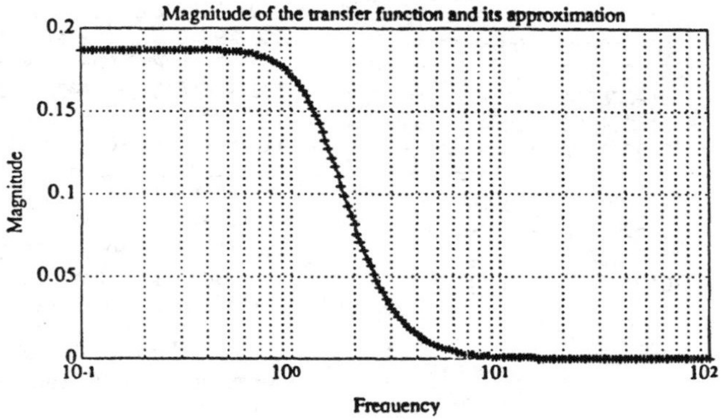


Figure 19

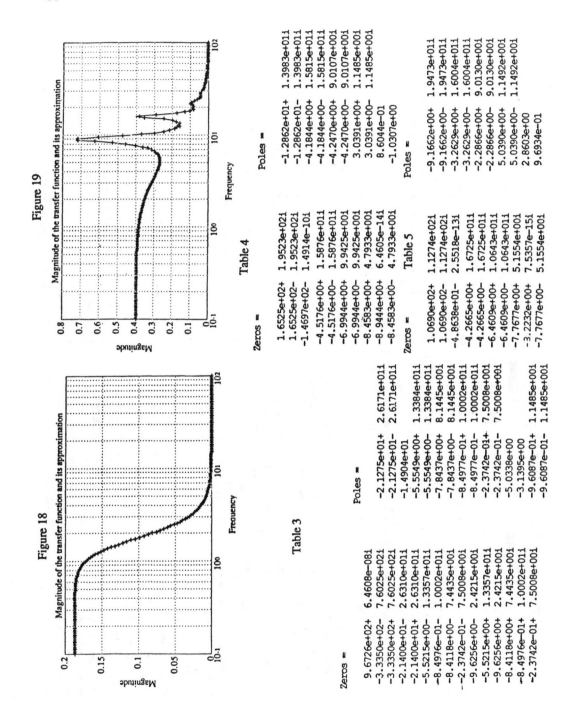

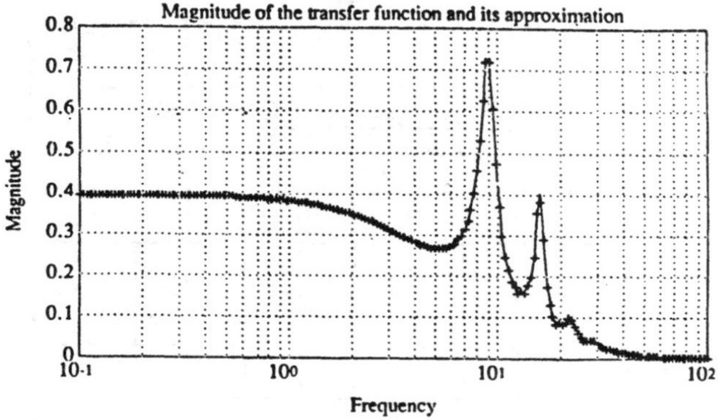


Table 3

| Zeros = | | Poles = | |
|---|---|---|---|
| 9.6726e+02+ | 6.4608e-08i | -2.1275e+01+ | 2.6171e+01i |
| -3.3350e+02- | 7.6025e+02i | -2.1275e+01- | 2.6171e+01i |
| -3.3350e+02+ | 7.6025e+02i | -1.4904e+01 | |
| -2.1400e+01- | 2.6310e+01i | -5.5549e+00+ | 1.3384e+01i |
| -2.1400e+01+ | 2.6310e+01i | -5.5549e+00- | 1.3384e+01i |
| -5.5215e+00- | 1.3357e+01i | -7.8437e+00+ | 8.1445e+00i |
| -8.4976e-01- | 1.0002e+01i | -7.8437e+00- | 8.1445e+00i |
| -8.4118e+00- | 7.4435e+00i | -8.4977e-01+ | 1.0002e+01i |
| -2.3742e-01- | 7.5008e+00i | -8.4977e-01- | 1.0002e+01i |
| -9.6256e+00- | 2.4215e+00i | -2.3742e-01+ | 7.5008e+00i |
| -5.5215e+00+ | 1.3357e+01i | -2.3742e-01- | 7.5008e+00i |
| -9.6256e+00+ | 2.4215e+00i | -5.0338e+00 | |
| -8.4118e+00+ | 7.4435e+00i | -3.1395e+00 | |
| -8.4976e-01+ | 1.0002e+01i | -9.6087e-01+ | 1.1485e+00i |
| -2.3742e-01+ | 7.5008e+00i | -9.6087e-01- | 1.1485e+00i |

Table 4

| Zeros = | | Poles = | |
|---|---|---|---|
| 1.6525e+02+ | 1.9523e+02i | -1.2862e+01+ | 1.3983e+01i |
| 1.6525e+02- | 1.9523e+02i | -1.2862e+01- | 1.3983e+01i |
| -1.4697e+02- | 1.4914e-10i | -4.1844e+00+ | 1.5815e+01i |
| -4.5176e+00+ | 1.5876e+01i | -4.1844e+00- | 1.5815e+01i |
| -4.5176e+00- | 1.5876e+01i | -4.2470e+00+ | 9.0107e+00i |
| -6.9944e+00+ | 9.9425e+00i | -4.2470e+00- | 9.0107e+00i |
| -6.9944e+00- | 9.9425e+00i | 3.0391e+00+ | 1.1485e+00i |
| -8.4583e+00+ | 4.7933e+00i | 3.0391e+00- | 1.1485e+00i |
| -8.9444e+00+ | 6.4605e-14i | 8.6044e-01 | |
| -8.4583e+00- | 4.7933e+00i | -1.0307e+00 | |

Table 5

| Zeros = | | Poles = | |
|---|---|---|---|
| 1.0690e+02+ | 1.1274e+02i | -9.1662e+00+ | 1.9473e+01i |
| 1.0690e+02- | 1.1274e+02i | -9.1662e+00- | 1.9473e+01i |
| -4.8638e+01- | 2.5518e-13i | -3.2629e+00+ | 1.6004e+01i |
| -4.2665e+00+ | 1.6725e+01i | -3.2629e+00- | 1.6004e+01i |
| -4.2665e+00- | 1.6725e+01i | -2.2866e+00+ | 9.0130e+00i |
| -6.4609e+00+ | 1.0643e+01i | -2.2866e+00- | 9.0130e+00i |
| -6.4609e+00- | 1.0643e+01i | 5.0390e+00+ | 1.1492e+00i |
| -7.7677e+00+ | 5.1554e+00i | 5.0390e+00- | 1.1492e+00i |
| -3.2232e+00+ | 7.5357e-15i | 2.8603e+00 | |
| -7.7677e+00- | 5.1554e+00i | 9.6934e-01 | |

Figure 20

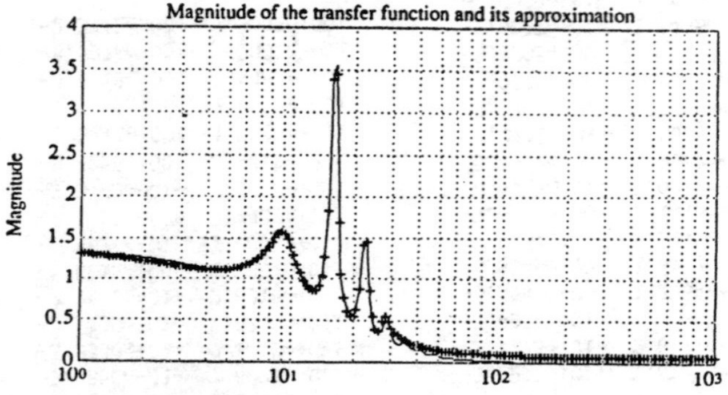


Table 6

| Zeros = | | Poles = | |
|---|---|---|---|
| 8.7406e+01+ | 7.4089e+01i | -1.6911e+00+ | 2.1435e+01i |
| 8.7406e+01- | 7.4089e+01i | -1.6911e+00- | 2.1435e+01i |
| -2.1736e+00+ | 1.9899e+01i | -6.9290e-01+ | 1.5808e+01i |
| -2.1736e+00- | 1.9899e+01i | -6.9290e-01- | 1.5808e+01i |
| -3.9681e+00+ | 1.3839e+01i | 7.0641e-01+ | 9.0098e+00i |
| -3.9681e+00- | 1.3839e+01i | 7.0641e-01- | 9.0098e+00i |
| -4.9386e+00+ | 8.2928e+00i | 8.1924e+00+ | 1.7771e+00i |
| -4.9386e+00- | 8.2928e+00i | 8.1924e+00- | 1.7771e+00i |
| -5.2553e+00+ | 2.7663e+00i | 5.3396e+00 | |
| -5.2553e+00- | 2.7663e+00i | 4.0649e+00 | |

Table 7

| Zeros = | | Poles = | |
|---|---|---|---|
| -6.6743e+01- | 1.6921e-13i | -1.4103e+00+ | 2.7232e+01i |
| 2.9396e+01+ | 8.2728e-15i | -1.4103e+00- | 2.7232e+01i |
| -1.3281e+00+ | 2.6104e+01i | -7.0556e-01+ | 2.2366e+01i |
| -1.3281e+00- | 2.6104e+01i | -7.0556e-01- | 2.2366e+01i |
| -2.8582e+00+ | 2.0191e+01i | 1.8193e+01 | |
| -2.8582e+00- | 2.0191e+01i | 3.6287e-01+ | 1.5813e+01i |
| -3.9347e+00+ | 1.4166e+01i | 3.6287e-01- | 1.5813e+01i |
| -3.9347e+00- | 1.4166e+01i | 1.7077e+00+ | 9.1703e+00i |
| -4.6100e+00+ | 8.4791e+00i | 1.7077e+00- | 9.1703e+00i |
| -4.6100e+00- | 8.4791e+00i | 4.5219e+00+ | 4.8952e+00i |
| -4.8614e+00+ | 2.8426e+00i | 4.5219e+00- | 4.8952e+00i |
| -4.8614e+00- | 2.8426e+00i | 2.9316e+00 | |

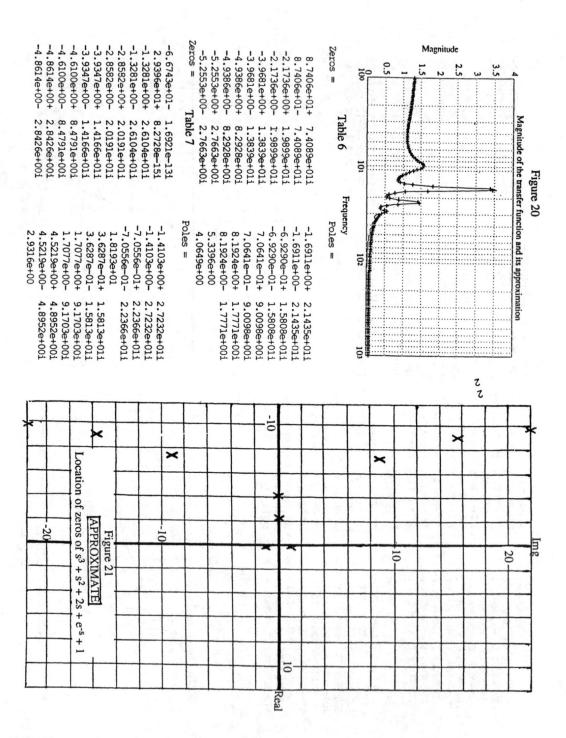

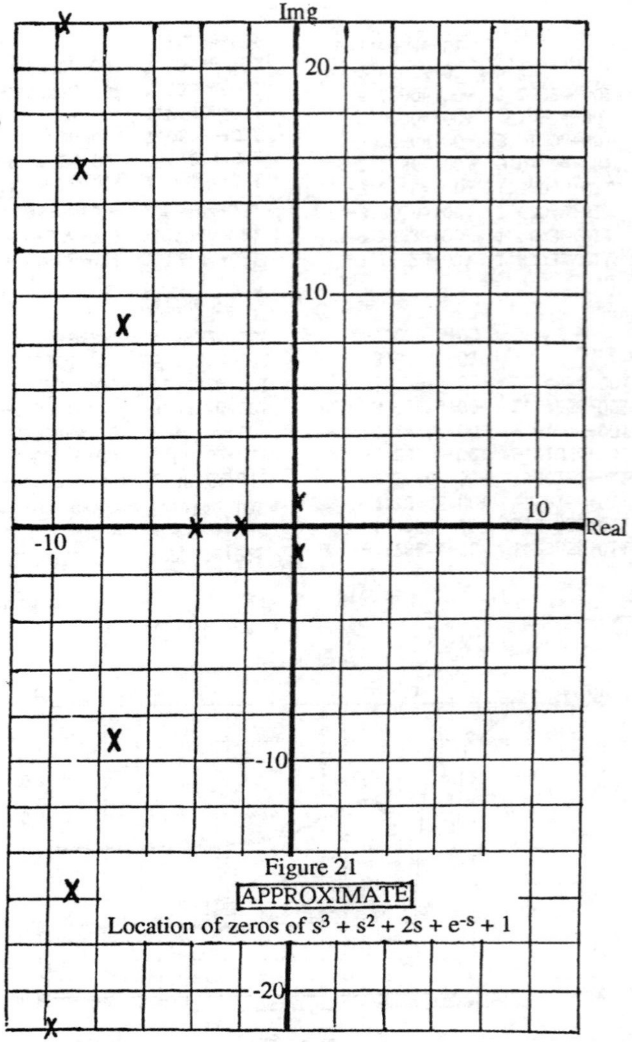


Figure 21

APPROXIMATE

Location of zeros of $s^3 + s^2 + 2s + e^{-s} + 1$

References


1. Alekal, Y., P. Brunovsky, D. H. Chyung and E. B. Lee, "The quadratic problem for systems with time delays," IEEE trans. on Automat. Contr. AC-16, (1971) 673-687.
2. Lee, E. B. and N. Eva Wu, "Sensitivity and optimal synthesis for a class of linear time-delay systems," In Control Prob. for Systems Described by PDE's (I. Lasiecka and R. Triggianni editors) Springer-Verlag, (1987) 282-298.
3. Alford, R. L. and E. B. Lee, "Sampled data hereditary systems: linear quadratic theory," IEEE Trans. on Automat. Contr., 31, (1986) 60-65.
4. Curtain, R. F. and K. Glover, "Robust stabilization of infinite dimensional systems by finite dimensional controllers," Syst. Contr. Lett. 7, (1986) 41-47.
5. Partington, J. R., K. Glover, H. J. Zwart and R. F. Curtain, "$L_\infty$ approximation and nuclearity of delay systems," Syst. Contr. Lett. 10, (1988) 59-65.
6. Partington, J. R., "Approximation of delay system by Fourier-Laguerre series," Automatica 27, (1991), 569-572.
7. Glader, C., G. Hognas, P. M. Makila and H.T. Toivonen, "Approximation of delay systems - a case study," Int. J. Contr. 53, (1991) 369-390.
8. Wahlberg, B., "System identification using Laguerre models," IEEE Trans. on Automat. Contr. 36, (1991) 551-562.
9. Makila, P. M., "Approximation of stable systems by Laguerre filters," Automatica 26, (1990) 333-345.
10. Moore, B. C., "Principal component analysis in linear systems: controllability, observability, and model reduction," IEEE Trans. on Automat. Contr. AC-26, (1981) 17-32.
11. Lu, W.-S. , E. B. Lee and Q.-T. Zhang, "Balanced approximation of two-dimensional and delay-differential systems," Int. J. Contr. 46, (1987) 2199-2218.
12. Premaratne, K., E. I. Jury and M. Mansour, "Model reduction for 2-D systems," IEEE Trans. on Circuits and systems, CAS-35, (1988) 100-115.
13. Gu, G., P. P. Khargonekar and E. B. Lee, "Approximation of infinite dimensional systems," IEEE Trans. on Automat. Contr., 34, (1989) 610-618.
14. Gu, G., P. P. Khargonekar, E. B. Lee, and P. Misra, "Finite-dimensional approximations of unstable infinite dimensional systems," SIAM J. Contr. and Opt., 30, (1992) 704-716.
15. Wu, N. Eva and E. B. Lee, "Feedback minimax synthesis for distributed systems," Proc. IEEE-CDC (1988) 492-496.
16. Partington, J. R., "Approximation of unstable infinite-dimensional systems using coprime factors," Syst. Contr. Lett., 16, (1991) 89-96.

17. Kamen, E. W., P. P. Khargonekar and A. Tannenbaum, "Stabilization of time-delay systems with finite-dimensional compensators," IEEE Trans. on Automat. Contr., 30 (1985) 75-78.

18. Naylor, A. W., and G. R. Sell, Linear Operator Theory in Engineering and Science, Holt, Rinehart and Winston, New York, 1971.

19. Sontag, E. D., "On first-order equations for multidimensional filters," IEEE Trans. Acoust., Speech, and Signal Processing, ASSP-26, (1978) 480-482.

20. Zak, S. H., E. B. Lee and W.-S. Lu, "Realizations of 2-D filters and time delay systems," IEEE Trans. on Circuits and Systems, CAS-33, (1986), 1241-1244.

21. Laub, A. J., M. J. Heath, C. C. Paige and R. C. Ward, "Computation of system balancing transformations" Proc. IEEE CDC, (1986) 548-553.

22. Driver, R. D., Ordinary and delay-differential equations, Springer-Verlag, New York, 1977.

23. Bellman, R. and K. L. Cooke, Differential-difference equations, Acad. Press, NY, 1963.

24. Wu, N. Eva, "A factorization approach to control synthesis of distributed linear systems", Ph.D. Thesis, Center for Control Science and Dynamical Systems, Univ. of Minn., 1987.

25. Fiagbedzi, Y. A., and A. E. Pearson, "Feedback stabilization of linear autonomous time lag systems," IEEE Trans. on Automat. Contr. AC-31, (1986) 847-854.

26. Cai, Manping and E. B. Lee, "Identification of linear systems using rational approximation techniques," In Stochastic Theory and Adaptive Control, (Edited by B. Pasik-Duncan and T. Duncan) Springer Verlag, Lecture notes in Control and Info. Sci, 186, New York (1993).

27. Manitius, A. and H. Tran, "Computation of closed-loop eigenvalues associated with the optimal regulator problem for functional differential equations," IEEE Trans. on Automat. Contr., AC-30, (1985) 1245-1248.

28. Brierley, S., "Stability and stabilization of generalized linear systems," Ph.D. Thesis, Center for Control Science and Dynamical Systems, Univ. of Minnesota, 1984.

29. Chiasson, J. N., "A method for computing the interval of delay values for which a differential-delay system is stable," IEEE Trans. on Automat. Contr., 33, (1988) 1176-1178.

30. Lee, E. B. and W-S Lu, "Feedback with delays: stabilization of linear time delay and two dimensional systems," In Signal Processing Part II, IMA Volume 23, Springer-Verlag, New York (1990) 155-195.

31. Mori, T. and H. Kokame, "Stability of $\dot{x}(t) = Ax(t) + Bx(t-\tau)$," IEEE Trans. on Automat. Contr., 34, (1989) 460-462.

32. Wang, S-S, "Further results on stability of $\dot{x}(t) = Ax(t) + Bx(t-\tau)$," Syst. and Contr. Lett., 19, (1992) 165-168.

# 43 Local Smoothing and Energy Decay for a Semi-Infinite Beam Pinned at Several Points, and Applications to Boundary Control

**Walter Littman** University of Minnesota, Minneapolis, Minnesota

**Stephen W. Taylor** Montana State University, Bozeman, Montana

ABSTRACT. The dynamics of a semi-infinite clamped Euler-Bernoulli beam, pinned at several points are investigated. Exact controllability for a finite such beam is derived, with Neumann controls and zero deflection of the non-clamped end.

## 0. Introduction.

In earlier papers [LM] [T], the authors have studied open loop boundary controllability for the "SCOLE Model", i.e., an Euler-Bernoulli beam, clamped at the left end, with a rigid body attached at the right end, subject to force and torque controls implemented at the right end. It was shown that the motion could be steered to zero in times of arbitrarily short duration via these controls. The controls turned out to be infinitely differentiable for $t > 0$. The method follows as a corollary to the careful study of the dynamics of the semi-infinite beam clamped at the left end.

In a related paper [LiTa] (see also [Lit]), a method briefly described as "local smoothing + reversibility + uniqueness $\Rightarrow$ controllability" is introduced to study boundary controllability for the Schrödinger equation with a nonsmooth potential.

The most popular competing method, that of a priori inequalities derived via "multipliers" (see, for example, [Lions 1,2], [LaTr1] (and the bibliography in [LaTr2]), [Z], has the advantage that, for the Euler-Bernoulli equation only one control is necessary. The original purpose of this paper was to show that our methods, suitably modified, could accomplish the same end. Thus a semi-infinite beam, clamped at $x = 0$, and pinned at $x = 1$, say, could serve as the first step in the controllability proof, as did the non-pinned semi-infinite beam on some earlier papers.

As the work progressed it became clear that a beam pinned at several points was of sufficient interest in its own right, and that is the spirit of this paper.

In a separate paper by the second author, the methods of this paper are used to show that indeed in the original SCOLE model boundary control can be achieved by controlling <u>either</u> the force <u>or</u> the torque at the non-clamped end, the other being zero.

Indeed the methods introduced here work with much more general boundary operators, and show promise of applicability in several space dimensions, for example plates.

## 1. The Dynamics of a Semi-Infinite Beam that is pinned at several points along its axis and clamped at its end.

### Description of the Physical Problem.

We consider a semi-infinite Euler-Bernoulli beam that is clamped at $x = 0$ and pinned (i.e. supported) at a finite number of locations $x = b_1$, $x = b_2, \ldots, x = b_n$ along its length. Let $w(x,t)$ denote the deflection of the beam from the non-negative $x$-axis at position $x$ and time $t$. Then, for $x \neq 0$, $b_1$, $b_2, \ldots, b_n$, $w(x,t)$ should satisfy the Euler-Bernoulli beam equation

$$\rho\frac{\partial^2 w}{\partial t^2} + EI\frac{\partial^4 w}{\partial x^4} = 0 \tag{1}$$

The "clamped" condition at $x = 0$ implies $w(0,t) = \frac{\partial w}{\partial x}(0,t) = 0$. The "pinned" conditions at $x = b_k$ imply $w(b_k,t) = 0$. Further, $\frac{\partial w}{\partial x}$ should be continuous at $x = b_k$. $\frac{\partial^2 w}{\partial x^2}$ should also be continuous at $x = b_k$ because the pin does not impose an external torque. However, the shearing force along the beam is proportional to $\frac{\partial^3 w}{\partial x^3}$ and because the pin at $x = b_k$ can impose an external shearing force at that point, we should expect a jump in $\frac{\partial^3 w}{\partial x^3}$ at $x = b_k$.

From physical reasoning, the energy $E(t)$ of the beam is given by

$$E(t) = \frac{1}{2}\int_0^\infty \rho\left(\frac{\partial w}{\partial t}\right)^2 + EI\left(\frac{\partial^2 w}{\partial x^2}\right)^2 dx. \tag{2}$$

A formal calculation shows that the energy is constant if $w$ satisfies the PDE (1) in each interval $(0,b_1)$, $(b_1,b_2),\ldots,(b_n,\infty)$ and the clamped and pinned conditions discussed above. In what follows this will be made rigorous. We will write Eq. (1) as an abstract O.D.E. in a Hilbert space whose norm is essentially given by Eq. (2). We will see that the solution operator is actually a strongly continuous unitary group on the Hilbert space (this accounts for the conservation of energy). In what follows, we assume that the variables have been scaled so that $\rho = EI = 1$.

### The Finite Energy Space and Abstract ODE.

**Definitions.** Let $\mathcal{C} = \{\phi \in C^\infty[0,\infty) : \phi''$ has compact support, $\phi(0) = \phi'(0) = 0$, $\phi(b_k) = 0$ for $k = 1,2,3,\ldots,n\}$. Let $H_2$ be the completion of $\mathcal{C}$ with the norm

$$\|\phi\|_2 = \int_0^\infty |\phi''(x)|^2 dx.$$

It is easy to see that $H_2$ is a Hilbert space with inner product

$$(\phi,\psi)_2 = \int_0^\infty \phi''(x)\bar{\psi}''(x)dx.$$

We define the finite energy space $\mathcal{H} = H_2 \times L^2$, where $L^2 = L^2[0,\infty)$. Clearly $\mathcal{H}$ is a Hilbert space with inner product

$$\left(\begin{bmatrix} w_1 \\ v_1 \end{bmatrix}, \begin{bmatrix} w_2 \\ v_2 \end{bmatrix}\right)_{\mathcal{H}} = \int_0^\infty w_1''(x)\bar{w}_2''(x) + v_1(x)\bar{v}_2(x)dx.$$

*Remark.* In previous considerations of infinite beam problems (see [LM], [T]) we have imposed the conditions $w(x,t) \to 0$, $v(x,t) \to 0$ as $x \to \infty$. Here, we essentially impose the conditions $v(x,t) \to 0$, $w(x,t) \to$ linear function of $x$ as $x \to \infty$. We will see later that the condition $w \to 0$ will be too restrictive for our applications to boundary control theory.

The previous considerations yield the following abstract o.d.e. system on $\mathcal{H}$:

$$\begin{aligned} \frac{\partial w}{\partial t} &= v \\ \frac{\partial v}{\partial t} &= -\frac{\partial^4 w}{\partial x^4} \end{aligned} \qquad \text{i.e. } \frac{d}{dt}\begin{pmatrix} w \\ v \end{pmatrix} = A\begin{pmatrix} w \\ v \end{pmatrix}$$

where $A = \begin{pmatrix} 0 & 1 \\ -\frac{\partial^4}{\partial x^4} & 0 \end{pmatrix}$. The domain of $A$ requires special attention. The $\frac{\partial^4}{\partial x^4}$ operator of the second component of $A\begin{pmatrix} w \\ v \end{pmatrix}$ should be understood to be acting only for $x$ in each of the intervals $(0,b_1)$, $(b_1,b_2), \ldots, (b_{n-1},b_n)$, $(b_n,\infty)$. The requirement that $A\begin{pmatrix} w \\ v \end{pmatrix}$ be in $\mathcal{H}$ leads to the definition of the domain of $A$, $\mathcal{D}(A)$.

**Definitions.** Let $M_0 = (0,b_1)$, $M_1 = (b_1,b_2), \ldots, M_{n-1} = (b_{n-1},b_n)$, $M_n = (b_n,\infty)$, $M = \bigcup_{k=0}^n M_k$. Then

$$\mathcal{D}(A) = (H_2 \cap \{f : f^{(iv)}\,|_{M_k} \in L^2(M_k), k = 0,1,2,\ldots,n\}) \times (H_2 \cap L^2)$$

It is easy to verify that $A$ is closed and densely defined. Further, one can show in the usual way that $iA$ is self-adjoint. Thus, by Stone's Theorem [S], we obtain the result.

**Theorem 1.1.** *$A$ is the infinitesimal generator of a strongly continuous group of unitary operators, $U(t)$ on the finite energy space $\mathcal{H}$.*

**Local Smoothing and Energy Decay Properties of $U(t)$.**

$U(t)$ is a unitary operator so one would not expect any smoothing or energy decay properties of $U(t)$. However, we see in this section that locally (in the space variable) $U(t)$ does have such properties. These are the properties that we will make use of in our control problems. Our results depend on the following Lemmas.

**Lemma 1.2.** *The point spectrum of $A$ is empty. i.e. $A$ has no eigenvalues.*

*Proof.* Suppose that $\lambda$ is an eigenvalue and that $A\binom{w}{v} = \lambda\binom{w}{v}$. From the definition of $A$, this implies that

$$v = \lambda w \quad \text{and} \quad w^{(iv)} + \lambda^2 w = 0 \quad \text{in} \quad M$$

Further, $w$ satisfies the conditions

$$w(0) = w'(0) = w(b_1) = w(b_2) = \cdots = w(b_n) = 0$$
$$w',\ w'' \text{ continuous at } b_1, b_2, \ldots b_n$$
$$w'',\ w^{(iv)} \in L^2(M_n)$$

Suppose $\lambda = 0$ is an eigenvalue. Then since $w'' \in L^2(M_n)$, we need $w$ linear for $x \geq b_n$. Hence $w''$ and $w'''$ vanish for $x \geq b_n$. Integrating by parts using the conditions at the pins and the clamped end condition gives

$$0 = \int \bar{w}w^{(iv)}dx = \int |w''|^2 dx.$$

Hence $w'' \equiv 0$. Thus $w$ is linear in each interval $M_k$. But, since $w(0) = w'(0) = 0$, $w$ vanishes in $M_0$. Also, since $w(b_k) = 0$ for each $k$, it follows that $w$ vanishes identically. Thus, 0 is not an eigenvalue.

Since $iA$ is self-adjoint, the spectrum of $A$ is purely imaginary. So if $\lambda \neq 0$ we can write

$$w^{(iv)} - \mu^4 w = 0$$

where $\mu > 0$ and $\lambda = \pm i\mu^2$. Since $\lambda w = v \in L^2$, it follows that $w = ce^{-\mu x}$ for $x \geq b_n$. But then $w(b_n) = 0 \Rightarrow c = 0$ so $w = 0$ for $x \geq b_n$. Since $w(b_n) = w'(b_n) = w''(b_n) = 0$, we obtain $w(x) = B(\sinh h[\mu(b_n - x)] - \sin[\mu(b_n - x)])$ for $b_{n-1} \leq x \leq b_n$. But the function sinh − sin is monotone increasing, so the only way to get $w(b_{n-1}) = 0$ is to have $B = 0$. Continuing in this manner, we see that $w \equiv 0$. Thus $A$ has no eigenvalues. □

**Definition.** Let $\mu \geq 0$. We say that $w$ satisfies the $\mu$-radiation condition if there exists $R \geq b_n$ such that for $x \geq R$

$$Pw(x) = w''(x) + (1+i)\mu w'(x) + i\mu^2 w(x) = 0$$

(equivalently, $w$ is a linear combination of $e^{-\mu x}$, $e^{-i\mu x}$ for $x \geq R$ if $\mu > 0$, or $w$ is linear if $\mu = 0$).

**Lemma 1.3.** *Let $w$ satisfy the $\mu$-radiation condition and*

(i) $w^{(iv)} - \mu^4 w = 0$ *in each interval* $M_k$

(ii) $w(0) = w'(0) = w(b_1) = w(b_2) = \cdots = w(b_n) = 0$

(iii) $w,\ w',\ w''$ *continuous at each* $x = b_k$.

*Then* $w = 0$.

*Proof.* If $\mu = 0$ then the $\mu$-radiation condition implies that $w$ is linear for $x \geq R$. Hence, if $w$ does not vanish, it is easily seen that $(w, 0)$ is an eigenfunction of $A$ with eigenvalue 0. But, by Lemma 1.2, this is impossible.

Suppose now that $\mu > 0$. The functions $w$ and $\bar{w}$ both satisfy the equation (i). Hence, in each interval $M_k$ we have

$$0 = w^{(iv)}\bar{w} - \bar{w}^{(iv)}w = \frac{d}{dx}(w'''\bar{w} - w\bar{w}''' - w''\bar{w}' + w'\bar{w}'') \tag{3}$$

Let $x \geq R$. Integrating Eq. (3) over each interval $M_k$ for $0 \leq k \leq n-1$ and over $(b_n, x)$ yields, with (ii) and (iii), the following identity which holds on $[R, \infty)$.

$$\begin{aligned} 0 &= w'''\bar{w} - w\bar{w}''' - w''\bar{w}' + w'\bar{w}'' \\ &= 2i\ \mathrm{Im}(w'''\bar{w} - w''\bar{w}'). \end{aligned}$$

But because $w$ satisfies the $\mu$-radiation condition we have $w = c_1 e^{-\mu x} + c_2 e^{-i\mu x}$ for $x \geq R$. Substituting this into the identity just obtained yields

$$\begin{aligned} 0 &= \mathrm{Im}(2i|c_2|^2 - (1+i)c_1\bar{c}_2 e^{-\mu(1+i)x} - (1-i)\bar{c}_1 c_2 e^{-\mu(1-i)x}) \\ &= 2|c_2|^2 \end{aligned}$$

Hence $c_2 = 0$. But if $w \neq 0$, this implies that $(w, i\mu^2 w)$ is an eigenvector of $A$ with eigenvalue $i\mu^2$. But, by Lemma 1.2, this is impossible, so $w = 0$. $\square$

Let $R_\lambda$ denote the resolvent operator of $A$. i.e. $R_\lambda = (\lambda - A)^{-1}$. One easily checks that $R_\lambda \begin{bmatrix} f_1 \\ f_2 \end{bmatrix} = \begin{bmatrix} w \\ v \end{bmatrix}$ if and only if $(w, v) \in \mathcal{D}(A)$ and

$$\left.\begin{aligned} w^{(iv)} + \lambda^2 w &= f_2 + \lambda f_1 \\ \lambda w - v &= f_1 \end{aligned}\right\} \text{ in each interval } M_j$$

To study these equations, we set $\lambda = k^2$ and consider the equation

$$w^{(iv)} + k^4 w = f$$

for $f \in L^2$ and $w \in H_2$ (see the definition of the finite energy space). We shall construct a Green's function $h(x,a,k)$ for the problem. $h = h(x,a,k)$ should satisfy

$$\left.\begin{aligned} &\frac{\partial^4 h}{\partial x^4} + k^4 h = \delta(x-a) \text{ on } M = \bigcup_{j=0}^{n} M_j \\ &h(0,a,k) = \frac{\partial h}{\partial x}(0,a,k) = h(b_j,a,k) = 0 \quad j = 1,2,\ldots,n \\ &h, \frac{\partial h}{\partial x}, \frac{\partial^2 h}{\partial x^2} \text{ continuous at each } x = b_j \\ &h(x,a,k) \to 0 \text{ as } x \to \infty \text{ if } \operatorname{Re}\lambda > 0 \text{ (i.e. if } |\arg k| < \frac{\pi}{4}) \end{aligned}\right\} \tag{4}$$

We will see that $h$ is closely related to the Green's function $g$ for the clamped beam with no pin conditions. This function can easily be found explicitly:

$$\begin{aligned} g(x,a,k) = \frac{1}{4k^3}\Big(& e^{\frac{\pi i}{4}} e^{-(\frac{1+i}{\sqrt{2}})k|x-a|} + e^{-\frac{\pi i}{4}} e^{-(\frac{1-i}{\sqrt{2}})k|x-a|} \\ &+ e^{-\frac{\pi i}{4}} e^{-\frac{k}{\sqrt{2}}(1+i)(x+a)} + e^{\frac{\pi i}{4}} e^{-\frac{k}{\sqrt{2}}(1-i)(x+a)} \\ &- \sqrt{2}\, e^{-\frac{k}{\sqrt{2}}(1+i)x - \frac{k}{\sqrt{2}}(1-i)a} - \sqrt{2}\, e^{-\frac{k}{\sqrt{2}}(1-i)x - \frac{k}{\sqrt{2}}(1+i)a}\Big). \end{aligned} \tag{5}$$

$g$ satisfies

$$\frac{\partial^4 g}{\partial x^4} + k^4 g = \delta(x-a)$$

$$g(0,a,k) = \frac{\partial g}{\partial x}(0,a,k) = 0$$

$$g(x,a,k) \to 0 \text{ as } x \to \infty \text{ if } |\arg k| < \frac{\pi}{4}.$$

$g$ also has a removable singularity at $k = 0$, so that if we define

$$g(x,a,0) = \frac{1}{12}(|x-a|^3 - x^3 - a^3 + 3x^2 a + 3a^2 x) \tag{6}$$

then $g$ is an entire function of $k$.

The Green's function $g$ is constructed in the usual way, so that $g(x,a,k)$, $\frac{\partial g}{\partial x}(x,a,k)$, $\frac{\partial^2 g}{\partial x^2}(x,a,k)$ are continuous at all points on the line $x = a$, but

$$\frac{\partial^3 g}{\partial x^3}(a^+,a,k) - \frac{\partial^3 g}{\partial x^3}(a^-,a,k) = 1.$$

We now define $h$ by

$$h(x,a,k) = g(x,a,k) - \sum_{j=1}^{n} c_j(a,k) g(x,b_j,k)$$

where the coefficients $c_j$ are found by imposing the conditions $h(b_j, a) = 0$ and using Cramer's Rule:

$$c_j = \frac{\det G_j(a,k)}{\det G(k)}$$

where $G(k) = (g(b_i, b_j, k))$ and $G_j$ is the matrix obtained by replacing the $j^{\text{th}}$ column of $G$ by the vector with $i^{\text{th}}$ component $g(b_i, a)$. Provided that $\det G \neq 0$, it is easy to check that $h$ is the desired Green's function (i.e. $h$ satisfies properties (4)).

**Lemma 1.4.** *$\det G(k) \neq 0$ for $k = 0$ and for all $k$ satisfying $|\arg k| \leq \frac{\pi}{4}$.*

*Proof.* Let $P(x, k)$ be the matrix function obtained by replacing the first row of $G$ by the vector with $j^{\text{th}}$ component $g(x, b_j, k)$, and let $p(x, k) = \det P(x, k)$. Then $p$ satisfies

(i) $p^{(iv)} + k^4 p = 0$ in each interval $M_i$
(ii) $p(0) = p'(0) = p(b_2) = p(b_3) = \cdots = p(b_n) = 0$
(iii) $p$, $p'$, $p''$ continuous at $b_1, \ldots, b_n$.

Suppose that $\det G(k) = 0$ for some $k = k_0$. Then, in addition to (i), (ii), (iii), we have $p(b_1, k) = 0$. If $|\arg k_0| < \frac{\pi}{4}$ then $p(x, k_0) \to 0$ as $x \to \infty$ so $(p, k_0^2 p)$ is either an eigenvector of $A$, or is identically zero. But by Lemma 1.2 we are forced to conclude that $p(x, k_0) \equiv 0$. Suppose now that $\arg k_0 = \frac{\pi}{4}$. Let $k_0 = (\frac{1+i}{\sqrt{2}})\mu$, where $\mu > 0$. One easily checks from Eq. (5) that $g(x, a, k_0)$ satisfies the $\mu$-radiation condition for $x > a$ when $k_0 = (\frac{1+i}{\sqrt{2}})\mu$. Hence $p$ satisfies the $\mu$-radiation condition for $x > b_n$. Thus, by Lemma 1.2, we again conclude that $p(x, k_0) \equiv 0$. We obtain the same conclusion if $\arg k_0 = \frac{-\pi}{4}$, for then $\overline{p(x, k_0)}$ will satisfy the $\mu$ radiation condition if $k_0 = (\frac{1-i}{\sqrt{2}})\mu$. Further, if $k_0 = 0$ then $p(x, 0)$ will satisfy the zero-radiation condition, so again $p = 0$.

But since $\frac{\partial^3 g}{\partial x^3}(a^+, a, k) - \frac{\partial^3 g}{\partial x^3}(a^-, a, k) = 1$, we obtain

$$p'''(b_1^+, k) - p'''(b_1^-, k) = \det G^{(2)}(k)$$

where $G^{(2)}(k)$ is the $1 - 1$ minor of $G(k)$, i.e. $G^{(2)}(k)$ is the matrix obtained by deleting the first row and first column of $G(k)$. But since $p(x, k_0) \equiv 0$, we see that $\det G^{(2)}(k_0) = 0$.

The matrix $G^{(2)}$ corresponds to the same physical problem without a pin located at $b_1$. Applying the same method to the new matrix $G^{(2)}$ in place of $G$, and continuing in this manner, we can reduce to the case for which there is only one pin located at $x = b_n$. The corresponding function $p(x, k_0)$ is easily calculated in this case

$$p(x, k_0) = g(x, b_n, k_0).$$

Clearly this is not identically zero. Hence, the assertion of the Lemma is correct. □

**Lemma 1.5.** *For each* $\delta \geq 0$ *there are positive constants* $M_1$, $M_2$ *depending on* $\delta$ *such that if* $k$ *is in the set*

$$\mathfrak{F}_\delta = \{k = z + \varepsilon : |\arg z| = \frac{\pi}{4}, \quad 0 \leq \varepsilon \leq \delta\}$$

*then* $M_2 \geq |\det 4(1 + |k|)^3 G| \geq M_1$.

**Note.** $\mathfrak{F}_\delta$ represents the set $\arg z = \pm\frac{\pi}{4}$, "smeared out" to the right by $\delta$ units.

*Proof.* We parametrize the upper part of $\mathfrak{F}_\delta$ by setting

$$k = \left(\frac{1+i}{\sqrt{2}}\right)\mu + \frac{(1-i)}{\sqrt{2}}\varepsilon \text{ for } \mu \geq 0, \quad 0 \leq \varepsilon \leq \frac{\delta}{\sqrt{2}}.$$

We note that $4k^3 G(k)$ is the matrix with elements

$$\begin{aligned} 4k^3 g_{mj} &= e^{\frac{\pi i}{4}} e^{|b_m - b_j|(-\mu i - \varepsilon)} + e^{-\frac{\pi i}{4}} e^{|b_m - b_j|(-\mu + i\varepsilon)} \\ &+ e^{\frac{\pi i}{4}} e^{(b_m + b_j)(-\mu i - \varepsilon)} + e^{\frac{\pi i}{4}} e^{(b_m + b_j)(-\mu + i\varepsilon)} \\ &- \sqrt{2}\, e^{b_m(-\mu i - \varepsilon) + b_j(-\mu + i\varepsilon)} - \sqrt{2}\, e^{b_j(-\mu i - \varepsilon) + b_m(-\mu + i\varepsilon)}. \end{aligned}$$

For $\mu$ large, $4k^3 G$ approaches a matrix $H$ with entries

$$\begin{aligned} h_{jj} &= \sqrt{2} + e^{-\frac{\pi i}{4}} e^{2b_j(-\mu i - \varepsilon)} \\ h_{mj} &= e^{\frac{\pi i}{4}} e^{|b_m - b_j|(-\mu i - \varepsilon)} + e^{-\frac{\pi i}{4}} e^{(b_m + b_j)(-\mu i - \varepsilon)} \quad (m \neq j). \end{aligned}$$

We claim that there are constants $c_1(n)$, $c_2(n)$ such that

$$0 < c_1 < |\det H| \leq c_2.$$

To see this, let $A_n$ be the determinant of the $n$ by $n$ matrix $H$ corresponding to the $n$ pins $b_1, b_2, \ldots, b_n$. Then $A_{n-1}$ is the determinant of the matrix obtained by deleting the last row and last column of $H$. On subtracting $e^{(b_n - b_{n-1})(-\mu i - \varepsilon)}$ times row $n-1$ from row $n$ in $H$, followed by the corresponding column operation, one sees that the sequence $A_n$ satisfies the recurrence relation

$$A_n = (\sqrt{2} - \sqrt{2}\, i\, e^{2(b_n - b_{n-1})(-\mu i - \varepsilon)}) A_{n-1} + i\, e^{2(b_n - b_{n-1})(-\mu i - \varepsilon)} A_{n-2}.$$

We set

$$\begin{aligned} A_n &= 2^{n/2} B_n, \\ \omega_1 &= -2\mu b_1 - \frac{\pi}{4} \\ \omega_n &= -2\mu(b_n - b_{n-1}) + \frac{\pi}{2}, \ n \geq 2 \\ \varepsilon_1 &= 2b_1\varepsilon \\ \varepsilon_n &= 2(b_n - b_{n-1})\varepsilon, \ n \geq 2 \end{aligned}$$

and find that

$$B_n = (1 - e^{i\omega_n - \varepsilon_n})B_{n-1} + \frac{1}{2}e^{i\omega_n - \varepsilon_n}B_{n-2} \tag{7}$$

An easy calculation yields

$$B_1 = 1 + \frac{1}{\sqrt{2}}e^{i\omega_1 - \varepsilon}$$
$$B_2 = (1 - e^{i\omega_2 - \varepsilon_2})(1 + \frac{1}{\sqrt{2}}e^{i\omega_1 - \varepsilon_1}) + \frac{1}{2}e^{i\omega_2 - \varepsilon_2}.$$

It is convenient to define

$$B_0 = 1,$$

for then the recurrence relation (7) is valid for $n \geq 2$.

For $n \geq 1$, we define $\rho_n = B_n/B_{n-1}$ so that

$$\rho_1 = B_1 = 1 + \frac{1}{\sqrt{2}}e^{i\omega_1 - \varepsilon_1} \text{ and}$$
$$\rho_n = 1 + e^{i\omega_n - \varepsilon_n}\left(\frac{1}{2\rho_{n-1}} - 1\right).$$

Clearly $\rho_1$ lies inside the closed disk $D$ centered at 1 with radius $\frac{1}{\sqrt{2}}$. Consider the transformation

$$f_{\varepsilon,\theta}(z) = 1 + e^{i\theta - \varepsilon}(\frac{1}{2z} - 1),$$

for $\varepsilon \geq 0$, $\theta \in \mathbb{R}$. The transformation $z \to \frac{1}{2z}$ maps $D$ onto $D$, so $f_{\varepsilon,\theta}$ maps $D$ into a disk with the same center as $D$, and with a radius no greater than that of $D$. i.e. $f(D) \subseteq D$. Hence $\rho_n \in D$ for each $n$, $\varepsilon \geq 0$, and we have the uniform bounds

$$1 - \frac{1}{\sqrt{2}} \leq |\rho_n| \leq 1 + \frac{1}{\sqrt{2}}.$$

This implies that

$$0 < c_1(n) < \det H < c_2(n)$$

for certain constants $c_1$ and $c_2$. Further, we can find $R > 0$ such that for $\mu \geq R$, $0 \leq \varepsilon \leq \frac{\delta}{\sqrt{2}}$,

$$|\det H - \det 4k^3G| < \frac{1}{2}\min(c_1,\, c_2 - c_1).$$

Hence, for $\mu \geq R$

$$\frac{1}{2}c_1 < |\det 4k^3G| < \frac{1}{2}(c_1 + c_2). \tag{8}$$

By Lemma 1.4, we can find positive constants $M_1$, $M_2$ such that for $0 \le \mu \le R$, $0 \le \varepsilon \le \frac{\delta}{\sqrt{2}}$ and $k = \frac{1+i}{\sqrt{2}}\varepsilon$,

$$M_1 \le |\det 4(1+k^3)G| \le M_2.$$

But because of estimate (8), an estimate of this form holds for all $\mu \ge 0$, $0 \le \varepsilon \le \frac{\delta}{\sqrt{2}}$. The proof for the lower part of $\mathfrak{F}_\delta$ is similar. □

The following theorem will allow us to construct an analytic continuation of the resolvent $R_\lambda$ into part of the left half $\lambda$-plane.

**Theorem 1.6.** *There exist positive constants $\eta$, $R_1$, $R_2$ such that for $k \in \mathcal{S}_\eta = \{k : |\arg(k+\eta)| \le \frac{\pi}{4}\}$ we have*

$$R_2 \ge |\det 4(1+|k|)^3 G| \ge R_1.$$

**Note.** $\mathcal{S}_\eta$ represents the quadrant $|\arg k| \le \frac{\pi}{4}$ shifted to the left by $\eta$ units.

*Proof.* Consider for $\varepsilon > 0$ the set $\mathcal{S}_{-\varepsilon}$. It is easy to check that for $k \in \mathcal{S}_{-\varepsilon}$ we have

$$|\sqrt{2}I - 4k^3G| \le \text{ constant } e^{-r\varepsilon}$$

for some $r > 0$.

Combining this with the result of Lemma 1.5 shows that the statement of Theorem 1.6 holds for $\eta = 0$. Further, it is easy to check that $\frac{d}{dk}(k^3G)$ is uniformly bounded in sets of the form $\mathcal{S}_\gamma$ for small positive constants $\gamma$. Hence we may invoke Lemma 1.5 once again to see that Theorem 1.6 must hold for some $\eta > 0$, and positive constants $R_1, R_2$. □

*Remark.* For the case $n = 1$ (one pin) one can easily show that the zeros of det $G$ occur near the boundary of such a region $\mathcal{S}_\eta$. Hence, in this sense, the region $\mathcal{S}_\eta$ is the best possible.

An immediate corollary to Theorem 1.6 is that the function $h$ is an analytic function of $k$ for $k \in \mathcal{S}_\eta$. It is convenient to define

$$\begin{aligned} g_2(x,a,k) = \frac{1}{4k^5}\Big(& e^{\frac{-\pi i}{4}} e^{-(\frac{1+i}{\sqrt{2}})k(a-x)} + e^{\frac{\pi i}{4}} e^{-(\frac{1-i}{\sqrt{2}})k(a-x)} \\ & - e^{\frac{\pi i}{4}} e^{-(\frac{1+i}{\sqrt{2}})k(x+a)} - e^{\frac{-\pi i}{4}} e^{-(\frac{1-i}{\sqrt{2}})k(x+a)} \\ & - \sqrt{2}\, i\, e^{-(\frac{1+i}{\sqrt{2}})kx - (\frac{1-i}{\sqrt{2}})ka} + \sqrt{2}\, i\, e^{-\frac{k}{\sqrt{2}}(1-i)x - \frac{k}{\sqrt{2}}(1+i)a}\Big) \end{aligned}$$

and $g_1(x,a,k) = \frac{\partial g_2}{\partial a}(x,a,k)$.

Then for $x < a$, $\frac{\partial^2 g_2}{\partial a^2} = \frac{\partial g_1}{\partial a} = g$. Similarly, we define

$$h_2(x,a,k) = g_2(x,a,k) - \sum_{j=1}^{n} c_j^{(2)}(a,k) g(x,b_j,k)$$

$$h_1(x,a,k) = g_1(x,a,k) - \sum_{j=1}^{n} c_j^{(1)}(a,k) g(x,b_j,k)$$

where $c_j^{(1)}$ (respectively $c_j^{(2)}$) is the determinant obtained by replacing the $j^{\text{th}}$ column of $G$ by the vector with $i^{\text{th}}$ component $g_1(b_i,a,k)$ (respectively $g_2(b_i,a,k)$). Comparing this with the definition of $h$ shows that

$$\frac{\partial h_2}{\partial a} = h_1 \text{ and for } x < a \text{ and } b_n \leq a, \ \frac{\partial^2 h_2}{\partial a^2} = \frac{\partial h_1}{\partial a} = h.$$

Unlike $g$, the functions $g_1$ and $g_2$ have poles at $k = 0$. One easily calculates that $g_1$ has a second order pole at $k = 0$, while $g_2$ has a third order pole. In fact, as $k \to 0$,

$$g_2 \sim \frac{\sqrt{2}\,x^2}{2k^3} - \frac{1}{6k^2} x^2(x+3a) + \frac{\sqrt{2}}{6k} x^3 a$$

$$g_1 \sim \frac{-x^2}{2k^2} + \frac{\sqrt{2}\,x^3}{6k}.$$

Consequently, $k^3 h_2$ and $k^2 h_1$ both have removable singularities at $k = 0$. The following theorem is an easy corollary of Theorem 1.6 and the definitions of $h$, $h_1$, $h_2$.

**Theorem 1.7.** *Let $\eta$ be the constant of Theorem 1.6 and let $0 \leq \varepsilon \leq \eta$. Then there are constants $c_1$, $c_2$ independent of $\varepsilon$ such that for $k$ satisfying $|\arg(k+\varepsilon)| = \frac{\pi}{4}$,*

$$\left| \frac{\partial^{i+j}}{\partial x^i \partial a^j} k^3 h_2(x,a,k) \right| \leq c_1\, e^{\varepsilon c_2 (x+a)} (1+|k|)^{i+j-2}$$

$$\left| \frac{\partial^{i+j}}{\partial x^i \partial a^j} k^2 h_1(x,a,k) \right| \leq c_1\, e^{\varepsilon c_2 (x+a)} (1+|k|)^{i+j-2}$$

*and for $x, a \in M = \bigcup_{j=0}^{n} M_j$, $\quad x \neq a$*

$$\left| \frac{\partial^{i+j}}{\partial x^i \partial a^j} h(x,a,k) \right| \leq c_1\, e^{\varepsilon c_2 (x+a)} (1+|k|)^{i+j-3} \qquad \square$$

**Theorem 1.8** (Local Smoothing Property). *Let $(f_1, f_2)$ be initial data in the finite energy space $\mathcal{H}$ and suppose that there is a constant $\alpha > 0$ such that $e^{\alpha x} f_1''(x) \in L^1(0,\infty)$ and $e^{\alpha x} f_2(x) \in L^1(0,\infty)$. Let*

$$\begin{pmatrix} w(\cdot,t) \\ v(\cdot,t) \end{pmatrix} = U(t) \begin{pmatrix} f_1 \\ f_2 \end{pmatrix}.$$

*Then both $w(x,t)$ and $v(x,t)$ are smooth functions of $(x,t)$ in the following sense:*

(i) *$w, v \in C^\infty(M \times (0,\infty))$, $v = \frac{\partial w}{\partial t}$ and for each bounded subset $K$ of $M$ and each interval $[t_0, t_1]$ with $t_0 > 0$ there exist constants $c, \theta$ such that for $(x,t) \in K \times [t_0, t_1]$,*

$$\left| \frac{\partial^{p+q} w}{\partial t^p \partial x^q} \right| \le c\theta^{2p+q}(2p+q)!$$

(ii) *Provided that $q \neq 3 \pmod 4$, we have for all $p$*

$$\frac{\partial^{p+q} w}{\partial t^p \partial x^q}(b_j^-, t) = \frac{\partial^{p+q} w}{\partial t^p \partial x^q}(b_j^+, t).$$

*Remarks.* 1. By estimate (i), for $t > 0$, $w(x,t)$ is a real analytic function of $x$ in $M$, and a Gevrey 2 function of $t$.

2. In previous investigations of beam problems (see [LM], [T]), it has been found that for each $\theta > 0$ a constant $c = c(\theta)$ can be found such that (i) holds. Here, there is a positive lower bound for $\theta$ due to the shape of $S_\eta$.

*Proof.*

We start with the standard identity (see [P], for example)

$$\int_0^t (t-s)U(s) \begin{pmatrix} f_1 \\ f_2 \end{pmatrix} ds = \frac{1}{2\pi i} \int_{\gamma - i\infty}^{\gamma + i\infty} \lambda^{-2} R_\lambda \begin{pmatrix} f_1 \\ f_2 \end{pmatrix} e^{\lambda t} d\lambda \qquad (9)$$

which is valid for all $t > 0$, $\gamma > 0$ and $(f_1, f_2) \in \mathcal{H}$. However, $R_\lambda$ is given explicitly by the formulas $R_\lambda \begin{pmatrix} f_1 \\ f_2 \end{pmatrix} = \begin{pmatrix} w_\lambda \\ v_\lambda \end{pmatrix}$, where

$$w_\lambda(x) = \int_0^\infty h(x, a, \lambda^{1/2})(f_2(a) + \lambda f_1(a))da$$
$$v_\lambda(x) = f_1(x) - \lambda w_\lambda(x).$$

We pick $R > b_n$ and, assuming $x < R$, Re $\lambda > 0$, integrate by parts to obtain

$$\begin{aligned} w_\lambda(x) = &\int_0^\infty h(x, a, \lambda^{1/2}) f_2(a)da + \int_0^R \lambda h(x, a, \lambda^{1/2}) f_1(a)da \\ &+ \lambda h_2(x, R, \lambda^{1/2}) f_1'(R) - \lambda h_1(x, R, \lambda^{1/2}) f_1(R) \\ &+ \int_R^\infty h_2(x, a, \lambda^{1/2}) \lambda f_1''(a)da. \end{aligned} \qquad (10)$$

Let $\Gamma_1$ be the image of the contour $(\gamma - i\infty, \gamma + i\infty)$ under the transformation $\lambda \to k = \lambda^{1/2}$. The square root is chosen so that $\Gamma_1$ lies in the set $|\arg k| < \frac{\pi}{4}$. Let $\kappa = \min(\frac{\alpha}{c_2}, \eta)$, where $c_2$ is the constant of Theorem 1.7, and $\eta$ is the constant of Theorem 1.6. Let $\Gamma_2$ be the contour consisting of the boundary of $S_\kappa$ with upward orientation (i.e. $\Gamma_2 = \{k : |\arg k + \kappa| = \frac{\pi}{4}\}$). If $\nu > 0$, let $\Gamma_3$ be the contour consisting of $\Gamma_2 \cap \{k : Re\, k \geq \nu\}$ and the line segment $\mathrm{Re}\, k = \nu$, $-\nu - \kappa \leq \mathrm{Im}\, k \leq \nu + \kappa$. Eq. (9) yields

$$\int_0^t (t-s)w(x,s)ds = \frac{1}{\pi i}\int_{\Gamma_1} k^{-3} w_{k^2}(x) e^{k^2 t} dk, \quad t > 0.$$

A standard deformation of contours argument, using the estimates of $h$, $h_1$, $h_2$ already derived, allows us to write

$$\int_0^t (t-s)w(x,s)ds = \frac{1}{\pi i}\int_{\Gamma_3} k^{-3} w_{k^2}(x) e^{k^2 t} dk, \qquad t > 0.$$

But the integrand on the right converges absolutely and is easily seen to be a $C^\infty$ function of $x, t$ for $t > 0$ and $x \neq b_j$, $j = 1, 2, 3, \ldots, n$. Moreover, differentiating twice with respect to $t$ yields

$$w(x,t) = \frac{1}{\pi i}\int_{\Gamma_3} k w_{k^2}(x) e^{k^2 t} dk.$$

But $k w_{k^2}$ has a removable singularity at $k = 0$, so we may write

$$w(x,t) = \frac{1}{\pi i}\int_{\Gamma_4} k w_{k^2}(x) e^{k^2 t} dk \tag{11}$$

where $\Gamma_4$ is defined as $\Gamma_3$ is defined, except with $\nu = 0$. We could even push the integration back further to the contour $\Gamma_2$, but $\Gamma_4$ yields better estimates.

Finally, the estimates (i) are obtained by parametrizing $\Gamma_4$ and estimating the integrand using Theorem 7. The resulting integrals are similar to the integral defining the Gamma Function. For details of similar estimates see [T].

*Remark.* It can also be shown that there exists a kernel function $K(x, a, t)$ which for $t > 0$, $x \neq b_j$, $a \neq b_j$ is real analytic in $(x, a)$ and Gevrey 2 in $t$, such that if $f_1$ and $f_2$ satisfy the conditions of Theorem 1.7

$$w(x,t) = \int_0^\infty \frac{\partial}{\partial t} K(x,a,t) f_1''(a) da + \int_0^\infty \frac{\partial^2 K}{\partial a^2}(x,a,t) f_2(a) da.$$

We omit the proof because we do not use this result. Again, details of a similar result may be found in [T].

**Definition.** We define for constants $R > 0$ and finite energy solutions $w, v$ i.e. $\begin{pmatrix} w(\cdot,t) \\ v(\cdot,t) \end{pmatrix} = U(t)\begin{pmatrix} f_1 \\ f_2 \end{pmatrix}$) the <u>local</u> <u>energy</u>

$$E_R(t) = \frac{1}{2}\int_0^R \left|\frac{\partial^2 w}{\partial x^2}(x,t)\right|^2 + |v(x,t)|^2 dx.$$

Clearly as $R \to \infty$, $E_R(t) \to$ total energy $= \frac{1}{2}\|(w,v)\|^2_{\mathcal{H}}$.

**Theorem 1.9** (Local Energy Decay). *Let $(f_1, f_2)$ and $(w, v)$ be as in Theorem 1.8. Then for each $t_0 > 0$, $R > 0$, there exists a constant $A = A(R, t_0, \alpha)$ such that for $t \geq t_0$, $0 \leq y \leq R$*

$$\left|\frac{\partial w}{\partial t}(y,t)\right| \leq A\,t^{-3/2}(\|f_1\|_2 + \|f_2 e^{\alpha x}\|_{L^1} + \|f_1'' e^{\alpha x}\|_{L^1})$$

$$\left|\frac{\partial^2 w}{\partial x^2}(y,t)\right| \leq A\,t^{-3/2}(\|f_1\|_2 + \|f_2 e^{\alpha x}\|_{L^1} + \|f_1'' e^{\alpha x}\|_{L^1})$$

*and thus there exists $B = B(R, t_0, \alpha)$ such that for $t \geq t_0$*

$$E_R(t) = Bt^{-3}(\|f_1\|_2^2 + \|f_2 e^{\alpha x}\|^2_{L^1} + \|f_1'' e^{\alpha x}\|^2_{L^1}).$$

*Remark.* We show later that $E_R(t) = \rho(t)E_\infty(0)$, where $\rho(t) < 1$ for all $t > 0$.

*Proof.* We estimate $v = \frac{\partial w}{\partial t}$ and $\frac{\partial^2 w}{\partial x^2}$ using the contour integral representation of Eq. (11). Using Eq. (10) and the estimates of Theorem 1.7, we find that for $k \in \Gamma_4$ and $\lambda = k^2$, there is a constant $C = C(\alpha, R)$ such that

$$\begin{aligned} |w_\lambda(x)| \leq\; & c_1[(1+|k|)^{-3}\|f_2 e^{\alpha x}\|_{L^1(0,\infty)} + |k|^2(1+|k|)^{-3}\|f_1\|_{L^1(0,R)} \\ & + |k|^{-1}(1+|k|)^{-2}|f_1'(R)| + (1+|k|)^{-2}|f_1(R)| \\ & + |k|^{-1}(1+|k|)^{-2}\|f_1'' e^{\alpha x}\|_{L^1(0,\infty)}] \\ \leq\; & c_2|k|^{-1}(\|f_1\|_2 + \|f_2 e^{\alpha x}\|_{L^1} + \|f_1'' e^{\alpha x}\|_{L^1}). \end{aligned}$$

Similarly, we may estimate $w_\lambda''$ and we find that both $k^2 w_{k^2}$ and $w_{k^2}''$ are bounded on $\Gamma_4$ by

$$c_3|k|(\|f_1\|_2 + \|f_2 e^{\alpha x}\|_{L^1} + \|f_1'' e^{\alpha x}\|_{L^1}).$$

Thus, we may estimate $\frac{\partial^2 w}{\partial x^2}$ and $\frac{\partial w}{\partial t}$ by estimating

$$\int_{\Gamma_4} |k|^2 |e^{k^2 t}||dk|.$$

We parametrize the three lines making up $\Gamma_4$ as follows:

$$\begin{aligned}
&\ell_1 : k = (1+i)\sigma + i\kappa && \sigma \geq 0 \\
&\ell_2 : k = i\sigma && -\kappa \leq \sigma \leq \kappa \\
&\ell_3 : k = (1-i)\sigma - i\kappa && \sigma \geq 0.
\end{aligned}$$

Hence

$$\begin{aligned}
\int_{\ell_2} |k|^2 |e^{k^2 t}|\, |dk| &\leq \int_{-\kappa}^{\kappa} \sigma^2 e^{-\sigma^2 t} d\sigma \\
&\leq t^{-3/2} \int_{-\infty}^{\infty} \tau^2 e^{-\tau^2} d\tau \\
&= c_4 t^{-3/2}
\end{aligned}$$

and for $j = 1, 3$

$$\begin{aligned}
\int_{\ell_j} |k|^2 |e^{k^2 t}\, |dk| &\leq e^{-\kappa^2 t} \int_0^{\infty} (2\sigma^2 + 2\sigma\kappa + \kappa^2) e^{-2\sigma\kappa t} d\sigma \\
&= \frac{1}{2} e^{-\kappa^2 t} (\kappa^{-3} t^{-3} + \kappa^{-1} t^{-2} + \kappa t^{-1}).
\end{aligned}$$

Thus, $\frac{\partial w}{\partial t}$ and $\frac{\partial^2 w}{\partial x^2}$ are both bounded by a constant times

$$(\|f_1\|_2 + \|f_2\, e^{\alpha x}\|_{L^1} + \|f_1'' e^{\alpha x}\|_{L^1})(t^{-3/2} + e^{-\kappa^2 t}[\kappa^{-3} t^{-3} + \kappa^{-1} t^{-2} + \kappa t^{-1}])$$

which proves the theorem. □

**2. Boundary Control of the finite pinned beam.**

In this section we use the smoothing property associated with the semi-infinite pinned beam of Section 1 to solve a control problem for the finite, pinned beam.

We assume that the beam has length $L$ and that it is clamped at $x = 0$ and pinned (i.e. supported) at $x = b_1, b_2, \ldots, b_n = L$. As in Section 1, we set $M_0 = (0, b_1)$, $M_1 = (b_1, b_2), \ldots, M_{n-1} = (b_{n-1}, b_n)$. The continuity properties of the displacement $w(x,t)$ of the beam in the vicinity of each pin $b_k$ $(1 \leq k \leq n-1)$ are just as described in Section 1 for the semi-infinite beam. We already have one boundary condition at $x = L = b_n$, i.e. $w(b_n, t) = 0$. We choose to control the beam by prescribing $\frac{\partial w}{\partial x}(b_n, t) = f(t)$, although it will be clear that the method we use below will apply for many other boundary control operators.

The method we use is similar to the method described in [LiTa]. Essentially, we take the initial data for the finite beam problem and extend it to the initial data for the semi-infinite beam problem. We show that the extension can be done in such a way that $w(x,T) = \frac{\partial w}{\partial t}(x,T) = 0$, $0 \leq x \leq L = b_n$. The corresponding control function is obtained from the trace: $f(t) = \frac{\partial w}{\partial x}(b_n, t)$. The primary ingredient for doing this is the smoothing property, Theorem 1.8. Theorem 1.8 allows us to prove that we can solve the control problem for arbitrary $T > 0$. Theorem 1.9 (local energy decay) can also be used if $T$ is sufficiently large. A similar method using local energy decay was used first by Russell [R] for the wave equation.

We solve the control problem for two classes of solutions: weak solutions of finite energy and strong solutions. These solutions are defined below. Specifically our finite pinned beam problem is

**Problem A.**

$$\frac{\partial^2 w}{\partial t^2} + \frac{\partial^4 w}{\partial x^4} = 0 \qquad x \in \tilde{M} = \bigcup_{j=0}^{n-1} M_j, \quad t \geq 0$$

$$w(x,0) = w_0(x), \quad \frac{\partial w}{\partial t}(x,0) = v_0(x)$$

$$w(0,t) = \frac{\partial w}{\partial x}(0,t) = 0$$

$$\begin{aligned} w(b_j,t) = 0, \quad w(b_j^+,t) - w(b_j^-,t) &= \frac{\partial w}{\partial x}(b_j^+,t) - \frac{\partial w}{\partial x}(b_j^-,t) \\ &= \frac{\partial^2 w}{\partial x^2}(b_j^+,t) - \frac{\partial^2 w}{\partial x^2}(b_j^-,t) = 0 \quad 1 \leq j \leq n-1 \end{aligned}$$

$$w(b_n,t) = 0, \quad \frac{\partial w}{\partial x}(b_n,t) = f(t)$$

The control problem for Problem A consists of finding a boundary control function $f$ that steers the solution with initial data $(w_0, v_0)$ to zero at time $T$. i.e. we want $f$ such that $\frac{\partial w}{\partial t}(x,T) = w(x,T) = 0$.

**Definitions.** 1. The finite energy space of problem A is $\mathcal{H}_A$, where

$$\mathcal{H}_A = \{(w,v) \in L^2(0,b_n) \times L^2(0,b_n) : \exists(\tilde{w},\tilde{v}) \in \mathcal{H} \text{ with } \tilde{w}\,|_{[0,b_n]} = w \text{ and } \tilde{v}\,|_{[0,b_n]} = v\}.$$

$\mathcal{H}_A$ is a Hilbert space with norm $\|\|_A$ and inner product $(\cdot)_A$ given by

$$\|(w,v)\|_A^2 = \int_0^{b_n} |w''|^2 + |v|^2 dx.$$

2. Problem B is precisely problem A with $f \equiv 0$. The relevant Hilbert space for this problem is

$$\mathcal{H}_B = \{(w,v) \in \mathcal{H}_A : \frac{\partial w}{\partial x}(b_n) = 0\}$$

with norm $\|(w,v)\|_B = \|(w,v)\|_A$. One can easily show that solutions of Problem B are given by a strongly continuous unitary group $U_B(t)$ whose infinitesimal generator $A_B$ is given by

$$A_B = \begin{pmatrix} 0 & 1 \\ \frac{-\partial^4}{\partial x^4} & 0 \end{pmatrix}$$

with domain

$$\mathcal{D}(A_B) = \{(w,v) \in \mathcal{H}_A : \exists(\tilde{w},\tilde{v}) \in \mathcal{D}_A \\ \text{with} \quad \tilde{w}\,|_{[0,b_n]} = w, \quad \tilde{v}\,|_{[0,b_n]} = v, \quad v'(b_n) = w'(b_n) = 0\}$$

3. We say that $w$ is a finite energy solution of Problem A if the mapping $t \to (w(\cdot,t),\ \frac{\partial w}{\partial t}(\cdot,t))$ is a continuous mapping from $[0,T]$ into $\mathcal{H}_A$ and if

$$\int_0^T \int_0^{b_n} \left(\frac{\partial^2 \phi}{\partial t^2} + \frac{\partial^4 \phi}{\partial x^4}\right) w\,dx\,dt + \int_0^T \frac{\partial^2 \phi}{\partial x^2}(b_n,t) f(t)dt \\ + \int_0^{b_n} w_0(x)\frac{\partial \phi}{\partial t}(x,0) - v_0(x)\phi(x,0)dx = 0 \qquad (1)$$

for all $\phi$ such that
(i) $\phi(x,T) = \frac{\partial \phi}{\partial t}(x,T) = 0$
(ii) The mapping $t \to (\phi(\cdot,t),\ \frac{\partial \phi}{\partial t}(\cdot,t))$ is a continuous mapping from $[0,T]$ into $\mathcal{D}(A_B)$, where $\mathcal{D}(A_B)$ is topologized by the graph norm $\|\|_{\mathcal{D}(A_B)}$ given by

$$\|(\phi,\psi)\|_{\mathcal{D}(A_B)} = \|(\phi,\psi)\|_B + \|A_B \begin{pmatrix} \phi \\ \psi \end{pmatrix}\|_B$$

4. We say that $w$ is a strong solution of Problem A if
(a) The mapping $t \to (w(\cdot,t),\ \frac{\partial w}{\partial t}(\cdot,t))$ is a continuous mapping from $[0,T]$ into $H^2[0,b_n] \times H^2[0,b_n]$.
(b) The mapping $t \to w(\cdot,t)$ is a continuous mapping from $[0,T]$ into $H^4(M_j)$ for $j = 0,1,2,\ldots,n-1$.
(c) The mapping $t \to \frac{\partial^2 w}{\partial t^2}(\cdot,t)$ is a continuous mapping from $[0,T]$ into $L^2[0,b_n]$.
(d) $w$ satisfies all of the equations listed in Problem A.

*Remarks.* 1. Obviously there are many other equivalent definitions of the spaces $\mathcal{H}_A$, $\mathcal{H}_B$. It is not necessary to define them in terms of $\mathcal{H}$.

2. It is easy to see that a strong solution of problem A is also a finite energy solution.

3. Finite energy solutions of problem A are unique, for if $w$ is such a solution with $w_0 = v_0 = f = 0$, then setting

$$\begin{pmatrix} \phi \\ \psi \end{pmatrix}(t) = -\int_t^T U_B(t-s)\begin{pmatrix} 0 \\ \bar{w}(\cdot, s) \end{pmatrix} ds$$

we find that $\begin{pmatrix} \phi \\ \psi \end{pmatrix}(t) \in \mathcal{D}(A_B)$ and $\frac{d}{dt}\begin{pmatrix} \phi \\ \psi \end{pmatrix} = A_B\begin{pmatrix} \phi \\ \psi \end{pmatrix} + \begin{pmatrix} 0 \\ \bar{w} \end{pmatrix}$. Hence $\frac{\partial^2\phi}{\partial t^2} + \frac{\partial^4\phi}{\partial x^4} = \bar{w}$. Substituting this into (1) shows that

$$\int_0^T \int_0^{b_n} |w(x,t)|^2 dx\, dt = 0 \Rightarrow w = 0.$$

4. If we define the energy of a strong solution by

$$E_B(t) = \frac{1}{2}\left\| \left( w, \frac{\partial w}{\partial t} \right) \right\|_B^2,$$

an easy computation shows that

$$\frac{dE_B}{dt} = \frac{\partial^2 w}{\partial x^2}(b_n, t) f'(t).$$

**Theorem 2.1.** *Let $T > 0$ and let $(w_0, v_0) \in \mathcal{H}_A$, the finite energy space. Then there exists a control function $f$ such that the finite energy solution of Problem A satisfies*

$$w(\cdot, T) = \frac{\partial w}{\partial t}(\cdot, T) = 0.$$

*Proof.* Consider the subspace $S$ of the semi-infinite problem's Hilbert space $\mathcal{H}$

$$S = \{(u,v) \in \mathcal{H} : u''(x) = v(x) = 0 \quad \text{for} \quad x \geq b_n\}.$$

Clearly $S$ is a closed subspace of $\mathcal{H}$ and the projection $P$ associated with $S$ is self-adjoint.

We extend our initial data to be in $S$. i.e. we define for $x \geq b_n$

$$w_0(x) = w_0'(b_n)(x - b_n)$$
$$v_0(x) = 0.$$

Recall that $U(t)$ is the unitary group associated with the semi-infinite beam problem. We define $U = U(T)$ and $g = (w_0, v_0)$. Consider the equation (in $\mathcal{H}$)

$$\tilde{h} - PU^{-1}PUP\tilde{h} = g.$$

If we can solve this, we set $h = Ph - U^{-1}PUPh$. Clearly we obtain $PUh = 0$ and $Ph = g$. i.e. $h$ will agree with the initial data $g$ for $0 \le x \le b_n$ and the solution $\begin{pmatrix} w(\cdot,t) \\ v(\cdot,t) \end{pmatrix} = U(t)h$ is zero for $0 \le x \le b_n$, $t = T$. Thus, a control function $f$ is obtained by setting $f(t) = \frac{\partial w}{\partial x}(b_n, t)$. Solving for $h$ in terms of $g$, we see that $h = Rg$ where

$$R = (P - U^{-1}PUP)(I - PU^{-1}PUP)^{-1}.$$

Thus, it is clear that this construction is possible if we can show that $PU^{-1}PUP$ is a contraction on $\mathcal{H}$. We do this in the following lemma. In fact, we show that $PUP$ is a contraction. □

**Lemma 2.2.** *PUP is a contraction.*

*Proof.* Set $M = PUP$. Clearly $\|M\| \le 1$. Suppose $\|M\| = 1$ and pick a sequence $\{z_n\}$ in $S$ such that

$$\|z_n\| = 1, \quad \lim_{n\to\infty} \|Mz_n\| = 1.$$

But, by the smoothing property (Theorem 1.8) and the compactness properties of Sobolev spaces on bounded domains, $PUP$ is compact. Hence, we might as well assume that $\{Mz_n\}$ is convergent. Let the limit $= g = \begin{bmatrix} w \\ v \end{bmatrix}$. Then $g \in S$ and $\|g\| = 1$. Put $z = U(-T)g$. $\|z\| = 1$ since $U$ is unitary. Also $g = Uz = PUz$.

$$\begin{aligned}
\|z - z_n\|^2 &= \|U(z - z_n)\|^2 \\
&= \|PU(z - z_n)\|^2 + \|(I - P)U(z - z_n)\|^2 \\
&= \|g - M\, z_n\|^2 + \|(I - P)U\, z_n\|^2 \\
&= \|g - M\, z_n\|^2 + \|U\, z_n\|^2 - \|PU\, z_n\|^2 \\
&= \|g - M\, z_n\|^2 + 1 - \|M\, z_n\|^2
\end{aligned}$$

But $M\, z_n \to g$ and $\|M\, z_n\| \to 1$ as $n \to \infty$. Hence $z_n \to z$ as $n \to \infty$. Thus, $z \in S$.

We have shown that if $\|M\| = 1$, we can find $z \in S$ such that $Uz \in S$. Thus,

$$V = \{z \in S : Uz \in S\}$$

is non-empty. Note that $V \subset \text{kernel}(I - PU^{-1}PUP)$ which is finite dimensional because $PUP$ is compact. Further, if $z \in V$ then $\begin{bmatrix} w_1 \\ v_1 \end{bmatrix} = Uz \in \mathcal{D}(A)$ (clearly by Theorem 1.8 $\begin{pmatrix} w_1 \\ v_1 \end{pmatrix}$ is smooth enough to be in $\mathcal{D}(A)$. But also $w_1''$ and $v_1$ have compact support, so $\begin{pmatrix} w_1 \\ v_1 \end{pmatrix}$ must be in $\mathcal{D}(A)$). But if $Uz \in \mathcal{D}(A)$ then $z \in \mathcal{D}(A)$. Hence $\mathcal{D}(A) \cap V = V$. Further, if $z \in \mathcal{D}(A) \cap V$ it is easy to check that $Az \in V$. Thus, $A$ is a bounded linear operator on the finite dimensional space $V$. Hence, $A$ must have an eigenvector in $V$. But this is impossible because $A$ has no eigenvectors (Lemma 1.2). □

*Remarks.* 1. The proof shows that there exists a constant $\rho = \rho(T, R) < 1$ such that the local energy satisfies

$$E_R(T) \le \rho E(0) \qquad (R = L = b_n)$$

for all solutions with initial data $(w_0, v_0)$ such that the supports of $w_0''$, $v_0$ lie inside $[0, R]$. It is clear that the proof will work for any $R > 0$. See the remark after the statement of Theorem 1.9.

2. One might wonder how robust this control method is, i.e. do small changes in the initial data cause small changes in the control function $f$? The answer is "yes" if $\|PUP\|$ is not close to 1. This is the case if $T$ is sufficiently large because by Theorem 1.9 $\|M\| = 0(T^{-3/2})$.

**Theorem 2.3.** *Let $T > 0$ and let $(w_0, v_0)$ be appropriate initial data for a strong solution (see definition 4). Then there exists a control function $f$ such that the strong solution of Problem A satisfies $w(\cdot, T) = \frac{\partial w}{\partial t}(\cdot, T) = 0$.*

*Proof* (Sketch). Consider again the unitary group $U(t)$ of the semi-infinite beam problem. $\mathfrak{D}(A)$ is itself a Hilbert space when topologized by the graph norm

$$\left\| \begin{pmatrix} w \\ v \end{pmatrix} \right\|^2_{\mathfrak{D}(A)} = 2\mu^4 \left\| \begin{pmatrix} w \\ v \end{pmatrix} \right\|^2_{\mathcal{H}} + \left\| A \begin{pmatrix} w \\ v \end{pmatrix} \right\|^2_{\mathcal{H}}$$

(Here we can take any $\mu > 0$, all such norms being equivalent). Further, the restriction of $U(t)$ to $\mathfrak{D}(A)$ is a strongly continuous unitary group with generator $\tilde{A}$ having $\mathfrak{D}(\tilde{A}) = \mathfrak{D}(A^2)$ and $\tilde{A}u = Au$ if $u \in \mathfrak{D}(\tilde{A})$.

We let $S$ be the closed subspace of $\mathfrak{D}(A)$ such that if $\begin{pmatrix} \tilde{w} \\ \tilde{v} \end{pmatrix} \in S$ then there are constants $c_1, c_2, c_3$ such that for $x \ge L = b_n$,

$$\tilde{v}(x) = c_1 e^{-\mu(x-L)} \sin \mu(x - L)$$
$$\tilde{w}(x) = c_2 e^{-\mu(x-L)} \sin \mu(x - L) + c_3(x - L)$$

and let $P$ be the projection operator associated with $S$. $S$ is chosen so that if $\begin{pmatrix} w \\ v \end{pmatrix} \in \mathfrak{D}(A)$ then $P\begin{pmatrix} w \\ v \end{pmatrix}$ agrees with $\begin{pmatrix} w \\ v \end{pmatrix}$ on the interval $0 \le x \le L$ and such that $P\begin{pmatrix} w \\ v \end{pmatrix}$ has minimal norm in $\mathfrak{D}(A)$. A simple application of the calculus of variations shows that the latter requirement implies that $\begin{pmatrix} \tilde{w} \\ \tilde{v} \end{pmatrix} = P\begin{pmatrix} w \\ v \end{pmatrix}$ satisfies for $x \ge b_n$ the equations

$$\tilde{v}^{(iv)} + 2\mu^4 \tilde{v} = 0$$
$$\tilde{w}^{(vi)} + 2\mu^4 \tilde{w}'' = 0$$

which explains the form of $\tilde{w}, \tilde{v}$ given above.

The rest of the proof is almost identical to that of Theorem 2.1. The main thing to check is that solutions with initial data in $S$ are smooth for $t > 0$. But, by Theorem 1.8, this is indeed the case. □

*Remark.* Strong solutions have the smoothness of $\mathcal{D}(A)$. We could look at solutions that have the smoothness of $\mathcal{D}(A^m)$ for any $m > 0$, and, by a technique similar to the one used above, prove the analogous controllability results.

## 3. Additional Remarks.

From the construction of the controllers it is clear that they are infinitely differentiable even of Gevrey class 2, for $0 < t < T$, and continuous for $0 \leq t \leq T$.

We now proceed with some thoughts about spaces of controls for Euler-Bernoulli beams, in general. In the literature, spaces of controls are usually taken to be $L^2$, if at all possible, since this creates an elegant mathematical theory. However from the physical point of view, $L^2$ may be somewhat narrow. After all, the principal physical consideration may be that imposition of the controls only involve expenditure of finite energy, and that the forces and torques (for example) do not destroy the beam. What better way to insure this than to choose forces and torques (for example) of magnitudes with which the right half of an infinite beam exerts "naturally" on the left half? This motivates the following question. Given a vibrating infinite beam of finite energy, (or other space) in what spaces (as a function of time) are the forces and torques at a fixed point in the beam? The following result is not difficult to prove, but because of space limitations will be presented elsewhere.

*Let $u(x,t)$ be a solution to the time dependent Schrödinger equation in $\mathbb{R}_1 \times \mathbb{R}_1$ with initial data in the Sobolev space $H^\alpha$ ($\alpha$ real). Then the restriction $(\partial_x^k \partial_t^\ell u)(0,t) = g(t)$ belongs to $H^\beta$ provided*

$$k + 2\ell + 2\beta - \tfrac{1}{2} = \alpha.$$

*As a consequence we see that for a beam of finite energy the vertical force = $u_{xxx}(0,t)$ is in $H^{-1/4}$, while the torque = $u_{xx}(0,t)$ is in $H^{1/4}$.*

*Acknowledgement.* The first author was partially supported by NSF grant DMS90-02919.

## BIBLIOGRAPHY

[LaLi] J. Lagnese and J. L. Lions, *Modelling, analysis and control of thin plates*, Masson, RMA6, Paris, 1988.

[LaTr1] I. Lasiecka and R. Triggiani, *Exact controllability of the Euler-Bernoulli equation with controls in the Dirichlet and Neumann boundary condition: A nonconservative case*, SIAM J. Control and Optim. **27** (2) (1989), 330–373.

[LaTr2] I. Lasiecka and R. Triggiani, *Lecture notes in control and Inf. Sci*, No. 164, Springer-Verlag, 1991.

[Lions1] J. L. Lions, *Exact controllability, stabilization and perturbations for distributed systems*, SIAM Rev **30** (1988), 1–68.

[Lions2] J. L. Lions, *Contrôlabilite exacte, perturbations et stabilization des systèmes distribués, Tome I*, Masson, RMA9, Paris, 1988.

[LM] W. Littman and L. Markus, *Exact boundary controllability of a hybrid system of elasticity*, Archive for Rational Mechanics and Analysis **103** (1988), 193–236.

[Lit] W. Littman, *Boundary controllability for polyhedral domains*, Springer Lecture Notes in Control and Information Sciences No 178, 1992.

[LiTa] W. Littman and S. Taylor, *Smoothing evolution equations and boundary control theory*, Journal d'Analyse Mathematique (to appear).

[P] A. Pazy, *Semigroups of linear operators and applications to partial differential equations*, Springer-Verlag, 1983.

[R] D. L. Russell, *A unified boundary controllability theory for hyperbolic and parabolic partial differential equations*, Studies in Applied Mathematics **52** (1973), 189–211.

[S] M. H. Stone, *On one parameter unitary groups in Hilbert space*, Ann. Math. **33** (1932), 643–648, (See also [P] for example).

[T] S. W. Taylor, *Gevrey solutions for equations of Euler-Bernoulli type*, J. Differential Equations **92** (2) (1991), 331–359.

[Z] E. Zuazua, *Contrôlabilité exacte en temps arbitrairement petit de quelques modèles de plaques*, Appendix I in [Lions 2].

# 44 Abnormal Sub-Riemannian Minimizers

**Wensheng Liu** University of Minnesota, Minneapolis, Minnesota

**Héctor J. Sussmann**[1] Rutgers University, New Brunswick, New Jersey

**ABSTRACT.** We give a new, simple proof of optimality of abnormal extremals, for an example similar to those recently considered by R. Montgomery and I. Kupka. Our optimality proof proceeds by directly establishing the desired inequality by elementary means, without making a detailed study of the Pontryagin extremals or using geometric arguments.

## §1. Introduction .

One of the most important questions in sub-Riemannian geometry is whether optimal abnormal extremals can exist (cf. [2], [7], [8]). In [7], a general statement was made that sub-Riemannian minimizers are necessarily smooth, and this was derived from the proposition that all minimizers are what we would now call "normal extremals." Subsequently, it was noticed that the proof of this assertion involved an incorrect application of the Pontryagin Maximum Principle, and that a truly correct analysis based on this result implied the possibility that a minimizer might be "abnormal." In [8], it was pointed out that the result of [7] remained valid for a very restrictive class of sub-Riemannian manifolds, namely, those that obey the "strong bracket-generating condition." Until recently, it was an open question whether optimal abnormal extremals can in fact occur. This was settled in recent work by R. Montgomery [5], who showed that there exist sub-Riemannian structures in $\mathbb{R}^3$, associated to a two-dimensional subbundle $E$ of the tangent bundle, for which there exist strictly abnormal extremals that are uniquely optimal. (We call an abnormal extremal *strictly abnormal* if it is not also a normal extremal. We call an admissible trajectory *optimal* if it minimizes length among all admissible trajectories with the same initial and terminal points, and *uniquely optimal* if it is the only optimal trajectory joining these two points, up to reparametrization of the time interval.) Montgomery's optimality proof is rather lengthy and involved, making it desirable to find simpler ways of establishing the

[1] Sussmann's work was supported in part by the National Science Foundation Grant DMS-8902994.

result. I. Kupka has provided in [3] a different proof, based on a detailed analysis of the solutions of the differential equation defining the normal extremals.

In this note, we present another, much simpler proof, for an example similar to those of Montgomery and Kupka. The optimality proof relies on an elementary inequality, and does not require any analysis of Pontryagin extremals. Moreover, we actually prove a slightly stronger optimality property, namely, that our abnormal extremal is optimal within the class of trajectories of a larger system, in which the control constraint $u^2 + v^2 \leq 1$ is replaced by the weaker restriction $|u| \leq 1$, $|v| \leq 1$.

**REMARK 1.** Our example involves a two-dimensional "nonregular distribution" $E$ in $\mathbb{R}^3$. The sections of the subbundle $E$ are globally spanned by two vector fields $f$ and $g$, and the vectors $f(p)$, $g(p)$, $[f,g](p)$ and $[f,[f,g]](p)$ span $\mathbb{R}^3$ for every $p$. However, the span of $f(p)$, $g(p)$ and $[f,g](p)$ is not always three-dimensional, since the dimension drops to 2 on the planes $x = 0$ and $x = 2$. So $E$ is not a "regular distribution." It is easy, however, to modify our example and produce a *regular* two-dimensional distribution $\tilde{E}$ on $\mathbb{R}^4$, with an associated metric, such that for the corresponding sub-Riemannian structure there exist uniquely optimal strictly abnormal extremals (cf. Remark 3 below).

**REMARK 2.** "Abnormal extremals" should in no way be confused with "singular extremals". For general optimal control problems abnormality and singularity are two unrelated notions, neither one of which implies the other. It is a fortuitous coincidence—which will be of no relevance to us here— that in the sub-Riemannian case these two concepts happen to equivalent. (For a detailed comparison of the two notions see the Appendix.)

## §2. Sub-Riemannian manifolds and abnormal extremals .

If $M$ is a $C^\infty$ manifold, and $p \in M$, we use $T_pM$ to denote the tangent space of $M$ at $p$, and $TM$ to denote the tangent bundle of $M$. If $E$ is any $C^\infty$ vector bundle over $M$, then $\Gamma(E)$ denotes the set of all $C^\infty$ sections of $E$. A subbundle $E$ of $TM$ is sometimes called a *distribution* on $M$. *A nonholonomic subbundle* (also known as a *bracket-generating distribution*) is a subbundle $E$ of $TM$ such that the Lie algebra $L(\Gamma(E))$ of vector fields generated by the sections of $E$ has the *full rank property*, i.e. satisfies $\{X(p) : X \in L(\Gamma(E))\} = T_pM$ for all $p \in M$. An *$E$-admissible arc* is an absolutely continuous curve $\gamma$ on $M$, defined on some compact interval $[a,b]$, such that $\dot{\gamma}(t) \in E(\gamma(t))$ for almost all $t \in [a,b]$. If $E$ is nonholonomic and $M$ is connected, then any two points in $M$ can be joined by an $E$-admissible arc.

A $C^\infty$ *Riemannian metric* on $E$ is a $C^\infty$ section $p \to G_p$ of the bundle $E^* \otimes E^*$ such that for each $p \in M$ the bilinear form $T_pM \times T_pM \ni (v,w) \to G_p(v,w) \in \mathbb{R}$ is symmetric and positive definite. A *sub-Riemannian structure* on a manifold $M$ is a pair $(E,G)$ where $E$ is a nonholonomic $C^\infty$ subbundle of $TM$ and $G$ is a $C^\infty$ Riemannian metric on $E$. A *sub-Riemannian manifold* is a triple $(M,E,G)$ such that $M$ is a $C^\infty$ manifold and $(E,G)$ is a sub-Riemannian structure on $M$. One can always construct a Riemannian metric on any subbundle $E$ of $TM$ by just taking a Riemannian metric on $TM$ and restricting it to $E$. If $p \in M$, $v \in E(p)$, then the *length* $||v||_G$ of $v$ is the number $G_p(v,v)^{1/2}$. The *length* $||\gamma||_G$ of an $E$-admissible arc $\gamma : [a,b] \to M$ is the integral $\int_a^b ||\dot{\gamma}(t)||_G dt$. If $p,q \in M$, then the infimum of the lengths of all the $E$-admissible curves $\gamma$ that go from $p$ to $q$ is the *distance* from $p$ to $q$, and is denoted by $d_G(p,q)$. If $M$ is connected and $E$ is nonholonomic, then $d_G(p,q) < \infty$ for all $p$, $q$, and $d_G : M \times M \to \mathbb{R}$ is a metric whose associated topology is the one of $M$. An $E$-admissible curve $\gamma : [a,b] \to M$ such that $d_G(\gamma(a),\gamma(b)) = ||\gamma||_G$ is called a *minimizer*.

An $E$-admissible curve $\gamma$ is *parametrized by arc length* if $||\dot{\gamma}(t)||_G = 1$ for almost all $t$ in the domain of $\gamma$. If $\gamma : [a,b] \to M$ is $E$-admissible, then we can define $\tau(t) = \int_a^t ||\dot{\gamma}(s)||_G ds$, so $\tau$ is a monotonically nondecreasing function on $[a,b]$ with range $[0, ||\gamma||_G]$. Moreover, if $t_1 < t_2$ but $\tau(t_1) = \tau(t_2)$, then $\gamma(t_2) = \gamma(t_1)$. So we can define $\tilde{\gamma} : [0, ||\gamma||_G] \to M$ by letting $\tilde{\gamma}(s) = \gamma(t)$ if $\tau(t) = s$. Then, if $s_1 < s_2$, and $s_i = \tau(t_i)$ for $i = 1,2$, the points $\tilde{\gamma}(s_1)$ and $\tilde{\gamma}(s_2)$ can be joined by the restriction of $\gamma$ to the interval $[t_1,t_2]$, whose $G$-length is $s_2 - s_1$. So $d_G(\tilde{\gamma}(s_1), \tilde{\gamma}(s_2)) \leq s_2 - s_1$. If $\hat{G}$ is a Riemannian metric on $M$ (i.e. a metric defined on the whole tangent bundle $TM$) that extends $G$, then the $\hat{G}$-distance $d_{\hat{G}}(\tilde{\gamma}(s_1), \tilde{\gamma}(s_2))$ is *a fortiori* $\leq s_2 - s_1$. So $\tilde{\gamma}$ is Lipschitz as a map into $(M, d_{\hat{G}})$. Clearly, $\gamma = \tilde{\gamma} \circ \tau$. Since $\tilde{\gamma}$ is Lipschitz and $\tau$ is integrable, we have $\int_0^s ||\dot{\tilde{\gamma}}(\sigma)||_G d\sigma = \int_0^t ||\dot{\gamma}(\theta)||_G d\theta = s$, if $s = \tau(t)$. So $||\dot{\tilde{\gamma}}(s)|| = 1$ for almost all $s$. Therefore $\tilde{\gamma}$ is parametrized by arc length.

In particular, every minimizer $\gamma$ is equivalent modulo reparametrization to an arc $\gamma^*$ which is parametrized by arc length and is optimal for the *minimum time problem* of joining two points $p$, $q$ by means of an $E$-admissible arc $\gamma$ that satisfies $||\dot{\gamma}(t)|| \leq 1$ for almost all $t$ and minimizes time among all such arcs. (It is clear that, if $\gamma$ is a minimizer, then the arc $\tilde{\gamma}$ constructed above is a solution of the minimum time problem.) Conversely, it is easy to see that, if $\gamma$ is a solution of the minimum time problem, then $\gamma$ is a minimizer parametrized by arc length. So the class of solutions of the minimum time problem coincides with the class of minimizers that are parametrized by arc length.

The solutions of the minimum time problem satisfy a necessary condition for optimality given by the Pontryagin Maximum Principle (cf. [1], [4], [6]). A trajectory that satisfies this condition is called a *Pontryagin extremal.* For our sub-Riemannian situation, the condition defining the Pontryagin extremals can be stated most succinctly as follows. We associate to our sub-Riemannian manifold $(M,E,G)$ a real-valued function $\mathcal{H}$ on the cotangent bundle $T^*M$ defined by $\mathcal{H}(\lambda) = -||\lambda||_{E,G}$, where the norm $||\lambda||_{E,G}$ of a covector $\lambda \in T_p^*M$ is the norm of the linear functional $\lambda_E$ obtained by restricting $\lambda$ to $E(p)$, and $E(p)$ is endowed with the norm arising from the quadratic form $G_p$. The function $\mathcal{H}$ is smooth on the set $\{\lambda \in T^*M : \mathcal{H}(\lambda) \neq 0\}$. However, *$\mathcal{H}$ is not $C^1$ on the set of zeros of $\mathcal{H}$.* Since $\mathcal{H}$ is locally Lipschitz, it has a well defined *generalized gradient* $\partial\mathcal{H}$, which is a set-valued function that assigns to each $\lambda \in T^*M$ a convex compact subset $\partial\mathcal{H}(\lambda)$ of the cotangent space $T^*_\lambda(T^*M)$. (By definition, $\partial\mathcal{H}(\lambda)$ is the closed convex hull of the set of all covectors $\Lambda \in T^*_\lambda(T^*M)$ such that $\Lambda$ is the limit of a sequence $\{\Lambda_j\}$ of covectors at points $\lambda_j \in T^*M$ such that $\mathcal{H}$ is differentiable at $\lambda_j$ and its differential at $\lambda_j$ is $\Lambda_j$.) The symplectic structure of $T^*M$ gives rise to linear maps $J_\lambda : T^*_\lambda(T^*M) \to T_\lambda(T^*M)$, for $\lambda \in T^*M$, depending smoothly on $\lambda$. So we can define a set-valued mapping $J\partial\mathcal{H}$, assigning to each $\lambda$ a compact convex subset of $T_\lambda T^*M$. A *Pontryagin extremal* is then, simply, an $E$-admissible arc $\gamma : [a,b] \to M$ which is the projection of a nontrivial solution $\Gamma : t \to (\gamma(t), \lambda(t))$ of the differential inclusion $\dot{\lambda} \in J\partial\mathcal{H}(\lambda)$. (This means that $\Gamma$ is absolutely continuous, $\dot{\Gamma}(t) \in J\partial\mathcal{H}(\Gamma(t))$ for almost all $t$, and $\lambda(t) \neq 0$ for all $t$.) As long as $\mathcal{H} \neq 0$, the set-valued map $J\partial\mathcal{H}$ is in fact single-valued, and is exactly the Hamilton vector field on $T^*M$ associated with $\mathcal{H}$. It follows in particular that, as long as $\mathcal{H} \neq 0$, a solution $\Gamma$ of the differential inclusion is in fact a solution in the ordinary Hamilton differential equations associated with $\mathcal{H}$. So $\mathcal{H}$ is constant along $\Gamma$. Therefore $\mathcal{H}$ is constant along *all* solutions of the differential inclusion.

The solutions $\Gamma$ of the differential inclusion $\dot{\lambda} \in J\partial\mathcal{H}(\lambda)$ along which $\mathcal{H} \neq 0$ (resp. $\mathcal{H} = 0$ ) are called *normal* (resp. *abnormal*) *solutions*, and a Pontryagin extremal that is the projection of a nontrivial normal (resp. abnormal) solution is called a *normal* (resp. *abnormal*) *extremal.* Since a curve $\gamma$ in $M$ can in principle be the projection of more than one solution of the differential inclusion, an extremal can be normal and abnormal at the same time. An extremal is ***strictly abnormal*** if it is abnormal and not normal. A normal solution $\Gamma$ of the differential inclusion is also a solution (up to a reparametrization $t \to \kappa t$ of time) of the Hamilton equations for the Hamiltonian $-\mathcal{H}^2$, which is everywhere smooth. In particular, a normal extremal is smooth. This probably explains the erroneous but widespread belief that all the optimal trajectories are projections of Hamilton trajectories of $-\mathcal{H}^2$. What makes the belief erroneous is precisely the abnormal extremals, for which no simple characterizations exist.

## §3. The example .

We now construct our example of a sub-Riemannian manifold for which there exist uniquely optimal strictly abnormal extremals. We let $M = \mathbb{R}^3$, with the usual coordinates $x$, $y$, $z$, and take $E$ to be the kernel of the 1-form

$$\omega = x^2 dy - (1-x) dz \ .$$

Since $\omega$ never vanishes, it is clear that $E$ is a smooth two-dimensional subbundle of the tangent bundle $TM$. The vector fields $f$, $g$, given by

$$f = \frac{\partial}{\partial x} \ , \quad g = (1-x)\frac{\partial}{\partial y} + x^2 \frac{\partial}{\partial z} \ ,$$

form a global basis of sections of $E$. The Lie brackets $[f,g]$, $[f,[f,g]]$ are easily computed, and turn out to be

$$[f,g] = -\frac{\partial}{\partial y} + 2x\frac{\partial}{\partial z} \ , \quad [f,[f,g]] = 2\frac{\partial}{\partial z} \ .$$

Moreover,

$$[g,[f,g]] = [f,[f,[f,g]]] = [g,[f,[f,g]]] = 0 \ .$$

This implies in particular that *the Lie algebra of vector fields generated by $f$ and $g$ is nilpotent.* Moreover, $f$, $g$ and $[f,g]$ are linearly independent everywhere except where $x = 0$ or $x = 2$, and $f$, $g$ and $[f,[f,g]]$ are independent everywhere except where $x = 1$. It follows that for every $p \in \mathbb{R}^3$ the values at $p$ of the four vector fields $f$, $g$, $[f,g]$ and $[f,[f,g]]$ span $\mathbb{R}^3$. Therefore *the subbundle $E$ is nonholonomic* or, equivalently, *the "distribution" $E$ is bracket-generating.*

We now define a metric $G$ on $E$ by

$$G = dx^2 + h(x)(dy^2 + dz^2)\ , \quad \text{where} \quad h(x) = \frac{1}{(1-x)^2 + x^4}\ .$$

Then the vector fields $f$ and $g$ form an orthonormal basis of sections of $E$.

The $E$-admissible arcs are the absolutely continuous curves $t \to (x(t), y(t), z(t)) = \gamma(t)$ that satisfy the differential equation $x^2\dot{y} = (1-x)\dot{z}$. If $\gamma : [a,b] \to M$ is such an arc, and $\gamma(t) = (x(t), y(t), z(t))$, then we let

$$u(t) = \dot{x}(t)\ , \quad v(t) = \frac{\dot{y}(t)}{1-x(t)} = \frac{\dot{z}(t)}{x(t)^2}\ .$$

(Notice that $v(t)$ is well defined and belongs to $L^1$, because the functions $1-x$ and $x^2$ never vanish simultaneously, and the two expressions defining $v$ agree when both $1-x$ and $x^2$ are $\neq 0$.) Equivalently, we can define $u$ and $v$ by expressing $\dot{\gamma}(t)$ as a linear combination $\dot{\gamma}(t) = u(t)f(\gamma(t)) + v(t)g(\gamma(t))$.

So $\gamma$ is a trajectory of the control system $\Sigma$ given by $\dot{p} = uf(p) + vg(p)$, i.e. by

$$\Sigma: \quad \dot{x} = u\ , \ \dot{y} = (1-x)v\ , \ \dot{z} = x^2 v\ ,$$

for a pair $(u,v)$ of integrable functions of $t$. Conversely, any absolutely continuous curve $\gamma : [a,b] \to M$ which is a solution of $\Sigma$ for some pair $(u,v)$ of integrable real-valued functions on $[a,b]$ is necessarily $E$-admissible. The length of $\gamma$ is then given by

$$||\gamma||_G = \int_a^b (u(t)^2 + v(t)^2)^{1/2} dt\ .$$

If $\gamma$ is parametrized by arc-length, i.e. $||\dot{\gamma}(t)|| = 1$ for almost all $t$, then the length of $\gamma$ is just the *duration* of $\gamma$, i.e. the number $b-a$.

The minimum time problem in our case is the problem of minimizing time in the class of trajectories of $\Sigma$ generated by control functions that satisfy the constraint $u^2 + v^2 \leq 1$, i.e. the class of trajectories of the system

$$\Sigma_1: \quad \dot{x} = u\ , \ \dot{y} = (1-x)v\ , \ \dot{z} = x^2 v\ , \ u^2 + v^2 \leq 1\ .$$

We now describe the Pontryagin extremals for this problem, and in particular determine the abnormal ones. In doing so, we revert to the terminology of optimal control theory, and use the control-theoretic Hamiltonian $H$, which is a smooth function on $T^*M$ times the set $U$ where the controls take values (i.e., in our case, the closed unit disc in the $(u,v)$ plane). The Hamiltonian $H$ of $\Sigma_1$ is given by

$$H(x,y,z,\xi,\eta,\zeta,u,v) = \xi u + \eta(1-x)v + \zeta x^2 v\ .$$

The adjoint equations say:

$$\dot{\xi} = (\eta - 2\zeta x)v\ , \quad \dot{\eta} = \dot{\zeta} = 0\ .$$

If $\gamma : [a,b] \to M$ is a time-optimal trajectory for $\Sigma_1$, whose components are functions $t \to x(t)$, $t \to y(t)$, $t \to z(t)$, such that $\gamma$ is generated by a control $t \to (u(t), v(t))$, then the Pontryagin Maximum Principle tells us that $\gamma$ is a Pontryagin extremal, that is, there exist three absolutely continuous functions $t \to \xi(t)$, $t \to \eta(t)$, $t \to \zeta(t)$ that satisfy the adjoint equations as well as the *nontriviality condition* (namely, $(\xi(t), \eta(t), \zeta(t)) \neq (0,0,0)$ for all $t$), and the *minimization property*, which in our case simply says that, if the vector $V(t) = \Big(\xi(t)\,,\, \eta(t)(1-x(t)) + \zeta(t)x(t)^2\Big)$ is $\neq (0,0)$, then the control vector $(u(t), v(t))$ is equal to $-\frac{V(t)}{\|V(t)\|}$. The adjoint equations then imply that $\eta$ and $\zeta$ are actually constant, and that the length of the vector $V(t)$ is constant as well.

An abnormal extremal is an extremal for which the two functions $t \to \xi(t)$ and $t \to \eta\Big((1-x(t)\Big)+\zeta x(t)^2$ vanish identically. The nontriviality condition then implies that $(\eta, \zeta) \neq (0,0)$. So, if $P(x)$ is the polynomial $\eta(1-x) + \zeta x^2$, then $P$ has at most two zeros, and therefore the function $t \to x(t)$ —whose values are zeros of $P$— is constant (because it is continuous). Therefore the derivative $u(t) = \dot{x}(t)$ vanishes identically. Since $\dot{\xi} = (\eta - 2\zeta x)v$, the constant value of $x$ must satisfy $\eta = 2\zeta x$, unless $v \equiv 0$. If $v \equiv 0$, then $\gamma$ is a constant trajectory, and so $\gamma$ is not time-optimal, unless $b = a$. So, if $b > a$, then $x$ satisfies the two equations $\eta(1-x) + \zeta x^2 = 0$, $-\eta + 2\zeta x = 0$. Therefore $\zeta \neq 0$ (for otherwise $\eta$ would be 0 as well, contradicting $(\eta, \zeta) \neq (0,0)$), and $x$ is a double root of $P$. So the discriminant of $P$ vanishes, that is, $\eta^2 = 4\eta\zeta$. Then either $\eta = 0$ or $\eta = 4\zeta$. In the former case, the equation $\eta = 2\zeta x$ yields $x = 0$. In the latter case, we get $x = 2$. We will only be interested in the case $x = 0$, which corresponds to $\eta = 0$. Clearly, when $x \equiv 0$, $z$ is constant, and $y$ satisfies $\dot{y} = v$. It is clear that such a trajectory cannot be optimal unless $v \equiv 1$ or $v \equiv -1$.

So we are left with the trajectories $\gamma$ along which $x \equiv 0$ (and therefore $u \equiv 0$) and $v \equiv 1$ or $v \equiv -1$. It is easy to see —by taking $\xi \equiv \eta \equiv 0$ and $\zeta \equiv 1$— that any such trajectory $\gamma$ is an abnormal extremal. We now show that $\gamma$ is strictly abnormal, i.e. that it is not a normal extremal. Indeed, for $\gamma$ to be a normal extremal we would have to be able to choose a function $t \to \xi(t)$ and constants $\eta$, $\zeta$ that satisfy the adjoint equations as well as the minimization and nontriviality conditions, and are such that the value of the Hamiltonian is $\neq 0$. We would still have to have $\xi \equiv 0$, because $u \equiv 0$. Then $(\eta, \zeta) \neq 0$ by nontriviality. The Hamiltonian would then have the value $\eta v(t)$, i.e. $-|\eta|$. But the $x$-derivative of $H$ at $x = 0$ is $-\eta v$, i.e. $|\eta|$. So $-|\eta| = \dot{\xi} = 0$, and then the value of the Hamiltonian would be 0, which is a contradiction.

**REMARK 3.** Recall that a subbundle $E$ is *regular* if, for each integer $k > 0$, the dimension of the subspace $E^{(k)}(p)$ is independent of $p$. (Here $E^{(k)}(p)$ is the linear span of all the vectors $X(p)$, for all vector fields $X$ that are Lie brackets of $k$ or fewer local sections of $E$ defined near $p$.) Clearly, the subbundle $E$ of our example is not regular. One can, however, construct a regular example by adding one more variable $w$, with the equation $\dot{w} = (x+y^2)v$. In other words, we let $\tilde{E}$ be the two-dimensional subbundle of $T\mathbb{R}^4$ spanned by the two vector fields $\tilde{f} = \frac{\partial}{\partial x}$, $\tilde{g} = (1-x)\frac{\partial}{\partial y} + x^2\frac{\partial}{\partial z} + (x+y^2)\frac{\partial}{\partial w}$. The metric on $\tilde{E}$ is defined by stipulating that $\tilde{f}$ and $\tilde{g}$ form an orthonormal basis of sections. The four vector fields $\tilde{f}$, $\tilde{g}$, $[\tilde{f}, \tilde{g}]$ and $[\tilde{f}, [\tilde{f}, \tilde{g}]]$ are linearly independent everywhere, so $\tilde{E}$ is regular.

Let $\gamma^* : [0, T] \to \mathbb{R}^3$ be one of our abnormal extremals for our three-dimensional problem, given by $\gamma^*(t) = (0, \bar{y} + t, \bar{z})$, and let $\tilde{\gamma}^* : [0, T] \to \mathbb{R}^4$ be a trajectory of the new system that projects down to $\gamma^*$. It is then easily shown that $\tilde{\gamma}^*$ is a strictly abnormal extremal. (Indeed, the Hamiltonian for our new problem is $\tilde{H} = \xi u + \Big(\eta(1-x) + \zeta x^2 + \mu(x + y^2)\Big)v$, where $\mu$ is a new adjoint variable dual to $w$. The adjoint equations say that $\zeta$ and $\mu$ are constant, and $\dot{\xi} = \eta - \mu - 2\zeta x$, $\dot{\eta} = -2\mu y$. Since $x(t) \equiv 0$ along $\tilde{\gamma}^*$ we can satisfy the adjoint equations by taking $\xi \equiv \eta \equiv \mu \equiv 0$, $\zeta = 1$. With this choice, the switching functions $\xi$ and $\eta(1-x) + \zeta x^2 + \mu y^2$ vanish identically, so $\tilde{\gamma}^*$ is an abnormal extremal. On the other hand, if $\tilde{\gamma}^*$ was a normal extremal as well, then there would exist a nontrivial solution of the adjoint equations that gives rise to a nonzero value of $\tilde{H}$ and satisfies the minimization condition. The latter condition implies that $\xi \equiv 0$. Since $\dot{\xi} = \eta - \mu - 2\zeta x = \eta - \mu$, we have $\eta = \mu$, so $\eta$ is a constant as well. Since $\dot{\eta} = -2\mu y$, and $y$ does not vanish identically, we have $\mu = 0$, and then $\eta = 0$. So the value of the Hamiltonian is $\zeta x^2 v$. i.e. 0. This is a contradiction.) Finally, it is obvious that if $\gamma^*$ is optimal then $\tilde{\gamma}^*$ is optimal as well. ∎

## §4. The optimality proof .

We now show that *every sufficiently short abnormal extremal along which $x \equiv 0$ is uniquely optimal.* Precisely, we will show that an abnormal extremal $\gamma^*$ of length $T$ along which $x \equiv 0$ and $v$ is either $\equiv 1$ or $\equiv -1$ is uniquely optimal, if $T \leq \frac{2}{3}$. We will just study the case $v \equiv 1$. (The other case is similar.) And we will always choose the starting time to be zero, so $\gamma^*$ is, for some $T > 0$, a function from $[0, T]$ to $\mathbb{R}^3$ given by $\gamma^*(t) = (0, \bar{y} + t, \bar{z})$. The optimality conclusion follows trivially from the following inequality:

LEMMA: *Let $\bar{\tau} = \frac{2}{3}$. Let $0 < \tau \leq \bar{\tau}$, and let $u$, $v : [0, \tau] \to \mathbb{R}$ be two measurable functions such that $|u(t)| \leq 1$ and $|v(t)| \leq 1$ for almost all $t$. Define $x : [0, \tau] \to \mathbb{R}$ by $x(t) = \int_0^t u(s)\,ds$. Assume that $x(\tau) = 0$ and $\int_0^\tau x(t)^2 v(t) dt = 0$. Then $\int_0^\tau \Big(1 - x(t)\Big)v(t) \leq \tau$, and equality holds if and only if $u(t) = 0$ and $v(t) = 1$ for almost all $t \in [0, \tau]$.*

To see that this lemma implies the optimality result, let $\gamma : t \to (x(t), y(t), z(t))$ be any trajectory of our system, defined on an interval $[0, \tau]$, that goes from $(0, \bar{y}, \bar{z})$ to $(0, \bar{y} + T, \bar{z})$, and corresponds to controls $u(\cdot)$, $v(\cdot)$. Then the functions $u(\cdot)$, $v(\cdot)$, $x(\cdot)$ satisfy the hypotheses of the lemma, and therefore the integral $\int_0^\tau \Big(1 - x(t)\Big)v(t)dt$, which is equal to $y(\tau) - y(0)$, i.e. to $T$, is $\leq \tau$, provided that $\tau \leq \frac{2}{3}$. Therefore $T \leq \tau$, if $T \leq \frac{2}{3}$. Moreover, the equality $T = \tau$ can only hold if $v \equiv 1$, $u \equiv 0$, i.e. if $\gamma$ coincides with $\gamma^*$. Therefore *the restriction of $\gamma^*$ to any interval of length $\leq \frac{2}{3}$ is uniquely optimal.*

PROOF OF THE LEMMA. Write $A = \int_0^\tau \Big(1 - x(t)\Big)v(t)dt$. Let $h(t) = \int_0^t v(s)ds$. Then $-\tau \leq h(\tau) \leq \tau$. Let $\alpha = \tau - h(\tau)$, $\beta = \sup\{|x(t)| : t \in [0, \tau]\}$. Since $|v(t)| \leq 1$, we have

$$\left|\int_0^\tau x(s)v(s)ds\right| \leq \beta\tau ,$$

so that

$$A \leq \int_0^\tau v(s)ds + \left|\int_0^\tau x(s)v(s)ds\right| \leq h(\tau) + \beta\tau = h(\tau) - \tau + \tau + \beta\tau = \tau - \alpha + \beta\tau.$$

Our conclusion will follow if we show that $\beta\tau \leq \alpha$. If we prove that $\beta \leq \frac{3}{2}\alpha$, then $\beta\tau \leq \frac{3}{2}\alpha\tau$, which implies $\beta\tau \leq \alpha$, since $\tau \leq \frac{2}{3}$.

To prove that $\beta \leq \frac{3}{2}\alpha$, we show that

$$\frac{2\beta^3}{3} \leq \int_0^\tau x(t)^2 dt \leq \beta^2\alpha \,.$$

The upper bound for $\int_0^\tau x(t)^2 dt$ follows by observing that, since $\int_0^\tau x(t)^2 v(t)dt = 0$, we have

$$\int_0^\tau x(t)^2 dt = \int_0^\tau x(t)^2(1 - v(t))dt \leq \beta^2 \int_0^\tau (1 - v(t))dt = \beta^2(\tau - h(\tau)) = \beta^2\alpha \,.$$

To prove the lower bound, we pick a point $a \in [0, \tau]$ such that $|x(a)| = \beta$. Then clearly $a \geq \beta$ and $\tau - a \geq \beta$, since $x(0) = 0$, $x(\tau) = 0$, and $|\dot{x}(t)| \leq 1$. So the intervals $I_1 = [a-\beta, a]$, $I_2 = [a, a+\beta]$ are entirely contained in $[0, \tau]$. On each of the two intervals $I_j$, the function $|x|$ is bounded below by the linear function $\varphi_j$ that takes the value $\beta$ at $a$ and is 0 at the other endpoint. Clearly, $\int_{I_j} \varphi_j^2 = \frac{\beta^3}{3}$. So $\int_{I_j} x(t)^2 dt \geq \frac{\beta^3}{3}$. Therefore

$$\int_0^\tau x(t)^2 dt \geq \int_{I_1} x(t)^2 dt + \int_{I_2} x(t)^2 dt \geq \frac{2\beta^3}{3} \,.$$

This concludes the proof. ▮

**REMARK 4.** The functions $u(\cdot)$, $v(\cdot)$ that correspond to a trajectory of $\Sigma_1$ satisfy $u^2 + v^2 \leq 1$. However, the only properties of these functions that are used in our optimality proof are the inequalities $|u| \leq 1$, $|v| \leq 1$. So we have in fact proved a stronger result, namely, that $\gamma^*$ *is time-optimal —if its length is* $\leq \frac{2}{3}$*— within a larger class of trajectories, namely, those of the control problem* $\Sigma_2$ *obtained from* $\Sigma$ *by imposing the new control constraint* $|u| \leq 1$, $|v| \leq 1$.

**REMARK 5.** The condition $x(\tau) = 0$ was only used to conclude that $a + \beta \leq \tau$, forcing the interval $I_2$ to be contained in $[0, \tau]$. If we do not assume that $x(\tau) = 0$, then all the other steps of the proof remain valid, except that now we can only bound $\int_0^\tau x^2$ from below by $\int_{I_1} x^2$, which is $\geq \frac{\beta^3}{3}$. So *the lemma is still valid even without the hypothesis* $x(\tau) = 0$, *provided that in this case we take* $\bar{\tau} = \frac{1}{3}$ *rather than* $\bar{\tau} = \frac{2}{3}$.

## Appendix: Abnormal and singular extremals.

For general optimal control problems the necessary conditions for optimality often give rise to the possibility of so-called "singular extremals" and "abnormal extremals." The purpose of this appendix is to explain that

(a) these two concepts are completely different in general;
(b) in the sub-Riemannian case they happen to be equivalent.

For simplicity, we will just consider minimum time control problems for systems described by an equation

$$\dot{x} = f(x, u), \tag{1}$$

where the map $f : \mathbb{R}^n \times U \to \mathbb{R}^n$ is smooth, and $U$ is a compact convex subset of $\mathbb{R}^m$. One version of the Maximum Principle says that if $\gamma : [a,b] \to \mathbb{R}^n$ is a time-optimal trajectory of (1) going from $\gamma(a)$ to $\gamma(b)$, corresponding to a control $u(\cdot)$, then the pair $(\gamma, u(\cdot))$ is an *extremal.* This means that there exists a pair $(\lambda, \lambda_0)$ such that:

(A) $\lambda = (\lambda_1, \ldots, \lambda_n) : [a,b] \to \mathbb{R}^n$ is a nowhere vanishing absolutely continuous vector-valued function,
(B) $\lambda_0$ is a nonnegative constant,
(C) the *adjoint equation*

$$\dot{\lambda}(t) = -\frac{\partial H}{\partial x}(\gamma(t), \lambda(t), u(t), \lambda_0)\,, \tag{2}$$

and the *minimization condition*

$$0 = H(\gamma(t), \lambda(t), u(t), \lambda_0) = \min_{v \in U} H(\gamma(t), \lambda(t), v, \lambda_0) \tag{3}$$

hold for almost every $t \in [a,b]$.

Here $H$ is the *Hamiltonian*, i.e. the real-valued function on $\mathbb{R}^n \times \mathbb{R}^n \times U \times \mathbb{R}$ given by

$$H(x, \lambda, u, \lambda_0) = \langle \lambda, f(x,u) \rangle + \lambda_0\,.$$

An absolutely continuous vector-valued function $\lambda$ on $[a,b]$ satisfying (2) for almost all $t$ is called an *adjoint vector* for $(\gamma, u(\cdot))$. A function $\lambda$ satisfying Conditions (A) and (C) for some nonnegative $\lambda_0$ is called a *minimizing adjoint vector* for $(\gamma, u(\cdot))$. If $\lambda$ actually satisfies (A) and (C) with $\lambda_0 = 0$, then $\lambda$ is called an *abnormal minimizing adjoint vector.* An *abnormal extremal* is an extremal for which there exists an abnormal minimizing adjoint vector.

Another version of the Maximum Principle states that a necessary condition for a trajectory $\gamma$ to be a boundary trajectory is that the pair $(\gamma, u(\cdot))$ satisfy (A), (B) and (C) with $\lambda_0 = 0$. i.e. that $(\gamma, u(\cdot))$ be an abnormal extremal. (We call $\gamma : [a,b] \to \mathbb{R}^n$ a *boundary trajectory* if $\gamma(b)$ belongs to the boundary of the set of points reachable from $\gamma(a)$.) This provides an easy way to produce plenty of abnormal extremals for general control problems: it suffices to consider problems where the reachable set from some point $p$ contains boundary points, and then any trajectory going from $p$ to a boundary point will be an abnormal extremal. *This method, however, does not work for the sub-Riemannian case, since in that case the reachable set from each point is the whole space, so boundary trajectories do not exist.*

A "singular extremal" is, roughly speaking, "an extremal for which the minimization condition (3) does not determine the control uniquely." As stated, this is not yet a precise definition. The control literature contains several different ways of making the definition precise, so that the definitions of "singular extremals" given by different authors (and sometimes even by the same author in different papers) are often not equivalent. In this paper, for simplicity, we choose to call a pair $(\gamma, u(\cdot))$ a *singular extremal* if there exists a pair $(\lambda, \lambda_0)$ that satisfies Conditions (A), (B), (C) and is such that

(SE) for almost every $t \in [a, b]$, the equation

$$H(\gamma(t), \lambda(t), v, \lambda_0) = 0 \tag{4}$$

does not have a unique solution. (In other words, (4) has at least one solution $v$ such that $v \neq u(t)$.)

As can be seen from the above discussion, the definitions of abnormal and singular extremals are in principle quite different. In fact, it is easy to exhibit examples showing that neither property implies the other.

**Example 1.** *(A singular extremal which is not abnormal.)* Consider the system $\Sigma$ given by

$$\dot{x} = 1 - y^2\,,$$
$$\dot{y} = u\,,$$

with control constraint $|u| \leq 1$. Let $\gamma : [0,1] \to \mathbb{R}^2$ be given by $\gamma(t) = (x(t), y(t)) = (t, 0)$. Then $\gamma$ is trajectory of $\Sigma$ going from $(0,0)$ to $(1,0)$ in time 1, and it is clear that no other trajectory of $\Sigma$ can reach $(1,0)$ from $(0,0)$ in time $\leq 1$. So $\gamma$ is time-optimal. The corresponding control $u(\cdot)$ is $u(t) \equiv 0$. Since $\gamma$ is time-optimal, it is an extremal, and since the Hamiltonian is linear with respect to the control variable $u$, the minimization condition can only be satisfied with $u(t) = 0$ if the coefficient of $u$ vanishes, in which case $H(\gamma(t), \lambda(t), v, \lambda_0)$ will be independent of $v$. So $H(\gamma(t), \lambda(t), v, \lambda_0) = 0$ for all $v \in [-1, 1]$, and therefore $(\gamma, u(\cdot))$ is a singular extremal. To see that $(\gamma, u(\cdot))$ is not an abnormal extremal, we write the Hamiltonian explicitly:

$$H(x, y, \lambda_1, \lambda_2, u, \lambda_0) = \lambda_1(1 - y^2) + \lambda_2 u + \lambda_0\,.$$

If the conditions of the Maximum Principle could be satisfied for $(\gamma, u(\cdot))$ with $\lambda_0 = 0$, then it would follow from the minimization condition that $\lambda_2 \equiv 0$. Since $\lambda_0 = 0$ and $y(t) \equiv 0$, the fact that $H(x(t), y(t), \lambda_1(t), \lambda_2(t), u(t), \lambda_0) = 0$ enables us to conclude that $\lambda_1(t) \equiv 0$, which contradicts the nontriviality of $(\lambda_1, \lambda_2)$. ∎

**Example 2.** *(An abnormal extremal which is not singular.)* Let $\Sigma$ be given by

$$\dot{x} = 2 + u\,,$$
$$\dot{y} = ux\,,$$

with $|u| \leq 1$. We consider the trajectory $\gamma(t) = (x(t), y(t))$ on $[0, 2]$ that starts at $p = (-3, 0)$ and corresponds to the control given by $u(t) = 1$ for $0 \leq t \leq 1$ and $u(t) = -1$ for $1 < t \leq 2$. An easy computation shows that

$$x(t) = -3 + 3t \text{ and } y(t) = -3t + \frac{3}{2}t^2 \text{ for } 0 \leq t \leq 1\,,$$

$$x(t) = t - 1 \text{ and } y(t) = -2 + t - \frac{t^2}{2} \text{ for } 1 \leq t \leq 2\,.$$

In particular, $\gamma(2) = q = (1, -2)$. It is clear that $\gamma$ is a boundary trajectory, so $(\gamma, u(\cdot))$ is an abnormal extremal. (One can actually show that $(\gamma, u(\cdot))$ is *strictly* abnormal, i.e. that it is not normal.) We now show that $(\gamma, u(\cdot))$ is not a singular extremal. The Hamiltonian is given by

$$H(x, y, \lambda_1, \lambda_2, u, \lambda_0) = \lambda_1(2 + u) + \lambda_2 ux + \lambda_0\,,$$

and the adjoint equations are

$$\dot{\lambda}_1 = -\lambda_2 u, \quad \dot{\lambda}_2 = 0\,.$$

The coefficient of $u$ in $H$ (i.e. the "switching function" for $u$) is $\lambda_1 + x\lambda_2$. For our trajectory to be singular, $\lambda_1(t) + x(t)\lambda_2(t)$ would have to vanish identically on $[0, 2]$. The adjoint equations imply that $\lambda_2$ is a constant. If $\lambda_1 + x\lambda_2 \equiv 0$, then $0 = \dot{\lambda}_1 + \dot{x}\lambda_2 = -\lambda_2 u + \lambda_2(2 + u) = 2\lambda_2$. Therefore $\lambda_2 = 0$. Then the identity $\lambda_1 + x\lambda_2 \equiv 0$ implies that $\lambda_1 \equiv 0$, which contradicts the nontriviality of $(\lambda_1, \lambda_2)$. ∎

Having established that the concepts of singularity and abnormality are different for general systems, we now analyze the situation for the sub-Riemannian case. For simplicity, we will only consider sub-Riemannian structures $(E, G)$ on $\mathbb{R}^n$ for which there is a global orthonormal basis $(f_1, \ldots, f_m)$ of sections of $E$. We associate to such a structure the system

$$\Sigma \;:\; \dot{x} = u_1 f_1(x) + \ldots + u_m f_m(x)\,,$$

with control constraint $u_1^2 + \ldots + u_m^2 \leq 1$, whose minimum time-trajectories are precisely the minimizers. System $\Sigma$ is of the general form $\dot{x} = f(x, u)$, but has in addition the special features that $f(x, u)$ is linear with respect to $u$, and the set of admissible control values is a ball centered at the origin. We now show that these two features imply that the singular extremals are exactly the abnormal ones. To prove this, we remark that the Hamiltonian $H$ is of the form $H = \tilde{H} + \lambda_0$, where

$$\tilde{H}(x, \lambda, u) = \sum_{k=1}^{m} \langle \lambda, f_k(x) \rangle u_k\,.$$

An abnormal extremal is a trajectory for which $\lambda$ and $\lambda_0$ can be chosen so that $\lambda_0 = 0$. On the other hand, the value of $H$ along such an extremal has to vanish, which means that $\tilde{H}$ must vanish. Since $\tilde{H}$ is linear as a function of the control $u$, and its value for $u = u(t)$ minimizes this linear function, it follows that $\langle \lambda(t), f_k(\gamma(t)) \rangle = 0$ for $k = 1, \ldots, m$, so our extremal is singular. Conversely, if an extremal $(\gamma, u(\cdot))$ is not abnormal, the minimum value of $\tilde{H}$ as a function of the control will be $-\lambda_0$, which is $\neq 0$. So the linear functional $u \to \tilde{H}(\gamma(t), \lambda(t), u)$ does not vanish identically, and therefore the problem of minimizing it over the unit ball has a unique solution. Hence $(\gamma, u(\cdot))$ is not singular.

## REFERENCES


[1] L.D. Berkovitz, *Optimal Control Theory*, Springer-Verlag, New York, 1974.

[2] R.W. Brockett, "Control Theory and Singular Riemannian Geometry," in *New Directions in Applied Mathematics* (P.J. Hilton and G.S. Young, eds.), Springer-Verlag, (1981).

[3] I. Kupka, "Abnormal extremals," preprint, 1992.

[4] E.B. Lee and L. Markus, *Foundations of Optimal Control Theory*, Wiley, New York, 1968.

[5] R. Montgomery, "Geodesics which do not Satisfy the Geodesic Equations," 1991 preprint.

[6] L.S. Pontryagin, V.G. Boltyanskii, R.V. Gamkrelidze and E.F. Mischenko, *The Mathematical Theory of Optimal Processes*, Wiley, New York, 1962.

[7] R. Strichartz, "Sub-Riemannian Geometry," *J. Diff. Geom.* 24, 221-263, (1986).

[8] R. Strichartz, "Corrections to 'Sub-Riemannian Geometry'," *J. Diff. Geom.* 30, no. 2, 595-596, (1989).

# 45 Some Algebraic Approaches for Stability Analysis of Two-Dimensional Systems and Digital Filters

**Wu-Sheng Lu** University of Victoria, Victoria, British Columbia, Canada


## Abstract

Stability analysis of 2-D systems and filters is a technically challenging problem as the zeros of the denominator polynomial of a general 2-D transfer function define algebraic curves as opposed to isolated points in the 1-D case. In the first part of the paper, Gutman's determinent condition and a technique of linearizing matrix polynomials are used to deduce a new stability criterion that reduces the 2-D stability verification problem to a generalized eigenvalue problem. The matrices involved in the generalized eigenvalue problem are of large size even for systems of moderate order, but they are very sparse. In the second part of the paper, two algorithms for testing 2-D stability are presented. The first algorithm is based on a singular-value perturbation analysis of a parameterized matrix, and the second is based on a 1-D robust stability analysis through a Lyapunov approach. It is shown that both algorithms can be considerably enhanced by a pre-processing procedure that reduces the norm of the system matrix. A direct method based on unconstrained optimization and an indirect method based on state-variable transformation are proposed for the norm reduction. A case study is given to illustrate the stability anslysis algorithms proposed and the critical role that the pre-processing for norm-reduction may play in the algorithms' implementation.


## I. Introduction

Recently several algebraic necessary and sufficient conditions for the stability of two-dimensional (2-D) sytems and digital filters have been proposed [4]–[8]. Using the root-clustering theory [1]–[3], Gutman[4] obtained a determinent type condition that ensures the stability of a matrix function of one complex variable $z$ on the unit circle[1]. Although the resulting stability test proposed there appears to be complicated to implement, this determinent condition is quite general and has found useful in developing feasible stability tests. Based on a 2-D version of the Schur-Cohn test, Gu and Lee [5] proposed a numerical algorithm for stability testing.. It is noted that the 2-D Schur-Cohn test is also a determinent type condition, where the matrix involved depends on a complex variable on the unit circle. In [6] and [7], Agathoklis, Jury and Mansour proposed several algebraic stability tests derived from a determinent type result similar to [4]. The resulting stability tests are, however, relatively easier to implement. In [8], Wang, Lee, and Boley proposed an algebraic stability test that utilizes the resultant matrix of the denominator polynomial and its reciprocal in conjunction with a linearization technique [9] for polynomial matrix to reduce the stability testing problem to a generalized eigenvalue problem for a pair of constant matrix of size $2n_1n_2 \times 2n_1n_2$ where $(n_1, n_2)$ is the order of the 2-D system.

In the first part (Section II) of this paper, we shall utilize the fact that the stability of a 2-D digital filter can be verified by computing the largest spectral radius of a *first-order* polynomial matrix when its complex parameter varies on the unit circle. This in conjunction with Gutman's determinent condition [4] and a linearization technique [10] leads to a generalized eigenvalue condition for 2-D stability testing. It is noted that the stability test obtained is similar in spirit to that of [8] but the matrices involved in the generalized eigenvalue problem are different. Nevertheless, as in [8] the matrices in our matrix pencil are very sparse so that special algorithms can be used to implement the test obtained [11][8]. In the second part of the paper, we present two new linear-algebraic type algorithms for testing stability of 2-D filters. The first algorithm, to be presented in Sec. III, is established based on the results obtained in Sec. II in conjunction with a singular-value perturbation analysis of the paramenterized matrix involved. The main idea in the proposed algorithm is that at each point on the unit circle where the stability is being checked, a "largest" possible step size is sought such that the filter is stable over the entire incremental arc along the unit circle. The second algorithm described in Sec. IV is similar in spirit to the first one except that the "largest" step size is obtained using a Lyapunov approach initiated in [13]. The case study presented in Section VI indicates that the algorithm is far more efficient than the existing algebraic algorithms [12] that was based on

---

[1]Throughout the paper, a square matrix is said to be stable if all its eigenvalues are in the open unit disk.

a stability-robust-analysis for a family of matrices with one complex parameter. In Section V two approaches to further improving the efficiency of both algorithms are discussed. These two approaches are based on unconstrained optimization and 2-D state-variable transform, respectively, to reduce the norm of the system matrix before stabilty test is applied. In Section VI we present a case study to demonstrate the feasibility of the algorithms proposed.

## II. A Stability Theorem

Consider a 2-D transfer function $H(z_1, z_2) = n(z_1, z_2)/d(z_1, z_2)$ of order $(n_1, n_2)$ and assume that $H(z_1, z_2)$ has no nonessential singularities of the second kind on the unit bicircle $T^2 = \{(z_1, z_2) : |z_1| = |z_2| = 1\}$. The bounded-input bounded-out (BIBO) stability of $H(z_1, z_2)$ is then equivalent to the BIBO stability of $\hat{H}(z_1, z_2) = 1/d(z_1, z_2)$. As the all-pole transfer function $\hat{H}(z_1, z_2)$ always admits a minimal realization [14], the stability analysis can be carried out in the Roesser state space where $\hat{H}(z_1, z_2)$ is realized by the state-space equation

$$\begin{bmatrix} x^h(i+1,j) \\ x^v(i,j+1) \end{bmatrix} = \begin{bmatrix} A_{11} & A_{12} \\ A_{21} & A_{22} \end{bmatrix} \begin{bmatrix} x^h(i,j) \\ x^v(i,j) \end{bmatrix} + \begin{bmatrix} b_1 \\ b_2 \end{bmatrix} u(i,j) \equiv Ax(i,j) + bu(i,j) \quad (1a)$$

$$y(i,j) = \begin{bmatrix} c_1 & c_2 \end{bmatrix} \begin{bmatrix} x^h(i,j) \\ x^v(i,j) \end{bmatrix} \equiv cx(i,j) \quad (1b)$$

with $A_{ii} \in R^{n_i \times n_i} \quad (i = 1, 2)$.

Defining

$$\begin{aligned}
A_{10} &= \begin{bmatrix} A_{11} & A_{12} \\ 0 & 0 \end{bmatrix} \equiv \begin{bmatrix} A_1 \\ 0 \end{bmatrix}, \\
A_{20} &= \begin{bmatrix} 0 & 0 \\ A_{21} & A_{22} \end{bmatrix} \equiv \begin{bmatrix} 0 \\ A_2 \end{bmatrix} \\
F(z) &= A_{10} + zA_{20} && (2) \\
\rho(M) &= \textit{the spectral redius of matrix } M \\
E_0 &= A_{10} \otimes A_{20} && (3) \\
E_1 &= A_{10} \otimes A_{10} + A_{20} \otimes A_{20} - I_N, \quad N = (n_1 + n_2)^2 && (4) \\
E_2 &= A_{20} \otimes A_{10} && (5)
\end{aligned}$$

where $\otimes$ denotes the Kronecker product. The following theorem reduces the 2-D stability testing problem to a 1-D stability testing problem and a generalized eigenvalue problem.

**Theorem 1** System (1) is stable if and only if

(i) Matrix $\mathbf{A}$ defined by (1a) is stable, and

(ii) There are no eigenvalues of the matrix pencil $L - \lambda K$ on the unit circle where

$$L = \begin{bmatrix} 0 & -I \\ E_0 & E_1 \end{bmatrix}, \quad K = \begin{bmatrix} -I & 0 \\ 0 & -E_2 \end{bmatrix} \tag{6}$$

**Remark** condition (i) of the theorem is a 1-D stability condition that is easy to test. Condition (ii) is a generalized eigenvalue problem for the matrix pencil $L - \lambda K$. Although the dimension of $L$ and $K$ is quite large, both $L$ and $K$ are very sparse. Special algorithms are available for efficient computation of the sparse generalized eigenvalue problem [15][11].
To prove Theorem 1, we recall the following lemmas.

**Lamma 1** [12] System (1) is stable if and only if

$$\max_{|z|=1} \rho(F(z)) < 1 \tag{7}$$

**Lemma2** [4] Let $Q(z)$ be an $N \times N$ matrix function in a complex variable $z$, continuous on $|z| = 1$. $Q(z)$ is stable on $|z| = 1$ if and only if

(i) $Q(1)$ is stable, and (8)

(ii) $\det[Q(z) \otimes \bar{Q}(z) - I] \neq 0$ on $|z| = 1$. (9)

**Proof of Theorem 1** By Lemma 1, the stability of (1) is equivalent to the stability of $F(z)$ on $|z| = 1$. Let $Q(z) = F(z)$ where $F(z)$ is defined by (2), Lemma 2 then implies that (1) is stable if and only if condition (i) holds and

$$\det[F(z) \otimes \bar{F}(z) - I] \neq 0 \quad on \quad |z| = 1 \tag{10}$$

As for $z$, $\bar{z} = z^{-1}$ on $|z| = 1$, it can readily be shown that

$$(10) \iff \det(z^{-1}E_0 + E_1 + zE_2) \neq 0 \qquad on \quad |z| = 1 \tag{11}$$

$$\iff \det(E_0 + zE_1 + z^2E_2) \neq 0 \qquad on \quad |z| = 1 \tag{12}$$

where $E_0$, $E_1$, and $E_2$ are defined by (3), (4), (5), respectively. Further notice that with $L$ and $\mathbf{K}$ defined by (6), we have

$$\begin{bmatrix} E_1 + zE_2 & I \\ -I & 0 \end{bmatrix} (L - zK) = \begin{bmatrix} E_0 + zE_1 + z^2E_2 & 0 \\ 0 & I \end{bmatrix} \begin{bmatrix} I & 0 \\ -zI & I \end{bmatrix}$$

Hence

$$\det\,(L - zK) = \det\,(E_0 + zE_1 + z^2E_2)$$

and it follows that

$$(12) \iff \det(L - zK) \neq 0 \quad on \quad |z| = 1 \tag{13}$$

$$\iff (ii) \tag{14}$$

□

**Collorary 1** System (1) is stable if and only if

(i) All eigenvalues of $\mathbf{A}$ are located in the open unit disk, and

(ii) There are no eigenvalues of the matrix pencil $\hat{L} - \lambda\hat{K}$ on the unit circle where

$$\hat{L} = \begin{bmatrix} 0 & -I \\ E_2 & E_1 \end{bmatrix}, \hat{K} = \begin{bmatrix} -I & 0 \\ 0 & -E_0 \end{bmatrix} \tag{13}$$

**Proof** By the proof of Theorem 1,

$$\begin{aligned} (10) &\iff \det(E_0 + zE_1 + z^2E_2 \neq 0 \quad on \quad |z| = 1 \\ &\iff \det(z^{-2}E_0 + z^{-1}E_1 + E_2) \neq 0 \quad on \quad |z| = 1 \\ &\iff \det(z^2E_0 + zE_1 + E_2) \neq 0 \quad on \quad |z| = 1 \end{aligned} \tag{14}$$

A similar linearization for polynomial matrix $z^2E_0 + zE_1 + E_2$ implies that

$$\begin{aligned} (14) &\iff \det(\hat{L} - z\hat{K}) \neq 0 \quad on \quad |z| = 1 \\ &\iff (ii). \end{aligned}$$

□

## III. A Stability Test Based On Singular-Value Perturbation Theory

The first algorithm described below is based on a singular-value perturbation analysis for matrix $z^{-1}E_0 + E_1 + zE_2$. Let $z = e^{j\theta}$ and define

$$\varepsilon(\theta) = e^{-j\theta}E_0 + E_1 + e^{j\theta}E_2 \qquad \theta \in [-\pi,\ \pi] \tag{15}$$

As a complex matrix is nonsingular if and only if its complex-conjugate is nonsingular, we conclude that

$$(10) \Longleftrightarrow \det \varepsilon(\theta) \neq 0 \quad for \quad \theta \in [0,\ \pi] \tag{16}$$

It is well known that a reliable way to test whether or not a matrix is nonsingular is to check its smallest singular value. Denote the smallest singular value of $\varepsilon(\theta)$ by $\underline{\sigma}(\theta)$ and assume $\varepsilon(\theta)$ for a fixed $\theta$ is nonsingular, it follows from [16, Ch. 6] that

$$\begin{aligned} |\underline{\sigma}(\theta+\Delta\theta) - \underline{\sigma}(\theta)| &\leq ||\varepsilon(\theta+\Delta\theta) - \varepsilon(\theta)|| \\ &= ||(e^{-j(\theta+\Delta\theta)} - e^{-j\theta})E_0 + (e^{j(\theta+\Delta\theta)} - e^{j\theta})E_2|| \\ &\leq \Delta\theta(||E_0|| + ||E_2||) \end{aligned}$$

Hence

$$\underline{\sigma}(\theta+\Delta\theta) \geq \underline{\sigma}(\theta) - \Delta\theta(||E_0|| + ||E_2||)$$

It follows that

$$\underline{\sigma}(\theta+\Delta\theta) > 0 \quad i.e. \quad \varepsilon(\theta+\Delta\theta) \quad is\ nonsingular$$

if

$$\Delta\theta < d(\theta) \tag{17}$$

where

$$d(\theta) = \frac{\underline{\sigma}(\theta)}{||E_0|| + ||E_2||} \tag{18}$$

In words, if the stability test starts at $\theta$, say $\theta = 0$, and finds $F(e^{j\theta})$ stable, then for any $\Delta\theta$ satisfying (17), $F(e^{j(\theta+\Delta\theta)})$ is also stable. Thus the next parameter value for stability testing is $\theta + d(\theta)$ where $d(\theta)$ is determined by (18). If $F(e^{j(\theta+d(\theta))})$ is unstable, then by Lemma 1 system (1) is unstable. Otherwise take $\theta + d(\theta)$ as a new $\theta$ and determine a new $d(\theta)$ by (18). This procedure continues until (a) a $d(\theta)$ at a certain value of $\theta$ ($0 \leq \theta \leq \pi$) is found such that $\theta + d(\theta) < \pi$ and $F(e^{j(\theta+d(\theta))})$ is unstable, then system (1) is deemed to be unstable, and the stability testing is complete; or (b) if a $d(\theta)$ at a certain value of $\theta$ is found such that $\theta + d(\theta) \geq \pi$, then system (1) is stable; or (c) if a $d(\theta)$ at a certain value of $\theta$ is found such that $\theta + d(\theta) < \pi$ and $d(\theta) < \delta$ where $\delta > 0$ is a prescribed margin, then the system is deemed to be *practically* unstable as even if the system might turn out to be stable, its stability margin is so small that the system should not be used in practice as a stable filter.

Concerning the computation of $d(\theta)$ defined by (18), we recall the following properties of Kronecker product [17]:

(i) $(P \otimes Q)^T = P^T \otimes Q^T$

(ii) $(P_1 \otimes Q_1)(P_2 \otimes Q_2) = P_1 P_2 \otimes Q_1 Q_2$

(iii) the eigenvalues of $P \otimes Q$ are $\{\lambda_i \mu_j,\ i = 1, \ldots, n \text{ and } j = 1, \ldots, n\}$ where $\lambda_i$ and $\mu_j$ are the $i$th eigenvalue of $\mathbf{P}$ and the $j$th eigenvalue of $\mathbf{Q}$, respectively.

It follows from (3) that

$$\begin{aligned}
||E_0|| &= \max[eig^{1/2}(E_0^T E_0)] \\
&= \max[eig^{1/2}(A_{10}^T A_{10} \otimes A_{20}^T A_{20})] \\
&= \max[eig^{1/2}(A_{10}^T A_{10})] \cdot \max[eig^{1/2}(A_{20}^T A_{20})] \\
&= ||A_{10}||\ ||A_{20}|| \\
&= ||A_1||\ ||A_2||
\end{aligned}$$

Similarly,

$$||E_2|| = ||A_1||\ ||A_2||$$

Therefore

$$d(\theta) = \frac{\underline{\sigma}(\theta)}{2||A_1||\ ||A_2||} \tag{19}$$

The above stability analysis can be summarized as an algorithm to test the stability of (1).

## Algorithm 1

**Step 1** Set $k = 0$ and $\theta_0 = 0$. Test stability of $F(e^{j\theta})$ at $\theta_0$. If stable, go to the next step. Otherwise stop and claim the instability of (1).

**Step 2** Compute $d(\theta_k)$ using (18).

**Step 3** Set $\theta_{k+1} = \theta_k + d(\theta_k)$. If $\theta_{k+1} \geq \pi$, stop and claim the stability of (1). Otherwise go to the next step.

**Step 4** If $d(\theta_k)$ is less than a prescribed margin $\delta$, then stop and claim that the filter is *practically* unstable. Otherwise update $k$ and go to Step 2.

We now make two remarks on (19) to conclude this section. First, by comparing (19) with (18), (19) is obviously more efficient to test the stability of a 2-D digital filter. Second, from (19) it is obvious that any pre-testing effort to reduce $||A_1||$ and $||A_2||$ will improve the computation efficiency of the stability test we just proposed. We will come back to this issue in some detail in Section V.

## IV. A Stability Test via a Lyapunov Approach

An immediate consequence of Lemma 1 is a Lyapunov type stability theorem stated as follows:

**Lemma 3** System (1) is stable if and only if for any $Q > 0$ there exists $P(\theta) > 0$, $\theta \in [0,\ \pi]$ such that

$$F^H(\theta)P(\theta)F(\theta) - P(\theta) = -Q \tag{20}$$

where $F(\theta)$ denotes $F(e^{j\theta})$ and $F^H(\theta)$ is the complex-conjugate transpose of $F(\theta)$.

Assume that $F(e^{j\theta})$ is stable at $\theta = \theta^*$, then the standard 1-D Lyapunov theory claims that there exists for $Q > 0$ a matrix $P(\theta^*) > 0$ such that

$$F^H(\theta^*)P(\theta^*)F(\theta^*) - P(\theta^*) = -Q \tag{21}$$

It is known [13] that $F(\theta^*) + \Delta F$ remains stable if

$$||\Delta F|| \leq \frac{\underline{\sigma}(Q)}{\bar{\sigma}(P(\theta^*)) + \bar{\sigma}^{1/2}(P(\theta^*)\bar{\sigma}^{1/2}(P(\theta^*) - Q)} \tag{22}$$

where $\bar{\sigma}(P)$ denotes the largest singular value of $P$ and that $||\Delta F||$ achieves its maximum when $Q = I$ is chosen.

Consider now matrix $F(\theta)$ at $\theta = \theta^* + \Delta\theta$ and write

$$F(\theta^* + \Delta\theta) = F(\theta^*) + \Delta F$$

with

$$\Delta F = [e^{j(\theta^*+\Delta\theta)} - e^{j\theta^*}]A_{20}$$

which implies that

$$||\Delta F|| \leq \Delta\theta ||A_{20}||$$

It follows that the largest possible step size $\Delta\theta$ which guarantees the stability of $F(\theta^* + \Delta\theta)$ based on the above Lyapunov approach is given by

$$\Delta\theta \leq \frac{1}{||A_2||[\bar{\sigma}(P(\theta^*)) + \bar{\sigma}^{1/2}(P(\theta^*))\bar{\sigma}^{1/2}(P(\theta^*) - I)]} \equiv \Delta\theta^* \tag{23}$$

where $P(\theta^*)$ is the solution of (21) with $Q = I$. The analysis given above may be summarized as the following algorithm to test the stability of (1).

### Algorithm 2

**Step 1** Set $k = 0$ and $\theta_0 = 0$. Test stability of $F(0)$. If stable, go to the next step. Otherwise stop and claim the instability of (1).

**Step 2** With $\theta^* = \theta_k$ and $Q = I$, solve (21) for $P(\theta^*)$ and compute $\Delta\theta^*$ defined in (23).

**Step 3** Set $\theta_{k+1} = \theta_k + \Delta\theta^*$. If $\theta_{k+1} \geq \pi$, stop and claim the stability of (1). Otherwise go to the next step.

**Step 4** If $\Delta\theta^*$ is less than a prescribed margin $\delta$, then stop and claim that the filter is *practically* unstable. Otherwise update $k$ and go to Step 2.

## V. Improvement of Computation Efficiency

As was pointed out earlier, an equivalent realization of (1) with smaller $||A_1||$ and $||A_2||$ will improve the convergence rate of the algorithm presented in Sec. III. Likewise, it follows from (23) that a smaller $||A_2||$ will improve the convergence rate of the algorithm in Sec. IV. In this section, we present two approaches to reducing the norm of $||A||$.

The first one is a direct approach that seeks a 2-D state-variable transformation $T = T_1 \oplus T_2$, where $T_1 \in R^{n_1 \times n_1}$ and $T_2 \in R^{n_2 \times n_2}$ are nonsingular and $\oplus$ denotes the direct sum, such that $||T^{-1}AT||$ is minimized (or reduced). To reduce the nmber of parameters involved in this minimization, a QR decomposition of $T$ may be used, i.e.,

$$T = T_1 \oplus T_2 = (R_1 \oplus R_2)(Q_1 \oplus Q_2) \equiv RQ$$

with $Q_i$ orthogonal and $R_i$ upper triangular ($i = 1, 2$). This leads to

$$||T^{-1}AT|| = ||R^{-1}AR|| = ||\hat{R}^{-1}A\hat{R}||$$

where $\hat{R} = \hat{R}_1 \oplus \hat{R}_2$, and

$$R_1 = \begin{bmatrix} r_{11} & & & \\ & r_{12} & & * \\ & & \ddots & \\ & 0 & & \\ & & & r_{1n_1} \end{bmatrix}, \quad R_2 = \begin{bmatrix} r_{21} & & & \\ & r_{22} & & * \\ & & \ddots & \\ & 0 & & \\ & & & r_{2n_2} \end{bmatrix}$$

$$\hat{R}_1 = \begin{bmatrix} 1 & & & \\ & r_{12}/r_{11} & & * \\ & & \ddots & \\ & 0 & & \\ & & & r_{1n_1}/r_{11} \end{bmatrix}, \quad \hat{R}_2 = \begin{bmatrix} r_{21}/r_{11} & & & \\ & r_{22}/r_{11} & & * \\ & & \ddots & \\ & 0 & & \\ & & & r_{2n_2}/r_{11} \end{bmatrix}$$

The optimization at hand is now to find $\hat{R}_1$ and $\hat{R}_2$ such that $f(\hat{R}) = ||\hat{R}^{-1}A\hat{R}||$ is minimized (or reduced), where the number of parameters involved is $0.5[n_1(1+n_1)+n_2(1+n_2)]-1$. Although there is a variety of reliable algorithms for unconstrained optimization [18], explicit, closed-form expression for the gradient of the objective function $f(\hat{R})$ with respect to the parameter vector is very difficult to find if not impossible. However, preliminary studies on this optimization approach have shown that the quasi-Newton algorithms such as Broyden-Fletcher-Goldfarb-Shanno and Davidon-Fletcher-Powell algorithms [18] which do not require the second-order information of the objective function can be used in conjunction with numerical differentiation to obtain a local minimum with fast convergence. This is especially true for 2-D filters of low order. For digital filters with moderate order, say $10 \leq n_1,\ n_2 \leq 15$, the dimension of the parameter space is quite high (between 110 and 240), and the conjugate gradient method with an objective-function-dependent restart procedure [20] is usually preferable.

Two remarks on the optimization approach are now in order. First, the optimization approach is a local approach in the sense that $||T^{-1}AT||$ can sometimes only be reduced slightly if the initial transformation $T$ is sufficiently close to a 'bad' local minimum, although the system is Q-stable [21] namely the global minimum of $||T^{-1}AT||$ is indeed less than unity [22]. Second, owing to the iterative nature of the approach, it is usually far more expensive in computation as compared to the algorithms proposed in the preceding sections. Consequently, an efficient stability test should combine a pre-processing for norm-reduction, which reduces $||A||$ to a value close to the unity, with one of the algorithms proposed in Sections III and IV.

The second approach for the norm reduction is based on the 2-D Lyapunov equation

$$A^TGA - G = -W \tag{24}$$

where $G = G_1 \oplus G_2$ and $W$ are positive definite. The filter (1) is said to be Q-stable [21] if there exists $G = G_1 \oplus G_2 > 0$ such that (24) holds for some $W > 0$. It has been known that the Q-stability of (1) implies its BIBO stability but not necessarily vice versa [23]. The connection of (24) to the norm reduction problem becomes obvious when setting $T = G^{-1/2}$ and writting (24) as

$$(T^{-1}AT)^T(T^{-1}AT) - I = -(T^TWT) < 0$$

which implies that

$$||T^{-1}AT|| < 1 \tag{25}$$

In other words, $||T^{-1}AT||$ can be reduced to a value less than unity at once if (24) is solved for some $W > 0$ with $G$ block-diagonal and positive definite. Unfortunately, solving (24) is not easy [24]. Here we seek a transformation $T = T_1 \oplus T_2$ that is "close" to $G^{-1/2}$ in a certain sense and is relatively easy to compute. Specifically, let us write (24) as

$$\begin{bmatrix} A_{11}^T G_1 A_{11} - G_1 + A_{21}^T G_2 A_{21} & A_{11}^T G_1 A_{12} + A_{21}^T G_2 A_{22} \\ A_{12}^T G_1 A_{11} + A_{22}^T G_2 A_{21} & A_{22}^T G_2 A_{22} - G_2 + A_{12}^T G_1 A_{12} \end{bmatrix} = - \begin{bmatrix} W_{11} & W_{12} \\ W_{12}^T & W_{22} \end{bmatrix} \tag{26}$$

and single out the two blocks along the diagonal, i.e.

$$A_{11}^T G_1 A_{11} - G_1 + A_{21}^T G_2 A_{21} = -W_{11} \tag{27a}$$

$$A_{22}^T G_2 A_{22} - G_2 + A_{12}^T G_1 A_{12} = -W_{22} \tag{27b}$$

Since (27) is a system of linear equations where the number of unknowns is equal to the number of equations, it is solvable under a typical rank condition. A more elegant solution approach to (27) will be given later in this section but for the moment let us assume that the solution of (27) for given $W_{11} > 0$ and $W_{22} > 0$ is available. Although matrix $G = G_1 \oplus G_2$ so obtained is not the solution to 2-D Lyapunov equation (26) in general, a state-variable transformation $T = G^{-1/2}$ can be used to *reduce* the norm of the system matrix. Indeed, if we set

$$\hat{A} = T^{-1}AT = \begin{bmatrix} \hat{A}_{11} & \hat{A}_{12} \\ \hat{A}_{21} & \hat{A}_{22} \end{bmatrix}$$

then (27) can be written as

$$\begin{bmatrix} \hat{A}_{11} \\ \hat{A}_{21} \end{bmatrix}^T \begin{bmatrix} \hat{A}_{11} \\ \hat{A}_{21} \end{bmatrix} - I = -\hat{W}_{11}$$

$$\begin{bmatrix} \hat{A}_{12} \\ \hat{A}_{22} \end{bmatrix}^T \begin{bmatrix} \hat{A}_{12} \\ \hat{A}_{22} \end{bmatrix} - I = -\hat{W}_{22}$$

where $\hat{W}_{ii} = G_i^{-1/2} W_{ii} G_i^{-1/2}$ $(i = 1, 2)$, i.e.,

$$\left\| \begin{bmatrix} \hat{A}_{11} \\ \hat{A}_{21} \end{bmatrix} \right\| < 1, \quad \left\| \begin{bmatrix} \hat{A}_{12} \\ \hat{A}_{21} \end{bmatrix} \right\| < 1 \tag{28}$$

which implies that

$$||\hat{A}|| < 2 \tag{29}$$

We see that the transformation $T = G^{-1/2}$ does reduce the norm of the system matrix of any dimension to satisfy (29) which is an acceptable bound for the purpose of executing the algorithms given in Sections III and IV more efficiently.

Concerning solving equations (27a) and (27b), let us consider a system of iterative Lyapunov equations

$$A_{11}^T G_1^{(n)} A_{11} - G_1^{(n)} + A_{21}^T G_2^{(n-1)} A_{21} = -W_{11} \tag{30$a$}$$

$$A_{22}^T G_2^{(n)} A_{22} - G_2^{(n)} + A_{12}^T G_1^{(n-1)} A_{12} = -W_{22} \tag{30$b$}$$

For given $W_{11} > 0$ and $W_{22} > 0$ and initial $G_1^{(0)} \geq 0$, $G_2^{(0)} \geq 0$, (30) becomes standard Lyapunov equations

$$A_{11}^T G_1^{(n)} A_{11} - G_1^{(n)} = -(W_{11} + A_{21}^T G_2^{(n-1)} A_{21}) \tag{31$a$}$$

$$A_{22}^T G_2^{(n)} A_{22} - G_2^{(n)} = -(W_{22} + A_{12}^T G_1^{(n-1)} A_{12}) \tag{31$b$}$$

and it is well-known that if the 2-D filter is stable, then matrices $A_{11}$ and $A_{22}$ are stable [26]. Thus $G_1^{(n)} > 0$ and $G_2^{(n)} > 0$, and if sequences $\{G_1^{(n)}\}$ and $\{G_2^{(n)}\}$ converge to $G_1$ and $G_2$ as $n \to \infty$, respectively, then $G_1 \geq 0$ and $G_2 \geq 0$ satisfy equations (27a) and (27b). This in turn implies that $G_1 > 0$ and $G_2 > 0$. To address the convergence issue, note that (31) gives

$$A_{11}^T (G_1^{(n+1)} - G_1^{(n)}) A_{11} - (G_1^{(n+1)} - G_1^{(n)}) = -A_{21}^T (G_2^{(n)} - G_2^{(n-1)}) A_{21} \tag{32$a$}$$

$$A_{22}^T (G_2^{(n+1)} - G_2^{(n)}) A_{22} - (G_2^{(n+1)} - G_2^{(n)}) = -A_{12}^T (G_1^{(n)} - G_1^{(n-1)}) A_{12} \tag{32$b$}$$

If one starts with $G_1^{(0)} = 0$ and $G_2^{(0)} = 0$, then it follows from (32a) and (32b) that $G_1^{(2)} - G_1^{(1)} > 0$ provided that $(A_{11},\ A_{21})$ is a 1-D observable pair [27] and that $G_2^{(2)} - G_2^{(1)} > 0$ if $(A_{22},\ A_{12})$ is a 1-D observable pair. Hence we obtain two *monotonically increasing* sequences of positive definite matrices $\{G_1^{(n)},\ n = 0, 1, 2, \ldots\}$

and $\{G_2^{(n)},\ n = 0, 1, 2, \ldots\}$ that satisfy (31a) and (31b). It follows that $G_1^{(n)} \to G_1$ and $G_2^{(n)} \to G_2$ for some $G_1 > 0$ and $G_2 > 0$ as $n \to \infty$ if $\{G_1^{(n)}\}$ and $\{G_2^{(n)}\}$ are bounded in norm [28]. Although the conditions, under which $||G_1^{(n)}||$ and $||G_2^{(n)}||$ are bounded, remain to be found, from our numerical experience on this matter these conditions have much to do with the Q-stability of the filter. As a matter of fact, for all Q-stable filters that we tested, the iterative Lyapunov equations (31) converge from any randomly selected non-negative $G_1^{(0)}$ and $G_2^{(0)}$, but (31) failed to converge for the stable but not Q-stable filter used in [29].

## VI. A Case Study

In this section we present a case study using three 2-D state-space filters to illustrate the stability analysis methods proposed. As stability is the only issue of concern, quantities other than system matrix will not be specified. For the sake of convenience, we call the algorithms proposed in Sections III and IV the K-test and L-test, respectively.

**6.1** We first consider a 2-D filter of order (2, 2) with

$$A = \begin{bmatrix} -0.5000 & -0.7500 & 0.3895 & 0.3895 \\ 0.0000 & 0.0000 & 0.0000 & 0.0000 \\ 0.1423 & 0.0000 & -0.4347 & 0.0981 \\ -0.0342 & 0.0000 & -0.1094 & -0.0903 \end{bmatrix}$$

With $n_1 = n_2 = 2$, the size of $E_0$, $E_1$ and $E_2$ is $16 \times 16$. By applying the K-test, we conclude that the system is stable after 8 iterations. The fast convergence of the algorithm is primarily due to the large gain factor $1/(||A_1||||A_2||)$ $(= 0.9990)$ in (19). Similarly, applying the L-test to the above filter leads to the same conclusion on stability after 8 iterations. However it is observed that the L-test is computationally more economical (177 Kflops) than the K-test (441 Kflops). Since $||A|| = 1.0786$ is already small enough for the use of these tests, the norm reduction technique will unlikely offer considerable improvement on the computation efficiency of the tests.

In an earlier work [12], it is shown that system (1) is stable if and only if there exist integers $L > 0$ and $K > 0$ such that

$$L > \frac{\pi||A_2||}{2(m_{LK} - \pi/K)} \quad \textit{and} \quad m_{LK} > \frac{\pi}{K}$$

where

$$\begin{aligned} m_{LK} &= \min_{0 \le l \le L} \min_{0 \le k \le K} \underline{\sigma}[\exp(j\omega_k I - F(\theta_l)] \\ \omega &= 2\pi k/K, \quad \theta_l = \pi l/L \end{aligned}$$

and $F(\theta_l)$ is 1-D stable for $l = 0, 1, \ldots, L$. This method as applied to the above filter requires $K = L = 32$ with 435 Kflops to achieve the same conclusion.

**6.2** Next we consider a 2-D filter of order (2, 2) with

$$A = \begin{bmatrix} 1.8890 & -0.9122 & -1.0000 & 0.0000 \\ 1.0000 & 0.0000 & 0.0000 & 0.0000 \\ 0.0277 & -0.0258 & 1.8890 & 1.0000 \\ -0.0258 & 0.0243 & -0.9122 & 0.0000 \end{bmatrix}$$

We compute $||A|| = 2.7584$ and begin the K-and L-test with $\theta = 0$. The average $\Delta\theta$ over first 800 iterations for the K- and L-test are $1.025 \times 10^{-7}$ and $2.509 \times 10^{-7}$, respectively, thus the implemention of either test is obviously not feasible. The iterative Lyapunov approach proposed in Sec. V is then used to reduced $||A||$. With $G_1^{(0)} = 0$ and $G_2^{(0)} = 0$, the iterative Lyapunov equations converge after 13 iterations to an equivalent system matrix

$$A_e = \begin{bmatrix} 0.9544 & -0.1873 & -0.2443 & 0.2088 \\ 0.1079 & 0.9346 & 0.1923 & -0.1644 \\ 0.0567 & -0.0370 & 0.9445 & 0.1924 \\ -0.0034 & 0.0375 & -0.1046 & 0.9445 \end{bmatrix}$$

whose norm is 1.1614. When the K-test is applied to the new sytem matrix $A_e$, the conclusion of $A_e$ being stable is achieved after 154 iterations with 9556 Kflops. The average $\Delta\theta$ per iteration is 0.0204, which is about $2 \times 10^5$ times larger than the average increment for the original system matrix. Similarly, applying the L-test to $A_e$ leads to the same stability conclusion after 181 iterations with 5074 Kflops. The average $\Delta\theta$ per iteration for the L-test is 0.0174. Again, we see that the Lyapunov approach is computationally more efficient than the K-test.

A remarkable fact associated with the Lyapunov method described in Sec. V to reduce the norm of **A** is that once a new but equivalent system matrix is obtained, the approach can be used again to obtain another equivalent system matrix with its norm being further reduced on most occasions. For the 2-D filter at hand, after the application of the Lyapunov type norm-reduction technique 6 times, we obtain a system matrix

$$A^* = \begin{bmatrix} 0.9421 & -0.1383 & -0.1116 & 0.0756 \\ 0.1455 & 0.9469 & 0.1209 & -0.0819 \\ 0.1260 & -0.0581 & 0.9419 & 0.1522 \\ -0.0041 & 0.0770 & -0.1322 & 0.9471 \end{bmatrix}$$

whose norm is 1.0049. The application of the K-test and L-test leads to the conclusion of stability with 91 iterations (5626 Kflops) and 98 iterations (2641 Kflops), respectively.

For the sake of comparison, the method of [12] turns out to be computationally not feasible even for $A_e$. When applying it to $A^*$, the method requires $K = L = 300$ with more than $1.564 \times 10^5$ Kflops to achieve the same conclusion the filter's stability.

**6.3** We now consider a lowpass 2-D filter of order (4, 8) with

$$A_{11} = \begin{bmatrix} 0.537000 & -0.06881 & 0.985510 & 0.503880 \\ 1 & 0 & 0 & 0 \\ 0 & 0 & 0.538819 & -0.066576 \\ 0 & 0 & 1 & 0 \end{bmatrix}$$

$$A_{12} = \begin{bmatrix} -1 & 0 & 0 & 0 & -1 & 0 & 1 & 0 \\ 0 & 0 & 0 & 0 & 0 & 0 & 0 & 0 \\ 0 & 0 & 0 & 0 & -1 & 0 & 0 & 0 \\ 0 & 0 & 0 & 0 & 0 & 0 & 0 & 0 \end{bmatrix}$$

$$A_{21} = \begin{bmatrix} -0.390721 & 0.244967 & -0.483603 & -0.248260 \\ 0.251223 & -0.145125 & 0.026974 & 0.013792 \\ 1.270478 & 1.106765 & 0.198140 & 0.101307 \\ 1.796448 & 0.421991 & 0.592117 & 0.302742 \\ 0 & 0 & -0.393352 & 0.242461 \\ 0 & 0 & 0.252014 & -0.145254 \\ 0 & 0 & 1.270845 & 1.107215 \\ 0 & 0 & 1.797547 & 0.423333 \end{bmatrix}$$

$$A_{22} = \begin{bmatrix} 0.490714 & 1 & 0 & 0 & 0.490714 & 0 & -0.490714 & 0 \\ -0.027371 & 0 & 0 & 0 & -0.027371 & 0 & 0.027371 & 0 \\ -0.201054 & 0 & 0 & 1 & -0.201054 & 0 & 0.201054 & 0 \\ -0.600823 & 0 & 0 & 0 & -0.600823 & 0 & 0.600823 & 0 \\ 0 & 0 & 0 & 0 & 0.491231 & 1 & 0 & 0 \\ 0 & 0 & 0 & 0 & -0.028179 & 0 & 0 & 0 \\ 0 & 0 & 0 & 0 & -0.201054 & 0 & 0 & 1 \\ 0 & 0 & 0 & 0 & -0.600823 & 0 & 0 & 0 \end{bmatrix}$$

We compute $||A|| = 3.8201$. Applying the K- and L-tests, it is found that the increment $\Delta\theta$ over first 100 iterations is $4.292 \times 10^{-5}$ and $9.860 \times 10^{-6}$, respectively, thus a pre-processing of the system matrix is necessary. By repeatedly applying the Lyapunov type norm-reduction technique 7 times, we obtained a new but equivalent system matrix $A_e$ with

$$A_{e11} = \begin{bmatrix} 0.3144 & 0.0127 & 0.0010 & 0.0644 \\ 0.3705 & 0.2348 & 0.0461 & -0.0012 \\ -0.1393 & 0.0584 & 0.3856 & -0.0395 \\ -0.0458 & 0.0999 & 0.2916 & 0.1411 \end{bmatrix}$$

$$A_{e12} = \begin{bmatrix} -0.3591 & 0.1057 & 0 & -0.0156 & -0.0748 & 0.0218 & 0.3591 & -0.0674 \\ 0.2802 & -0.0810 & 0 & 0.0121 & 0.0429 & -0.0127 & -0.2802 & 0.0512 \\ 0.0643 & 0.0465 & 0 & 0.0001 & -0.6710 & 0.1871 & -0.0643 & -0.0499 \\ -0.1435 & 0.0164 & 0 & -0.0052 & 0.2409 & -0.0668 & 0.1435 & -0.0024 \end{bmatrix}$$

$$A_{e21} = \begin{bmatrix} -0.1031 & 0.3081 & -0.1758 & -0.3058 \\ 0.1504 & -0.3040 & -0.1743 & -0.1062 \\ 0.0671 & 0.0593 & 0.0032 & 0.0053 \\ 0.1715 & 0.1217 & 0.0094 & 0.0075 \\ -0.0742 & 0.0985 & -0.4069 & 0.4585 \\ 0.1085 & -0.1479 & 0.0579 & -0.3627 \\ 0.1008 & -0.3014 & 0.1756 & 0.3066 \\ -0.1325 & 0.2651 & 0.1914 & 0.1334 \end{bmatrix}$$

$$A_{e22} = \begin{bmatrix} 0.0878 & 0.4759 & 0 & 0.0008 & 0.0450 & -0.0224 & -0.0878 & -0.1446 \\ 0.0101 & 0.1493 & 0 & -0.0007 & 0.0354 & -0.0312 & -0.0101 & -0.1473 \\ -0.0069 & 0.0017 & 0 & 0.3332 & -0.0010 & 0.0003 & 0.0069 & -0.0013 \\ -0.0088 & 0.0130 & 0 & -0.0005 & -0.0031 & 0.0016 & 0.0088 & -0.0118 \\ -0.0042 & -0.0244 & 0 & 0 & 0.1944 & 0.6701 & 0.0042 & 0.0244 \\ -0.0031 & -0.0012 & 0 & -0.0002 & -.0599 & 0.3094 & 0.0031 & 0.0016 \\ -0.0706 & -0.1466 & 0 & -0.0005 & -0.0440 & 0.0220 & 0.0706 & 0.4811 \\ -0.0169 & -0.1726 & 0 & 0.0007 & -0.0392 & 0.0331 & 0.0169 & 0.1709 \end{bmatrix}$$

for which $||A_e|| = 1.0116$. The K-test as applied to the new system matrix $A_e$ leads to the conclusion of $A_e$ being stable after 18 iterations with $5.7 \times 10^5$ Kflops, while the use of the L-test reaches the same conclusion after 24 iterations with $1.5 \times 10^4$ Kflops. We see that as filter order increases, the superiory of the L-test to the K-test in computation efficiency becomes more apparent.

It should be mentioned that the matrix $A_e$ given above is not the best that can be obtained by the Lyapunov type norm-reduction technique. As a matter of fact, one can obtain an equivalent system matrix $A^*$ after applying 15 times of the reduction technique to obtain an $A^*$ with $||A^*|| = 0.9335$. Thus we conclude that the system is stable [22] without using the K- or L-test.

Although it is quite obvious from the numerical results in Sec. 6.2 and 6.3, we would like to emphasize one point as our final remark on this case study: reducing the system matrix norm improves not only the gain factor $1/(2||A_1||||A_2||)$ in (19) and $1/||A_2||$ in (23) but the "structure" of the filter as well that in turn increases $\underline{\sigma}(\theta)$ in (19) and reduces the rest of denominator in (23).

## References


[1] S. Gutman, "Root-clustering of a real matrix in an algebraic region," *Int. J. Control*, vol. 29, pp. 871-880, 1979.

[2] S. Gutman and E. I. Jury, "A general theory of matrix root-clustering in subregions of the complex plane," *IEEE Trans. Automat. Contr.*, vol. 26, pp. 853-863, Aug. 1981.

[3] S. Gutman, "Matrix root-clustering in algebraic regions," *Int. J. Control*, vol. 39, pp. 773-778, 1984.

[4] S. Gutman, "State-space stability of two-dimensional systems," *IMA J. Math. Contr. Info.*, vol. 4, pp. 55-63, 1987.

[5] G. Gu and E. B. Lee, "A numeriral algorithm for stability testing of 2-D recursive digital filters," *IEEE Trans. Circuits Syst.*, vol. 37. pp. 135-138, Jan. 1990.

[6] P. Agathoklis, E. I. Jury, and M. Mansour, "An algebraic test for internal stability of 2-D discrete systems," in *Proc. Int. Symp. MTNS 89*, vol. 1, pp. 303-310, 1990.

[7] P. Agathoklis, E. I. Jury, and M. Mansour, "Algebraic necessary and sufficient conditions for the stability of 2-D discrete systems," in *Proc. IEEE Int. Symp. Circuits Syst.*, vol. 1, pp. 610-613, June 1991.

[8] M. Wang, E. B. Lee, and D. Boley, "A simple method to determine the stability and margin of stability of 2-D recursive filters," *IEEE Trans. Circuits Syst.*, vol. 39, pp. 237-239, March 1992.

[9] T. Beelen and P. Van Dooren, "A pencil approach for embedding a polynomial matrix into a unimodular matrix," *SIAM J. Matrix Appli.*, vol. 9, pp. 77-89, 1988.

[10] I. Gohberg, P. Lancaster, and L. Rodman, *Matrix Polynomials*, New York: Accedemic Press, 1982.

[11] J. Cullum, W. Kerner, and R. Willoughby, "A generalized nonsymmetric Lanczos procedure," *Comput. Phys. Comm.*, pp. 19-48, 1989.

[12] W.-S. Lu, "2-D stability test via 1-D stability robust analysis," *Int. J. Control*, vol. 48, no. 4, pp. 1735-1741, 1988.

[13] M. E. Sezer and D. D. Saljak, "Robust stability of discrete systems," Int. J. Control, vol. 48, pp. 2055-2063, 1988.

[14] S. Y. Kung, B. Levy, M. Morf, and T. Kailath, "New results in 2-D system theory, Part II: State-space models, realization and notions of controllability, observability and minimality," *Proc. IEEE*, vol. 65, pp. 945-961, June 1977.

[15] S. Pissanetzky, *Sparse Matrix Technology*, London: Academic Press, 1984.

[16] G. W. Stewart, *Introduction to Matrix Computations*, New York: Academic Press, 1973.

[17] S. Barnett, *Polynomials and Linear Control Systems*, New York: Marcel Dekker, 1983.

[18] R. Fletcher, *Practical Methods of Optimization*, Chichester, UK: John Wiley, 1987.

[19] Q.-H. Meng, A. Hasan, W.-S. Lu, and V. K. Bhargava, "On the minimization of system matrix for 2-D digital filters," in *Proc. IEEE Pacific Rim Conf.*, pp. 396-399, May 1991.

[20] M. J. D. Powell, "Restart procedures for the conjugate gradient method," *Math. Programming*, vol. 12, pp. 241-254, 1977.

[21] W. Wang, J. Doyle, C. Beck, and K. Glover, "Model reduction of LFT systems," in *Proc. IEEE Conf. Decision and Control*, pp. 1233-1238, Brighton, UK, Dec. 1991.

[22] W.-S. Lu, A. Antoniou, and P. Agathoklis, "Stability of 2-D digital filters under parameter variations," *IEEE Trans. Circuits Syst.*, vol. 33, pp. 476-482, May 1986.

[23] B. D. O. Anderson, P. Agathoklis, E. I. Jury, and M. Mansour, "Stability and the matrix Lyapunov equation for discrete 2-D systems," *IEEE Trans. Circuits Syst.*, vol. 33, pp. 261-266, March 1986.

[24] P. Agathoklis, E. I. Jury, and M. Mansour, "The discrete-time strictly bounded-real lemma and its computation of positive definite solutions to the 2-D Lyapunov equations," *IEEE Trans. Circuits Syst.*, vol. 36, pp. 830-837, June 1989.

[25] K. Zhou, J. L. Aravena, G. Gu, and D. Xiong, "Two-dimensional system model reduction by quasi-balanced truncation and singular perturbation," Research Report, Dept. of Electrical and Computer Engineering, Louisiana State University, Oct. 1991.

[26] W.-S. Lu and E. B. Lee, "Stability analysis for 2-D systems," *IEEE Trans. Circuits Syst.*, vol. 30, pp. 455-461, July 1983.

[27] T. Kailath, *Linear Systems*, Englewood Cliffs, NJ: Prentice-Hall, 1980.

[28] L. V. Kantorovich and G. P. Akilov, *Functional Analysis in Normed Space*, Oxford: Pergamon Press, 1964.

[29] B. D. O. Anderson, P. Agathoklis, E. I. Jury, and M. Mansour, "Stability and the matrix Lyapunov equation for discrete 2-Dimensional systems," *IEEE Trans. Circuits Syst.*, vol. 33, pp. 261-267, March 1986.

# 46 A Control-Theoretic Banach Lie Group $G_{A,B}$—The Stability Group

**Dahlard L. Lukes** University of Virginia, Charlottesville, Virginia

## 1. Introduction.

This paper studies a group of symmetries, $\mathbf{G}_{A,B}$ , associated with the differential control equation

$$\dot{x} = Ax + Bu \tag{1}$$

having state variable $x \in R^n$, control variable $u \in R^p$ and real coefficient matrices $A \in \mathbf{M}_n(R)$, $B \in \mathbf{M}_{n,p}(R)$ . An interval $I = [t_0 , t_1]$ is prescribed. To each initial state $x^0 \in R^n$ and integrable control function $u(\cdot)$ on the interval there exists a unique solution $x(\cdot)$ to (1) satisfying the initial condition $x(t_0) = x^0$. That solution is called the *response* of (1) to the input $(x^0, u(\cdot))$.

Consider any matrix functions, $P(\cdot): I \to GL(n,R)$ and $Q(\cdot): I \to GL(p,R)$ absolutely continuous and $K(\cdot): I \to \mathbf{M}_{p,n}$ integrable. Let $\tilde{x}(\cdot)$ be the response of (1) to the initial state $\tilde{x}^0 = P(t_0)x^0$ and control function $\tilde{u} = K(\cdot)x(\cdot) + Q(\cdot)u(\cdot)$. This article studies the induced constraints on $P(\cdot)$, $Q(\cdot)$ and $K(\cdot)$ resulting from the invariance requirement that $\tilde{x}(\cdot) = P(\cdot)x(\cdot)$ for all $x^0$ and $u(\cdot)$.

To keep this article's length within limits, $rank(B) = p$ is assumed, henceforth, starting with Theorem 2.2.1 (where it is used only in showing $\mathbf{G}_{A,B}$ to be a group). Note for later reference that invertibility of $B^*B$ and $Q$, the solution to $PB = BQ$ for $P$ invertible, are two transparent consequences. Later, in addition, and closer to its application, $(A,B)$ is assumed controllable. Thus the (generic) case of primary control-theoretic interest is covered.

The literature contains a number of theories of (1) whose starting point is changes of state and control coordinates and a feedback equation via $P(\cdot)$, $Q(\cdot)$ and $K(\cdot)$, as appear in the previous paragraphs, mainly for constant matrix functions and generally oriented toward changing the form of the model equation (1) or its dynamics. Notable among such results are the Brunovsky form of (1) and Heymann's control variable dimension reduction. This article makes application of the former. (See Markus[1] for earlier studies of uncontrolled equations.)

However, the opposite tack is taken in this paper wherein the original system (1) is embedded into a feedback system, along with a maximal set of parameters, allowing some control over the inputs experienced by the original component of the feedback system while the ambient feedback system is required to maintain the same input-output characteristics as the original system. The author's interest in such invariance was generated by problems of software fault tolerance. Applications of the symmetries found here are deferred to other articles.

## 2. The Lie Algebra $\mathfrak{g}_{A,B}$ and Group $\mathcal{G}_{A,B}$

### 2.1 The Equations for Invariance

Imposition of the invariance defined in the *Introduction* easily translates into the equivalent requirement that $P(\cdot)$, $Q(\cdot)$ and $K(\cdot)$ satisfy the equations

$$\dot{P}(t) + P(t)A - AP(t) = BK(t) \ , \ a.e. \ t \in I \tag{1}$$

$$P(t)B = BQ(t) \ , \ t \in I \tag{2}$$

and that $P(t)$ and $Q(t)$ be invertible at each $t$ in that interval. (Hereafter the "a.e." where omitted is understood.)

### 2.2 Definitions of $\mathfrak{g}_{A,B}$ and $\mathcal{G}_{A,B}$

Equations (2.1/1) and (2.1/2) are the basis for the following definitions:

$$\mathfrak{g}_{A,B} = \{P(\cdot) \mid \ (2.1/1) - (2.1/2) \ \textit{hold for some} \ Q(\cdot), \ K(\cdot) \ \}$$

$$\mathcal{G}_{A,B} = \{P(\cdot) \mid \ P(\cdot) \in \mathfrak{g}_{A,B} \ , \ det(P(t)) \neq 0 \ , \ t \in I \ \}$$

*in which $P(\cdot)$ and $Q(\cdot)$ are absolutely continuous and $K(\cdot)$ is integrable on I.*

The above definitions are meant to entail the conspicuous linear space structure of $\mathfrak{g}_{A,B}$ over the field $R$ resulting from pointwise matrix addition and scalar multiplication of its elements and on $\mathcal{G}_{A,B}$ the pointwise matrix product. Moreover, $\mathfrak{g}_{A,B}$ is endowed with the pointwise matrix commutator product.

**Theorem 2.2.1** *$\mathfrak{g}_{A,B}$ is a Lie algebra and $\mathcal{G}_{A,B}$ is a group.*

Proof. Let $P_1$, and $P_2$ be arbitrary elements of $\mathfrak{g}_{A,B}$ ; hence they satisfy (2.1/1) - (2.1/2) for respective integrable $K_1$ , $K_2$ and absolutely continuous $Q_1$, $Q_2$. But $K_1 + K_2$ is integrable and $Q_1 + Q_2$ is absolutely continuous. With this and the linearity of (2.1/1), (2.1/2) it is evident that $\mathfrak{g}_{A,B}$ is closed under addition and moreover under scalar multiplication. From the fact that $P_1$ and $P_2$ each satisfy an equation of type (2.1/1) with their respective $K$'s, differentiation of the product

$$P = [P_1, P_2] = P_1P_2 - P_2P_1$$

gives

$$\dot{P} = -PA + AP + B(K_1P_2 - K_2P_1) + P_1BK_2 - P_2BK_1 \, .$$

Since $P_1B = BQ_1$ and $P_2B = BQ_2$ it is evident that $P$ satisfies (2.1/1) for $K = K_1P_2 - K_2P_1 + Q_1K_2 - Q_2K_1$ which is integrable. The calculation

$$PB = [P_1, P_2]B = P_1P_2B - P_2P_1B = P_1BQ_2 - P_2BQ_1 = BQ_1Q_2 - BQ_2Q_1 = BQ,$$

with $Q = [Q_1, Q_2]$ noted to be absolutely continuous, finishes the verification that $\mathcal{g}_{A,B}$ is a Lie algebra.

For $P_1, P_2$ as above $P = P_1P_2$ is absolutely continuous and satisfies (2.1/1) with $K = Q_1K_2 + K_1P_2$. Furthermore, (2.1/2) is satisfied by $P$ for $Q = Q_1Q_2$. Thus $\mathcal{g}_{A,B}$ is closed under the standard as well as commutator (pointwise) product and clearly $\mathcal{G}_{A,B}$ is then closed under the standard product as well.

For any $P \in \mathcal{G}_{A,B}$, $PB = BQ$ and thus $P^{-1}B = BQ^{-1}$. Since $det(Q(\cdot))$ is uniformly bounded away from zero on $I$ it is evident that $Q^{-1}$ is absolutely continuous on that interval. Using (2.1/1) and (2.1/2) it follows that $\frac{d}{dt}P^{-1} = -P^{-1}A + AP^{-1} + B\tilde{K}$ where $\tilde{K} = Q^{-1}KP^{-1}$. At this point it is clear that $\mathcal{G}_{A,B}$ is closed under pointwise matrix inversion and, but for a few trivial details, $\mathcal{G}_{A,B}$ has been shown to be a group. The proof of Theorem 2.2.1 is finished.

## 2.3 An Alternative Characterization of $\mathcal{g}_{A,B}$ and $\mathcal{G}_{A,B}$

On the Lie algebra of absolutely continuous matrix functions $P(\cdot) : I \to \mathbf{M}_n$ define $f = f_1 \otimes f_2$ by the formulas

$$f_1(P) = [I_n - B(B^*B)^{-1}B^*](\dot{P} + PA - AP) \tag{1}$$

$$f_2(P) = [I_n - B(B^*B)^{-1}B^*]PB. \tag{2}$$

(Since $B^*B$ is invertible (2.3/1) and (2.3/2) make sense and $f(P) : I \to \mathbf{M}_n \times \mathbf{M}_{n,p}$ is integrable.)

**Proposition 2.3.1**

$$f^{-1}(0) = \mathcal{g}_{A,B}.$$

Proof. For $P \in f^{-1}(0)$ the equation $f(P) = 0$ can be written as the pair (2.1/1), (2.1/2) for the choice

$$K = (B^*B)^{-1}B^*[\dot{P} + PA - AP] \tag{3}$$

$$Q = (B^*B)^{-1}B^*PB \tag{4}$$

This shows that $f^{-1}(0) \subseteq \mathcal{g}_{A,B}$. The reverse inclusion also holds since each $P \in \mathcal{g}_{A,B}$ satisfies (2.1/1) and (2.1/2) for some $K$ and $Q$ and (3), (4) follow.

## 3. The Algebraic Structure of $\mathcal{G}_{A,B}$

3.1 Canonical Case : $A = A^o, B = B^o$

Analysis of $\mathcal{G}_{A,B}$ is made first for the special coefficient matrices

$$A^o = diag\,[A_1, A_2, \cdots, A_p] \tag{1}$$

in which

$$A_i = \begin{bmatrix} 0 & 1 & 0 & \cdots & 0 \\ 0 & 0 & 1 & & \cdot \\ \cdot & & & \cdot & \cdot \\ \cdot & & & & \cdot \\ \cdot & & & & 1 \\ 0 & 0 & \cdots & \cdots & 0 \end{bmatrix}$$

is $r_i \times r_i$, $(i = 1, 2, \cdots, p)$, with $r_1 \geq r_2 \geq \cdots \geq r_p \geq 1$, $r_1 + r_2 + \cdots + r_p = n$, and

$$B^o = [e_{r_1}, e_{r_1+r_2}, \cdots, e_n] \tag{2}$$

in which $e_m$ denotes the mth column of the $n \times n$ identity matrix $I_n$.

Define $\Lambda = [I_n - B(B^*B)^{-1}B^*]$ which is easily computed to be

$$\Lambda^o = diag\,[\Lambda_1, \Lambda_2, \cdots, \Lambda_p] \tag{3}$$

for $B = B^o$ where $\Lambda_i = diag\,[1, 1, \cdots, 1, 0]$ is $r_i \times r_i$. The problem is to solve

$$\Lambda^o P B^o = 0 \tag{4}$$

$$\Lambda^o[\dot{P} + PA - AP] = 0 \tag{5}$$

for $P$. It is dealt with in block form using $P = (P^{ij})$, where $P^{ij}$ is of size $r_i \times r_j$, $(i = 1, 2, \cdots, p)$, $(j = 1, 2, \cdots, p)$. (For notational simplicity the naught superscript is dropped from $A^o$, $B^o$ and $\Lambda^o$.)

If $r_1 = 1$ then $\Lambda = 0$ and all absolutely continuous $P$ are solutions.

Now consider the complementary case where $r_k \geq 2$ for largest $k = p_*$.

The first blocks to be examined are the $r_i \times 1$ blocks $P^{ij}$, $(i = 1, 2, \cdots, p_*)$, $(j = p_* + 1, p_* + 2, \cdots, p)$. The claim made is that each must be zero. If $p = p_*$ the argument is vacuous. Suppose $p_* < p$. The submatrix of $B$ consisting of its $q = n - (r_1 + r_2 \cdots + r_{p_*})$ rightmost columns has $I_q$ at its bottom with the other elements zero. In view of this and the diagonal nature of $\Lambda$ it is evident that (4) requires all elements of the $P^{ij}$ under consideration to be zero with the exception of those elements occurring in the rows of $P$ numbered $r_1, r_1 + r_2, \cdots, r_1 + r_2 + \cdots + r_{p_*}$ upon which (4) imposes no constraint. However, (5) forces those elements to be zero too. The validity of this last statement can be seen by observing that the last $q$ column equations of (5) may be written as

$$\Lambda[\dot{P}_q - AP_q] = 0 \tag{6}$$

in which $P_q$ is the sub-block of $P$ consisting of the last $q$ columns of $P$. (Those columns of $A$ are zero.) But multiplication raises the rows under examination one level and hence (6) forces them to be zero.

Since the bottom-most $q$ rows of $\Lambda$ are zero, as are its right-most $q$ columns, the remaining equations of (4)-(5) under analysis constitute a problem of the original type but with $n$ now replaced by $n-q$ and $p = p*$. Once the solution to this reduce problem is found, the solution to the original problem can be obtained by simply adjoining $q$ rows of length $n$ to the bottom, whose elements are arbitrary absolutely continuous functions on $I$ and filling in the remaining $(n-q)\times q$ block with zeros.

Thus the analysis of (4)-(5) commences once again, this time under the assumption that $p = p*$. Such a system can be written as

$$\Lambda_i[P^{i,1}_{r_1}, P^{i,2}_{r_2} \cdots, P^{i,p}_{r_p}] = 0 \tag{7}$$

$$\Lambda_i[\dot{P}^{i,j} + P^{i,j}A_i - A_iP^{i,j}] = 0, \tag{8}$$

$(i = 1,, 2, \cdots, p)$, $(j = 1, 2, \cdots, p)$ in which $P_k{}^{i,j}$ denotes the kth column of $P^{i,j}$. The problem breaks up into three cases:

*Case* $r_i = r_j$. In terms of its columns $P^{i,j} = [c^1, c^2, \cdots, c^{r_j}]$, equation (8) can be written as

$$\Lambda_i\{D[c^1, c^2, \cdots, c^{r_j}] + [0, c^1, c^2, \cdots c^{r_j-1}] - A_i[c^1, c^2, \cdots, c^{r_j}]\} = 0 \tag{9}$$

in which $D = \dfrac{d}{dt}$. This in turn is equivalent to

$$A_ic^1 = Dc^1, A_ic^2 = \Lambda_i(Dc^2 + c^1), \cdots, A_ic^{r_j} = \Lambda_i(Dc^{r_j} + c^{r_j-1}). \tag{10}$$

Solution of the first equation of (10) and the ith equation of (7) gives

$$c^1 = [c_1^1, Dc_1^1, \cdots, D^{r_i-1}c_1^1]^* \quad \text{and} \quad c^{r_j} = [0, 0, \cdots 0, c_{r_i}^1]^*.$$

With these "boundary conditions" the system (10) can be solved, first for the elements above the diagonal which all turn out to be zero, then for the diagonal terms which all must equal $c_1^1$ and then for the remaining elements below the diagonal using the rule

$\alpha_{s+1,k+1} = \alpha_{s,k} + \alpha_{s,k+1}$ to generate the remaining coefficients in the solution

$$P^{ij} = \begin{bmatrix} c_1^1 & 0 & \cdots & \cdot & 0 \\ 1Dc_1^1 & c_1^1 & & & \cdot \\ 1D^2c_1^1 & \alpha_{3,2}Dc_1^1 & & & \cdot \\ \cdot & \cdot & \cdot & & \cdot \\ \cdot & \cdot & & \cdot & \cdot \\ \cdot & \cdot & & c_1^1 & 0 \\ 1D^{r_i-1}c_1^1 & \alpha_{r_i,2}D^{r_i-2}c_1^1 & \cdots & \alpha_{r_i,r_i-1}Dc_1^1 & c_1^1 \end{bmatrix}. \tag{11}$$

*Case* $r_i < r_j$. Consider any $1 \le i \le p$, $1 \le j \le p$ for which $r_i < r_j$. This necessitates $i > j$. Solution of the first equation of system (10) and the jth column equation of (7) provide

$$c^1 = [c_1^1, Dc_1^1, \cdots, D^{r_i-1}c_1^1]^* \qquad \text{and} \qquad c^{r_j} = [0, 0, \cdots 0, c_{r_i}^{r_j}]^*$$

which act as boundary conditions to the remaining $r_j-1$ equations of (10). One can easily verify that the solution obtained is

$$P^{ij} = \begin{bmatrix} c_1^1 & c_1^2 & \cdots & c_1^{r_j-r_i+1} & 0 & \cdot & \cdot & \cdot & 0 \\ Dc_1^1 & \beta_{2,2} & \cdots & \beta_{2,r_j-r_i} & c_1^{r_j-r_i+1} & & & & \cdot \\ \cdot & \cdot & & & & \cdot & & & \cdot \\ \cdot & \cdot & & & & & \cdot & & \cdot \\ \cdot & \cdot & & & & & & \cdot & 0 \\ D^{r_i-1}c_1^1 & \beta_{r_j-1,2} & \cdots & \beta_{r_j-1,r_j-r_i} & & \cdot & \cdot & \cdot\ \beta_{r_j-1,r_j-1} & c_1^{r_j-r_i+1} \end{bmatrix} \tag{12}$$

in which the remaining $\beta_{i,j}$ are filled in according to the scheme $\beta_{s+1,k+1} = \beta_{s,k} + D\beta_{s,k+1}$.

*Case* $r_i > r_j$. This occurs only if $i < j$. First solve the jth equation of (7) to get

$$c^{r_j} = [0, 0, \cdots 0, c_{r_i}^{r_j}]^*$$

in which $c_{r_i}^{r_j}$ is unconstrained. By considering the problem of solving the system (10) in reverse order, using the starting value $c^{r_j}$ just found, it is clear that due to the facts that each equation has $A_i$ as a coefficient and that $r_i > r_j$, it is necessary that $c_1^j = 0$ for $j = r_j, r_j-1, \cdots, 1$. Now since the first equation in (10) has a solution of the form

$$c^1 = [c_1^1, Dc_1^1, \cdots, D^{r_i-1}c_1^1]^*$$

it follows that $c^1 = 0$. But the solution of the second equation then has the similar form

$$c^2 = [c_1^2, Dc_1^2 \cdots, D^{r_i-1}c_1^2]^*$$

and it likewise must be zero. By obvious induction $c^j = 0$, $(j = 1, 2, \cdots, r_j)$. Our conclusion is that for the case in which $r_i > r_j$ the only solution is $P^{ij} = 0$.

The results of the above computations are summarized by the following theorem.

**Theorem 3.1.1** *For coefficients $A^0$, $B^0$ and $\Lambda^0$ of the type given by (1), (2) and (3), respectively, the pair of matrix equations (4)-(5) has as its absolutely continuous solutions all real $n \times n$ matrix functions $P(t) = (P^{ij}(t))$ having elements of the block $P^{ij}(t) = (\lambda_{ij}^k(t))$ absolutely continuous but otherwise arbitrary, $(i = p_*+1, p_*+2, \cdots, p)$, $(j = 1, 2, \cdots, p)$, in which $p_*$ is the largest integer $k$ having $r_k \geq 2$;*

$$P^{ij}(t) = \begin{bmatrix} \lambda_{ij}^1(t) & \lambda_{ij}^2(t) & \cdots & \lambda_{ij}^{r_j-r_i+1}(t) & 0 & \cdot & \cdot & \cdot & 0 \\ D\lambda_{ij}^1(t) & \beta_{22}(t) & \cdots & \beta_{2,r_j-r_i}(t) & \lambda_{ij}^{r_j-r_i+1}(t) & & & & \cdot \\ \cdot & \cdot & & & & & \cdot & & \cdot \\ \cdot & \cdot & & & & & & \cdot & \cdot \\ \cdot & \cdot & & & & & & \cdot & 0 \\ D^{r_i-1}\lambda_{ij}^1(t) & \beta_{r_j-1,2}(t) & \cdot & \cdot & \cdot & \cdot & \cdot & \beta_{r_j-1,r_j-1}(t) & \lambda_{ij}^{r_j-r_i+1}(t) \end{bmatrix}$$

*for $r_i \leq r_j$, $(i = 1, 2, \cdots, p_*)$, $(j = 1, 2, \cdots, p_*)$, with $\lambda_{ij}^k(t)$ having $D^{r_i-1}\lambda_{ij}^k$ absolutely continuous but otherwise arbitrary, determining the $\beta_{k,l}$'s by the relation $\beta_{k+1,l+1} = \beta_{k,l} + D\beta_{k,l+1}$; and $P^{ij} = 0_{r_i,r_j}$ zero otherwise. Those solutions $P(t)$ that are invertible on $I$ constitute the group $\boldsymbol{G}_{A^o,B^o}$.*

**Example 3.1.2** Consider $A^o$, $B^o$ having $n = 8$, $p = 4$ and $r_1 = 3$, $r_2 = 2$, $r_3 = 2$, $r_4 = 1$. According to Theorem 3.1.1 the elements $P \in \boldsymbol{g}_{A^o,B^o}$ are those of the form

$$\begin{bmatrix} \alpha & 0 & 0 & 0 & 0 & 0 & 0 & 0 \\ D\alpha & \alpha & 0 & 0 & 0 & 0 & 0 & 0 \\ D^2\alpha & 2D\alpha & \alpha & 0 & 0 & 0 & 0 & 0 \\ a & b & 0 & \beta_1 & 0 & \beta_2 & 0 & 0 \\ Da & a+Db & b & D\beta_1 & \beta_1 & D\beta_2 & \beta_2 & 0 \\ c & d & 0 & \beta_3 & 0 & \beta_4 & 0 & 0 \\ Dc & c+Dd & d & D\beta_3 & \beta_3 & D\beta_4 & \beta_4 & 0 \\ e_1 & e_2 & e_3 & e_4 & e_5 & e_6 & e_7 & \gamma \end{bmatrix}$$

in which $D^3\alpha$, $D^2\beta_1$, $D^2\beta_2$, $D^2\beta_3$, $D^2\beta_4$, $D\gamma$, $D^2a$, $D^2b$, $D^2c$, $D^2d$, $De_1$, $De_2$, $De_3$, $De_4$, $De_5$, $De_6$, and $De_7$ are integrable on $I$ but otherwise the parameter functions are arbitrary. For this example $p_* = 3$.

3.2 The Finite-Dimensional Lie Algebra $\mathcal{g}^o_{A,B}$ and Lie Group $\mathcal{G}^o_{A,B}$ -- Constant Matrices

The preceding analysis can be restricted to solutions $P$ of (2.1/1)-(2.1/2) wherein $P$, $Q$ and $K$ are independent of $t \in I$. These constant elements of $\mathcal{g}_{A,B}$ constitute a finite-dimensional Lie algebra, to be denoted by $\mathcal{g}^o_{A,B}$, and similarly the constant elements of $\mathcal{G}_{A,B}$ are a subgroup of $\mathcal{G}_{A,B}$, to be denoted by $\mathcal{G}^o_{A,B}$. This subgroup is actually a finite-dimensional Lie group since it occurs as a closed subgroup of $GL(n,R)$, the $n \times n$ real non-singular matrices. (See Proposition 2.3.1.)

In the proof of Theorem 2.2.1 it was shown that $\mathcal{g}_{A,B}$ is not only a Lie algebra but is also closed under standard pointwise matrix multiplication and thus at the same time is an associative algebra. In particular, $\mathcal{g}^o_{A,B}$ can be normed to become a (finite dimensional, hence Banach) normed algebra. For example $|P| = \max \sum_{j=1}^{n} |P_{ij}|$, where the max is over $i \in \{1, 2, \cdots, n\}$, provides one choice of norm ensuring the requisite inequality $|P_1P_2| \leq |P_1||P_2|$ for $P_i \in \mathcal{g}^o_{A,B}$, $(i = 1, 2)$. (See Lukes[1,p.66].)

The elements of $\mathcal{g}^o_{A^o,B^o}$ could be viewed as a matrix with real parameters. (See Theorem 3.1.1 and look at Example 3.1.2. The matrix elements containing $D$ will be zero.) The parameters provide coordinates on $\mathcal{G}^o_{A^o,B^o}$. Those parameters could be carried over to $\mathcal{g}^o_{A,B}$ via a nonsingular linear transformation $h : \mathcal{g}^o_{A^o,B^o} \to \mathcal{g}^o_{A,B}$ to be introduced by Theorem 3.3.1. Whereas these coordinates were not essential to showing $\mathcal{G}^o_{A,B}$ to be a finite-dimensional Lie group, (that was done another way using Proposition 2.3.1 explained above) the transformation $h$ will be an essential ingredient of the proof that $\mathcal{G}_{A,B}$ is an (infinite-dimensional) Lie group. A global coordinate system for $\mathcal{G}_{A,B}$ will be obtained, based in a collection of maps of $I$ into $\mathcal{g}^o_{A,B}$, actually a Banach Lie algebra. Use will be made of the following theorem. (It continues to assume that $rank(B) = p$.)

**Theorem 3.2.1** *(Brunovsky form) For $(A,B)$ a controllable pair of matrices there exists a matrix $K_o \in \mathbf{M}_{p,n}$ and invertible matrices $P_o \in \mathbf{M}_n$ and $Q_o \in \mathbf{M}_p$ such that for some unique choice of dimension numbers $r_1 \geq r_2 \geq \cdots \geq r_p$ defining $A^o$ and $B^o$, (see (3.1/1), (3.1/2)), the equations*

$$P_o(A + BK_o)P_o^{-1} = A^o \tag{1}$$

$$P_oBQ_o = B^o \tag{2}$$

*are satisfied.*

### 3.3 Linear Isomorphisms $h$

The following theorem assigns to each element of the Lie algebra $\mathfrak{g}_{A,B}$ and to each element of the group $\mathcal{G}_{A,B}$ a canonical form.

**Theorem 3.3.1** *For $(A,B)$ a controllable pair of matrices the maps*

$$\mathfrak{g}_{A^o,B^o} \to \mathfrak{g}_{A,B} \tag{1}$$

$$\mathcal{G}_{A^o,B^o} \to \mathcal{G}_{A,B} \tag{2}$$

*defined by the formula $h(P) = P_o^{-1}PP_o$ are Lie algebra and group isomorphisms, respectively. ($P_o$ is defined by Theorem 3.2.1.)*

Proof. To show that $h$ maps $\mathfrak{g}_{A^o,B^o}$ into $\mathfrak{g}_{A,B}$ choose arbitrary $P$ in the former, which necessarily satisfies

$$\dot{P} + PA^o - A^oP = B^oK \tag{3}$$

$$PB^o = B^oQ \tag{4}$$

for some $K$ and $Q$. Let $P_1$ denote $h(P) = P_o^{-1}PP_o$. From (3), for $P$ in terms of $P_1$,

$$\dot{P}_1 + P_1P_o^{-1}A^oP_o - P_o^{-1}A^oP_oP_1 = P_o^{-1}B^oKP_o. \tag{5}$$

Using $P_o^{-1}A^oP_o = A + BK_o$ which follows from (3.2/1), rewrite (5) as

$$\dot{P}_1 + P_1A - AP_1 = BK_oP_1 + P_o^{-1}B^oKP_o - P_1BK_o = BK_oP_1 + BQ_oKP_o - P_1BK_o \tag{6}$$

where the last equality uses (3.2/2). But

$$P_1BK_o = P_o^{-1}PP_oBK_o = P_o^{-1}PB^oQ_o^{-1}K_o = P_o^{-1}B^oQQ_o^{-1}K_o = BQ_oQQ_o^{-1}K_o$$

because of the definition of $P_1$, (3.2/2), (4), then (3.2/2) again. Applied to (6) this shows that

$$\dot{P}_1 + P_1A - AP_1 = BK_1 \tag{7}$$

where $K_1 = K_oP_1 + Q_oKP_o - Q_oQQ_o^{-1}K_o$. Moreover,

$$P_1B = BQ_1 \tag{8}$$

for $Q_1 = Q_oQQ_o^{-1}$ as seen from the calculation using (3.2/2) and (4),

$$P_1B = P_o^{-1}PP_oB = P_o^{-1}PB_oQ_o^{-1} = P_o^{-1}B^oQQ_o^{-1} = BQ_oQQ_o^{-1}.$$

Equations (7) and (8) show $P_1 \in \mathfrak{g}_{A,B}$ and thus $h$ is into $\mathfrak{g}_{A,B}$.

The required surjectivity would follow if for each $P_1 \in \mathfrak{g}_{A,B}$ there were a $P \in \mathfrak{g}_{A^o,B^o}$ for which $P_o^{-1}PP_o = P_1$. The candidate is $P = P_oP_1P_o^{-1}$. It suffices to show that $P \in \mathfrak{g}_{A^o,B^o}$.

Thus it must be shown that

$$\dot{P} + PA^o - A^oP = B^oK_2 \tag{9}$$

$$PB^o = B^oQ_2 \tag{10}$$

for some $K_2$ and $Q_2$. The by now routine calculations, using (3.2/1)-(3.2/2) and the equations satisfied by $P_1$ by being an element of $\mathcal{G}_{A,B}$, that lead to such $K_2$ and $Q_2$ are not detailed here. Clearly $h$ preserves commutator products and is injective. Verification of the isomorphism of $\mathcal{G}_{A^o,B^o}$ with $\mathcal{G}_{A,B}$ as groups is omitted.

For $(A,B)$ controllable, as assumed in Theorem 3.2.1, let $\rho_1 > \rho_2 > \cdots > \rho_s$ be the distinct values of the dimension numbers $r_1 \geq r_2 \geq \cdots \geq r_p$ of $(A,B)$, i.e., of $(A^o,B^o)$ and let $m_k$ denote the multiplicity of $\rho_k$. Hence, $n = m_1\rho_1 + m_2\rho_2 + \cdots + m_s\rho_s$. Choose arbitrary $P \in \mathfrak{g}_{A^o,B^o}$. Its form is described by Theorem 3.1.1. Let $\hat{P}$ be that matrix obtained from $P$ by replacing each element of $P$ in which $D$ appears by zero. Observe that $\hat{P}$ is of block lower triangular form with $s$ square block matrices down the diagonal with the kth one of the block form $(\mu_{ij}^k I_{\rho_k})$, $(i = 1, 2, \cdots, m_k)$, $(j = 1, 2, \cdots, m_k)$ in which $\mu_{ij}^k$ has $D^{\rho_k - 1}\mu_{ij}^k$ integrable on $I = [t_o, t_1]$ but otherwise arbitrary.

**Corollary 3.3.2** *For $P \in \mathfrak{g}_{A^o,B^o}$ as discussed above,*

$$det(P) = det(h(P)) = \prod_{k=1}^{s} det(\mu_{ij}^k I_{\rho_k}) = \prod_{k=1}^{s} [det(\mu_{ij}^k)]^{\rho_k} . \tag{11}$$

Proof. Start with showing $det(P) = det(\hat{P})$ and the fact that only the diagonal blocks contribute as the product of their determinants. Details are omitted.

**Example 3.3.3** For Example 3.1.2, $\rho_1 = 3$, $\rho_2 = 2$, $\rho_3 = 1$, $m_1 = 1$, $m_2 = 2$, $m_3 = 1$. Thus (11) gives

$$det(P) = \alpha^3 \begin{vmatrix} \beta_1 & \beta_2 \\ \beta_3 & \beta_4 \end{vmatrix}^2 \gamma = \alpha^3(\beta_1\beta_4 - \beta_2\beta_3)^2\gamma .$$

## 4. $\mathcal{G}_{A,B}$ as a Lie Group

### 4.1 Infinite Dimensional Lie Groups

The parameters entering the elements of the matrices constituting $\mathcal{G}_{A,B}^o$ admitted an interpretation as coordinates on a manifold and $\mathcal{G}_{A,B}^o$ was found to be a Lie group. (See 3.3.) The task of showing that the Lie group property extends in a natural way to $\mathcal{G}_{A,B}$ begins with a recollection of the necessary notion of infinite dimensional Lie group.

A Lie group encompasses both the group and manifold structures. The axioms of a finite dimensional manifold generalize in a rather direct way to manifolds modeled on Banach spaces. (See Lang [1].) Having made that generalization the subject adds the group structure and, as in the finite dimensional setting, requires smoothness of multiplication and inversion to complete the axiomatic definition of an infinite dimensional Lie group. To begin it is necessary to single out appropriate Banach spaces in which to base the coordinate systems.

## 4.2 Selection of Coordinate Spaces

Recall that $BV(I)$, the linear space of real valued functions of bounded variation on $I = [t_0, t_1]$, is a Banach space relative to the norm $|\lambda| = |\lambda(t_0)| + V[\lambda]$ in which $V[\lambda]$ is the total variation of $\lambda$ on $I$. Moreover, it is a Banach algebra relative to pointwise multiplication and thus $|\lambda_1\lambda_2| \leq |\lambda_1||\lambda_2|$ for $\lambda_i \in BV(I)$, $(i = 1, 2)$. (See Kuller [1, pp.129,132].) It will prove useful to observe that $|\lambda|^s \leq |\lambda|$ where $|\lambda|^s = \sup_t |\lambda(t)|$ and consequently any sequence convergent in $BV(I)$ is uniformly convergent on $I$.

The analysis takes place within $AC(I)$, the absolutely continuous functions on $I$, which is easily shown to be a closed linear subspace of $BV(I)$ and hence is a Banach subalgebra. A nested sequence of subspaces of $AC(I)$ will play a significant role in that analysis. In defining that sequence it is convenient to introduce the notations $AC^o(I)$ for $AC(I)$ and $|\lambda|_o$ for $|\lambda|$.

**Definition 4.2.1**

$$AC^r(I) = \{\lambda \mid \lambda \in AC^{r-1}(I),\ D\lambda \in AC^{r-1}(I)\}$$

*with norm* $|\lambda|_r = |\lambda(t_0)| + |D\lambda|_{r-1}$ , $(r = 1, 2, \cdots)$.

Note that

$$|\lambda|_r = |\lambda(t_0)| + |D\lambda(t_0)| + \cdots + |D^r\lambda(t_0)| + V[D^r\lambda]\ ,\ (r = 0, 1, 2, \cdots).$$

**Theorem 4.2.2** *$AC^r(I)$ is a Banach space, $(r = 0, 1, 2, \cdots)$. For any integers $r_1 \geq r_2 \geq 0$,*

$$|h_2 D^s h_1|_{r_2} \leq [3\cdot 2^{r_2} - 2]\,|h_1|_{r_1}|h_2|_{r_2}$$

*for $(s = 0, 1, 2, \cdots, r_1 - r_2)$ and all $h_i \in AC^{r_i}(I)$, $(i = 1, 2)$.*

Proof. The conclusion of Theorem 4.2.2 for $r = 0$ is true by the previous remarks. To prove by induction that each $AC^r(I)$ is a Banach space assume that the claim is true for an arbitrary fixed integer $r \geq 0$. It is evident that $AC^{r+1}(I)$ is a normed linear space. To prove its completeness let $\{\lambda_n\}_1^\infty$ be any Cauchy sequence in $AC^{r+1}(I)$ The problem is to prove that $\lambda_n \to \lambda_\infty$ as $n \to \infty$ for some $\lambda_\infty \in AC^{r+1}(I)$. Since the sequence is Cauchy, for each $\varepsilon > 0$ there exists a $N(\varepsilon)$ such that

$$|\lambda_n - \lambda_m|_{r+1} = |\lambda_n(t_0) - \lambda_m(t_0)| + |\omega_n - \omega_m|_r \leq \varepsilon \qquad (1)$$

for all $n \geq N(\varepsilon)$ and $m \geq N(\varepsilon)$ where by definition $\omega_k = D\lambda_k$. Clearly this implies that

$\{\omega_n\}_1^\infty$ is Cauchy in $AC^r(I)$, hence by the inductive hypothesis there exists an $\omega_\infty \in AC^r(I)$ such that $\omega_n \to \omega_\infty$ as $n \to \infty$ and in particular the convergence is uniform on $I$. Since each $\lambda_n \in AC^o(I)$,

$$\lambda_n(t) = \lambda_n(t_0) + \int_{t_0}^{t} D\lambda_n(\sigma)d\sigma = \lambda_n(t_0) + \int_{t_0}^{t} \omega_n(\sigma)d\sigma. \tag{2}$$

Using the fact that $\lim_{n \to \infty} \lambda_n(t_0)$ exists, which is evident from (1) and the completeness of $R$, together with the uniform convergence mentioned it is possible to define the pointwise limit

$$\lambda_\infty(t) = \lim_{n \to \infty} [\lambda_n(t_0) + \int_{t_0}^{t} D\lambda_n(\sigma)d\sigma] = \lim_{n \to \infty} \lambda_n(t_0) + \int_{t_0}^{t} \omega_\infty(\sigma)d\sigma = \lambda_\infty(t_0) + \int_{t_0}^{t} \omega_\infty(\sigma)d\sigma. \tag{3}$$

Then $D\lambda_\infty \in AC^o(I)$ since $D\lambda_\infty = \omega_\infty \in AC^o(I)$, a consequence of (3). This shows that $\lambda_\infty \in AC^{r+1}(I)$. Now it is apparent that

$$|\lambda_n - \lambda_\infty|_{r+1} = |\lambda_n(t_0) - \lambda_\infty(t_0)| + |\omega_n - \omega_\infty|_r \to 0 \tag{4}$$

as $n \to \infty$ and the first part of Theorem 4.2.2 is proved.

Assume the hypothesis concerning $r_1, r_2, h_1, h_2$ and $s$. By the formula for the $r_2$-norm, the Leibniz differentiation formula for a product and standard properties of total variation, (with $C_k^n$ denoting the binomial coefficient $\frac{n!}{k!(n-k)!}$),

$$|h_2 D^s h_1| = \sum_{j=0}^{r_2} |D^j(h_2 D^s h_1)(t_0)| + V[D^{r^2}(h_2 D^s h_1)] =$$

$$\sum_{j=0}^{r_2} \Big| \sum_{k=0}^{j} C_k^j (D^k h_2) D^{j-k}(D^s h_1)(t_0) \Big| + V[\sum_{k=0}^{r_2} C_k^{r_2}(D^k h_2) D^{r_2-k}(D^s h_1)]$$

$$\leq \sum_{j=0}^{r_2} \sum_{k=0}^{j} C_k^j |D^k h_2(t_0)| |D^{j-k+s} h_1(t_0)| + \sum_{k=0}^{r_2} C_k^{r_2} V[D^k h_2] V[D^{r_2-h+s} h_1] \tag{5}$$

$$\leq |h_1|_{r_1} \left\{ \sum_{j=0}^{r_2} \sum_{k=0}^{j} C_k^j |D^k h_2(t_0)| + \sum_{k=0}^{r_2} C_k^{r_2} V[D^k h_2] \right\}$$

$$= |h_1|_{r_1} \left\{ \sum_{k=0}^{r_2} C_k^{r_2} |D^k h_2(t_0)| + C_{r_2}^{r_2} V[D^{r_2} h_2] + \sum_{j=0}^{r_2-1} \sum_{k=0}^{j} C_k^j |D^k h_2(t_0)| + \sum_{k=0}^{r_2-1} C_k^{r_2} V[D^k h_2] \right\}.$$

The parts can now be estimated.

$$\sum_{k=0}^{r_2} C_k^{r_2}|D^k h_2(t_0)|+C_{r_2}^{r_2}V[D^{r_2}h_2]\leq\sum_{k=0}^{r_2} C_k^{r_2}\{\sum_{j=0}^{r_2}|D^j h_2(t_0)|+V[D^{r_2}h_2]\}=2^{r_2}|h_2|_{r_2} \cdot \quad (6)$$

$$\sum_{j=0}^{r_2-1}\sum_{k=0}^{j} C_k^j|D^k h_2(t_0)| + \sum_{k=0}^{r_2-1} C_k^{r_2}V[D^k h_2] \leq |h_2|_{r_2}\left\{\sum_{j=0}^{r_2-1}\sum_{k=0}^{j} C_k^j + \sum_{k=0}^{r_2-1} C_k^{r_2}\right\} \quad (7)$$

$$= |h_2|_{r_2}\left\{\sum_{j=0}^{r_2-1} 2^j + \sum_{k=0}^{r_2-1}\frac{r_2}{r_2-k}C_k^{r_2-1}\right\} = |h_2|_{r_2}\{(2^{r_2}-1)+(2^{r_2}-1)\} = 2(2^{r_2}-1)|h_2|_{r_2} \cdot$$

Combining the results of (5)-(7) establishes the estimate in Theorem 4.2.2. (The last sum in (7) can be evaluated by integration of the binomial expansion of $(x+y)^n$ on one of the variables and evaluation at $x=y=1$.) This concludes the proof of Theorem 4.2.2.

**Corollary 4.2.3** $AC^r(I)$ is a Banach algebra relative to the (equivalent) norm $|\lambda|_r' = c_r|\lambda|_r$ in which $c_r = 3{\cdot}2^r - 2$, $(r=0, 1, 2, \cdots)$. (Note: $c_0 = 1$.)

4.3 $G_{A,B}$ as a Lie Group Modeled on a Banach Algebra with Banach Lie Algebra $g_{A,B}$

**Theorem 4.3.1** *If $(A,B)$ is controllable then $G_{A,B}$ is a Lie group, whose manifold is infinite dimensional and modeled on a Banach algebra, and whose Lie algebra is $g_{A,B}$. The maps (see Theorem 3.3.1)*

$$g_{A^o,B^o} \rightarrow g_{A,B}$$

$$G_{A^o,B^o} \rightarrow G_{A,B}$$

*defined by $h(P) = P_o^{-1}PP_o$ are Lie algebra and Lie group isomorphisms, respectively.*

Proof. First consider $A = A^o$, $B = B^o$ canonical. To define a manifold structure on it $G_{A,B}$ must be provided with a $C^\omega$ atlas. This can be done with one chart as follows: According to Theorem 3.1.1 there is a bijection of $g_{A,B}$ onto a Banach space $E$ consisting of a product of a finite number of Banach spaces of the types $AC^r(I)$, $(r=0, r_1, r_2, \cdots, r_p)$, where the $r_i$ are the canonical dimension numbers of $(A,B)$. Therefore $g_{A,B}$ can be given the topology of $E$ by regarding its elements as a function $P[\lambda]$ of $\lambda \in E$ where $\lambda$ has coordinates $\lambda_{ij}^k$ described in Theorem 3.1.1. Convergence in that topology implies uniform convergence of the matrix functions $P[\lambda](\cdot)$ on $I$ since that was the case for the $AC^r(I)$ spaces. Passage from $g_{A,B}$ to $G_{A,B}$ deletes the set of those $P \in g_{A,B}$ failing to satisfy the condition that $det(P) \neq 0$ on $I$. By using the uniform convergence mentioned it is easy to check that the set is closed. This leaves an open set $U \subset E$ and the resultant identification of $G_{A,B}$ with $U$ provides a one chart atlas; i.e., the $\lambda_{ij}^k \in AC^{r_i}(I)$ can be regarded as coordinate variables on $G_{A,B}$ as a manifold.

For $G_{A,B}$ to qualify as a Lie group its multiplication and inversion operations must be $C^\omega$ maps. It is sufficient to prove that they are $C^1$ maps. (See Varadarajan[1].)

Consider multiplication first. From Theorem 3.1.1 we see that if $(\lambda^1, \lambda^2)$ are coordinate maps on $\boldsymbol{G}_{A,B}\times\boldsymbol{G}_{A,B}$ into $\boldsymbol{E}\times\boldsymbol{E}$ then each coordinate component of $P[\lambda^1]P[\lambda^2]$ is a finite linear combination of terms of the type $\lambda_2 D^s\lambda_1$ with $\lambda_i \in AC^{r_i}(I)$, $(i = 1, 2)$, for $0 \le s \le r_1 - r_2$ where $r_1 \ge r_2$ are canonical dimension numbers ( not necessarily the first two) of $(A,B)$. (Note that the only elements of the product matrix of concern are those in the position of the coordinate components.) Therefore it is sufficient to prove that the map from $AC^{r_1}(I)\times AC^{r_2}(I)$ into $AC^{r_2}(I)$ sending $(\lambda_1, \lambda_2)$ to $\lambda_2 D^s\lambda_1$ is once continuously differentiable. The candidate for the derivative is the map sending $(\lambda_1, \lambda_2)$ to the bounded linear map on $AC^{r_1}(I)\times AC^{r_2}(I)$ into $AC^{r_2}(I)$ with value determined by $h_2 D^s\lambda_1 + \lambda_2 D^s h_1$. Certainly it is linear in $h = (h_1, h_2)$ and with the norm on $AC^{r_1}(I)\times AC^{r_2}(I)$ being given by $|h| = |h_1|_{r_1} + |h_2|_{r_2}$ it is bounded since according to Theorem 4.2.2,

$$|h_2 D^s\lambda_1 + \lambda_2 D^s h_1|_{r_2} \le |h_2 D^s\lambda_1|_{r_2} + |\lambda_2 D^s h_1|_{r_2} \le 2(3\cdot 2^{r_2} - 2)|h_1|_{r_1}|h_2|_{r_2} \quad (1)$$

$$\le (3\cdot 2^{r_2} - 2)(|h_1|_{r_1} + |h_2|_{r_2})^2 \le (3\cdot 2^{r_2} - 2)(|h_1|_{r_1} + |h_2|_{r_2}) = (3\cdot 2^{r_2} - 2)|h|$$

for $|h| \le 1$. To verify that the candidate is indeed the derivative, again employ Theorem 4.2.2 to do the estimation

$$|(\lambda_2 + h_2)D^s(\lambda_1 + h_1) - \lambda_2 D^s\lambda_1 - (h_2 D^s\lambda_1 + \lambda_2 D^s h_1)|_{r_2} = |h_2 D^s h_1|_{r_2} \le \quad (2)$$

$$(3\cdot 2^{r_2} - 2)|h_1|_{r_1}|h_2|_{r_2} \le (3\cdot 2^{r_2 - 1} - 1)(|h_1|_{r_1} + |h_2|_{r_2})^2 = (3\cdot 2^{r_2 - 1} - 1)|h|^2 .$$

This establishes the differentiation. The continuity of the derivative at $\lambda$, $(\cdot)D^s\lambda_1 + \lambda_2 D^s(\cdot)$, follows from the inequality involving operator norms,

$$\Big|\, |(\cdot)D^s(\lambda_1 + \delta_1) + (\lambda_2 + \delta_2)D^s(\cdot)|_{r_2} - |(\cdot)D^s\lambda_1 + \lambda_2 D^s(\cdot)|_{r_2} \Big| \le \quad (3)$$

$$2(3\cdot 2^{r_2} - 2)(|\delta_1|_{r_1} + |\delta_2|_{r_2})$$

which can be derived for all $\delta = (\delta_1, \delta_2)$ in a neighborhood of the origin in $AC^{r_1}(I)\times AC^{r_2}(I)$ by using Theorem 4.2.2 along with standard estimation procedures.

In preparation for the proof that inversion on $\boldsymbol{G}_{A^o,B^o}$ is a $C^1$ map consider the factorization $P[\lambda] = L[\lambda]R[\lambda]$ in which $\lambda = (\lambda_{ij}^k)$ is the global coordinate system which first appeared in this proof of Theorem 4.3.1. In the factorization $L[\lambda]$ is the block diagonal part of $P[\lambda]$. That is, $L[\lambda]$ is the block diagonal matrix consisting of only the blocks $P^{ij}$ of Theorem 3.1.1 having size $r_i\times r_j$ with $r_i = r_j$, $(i = 1, 2, \cdots, p_*)$, $(j = 1, 2, \cdots, p_*)$ together with the bottom right corner block of size $m_s\times m_s$ which occurs if the distinct dimension numbers $\rho_1 > \rho_2 > \cdots > \rho_s$ have $\rho_s = 1$ and then $m_s$ is the multiplicity of $\rho_s$. From its form it is clear that $L[\lambda]$ maps $\boldsymbol{G}_{A^o,B^o}$ into itself and since $\boldsymbol{G}_{A^o,B^o}$ is a group, defining $R[\lambda] = L^{-1}[\lambda]P[\lambda]$ results in $R[\lambda]$ likewise mapping the group back into itself. Whereas $P^{-1}[\lambda] = R^{-1}[\lambda]L^{-1}[\lambda]$ and multiplication in $\boldsymbol{G}_{A^o,B^o}$ has already been proved to be $C^1$ it is sufficient to show that each of $R^{-1}[\lambda]$ and $L^{-1}[\lambda]$
is $C^1$.

Since $L[\lambda]$ is block diagonal, it is sufficient to prove that inversion of each block is $C^1$. As indicated in the definition of $L[\lambda]$ there are two types of diagonal blocks to consider.

The first type is a square matrix of blocks in which the blocks are of the form

$$\begin{bmatrix} \lambda_{ij} & 0 & \cdots & \cdot & 0 \\ D\lambda_{ij} & \lambda_{ij} & \cdots & \cdot & \cdot \\ D^2\lambda_{ij} & \alpha_{3,2}D\lambda_{ij} & \cdots & \cdot & \cdot \\ \cdot & \cdot & \cdots & \cdot & \cdot \\ \cdot & \cdot & \cdots & \cdot & \cdot \\ \cdot & \cdot & \cdots & \lambda_{ij} & 0 \\ D^{r-1}\lambda_{ij} & \alpha_{r,2}D^{r-2}\lambda_{ij} & \cdots & \alpha_{r,r-1}D\lambda_{ij} & \lambda_{ij} \end{bmatrix} \tag{4}$$

$(i = 1, 2, \cdots, r)$, $(j = 1, 2, \cdots, r)$. The $\lambda_{ij}$ is a component of $\lambda$. (The values of the scalars $\alpha_{k,l}$ also depend on $i$, $j$, etc., but are of no particular interest in the present discussion. See Theorem 3.1.1.) The important point is that the inversion of $L[\lambda]$ transforms the coordinate $\lambda_{ij}$ into $[(\lambda_{k,l})^{-1}]_{i,j}$ in which the $\lambda_{k,l}$'s are the $r^2$ coordinate variables of $L[\lambda]$ appearing in the block under discussion, one from each sub-block of the type pictured in (4). That is, $\lambda_{ij}$ is replace by a rational function of the $\lambda_{k,l}$'s (the denominator being the determinant of the block, pointed out earlier by Corollary 3.3.2 to be independent of the differentiated terms in the block). The $\lambda_{k,l}$ are elements of $AC^r(I)$. The previous part of this proof of Theorem 4.3.1 included proof that multiplication in $AC^r(I)$ is a $C^1$ function and certainly addition is likewise $C^1$. Since composition of $C^1$ maps produces $C^1$ maps it is sufficient to prove that inversion is $C^1$ on the (open) subset of $AC^r(I)$ where it is defined. The proof of this inversion is left to the reader as an exercise in applying the inequality of Theorem 4.2.2.

The second type of block described in the definition of $L[\lambda]$, which occurs if $\rho_s = 1$, contains only elements that are components of $\lambda$ and are points in $AC^o(I)$; no differentiated terms are present. Thus, the inverted matrix has elements which are rational functions in the components of $\lambda$ and the problem again reduces to proving inversion in $AC^o(I)$ to be $C^1$. The proof, using Theorem 4.2.2 once again, is omitted. This finishes the demonstration that inversion of $L[\lambda]$ is $C^1$.

The task has been reduced to showing that $R[\lambda] = L^{-1}[\lambda]P[\lambda]$ is likewise $C^1$. It is made easy by the observation that $R^{-1}[\lambda] = 2I - R[\lambda]$. In left multiplication of $P[\lambda]$ by $L^{-1}[\lambda]$ the rational functions of the $\lambda_{k,l}$'s that appear on the diagonal of the blocks of $L^{-1}[\lambda]$ are found to multiply those sub-diagonal row segments of $P[\lambda]$ containing the derivative free coordinate functions. In summary, the coordinate elements of $R[\lambda]$ are each rational functions of the coordinate components of $\lambda$. Hence, the same is true of $R^{-1}[\lambda]$ and once again, the proof reduces to that of the $C^1$ differentiability of rational functions which was already argued. Therefore, the proof that $\boldsymbol{G}_{A^o,B^o}$ is a Lie group is finished. As concluded already by Theorem 3.3.1, $h$ provides an isomorphism of $\boldsymbol{G}_{A^o,B^o}$

with $\mathcal{G}_{A,B}$ as groups. The easy details showing that $h$ moreover is an isomorphism between the underlying manifolds can be supplied to conclude the proof that $\mathcal{G}_{A^o,B^o}$ and $\mathcal{G}_{A,B}$ are isomorphic Lie groups.

A $C^1$ curve $P_{(\cdot)}$ in $\mathcal{G}_{A,B}$ identifies to a $C^1$ curve in $\mathfrak{g}_{A,B}$. (See Theorem 3.1.1.) Thus the derivative $P'_{(0)}$ is an element of $\mathfrak{g}_{A,B}$ for each curve having $P_{(0)}$ the identity. By obvious choice of $P_{(\cdot)}$ each element of $\mathfrak{g}_{A,B}$ can be realized in this manner. Thus $\mathfrak{g}_{A,B}$ constitutes the tangent space to $\mathcal{G}_{A,B}$ at the identity. This concludes the proof of Theorem 4.3.1.

## 5. $\mathcal{G}_{A,B}$ as a Semi-Direct Product

To summarize the manifold structure of $\mathcal{G}_{A,B}$ it is convenient to let $GL(m,AC^{\rho})$ denote the $m{\times}m$ matrices $M$ with elements in $AC^{\rho}(I)$ and satisfying the condition that $det(M) \neq 0$ on $I$. It is an open submanifold of $[AC^{\rho}]^{m^2}$.

**Theorem 5.1** *Let $(A,B)$ be controllable with $n = m_1\rho_1 + m_2\rho_2 + \cdots + m_s\rho_s$ as described in 3.3. Then there exists a manifold isomorphism*

$$\mathcal{G}_{A,B} \cong \prod_{k=1}^{s} GL(m_k\,, AC^{\rho_k}) \times \prod_{i=1}^{s_*-1} [AC^{\rho_i}]^{\alpha_i} \times [AC^{o}]^{(n-q)q}$$

*in which*

$$s_* = \text{the largest } k \text{ for which } \rho_k \geq 2$$

$$\alpha_i = m_{i+1} \sum_{k=1}^{i} (\rho_i - \rho_{i+k} + 1) m_k$$

$$q = \begin{cases} 0 & \text{if } \rho_s > 1 \\ m_s & \text{if } \rho_s = 1\,. \end{cases}$$

Proof. The required counting can be done using Theorem 3.1.1.

The proof of Theorem 4.3.1 factored each element $P$ of $\mathcal{G}_{A^o,B^o}$ uniquely as a product in that group, $P = LR$, in which $L$ is the diagonal part of $P$ and $R$ is the lower triangular part defined as $R = L^{-1}P$. By defining $\mathcal{D}_{A^o,B^o}$ to be the collection of diagonal parts and $\mathcal{L}_{A^o,B^o}$ to be the collection of lower triangular parts it is apparent from the proof of Theorem 4.3.1 that each is a Lie subgroup of $\mathcal{G}_{A^o,B^o}$ and thus

$$\mathcal{G}_{A^o,B^o} = \mathcal{D}_{A^o,B^o} \cdot \mathcal{L}_{A^o,B^o} \tag{1}$$

Each $R \in \mathcal{L}_{A^o,B^o}$ can be factored further as $R = R_1R_2$ uniquely in $\mathcal{L}_{A^o,B^o}$ where $R_2$ is the matrix obtained from $R$ by setting to zero all coordinate functions occupying the bottom $q$ rows ($q$ as defined in Theorem 5.1) which occur when $\rho_s = 1$ and $R_1$ is obtained by setting the complementary set of coordinate functions in $R$ to zero.

This leads to the factorization

$$\mathcal{L}_{A^o,B^o} = \mathcal{L}^1_{A^o,B^o} \cdot \mathcal{L}^2_{A^o,B^o} \tag{2}$$

as the product of the two respective Lie subgroups. The next theorem summarizes major aspects of the structure of $\mathcal{G}_{A,B}$.

**Theorem 5.2** *Assume that* $(A,B)$ *is controllable. Let* $\sigma$ *be the homomorphism of* $\mathcal{D}_{A^o,B^o}$ *into the group of automorphisms of* $\mathcal{L}_{A^o,B^o}$ *defined by* $\sigma_L(\cdot) = L^{-1}(\cdot)L$. *There exists a Lie group isomorphism,*

$$\mathcal{G}_{A,B} \cong \mathcal{D}_{A^o,B^o} \times_\sigma \mathcal{L}_{A^o,B^o} \tag{3}$$

*in which* $\times_\sigma$ *denotes the semi-direct product relative to* $\sigma$ *defined by* $(L, R)(\tilde{L}, \tilde{R}) = (L\tilde{L}, \sigma_{\tilde{L}}(R)\tilde{R})$ *and* $(A^o, B^o)$ *is the canonical form of* $(A, B)$. *It is possible to factor, as a semi-direct product,*

$$\mathcal{L}_{A^o,B^o} \cong \mathcal{L}^1_{A^o,B^o} \times_\tau \mathcal{L}^2_{A^o,B^o} \tag{4}$$

*in which* $\tau_{R_1}(\cdot) = R_1^{-1}(\cdot)R_1$ *for* $R_1 \in \mathcal{L}^1_{A^o,B^o}$. *Moreover, there are further Lie group isomorphisms*

$$\mathcal{D}_{A^o,B^o} \cong \prod_{k=1}^{s} GL(m_k, AC^{\rho_k}), \tag{5}$$

*as a direct product in which* $\rho_1 > \rho_2 > \cdots > \rho_s$ *are the distinct canonical dimension numbers of* $(A, B)$ *and* $\rho_k$ *has multiplicity* $m_k$ ;

$$\mathcal{L}^1_{A^o,B^o} \cong \sum_{i=1}^{s_*-1} \sum_{j=1}^{\alpha_i} \oplus AC^{\rho_i} \tag{6}$$

*in which*

$$\alpha_i = m_{i+1} \sum_{k=1}^{i} (\rho_k - \rho_{i+k} + 1)m_k \tag{7}$$

*and*

$$\mathcal{L}^2_{A^o,B^o} \cong \sum_{i=1}^{(n-q)q} \oplus AC^o . \tag{8}$$

*For* $R_i \in \mathcal{L}^i_{A^o,B^o}$, $R_i^{-1} = 2I - R_i$, $(i = 1, 2)$.

Proof. Any pair $P$, $\tilde{P}$ in $\mathcal{G}_{A^o,B^o}$ can be factored uniquely as $P = LR$, $\tilde{P} = \tilde{L}\tilde{R}$ with the $L$, $\tilde{L}$
in $\mathcal{D}_{A^o,B^o}$ and the $R$, $\tilde{R}$ in $\mathcal{L}_{A^o,B^o}$. It can be verified that $\tilde{L}^{-1}R\tilde{L} \in \mathcal{L}_{A^o,B^o}$. For canonical $A = A^o$, $B = B^o$ the fact that the product is semi-direct and that the map $P = LR \to (L, R)$ is a group isomorphism is evident from the calculations

$$P\tilde{P} = LR\tilde{L}\tilde{R} = (L\tilde{L})(\tilde{L}^{-1}RL)\tilde{R}, \tag{9}$$

$P^{-1} = L^{-1}(LR^{-1}L^{-1})$ and the uniqueness of factorization. The $C^1$ character of the semi-direct multiplication and inversion is a consequence of that of $\mathcal{G}_{A^o,B^o}$ and hence of $\mathcal{D}_{A^o,B^o}$ and of $\mathcal{L}_{A^o,B^o}$. Now (3) follows from application of Theorem 4.3.1.

$L[\lambda] \in \mathcal{D}_{A^o,B^o}$ is of block diagonal type with at most two kinds of diagonal blocks that can occur as discussed in the proof of Theorem 4.3.1. It was pointed out that each block can be identified with a matrix of coordinate components of $\lambda$ that multiply in response to multiplication in $\mathcal{D}_{A^o,B^o}$. The identification mentioned provides the isomorphism (5). The isomorphisms (6) and (8) are apparent from checking that matrix multiplication in each of $\mathcal{L}^1_{A^o,B^o}$ and $\mathcal{L}^2_{A^o,B^o}$ adds the coordinate functions. In each of these commutative groups the fact that $R[\lambda]R[\tilde{\lambda}] = R[\lambda+\tilde{\lambda}]$ and $R[0] = I$ implies that $R^{-1}[\lambda] = R[-\lambda]$ and thus $R^{-1}[\lambda] = 2I - R[\lambda]$ as claimed. This concludes the proof of Theorem 5.2.

The results of Theorem 5.2 can be more briefly stated as Theorem 5.3.

**Theorem 5.3** *For (A,B) controllable $\mathcal{G}_{A,B}$ is a Lie group with Lie algebra $\mathfrak{g}_{A,B}$. There are semi-direct products $\times_\sigma$ and $\times_\tau$ for which there is a Lie group isomorphism*

$$\mathcal{G}_{A,B} \cong \prod_{k=1}^{s} GL(m_k, AC^{\rho_k}) \times_\sigma \left[ \sum_{i=1}^{s_*-1} \sum_{j=1}^{\alpha_i} \oplus AC^{\rho_i} \times_\tau \sum_{i=1}^{(n-q)q} \oplus AC^{o} \right]. \tag{10}$$

*The norm for the Banach space $\mathfrak{g}_{A,B}$ in which the coordinates on $\mathcal{G}_{A,B}$ are based can be taken to be one relative to which $\mathfrak{g}_{A,B}$ is both a Lie algebra and a Banach algebra.*

Proof. Almost everything has been proved. For $P \in \mathfrak{g}_{A,B}$ let $\hat{P}$ be the projection of its pre-image in $\mathfrak{g}_{A^o,B^o}$ under $h$ onto the block matrix of coordinate functions used on $\mathcal{G}_{A^o,B^o}$.
(See Theorem 3.1.1 and the discussion in 3.2.) The coordinate functions $\lambda^k_{ij}$ in block $\hat{P}^{ij}$ lie in a Banach space $AC^{\rho_{i,j}}$ associated with that block. ($\rho_{i,j} \in \{\rho_1, \rho_2, \cdots, \rho_s\}$.) Define $|\hat{P}^{ij}|_{i,j} = \sum_k |\lambda^k{}_{ij}|_{\rho_{i,j}}$ and finally $|P|^o = \sum_{i,j} |\hat{P}^{ij}|_{i,j}$. This is the norm on $\mathfrak{g}_{A^o,B^o}$, consistent with the coordinatization of $\mathcal{G}_{A^o,B^o}$ and, moreover, which is compatible with $\mathfrak{g}_{A^o,B^o}$ both as a Lie algebra and Banach algebra. Finally the induced norm on $\mathfrak{g}_{A,B}$ given by $|P| = |h^{-1}(P)|^o = |P_o P P_o^{-1}|^o$ satisfies the requirements of Theorem 5.3.

The finite dimensional Lie subgroup of $\mathcal{G}_{A,B}$ discussed in 3.2 can be similarly described as a semi-direct product, as stated in Theorem 5.4.

**Theorem 5.4** *For (A,B) controllable $\mathcal{G}^o_{A^o,B^o}$ is a finite dimensional subgroup of $\mathcal{G}_{A,B}$. There are semi-direct products $\times_\sigma$ and $\times_\tau$ for which there is a Lie group isomorphism*

$$\mathcal{G}^o_{A,B} \cong \prod_{k=1}^{s} GL(m_k, R) \times_\sigma \left[ R^{\alpha_*} \times_\tau R^{(n-q)q} \right] \tag{11}$$

*where* $\alpha_* = \sum_{i=1}^{s_*-1} \alpha_i$ .

## REFERENCES

Kuller, R.G.
*Topics in Modern Analysis* Prentice Hall, Englewoods Cliffs, N.J., 1969

Lang, S.
*Introduction to Differentiable Manifolds* Interscience, New York, 1967

Lukes, D. L.
*Differential Equations: Classical to Controlled,* Academic Press, New York, 1982

Markus, L.
Continuous matrices and stability of differential systems, *Math. Zeitschr.,* vol. 62, 310-319, 1955

Varadarajan, V. S.
*Lie Groups, Lie Algebras and their Representations,* Springer-Verlag, New York, 1984

# 47 Min-Max Game Theory and Algebraic Riccati Equations for Boundary Control Problems with Analytic Semigroups: The Stable Case

**Christine A. McMillan**[1] University of Virginia, Charlottesville, Virginia

**Roberto Triggiani**[2] University of Virginia, Charlottesville, Virginia

**Abstract**

We consider a min-max game theory problem for an abstract model [L-T.1, class (H.1)], which covers in particular parabolic and parabolic-like partial differential equations in a general bounded domain with both control function ("good" player) and deterministic disturbance ("bad" player) acting on the boundary of the spatial domain. The case of point control and point disturbance is also included. Specific examples encompass not only mixed problems for heat/diffusion equations, but also for wave/plate equations with a sufficiently high degree of internal damping [L-T.1, section 6].

The present paper treats the case where the original free dynamics is stable. Here, a direct treatment may be given which provides an explicit solution of all the relevant quantities directly in terms of the data of the problem. The optimal control and the worst disturbance are both synthesized in a feedback form, pointwise in time, in terms of the unique solution of an algebraic Riccati operator equation. The overall proof is a fusion of the strategy devised in [M-T.1] for hyperbolic/plate-like or Schroedinger equations with the technicalities of the corresponding linear quadratic regulator problem for boundary control parabolic equations, where the disturbance $w \equiv 0$ [L-T.1] – [L-T.3], [D-I.1], [F.1].

[1]Research partially supported by an IBM Graduate Student Fellowship

[2]Research partially supported by the National Science Foundation under Grant NSF-DMS-8902811-01

# 1 Introduction

## 1.1 Problem setting

Let $U$ (control) and $Y$ (state) be separable Hilbert spaces. We introduce the following abstract state equation

$$\dot{y}(t) = Ay(t) + Bu(t) + Gw(t) \quad \text{in } [D(A^*)]'; \; y(0) = y_0 \in Y \tag{1.1.1}$$

Here, the function $u \in L_2(0,\infty;U)$ is the control and $w \in L_2(0,\infty;Y)$ is a deterministic disturbance. The dynamics (1.1.1), (1.1.2) is subject to the following assumptions, which will be maintained throughout the paper [L-T.1, class (H.1)]:

(H.1) $A : Y \supset D(A) \longrightarrow Y$ is the infinitesimal generator of a strongly continuous (s.c.) analytic semigroup $e^{At}$ on the Hilbert space $Y$.

(H.2) $B$ is a linear continuous operator $U \longrightarrow [D(A^*)]'$, the dual space of $D(A^*)$, $A^*$ being the adjoint of $A$ in $Y$, such that

$$A^{-\delta}B \in L(U;Y) \quad \textit{for some fixed constant } \delta < 1 \tag{1.1.2a}$$

$$\|A^{-\delta}B\|_{L(U;Y)} = \|B^*A^{*-\delta}\|_{L(Y;U)} = c_\delta < \infty \tag{1.1.2b}$$

(H.3) $G$ is a linear continuous operator $V \longrightarrow [D(A^*)]'$ of the same class as $B$, i.e. such that

$$A^{-\rho}G \in L(V;Y) \quad \textit{for some fixed constant } \rho < 1 \tag{1.1.3a}$$

$$\|A^{-\rho}G\|_{L(V;Y)} = \|G^*A^{*-\rho}\|_{L(Y;V)} = c_\rho < \infty \tag{1.1.3b}$$

where $V$ is, possibly, another Hilbert space. To simplify notation, we shall henceforth take $V$ equal to $U$: $U \equiv V$.

Remark 1.1.1 The above abstract setting includes partial differential equations of parabolic type with boundary/point control $u$ and boundary/point disturbance $w$. See specific classes of P.D.E. examples in [L-T.1, section 6]; these encompass not only mixed problems for heat/diffusion equations but also for wave/plate equations with a sufficiently high degree of internal damping. □

In this paper, we shall study the min-max game theoretic problem with indefinite cost described below in section 1.2 for the above abstract dynamics, where the observation operator $R$ in (1.2.1) below is subject to the assumption (see Remark 1.2.1 below)

(H.4) $$R \in \mathcal{L}(Y). \tag{1.1.4}$$

The same min-max game problem, however for an abstract model which encompasses hyperbolic, or plate-like, or Schroedinger, partial differential equations with boundary/point control is studied in references [M-T.1], [M-T.2], which provide a rather complete and comprehensive theory in terms of an algebraic Riccati equation by falling into the linear quadratic regulator treatment of [F-L-T.1]. A different approach which likewise ultimately and critically relies on [F-L-T.1] is given in [B.1]. In the present paper, the emphasis is on the "abstract boundary control parabolic" case. In line with the strategy followed in the hyperbolic/plate case, we find it both useful and convenient to divide our efforts into two parts. In the present Part I, we shall add a stability assumption on the free system, i.e.:

(H.5) There exist constants $M \geq 1$ and $\omega > 0$ such that

$$\|e^{At}\|_{\mathcal{L}(Y)} \leq Me^{-\omega t}, \qquad t \geq 0 \tag{1.1.5}$$

Many of the canonical P.D.E. parabolic problems do satisfy assumption (H.5) = (1.1.5). In this case, then, it is possible to give a simplified, short-cut treatment – which is also more informative and fully explicit – of the min-max game problem. Our main result is contained in Theorem 1.3.1 below, which provides the explicit characterizations of the relevant optimal quantities which are not possible when (H.5) is omitted. This theorem is obtained by combining the strategy of [M-T.1] of the hyperbolic/plate-like stable case with the technicalities of the parabolic case, as in the corresponding linear quadratic regulator problem where the disturbance $w \equiv 0$ [L-T.2], [L-T.3, Chapter 3], [D-I.1], [F.1]. The more general case, where assumption (H.5) = (1.1.5) is dispensed with, is more complicated and requires the additional assumptions of "Finite Cost Condition" or "Stabilizability", and of "Detectability" (the former for existence, the latter for uniqueness of the Riccati operator), which are automatically satisfied under assumption (H.5). A treatment thereof is given in the companion paper [M-T.3]: unlike the present paper, [M-T.3] critically uses the corresponding Riccati operator when the disturbance $w \equiv 0$, as is the case in the general hyperbolic treatment of [M-T.2], of which [M-T.3] is the parabolic counterpart.

The solution to the state equation (1.1.1) is given explicitly by

$$y(t) = y(t; y_0) = e^{At}y_0 + (Lu)(t) + (Ww)(t) \tag{1.1.6}$$

where, the operators $L$, $W$ (under assumption (H.5) = (1.1.5)) and their regularity properties are given below in (2.1.2), (2.1.5). It will be freely used below, that as a consequence of analyticity in (H.1) and of (H.5), the fractional powers $(-A)^\theta$ of $(-A)$ are well-defined, $0 < \theta < 1$.

## 1.2 Game Theory Problem

For a fixed $\gamma > 0$, we associate with (1.1.1) or (1.1.6) the cost functional

$$J(u, w) = J(u, w, y(u, w)) = \int_0^\infty [\|Ry(t)\|_Y^2 + \|u(t)\|_U^2 - \gamma^2\|w(t)\|_Y^2]dt \tag{1.2.1}$$

where $y(t) = y(t; y_0)$ is given by (1.1.6). The aim of this paper, is to study the following game-theory problem:

$$\sup_w \inf_u J(u, w) \tag{1.2.2}$$

where the infimum is taken over all $u \in L_2(0, \infty; U)$, for $w$ fixed, and the supremum is taken over all $w \in L_2(0, \infty; Y)$.

Remark 1.2.1 As remarked in [B.1], a more general "output" equation may be likewise treated at essentially no additional difficulty. □

## 1.3 Statement of main results

**Main Theorem 1.3.1** *Assume (H.1) - (H.5). Then there exists a (critical) value $\gamma_c > 0$ defined explicitly in terms of the problem data by Eq. (2.2.1) below such that:*

*(a) if $0 < \gamma < \gamma_c$, then taking the supremum in $w$ as in (1.2.2) leads to $+\infty$; i.e. there is no finite solution of the game theory problem (1.2.2) (see Theorem 3.1 (iii));*

*(b) if $\gamma > \gamma_c$, then:*

*(i) there exists a unique solution $\{u^*(\cdot\,; y_0); w^*(\cdot\,; y_0); y^*(\cdot\,; y_0)\}$ of the game theory problem (1.2.2) (see Theorem 3.1 (ii));*

*(ii) there exists a unique bounded, nonnegative self-adjoint operator, $P = P^* \in \mathcal{L}(Y)$, which satisfies the following Algebraic Riccati Equation $ARE_\gamma$, for all $x,\ z \in D((-A)^\epsilon)$, $\forall \epsilon > 0$:*

$$(PAx, z)_Y + (Px, Az)_Y + (Rx, Rz)_Y = (B^*Px, B^*Pz)_U - \gamma^{-2}(G^*Px, G^*Pz)_U \quad (1.3.1)$$

*(see Theorem 7.2.1 below), with the properties (see (7.1.2) – (7.1.3) below)*

$$(-A)^\theta P \in \mathcal{L}(Y) \quad 0 \leq \theta < 1 \quad (1.3.2)$$

$$B^*P \in \mathcal{L}(Y;U); \quad G^*P \in \mathcal{L}(Y;U) \quad (1.3.3)$$

*(iii) the following pointwise feedback relations hold*

$$u^*(t;y_0) = -B^*Py^*(t;y_0) \in L_2(0,\infty;U) \cap C([0,\infty];U) \quad (1.3.4)$$

$$\gamma^2 w^*(t;y_0) = G^*Py^*(t;y_0) \in L_2(0,\infty;U) \cap C([0,\infty];U) \quad (1.3.5)$$

*(see Corollary 7.1.1 below)*

*(iv) the operator ($F$ stands for "feedback") with maximal domain*

$$A_F = A - BB^*P + \gamma^{-2}GG^*P \quad (1.3.6)$$

*is the generator of a s.c. semigroup $e^{A_F t}$ on $Y$ which is, moreover, analytic for $t > 0$ (see Lemma 7.2.2 below) and, in fact, for $y_0 \in Y$ (see (7.2.6) below):*

$$y^*(t;y_0) = e^{A_F t}y_0 = e^{(A-BB^*P+\gamma^{-2}GG^*P)t}y_0 \in L_2(0,\infty;Y) \cap C([0,\infty];Y) \quad (1.3.7)$$

*Moreover, the semigroup $e^{A_F t}$ is uniformly stable in $Y$ (see Eq. (6.1.2));*

*(v) for any $y_0 \in Y$ (see (7.1.5) below)*

$$(Py_0, y_0) = J^*(y_0) \equiv J(u^*(\cdot\,;y_0), w^*(\cdot\,;y_0), y^*(\cdot\,;y_0)) = \sup_w \inf_u J(u, w, y(\cdot\,;y_0)) \quad \square(1.3.8)$$

*Conversely, suppose that $P = P^* \geq 0$ is an operator in $\mathcal{L}(Y)$ such that:*

*(a) the operator $A_F = A - BB^*P + \gamma^{-2}GG^*P$ is the generator of a s.c. uniformly stable semigroup on $Y$ for some $\gamma > 0$;*

*(b) $B^*P \in \mathcal{L}(U;Y)$ and $G^*P \in \mathcal{L}(Y)$; and*

*(c) $P$ is a solution of the corresponding $ARE_\gamma$ in (1.3.1), $\forall x, z \in D((-A)^\epsilon)$ $\forall \epsilon > 0$.*

*Then, the operator $(A-BB^*P)$ is likewise the generator of a s.c. uniformly stable semigroup on $Y$ and, moreover, the game problem (1.2.2) has a finite optimal cost functional for all $y_0 \in Y$, so that then $\gamma \geq \gamma_c$.* □

Additional results are given in the treatment below. In short, our proof of Theorem 1.3.1 here is a fusion of the strategy for the differential game problem (1.2.2) devised in [M-T.1] for hyperbolic/plate-like equations, combined with the technicalities of the parabolic dynamics which are already critical in the case of the corresponding linear quadratic regulator problem with no disturbance ($w \equiv 0$) as e.g. in [L-T.1] - [L-T.3]. A major difference of the present parabolic case over the hyperbolic/plate-like equation in [M-T.1] is the issue of the continuity in time (or, more properly, lack thereof) of the solutions $y$ of the parabolic dynamics (1.1.1) in (1.1.6) for which we refer to Remark 4.2 below for a more technical comment. This issue is settled in section 5.

# 2 Minimization of $J$ over $u$ for $w$ fixed

## 2.1 Existence of a unique optimal pair and optimality conditions

We return to the functional cost $J$ in (1.2.1). In this section we consider the following problem: given a fixed but arbitrary $w \in L_2(0,\infty;Y)$: minimize

$$J(u,w,y_0) = \int_0^\infty [\|Ry(t)\|_Y^2 + \|u(t)\|_U^2 - \gamma^2\|w(t)\|_Y^2]dt \tag{2.1.1}$$

over all $u \in L_2(0,\infty;U)$, where $y(t) = y(t;y_0)$ is the corresponding solution of (1.1.1) given explicitly by (1.1.6). The advantages of assumption (H.5) = (1.1.5) of the present Part I are reaped at the very outset of the analysis. The operators $L$ and $W$ in Eq. (1.1.6) are given by

$$(Lu)(t) = \int_0^t e^{A(t-\tau)}Bu(\tau)d\tau \tag{2.1.2}$$

$$: \textit{continuous } L_2(0,\infty;U) \longrightarrow L_2(0,\infty;Y) \tag{2.1.3}$$

$$(Wu)(t) = \int_0^t e^{A(t-\tau)}Gw(\tau)d\tau \tag{2.1.4}$$

$$: \textit{continuous } L_2(0,\infty;Y) \longrightarrow L_2(0,\infty;Y) \tag{2.1.5}$$

The dual versions of (2.1.2)-(2.1.5) are

$$(L^*f)(t) = B^* \int_t^\infty e^{A^*(\tau - t)} f(\tau) d\tau \tag{2.1.6}$$

$$: \quad continuous \quad L_2(0,\infty;Y) \longrightarrow L_2(0,\infty;U) \tag{2.1.7}$$

$$(W^*v)(t) = G^* \int_t^\infty e^{A^*(\tau - t)} v(\tau) d\tau \tag{2.1.8}$$

$$: \quad continuous \quad L_2(0,\infty;Y) \longrightarrow L_2(0,\infty;Y) \tag{2.1.9}$$

Remark 2.1.1 The above regularity results are conservative. Indeed, a *key* feature of the present analytic semigroup case is that $L$, $W$, $L^*$ and $W^*$ are *smoothing* or *regularizing* operators. This will be noted and used at the appropriate spot in our analysis below, see section 5. □

**Theorem 2.1.1** *(i) With reference to the minimization problem (2.1.1), there exists a unique optimal pair denoted by* $\{u_w^0(\cdot\,; y_0), y_w^0(\cdot\,; y_0)\}$, *with corresponding optimal cost denoted by*

$$\begin{aligned} J_w^0(y_0) &= J(u_w^0(\cdot\,; y_0), y_w^0(\cdot\,; y_0)) \\ &= \int_0^\infty [\|Ry_w^0(t; y_0)\|_Y^2 + \|u_w^0(t; y_0)\|_U^2 - \gamma^2\|w(t)\|_Y^2]dt \end{aligned} \tag{2.1.10}$$

*This statement does not require the stability hypothesis (1.1.5).*

*(ii) The optimal pair is related by*

$$u_w^0(\cdot\,; y_0) = -L^*R^*Ry_w^0(\cdot\,; y_0) \tag{2.1.11}$$

*and is explicitly given in terms of the problem data by the following formulas:*

$$\begin{aligned} -u_w^0(\cdot\,; y_0) &= [I + L^*R^*RL]^{-1}L^*R^*R[e^{A\cdot}\, y_0 + Ww] \in L_2(0,\infty;U) \\ &= -u_{w=0}^0(\cdot\,; y_0) - u_w^0(\cdot\,; y_0 = 0) \end{aligned} \tag{2.1.12}$$

$$y_w^0(\cdot\,; y_0) = [I + LL^*R^*R]^{-1}[e^{A\cdot}\, y_0 + Ww] \in L_2(0,\infty;Y) \tag{2.1.13}$$

$$= y_{w=0}^0(\cdot\,; y_0) + y_w^0(\cdot\,; y_0 = 0)$$

*where both inverse operators in (2.1.12) and (2.1.13) are well defined as bounded operators on all of* $L_2(0,\infty;U)$ *and* $L_2(0,\infty;Y)$ *respectively (for* $[I + LL^*R^*R]^{-1}$ *see [L-T.4 p.891, below Eq. (2.8e) ]). Moreover, the optimal dynamics is, of course,*

$$y_w^0(t; y_0) = e^{At}y_0 + \{Lu_w^0(\cdot\,; y_0)\}(t) + \{Ww(\cdot)\}(t) \in L_2(0,\infty;Y) \tag{2.1.14}$$

*(iii) The optimal cost $J^0_w(y_0)$ in (2.1.10) is given explicitly in terms of the data by the following formulas:*

$$J^0_w(y_0) = (e^{A\cdot}\, y_0 + Ww, R^*R[I + LL^*R^*R]^{-1}[e^{A\cdot}\, y_0 + Ww])_{L_2(0,\infty;Y)} - \gamma^2(w,w)_{L_2(0,\infty;Y)} \quad (2.1.15a)$$

$$= J^0_{w=0}(y_0) + J^0_w(y_0 = 0) + \chi_{y_0,w} \quad (2.1.15b)$$

$$J^0_{w=0}(y_0) = (e^{A\cdot}\, y_0, R^*R[I + LL^*R^*R]^{-1}(e^{A\cdot}\, y_0))_{L_2(0,\infty;Y)} \quad (2.1.16a)$$

$$= \|[I + (R^*R)^{1/2}LL^*(R^*R)^{1/2}]^{-1/2}(R^*R)^{1/2}(e^{A\cdot}\, y_0)\|^2_{L_2(0,\infty;Y)} \quad (2.1.16b)$$

$$J^0_w(y_0 = 0) = (Ww, R^*R[I + LL^*R^*R]^{-1}Ww)_{L_2(0,\infty;Y)} - \gamma^2(w,w)_{L_2(0,\infty;Y)} \quad (2.1.17a)$$

$$= -(w, E_\gamma w)_{L_2(0,\infty;Y)} \quad (2.1.17b)$$

$$= \|[I + (R^*R)^{1/2}LL^*(R^*R)^{1/2}]^{-1/2}(R^*R)^{1/2}Ww\|^2_{L_2(0,\infty;Y)} - \gamma^2\|w\|^2_{L_2(0,\infty;Y)} \quad (2.1.17c)$$

*where*

$$E_\gamma = \gamma^2 I - W^*R^*R[I + LL^*R^*R]^{-1}W \quad (2.1.18a)$$

$$= \gamma^2 I - W^*(R^*R)^{1/2}[I + (R^*R)^{1/2}LL^*(R^*R)^{1/2}]^{-1}(R^*R)^{1/2}W \quad (2.1.18b)$$

$$= \gamma^2 I - S;\ S \text{ nonnegative self}-\text{adjoint operator in } \mathcal{L}(L_2(0,\infty;Y)) \quad (2.1.18c)$$

$$S = W^*R^*R[I + LL^*R^*R]^{-1}W \quad (2.1.18d)$$

*The cross terms in (2.1.15b) are linear in $w$:*

$$\chi_{y_0,w} = (e^{A\cdot}\, y_0, R^*R[I + LL^*R^*R]^{-1}Ww)_{L_2(0,\infty;Y)} + (Ww, R^*R[I + LL^*R^*R]^{-1}e^{A\cdot}\, y_0)_{L_2(0,\infty;Y)} \quad (2.1.19a)$$

$$= 2(e^{A\cdot}\, y_0, R^*R[I + LL^*R^*R]^{-1}Ww)_{L_2(0,\infty;Y)} \quad (2.1.19b)$$

$$= 2(e^{A\cdot}\, y_0, (R^*R)^{1/2}[I + (R^*R)^{1/2}LL^*(R^*R)^{1/2}]^{-1}(R^*R)^{1/2}Ww)_{L_2(0,\infty;Y)} \quad (2.1.19c)$$

*In going from (2.1.16a) to (2.1.16b), as well as from (2.1.17a) to (2.1.17b), and from (2.1.19b) to (2.1.19c) we have used the identity*

$$(R^*R)^{1/2}[I + LL^*R^*R]^{-1} = [I + (R^*R)^{1/2}LL^*(R^*R)^{1/2}]^{-1}(R^*R)^{1/2} \in L(L_2(0,\infty;Y)) \tag{2.1.20}$$

*so that*

$$\begin{aligned} R^*R[I + LL^*R^*R]^{-1} &= (R^*R)^{1/2}[I + (R^*R)^{1/2}LL^*(R^*R)^{1/2}]^{-1}(R^*R)^{1/2} \\ &= \text{self} - \text{adjoint operator in } L(L_2(0,\infty;Y)) \end{aligned} \tag{2.1.21}$$

**Proof:** A proof which uses Liusternik's Lagrange Multiplier Theorem is given in [M-T.1].

□

Remark 2.1.2 One could give better regularity results for the intermediate quantities $\{u_w^0, y_w^0\}$ using the subsequent Theorem 5.1: however, we shall do this for the final quantities $\{u^*, y^*, w^*\}$ of interest in section 5. □

# 3 Minimization of $J$ over $u$ for $w$ fixed

## 3.1 Existence of a unique optimal pair and optimality conditions

We return to the functional cost $J$ in (1.2.1). In this section we consider the following problem: given a fixed but arbitrary $w \in L_2(0,\infty;Y)$: minimize

$$J(u, w, y_0) = \int_0^\infty [\|Ry(t)\|_Y^2 + \|u(t)\|_U^2 - \gamma^2\|w(t)\|_Y^2]dt \tag{3.1.1}$$

over all $u \in L_2(0,\infty;U)$, where $y(t) = y(t;y_0)$ is the corresponding solution of (1.1.1) given explicitly by (1.1.6). The advantages of assumption (H.5) = (1.1.5) of the present Part I are reaped at the very outset of the analysis. The operators $L$ and $W$ in Eq. (1.1.6) are given by

$$(Lu)(t) = \int_0^t e^{A(t-\tau)}Bu(\tau)d\tau \tag{3.1.2}$$

$$: \textit{continuous } L_2(0,\infty;U) \longrightarrow L_2(0,\infty;Y) \tag{3.1.3}$$

$$(Wu)(t) = \int_0^t e^{A(t-\tau)}Gw(\tau)d\tau \tag{3.1.4}$$

$$: \textit{continuous } L_2(0,\infty;Y) \longrightarrow L_2(0,\infty;Y) \tag{3.1.5}$$

The dual versions of (2.1.2)-(2.1.5) are

$$(L^*f)(t) = B^* \int_t^\infty e^{A^*(\tau-t)} f(\tau) d\tau \tag{3.1.6}$$
$$: \ continuous \ L_2(0,\infty;Y) \longrightarrow L_2(0,\infty;U) \tag{3.1.7}$$

$$(W^*v)(t) = G^* \int_t^\infty e^{A^*(\tau-t)} v(\tau) d\tau \tag{3.1.8}$$
$$: \ continuous \ L_2(0,\infty;Y) \longrightarrow L_2(0,\infty;Y) \tag{3.1.9}$$

Remark 2.1.1 The above regularity results are conservative. Indeed, a *key* feature of the present analytic semigroup case is that $L$, $W$, $L^*$ and $W^*$ are *smoothing* or *regularizing* operators. This will be noted and used at the appropriate spot in our analysis below, see section 5. □

**Theorem 3.1.1** *(i) With reference to the minimization problem (2.1.1), there exists a unique optimal pair denoted by* $\{u_w^0(\cdot\,;y_0), y_w^0(\cdot\,;y_0)\}$*, with corresponding optimal cost denoted by*

$$\begin{aligned} J_w^0(y_0) &= J(u_w^0(\cdot\,;y_0), y_w^0(\cdot\,;y_0)) \\ &= \int_0^\infty [\|Ry_w^0(t;y_0)\|_Y^2 + \|u_w^0(t;y_0)\|_U^2 - \gamma^2\|w(t)\|_Y^2]dt \end{aligned} \tag{3.1.10}$$

*This statement does not require the stability hypothesis (1.1.5).*

*(ii) The optimal pair is related by*

$$u_w^0(\cdot\,;y_0) = -L^*R^*Ry_w^0(\cdot\,;y_0) \tag{3.1.11}$$

*and is explicitly given in terms of the problem data by the following formulas:*

$$\begin{aligned} -u_w^0(\cdot\,;y_0) &= [I + L^*R^*RL]^{-1}L^*R^*R[e^{A\cdot}\,y_0 + Ww] \in L_2(0,\infty;U) \\ &= -u_{w=0}^0(\cdot\,;y_0) - u_w^0(\cdot\,;y_0 = 0) \end{aligned} \tag{3.1.12}$$

$$\begin{aligned} y_w^0(\cdot\,;y_0) &= [I + LL^*R^*R]^{-1}[e^{A\cdot}\,y_0 + Ww] \in L_2(0,\infty;Y) \\ &= y_{w=0}^0(\cdot\,;y_0) + y_w^0(\cdot\,;y_0 = 0) \end{aligned} \tag{3.1.13}$$

*where both inverse operators in (2.1.12) and (2.1.13) are well defined as bounded operators on all of* $L_2(0,\infty;U)$ *and* $L_2(0,\infty;Y)$ *respectively (for* $[I + LL^*R^*R]^{-1}$*see [L-T.4 p.891, below Eq. (2.8e) ]). Moreover, the optimal dynamics is, of course,*

$$y_w^0(t;y_0) = e^{At}y_0 + \{Lu_w^0(\cdot\,;y_0)\}(t) + \{Ww(\cdot)\}(t) \in L_2(0,\infty;Y) \tag{3.1.14}$$

*(iii) The optimal cost $J_w^0(y_0)$ in (2.1.10) is given explicitly in terms of the data by the following formulas:*

$$J_w^0(y_0) = (e^{A\cdot}\, y_0 + Ww, R^*R[I + LL^*R^*R]^{-1}[e^{A\cdot}\, y_0 + Ww])_{L_2(0,\infty;Y)} - \gamma^2(w,w)_{L_2(0,\infty;Y)} \quad (3.1.15a)$$

$$= J_{w=0}^0(y_0) + J_w^0(y_0 = 0) + \chi_{y_0,w} \quad (3.1.15b)$$

$$J_{w=0}^0(y_0) = (e^{A\cdot}\, y_0, R^*R[I + LL^*R^*R]^{-1}(e^{A\cdot}\, y_0))_{L_2(0,\infty;Y)} \quad (3.1.16a)$$

$$= \|[I + (R^*R)^{1/2}LL^*(R^*R)^{1/2}]^{-1/2}(R^*R)^{1/2}(e^{A\cdot}\, y_0)\|^2_{L_2(0,\infty;Y)} \quad (3.1.16b)$$

$$J_w^0(y_0 = 0) = (Ww, R^*R[I + LL^*R^*R]^{-1}Ww)_{L_2(0,\infty;Y)} - \gamma^2(w,w)_{L_2(0,\infty;Y)} \quad (3.1.17a)$$

$$= -(w, E_\gamma w)_{L_2(0,\infty;Y)} \quad (3.1.17b)$$

$$= \|[I + (R^*R)^{1/2}LL^*(R^*R)^{1/2}]^{-1/2}(R^*R)^{1/2}Ww\|^2_{L_2(0,\infty;Y)} - \gamma^2\|w\|^2_{L_2(0,\infty;Y)} \quad (3.1.17c)$$

*where*

$$E_\gamma = \gamma^2 I - W^*R^*R[I + LL^*R^*R]^{-1}W \quad (3.1.18a)$$

$$= \gamma^2 I - W^*(R^*R)^{1/2}[I + (R^*R)^{1/2}LL^*(R^*R)^{1/2}]^{-1}(R^*R)^{1/2}W \quad (3.1.18b)$$

$$= \gamma^2 I - S;\ S \text{ nonnegative self-adjoint operator in } \mathcal{L}(L_2(0,\infty;Y)) \quad (3.1.18c)$$

$$S = W^*R^*R[I + LL^*R^*R]^{-1}W \quad (3.1.18d)$$

*The cross terms in (2.1.15b) are linear in w:*

$$\chi_{y_0,w} = (e^{A\cdot}\, y_0, R^*R[I + LL^*R^*R]^{-1}Ww)_{L_2(0,\infty;Y)} + (Ww, R^*R[I + LL^*R^*R]^{-1}e^{A\cdot}\, y_0)_{L_2(0,\infty;Y)} \quad (3.1.19a)$$

$$= 2(e^{A\cdot}\, y_0, R^*R[I + LL^*R^*R]^{-1}Ww)_{L_2(0,\infty;Y)} \quad (3.1.19b)$$

$$= 2(e^{A\cdot}\, y_0, (R^*R)^{1/2}[I + (R^*R)^{1/2}LL^*(R^*R)^{1/2}]^{-1}(R^*R)^{1/2}Ww)_{L_2(0,\infty;Y)} \quad (3.1.19c)$$

*In going from (2.1.16a) to (2.1.16b), as well as from (2.1.17a) to (2.1.17b), and from (2.1.19b) to (2.1.19c) we have used the identity*

$$(R^*R)^{1/2}[I+LL^*R^*R]^{-1} = [I+(R^*R)^{1/2}LL^*(R^*R)^{1/2}]^{-1}(R^*R)^{1/2} \in \mathcal{L}(L_2(0,\infty;Y)) \tag{3.1.20}$$

*so that*

$$\begin{aligned} R^*R[I+LL^*R^*R]^{-1} &= (R^*R)^{1/2}[I+(R^*R)^{1/2}LL^*(R^*R)^{1/2}]^{-1}(R^*R)^{1/2} \\ &= \text{self-adjoint operator in } \mathcal{L}(L_2(0,\infty;Y)) \end{aligned} \tag{3.1.21}$$

**Proof:** A proof which uses Liusternik's Lagrange Multiplier Theorem is given in [M-T.1]. □

Remark 2.1.2 One could give better regularity results for the intermediate quantities $\{u^0_w, y^0_w\}$ using the subsequent Theorem 5.1: however, we shall do this for the final quantities $\{u^*, y^*, w^*\}$ of interest in section 5. □

# 4 Maximization of $J^0_w(y_0)$ over $w$: existence of a unique optimal $w^*$ .

In this section we return to the optimal $J^0_w(y_0)$ in (2.1.10) for $w \in L_2(0,\infty;Y)$ fixed and consider the problem:

$$\text{maximize } J^0_{w,\infty}(y_0), \text{ equivalently, minimize } -J^0_{w,\infty}(y_0), \text{ over all } w \in L_2(0,\infty;Y). \tag{4.1}$$

**Theorem 4.1** *(i) For $\gamma > \gamma_c$ (defined in (2.2.1)), the following estimate holds true for any $\epsilon > 0$ and every $w \in L_2(0,\infty;Y)$:*

$$-J^0_{w,\infty}(y_0) \geq [\gamma^2 - ({\gamma_c}^2+\epsilon)]\|w\|^2_{L_2(0,\infty;Y)} - J^0_{w=0}(y_0) - C_\epsilon\|y_0\|^2_Y \tag{4.2}$$

*(ii) For $\gamma > \gamma_c$ (defined in (2.2.1)), there exists a unique optimal solution $w^*(\cdot\,;y_0) \in L_2(0,\infty;Y)$ for the optimal problem (3.1):*

$$\max_{w\in L_2(0,\infty;Y)} J^0_w(y_0) \equiv J^0_{w=w^*}(y_0) \equiv J^*(y_0). \tag{4.3}$$

*(iii) If $0 < \gamma < \gamma_c$, then* $\sup_w J^0_{w,\infty}(y_0) = +\infty$. □

Proof: (i) We return to (2.1.15b) which gives $-J^0_{w,\infty}(y_0)$ as the sum of three contributions: a quadratic term in $w$, given by $-J^0_w(y_0 = 0) = (w, E_\gamma w)$ in (2.1.17b), which satisfies (2.2.3); a linear term in $w$, given by $-\chi_{y_0,w}$ in (2.1.19b), and satisfying with $V = W^*[I + R^*RLL^*]^{-1}R^*Re^{A\cdot}$

$$\begin{aligned} |\chi_{y_0,w}| &\le 2\|w\|_{L_2(0,\infty;Y)}\|Vy_0\|_{L_2(0,\infty;Y)} \\ &\le \epsilon\|w\|^2_{L_2(0,\infty;Y)} + \epsilon^{-1}\|V\|^2\|y_0\|^2_Y \end{aligned} \tag{4.4}$$

where the norm of $V$ is in $L(Y; L_2(0,\infty;Y))$; finally a constant term in $w$ given by $-J^0_{w=0}(y_0)$. Thus, (3.2) follows with $C_\epsilon = \epsilon^{-1}\|V\|^2$.

(ii) The expression of $-J^0_{w,\infty}(y_0)$ given by (2.1.15b) as a quadratic functional, bounded below by part (i), guarantees that there exists a unique optimal solution $w^*$ in $L_2(0,\infty;Y)$.

(iii) If $0 < \gamma < \gamma_c$, then

$$\inf_{\|w\|=1} (E_\gamma w, w)_{L_2(0,\infty;Y)} = m < 0$$

Here, for $\epsilon > 0$ sufficiently small, there exists $w_\epsilon$, $\|w_\epsilon\| = 1$ such that $(E_\gamma w_\epsilon, w_\epsilon)_{L_2(0,\infty;Y)} < m + \epsilon < 0$. Then, define $w_k = kw_\epsilon \in L_2(0,\infty;Y)$, for a real constant $k$. From (2.1.15b), (2.1.16a), (2.1.17b), we have

$$\begin{aligned} -J^0_{w_k}(y_0) &= k^2(E_\gamma w_\epsilon, w_\epsilon)_{L_2(0,\infty;Y)} - 2k(w_\epsilon, Vy_0) - J^0_{w=0}(y_0) \\ &\longrightarrow -\infty \quad as \quad k \longrightarrow \infty \end{aligned}$$

as desired, since $(E_\gamma w_\epsilon, w_\epsilon) < 0$. □

With the optimal $w^*$ provided by Theorem 3.1(ii), we return to the optimal pair $\{u^0_w, y^0_w\}$ over $u$ of Theorem 2.1.1. and set, along with (3.3):

$$u^*(\cdot\,; y_0) \equiv u^0_{w=w^*}(\cdot\,; y_0) \in L_2(0,\infty;U); \qquad y^*(\cdot\,; y_0) \equiv y^0_{w=w^*}(\cdot\,; y_0) \in L_2(0,\infty;Y); \tag{4.5}$$

**Theorem 4.2** *(i) The unique optimal $w^*(\cdot\,; y_0)$ provided by Theorem 3.1(ii) is given explicitly in terms of the problem data by (see (2.1.8) and (3.5)):*

$$\gamma^2 w^*(\cdot\,; y_0) = W^*R^*Ry^*(\cdot\,; y_0) \in L_2(0,\infty;Y), \quad \gamma > \gamma_c \tag{4.6}$$

*(ii) Thus, for $\gamma > \gamma_c$ (defined by (2.2.1)), the original minimax problem (1.2.2) has a unique solution $\{u^*(\cdot\,;y_0), y^*(\cdot\,;y_0), w^*(\cdot\,;y_0)\}$ satisfying (3.6) and given by*

$$-u^*(\cdot\,;y_0) = [I + L^*R^*RL]^{-1}L^*R^*R\,[e^{A\cdot}\,y_0 + Ww^*(\cdot\,;y_0)] \in L_2(0,\infty;U) \tag{4.7}$$

$$y^*(\cdot\,;y_0) = [I + LL^*R^*R]^{-1}[e^{A\cdot}\,y_0 + Ww^*(\cdot\,;y_0)] \in L_2(0,\infty;Y) \tag{4.8}$$

$$u^*(\cdot\,;y_0) = -L^*R^*Ry^*(\cdot\,;y_0) \tag{4.9}$$

*with optimal dynamics*

$$y^*(t;y_0) = e^{At}y_0 + \{Lu^*(\cdot\,;y_0)\}(t) + \{Ww^*(\cdot\,;y_0)\}(t) \tag{4.10}$$

*which therefore satisfies*

$$\{[I + LL^*R^*R - \gamma^{-2}WW^*R^*R]y^*(\cdot\,;y_0)\}(t) = e^{At}y_0 \tag{4.11}$$

**Proof:** Again see [M-T.1] where a Lagrange multiplier proof is given as well as a proof on setting to zero the variation in $w$. □

Remark 3.1 We are not authorized to boundedly invert on $L_2(0,\infty;Y)$ the operator $[I + LL^*R^*R - \gamma^{-2}WW^*R^*R]$ for all $\gamma > \gamma_c$; only for $\gamma$ sufficiently large, in fact $\gamma^2 > \gamma_1^2 = \|I + LL^*R^*R\|/\|WW^*R^*R\|$, in $L_2(0,\infty;Y)$-norms. Thus, a modified argument will be indicated in Proposition 4.1 below; see also Remark 4.1. □

# 5 Explicit expressions of $\{u^*, y^*, w^*\}$ for $\gamma > \gamma_c$ in terms of the data via $E_\gamma^{-1}$.

We begin with an explicit expression of $w^*(\cdot\,;y_0)$ for $\gamma > \gamma_c$ in terms of the data via $E_\gamma^{-1}$.

**Proposition 5.1** *For $\gamma > \gamma_c$ (defined by (2.2.1)), we have*

$$w^*(\cdot\,;y_0) = E_\gamma^{-1}W^*R^*R[I + LL^*R^*R]^{-1}(e^{A\cdot}\,y_0) \in L_2(0,\infty;Y) \tag{5.1}$$

Proof: We insert (3.8) into (3.6) thereby obtaining

$$[\,\gamma^2 I - W^*R^*R[I + LL^*R^*R]^{-1}W]w^*(\cdot\,;y_0) = W^*R^*R[I + LL^*R^*R]^{-1}(e^{A\cdot}\,y_0) \qquad (5.2)$$

Recalling the definition (2.1.18) of $E_\gamma$ , we rewrite (4.1) as

$$E_\gamma w^*(\cdot\,;y_0) = W^*R^*R[I + LL^*R^*R]^{-1}(e^{A\cdot}\,y_0) \qquad (5.3)$$

from which (4.1) follows for $\gamma > \gamma_c$ by Corollary 2.2.1. □

Remark 4.1 If we apply $W^*R^*R[I + LL^*R^*R]^{-1}$ to Eq. (3.11), we get

$$W^*R^*R[I + LL^*R^*R]^{-1}\{[I + LL^*R^*R]y^*(\cdot\,;y_0) \;\; -\gamma^{-2}WW^*R^*Ry^*(\cdot\,;y_0)\} = W^*R^*R[I + LL^*R^*R]^{-1}(e^{A\cdot}\,y_0) \qquad (5.4)$$

We then use (3.6) to obtain Eq. (4.2). But, Eq. (4.2) is solvable for $w^*$ for $\gamma > \gamma_c$, while Eq. (3.11) is solvable for $y^*$ for $\gamma$ sufficiently large, see Remark 3.1. On the other hand, inserting $w^*$ given by (4.1) into the right hand sides of (3.7) and (3.8), produces explicit expressions for $u^*$ and $y^*$ for all $\gamma > \gamma_c$ in terms of the problem data, which we omit writing explicitly. □

Remark 4.2 So far our treatment has followed closely the development for hyperbolic/plate-like equations in [M-T.1] with one important exception. In the abstract model of [M-T.1] – which captures intrinsic properties of hyperbolic equations, plate-like equations, Schroedinger equations, etc. [L-T.1] – the *continuity in time* $y \in C([0,T];Y)$ of *any* solution $y$ is guaranteed by the model. In the present "abstract parabolic case" (H.1) – (H.3), the situation is drastically different. Continuity in time $y \in C([0,T];Y)$ of all solutions $y$ (with $Y$ a correctly chosen, sharp space of parabolic regularity in mixed problems) is *false*. However, in the mildly unbounded case, where the constants $\delta$ and $\rho$ in (1.1.2) and (1.1.3) are $< 1/2$, one obtains the desired regularity for the *optimal* solutions $y^0_w(t;y_0) \in C([0,\infty];Y)$, (as opposed to a general solution), in particular, $y^*(t;y_0) \in C([0,\infty];Y)$, and $w^*(t;y_0) \in C([0,\infty];Y)$, as it follows directly from Theorem 5.1 (i) below. Instead, in the general case where $1/2 < \delta,\ \rho < 1$, the continuity in time of the *optimal* solutions $y^*(t;y_0)$, $w^*(t;y_0)$ and $u^*(t;y_0)$ are still true – and this is a key achievement of the theory – but it requires a much more complicated boot-strap argument (as in the corresponding linear quadratic regulator problem where the disturbance

$w \equiv 0$). This will be our next objective, a point of departure from the hyperbolic/plate-like equations of [M-T.1], and a point of convergence with the parabolic theory with no disturbance, $w \equiv 0$, [L-T.1], [L-T.2], [L-T.3], [D-I.1], [F.1].

# 6 Smoothing properties of the operators $L$, $L^*$, $W$, $W^*$: the optimal $u^*$, $y^*$, $w^*$ are continuous in time

The key feature in the present analytic case is the following theorem whose proof may be found in [L-T.2, Theorem 2.5], or [L-T.3, Chapter 3].

**Theorem 6.1** *(a) With reference to the operators $L$ and $L^*$ defined by (2.1.2) and (2.1.6) we have:*

*(i)*
$$L: \text{ continuous } L_2(0,\infty;U) \longrightarrow L_{l_1}(0,\infty;Y), \tag{6.1}$$

*where $l_1$ is an arbitrary positive number satisfying $l_1 < 2/(2\delta-1)$; here $2/(2\delta-1) > 2$ for $1/2 < \delta < 1$, while for $0 \le \delta \le 1/2$ we may take $l_1 = \infty$, and indeed replace $L_\infty(0,\infty;\cdot\,)$ with $C([0,\infty];\cdot\,)$.*

*(ii)*
$$L^*: \text{ continuous } L_{l_1}(0,\infty;Y) \longrightarrow L_{l_2}(0,\infty;U) \tag{6.2}$$

*where $l_1$ is as in (i), and $l_2$ is positive number satisfying $l_2 < 2/(4\delta-3)$; here $2/(4\delta-3) > l_1$ for $3/4 < \delta < 1$; for $0 < \delta < 3/4$ we may take $l_2 = \infty$ and indeed replace $L_\infty(0,\infty;\cdot\,)$ with $C([0,\infty];\cdot\,)$.*

*(iii) Generally, let $l_0 = 2$ and let $l_n$, $1,2,\ldots$ be arbitrary positive numbers such that*

$$l_n < \frac{2}{2n\delta-(2n-1)}, \quad n = 1,2,\ldots\,. \tag{6.3}$$

*Then, for $n = 0,2,4,\ldots$ we have that*

$$L: \text{ continuous} L_{l_n}(0,\infty;U) \longrightarrow L_{l_{n+1}}(0,\infty;Y) \tag{6.4}$$

*where for $0 \le \delta \le \frac{2(n+1)-1}{2(n+1)}$ we may take $l_{n+1} = \infty$ and indeed replace $L_\infty(0,\infty;\cdot\,)$ with $C([0,\infty];\cdot\,)$, and moreover,*

$$L^*: \text{ continuous } L_{l_{n+1}}(0,\infty;Y) \longrightarrow L_{l_{n+2}}(0,\infty;U), \tag{6.5}$$

*where $l_{n+1}$ in (5.5) is the same as in (5.4), and where for* $0 \leq \delta \leq \frac{2(n+1)-1}{2(n+2)}$ *we may take* $l_{n+2} = \infty$, *and indeed replace* $L_\infty(0,\infty;\cdot\,)$ *with* $C([0,\infty];\cdot\,)$.

(*iv*) *For* $p > 1/(1-\delta)$,

$$L: \quad \textit{continuous } L_p(0,\infty;U) \longrightarrow C([0,\infty];Y). \tag{6.6}$$

(*v*)

$$L^*: \quad \textit{continuous } C([0,\infty];Y) \longrightarrow C([0,\infty];U) \tag{6.7}$$

(*b*) *Similar results apply for the operators W and $W^*$ mutatis mutandis, i.e. replacing δ in (1.1.2) with ρ in (1.1.3).* □

As a corollary, we then obtain the desired time regularity property property in $C([0,\infty];\cdot\,)$ of the optimal quantities.

**Corollary 5.2** *With reference to the optimal solutions $\{u^*, w^*, y^*\}$, we have for any $y_0 \in Y$:*

$$u^*(\cdot\,;y_0) \in C([0,\infty];U); \quad y^*(\cdot\,;y_0) \in C([0,\infty];Y); \quad w^*(\cdot\,;y_0) \in C([0,\infty];U); \tag{6.8}$$

**Proof:** By a boot-strap argument as in [L-T.2], [L-T.3] playing alternatively between (3.10) on the one hand, and (3.6) and (3.9) on the other. To begin with, we already have from Theorem 3.2 that

$$u^*(\cdot\,;y_0) \;\; \textit{and} \;\; w^*(\cdot\,;y_0) \in L_2(0,\infty;U); \quad y^*(\cdot\,;y_0) \in L_2(0,\infty;Y) \tag{6.9}$$

It is not restrictive to assume that $\delta = \rho$ for the purposes of this proof and thus use the same regularity results for $L$ and $W$, and for $L^*$ and $W^*$ from Theorem 5.1. Thus, invoking (5.1) for $L$ and $W$ we obtain that $Lu^*, Ww^* \in L_{l_1}(0,\infty;Y)$. Since $e^{At}$ is exponentially stable as in (1.1.5), we then obtain from the optimal dynamics (3.10) that $y^* \in L_{l_1}(0,\infty;Y)$ as well. Next, we recall (3.6) and (3.9) and apply (5.2) for $L^*$ and $W^*$ to obtain that $w^*$, $u^* \in L_{l_2}(0,\infty;U)$. We then repeat this boot-strap argument on $L$ and $W$ in (3.10), invoking (5.4), and on $L^*$ and $W^*$ in (3.6), (3.9), invoking (5.5), and we obtain the desired time continuity for $u^*(\cdot\,;y_0)$; $w^*(\cdot\,;y_0)$, $y^*(\cdot\,;y_0)$, by Theorem 5.1 (iv), (v). □

Having obtained the desired time continuity for $u^*(\cdot\,;y_0), w^*(\cdot\,;y_0), y^*(\cdot\,;y_0)$, we can next proceed as in [M-T.1] to obtain the semigroup property of $y^*(\cdot\,;y_0)$.

# 7 A transition property for $w^*$ and the semigroup property for $y^*$ for $\gamma > \gamma_c$

We have the following important property:

**Theorem 7.1** *For $\gamma > \gamma_c$ (defined in (2.2.1), we have:*

$$w^*(t+\sigma; y_0) = w^*(\sigma; y^*(t; y_0)) \qquad \forall t, \sigma > 0 \tag{7.1}$$

*for t fixed, the equality being intended in $C([0,\infty]; Y)$ in $\sigma$* □.

**Proof:** see [M-T.1, Theorem 4.2.1]. Because of its importance, we provide here a sketch. This is based on (3.11) to obtain first that for $\gamma > \gamma_c$ (defined in (2.2.1), we have:

$$[I + LL^*R^*R - \gamma^{-2}WW^*R^*R][y^*(t+\cdot\,; y_0) - y^*(\cdot\,; y^*(t; y_0))] = 0 \tag{7.2}$$

Next, starting from (6.2) and applying to it the operator $W^*R^*R[I + LL^*R^*R]^{-1}$ as in Remark 4.1, we obtain

$$\begin{aligned} W^*R^*R \quad & [y^*(t+\cdot\,; y_0) - y^*(\cdot\,; y^*(t; y_0))] \\ & -\gamma^{-2}W^*R^*R[I + LL^*R^*R]^{-1}WW^*R^*R[y^*(t+\cdot\,; y_0) - y^*(\cdot\,; y^*(t; y_0)] = 0 \end{aligned} \tag{7.3}$$

Furthermore, one verifies that: for $\gamma > \gamma_c$, (defined in (2.2.1)) and with reference to (3.6) we have:

$$\gamma^2 w^*(t+\sigma; y_0) = \{W^*R^*Ry^*(t+\cdot\,; y_0)\}(\sigma) \tag{7.4}$$

Using (6.4) and (3.6) in (6.3), we rewrite (6.3) as

$$\begin{aligned} \gamma^2[w^*(t+\sigma; y_0) \quad & -w^*(\sigma; y^*(t; y_0))] \\ & -W^*R^*R[I + LL^*R^*R]^{-1}W[w^*(t+\cdot\,; y_0) - w^*(\cdot\,; y^*(t; y_0))] = 0 \end{aligned} \tag{7.5}$$

or, recalling the definition of $E_\gamma$ in (2.1.18):

$$E_\gamma[w^*(t+\cdot\,; y_0) - w^*(\cdot\,; y^*(t; y_0))] = 0 \tag{7.6}$$

Thus, by Corollary 2.2.1, if $\gamma > \gamma_c$, then $E_\gamma{}^{-1} \in L_2(0,\infty; Y)$, and so by (6.6) we obtain

$$w^*(t+\cdot\,; y_0) - w^*(\cdot\,; y^*(t; y_0)) = 0 \tag{7.7}$$

first in $L_2(0,\infty;Y)$, next in $C([0,\infty];Y)$ as desired. Theorem 6.1 is proved. $\square$

Defining, the operator $\Phi(t)$ (which depends on $\gamma$) by

$$y^*(t;x) = \Phi(t)x \ \in C([0,\infty];Y), \quad \forall\, x \in Y \tag{7.8}$$

we obtain the semigroup property:

**Theorem 7.2** *For $\gamma > \gamma_c$ (see (2.2.1)), $y_0 \in Y$, and $t,\ \sigma > 0$ we have:*

$$y^*(t+\sigma;y_0) = y^*(\sigma;y^*(t;y_0)) \ \in C([0,\infty];Y) \tag{7.9}$$

*so that $\Phi(t)$ is a s.c. semigroup on $Y$.* $\square$

**Proof:** It is a consequence of Theorem 6.1, see [M-T.1, Theorem 4.3.1], to obtain

$$[I + LL^*R^*R][y^*(t+\cdot\,;y_0) - y^*(\cdot\,;y^*(t;y_0))] = 0 \tag{7.10}$$

Since $[I + LL^*R^*R]^{-1} \in \mathcal{L}(L_2(0,\infty;Y))$, see [L-T.2], [F-L-T.1] , we obtain from (4.3.5)

$$y^*(t+\sigma;y_0) - y^*(\sigma;y^*(t;y_0)) = 0, \tag{7.11}$$

first in $L_2(0,\infty;Y)$, and then, by the regularity of $y^*$, in $C([0,\infty];Y)$. $\square$

**Corollary 6.3** *The s.c. semigroup $\Phi(t)$ in (6.8), guaranteed by Theorem 6.2, is exponentially stable: there exists $C \geq 1,\ k > 0$ such that*

$$\|\Phi(t)\|_{\mathcal{L}(Y)} \leq Ce^{-kt}, \quad t \geq 0 \tag{7.12}$$

Proof: The s.c. semigroup $\Phi(t)$ satisfies $y^*(t;x) = \Phi(t)x \in L_2(0,\infty;Y)$ for all $x \in Y$ by optimality (3.5) and then conclusion (6.12) follows via a known theorem in [D.1]. $\square$

# 8 The Riccati operator, $P$, for $\gamma > \gamma_c$

The property of $\Phi(t)$ in (6.8) as a s.c. uniformly stable semigroup on $Y$ – as guaranteed by section 6 – allows us to readily fall into e.g. the abstract treatment of [L-T.2], [L-T.3, Chapter 3]. Accordingly, we shall first define (in terms of the problem data) an operator $P \in \mathcal{L}(Y)$ for $\gamma > \gamma_c$, and we shall next show that $P$ is, in fact, a solution of an algebraic Riccati (operator) equation and indeed the unique such solution (within a certain class).

## 8.1 Definition of $P$ and its preliminary properties

For $\gamma > \gamma_c$, and recalling (1.1.5), we define the operator $P \in \mathcal{L}(Y)$ by (see (6.8)):

$$Px = \int_0^\infty e^{A^*\sigma} R^* R y^*(\sigma; x) d\sigma = \int_0^\infty e^{A^*\sigma} R^* R \Phi(\sigma) x d\sigma \tag{8.1.1a}$$

$$= \int_t^\infty e^{A^*(\tau - t)} R^* R y^*(\tau - t; x) d\tau = \int_t^\infty e^{A^*(\tau-t)} R^* R \Phi(\tau - t) x d\tau \tag{8.1.1b}$$

We now collect preliminary properties of $P$.

**Proposition 8.1.1** *With reference to (7.1.1) we have for $\gamma > \gamma_c$:*

(*i*)
$$\text{range of } P = PY \subset D((-A^*)^\theta) \quad \forall\ 0 \le \theta < 1 \tag{8.1.2a}$$

$$(-A^*)^\theta P \in \mathcal{L}(Y) \tag{8.1.2b}$$

(*ii*)
$$B^* P \in \mathcal{L}(Y); \qquad G^* P \in \mathcal{L}(Y) \tag{8.1.3}$$

**Proof:** As usual, applying $(-A^*)^\theta$, $0 \le \theta < 1$ to (7.11) and using the analytic semigroup bound (see (1.1.5)):

$$\|(-A^*)^\theta e^{A^* t}\| \le \frac{Ce^{-\omega t}}{t^\theta}, \quad t > 0 \tag{8.1.4}$$

yields readily (7.1.2) via, say, (6.1.2). then (7.1.2) with $\theta = \delta < 1$ implies (7.1.3) by writing $B^*P = B^*(-A^*)^{-\delta}(-A^*)^\delta P$ and using (1.1.2). Similarly, we can prove the desired result for $G^*P$, recalling (1.1.3) with $\rho < 1$. □

**Corollary 7.1.2** *With reference to (7.1.1), we have for $\gamma > \gamma_c$:*

(*i*)
$$u^*(t; y_0) = -B^* P y^*(t; y_0) = -B^* P \Phi(t) y_0 \quad \text{a.e. in } t;\ y_0 \in Y \tag{8.1.5a}$$

$$B^* P \Phi(t): \ \text{continuous } Y \longrightarrow L_2(0, \infty; U) \cap C([0, \infty]; U) \tag{8.1.5b}$$

(*ii*)
$$\gamma^2 w^*(t; y_0) = G^* P y^*(t; y_0) = G^* P \Phi(t) y_0 \tag{8.1.6a}$$

$$G^* P \Phi(t): \ \text{continuous } Y \longrightarrow L_2(0, \infty; U) \cap C([0, \infty]; Y) \tag{8.1.6b}$$

(*iii*) *the operator $P \in \mathcal{L}(Y)$ satisfies the symmetric relation for $x_1, x_2 \in Y$:*

$$(Px_1, x_2) = \int_0^\infty [(Ry^*(t; x_1), Ry^*(t; x_2))_Y + (u^*(t; x_1), u^*(t; x_2))_U - \gamma^2 (w^*(t; x_1), w^*(t; x_2))_Y] dt \tag{8.1.7}$$

*from which it follows that $P$ is a nonnegative self-adjoint operator: $P = P^* \geq 0$ on $Y$ and that the optimal cost of problem (2.1.1) is*

$$\begin{aligned}(Py_0, y_0)_Y &= J^*(y_0) \quad \text{[optimal cost in (3.3)]} \\ &= J(u^*(\cdot\,;y_0), y^*(\cdot\,;y_0), w^*(\cdot\,;y_0))\end{aligned} \tag{8.1.8}$$

**Proof:** (i) One applies $B^*$ to (7.1.1b) with $x$ replaced by $y^*(t;y_0)$ and obtains (7.1.5) via the semigroup property $\Phi(\tau - t)\Phi(t) = \Phi(\tau)$ as well as (2.1.6) and (3.9). The proof for (ii) is similar, this time by use of (2.1.8) and (3.6). The indicated regularity follows from (7.1.3).

(ii) For $x_1, x_2 \in Y$ we write from (7.1.1) using $e^{A\cdot}\, x_2$ from (3.10) :

$$\begin{aligned}(Px_1, x_2)_Y &= (R^*Ry^*(\cdot\,;x_1), e^{A\cdot}\, x_2)_{L_2(0,\infty;Y)} \\ &= (Ry^*(\cdot\,;x_1), Ry^*(\cdot\,;x_2)) - (L^*R^*Ry^*(\cdot\,;x_1), u^*(\cdot\,;x_2))_{L_2(0,\infty;U)} \\ &\quad -(W^*R^*Ry^*(\cdot\,;x_1), w^*(\cdot\,;x_2))_{L_2(0,\infty;Y)}\end{aligned} \tag{8.1.9}$$

and (7.1.7) follows from (7.1.9) recalling (3.9) and (3.6). □

# 9 The Riccati operator, $P$, for $\gamma > \gamma_c$

The property of $\Phi(t)$ in (6.8) as a s.c. uniformly stable semigroup on $Y$ – as guaranteed by section 6 – allows us to readily fall into e.g. the abstract treatment of [L-T.2], [L-T.3, Chapter 3]. Accordingly, we shall first define (in terms of the problem data) an operator $P \in L(Y)$ for $\gamma > \gamma_c$, and we shall next show that $P$ is, in fact, a solution of an algebraic Riccati (operator) equation and indeed the unique such solution (within a certain class).

## 9.1 Definition of $P$ and its preliminary properties

For $\gamma > \gamma_c$, and recalling (1.1.5), we define the operator $P \in L(Y)$ by (see (6.8)):

$$Px = \int_0^\infty e^{A^*\sigma}R^*Ry^*(\sigma;x)d\sigma = \int_0^\infty e^{A^*\sigma}R^*R\Phi(\sigma)x d\sigma \tag{9.1.1a}$$

$$= \int_t^\infty e^{A^*(\tau-t)}R^*Ry^*(\tau - t;x)d\tau = \int_t^\infty e^{A^*(\tau-t)}R^*R\Phi(\tau - t)x d\tau \tag{9.1.1b}$$

We now collect preliminary properties of $P$.

**Proposition 9.1.1** *With reference to (7.1.1) we have for $\gamma > \gamma_c$:*

(*i*)
$$range\ of\ P = PY \subset D((-A^*)^\theta) \quad \forall\ 0 \leq \theta < 1 \tag{9.1.2a}$$
$$(-A^*)^\theta P \in \mathcal{L}(Y) \tag{9.1.2b}$$

(*ii*)
$$B^*P \in \mathcal{L}(Y); \qquad G^*P \in \mathcal{L}(Y) \tag{9.1.3}$$

**Proof:** As usual, applying $(-A^*)^\theta$, $0 \leq \theta < 1$ to (7.11) and using the analytic semigroup bound (see (1.1.5)):

$$\|(-A^*)^\theta e^{A^*t}\| \leq \frac{Ce^{-\omega t}}{t^\theta}, \quad t > 0 \tag{9.1.4}$$

yields readily (7.1.2) via, say, (6.1.2). then (7.1.2) with $\theta = \delta < 1$ implies (7.1.3) by writing $B^*P = B^*(-A^*)^{-\delta}(-A^*)^\delta P$ and using (1.1.2). Similarly, we can prove the desired result for $G^*P$, recalling (1.1.3) with $\rho < 1$. □

**Corollary 7.1.2** *With reference to (7.1.1), we have for $\gamma > \gamma_c$:*

(*i*)
$$u^*(t; y_0) = -B^*Py^*(t; y_0) = -B^*P\Phi(t)y_0 \quad \text{a.e. in } t;\ y_0 \in Y \tag{9.1.5a}$$
$$B^*P\Phi(t): \ \text{continuous } Y \longrightarrow L_2(0, \infty; U) \cap C([0, \infty]; U) \tag{9.1.5b}$$

(*ii*)
$$\gamma^2 w^*(t; y_0) = G^*Py^*(t; y_0) = G^*P\Phi(t)y_0 \tag{9.1.6a}$$
$$G^*P\Phi(t): \ \text{continuous } Y \longrightarrow L_2(0, \infty; U) \cap C([0, \infty]; Y) \tag{9.1.6b}$$

(*iii*) *the operator $P \in \mathcal{L}(Y)$ satisfies the symmetric relation for $x_1, x_2 \in Y$:*

$$(Px_1, x_2) = \int_0^\infty [(Ry^*(t; x_1), Ry^*(t; x_2))_Y + (u^*(t; x_1), u^*(t; x_2))_U - \gamma^2(w^*(t; x_1), w^*(t; x_2))_Y]dt \tag{9.1.7}$$

*from which it follows that $P$ is a nonnegative self-adjoint operator: $P = P^* \geq 0$ on $Y$ and that the optimal cost of problem (2.1.1) is*

$$\begin{aligned}(Py_0, y_0)_Y &= J^*(y_0) \quad \text{[optimal cost in (3.3)]} \\ &= J(u^*(\cdot\ ; y_0), y^*(\cdot\ ; y_0), w^*(\cdot\ ; y_0))\end{aligned} \tag{9.1.8}$$

**Proof:** (i) One applies $B^*$ to (7.1.1b) with $x$ replaced by $y^*(t; y_0)$ and obtains (7.1.5) via the semigroup property $\Phi(\tau - t)\Phi(t) = \Phi(\tau)$ as well as (2.1.6) and (3.9). The proof for (ii) is similar, this time by use of (2.1.8) and (3.6). The indicated regularity follows from (7.1.3).

(ii) For $x_1, x_2 \in Y$ we write from (7.1.1) using $e^{A\cdot}\, x_2$ from (3.10) :

$$\begin{aligned}(Px_1, x_2)_Y &= (R^*Ry^*(\cdot\,; x_1), e^{A\cdot}\, x_2)_{L_2(0,\infty;Y)} \\ &= (Ry^*(\cdot\,;x_1), Ry^*(\cdot\,;x_2)) - (L^*R^*Ry^*(\cdot\,;x_1), u^*(\cdot\,;x_2))_{L_2(0,\infty;U)} \\ &\qquad -(W^*R^*Ry^*(\cdot\,;x_1), w^*(\cdot\,;x_2))_{L_2(0,\infty;Y)} \qquad (9.1.9)\end{aligned}$$

and (7.1.7) follows from (7.1.9) recalling (3.9) and (3.6). □

# 10 References


[B.1 ] V. Barbu, $H^\infty$-Boundary Control with State Feedback; the Hyperbolic Case, preprint 1992.

[B-B.1 ] T. Başar and P. Bernhard. *$H^\infty$-Optimal Control and Related Minimax Design Problems. A Dynamic Game Approach*, Birkhaüser Boston (1991).

[D.1 ] R. Datko. Extending a Theorem of Liapunov to Hilbert Space, *J. Math. Anal. Appl.*, **32** (1970), pp. 610-616.

[D-I.1 ] G. Da Prato and A. Ichikawa. Riccati Equations with Unbounded Coefficients, *Annali di Matem. Pura e Applic.*, **140** (1985), pp. 209-221.

[F.1 ] F. Flandoli. Algebraic Riccati Equations Arising in Boundary Control Problems, *SIAM J. Control and Optim.*, **25** (1987), pp. 612 -636.

[L-T.1 ] I. Lasiecka and R. Triggiani. *Differential and Algebraic Riccati Equations with Application to Boundary/Point Control Problems: Continuous Theory and Approximation Theory*, Volume # 164 in the Springer-Verlag Lectures Notes LNCIS series (1991), pp. 160.

[L-T.2 ] I. Lasiecka and R. Triggiani. The Regulator Problem for Parabolic Equations with Dirichlet Boundary Control. Part I: Riccati's Feedback Synthesis, and Regularity of the Optimal Solutions, *Appl. Math. and Optimiz.*, **15** (1987), pp. 147-168.

[L-T.3 ] I. Lasiecka and R. Triggiani. Monograph for Encyclopedia of Mathematics, Cambridge University Press, to appear.

[L-T.4 ] I. Lasiecka and R. Triggiani. Riccati Equations for Hyperbolic Partial Differential Equations with $L_2(0, T; L_2(\Gamma))$-Dirichlet Boundary Terms, *SIAM J. Control Optimiz.*, **24** (1986), pp. 884-924.

[M.1 ] C. McMillan. Ph.D. dissertation, University of Virginia (1993).

[M-T.1 ] C. McMillan and R. Triggiani. Min-Max Game Theory and Algebraic Riccati Equations for Boundary Control Problems with Continuous Input-Solution Map. Part I: the Stable Case, University of Virginia preprint 1992.

[M-T.2 ] C. McMillan and R. Triggiani. Min-Max Game Theory and Algebraic Riccati Equations for Boundary Control Problems with Continuous Input-Solution Map. Part II: the General Case, preprint 1992. Presented at the International Conference on Mathematical Physics and Differential Equations held at the Georgia Institute of Technology, March 1992.

[M-T.3 ] C. McMillan and R. Triggiani. Min-Max Game Theory and Algebraic Riccati Equations for Boundary Control Problems with Analytic Semigroups. Part II: the General Case, University of Virginia preprint 1992. Presented at the SIAM Conference on Control and its Applications held at Minneapolis, September 1992.

# 48 Approximate Controllability of Linear Functional-Differential Systems: A State Space Independent Approach

**Andrzej W. Olbrot** Wayne State University, Detroit, Michigan

**Abstract**

Approximate versions of controllability, reachability, and null controllability for linear functional-differential systems of retarded type are considered in relation to spectral controllability and stabilizability in general function spaces not necessarily generating a $C_0$ - semigroup in the state space. It is shown that spectral controllability is necessary for approximate controllability and reachability in a large variety of state spaces. Spectral controllability is also necessary for approximate null controllability on a fixed time interval. On the other hand, if the space of control functions is closed under right shifts, uniform approximate controllability implies spectral controllability. Also, open loop stabilizability is necessary for approximate null controllability. Finally, we show that three large families of concrete function spaces, $C^{(k)}$, $M^p$, and $W^{(k,p)}$, satisfy all our general assumptions.

## 1. INTRODUCTION

Our aim is to examine various approximate controllability and related stabilizability properties of linear systems with time delays described by functional-differential equations of the form

(1) $d/dt\, x(t) = A\, x_t + B\, u(t)$

with an initial condition

(2) $x_0 = \phi$

where the state $x_t$ is a piece of past trajectory of the solution vector, $x_t(r) = x(t + r) \in R^n$, $r \in [-h, 0]$, $B$ is a real matrix, and $A$ is a linear convolution operator, $A\, x_t = (a * x)(t)$, where the matrix $a(.)$ has support in $[0, h]$ and its atomic part is finite. More specifically, we assume the following representation for $A$

$$A\varphi = \sum_{i=0}^{N} A_i \varphi(-h_i) + \int_{-h}^{0} \alpha(r)\varphi(r)dr$$

where $0 = h_0 < h_1 < \dots < h_N = h$, the matrices $A_i$ are real and $\alpha(.)$ is a real matrix valued function assumed to be Lebesgue integrable in $[-h,0]$.

The literature on controllability of delay systems is quite rich; the reader is referred to [1] and [2] for basic references. In early works the research concentrated mainly on two problems: the reachability of the trajectory endpoint $x(T)$ for some final time $T$ where various forms of complete solutions were obtained [3], [4], and the controllability to equilibrium (to the zero final state $x_T = 0$) where the first complete solution was obtained in [5] in 1973 by reducing the problem to geometric problems of minimal and maximal controllability subspaces in finite dimension and later algebraic solutions over specific subrings of entire functions as Laplace transforms were obtained, [6], [7]. Next, some authors tried to examine reachability of arbitrary final states $x_T$ in some function spaces. It soon appeared that this concept is much too strong to be useful in control theory since it typically required rank $B = n$, thus excluding systems with a single control and more than one state variable. It occurs that the concept of approximate controllability is much less restrictive, still appealing from the application point of view, and strongly related to other fundamental concepts like stabilizability and spectral controllability.

First complete results on approximate controllability were derived in [1] in 1977 where, for the state spaces $L^p$, $M^p$, $C$, and $W^{(1,p)}$, this property was characterized by dual observability problems for a dual (transposed) system. It was also shown that multipoint controllability and spectral controllability are necessary for approximate controllability in all these spaces except $L^p$. For the case of commensurable delays, the complete set of necessary and sufficient conditions for approximate controllability on a fixed time interval was derived. A sequence of other results followed, [8], [9], [10], [2]. Manitius and Triggiani, [8], examined the case with one delay and the state spaces $L^2$ and $M^2$. Marchenko, [9], proved that approximate null controllability, with finite time dependent on an initial state, is equivalent to feedback stabilizability. Manitius, [10], proved that approximate controllability in the space $M^2$ is equivalent to spectral controllability and complettability where the latter condition means the existence of a linear feedback which transforms the original system into a system with a complete set of generalized eigenfunctions. The rank condition equivalent to complettability appeared earlier in [1]. Salamon, [2], demonstrated that approximate null controllability on a sufficiently long time interval is equivalent to spectral controllability. For systems with neutral terms, criteria for approximate controllability in $W^{(1,p)}$ were given in [11] (see also [2] for more results and references on neutral systems).

Summarizing this short overview of literature, we note that the results were obtained for very specific function spaces (mostly $M^2$ and $W^{(1,p)}$) using the methods of the $C_0$ - semigroup theory and, typically, for systems with one or several discrete time delays.

In this paper, we extend the above mentioned results in several ways. We will consider systems with both discrete and distributed delays. The main feature of our approach is that we are not restricted to a specific function space as opposed to most previous papers and we do not need the $C_0$ - semigroup properties.

In Section 2, we recall definitions of various approximate controllability notions and show basic interrelationships between them due to system linearity and time invariance. In Section 3, we prove that spectral controllability is necessary for approximate controllability on both fixed and state dependent intervals, as well as for approximate null controllability on a fixed interval, and this holds in general function spaces satisfying some natural mild axioms. Next result is that the open loop stabilizability is necessary for approximate null controllability (on intervals dependent on initial states). Finally, we show that three large families of concrete function spaces, $X = C^{(k)}$, $M^p$, and $W^{(k,p)}$ satisfy all our general assumptions.

In another paper, we intent to generalize the notions of a dual transposed system and dual observability concepts, developed in [1], corresponding to approximate controllability concepts. This enables completion of our general results by proving that spectral controllability is sufficient for approximate null controllability and by deriving testable criteria from the dual problems. Such criteria assume an especially simple form in case of discrete delays where connections to controllability over the ring of polynomials in a delay operator can be established.

## 2. CONTROLLABILITY DEFINITIONS

Consider a control interval $[0, T]$, some linear function space $X$ of states with a topology, and some space $U$ of control functions on $[0, T]$. We assume that the pair $(U,X)$ is compatible in the sense that $U$ is a subset of integrable functions, X is a subset of essentially bounded functions, and for any initial condition $\phi$ in X and any control u in U the corresponding states $x_t$ are in X for all $t \in [0, T]$. The solution to the system equations (1),(2) can be represented as

$$(3) \quad x_t = S(t)\phi + C(t)u$$

where S(t) and C(t) are linear operators defined by the variation of constants formula with the fundamental matrix solution $\Xi(t)$ (see Hale, [12])

$$(4) \quad x(t) = \Xi(t)\varphi(0)$$
$$+ \int_0^h \Xi(t-\tau)\{\sum_{i=0}^{N} A_i\Phi(\tau - h_i) + \int_{-h}^{0} \alpha(r)\Phi(\tau + r)dr\}d\tau + \int_0^t \Xi(t-\tau)Bu(\tau)d\tau$$

where $\Phi(\tau) = \phi(\tau)$ for $\tau \in [-h, 0]$ and $\Phi(\tau) = 0$ otherwise.

The closure of the reachable subspace $R(t) = \text{cl } C(t)U$ will play an essential role in our considerations. The following are basic definitions of controllability for system (1),(2).

**Definition 2.1 (Approximate controllability)** System (1),(2) is approximately controllable if for any initial condition in X, any $\psi$ in X, and any neighborhood $N_\psi$ of $\psi$ there is a final time T and a control u in U such that the final state $x_T$ of the corresponding solution x(.) to (1),(2) belongs to $N_\psi$. This system is approximately controllable on an interval [0, T] if the above property holds with a fixed value of T and it is uniformly approximately controllable if there is a T such that it is approximately controllable on the interval [0, T].

**Definition 2.2 (Approximate null controllability)** System (1),(2) is approximately null controllable if for any initial condition in X, and any neighborhood N of zero there is a final time T and a control u in U such that the final state $x_T$ of the corresponding solution x(.) to (1),(2) belongs to N. This system is approximately null controllable on the interval [0, T] if the above property holds with the fixed value of T and it is uniformly approximately null controllable if there is a T such that it is approximately null controllable on the interval [0, T].

**Definition 2.3 (Approximate reachability)** If system (1),(2) satisfies the properties of Definition 1, except that a fixed zero initial state (2) is assumed, then it is, respectively, approximately reachable, approximately reachable on [0, T], and uniformly approximately reachable.

Clearly, our definitions make sense only if $T > h$. Some of the above properties are obviously interrelated by definition, others are related thanks to the linearity and time invariance of eq. (1). To the latter category we include

**Proposition 2.4** The following implications hold:
(i) Approximate controllability on [0,T] implies approximate null controllability on [0,T].
(ii) Uniform approximate controllability implies uniform approximate null controllability.
(iii) If U is closed under right shifts in time then approximate (null) controllability on $[0,T_1]$ implies approximate (null) controllability on $[0,T_2]$ for $T_2 > T_1 > 0$. A similar implication holds for reachability.

**Proof**: To prove (i), take $\psi = -S(T)\phi$. If the system is approximately controllable on [0,T] then, starting from a zero state, we can control the system to some state $C(T)u$ in a neighborhood $N_\psi$ of $\psi$, that is, $C(T)u - \psi = C(T)u + S(T)\phi$ belongs to the neighborhood of zero $N = N_\psi - \psi$. Hence, by formula (3), for a given initial state $x_0 = \phi$, control u brings the final state in a neighborhood of the zero state which means null controllability since $\phi$ is arbitrary.

Property (ii) follows directly from (i). To prove (iii), we can consider, by the right shift property, controls u on $[0,T_2]$ with $u(t) = 0$ on $[0, T_2 - T_1]$ and choose $\tau = T_2 - T_1$ as an initial time. If $x_0 = 0$ then $x_\tau = 0$ and the reachability property on the shorter interval $[\tau, T_2]$, being equivalent (by time-invariance of system equations) to reachability on $[0, T_1]$, implies reachability on $[0,T_2]$. If $x_0$ is an arbitrary state in X then, by the compatibility assumption above, the state $x_\tau$ is in X and controllability on $[0,T_2]$ follows from controllability on $[0, T_1]$. Also, if any initial state $x_0$ in X can be brought to any neighborhood of zero at time $T_1$ then the same holds true for all states of form $x_0 = S(\tau)\phi$ since $S(\tau)X$ is a subset of X. Utilizing time-invariance again we complete the proof. QED

## 3. CONNECTIONS WITH SPECTRAL CONTROLLABILITY AND STABILIZABILITY

In this section we will examine necessary conditions for approximate controllability in terms of stabilizability and spectral controllability which are known to be algebraic properties in the sense that they do not depend on the state space, they are rather characterized by algebraic rank conditions involving system matrices, [13]. We will show that spectral controllability is necessary for both approximate controllability and uniform approximate null controllability and stabilizability is necessary for approximate null controllability and these facts hold true independent of the set of controls and, under some mild assumptions, independent of the state space X.

Let us recall the definitions of spectral controllability and stabilizability. Note that we use "open loop" definitions, no feedback control is involved contrary to most literature. Although the feedback versions are fully equivalent, under some restrictions on feedback, as demonstrated by the theory developed in [13], we find the open loop versions more natural and easier to apply in considerations concerning controllability properties.

**Definition 3.1 (Stabilizability and spectral controllability)** System (1),(2) is $\gamma$-stabilizable, for a given real number $\gamma$, if for any $x_0$ in X there exists a control u in $L^1_{loc}(0, \infty)$ such that both u and the corresponding solution x behaves asymptotically as $O(e^{-\gamma t})$. The system is called stabilizable if it is $\gamma$-stabilizable with $\gamma = 0$, and it is spectrally controllable if it is $\gamma$-stabilizable for any $\gamma$.

It follows from [13] that $\gamma$-stabilizability is equivalent to

$$(5) \qquad \operatorname{rank}\,[sI - \mathbb{A}(s), B] = n$$

for any complex s with $re(s) \geq \gamma$ and spectral controllability holds if and only if (5) holds for any complex s where

$$(6) \qquad \mathbb{A}(s) = \sum_{i=0}^{N} e^{-sh_i} A_i + \int_{-h}^{0} e^{sr} \alpha(r)\,dr$$

Now, we are in a position to prove the main result of the paper.

**Theorem 3.2** Assume that X is a normed state space such that all functionals F of the form

$$(7) \qquad F(\phi) = \langle q, \phi(0) \rangle + \int_{-h}^{0} \langle f(r), \phi(r) \rangle \, dr$$

where $q \in R^n$, $f$ is a sum of a piecewise analytic and a $W^{(1,\infty)}$ function, are continuous and $F$ is nontrivial if $q$ is nonzero. Then spectral controllability is necessary for approximate reachability (and therefore approximate controllability). If, additionally, the function space of control inputs is closed under right shifts and, for some T, all final states $x_T$ generated by zero controls, zero initial functions, and nonzero $x(0)$ belong to the state space $X$ then uniform approximate null controllability implies spectral controllability.

**Proof**: Suppose spectral controllability does not hold and thus the rank condition (5) is not valid for some complex $s_0$. Then there exists a nonzero complex row vector $q$ such that $q\mathbb{A}(s_0) = s_0 q$ and $qB = 0$. Without loss of generality we can consider all control functions to have Laplace transforms $\mathbb{U}(s)$ since by replacing the control values by zero after the final time T we do not change the states on [0,T]. Thus, we can use Laplace transforms $\mathbb{X}(s)$ of x(.) as well, where the existence is guaranteed by the exponential boundedness of solutions. Taking the Laplace transforms of (1) we get

$$[sI - \mathbb{A}(s)]\,\mathbb{X}(s) = x(0) + \mathbb{Y}(s) + B\,\mathbb{U}(s)$$

where $\mathbb{Y}(s)$ is the Laplace transform of $y(t) = Ax_{0t}$ and $x_{0t}$ is the zero extension of the initial function , $x_{0t}(r) = 0$ if $t+r > 0$ and $x_{0t}(r) = x_0(t+r)$ otherwise (with $t$ nonegative and $r$ in $[-h,0]$ ). Multiplying this inequality by $q$ and using the above equations we obtain

$$q[(s-s_0)I + \mathbb{A}(s) - \mathbb{A}(s_0)]\,\mathbb{X}(s) = q[x(0) + \mathbb{Y}(s)]$$

or, equivalently,

$$(8) \qquad [q + (s-s_0)^{-1} q(\mathbb{A}(s) - \mathbb{A}(s_0))]\,\mathbb{X}(s) = (s-s_0)^{-1} q[x(0) + \mathbb{Y}(s)].$$

Note that $(s-s_0)^{-1}(\mathbb{A}(s) - \mathbb{A}(s_0))$ is the Laplace transform of the function $H(t) = AG_t$ where the matrix valued function $G_t(r) = G(t+r)$, $r \in [-h, 0]$, and $G(t) = \exp(ts_0)I$ for $t \in [-h, 0]$ and $G(t) = 0$ for $t > 0$. Hence, $H(t) = 0$ for $t > h$. Thus, the left hand side of the equation above can be written in the time domain, for $t > h$, as

$$qx(t) + \int_0^h qH(r)x(t-r)dr$$

or

$$qx_t(0) + \int_{-h}^0 qH(-r)x_t(r)dr$$

which defines a functional $F(x_t)$. This functional belongs to the topological adjoint $X^*$ of the state space $X$ since it satisfies the hypotheses of the theorem (due to the assumed form of $A$, the function $H$ is a sum of a piecewise analytic function and a function from $W^{(1,\infty)}$ ).

Now, consider approximate reachability (and controllability). Setting $x_0 = 0$ in (8) yields $F(x_t) = 0$ for any $t > 0$ and any control $u(.)$. Hence

$$\|F(\psi)\| = \|F(\psi - x_t)\| \leq \|F\| \, \|(\psi - x_t)\|$$

and therefore

$$\|F(\psi)\| / \|F\| \leq \|\psi - x_t\|$$

which shows that the system is not approximately reachable (controllable) since $\|\psi - x_t\|$ cannot be made arbitrarily small if $\psi$ is not in the kernel of $F$. Now, $\ker F$ cannot be the whole of $X$ since $q$ is nonzero and therefore $F$ is a nonzero functional. This completes the proof for the first statement of the theorem (An obvious cosmetic change is needed if $q$ is not a real vector: Either the real or imaginary part of $F$ should be taken into consideration whichever is nonzero)

Next, to prove that uniform approximate null controllability implies spectral controllability, chose an initial function $x_0(.)$ such that $x_0(r) = 0$ for $r<0$ and $qx(0) \neq 0$. This form of $x_0(.)$ implies that $Y(s) = 0$ in (8). If $x_0$ is not in $X$ we use the assumption on the existence of time $T$, put $u(t) = 0$ for $t < T$ and consider the corresponding state $x_T$ as the initial state for the interval $[T, T_1]$, for some $T_1 > T$. Then, from

(8), $\exp(s_0 t)qx(0) = F(x_t)$ and thus

(9) $\exp(re(s_0 t))\,|qx(0)| = |F(x_t)| \le \|F\|\,\|x_t\|$

This shows that $\|x_t\|$ cannot be made arbitrarily small since $qx(0) \neq 0$ and $F$ is nontrivial. QED

**Remark 3.3** By Proposition 2.1 and definitions, spectral controllability is also necessary for approximate controllability and approximate null controllability on a given interval as well as for uniform approximate controllability but not for approximate null controllability. The latter is obvious since some systems without control, like $d/dt\ x(t) = -x(t)$, are formally approximately null controllable according to Definition 2.2. However, in general, we have

**Theorem 3.4** Let all assumptions of Theorem 3.2 be satisfied. Then approximate null controllability implies stabilizability.

**Proof:** As in the proof of Theorem 3.2, assuming that the rank condition (5) does not hold for some $s_0$ with $re(s_0) \ge 0$, we arrive at (9). Now, observe that if $re(s_0) \ge 0$ then, for the initial conditions generated by $x(0) = q'$, the transpose of $q$, if $q$ is real, or by the transpose of $im(q)$ otherwise, the state $x_t$ is uniformly (for all $t$) bounded away from zero which proves that approximate null controllability cannot be achieved if stabilizability does not hold. QED

**Remark 3.5** The results of Theorems 3.2 and 3.4 can be strengthen by weakening the topology in $X$. In fact, it follows from the proofs that, in the weak topology in $X$, spectral controllability is necessary for approximate controllability and uniform approximate null controllability while stabilizability is necessary for approximate null controllability. Moreover, even a weaker topology generated by all functionals of type (7) can be taken without losing validity of Theorems 3.2 and 3.4.

In the theory of delay systems it is customary to work with specific function spaces as state spaces $X$. Below, we show that a large family of concrete function spaces satisfies the assumptions of Theorems 3.2 and 3.4.

**Theorem 3.6** The following function spaces satisfy the assumptions for the state space $X$ of Theorems 3.2 and 3.4.

(a) The space $X = C^{(k)}$ of $R^n$ - valued functions on $[-h,0]$ with continuous k-th derivative and the norm

$$\|x\| = \max\{ |x^{(i)}(0)|,\ 0 \le i \le k\} + \sup\{ |x^{(k)}(t)|,\ t \in [-h, 0] \}.$$

(b) The space $X = M^p$ of $R^n$ - valued p-integrable functions on $[-h,0]$ with the norm

$$\|x\| = |x(0)| + \left(\int_{-h}^{0} |x(t)|^p dt\right)^{1/p}$$

for $1 \le p \le \infty$ and

$$\|x\| = |x(0)| + \text{ess sup}\{ |x(t)|,\ t \in [-h, 0] \}$$

for $p = \infty$.

(c) The space $X = W^{(k,p)}$ of $R^n$ - valued functions on $[-h,0]$ with p-integrable k-th derivative and the norm

$$\|x\| = \sum_{i=0}^{k-1} |x^{(i)}(0)| + \left(\int_{-h}^{0} |x^{(k)}(t)|^p dt\right)^{1/p}$$

for $1 \le p \le \infty$ and

$$\|x\| = \sum_{i=0}^{k-1} |x^{(i)}(0)| + \text{ess sup}\{ |x(t)|,\ t \in [-h, 0] \}$$

for $p = \infty$.

**Proof:** It is evident that the functionals of form (7) are nonzero on $X$ if $q \neq 0$. It remains to prove that they are bounded.

(a) Evaluate (7) as follows

$$|F(\varphi)| \le |q|\,|\varphi(0)| + h \sup\{ |f(t)|,\ t \in [-h, 0] \} \sup\{ |\varphi(t)|,\ t \in [-h, 0] \}$$

$$\le L_0( |\varphi(0)| + \sup\{ |\varphi(t)|,\ t \in [-h, 0] \} )$$

where $L_0 = \max\{ |q|, \sup\{ |f(t)|,\ t \in [-h, 0] \}\}$. This proves the case $k = 0$. For $k > 0$, we use k-times the equality

(10) $$\varphi(t) = \varphi(0) - \int_t^0 \varphi^{(1)}(r)dr$$

to obtain

$$|\varphi(0)| + \sup\{ |\varphi(t)|, t \in [-h,0] \}$$

$$\leq 2|\varphi(0)| + h|\varphi^{(1)}(0)| + \dots + h^k|\varphi^{(k)}(0)| + h^k \sup\{ |\varphi^{(k)}(t)|, t \in [-h,0] \}$$

$$\leq \max\{2, h, h^k\} ( \max\{|\varphi^{(i)}(0)|, i=0,1,\dots,k\} + \sup\{ |\varphi^{(k)}(t)|, t \in [-h,0] \} )$$

and to get finally

$$|F(\varphi)| \leq L_0 \max\{2, h, h^k\} \|\varphi\| .$$

This completes the proof in case (a).

(b) We have

$$|F(\varphi)| \leq |q|\,|\varphi(0)| + \sup\{ |f(t)|, t \in [-h,0] \} \int_{-h}^0 |\varphi(r)|\, dr$$

$$\leq ( |q| + \sup\{ |f(t)|, t \in [-h,0] \} ) ( |\varphi(0)| + \int_{-h}^0 |\varphi(r)|\, dr )$$

which proves the case $p = 1$. The Hoelder inequality implies that

(11) $$\int_{-h}^0 |\varphi(r)|\, dr \leq h^{p/(p-1)} ( \int_{-h}^0 |\varphi(r)|^p dr)^{1/p}.$$

and this, combined with the above, proves the case $1 < p < \infty$. The case $p = \infty$ follows from the first inequality in case (a) where sup can be replaced by ess sup.

(c) The final inequality in case (a) implies

$$|F(\varphi)| \leq (k-1)L_{k-1} ( \sum_{i=0}^{k-1} |\varphi^{(i)}(0)| + \sup\{ |\varphi^{(k-1)}(t)|, t \in [-h,0] \} ).$$

Applying (10), with $\varphi$ replaced by $\varphi^{(k-1)}$, we get

$$|F(\varphi)| \le (k-1)L_{k-1} \left( \sum_{i=0}^{k-1} |\varphi^{(i)}(0)| + \int_{-h}^{0} |\varphi^{(k)}(r)| dr \right)$$

which proves the case $p=1$ and implies immediately the case $p=\infty$. For $1 < p < \infty$, we apply (11) with $\varphi$ replaced by $\varphi^{(k)}$.

## REFERENCES

[1]. A. W. Olbrot, Control of retarded systems with function space constraints. Part 2: Approximate controllability, Control & Cybernetics, Vol. 6, No. 2, pp. 17-69, 1977
[2]. D. Salamon, Control and Observation of Neutral Systems, Pitman, 1984
[3]. R. V. Gabasov and F. M. Kirillova, Qualitative Theory of Optimal Processes, Nauka, Moscow, 1971
[4]. A. Manitius and A. W. Olbrot, Controllability conditions for linear systems with delayed state and control, Archiv. Autom. Telemech., Vol. 17, pp. 119-131, 1972.
[5]. A. W. Olbrot, Algebraic criteria of controllability to zero function for linear constant time-lag systems, Control & Cybernetics, Vol. 2, No. 1/2, pp. 59-77, 1973
[6]. A. W. Olbrot, Control to equilibrium of linear delay-diffeential systems, IEEE Trans. Autom. Control, Vol. AC-28, No. 4, pp. 521-523, 1983
[7] A. W. Olbrot and L. Pandolfi, Null controllability of a class of functional differential systems, Int. J. Control, Vol. 43, No. 1, pp. 193-208, 1988
[8]. A. Manitius and R. Triggiani, Function space controllability of linear retarded systems: A derivation from abstract operator conditions, SIAM J. Control & Optim., Vol. 16, pp. 599-645, 1978
[9]. V. M. Marchenko, Quasicontrollability of linear systems with aftereffect (in Russian), Avtomatika Telemech., Vol. 18, No. 3, pp. 18-22, 1979 (English transl. in Automat. Remote Control, Vol. 40, No. 3, part 1, pp. 335-339, 1979
[10]. A. Manitius, Necessary and sufficient conditions of approximate controllability for general linear retarded systems, SIAM J. Control & Optim., Vol. 19, pp. 516-632, 1981
[11]. D. A. O'Connor and T. J. Tarn, On the function space controllability of linear neutral systems, SIAM J. Control & Optim., Vol. 21, pp. 306-329, 1983
[12]. J. Hale, Theory of Functional Differential Equations, Springer-Verlag, 1977
[13]. A. W. Olbrot, Stabilizability, detectability, and spectrum assignment for linear systems with general time delays, IEEE Trans. Autom. Control, Vol. AC-23, No. 5, pp. 887-890, 1978

# 49 The Attainability Order in Control Systems

**Emilio O. Roxin** Department of Mathematics, University of Rhode Island, Kingston, Rhode Island

## 1. Control Systems and Attainable Set

Let $X$, $Y$ and $U$ be Banach spaces and let $T$ be a given interval of $\mathbb{R}$ (continuous case) or $I$ (discrete case, $I$ = the integers). A control system is usually given in one of the following two ways, which are to some extent equivalent:

a) The classical control formulation, where for each "admissible control function" $u(t)$, (an element of a prescribed class of functions $u: T \to U$) and each "initial condition" $x_0 \in X$, there is a well determined "evolution" of the system, given by a function $\phi: T \to X$. In many cases this evolution is determined by a differential or difference equation

$$\text{(a)} \quad \dot{x} = f(x, u) \qquad \text{or} \qquad \text{(b)} \quad x_{k+1} = f(x_k, u_k). \tag{1}$$

$$\text{(a}'\text{)} \quad u(t) \in U \qquad \text{or} \qquad \text{(b}'\text{)} \quad u_k \in U \tag{2}$$

The interpretation of equations (1) is as follows. The independent variable $t$ is the "time" and the variable $x = x(t)$ is the "state" describing the system under consideration. The variable $u = u(t)$ is the "control," which is some input into the system, chosen "arbitrarily" from the set of "admissible controls," usually the measurable functions with values in a given set $U$ specified in (2), and sometimes subjected to other additional restrictions. This setting makes it possible to state interesting problems, like for example optimization problems over the class of all admissible controls.

The function $f(x, u)$ in (1) is assumed to satisfy some conditions (usually the classical Carathéodory conditions) guaranteeing existence and uniqueness of the solutions of (1) for any admissible control $u(t)$. In many cases also a growth condition in $x$ is assumed, in order to avoid escape to infinity in finite time of any solution corresponding to a bounded control. In case of the difference equation (1-b), local existence is automatically insured and the growth condition is the only one to be possibly added.

Equation (1) with condition (2) can be replaced by the "differential inclusion"

$$\text{(a)} \quad \dot{x} \in F(x), \qquad \text{or} \qquad \text{(b)} \quad x_{k+1} \in F(x_k) \tag{3}$$

where now $F(x)$ is a "set valued" (or "multi-valued") function, a mapping from $X$ (or a suitable subset of $X$) into the set of subsets of $X$. To do this we just let $F(x) = f(x, U)$, the set of values of $f(x, u)$ for $u \in U$. A solution of (3-a) is, of course, an absolutely continuous function $x(t)$ satisfying (3-a) a.e.

When the control $u$ is absent in (1) (or when $U$ is a single point), the control system reduces to a differential equation (dynamical system). In the form (3) this means that the set $F(x)$ is a single point. Hence a control system is a generalization of a dynamical system and it should be possible to generalize the properties known from dynamical systems to control systems. This has been done in [3,4,5,6,7] for invariance and stability, in both a weak and strong form.

A fundamental concept related to the description of a control system is the "attainable set" $\mathcal{A}(t, t_0, x_0)$ (see [8]), defined by

$$\mathcal{A}(t, t_0, x_0) = \{x(t) \in X \mid x(\cdot) \text{ is a solution of the control system with } x(t_0) = x_0\} \tag{4}$$

In the case of an autonomous control system, which is our main interest in this paper, one can write

$$\mathcal{A}(t - t_0, x_0), \tag{4'}$$

since this set is translation invariant in $t$.

The most important property of this attainable set is the "semigroup" property

$$\mathcal{A}(t_3, x_0) = \bigcup \mathcal{A}(t_2, x_1) \text{ for all } x_1 \in \mathcal{A}(t_1, x_0), \text{ with either } 0 \le t_1 \le t_3 = t_1 + t_2 \text{ or } 0 \ge t_1 \ge t_3 = t_1 + t_2. \tag{5}$$

Related is the concept of "reachable set"

$$\mathcal{R}^+(t_0, x_0) = \bigcup \mathcal{A}(t, t_0, x_0) \text{ for all } t > t_0, \tag{6}$$

which in the autonomous case becomes simply $\mathcal{R}^+(x_0)$. Similarly,

$$\mathcal{R}^-(t_0, x_0) = \bigcup \mathcal{A}(t, t_0, x_0) \text{ for all } t < t_0, \tag{7}$$

which in the autonomous case is $\mathcal{R}^-(x_0)$.

Property (5) implies

$$\text{if } z \in \mathcal{R}^+(y) \text{ and } y \in \mathcal{R}^+(x), \text{ then } z \in \mathcal{R}^+(x), \tag{8}$$

$$\text{if } z \in \mathcal{R}^-(y) \text{ and } y \in \mathcal{R}^-(x), \text{ then } z \in \mathcal{R}^-(x). \tag{9}$$

Finally, another useful concept is the "holding set from $x_0$," defined by

$$\mathcal{H}(x_0) = \mathcal{R}^+(x_0) \cap \mathcal{R}^-(x_0). \tag{10}$$

This set has been introduced by the author [9,10] and others [1,2]. It may be empty, and is the set of points $y$ for which it is possible to go from $x_0$ to $y$ and come back to $x_0$.

## 2. The Natural Pre-Order Defined by a Control System

**Definitions.** Given an autonomous control system as described above, we say that state $x_1$ is *attainable* from state $x_0$ in time $\tau$, if there exists an admissible control $u(\cdot)$ such that the corresponding solution of (1) with condition (2) and $x(0) = x_0$ lead to $x(\tau) = x_1$.

We will furthermore say that $x_1$ is (*positively*) *reachable* from $x_0$, if there is a value of $\tau > 0$ such that $x_1$ is attainable from $x_0$ at time $\tau$.

We will say that $x_2$ is *negatively reachable* from $x_0$ if the same is true for some value of $\tau < 0$.

**Notation.** Obviously, if $x_1$ is positively reachable from $x_0$, then $x_0$ is negatively reachable from $x_1$. We will denote this fact by writing

$$x_0 \prec x_1 \qquad \text{or, equivalently,} \qquad x_1 \succ x_0. \tag{11}$$

This relation of reachability is *transitive*:

$$x_1 \prec x_2 \;\text{ and }\; x_2 \prec x_3 \;\text{ imply }\; x_1 \prec x_3. \tag{12}$$

The proof is immediate but it should be noted that it works only in the case of an autonomous control system.

It is not necessarily true that for every $x \in X$, $x \prec x$, since the $\tau$ in the definition of reachable points is not allowed to be zero. The "points" $x$ for which $x \prec x$ is not true will, indeed, constitute a special subset of $X$.

Note also that if $x \prec x$ is true, this implies the existence of some periodic solution of the control system, passing through $x$.

The definitions of reachable sets and of holding sets given above, can now be rephrased in the following way:

The POSITIVELY REACHABLE SET from $x_0$ is defined by

$$\mathcal{R}^+(x_0) = \{x \in X;\; x_0 \prec x\}. \tag{13}$$

The NEGATIVELY REACHABLE SET from $x_0$ is defined by

$$\mathcal{R}^-(x_0) = \{x \in X;\; x \prec x_0\}. \tag{14}$$

The HOLDING SET from $x_0$ is defined by

$$\mathcal{H}(x_0) = \mathcal{R}^+(x_0) \cap \mathcal{R}^-(x_0) = \{x \in X;\; x \prec x_0 \prec x\}. \tag{15}$$

**Theorem.** *Let $y \in \mathcal{H}(x)$. Then $\mathcal{H}(y) = \mathcal{H}(x)$.*

The proof is immediate, since the assumption implies that

$$y \prec x \prec y, \text{ hence } x \prec y \prec x.$$

**Theorem.** *The relation*

$$x \sim y \text{ if } y \in \mathcal{H}(x) \tag{16}$$

*is an equivalence relation.*

Indeed, $x \sim y$ and $y \sim z$ imply

$$x \prec y \prec z \text{ and also } z \prec y \prec x.$$

Therefore the state space $X$ gets subdivided into equivalence classes $\mathcal{H}_\alpha$, where $\alpha$ is in some index set $\Lambda$. There is a remainder set of points $x$ with empty holding set:

$$\mathcal{T} = \{x \in X;\ \mathcal{H}(x) = \emptyset\}. \tag{17}$$

This set $\mathcal{T}$ is called *transient set*; it is the set of points for which no admissible control can hold $x(t)$ = constant or produce a periodic solution returning to $x$. If a holding set $\mathcal{H}(x_0) \neq \emptyset$, then $x_0 \in \mathcal{H}(x_0)$.

**Some Examples**

a) $x \in \mathbb{R};\ \dot{x} = u,\ |u| \leq 1$. The whole real line is just one holding set.

b) $x \in \mathbb{R};\ \dot{x} = u,\ 0 \leq u \leq 1$. $\mathcal{H}(x) = \{x\}$ for every $x \in \mathbb{R}$.

c) $x \in \mathbb{R};\ \dot{x} = u,\ 1 \leq u \leq 2$. $\mathcal{H}(x) = \emptyset$ for every $x \in \mathbb{R}$, hence $\mathcal{T} = \mathbb{R}$.

## 3. Partial Order Among the Holding Sets

The relation "$\prec$" defines a preorder on the set $X \setminus \mathcal{T}$. The equivalence relation "$\sim$" defined above not only induces a partition of the state space $X$ into equivalence classes $\mathcal{H}_\alpha$ (the holding sets), with a remainder set $\mathcal{T}$ (the transient set), as seen above, but also induces a partial order among the holding sets themselves:

$$\mathcal{H}_\alpha \prec \mathcal{H}_\beta \text{ if, for any } x \in \mathcal{H}_\alpha,\ y \in \mathcal{H}_\beta,\ x \prec y. \tag{18}$$

Indeed, definition (18) is independent of the particular choice of $x$, $y$ within their holding sets. Furthermore,

$$\mathcal{H}_\alpha \prec \mathcal{H}_\alpha \text{ for every } \alpha \text{ in the subindex set } \Lambda. \tag{19}$$

Hence (18) defines a partial order in $\Lambda$.

We can define a "*reduced state space*" $\tilde{H}$, whose elements are all the transient points $x \in \mathcal{T}$, plus the holding sets $\mathcal{H}_\alpha$. Among all of these elements, the relation $\prec$, defined originally for points in $X$, induces a similar relation. We may therefore consider this space as a model of some generalized kind, related to the control system, in the sense that the reachability relation of the given control system is preserved, but the independent variable "$t$" is lost: it does not make sense (since it is not properly defined) to ask what time interval $t_1 - t_0$ is needed "to move $x(t)$ from $\mathcal{H}_\alpha$ to $\mathcal{H}_\beta$."

This "reduced state space" $\tilde{H}$ could be interpreted more as a *directed graph* representing the possible evolutions of the original control system. This opens the possibility of studying control systems using all the techniques developed in graph theory.

We end by giving a few more simple illustrative examples. More elaborate examples can also be constructed (see [9,10]).

## More Examples

a) $x \in \mathbb{R}$; $\dot{x} = 1 + u \cos \pi x$, $|u| \leq 1$. Here $\Lambda = \{n \text{ integer}\}$, $\mathcal{H}_n = \{n\}$, $\mathcal{T} = \{\text{all non-integer reals}\}$, and $x \prec y$ if $x < y$.

b) $x \in \mathbb{R}$; $\dot{x} = 1 + u \cos \pi x$, $|u| \leq 2$.
Now we get a sequence of holding sets $\mathcal{H}_n = (n - (1/6), n + (1/6))$ for $n =$ integer; the boundary points $x = m \pm (1/6)$, $m$ integer, are one-point holding sets, all remaining points constitute $\mathcal{T}$.

c) From the above, a two-dimensional example can easily be constructed: $x \in \mathbb{R}^2$, $u \in \mathbb{R}^2$; $\dot{x}_i = 1 + u_i \cos \pi x_i$, $|u_i| \leq 1$, $i = 1, 2$.
Here the only holding sets are the single points of coordinates $(n, m)$ with $n, m$ integers; the reachability relation is generated by $(n, m) \prec (n + 1, m)$ and $(n, m) \prec (n, m + 1)$ (motion to right and up).

## References

[1] Gayek, J.E. and Vincent, T.L., On the intersection of controllable and reachable sets, *J. Optim. Theory Appl.*, Vol. 50, No. 23, 1986, 267–278.

[2] Panasyuk, A.I., Dynamics of sets defined by differential inclusions, Siberian Math. J. (Russian, English transl. available), Vol. 27, No. 5, 1986, 757–765.

[3] E. Roxin, Stability in general control systems, J. Differential Equa. 1, 1965, 115–150.

[4] E. Roxin, On generalized dynamical systems defined by contingent equations, J. Differential Equa. 1, 1965, 188–205.

[5] E. Roxin, On stability in control systems, SIAM J. Control 3, 1966, 357–372.

[6] E. Roxin, On asymptotic stability in control systems, Rendiconti Circolo Mat. Palermo II, 15, 1966, 193–207.

[7] E. Roxin, On finite stability in control systems, Rendiconti Circolo Mat. Palermo II, 15, 1966, 273–282.

[8] E. Roxin, Problems about the set of attainability, Lecture Notes Summer Course 1966 CIME, Edizione Cremonese, 1967, 241–369.

[9] E. Roxin, Limit sets in infinite horizon optimal control systems, Proc. VI Int. Conference on Trends in Theory and Practice in Nonlinear Analysis, Arlington, Texas, 1984, North-Holland, 401–407.

[10] Reachable sets, limit sets and holding sets in control systems, Proc. VII Int. Conference on Nonlinear Analysis, Arlington, Texas, 1986, 533–540.

# 50 Extending Linear-Quadratic Optimal Control Laws to Nonlinear Systems and/or Nonquadratic Cost Criteria

**D. L. Russell** Virginia Polytechnic Institute and State University, Blacksburg, Virginia

**Xiaohong Zhang** Virginia Polytechnic Institute and State University, Blacksburg, Virginia

## 1. Background on Nonlinear / Nonquadratic Optimal Control Problems

The linear-quadratic optimal control theory, introduced in 1960 by R.E. Kalman and R. S. Bucy [ 8 ], [ 9 ], has since that time enjoyed wide popularity as a mathematical framework in terms of which a wide variety of design objectives can be expressed and corresponding design techniques developed. This theory, further developed in an enormous number of journal articles and various books (see, e.g., [ 1 ],[ 10 ],[ 19 ]) is so familiar that we need not give a systematic description of its contents in this paper. Rather, we will cite results from that theory as required to support our basic objective, the extension of that theory to nonlinear systems and to non-quadratic cost functionals in order to achieve improved performance and larger regions of asymptotic stability in the nonlinear system context. This work will be carried out primarily in neighborhoods of invariant sets of such systems, notably critical points and periodic solutions. In the process we will have reason to develop certain variations of the stable manifold theorem [ 4 ],[ 5 ] of nonlinear differential equations theory, including some new approaches to proving such theorems.

Henceforth in this paper we refer to linear-quadratic theory, in whatever context, as LQ theory (not LQG because we do not consider any connections with stochastic control theory in this article). We will refer to extensions of that theory to cover nonlinear systems of ordinary differential equations, or discrete recursion equations, together with non-quadratic cost criteria, as NLNQ (nonlinear, nonquadratic) theory. To the extent that the theory is developed as a perturbation of LQ theory in small neighborhoods of invariant sets we might also want to interpret NLNQ as meaning nearly linear, nearly quadratic.

Only part of what we address here is new with this paper. In particular, we should acknowledge at the outset that the whole subject really derives from the original work of D.L. Lukes [ 15 ],[ 16 ] carried out in the context of autonomous nonlinear systems in the neighborhood of a critical point. Much of what we do consists of variations on his original theme.

We begin in §2 by considering the form which NLNQ theory takes in a setting corresponding to the finite interval LQ case. Here there are no stability criteria as such; just optimality. We will see that the extension from LQ to NLNQ simply involves the implicit function theorem in a quite specific setting, the desired nonsingular Jacobian being derived from the LQ theory. In §3 we review Lukes' development of NLNQ theory for nonlinear autonomous systems in the neighborhood of a critical point; our main contribution will be the development of an alternative setting within which the stable manifold theorem of nonlinear ordinary differential equations theory can be established. This alternative approach appears to be particularly well adapted to allow numerical approximation of NLNQ feedback control laws by spline, or other finite element, methods. In §4 we discuss the counterpart of this theory for nonlinear recursion equations and the pertinent formulation, proof and application of the stable manifold theorem in this context.

Our main goal here is the development of NLNQ theory as it relates to nonautonomous systems of a particular type; specifically periodic systems, and to the development of NLNQ theory in the context of stabilization of autonomous systems with reference to invariant sets other than critical points; e.g., periodic solutions. The NLNQ theory for stabilization of periodic ODE systems in the neighborhood of a (time invariant) critical point is considered in §5 and related to the recursion equation developments of §4 and the finite interval theory of §2. In §6 we consider NLNQ theory in the context of stabilization of periodic solutions of autonomous systems of ordinary differential equations; e.g., those arising in the study of self-excited oscillations. At the end of §6 we present some conjectures and speculations as to the form which NLNQ theory may take with reference to arbitary compact invariant sets of an autonomous system of differential equations - but these are, indeed, speculative.

Our purpose in this article is primarily descriptive in character; it will be easy to point out cases we have not covered and further precisions which might be introduced into our analysis. In this last connection we note that some of the detailed calculations for §4 and §5 already appear in the second named author's thesis [ 20 ]. The material presented in §2,3 is either sufficiently elementary or sufficiently developed in the literature already that description is really all that is required here. If in writing this article we succeed in encouraging others to improve upon and extend our preliminary work here we will have achieved the main purpose of present efforts.

## 2. Finite Interval NLNQ Theory

Let us consider a system of differential equations, with $\dot{} \equiv \frac{d}{dt}$,

$$\dot{x} = F(t,x,u),\ t \in [t_0,t_1],\ x \in R^n,\ u \in R^m, \tag{2.01}$$

where $f:R^{1+n+m} \to R^n$ is continuous with respect to all its arguments and twice continuously differentiable with respect to x and u in the region of interest. Let us suppose that $\bar{x}(t)$ is a solution of (2.01) corresponding to an initial state

$$\bar{x}(t_0) = \bar{x}_0, \tag{2.02}$$

and a continuous control input function $\bar{u}(t)$ defined for $t \in [t_0,t_1]$. If $\bar{x}(t)$, $\bar{u}(t)$ constitutes the desired system trajectory, it may be desirable to counter the effect of possible disturbances (not modelled in our system representation) by requiring that perturbations from the trajectory $\bar{x}(t)$ to nearby trajectories $\tilde{x}(t)$ should invoke modified controls $\tilde{u}(t)$, lying near $\bar{u}(t)$, in such a way as to minimize an appropriate cost functional. Redefining $x(t) = \tilde{x}(t) - \bar{x}(t)$, $u(t) = \tilde{u}(t) - \bar{u}(t)$ the base trajectory becomes $x(t) \equiv 0$, $u(t) \equiv 0$ and, with appropriate modification of F in (2.01), we may assume that we have that equation with (0,0) as the (desired) equilibrium point, i.e.,

$$F(t,0,0) \equiv 0,\ t \in [t_0,t_1]\ . \tag{2.03}$$

In this context we will assume that the cost functional has the form

$$J(x_0,u) = \int_{t_0}^{t_1} G(t,x(t),u(t))\,dt + V(x(t_1)), \tag{2.04}$$

where $G:R^{1+n+m} \to R^{1+n+m}$ is continuous, twice continuously differentiable with respect to x and u, with

$$G(t,0,0) \equiv 0,\ \frac{\partial G}{\partial x}(t,0,0) \equiv 0,\ \frac{\partial G}{\partial u}(t,0,0) \equiv 0,\ t \in [t_0,t_1]. \tag{2.05}$$

Further, $V:R^n \rightarrow R^n$ is twice continuously differentiable with

$$V(0) = 0, \; \frac{\partial V}{\partial x}(0) = 0 \; . \tag{2.06}$$

In (2.05) and (2.06) and subsequently the indicated partial derivatives are, of course, Jacobian matrices (gradient vectors) of appropriate dimension.

We may suppose then, with

$$A(t) = \frac{\partial F}{\partial x}(t,0,0), \; B(t) = \frac{\partial F}{\partial u}(t,0,0), \; t \in [t_0,t_1], \tag{2.07}$$

that (2.01) takes the form

$$\dot{x} = A(t)\, x + B(t)\, u + f(t,x,u), \; x \in R^n, \; u \in R^m, \tag{2.08}$$

where f has the same properties as originally stated for F but now

$$f(t,0,0) \equiv 0, \; \frac{\partial f}{\partial x}(t,0,0) \equiv 0, \; \frac{\partial f}{\partial u}(t,0,0) \equiv 0, \; t \in [t_0,t_1]. \tag{2.09}$$

Correspondingly, we suppose that

$$G(t,x,u) = (x^* \; u^*)\begin{pmatrix} W(t) & R(t) \\ R(t)^* & U(t) \end{pmatrix}\begin{pmatrix} x \\ u \end{pmatrix} + g(t,x,u), \; t \in [t_0,t_1], \tag{2.10}$$

$$V(x) = x^*P\, x + v(x) \; , \tag{2.11}$$

wherein

$$g(t,0,0) \equiv 0, \; \frac{\partial g}{\partial(x,u)}(t,0,0) \equiv 0, \; \frac{\partial^2 g}{\partial(x,u)^2}(t,0,0) \equiv 0, \; t \in [t_0,t_1], \tag{2.12}$$

$$v(0) = 0, \; \frac{\partial v}{\partial x}(0) = 0, \; \frac{\partial^2 v}{\partial x^2}(0) = 0 \; . \tag{2.13}$$

We will further suppose that the composite matrix displayed in (2.10) is symmetric and non-negative for $t \in [t_0,t_1]$, with U(t) uniformly positive definite there, while P is symmetric and non-negative.

We pose then the optimal control problem of minimizing the cost (2.04), subject to x(t) and u(t) satisfying (2.01), modified, as described, so that (2.03) holds, with

$$x(t_0) = x_0 \in R^n . \tag{2.14}$$

Application of the Pontryagin Maximum Principle [ 18 ],[ 11 ] allows us to characterize the optimal control $\hat{u}(t)$ and the corresponding optimal state trajectory $\hat{x}(t)$ in terms of a solution pair $\hat{x}(t)$, $\lambda(t)$ of the two point boundary value problem consisting of the differential equations and boundary conditions

$$\dot{\hat{x}}(t) = F(t,\hat{x}(t),\hat{u}(t)), \quad \hat{x}(t_0) = x_0, \tag{2.15}$$

$$\dot{\lambda}(t) = - \hat{A}(t)^*\lambda(t) - \frac{\partial G}{\partial x}(t,\hat{x}(t),\hat{u}(t)), \quad \lambda(t_1) = \frac{\partial V}{\partial x}(\hat{x}(t_1))^*, \tag{2.16}$$

wherein

$$\hat{A}(t) = \frac{\partial F}{\partial x}(t,\hat{x}(t),\hat{u}(t)), \quad t \in [t_0,t_1] ,$$

further coupled through the condition

$$\lambda(t)^*\frac{\partial F}{\partial x}(t,\hat{x}(t),\hat{u}(t)) + \frac{\partial G}{\partial u}(t,\hat{x}(t),\hat{u}(t)) = 0 . \tag{2.17}$$

Since

$$\frac{\partial^2 G}{\partial u^2}(t,0,0) = U(t) > 0$$

the implicit function theorem yields a unique solution of (2.17) in the form

$$\hat{u}(t) = \mathcal{K}(t,\hat{x}(t),\lambda(t)) \tag{2.18}$$

near u = 0 in $R^m$ for $\hat{x}(t),\lambda(t)$ near $(x,\lambda) = (0,0)$ in $R^{2n}$, $\mathcal{K}(t,x,\lambda)$ being continuously differentiable there. This allows elimination of

$\hat{u}(t)$ from (2.15) and (2.16), resulting in the more straightforward two point boundary value problem

$$\dot{\hat{x}}(t) = F(t,\hat{x}(t),\mathfrak{K}(t,\hat{x}(t),\lambda(t))), \quad \hat{x}(t_0) = x_0, \qquad (2.19)$$

$$\dot{\lambda}(t) = -A_{\mathfrak{K}}(t)^*\lambda(t) - \frac{\partial G}{\partial x}(t,\hat{x}(t),\mathfrak{K}(t,\hat{x}(t),\lambda(t))), \quad \lambda(t_1)^* = \frac{\partial V}{\partial x}(\hat{x}(t_1)), \qquad (2.20)$$

in which

$$A_{\mathfrak{K}}(t) \equiv \frac{\partial F}{\partial x}(t,\hat{x}(t),\mathfrak{K}(t,\hat{x}(t),\lambda(t))) , \quad t \in [t_0,t_1].$$

The question then arises as to whether $\hat{u}(t)$ can be further character- in time-varying linear feedback form

$$\hat{u}(t) = \mathfrak{K}(t,\hat{x}(t)), \qquad (2.21)$$

so that (2.19) takes the synthesized form, decoupled from (2.20),

$$\dot{\hat{x}}(t) = F(t,\hat{x}(t),\mathfrak{K}(t,\hat{x}(t))), \quad \hat{x}(t_0) = x_0. \qquad (2.22)$$

We will see that this is, indeed, the case and is, in fact, a quite direct consequence of the implicit function theorem, with the relevant nonsingular Jacobian coming out of the corresponding linear quadratic optimal control problem for which, in (2.01) and (2.04), respectively, replacing the variables $\hat{x}(t)$, $\hat{u}(t)$ and $\lambda(t)$ of the non-linear case by $\hat{y}(t)$, $\hat{v}(t)$ and $p(t)$ for the linear case,

$$f(t,y,v) \equiv 0, \quad g(t,y,v) \equiv 0 .$$

For this problem the equation corresponding to (2.17), i.e.,

$$p(t)^*B(t) + 2\,\hat{y}(t)^*R(t) + 2\,\hat{v}(t)^*U(t) = 0 ,$$

is directly solvable, yielding $\hat{v}(t)$ as a linear function of $\hat{y}(t)$ and $p(t)$,

$$\hat{v}(t) = -\frac{1}{2}\,U(t)^{-1}\left[B(t)^*p(t) + 2\,R(t)^*\hat{y}(t)\right] \equiv \mathfrak{K}_0(t,\hat{y}(t),p(t)). \quad (2.23)$$

With substitution of the relation (2.23) into the linear two point boundary value problem corresponding to (2.15), (2.16), we obtain

$$\dot{\hat{y}}(t) = A(t)\,\hat{y}(t) + B(t)\,\mathcal{K}_0(t,\hat{y}(t),p(t))$$

$$= \left[A(t)-B(t)U(t)^{-1}R(t)^*\right]\hat{y}(t) - \frac{1}{2}\,B(t)U(t)^{-1}B(t)^*p(t) \; ,$$
$$\hat{y}(t_0) = y_0 \; , \qquad (2.24)$$

$$\dot{p}(t) = -\,A(t)^*p(t) - 2\,W(t)\hat{y}(t) - 2\,R(t)\hat{v}(t)$$

$$= -\,A(t)^*p(t) - 2\,W(t)\hat{y}(t) + R(t)U(t)^{-1}\left[B(t)^*p(t) + 2\,R(t)^*\hat{y}(t)\right]$$

$$= -\left[A(t)-B(t)U(t)^{-1}R(t)^*\right]^*p(t) - 2\,W(t)\hat{y}(t) + 2\,R(t)U(t)^{-1}R(t)^*\hat{y}(t) \; ,$$
$$p(t_1) = 2\,P\,\hat{y}(t_1) \; . \qquad (2.25)$$

is obtained. Then defining

$$\tilde{A}(t) = A(t) - B(t)U(t)^{-1}R(t)^*, \qquad D(t) = -\,\frac{1}{2}\,B(t)U(t)^{-1}B(t)^*,$$

$$E(t) = 2\,R(t)U(t)^{-1}R(t)^* - 2\,W(t) \; ,$$

the differential equations in (2.24),(2.25) are equivalent to the system

$$\begin{bmatrix} \dot{\hat{y}} \\ \dot{p} \end{bmatrix} = \begin{bmatrix} \tilde{A}(t) & D(t) \\ E(t) & -\tilde{A}(t)^* \end{bmatrix} \begin{bmatrix} \hat{y} \\ p \end{bmatrix} . \qquad (2.26)$$

With a modest amount of computation one can verify that (2.26) is the variational equation for the system (2.19), (2.20) based on the solution $\hat{x}(t) \equiv 0$, $\lambda(t) \equiv 0$ corresponding to $x_0 = 0$.

Let $\tau \in [t_0,t_1]$. It is natural to conjecture there should be a functional relationship

$$\lambda(\tau) = \mathcal{Q}(\tau,\hat{x}(\tau)) \qquad (2.27)$$

such that solutions of the differential equations in (2.19),(2.20) satisfying this relationship at $t = \tau$ will, at $t = t_1$, satisfy the terminal condition

$$\lambda(t_1)^* = \frac{\partial V}{\partial x}(\hat{x}(t_1)) . \tag{2.28}$$

At $\tau = t_1$ the condition for agreement between (2.27) and (2.28) is

$$\mathcal{Q}(t_1,\hat{x}(t_1))^* = \frac{\partial V}{\partial x}(\hat{x}(t_1)) . \tag{2.29}$$

Given conditions at $t = \tau$:

$$\hat{x}(\tau) = x_\tau, \quad \lambda(\tau) = \lambda_\tau, \tag{2.30}$$

let us designate the resulting solution of the differential equations (2.19),(2.20) by

$$\hat{x}(t,x_\tau,\lambda_\tau), \ \lambda(t,x_\tau,\lambda_\tau), \ t \in [\tau,t_1]. \tag{2.31}$$

This pair satisfies (2.28) just in case

$$\mathcal{V}(x_\tau,\lambda_\tau) \equiv \lambda(t_1,x_\tau,\lambda_\tau) - \frac{\partial V}{\partial x}(\hat{x}(t_1,x_\tau,\lambda_\tau)) = 0 . \tag{2.32}$$

Since the equation is clearly satisfied when $x_\tau = 0$, $\lambda_\tau = 0$, and the functions involved are of class $C^1$, by virtue of standard regularity results for ordinary differential equations, the implicit function theorem guarantees a solution of (2.32) in the form (2.27) if the partial Jacobian matrix

$$\frac{\partial \mathcal{V}}{\partial \lambda_\tau}(0,0) = \frac{\partial \lambda}{\partial \lambda_\tau}(t_1,0,0) - \frac{\partial^2 V}{\partial x^2}(0)\frac{\partial \hat{x}}{\partial \lambda_\tau}(t_1,0,0) = \frac{\partial \lambda}{\partial \lambda_\tau}(t_1,0,0) - 2P\frac{\partial \hat{x}}{\partial \lambda_\tau}(t_1,0,0) \tag{2.33}$$

is nonsingular. Thus our next task will be to identify this Jacobian.

It is familiar (see, e.g., [ 14 ]) that if we make the change of variable in (2.26)

$$p(t) = q(t) + 2\,Q(t)\,\hat{y}(t) , \tag{2.34}$$

with Q(t) the solution of the matrix Riccati differential equation

$$\dot{Q}(t)+A(t)^*Q(t)+Q(t)A(t)+W(t)-(Q(t)B(t)+R(t))^*U(t)^{-1}(B(t)^*Q(t)+R(t))=0 \tag{2.35}$$

satisfying (cf.(2.25)) the terminal condition

$$Q(t_1) = P , \tag{2.36}$$

we obtain the decoupled system

$$\begin{pmatrix} \dot{\hat{y}} \\ \dot{q} \end{pmatrix} = \begin{pmatrix} \tilde{C}(t) & D(t) \\ 0 & -\tilde{C}(t)^* \end{pmatrix} \begin{pmatrix} \hat{y} \\ q \end{pmatrix} ,$$

wherein

$$\tilde{C}(t) = C(t) + D(t)Q(t) .$$

Given (cf.(2.30)) transformed initial conditions at $t = \tau$:

$$\hat{y}(\tau) = y_\tau, \ q(\tau) = q_\tau ,$$

and, taking $\Psi(t,s)$ and $\Phi(t,s)$ to be the fundamental matrix solutions of

$$\frac{\partial\Psi}{\partial t} = C(t)\ \Psi , \quad \frac{\partial\Phi}{\partial t} = - C(t)^*\Phi , \tag{2.37}$$

reducing to the identity when $t = s$, we have,

$$q(t) = \Phi(t,\tau)\ q_\tau , \tag{2.38}$$

$$\hat{y}(t) = \Psi(t,\tau)\ y_\tau + \int_\tau^t \Psi(t,s)D(s)q(s)\ ds$$

$$= \Psi(t,\tau)\ y_\tau + \int_\tau^t \Psi(t,s)D(s)\Phi(s,\tau)\ ds\ q_\tau \equiv \Psi(t,\tau)\ y_\tau + \Delta(t,\tau)\ q_\tau . \tag{2.39}$$

Substituting (2.38) into (2.34) and setting $t = t_1$ we have

$$p(t_1) = q(t_1) + 2\, Q(t_1)\, \hat{y}(t_1) = (\text{cf.}(2.36))$$

$$= \Phi(t_1,\tau)\, q_\tau + 2\, P\, \hat{y}(t_1) = \Phi(t_1,\tau)\left[p_\tau - Q(\tau)\, y_\tau\right] + 2\, P\, \hat{y}(t_1)$$

$$= \Phi(t_1,\tau)\, p_\tau + 2\, P\, \hat{y}(t_1) - \Phi(t_1,\tau)Q(\tau)\, y_\tau\,. \tag{2.40}$$

Then substituting (2.33) (for $t = \tau$) into (2.38) we have

$$\hat{y}(t_1) = \Psi(t_1,\tau)\, y_\tau + \Delta(t_1,\tau)\left[p_\tau - Q(\tau)\, y_\tau\right]$$

$$= \left[\Psi(t_1,\tau) - \Delta(t_1,\tau)Q(\tau)\right] y_\tau + \Delta(t_1,\tau)\, p_\tau. \tag{2.41}$$

Further substitution of (2.40) into (2.39) then yields

$$p(t_1) = \Phi(t_1,\tau)p_\tau + 2P\left[\left[\Psi(t_1,\tau) - \Delta(t_1,\tau)Q(\tau)\right]y_\tau + \Delta(t_1,\tau)p_\tau\right] - \Phi(t_1,\tau)Q(\tau)y_\tau$$

$$= \left[\Phi(t_1,\tau) + 2P\Delta(t_1,\tau)\right]p_\tau + \left[2P\left[\Psi(t_1,\tau) - \Delta(t_1,\tau)Q(\tau)\right] - \Phi(t_1,\tau)Q(\tau)\right]y_\tau.$$

Since (2.26) is the variational system for (2.19),(2.20) based on the zero solution we conclude that

$$\frac{\partial\lambda}{\partial\lambda_\tau}(t_1,0,0) = \Phi(t_1,\tau) + 2\, P\, \Delta(t_1,\tau)\,. \tag{2.42}$$

On the other hand, (2.41) implies that

$$\frac{\partial\hat{x}}{\partial\lambda_\tau}(t_1,0,0) = \Delta(t_1,\tau)\,. \tag{2.43}$$

Substituting (2.42) and (2.43) into (2.33) we have

$$\frac{\partial\nu}{\partial\lambda_\tau}(0,0) = \frac{\partial\lambda}{\partial\lambda_\tau}(t_1,0,0) - 2\, P\, \frac{\partial\hat{x}}{\partial\lambda_\tau}(t_1,0,0)$$

$$= \Phi(t_1,\tau) + 2\, P\, \Delta(t_1,\tau) - 2\, P\, \Delta(t_1,\tau) = \Phi(t_1,\tau)\,,$$

which is nonsingular by virtue of being the value at $t = t_1$ of the fundamental matrix solution $\Phi(t,\tau)$ of the second equation in (2.37).

It thus follows that (2.32) has a solution of the form (2.27) so that (2.18) yields

$$\hat{u}(t) = \mathcal{K}(t,\hat{x}(t),\mathcal{Q}(t,\hat{x}(t)) \equiv \mathfrak{K}(t,\hat{x}(t)) .$$

From the twice continuous differentiability of V in (2.32) and the standard regularity results for solutions of (2.19),(2.20) it follows that $\mathcal{Q}(t,x)$ is continuously differentiable with respect to x near the origin in $R^n$, uniformly for $t\in[t_0,t_1]$. This verifies our conjecture (2.27) and demonstrates that the optimal control $\hat{u}(t)$ is synthesized by the feedback relation (2.21), with the optimal state trajectory $\hat{x}(t)$ satisfying (2.22). Moreover, $\mathfrak{K}(t,x)$ is continuously differentiable with respect to x near the origin in $R^n$, uniformly for $t\in[t_0,t_1]$.

The optimal value function $\hat{J}(\tau,x)$ can now be defined for $x_\tau$ near the origin in $R^n$ and for $\tau \in [t_0,t_1]$ by the formula (cf.(2.04))

$$\hat{J}(\tau,x_\tau) = \int_\tau^{t_1} G(t,\hat{x}(t),\mathfrak{K}(t,\hat{x}(t)))\, dt + V(\hat{x}(t_1)),$$

it being understood that $\hat{x}(t)$ satisfies (2.22) with $\hat{x}(\tau) = x_\tau$. The standard regularity results for solutions of (2.22) show $\hat{J}(t,x)$ to be continuously differentiable with respect to x near the origin in $R^n$, uniformly for $t \in [t_0,t_1]$. A standard computation (see [ 16 ],p. 92, e.g.) then shows that

$$\frac{\partial \hat{J}}{\partial x}(t,x)^* = \mathcal{Q}(t,x) .$$

We have already noted that $\mathcal{Q}(t,x)$ is continuously differentiable with respect to x, so it follows that $\hat{J}(t,x)$ is, in fact, twice continuously differentiable with respect to x. Then it is quite straightforward to see that with Q(t) defined by (2.35),(2.36) we have

$$\hat{J}(t,0) = 0, \ \frac{\partial \hat{J}}{\partial x}(t,0) = 0, \ \frac{\partial^2 \hat{J}}{\partial x^2}(t,0) = 2\, Q(t), \ t \in [t_0,t_1], \quad (2.44)$$

so that

$$\hat{J}(t,x) = x^* Q(t)x + \hat{j}(x), \ \hat{j}(x) = o(\|x\|^2), \ \|x\| \to 0, \quad (2.45)$$

uniformly for $t \in [t_0,t_1]$. The optimal value function $\hat{J}(t,x)$ is, of course, the unique solution of the Hamilton-Jacobi partial differential equation

$$\frac{\partial \hat{J}}{\partial t}(t,x) + \frac{\partial \hat{J}}{\partial x}(t,x)F(t,x,\hat{\mathfrak{K}}(t,x)) + G(t,x,\hat{\mathfrak{K}}(t,x)) = 0 \quad (2.46)$$

satisfying the conditions (cf.(2.04)

$$\hat{J}(t_1,x) = V(x), \ \hat{J}(t,0) \equiv 0 . \quad (2.47)$$

## 3. Remarks on D. L. Lukes' Infinite Interval NLNQ Theory

All of the various parts of the present paper amount to variations on results proved by D. L. Lukes in his 1967 University of Minnesota thesis, supervised by L. W. Markus, subsequently published in [ 15 ], [ 16 ]. That theory treats an autonomous version of (2.01), for which we retain the assumptions of §2, specialized as appropriate with some additions which we will introduce below. Thus we are concerned with

$$\dot{x} = F(x,u), \quad x \in R^n, \quad u \in R^m, \tag{3.01}$$

satisfying

$$F(0,0) \equiv 0 \; .$$

The problem differs from that of the previous section in that the cost functional involves an infinite integral

$$J(x_0,x,u) = \int_0^{\infty} G(x(t),u(t))\; dt \tag{3.02}$$

in which $x(t)$ is the unique solution of (2.01) with $x(0) = x_0$. We continue to assume that

$$G(0,0) \equiv 0, \quad \frac{\partial G}{\partial(x,u)}(0,0) \equiv 0,$$

and thus have

$$G(x,u) = (x^* \; u^*)\begin{bmatrix} W & R \\ R^* & U \end{bmatrix}\begin{bmatrix} x \\ u \end{bmatrix} + g(x,u) \; , \tag{3.03}$$

with

$$g(0,0) = 0, \quad \frac{\partial g}{\partial(x,u)}(0,0) = 0, \quad \frac{\partial^2 g}{\partial^2(x,u)}(0,0) = 0 \; . \tag{3.04}$$

With

$$A = \frac{\partial F}{\partial x}(0,0), \quad B = \frac{\partial F}{\partial u}(0,0) \ , \tag{3.05}$$

(3.01) takes the form

$$\dot{x} = A\,x + B\,u + f(x,u), \quad x \in R^n, \ u \in R^m, \tag{3.06}$$

with

$$f(0,0) = 0, \quad \frac{\partial f}{\partial (x,u)}(0,0) = 0 \ . \tag{3.07}$$

Beyond the non-negativity for the composite matrix in (3.03) and non-negativity of U coming from the assumptions of §2, we assume (see [DLR],e.g.) that the pair A,B is stabilizable and the pair W,A is detectable.

In the corresponding LQ problem wherein $f,g \equiv 0$ the classical ([ 8 ],[ 9 ]) result is that the quadratic cost corresponding to $g \equiv 0$ in (3.03) is uniquely solved by the synthesized control-trajectory pair

$$\hat{u}(t) \equiv -\,U^{-1}(B^*Q + R^*)\,\hat{x}(t) \equiv \hat{K}\,\hat{x}(t) \ , \tag{3.08}$$

$$\dot{\hat{x}}(t) = \left[A + B\hat{K}\right]\hat{x}(t) \ , \quad \hat{x}(0) = x_0 \ , \tag{3.09}$$

wherein Q is the unique symmetric positive definite solution of the algebraic matrix Riccati equation

$$A^*Q + Q\,A + W - (QB+R)U^{-1}(B^*Q+R^*) = 0 \ . \tag{3.10}$$

Lukes showed that this result has an NLNQ extension to the problem of minimizing a general cost (3.03) subject to x and u satisfying (3.01) with $x(0) = x_0$ if $x_0$ is sufficiently small (or, more generally, lies in the set of initial states which can be steered to the origin with finite cost (3.03)). To facilitate our re-statement of his theorem, we define the set of admissible controls, $\mathcal{U}_\infty$, in the following way. The space $\mathcal{U}_\infty$ consists of all continuous m-vector functions u(t) defined for $t \in [0,\infty)$ such that, given any $T > 0$,

$$\sum_{k=0}^{\infty} \|u_{T,k}\|^2_{C^0[kT,(k+1)T]} < \infty ,$$

$u_{T,k}$ being the restriction of u to [kT,(k+1)T].

<u>Theorem 3.1</u> *With the foregoing assumptions the problem*

$$\min_{u \in \mathfrak{U}_\infty} \int_0^\infty G(x(t),u(t))\, dt ,$$

*subject to* $\dot{x}(t) = F(x(t),u(t))$, $x(0) = x_0$, *has, for* $\|x_0\|$ *sufficiently small, a unique solution* $\hat{u} \in \mathfrak{U}_\infty$. *Moreover,* $\hat{u}(t)$ *is generated by a nonlinear feedback relation*

$$\hat{u}(t) = \mathfrak{K}(\hat{x}(t)) = \hat{K}\, \hat{x}(t) + k(x(t)) , \tag{3.11}$$

*wherein* $\hat{x}(t)$ *satisfies the closed loop equation and initial condition*

$$\dot{\hat{x}}(t) = F(\hat{x}(t),\mathfrak{K}(\hat{x}(t))),\ \hat{x}(0) = x_0 . \tag{3.12}$$

*In* (3.11) $\hat{K}$ *is defined as in* (3.08) *and* k(x) *is a continuously differentiable function defined for* $\|x\|$ *sufficiently small with*

$$k(0) = 0,\ \frac{\partial k}{\partial x}(0) = 0 .$$

*The optimally controlled solutions* $\hat{x}(t)$ *have the uniform exponential decay property, valid uniformly for* $x_0$ *in sufficiently small neighborhoods of the origin in* $R^n$,

$$\|\hat{x}(t)\| \leq M\, e^{-\gamma t}\, \|x_0\| ,\ t \geq 0 , \tag{3.13}$$

*where* M *and* $\gamma$ *are fixed positive numbers.*

We do not intend to repeat the proof of this theorem here; the details appear in [ 16 ] and are, in any case, repeated in slightly different contexts in the subsequent sections of this paper. However, in order to provide a background for the new material introduced at the end of this section we will review, in broad outline, the main points of the proof.

Applying the Pontryagin Maximum Principle ([ 11 ],[ 13 ],[ 18 ]) the optimal control $\hat{u}(t)$ is characterized as the unique solution near $u = 0$ of the equation

$$\lambda(t)^* \frac{\partial F}{\partial u}(\hat{x}(t),\hat{u}(t)) + \frac{\partial G}{\partial u}(\hat{x}(t),\hat{u}(t)) = 0 , \qquad (3.14)$$

with $\hat{x}(t)$, $\hat{u}(t)$ together satisfying (3.01) and $\hat{x}(0) = x_0$, while the Lagrange multiplier function $\lambda(t)$ satisfies the adjoint system

$$\dot{\lambda}(t)^* = - \lambda(t)^* \frac{\partial F}{\partial x}(\hat{x}(t),\hat{u}(t)) - \frac{\partial G}{\partial x}(\hat{x}(t),\hat{u}(t)) \qquad (3.15)$$

with

$$\lim_{t\to\infty} \lambda(t) = 0 .$$

For $\hat{u}(t)$ near $u = 0$ the nonsingularity of $U = \frac{\partial^2 G}{\partial x^2}(0,0)$ together with the implicit function theorem yields the existence of a continuously differentiable function $\mathcal{K}(x,\lambda)$ near the origin in $R^{2n}$ such that

$$\hat{u}(t) = \mathcal{K}(\hat{x}(t),\lambda(t)) = - U^{-1}(B^*\lambda(t) + R^*\hat{x}(t)) + \ell(\hat{x}(t),\lambda(t)) \ ,(3.16)$$

$$\ell(0,0) = 0, \ \frac{\partial \ell}{\partial(x,\lambda)}(0,0) = 0 .$$

Thus the optimal trajectory $\hat{x}(t)$ and the Lagrange multiplier $\lambda(t)$ together satisfy (cf.(3.12),(3.15))

$$\dot{\hat{x}} = F(\hat{x}(t),\mathcal{K}(\hat{x}(t),\lambda(t))) \ , \ \hat{x}(0) = x_0, \qquad (3.17)$$

$$\dot{\lambda}(t)^* = - \lambda(t)^* \frac{\partial F}{\partial x}(\hat{x}(t),\hat{u}(t)) - \frac{\partial G}{\partial x}(\hat{x}(t),\hat{u}(t))$$

$$= - \lambda(t)^* \frac{\partial F}{\partial x}(\hat{x}(t), \mathcal{K}(\hat{x}(t), \lambda(t))) - \frac{\partial G}{\partial x}(\hat{x}(t), \mathcal{K}(\hat{x}(t), \lambda(t))), \quad \lim_{t\to\infty} \lambda(t) = 0. \quad (3.18)$$

Solution of the corresponding boundary value problem in the LQ case is enabled by the transformation, with Q the indicated solution of (3.10),

$$\lambda(t) = 2Q\,\hat{x}(t) + \eta(t) . \quad (3.19)$$

The solutions of the system corresponding to (3.17),(3.18) in that restricted case satisfying the indicated boundary conditions are obtained by setting $\eta(t) \equiv 0$. Equivalently, one demonstrates that if we replace $\lambda(t)$ by $2Q\,\hat{x}(t)$ in the linear version of (3.17) and solve the resulting initial value problem for $\hat{x}(t)$, then $\hat{x}(t)$, $\lambda(t) \equiv Q\,\hat{x}(t)$ together solve the linear counterpart of the system (3.17),(3.18).

If (3.19) is applied to the system (3.17),(3.18), followed by some algebraic manipulations, then with $\hat{A}$ = (cf.(3.08)) = $A + B\hat{K}$, which is known from the standard LQ theory to be a stability matrix, there results a system in the form

$$\dot{\hat{x}}(t) = \hat{A}\,\hat{x}(t) - BU^{-1}B^*\eta(t) + f_1(\hat{x}(t), \eta(t)) , \quad (3.20)$$

$$\dot{\eta}(t) = - \hat{A}^*\,\eta(t) - f_2(\hat{x}(t), \eta(t)) , \quad (3.21)$$

wherein $f_1$, $f_2$ are continuously differentiable near the origin in $R^{2n}$ and

$$f_i(0,0) = 0, \quad \frac{\partial f_i}{\partial (x,\eta)}(0,0) = 0 , \quad i = 1,2 . \quad (3.22)$$

The next stage in the NLNQ case is to seek for solutions of the system (3.20),(3.21) in the form

$$\eta(t) = \mathcal{Q}(\hat{x}(t)) . \quad (3.23)$$

Evidently this project succeeds if and only if

$$- \hat{A}^* \mathcal{Q}(\hat{x}(t)) - f_2(\hat{x}(t),\mathcal{Q}(\hat{x}(t))) = \dot{\eta}(t) = \frac{\partial \mathcal{Q}}{\partial x}(\hat{x}(t)) \dot{\hat{x}}(t)$$

$$= \frac{\partial \mathcal{Q}}{\partial x}(\hat{x}(t))\left[\hat{A} \hat{x}(t) - BU^{-1}B^*\mathcal{Q}(\hat{x}(t)) + f_1(\hat{x}(t),\mathcal{Q}(\hat{x}(t)))\right] .$$

Taking account of all possible values of $\hat{x}(t)$ under consideration we see that this is the case if and only if $\mathcal{Q}(x)$ satisfies the partial differential equation

$$\hat{A}^*\mathcal{Q}(x) + f_2(x,\mathcal{Q}(x)) + \frac{\partial \mathcal{Q}}{\partial x}(x)\left[\hat{A}x - BU^{-1}B^*\mathcal{Q}(x) + f_1(x,\mathcal{Q}(x))\right] = 0. \quad (3.24)$$

The corresponding boundary condition is

$$\mathcal{Q}(0) = 0 . \quad (3.25)$$

Since it is familiar from the Hamilton-Jacobi theory that

$$\lambda(t)^* = \frac{\partial \hat{J}}{\partial x}(\hat{x}(t)) ,$$

where $\hat{J}$ is the optimal value function, it must be true that

$$2x^*Q + \mathcal{Q}(x)^* \equiv \frac{\partial \hat{J}}{\partial x}$$

and therefore

$$\frac{\partial \mathcal{Q}}{\partial x} = \frac{\partial}{\partial x}\left(\frac{\partial \hat{J}^*}{\partial x}\right) - 2Q = \frac{\partial^2 \hat{J}}{\partial x^2} - 2Q,$$

must be symmetric. This is not directly evident from (3.24), however.

Solution of the partial differential equation (3.24) yields the optimal control for the nonlinear system in the form (cf.(3.16))

$$\hat{u}(t) = \mathcal{R}(\hat{x}(t)) = \mathcal{K}(\hat{x}(t),2Q\hat{x}(t)+\mathcal{Q}(\hat{x}(t))) . \quad (3.26)$$

For a given $x_0$ this control is obtained by substituting (3.26) into (3.17) to obtain

$$\dot{\hat{x}}(t) = F(\hat{x}(t),\mathfrak{K}(\hat{x}(t))) \equiv \hat{F}(\hat{x}(t)) \qquad (3.27)$$

and then integrating this equation to give $\hat{x}(t)$. Then $\hat{u}(t)$ may be computed from (3.26). The solution $\mathcal{Q}(x)$ of (3.24),(3.25) also yields the stable manifold for the system (3.17),(3.18) in the x-parametrized form

$$\mathcal{S} = \left\{(x,\lambda)\,\middle|\,\lambda = 2Qx+\mathcal{Q}(x)\right\} .$$

It is clear, therefore, that methods for solution of (3.24),(3.25) are of considerable interest.

The partial differential equation (3.24) is a nonlinear first order hyperbolic system. Indeed, the characteristic curves are precisely curves in $R^n$ corresponding to solutions of (3.27). To see this it is sufficient to note that

$$F(x,\mathfrak{K}(x)) \equiv \hat{A}x - BU^{-1}B^*\mathcal{Q}(x) + f_1(x,\mathcal{Q}(x)) \qquad (3.28)$$

so that along solutions $\hat{x}(t)$ of (3.27) we have

$$\hat{A}^* \mathcal{Q}(\hat{x}(t)) + f_2(\hat{x}(t),\mathcal{Q}(\hat{x}(t))) + \frac{\partial\mathcal{Q}}{\partial x}(\hat{x}(t))\,\dot{\hat{x}}(t) = 0 .$$

Then, again along such trajectories, we have

$$\hat{A}^* \mathcal{Q}(\hat{x}(t)) + f_2(\hat{x}(t),\mathcal{Q}(\hat{x}(t))) + \frac{d}{dt}\,\mathcal{Q}(\hat{x}(t)) = 0 . \qquad (3.29)$$

This is, in fact, just a restatement of (3.21). In principle the equations (3.27),(3.29) can be integrated together along with the condition (3.25) to give values of $\mathcal{Q}$ along trajectories. Aside from the obvious difficulty associated with the fact that the interval of integration is infinite, so that the boundary condition, in fact, is $\mathcal{Q}(\hat{x}(-\infty)) = 0$, information given along a selected set of curves is of limited use in implementation of the optimal feedback law (3.11); there remains the matter of interpolation/extrapolation of the available information so that estimates of $\mathcal{Q}(x)$, hence of $\mathfrak{K}(x)$, can be obtained at arbitrary points x in the region of interest.

Now, in fact, existence and uniqueness of solutions of (3.24) can be obtained in the following way. Let $\mathfrak{X}$ be a region of $R^n$ containing the origin and let $C^1_{0,\varepsilon}(\mathfrak{X})$ be the collection of continuously differentiable n-vector functions q(x) defined on $\mathfrak{X}$ satisfying

$$q(0) = 0, \quad \frac{\partial q}{\partial x}(0) = 0, \quad \left\|\frac{\partial q}{\partial x}(x)\right\| < \varepsilon, \quad x \in \mathfrak{X} .$$

Let $T > 0$ be fixed. Given a point $x_0 \in \mathfrak{X}$, the differential equation (3.27) can be solved on [0,T] to obtain a solution $\hat{x}(t,x_0)$ satisfying the initial condition

$$\hat{x}(0,x_0) = x_0 .$$

At $t = T$ we obtain a point

$$x_1 = \hat{x}(T,x_1) \equiv x_1(x_0) .$$

Clearly (3.29) implies that

$$\mathcal{Q}(x_0) = \mathcal{Q}(x_1(x_0)) + \int_0^T \left[\hat{A}^*\mathcal{Q}(\hat{x}(t,x_0)) + f_2(\hat{x}(t,x_0),\mathcal{Q}(\hat{x}(t,x_0)))\right]dt . \tag{3.30}$$

One can see that satisfaction of this functional-integral equation is equivalent to the condition that $\mathcal{Q}(x)$ should be a solution of the partial differential equation (3.24) - provided that the region $\mathfrak{X}$ has the property of being invariant under (forward) solution of (3.27), so that

$$x_0 \in \mathfrak{X} \to x(t,x_0) \in \mathfrak{X}, \quad t \in [0,T].$$

Since the matrix Q of (3.10) may be seen to satisfy (cf.(3.08))

$$(A+B\hat{K})^*Q + Q(A+B\hat{K}) + W + R\hat{K} + \hat{K}^*R^* + \hat{K}^*U\hat{K} - 0 ,$$

defining, for $r > 0$,

$$\mathfrak{X}_r = \left\{ x \in R^n \;\middle|\; x^*Qx \leqq r \right\},$$

it is not hard to see from the earlier indicated higher order character of $f_2$ that this invariance obtains if $\mathcal{Q} \in C^1_{0,\varepsilon}(\mathfrak{X}_r)$ and r and $\varepsilon$ are sufficiently small. From this one can see that for r and $\varepsilon$ sufficiently small the map

$$\tilde{q} = \mathcal{T}(q)$$

defined by

$$\tilde{q}(x_0) = q(x_1(x_0)) + \int_0^T \left[\hat{A}^* q(\hat{x}(t,x_0)) + f_2(\hat{x}(t,x_0), q(\hat{x}(t,x_0)))\right] dt \tag{3.31}$$

with q an arbitrary element of $C^1_{0,\varepsilon}(\mathfrak{X}_r)$ and $\hat{x}(t,x_0)$ constructed as indicated, but now a solution of

$$\dot{\hat{x}}(t) = \hat{A}\,\hat{x}(t) - BU^{-1}B^* q(\hat{x}(t)) + f_1(x, q(x(t))) ,$$

will be a contraction mapping on $C^1_{0,\varepsilon}(\mathfrak{X}_r)$ if r and $\varepsilon$ are sufficiently small. The unique fixed point is then the desired solution of (3.30), hence of (3.24).

What is notable about the method is its adaptability to various types of finite element (e.g., spline) computational realization. The space $C^1_{0,\varepsilon}$ can be replaced be a space of appropriate finite element functions, each function of which is determined by its values on a finite set of points X. Given q(x) a member of this space, for $x_0 \in X$ $\tilde{q}(x_0)$ can be computed from q(x) using (3.31) and then, using the rule for construction of the finite element function $\tilde{q}(x)$ from its values $\tilde{q}(x_0)$, $x_0 \in X$, one obtains the finite element function $\tilde{q}(x)$. Using this iteration procedure with starting function $q_0(x) \equiv 0$, one can construct successively a sequence of finite element functions $q_k(x)$ converging to an approximation q(x) to the exact solution $\mathcal{Q}(x)$ of the partial differential equation (3.24). An approximation to the optimal control law $\mathfrak{K}(x)$ can then be constructed from $q(x)$ in the same way as $\mathfrak{K}(x)$ itself is constructed from $\mathfrak{Q}(x)$. Computational tests of this procedure should show how it compares to the power series method of [ 15 ], [ 16 ] for practical nonlinear control implementation.

4. NLNQ Theory for General Discrete Systems on an Infinite Interval

Let us consider a general discrete (recursion) system with state $x \in R^n$ and control $u \in R^m$:

$$x_{k+1} = F(x_k,u_k),\ k = 0,1,2,\ldots, \tag{4.01}$$

and a corresponding cost, depending on the initial state $x_0$ and the applied control sequence $u = \{u_k\}$,

$$J(x_0,u) = \sum_{k=0}^{\infty} G(x_k,u_k)\ . \tag{4.02}$$

In (4.01) and (4.02) we assume that $F:R^{n+m} \to R^n$ and $G:R^{n+m} \to R^1$ are at least twice continuously differentiable in some region containing the origin in $R^{n+m}$ with

$$F(0,0) = 0,\ G(0,0) = 0,\ \frac{\partial G}{\partial(x,u)}(0,0) = 0\ . \tag{4.03}$$

We may consequently rewrite (4.01), (4.02) in the form

$$x_{k+1} = A\,x_k + B\,u_k + f(x_k,u_k),\ k = 0,1,2,\ \ldots\ , \tag{4.04}$$

$$J(x_0,\{u\}) = \sum_{k=0}^{\infty}\left[(x_k^*\ u_k^*)\begin{pmatrix} W & R \\ R^* & U \end{pmatrix}\begin{pmatrix} x_k \\ u_k \end{pmatrix} + g(x_k,u_k)\right]\ , \tag{4.05}$$

with $f(x,u)$, $g(x,u)$ at least twice continuously differentiable near the origin in $R^{n+m}$ and, in addition to the conditions

$$f(0,0) = 0,\ g(0,0) = 0,\ \frac{\partial g}{\partial(x,u)}(0,0) = 0 \tag{4.06}$$

implied by (4.03),

$$\frac{\partial f}{\partial(x,u)}(0,0) = 0\ ,\ \frac{\partial^2 g}{\partial(x,u)^2}(0,0) = 0\ . \tag{4.07}$$

We further suppose, for simplicity in our current presentation, that the pair A,B is (discrete) stabilizable and the $(n+m) \times (n+m)$ matrix in the quadratic term of (4.05) is positive definite (these conditions can be weakened along the lines already indicated for the continuous autonomous system in the preceding section). This corresponds to the condition on the Hessian

$$\frac{\partial^2 G}{\partial (x,u)^2}(0,0) > 0 . \qquad (4.08)$$

Much as before, we define $\mathcal{U}_\infty = \mathcal{U}_\infty(x_0)$, for a given $x_0 \in R^n$, to be the set of control sequences

$$\{u\} = \left\{ u_k \;\middle|\; k = 0,1,2,..., \right\} \in \ell^2_m$$

for which the cost (4.02), with $x_k$, $u_k$ assumed to satisfy (4.02) for $k = 0,1,2,...,$ is finite. Since it is straightforward to see that control sequences $\{u\}$ generated by $u_k = K\, x_k$, $k = 0,1,2,...,$ with $A + BK$ a discrete stability matrix, lie in $\mathcal{U}_\infty(x_0)$ for $\|x_0\|$ sufficiently small, it is reasonable to pose the optimal control problem of minimizing $J(x_0,\{u\})$ subject to the constraint that $\{x\}$ and $\{u\}$ together satisfy (4.01).

Standard arguments, which are detailed in [XZh], show that if we introduce the adjoint system based on a state-control pair $\{\hat{x}\}$, $\{\hat{u}\}$,

$$\lambda_{k+1}^* \frac{\partial F}{\partial x}(\hat{x}_k,\hat{u}_k) - \lambda_k^* + \frac{\partial G}{\partial x}(\hat{x}_k,\hat{u}_k) = 0 , \qquad (4.09)$$

with the condition

$$\lim_{k\to\infty} \lambda_k = 0 . \qquad (4.10)$$

For $\{\hat{u}\} \in \mathcal{U}_\infty$ and $x_0$ sufficiently small one can show that (4.09), (4.10) has a unique solution $\left\{ \lambda_k \;\middle|\; k = 0,1,2,..., \right\}$. Then a necessary condition for optimality of the pair $\{\hat{x}\}$, $\{\hat{u}\}$ is that for each $k = 0,1,2,...,$ we should have

$$\lambda_{k+1}{}^{*}\frac{\partial F}{\partial u}(\hat{x}_k,\hat{u}_k) + \frac{\partial G}{\partial u}(\hat{x}_k,\hat{u}_k) = 0\ . \tag{4.11}$$

Using the assumed twice continuous differentiability of F and G and (4.11), the implicit function theorem allows us to see that for $(x,\lambda)$ near the origin in $R^{2n}$ the equation

$$H(x,\lambda,u) \equiv \lambda^{*}\frac{\partial F}{\partial u}(x,u) + \frac{\partial G}{\partial u}(x,u) = 0 \tag{4.12}$$

has a unique solution u near the origin in $R^m$ taking the form

$$u = \mathcal{K}(x,\lambda) \tag{4.13}$$

with $\mathcal{K}$ a continuously differentiable function of x and $\lambda$. Thus the combined conditions (4.01),(4.09),(4.10) and (4.11) are satisfied if the two point boundary value problem obtained by substituting the relationship $u_k = \mathcal{K}(\hat{x}_k,\lambda_{k+1})$ into (4.01) (with $u_k \equiv \hat{u}_k$) and (4.09), namely

$$\hat{x}_{k+1} = F(\hat{x}_k,\mathcal{K}(\hat{x}_k,\lambda_k)),\ k = 0,1,2,\ldots,\ x_0 \text{ given}, \tag{4.14}$$

$$\lambda_{k+1}{}^{*}\frac{\partial F}{\partial x}(\hat{x}_k,\mathcal{K}(\hat{x}_k,\lambda_{k+1})) - \lambda_k{}^{*} + \frac{\partial G}{\partial x}(\hat{x}_k,\mathcal{K}(x_k,\lambda_{k+1})) = 0\ , \tag{4.15}$$

$$\lim_{k\to\infty}\lambda_k = 0\ ,$$

can be shown to have a unique solution for $x_0$ near the origin in $R^n$.

Careful analysis, using the form of (4.11), shows the system (4.14),(4.15) to have the form (transposing the second equation now)

$$\hat{x}_{k+1} = (A-BU^{-1}R^{*})\hat{x}_k - \tfrac{1}{2}BU^{-1}B^{*}\lambda_{k+1} + r_1(\hat{x}_k,\lambda_{k+1})\ , \tag{4.16}$$

$$(A-BU^{-1}R^{*})^{*}\lambda_{k+1} - \lambda_k + 2(W-RU^{-1}R^{*})\hat{x}_k + r_2(\hat{x}_k,\lambda_{k+1}) = 0\ . \tag{4.17}$$

The remainder terms are of at least second order; the linear terms shown are all terms present in the corresponding equations for the linear quadratic case wherein $g(x,u) \equiv 0$ in (4.05). The theory of the linear quadratic problem is, of course, long since complete (cf. [ 1 ],[ 2 ],[ 3 ]). From that theory we know that there is a unique $n \times n$ positive definite symmetric solution matrix P, and associated $m \times n$ solution matrix $\hat{K}$, for the system of equations

$$P = W + R\hat{K} + \hat{K}^*R^* + \hat{K}^*U\hat{K} + (A+B\hat{K})^*P(A+B\hat{K}) \ , \tag{4.18}$$

$$\hat{K} = -\ (U + B^*PB)^{-1}(R^* + B^*PA) \ , \tag{4.19}$$

and it can be seen that the transformation

$$\lambda = \eta + 2Px \tag{4.20}$$

carries (4.16),(4.17) into the linearly decoupled form

$$\hat{x}_{k+1} = (A+B\hat{K})\hat{x}_k + V\left[(A+B\hat{K})^{-1}\right]^*\eta_k + f_1(\hat{x}_k,\eta_k) \ , \tag{4.21}$$

$$\eta_{k+1} = \left[(A+B\hat{K})^{-1}\right]^*\eta_k + f_2(\hat{x}_k,\eta_k) \ . \tag{4.22}$$

Here

$$V = -\ \frac{1}{2}(I + BU^{-1}B^*P)^{-1}BU^{-1}B^* \ .$$

The matrix $A + B\hat{K}$ is a discrete stability matrix with eigenvalues in the interior of the unit disc in the complex plan while $\left[(A+B\hat{K})^{-1}\right]^*$ clearly has eigenvlues in the exterior of the unit disc. The functions $f_1$, $f_2$ are of at least second order in $x,\eta$ near the origin in $R^{2n}$.

For ease of manipulation, let us abbreviate (4.21),(4.22) to

$$\hat{x}_{k+1} = \hat{A}\ \hat{x}_k + C\ \eta_k + f_1(\hat{x}_k,\eta_k) \ , \tag{4.23}$$

$$\eta_{k+1} = \hat{A}^{-*}\eta_k + f_2(\hat{x}_k,\eta_k) \ , \tag{4.24}$$

wherein

$$\hat{A} = A + BK \ , \ \hat{A}^{-*} = (\hat{A}^{-1})^*, \ C = V\hat{A}^{-*} \ .$$

By making an appropriate change of variable, $\hat{x}_k = S\,\hat{\xi}_k$, $\eta_k = S^*\vartheta_k$, which we will not explicitly carry out here, we may assume the matrices $\hat{A}$ and $\hat{A}^*$ to be strict contractions, i.e.,

A

$$\|\hat{A}\| = \|\hat{A}^*\| < 1\,. \tag{4.25}$$

As in the continuous framework of the preceding section, in order to convert (4.13) into a true nonlinear state feedback control law, we need to find a continuously differentiable functional relationship

$$\eta = \mathcal{P}(x), \quad \mathcal{P}(0) = 0, \quad \frac{\partial \mathcal{P}}{\partial x}(0) = 0, \tag{4.26}$$

so that if the first equation of (4.26) is substituted into (4.23) to yield

$$\hat{x}_{k+1} = \hat{A}\,\hat{x}_k + C\,\mathcal{P}(\hat{x}_k) + f_1(\hat{x}_k, \mathcal{P}(\hat{x}_k))\,,$$

then $\hat{x}_k$, $\eta_k = \mathcal{P}(\hat{x}_k)$ will satisfy (4.23), (4.24). Clearly, for this to be the case we must have

$$\hat{A}^{-*}\mathcal{P}(\hat{x}_k) + f_2(\hat{x}_k, \mathcal{P}(\hat{x}_k)) = \eta_{k+1} = \mathcal{P}(\hat{x}_{k+1})$$

$$= \mathcal{P}\left[\hat{A}\,\hat{x}_k + C\,\mathcal{P}(\hat{x}_k) + f_1(\hat{x}_k, \mathcal{P}(\hat{x}_k))\right]\,.$$

Multiplying on the left by $\hat{A}^*$ we have

$$\mathcal{P}(\hat{x}_k) = \hat{A}^*\mathcal{P}\left[\hat{A}\,\hat{x}_k + C\,\mathcal{P}(\hat{x}_k) + f_1(\hat{x}_k, \mathcal{P}(\hat{x}_k))\right] - \hat{A}^* f_2(\hat{x}_k, \mathcal{P}(\hat{x}_k))\,.$$

Recognizing that this should be an identity over the entire range of $\hat{x}_k$ involved, we arrive at the functional equation

$$\mathcal{P}(x) = \hat{A}^*\mathcal{P}\left[\hat{A}\,x + C\,\mathcal{P}(x) + f_1(x, \mathcal{P}(x))\right] - \hat{A}^* f_2(x, \mathcal{P}(x))\,. \tag{4.27}$$

Using (4.25) it is not difficult to show the existence of a unique small solution of (4.27) by the contraction fixed point theorem, provided the region $\mathfrak{X}$ to which x is restricted is appropriately specified. Writing (4.27) as

$$\mathcal{P} = \mathfrak{Q}(\mathcal{P}),\ \mathcal{P} \in C^1_{0,\varepsilon}(\mathfrak{X}), \tag{4.28}$$

where $\mathfrak{X}$ is a small region containing the origin in $R^n$ and $C^1_{0,\varepsilon}(\mathfrak{X})$ denotes the metric space of n - dimensional C functions defined on $\mathfrak{X}$ vanishing, together with the first order partial derivatives, at x = 0 with the Jacobian bounded in norm by $\varepsilon$ on $\mathfrak{X}$. The contraction property and $\mathfrak{Q}: C^1_{0,\varepsilon}(\mathfrak{X}) \to C^1_{0,\varepsilon}(\mathfrak{X})$ follow easily if $\mathfrak{X}$ is sufficiently small and further specified so that

$$x \in \mathfrak{X}\ ,\ \mathcal{P} \in C^1_{0,\varepsilon}(\mathfrak{X}) \to \hat{A}\, x + C\, \mathcal{P}(x) + f_1(x,\mathcal{P}(x)) \in \mathfrak{X}\ . \tag{4.29}$$

Since (4.18) has the form

$$P = \hat{A}^* P \hat{A} + \hat{W} \tag{4.30}$$

with both P and $\hat{W}$ positive definite, defining

$$\mathfrak{X}_r = \left\{ x \in R^n \middle|\ x^* P x \leqq r \right\}\ ,\ r > 0\ ,$$

it is easy to see from (4.30) that there is a positive number $\rho$ with $0 < \rho < r$ such that

$$x \in \mathfrak{X}_r \to \hat{A}\, x \in \mathfrak{X}_\rho\ .$$

It is then straightforward to see that (4.29) will be true if $\varepsilon$ is chosen sufficiently small and $\mathfrak{X}$ is taken equal to $\mathfrak{X}_r$ with r sufficiently small. The nonlinear feedback law synthesizing the optimal control then takes the form (cf. (4.13),(4.20),(4.26))

$$u = \mathfrak{K}(x,\ 2Px + \mathcal{P}(x)) \equiv \hat{\mathfrak{K}}(x) = \hat{K}\, x + o(x),\ \|x\| \to 0\ . \tag{4.31}$$

The analyticity of $\hat{\mathfrak{K}}(x)$ in $\mathfrak{X}$, when F(x,u) and G(x,u) are analytic functions of x and u near the origin in $R^{n+m}$ also follows from the indicated iteration solution of (4.27). Starting the iteration with $\mathcal{P}_0(x) \equiv 0$, one readily verifies with the assumed analyticity that all of the later iterates $\mathcal{P}_k(x)$ will be analytic functions of x in $\mathfrak{X}$ and the analyticity of $\mathcal{P}(x)$ then follows from the uniform convergence of the $\mathcal{P}_k(x)$ to $\mathcal{P}(x)$ in $\mathfrak{X}$. Since the implicit function theorem gives the analyticity of $\mathcal{K}(x,\lambda)$ with respect to $x,\lambda$ near the origin in $R^{2n}$, the result follows. It should be clear that approximations to $\mathcal{P}(x)$ and the corresponding optimal control law $\mathfrak{K}(x)$ can be obtained by applying the appropriately modified map (4.28) to a space of finite element functions as described for the ordinary differential case in the preceding section.

It will be recognized that solution of (4.27) provides a proof of the stable manifold theorem for discrete systems different from the standard adaptation to discrete systems of the integral equation argument generally used (see [ 4 ],[ 5 ]) to prove the stable manifold theorem in the ordinary differential equations context.

We conclude this section with two remarks based on work in the thesis [ 20 ] of the second author. There it is shown that all of these results remain substantially unchanged if we replace the requirement $u \in R^m$ with a more general specification $u \in \mathfrak{B}$, where $\mathfrak{B}$ is a general (infinite dimensional) Banach space, provided the continuity, twice continuous differentiability, etc., are maintained (in the linear case this would correspond to requiring that the control operator $B:\mathfrak{B} \to R^n$ should be a bounded operator) and provided an appropriate sense can be given to the forms $u^*Uu$ and $u^*Rx$ appearing in the formula (4.05). It is further shown in [ 20 ] that

$$\lambda_0 = 2P\,x_0 + \mathcal{P}(x_0) = \frac{\partial \hat{J}}{\partial x}(x_0)^*, \quad \mathcal{P}(x) = o(\|x\|), \ \|x\| \to 0 \ , \quad (4.32)$$

where $\hat{J}(x_0)$ is the optimal value function; the minimal value of the cost (4.02) achieved through use of the optimal control synthesized by (4.31). We will see in the next section, which applies the results of the present section to the problem of NLNQ control of nonlinear periodic systems near a critical point, that the existence and properties of the optimal value function in the present discrete context serves as an essential starting point for that theory.

## 5. NLNQ Theory for Nonlinear Periodic Systems in the Neighborhood of a Critical Point

Let us consider a system of ordinary differential equations of the form

$$\dot{x} = F(t,x,u) \ , \ t \in [0,\infty), \ x \in R^n, \ u \in R^m \ . \tag{5.01}$$

We assume that F and its first and second order partial derivatives with respect to x and u are continuous functions of (t,x,u) for t as indicated and for (x,u) in a neighborhood of the origin in $R^{n+m}$. We will further suppose that there is a positive number $T > 0$ such that the system is periodic in t with period T; i.e.,

$$F(t,x,u) \equiv F(t+T,x,u) \tag{5.02}$$

for all values of (t,x,u) under consideration. We do not necessarily assume that T is minimal in this respect. We will suppose the system has $x = 0$, $u = 0$ as a critical point; i.e.,

$$F(t,0,0) = 0 \ , \ t \in [0,\infty) \ . \tag{5.03}$$

We let

$$A(t) = \frac{\partial F}{\partial x}(t,0,0), \ B(t) = \frac{\partial F}{\partial u}(t,0,0), \ t \in [0,\infty); \tag{5.04}$$

thus the linear variational system about the origin in $R^{n+m}$ is

$$\dot{x}(t) = A(t)\ x + B(t)\ u \ ; \tag{5.05}$$

this is a linear T-periodic system; i.e.,

$$A(t) = A(t+T), \ B(t) = B(t+T). \tag{5.06}$$

With this system we consider a cost functional of the form

$$J(x_0,u) = \int_0^\infty G(t,x(t),u(t))\ dt \ , \tag{5.07}$$

where G(t,x,u) and its first and second order partial derivatives with respect to x and u are continuous functions of (t,x,u) in the region of interest and G is T-periodic in t; i.e.,

$$G(t,x,u) \equiv G(t+T,x,u) \ . \tag{5.08}$$

We further suppose that G(t,x,u) takes the form

$$G(t,x,u) = (x^* \ u^*)\begin{bmatrix} W(t) & R(t) \\ R(t)^* & U(t) \end{bmatrix}\begin{bmatrix} x \\ u \end{bmatrix} + g(t,x,u) \ , \tag{5.09}$$

where, uniformly in t,

$$g(t,x,u) = o(\|x\|^2 + \|u\|^2), \text{ as } \|(x,u)\| \to 0 \text{ in } R^{n+m} \ . \tag{5.10}$$

Clearly the matrix in (5.09) is also T-periodic; we will assume in addition that it is positive definite, uniformly with respect to t (as before, recognizing the possibility of weakening this condition). From the periodicity property it is enough to assume these uniformity properties are valid for $t \in [0,T]$ in order to conclude that they are also valid for $t \in [0,\infty)$.

We will suppose that the linearized system (5.05) is T-*periodically stabilizable,* by which we mean that there exists a continuous T-periodic $m \times n$ matrix function

$$K(t) = K(t+T)$$

such that the T-periodic linear feedback law

$$u(t) = K(t)\ x(t) \tag{5.11}$$

yields a synethesized system

$$\dot{x}(t) = \left[A(t) + B(t)\ K(t)\right]\ x(t) \tag{5.12}$$

which is asymptotically stable. This means that if $\Phi_K(t)$ is the fundamental matrix solution reducing to the identity when t = 0, then

$$\|\Phi_K(nT)\| = \|\Phi_K(t)^n\| \leqq M\ \gamma^n \ , \ n = 0,1,2,\ldots \ , \tag{5.13}$$

for some positive $M \geqq 1$ and some positive $\gamma$ with $0 < \gamma < 1$.

We define the set of admissible controls for the linear system to be the set, $\mathcal{U}_\infty$, of controls $u \in L^2_m[0,\infty)$ for which the quadratic cost corresponding to (5.09), obtained by replacing $g(t,x,u)$ with 0 in that formula, is finite. The T-periodic linear quadratic control problem may then be posed, for a given initial state $x_0 \in R^n$, as the problem of finding, if possible, a control function $\hat{u} \in \mathcal{U}_\infty$ for which

$$J(x_0,\hat{u}) \leqq J(x_0,u), \quad u \in L^2_m[0,\infty) . \tag{5.14}$$

While a number of treatments of periodic optimal control problems appear in the literature (cf.[ 7 ],[ 17 ],e.g.), we will present here a simplified argument communicated to the first author by Prof. Georg Schmidt of McGill University. Then we will indicate how that argument can be modified to treat the corresponding NLNQ optimal control problem.

Let $\Phi(t,\tau)$ denote the fundamental matrix solution of the homogeneous version of (5.05) reducing to the identity when $t = \tau$. Then we have, for solutions of (5.05) itself, the familiar variation of parameters formula

$$x(t) = \Phi(t,\tau)\, x(\tau) + \int_\tau^t \Phi(t,s)\, B(s)\, u(s)\, ds . \tag{5.15}$$

If we let

$$x_k = x(kT), \quad k = 0,1,2,\ldots, \tag{5.16}$$

then, using the periodicity of (5.05) and the corresponding property of the fundamental solutions, we have

$$x_{k+1} = \Phi((k+1)T,kT)\, x_k + \int_{kT}^{(k+1)T} \Phi((k+1)T,s)\, B(s)\, u(s)\, ds$$

$$= \Phi(T,0)\, x_k + \int_0^T \Phi(T,s)\, B(s)\, u(s+kT)\, ds \equiv \mathbb{A}\, x_k + \mathbb{B}\, u_k , \tag{5.17}$$

wherein

$$\mathbb{A} = \Phi(T,0), \quad u_k(s) = u(s+kT), \quad s \in [0,T], \tag{5.18}$$

and $\mathbb{B}:L^2_m[0,T] \to R^n$ is the bounded operator defined by

$$\mathbb{B}\, u_k = \int_0^T \Phi(T,s)\, B(s)\, u_k(s)\, ds .$$

This gives us a discrete linear system

$$x_{k+1} = \mathbb{A}\, x_k + \mathbb{B}\, u_k \,, \quad k = 0,1,2,\ldots, \tag{5.19}$$

with state in $R^n$ and control in $L^2_m[0,T]$. This system is stabilizable in the discrete sense because, if $K(t)$ is the periodic stabilizing feedback matrix (5.11) and $\Phi_K(t)$ is the fundamental matrix for the synthesized system (5.12), use of that feedback relation in (5.05) gives

$$u_k(s) = K(s)\, x(s+kT) = K(s)\, \Phi_K(s)\, x_k \equiv (\mathbb{K}\, x_k)(s) \,, \tag{5.20}$$

which is readily seen to be a stabilizing feedback control law for the discrete linear system (5.19); the synthesized system is

$$x_{k+1} = \left(\mathbb{A} + \mathbb{B}\,\mathbb{K}\right) x_k \,, \quad k = 0,1,2,\ldots. \tag{5.21}$$

Since $\mathbb{K}:R^n \to C_m[0,T] \subset L^2_m[0,T]$ and $\mathbb{B}:L^2_m[0,T] \to R^n$, $\mathbb{B}\,\mathbb{K}$ is an ordinary $n \times n$ matrix. We observe that the discrete systems (5.19), (5.21) are autonomous linear discrete systems - a consequence of the periodicity of the original ODE system.

The quadratic cost obtained by replacing $g(t,x,u)$ by zero in the formula (5.09) can be written as

$$J_0(x_0,u) = \sum_{k=0}^{\infty} \int_{kT}^{(k+1)T} (x(s)^* \;\; u(s)^*) \begin{pmatrix} W(s) & R(s) \\ R(s)^* & U(s) \end{pmatrix} \begin{pmatrix} x(s) \\ u(s) \end{pmatrix} ds$$

$$= \sum_{k=0}^{\infty} \int_0^T (x(s+kT)^* \ u(s+kT)^*) \begin{pmatrix} W(s+kT) & R(s+kT) \\ R(s+kT)^* & U(s+kT) \end{pmatrix} \begin{pmatrix} x(s+kT) \\ u(s+kT) \end{pmatrix} ds$$

$$= \sum_{k=0}^{\infty} \int_0^T (x(s+kT)^* \ u(s+kT)^*) \begin{pmatrix} W(s) & R(s) \\ R(s)^* & U(s) \end{pmatrix} \begin{pmatrix} x(s+kT) \\ u(s+kT) \end{pmatrix} ds \ ; \quad (5.21)$$

the last equality a consequence of the periodicity of the matrix in (5.09). If we now use (5.16) in the last expression here and again define $u_k$ as in (5.18) we can see, with a certain amount of calculation, that the quadratic cost (5.21) can be expressed as

$$J(x_0,\{u_k\}) = \sum_{k=0}^{\infty} (x_k^* \ u_k^*) \begin{pmatrix} \mathbb{W} & \mathbb{R} \\ \mathbb{R}^* & \mathbb{U} \end{pmatrix} \begin{pmatrix} x_k \\ u_k \end{pmatrix} \quad (5.22)$$

wherein the matrix operator

$$\begin{pmatrix} \mathbb{W} & \mathbb{R} \\ \mathbb{R}^* & \mathbb{U} \end{pmatrix}$$

is a bounded, self-adjoint positive definite operator on $R^n \times L_m^2[0,T]$. We are thus set up to use the standard linear quadratic optimal control theory for discrete systems; since all involved operators from $L_m^2[0,T]$ to $R^n$ are bounded there is no change in the theory. We should remark that, for any vector $v \in L_m^2[0,T]$, $u_k^* v$ means $(v,u_k)_{L_m^2[0,T]}$.

It follows then, from the standard linear quadratic theory for autonomous discrete linear systems on an infinite interval, which we have already described in the preceding section, that the optimal control is given by

$$\hat{u}_k = \hat{\mathbb{K}} x_k \ , \ k = 0,1,2,\ldots, \quad (5.23)$$

with the optimal state trajectory, $\hat{x}_k$, satisfying (5.21) with $\mathbb{K}$ replaced by $\hat{\mathbb{K}}$ and the optimal (i.e., minimal) cost is given by

$$J(x_0,\{\hat{u}_k\}) = x_0{}^*\mathbb{P}\, x_0 \ , \quad (5.24)$$

where $\mathbb{P}$ is the unique positive definite solution, with associated $\hat{\mathbb{K}}$, of the system of matrix equations (cf.(4.18),(4.19))

$$\mathbb{P} = \mathbb{W} + \mathbb{R}\hat{\mathbb{K}} + \hat{\mathbb{K}}^*\mathbb{R}^* + \hat{\mathbb{K}}^*\mathbb{U}\hat{\mathbb{K}} + (\mathbb{A}+\mathbb{B}\hat{\mathbb{K}})^*\mathbb{P}(\mathbb{A}+\mathbb{B}\hat{\mathbb{K}}) , \tag{5.25}$$

$$\hat{\mathbb{K}} = - (\mathbb{U} + \mathbb{B}^*\mathbb{P}\mathbb{B})^{-1}(\mathbb{R}^* + \mathbb{B}^*\mathbb{P}\mathbb{A}) . \tag{5.26}$$

Now we can return to the linear periodic system (5.05) with initial state $x(0) = x_0$. Since the discrete system just discussed is entirely equivalent, the minimal cost is still (5.24) and, since the system (5.05) and the matrix determining the quadratic cost in (5.09) are both T-periodic, we can see that if the initial state were specified to be $x_0$, but at time T, and if the cost were re-defined to be

$$\int_T^\infty (x(s)^* \ u(s)^*)\begin{pmatrix} W(s) & R(s) \\ R(s)^* & U(s)\end{pmatrix}\begin{pmatrix} x(s) \\ u(s)\end{pmatrix} ds ,$$

the new minimal cost would still be given by (5.24). Applying the principle of optimality, the originally posed linear quadratic optimal control problem of minimizing $J_0(x_0,u)$ can be replaced by

$$\min_u J_{T,\mathbb{P}}(x_0,u) \tag{5.27}$$

with

$$J_{T,\mathbb{P}}(x_0,u) = \int_0^T (x(s)^* \ u(s)^*)\begin{pmatrix} W(s) & R(s) \\ R(s)^* & U(s)\end{pmatrix}\begin{pmatrix} x(s) \\ u(s)\end{pmatrix} ds + x(T)^*\mathbb{P}\, x(T). \tag{5.28}$$

This, of course, is just the standard finite interval ODE linear quadratic optimal control problem, so we know the optimal control is given on [0,T] by

$$\hat{u}(t) = \hat{K}(t)\, \hat{x}(t) = - U(t)^{-1}\left[B(t)^*Q(t) + R(t)^*\right] \hat{x}(t) , \tag{5.29}$$

where $\hat{x}(t)$ satisfies (5.05) with $u(t)$ replaced by $\hat{K}(t)\,\hat{x}(t)$ with $\hat{x}(0) = x_0$, $Q(t)$ being the solution of the familiar Riccati matrix differential equation (2.35) with terminal condition (cf.(5.28))

$$Q(T) = \mathbb{P} . \tag{5.30}$$

For the problem thus posed we know that the minimal cost is given by

$$J_{T,\mathbb{P}}(x_0,\hat{u}) = x_0{}^*Q(0)\, x_0 .$$

But from the discrete formulation the same minimal cost must be given by (5.24). We conclude therefore that, in addition to (2.35), $Q(t)$ satisfies

$$Q(0) = \mathbb{P} .$$

The matrix function $Q(t)$ thus extends to $[0,\infty)$ as the unique T-periodic symmetric positive definite solution of the Riccati differential equation (2.35). From the T-periodicity of the problem it is easy to see that the optimal control $\hat{u}(t)$ must then be given on $[0,\infty)$ by (5.29), $\hat{K}(t)$ also being extended to $[0,\infty)$ by T-periodicity. From

$$\lim_{k\to\infty} \hat{x}_k = 0 ,$$

a conclusion of the corresponding discrete theory, we readily conclude that

$$\lim_{t\to\infty} x(t) = 0$$

and it is easy to see that this decay is uniformly exponential in t. There follows

<u>Theorem 5.1</u> *For the* T*-periodic linear quadratic control problem consisting of the linear system* (5.05) *with initial state* $x_0$ *and quadratic cost* (5.21) *the optimal control is given by the formula* (5.29), $t \in [0,\infty)$, *the optimal state* $\hat{x}(t)$ *satisfying* (5.05) *with* $u(t)$ *replaced by* $\hat{K}(t)\ \hat{x}(t)$ *on* $[0,\infty)$, *and with* $Q(t)$ *in* (5.29) *being the unique positive definite symmetric* T*-periodic solution of the Riccati matrix differential equation* (2.35) *on* $[0,\infty)$. *Thus the optimal feedback matrix* $\hat{K}(t)$ *is also* T*-periodic. The fundamental solution matrix* $\Phi_{\hat{K}}(t)$ *satisfies*

$$\|\Phi_{\hat{K}}(kT)\| \leqq M\,\gamma^k\ ,\ k = 0,1,2,\ldots,$$

*from which we conclude that the solutions* $\hat{x}(t)$ *of the optimal synthesized* T*-periodic system have uniform exponential decay to the origin in* $R^n$ *as* $t \to \infty$.

The previous analysis is considerably facilitated by applying the results from the discrete formulation; without that considerable work is required to establish the existence of the desired positive definite symmetric periodic solution of the Riccati equation. We will see that this continues to be the case for the nonlinear periodic system as our analysis proceeds.

We define the space of admissible controls for the NLNQ periodic optimal control problem to be $\mathfrak{U}_\infty$ as in §3; it will be adequate to fix T introduced there to the value corresponding to the period T of the present section. We may then pose our NLNQ periodic optimal control problem as that of finding $\hat{u} \in \mathfrak{U}_\infty$ such that, with $J(x_0,u)$ defined as in (5.07),

$$J(x_0,\hat{u}) \leqq J(x_0,u),\ u \in \mathfrak{U}_\infty.$$

Using the standard existence, uniqueness and regularity properties of ordinary differential equations, as developed, e.g., in [ 4 ], one sees that for solutions $x(t)$ of (5.01) lying sufficiently near the origin in $R^n$, and corresponding to continuous control functions $u(t)$ lying sufficiently near the origin in $R^m$, we have

$$x_{k+1} = \mathbf{F}(x_k,u_k),\ k = 0,1,2,\ldots, \qquad (5.31)$$

where, as previously, $x_k = x(kT)$ and $u_k$ is the restriction of u to the interval $[kT,(k+1)T]$. Moreover, with our differentiability assumptions on $F(x,u)$ of (5.01),

$$\mathbf{F}: R^n \times C^0[0,T] \to R^n$$

is twice continuously differentiable as a function of $x \in R^n$, $u \in C^0[0,T]$ sufficiently small. Further, the linear variational system for (5.31), based on the $x_k \equiv 0$, $u_k \equiv 0$ equilibrium solution of that system, is easily seen to be the linear system (5.19). From this it is quite direct to see that a linear feedback control $u_k = \mathbb{K}\, x_k$, constructed as in (5.26), yields solutions $x_k$ of the nonlinear system with uniform exponential decay to zero as $k \to \infty$, provided $\|x_0\|$ is sufficiently small in $R^n$. In much the same way the cost (5.07) can be re-expressed in the form

$$J(x_0,\{u_k\}) = \sum_{k=0}^{\infty} \mathbb{G}(x_k,u_k) \ , \quad \mathbb{G}: R^n \times C^0[0,T] \to R^1 \ , \tag{5.32}$$

with $\mathbb{G}(x_k,u_k)$ twice continuously differentiable in the same sense as described earlier for $\mathbb{F}(x,u)$ and (cf.(5.22))

$$\mathbb{G}(x,u) = (x^* \ u^*)\begin{pmatrix} \mathbb{W} & \mathbb{R} \\ \mathbb{R}^* & \mathbb{U} \end{pmatrix}\begin{pmatrix} x \\ u \end{pmatrix} + o(\|x\|^2 + \|u\|^2) \tag{5.33}$$

as $\|x\|$ and $\|u\|$ tend to 0 in $R^n$, $C^0[0,T]$, respectively.

The foregoing puts our problem precisely in the framework of §4, with an infinite dimensional control as discussed at the end there. From the work there on the discrete version of our problem we know that the optimal value function $\hat{\mathbb{J}}(x_0)$ is defined as a twice continuously differentiable function for $x_0$ near the origin in $R^n$ and, with $\mathbb{P}$ as defined above and $\mathcal{P}(x) = o(\|x\|^2)$ as $\|x\| \to 0$,

$$\hat{\mathbb{J}}(x_0) = x_0^* \mathbb{P}\, x_0 + \mathcal{P}(x_0). \tag{5.34}$$

From the T-periodicity of the problem, it is clear that if the initial state $x_0$ were given at time T instead of 0 and if the cost were redefined to be

$$\int_T^{\infty} G(t,x(t),u(t))\, dt \ ,$$

the expression for the minimal cost would still be (5.34). Citing the Principle of Optimality again, the periodic NLNQ optimal control problem can be restated, at least as far as the restriction to [0,T] is concerned, as

$$\min_{u\in C[0,T]}\left\{\int_0^T G(t,x(t),u(t))\,dt + \hat{\mathcal{J}}(\hat{x}(T))\right\}. \tag{5.35}$$

With the reformulation (5.35), we are now in a position to apply the results of §2. The notation of (5.01) and (5.07) is the same as that of §2, if we agree to identify $V(x)$ with $\hat{\mathcal{J}}(x)$, and we will use the same symbols $\hat{u}(t), \hat{x}(t), \lambda(t), \mathcal{Q}(t,x), \mathcal{K}(t,x)$ and $\hat{J}(t,x)$ used there to describe the results.

Accordingly, the optimal value function $\hat{J}(t,x)$, which satisfies the Hamilton-Jacobi partial differential equation (2.46), must agree with $\hat{\mathcal{J}}(x)$ of (5.34) both when $t = 0$ and $t = T$, so that $\hat{J}(t,x)$ extends to a T-periodic solution of (2.46), (2.47) for $t \in [0,\infty)$. The optimal value function $\hat{J}(t,x)$ is defined, continuous with respect to t and twice continuously differentiable with respect to x near the origin in $R^n$, uniformly for $t \in [0,\infty)$, with (cf.(2.44),(2.45),(5.34))

$$\hat{J}(t,0) = 0,\quad \frac{\partial \hat{J}}{\partial x}(t,0) = 0,\quad \frac{\partial^2 \hat{J}}{\partial x^2}(t,0) = 2\,Q(t),\quad t \in [0,\infty),$$

$$Q(kT) = \mathbb{P},\quad k = 0,1,2,\ldots,\,. \tag{5.36}$$

where $Q(t)$ is the unique periodic solution of the Riccati equation described earlier in this section and in §2. The optimal solutions $\hat{x}(t)$ and corresponding solutions $\lambda(t)$ of (2.16) are defined on $[0,\infty)$ and are related by the (now T-periodic) relationship

$$\lambda(t) = \mathcal{Q}(t,\hat{x}(t)) = \frac{\partial J}{\partial x}(t,\hat{x}(t))^*$$

described in §2. The optimal controls $\hat{u}(t)$ are synthesized by a non-linear, T-periodic control law

$$\hat{u}(t) = \mathcal{K}(t,\hat{x}(t),\mathcal{Q}(t,\hat{x}(t))) = \mathcal{K}(t,\hat{x}(t)) = \hat{K}(t)\hat{x}(t) + \ell(t,\hat{x}(t)), \tag{5.37}$$

where $\hat{K}(t)$ is the periodic optimal feedback matrix for the T-periodic LQ problem described earlier in this section and

$$\ell(t,x) = \ell(t+T,x), \; \ell(t,x) = o(\|x\|), \; \|x\| \to 0,$$

uniformly for $t \in [0,\infty)$. Again the role of the discrete formulation (5.31),(5.33) is to provide values for the optimal value function at $0, kT$, $k = 0,1,2,\ldots$, serving as the basis for very direct development of the rest of the theory from the finite interval NLNQ theory of §2.

Although we do not pursue this here, in the case where $F(t,x,u)$ and $G(t,x,u)$ are real analytic functions of $x$ and $u$ it should be possible to express the optimal nonlinear feedback function $\mathfrak{K}(t,x)$ as a power series in $x$ whose coefficients are T-periodic functions of $t$ and thus developable in T-periodic Fourier series, or, equivalently, as a T-periodic Fourier series whose coefficients are power series in $x$. It is intriguing to speculate, in the latter case, that the coefficients, as functions of $x$, might well satisfy a functional equation related to (3.24) and/or (4.27), again permitting expression in terms of spline or other finite element functions.

## 6. NLNQ Theory in the Neighborhood of a Periodic Solution of an Autonomous Nonlinear System

Let us consider an autonomous system of nonlinear differential equations

$$\dot{x} = F(x,u) \ , \qquad (6.01)$$

where $F(x,u)$ is defined and twice continuously differentiable in some region of $R^{n+m}$. We will suppose that when $u = 0$ the (uncontrolled) system

$$\dot{x} = F(x,0) \qquad (6.02)$$

has a nontrivial periodic solution $\tilde{x}(t)$ with least positive period T. The orbit $\mathcal{O}(\tilde{x})$ of $\tilde{x}$ then forms a simple closed curve in $R^n$ which is of class $C^3$, at least. For each $x \in \mathcal{O}(\tilde{x})$ we define $\hat{x}$ by (6.02). We select a point (any point) $\tilde{x}_0 \in \mathcal{O}(\tilde{x})$ as our starting point, so that we may consider $\tilde{x}(t)$ to be the solution of (6.02) with initial state

$$\tilde{x}(0) = \tilde{x}_0 \ . \qquad (6.03)$$

Given an arbitrary point $x \in \mathcal{O}(\tilde{x})$ there will be a least non-negative value of $\tau$ such that $\tilde{x}(\tau) = x$; then we will have

$$\tilde{x}(\tau+kT) = x, \ k = 0,1,2,\ldots, \ . \qquad (6.04)$$

We let $\xi_1(\tau)$, $\xi_2(\tau)$, ... , $\xi_n(\tau)$ be twice continuously differentiable functions of $\tau$ forming, for each $\tau$, an orthonormal basis for $R^n$ and such that

$$\xi_1(\tau) \equiv \frac{\hat{x}(\tau)}{\|\hat{x}(\tau)\|} \ ,$$

all periodic with period T. We assume this system is such that the matrix relating the vectors $\xi_i(t)$ to the vectors $\xi_i(\sigma)$, $\sigma \neq \tau$, has positive determinant for all such pairs $\tau$, $\sigma$. It is then possible to demonstrate that there is a neighborhood $\mathfrak{N}$ of $\mathcal{O}(\tilde{x})$ in $R^n$ with the property that each $x \in \mathfrak{N}$ lies on exactly one of the surfaces $\Sigma(\tau)$ generated parametrically by the expression

$$\tilde{x}(\tau) + \sum_{k=2}^{n} z_k \xi_k(\tau) .$$

This means that each $x \in \mathfrak{N}$ has a unique representation

$$x = \tilde{x}(\tau) + \sum_{k=2}^{n} z_k(\tau) \xi_k(\tau), \quad \tau \in [0,T). \tag{6.05}$$

This representation can clearly be extended by periodicity so that we also have

$$x = \tilde{x}(\tau+\ell T) + \sum_{k=2}^{n} z_k(\tau+\ell T) \xi_k(t+\ell T), \quad \ell = 0,1,2,\dots .$$

Indeed, we may suppose $\mathfrak{N}$ chosen so that the correspondence

$$x \leftrightarrow \begin{bmatrix} \tau \\ z \end{bmatrix}$$

is a $C^2$, locally one to one map between $\mathfrak{N}$ and $[0,\infty) \times \mathfrak{M}$, where $\mathfrak{M}$ is a fixed neighborhood of the origin in $R^{n-1}$; thus $x = x(\tau,z)$ is defined on $[0,\infty) \times \mathfrak{M}$ and is T-periodic with respect to $\tau$, while $\tau = \tau(x)$, $z = z(x)$, the inverse map, is multi-valued as far as $\tau$ is concerned, the possible values of $\tau(x)$ differing by multiples of T. For convenience we will re-write (6.05) in the form

$$x(\tau,z) = \tilde{x}(\tau) + \Xi(\tau)z \tag{6.06}$$

where $\Xi(\tau)$ is the $n \times n-1$ matrix whose columns are the vectors $\xi_k(\tau)$, $k = 2,3,\dots,n$. Then

$$F(x,u) = F(\tilde{x}(\tau)+\Xi(\tau)z,u) \equiv H(\tau,z,u)$$

is defined and twice continuously differentiable on $[0,\infty) \times \mathfrak{M}$, T - periodic with respect to $\tau$ and, with $\dot{}$ still representing the derivative with respect to t, the system (6.06) has, in the $\tau$,z coordinates, the twice continuously differentiable representation

$$\dot{\tau} = \frac{\partial\tau}{\partial x}(x(\tau))H(\tau,z,u) \equiv 1 + h_\tau(\tau,z,u), \tag{6.07}$$

$$\dot{z} = \frac{\partial z}{\partial x}(x(\tau))H(\tau,z,u) \equiv h_z(\tau,z,u) , \tag{6.08}$$

wherein, as a consequence of the fact that the periodic solution $\tilde{x}(t)$ corresponds to $\tau \equiv t$, $z \equiv 0$,

$$h_\tau(\tau,0,0) \equiv 0,\ h_z(\tau,0,0) \equiv 0,\ \frac{\partial h_\tau}{\partial\tau}(\tau,0,0) \equiv 0,\ \frac{\partial h_z}{\partial\tau}(\tau,0,0) \equiv 0, \tag{6.09}$$

and with

$$a(t)^* = \frac{\partial h_\tau}{\partial z}(t,0,0) \in \left(R^{n-1}\right)^*,\ A(t) = \frac{\partial h_z}{\partial z}(t,0,0),\ n-1 \times n-1, \tag{6.10}$$

$$B_\tau(t) = \frac{\partial h_\tau}{\partial u}(t,0,0),\ B_z(t) = \frac{\partial h_z}{\partial u}(t,0,0),\ B(t) = \begin{bmatrix} B_\tau(t) \\ B_z(t) \end{bmatrix} , \tag{6.11}$$

the linear variational system based on $\tau \equiv t$, $z \equiv 0$, corresponding to the original periodic solution $\tilde{x}(t)$, takes the form

$$\begin{bmatrix} \delta\dot{\tau} \\ \delta\dot{z} \end{bmatrix} = \begin{bmatrix} 0 & a(t)^* \\ 0 & A(t) \end{bmatrix}\begin{bmatrix} \delta\tau \\ \delta z \end{bmatrix} + B(t)\delta u . \tag{6.12}$$

The n-1 dimensional row vector $a(t)^*$, the $n-1 \times n-1$ matrix A(t) and the $n \times m$ matrix B(t) are, of course, T-periodic and twice continuously differentiable with respect to t.

Now let us turn to the cost functional for our optimal control problem. Assuming that our control objective is either orbital stabilization or enhancement of existing orbital stability [ 5 ] of the periodic solution $\tilde{x}(t)$, we select a cost integrand G(x,u) which is twice continuously differentiable for $x \in \Omega$ and u near the origin in $R^m$ and such that

$$G(x,0) \equiv 0,\ x \in \mathcal{O}(\tilde{x}),\ G(x,u) > 0,\ (x,u) \notin \mathcal{O}(\tilde{x}) \times \{0\}. \tag{6.13}$$

Further assumptions will be made concerning the second order partial derivatives, but these are easier to state in terms of the $\tau$,z variables, with respect to which we have

$$G(x,u) = G(\tilde{x}(\tau)+\Xi(\tau)z,u) = \Gamma(\tau,z,u). \tag{6.14}$$

Clearly

$$\Gamma(\tau,0,0) \equiv 0,\ \Gamma(\tau,z,u) > 0,\ (z,u) \neq (0,0) \text{ in } R^{n+m-1}, \tag{6.15}$$

and the assumed smoothness of G, and hence $\Gamma$, together with the second condition of (6.15) (equiv.(6.13)) implies that

$$\frac{\partial\Gamma}{\partial z}(\tau,0,0) \equiv 0,\ \frac{\partial\Gamma}{\partial u}(\tau,0,0) \equiv 0\ . \tag{6.16}$$

Therefore

$$\frac{\partial^2\Gamma}{\partial\tau\partial z}(\tau,0,0) \equiv 0,\ \frac{\partial^2\Gamma}{\partial\tau\partial u}(\tau,0,0) \equiv 0$$

and it follows that the Hessian matrix of $\Gamma$ at z = 0, u = 0 takes the form

$$\frac{\partial^2\Gamma}{\partial(\tau,z,u)^2} = \begin{bmatrix} 0 & 0 & 0 \\ 0 & W(\tau) & R(\tau) \\ 0 & R(\tau)^* & U(\tau) \end{bmatrix}; \tag{6.17}$$

continuous and T-periodic in $\tau$. We will assume that the indicated n+m-1 $\times$ n+m-1 matrix in the lower right is symmetric and uniformly positive definite; as in earlier sections this requirement can be weakened. Thus we have

$$\Gamma(\tau,z,u) = (z^*\ u^*)\begin{pmatrix} W(\tau) & R(\tau) \\ R(\tau)^* & U(\tau) \end{pmatrix}\begin{pmatrix} z \\ u \end{pmatrix} + g(\tau,z,u), \tag{6.18}$$

wherein, uniformly in $\tau$,

$$g(\tau,z,u) = o\left(\|z\|^2+\|u\|^2\right)\ ,\ \|z\|,\|u\| \to 0,0\ . \tag{6.19}$$

An example of cost integrands G, $\Gamma$ with these properties can be obtained, for x sufficiently near $\mathcal{O}(\tilde{x})$, by choosing a symmetric positive definite $n \times n$ matrix $\mathbb{W}$ and a symmetric positive definite $m \times m$ matrix $\mathbb{U}$ and then defining

$$G(x,u) = \min_{\xi \in \mathcal{O}(\tilde{x})} \left\{(x-\xi)^* \mathbb{W}(x-\xi)\right\} + u^* \mathbb{U} u .$$

We define the space of admissible controls $\mathcal{U}_\infty$ as in §5 and pose the optimal control problem

$$\min_{u \in \mathcal{U}_\infty} \int_0^\infty G(x(t),u(t))\, dt \tag{6.20}$$

subject to the constraint that x(t) and u(t) should together satisfy (6.01) with $x(0) = x_0$ sufficiently close to $\tilde{x}_0$ (cf.(6.03)). The corresponding problem in the $\tau$,z variables is clearly

$$\min_{u \in \mathcal{U}_\infty} \int_0^\infty \Gamma(\tau(t),z(t),u(t))\, dt \tag{6.21}$$

with $\tau(t),z(t)$ satisfying (6.07),(6.08) and $\tau(0),z(0)$ sufficiently close to $\tau(\tilde{x}_0) = 0$, $z(\tilde{x}_0) = 0$.

The subspace $\delta z = 0$ is clearly an invariant subspace for (6.12) when $u = 0$. If we can find a continuous T-periodic $m \times n-1$ matrix K(t) such that the periodic linear feedback relation

$$\delta u(t) = K(t)\, \delta z(t) \tag{6.22}$$

yields, in place of the second equation of (6.12) (cf.(6.11)),

$$\delta \dot{z} = \left[A(t) + B_z(t)K(t)\right] \delta z \tag{6.23}$$

as an asymptotically stable linear T-periodic system; thus the corre sponding Floquet multipliers (exponents)([ 12 ]) have absolute value less than unity (real part less than zero); solutions $\delta z(t)$ enjoy a a uniform exponential decay property as $t \to 0$ and the corresponding quadratic cost (cf.(6.18), noting that $\tau \equiv t$ when $z = 0$, $u = 0$)

$$\int_0^\infty (\delta z(t)^* \;\; \delta u(t)^*) \begin{pmatrix} W(t) & R(t) \\ R(t)^* & U(t) \end{pmatrix} \begin{pmatrix} \delta z(t) \\ \delta u(t) \end{pmatrix} dt \tag{6.24}$$

will be finite. It should be noted that for $\delta z(0) = 0$ we have $\delta z(t) \equiv 0$ and $\delta\tau(t)$ is constant. For $\delta z(0) \neq 0$ the indicated exponential decay of $\delta z(t)$ guarantees that the integral representation

$$\delta\tau(t) = \int_0^t \left[a(s)^* + B_\tau(s)K(s)\right] \delta z(s)\, ds + \delta\tau(0)$$

converges as $t \to \infty$ but does not necessarily remain constant; there is an asymptotic phase difference

$$\lim_{t\to\infty} \left[\delta\tau(t) - \delta\tau(0)\right] = \int_0^\infty \left[a(s)^* + B_\tau(s)\right] \delta z(s)\, ds\ . \tag{6.25}$$

This clearly does not affect the cost (6.24). We will assume that the linearized system (6.12) is stabilizable in this sense. Standard results then show that if the feedback relation

$$u(t) = K(\tau(t))\ z(t) \tag{6.26}$$

is used in the nonlinear system (6.07),(6.08) the corresponding cost of the form displayed in (6.21) will be finite, provided the initial state $\tau(0)$, $z(0)$ is sufficiently small; if this relationship is mapped back to the original system (6.01) the periodic solution $\tilde{x}(t)$ is seen to be exponentially orbitally stable.

Let us first examine the LQ optimal control problem for the linear system (6.12) and the quadratic cost (6.24). Denoting the first component of the adjoint variable $\tilde{\lambda}$ for this linear quadratic case by by $\tilde{\lambda}_\tau$ and the last n-1 components collectively by $\tilde{\lambda}_z$, that equation may be seen to take the form

$$\dot{\tilde{\lambda}}_\tau = 0\ , \tag{6.27}$$

$$\dot{\tilde{\lambda}}_z = -a(t)\tilde{\lambda}_\tau - A(t)^*\tilde{\lambda}_z - 2W(t)\delta z - 2R(t)\delta u\ . \tag{6.28}$$

If we select the solution $\tilde{\lambda}_\tau(t) \equiv 0$ of (6.27), (6.28) becomes

$$\dot{\tilde{\lambda}}_z = - A(t)^*\tilde{\lambda}_z - 2W(t)\delta z - 2R(t)\delta u . \qquad (6.29)$$

The coefficients here are T-periodic and we can see that the periodic LQ problem of minimizing (6.24) subject to $\delta z(t)$ satisfying (6.12) is basically the same LQ problem as we considered in §5, except that now $\delta z(t)$ and $\tilde{\lambda}_z(t)$ are of dimension n-1. Thus it follows that the optimal controls for this LQ problem are characterized by the relationship

$$\delta\hat{u}(t) = - \tfrac{1}{2}U(t)^{-1}\left[B_\tau(t)^*\tilde{\lambda}_\tau(t)+B_z(t)^*\tilde{\lambda}_z(t)+2R(t)^*\delta\hat{z}(t)\right], \qquad (6.30)$$

which reduces, when we specify $\tilde{\lambda}_\tau(t) \equiv 0$, to the simpler form

$$\delta\hat{u}(t)=- \tfrac{1}{2}U(t)^{-1}\left[B_z(t)^*\tilde{\lambda}_z(t)+2R(t)^*\delta\hat{z}(t)\right] \equiv \tilde{K}(t,\delta\hat{z}(t),\tilde{\lambda}_z(t)). \qquad (6.31)$$

In (6.29) and (6.30) $\delta\tau(t)$, $\delta\hat{z}(t)$, $\tilde{\lambda}_\tau(t)$ and $\tilde{\lambda}_z(t)$ satisfy the coupled system of ordinary differential equations

$$\delta\dot{\tau}(t) = \tilde{a}(t)^*\delta\hat{z}(t) - \tfrac{1}{2} B_\tau(t)U(t)^{-1}\left[B_\tau(t)^*\tilde{\lambda}_\tau(t)+B_z(t)^*\tilde{\lambda}_z(t)\right], \qquad (6.32)$$

$$\delta\dot{\hat{z}}(t) = \tilde{A}(t)\delta\hat{z}(t) - \tfrac{1}{2} B_z(t)U(t)^{-1}\left[B_\tau(t)^*\tilde{\lambda}_\tau(t)+B_z(t)^*\tilde{\lambda}_z(t)\right], \qquad (6.33)$$

$$\dot{\tilde{\lambda}}_\tau(t) \equiv 0 , \qquad (6.34)$$

$$\dot{\tilde{\lambda}}_z(t) = -\tilde{a}(t)\tilde{\lambda}_\tau(t)-\tilde{A}(t)^*\tilde{\lambda}_z(t)-\left[2W(t)-R(t)U(t)^{-1}R(t)^*\right]\delta\hat{z}(t), \qquad (6.35)$$

with

$$\tilde{A}(t) = A(t)-B_z(t)U(t)^{-1}R(t)^*, \quad \tilde{a}(t) = a(t)-R(t)U(t)^{-1}B_\tau(t)^*. \qquad (6.36)$$

Since the $\delta\hat{z}$ and $\tilde{\lambda}_z$ equations do not depend on $\tau$, when we set $\tilde{\lambda}_\tau(t)$ identically equal to zero we obtain a simpler system of equations involving only $\delta\hat{z}$ and $\tilde{\lambda}_z$. If we let Q(t) be the unique T - periodic symmetric positive definite solution of the Riccati equation (2.35), as discussed in §4, we see that with the change of variable

$$\tilde{\lambda}_z(t) = \tilde{\mu}_z(t) + 2\,Q(t)\delta\hat{z}(t) \tag{6.37}$$

we obtain the decoupled system

$$\begin{bmatrix} \delta\dot{\hat{z}} \\ \dot{\tilde{\lambda}}_z \end{bmatrix} = \begin{bmatrix} \hat{A}(t) & \hat{D}(t) \\ 0 & -\hat{A}(t)^* \end{bmatrix} \begin{bmatrix} \delta\hat{z} \\ \tilde{\lambda}_z \end{bmatrix}, \tag{6.38}$$

wherein

$$\hat{A}(t) = \tilde{A}(t) + \hat{D}(t)Q(t) = A(t) + B(t)\hat{K}(t), \tag{6.39}$$

$$\hat{K}(t) = -U(t)^{-1}\left[B(t)^*Q(t)+R(t)^*\right], \quad \hat{D}(t) = -\tfrac{1}{2}\,B_z(t)U(t)^{-1}B_z(t)^*. \tag{6.40}$$

If the transformation (6.37) is made in the complete system (6.32) - (6.35) (not assuming $\tilde{\lambda}_\tau(t) = 0$), we obtain the system of equations

$$\begin{aligned} \delta\dot{\tau} &= (\tilde{a}(t)^* + B_\tau(t)\hat{K}(t))\delta\hat{z} - \tfrac{1}{2}\,B_\tau(t)U(t)^{-1}\left[B_\tau(t)^*\tilde{\lambda}_\tau + B_z(t)^*\tilde{\mu}_z\right] \\ &\equiv \hat{a}(t)^*\delta\hat{z} + b_1(t)\tilde{\lambda}_\tau + b_2(t)^*\tilde{\mu}_z, \end{aligned} \tag{6.41}$$

$$\delta\dot{\hat{z}} = \hat{A}(t)\delta\hat{z} - \tfrac{1}{2}\,B_z(t)U(t)^{-1}B_\tau(t)^*\tilde{\lambda}_\tau + \hat{D}(t)\tilde{\mu}_z \equiv \hat{A}(t)\delta\hat{z} + \tilde{d}(t)\tilde{\lambda}_\tau + \hat{D}(t)\tilde{\mu}_z \tag{6.42}$$

$$\dot{\tilde{\lambda}}_\tau = 0, \tag{6.43}$$

$$\dot{\tilde{\mu}}_z = -\hat{a}(t)\tilde{\lambda}_\tau - \hat{A}(t)^*\tilde{\mu}_z, \tag{6.44}$$

wherein all time-varying coefficient matrices shown are T-periodic. Finally, setting

$$\delta\tau(t) = \delta\sigma(t) + q(t)^*\delta\hat{z}(t), \tag{6.45}$$

$$\tilde{\mu}_z(t) = \tilde{\nu}_z(t) + p(t)\tilde{\lambda}_\tau(t), \tag{6.46}$$

where $q(t)^*$ and $p(t)$ are the unique T-periodic solutions of the (n-1) al row and column vector differential equations, respectively,

$$\dot{q}(t)^* = \hat{a}(t)^* - q(t)^*\hat{A}(t),$$

$$\dot{p}(t) = \hat{a}(t) + \hat{A}(t)^* p(t),$$

whose existence is assured because all of the Floquet multipliers associated with $-\hat{A}(t)^*$ have modulus greater than unity and all of those associated with $\hat{A}(t)^*$ have modulus less than unity, we arrive at a system, which we show in matrix form,

$$\begin{bmatrix} \delta\dot{\sigma} \\ \delta\dot{\hat{z}} \\ \dot{\tilde{\lambda}}_\tau \\ \dot{\tilde{\nu}}_z \end{bmatrix} = \begin{bmatrix} 0 & 0 & B_{\sigma\lambda}(t) & B_{\sigma\nu}(t) \\ 0 & \hat{A}(t) & B_{z\lambda}(t) & B_{z\nu}(t) \\ 0 & 0 & 0 & 0 \\ 0 & 0 & 0 & -\hat{A}(t)^* \end{bmatrix} \begin{bmatrix} \delta\sigma \\ \delta\hat{z} \\ \tilde{\lambda}_\tau \\ \tilde{\nu}_z \end{bmatrix}, \tag{6.47}$$

wherein all of the entries of the matrix shown are T-periodic in t. From this it is clear that the stable subspace for (6.47) is given by

$$\delta\sigma(t) \equiv 0, \text{ i.e., } \delta\tau(t) \equiv q(t)^*\delta\hat{z}(t), \tag{6.48}$$

$$\tilde{\lambda}_\tau(t) \equiv 0,\ \tilde{\nu}_z(t) \equiv 0 \rightarrow \tilde{\lambda}_\tau(t) \equiv 0,\ \tilde{\mu}_z(t) \equiv 0 . \tag{6.49}$$

Since $q(t)$ is T-periodic, for $t = kT$, $k = 0,1,2,\ldots,$, (6.48) gives

$$\delta\tau(kT) = q(kT)^*\delta\hat{z}(kT) = q(0)^*\delta\hat{z}(kT) .$$

For $k = 0$ this describes the relationship which must hold between $\delta\tau(0)$ and $\delta\hat{z}(0)$ in order for us to have $\lim_{t\to\infty} \delta\tau(t) = 0$; if this relationship does not hold we will have (cf.(6.25)) the more general relationship

$$\lim_{t\to\infty} \delta\tau(t) = \delta\tau(0) + \int_0^\infty \hat{a}(s)^* \delta\hat{z}(s)\, ds .$$

From the form of (6.47) we see that if we let

$$\delta\sigma_k = \delta\sigma(kT),\ \delta\hat{z}_k = \delta\hat{z}(kT),\ \tilde{\lambda}_{\tau,k} = \tilde{\lambda}_\tau(kT),\ \tilde{\nu}_{z,k} = \tilde{\nu}_z(kT),\ k = 0,1,2,\ldots,$$

then with

$$\mathbb{A} = \hat{\Phi}(T)$$

the value at T of the fundamental solution matrix for the T-periodic linear homogeneous equation

$$\dot{w} = \hat{A}(t)w$$

reducing to the identity at $t = 0$, we have a discrete linear recursion equation of the form

$$\begin{bmatrix} \delta\sigma_{k+1} \\ \delta\hat{z}_{k+1} \\ \tilde{\lambda}_{\tau,k+1} \\ \tilde{\nu}_{z,k+1} \end{bmatrix} = \begin{bmatrix} 1 & 0 & \Phi_{\sigma,\lambda} & \Phi_{\sigma,\nu} \\ 0 & \mathbb{A} & \hat{\Phi}_{z,\lambda} & \Phi_{z,\nu} \\ 0 & 0 & 1 & 0 \\ 0 & 0 & 0 & (\mathbb{A}^*)^{-1} \end{bmatrix} \begin{bmatrix} \delta\sigma_k \\ \delta\hat{z}_k \\ \tilde{\lambda}_{\tau,k} \\ \tilde{\nu}_{z,k} \end{bmatrix}, \tag{6.50}$$

for which the stable manifold is given by the identities (6.48) and (6.49), appropriately restated for the discrete situation. The optimal cost for an initial state $\delta\tau_0$, $\delta z_0$ given at $t = 0$ is given by

$$\delta z_0^* \, P \, \delta z_0 , \tag{6.51}$$

with the $n-1 \times n-1$ matrix P satisfying an equation of the form (4.30) with $\hat{A}$ in that formula replaced by $\mathbb{A}$ appearing here. Moreover, the matrix P agrees with the value Q(0) of the T - periodic matrix function Q(t) in (6.37).

Now if we set up the NLNQ problem for the original nonlinear system (6.01) and cost function

$$\int_0^\infty G(x(t),u(t))\,dt\ ,$$

equivalently for the system (6.07),(6.08) with cost functional

$$\int_0^\infty \Gamma(\tau(t),z(t),u(t))\,dt\ , \tag{6.52}$$

with $\Gamma(\tau,z,u)$ given by (6.18), and if we subsequently carry out the same transformations as we did in the linear case except that (6.45) is replaced, in view of the constant term in (6.07), by

$$\tau(t) = t + \sigma(t) + q(t)^*\hat{z}(t)\ , \tag{6.53}$$

we obtain necessary conditions for optimality in the form of combined state and adjoint equations analogous to (2.19),(2.20) and a formula for the optimal control in the form

$$\hat{u}(t) = \mathfrak{K}(\tau(t),\hat{z}(t),\lambda_\tau(t),\lambda_z(t)) =$$

$$\hat{\mathfrak{K}}(\sigma(t),\hat{z}(t),\lambda_\tau(t),\nu_z(t)) = \hat{K}(t)\hat{z}(t)+\ell(\sigma(t),\hat{z}(t),\lambda_\tau(t),\nu_z(t)), \tag{6.54}$$

the last two identities following from the transformations indicated above for the LQ case case. In the last of these expressions the function $\ell$ and its first order partial derivatives vanish at (0,0,0,0). If this relationship is substituted into the combined state/adjoint system and if we then construct the corresponding period map, we obtain a discrete system of the form

$$\begin{bmatrix} \sigma_{k+1} \\ \hat{z}_{k+1} \\ \lambda_{\tau,k+1} \\ \nu_{z,k+1} \end{bmatrix} = \begin{bmatrix} 1 & 0 & \Phi_{\sigma,\lambda} & \Phi_{\sigma,\nu} \\ 0 & \mathbb{A} & \Phi_{z,\lambda} & \Phi_{z,\nu} \\ 0 & 0 & 1 & 0 \\ 0 & 0 & 0 & (\mathbb{A}^*)^{-1} \end{bmatrix} \begin{bmatrix} \sigma_k \\ \hat{z}_k \\ \lambda_{\tau,k} \\ \nu_{z,k} \end{bmatrix} + \begin{bmatrix} \varphi_\sigma(\sigma_k,\hat{z}_k,\lambda_{\tau,k},\nu_{z,k}) \\ \varphi_z(\sigma_k,\hat{z}_k,\lambda_{\tau,k},\nu_{z,k}) \\ \varphi_\lambda(\sigma_k,\hat{z}_k,\lambda_{\tau,k},\nu_{z,k}) \\ \varphi_\nu(\sigma_k,\hat{z}_k,\lambda_{\tau,k},\nu_{z,k}) \end{bmatrix}, \tag{6.55}$$

the functions appearing as components of the vector at the extreme right reducing to zero and having zero first order partial derivatives with respect to all of their arguments at (0,0,0,0).

The next task is to obtain the stable manifold for this system.

To this end we interchange the first two equations in (6.55), collectively designate the other variables by w, i.e., $w^* = (\sigma, \lambda_T, \nu^*) \in R^{n+1}$, so that we have a system of the form in $R^{n-1} \times R^{n+1}$,

$$\hat{z}_{k+1} = \mathbb{A}\, \hat{z}_k + \hat{\Phi}\, w_k + \hat{\varphi}_z(\hat{z}_k, w_k) \ , \tag{6.56}$$

$$w_{k+1} = \tilde{\mathbb{A}}\, w_k + \hat{\varphi}_w(\hat{z}_k, w_k) \ , \tag{6.57}$$

wherein the definitions of $\hat{\Phi}$, $\hat{\varphi}_z$, $\hat{\varphi}_w$ should be clear and

$$\tilde{\mathbb{A}} = \begin{pmatrix} 1 & \Phi_{\sigma,\lambda} & \Phi_{\sigma,\nu} \\ 0 & 1 & 0 \\ 0 & 0 & (\mathbb{A}^*)^{-1} \end{pmatrix} . \tag{6.58}$$

We want to find a continuously differentiable functional relationship

$$w = \mathcal{W}(z), \ \mathcal{W}(0) = 0,$$

so that if substitute $w = \mathcal{W}(z)$ into (6.56) to obtain

$$\hat{z}_{k+1} = \mathbb{A}\, \hat{z}_k + \hat{\Phi}_w\, \mathcal{W}(z_k) + \hat{\varphi}_z(\hat{z}_k, \mathcal{W}(z_k)) \ ,$$

then $\hat{z}_k$, $w_k = \mathcal{W}(\hat{z}_k)$ will be a solution of the coupled system (6.56), (6.57). Just as in §4, this is true just in case

$$\tilde{\mathbb{A}}\, \mathcal{W}(\hat{z}_k) + \varphi_w(\hat{z}_k, \mathcal{W}(\hat{z}_k)) = w_{k+1} = \mathcal{W}(\hat{z}_{k+1})$$

$$= \mathcal{W}\left[\mathbb{A}\, \hat{z}_k + \hat{\Phi}\, \mathcal{W}(\hat{z}_k) + \varphi_z(\hat{z}_k, \mathcal{W}(\hat{z}_k))\right] .$$

Multiplying on the left by $\hat{\mathbb{A}} = \tilde{\mathbb{A}}^{-1}$ we see, as in §4, that $\mathcal{W}(z)$ should obey the functional equation

$$\mathcal{W}(z) = \hat{\mathbb{A}}\, \mathcal{W}\left[\mathbb{A}\, z + \hat{\Phi}\, \mathcal{W}(z) + \varphi_z(z, \mathcal{W}(z))\right] - \hat{\mathbb{A}}\, \varphi_w(z, \mathcal{W}(z)) \ . \tag{6.59}$$

By making transformations as in §4, which we again do not carry out explicitly, we can assume $\mathbb{A}$ and $\mathbb{A}^*$ to be strict contractions. Application of a linear "shearing" transformation permits us to suppose $\|\Phi_{\sigma,\lambda}\|$ and $\|\Phi_{\sigma,\nu}\|$ in (6.59) to be as small as we please. Taking account of the "1"'s in (6.58), we see that we may suppose

$$\|\mathbb{A}\| = \|\mathbb{A}^*\| < 1, \ \|\hat{\mathbb{A}}\| \leq 1 + \delta, \tag{6.60}$$

where $\delta > 0$ is as small as desired. The first of these inequalities enables us to determine an invariant region $\mathfrak{Z} \subset R^{n-1}$ for the expres- $\mathbb{A}\, z + \hat{\Phi}\, \mathcal{W}(z) + \varphi_z(z,\mathcal{W}(z))$ when $\mathcal{W} \in C^1_{0,\varepsilon}$, as described earlier in §4. Because the second inequality in (6.58) does not express a strict contraction property, however, we cannot be quite as cavalier as we were in §4 in regard to the contraction property in the metric space $C^1_{0,\varepsilon}$ for $\varepsilon$ sufficiently small; we elaborate on this a little bit.

It will be convenient to use, in the space $C^1_0$, the norm

$$\|\mathcal{W}\|_{C^1_0} = \sup_{z\in\mathfrak{Z}} \left\| \frac{\partial \mathcal{W}}{\partial z} \right\| ,$$

relying on the Poincare' inequality ([Poi]) and the requirement $\mathcal{W}(0) = 0$ in the definition of $C^1_0$, to bound the values of $\mathcal{W}(z)$ itself. If we then have two functions $\mathcal{W}(z)$ and $\tilde{\mathcal{W}}(z)$ in $C^1_0$ we can estimate

$$\left\|\hat{\mathbb{A}} \left[ \frac{\partial}{\partial z}(\mathcal{W}(\mathbb{A}z)) - \frac{\partial}{\partial z}(\tilde{\mathcal{W}}(\mathbb{A}z)) \right]\right\| = \left\|\hat{\mathbb{A}} \left[\frac{\partial \mathcal{W}}{\partial z}(\mathbb{A}z) - \frac{\partial \tilde{\mathcal{W}}}{\partial z}(\mathbb{A}z)\right]\mathbb{A} \right\|$$

$$\leq \|\hat{\mathbb{A}}\|\|\mathbb{A}\| \left\| \frac{\partial \mathcal{W}}{\partial z}(\mathbb{A}z) - \frac{\partial \tilde{\mathcal{W}}}{\partial z}(\mathbb{A}z) \right\| \leq d \left\| \frac{\partial \mathcal{W}}{\partial z}(\mathbb{A}z) - \frac{\partial \tilde{\mathcal{W}}}{\partial z}(\mathbb{A}z) \right\|$$

with $0 < d < 1$, using (6.58) with $\delta$ sufficiently small. Then, using the remark immediately following (6.53), and increasing d slightly to some value $\tilde{d}$, still less than 1, if necessary, we can obtain the required contraction property, as described in §4, in $C^1_{0,\varepsilon}$ for $\varepsilon$ sufficiently small. The contraction fixed point theorem is applied and we obtain the specified continuously differentiable $\mathcal{W}(z)$.

Once we have $w = \hat{\mathcal{W}}(z)$, which we may further detail as

$$\sigma = \mathcal{W}_\sigma(\hat{z}),\ \lambda_\tau = \mathcal{W}_\lambda(\hat{z}),\ \nu_z = \mathcal{W}_\nu(\hat{z}), \tag{6.61}$$

the stable manifold for the discrete system (6.55) is seen to consist of initial states

$$\sigma_0 = \mathcal{W}_\sigma(z_0),\ \lambda_{\tau,0} = \mathcal{W}_{\lambda,0}(z_0),\ \nu_{z,0} = \mathcal{W}_\nu(z_0).$$

This manifold is invariant under (6.55), so we also have

$$\sigma_k = \mathcal{W}_\sigma(\hat{z}_k),\ \lambda_{\tau,k} = \mathcal{W}_{\lambda,k}(\hat{z}_k),\ \nu_{z,k} = \mathcal{W}_\nu(\hat{z}_k),\ k = 1,2,\ldots\ . \tag{6.62}$$

Moreover, the optimal cost, for $\sigma_0$, $z_0$ related by the first equation in (6.61), takes the form of a twice continuously differentiable function of z of the form (cf.(6.51))

$$\hat{\mathcal{J}}(z_0) = z_0{}^*P\, z_0 + \mathcal{P}(z_0),\ \mathcal{P}(z) = o(\|z\|^2),\ \|z\| \to 0 \text{ in } R^{n-1}.$$

Now we can return to the continuous formulation again. Here we already know that the optimal control is given, in terms of $\sigma$, $\hat{z}$, $\lambda_\tau$ and $\nu_z$ by the last formula in (6.54). Proceeding as in §2 and §5, for $t \in [0,T]$ we can find a relationship of the same form as (6.62), which we now write as

$$\sigma(t) = \mathcal{W}_\sigma(t,\hat{z}(t),\ \lambda_\tau(t) = \mathcal{W}_\lambda(t,\hat{z}(t)),\ \nu_z(t) = \mathcal{W}_\nu(t,\hat{z}(t))\ , \tag{6.63}$$

such that solutions of the combined state/adjoint system initiating on the manifold described by these equations lie on the manifold (6.61) for t = T. When these relationships are substituted into the formula (6.54) for the optimal control, we obtain a relationship

$$\hat{u}(t)) = \hat{K}(t)\hat{z}(t)+\ell(\mathcal{W}_\sigma(t,\hat{z}(t),\hat{z}(t),\mathcal{W}_\lambda(t,\hat{z}(t),\mathcal{W}_\nu(t,\hat{z}(t)))\ .$$

Then, taking the relationship (6.53) into account, as is also done in (6.54), we obtain a continuously differentiable time varying feedback law for the optimal control in the form

$$\hat{u}(t) = \hat{K}(t)\hat{z}(t)+\mathfrak{L}(t,\tau(t),\hat{z}(t)),\ \mathfrak{L}(t,\tau,z) = o(|\tau|+\|z\|),\ |\tau|+\|z\| \to 0, \tag{6.64}$$

valid for solutions lying for all t on the manifolds described by (6.63) and its expression, via (6.53), in terms of $\tau$ and z. As these manifolds cover a neighborhood of the $\tau$ axis as $\tau$ evolves through its values, all values of $\tau$ and z are covered with the additional feature that the feedback law in question guarantees that the optimally controlled solutions will have the property that

$$\lim_{t\to\infty} |\tau(t) - t| = 0,$$

rather than there being some asymptotic phase difference between the two.

Now a very natural question arises: since the original system (6.01) is autonomous, the optimal control law should not, in fact, explicitly depend on t as we have indicated in (6.64). This matter is quite easily resolved. From (6.53), taking the first equation of (6.63) into account, we obtain the identity

$$\mathfrak{F}(t,\tau,\hat{z}) \equiv \tau - t - \mathcal{W}_\sigma(t,\hat{z}) - q(t)^*\hat{z} \equiv 0 .$$

Evidently

$$\frac{\partial\mathfrak{F}}{\partial t}(t,\tau,0) \equiv -1 \longrightarrow \frac{\partial\mathfrak{F}}{\partial t}(\tau,\tau,0) \equiv -1 .$$

Since, from our earlier development, we have $\tau(t) \equiv t$ when $z = 0$, application of the implicit function theorem shows that there is a continuously differentiable function $t = \mathfrak{T}(\tau,z)$ with

$$\mathfrak{T}(\tau,0) \equiv \tau,\ \frac{\partial\mathfrak{T}}{\partial z}(\tau,0) = -\ q(\tau)^*,$$

such that

$$\mathfrak{F}(\mathfrak{T}(\tau,\hat{z},\tau,\hat{z}) \equiv 0 ;$$

from periodicity we can see that this is true for $\hat{z}$ sufficiently small, uniformly with respect to $\tau$. Then (6.64) yields

$$\hat{u}(t) = \hat{u}(\tau,\hat{z}) = \hat{K}(\mathfrak{T}(\tau,\hat{z}))\hat{z} + \mathfrak{L}(\mathfrak{T}(\tau,z),\tau,\hat{z})$$

$$= \left[\hat{K}(\tau)-\hat{K}(\tau)q(\tau)^*\right]\hat{z} + \tilde{\mathfrak{L}}(\tau,\hat{z}) \equiv \hat{K}_1(\tau)\hat{z} + \tilde{\mathfrak{L}}(\tau,\hat{z}) , \qquad (6.65)$$

with $\tilde{\mathfrak{L}}$ and its first order partial derivatives vanishing identically when $\hat{z} = 0$. Taking (6.06) into account we obtain, in terms of the original x coordinates, an optimal control law of the form

$$\hat{u}(t) = \hat{u}(\hat{x}) = \mathfrak{K}(\hat{x}), \qquad (6.66)$$

expressing the optimal control in terms of the optimally controlled solution. We can summarize all of this in

Theorem 6.1 *The optimal control problem for the system* (6.01) *with cost integrand described by* (6.13) - (6.17) *has the* (*unique*) *solution expressed by* (6.66) *in a neighborhood of the stabilizable periodic orbit* $\mathcal{O}(\tilde{x})$. *The nonlinear optimal feedback function,* $\mathfrak{K}(x)$, *is identically equal to zero on* $\mathcal{O}(\tilde{x})$ *and is continuously differentiable in a nieghborhood of* $\mathcal{O}(\tilde{x})$ *with first order behavior in directions orthogonal to* $\tilde{x}$ *being described by the matrix* $\hat{K}_1$ *of* (6.65). *The synthesized system*

$$\dot{x} = F(x,\mathfrak{K}(x)) \qquad (6.67)$$

*obtained with the use of this control has* $\mathcal{O}(\tilde{x})$ *as an exponentially orbitally stable invariant set, still a* T - *periodic solution of the system* (6.67).

The claimed uniqueness is not proved here. That proof, along with other desirable features such as sufficiency of the necessary conditions, etc., follows much the same lines as originally given by Lukes in [ 16 ]; it will appear in an extended version of this work to appear separately, which will include more detailed development of the proof process which we have only sketched here.

We believe that this result can be extended to other types of compact invariant sets for an autonomous system as described here. Nevertheless we should be frank in admitting that, at this writing, it is not at all clear just how this result could be proved. The present proof for periodic orbits is very much dependent on use of the autonomous discrete formulation made possible by the periodic structure of the flow on $\mathcal{O}(\tilde{x})$. For a more general invariant manifold, such as an invariant torus with non-periodic flow, this type of approach appears to have no counterpart. It may be that the first clues as to how to approach this general problem may be obtained by devising a proof of the present result which does not rely on the discrete formulation of the problem in any way.

## References


[ 1 ] Anderson, B.D.O., and J.B. Moore: *Linear Optimal Control*, Prentice-Hall, Inc., Englewood Cliffs, NJ, 1971

[ 2 ] Aoki, M. *Optimal Control and System Theory in Dynamic Economic Analysis*, North-Holland Pub. Co., New York, Oxford, Amsterdam, 1976

[ 3 ] Chow, G. *Analysis and Control of Dynamic Economic Systems*, Wiley Series in Probability, John Wiley & Sons, Inc., New York, London, Sydney, Toronto, 1975

[ 4 ] Coddington, E.A., and N. Levinson: *Theory of Ordinary Differential Equations*, McGraw-Hill Book Co., New York, 1955

[ 5 ] Cronin, J. *Differential Equations: Introduction and Qualitative Theory*, Vol. 54, Pure and Applied Mathematics, Marcel Dekker, Inc., New York, 1980

[ 6 ] Garabedian, P.R.: *Partial Differential Equations*, John Wiley & Sons, Inc., New York, 1964

[ 7 ] Guardabassi, G., A. Locatelli and S. Rinaldi: *Status of periodic optimization of dynamical systems*, J. Optim. Theory Appl., 14 (1974), pp. 1 - 20

[ 8 ] Kalman, R.E.: *Contributions to the theory of optimal control*, Bol. Soc. Mat. Mexicana, 5 (1960), pp. 102 - 119

[ 9 ] Kalman, R.E., and R.S. Bucy: *New results in linear prediction and filtering theory*, J. Basic Engr. (Trans. ASME, Ser. D) 83D (1961), pp. 95 - 100

[ 10 ] Kalman, R.E., P.L. Falb and M.A. Arbib: *Topics in Mathematical System Theory*, McGraw-Hill Book Co., New York, 1969

[ 11 ] Lee, E.B., and L.W. Markus: *Foundations of Optimal Control Theory*, John Wiley & Sons, Inc., 1967

[ 12 ] Leipholz, H.: *Stability Theory*, Academic Press, New York, 1970

[ 13 ] Leitmann, G.: *Introduction to the Calculus of Variations and*

*Optimal Control,* Plenum Press, New York

[ 14 ] Lions, J.L.: *Optimal Control of Systems Governed by Partial Differential Equations* (Translated by S.K. Mitter), Springer-Verlag, New York, Heidelberg, Berlin, 1971

[ 15 ] Lukes, D.L.: *Optimal regulation of nonlinear dynamical systems,* SIAM J. Control 7 (1969), pp. 75 - 100

[ 16 ] Lukes, D.L.: *Stabilizability and optimal control,* Funkcialaj Ekvacioj, 11 (1968), pp. 39 - 50

[ 17 ] Noldus, E.: *A survey of optimal periodic control of continuous systems,* (ASME) Journal A, 16 (1975), pp. 11 - 16

[ 18 ] Pontryagin, L.S., V.G. Boltyanskii, R.V. Gamkrelidze and E.F. Mishchenko: *The Mathematical Theory of Optimal Processes* (Trans. by K.N. Trirogoff, Ed. by L.W. Neustadt), Interscience Pub., John Wiley & Sons, Inc., New York and London, 1962

[ 19 ] Russell, D.L.: *Mathematics of Finite Dimensional Control Systems,* Marcel Dekker, Inc., New York, 1979

[ 20 ] Zhang,X.: Thesis, Department of Mathematics, Virginia Polytechnic Institute and State University, 1993

# 51 Existence of Optimal Controls for a Free Boundary Problem

**Thomas I. Seidman** University of Maryland, Baltimore County, Baltimore, Maryland

ABSTRACT: We consider a model for the radially symmetric evolution (by accretion or dissolution) of the interface between a 'pure' solid grain and a liquid solution of the same substance with diffusion both of material and heat. After analyzing the direct problem, we show existence of optimal controls.

## 1. Introduction

We take up a 'promise' from [4] to consider the controlled evolution (growth/dissolution) of a grain in a diffusive medium with temperature coupling, using a heat source as control.

Some of the technical difficulties addressed in [4], [3], [5] were due to the necessary consideration of possibly rough boundary data for the concentration as control which now become unnecessary; compare [1]. On the other hand, the setting now becomes a system of partial differential equations whose coupling involves new technical difficulties. We also address an implicit consideration of controlability to verify that there may be *any* admissible control for the optimization since the existence theory for the direct problem need not ensure that the grain 'survives' throughout the relevant time interval. We defer any characterization of optimal controls in terms of first order necessary conditions for optimality but anticipate that they would be rather comparable to those obtained in [3].

We continue to consider the problem only in the simplified context of radial symmetry. We are considering the problem in a fixed region $\Omega = \{|x| < L\}$ where we have scaled so $L = 1$; the nature of the 'outer' boundary condition at $r = L$ is somewhat arbitrary and the problem could alternatively have been considered in an unbounded region $L = \infty$ as in [2].

The physical setting for the problem begins with consideration of diffusion of some substance in a liquid whose concentration $C$ satisfies

$$(1.1) \qquad \dot{C} = \Delta C = r^{1-d}\left(r^{d-1}C_r\right)_r - F(C,\Theta) \qquad \text{on } \boldsymbol{Q}$$

where, with $\mathcal{I} := (0,T)$,

$$(1.2) \qquad \boldsymbol{Q} := \{(t,x) \in \mathcal{I} \times \Omega : R(t) < r := |x|\} \quad \Omega := \{x \in \mathbb{R}^d : |x| < 1\}.$$

Here $R(t) > 0$ is the radius of a solid grain[1] of the substance and at this solid/liquid interface conservation gives

(1.3) $$C_r = \dot{R}[1 - C] \qquad \text{at } r = R(t)+$$

where we have normalized the 'pure' concentration in the grain to 1, just as in (1.1) we chose units to make the diffusion coefficient 1. Note that the concentration $C$ must physically take values in $[0, 1]$. Note also that (1.3) only applies when $R(t) > 0$, as otherwise the grain has (permanently) 'disappeared' and we would shift to using (1.1) on all of $\Omega$.

This problem is a *free boundary problem* because the time-varying grain radius $R(\cdot)$, defining the boundary of $\boldsymbol{Q}$, is to be determined as part of the problem. We have an ordinary differential equation

(1.4) $$\dot{R} = h(t) \qquad \text{on } \mathcal{I}$$

— with the constraint that

(1.5) $$R(t_*) = 0 \qquad \Rightarrow \qquad R(t) \equiv 0 \text{ for } t_* \leq t \leq T$$

since, once the grain is completely dissolved, we have no provision for its resurrection — and otherwise (when $R > 0$) we assume for growth/dissolution a constitutive law of the form

(1.6) $$h := K(\vartheta)[\omega - G(\kappa, \vartheta)] =: H(\kappa, \omega, \vartheta)$$

where $\kappa := 1/R$ is the *curvature* of the interface, $\omega(t) := C(t, r)\big|_{r=R(t)+}$ is the local concentration (in the liquid next to the grain boundary), and $\vartheta(t) := \Theta(t, R(t))$ is the temperature there. Here $K$ is a 'reaction rate' coefficient and $G$ is the 'critical concentration', determining whether one has accretion (growth of the grain if $C > G$ at the interface) or dissolution of grain material into the liquid (if $C < G$). Of course, $0 < K$ and $0 < G < 1$. The curvature dependence in $G$ comes from a surface component in the Gibbs Free Energy (surface tension) so, plausibly, $G$ increases with (positive) curvature — i.e., decreases with increasing $R(t)$. We model the surface reaction of adsorption as endothermic, making this 'easier' for higher temperatures, so plausibly $G$ decreases (and $K$ increases) with increasing $\vartheta$. In the same spirit, we consider an endothermic reaction in the liquid so $F(C, \Theta)$ would plausibly be an increasing function of each of its arguments

For the temperature we have

(1.7) $$\dot{\Theta} = \nabla \cdot D\nabla\Theta - \alpha_0 F(C, \Theta) + S \qquad \text{for } 0 \leq r < 1 \text{ with } r \neq R(t)$$

where we permit different diffusion coefficients within the grain ($D = D_1$) and in the liquid ($D = D_0$). The term $\alpha_0 F$ represents the heat absorbed by the reaction included in (1.1). It is precisely the exogenous source term $S$ in (1.7) which will involve the (scalar) control function $u(\cdot)$: we model[2] this as

(1.8) $$S = u(t)\beta \qquad \text{with } \beta = \beta(t, r) := \begin{cases} \beta_1(r) & \text{for } 0 \leq r < R(t) \\ \beta_0(r) & \text{for } R(t) < r < 1 \end{cases}.$$

---

[1]Physically, $d = 2$ or $3$ with $d = 2$ corresponding, e.g., to a thin surface film. The problem as described will actually be a physically somewhat implausible 'mixture' since we take $d = 2$ for descriptive convenience and adopt a heat source model somewhat appropriate to a planar problem whereas we continue to speak of a 'grain', which is more appropriate to $d = 3$.

[2]This might, for example, correspond to the effect of a heatlamp with adjustable intensity $u(t)$ whose radiation has a radial spatial pattern $\beta_*(r)$ with possibly different absorption coefficients $\tilde{\beta}_j$ for the two phases so $\beta = \beta_*(r)\tilde{\beta}_j$ with $j = 0, 1$ for the liquid *vs.* solid phases. Alternatively, one could consider $\tilde{\beta} = \tilde{\beta}(C)$, so $\tilde{\beta}_1 = \tilde{\beta}(1)$ and $\tilde{\beta}_0 \approx \tilde{\beta}(C)$ for $C << 1$, just as one might take $D = D(C)$ in (1.7).

Since (1.7) holds separately in the two phases, we must balance the heat transport mechanisms across $r = R$; we simplify this by assuming the heat content is the same in both phases so material transport is temperature–neutral. If we assume that the heat absorbed by the surface reaction is simply proportional to the reaction, we adjoin to (1.7) the *jump condition*

$$D\Theta_r|_{r=R(t)-}^{r=R(t)+} =: [D\nabla\Theta] = \alpha\dot{R} \tag{1.9}$$

with the constant $\alpha$ positive, corresponding to the physical assumption that adsorption is endothermic. Formally, we may write (1.7)+(1.9) as

$$\Theta_t = r^{-1}(rD\Theta_r)_r - \alpha_0 F + u\beta - \alpha h\delta_R \tag{1.10}$$

where $h = \dot{R}$ as in (1.4) and $\delta_R = \delta(r - R(t))$ — a $\delta$-function with respect to the variable $r$ (using the measure $r\,dr$ to correspond to integration over $\Omega$).

The 'direct problem' consists of the coupled system of differential equations (1.4), (1.1)+(1.3), (1.10), together with boundary conditions at $r = 1$ and initial conditions. The 'control problem' then comes from the attempt to specify $u(\cdot)$ optimally in (1.8) so as to minimize a *cost functional* $\mathcal{J}$. Our objectives in this paper are to verify solvability for the direct problem for 'general' $u(\cdot)$ and then to demonstrate existence of an optimal control.

## 2. Formulation

We begin by stating the relevant hypotheses for the constitutive functions $F$ in (1.1), (1.7) and $H = K[C - G]$ in (1.6):

**(H-1)** $K = K(\vartheta)$ is positive, increasing, and continuous on $\mathbb{R}$ with at most linear growth as $\vartheta \to \infty$

**(H-2)** $G = G(\kappa, \vartheta)$ is continuous with values in $(0, 1)$

**(H-3)** $G(1, \gamma_1) > \gamma_0$

**(H-4)** $\vartheta G(\kappa, \vartheta)$ is bounded as $\vartheta \to \infty$, uniformly on bounded $\kappa$-intervals

**(H-5)** we extend $H$ outside the physical range[3] by setting

$$H(\kappa, C, \vartheta) := \begin{cases} H(\kappa, 0, \vartheta) & \text{if } C < 0, \qquad H(\kappa, 1, \vartheta) \quad \text{if } C > 1 \\ H(1, C, \vartheta) & \text{if } \kappa < 1. \end{cases}$$

**(H-6)** $F = F(C, \Theta)$ is uniformly Lipschitzian and is increasing in each argument, extended so

$$F(C, \Theta) = F(0, \Theta) = 0 \text{ if } C < 0, \qquad F(C, \Theta) = F(1, \Theta) \text{ if } C > 1.$$

As in [1], etc., we extend (1.3) outside the physical range as

$$C_r = \begin{cases} \dot{R}[1 - C] & \text{if } C \le 1 \\ \dot{R}[1 - C]/2 & \text{if } C \ge 1 \end{cases} \qquad \text{at } r = R(t)+ \tag{2.1}$$

[3] I.e., $0 \le C < 1$ and $0 < R < 1$ so $\kappa \ge 1$ — noting that $R = 0$ would mean there is no grain at all.

and then make a change of variables to the fixed domain $(t,y) \in \mathcal{Q} := \mathcal{I} \times (0,1)$ with

$$y = \rho(1-r), \qquad r := |x|, \qquad \rho := \frac{1}{1-R(t)}. \tag{2.2}$$

In this setting — abusing notation slightly more by letting $C$ denote not only $C(t,x)$ and $C(t,r)$, as earlier, but also $C(t,y)$ — (1.1), with boundary conditions and initial conditions, then becomes

$$\begin{array}{rcl} C_t = \rho^2 C_{yy} - \psi C_y - F(C,\Theta) & & \text{on } \mathcal{Q} \\ C(\cdot,0) & = & \gamma_0 \\ -\rho C_y(\cdot,1) & = & \begin{cases} h(\cdot)[1-C] & \text{if } C \le 1 \\ h(\cdot)[1-C]/2 & \text{if } C \ge 1 \end{cases} \end{array} \Bigg\} \quad \text{on } \mathcal{I};\ C = C(\cdot,1) \tag{2.3}$$

with

$$\psi(t,y) := \rho(t)\left[\frac{(d-1)\rho(t)}{\rho(t)-y} + yh(t)\right]. \tag{2.4}$$

The function $h(\cdot)$ is, for the moment, a specified function of $t$ as in (1.4) — only later to be determined as in (1.6). The weak formulation of (2.3) is standard, taking test functions in

$$\begin{array}{rcl} \boldsymbol{\mathcal{U}} & := & L^2(\mathcal{I} \to \mathcal{U}) \cap H^1(\mathcal{I} \to \mathcal{U}^*) \text{ with} \\ \mathcal{U} & := & \{w \in H^1(0,1) : w(0) = 0\}; \end{array} \tag{2.5}$$

requiring that $[C(t) - \gamma_0] \in \boldsymbol{\mathcal{U}}$ imposes the Dirichlet boundary condition at $y = 0$.

The equation (1.7) for $\Theta$ is already given on a fixed domain $\hat{\mathcal{Q}} := \mathcal{I} \times \Omega$ and, noting that this is purely radial, we again abuse notation slightly to obtain the weak formulation of (1.10):

$$\int_0^1 [\eta\dot{\Theta} + \eta_r\Theta_r]\, r\, dr = \int_0^1 \eta[\alpha_0 F + S]\, r\, dr - \alpha h(t)\eta(t, R(t)) \tag{2.6}$$

for test functions $\eta = \eta(t,r) \in \boldsymbol{\mathcal{V}} := L^2(\mathcal{I} \to \mathcal{V}) \cap H^1(\mathcal{I} \to \mathcal{V}^*)$ where $\mathcal{V}$ is the space of functions on $[0,1)$ corresponding to consideration of the purely radial functions in $H_0^1(\Omega)$. The Dirichlet boundary condition

$$\Theta = \gamma_1 \qquad \text{at } r = 1- \tag{2.7}$$

is imposed by requiring that $[\Theta(t,\cdot) - \gamma_1] \in \mathcal{V}$.

Finally, we will impose the initial conditions

$$\left.\begin{array}{l} C(0,\cdot) = C_0(\cdot) \\ \Theta(0,\cdot) = \Theta_0(\cdot) \end{array}\right\} \text{on } \Omega \qquad R(0) = R_0, \tag{2.8}$$

subject to the requirements that $C_0$ and $\Theta_0$ are in $L^2(\Omega)$ and that $0 < C_0 < 1$, $0 < R_0 < 1$.

## 3. Existence

The existence and compactness arguments for the direct problem — i.e., given a fixed 'control function' $u$ in (1.8) for which we assume

$$\int_0^t u(s)\, ds = U(t) < \infty \qquad \text{with } U \text{ increasing } (u \ge 0) \tag{3.1}$$

— are similar enough to the arguments of [1], etc., that it suffices merely to sketch them, with particular attention to the modification due to the coupling with (1.10). As there, we initially obtain a semi-local existence result: on an interval $\mathcal{I}_\tau := [0, \tau]$. We note one difference: that *a priori* we no longer can expect bounded $h(\cdot)$ as in [1], etc., but instead will only have $h(\cdot) \in L^2(\mathcal{I}_\tau)$. We will be using a fixpoint argument for existence and a further difference is related to that between the approaches of [1] and of [4]: while we could have used a Contraction Mapping argument and gotten uniqueness at the same time as in [1], we actually use a Schauder Theorem argument since the estimates are slightly simpler and, in any case, we do not need uniqueness but do need the compactness obtained in order to consider optimal controls later, as in [4]. Thus, this section is the technical heart of the paper.

We wish to construct a map $\boldsymbol{\mathcal{F}} : [C, h] \mapsto [\hat{C}, \hat{h}]$ whose fixpoint will provide the desired solution of the system, at least on some interval $\mathcal{I}_\tau$. We start with $C$ on $\mathcal{Q}$ such that $0 \le C \le 1$ and $h \in L^2$ with an *a priori* bound $M_*$ to be specified later

$$h \in \mathcal{X}_* := \{h \in L^2(\mathcal{I}) : \|h\| \le M_*\} \tag{3.2}$$

and proceed as follows:

**Step 1** Solve (1.4) for $R(\cdot)$, stopping[4] at $t_*$ if there is a $t_*$ giving $R(t_*) = 0, 1$; else $t_* = T$.

**Step 2** Use $R, h$ to define $\rho, \psi$ by (2.2), (2.4) for $0 \le t < t_*$ and solve (2.6), (2.7), (2.8) to obtain $\Theta$.

**Step 3** Use (2.3) to obtain $\tilde{C}$. Let $\hat{C}$ be the truncation of $\tilde{C}$ to take values in $[0, 1]$.

**Step 4** Obtain the traces at the interface:

$$\omega := \hat{C}(\cdot, R(\cdot)), \qquad \vartheta := \Theta(\cdot, R(\cdot)) \tag{3.3}$$

and define $\hat{h} \in \mathcal{X}_*$ by setting

$$\hat{h} := \begin{cases} \tilde{h} & \text{if } \|\tilde{h}\| \le M_* \\ M_* \tilde{h}/\|\tilde{h}\| & \text{else} \end{cases} \quad \text{with } \tilde{h} := H(R, \omega, \vartheta). \tag{3.4}$$

We accept, here, the solvability of the equations in Steps 2,3; note that $t_* \ge \min\{T, [\min\{R_0, 1-R_0\}/M_*]^2\} =: \tau$ in Step 1. A standard estimate, taking $[C - \gamma_0]$ as test function in the weak formulation of (2.3) and using Gronwall (compare [1], etc.), gives a bound on $C - \gamma_0$ in $\boldsymbol{\mathcal{U}}_\tau$ so the trace $\omega$ is well defined in, say, $L^2(\mathcal{I}_\tau)$. Indeed, the $\boldsymbol{\mathcal{U}}$-bound gives $C$ in a compact subset of, say, $L^2(\mathcal{I}_\tau \to H^s(0, 1)$ for any $s < 1$ by the Aubin Compactness Theorem and taking $s > \frac{1}{2}$ gives $\omega$ in a compact subset of $L^2(\mathcal{I}_\tau)$ by trace theory. We will also, somewhat similarly, obtain a $\boldsymbol{\mathcal{V}}$-bound for $\Theta - \gamma_1$ and the corresponding argument then gives $\vartheta$ in a compact subset of $L^2(\mathcal{I}_\tau)$. The hypotheses (H-1,2,5) ensure that $\hat{h}$ depends continuously on $\omega, \vartheta$ and is also in a compact subset of $L^2(\mathcal{I}_\tau)$. It is not difficult to verify the continuity of $\boldsymbol{\mathcal{F}}$ on $\mathcal{X}_*$ and the Schauder Theorem then gives existence of a fixpoint, determining $C, \Theta$.

[4]We will not be interested in situations in which the grain disappears ($R = 0$) and will show that $R = 1$ cannot occur for the solution.

The precise extension we made in (2.1) just permits a weak Maximum Principle argument[5] (using $C_-$ and $(C-1)_+$ as test functions) to show that $0 \le C \le 1$; cf., [1], etc.

We return to the estimation of $\Theta$ to show, in particular, that we may choose $M_*$ so as to have (3.4) give (1.6) at the solution. Using $\eta = 2[\Theta - \gamma_1]$ in (2.6), we first note that

$$|\vartheta - \gamma_1|^2 \le \delta\|\Theta_r\|^2 + C_\delta\|\Theta - \gamma_1\|^2$$

for any $\delta > 0$ so, with $\delta > 0$ a lower bound for $D(r)$, we use (H-6), etc., to get

$$\|\Theta - \gamma_1\|^2 + \delta \int_0^t \|\Theta_r\|^2\, dt' \le M\left[M_0 + C_\delta M_*^2 + \int_0^t [1 + u(t)]\|\Theta - \gamma_1\|^2\, dt'\right]$$

and Gronwall gives a $\mathcal{V}$-bound, here depending on $M_*$ but adequate to justify our previous argument. However, at the solution, where we have $\omega \in [0,1]$ and (3.4), we note that the last term in (2.6) gives $-(\vartheta - \gamma_1)\tilde{\alpha}\vartheta$ with $\tilde{\alpha} = \alpha$ when $\|\vartheta\| \le M_*$ and $\tilde{\alpha} = \alpha M_*/\|\vartheta\| < \alpha$ else; using (H-1,2,4), one has

$$\begin{aligned} -(\vartheta - \gamma_1)\tilde{\alpha}\vartheta &\le \begin{cases} \alpha[\gamma_1 + \vartheta G(\kappa, \vartheta)]K(\vartheta) & \text{if } \vartheta \ge 0 \\ \alpha[\gamma_1 - \vartheta]K(\gamma_1) & \text{if } \vartheta \le 0 \end{cases} \\ &\le M_1 + M_2|\vartheta - \gamma_1|^2 \end{aligned}$$

where it is important to note that the constants $M_1, M_2$ are independent of $M_*$ and of $u(\cdot)$. We now have (with a new $M$)

$$\|\Theta - \gamma_1\|^2 + \delta \int_0^t \|\Theta_r\|^2\, dt' \le M\left[M_0 + \int_0^t [u(t) + C_\delta]\|\Theta - \gamma_1\|^2\, dt'\right]$$

and Gronwall now gives a $\mathcal{V}$-bound on $[\Theta - \gamma_1]$ which depends only on the data (and on an integral bound for the control $u$) and so an $L^2(\mathcal{I}_\tau)$-bound for the trace $\vartheta$. The linear growth of $K(\cdot)$ then gives a corresponding bound for $\tilde{h}$ in (3.4) and this gives (1.6) if we have chosen $M_*$ at least equal to that.

We note that (H-3) ensures — together with the Dirichlet conditions for $C, \Theta$ at $r = 1-$ — that $H(R, C(\cdot, R), \Theta(\cdot, R))$ would become negative if we were to have $R \to 1$. Hence, (1.4)+(1.6) ensures that $R$ can never reach 1 for a solution of the direct problem. One consequence of this is that we need never contend with a possible degeneracy of having $\rho \to 0$ in (2.3).

This procedure gives such a solution on the interval $\mathcal{I}_\tau$ with $\tau$ as noted above. Obviously, if we do not actually have $R(\tau) = 0$ then we can restart the problem and continue the solution to a maximal interval of existence, again denoted by $\mathcal{I}_\tau$ with either $\tau = T$ or $R(\tau) = 0$; one sees as in [1] that a solution cannot become non-continuable for any other reason so, now with $\mathcal{I}_\tau$ the maximal interval of existence, we could only have $\tau < T$ if the grain were to dissolve completely: $R(\tau) = 0$.

## 4. Optimal Controls

We will consider as admissible controls only those satisfying (3.1) which provide a solution with $R > 0$ on all of $\mathcal{I}$. We will simply assume the existence of such controls but do observe

[5]Since the direction of motion of the interface is not fixed, this argument is necessarily a bit more delicate than, e.g., for the one-phase Stefan problem.

heuristically that, if we have no upper constraint on $u(\cdot)$, it should always be possible to pump enough heat into the system to make $\vartheta$ large so, using (H-4), $G(\kappa,\vartheta) \leq \omega$ and $\dot{R} \geq 0$ — noting that we cannot be prevented from this by having $\dot{R} < 0$ with $\omega = 0$ since (1.3) with $\dot{R} < 0$ implies $C_r\big|_{r=R} < 0$ which would be impossible for $\omega = 0$, which shows that the possible effect of increasing $\Theta$ to increase $F$ and so decrease $C$ must be overbalanced at the interface, itself, by the surface reaction.

Somewhat arbitrarily, we may consider $\mathcal{J}$ as having the form

$$\mathcal{J} := \int_0^T |\kappa(t) - \kappa_*(t)|^2 \, dt + \lambda U(T) \tag{4.1}$$

with $U$ as in (3.1); here $\kappa_*(\cdot)$ corresponds to some specified desirable 'trajectory' for the free boundary. One significance of using this particular form for $\mathcal{J}$ is that we will have $\mathcal{J} = \infty$ if the grain is not there for the full interval $\mathcal{I} = [0,T]$ — so the solution gives $R(\tau) = 0$ for some $\tau < T$ and so, by our interpretation of (1.4), gives $\kappa(t) \equiv \infty$ on $[\tau, T]$. Indeed, if we were to set

$$\mathcal{J}_\tau := \int_0^\tau [|\kappa(t) - \kappa_*(t)|^2 + \lambda u(t)] \, dt,$$

then $R(\tau) = 0$ would give

$$\begin{gathered} R(\tau - s) = -\textstyle\int_{\tau-s}^{\tau} h(\sigma)\, d\sigma \leq \sqrt{s}\|h\| \\ \textstyle\int_{\tau-s}^{\tau} \kappa^{-2}\, d\sigma \geq \|h\|^{-2} \int_{\tau-s}^{\tau} \frac{1}{\sigma}\, d\sigma \end{gathered}$$

and this would make $\mathcal{J}_\tau = \infty$. Thus, any control $u$ for which $\mathcal{J}$ or $\mathcal{J}_\tau$ would be finite necessarily implies a solution existing on the full interval $\mathcal{I}$ with $R(t)$ bounded away from $0, 1$ and, in fact, a bound on the control will imply a uniform such bound for $R(\cdot)$.

We will permit the inclusion of positive measures as controls so controls are taken in the dual space $[\mathcal{C}(\mathcal{I})]^*$; another consequence of the use of (4.1) is that any minimizing sequence $\{u_k\}$ for $\mathcal{J}$ is then necessarily bounded and so, possibly extracting a subsequence, we have weak-$*$ convergence $u_k \rightharpoonup \bar{u}$ (whence strong convergence $u_k \to \bar{u}$ in, say, $H^{-1}(\mathcal{I})$). By the discussion in the previous section, the corresponding $\{\vartheta_k, \omega_k, h_k\}$ lie in a compact subset of $L^2(\mathcal{I})$ and the functions $R_k(\cdot)$ are uniformly bounded away from $0, 1$. Again taking a subsequence, we may assume that that $\vartheta_k \to \bar{\vartheta}$, etc. By Krasnosel'skĭi's Theorem on the continuity of Nemitsky operators, one has (1.6) in the limit for $\bar{h}$, etc., and, using smooth test functions (which are dense) one easily verifies that $\bar{u}$ is an admissible control with uniform convergence $R_k \to \bar{R}$. Since the last term in $\mathcal{J}$ is actually continuous in $u$ with respect to weak-$*$ convergence (noting the positivity of the controls), this shows that $\bar{u}$ is an *optimal control*, as desired.

An essentially similar argument — indeed, technically slightly simpler — would give the same result for, e.g., a cost functional of the form

$$\mathcal{J} := \int_0^T |\kappa(t) - \kappa_*(t)|^2 \, dt + \lambda \int_0^T |u(t)|^2 \, dt \tag{4.2}$$

and it is for this that one might expect a derivation of the first order necessary conditions along the lines of [3]. On the other hand, if we were to try to use, say,

$$\mathcal{J} := \int_0^T |R(t) - R_*(t)|^2 \, dt + \lambda U(T), \tag{4.3}$$

then we do not have a similar argument that $\bar{R}(\cdot)$ never vanishes, which was needed to have $\bar{u}$ admissible. It is likely, in that case, that the existence of an optimal control would depend on the particular initial conditions. Finally, one can also obtain, by now-standard arguments, existence of time-optimal controls: minimizing the time $T$ at which one would have $R(t) = R_* < R_0$ — where the requirement that $0 < R_* < R_0$ ensures that there can be no intermediate difficulties with disappearance of the grain.

Note that we could obtain the existence results for optimal controls as rapidly as we have only because the relevant compactness results for $\{h_k\}$, etc., were obtained in the course of the existence arguments in the previous section.

# References


[1] F. Conrad, D. Hilhorst, and T.I. Seidman, *Well-posedness of a moving boundary problem arising in a dissolution–growth process,* Nonlinear Anal-TMA **15**, pp. 445–465 (1990).

[2] D. Hilhorst, F. Issard-Roch, and T.I. Seidman, *On a free boundary problem on an unbounded domain arising in a dissolution growth process*, in *Proc. 5$^{th}$ International Conf. on Free Boundary Problems*, (Montreal, 1990), to appear [Longman, London, 1993].

[3] P. Neittaanmäki and T.I. Seidman, *Optimal solutions for a free boundary problem for crystal growth,* in *Control and Estimation of Distributed Parameter Systems* (Vorau, 1988) (F. Kappel, K. Kunisch, W. Schappacher, eds.), pp. 323–334, Birkhäuser Verlag, Basel (1989).

[4] T.I. Seidman, *Some control-theoretic questions for a free boundary problem,* in *Control of Partial Differential Equations* (LNCIS #114; A. Bermúdez, ed.), Springer-Verlag, New York, pp. 265–276, (1989).

[5] T.I. Seidman, *Optimal control and well-posedness for a free boundary problem*, in *Proc. 5$^{th}$ International Conf. on Free Boundary Problems*, (Montreal, 1990), to appear [Longman, London, 1993].

# 52 Maximum Principle for Optimal Control of Distributed Parameter Stochastic Systems with Random Jumps

**Shanjian Tang** Fudan University, Shanghai, China
**Xun-Jing Li** Fudan University, Shanghai, China

**Abstract.** The optimal control is considered for distributed parameter stochastic systems with random jumps. The control is allowed to enter into both diffusion and jump terms, and a fairly general end constraint is taken into consideration. Under the assumption of a finite-codimension condition, which is always satisfied when the state space is of finite dimension, a maximum principle is proved for the optimal control. When calculating the variation of the cost, we use the vector-valued measure theory. As a matter of fact, we use only the scalar case of this theory, although the optimal control problem involves infinite-dimensional state space. The study unifies the Pontryagin-type maximum principle for continuous, discontinuous, deterministic, stochastic, finite-dimensional, and infinite-dimensional optimal controls.

**Keywords.** maximum principle, optimal stochastic control, Poisson point process, Lebesgue integral, distributed parameter stochastic system
**AMS (MOS) subject classification.** 49K27,49K45,93C25,93E20.

## §2. Introduction.

The idea of optimal control is one of the central themes in control theory. Pontryagin et al. [27] derived the Maximum Principle as a necessary condition of optimal control for deterministic finite-dimensional systems. The book [20] written by E.B. Lee and L. Markus is a treatise on optimal control. Following that there is much literature on this subject.

This paper considers the optimal control problem for distributed parameter stochastic systems with random jumps, and is dedicated to the seventieth birthday of Professor L. Markus.

In optimal stochastic control theory [2,6,11,13,19,21,26,28] were for the finite-dimensional case and [14,16] were for the infinite-dimensional case, and are concerned with continuous stochastic processes. The optimal control problem of jump processes was discussed in [28] for the finite-dimensional case, but under the assumption that no control

This work was partially supported by NSF of China and the Chinese State Education Commission Science Foundation

variable entered into the jump term. In [29], we considered the optimal control problem for finite-dimensional stochastic systems with jumps, allowing the control variable to enter into both diffusion and jump terms. This paper is an extension of the results of [29] to the infinite-dimensional case.

When deriving the maximum principle , we use the needle-variation method and the vector-valued measure theory ( **VVMT** for shorthand ). They have been known to be useful in deriving the Pontryagin-type maximum principle ( cf. [13,22,23] ). Here,however, we provide a routine of applying the VVMT, different from the traditional one. In the traditional routine,the VVMT was applied in a vector space of as many dimensions as the system state (note that the state of a stochastic system is considered here to be of infinite dimensions. ). While in the routine put forward here, the VVMT is applied in the real number space **R**, which is of only one dimension. Then if we use the latter routine, we only need the following property of a real Lebesgue-integrable function $l(\cdot)$ :

$$\rho\int_0^1 l(s)\,ds = \int_{I_\rho} l(s)\,ds + o(\rho),$$
$$\text{for some } I_\rho \subset [0,1] \text{ such that } |I_\rho| = \rho,$$

whether the control system is deterministic or stochastic, finite or infinite-dimensional. This not only overcomes the diffficulty caused by the appearance of the control in diffusion and/or jump terms, but also unifies the study of the Pontryagin-type maximum principle for continuous, discontinuous, deterministic, stochastic, finite-dimensional, and infinite-dimensional optimal controls.

The rest of the paper is organized as follows. Section 2 is the formulation of the optimal control problem discussed in this paper. Section 3 is the formulation of the main result. The proof of the main results are in Sections 4 and 5. Section 6 is a conclusion of the paper.

## §1. Preliminaries and the problem.

Let $\mathbf{D}, \mathbb{D}_1, \mathbb{D}_2, \mathbf{E}, \mathbf{F}, \mathbf{H}$ be real separable Hilbert spaces, **R** be the real number space,and **K** be some nonempty open subset of a finite-dimensional Euclidean space. Let $\langle , \rangle$ stand for inner products in Hilbert spaces and $\|\cdot\|$ for norms of vectors and operators. We will use subscripts to specify the spaces (such as $\langle , \rangle_{\mathbb{H}}, \|\cdot\|_{\mathbb{H}}$) when necessary. Let $L(\mathbb{E},\mathbb{H}), \mathcal{L}^2(\mathbf{E},\mathbf{H}), L(\mathbb{H})$ and $\mathcal{L}^2(\mathbf{H})$ be defined as in Chapter 1 [32]. In any given topological space $\mathbb{G}$,let $\mathcal{B}(\mathbb{G})$ be its topological $\sigma$-field.

Let $(\Omega, \mathcal{F}, \nu)$ be a complete probability space with $\nu$-completed right continuous filtration $\mathcal{F}_t$. Consider a stationary $(\mathcal{F}_t)$-Poisson point process $k(\cdot)$ on **K** with the characteristic measure $\pi(dz)$. Its counting measure is denoted by $N_k(dzdt)$. Let $\widetilde{N}_k(dzdt) = N_k(dzdt) - \pi(dz)dt$. Let $w(\cdot)$ be an $\mathbb{E}$-valued $(\mathcal{F}_t)$-adapted Wiener process with nuclear operator $W$. We always assume that

$$\mathcal{F}_t = \mathcal{F}_t^w \bigvee \mathcal{F}_t^k$$

with $\mathcal{F}_t^w$ and $\mathcal{F}_t^k$ being defined as

$$\mathcal{F}_t^w =: \sigma\Big[\langle w(s), e\rangle; s \le t, e \in \mathbb{E}\Big] \bigvee \mathcal{N}$$
$$\mathcal{F}_t^k =: \bigcap_{\epsilon>0} \sigma\Big[N_k(U, (0, s]); s \le t+\epsilon, U \in \mathcal{B}(\mathbf{K})\Big] \bigvee \mathcal{N}$$
$$\mathcal{F}_t^* =: \mathcal{F}_t^w \bigvee \mathcal{F}_t^k.$$

For $g \in$ $\mathbf{H}$ and $h \in$ $\mathbf{E}$,we define $g \otimes h$ as an element of $\mathcal{L}^2(\mathbf{E}, \mathbf{H})$ in the following way:

$$(g \otimes h)k = g\langle h, k\rangle, \quad \forall\, k \in \mathbb{E}$$

and $\mathbf{H} \otimes \mathbb{E}$ as

$$\mathbf{H} \otimes \mathbf{E} =: \left\{ z = \sum_{i=1}^{m} x_i \otimes y_i \,\middle|\, x_i \in \mathbf{H}, y_i \in \mathbb{E}, i = 1, \ldots, m;\ m \text{ is some finite integer} \right\}.$$

We can prove that $\mathcal{L}^2(\mathbb{E}, \mathbf{H})$ is the completion of $\mathbb{H} \otimes \mathbf{E}$ for the norm associated with the following scalar product

$$\langle z, \bar{z}\rangle =: \sum_{i=1}^{m}\sum_{j=1}^{n} \langle x_i, \bar{x}_j\rangle\langle y_i, \bar{y}_j\rangle, \quad \forall\, z = \sum_{i=1}^{m} x_i \otimes y_i \text{ and } \bar{z} =: \sum_{j=1}^{n} \bar{x}_i \otimes \bar{y}_i.$$

For $q_1, q_2 \in L(\mathbf{E}, \mathbf{H})$, we define $q_1 \otimes q_2$ as the elements of $L(\mathcal{L}^2(\mathbf{E}), \mathcal{L}^2(\mathbf{H}))$ which is the unique continuous linear extension of $x \otimes y \mapsto q_1(x) \otimes q_2(y)$.

**Definition 1.1.** A function $f$: $[0,1] \times \mathbb{K} \times \Omega \to \mathbf{H}$ is called $(\mathcal{F}_t)$-predictable if the mapping $(t, z, \omega) \mapsto f(t, z, \omega)$ is strongly measurable w.r.t. $\varphi$, the smallest $\sigma$-field on $[0,1] \times \mathbf{K} \times \Omega$ w.r.t. which all $g$: $[0,1] \times \mathbb{K} \times \Omega \to \mathbf{H}$, having the following properties, are strongly measurable:

1). for each $t \in [0,1], (z, \omega) \mapsto g(t, z, \omega)$ is strongly measurable w.r.t. $\mathcal{B}(\mathbb{K}) \times \mathcal{F}_t$;

2). for each $(z, \omega)$, $t \mapsto g(t, z, \omega)$ is strongly left-continuous.

**Remark 1.1.** Let $\tilde{\varphi}$ denote the smallest $\sigma$-field on $[0,1] \times \mathbf{K} \times \Omega$ generated by all $\tilde{g}$: $[0,1] \times \mathbf{K} \times \Omega \to \mathbf{R}$ satisfying the following properties:

1)'. for each $t \in [0,1], (z, \omega) \mapsto \tilde{g}(t, z, \omega)$ is $\mathcal{B}(\mathbf{K}) \times \mathcal{F}_t$-measurable;

2)'. for each $(z, \omega)$, $t \mapsto \tilde{g}(t, z, \omega)$ is left-continuous.

Then $\tilde{\varphi} = \varphi$.

Let $F^2(\mathbf{K}, \mathcal{B}(\mathbf{K}); \mathbb{H})$ be the Hilbert space of measurable functions $f(\cdot) : \mathbf{K} \to \mathbf{H}$ such that

$$\int_{\mathbb{K}} \|f(z)\|^2 \pi(dz) < \infty$$

with the inner product being defined as

$$\langle f,g\rangle =: \int_{\mathbf{K}} \Big\langle f(z),g(z)\Big\rangle\, \pi(dz).$$

Let $F^2_{\mathcal{F},k,p}(0,1;\mathbf{H})$ denote the Hilbert space of $\mathbf{H}$-valued $(\mathcal{F}_t)$-predictable functions $f(\cdot,\cdot,\cdot)$ defined on $\mathbf{K}\times[0,1]\times\Omega$ such that

$$\iint_{\mathbf{K}\times[0,1]} E\Big\|f(z,t,\cdot)\Big\|^2\, \pi(dz)\,dt < \infty$$

with the inner product being defined as

$$\langle f,g\rangle =: \iint_{\mathbf{K}\times[0,1]} E\Big\langle f(z,t,\cdot),g(z,t,\cdot)\Big\rangle_{\mathbf{H}} \pi(dz)\,dt.$$

For $\hat{f}(\cdot,\cdot,\cdot)\in F^2_{\mathcal{F},k,p}(0,1;\mathbf{H})$, define

$$\begin{aligned}&\iint_{\mathbf{K}\times(s,t]} \hat{f}(z,\tau,\omega)\,\widetilde{N}_k(dz\,d\tau)\\ &=: \iint_{\mathbf{K}\times(0,t]} \hat{f}(z,\tau,\omega)\,\widetilde{N}_k(dz\,d\tau) - \iint_{\mathbf{K}\times(0,s]} \hat{f}(z,\tau,\omega)\,\widetilde{N}_k(dz\,d\tau)\end{aligned}$$

for $s\le t$. We easily varify that for $s\le t$, the process

$$\iint_{\mathbf{K}\times(s,t]} \hat{f}(z,\tau,\omega)\,\widetilde{N}_k(dz\,d\tau)$$

is an $(\mathcal{F}_t)$-martingale with the first increasing process being

$$\iint_{\mathbf{K}\times(s,t]} \hat{f}(z,\tau,\omega)\,\pi(dz)\,d\tau.$$

Let $y\in L^2(\Omega,\mathcal{F};\mathbf{E})$. The covariance operator of $y$ is

$$\mathrm{Cov}[y] = E\Big[(y-Ey)\otimes(y-Ey)\Big].$$

Then Cov[$y$] is a selfadjoint-nonnegative nuclear operator.

**Definition 1.2.** Let $W$ be a nonnegative self-adjoint nuclear operator on Hilbert space $\mathbf{E}$. An $\mathbf{E}$-valued process is called $(\mathcal{F}_t)$-Wiener process with covariance $W$ if it satisfies

(1) for every $s<t, w(t)-w(s)$ is independent of $\mathcal{F}_s$;

(2) for every $s < t$ and $h \in \mathbf{H}$, the real random variable $\langle w(t) - w(s), h\rangle$ is Gaussian centered with variance $(t-s)\langle W, h \otimes h\rangle$.

**Definition 1.3.** Let $\widetilde{W} \in \mathcal{L}^1(\mathbf{E}, \mathbf{H})$ be symmetric and nonnegative. $\mathcal{L}^2(\mathbf{E}, \mathbf{H}; \widetilde{W})$ is defined as the Hilbert space of linear operators $f$ from $\mathbf{E}$ into $\mathbf{H}$ satisfying the following

(1) the domain of $f$ contains the range of $\widetilde{W}^{\frac{1}{2}}$;

(2) $f \circ \widetilde{W}^{\frac{1}{2}} \in \mathcal{L}^2(\mathbf{E}, \mathbf{H})$ with the inner product being defined by

$$\langle f, g\rangle = \text{trace } f \circ \widetilde{W} \circ g^*,$$

where the symbol $\circ$ stresses the operation of operators.

We shall also denote by $L^2_{\mathcal{F}}(0,1;\mathbf{H})$ the space of $\mathbf{H}$-valued square integrable $(\mathcal{F}_t)$-adapted process, $L^2_{\mathcal{F},p}(0,1;\mathbf{H})$ the space of $(\mathcal{F}_t)$-predictable versions of equivalent classes in $L^2_{\mathcal{F}}(0,1;\mathbf{H})$. Then $L^2_{\mathcal{F},p}(0,1;\mathcal{L}^2(\mathbf{E},\mathbf{H};W))$ is the Hilbert space of $\mathcal{L}^2(\mathbf{E},\mathbf{H};W)$-valued processes such that

$$\int_0^1 E\left\|f(t)\right\|^2 dt < \infty$$

and for any $h \in \mathbf{E}$, $f \circ W^{\frac{1}{2}}(h)$ is $(\mathcal{F}_t)$-predictable. For $f \in L^2_{\mathcal{F},p}(0,1;\mathcal{L}^2(\mathbf{E},\mathbf{H};W))$, we can define the stochastic integral [24]

$$\int_0^t f(s) \circ dw(s).$$

**Definition 1.4.** Let $\mathbf{H}$ be a separable Hilbert space. $T(\cdot,\cdot)$: $\{(t,s): 0 \le s \le t \le 1\} := \bar{\Delta} \to L(\mathbf{H})$ is called an evolution operator on $\mathbf{H}$ defined on $\bar{\Delta}$, if

1). $T(t,t) = I$ (Identity),

2). $T(t,\tau)T(\tau,s) = T(t,s)$, $\quad 0 \le s \le \tau \le t \le 1$,

3). $T(t,s)$ is strongly continuous in $s$ on $[0,t]$ and strongly continuous in $t$ on $[s,1]$.

Let $T(\cdot,\cdot) : \bar{\Delta} =: \{(t,s) : 0 \le s \le t \le 1\} \to L(\mathbf{H})$ be an evolution operator on $\mathbf{H}$, and $T^*(\cdot,\cdot)$ be strongly continuous in $s$ on $[0,t]$ and strongly continuous in $t$ on $[s,1]$ Then consider the following stochastic evolution equation

$$\begin{aligned} x(t) = &T(t,0)x_0 + \int_{(0,t]} T(t,s)A(x(s),s)\,ds + \int_{(0,t]} T(t,s)B(x(s),s) \circ dw(s) + \\ &+ \iint_{\mathbb{K}\times(0,t]} T(t,s)C(x(s-),s,z)\widetilde{N}_k(dz\,ds). \end{aligned} \tag{2.1}$$

An **H**-valued process $x(\cdot)$ is called a solution of (2.1) if it is an $(\mathcal{F}_t)$-adapted cadlag (i.e.,right-continuous with left-hand limits) process such that (2.1) holds.

Let $x_0$ be $\mathcal{F}_0$-measurable with $E|x_0|^2 < \infty$, and the coefficients of (2.1) satisfy Lipschitz conditions ( cf. [17] for the detailed statement). Then following the routine of [16], we easily obtain the existence and uniqueness of solution for (2.1) in the space $D(0,1;\mathbf{H})$ which is defined as the space of **H**-valued $(\mathcal{F}_t)$-adapted cadlag processes $x(\cdot)$ such that

$$E \sup_{0\le t\le 1} \left\| x(t) \right\|^2 \le \infty.$$

The distributed parameter stochastic control system is govened by

$$\begin{aligned} x(t) =& T(t,0)x_0 + \int_{(0,t]} T(t,s)a(x(s),v(s))\,ds + \int_{(0,t]} [T(t,s)b(x(s),v(s))] \circ dw(s) + \\ &+ \iint_{\mathbb{K}\times(0,t]} T(t,s)c(x(s-),v(s),z)\,\tilde{N}_k(dz\,ds). \end{aligned} \tag{2.2}$$

An admissible control $v(\cdot)$ is defined as a $U$-valued $(\mathcal{F}_t)$-predictable process such that

$$\sup_{0\le t\le 1} E\left\| v(t) \right\|^8 < \infty \tag{2.3}$$

with $U$ being a nonempty subset of some normed linear space $\mathbb{U}$. The collective of admissible controls $v(\cdot)$ is denoted by $U_{ad}$. The end constraint is

$$Ef(x_0, x(1)) \in Q \subset \mathbb{F}. \tag{2.4}$$

The index functional is

$$J(v(\cdot), x_0) = \Gamma(E\int_0^1 g(x(s),v(s))ds, Eh(x_0, x(1))). \tag{2.5}$$

In the above statement, we require that $a(\cdot,\cdot) : \mathbf{H}\times U \to \mathbf{H}$, $b(\cdot,\cdot) : \mathbf{H}\times U \to \mathcal{L}^2(\mathbf{E},\mathbf{H};W)$, $c(\cdot,\cdot,\cdot) : \mathbf{H}\times U\times \mathbf{K} \to \mathbf{H}$, $f(\cdot,\cdot) : \mathbf{H}\times\mathbf{H} \to \mathbf{F}$, $g(\cdot,\cdot) : \mathbf{H}\times U \to \mathbf{D}_1$, $h(\cdot,\cdot) : \mathbf{H}\times\mathbf{H} \to \mathbf{D}_2$, and $\Gamma(\cdot,\cdot) : \mathbf{D}_1 \times \mathbf{D}_2 \to \mathbf{R}$. The optimal control problem is to find a pair $(y_0, u(\cdot)) \in \mathbf{H}\times U_{ad}$ such that (2.2) and (2.4) are satisfied and (2.5) minimized.

Throughout the chapter we will make the following assumptions:

$(\mathcal{A}1)$ The vector functions $l(x,y,v,z) =: a(x,v)$, $b(x,v)\circ W^{\frac{1}{2}}$, $c(x,v,z)$, $f(y,x)$, $g(x,v)$, $h(y,x)$, are Borel measurable in all arguments, differentiable in $y$, and are twice Fréchet differentiable in. Also $l_x(x,y,v,z), l_y(x,y,v,z) \in L(\mathbf{H},\mathbf{B})$ and $l_{xx}(x,y,v,z) \in L(\mathcal{L}^2(\mathbf{H}),\mathbf{B})$ are uniformly continuous in $(x,y,v)$ for all $x,y\in\mathbf{H}, v\in U, z\in\mathbb{K}$. Here **B** is the space where $l(x,y,v,z)$ takes its values. $a(x,v), b(x,v)\circ W^{\frac{1}{2}}$,

$$\left[\int_{\mathbb{K}} \left\| c(x,v,z) \right\|^{2k} \pi(dz)\right]^{\frac{1}{2k}}, \quad k = 1,2,$$

$f_y(y,x), f_x(y,x), g_x(x,v), h_y(y,x)$,and $h_x(y,x)$ are bounded by $(1+\|x\|+\|y\|+\|v\|)$. $f(y,x), g(x,v), h(y,x)$ are bounded by $(1+\|x\|^2+\|y\|^2+\|v\|^2)$. $a_x(x,v), a_{xx}(x,v), b_x(x,v)$, $b_{xx}(x,v)$,

$$\int_{\mathbb{K}} \left\|c_x(x,v,z)\right\|^{2k} \pi(dz), \quad k=1,2, \quad \int_{\mathbb{K}} \left\|c_{xx}(x,v,z)\right\|^2 \pi(dz),$$

$f_{xx}(y,x), g_{xx}(x,v)), h_{xx}(y,x)$ are bounded;

($\mathcal{A}2$) $T(t,s)$ defined on $\bar{\Delta}$ is an evolution operator on $\mathbf{H}$;

($\mathcal{A}3$) The set Q is closed and convex;

($\mathcal{A}4$) $\Gamma(g,h)$ is continuously Fréchet differentiable with respect to its arguments,with its derivatives being bounded.

Under Assumption ($\mathcal{A}1$),(2.2) admits unique solution for given $x_0 \in \mathbb{H}$ and $v(\cdot) \in U_{ad}$ (cf. [21]) and the above formulation of the optimal control problem is well-defined.

## §3. Main result and auxiliary lemmas.

The Hamiltonian is defined as

$$\begin{aligned} H(x,v,\lambda,p,q,r(\cdot)) =& \lambda g(x,v) + \Big\langle p, a(x,v)\Big\rangle + \Big\langle q, b(x,v)\Big\rangle + \\ &+ \Big\langle r(\cdot), c(x,v,\cdot)\Big\rangle, \end{aligned} \tag{3.1}$$

being a map from $\mathbf{H} \times U \times \mathbf{D}_1 \times \mathbf{H} \times \mathcal{L}^2(\mathbf{E},\mathbf{H};W) \times F^2(\mathbf{K},\mathcal{B}(\mathbf{K});\mathbf{H})$ to $\mathbb{R}$.

For $\rho \in (0,1]$, $I_\rho \subset [0,1]$,and $v(\cdot) \in U_{ad}$, define

$$\begin{aligned} |I_\rho| &=: \text{the Lebesgue measure of set } I_\rho, \\ u^\rho(s) &=: u(s)1_{[0,1]\setminus I_\rho}(s) + v(s)1_{I_\rho}(s), \qquad s \in [0,1], \\ y_0^\rho &=: y_0 + |I_\rho|\eta, \ \ \eta \in \mathbf{H}, \end{aligned} \tag{3.2}$$

with $1_A(\cdot)$ standing for the indicator function of some set $A \subset [0,1]$. It's easily proved that $u^\rho(\cdot) \in U_{ad}$.

Let $y_0 \in \mathbb{H}$ and $u(\cdot) \in U_{ad}$ be given,and $y(\cdot)$ denote the solution of (2.1) corresponding to $(y_0, u(\cdot))$. For each $v \in U$, we introduce the following simplified notation

$$\begin{aligned} \Delta m(s;v) &=: m(y(s-),v) - m(y(s-),u(s)), \\ m(s) &=: m(y(s-),u(s)), \\ \Delta n(s,z;v) &=: n(y(s-),v,z) - n(y(s-),u(s),z), \\ n(s,z) &=: n(y(s-),u(s),z), \end{aligned}$$

with $m$ standing for $a, b, g$ and all their (up to second-) derivatives in $x$,and $n$ for $c$ and its (up to second-) derivatives in $x$.

For given $p(1) \in L^2(\Omega, \mathcal{F}_1; \mathbb{H})$ and $P(1) \in L^2(\Omega, \mathcal{F}_1; \mathcal{L}^2(\mathbb{H}))$, we see from [ Theorem 3.1 of Chapter 1, 32] that the Itô-type adjoint equations

$$\begin{aligned} p(t) =& T^*(1,t)p(1) + \int_{(t,1]} T^*(s,t)H_x(y(s),u(s),\lambda\Gamma_g,p(s),q(s),r_\cdot(s))\,ds- \\ &- \int_{(t,1]} T^*(s,t)q(s)\circ dw(s) - \iint_{\mathbb{K}\times(t,1]} T^*(s,t)r_s(z)\,\widetilde{N}_k(dz\,ds) \end{aligned} \tag{3.3}$$

and

$$\begin{aligned} P(t) = & T^*(1,t)P(1)T(1,t) + \int_{(t,1]} T^*(s,t)\Big\{a_x^*(s)P(s) + P(s)a_x(s)+ \\ &+ \Big[b_x(s)\circ W^{\frac{1}{2}}\Big]^* P(s)\Big[b_x(s)\circ W^{\frac{1}{2}}\Big] + \int_{\mathbb{K}} c_x^*(s,z)P(s)c_x(s,z)\,\pi(dz)+ \\ &+ \Big[(b_x(s)\circ W^{\frac{1}{2}}\Big]^* Q(s)\circ W^{\frac{1}{2}} + \Big[Q(s)\circ W^{\frac{1}{2}}\Big]\Big[b_x(s)\circ W^{\frac{1}{2}}\Big]+ \\ &+ \int_{\mathbb{K}} \Big[c_x^*(s,z)R(s)c_x(s,z) + c_x^*(s,z)R(s) + R(s)c_x(s,z)\Big]\,\pi(dz)+ \\ &+ H_{xx}(y(s),u(s),\lambda\Gamma_g,p(s),q(s),r_\cdot(s))\Big\}T(s,t)\,ds- \\ &- \int_{(t,1]} T^*(s,t)\Big[Q(s)\circ dw(s)\Big]T(s,t)- \\ &- \iint_{\mathbb{K}\times(t,1]} T^*(s,t)R_s(z)T(s,t)\,\widetilde{N}_k(dz\,ds) \end{aligned} \tag{3.4}$$

admit unique solutions

$$(p(\cdot),q(\cdot),r_\cdot(\cdot)) \in L^2_{\mathcal{F}}(0,1;\mathbb{H}) \times L^2_{\mathcal{F},p}(0,1;\mathcal{L}^2(\mathbb{E},\mathbb{H};W)) \times F^2_{\mathcal{F},k,p}(0,1;\mathbb{H})$$

and

$$(P(\cdot),Q(\cdot),R_\cdot(\cdot)) \in L^2_{\mathcal{F}}(0,1;\mathcal{L}^2(\mathbb{H})) \times L^2_{\mathcal{F},p}(0,1;\mathcal{L}^2(\mathbb{E},\mathcal{L}^2(\mathbb{H});W)) \times F^2_{\mathcal{F},k,p}(0,1;\mathcal{L}^2(\mathbb{H}))$$

respectively,with $p(\cdot)$ and $P(\cdot)$ being cadlag processes.

We also define for each $v(\cdot) \in U_{ad}$, $y_1(\cdot)$ and $y_2(\cdot)$ respectively as the solution of

$$\begin{aligned} y_1(t) = & \int_{(0,t]} T(t,s)a_x(s)y_1(s)\,ds+ \\ &+ \int_{(0,t]} T(t,s)[b_x(s)y_1(s) + \Delta b(s;v(s)]\circ dw(s)+ \\ &+ \iint_{\mathbb{K}\times(0,t]} T(t,s)\Big[c_x(s,z)y_1(s-) + \Delta c(s,z;v(s))\Big]\,\widetilde{N}_k(dz\,ds) \end{aligned} \tag{3.5}$$

and the second-order variational process as the solution of

$$
\begin{aligned}
y_2(t) =& T(t,0)|I_\rho|\eta + \\
&+ \int_{(0,t]} T(t,s)\Big[a_x(s)y_2(s) + \Delta a(s;v(s)) + \frac{1}{2}a_{xx}(s)(y_1(s)\otimes y_1(s))\Big]\,ds + \\
&+ \int_{(0,t]} T(t,s)\Big[b_x(s)dw(s)y_2(s) + \Delta b_x(s;v(s))dw(s)y_1(s) + \\
&+ \frac{1}{2}b_{xx}(s)dw(s)(y_1(s)\otimes y_1(s))\Big] + \\
&+ \iint_{\mathbb{K}\times(0,t]} T(t,s)[c_x(s,z)y_2(s-) + \Delta c_x(s,z;v(s))y_1(s-) + \\
&+ \frac{1}{2}c_{xx}(s,z)\Big(y_1(s-)\otimes y_1(s-)\Big)\Big]\,\widetilde{N}_k(dz\,ds).
\end{aligned}
\tag{3.6}
$$

When necessary,we use $y_{1v(\cdot)}(\cdot), y_{2v(\cdot)}(\cdot)$ to indicate the dependence of $y_1(\cdot), y_2(\cdot)$ on $v(\cdot)$. When $|I_\rho| = 1$,we set

$$\tilde{y}_i(\cdot) =: y_i(\cdot),\ i = 1,2.$$

We introduce the set defined by

$$
\begin{aligned}
\widetilde{Q} =:& \Big\{z - Ef(y_0,y(1)) - Ef_y(y_0,y(1))\eta - Ef_x(y_0,y(1))\Big(\tilde{y}_1(1)+\tilde{y}_2(1)\Big) - \\
&- \frac{1}{2}Ef_{xx}(y_0,y(1))\Big(\tilde{y}_1(1)\otimes\tilde{y}_1(1)\Big)\Big|\ \forall\, z\in Q, \eta\in\mathbb{H},\ v(\cdot)\in U_{ad}\Big\}.
\end{aligned}
\tag{3.7}
$$

Our main result is stated as

**Theorem 3.1.** Let Assumptions ($\mathcal{A}1$)-($\mathcal{A}3$) be satisfied,and $\widetilde{Q}$ be of finite codimension in $\mathbf{F}$. Assume that $(y_0, y(\cdot), u(\cdot))$ is an optimal triplet. Then there exist $0 \le \lambda \in \mathbf{R}, \gamma \in \mathbf{F}$, $(p(\cdot), q(\cdot), r.(\cdot)) \in L^2_{\mathcal{F}}(0,1;\mathbf{H}) \times L^2_{\mathcal{F},p}(0,1;\mathcal{L}^2(\mathbf{E},\mathbf{H};W)) \times F^2_{\mathcal{F},k,p}(0,1;\mathbf{H})$,and $(P(\cdot), Q(\cdot), R.(\cdot)) \in L^2_{\mathcal{F}}(0,1;\mathcal{L}^2(\mathbf{H})) \times L^2_{\mathcal{F},p}(0,1;\mathcal{L}^2(\mathbf{E},\mathcal{L}^2(\mathbf{H});W)) \times F^2_{\mathcal{F},k,p}(0,1;\mathcal{L}^2(\mathbf{H}))$ such that

1) the nontrivial condition

$$|\lambda|^2 + \|\gamma\|^2 > 0; \tag{3.8}$$

2) (3.3),(3,4) and

$$
\begin{aligned}
p(1) =& \lambda\Gamma_h h_x(y_0,y(1)) + \gamma f_x(y_0,y(1)), \\
p(0) =& -\lambda\Gamma_h h_y(y_0,y(1)) - \gamma f_y(y_0,y(1)), \\
P(1) =& \lambda\Gamma_h h_{xx}(y_0,y(1)) + \gamma f_{xx}(y_0,y(1)),
\end{aligned}
\tag{3.9}
$$

with $p(\cdot)$ and $P(\cdot)$ being cadlag processes;

3) the maximum condition

$$H\big(y(s-),v,\lambda,p(s-),q(s),r.(s)\big)-H\big(y(s-),u(s),\lambda\Gamma_g,p(s-),q(s),r.(s)\big)+$$
$$+\frac{1}{2}\Big\langle P(s-),\Delta b(s;v(s))\circ W\circ\Big(\Delta b(s;v(s))\Big)^*+\int_{\mathbb{K}}\Delta c(s,z;v(s))\otimes\Delta c(s,z;v(s))\pi(dz)\Big\rangle$$
$$+\frac{1}{2}\int_{\mathbb{K}}\Big\langle R_s(z),\Delta c(s,z;v(s))\otimes\Delta c(s,z;v(s))\Big\rangle\pi(dz)\geq 0,\quad\forall\, v\in U, a.e.a.s.; \tag{3.10}$$

4) the transversality condition

$$\Big\langle\gamma,z-Ef(y_0,y(1))\Big\rangle\leq 0,\ \forall\ z\in Q \tag{3.11}$$

are all satisfied.

First we are to give some lemmas.

The following estimates play an elementary role in calculating the variation of the index.

**Lemma 3.1.** Let $(\mathcal{A}1)$ be satisfied. Then for fixed $v(\cdot)\in U_{ad}$, we have

$$\begin{aligned}\sup_{0\leq t\leq 1}E\Big\|y_{1u^\rho(\cdot)}(t)\Big\|^8&=O(|I_\rho|^4),\\ \sup_{0\leq t\leq 1}E\Big\|y_{2u^\rho(\cdot)}(t)\Big\|^4&=O(|I_\rho|^4),\\ \sup_{0\leq t\leq 1}E\Big\|y^\rho(t)-y(t)-y_{1u^\rho(\cdot)}(t)-y_{2u^\rho(\cdot)}(t)\Big\|^2&=o(|I_\rho|^2),\end{aligned} \tag{3.12}$$

as $|I_\rho|\to 0$. Here $y^\rho(\cdot)$ stands for the solution of (2.1) corresponding to $(y_0^\rho,u^\rho(\cdot))\in\mathbf{H}\times U_{ad}$.

*Proof.* The first two estimates are easily proved by using the familiar elementary inequalities

$$(m_1+m_2+m_3)^i\leq C(|m_1|^i+|m_2|^i+|m_3|^i),\qquad i=4,8,$$

the Burkholder's inequality (in essential,the transformation formula),and the well-known Grownwall's inequality. The proof for the last estimate is easily concluded by an adaptation of that in [26],although the definitions for $y_1(\cdot)$ and $y_2(\cdot)$ are slightly different from those in [26]. □

**Lemma 3.2.** Assume that $l(\cdot)$ is a scalar-valued Lebesgue integrable function defined on $[0,1]$. Then for $\rho \in (0,1]$, there exists a measurable subset $I_\rho \subset [0,1]$, such that

$$|I_\rho| = \rho,$$
$$\int_{I_\rho} l(s)\,ds = \rho \int_{[0,1]} l(s)\,ds + o(\rho) \qquad \text{as } \rho \to 0. \tag{3.13}$$

The proof is quite elementary and the reader is referred to [22]. We would like to mention that a stronger statement (more precisely,this lemma with the term $o(\rho)$ vanishing) is available via an application of the well-known Liapunov convexity theorem to the $\mathbb{R}^2$-valued integrable vector function $(l(\cdot),1)$. However,this lemma is enough for our argument below.

Define the following function

$$\Phi(s,z;\varepsilon) =: \inf_{(\bar{z},t)\in Q\times(-\infty, J(u(\cdot),y_0)-\varepsilon]} \sqrt{\|\bar{z}-z\|^2 + |s-t|^2}. \tag{3.14}$$

**Lemma 3.3.** For given $\varepsilon > 0$, the function $\Phi(s,z;\varepsilon)$ is continuously differentiable on the open set $\widehat{Q} =: \{(s,z) : \Phi(s,z;\varepsilon) > 0\}$. Moreover,when $\Phi(s,z;\varepsilon) > 0$, we have

$$\begin{aligned} &\Big\langle \Phi_z(s,z;\varepsilon), \hat{z} - z \Big\rangle \le 0, \quad \forall\, \hat{z} \in Q, \\ &\Phi_s(s,z;\varepsilon) \ge 0, \\ &\Big|\Phi_s(s,z;\varepsilon)\Big|^2 + \Big\|\Phi_z(s,z;\varepsilon)\Big\|^2 = 1. \end{aligned} \tag{3.15}$$

*Proof.* We easily see that the distance function $\Phi(\cdot,\cdot;\varepsilon)$ is Lipschitz with Lipschitz modulus 1 and we easily derive from Lemma 3.4 and Corollary 3.5 of [23] that $\Phi(s,z;\varepsilon)$ is continuously differentiable at $(s,z)$, and further

$$\partial\Phi(s,z;\varepsilon) = \big\{D\Phi(s,z;\varepsilon)\big\}, \tag{3.16}$$

$$\Big\|D\Phi(s,z;\varepsilon)\Big\|^2_{\mathbb{R}\times\mathbb{R}^k} = 1 \tag{3.17}$$

whenever $\Phi(s,z;\varepsilon) > 0$. From the definition of $\partial\Phi(s,z;\varepsilon)$ ( cf. [23]) and (3.16), we have

$$\Big\langle D\Phi(s,z;\varepsilon), (\hat{s},\hat{z}) - (s,z)\Big\rangle \le 0, \quad \forall(\hat{s},\hat{z}) \in (-\infty, J(u(\cdot),y_0) - \varepsilon] \times Q. \tag{3.18}$$

This implies

$$\Big\langle \Phi_s(s,z;\varepsilon), \hat{s} - s\Big\rangle \le 0, \quad \forall \hat{s} \in (-\infty, J(u(\cdot),y_0) - \varepsilon] \tag{3.19}$$

and

$$\left\langle \Phi_z(s, z; \varepsilon), \hat{z} - z \right\rangle \leq 0, \ \forall \hat{z} \in Q. \tag{3.20}$$

The last two relations of (3.15) follow respectively from (3.19) and (3.17). □

Define the smooth approximation $\Psi(s, z; \varepsilon, \delta)$ of $\Phi(\cdot, \cdot; \varepsilon)$ as

$$\Psi(s, z; \varepsilon, \delta) =: \min_{\substack{\bar{s} \in \mathbb{R} \\ \bar{z} \in \mathbb{F}}} \left\{ \Phi(\bar{s}, \bar{z}; \varepsilon) + \frac{1}{2\delta} \left( \left| \bar{s} - s \right|^2 + \left\| \bar{z} - z \right\|^2 \right) \right\}. \tag{3.21}$$

Then we easily have

$$0 \leq \Psi(s, z; \varepsilon, \delta) \leq \Phi(s, z; \varepsilon).$$

Moreover, we can prove the following

**Lemma 3.4.** For $\widehat{Q}$ defined in Lemma 3.2,we have for $(s, z) \in \widehat{Q}$,

$$\begin{aligned} \lim_{\delta \to 0+} \Psi_s(s, z; \varepsilon, \delta) &= \Phi_s(s, z; \varepsilon), \\ \lim_{\delta \to 0+} \Psi_z(s, z; \varepsilon, \delta) &= \Phi_z(s, z; \varepsilon). \end{aligned} \tag{3.22}$$

*Proof.* Observe (cf. [7]) that

$$D\Psi(s, z; \varepsilon, \delta) = D\Phi(s - \delta\Psi_s(s, z; \varepsilon, \delta), z - \delta\Psi_z(s, z; \varepsilon, \delta); \varepsilon)$$

and

$$\Psi(s, z; \varepsilon, \delta) - \frac{1}{2}\delta |D\Psi(s, z; \varepsilon, \delta)|^2 = \Phi(s - \delta\Psi_s(s, z; \varepsilon, \delta), z - \delta\Psi_z(s, z; \varepsilon, \delta); \varepsilon),$$

and we easily finish the proof. □

## §4. A basic case.

Now we assume that

$$Q = \mathbb{F}$$

and that the operator function $\Gamma(\cdot, \cdot)$ is twice differentiable with respect to its arguments.

Choose

$$\lambda = 1, \quad \gamma = 0.$$

Then (3.8) and (3.11) hold.

To avoid notational complexity,we use the short forms $y_1(\cdot), y_2(\cdot)$ of $y_{1u^\rho}(\cdot), y_{2u^\rho}(\cdot)$.

Using Taylor's formula, we have

$$\begin{aligned}
&J(u^\rho(\cdot), y_0 + |I_\rho|\eta) - J(u(\cdot), y_0) \\
&= \Gamma(E\int_0^1 g(y^\rho(t), u^\rho(t))dt, Eh(y_0^\rho, y^\rho(1))) - \Gamma(E\int_0^1 g(y(t), u(t))dt, h(y_0, y(1))) \\
&= \Gamma_g E\int_0^1 [g(y^\rho(t), u^\rho(t)) - g(y(t), u(t))]dt + \Gamma_h[Eh(y_0^\rho, y^\rho(1)) - Eh(y_0, y(1))] + \\
&+ O(\|E\int_0^1 [g(y^\rho(t), u^\rho(t)) - g(y(t), u(t))]dt\|^2 + \|Eh(y_0^\rho, y^\rho(1)) - Eh(y_0, y(1))\|^2).
\end{aligned} \tag{4.1}$$

Moreover from Lemma 3.1, we have

$$\begin{aligned}
&\Gamma_g E\int_0^1 \Big[g(y^\rho(t), u^\rho(t)) - g(y(t), u(t))\Big]\,dt + \Gamma_h\Big[Eh(y_0^\rho, y^\rho(1)) - Eh(y_0, y(1))\Big] \\
=&\Gamma_g E\int_0^1 \Big[g(y(t) + y_1(t) + y_2(t), u^\rho(t)) - g(y(t), u(t))\Big]\,dt + \\
&+ \Gamma_h E\Big[h(y_0^\rho, y(1) + y_1(1) + y_2(1)) - h(y_0, y(1))\Big] + o(|I_\rho|) \\
=&\Gamma_g E\int_0^1 \Big[g(y(t) + y_1(t) + y_2(t), u(t)) - g(y(t), u(t))\Big]\,dt + \\
&+ \Gamma_h E\Big[h(y_0^\rho, y(1) + y_1(1) + y_2(1)) - h(y_0, y(1))\Big] + \\
&+ \Gamma_g E\int_0^1 \Big[g(y(t) + y_1(t) + y_2(t), u^\rho(t)) - g(y(t) + y_1(t) + y_2(t), u(t))\Big] + o(|I_\rho|) \\
=&|I_\rho|\Gamma_h\Big\langle Eh_y(y_0, y(1)), \eta\Big\rangle + \Gamma_h E\langle h_x(y_0, y(1)), y_1(1) + y_2(1)\rangle + \\
&+ \frac{1}{2}\Gamma_h Eh_{xx}(y_0, y(1))\Big(y_1(1)\otimes y_1(1)\Big) + \\
&+ \Gamma_g E\int_0^1 g_x(s)\Big[y_1(s) + y_2(s)\Big]\,ds + \frac{1}{2}\Gamma_g E\int_0^1 g_{xx}(s)\Big(y_1(s)\otimes y_1(s)\Big)\,ds + \\
&+ E\int_0^1 \Gamma_g \Delta g\big(s; u^\rho(s)\big)\,ds + o(|I_\rho|),
\end{aligned} \tag{4.2}$$

and

$$\begin{aligned}
&\Big\|E\int_0^1 \Big[g(y^\rho(t), u^\rho(t)) - g(y(t), u(t))\Big]\,dt\Big\|^2 + \Big\|Eh(y_0^\rho, y^\rho(1)) - Eh(y_0, y(1))\Big\|^2 \\
=&\Big\|E\int_0^1 g_x(s)\Big[y_1(s) + y_2(s)\Big]\,ds + \frac{1}{2}E\int_0^1 g_{xx}(s)\Big(y_1(s)\otimes y_1(s)\Big)\,ds + \\
&+ E\int_0^1 \Delta g\big(s; u^\rho(s)\big)\,ds\Big\|^2 + \Big\|E\Big\langle h_x(y_0, y(1)), y_1(1) + y_2(1)\Big\rangle + \\
&+ \frac{1}{2}Eh_{xx}(y_0, y(1))\Big(y_1(1)\otimes y_1(1)\Big)\Big\|^2 + o(|I_\rho|) = o(|I_\rho|).
\end{aligned} \tag{4.3}$$

From the optimality of $(y_0, u(\cdot))$, we conclude that

$$
\begin{aligned}
&|I_\rho|\Gamma_h\Big\langle Eh_y(y_0, y(1)), \eta\Big\rangle + \Gamma_h E\Big\langle h_x(y_0, y(1)), y_1(1) + y_2(1)\Big\rangle + \\
&+ \frac{1}{2}\Gamma_h E h_{xx}(y_0, y(1))\Big(y_1(1) \otimes y_1(1)\Big) + \\
&+ \Gamma_g E \int_0^1 g_x(s)\Big[y_1(s) + y_2(s)\Big]\, ds + \\
&+ \frac{1}{2}\Gamma_g E \int_0^1 g_{xx}(s)\Big(y_1(s) \otimes y_1(s)\Big)\, ds + \\
&+ E\int_0^1 \Gamma_g \Delta g(s; u^\rho(s))\, ds + o(|I_\rho|) \leq 0.
\end{aligned} \tag{4.4}
$$

From Theorem 3.1 of [ Chapter 1, 32 ], we see that there exist $(p(\cdot), q(\cdot), r.(\cdot))$ ,and $(P(\cdot), Q(\cdot), R.(\cdot))$ such that they satisfy equations (3.3), $(3.9)_1$ and (3.4), $(3.9)_3$ respectively.

Using Theorems 2.2 and 2.3 of [ Chapter 1, 32 ] we have from (3.3) – (3.6) and (3.9) that

$$
\begin{aligned}
&E\Big\langle \Gamma_h h_x(y_0, y(1)), y_1(1) + y_2(1)\Big\rangle \\
=&E\Big\langle p(1), y_1(1) + y_2(1)\Big\rangle \\
=&\Big\langle p(0), \eta\Big\rangle |I_\rho| + E\int_0^1 \Big[H\big(y(s), u^\rho(s), 0, p(s), q(s), r_s(\cdot)\big) - \\
&- H\big(y(s), u(s), 0, p(s), q(s), r_s(\cdot)\big)\Big]\, ds + \\
&+ \frac{1}{2}E\int_0^1 y_1(s) H_{xx}\big(y(s), u(s), 0, p(s), q(s), r_s(\cdot)\big)\Big(y_1(s) \otimes y_1(s)\Big)\, ds + \\
&+ E\int_0^1 \Big\langle q, \Delta b_x(s; u^\rho(s)) y_1(s)\Big\rangle\, ds + \\
&+ E\iint_{\mathbb{K}\times(0,1]} \Big\langle r_s(z), \Delta c_x(s; u\rho(s)) y_1(s)\Big\rangle\, \pi(dz)\, ds,
\end{aligned} \tag{4.5}
$$

$$
\begin{aligned}
& y_1(t) \otimes y_1(t) \\
&= \int_{(0,t]} T(t,s)\Big\{ a_x(s) y_1(s) \otimes y_1(s) + y_1(s) \otimes y_1(s) a_x^*(s) + \\
&+ \Big[ b_x(s) \circ W^{\frac{1}{2}} \Big] y_1(s) \otimes y_1(s) \Big[ b_x(s) \circ W^{\frac{1}{2}} \Big]^* + \\
&+ \int_{\mathbb{K}} c_x(s,z) y_1(s) \otimes y_1(s) c_x^*(s,z)\, \pi(dz) + \\
&+ \Big[ b_x(s) \circ W^{\frac{1}{2}} \Big] y_1(s) \otimes \Big[ \Delta b(s; u^\rho(s)) \circ W^{\frac{1}{2}} \Big] + \\
&+ \Big[ \Delta b(s; u^\rho(s)) \circ W^{\frac{1}{2}} \Big] \otimes y_1(s) \Big[ b_x(s) \circ W^{\frac{1}{2}} \Big]^* + \\
&+ \Big[ \Delta b(s; u^\rho(s)) \circ W^{\frac{1}{2}} \Big] \otimes \Big[ \Delta b(s; u^\rho(s)) \circ W^{\frac{1}{2}} \Big] + \\
&+ \int_{\mathbb{K}} \Big[ c_x(s,z) y_1(s) \otimes \Delta c(s,z; u^\rho(s)) + \Delta c(s,z; u^\rho(s)) \otimes y_1(s) c_x^*(s,z) + \\
&+ \Delta c(s,z; u^\rho(s)) \otimes \Delta c(s,z; u^\rho(s)) \Big]\, \pi(dz) \Big\} T^*(t,s)\, ds + \\
&+ \int_{(0,t]} T(t,s) \Big\{ \Big[ b_x(s) \circ dw(s) \Big] y_1(s) \otimes y_1(s) + y_1(s) \otimes y_1(s) \Big[ b_x(s) \circ dw(s) \Big]^* + \\
&+ y_1(s) \otimes \Big[ \Delta b(s; u^\rho(s)) \circ dw(s) \Big] + \Big[ \Delta b(s; u^\rho(s)) \circ dw(s) \Big] \otimes y_1(s) \Big\} T^*(t,s) + \\
&+ \iint_{\mathbb{K}\times(0,t]} T(t,s) \Big[ c_x(s,z) y_1(s-) \otimes y_1(s-) + y_1(s-) \otimes y_1(s-) c_x^*(s,z) + \\
&+ c_x(s,z) y_1(s-) \otimes y_1(s-) c_x^*(s,z) + \\
&+ c_x(s,z) y_1(s-) \otimes \Delta c(s,z; u^\rho(s)) + \Delta c(s,z; u^\rho(s)) \otimes y_1(s-) c_x^*(s,z) + \\
&+ \Delta c(s,z; u^\rho(s)) \otimes \Delta c(s,z; u^\rho(s)) + \\
&+ y_1(s-) \otimes \Delta c(s,z; u^\rho(s)) + \Delta c(s,z; u^\rho(s)) \otimes y_1(s-) \Big] T^*(t,s)\, \tilde{N}_k(dz\, ds),
\end{aligned}
\tag{4.6}
$$

and

$$
\begin{aligned}
& \Gamma_h E h_{xx}(y_0, y(1)) \Big( y_1 \otimes y_1(1) \Big) \\
&= E \Big\langle P(1), y_1(1) \otimes y_1(1) \Big\rangle \\
&= E \int_0^1 \Big\{ H_{xx}(y(s), u(s), \Gamma_g, p(s), q(s), r_s(\cdot)) \Big( y_1(s) \otimes y_1(s) \Big) ds + \\
&\quad + \Big\langle P(s), \Big[ \Delta b(s; u^\rho(s)) \circ W^{\frac{1}{2}} \Big] \otimes \Big[ \Delta b(s; u^\rho(s)) \circ W^{\frac{1}{2}} \Big] \Big\rangle + \\
&\quad + \Big\langle P(s), \int_{\mathbb{K}} \Delta c(s,z; u^\rho(s)) \otimes \Delta c(s,z; u^\rho(s))\, \pi(dz) \Big\rangle + \\
&\quad + \int_{\mathbb{K}} \Big\langle R_s(z), \Delta c(s,z; u^\rho(s)) \otimes \Delta c(s,z; u^\rho(s)) \Big\rangle\, \pi(dz) \Big\} ds + \\
&\quad + 2E \int_0^1 \Big\{ \Big\langle P(s), \Big[ b_x(s) \circ W^{\frac{1}{2}} \Big] y_1(s) \otimes \Big[ \Delta b(s; u^\rho(s)) \circ W^{\frac{1}{2}} \Big] \Big\rangle +
\end{aligned}
\tag{4.7}
$$

$$
\begin{aligned}
&+\Big\langle P(s), \int_{\mathbb{K}} c_x(s,z)y_1(s)\otimes \Delta c(s,z;u^\rho(s))\,\pi(dz)\Big\rangle +\\
&+\Big\langle Q(s), y_1(s)\otimes \Delta b(s;u^\rho(s))\Big\rangle + \\
&+\int_{\mathbb{K}}\Big\langle R_s(z), c_x(s,z)y_1(s)\otimes \Delta c(s,z;u^\rho(s)) + y_1(s)\otimes \Delta c(s,z;u^\rho(s))\Big\rangle\,\pi(dz)\Big\}\,ds.
\end{aligned}
\tag{4.7}
$$

Noting the estimates (3.12), we conclude from (4.1) – (4.7) that

$$
\begin{aligned}
&\Big\langle \Gamma_h E h_y(y_0^\epsilon, y^\epsilon(1)) + p(0), \eta\Big\rangle |I_\rho| + \\
&+\Gamma_g \int_0^1 l(s;u^\rho(\cdot))\,ds + o(|I_\rho|) \geq 0,
\end{aligned}
\tag{4.8}
$$

where $l(\cdot;v)$ is defined by

$$
\begin{aligned}
l(s;v) =: E\Big\{ &H\big(y(s-), v, \Gamma_g, p(s-), q(s), r_s(\cdot)\big) - \\
&- H\big(y(s-), u(s), \Gamma_g, p(s-), q(s), r_s(\cdot)\big) + \\
&+ \frac{1}{2}\Big\langle P(s-), \Delta b(s;v)\circ W \circ \Delta b^*(s;v)\Big\rangle + \\
&+ \frac{1}{2}\Big\langle P(s-), \int_{\mathbb{K}} \Delta c(s,z;v)\otimes \Delta c(s,z;v)\,\pi(dz)\Big\rangle + \\
&+ \frac{1}{2}\int_{\mathbb{K}} \Big\langle R_s(z), \Delta c(s,z;v)\otimes \Delta c(s,z;v)\Big\rangle\,\pi(dz)\Big\}.
\end{aligned}
\tag{4.9}
$$

Then for given $v(\cdot) \in U_{ad}$, applying Lemma 3.2 to the real number valued Lebesgue integrable function $l(\cdot;v(\cdot))$, we know that there exists $I_\rho \subset [0,1]$ such that

$$
\begin{aligned}
&|I_\rho| = \rho,\\
&\int_{I_\rho} l(s;v(s))\,ds = \rho\int_0^1 l(s;v(s))\,ds + o(\rho), \qquad \text{as } \rho \to 0.
\end{aligned}
\tag{4.10}
$$

Next choose the above $I_\rho$ in (3.2), and we have

$$
\int_{I_\rho} l(s;v(s))\,ds = \int_0^1 l(s;u^\rho(s))\,ds. \tag{4.11}
$$

From (4.8) – (4.11),we conclude for given $v(\cdot) \in U_{ad}$ that

$$
\begin{aligned}
&\rho\Big\langle \Gamma_h E h_y(y_0, y(1)) + p(0), \eta\Big\rangle + \rho\int_0^1 l(s;v(s))\,ds \\
&\geq o(\rho), \text{ as } \rho \to 0.
\end{aligned}
\tag{4.12}
$$

So we have

$$\Big\langle \Gamma_h E h_y(y_0, y(1)) + p(0), \eta \Big\rangle + \int_0^1 l(s; v(s))\, ds \\ \geq 0, \ \forall\, \eta \in \mathbf{R}^n \text{ and } \forall\, v(\cdot) \in U_{ad}, \tag{4.13}$$

which implies $(3.9)_2$ and the following

$$\int_0^1 E\Big\{ H(y(s), v(s), \lambda\Gamma_g, p(s), q(s), r_\cdot(s)) - H(y(s), u(s), \lambda, p(s), q(s), r_\cdot(s)) + \\ + \frac{1}{2}\Big\langle P(s), \Delta b(s; v(s)) \circ W \circ \Big(\Delta b(s; v(s))\Big)^* + \int_{\mathbb{K}} \Delta c(s, z; v(s)) \otimes \Delta c(s, z; v(s))\, \pi(dz) \Big\rangle \\ + \frac{1}{2} \int_{\mathbb{K}} \Big\langle R_s(z), \Delta c(s, z; v(s)) \otimes \Delta c(s, z; v(s)) \Big\rangle\, \pi(dz) \Big\}\, ds \geq 0, \\ \forall v(\cdot) \in U_{ad}. \tag{4.14}$$

Therefore (3.10) easily follows from (4.14).

## §5. The proof of Theorem 3.1 for the general case.

Now we begin proving Theorem 3.1.

Step 1. Application of Ekeland's variational principle leads to the reduction of general end-constraint problem to a family of free end problems.

Define the following auxiliary function

$$J(v(\cdot), x_0; \varepsilon, \delta) = \Psi(J(v(\cdot), x_0), Ef(x_0, x(1)); \varepsilon, \delta) \tag{5.1}$$

with $\Psi(\cdot, \cdot; \varepsilon, \delta)$ being defined as in §4.

Using the technique in [14],we may assume without loss of generality that $U$ is bounded. Then consider the metric space $(\mathbf{H} \times U_{ad}, d)$ with the distance $d$ defined by

$$d((x_1, v_1(\cdot)), (x_2, v_2(\cdot))) = \sqrt{\|x_1 - x_2\|^2 + \hat{d}^2(v_1(\cdot), v_2(\cdot))}, \\ \hat{d}(v_1(\cdot), v_2(\cdot)) =: \Big|\Big\{ t \in [0,1]; E\|v_1(t) - v_2(t)\|^2 > 0 \Big\}\Big|. \tag{5.2}$$

We can easily prove that $(\mathbf{H} \times U_{ad}, d)$ is complete and $J(v(\cdot), x_0; \varepsilon)$ is continuous and bounded. Also we have for any given $\varepsilon > 0$,

$$\Phi(J(v(\cdot), x_0), Ef(x_0, x(1)); \varepsilon) > 0, \qquad \forall\, (x_0, v(\cdot)) \in \mathbf{H} \times U_{ad}; \\ \Phi(J(u(\cdot), y_0), Ef(y_0, y(1)); \varepsilon) = \varepsilon; \\ J(v(\cdot), x_0; \varepsilon, \delta) > 0, \quad \forall\, (x_0, v(\cdot)) \in \mathbf{H} \times U_{ad} \text{ for sufficiently small } \delta > 0; \\ J(u(\cdot), y_0; \varepsilon, \delta) \leq \varepsilon + \inf_{(x_0, v(\cdot))} J(v(\cdot), x_0; \varepsilon, \delta). \tag{5.3}$$

Therefore we can apply Ekeland's variational principle and conclude that there exist $u^{\varepsilon\delta}(\cdot) \in U_{ad}$ and $y_0^{\varepsilon\delta} \in \mathbf{H}$ such that

1) $J(u^{\varepsilon\delta}(\cdot), y_0^{\varepsilon\delta}; \varepsilon, \delta) \leq \varepsilon$;
2) $d((y_0^{\varepsilon\delta}, u^{\varepsilon\delta}(\cdot)), (y_0, u(\cdot))) \leq \sqrt{\varepsilon}$;
3) $\bar{J}(v(\cdot), x_0; \varepsilon, \delta) =: J(v(\cdot), x_0; \varepsilon, \delta) + \sqrt{\varepsilon} d((x_0, v(\cdot)), (y_0^{\varepsilon\delta}, u^{\varepsilon\delta}(\cdot)))$
$\geq J(u^{\varepsilon\delta}(\cdot), y_0^{\varepsilon\delta}), \quad \forall\, (x_0, v(\cdot)) \in \mathbf{H} \times U_{ad}$.

Set

$$\begin{aligned}
\lambda^{\varepsilon\delta} &=: \Psi_s(J(u^{\varepsilon\delta}(\cdot), y_0^{\varepsilon\delta}), Ef(y_0^{\varepsilon\delta}, y^{\varepsilon\delta}(1)); \varepsilon, \delta),\\
\gamma^{\varepsilon\delta} &=: \Psi_z(J(u^{\varepsilon\delta}(\cdot), y_0^{\varepsilon\delta}), Ef(y_0^{\varepsilon\delta}, y^{\varepsilon\delta}(1)); \varepsilon, \delta),\\
\hat{\lambda}^{\varepsilon\delta} &=: \Phi_s(J(u^{\varepsilon\delta}(\cdot), y_0^{\varepsilon\delta}), Ef(y_0^{\varepsilon\delta}, y^{\varepsilon\delta}(1)); \varepsilon),\\
\hat{\gamma}^{\varepsilon\delta} &=: \Phi_z(J(u^{\varepsilon\delta}(\cdot), y_0^{\varepsilon\delta}), Ef(y_0^{\varepsilon\delta}, y^{\varepsilon\delta}(1)); \varepsilon).
\end{aligned} \tag{5.4}$$

From the first relation (5.3) and Lemmas 3.3-3.4, we have for each sufficiently small $\varepsilon > 0$,

$$\lim_{\delta\to 0+} (\lambda^{\varepsilon\delta} - \hat{\lambda}^{\varepsilon\delta}) = 0,$$

$$\lim_{\delta\to 0+} (\gamma^{\varepsilon\delta} - \hat{\gamma}^{\varepsilon\delta}) = 0,$$

$$|\hat{\lambda}^{\varepsilon\delta}|^2 + \|\hat{\gamma}^{\varepsilon\delta}\|^2 = 1.$$

Therefore, for each sufficiently small $\varepsilon > 0$, we can choose $\delta(\varepsilon) > 0$, such that

$$\begin{aligned}
&J(u^{\varepsilon\delta(\varepsilon)}(\cdot), y_0^{\varepsilon\delta(\varepsilon)}; \varepsilon, \delta(\varepsilon)) > 0,\\
&\delta(\varepsilon) \leq \varepsilon,\\
&|\lambda^{\varepsilon\delta(\varepsilon)}|^2 + \|\gamma^{\varepsilon\delta(\varepsilon)}\|^2 = 1 + o(1), \text{ as } \varepsilon \to 0.
\end{aligned} \tag{5.5}$$

Set

$$\lambda^{\varepsilon} =: \lambda^{\varepsilon\delta(\varepsilon)}, \gamma^{\varepsilon} =: \gamma^{\varepsilon\delta(\varepsilon)},$$

$$y_0^{\varepsilon} =: y_0^{\varepsilon\delta(\varepsilon)}, u^{\varepsilon}(\cdot) =: u^{\varepsilon\delta(\varepsilon)}(\cdot).$$

Step 2. Necessary conditions for minimization of $\bar{J}(v(\cdot), x_0; \varepsilon, \delta(\varepsilon))$ at $(y_0^{\varepsilon}, u^{\varepsilon}(\cdot))$.

Let $y^{\varepsilon}(\cdot)$ denote the solution of (2.1) corresponding to $(y_0, u(\cdot))$. For each $v \in U$, we introduce the following simplified notations

$$\begin{aligned}
\Delta m^{\varepsilon}(s; v) &=: m(y^{\varepsilon}(s-), v) - m(y^{\varepsilon}(s-), u^{\varepsilon}(s)),\\
m^{\varepsilon}(s) &=: m(y^{\varepsilon}(s-), u^{\varepsilon}(s)),\\
\Delta n^{\varepsilon}(s, z; v) &=: n(y^{\varepsilon}(s-), v, z) - n(y^{\varepsilon}(s-), u^{\varepsilon}(s), z),\\
n^{\varepsilon}(s, z) &=: n(y^{\varepsilon}(s-), u^{\varepsilon}(s), z),
\end{aligned}$$

with $m$ standing for $a, b, g$ and all their (up to second-) derivatives in $x$,and $n$ for $c$ and its (up to second-) derivatives in $x$.

From Theorem 3.1 of [ Chapter 1,32 ], we see that the following two equations

$$\begin{cases} p^\varepsilon(t) = p^\varepsilon(1) + \int_{(t,1]} T^*(s,t) H_x(y^\varepsilon(s), u^\varepsilon(s), \lambda^\varepsilon \Gamma_g^\varepsilon, p^\varepsilon(s), q^\varepsilon(s), r_s^\varepsilon(\cdot)) T(s,t)\, ds - \\ \quad - \int_{(t,1]} T^*(s,t) q^\varepsilon(s) dw(s) T(s,t) - \\ \quad - \iint_{\mathbb{K}\times(t,1]} T^*(s,t) r_s^\varepsilon(z) T(s,t) \widetilde{N}_k(dz\, ds), \\ p^\varepsilon(1) = \lambda^\varepsilon \Gamma_h^\varepsilon h_x(y_0^\varepsilon, y^\varepsilon(1)) + \gamma^\varepsilon f_x(y_0^\varepsilon, y^\varepsilon(1)) \end{cases} \tag{5.6}$$

and

$$\begin{cases} P^\varepsilon(t) = T^*(1,t) P^\varepsilon(1) T(1,t) + \int_{(t,1]} T^*(s,t) \Big\{ a_x^{\varepsilon *}(s) P^\varepsilon(s) + P^\varepsilon(s) a_x^\varepsilon(s) + \\ \quad + \left[b_x^\varepsilon(s) \circ W^{\frac{1}{2}}\right]^* P^\varepsilon(s) \left[b_x^\varepsilon(s) \circ W^{\frac{1}{2}}\right] + \int_{\mathbb{K}} c_x^{\varepsilon *}(s,z) P(s) c_x^\varepsilon(s,z)\, \pi(dz) + \\ \quad + \left[b_x^\varepsilon(s) \circ W^{\frac{1}{2}}\right]^* \left[Q^\varepsilon(s) \circ W^{\frac{1}{2}}\right] + \left[Q^\varepsilon(s) \circ W^{\frac{1}{2}}\right] \left[b_x^\varepsilon(s) \circ W^{\frac{1}{2}}\right] + \\ \quad + \int_{\mathbb{K}} \left[ c_x^{\varepsilon *}(s,z) R_s^\varepsilon(z) c_x^\varepsilon(s,z) + c_x^{\varepsilon *}(s,z) R_s^\varepsilon(z) + R_s^\varepsilon(z) c_x^\varepsilon(s,z) \right] \pi(dz) + \\ \quad + H_{xx}(y^\varepsilon(s), u^\varepsilon(s), \lambda^\varepsilon \Gamma_g^\varepsilon, p^\varepsilon(s), q^\varepsilon(s), r_s^\varepsilon(\cdot)) \Big\} T(s,t)\, ds - \\ \quad - \int_{(t,1]} T^*(s,t) \left[ Q^\varepsilon(s) \circ dw(s) \right] T(s,t) - \\ \quad - \iint_{\mathbb{K}\times(t,1]} T^*(s,t) R_s^\varepsilon(z) T(s,t) \widetilde{N}_k(dz\, ds), \\ P^\varepsilon(1) = \lambda^\varepsilon \Gamma_h^\varepsilon h_{xx}(y_0^\varepsilon, y^\varepsilon(1)) + \gamma^\varepsilon f_{xx}(y_0^\varepsilon, y^\varepsilon(1)) \end{cases} \tag{5.7}$$

have unique solutions $(p^\varepsilon(\cdot), q^\varepsilon(\cdot), r_\cdot^\varepsilon(\cdot))$ and $(P^\varepsilon(\cdot), Q^\varepsilon(\cdot), R_\cdot^\varepsilon(\cdot))$ respectively,with $p^\varepsilon(\cdot)$ and $P^\varepsilon(\cdot)$ being cadlag processes.

Proceeding similarly as in §4, we easily conclude from 3) of Step 1 that

$$\begin{aligned} &\left\langle \lambda^\varepsilon \Gamma_h^\varepsilon E h_y(y_0^\varepsilon, y^\varepsilon(1)) + \gamma^\varepsilon E f_y(y_0^\varepsilon, y^\varepsilon(1)) + p^\varepsilon(0), \eta \right\rangle |I_\rho| + \\ &\quad + \int_0^1 l^\varepsilon(s; u^{\varepsilon\rho}(\cdot))\, ds + o(|I_\rho|) \geq -|I_\rho| \sqrt{\varepsilon} \sqrt{1 + |\eta|^2}, \end{aligned} \tag{5.8}$$

where $l^\varepsilon(\cdot;v)$ is defined by

$$\begin{aligned} l^\varepsilon(s;v) =:E\Big\{&H(y^\varepsilon(s-),v,\lambda^\varepsilon\Gamma_g^\varepsilon,p^\varepsilon(s-),q^\varepsilon(s),r_s^\varepsilon(\cdot))- \\ &-H(y^\varepsilon(s-),u^\varepsilon(s),\lambda^\varepsilon\Gamma_g^\varepsilon,p^\varepsilon(s-),q^\varepsilon(s),r_s^\varepsilon(\cdot))+ \\ &+\frac{1}{2}\Big\langle P^\varepsilon(s-),\Delta b^\varepsilon(s;v)\circ W\circ\Delta b^{\varepsilon *}(s;v)\Big\rangle+ \\ &+\frac{1}{2}\Big\langle P^\varepsilon(s-),\int_{\mathbb{K}}\Delta c^\varepsilon(s,z;v)\otimes\Delta c^\varepsilon(s,z;v)\,\pi(dz)\Big\rangle+ \\ &+\frac{1}{2}\int_{\mathbb{K}}\Big\langle R_s(z),\Delta c^\varepsilon(s,z;v)\otimes\Delta c^\varepsilon(s,z;v)\Big\rangle\,\pi(dz)\Big\}. \end{aligned} \tag{5.9}$$

Then for given $v(\cdot)\in U_{ad}$, applying Lemma 2.2 of Chapter 2 [32] to this real number valued Lebesgue integrable function, we know that there exists $I_\rho\subset[0,1]$ such that

$$\begin{aligned} &|I_\rho|=\rho, \\ &\int_{I_\rho} l^\varepsilon(s;v(s))\,ds=\rho\int_0^1 l^\varepsilon(s;v(s))\,ds+o(\rho), \qquad \text{as } \rho\to 0. \end{aligned} \tag{5.10}$$

Next choose the above $I_\rho$ in (3.2), and we have

$$\int_{I_\rho} l^\varepsilon(s;v(s))\,ds=\int_0^1 l^\varepsilon(s;u^{\varepsilon\rho}(s))\,ds. \tag{5.11}$$

From (5.8) – (5.11), we conclude for given $v(\cdot)\in U_{ad}$ that

$$\begin{aligned} &\rho\Big\langle \lambda^\varepsilon\Gamma_h^\varepsilon Eh_y(y_0^\varepsilon,y^\varepsilon(1))+\gamma^\varepsilon Ef_y(y_0^\varepsilon,y^\varepsilon(1))+p^\varepsilon(0),\eta\Big\rangle+\rho\int_0^1 l^\varepsilon(s;v(s))\,ds \\ &\qquad\geq-\rho\sqrt{\varepsilon}\sqrt{1+|\eta|^2}+o(\rho), \text{ as } \rho\to 0. \end{aligned} \tag{5.12}$$

So we have

$$\begin{aligned} &\Big\langle \lambda^\varepsilon\Gamma_h^\varepsilon Eh_y(y_0^\varepsilon,y^\varepsilon(1))+\gamma^\varepsilon Ef_y(y_0^\varepsilon,y^\varepsilon(1))+p^\varepsilon(0),\eta\Big\rangle+\int_0^1 l^\varepsilon(s;v(s))\,ds \\ &\qquad\geq-\sqrt{\varepsilon}\sqrt{1+|\eta|^2}, \quad \forall\,\eta\in\mathbf{R}^n \text{ and } \forall\, v(\cdot)\in U_{ad}, \end{aligned} \tag{5.13}$$

which implies that

$$\begin{aligned} &\Big\|p^\varepsilon(0)+\lambda^\varepsilon\Gamma_h^\varepsilon Eh_y(y_0^\varepsilon,y^\varepsilon(1))+\gamma^\varepsilon Ef_y(y_0^\varepsilon,y^\varepsilon(1))\Big\|\leq C\sqrt{\varepsilon}, \\ &\int_0^1 l^\varepsilon(s;v(s))\,ds\geq-\sqrt{\varepsilon}, \quad \forall\, v(\cdot)\in U_{ad}. \end{aligned} \tag{5.14}$$

Applying Lemmas 3.2-3.3, we easily show that $\lambda^\varepsilon \geq 0$ and $u^\varepsilon \in \mathbf{F}$ satisfy the following properties

$$\left\langle \gamma^\varepsilon, z - Ef(y_0^\varepsilon, y^\varepsilon(1)) \right\rangle \leq \delta(\varepsilon) \leq \varepsilon, \quad \forall\, z \in Q; \tag{5.15}$$

$$|\lambda^\varepsilon|^2 + \|\gamma^\varepsilon\|^2 = 1 + o(1). \tag{5.16}$$

Step 3. Passing to the limit $\varepsilon \to 0$, we conclude the proof.

The key point is to show that there exists some weak sub-limit $(\lambda, \gamma)$ of $(\lambda^\varepsilon, \gamma^\varepsilon)$ as $\varepsilon \to 0+$, not identically zero. We will see that the nontriviality property follows from the condition of finite-codimension and (5.16).

Let $\lambda^\varepsilon \to 0$ (or the nontriviality property is obtained) as $\varepsilon \to 0+$. Then from (5.13) and (5.16), we have

$$\begin{aligned} &\left\langle \gamma^\varepsilon, z - Ef(y_0^\varepsilon, y^\varepsilon(1)) - Ef_y(y_0^\varepsilon, y^\varepsilon(1))\eta \right\rangle - \langle p^\varepsilon(0), \eta\rangle - \\ &\quad - \int_0^1 l^\varepsilon(s; v(\cdot))\,ds \leq o(1), \ \forall\, z \in Q \ \text{ as } \varepsilon \to 0+. \end{aligned} \tag{5.17}$$

Compute $\langle p^\varepsilon(1), \tilde{y}_1(1) + \tilde{y}_2(1)\rangle$, $\langle P^\varepsilon(1), \tilde{y}_1(1) \otimes \tilde{y}_1(1)\rangle$ and combine them with (5.17), and we get

$$\begin{aligned} &\left\langle \gamma^\varepsilon, z - Ef(y_0^\varepsilon, y^\varepsilon(1)) - Ef_y(y_0^\varepsilon, y^\varepsilon(1))\eta - Ef_x(y_0^\varepsilon, y^\varepsilon(1))(\tilde{y}_1(1) + \tilde{y}_2(1)) - \right. \\ &\quad \left. - \frac{1}{2} Ef_{xx}(y_0^\varepsilon, y^\varepsilon(1))(\tilde{y}_1(1) \otimes \tilde{y}_1(1)) \right\rangle \leq o(1), \\ &\quad \forall\, z \in Q, \eta \in \mathbf{H}, \ v(\cdot) \in U_{ad} \ \text{ as } \varepsilon \to 0+, \end{aligned}$$

which implies that

$$\begin{aligned} &\left\langle \gamma^\varepsilon, z - Ef(y_0, y(1)) - Ef_y(y_0, y(1))\eta - Ef_x(y_0, y(1))(\tilde{y}_1(1) + \tilde{y}_2(1)) - \right. \\ &\quad \left. - \frac{1}{2} Ef_{xx}(y_0, y(1))(\tilde{y}_1(1) \otimes \tilde{y}_1(1)) \right\rangle \leq o(1), \\ &\quad \forall\, z \in Q, \ \eta \in \mathbf{H}, \ v(\cdot) \in U_{ad}, \ \text{as } \varepsilon \to 0+. \end{aligned} \tag{5.19}$$

Now by Lemma 3.2 of [23], we may assume without loss of generality that

$$\lim_{\varepsilon \to 0+} \lambda^\varepsilon = \lambda, \quad \lim_{\varepsilon \to 0+} \gamma^\varepsilon = \gamma,$$

with (3.8) being satisfied. The remainder of the proof is referred to the proof of Theorem 2.1 of [ Chapter 2, 32 ].

## §6. Conclusion.

The stochastic maximum principle with random jumps is quite different from the one corresponding to a pure diffusion system. In calculating the variation for the index functional, we use only a property of Lebesgue integrals in the real number space **R**. Thus the study of maximum principles can be unified for finite-dimensional,infinite-dimensional,deterministic and stochastic optimal controls.

**Acknowledgement**. The authors would like to thank Professor Yong Jiongmin for useful conversations related to this subject.

## References

[1] A. Bensoussan, *Stochastic Control by Functional Analysis Methods,* North-Holland, Amsterdam, 1982.

[2] A. Bensoussan, *Lectures on Stochastic Control, Lecture Notes in Mathematics,* Vol. 972, *Nonlinear Filtering and Stochastic Control,* Proceedings, Cortona, 1981.

[3] A. Bensoussan, *Perturbation Methods in Optimal Control,* Dunod, Gauthier-Villars, 1988.

[4] A. Bensoussan, *Stochastic Maximum Principle for Distributed Parameter System, J. of the Francline Institute,* 315 (1983), 387–406.

[5] A. Bensoussan, *Maximum Principle and Dynamic Programming Approaches of the Optimal Control of Partially Observed Diffusions, Stochastics,* 9 (1983), 169–222.

[6] J. M. Bismut, *An Introductary Approach to Duality in Optimal Stochastic control, SIAM Rev.,* 20 (1978), 62–78.

[7] M. G. Crandall and H. Ishii, *The Maximum Principle for Semicontinuous Functions, Diff. Int. Eqn.,* 3 (1990), 1001–1014.

[8] I. Ekeland, *Sur les Problems Variationels, Acad. Sci. Paris,* 275 (1972), 1057–1059.

[9] I. Ekeland, *Nonconvex Minimization Problems, Bull. Amer. Math. Soc. (NS),* 1 (1979), 443–474.

[10] W. H. Fleming, *Optimal Continuous-parameter Stochastic Control, SIAM Rev.,* 11 (1969), 470–509.

[11] U. G. Haussmann, *General Necessary Conditions for Optimal Control of Stochastic System, Math. Programming Study,* 6 (1976), 34–58.

[12] E. Hille and R. Phillips, *Functional Analysis and Semigroups*, Vol.31, *American Mathematical Society*, Colloquium Publication, Providence, R. I., 1957.

[13] Y. Hu, *Maximum Principle of Optimal Control for Markov Processes*, Acta *Mathematica Sinica*, 33 (1990), 43–56.

[14] Y. Hu and S. Peng, *Maximum Principle for Semilinear Stochastic Evolution Control Systems*, *Stochastics and Stochastics Reports*, 33 (1990), 159–180.

[15] Y. Hu and S. Peng, *Adapted Solution of a Backward Semilinear Stochastic Evolution Equation*, *Stochastic Analysis and Applications*, 9 (1991), 445–459.

[16] A. Ichikawa, *Stability of Semilinear Stochastic Evolution Equations*, *J. Math. Anal. Appl.*, 90 (1982), 12–44.

[17] N. Ikeda and S. Watanabe, *Stochastic Differential Equatiopns and Diffusion Processes*,North-Holland, Kodansha, 1989.

[18] H. Kunita and S. Watanabe, *On Square Integrable Martingales*, *Nagayo Math. Journal*, 30 (1967), 209–245.

[19] H. J. Kushner, *Necessary Conditions for Continuous Parameter Stochastic Optimization Problems*, *SIAM J. Control*, 10 (1972), 550–565.

[20] E. B. Lee and L. Markus, *Foundations of Optimal Control Theory*, N.Y., Wiley, 1967.

[21] Xiaojun Li, *Optimal Stochastic Control Problems in Hilbert Spaces*, Ph D. Thesis, Zhongshan University, Guangzhou, Guangdong Province.

[22] Xunjing Li and Yunlong Yao, *Maximum Principle of Distributed Parameter Systems with Time Lags*, *Distributed Parameter Systems, Lecture Notes in Control and Information* Sciences 75, Springer-Verlag, NY, 1985, 410–427.

[23] Xunjing Li and Jiongmin Yong, *Necessary Conditions for Optimal Control of Distributed Parameter Systems*, *SIAM J. Control* and *Optimization*, 29 (1991), 895–908.

[24] M. Métivier, *Semimartingales*, de Gruyter, 1982.

[25] E. Pardoux and S. G. Peng, *Adapted Solution of Backward Stochastic Equation*, *Systems and Control Letters*, 14 (1990), 55–61.

[26] S. G. Peng, *A General Stochastic Maximum Principle for Optimal Control Problems*, *SIAM J. Control*, 28 (1990), 966–979.

[27] L. S. Pontryagin, *The Mathematical Theory of Optimal Processes*, *L. S. Pontryagin SELECTED WORKS*, *Vol. 4*, Gordon and Breach Science Publications.

[28] R. Situ, *A Maximum Principle for Optimal Controls of Stochastic Systems with Random Jumps*, *Proceedings of National Conference on Control Theory and Its Applications*, 1991.

[29] S. J. Tang and X. J. Li, *Necessary Conditions for Optimal Control of Stochastic Systems with Random Jumps*, preprint.

[30] X. Y. Zhou, *The Connection between the Maximum Principle and Dynamic Programming in Stochastic Control*, Stochastics and Stochastics Reports, 1990, 1–13.

[31] X. Y. Zhou, *A Unified Treatment of Maximum Principle and Dynamic Programming in Stochastic Controls*, Stochastics and Stochastics Reports, 36 (1991), 137–164.

[32] Shanjian Tang, *Optimal Control of Stochastic Systems with Random Jumps*, Ph D. thesis, Fudan University, 1992.

# 53 Spillover Problem and Global Dynamics of Nonlinear Beam Equations

**Yuncheng You** University of South Florida, Tampa, Florida

**Abstract.** In this paper a nonlinear extensible and elastic beam equation with structural damping and Balakrishnan-Taylor damping is considered. The "spillover" problem in stabilization with feedback controllers involving only a finite number of modes is resolved in this paper by an approach of showing the existence of inertial manifolds for the uncontrolled equation. Based upon that existence, a stabilization of the system at a uniform exponential rate is achieved by a linear finite-dimensional feedback control which is robust with respect to the uncertainty of parameters.

## § 1. Introduction

The objective of this paper is to study the following initial-boundary value problem of a nonlinear beam equation, cf. Bass and Zes (1991),

$$(1.1)\qquad u_{tt} + \alpha u_{xxxx} - \delta u_{xxt} - \left[a + b\int_0^1 |u_x(t,\xi)|^2 d\xi + q\left(\int_0^1 (u_x u_{xt})(t,\xi)d\xi\right)^{2(n+\beta)+1}\right]u_{xx} = f,$$

for $t > 0$, and $x \in (0, 1)$, and

$$(1.2)\qquad \begin{aligned} &u(t, 0) = u_{xx}(t, 0) = u(t, 1) = u_{xx}(t, 1) = 0, && \text{for } t \geq 0, \\ &u(0, x) = u_o(x), \qquad u_t(0, x) = u_1(x), && \text{for } x \in [0,1]. \end{aligned}$$

Here u(t, x) is the dynamical transverse deflection of the beam. Assume that $\alpha = EI/\rho$ with $\rho$ the density, E the Young modulus of elasticity, I the cross-sectional moment of inertia, $\delta = C/\rho$ with C the coefficient of viscous structural damping, $a = H/\rho$ with H the axial force (tension or compression) per unit length, $b = EA_c/2$ with $A_c$ the area of the uniform cross-section, and $q = \Gamma/\rho$ with $\Gamma$ the coefficient of Balakrishnan damping. All these parameters are assumed to be positive constants except that $a \in \mathbf{R}$. The term $-\delta u_{xxt}$ represents the *structural damping,* $[a + b\|u_x\|^2]u_{xx}$ is the tension from the extensibility, and

$$(1.3)\qquad u_{xx}\left(\int_0^1 (u_x u_{xt})dx\right)^{2(n+\beta)+1}, \text{ with } 0 \leq \beta < 1/2,\ n \geq 0 \text{ integer},$$

stands for Balakrishnan-Taylor damping, and $f = f(t, x)$ is an external input which in this paper is a control function. In this work we consider the hinged boundary condition (1.2) at two endpoints. We also make a simplification assumption that in Balakrishnan damping term (1.3), $\beta = n = 0$ so that the exponent is 1. The case with full exponents $2(n+\beta) + 1$ will be treated in a separate future paper.

As a significant mathematical model of (small-amplitude) deflection and control of nonlinear flexural aerospace structures, there is a "spillover" problem which was initially put forward by Balakrishnan (1981). The problem concerns whether it is possible and how to design a control involving only finitely many modes and achieving a high performance, for instance, in terms of stabilizing the system of some parameter uncertainty at a uniform rate. In the Technical Report by Bass and Zes (1991), only the Lyapunov-Schimidt bifurcation method was used to prove the global existence of solutions for the uncontrolled equation under the additional conditions that $a > 0$ and that initial data be small in certain sense.

In this work we deal with the spillover problem of stabilization in connection with the global dynamics of the uncontrolled equation. Taking this viewpoint, first we investigate the dissipative property of the uncontrolled equation, take advantage of the nonlocal nonlinearity to prove the existence of flat inertial manifolds by direct energy estimates, then based upon the obtained inertial form, which provides a precise attracting subflow governed by a system of ordinary differential equations, we construct a robust stabilizing feedback control linearly involving the first finite eigen-modes. This approach is expected to go beyond this concrete model and lead to more results on the high performance control and its computational implementation of aerospace structure systems and fluid flow.

We refer the concepts and theory of inertial manifolds to the founding paper, Foias-Sell-Temam (1988). The rest of the paper is organized as follows. In § 2, the initial-boundary value problem (1.1)-(1.2) without control is formulated as an abstract evolution equation in a Hilbert space which is physically the energy space.The properties of the linear operator and the local existence of solutions along with the regularity are studied. In §3 the global existence of solutions and the absorbing property are proved by *a priori* estimates on the approriately chosen quasi-energy functionals. In § 4, by making a spectral decomposition, we prove the existence of flat inertial manifolds and estimate its dimension in terms of the parameters. In § 5, based upon the obtained inertial form, the spillover problem of stabilization is resolved, and the implementable feedback control which involves only finitely many master modes is robust with respect to the parameter uncertainty.

## § 2. Formulation as Abstract Evolution Equation

To study the dynamics and control of the nonlinear partial differential equation (1.1) with the boundary and initial conditions (1.2) under the simplification assumption, in this section we first formulate the following uncontrolled equation

(2.1) $$u_{tt} + \alpha u_{xxxx} - \delta u_{xxt} - [a + b \int_0^1 |u_x(t,\xi)|^2 d\xi + q( \int_0^1 (u_x u_{xt})(t,\xi) d\xi ) ] u_{xx} = 0,$$

$$\text{for } t > 0, \text{ and } x \in (0, 1),$$

$$u(t, 0) = u_{xx}(t, 0) = u(t, 1) = u_{xx}(t, 1) = 0, \quad \text{for } t \geq 0,$$

$$u(0, x) = u_0(x), \qquad u_t(0, x) = u_1(x), \qquad \text{for } x \in [0,1],$$

as an abstract semilinear evolution equation and consider the existence and properties of local solutions. Denote by $H = L^2(0, 1)$ with its norm $|\,.\,|$ and inner-product $\langle \, , \, \rangle$. Define a linear operator $A: D(A) \to H$ by

(2.2) $$A\varphi = \frac{d^4\varphi}{dx^4} \text{ (in the distribution sense)}, \quad \forall \varphi \in D(A),$$

$$D(A) = \{\varphi \in H^4(0, 1): \varphi(0) = \varphi''(0) = \varphi(1) = \varphi''(1) = 0\}.$$

In this paper, we shall denote by BL(W) the space formed by all the bounded linear operators from a Banach space W to itself.

**Lemma 2.1** *The operator* A *is densely defined, self-adjoint, and coercively positive, with compact resolvent* $A^{-1}$. *The spectrum* $\sigma(A)$ *consists of eigenvalues* $\{\lambda_k = k^4\pi^4 : k = 1, 2, \ldots\}$ *of multiplicity one, with the eigenvectors* $\{e_k = \sqrt{2}\sin(k\pi x): k = 1, 2, \ldots\}$.

The proof of Lemma 2.1 is omitted. In terms of the eigen-expansion with respect to the orthonormal basis consisting of $\{e_k\}_{k=1}^{\infty}$, the fractional power operator $A^{1/2}$ is given by

$$(2.3) \qquad A^{1/2}\varphi = - d^2\varphi/dx^2, \quad D(A^{1/2}) = \{\varphi = \sum \varphi_k e_k \in H^2(0, 1): \sum |\varphi_k|^2(k\pi)^4 < \infty\}.$$

Note that for a function $\varphi(x) \in D(A^{1/2})$, the boundary conditions are interpreted as $\varphi(0) = \varphi(1) = 0$ in the usual sense, but $\varphi_{xx}(0) = \varphi_{xx}(1) = 0$ holds in the Sobolev space $H^{-(1/2+\varepsilon)}$ for any small $\varepsilon > 0$. Another observation is that even though $A^{1/4}\varphi \neq \varphi_x$, but it is true that

$$(2.4) \qquad |\varphi_x|^2 = |A^{1/4}\varphi|^2 \qquad \text{for } \varphi \in D(A^{1/4}).$$

This can be verified by the eigen-expansion together with the fact that $M(\varphi_x) = \int_0^1 \varphi_x \, dx = 0$.

Thus the original equation (2.1) can be formulated as a second-order evolution equation

$$(2.5) \qquad \frac{d^2u}{dt^2} + \alpha Au + \delta A^{1/2}\frac{du}{dt} + [a + b|A^{1/4}u|^2 + q\langle A^{1/2}u, u_t\rangle]A^{1/2}u = 0, \quad t > 0,$$

$$u(0) = u_o, \quad \frac{du}{dt}(0) = u_1,$$

where we denote by u(t) a time-dependent abstract function u(t, .) valued in a Hilbert space which will be specified below along with the initial data, according to our regularity setting for the solutions.Denote by $V = D(A^{1/2})$ with the norm $\|v\| = |A^{1/2}v|$, which is equivalent to the graph norm or to the norm of Sobolev space $H^2(0,1)$. Define a product real Hilbert space $E = V \times H$ with the norm $\|(v, h)\|_E = [\|v\|^2 + |h|^2]^{1/2}$. Basically we assume that $(u_o,u_1) \in E$, since $(1/2)\|(u, u_t)\|_E^2$ represents the total energy (the potential energy plus the kinetic energy) of the physical status $(u, u_t)$ at any time instant, the space E can be interpreted as the energy space. Furthermore define a comprehensive linear operator

$$(2.6) \qquad G = \begin{pmatrix} 0 & I_V \\ -\alpha A & -\delta A^{1/2} \end{pmatrix} : D(G)\ (= D(A) \times D(A^{1/2})) \rightarrow E,$$

in which $I_V$ is the identity on V. Also define a nonlinear mapping g by

$$(2.7) \qquad g\begin{pmatrix} \varphi \\ \psi \end{pmatrix} = \begin{pmatrix} 0 \\ -[a + b|A^{1/4}\varphi|^2 + q\langle A^{1/4}\varphi\ A^{1/4}\psi\rangle]A^{1/2}\varphi \end{pmatrix}.$$

Then the equation (2.7) can be formulated as a first-order semilinear evolution equstion:

$$(2.8) \qquad \frac{d}{dt}\begin{pmatrix} u \\ v \end{pmatrix} = G\begin{pmatrix} u(t) \\ v(t) \end{pmatrix} + g\begin{pmatrix} u(t) \\ v(t) \end{pmatrix}, \quad t \geq 0, \quad \begin{pmatrix} u(0) \\ v(0) \end{pmatrix} = \begin{pmatrix} u_o \\ v_o \end{pmatrix} \in E,$$

or, let $w(t) = \begin{pmatrix} u(t) \\ v(t) \end{pmatrix}$ and $w_o = \begin{pmatrix} u_o \\ v_o \end{pmatrix}$, it is written as

$$(2.9) \qquad \frac{d}{dt} w = Gw + g(w), \quad t \geq 0, \quad w_o \in E.$$

**Lemma 2.2** *The operator* - G *is sectorial,* G *generates an analytic semigroup of contraction, denoted by* $\{T(t), t \geq 0\}$, *and* G *has compact resolvent* $G^{-1} \in BL(E)$. *Besides its spectrum* $\sigma(G)$ *consists of eigenvalues given by*

$$\sigma(G) = \{\mu_k^+ = \frac{\delta k^2\pi^2}{2}(-1 + \sqrt{1 - 4\alpha/\delta^2}) \text{ and } \mu_k^- = \frac{\delta k^2\pi^2}{2}(-1 - \sqrt{1 - 4\alpha/\delta^2})\}_{k=1}^{\infty},$$

*with the corresponding normalized eigenvectors*

$$\{(k^4\pi^4 + |\mu_k^+|^2)^{-1/2}\begin{pmatrix} e_k \\ \mu_k^+ e_k \end{pmatrix} \text{ and } (k^4\pi^4 + |\mu_k^-|^2)^{-1/2}\begin{pmatrix} e_k \\ \mu_k^- e_k \end{pmatrix}\}_{k=1}^{\infty}.$$

Proof. That G is a sectorial operator and generates an analytic semigroup of contraction in this case with the structural damping was proved in Chen and Triggiani (1989). Also G is dissipative so that its generated semigroup T(t) is contraction for $t \geq 0$. As to the eigenvalues and eigenvectors, we can calculate them directly from the fact that $\mu \in \sigma(G)$ if and only if there is nonzero $\varphi \in D(A)$, such that $\mu^2\varphi + \mu\delta A^{1/2}\varphi + \alpha A\varphi = 0$. ¶

Denote by $E^1$ = the Hilbert space $D((-G)^{1/2})$ with the graph norm, and $E^2$ = the Hilbert space $D(A) \times D(A^{1/2})$ with the graph norm.

**Lemma 2.3** *The nonlinear mapping* $g: E \to E$ (resp. $g: E^1 \to E^1$) *is locally Lipschitz continuous and maps any bounded set of* E *to a bounded set of* E (*resp. for* $E^1$).

Proof. Note that (2.7) can be written as

$$g\begin{pmatrix} \varphi \\ \psi \end{pmatrix} = \begin{pmatrix} 0 \\ [a + b|\varphi_x|^2 - q\langle\varphi_{xx}\, \psi\rangle]\varphi_{xx} \end{pmatrix}.$$

Therefore it is a mapping from E to E and satisfies the local Lipschitz continuity and boundedness as described. The same argument also goes through regarding $g: E^1 \to E^1$ .¶

An E-valued function $w(t) = (u(t), v(t))$ is called a mild solution of the semilinear evolution equation (2.9) if $w \in C([0, \tau]; E)$ for some $\tau > 0$ and it satisfies the following integral equation

$$w(t) = T(t)w_0 + \int_0^t T(t - s)g(w(s))ds\,, \qquad \text{for } t \in [0,\tau]. \tag{2.10}$$

We refer the definitions of *strong* solution and *classical* solution of evolution equations to Pazy (1983).

**Lemma 2.4** *For any* $w_0 \in E$, *there is a* $\tau = \tau(w_0) > 0$ *such that the mild solution of the equation* (2.9) *with the initial condition* $w(0) = w_0$ *exists uniquely for* $t \in [0,\tau]$, *and* $w \in C([0,\tau]; E)) \cap C^1((0,\tau); E) \cap C((0,\tau); E^2)$. *If* $w_0 \in E^2$, *then this mild solution is a classical solution of* (2.9) *for* $t \in [0,\tau]$.

Proof. These are the consequences of the basic results, cf. Theorem 6.3.1 and Theorem 4.3.5 of Pazy (1983). It should be mentioned that this mild solution $w(t)$, $t \in [0,\tau]$, is actually (locally) Holder continuous with an exponent $0 < \gamma < 1/2$. ¶

## § 3. Dissipation of the Semiflow

In this section we will prove the global existence of mild solutions of the equation (2.9) and the dissipation property of the semiflow generated by the solution semigroup in terms of the existence of absorbing sets in E.

To introduce the terminology, suppose that the mild solution $w(t; w_0)$ of (2.9) with $w(0; w_0) = w_0 \in E$ exists uniquely on $[0,\infty)$, with the continuous dependence on $(t,w_0) \in \mathbf{R}^+ \times E$, then the mapping $S(t)$: $w_0 \to w(t; w_0)$ is called the *solution semigroup*, and the mapping $\vartheta(t, w_0) = S(t)w_0 : \mathbf{R}^+ \times E \to E$ is called the *semiflow* generated by the evolution equation (2.9). A subset N of E is called an *absorbing set* for the semiflow $\vartheta$ if for any bounded set Z of E, there exists a $t_0 = t_0(Z)$ such that, after the transient period $[0, t_0]$, the trajectory $\{S(t)w_0 : t \geq t_0\} \subset N$, for all $w_0 \in Z$. The semiflow $\vartheta$ is *dissipative* if there exists an absorbing set in E.

**Lemma 3.1** *For* $w_0 \in E$, *the mild solution of the equation* (2.9) *satisfies the following inequality for* $t \in I_{max}$ (*the maximal interval* $[0, \tau_{max})$ *of existence*),

$$(3.1) \qquad |u_t|^2 + \alpha |u_{xx}|^2 + \frac{1}{2b}(a + b|u_x|^2)^2 + \varepsilon\langle u_t, u\rangle + \frac{\varepsilon\delta}{2}|u_x|^2 + \frac{\varepsilon q}{4}|u_x|^4$$

$$\leq \exp(-\frac{\varepsilon}{2}t)\{|u_1|^2 + \alpha|u_{oxx}|^2 + \frac{1}{2b}(a + b|u_{ox}|^2)^2 + \varepsilon\langle u_1,u_o\rangle + \frac{\varepsilon\delta}{2}|u_{ox}|^2 + \frac{\varepsilon q}{4}|u_{ox}|^4\} + \frac{a^2}{b},$$

*where* $\varepsilon > 0$ *is a sufficiently small and uniform constant.*

Proof. As a first step, assume that $w_0 \in E^2$ so that the mild solution $w(t)$ of the equation (2.9) with such an initial point is a classical solution over $I_{max}$, which means that the first component function $u(t) \in D(A)$ of the solution $w(t)$ satisfies the differential equation (2.1) for all $t \in I_{max}$. Take the inner-product in H of the equation (2.1) with $2u_t$, we have

$$(3.2) \qquad \frac{d}{dt}(|u_t|^2 + \alpha|u_{xx}|^2) + 2\delta|u_{xt}|^2 + [a + b|u_x|^2 + (q/2)\frac{d}{dt}|u_x|^2]\frac{d}{dt}|u_x|^2$$

$$= \frac{d}{dt}[|u_t|^2 + \alpha|u_{xx}|^2 + \frac{1}{2b}(a + b|u_x|^2)^2] + 2\delta|u_{xt}|^2 + (q/2)(\frac{d}{dt}|u_x|^2)^2 = 0.$$

Then take the inner-product in H of the equation (2.1) with $\varepsilon u$, where $\varepsilon > 0$ is an undetermined constant, we have

$$(3.3) \quad \frac{d}{dt}(\varepsilon\langle u_t, u\rangle + \frac{\varepsilon\delta}{2}|u_x|^2) + \varepsilon\alpha|u_{xx}|^2 - \varepsilon|u_t|^2 + \varepsilon[a + b|u_x|^2 + (q/2)\frac{d}{dt}|u_x|^2]|u_x|^2$$

$$= \frac{d}{dt}(\varepsilon\langle u_t, u\rangle + \frac{\varepsilon\delta}{2}|u_x|^2) + \varepsilon\alpha|u_{xx}|^2 - \varepsilon|u_t|^2 + \varepsilon[\frac{1}{2}ab^{-1/2} + b^{1/2}|u_x|^2]^2 - \frac{1}{4}\varepsilon a^2 b^{-1} + \frac{\varepsilon q}{4}\frac{d}{dt}|u_x|^4 = 0.$$

Add up (3.2) and (3.3), we obtain

$$\frac{d}{dt}[\,|u_t|^2 + \alpha|u_{xx}|^2 + \frac{1}{2b}(a + b|u_x|^2)^2 + \varepsilon\langle u_t, u\rangle + \frac{\varepsilon\delta}{2}|u_x|^2 + \frac{\varepsilon q}{4}|u_x|^4\,] \tag{3.4}$$

$$+ \{2\delta|u_{xt}|^2 + (q/2)(\frac{d}{dt}|u_x|^2)^2 + \varepsilon\alpha|u_{xx}|^2 - \varepsilon|u_t|^2 + \varepsilon[\frac{1}{2}ab^{-1/2} + b^{1/2}|u_x|^2]^2\} = \frac{1}{4}\varepsilon a^2 b^{-1}.$$

Let

$$N(t) = 2\delta|u_{xt}|^2 + \varepsilon\alpha|u_{xx}|^2 - \varepsilon|u_t|^2 + \varepsilon[\frac{1}{2}ab^{-1/2} + b^{1/2}|u_x|^2]^2$$

$$= \varepsilon\{2\delta\varepsilon^{-1}|u_{xt}|^2 + \alpha|u_{xx}|^2 - |u_t|^2 + [\frac{1}{2}ab^{-1/2} + b^{1/2}|u_x|^2]^2\},$$

and

$$L(t) = |u_t|^2 + \alpha|u_{xx}|^2 + \frac{1}{2b}(a + b|u_x|^2)^2 + \varepsilon\langle u_t, u\rangle + \frac{\varepsilon\delta}{2}|u_x|^2 + \frac{\varepsilon q}{4}|u_x|^4.$$

Since $|u_{xt}| \geq |u_t|$ and $|u_{xx}| \geq |u_x| \geq |u|$, we get

$$2\varepsilon^{-1}N(t) - L(t) \geq (\frac{4\delta}{\varepsilon} - 3 - \varepsilon)|u_{xt}|^2 + (\alpha - \varepsilon - \frac{\varepsilon\delta}{2})|u_{xx}|^2 \tag{3.5}$$

$$+ \frac{1}{2b}[a + 2b|u_x|^2]^2 - \frac{1}{2b}(a + b|u_x|^2)^2 - \frac{\varepsilon q}{4}|u_x|^4$$

$$= (\frac{4\delta}{\varepsilon} - 3 - \varepsilon)|u_{xt}|^2 + (\alpha - \varepsilon - \frac{\varepsilon\delta}{2})|u_{xx}|^2 + \frac{1}{2}|u_x|^2(2a + 3b|u_x|^2) - \frac{\varepsilon q}{4}|u_x|^4$$

$$= (\frac{4\delta}{\varepsilon} - 3 - \varepsilon)|u_{xt}|^2 + (\alpha - \varepsilon - \frac{\varepsilon\delta}{2})|u_{xx}|^2 + (\frac{3b}{2} - \frac{\varepsilon q}{4})|u_x|^4 + a|u_x|^2.$$

We can choose $\varepsilon > 0$ sufficiently small such that

$$\frac{4\delta}{\varepsilon} - 3 - \varepsilon \geq 0, \quad \alpha - \varepsilon - \frac{\varepsilon\delta}{2} \geq 0, \text{ and } \frac{3b}{2} - \frac{\varepsilon q}{4} \geq b. \tag{3.6}$$

Note that such a choice is independent of specifical initial data and is uniform. Then from (3.5) it follows that

(3.7) $$2\varepsilon^{-1}N(t) - L(t) \geq -\frac{a^2}{4b}.$$

Substitute (3.9) into (3.4), we obtain the following differential inequality in L(t),

(3.8) $$\frac{d}{dt}L(t) + \frac{\varepsilon}{2}L(t) \leq \frac{1}{4}\varepsilon a^2 b^{-1} + \frac{\varepsilon}{2}\left(\frac{a^2}{4b}\right) \leq \frac{\varepsilon a^2}{2b}, \quad \text{for } t \in I_{max}.$$

Solve (3.8), we get

(3.9) $$L(t) \leq L(0)\exp\left(-\frac{\varepsilon}{2}t\right) + \frac{a^2}{b}, \quad \text{for } t \in I_{max}.$$

Then for any intial data $w_0 \in E$, due to the denseness of $E^2$ in E, there is a sequence $\{w_0^n\}_{n=1}^{\infty}$ in $E^2$ such that $w_0^n \to w_0$ as $n \to \infty$. By the continuous dependence in the sense that

$\| u(t; w_0^n) - u(t; w_0) \|^2 + |u_t(t; w_0^n) - u_t(t; w_0)|^2 \to 0$, as $n \to \infty$, (3.11) remains valid for $u(t; w_0)$

for any $w_0 \in E$. ¶

**Theorem 3.2** *For any* $w_0 \in E$, *there exists a unique global mild solution* w(t), $t \in [0,\infty)$, *of the equation* (2.9), *which has the regularity as described in Lemma* 2.5. *Moreover, the semiflow* $\vartheta$ *generated by the equation* (2.9) *is dissipative, i.e. absorbing sets exist.*

Proof. We have shown that

(3.10) $$L(t) = |u_t|^2 + \alpha|u_{xx}|^2 + \frac{1}{2b}(a + b|u_x|^2)^2 + \varepsilon\langle u_t, u\rangle + \frac{\varepsilon\delta}{2}|u_x|^2 + \frac{\varepsilon q}{4}|u_x|^4$$

$$\geq (1-\varepsilon)|u_t|^2 + (\alpha - \varepsilon)|u_{xx}|^2 \geq \frac{1}{2}\min\{1,\alpha\}\|(u(t), u_t(t))\|_E^2.$$

Combine (3.1) or (3.9) with (3.10), we have

(3.11) $$\frac{1}{2}\min\{1,\alpha\}\|(u(t), u_t(t))\|_E^2 \leq L(0)\exp\left(-\frac{\varepsilon}{2}t\right) + \frac{a^2}{b}, \quad \text{for } t \in I_{max}.$$

This inequality shows that the mild solution $w(t) = \begin{pmatrix} u(t) \\ u_t(t) \end{pmatrix}$ does not blow up in any finite time interval so that it exists globally for $t \in [0,\infty)$. Besides (3.11) shows that

(3.12) $$\limsup_{t\to\infty} \| w(t) \|^2 \le 2a^2b^{-1}\min\{1,\alpha\}^{-1} .$$

Therefore the closed bounded ball

(3.13) $$B_R = \{y\in E: \|y\|_E \le R\}, \text{ with } R = [2a^2b^{-1}\min\{1, \alpha\}^{-1} + 1]^{1/2},$$

is an absorbing set for the semiflow $\vartheta$. ¶

## § 4. The existence of Inertial Manifolds

First introduce the terminology in this section. For a given semiflow $\sigma$ on Hilbert space W and denoted its associated semigroup by $\Sigma(t)$, a set $M \subset W$ is called an *inertial manifold* of $\sigma$, if the following three conditions are all satisfied:

1) M is a Lipschitz continuous manifold of finite dimension.

2) M is positively invariant under the semiflow $\sigma$.

3) M attracts exponentially all the trajectories in the sense that there exists a uniform constant $\nu > 0$, for any bounded set B of W, there is a constant C(B) such that

$$\mathrm{dist}_W(\Sigma(t)w_0, M) \le C(B)\exp(-\nu t), \quad \text{for } t \ge 0, \text{ and any } w_0 \in B.$$

In this section we shall prove the existence of inertial manifolds of our concerned semiflow $\vartheta$, and estimate their dimensions.

Let the Hilbert space H and $V = D(A^{1/2})$ be decomposed into two orthogonal subspaces as follows. Observe that the complete normalized eigenvectors $\{e_k\}_{k=1}^{\infty}$ form an orthonormal basis for H and an orthogonal basis for V. The corresponding eigenvalues $\{\lambda_k\}_{k=1}^{\infty}$ satisfies the order $\lambda_1 < \lambda_2 < \dots$ . Let $H_m$ be defined by $H_m = \mathrm{Span}\{e_1, \dots, e_m\}$. Denote by $P_m: H \to H_m$ the orthogonal projection from H onto $H_m$, and $Q_m = I_H - P_m$. Denote by $\Pi_m = \begin{pmatrix} P_m & 0 \\ 0 & P_m \end{pmatrix}: E \to H_m \times H_m$ and $\Theta_m = I_E - \Pi_m$. Thus we have decompositions

$$H = P_mH \oplus Q_mH \qquad \text{and} \qquad E = (\Pi_mE) \oplus (\Theta_mE).$$

The H-valued function u(t) has a corresponding orthogonal decomposition $u(t) = p(t) + h(t)$, with $p(t) = P_mu(t)$ and $h(t) = Q_mu(t)$. The E-valued function $w(t) = (u(t), v(t))$= (in the strong or classical solution case) $(u(t), u_t(t))$ has also a corresponding orthogonal decomposition $w(t) =$

$\pi(t) + \theta(t)$, with $\pi(t) = \Pi_m w(t)$ and $\theta(t) = \Theta_m w(t)$. Thanks to the *commutivity* between $A^{1/2}$ and $P_m$, the second-order evolution equation (2.5) is decomposed into the following coupled equations:

$$(4.1)_p \qquad \frac{d^2p}{dt^2} + \alpha Ap + \delta A^{1/2}\frac{dp}{dt} + [a + b|A^{1/4}u|^2 + q\langle A^{1/2}u, u_t\rangle]A^{1/2}p = 0,$$

$$(4.1)_h \qquad \frac{d^2h}{dt^2} + \alpha Ah + \delta A^{1/2}\frac{dh}{dt} + [a + b|A^{1/4}u|^2 + q\langle A^{1/2}u, u_t\rangle]A^{1/2}h = 0,$$

with initial conditions $p(0) = P_m u_o$, $p_t(0) = P_m u_1$, and $h(0) = Q_m u_o$, $h_t(0) = Q_m u_1$, respectively. For convenience denote by

$$(4.2) \qquad J_u(t) = a + b|A^{1/4}u|^2 + q\langle A^{1/2}u, u_t\rangle.$$

**Theorem 5.1** *There exists a flat inertial manifold* $\mathcal{M}$ *in* E,

$$(4.3) \qquad \mathcal{M} = H_m \times H_m,$$

*for the semiflow* $\vartheta$ *generated by the equation* (2.9), *where* $m > 0$ *is a suitably large integer.*

Proof. First it is obvious that $\mathcal{M}$ given by (4.3) is a finite-dimensional subspace in E and that $\mathcal{M}$ is a Lipschitz continuous (actually linear) manifold. Secondly, this $\mathcal{M}$ is positively invariant under the semiflow $\vartheta$. This can be easily shown essentially due to the commutivity between $A^{1/2}$ and $P_m$. In fact, if $w_o = (u_o, u_1) \in \mathcal{M} \subset D(G)$, the mild solution $w(t) = (u(t), u_t(t))$ of the equation (2.9) is a classical solution as we have shown, so that the first component $u(t) = p(t) + h(t)$ satisfies the equation (2.5), and the functions $p(t)$ and $h(t)$ are respectively classical solutions of the equations $(4.1)_p$ and $(4.1)_h$, with $h(0) = h_t(0) = 0$. By the uniqueness of solutions of the equation $(4.1)_h$, it must be such that $h(t) \equiv 0$, for $t \geq 0$, and $p(t)$ is the solution of the following equsation,

$$(4.4) \qquad \frac{d^2p}{dt^2} + \alpha Ap + \delta A^{1/2}\frac{dp}{dt} + [a + b|A^{1/4}p|^2 + q\langle A^{1/2}p, p_t\rangle]A^{1/2}p = 0,$$

$$p(0) = u_o \in H_m, \qquad p_t(0) = u_1 \in H_m.$$

It means that for $w_o \in \mathcal{M}$, $S(t)w_o \in \mathcal{M}$ for all $t \geq 0$. Namely, $\mathcal{M}$ has the positive invariance.

It remains to prove that this M has the exponential attraction property required in the definition of inertial manifolds. Now, as usual, we will conduct *a priori* estimates in such a way that first it is for initial data in D(G) so that the solution has the sufficient regularity as we proceed, and then the integrated inequality results can be extended to solutions with any intial data in E by an approximation and the continuous dependence property. Take the inner-product in H of the equation $(4.1)_h$ with $2h_t + \xi h$, where $\xi$ is an undetermined constant, we have

$$(4.5) \qquad \frac{d}{dt}\{ |h_t|^2 + \alpha |h_{xx}|^2 + \xi\langle h_t, h\rangle + (\xi\delta/2)|h_x|^2 \}$$

$$+ \{2\delta |h_{xt}|^2 - \xi |h_t|^2 + \xi\alpha |h_{xx}|^2\} + \{2J_u(t)\langle h_x, h_{xt}\rangle + \xi J_u(t)|h_x|^2\} = 0.$$

By Theorem 3.2 and its proof, it is known that for every given bounded set Z in E and for any initial point $w_o \in Z$, the solution trajectory $w(t; w_o)$ will enter a fixed absorbing ball $B_R$ in E (and stay in forever) at a universal exponential decaying rate $\varepsilon/2$ (which is independent of specific Z) and after a uniform transient period $[0, t_o]$ with $t_o=t_o(Z)$ depending only on Z. Thanks to this observation, here we consider the trajectories already within the fixed absorbing ball $B_R$ after the indicated transient period. We have

$$(4.6) \qquad |J_u(t)| \le |a + b|u_x|^2 - q\langle u_{xx}, u_t\rangle| \le |a| + (b+q)\|w(t; w_o)\|_E^2$$

$$\le |a| + (b+q)R^2, \quad \text{for } t \ge t_o,$$

where R is the radius of the fixed absorbing ball $B_R$. Use (4.6) we can estimate the last two terms in (4.5) as follows,

$$(4.7) \qquad |2J_u(t)\langle h_x, h_{xt}\rangle + \xi J_u(t)|h_x|^2|$$

$$\le 2(|a| + (b+q)R^2)|h_x||h_{xt}| + \xi(|a| + (b+q)R^2)|h_x|^2$$

$$\le (|a| + (b+q)R^2)^2\delta^{-1}|h_x|^2 + \delta|h_{xt}|^2 + \xi(|a| + (b+q)R^2)|h_x|^2$$

$$\le [(|a| + (b+q)R^2)^2\delta^{-1} + \xi(|a| + (b+q)R^2)]|h_x|^2 + \delta|h_{xt}|^2$$

$$\le \frac{(|a| + (b+q)R^2)^2\delta^{-1} + \xi(|a| + (b+q)R^2)}{\sqrt{\lambda_{m+1}}}|h_{xx}|^2 + \delta|h_{xt}|^2$$

$$\leq K(R,\xi)\,(m+1)^{-2}\pi^{-2}\,|h_{xx}|^2 + \delta|h_{xt}|^2.$$

where

(4.8) $$K(R,\xi) = (|a| + (b+q)R^2)^2\delta^{-1} + \xi(|a| + (b+q)R^2).$$

Substitute (4.7) into (4.5), to obtain

(4.9) $$\frac{d}{dt}\{\,|h_t|^2 + \alpha|h_{xx}|^2 + \xi\langle h_t, h\rangle + (\xi\delta/2)|h_x|^2\,\}$$

$$+ \{\delta|h_{xt}|^2 - \xi|h_t|^2 + \xi\alpha|h_{xx}|^2 - K(R,\xi)\,(m+1)^{-2}\pi^{-2}|h_{xx}|^2\} \leq 0.$$

Define

(4.10) $$Y(t) = |h_t|^2 + \alpha|h_{xx}|^2 + \xi\langle h_t, h\rangle + (\xi\delta/2)|h_x|^2,$$

$$V(t) = \delta|h_{xt}|^2 - \xi|h_t|^2 + \xi\alpha|h_{xx}|^2 - K(R,\xi)\,(m+1)^{-2}\pi^{-2}|h_{xx}|^2$$

Then we have

(4.11) $$V(t) - (\xi/2)Y(t)$$

$$\geq \delta|h_{xt}|^2 - (3\xi/2)|h_t|^2 + (\xi\alpha/2)|h_{xx}|^2 - K(R,\xi)\,(m+1)^{-2}\pi^{-2}|h_{xx}|^2 - (\xi/2)\langle h_t, h\rangle$$

$$\geq (\delta - 2\xi)|h_{xt}|^2 + [(\xi\alpha/2) - ((\xi/2) + K(R,\xi))\,(m+1)^{-2}\pi^{-2}]\,|h_{xx}|^2 \geq 0,$$

by choosing and fixing $\xi$ satisfying

(4.12) $$0 < \xi \leq \min\{1, \alpha(1+\delta)^{-1}, \delta/2\},$$

and then letting the positive integer m be large enough such that

$$(m+1)^2 \geq (\xi\alpha\pi^2)^{-1}(\xi + 2K(R,\xi)),$$

or

(4.13) $$m \geq -1 + \sqrt{(\xi\alpha\pi^2)^{-1}(\xi + 2K(R,\xi))}\,.$$

From (4.7) and (4.11) we obtain that if m satisfies (4.13), then

$$\frac{d}{dt}Y(t) + \frac{\xi}{2}Y(t) \leq 0, \quad \text{for } t \geq t_o, \tag{4.14}$$

so that

$$Y(t) \leq Y(t_o)\exp(-\frac{\xi}{2}(t - t_o)), \quad \text{for } t \geq t_o. \tag{4.15}$$

Thanks to the choice of $\xi$ satisfying (4.12), $Y(t)$ satisfies

$$\begin{aligned} Y(t) &= |h_t|^2 + \alpha|h_{xx}|^2 + \xi\langle h_t, h\rangle + (\xi\delta/2)|h_x|^2 \\ &\geq \frac{1}{2}\min\{1,\alpha\}(|h_t|^2 + |h_{xx}|^2) + \frac{1-\xi}{2}|h_t|^2 + \frac{1}{2}(\alpha - \xi(1+\delta))|h_{xx}|^2 \\ &\geq \frac{1}{2}\min\{1,\alpha\}(|h_t|^2 + |h_{xx}|^2) = \frac{1}{2}\min\{1,\alpha\}\left\|\begin{pmatrix} h(t) \\ h_t(t)\end{pmatrix}\right\|_E^2 . \end{aligned} \tag{4.16}$$

On the other hand, we have

$$\begin{aligned} Y(t_o) &\leq [1 + \alpha + \xi + (\xi\delta/2)]\left\|\begin{pmatrix} h(t_o) \\ h_t(t_o)\end{pmatrix}\right\|_E^2 \\ &\leq (2 + \alpha + \delta)\left\|\begin{pmatrix} u(t_o) \\ u_t(t_o)\end{pmatrix}\right\|_E^2 \leq (2 + \alpha + \delta)R^2. \end{aligned} \tag{4.17}$$

Now combine (4.15) with (4.16) and (4.17), we obtain

$$\begin{aligned} \left\|\begin{pmatrix} h(t) \\ h_t(t)\end{pmatrix}\right\|_E^2 &\leq 2\min\{1,\alpha\}^{-1}(2 + \alpha + \delta)\left\|\begin{pmatrix} u(t_o) \\ u_t(t_o)\end{pmatrix}\right\|_E^2 \exp(-\frac{\xi}{2}(t - t_o)) \\ &\leq 2\min\{1,\alpha\}^{-1}(2 + \alpha + \delta)R^2\exp(-\frac{\xi}{2}(t - t_o)), \quad \text{for } t \geq t_o. \end{aligned} \tag{4.18}$$

According to our remark at the beginning on the *a priori* estimates, the result (4.18) remains valid for all the solutions with any initial data in E, in the sense that

$$\begin{aligned} \|\theta(t)\|_E^2 = \|\Theta_m w(t)\|_E^2 &\leq 2\min\{1,\alpha\}^{-1}(2 + \alpha + \delta)\|w(t_o)\|_E^2 \exp(-\frac{\xi}{2}(t - t_o)) \\ &\leq 2\min\{1,\alpha\}^{-1}(2 + \alpha + \delta)R^2\exp(-\frac{\xi}{2}(t - t_o)), \quad \text{for } t \geq t_o. \end{aligned} \tag{4.19}$$

Finally, (4.19) impies that

$$\text{dist}_E(S(t)w_o, M) = \text{dist}_E(\pi(t) + \theta(t), M) \quad (\text{cf. } (5.4)) \tag{4.20}$$

$$\leq \| \theta(t) \|_E \leq 2 \min\{1,\alpha\}^{-1}(2 + \alpha + \delta) \|w(t_0)\|_E^2 \exp(-\frac{\xi}{2}(t - t_0)), \qquad \text{for } t \geq t_0.$$

We also have the uniform exponential estimate of solutions during the transient period $[0, t_0]$, cf. (3.11),

$$\|w(t_0)\|_E^2 \leq 2 \min\{1,\alpha\}^{-1}\{L(0) \exp(-\frac{\varepsilon}{2} t_0) + \frac{a^2}{b}\},$$

where L(0) is a functional of $w_0$. For any given bounded set Z in E, let

$$K_1(Z) = \sup \{L(0): w_0 \in Z\} \tag{4.21}$$

Then it follows that

$$\|w(t_0)\|_E^2 \leq 2 \min\{1,\alpha\}^{-1}\{K_1(Z) \exp(-\frac{\varepsilon}{2} t_0) + \frac{a^2}{b}\}. \tag{4.22}$$

Substitute (4.22) into (4.20), we get

$$\text{dist}_E (S(t)w_0, M) \tag{4.23}$$

$$\leq 4 \min\{1,\alpha\}^{-2}(2 + \alpha + \delta) \{K_1(Z) \exp(-\frac{\varepsilon}{2} t_0) + \frac{a^2}{b}\} \exp(-\frac{\xi}{2}(t - t_0)), \quad t \geq t_0.$$

To include the behavior of solutions in the transient period, denote by

$$\nu = \frac{1}{2} \min\{\varepsilon, \xi\}, \tag{4.24}$$

$$K_2(Z, t_0(Z)) = 4 \min\{1,\alpha\}^{-2}(2 + \alpha + \delta) \{K_1(Z) + \frac{a^2}{b} \exp(\frac{\varepsilon}{2} t_o)\}. \tag{4.25}$$

Then we achieve the following unified exponential attraction expression,

$$\text{dist}_E (S(t)w_0, M) \leq K_2(Z, t_0(Z)) \exp(-\nu t), \quad \text{for } t \geq 0, \tag{4.26}$$

in which the constant $K_2$ only depends on the given bounded set Z containing the initial data. Thus by definition, the set $M = H_m \times H_m$ is an inertial manifold for the concerned semiflow $\vartheta$. The proof of Theorem 5.1 is completed. ¶

We can make an estimate of the bound of the dimension of the inertial manifold shown in Theorem 5.1. Note that

$$(4.27) \qquad \dim \mathcal{M} = 2m,$$

where m is an integer satisfying the condition (4.13) in which $\xi$ satisfies (4.12).

**Corollary 5.2** Let m be the smallest positive integer which satisfies

$$(4.28) \quad m > -1 + \frac{1}{\pi\alpha^{1/2}}\sqrt{1 + 2\rho(\alpha,a,b,q)\,[1 + \delta^{-1}\rho(\alpha,a,b,q)\max\{1\ \alpha^{-1}(1+\delta)\ 2\delta^{-1}\}]}\ ,$$

where

$$(4.29) \qquad \rho(\alpha, a, b, q) = |a| + 2a^2(1+ q/b)\max\{1, \alpha^{-1}\},$$

then there exists an inertial manifold $\mathcal{M}$ given by (5.7) with $\dim \mathcal{M} = 2m$.

Proof. Substitude $R^2 = 2a^2b^{-1}\max\{1, \alpha^{-1}\}$ and $\xi = \min\{1, \alpha(1+\delta)^{-1}, \delta/2\}$ into (4.8), then substitute (4.8) with the same $\xi$ into (4.13), we get

$$m \geq -1 + \frac{1}{\pi\alpha^{1/2}}\sqrt{(1 + 2\xi^{-1}K(R,\xi))}\ ,$$

Then the result (4.28) with (4.29) follows. ¶

**Remark 5.3** From (4.28) one can see that when the coefficient $\delta$ of the structural damping becomes smaller, the lower bound of the dimension of inertial manifold will increase, and that its growth rate is roughly proportional to $\delta^{-1}$, provided there are no dramatic changes for the other parameters.

More important, if one does not know each parameter exactly but in some moderate range, then use this formula (4.28) one can at least acquire a conservative estimate of the lower bound of the dimension of the inertial manofold. It, together with the result in the next section, will provide a *robust stabilization with respect to the parameter uncertainty* by a finite-modes feedback control. ¶

The governing equation of the subflows on an inertial manifold is called an *inertial form*, which is indeed a system of ordinary differential equations.

**Corollary 5.4** *For the inertial manifold* $\mathcal{M} = H_m \times H_m$, *the inertial form is the following equation in the subspace* $H_m$,

$$\frac{d^2p}{dt^2} + \alpha Ap(t) + \delta A^{1/2}\frac{dp}{dt} + [a + b\,|A^{1/4}p(t)|^2 + q\langle A^{1/2}p(t), \frac{dp}{dt}\rangle]A^{1/2}p(t) = 0, \quad t \geq 0, \tag{4.30}$$

$$p(0) = p_o \in H_m, \qquad p_t(0) = p_1 \in H_m.$$

*Here* p(t) *has finite components and accordingly* (4.30) *can be regarded as a system of ODEs.*

Proof. Since on the inertial manifold $M = H_m \times H_m$ the component $h(t) \equiv 0$ for $t \geq 0$, so that the solution must be a classical solution, and the governing equation on $M$ is given by $(4.1)_p$ with the change $J_u(t) = J_p(t)$. Thus we obtain the inertial form (4.30). ¶

## § 5. Robust Stabilization by Finite-Modes Feedback Control

In this section we shall resolve the *spillover problem.* The problem will be solved based on the existence of inertial manifolds. This approach of stabilization and its adaptions are bebieved to be useful to many distributed parameter control systems if the nonlinearity has the nonlocal feature.

Now consider the full equation (1.1) with control function f(t,x) on the right-hand side. The assumtion $n = \beta = 0$ is still valid. This equation is written here:

$$u_{tt} + \alpha u_{xxxx} - \delta u_{xxt} - [a + b\int_0^1 |u_x(t,\xi)|^2 d\xi + q(\int_0^1 (u_x u_{xt})(t,\xi)d\xi)]u_{xx} = f, \tag{5.1}$$

the boundary condition is still (1.2), and we assume that $(u_o, u_1) \in E$.

The stabilization of the system (5.1) is stated as: to find a linear or nonlinear feedback operator $F: E \rightarrow H$, such that the feedback control

$$f(t) = f(t, .) = F(u(t), u_t(t)), \quad t \geq 0, \tag{5.2}$$

makes the *closed-loop system* stable in the sense that all the mild solutions w(t) of the closed-loop equation converge to zero in E as $t \rightarrow \infty$. If so, (5.1) ia called *strongly stabilizable* by the indicated feedback control (5.2). If in addition the convergence is at a uniformly exponential rate, then (5.1) is called *exponentially stabilizable* by (5.2).

**Theorem 5.1** *The control system* (5.1) *is exponentially stabilizable by a finite-dimensional linear feedback control*

$$f(t) = a\,A^{1/2}P_m u(t), \quad t \geq 0, \tag{5.3}$$

*where* $P_m: H \rightarrow H_m$ *is the orthogonal projection, and* $H_m$ *is the factor subspace associated with the inertial manifold* $M = H_m \times H_m$ *for the uncontrolled equation* (2.9).

Proof. Apply this feedback control in the equation (5.1) and decompose it into two component equations in accordance with the decomposition of $H = H_m \oplus Q_m H$, we get

$$(5.4)_p \qquad p_{tt} + \alpha p_{xxxx} - \delta p_{xxt} - (J_u(t) - a)p_{xx} = 0,$$

$$(5.4)_h \qquad h_{tt} + \alpha h_{xxxx} - \delta h_{xxt} - J_u(t)h_{xx} = 0,$$

where $J_u(t)$ is defined by (4.2)., and $u(t) = p(t) + h(t)$ is a solution of the closed-loop equation

$$u_{tt} + \alpha Au + \delta A^{1/2}u_t + [a + b|A^{1/4}u|^2 + q\langle A^{1/2}u, u_t\rangle]A^{1/2}u = a\,A^{1/2}P_m u(t), \quad t > 0, \tag{5.5}$$

$$u(0) = u_o\,, \; u_t(0) = u_1.$$

Since using this linear feedback (5.3) only partially cancels a term " $- a\, p_{xx}$ " on the left-hand side of the equation (5.5) or (5.4), an easy adaption in the proof of Lemma 3.1 ensures that Theorem 3.2 remains valid and the ball $B_R$ in (3.13) is still an absorbing set for the new closed-loop equation (5.5) or (5.4).

Since this linear feedback (5.3) does not change the h- component equation at all, the same argument in the proof of Theorem 4.1 in showing the exponential attraction (especially within the absorbing ball $B_R$) of the manifold $M$ remains true without even any change in constants. Hence, it is true that

$$\left\|\begin{pmatrix} h(t) \\ h_t(t) \end{pmatrix}\right\|_E^2 \leq K_2(Z, t_o(Z))\exp(-\nu t), \qquad t \geq 0, \tag{5.6}$$

where $h(t) = Q_m u(t)$, and the constants $K_2$ and $\nu$ are the same as defined by (4.25) and (4.24).

It remains to handle the p-component equation $(5.4)_p$. We want to prove that the component $p(t) = P_m u(t)$ of the solution $u(t)$ of the closed-loop equation (5.5) also converges to zero at a uniform exponential decay rate. Taking the inner-product of the equation $(5.4)_p$ in H with $2p_t + \kappa p$, with $\kappa$ an undetermined constant, we have

(5.7) $$\frac{d}{dt}\{|p_t|^2 + \alpha|p_{xx}|^2 + \kappa\langle p_t, p\rangle + (\kappa\delta/2)|p_x|^2 + (b/2)|p_x|^4 + (\kappa q/4)|p_x|^4\}$$

$$+ \{ 2\delta|p_{xt}|^2 - \kappa|p_t|^2 + 2q\left|\langle p_x, p_{xt}\rangle\right|^2 + \kappa\alpha|p_{xx}|^2 + \kappa b|p_x|^4\}$$

$$+ \{ 2b\langle p_{xx}, p_t\rangle |h_x|^2 + 2q\langle p_{xx}, p_t\rangle\langle h_{xx}, h_t\rangle + \kappa b|p_x|^2|h_x|^2 - \kappa q|p_x|^2\langle h_{xx}, h_t\rangle\} = 0.$$

Denote by

(5.8) $$\Gamma(t) = |p_t|^2 + \alpha|p_{xx}|^2 + \kappa\langle p_t, p\rangle + (\kappa\delta/2)|p_x|^2 + [(b/2) + (\kappa q/4)]|p_x|^4,$$

(5.9) $$\Delta(t) = 2\delta|p_{xt}|^2 - \kappa|p_t|^2 + 2q\left|\langle p_x, p_{xt}\rangle\right|^2 + \kappa\alpha|p_{xx}|^2 + \kappa b|p_x|^4.$$

Then we have

(5.10) $$\Delta(t) - \frac{\kappa}{2}\Gamma(t) \geq (2\delta - \frac{3\kappa}{2} - \frac{\kappa^2}{2})|p_t|^2 + \kappa(\frac{\alpha}{2} - \frac{\kappa}{2} - \frac{\kappa\delta}{4})|p_{xx}|^2$$

$$+ \frac{\kappa}{4}( 3b - \frac{\kappa q}{2} )|p_x|^4 \geq 0, \qquad t \geq 0,$$

if we choose $\kappa > 0$ small enough to satisfy

(5.11) $$2\delta - \frac{3\kappa}{2} - \frac{\kappa^2}{2} \geq 0, \qquad \alpha - \kappa - \frac{\kappa\delta}{2} \geq 0, \qquad 3b - \frac{\kappa q}{2} \geq 0, \qquad \min\{1,\alpha\} - \kappa > 0.$$

On the other hand, for the p-h mixed terms in (5.7), we have the following estimate which is valid within the absorbing ball $B_R$ given by (3.13).

(5.12) $$\left| 2b\langle p_{xx}, p_t\rangle |h_x|^2 + 2q\langle p_{xx}, p_t\rangle\langle h_{xx}, h_t\rangle + \kappa b|p_x|^2|h_x|^2 - \kappa q|p_x|^2\langle h_{xx}, h_t\rangle\right|$$

$$\leq (2 + \kappa)(b + q)R^2 (|h_x|^2 + |h_{xx}|\,|h_t|) \leq (2 + \kappa)(b + q)R^2 K_2(Z, t_0(Z)) \exp(- \nu t), \quad \text{for } t \geq t_0,$$

in which we used the fact $|p_{xx}|, |p_x|, |p_t| \le R$ in that absorbing ball, and in the last step $|h_x|^2 + |h_{xx}||h_t|$ is substituted by the esimate (5.6) for h- component.

Now substitute (5.10) and (5.12) into (5.7), we get

$$\frac{d}{dt}\Gamma(t) + \frac{\kappa}{2}\Gamma(t) \le K_3(Z, t_o(Z)) \exp(-\nu t), \quad t \ge t_o, \tag{5.13}$$

where

$$K_3(Z, t_o(Z)) = (2 + \kappa)(b + q)R^2 K_2(Z, t_o(Z)), \tag{5.14}$$

with $\kappa$ satisfying (5.11) and fixed. Multiply this differential inequality (5.13) by the factor $\exp(\kappa t/2)$ and integrate it over $[t_o, t]$, we obtain

$$\Gamma(t) \le \Gamma(t_o) \exp(-\frac{\kappa}{2}(t - t_o)) + \frac{K_3(Z,t_o(Z))}{|\frac{\kappa}{2} - \nu|} \exp(-\min\{\frac{\kappa}{2}, \nu\}(t - t_o)), \quad t \ge t_o. \tag{5.15}$$

By (4.24), $\nu = (1/2)\min\{\varepsilon, \xi\}$. Denote by

$$\mu = \frac{1}{2}\min\{\kappa, \varepsilon, \xi\}. \tag{5.16}$$

On the other hand, from (5.8) and the last inequality of (5.11), we have

$$\Gamma(t) \ge \frac{1}{2}\min\{1,\alpha\}(|p_t|^2 + |p_{xx}|^2) = \frac{1}{2}\min\{1,\alpha\}\|\binom{p(t)}{p_t(t)}\|_E^2, \tag{5.17}$$

$$\Gamma(t_o) \le (3 + \frac{\delta}{2})R^2 + (\frac{b}{2} + \frac{q}{4})R^4, \tag{5.18}$$

where $R^2 = 2a^2b^{-1}\min\{1, \alpha\}^{-1} + 1$ as shown in (3.13). Denote by

$$K_4(Z, t_o(Z)) = 2\max\{1,\alpha^{-1}\}[(3 + \frac{\delta}{2})R^2 + (\frac{b}{2} + \frac{q}{4})R^4 + \frac{K_3(Z,t_o(Z))}{|\frac{\kappa}{2} - \nu|}] \exp(-\mu t_o). \tag{5.19}$$

Then from (5.15) it follows that

$$\|\binom{p(t)}{p_t(t)}\|_E^2 \le K_4(Z, t_o(Z)) \exp(-\mu t), \qquad t \ge t_o, \tag{5.20}$$

where the constant $K_4$ depends on the specific bounded set Z in which the initial point $w_0$ lies. Combine this result with the exponential decay during the transient period $[0, t_o]$, i.e.

(5.21) $$\|\binom{p(t)}{p_t(t)}\|_E^2 \le \|\binom{u(t)}{u_t(t)}\|_E^2 \le 2\max\{1,\alpha^{-1}\}[L(0)\exp(-\frac{\varepsilon}{2}t) + \frac{a^2}{b}]$$

$$\le 2\max\{1,\alpha^{-1}\}[K_1(Z) + \frac{a^2}{b}\exp(\frac{\varepsilon}{2}t_o)]\exp(-\frac{\varepsilon}{2}t),$$

$$\le K_5(Z, t_o(Z))\exp(-\frac{\varepsilon}{2}t) \le K_5(Z, t_o(Z))\exp(-\mu t), \quad \text{for } t\in[0,t_o],$$

we obtain the uniform exponential decay of the p-component of the closed-loop equstion, namely,

(5.22) $$\|\binom{p(t)}{p_t(t)}\|_E^2 \le (K_4(Z,t_o(Z)) + K_5(Z,t_o(Z))\exp(-\mu t), \qquad t \ge 0.$$

Finally, (5.6) and (5.22) imply that the solutions of the closed-loop equation (5.5) satisfies

(5.23) $$\|\binom{u(t)}{u_t(t)}\|_E^2 \le K_6(Z,t_o(Z))\exp(-\mu t), \qquad t \ge 0,$$

where $K_6(Z,t_o(Z)) = \max\{K_2(Z,t_o(Z)), K_4(Z,t_o(Z)) + K_5(Z,t_o(Z)\}$. The proof of Theorem 5.1 is completed. ¶

**Remark 5.2** First of all, Theorem 5.1 shows that using this finite-dimensional linear feedback, the exponential decay rate is uniform by (5.16) and can be estimated in terms of the pamameters. The size m of finitely many modes $\{e_1, \dots, e_m\}$ involved in this stabilizing feedback can be estimated by Corollary 4.2. Note that m increases (roughly propotional to $1/\delta$) as the structural damping coefficient $\delta$ decreases.

Most importantly, if we simply replace the parameters appearing in our results, say in the dimension bound formula (4.28), by their conservative bounds of uncertainty, then the same results Theorem 4.1 and Theorem 5.1 become the robust existence of inertial manifolds and the robust stabilization, respectively. This means that as the spillover problem is resolved by this approach, the solution will be *robust* with respect to the structure parameter uncertainty.

## References

[1] A.V. Balakrishnan (1981): *Applied Functional Analysis*, Springer-Verlag, New York.

[2] A.V. Balacrishnan and L.W. Taylor, Jr. (1989): Distributed parameter nonlinear damping models for flight structures, *Proceedings "Damping '89"*, Flight Dynamics Lab and Air Force Wright Aeronautical Labs, WPAFB.

[3] R.W. Bass and D. Zes (1991): Spillover, nonlinearity, and flexible structures, *The 4th NASA Workshop on Computational Control of Flexible Aerospace Systems*, NASA Conference Publication 10065, compiled by L.W. Taylor, Jr., Hampton, VA, pp. 1 - 14.

[4] S. Chen and R. Triggiani (1989): Proof of extensions of two conjectures on structural damping for elastic systems, *Pacific Journal of Math*, Vol.136, No.1, pp. 15-55.

[5] C. Foias, G.R. Sell and R. Temam (1988): Inertial manifolds for evolutionary equations, *J. Diff. Eqns.*, Vol. 73, pp. 309-353.

[6] X. Mora (1983): Finite-dimentional attracting invariant manifolds for damped semilinear wave equations, in *Contributions to Nonlinear Partial Differential Equations*, Vol.2, J.I. Diaz & P.L. Lions, edit., Pitman Res. Notes on Math, Vol.155, London, pp. 172-183.

[7] A. Pazy (1983): *Semigroups of Linear Operators and Applications to Partial Differential Equations*, Springer-Verlag, New York.

[8] M. Taboada and Y. You (1991): Global attractor, inertial manifolds and stabilization of nonlinear damped beam equations, to appear in *Communications in Partial Diff. Eqns.*

[9] R. Temam (1988): *Infinite Dimensional Dynamical systems in Mechanics and Physics*, Springer-Verlag, New York.

# 54 A Dynamical Systems Approach to Solving Linear Programming Problems

**Stanislaw H. Żak** Purdue University, West Lafayette, Indiana

**Viriya Upatising** Purdue University, West Lafayette, Indiana

**Walter E. Lillo** The Aerospace Corporation, Los Angeles, California

**Stefen Hui** San Diego State University, San Diego, California

**Abstract**

We propose and analyze dynamic linear programming problem solvers. We use a penalty function method and variable structure systems approach to solve linear programming problems. We introduce a family of penalty functions which allows us to transform linear programming problems into unconstrained optimization problems. Bounds on the weight parameters of the penalty functions are derived, using a method from variable structure systems theory, for which the given linear program and the associated unconstrained optimization problem have the same solution. In addition, we have combined gradient projection and the penalty function methods for solving linear programming problems. Bounds on the weight parameters of the penalty functions for this method resulting in the exact solution are also given. Both proposed methods for solving linear programming problems can be interpreted from the variable structure systems viewpoint. Simulation examples are given to illustrate the results obtained.

# 1 Introduction

Linear programming (LP) plays a fundamental role in many disciplines such as economics, strategic planning, analysis of algorithms, combinatorial problems, etc. In 1947, Dantzig ([1]) developed a method for solving linear programs which is known today as the simplex method. In 1951, Brown and Koopmans ([2]) described the first of a series of interior point methods for solving linear programs. Khachiyan ([3]) showed that linear programming problems can be solved in polynomial time using his ellipsoid method (see [4] for a discussion of this technique). However, computational experience with the ellipsoid algorithm has shown that it is not a practical alternative to the simplex method ([5, 6]). Recently, Karmarkar ([7]) (see also Strang ([8]) or Schrijver ([4]) for more details of this method) developed a linear programming algorithm which solves some complicated real-world problems of scheduling, routing and planning more efficiently than the simplex method. The simplex method is classified as an exterior point method and Karmarkar's method is classified as an interior point method ([9]). Recently, Gill et al. ([6]) showed that the whole family of the interior point methods can be derived from some classical results from the field of nonlinear programming. The classical methods of nonlinear programming include the penalty function method ([10]) and the barrier method ([11]). Conn ([12]) developed alternative methods for solving linear programs utilizing unconstrained optimization with penalty function methods. Another approach to solving LP problems, using interconnected networks of simple analog processors, was proposed by Pyne ([13]) and later studied by Rybashov ([14, 15]), Karpinskaya ([16]), and others. Since then various dynamic LP problem solvers have been proposed - see e.g. Tank and Hopfield ([17]), Kennedy and Chua ([18]), Rodríguez-Vázquez et al. ([19]), and

Cichocki and Unbehauen ([20]).

For the past thirty years, considerable attention has been given to devising penalty function methods which solve a constrained minimization problem by means of a single unconstrained minimization ([12, 21, 22, 23]). The penalty method transforms the following constrained optimization problem,

$$\textit{minimize} \;\; f(\mathbf{x})$$

$$\textit{subject to} \;\; \mathbf{x} \in S,$$

into an unconstrained optimization problem of the form

$$\textit{minimize} \;\; (f(\mathbf{x}) + c\mathcal{P}(\mathbf{x})),$$

where $f$ is a continuous function on $\mathbb{R}^n$, $S$ is a constraint set in $\mathbb{R}^n$, $c$ is a positive constant and $\mathcal{P}(\mathbf{x})$ is a penalty function on $\mathbb{R}^n$ satisfying: (i) $\mathcal{P}(\mathbf{x})$ is continuous, (ii) $\mathcal{P}(\mathbf{x}) \geq 0$ for all $\mathbf{x} \in \mathbb{R}^n$, and (iii) $\mathcal{P}(\mathbf{x}) = 0$ if and only if $\mathbf{x} \in S$ (see Luenberger ([24]) or Peressini et al. ([25]) for more details on the penalty method). When the solution of the associated penalty problem yields the exact solution to the original problem for a finite value of the penalty parameter $c$, the penalty function is called an exact penalty function. Bertsekas ([26]) has shown that, except for trivial cases, nondifferentiable penalty functions are necessary for the penalty method to yield the solution of the original constrained problem in a single unconstrained minimization.

In this chapter, we use nondifferentiable penalty functions to transform linear programming problems into equivalent unconstrained optimization problems. We then use techniques from variable structure systems theory to derive sufficient conditions for our penalty functions to be exact. Variable structure systems control, the control of dynamic systems with discontinuous feedback controllers, has been developed over the last 35 years. The theory of variable structure control rests on the concept of changing the structure of the controller in response to the changing states of the system to obtain a desired response. This is accomplished by the use of a discontinuous (switching) control law which forces the trajectories of the system onto a chosen manifold, where they are maintained thereafter. The motion of the trajectory while on the manifold is referred to as sliding. The reader is referred to Utkin ([27, 28, 29]) for more information on the subject. The variable structure systems approach proved to be very effective in the synthesis of robust control systems. However, variable structure systems theory can also be used to solve optimization problems as indicated by Utkin ([28], Chp.XII, or [29]). In this chapter we combine penalty methods and a variable structure systems approach in the synthesis of the proposed dynamic LP problem solvers.

The chapter is organized as follows. In Section 2, we use penalty functions to transform linear programs into unconstrained minimization problems. Subsequently, in Section 3 and 4 we give sufficient conditions for the proposed penalty functions to be exact, and we use a combined gradient projection method and the penalty function method to synthesize LP problem solvers. We present examples and simulation results to illustrate the results obtained. We offer concluding remarks in Section 5.

# 2 Converting a Linear Programming Problem into an Unconstrained Optimization Problem

The penalty function method allows one to transform a constrained optimization problem into an unconstrained optimization problem. Recall that the linear programming problem in the standard form is defined as

$$\min \ \mathbf{c}^T\mathbf{z}$$

subject to

$$\mathbf{Az} = \mathbf{b}, \quad \mathbf{z} \geq \mathbf{0},$$

where

$$\mathbf{A} \in \mathbb{R}^{m \times n}, \ m < n, \ rank \ \mathbf{A} = m.$$

Any $\mathbf{z} \in \mathbb{R}^n$ which minimizes the objective function over the given constraint set is called a minimizer. We call two problems equivalent if they have the same minima and minimizers. In this chapter, we show that, for a proper choice of the scalars $\rho$ and $\gamma$, the problem:

$$\min \ \mathbf{c}^T\mathbf{z} \text{ subject to } \mathbf{Az} = \mathbf{b}, \ \mathbf{z} \geq \mathbf{0}$$

is equivalent to the unconstrained problem

$$\min \ (\mathbf{c}^T\mathbf{z} + \rho\|\mathbf{Az} - \mathbf{b}\|_p + \gamma \sum_{i=1}^{n} (\mathbf{z})_i^-),$$

where

$$1 \leq p \leq \infty \ \text{ and } \ (\mathbf{z})_i^- = \begin{cases} -z_i & \text{if } z_i < 0 \\ 0 & \text{if } z_i \geq 0. \end{cases}$$

We first introduce a simple change of variables which makes the development and the proof of Theorem 1 notationally simpler. Let $\mathbf{r} \in \mathbb{R}^n$ be such that $\mathbf{Ar} = -\mathbf{b}$ and let $\mathbf{x} = \mathbf{z}+\mathbf{r}$. Note that one such $\mathbf{r}$ is $\mathbf{r} = -\mathbf{A}^T(\mathbf{AA}^T)^{-1}\mathbf{b}$. Then the linear programming problem has the form

$$\min \quad \mathbf{c}^T(\mathbf{x}-\mathbf{r})$$

subject to

$$\mathbf{Ax} = \mathbf{0}, \quad \mathbf{x}-\mathbf{r} \geq \mathbf{0}.$$

The associated unconstrained optimization problem has the form

$$\min \ E_{p,\rho,\gamma}(\mathbf{x}) = \min \ \left(\mathbf{c}^T(\mathbf{x}-\mathbf{r}) + \rho\|\mathbf{Ax}\|_p + \gamma\sum_{i=1}^{n}(\mathbf{x}-\mathbf{r})_i^-\right).$$

When there exist finite constants $\rho$ and $\gamma$, for a given $p$, such that the constrained and unconstrained optimization problem are equivalent then the penalty functions are called exact penalty functions. Note that the derivative of $\sum_{i=1}^{n}(\mathbf{x}-\mathbf{r})_i^-$ is not well defined at any point where $x_i = r_i$. However, it was observed that, even for smooth, well-posed problems, methods based on a differentiable penalty function suffer from inevitable ill-conditioning and the need to solve a sequence of subproblems ([30]). In addition Bertsekas ([26]) has shown that, except for trivial cases, nondifferentiable penalty functions are necessary for the method to yield an optimal solution in a single unconstrained minimization.

# 3 Equivalency of a LP Problem and Associated Unconstrained Optimization Problem

In this chapter we use the continuous gradient descent method to solve the unconstrained optimization problem. We would like to note that continuous gradient descent has many applications in optimal control problems - see e.g. Lee and Markus ([31]).

The trajectories of the continuous gradient descent, in our case, are governed by:

$$\dot{\mathbf{x}} = \frac{d\mathbf{x}}{dt} = -\nabla E_{p,\rho,\gamma}(\mathbf{x}), \quad \mathbf{x}_0 = \mathbf{x}(0).$$

The gradient of the objective function, $\nabla E_{p,\rho,\gamma}(\mathbf{x})$, exists except at the points of the set $\{\mathbf{x}|x_i = r_i \text{ for some } 1 \leq i \leq n\} \cup N(\mathbf{A})$, where $N(A) = \{\mathbf{x}|\mathbf{A}\mathbf{x} = \mathbf{0}\}$. For j $= 1, \ldots, m$, let $\mathbf{a}_j$ be the j-th row of $\mathbf{A}$. Where it is defined,

$$\nabla E_{p,\rho,\gamma}(\mathbf{x}) = \mathbf{c} + \frac{\rho \mathbf{A}^T}{\|\mathbf{A}\mathbf{x}\|_p^{p-1}} \begin{bmatrix} sgn(\mathbf{a}_1\mathbf{x})|\mathbf{a}_1\mathbf{x}|^{p-1} \\ \vdots \\ sgn(\mathbf{a}_m\mathbf{x})|\mathbf{a}_m\mathbf{x}|^{p-1} \end{bmatrix} + \gamma \begin{bmatrix} \frac{\partial(\mathbf{x}-\mathbf{r})_1^-}{\partial x_1} \\ \vdots \\ \frac{\partial(\mathbf{x}-\mathbf{r})_n^-}{\partial x_n} \end{bmatrix}.$$

Note that

$$\frac{\partial(\mathbf{x}-\mathbf{r})_i^-}{\partial x_i} = \begin{cases} -1 & \text{if } (\mathbf{x}-\mathbf{r})_i < 0 \\ 0 & \text{if } (\mathbf{x}-\mathbf{r})_i > 0. \end{cases}$$

We extend the definition of $\nabla E_{p,\rho,\gamma}(\mathbf{x})$ to all of $\mathbb{R}^n$ by defining

$$\frac{\partial(\mathbf{x}-\mathbf{r})_i^-}{\partial x_i} = 0 \quad \text{if } (\mathbf{x}-\mathbf{r})_i = 0$$

and

$$\frac{\mathbf{A}^T}{\|\mathbf{A}\mathbf{x}\|_p^{p-1}} \begin{bmatrix} sgn(\mathbf{a}_1\mathbf{x})|\mathbf{a}_1\mathbf{x}|^{p-1} \\ \vdots \\ sgn(\mathbf{a}_m\mathbf{x})|\mathbf{a}_m\mathbf{x}|^{p-1} \end{bmatrix} = \mathbf{0} \quad \text{if } \mathbf{A}\mathbf{x} = \mathbf{0}.$$

**Remark 1** Note that we extended the definition of $\nabla E_{p,\rho,\gamma}(\mathbf{x})$ to all of $\mathbb{R}^n$. The reason for this is our goal in synthesizing physically realizable dynamic LP problem solvers. Practical implementation requires that the synthesized system be well defined.

We shall now show that given $p$, we can find sufficiently large $\rho$ and $\gamma$ so that for any initial condition $\mathbf{x}_0 = \mathbf{x}(0)$, the trajectory of $\dot{\mathbf{x}} = -\nabla E_{p,\rho,\gamma}(\mathbf{x})$ will converge to the solution of the constrained optimization problem. In proving the above statement we use arguments from theory of variable structure systems. Specifically, we will treat the proposed LP problems solver as a control system of the form

$$\dot{\mathbf{x}} = -\mathbf{c} + \mathbf{A}^T\mathbf{u}^{(1)} + \mathbf{u}^{(2)},$$

where

$$\mathbf{u}^{(1)} = \mathbf{u}^{(1)}(\mathbf{x}) = -\rho\nabla_\sigma\|\mathbf{A}\mathbf{x}\|_p, \quad \mathbf{u}^{(2)} = \mathbf{u}^{(2)}(\mathbf{x}) = -\gamma\nabla_{\mathbf{x}}(\sum_{i=1}^{n}(\mathbf{x}-\mathbf{r})_i^-), \qquad \sigma = \mathbf{A}\mathbf{x}.$$

We define $\mathbf{u}^{(1)}(\mathbf{x}) = \mathbf{0}$ if $\mathbf{A}\mathbf{x} = \mathbf{0}$ and $\mathbf{u}^{(2)}(\mathbf{x}) = \mathbf{0}$ if $\mathbf{x} - \mathbf{r} \geq \mathbf{0}$.

The role of the control laws $\mathbf{u}^{(1)}$ and $\mathbf{u}^{(2)}$ is to force system trajectories into the feasible region and maintain it there while moving in the direction which decreases the cost $\mathbf{c}^T(\mathbf{x}-\mathbf{r})$. Let

$$\nabla(\sum_{i=1}^{n}(\mathbf{x}-\mathbf{r})_i^-) = \begin{bmatrix} \frac{\partial(\mathbf{x}-\mathbf{r})_1^-}{\partial x_1} \\ \vdots \\ \frac{\partial(\mathbf{x}-\mathbf{r})_n^-}{\partial x_n} \end{bmatrix},$$

and let $s$ be the conjugate exponent of $p$, that is,

$$\frac{1}{p} + \frac{1}{s} = 1.$$

Let

$$\mathbf{P} = \mathbf{I}_n - \mathbf{A}^T(\mathbf{A}\mathbf{A}^T)^{-1}\mathbf{A}.$$

**Lemma 1** *Suppose* $\{\mathbf{x}|\mathbf{Ax} = \mathbf{0}\} \cap \{\mathbf{x}|\mathbf{x} - \mathbf{r} \geq \mathbf{0}\} \neq \phi$, *i.e. the constraints of the linear programming problem are consistent. Then :*

*(i) If* $\mathbf{Ax} = \mathbf{0}$ *and* $(\mathbf{x} - \mathbf{r})_i < 0$ *for some* $i$, *then* $\mathbf{P}\nabla(\sum_{i=1}^n (\mathbf{x} - \mathbf{r})_i^-) \neq \mathbf{0}$.

*(ii)* $\inf\{\|\mathbf{P}\nabla(\sum_{i=1}^n (\mathbf{x} - \mathbf{r})_i^-)\|_2 | \mathbf{Ax} = \mathbf{0}, (\mathbf{x} - \mathbf{r})_i < 0$ *for some* $i\} > 0$.

*(iii)* $\inf\{\nabla^T(\sum_{i=1}^n (\mathbf{x} - \mathbf{r})_i^-)\mathbf{P}\nabla(\sum_{i=1}^n (\mathbf{x} - \mathbf{r})_i^-) | \mathbf{Ax} = \mathbf{0}, (\mathbf{x} - \mathbf{r})_i < 0$ *for some* $i\} > 0$.

Proof: (i) We will use proof by contradiction. Suppose $\mathbf{Ax} = \mathbf{0}$, $(\mathbf{x} - \mathbf{r})_i < 0$ for some $i$ and $\mathbf{P}\nabla(\sum_{i=1}^n (\mathbf{x} - \mathbf{r})_i^-) = \mathbf{0}$. Without loss of generality, let $(\mathbf{x} - \mathbf{r})_i < 0$ for $1 \leq i \leq m$ and $(\mathbf{x} - \mathbf{r})_i \geq 0$ for $m + 1 \leq i \leq n$. Then

$$\nabla(\sum_{i=1}^n (\mathbf{x} - \mathbf{r})_i^-) = [\underbrace{-1, -1, \cdots, -1}_{m}, \underbrace{0, \cdots, 0}_{n-m}]^T.$$

Since $\mathbf{P}\nabla(\sum_{i=1}^n (\mathbf{x} - \mathbf{r})_i^-) = \mathbf{0}$, we have for $\mathbf{z} \in N(\mathbf{A})$,

$$\begin{aligned} [\nabla(\sum_{i=1}^n (\mathbf{x} - \mathbf{r})_i^-)]^T \mathbf{z} &= [\mathbf{A}^T(\mathbf{AA}^T)^{-1}\mathbf{A}\nabla(\sum_{i=1}^n (\mathbf{x} - \mathbf{r})_i^-)]^T \mathbf{z} \\ &= [(\mathbf{AA}^T)^{-1}\mathbf{A}\nabla(\sum_{i=1}^n (\mathbf{x} - \mathbf{r})_i^-)]^T \mathbf{Az} \\ &= 0. \end{aligned}$$

Thus $z_1 + \cdots + z_m = 0$ for $\mathbf{z} \in N(\mathbf{A})$. By consistency, there exist $\mathbf{z}$ such that $\mathbf{Az} = \mathbf{0}$ and $\mathbf{z} - \mathbf{r} \geq \mathbf{0}$. We can conclude that

$$r_1 + \cdots + r_m \leq z_1 + \cdots + z_m = 0.$$

On the other hand, we have by assumption that $\mathbf{Ax} = \mathbf{0}$ and $x_i - r_i < 0$ for $i = 1, \cdots, m$. Hence $0 = x_1 + \cdots + x_m < r_1 + \cdots + r_m$. We arrive at a contradiction and the proof is complete.

(ii) Observe that for $\mathbf{x} \in \mathbb{R}^n$, there are only $2^n$ possible $\nabla(\sum_{i=1}^n (\mathbf{x}-\mathbf{r})_i^-)$:

$$\nabla(\sum_{i=1}^n (\mathbf{x}-\mathbf{r})_i^-) \in \{\mathbf{z} \in \mathbb{R}^n | \mathbf{z}_i = 0 \ or \ -1 \ for \ i = 1, \cdots, n\}.$$

By part (i), if $(\mathbf{x}-\mathbf{r})_i < 0$ for some $i$ and $\mathbf{Ax} = \mathbf{0}$, we have $\mathbf{P}\nabla(\sum_{i=1}^n (\mathbf{x}-\mathbf{r})_i^-) \neq \mathbf{0}$. Therefore,

$$\min\{\|\mathbf{P}\nabla(\sum_{i=1}^n (\mathbf{x}-\mathbf{r})_i^-)\|_2 | \mathbf{Ax} = \mathbf{0}, (\mathbf{x}-\mathbf{r})_i < 0 \ for \ some \ i = 1, \cdots, n\} > 0.$$

(iii) Since we have $\mathbf{P} = \mathbf{P}^T = \mathbf{P}^2$,

$$\begin{aligned} \nabla^T(\sum_{i=1}^n (\mathbf{x}-\mathbf{r})_i^-)\mathbf{P}\nabla(\sum_{i=1}^n (\mathbf{x}-\mathbf{r})_i^-) &= \nabla^T(\sum_{i=1}^n (\mathbf{x}-\mathbf{r})_i^-)\mathbf{P}^T\mathbf{P}\nabla(\sum_{i=1}^n (\mathbf{x}-\mathbf{r})_i^-) \\ &= \|\mathbf{P}\nabla(\sum_{i=1}^n (\mathbf{x}-\mathbf{r})_i^-)\|_2^2. \end{aligned}$$

The desired inequality now follows from (ii). □

Let

$$\beta = \inf\{\|\mathbf{P}\nabla(\sum_{i=1}^n (\mathbf{x}-\mathbf{r})_i^-)\|_2 | \mathbf{Ax} = \mathbf{0}, (\mathbf{x}-\mathbf{r})_i < 0 \ for \ some \ i\},$$

$$\tau = \frac{\|\mathbf{c}\|_2}{\beta}.$$

Note that by Lemma 1, we have $\beta > 0$ and $0 < \tau < \infty$ since $0 < \|\mathbf{c}\|_2 < \infty$.

We can now state the main result of this section.

**Theorem 1** *Let* $1 \leq p \leq \infty$ *and let* $\tau$ *be defined as above. Then for* $\gamma > \tau$ *and* $\rho > \|\mathbf{c}^T\mathbf{A}^T(\mathbf{AA}^T)^{-1}\|_s + \gamma n^{\frac{1}{s}}\|\mathbf{A}^T(\mathbf{AA}^T)^{-1}\|_s$, *the constrained problem*

$$\min \quad \mathbf{c}^T(\mathbf{x}-\mathbf{r}) \quad \text{subject to } \mathbf{Ax} = \mathbf{0}, \ \mathbf{x}-\mathbf{r} \geq \mathbf{0}$$

*is equivalent to the unconstrained optimization problem*

$$\min \quad (\mathbf{c}^T(\mathbf{x}-\mathbf{r}) + \rho\|\mathbf{Ax}\|_p + \gamma\sum_{i=1}^n (\mathbf{x}-\mathbf{r})_i^-).$$

Before proving the above theorem we first discuss some necessary technical results.

**Lemma 2** *If for some* $k > 0$, *we have*

$$-\mathbf{x}^T\mathbf{A}^T(\mathbf{A}\mathbf{A}^T)^{-1}\nabla E_{p,\rho,\gamma}(\mathbf{x}) \leq -k\|\mathbf{A}\mathbf{x}\|_p \ \forall \ \mathbf{x} \notin N(\mathbf{A}),$$

*then the trajectories of* $\dot{\mathbf{x}} = -\nabla E_{p,\rho,\gamma}(\mathbf{x}), \mathbf{x}_0 = \mathbf{x}(0)$, *reach the manifold* $N(\mathbf{A}) = \{\mathbf{x}|\mathbf{A}\mathbf{x} = \mathbf{0}\}$ *in finite time and stay on the manifold thereafter.*

Proof: Observe that the trajectories of $\dot{\mathbf{x}} = -\nabla E_{p,\rho,\gamma}(\mathbf{x})$ are by definition absolutely continuous. Note that each $\mathbf{x} \in \mathbb{R}^n$ can be orthogonally decomposed as

$$\mathbf{x} = \mathbf{P}\mathbf{x} + (\mathbf{I}_n - \mathbf{P})\mathbf{x}.$$

with $\mathbf{P}\mathbf{x} \in N(\mathbf{A})$. Therefore the Euclidean distance from $\mathbf{x}$ to $N(\mathbf{A})$ is $\|(\mathbf{I}_n - \mathbf{P})\mathbf{x}\|_2 = \|\mathbf{A}^T(\mathbf{A}\mathbf{A}^T)^{-1}\mathbf{A}\mathbf{x}\|_2$. We have for $\mathbf{x} \notin N(\mathbf{A})$,

$$\begin{aligned} \frac{1}{2}\frac{d}{dt}\|\mathbf{A}^T(\mathbf{A}\mathbf{A}^T)^{-1}\mathbf{A}\mathbf{x}\|_2^2 &= \mathbf{x}^T\mathbf{A}^T(\mathbf{A}\mathbf{A}^T)^{-1}\mathbf{A}\dot{\mathbf{x}} \\ &= -\mathbf{x}^T\mathbf{A}^T(\mathbf{A}\mathbf{A}^T)^{-1}\mathbf{A}\nabla E_{p,\rho,\gamma}(\mathbf{x}). \end{aligned}$$

Thus by assumption,

$$\frac{1}{2}\frac{d}{dt}\|\mathbf{A}^T(\mathbf{A}\mathbf{A}^T)^{-1}\mathbf{A}\mathbf{x}\|_2^2 \leq -k\|\mathbf{A}\mathbf{x}\|_p \quad \forall \ \ \mathbf{x} \notin N(\mathbf{A}).$$

Since all norms on $\mathbb{R}^n$ are equivalent, there exists $\alpha_p > 0$ such that $\alpha_p\|\mathbf{A}\mathbf{x}\|_2 \leq \|\mathbf{A}\mathbf{x}\|_p$. We have then

$$\begin{aligned} \frac{1}{2}\frac{d}{dt}\|\mathbf{A}^T(\mathbf{A}\mathbf{A}^T)^{-1}\mathbf{A}\mathbf{x}\|_2^2 &\leq -k\alpha_p\|\mathbf{A}\mathbf{x}\|_2 \\ &\leq \frac{-k\alpha_p}{\|\mathbf{A}^T(\mathbf{A}\mathbf{A}^T)^{-1}\|_2}\|\mathbf{A}^T(\mathbf{A}\mathbf{A}^T)^{-1}\mathbf{A}\mathbf{x}\|_2, \end{aligned}$$

and it follows that

$$\frac{d}{dt}\|\mathbf{A}^T(\mathbf{A}\mathbf{A}^T)^{-1}\mathbf{A}\mathbf{x}\|_2 \leq \frac{-k\alpha_p}{\|\mathbf{A}^T(\mathbf{A}\mathbf{A}^T)^{-1}\|_2},$$

a constant independent of $t$. Thus $\|\mathbf{A}^T(\mathbf{A}\mathbf{A}^T)^{-1}\mathbf{A}\mathbf{x}\|_2$ is a decreasing function of time and that for $\mathbf{x}(t) \notin N(\mathbf{A})$,

$$\|\mathbf{A}^T(\mathbf{A}\mathbf{A}^T)^{-1}\mathbf{A}\mathbf{x}(t)\|_2 \leq \|\mathbf{A}^T(\mathbf{A}\mathbf{A}^T)^{-1}\mathbf{A}\mathbf{x}_0\| - \frac{k\alpha_p}{\|\mathbf{A}^T(\mathbf{A}\mathbf{A}^T)^{-1}\|_2}t.$$

Therefore the trajectories reach $N(\mathbf{A})$ in finite time and stay there thereafter. □

**Lemma 3** *The motion of* $\dot{\mathbf{x}} = -\nabla E_{p,\rho,\gamma}(\mathbf{x})$ *when restricted to* $N(\mathbf{A}) = \{\mathbf{x}|\mathbf{A}\mathbf{x} = \mathbf{0}\}$ *is described by the equations*

$$\begin{cases} \dot{\mathbf{x}} = -\mathbf{P}\nabla E_{p,\rho,\gamma}(\mathbf{x}) \\ \mathbf{A}\mathbf{x} = \mathbf{0}. \end{cases}$$

Proof: The above result follows from theory of sliding modes - see Utkin ([28], pp. 43,44) or Żak et al. ([32], pp. 110,111). We proceed as follows. We treat the system $\dot{\mathbf{x}} = -\nabla E_{p,\rho,\gamma}(\mathbf{x})$ as a control system

$$\dot{\mathbf{x}} = -\mathbf{c} + \mathbf{A}^T\mathbf{u}^{(1)} + \mathbf{u}^{(2)},$$

where $\mathbf{u}^{(1)}$ and $\mathbf{u}^{(2)}$ are as previously defined. While sliding along $N(\mathbf{A})$, the system can be described by the equations $\mathbf{A}\mathbf{x} = \mathbf{0}$, $\mathbf{A}\dot{\mathbf{x}} = \mathbf{0}$. The relation $\mathbf{A}\dot{\mathbf{x}} = \mathbf{0}$ can be expressed in an alternative form using the method of equivalent control. We compute the equivalent control $\mathbf{u}_{eq}^{(1)}$ by solving the algebraic equation in $\mathbf{u}^{(1)}$

$$\mathbf{A}\dot{\mathbf{x}} = -\mathbf{A}\mathbf{c} + \mathbf{A}\mathbf{A}^T\mathbf{u}^{(1)} + \mathbf{A}\mathbf{u}^{(2)} = \mathbf{0}.$$

We obtain the equivalent control by solving for $\mathbf{u}^{(1)}$ :

$$\mathbf{u}_{eq}^{(1)} = (\mathbf{A}\mathbf{A}^T)^{-1}\mathbf{A}\mathbf{c} - (\mathbf{A}\mathbf{A}^T)^{-1}\mathbf{A}\mathbf{u}^{(2)}$$

and substitute this into the orginal dynamic equation to obtain

$$\dot{\mathbf{x}} = (\mathbf{I}_n - \mathbf{A}^T(\mathbf{A}\mathbf{A}^T)^{-1}\mathbf{A})(-\mathbf{c} + \mathbf{u}^{(2)}) = \mathbf{P}(-\mathbf{c} + \mathbf{u}^{(2)}).$$

Hence, in sliding mode along the manifold $N(\mathbf{A})$, the system is described by

$$\begin{cases} \dot{\mathbf{x}} = -\mathbf{P}\nabla E_{p,\rho,\gamma}(\mathbf{x}) \\ \mathbf{A}\mathbf{x} = \mathbf{0}. \end{cases} \qquad \square$$

**Remark 2** In proving Theorem 1 we employ the method of control hierarchy from variable structure systems theory - see Utkin ([28], pp. 101-104) for a description of this method. Specifically, we will view $\rho$ and $\gamma$ as the controllers gains. We then establish a hierarchy of " controls " so that the first controller should drive the system trajectory onto the surface $N(\mathbf{A})$. The second control should then drive the system trajectory onto the intersection of $N(\mathbf{A})$ and $\{\mathbf{x}|\mathbf{x} - \mathbf{r} \geq \mathbf{0}\}$ while the first control maintains a sliding mode on $N(\mathbf{A})$. Note that while choosing the gain $\gamma$ for the second controller we assume that the first controller guarantees sliding mode along $N(\mathbf{A})$ for any possible values of $\gamma$. We also assume knowledge of the system structure in sliding along the manifold $N(\mathbf{A})$ - see Lemma 2.

Proof of Theorem 1: Suppose that the trajectories are restricted to $N(\mathbf{A})$. Note that $\mathbf{P}$ is the orthogonal projection onto $N(\mathbf{A})$. The dynamics of $\dot{\mathbf{x}} = -\nabla E_{p,\rho,\gamma}(\mathbf{x})$ on $N(\mathbf{A})$ are given by (see Lemma 3) :

$$\dot{\mathbf{x}} = -\mathbf{P}\mathbf{c} - \gamma\mathbf{P}\nabla(\sum_{i=1}^{n}(\mathbf{x} - \mathbf{r})_i^-).$$

Consider the case when some inequality constraints are violated, i.e. when $(\mathbf{x} - \mathbf{r})_i^- > 0$ for some $i$. We would like the representative point while sliding along $N(\mathbf{A})$ to move in the direction of the set $\{\mathbf{x}|\mathbf{x} - \mathbf{r} \geq \mathbf{0}\}$. Consider the function

$$h(\mathbf{x}) = \sum_{i=1}^{n}(\mathbf{x} - \mathbf{r})_i^-.$$

Note that $h(\mathbf{x})$ gives the distance of $\mathbf{x}$ to $\{\mathbf{x}-\mathbf{r}\geq \mathbf{0}\}$ in the $l^1$-norm. Since all norms on $\mathbb{R}^n$ are equivalent, if we show that $h(\mathbf{x}(t))\to 0$, then the Euclidean distance of $\mathbf{x}(t)$ to $\{\mathbf{x}-\mathbf{r}\geq \mathbf{0}\}$ will also tend to 0. Taking into account the fact that $\mathbf{P}=\mathbf{P}^T=\mathbf{P}^2$ and Lemma 1, we have everywhere outside the feasible region :

$$
\begin{aligned}
\frac{d}{dt}h(\mathbf{x}(t)) &= \nabla^T(\sum_{i=1}^{n}(\mathbf{x}-\mathbf{r})_i^-)\,\dot{\mathbf{x}} \\
&= \nabla^T(\sum_{i=1}^{n}(\mathbf{x}-\mathbf{r})_i^-)[-\mathbf{P}\mathbf{c}-\gamma\mathbf{P}\nabla(\sum_{i=1}^{n}(\mathbf{x}-\mathbf{r})_i^-)] \\
&\leq |\nabla^T(\sum_{i=1}^{n}(\mathbf{x}-\mathbf{r})_i^-)\,\mathbf{P}\mathbf{c}| - \gamma\nabla^T(\sum_{i=1}^{n}(\mathbf{x}-\mathbf{r})_i^-)\,\mathbf{P}\nabla(\sum_{i=1}^{n}(\mathbf{x}-\mathbf{r})_i^-) \\
&\leq \|\mathbf{c}\|_2\|\mathbf{P}\nabla(\sum_{i=1}^{n}(\mathbf{x}-\mathbf{r})_i^-)\|_2 - \gamma\|\mathbf{P}\nabla(\sum_{i=1}^{n}(\mathbf{x}-\mathbf{r})_i^-)\|_2^2 \\
&= (\|\mathbf{c}\|_2-\gamma\|\mathbf{P}\nabla(\sum_{i=1}^{n}(\mathbf{x}-\mathbf{r})_i^-)\|_2)\|\mathbf{P}\nabla(\sum_{i=1}^{n}(\mathbf{x}-\mathbf{r})_i^-)\|_2 \\
&< 0
\end{aligned}
$$

if $\gamma>\tau$. Since there are only finitely many $\nabla(\sum_{i=1}^{n}(\mathbf{x}-\mathbf{r})_i^-)$'s and $\mathbf{c}$ is fixed, there exists $\epsilon>0$ such that

$$\frac{d}{dt}h(\mathbf{x}(t)) < -\epsilon$$

if $\gamma>\tau$. Hence we conclude that $h(\mathbf{x}(t))$ decreases with $t$ and $\mathbf{x}(t)$ will reach $\{\mathbf{x}-\mathbf{r}\geq\mathbf{0}\}$ in finite time and stay there thereafter.

We now establish a bound on $\rho=\rho(\gamma)$ for which the trajectories of $\dot{\mathbf{x}}=-\nabla E_{p,\rho,\gamma}(\mathbf{x})$ are guaranteed to converge to and remain on the manifold $N(\mathbf{A})$. We have

$$
\begin{aligned}
&-\mathbf{x}^T\mathbf{A}^T(\mathbf{A}\mathbf{A}^T)^{-1}\mathbf{A}\nabla E_{p,\rho,\gamma}(\mathbf{x}) \\
&= -\mathbf{x}^T\mathbf{A}^T(\mathbf{A}\mathbf{A}^T)^{-1}\mathbf{A}(\mathbf{c}+\frac{\rho\mathbf{A}^T}{\|\mathbf{A}\mathbf{x}\|_p^{p-1}}\begin{bmatrix} sgn(\mathbf{a}_1\mathbf{x})|\mathbf{a}_1\mathbf{x}|^{p-1} \\ \vdots \\ sgn(\mathbf{a}_m\mathbf{x})|\mathbf{a}_m\mathbf{x}|^{p-1}\end{bmatrix}+\gamma\nabla(\textstyle\sum_{i=1}^{n}(\mathbf{x}-\mathbf{r})_i^-)).
\end{aligned}
$$

We will now show that $-\mathbf{x}^T\mathbf{A}^T(\mathbf{A}\mathbf{A}^T)^{-1}\mathbf{A}\nabla E_{p,\rho,\gamma}(\mathbf{x}) < 0$ for $\rho > \|\mathbf{c}^T\mathbf{A}^T(\mathbf{A}\mathbf{A}^T)^{-1}\|_s + \gamma n^{\frac{1}{s}}\|\mathbf{A}(\mathbf{A}\mathbf{A}^T)^{-1}\|_s$. Consider the second term in the product $[\mathbf{x}^T\mathbf{A}^T(\mathbf{A}\mathbf{A}^T)^{-1}\mathbf{A}]\nabla E_{p,\rho,\gamma}(\mathbf{x})$. We have

$$\begin{aligned}
II &= \frac{\rho\mathbf{x}^T\mathbf{A}^T(\mathbf{A}\mathbf{A}^T)^{-1}\mathbf{A}\mathbf{A}^T}{\|\mathbf{A}\mathbf{x}\|_p^{p-1}}\begin{bmatrix} sgn(\mathbf{a}_1\mathbf{x})|\mathbf{a}_1\mathbf{x}|^{p-1} \\ \vdots \\ sgn(\mathbf{a}_m\mathbf{x})|\mathbf{a}_m\mathbf{x}|^{p-1} \end{bmatrix} \\
&= \frac{\rho\mathbf{x}^T\mathbf{A}^T}{\|\mathbf{A}\mathbf{x}\|_p^{p-1}}\begin{bmatrix} sgn(\mathbf{a}_1\mathbf{x})|\mathbf{a}_1\mathbf{x}|^{p-1} \\ \vdots \\ sgn(\mathbf{a}_m\mathbf{x})|\mathbf{a}_m\mathbf{x}|^{p-1} \end{bmatrix} \\
&= \frac{\rho}{\|\mathbf{A}\mathbf{x}\|_p^{p-1}}\sum_{j=1}^{m}|\mathbf{a}_j x|^p \\
&= \rho\|\mathbf{A}\mathbf{x}\|_p.
\end{aligned}$$

We now estimate the first term in the product $[\mathbf{x}^T\mathbf{A}^T(\mathbf{A}\mathbf{A}^T)^{-1}\mathbf{A}]\nabla E_{p,\rho,\gamma}(\mathbf{x})$. By Hölder's inequality:

$$\begin{aligned}
|I| &= |\mathbf{x}^T\mathbf{A}^T(\mathbf{A}\mathbf{A}^T)^{-1}\mathbf{A}\mathbf{c}| \\
&\leq \|\mathbf{c}^T\mathbf{A}^T(\mathbf{A}\mathbf{A}^T)^{-1}\|_s\|\mathbf{A}\mathbf{x}\|_p, \quad where \ \frac{1}{s}+\frac{1}{p}=1.
\end{aligned}$$

We now estimate the third term in the product $[\mathbf{x}^T\mathbf{A}^T(\mathbf{A}\mathbf{A}^T)^{-1}\mathbf{A}]\nabla E_{p,\rho,\gamma}(\mathbf{x})$. By Hölder's inequality:

$$\begin{aligned}
|III| &= |\gamma\mathbf{x}^T\mathbf{A}^T(\mathbf{A}\mathbf{A}^T)^{-1}\mathbf{A}\nabla(\sum_{i=1}^{n}(\mathbf{x}-\mathbf{r})_i^-)| \\
&\leq \gamma\|\nabla^T(\sum_{i=1}^{n}(\mathbf{x}-\mathbf{r})_i^-)\mathbf{A}^T(\mathbf{A}\mathbf{A}^T)^{-1}\|_s\|\mathbf{A}\mathbf{x}\|_p \\
&\leq \gamma\|\nabla(\sum_{i=1}^{n}(\mathbf{x}-\mathbf{r})_i^-)\|_s\|\mathbf{A}^T(\mathbf{A}\mathbf{A}^T)^{-1}\|_s\|\mathbf{A}\mathbf{x}\|_p \\
&\leq \gamma n^{\frac{1}{s}}\|\mathbf{A}^T(\mathbf{A}\mathbf{A}^T)^{-1}\|_s\|\mathbf{A}\mathbf{x}\|_p, \quad where \ \frac{1}{s}+\frac{1}{p}=1.
\end{aligned}$$

Therefore

$$\begin{aligned} & \mathbf{x}^T\mathbf{A}^T(\mathbf{A}\mathbf{A}^T)^{-1}\mathbf{A}\nabla E_{p,\rho,\gamma}(\mathbf{x}) \\ = \; & I + II + III \\ \geq \; & II - |I| - |III| \\ \geq \; & (\rho - \|\mathbf{c}^T\mathbf{A}^T(\mathbf{A}\mathbf{A}^T)^{-1}\|_s - \gamma n^{\frac{1}{s}}\|\mathbf{A}^T(\mathbf{A}\mathbf{A}^T)^{-1}\|_s)\|\mathbf{A}\mathbf{x}\|_p. \end{aligned}$$

Hence if $\rho > \|\mathbf{c}^T\mathbf{A}^T(\mathbf{A}\mathbf{A}^T)^{-1}\|_s + \gamma n^{\frac{1}{s}}\|\mathbf{A}^T(\mathbf{A}\mathbf{A}^T)^{-1}\|_s$ then the condition $-\mathbf{x}^T\mathbf{A}^T(\mathbf{A}\mathbf{A}^T)^{-1}$ $\mathbf{A}\nabla E_{p,\rho,\gamma}(\mathbf{x}) \; < -k\|\mathbf{A}\mathbf{x}\|_p \; \forall \; \mathbf{x} \notin N(\mathbf{A})$ is satisfied and hence by Lemma 2 the trajectories of $\dot{\mathbf{x}} = -\nabla E_{p,\rho,\gamma}(\mathbf{x})$ will be forced onto the manifold $N(\mathbf{A})$.

Since the penalized function is convex and the gradient descent always reaches the feasible region, the minimizers must be in the feasible region. In the feasible region, the two problems are identical and so they must have the same minimizers. □

While on the surface $N(\mathbf{A})$ the dynamics of $\dot{\mathbf{x}} = -\nabla E_{p,\rho,\gamma}(\mathbf{x})$ are described by

$$\dot{\mathbf{x}} = \mathbf{P}(-\mathbf{c} + \mathbf{u}^{(2)}) = \mathbf{P}(-\mathbf{c} - \gamma\nabla(\sum_{i=1}^{n}(\mathbf{x}-\mathbf{r})_i^-)).$$

If $\mathbf{x} - \mathbf{r} \geq \mathbf{0}$ then we set $\mathbf{u}^{(2)} = \mathbf{0}$. On the other hand, one can show by contradiction, using arguments similar to those employed in the proof of Lemma 1(i), that for consistent constraints if $\mathbf{x} \in \mathbb{R}^n$ satisfies $\mathbf{A}\mathbf{x} = \mathbf{0}$ and $\mathbf{P}\nabla(\sum_{i=1}^{n}(\mathbf{x}-\mathbf{r})_i^-) = \mathbf{0}$ then, $\mathbf{x} - \mathbf{r} \geq \mathbf{0}$.

Using Theorem 1, we can now propose a dynamic LP problem solver. The proposed dynamic LP problem solver can be viewed as a variable structure feedback system. It has the form

$$\dot{\mathbf{x}} = -\mathbf{c} + \mathbf{A}^T\mathbf{u}^{(1)} + \mathbf{u}^{(2)}, \tag{1}$$

where

$$\mathbf{u}^{(1)} = -\rho\nabla_{\sigma}\|\mathbf{Ax}\|_p, \quad \sigma = \mathbf{Ax},$$

$$\rho = \|\mathbf{c}^T\mathbf{A}^T(\mathbf{AA}^T)^{-1}\|_s + \frac{\|\mathbf{c}\|_2}{\|\mathbf{P}\nabla(\sum_{i=1}^n(\mathbf{x}-\mathbf{r})_i^-)\|_2} n^{\frac{1}{s}}\|\mathbf{A}^T(\mathbf{AA}^T)^{-1}\|_s + \varepsilon_1,$$

$$\mathbf{u}^{(2)} = -\gamma\nabla_{\mathbf{x}}(\sum_{i=1}^n(\mathbf{x}-\mathbf{r})_i^-),$$

$$\gamma = \frac{\|\mathbf{c}\|_2}{\|\mathbf{P}\nabla(\sum_{i=1}^n(\mathbf{x}-\mathbf{r})_i^-)\|_2} + \varepsilon_2,$$

and $\varepsilon_1 > 0, \varepsilon_2 > 0$ are any positive constants.

**Remark 3** Note that the proposed dynamic LP problem solver does not require that the initial condition be in the feasible set of the original LP problem.

We will now illustrate the effectiveness of the dynamic LP problem solver on the following example.

**Example 1** Consider the following LP problem:

$$\max 3z_1 + 5z_2$$

subject to

$$\begin{aligned} z_1 \quad +z_2 \quad &\leq 4 \\ 5z_1 \quad +3z_2 \quad &\geq 8 \\ z_1 \geq 0 \text{ and } z_2 &\geq 0. \end{aligned}$$

We transform the above problem into standard form. First, we change the maximization to minimization by multiplying the objective function by minus one. We then transform the first and second inequality constraints to the corresponding equality constraints by introducing nonnegative slack and surplus variables denoted by $z_3$ and $z_4$, respectively. The above LP problem in standard form is

$$\min -3z_1 - 5z_2$$

subject to

$$\begin{array}{lllll} z_1 & +z_2 & +z_3 & & = 4 \\ 5z_1 & +3z_2 & & -z_4 & = 8 \end{array}$$

$$z_1 \geq 0, z_2 \geq 0, z_3 \geq 0, \text{ and } z_4 \geq 0.$$

One can check that the solution of the above LP problem is

$$\mathbf{z}^T = [0, 4, 0, 4].$$

For this problem

$$\mathbf{A} = \begin{bmatrix} 1 & 1 & 1 & 0 \\ 5 & 3 & 0 & -1 \end{bmatrix}, \quad \mathbf{b}^T = [4, 8], \quad \mathbf{c}^T = [-3, -5, 0, 0],$$

and

$$\mathbf{P} = [\mathbf{I}_2 - \mathbf{A}^T(\mathbf{A}\mathbf{A}^T)^{-1}\mathbf{A})] = \begin{bmatrix} 0.2683 & -0.3902 & 0.1220 & 0.1707 \\ -0.3902 & 0.6585 & -0.2683 & 0.0244 \\ 0.1220 & -0.2683 & 0.1463 & -0.1951 \\ 0.1707 & 0.0244 & -0.1951 & 0.9268 \end{bmatrix}.$$

We will find the solution of the above LP problem utilizing the proposed dynamic program solver. We select the following initial point

$$\mathbf{z}^T(0) = [z_1(0), z_2(0), z_3(0), z_4(0)] = [0, 0, 0, 0].$$

Note that the initial point can be any arbitrary point in $\mathbb{R}^n$, (in this example n=4) and need not be in the feasible region. In fact our initial point is not in the feasible region. We first form the associated objective function

$$E_{p,\rho,\gamma} = \mathbf{c}^T\mathbf{z} + \rho\|\mathbf{A}\mathbf{z} - \mathbf{b}\|_p + \gamma(\sum_{i=1}^{4} z_i^-)$$

We will synthesize three different dynamic LP problem solvers. Specifically we consider cases for $p = 1, 2,$ and $\infty$.

For $p = 1$ the dynamic LP problem solver has the form

$$\dot{\mathbf{z}} = -\mathbf{c} - \rho \mathbf{A}^T \begin{bmatrix} sgn(\mathbf{a}_1 \mathbf{z} - b_1) \\ sgn(\mathbf{a}_2 \mathbf{z} - b_2) \end{bmatrix} - \gamma \nabla (\sum_{i=1}^{n} (\mathbf{z})_i^-),$$

where

$$\nabla (\sum_{i=1}^{n} (\mathbf{z})_i^-) = [\frac{\partial z_1^-}{\partial z_1}, \cdots, \frac{\partial z_4^-}{\partial z_4}]^T,$$

$$\gamma = \frac{6}{\|\mathbf{P}\nabla(\sum_{i=1}^{n}(\mathbf{z})_i^-)\|_2} > \frac{\|\mathbf{c}\|_2}{\|\mathbf{P}\nabla(\sum_{i=1}^{n}(\mathbf{z})_i^-)\|_2},$$

$$\rho = 1 + 11\gamma > \|\mathbf{c}^T \mathbf{A}^T (\mathbf{A}\mathbf{A}^T)^{-1}\|_\infty + \gamma \|\mathbf{A}^T (\mathbf{A}\mathbf{A}^T)^{-1}\|_\infty.$$

In Figure 1, we depicted the evolution of the variables $z_i$, $i = 1, 2, 3, 4$ versus time.

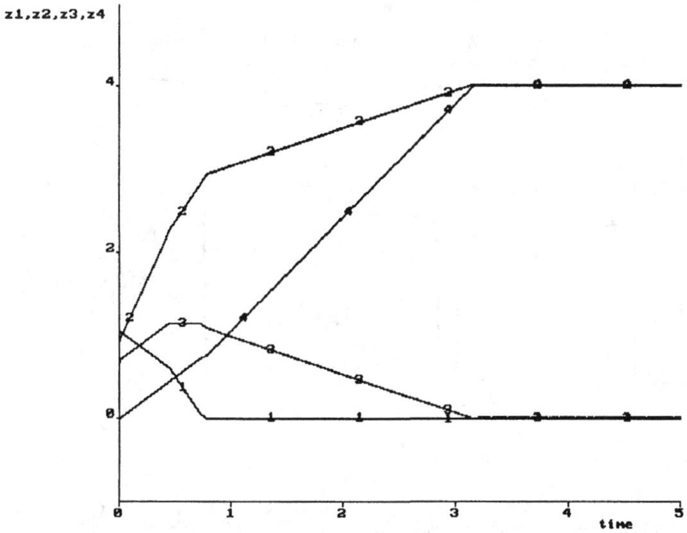


Figure 1: Trajectories of $z_1, z_2, z_3$, and $z_4$ versus time for $p = 1$.

In Figure 2, the projection of the solution on the $(z_1, z_2)$-plane is shown.

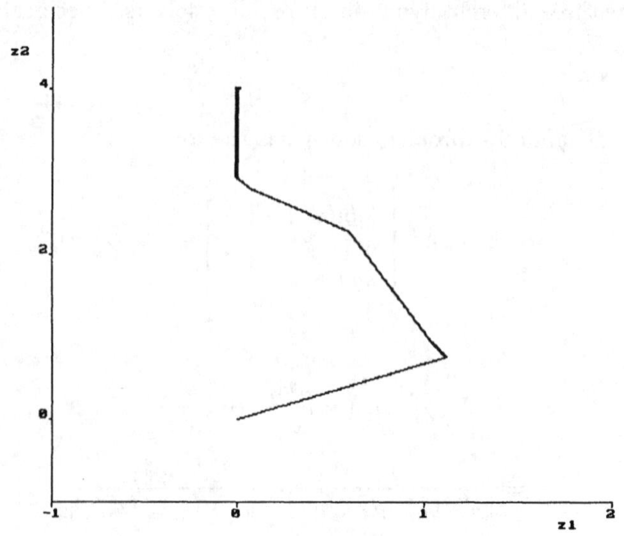


Figure 2: Trajectory $z_2$ versus $z_1$ for $p = 1$.

For $p = 2$, the dynamic LP problem solver has the form

$$\dot{\mathbf{z}} = -\mathbf{c} - \rho \frac{\mathbf{A}^T}{\|\mathbf{A}\mathbf{z} - \mathbf{b}\|_2} \begin{bmatrix} \mathbf{a}_1\mathbf{z} - b_1 \\ \mathbf{a}_2\mathbf{z} - b_2 \end{bmatrix} - \gamma \nabla \left(\sum_{i=1}^{n} (\mathbf{z})_i^-\right).$$

$$= \begin{bmatrix} 3 \\ 5 \\ 0 \\ 0 \end{bmatrix} - \frac{\rho}{\omega} \begin{bmatrix} 1 & 5 \\ 1 & 3 \\ 1 & 0 \\ 0 & -1 \end{bmatrix} \begin{bmatrix} z_1 + z_2 + z_3 - 4 \\ 5z_1 + 3z_2 - z_4 - 8 \end{bmatrix} - \gamma \nabla \left(\sum_{i=1}^{n} (\mathbf{z})_i^-\right),$$

where $\gamma$ is the same as before and

$$\rho = 2 + 11\gamma > \|\mathbf{c}^T \mathbf{A}^T (\mathbf{A}\mathbf{A}^T)^{-1}\|_2 + 2\gamma \|\mathbf{A}^T (\mathbf{A}\mathbf{A}^T)^{-1}\|_2,$$

$$\omega = \sqrt{(z_1 + z_2 + z_3 - 4)^2 + (5z_1 + 3z_2 - z_4 - 8)^2}.$$

In Figure 3, we depicted the evolution of the variables $z_i$, $i = 1, 2, 3, 4$ versus time.

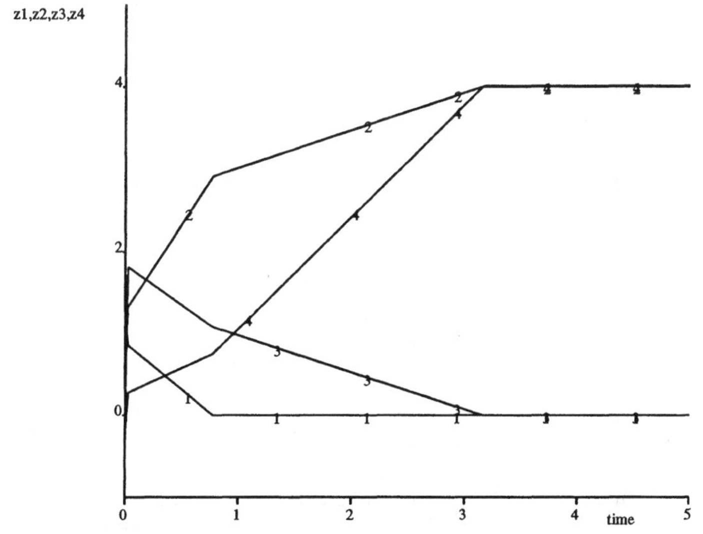


Figure 3: Trajectories of $z_1, z_2, z_3$, and $z_4$ versus time for $p = 2$.

In Figure 4, the projection of the solution on the $(z_1, z_2)$-plane is shown.

For $p = \infty$, the dynamic LP problem solver for the given problem has the form

$$\dot{z} = \begin{bmatrix} 3 \\ 5 \\ 0 \\ 0 \end{bmatrix} - \rho \begin{bmatrix} 1 & 5 \\ 1 & 3 \\ 1 & 0 \\ 0 & -1 \end{bmatrix} \begin{bmatrix} v_1 \\ v_2 \end{bmatrix} - \gamma \nabla (\sum_{i=1}^{n} (\mathbf{z})_i^-),$$

where

$$v1 = \text{if } max(|\mathbf{a}_1\mathbf{z} - b_1|, |\mathbf{a}_2\mathbf{z} - b_2|) > |\mathbf{a}_1\mathbf{z} - b_1| \text{ then } 0 \text{ else } sgn(\mathbf{a}_1\mathbf{z} - b_1),$$

$$v2 = \text{if } max(|\mathbf{a}_1\mathbf{z} - b_1|, |\mathbf{a}_2\mathbf{z} - b_2|) > |\mathbf{a}_2\mathbf{z} - b_2| \text{ then } 0 \text{ else } sgn(\mathbf{a}_2\mathbf{z} - b_2),$$

$\gamma$ is the same as before, and

$$\rho = 2 + 26\gamma > \|\mathbf{c}^T\mathbf{A}^T(\mathbf{A}\mathbf{A}^T)^{-1}\|_1 + 4\gamma\|\mathbf{A}^T(\mathbf{A}\mathbf{A}^T)^{-1}\|_1.$$

In Figure 5, we depicted the evolution of the variables $z_i$, $i = 1, 2, 3, 4$ versus time.

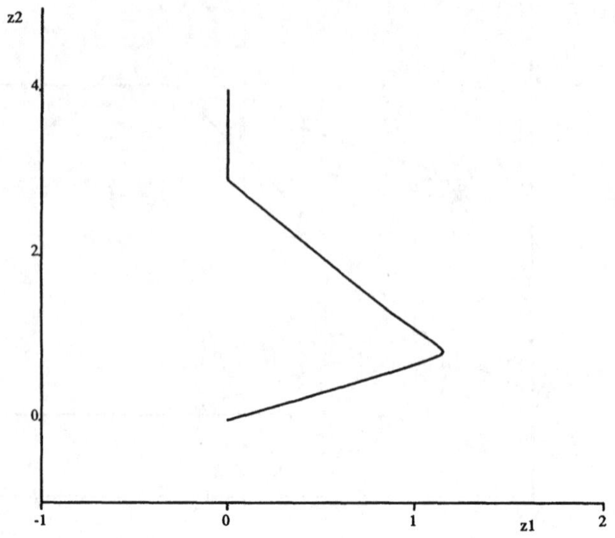


Figure 4: Trajectory $z_2$ versus $z_1$ for $p = 2$.

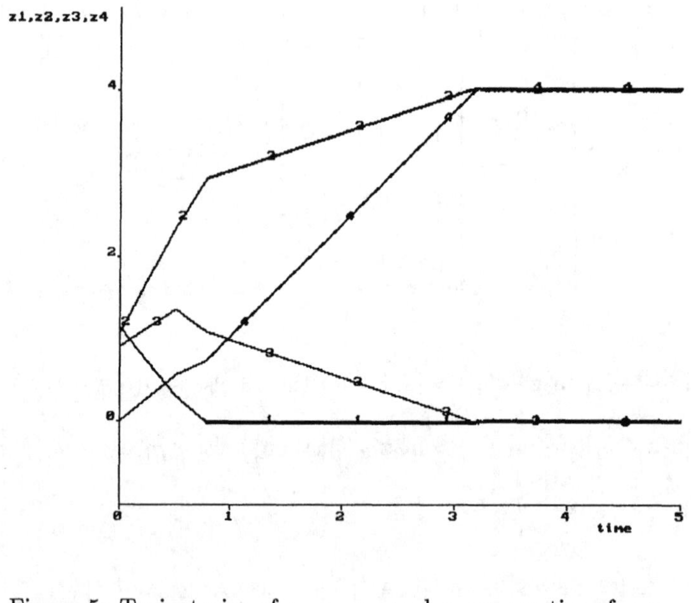


Figure 5: Trajectories of $z_1, z_2, z_3$, and $z_4$ versus time for $p = \infty$.

In Figure 6, the projection of the solution on the $(z_1, z_2)$-plane is shown.

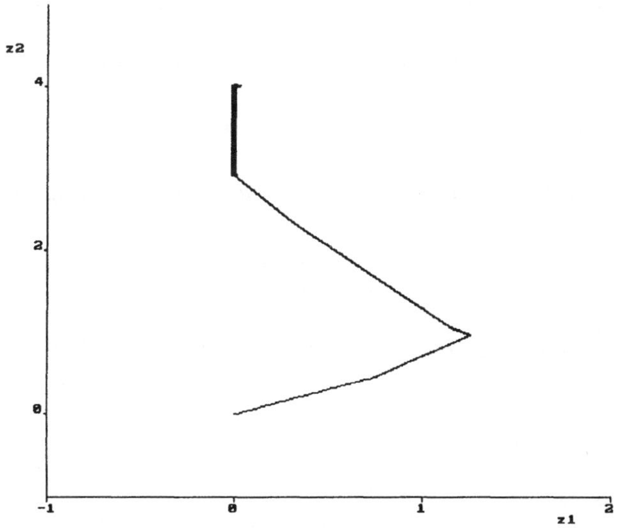


Figure 6: Trajectory $z_2$ versus $z_1$ for $p = \infty$.

We used the SIMNON package to perform our simulations. In our simulations we called the Euler method package to solve the differential equations. We chose the step size h=$10^{-5}$ second.

# 4 The Combined Gradient Projection and Penalty Methods for Solving LP Problems

In 1960, Rosen [33, 34] developed the gradient projection method for solving constrained optimization problems. The gradient projection method is motivated by the method of steepest descent for unconstrained problems. The negative gradient is projected onto the working surface defining the direction of movement.

In our proposed dynamic LP problem solvers, we use the gradient projection method for maintaining the trajectories of the solution on the surface $\mathbf{Az} = \mathbf{b}$. Moreover, we use a penalty function method and techniques from variable structure systems to ensure that trajectories of the proposed LP problem solver remain on the feasible surface $\mathbf{z} \geq \mathbf{0}$.

Before proceeding with a discussion of another dynamic LP problem solver, we first present the following lemma which we will use in our discussion.

**Lemma 4** *For the linear program in standard form, the Kuhn-Tucker conditions are*

$(i)$ $\mathbf{c}^T + \lambda^T \mathbf{A} + \mu^T = \mathbf{0}$,

$(ii)$ $\mu^T \mathbf{z} = 0$,

$(iii)$ $\mu \leq \mathbf{0}$,

*where* $\lambda \in I\!R^m$ *is the vector of Lagrange multipliers and* $\mu \in I\!R^n$ *is the vector of Kuhn-Tucker variables.*

Observe that $(i)$ implies that $\mathbf{c} + \mu \in R(\mathbf{A}^T)$.

Note that the Kuhn-Tucker conditions are also sufficient for a point to be a LP problem minimizer. This is because the LP problems are convex optimization problems.

It follows from the above lemma that if the inequality constraints are all inactive at a minimizer of the given LP problem or all Kuhn-Tucker variables $\mu_i$ are zero, then $\mathbf{c} \in R(\mathbf{A}^T)$.

We now describe the proposed LP problem solver. First, we start at any initial point satisfying $\mathbf{Az} = \mathbf{b}$. This initial point is very easy to calculate, e.g. a possible initial point is $\mathbf{A}^T(\mathbf{AA}^T)^{-1}\mathbf{b}$. We use the continuous steepest descent technique combined with the penalty method and a variable structure systems approach to synthesize a LP problem solver of the form

$$\dot{\mathbf{z}} = -\mathbf{P}(\nabla(\mathbf{c}^T\mathbf{z} + \eta \sum_{i=1}^{n}(\mathbf{z})_i^-)), \quad \mathbf{Az}(0) = \mathbf{b}.$$

Let

$$\beta = \min\{\|\mathbf{P}\nabla(\sum_{i=1}^{n}(\mathbf{z})_i^-)\|_2 | \mathbf{Az} = \mathbf{b}, \mathbf{z}_i < 0 \ \ for\ some \ \ i = 1, \cdots, n\},$$

and

$$\tau = \frac{\|\mathbf{c}\|_2}{\beta}.$$

By Lemma 1, $\beta > 0$, and hence $0 < \tau < \infty$.

**Theorem 2** *Trajectories of the dynamic system*

$$\dot{\mathbf{z}} = -\mathbf{P}(\nabla(\mathbf{c}^T\mathbf{z} + \eta \sum_{i=1}^{n}(\mathbf{z})_i^-)) = -\mathbf{P}(\mathbf{c} + \eta\nabla(\sum_{i=1}^{n}(\mathbf{z})_i^-)), \quad \mathbf{Az}(0) = \mathbf{b}, \tag{2}$$

*for $\eta > \tau$ converge to the feasible set of the corresponding linear programming problem.*

Proof:

The initial condition satisfies the equality constraints and the system trajectories are restricted to $\{\mathbf{z}|\mathbf{Az} = \mathbf{b}\}$.

We will now show that for $\eta > \tau$, the representative point of the above dynamic LP problem solver will be moving toward $\mathbf{z} \geq \mathbf{0}$ while sliding along $\{\mathbf{z}|\mathbf{Az} = \mathbf{b}\}$. We only need to consider the case when some inequality constraints are violated, i.e. when $z_i^- > 0$ for some i, and we would like a representative point to move in the direction of the set $\{\mathbf{z}|\mathbf{z} \geq \mathbf{0}\}$. Consider the function

$$h(\mathbf{z}) = \sum_{i=1}^{n}(\mathbf{z})^-.$$

Note that $h(\mathbf{z})$ gives the distance of $\mathbf{z}$ to $\{\mathbf{z} \geq \mathbf{0}\}$ in the $l^1$-norm. Since all norms on $\mathbb{R}^n$ are equivalent, if we show that $h(\mathbf{z}(t)) \to 0$, then the Euclidean distance of $\mathbf{z}(t)$ to $\{\mathbf{z} \geq \mathbf{0}\}$

will also tend to 0. Taking into account Lemma 1 and using arguments similar to those in proving Theorem 1, we obtain

$$\begin{aligned}
\frac{d}{dt}h(\mathbf{z}(t)) &= \nabla^T(\sum_{i=1}^{n}(\mathbf{z})_i^-)\ \dot{\mathbf{z}} \\
&= \nabla^T(\sum_{i=1}^{n}(\mathbf{z})_i^-)[-\mathbf{Pc} - \gamma\mathbf{P}\nabla(\sum_{i=1}^{n}(\mathbf{z})_i^-)] \\
&\leq |\nabla^T(\sum_{i=1}^{n}(\mathbf{z})_i^-)\ \mathbf{Pc}| - \gamma\nabla^T(\sum_{i=1}^{n}(\mathbf{z})_i^-)\ \mathbf{P}\nabla(\sum_{i=1}^{n}(\mathbf{z})_i^-) \\
&< 0
\end{aligned}$$

if $\eta > \tau$. Since there are only finitely many $\nabla(\sum_{i=1}^{n}(\mathbf{z})_i^-)$'s and $c$ is fixed, there exists $\epsilon > 0$ such that

$$\frac{d}{dt}h(\mathbf{z}(t)) < -\epsilon$$

if $\eta > \tau$. As in the proof of Lemma 2, we conclude that $h(\mathbf{z}(t))$ decreases with $t$ and $\mathbf{z}(t)$ will reach $\{\mathbf{z} \geq \mathbf{0}\}$ in finite time. □

Once in the feasible region, the trajectories of (2) will converge to the solution of the corresponding LP problem. This results from the following Propositions.

Let

$$F(\mathbf{z}) = \mathbf{c}^T\mathbf{z} + \eta\sum_{i=1}^{n}(\mathbf{z})_i^-.$$

Then we have :

**Proposition 1** *The function $F(\mathbf{z})$ is a decreasing time function on the trajectories of (2).*

Proof: We have

$$\begin{aligned}
\frac{d}{dt}F(\mathbf{z}) &= \nabla^T F\dot{\mathbf{z}} \\
&= -\nabla^T F\mathbf{P}\nabla F
\end{aligned}$$

$$= -\|\mathbf{P}\nabla(\mathbf{c}^T\mathbf{z} + \eta\sum_{i=1}^{n}(\mathbf{z})_i^-)\|_2^2$$
$$\leq 0. \qquad \square$$

**Proposition 2** *The time derivative of $F(\mathbf{z})$ is zero only at the equilibrium points of the system (2).*

Proof: The result follows from Proposition (1). $\square$

Observe that (2) does not have equilibrium points outside the feasible region. For if it had, then we would have

$$\mathbf{P}\nabla(\mathbf{c}^T\mathbf{z} + \eta\sum_{i=1}^{n}(\mathbf{z})_i^-) = \mathbf{0},$$

where for some $i$, $z_i < 0$. However, if $\eta > \tau$ then there is no solution to the above equation if $z_i < 0$ for some $i$. This follows from the fact that $\frac{d}{dt}h(\mathbf{z}(t)) < 0$.

**Proposition 3** *The equilibrium points of (2) satisfy the Kuhn-Tucker conditions for optimality.*

Proof: Recall that a point is an equilibrium point of (2) if

$$\mathbf{P}(\mathbf{c} + \eta\nabla(\sum_{i=1}^{n}(\mathbf{z})_i^-) = \mathbf{0},$$

that is, there is a vector $\lambda \in \mathbb{R}^m$ such that

$$\mathbf{c} + \eta\nabla(\sum_{i=1}^{n}(\mathbf{z})_i^-) = \mathbf{A}^T\lambda.$$

The above equation can be rewritten as

$$\mathbf{c} + \mu = \mathbf{A}^T\lambda,$$

where $\mu = \eta\nabla(\sum_{i=1}^{n}(\mathbf{z})_i^-) \leq \mathbf{0}$, since $\eta > 0$. The above equation is equivalent to (i) and (iii) in Lemma 4. To show (ii), note that $\mu_i = \eta\frac{\partial z_i^-}{\partial z_i}$ and $\mu_i = 0$ if $z_i \geq 0$. Hence $\mu^T\mathbf{z} = 0$ for $\mathbf{z} \geq \mathbf{0}$. The proof is complete, since (2) does not have any equilibrium points outside the feasible region. □

**Remark 4** Note that (2) may not have any equilibrium points. This is because for the equilibrium points of (2) we have to have $\mathbf{c} \in R(\mathbf{A}^T)$. Thus, for example, if for an LP problem $\mu \neq \mathbf{0}$ then (2) corresponding to this problem will not have any equilibrium points. The trajectories of (2), however, will converge to the solution of the LP problem, and then will "chatter" about this point. This phenomenon can be observed in our simulation examples. The above Propositions and observation also apply to the LP problem solver (1).

As in the previous sections, the proposed LP problem solver can also be viewed as a variable structure feedback system. It has the form :

$$\dot{\mathbf{z}} = -\mathbf{P}(\mathbf{c} - \mathbf{u}), \quad \mathbf{A}\mathbf{z}(0) = \mathbf{b},$$

where

$$\mathbf{u} = -\eta\nabla_{\mathbf{z}}(\sum_{i=1}^{n} z_i^-), \quad \eta = \frac{\|\mathbf{c}\|_2}{\|\mathbf{P}\nabla(\sum_{i=1}^{n}(\mathbf{z})_i^-)\|_2} + \varepsilon,$$

and $\varepsilon > 0$ is any positive constant. Note that unlike in the previous LP solver, the initial condition for this system must satisfy the relation $\mathbf{A}\mathbf{z}(0) = \mathbf{b}$.

We will now illustrate the behavior of this dynamic LP problem solver on the following example.

**Example 2**

Consider the LP problem described in Example 1. We now use the second dynamic LP problem solver to find the solution of this LP problem. However, we now must select the initial point satisfying $\mathbf{Az}(0) = \mathbf{b}$. Here, we select

$$\mathbf{z}(0) = [0, 0, 4, -8]^T.$$

We choose

$$\eta = \frac{6}{\|\mathbf{P}\nabla(\sum_{i=1}^{n}(\mathbf{z})_i^-)\|_2} > \frac{\|\mathbf{c}\|_2}{\|\mathbf{P}\nabla(\sum_{i=1}^{n}(\mathbf{z})_i^-)\|_2}.$$

In Figure 7, we depicted the evolution of the variables $z_i$, $i = 1, 2, 3, 4$ versus time.

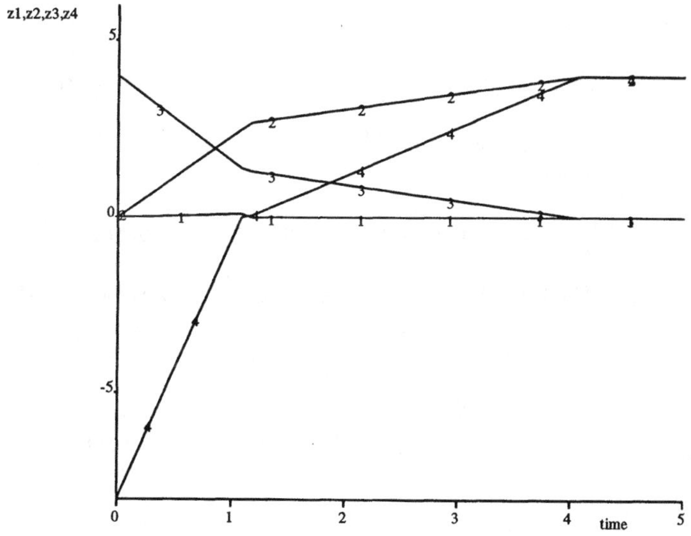


Figure 7: Trajectories of $z_1, z_2, z_3$, and $z_4$ versus time.

In this section, we also used the Euler method with step size h=$10^{-5}$ to solve the differential equation.

# 5 Conclusions

In this paper, we proposed and analyzed two dynamic systems for solving LP problems. The connections of both dynamic LP solvers to variable structure systems were discussed. We included numerical examples to illustrate the results obtained.

## References

[1] G. B. Dantzig, *Linear Programming and Extensions.* Princeton, New Jersey: Princeton University Press, 1963.

[2] G. W. Brown and T. C. Koopmans, "Computation suggestions for maximizing a linear function subject to linear inequalities," in *Activity Analysis of Production and Allocation* (T. Koopmans, ed.), pp. 377–380, New York: J. Wiley, 1951.

[3] L. G. Khachiyan, "A polynomial algorithm in linear programmig," *Dokl. Akad. Nauk SSSR*, no. 244, pp. 1093–1096, 1979. (English Translation: Soviet Math. Dokl. 20, 191-194).

[4] A. Schrijver, *Theory of Linear and Integer Programming.* New York: J. Wiley and Sons, 1986.

[5] R. G. Bland, D. Goldfarb, and M. J. Todd, "The ellipsoid method: a survey," *Oper. Research*, vol. 29, pp. 1039–1091, 1981.

[6] P. E. Gill, W. Murray, M. A. Saunders, J. A. Tomlin, and M. H. Wright, "On projected Newton barrier method for LP and an equivalent to Karmarkar's projected method," *Mathematical Programming*, vol. 36, pp. 183–209, Nov 1986.

[7] N. K. Karmarkar, "A new polynomial-time algorithm for linear programming," *Combinatorica*, vol. 4, pp. 373–395, Dec 1984.

[8] G. Strang, *Linear Algebra and Its Applications*, ch. 8, pp. 405–410. San Diego: Harcourt Brace Jovanovich, third ed., 1988.

[9] A. Ravindran and R. Marsten, "Linear Programming," in *Handbook of Industrial Engineering* (G. Salvendy, ed.), ch. 103, pp. 2597–2612, New York: J. Wiley and Sons, second ed., 1992.

[10] R. Courant, "Calculus of Variations and Supplementary Notes and Exercises," 1962. (mimeographed notes), supplementary notes by M. Kruskal and H. Rubin, revised and amended by J. Moser, New York University.

[11] C. W. Carroll, "The created response surface technique for optimizing non-linear restrained systems," *Operations Research*, vol. 9, no. 12, pp. 169–184, 1961.

[12] A. R. Conn, "Linear programming via a nondifferentiable penalty function," *SIAM J. Numerical Analysis*, vol. 13, pp. 145–154, 1976.

[13] I. B. Pyne, "Linear programming on an electronic analogue computer," *Transactions of the American Institute of Electrical Engineers*, vol. 75, pp. 139–143, May 1956.

[14] M. V. Rybashov, "The gradient method of solving convex programming problems on electronic analog computers," *Automation and Remote Control*, vol. 26, no. 11, pp. 1886–1898, 1965.

[15] M. V. Rybashov, "Gradient method of solving linear and quadratic programming problems on electronic analog computers," *Automation and Remote Control*, vol. 26, no. 12, pp. 2079–2089, 1965.

[16] N. N. Karpinskaya, "Method of "penalty" functions and the foundations of Pyne's method," *Automation and Remote Control*, vol. 28, no. 1, pp. 124–129, 1967.

[17] D. W. Tank and J. J. Hopfield, "Simple "neural" optimization networks: An A/D converter, signal decision circuit, and a linear programming circuit," *IEEE Transactions on Circuits and Systems*, vol. CAS-33, pp. 533–541, May 1986.

[18] M. P. Kennedy and L. O. Chua, "Neural networks for nonlinear programming," *IEEE Transactions on Circuits and Systems*, vol. 35, pp. 554–562, May 1988.

[19] A. Rodríguez-Vázquez, R. Domínguez-Castro, A. Rueda, J. L. Huertas, and E. Sánchez-Sinencio, "Nonlinear switched-capacitor "neural" networks for optimization problems," *IEEE Transactions on Circuits and Systems*, vol. 37, pp. 384–397, Mar 1990.

[20] A. Cichocki and R. Unbehauen, "Neural networks for solving systems of linear equations and related problems," *IEEE Transactions on Circuits and Systems-I:Fundamental theory and applications*, vol. 39, pp. 124–138, Feb. 1992.

[21] S. Hui, W. E. Lillo, and S. H. Żak, "Minimizing quotient space norms using penalty functions," Tech. Rep. TR-EE 91-48, Purdue University, W. Lafayette, IN 47907, Nov 1991.

[22] J. V. Burke, "An exact penalization viewpoint of constrained optimization," *SIAM J. Control and Optimization*, vol. 29, pp. 968–998, 1991.

[23] A. R. Conn, "Constrained optimization using a nondifferential penalty function," *SIAM J. Numerical Analysis*, vol. 10, pp. 760–783, 1973.

[24] D. G. Luenberger, *Linear and Nonlinear Programming.* Reading, Massachusetts: Addison-Wesley, second ed., 1984.

[25] A. L. Peressini, F. E. Sullivan, and J. J. Uhl, Jr., *The Mathematics of Nonlinear Programming.* New York: Springer-Verlag, 1988.

[26] D. P. Bertsekas, "Necessary and sufficient conditions for a penalty method to be exact," *Mathematical Programming*, vol. 9, pp. 87–99, 1975.

[27] V. I. Utkin, "Variable structure systems with sliding modes," *IEEE Transactions on Automatic Control*, vol. AC-22, no. 2, pp. 212–222, 1977.

[28] V. I. Utkin, *Sliding Modes and Their Application in Variable Structure Systems.* Moscow: Mir Publishers, 1978.

[29] V. I. Utkin, *Sliding Modes in Control and Optimization.* Berlin, Heidelberg: Springer-Verlag, 1992.

[30] P. E. Gill, W. Murray, and M. H. Wright, *Practical Optimization.* London, England: Academic Press, 1981.

[31] E. B. Lee and L. Markus, *Foundations of Optimal Control Theory.* New York: J. Wiley and Sons, 1967.

[32] S. H. Żak, S. M. Madani-Esfahani, and S. Hui, "Control of dynamic systems via neural networks," in *Neural Networks: Concepts, Applications, and Implementations* (P. Antognetti and V. Milutinović, eds.), vol. 2, ch. 5, pp. 101–143, Prentice Hall, 1991.

[33] J. B. Rosen, "The gradient projection method for nonlinear programming, I. Linear constraints," *J. SIAM*, vol. 8, pp. 181–217, 1960.

[34] J. B. Rosen, "The gradient projection method for nonlinear programming, II. Nonlinear constraints," *J. SIAM*, vol. 9, pp. 514–532, 1961.

ISBN 0-8247-8904-0
90000>
9 780824 789046